Biotechnology, Multiple Omics, and Precision Breeding in Medicinal Plants

Biotechnology, Multiple Omics, and Precision Breeding in Medicinal Plants explores the various methods for advancing medicinal plant research. It covers a wide range of approaches, including integrated and advanced plant biotechnology, mutagenesis, nanotechnology, genome-wide association studies, multiple omics tools, and high-throughput technologies. The book highlights the significant impact of combining pan-genomics with metabolomics in medicinal plant research, particularly in understanding how genetic diversity influences the profiles of secondary metabolites and the therapeutic potential of these plants.

FEATURES:

- Explores ways to improve the production of secondary metabolites and bioactive compounds in key medicinal plants.
- Features information on bioinformatics, artificial intelligence models, molecular markers, and genome editing techniques such as CRISPR-assisted precision breeding.
- Promotes specific prebiotic formulas to ward off adverse effects of antibiotics.
- Covers information on epigenetic regulation in boosting secondary metabolite production and the use of speed breeding combined with high-throughput technologies.

Proposing a multitude of technologies and methodologies in plant biotechnology with focus on enhancing the production of secondary metabolites and bioactive compounds from medicinal plants, this book is an ideal resource for researchers and academia in plant sciences/breeding, agriculture, and horticulture industries.

Biotechnology, Multiple Omics, and Precision Breeding in Medicinal Plants

Edited by Jen-Tsung Chen

CRC Press
Taylor & Francis Group
Boca Raton London New York

CRC Press is an imprint of the
Taylor & Francis Group, an **informa** business

Designed cover image: Irina Vaneeva\shutterstock

First edition published 2025
by CRC Press
2385 NW Executive Center Drive, Suite 320, Boca Raton FL 33431

and by CRC Press
4 Park Square, Milton Park, Abingdon, Oxon, OX14 4RN

CRC Press is an imprint of Taylor & Francis Group, LLC

ISBN: 978-1-032-75467-3 (hbk)
ISBN: 978-1-032-75751-3 (pbk)
ISBN: 978-1-003-47549-1 (ebk)

DOI: 10.1201/9781003475491

Typeset in Times
by Apex CoVantage, LLC

Contents

About the Editor

Dr. Jen-Tsung Chen is a professor of cell biology at the National University of Kaohsiung in Taiwan. He also teaches genomics, proteomics, plant physiology, and plant biotechnology. Dr. Chen's research interests include bioactive compounds, chromatography techniques, plant molecular biology, plant biotechnology, bioinformatics, and systems pharmacology. He is an active editor of academic books and journals to advance the exploration of multidisciplinary knowledge involving plant physiology, plant biotechnology, nanotechnology, ethnopharmacology, systems biology, and drug discovery. He serves as an editorial board member and a guest editor in several reputed journals. Dr. Chen published books in collaboration with international publishers on diverse topics such as drug discovery, herbal medicine, medicinal biotechnology, nanotechnology, bioengineering, plant functional genomics, plant speed breeding, CRISPR-based plant genome editing, and artificial intelligence. In 2023 and 2024, Dr. Chen was included in the "World's Top 2% Scientists" by Stanford University.

Contributors

Malika Abid
Laboratory for Agricultural Productions Improvement, Biotechnology, and Environment
University Mohammed First
Oujda, Morocco

Mohamed Addi
Laboratory for Agricultural Productions Improvement, Biotechnology, and Environment
University Mohammed First
Oujda, Morocco

Sujaan Aehsas
Faculty of Agriculture and Allied Sciences
C.V. Raman Global University
Bhubaneswar, India

Mostafa Ahmadizadeh
Minab Higher Education Center
University of Hormozgan
Bandar Abbas, Iran

H.S. Likitha Aishwayra
Department of Plant Sciences
Manipal Academy of Higher Education
Manipal, Karnataka, India

Arumugam Vijaya Anand
Department of Human Genetics and Molecular Biology
Bharathiar University
Coimbatore, Tamil Nadu, India

Muthukrishnan Anand
Department of English
Ramco Institute of Technology
Rajapalayam, Tamil Nadu, India

Ashika Ansari
Department of Botany and Plant Physiology
Chaudhary Charan Singh Haryana Agricultural University
Hisar, India

Muthukrishnan Arun
Department of Biotechnology
Bharathiar University
Coimbatore, Tamil Nadu, India

Santhanalakshmi Balasubramaniam
Translational Plant Research Laboratory, Department of Microbial Biotechnology
Bharathiar University
Coimbatore, Tamil Nadu, India

Sampurna Banerjee
Faculty of Agriculture
Bidhan Chandra Krishi Viswavidyalaya
Mohanpur, Nadia, India

Swastika Banerjee
Department of Science
Alliance University
Bangalore, Karnataka, India

Piyush Bania
Palli Siksha Bhavana (Institute of Agriculture)
Visva-Bharati
Sriniketan, West Bengal, India

Ahamed Kabeer Umera Begam
Department of Biochemistry
Holy Cross College (Autonomous)
Tiruchirappalli, Tamil Nadu, India

Tajudheen Nargis Begum
Department of Biotechnology
Jamal Mohamed College (Autonomous)
Tiruchirappalli, Tamil Nadu, India

Khadija Benamar
Phytopathology Unit, Department of Plant Protection
Ecole Nationale d'Agriculture de Meknès
Meknès, Morocco

Adalberto Benavides-Mendoza
Departamento de Horticultura
Universidad Autónoma Agraria Antonio Narro
Saltillo, Mexico

Sourav Bhattacharjee
Department of Botany
Pandit Deendayal Upadhyaya Adarsha Mahavidyalaya Tulungia
Bongaigaon, Assam, India

Jen-Tsung Chen
Department of Life Sciences
National University of Kaohsiung
Kaohsiung, Taiwan

Privilege Chikove
Faculty of Agriculture and Allied Sciences
C.V. Raman Global University
Bhubaneswar, Odisha, India

Dyuti Ballav Chongder
Palli Siksha Bhavana (Institute of Agriculture)
Visva-Bharati
Sriniketan, West Bengal, India

Abhra Abir Chowdhury
PG Department of Botany
Darjeeling Government College
Darjeeling, West Bengal, India

Ankan Das
Palli Siksha Bhavana (Institute of Agriculture)
Visva-Bharati
Sriniketan, West Bengal, India

Samrat Das
Division of Plant Physiology
Indian Agricultural Research Institute
New Delhi, India

Sandip Debnath
Department of Genetics and Plant Breeding
Palli Siksha Bhavana (Institute of Agriculture)
Visva-Bharati
Sriniketan, West Bengal, India

Aniket Dhariwal
Department of Botany and Plant Physiology
Chaudhary Charan Singh Haryana Agricultural University
Hisar, India

Jayakumar Dharshini
Department of Human Genetics and Molecular Genetics
Bharathiar University
Coimbatore, Tamil Nadu, India

Hien Thi Thu Do
Research Center for Infectious Diseases
Vietnam National University
Ho Chi Minh City, Vietnam

Shreyasi Dutta
Research Center for Natural and Applied Sciences
Raja Narendralal Khan Women's College (Autonomous)
Midnapur, India

Tuward J. Dweh
Faculty of Agriculture and Allied Sciences
C.V. Raman Global University
Bhubaneswar, Odisha, India

Ghizlane Echchgadda
Agroecology Unit, Department of Plant Protection
Ecole Nationale d'Agriculture de Meknès
Meknès, Morocco

M.R. Elakkiya
Department of Botany
PSGR Krishnammal College for Women
Coimbatore, Tamil Nadu, India

Suriya Prakash Eswaran
Translational Plant Research Laboratory, Department of Microbial Biotechnology
Bharathiar University
Coimbatore, Tamil Nadu, India

Rachid Ezzougari
Phytopathology Unit, Department of Plant Protection
Ecole Nationale d'Agriculture de Meknès
Meknès, Morocco

Abdelaaziz Farhaoui
Phytopathology Unit, Department of Plant Protection
Ecole Nationale d'Agriculture de Meknès
Meknès, Morocco

Paola Fincheira
Centro de Excelencia en Investigación Biotecnológica Aplicada al Medio Ambiente (CIBAMA)
Universidad de La Frontera
Temuco, Chile

Laura Olivia Fuentes-Lara
Departamento de Nutrición Animal
Universidad Autónoma Agraria Antonio Narro
Saltillo, Mexico

Arindam Ganguly
Department of Microbiology
Bankura Sammilani College
Bankura, West Bengal, India

Shanmugam Gavaskar
Department of Computer Application
Bharathiar University
Coimbatore, Tamil Nadu, India

Banita Gawaria
Department of Botany and Plant Physiology
Chaudhary Charan Singh Haryana Agricultural University
Hisar, India

Kapildev Gnanajothi
Translational Plant Research Laboratory, Department of Microbial Biotechnology
Bharathiar University
Coimbatore, Tamil Nadu, India

M. Gnanaraj
Department of Biotechnology, School of Life Sciences
St Joseph's University
Bengaluru, Karnataka, India

Susana González-Morales
CONAHCYT-UAAAN
Saltillo, Mexico

Khadija Goura
Phytopathology Unit, Department of Plant Protection
Ecole Nationale d'Agriculture de Meknès
Meknès, Morocco

Jayachandran Halka
Department of Biotechnology
Bharathiar University
Coimbatore, Tamil Nadu, India

Yousra Hammouti
Laboratory for Agricultural Productions Improvement, Biotechnology, and Environment
University Mohammed First
Oujda, Morocco

Parviz Heidari
Faculty of Agriculture
Shahrood University of Technology
Shahrood, Iran

T. Jebastin
Department of Bioinformatics
Bishop Heber College (Autonomous)
Trichy, Tamil Nadu, India

Angeline Jurry
Faculty of Agriculture and Allied Sciences
C.V. Raman Global University
Bhubaneswar, Odisha, India

Indrajit Kalita
Department of Botany
Gauhati University
Guwahati, Assam, India

Najwa Seddiqi Kallali
Phytopathology Unit, Department of Plant Protection
Ecole Nationale d'Agriculture de Meknès
Meknès, Morocco

Greeshma P. Kammar
Department of Plant Sciences
Manipal School of Life Sciences, Manipal Academy of Higher Education
Manipal, Karnataka, India

Arya Kaniyassery
Department of Plant Sciences
Manipal School of Life Sciences, Manipal Academy of Higher Education
Manipal, Karnataka, India

Zoofishan Kazi
Department of Botany
St. Xavier's College (Autonomous)
Mumbai, Maharashtra, India

Mohammed Khadiri
Phytopathology Unit, Department of Plant Protection
Ecole Nationale d'Agriculture de Meknès
Meknès, Morocco

Kiran
Department of Botany and Plant Physiology
Chaudhary Charan Singh Haryana Agricultural University
Hisar, India

Kumaresan Kowsalya
Department of Biotechnology
Bharathiar University
Coimbatore, Tamil Nadu, India

M. Krishnasreya
Department of Botany
PSGR Krishnammal College for Women
Coimbatore, Tamil Nadu, India

Manoj Kumar
Department of Plant Sciences
Manipal School of Life Sciences, Manipal Academy of Higher Education
Manipal, Karnataka, India

Neeraj Kumar
Department of Botany and Plant Physiology
Chaudhary Charan Singh Haryana Agricultural University
Hisar, India

Rahul Kumar
Department of Zoology, College of Basic Sciences and Humanities
Chaudhary Charan Singh Haryana Agricultural University
Hisar, India

Rachid Lahlali
Phytopathology Unit, Department of Plant Protection
Ecole Nationale d'Agriculture de Meknès
Meknès, Morocco

Haitam Lahmamsi
Phytopathology Unit, Department of Plant Protection
Ecole Nationale d'Agriculture de Meknès
Meknès, Morocco

Huy Phu Quang Le
School of Biotechnology
International University
Ho Chi Minh City, Vietnam

Salah-Eddine Laasli
Phytopathology Unit, Department of Plant Protection
Ecole Nationale d'Agriculture de Meknès
Meknès, Morocco

Manaswini Mahapatra
Faculty of Agriculture and Allied Sciences
C.V. Raman Global University
Bhubaneswar, Odisha, India

Aqsa Majgaonkar
Department of Botany
St. Xavier's College (Autonomous)
Mumbai, Maharashtra, India

R. Manikandan
Department of Microbiology
Padmavati Arts & Science College for Women
Salem, Tamil Nadu, India

Debdut Manna
Division of Biochemistry
Indian Agricultural Research Institute
New Delhi, India

Mani Manoj
Department of Human Genetics and Molecular Genetics
Bharathiar University
Coimbatore, Tamil Nadu, India

Arockiya Anita Margret
Department of Biotechnology and Bioinformatics
Bishop Heber College (Autonomous)
Tiruchirappalli, Tamil Nadu, India

Ahmed Marhri
Laboratory for Agricultural Productions Improvement, Biotechnology, and Environment
University Mohammed First
Oujda, Morocco

Arun Kumar Maurya
Department of Botany
Multanimal Modi College
Modinagar, Ghaziabad, Uttar Pradesh, India

Julia Medrano-Macías
Universidad Autónoma Agraria Antonio Narro
Departamento de Horticultura
Saltillo, Mexico

Sarmad Moin
School of Applied Sciences
Suresh Gyan Vihar University
Jaipur, Rajasthan, India

Partha Mondal
Department of Agricultural Biochemistry
Bidhan Chandra Krishi Viswavidyalaya
West Bengal, India

Tanmoy Mondal
Assistant Professor
Department of Horticulture
MITS Institute of Professional Studies
Rayagada, Odisha

Rashmi Muherjee
Research Center for Natural and Applied Sciences, Department of Botany (UG & PG)
Raja Narendralal Khan Women's College (Autonomous)
Midnapur, India

Shahla Haseena Muthalib
Department of Plant Sciences
Manipal School of Life Sciences, Manipal Academy of Higher Education
Manipal, Karnataka, India

V.N. Mutharaian
Department of Botany
Gobi Arts & Science College
Gobichettipalayam, Tamil Nadu, India

Annamalai Muthusamy
Department of Plant Sciences
Manipal School of Life Sciences, Manipal Academy of Higher Education
Manipal, Karnataka, India

Prayasi Nayak
Faculty of Agriculture and Allied Sciences
C.V. Raman Global University
Bhubaneswar, Odisha, India

Ngoc Diem Thuy Nguyen
School of Biotechnology
International University
Ho Chi Minh City, Vietnam

Phuong-Hien Nguyen
University of Cologne, Faculty of Medicine
Cologne, Germany

Nguyen Phuong Thao
Research Center for Infectious Diseases
International University
Ho Chi Minh City, Vietnam

Tram Huyen Ngoc Nguyen
School of Biotechnology
International University
Ho Chi Minh City, Vietnam

Lam Nguyen Ngoc Pham
School of Biotechnology
International University
Ho Chi Minh City, Vietnam

Madhuri Pradhan
Faculty of Agriculture and Allied Sciences
C.V. Raman Global University
Bhubaneswar, Odisha, India

Ritashree Pramanik
Palli Siksha Bhavana (Institute of Agriculture)
Visva-Bharati
Sriniketan, West Bengal, India

Ramesh Preyenga
Department of Biotechnology and Bioinformatics
Bishop Heber College (Autonomous)
Tiruchirappalli, Tamil Nadu, India

Mohammed Radi
Phytopathology Unit, Department of Plant Protection
Ecole Nationale d'Agriculture de Meknès
Meknès, Morocco

M. Rajadurai
Department of Bioinformatics
Bishop Heber College (Autonomous)
Trichy, Tamil Nadu, India

D. Rajasudhakar
Department of Bioinformatics
Bishop Heber College (Autonomous)
Trichy, Tamil Nadu, India

Karuppiah Ramalashmi
Department of Biotechnology
Bharathiar University
Coimbatore, Tamil Nadu, India

Nikhil Kumar Ramesha
Department of Plant Sciences
Manipal School of Life Sciences, Manipal Academy of Higher Education
Manipal, Karnataka, India

Shital Rani
Department of Biophysics
Panjab University
Chandigarh, India

Balasubramaniam Rashmika
Department of Human Genetics and Molecular Genetics
Bharathiar University
Coimbatore, Tamil Nadu, India

Muthuswamy Razia
Department of Biotechnology
Mother Teresa Women's University
Kodaikanal, Tamil Nadu, India

Anna Rera
Faculty of Agriculture and Allied Sciences
C.V. Raman Global University
Bhubaneswar, Odisha, India

Asirvatham Alwin Robert
Department of Endocrinology and Diabetes
Prince Sultan Military Medical City, Ministry of Defense
Riyadh, Kingdom of Saudi Arabia

Arokiaraj Sherlin Rosita
Department of Bioinformatics
Bishop Heber College (Autonomous)
Tiruchirappalli, Tamil Nadu, India

Rajaswini Rupali
Department of Biotechnology
Bharathiar University
Coimbatore, Tamil Nadu, India

Manalisha Saharia
Department of Botany
Madhab Choudhury College
Barpeta, Assam, India

Subhashree Sahu
Faculty of Agriculture and Allied Sciences
C.V. Raman Global University
Bhubaneswar, Odisha, India

Jyoti Prakash Sahoo
Faculty of Agriculture and Allied Sciences
C.V. Raman Global University
Bhubaneswar, Odisha, India

Aarsha Saji
Department of Bioinformatics
Bishop Heber College (Autonomous)
Tiruchirappalli, Tamil Nadu, India

Anitha Sakthivel
Translational Plant Research Laboratory, Department of Microbial Biotechnology
Bharathiar University
Coimbatore, Tamil Nadu, India

Manojkumar Sampath
Translational Plant Research Laboratory, Department of Microbial Biotechnology
Bharathiar University
Coimbatore, Tamil Nadu, India

P. Samuel
Department of Biotechnology
Bishop Heber College (Autonomous)
Trichy, Tamil Nadu, India

Fatima-Ezzahra Saouab
Laboratory of Biotechnology, Environment, Food and Health (LBEFH), Department of Biology
Faculty of Sciences Dhar el Mahraz, Sidi Mohammed Ben Abdellah University
Atlas, Fez, Morocco

Balasundaram Saraswathy Chithra Devi
Department of Botany
PSGR Krishnammal College for Women
Coimbatore, Tamil Nadu, India

Arumugam Sembiyaa
Department of Human Genetics and Molecular Genetics
Bharathiar University
Coimbatore, Tamil Nadu, India

Pankaj Sharma
Department of Ophthalmology & Visual Sciences
University of Illinois Chicago
Chicago, IL, United States

Shweta Sharma
Department of Physiology
AIIMS
New Delhi, India

Nyashadzashaishe Shengezi
Faculty of Agriculture and Allied Sciences
C.V. Raman Global University
Bhubaneswar, Odisha, India

Reena Sheoran
Department of Zoology
Maharshi Dayanand University
Rohtak, Haryana, India

Jambulingam Shivani
Department of Bioinformatics
Bishop Heber College (Autonomous)
Tiruchirappalli, Tamil Nadu, India

Mathivanan Jay Shree
Department of Bioinformatics
Bishop Heber College (Autonomous)
Tiruchirappalli, Tamil Nadu, India

Sudeep Sindhu
Department of Biochemistry
College of Basic Sciences and Humanities, Chaudhary Charan Singh Haryana Agricultural University
Hisar, India

Devendra Singh
Faculty of Biotechnology
Institute of Biosciences & Technology, Shri Ramswaroop Memorial University
Lucknow-Deva Road, Barabanki, Uttar Pradesh, India

Dilbag Singh
School of Agriculture
Galgotias University
Greater Noida, Uttar Pradesh, India

Veena Singh
Faculty of Commerce and Economics
Institute of Management of Commerce & Economics, Shri Ramswaroop Memorial University
Lucknow-Deva Road, Barabanki, Uttar Pradesh, India

Dwaipayan Sinha
Department of Botany
Government General Degree College
Mohanpur, Paschim Medinipur, West Bengal, India

Ganeshan Sivanandhan
Department of Horticulture
Chungnam National University
Daejeon, South Korea

P.V. Sreelakshmi
Department of Plant Sciences
Manipal School of Life Sciences, Manipal Academy of Higher Education
Manipal, Karnataka, India

Stuti Srivastava
Faculty of Biosciences
Institute of Biosciences & Technology, Shri Ramswaroop Memorial University
Lucknow-Deva Road, Barabanki, Uttar Pradesh, India

Mayavan Subramani
Plant Molecular Genetics and Epigenomics
Delaware State University
Dover, Delaware, USA

Alagarsamy Sumithra
Department of Computer Science and Engineering
SNS College of Technology
Coimbatore, Tamil Nadu, India

Muthukumar Bala Sundar
Department of Bioinformatics
Bishop Heber College (Autonomous)
Tiruchirappalli, Tamil Nadu, India

S. Tharani
GRG BIRAC EYUVA Centre
PSGR Krishnammal College for Women
Coimbatore, Tamil Nadu, India

Gangadharan Thilagavathi
Department of Information Technology
Coimbatore Institute of Technology
Coimbatore, Tamil Nadu, India

R. Thirugnanasampandan
Department of Botany, Assistant Professor of Botany
Kongunadu Arts & Science College
Coimbatore, Tamil Nadu, India

Sachin Ashok Thorat
Department of Plant Sciences
Manipal School of Life Sciences, Manipal Academy of Higher Education
Manipal, Karnataka, India

Gonzalo Tortella
Centro de Excelencia en Investigación Biotecnológica Aplicada al Medio Ambiente (CIBAMA), Departamento de Ingeniería Química
Universidad de La Frontera
Temuco, Chile

Srivatsa Udupa
Department of Plant Sciences
Manipal School of Life Sciences, Manipal Academy of Higher Education
Manipal, Karnataka, India

Shanmugam Velayuthaprabhu
Department of Biotechnology
Bharathiar University
Coimbatore, Tamil Nadu, India

Sunil Kumar Verma
Faculty of Biotechnology
Institute of Biosciences & Technology, Shri Ramswaroop Memorial University
Barabanki, Uttar Pradesh, India

Nandakumar Vidya
Department of Biotechnology
Bharathiar University
Coimbatore, Tamil Nadu, India

P.T. Vismaya
Department of Plant Sciences
Manipal School of Life Sciences, Manipal Academy of Higher Education
Manipal, Karnataka, India

P. Servin Wesley
Department of Biotechnology
SAFI Institute of Advanced Study
Malappuram, Kerala, India

Preface

Plants synthesize a wide range of secondary metabolites, such as alkaloids, flavonoids, terpenoids, and so on, to support their reproduction, competition, and defense mechanisms, and many of them have been well-recognized to act as bioactive compounds for offering health benefits and mitigating human diseases. These plants were defined as 'medicinal plants'. They always got great attention from scientists to develop technologies for enhancing the production of secondary metabolites/bioactive compounds by advancing in vitro culture/bioreactor systems or field-grown plants. Also, the breeding of medicinal plants has a high demand to develop high-quality and yield future elites.

This book summarizes current technologies, including: (1) integrated and advanced plant biotechnology, such as bioreactors and genetic mutagenesis, nanotechnology, GWAS (Genome-Wide Association Study); (2) integrated omics tools; (3) high-throughput technology; (4) bioinformatics such as artificial intelligence (AI) approaches; (5) molecular-marker and genome editing such as CRISPR (Clustered Regularly Interspaced Short Palindromic Repeats)-assisted precision breeding; (6) plant biotechnology for advancing in vitro systems of the plant cell, tissue, and organ culture, suspension culture, and bioreactor-based culture and obtaining genetically engineered plants for the enhanced and smarter production of secondary metabolites/bioactive compounds in critical medicinal plants including traditional Chinese medicine (TCM) and some important pharmaceutical valuable species.

Among omics tools, this book highlights the impact of pan-genomics combined with metabolomics on medicinal plant research, such as to advance the understanding of how genetic diversity affects the profiles of secondary metabolites and the therapeutic potential of medicinal plants. In recent years, another hot subtopic involves epigenetics, and the use of epigenetic regulation in enhancing secondary metabolite production is included in this book. Also, the applications of epigenetic modulation for advancing herbal medicines, such as TCM, are comprehensively discussed. Speed breeding, combined with high-throughput technologies for medicinal plant research, is presented in this book, including the implications of environmental condition surveillance, light regime optimization, temperature modulation, humidity control, genomic and transcriptomic analyses, and genetic up-regulation in biosynthetic pathways. In the part of advanced genetic and molecular tool-assisted speed breeding for improving secondary metabolite production, ways to conduct gene insertion and overexpression, CRISPR/Cas9-mediated genome editing, transcription factor engineering, and metabolic engineering are presented.

Selected modern plant biotechnological approaches for medicinal plants have been included in this book to present their fundamentals and applications, including cell and tissue culture, gene mapping, genomic selection, marker-assisted selection, metabolic engineering, phytochemical analysis, and biopharming. Additionally, bioinformatics including artificial intelligence, together with systems biology/network pharmacology and synthetic biology are promising tools in medical plant research.

Selected anti-cancer compounds are introduced and comprehensive strategies for enhanced production of anti-cancer compounds are provided in this book. Methodologies are chiefly based on optimizing in vitro culture systems and applying metabolic engineering. In addition, omics tools such as genomics, transcriptomics, proteomics, and metabolomics, can lead to the elucidation of metabolic pathways, and then move forward to genetic engineering and targeted breeding. In recent years, genome editing technologies have been applied to medical plant research; among them is the CRISPR/Cas9-mediated system, presented in this book for the application in promoting anti-cancer compound production using strategies such as CRISPR/Cas9 mediated gene knockout, multiplex pYLCRISPR/Cas9 system-based gene knockout, site-specific mutagenesis leading to overexpression, triple mutation leading to overexpression, etc.

The readers can efficiently have an overview knowledge of modern medical plant research from this book, and hopefully, can inspire their ideas to design future research engaging in enhancing plant secondary metabolite production and therapeutic applications.

The book editor, Dr. Jen-Tsung Chen, would like to thank all the contributors for their innovative chapters and also appreciates the staff of the publisher, CRC Press, Taylor & Francis Group, for their valuable instructions.

1 Breeding Strategies for Stress-Tolerant and Climate-Smart Medicinal Plants

Rachid Ezzougari, Fatima-Ezzahra Saouab, Salah-Eddine Laasli, Khadija Benamar, Mohammed Khadiri, Abdelaaziz Farhaoui, Ghizlane Echchgadda, and Rachid Lahlali

1.1 INTRODUCTION

In 1992, the scientific community issued its initial warning to humanity, emphasizing the perilous consequences of unsustainable practices across various ecosystems. This warning underscored the alarming rates at which human activities were causing damage to the atmosphere, topsoil, forests, freshwater, and ocean reserves, as well as overall biodiversity (Kendall, 1992). The subsequent Second Warning to Humanity provided updated evidence of the ongoing unsustainable depletion of crucial environmental resources essential for human well-being (Ripple et al., 2017). Applequist et al. (2020), among others, highlighted the adverse impacts of climate change on medicinal plants and their users, including diminished productivity, dwindling availability, and the potential extinction of species. Importantly, climate change may also result in the displacement of human populations from their traditional habitats, rendering them devoid of access to medicinal plants as they become climate refugees. Additionally, climate change poses a significant concern regarding the alteration of phytochemical content in surviving plant populations, impacting their pharmaceutical properties.

Climate change is anticipated to modify environmental conditions in various regions, rendering them unsuitable or inhospitable for certain species currently inhabiting them. Projections suggest that the viable habitat range for numerous species, including medicinal plants, will either shrink or shift significantly due to anticipated climate shifts (Shafer et al., 2001; Zhao et al., 2018), although some species may experience range expansions. Already, there is observable rapid migration of many organisms toward higher latitudes or elevations (Lamprecht et al., 2018; Lenoir & Svenning, 2013), intensifying competitive pressures on resident species within these areas. Fragmentation of habitats escalates the risk of species being unable to migrate and facing extinction. Species dwelling at higher elevations are particularly vulnerable to extinction, as they may be displaced by lower-elevation species expanding their ranges upward, leaving them with limited options for survival (Salick et al., 2014). For instance, species inhabiting montane ecosystems, particularly nival or subnival species, face heightened threats of habitat loss, with future climate projections indicating severe impacts in northern latitude mountainous regions (Lamprecht et al., 2018; Nogués-Bravo et al., 2007). Alpine meadows, among the plant communities most vulnerable to environmental changes, are diminishing due to the upward encroachment of shrubs influenced by warming temperatures (Brandt et al., 2013). Additionally, medicinal plants in arid zones face heightened risks. Deserts and arid shrublands are expected to experience some of the swiftest rates of climate change, hindering the ability of plants to migrate to more suitable habitats (Loarie et al., 2009). A recent study revealed that nearly 600 plant species have become extinct in the last 250 years (Humphreys et al., 2019). Even if a species manages to survive by spreading into new habitats, the loss of medicinally or economically valuable plants from locally accessible areas would still adversely affect human populations (Khanum et al., 2013). The combination of overharvesting for global markets and climate change poses a significant threat. For example, in North America, the extinction risk for American ginseng was estimated to be 8% over 70 years due to harvesting alone, 6% due to climate change alone, but a staggering 65% when both factors were considered together (Souther & McGraw, 2014). Other contributing factors to past population declines include habitat loss from agricultural clearing and inadequate fire management. Consequently, the interplay of climate change, harvesting, habitat loss, logging, grazing, and fire exacerbates the detrimental impacts on vulnerable species in seasonally dry African woodlands.

Abiotic stress is anticipated to rise in numerous crop-growing regions worldwide. While heat and drought stress have garnered the most attention, other forms of abiotic stress that may become more prevalent include flooding, salinity, ozone exposure, and unseasonably cold temperatures (Zandalinas et al., 2022). Despite the significant economic and social benefits provided by medicinal plants, their productivity is globally challenged by such abiotic stresses. The unique value of medicinal plants lies in their active compounds. However, these abiotic stresses can alter the morphology, physiology, and metabolism of medicinal plants, in response to these challenges, medicinal plants employ various protective mechanisms, including the preservation of cell wall architecture, accumulation of osmolytes, production of antioxidants, initiation of stomatal closure, and enhancement of secondary metabolite levels (Nazari et al., 2023). As well as certain genetic changes or modifications may aid in mitigating or adapting to changing environmental conditions (Gupta

DOI: 10.1201/9781003475491-1

et al., 2019). The most efficient and expeditious approach to bridging the gap between the supply and demand of medicinal plants is through large-scale cultivation. Achieving such a scale requires concerted efforts in breeding studies to advance medicinal plant cultivation. Given the prevailing trends of yield, burgeoning population growth, and the adverse impacts of climate change, prioritizing traits such as tolerance to abiotic stress, robust disease resistance, and improved efficiency in nutrient and water utilization is imperative (Mackill et al., 1999; Slafer et al., 2005). Despite advancements in breeding some medicinal plants, numerous challenges persist, stemming from the specificity and complexity involved in determining breeding targets. The prevalence of biotic and abiotic stresses, along with climate change, population growth, and the depletion of water for agricultural purposes, poses significant hurdles for breeders striving to develop enhanced crop varieties with greater yield potential (Lateef, 2015). To overcome these obstacles, there is an urgent need to embrace new technologies. Molecular breeding approaches, including marker-assisted selection (MAS), genomic selection (GS or GWS), gene editing (CRISPR-Cas9), and speed breeding, offer a platform for expediting progress in selecting stress-tolerant crop plants while enhancing the precision of selection (Devi et al., 2017). In this chapter, we summarize modern breeding strategies used in medicinal plants to improve their resistance to environmental stress as well as their productivity in terms of quality and quantity.

1.2 UNDERSTANDING CLIMATIC STRESSES AND CHALLENGES FOR MEDICINAL PLANTS

Medicinal plants, as living organisms, are subject to a diverse range of stresses, both biotic and abiotic, which can have significant implications for their health and productivity. Biotic stresses arise from interactions with living organisms such as pathogens, including bacteria, fungi, nematodes, and insects. These pathogens can trigger infectious diseases, rot, or infestations, thereby weakening the plant and reducing its ability to synthesize beneficial medicinal compounds (Gull et al., 2019). On the other hand, abiotic stresses are non-biological factors that negatively affect medicinal plants. These factors include soil salinity, heavy metal presence, nutrient deficiencies, UV radiation exposure, atmospheric pollution, extreme temperatures, flooding, drought, etc. (Rahman et al., 2023) (Figure 1.1).

For instance, increased soil salinity can disrupt the osmotic balance of plants, while heavy metal presence can inhibit root growth and hinder the absorption of essential nutrients. However, among all these stresses, climate change represents a major challenge for medicinal plants. Extreme weather events, such as high temperatures, prolonged droughts, and flooding episodes, can have a devastating impact on medicinal plant crops. For instance, studies have shown that rainstorms can damage crops such as cumin (*Cuminum cyminum*) and wild mint (*Mentha arvensis*), while periods of drought can lead to a complete absence of yield for plants like fennel (*Foeniculum vulgare*) in Bulgaria (Das et al., 2016). The adverse effects of climate change on medicinal plants can result from both direct physical damage, such as that caused by flooding, and its impact on plant metabolism. The latter can lead to a decrease in plant yield and an alteration of their medicinal properties. For example, abiotic stress related to climate change can modify plant growth and physiology, reduce biomass production, and influence the production of secondary metabolites and active phytochemical compounds (Patni et al., 2022).

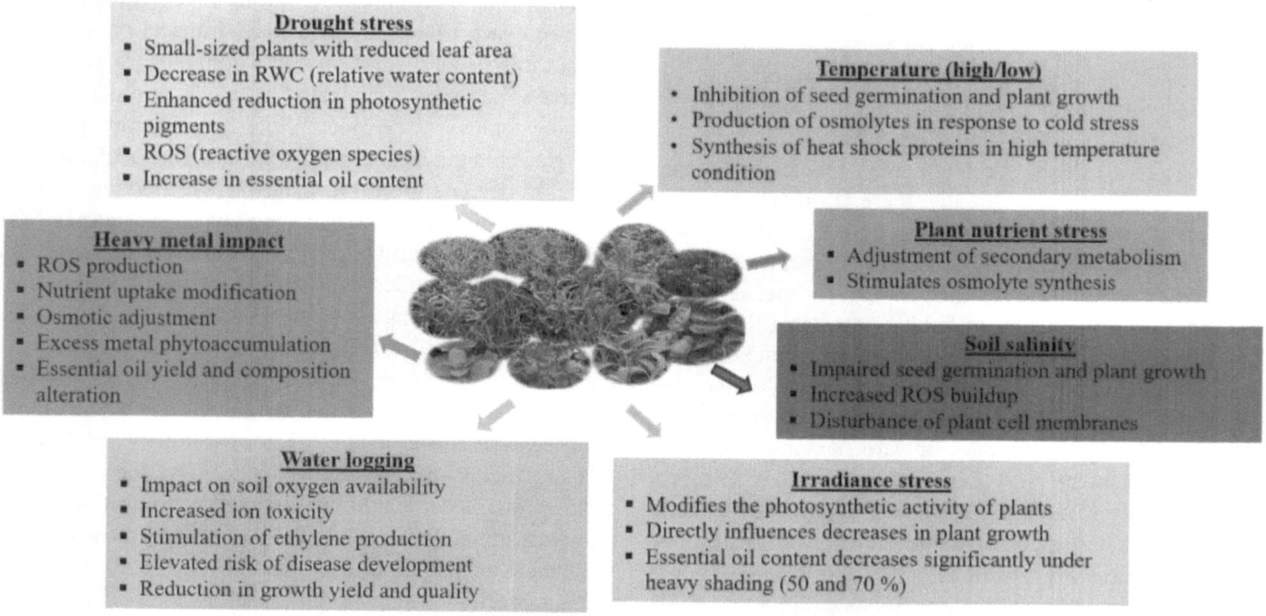

FIGURE 1.1 Abiotic stressors affecting medicinal plants.

Medicinal plants possess specific requirements that enable them to adapt to fluctuations in temperature and climatic stress. Therefore, creating a suitable environment with favorable soil texture, adequate water availability, and appropriate salinity levels is essential for plants to withstand the adverse conditions associated with climate change (Pourmeidani et al., 2020). For instance, research conducted across different climatic zones has revealed that the medicinal plant coleus (*Plectranthus barbatus*) thrives in soils with varying characteristics, including differences in clay content, organic carbon, nitrogen, phosphorus, potassium, and other micronutrient availability (Balasubramanian et al., 2020). Moreover, since climate changes can lead to an increase in soil salinity, plants may require support from organic amendments such as waste compost and inorganic amendments like phosphogypsum. These amendments serve to enhance soil density, and hydraulic conductivity, and adjust soil pH in areas affected by salinity resulting from climatic changes. Furthermore, in regions experiencing stress induced by drought in arid and semiarid conditions, plants may need irrigation to mitigate the adverse effects of water deficiency. Conversely, in cases of water excess, proper drainage is essential to remove excessive water accumulation resulting from high precipitation events (Mukhopadhyay et al., 2021). By addressing these specific needs and employing appropriate soil management techniques, medicinal plants can better adapt to the challenges posed by climate change. This proactive approach can help optimize their growth, productivity, and resilience in the face of changing environmental conditions.

1.3 CLIMATE-SMART TRAITS

Now that we have a comprehensive understanding of the importance of developing climate-smart traits in medicinal plants, let's delve into the specific strategies employed by these plants to effectively cope with various environmental stresses. These strategies encompass a wide range of adaptive responses to challenges, such as drought, cold, salinity, fluctuations in light intensity, and elevated levels of atmospheric carbon dioxide. By gaining insight into how medicinal plants navigate and respond to these stressors, researchers can pinpoint key pathways for enhancing their resilience and productivity in the ever-changing landscape of climate change.

1.3.1 Drought Stress Tolerance

Under conditions of water stress, plant growth diminishes while the content of secondary metabolites tends to increase. This rise in secondary metabolites is attributed to a decrease in biomass production, rather than an escalation in synthesis rates. Consequently, the overall concentration of metabolites increases relative to either fresh or dry weight (Selmar & Kleinwächter, 2013). Nowak et al. (2010) observed that *S. officinalis*, traditionally used for treating digestive and stomach ailments, experiences heightened levels of monoterpenes when subjected to drought. Similarly, Jaafar et al. (2012) noted that *Labisia pumila*, utilized for cardiovascular protection and addressing osteoporosis, exhibits elevated concentrations and enhanced production of total phenolics and flavonoids during drought stress.

Plants possess numerous adaptive strategies to endure periods of drought or water scarcity. It is crucial to comprehend these strategies in order to cultivate crop varieties that can thrive under such challenging conditions. In a study by Quéro et al. (2015), the mechanisms behind the induction of drought tolerance in plants through b aminobutyric acid (BABA) were investigated. Analysis of metabolomic and ionomic profiles in flax leaves revealed that BABA triggers a restructuring of solute composition, resulting in enhanced accumulation of nonstructural carbohydrates and proline, while reducing levels of inorganic solutes. Consequently, BABA treatment was shown to instigate changes that bolstered the resilience of flax plants against drought stress.

1.3.2 Cold Stress Tolerance

Low temperatures alter the fluidity of cell membranes by impacting their concentration (Sevillano et al., 2009; Upchurch, 2008). Furthermore, cold temperatures prompt the production of free oxygen radicals, elevating stress levels within cells, and necessitating the activation of antioxidants by plant cells to mitigate this effect (Ruelland et al., 2009). Extensive research has delved into the adverse impacts of cold on medicinal plants across various species such as *Withania somnifera* (Mir et al., 2015), *A. thaliana* (Mishra et al., 2013), *Salvia sclarea* (T. Kaur et al., 2015), *Teucrium polium* (Lianopoulou, Patakas et al., 2014), and *Thymus sibthorpii* (Lianopoulou, Bosabalidis et al., 2014). Indeed, to endure cold stress, plants adapt their survival strategies by redistributing resources and reducing growth (Eremina et al., 2016). The dimorphism observed in these plants is adapted by various defensive mechanisms triggered by different hormones (Gupta et al., 2019).

1.3.3 Salinity Stress Tolerance

Salinity poses a significant constraint to rice cultivation in the context of climate change. Hokmalipour (2015) assessed varying salinity levels on the germination and seed vigor of three medicinal plants: chicory (*Cichorium intybus* L.), cumin (*Cuminum cyminum* L.), and fennel (*Foeniculum vulgare*). The findings revealed a reduction in seedling length, aerial part length, root length, germination percentage, germination rate, as well as seedling dry and fresh mass due to salinity. Similarly, Saberali and Moradi (2019) investigated the effects of NaCl concentrations on seeds of fenugreek (*Trigonella foenumgraecum* L.), dragonhead (*Dracocephalum moldavica* L.), savory (*Satureja hortensis* L.), and dill (*Anethum graveolens* L.), all common medicinal species in the Middle East, observing a consistent decrease in germination and seedling growth with increasing NaCl concentration across all species. Moreover, Rahimi-Dehgolan et al. (2012) found that aloe vera exhibited inhibited plant growth, altered morphological characteristics, and reduced biomass when subjected to different salinity levels. Yu et al. (2014) analyzed the digital gene expression of flax plants under alkaline stress, alkaline–salt stress, and neutral salt stress, noting significant impacts on photosynthesis and response to biotic stimulus under alkaline–salt stress, and carbohydrate metabolism under neutral salt stress. They identified key regulators involved in abiotic stress responses such as mitogen-activated protein kinase kinase (MAPKK), WRKY, abscisic acid (ABA), and ion channels, which were differentially expressed. Notably, alkaline–salt stress induced a higher number of differentially expressed genes compared to neutral salt stress and alkaline stress, indicating the involvement of a broader set of genes in regulating the alkaline–salt stress pathway.

The impact of NaCl salinity was investigated in *Cymbopogon nardus*. The plants were subjected to varying concentrations of NaCl (25, 50, 100, 200, and 300 mM). At higher salt levels, a reduction in chl a, chl b, and carotenoids was noted. Conversely, the concentration of polyphenols increased with rising salt concentrations, peaking at 300 mM. These findings imply that heightened polyphenol levels under increased salinity contribute to the accumulation of secondary metabolites in *C. nardus*, aiding its tolerance to high salt stress (Mane et al., 2010). In the case of the salt-tolerant plant *Artemisia annua*, there was a decline in total chlorophyll content, diminished activity of carbonic anhydrase and nitrate reductase, elevated proline content, and increased antioxidant enzyme activity, along with enhanced artemisinin content at specific salt concentrations (Aftab & Hakeem, 2021). Similarly, *Matricaria chamomilla* demonstrated an increase in proline content and concentrations of $Na+$ and Fe as part of its tolerance mechanisms (Heidari & Sarani, 2012).

1.3.4 Light Intensity Tolerance

Zhao et al. (2023) suggest that light intensity plays a regulatory role in promoting the accumulation of flavonoids, enabling the alpine herbal plant *Sinopodophyllum hexandrum* to adjust to the heightened altitudes associated with intense light. Furthermore, it has been observed that increased light intensities are linked to higher levels of flavonol, flavonoid, and anthocyanin compounds, along with elevated ratios of anthocyanin to total flavonoid and anthocyanin to total flavonol.

1.3.5 Carbon Sequestration

Anthropogenic activities have led to a significant increase in CO2emissions over the past four decades since 1750 (Field et al., 2014). Medicinal plants possess an inherent ability to adapt to changing environmental conditions. Their metabolic pathways exhibit flexibility due to secondary metabolites, although this adaptability may potentially impact the production of metabolites crucial for their medicinal properties (Mishra, 2016). *Ginkgo biloba*, utilized for treating Alzheimer's disease, vascular or mixed dementia, demonstrates altered terpenoid content when exposed to elevated CO2 levels, showing a 15% increase in quercetin aglycon and a 10% decrease in kaempferol aglycon, along with changes in isorhamnetin and bilobalide levels to some extent (Weinmann et al., 2010). Similarly, *Catharanthus roseus*, known for its anticancer, antiviral, and diuretic properties, experiences reductions in various plant secondary metabolites (flavonoids, phenolics, alkaloids, and tannins) when subjected to high CO2 levels. *Pseudotsuga menziesii*, utilized in treating kidney and bladder problems, as well as venereal diseases (Rahimtoola, 2004), demonstrates a significant decrease in terpenes, particularly monoterpenes, under high-CO2 conditions. The rise in atmospheric CO2 concentration contributes significantly to global warming. Plant systems offer efficient solutions, including carbon sequestration in biomass and soil, which has been proposed as a potential remedy (Smith, 2008). Van Noorden (2010) investigated the carbon sequestration potential of vetiver (*Vetiveria zizanioides* L.) in field plots in Bangalore, India. Vetiver is a perennial, economically viable crop thriving in tropical and subtropical regions, possessing medicinal and aromatic properties. He estimated that vetiver cultivation could sequester 150 Tg per year in India, nearly 46% of the country's carbon emissions. Although land-based carbon capture and storage (CCS), involving the capture and underground storage of CO2 in rock formations, have been proposed, their future remains debatable due to high costs.

1.3.6 Rapid Growth and Regeneration

Abiotic stressors like drought, salinity, high light exposure, extreme temperatures, and others can diminish the quality and yield of medicinal plants by disrupting their biochemical, metabolic, and physiological functions (Dobrikova, 2023). Early crop maturation represents a crucial agronomic characteristic that enables plants to withstand various abiotic and biotic stresses (Shivaraj et al., 2018; Tyagi et al., 2018). For instance, the typical life cycle of a flax plant spans 90–150 days, with approximately 45–60 days devoted to the vegetative phase, followed by a flowering period lasting 15–25 days, and concluding with a maturation period lasting about 30–40 days (Muir & Westcott, 2003).

1.4 BREEDING APPROACHES FOR STRESS-RESILIENT AND CLIMATE-ADAPTIVE MEDICINAL PLANTS

Particular attention has been given to improving the stress resistance and tolerance capacity of medicinal plants, as well as their suitability for phytoremediation, through a variety of means and techniques. Unlike commercial crops, most medicinal plants originate from wild environments. However, given the increasing demand for their products, their cultivation and selection have become essential to ensure high-quality, profitable, and sustainable production.

As a result, several medicinal plants have been the subject of research efforts aimed at understanding their genomic composition. Reports on these genomes are available, For example, ginseng (*Panax ginseng*), feathered poppies (*Maclaeya cordata*), and garlic (*Allium sativum*) (Kim et al., 2018; Liu et al., 2017; Sun et al., 2020) are ready to be explored in future genetic studies and selection processes. However, these technological advancements are still underutilized, especially for plants that have only therapeutic value. Traditional approaches for cultivating stress-tolerant medicinal plants may still prevail.

Different traditional approaches have been adopted in order to cultivate medicinal plants resistant to climate change and stress. These approaches are mainly based on the use of conventional breeding techniques and some suitable agricultural practices to overcome the bad effects of climate change on medicinal plants. These techniques aim to obtain not only medicinal plants resilient to climate change, associated with good growth, yield, quality, resistance to pests, and high content in secondary metabolites and bioactive compounds.

1.4.1 Conventional Breeding

Conventional breeding consists of the development of novel plant varieties through the use of traditional processes that rely on target features pre-existing in the genetic material of the species without the introduction of new genes (Sisea, 2024). Indeed, traditional breeding depends on the experience of breeders and is based on the visible plant phenotypes which could be influenced by several factors, such as the genotype, plant age, environment, and field (Wang et al., 2020). Selective breeding and hybrid breeding represent the basic conventional breeding techniques. Indeed, selective breeding refers to the selection from mixed populations, of certain varieties which represent specific desired traits, whereas hybridization indicates the crossing of parents for the creation of new cultivars and it involves the application of controlled pollination (Acquaah, 2015; Wang et al., 2020). These conventional breeding plant techniques present some advantages as well as some limitations. Thus, selective breeding is characterized by simplicity and doesn't need sophisticated equipment; however, the improvement of the desired traits depends on the natural variation of the selected population with a low probability of the occurrence of favorable mutation. As regard hybrid breeding, it represents the advantage of heterosis or hybrid vigor, but it is still in the primary stage since several medicinal plants are heterozygous (Wang et al., 2021). Furthermore, these conventional techniques have been successfully applied to achieve different goals including the obtention of medicinal plants resistant to stressful climate conditions. Thus, for instance, Gholizadeh et al. (2018) has used selective breeding for the selection of drought-tolerant genotypes of coriander (*Coriandrum sativum* L.). The resilience to drought stress of *Coriandrum sativum* genotypes was associated with their high production of essential oils. Moreover, Mirmiran et al. (2021) have also used this conventional technique for the selection of fenugreek (*Trigonella foenum graecum*) landraces resistant to freezing conditions and characterized by high weight and yield. Similarly, Zarghami Moghadam et al. (2021) selected from 13 *Calendula officinalis* cultivars, the cultivars that were tolerant to drought stress, and contained high flower number and proline levels.

1.4.2 Agricultural Practices

Several agricultural practices have been used in order to allow the adaptation of medicinal plants to climate change-related stress. These methods include management of land use, irrigation, and drainage techniques. The management of land use relies on the change of land use according to environmental conditions (Zulkarnain et al., 2015). Thus, under drought conditions, medicinal plants tolerant to high temperatures and needing a low amount of water could be sown instead of medicinal plants sensitive to such conditions. The same approach could be adopted for medicinal plants tolerant to low temperatures in the case of the intervention of freezing conditions. Besides, in drought conditions, irrigation systems have been used in order to irrigate including sprinkler irrigation, surface irrigation, and micro-irrigation (Eisenhauer et al., 2021). These systems provide plants with appropriate amounts of water under drought conditions, thus, vue their importance, use, and

TABLE 1.1
Advantages and Limitations of Conventional Plant Breeding Techniques

Conventional methods	Characteristics	Advantages	Limitations
Selective breeding	Direct use of the existing plant resources	- Simplicity - Doesn't need sophisticated equipment	- The improvement of the desired traits depends on the natural variation of the selected population. - Low probability of the occurrence of favorable mutation
Hybrid breeding	Forming different varieties based on gene recombination of parents.	- Hybrid vigor - Obtention of rich materials for plant breeding	- Hybrid breeding is still in the primary stage since several medicinal plants are heterozygous

development continue to be of great importance to researchers. (Rodríguez-Ortega et al., 2018) have demonstrated the efficiency of smart irrigation systems in the management of *Solanum lycopersicum* growth under different temperature regimes. Furthermore, drought conditions have been revealed to be overcome also by using other agricultural practices. For instance, the exogenous application of phytohormones, minerals nutrients, and organic manures, in addition to seed priming was found to be effective in the management of drought stress in sunflowers (Swain et al., 2023). Regarding flooding conditions, agricultural drainage allows removing the excess of water and therefore provides plants with good conditions for their survival. Drainage systems include surface and subsurface tile drains (Sheler, 2013). The surface drainage permits the prevention of water ponding surface formation and the control of runoff without inducing erosion. As for subsurface drainage, it consists of the elimination of water from the soil profile (Skaggs & Van Schilfgaarde, 2015). In addition to artificial drainage systems, bio-drainage has also been successful in the management of water excess in flooding conditions. Indeed, this technique relies on the use of tree species to reduce water excess through transpiration (Mukhopadhyay et al., 2021). In this context, it was found that *Eucalyptus* species significantly lowered water levels in waterlogged soils (Dagar et al., 2016). Other plant species, including populus (*Populus deltoides*), leptochloa (*Leptochloa fusca*), and panicum (*Panicum maximum*), were recommended for this purpose due to their high water demand (Dagar, 2014).

1.4.3 Implications and Effective Ways for the Culture and Preservation of Medicinal Plants Adapted to Stress Induced by Climate Changes

Medicinal plants are threatened by climate change. Knowledge of the impact of these conditions on plants and their secondary metabolites is therefore extremely important. It is essential to step up studies into the conservation of medicinal plants under conditions of climate change and stress, as these plants represent an important source of bioactive and neutraceutical compounds (Sharma et al., 2020). Modern cultivation techniques introducing new selection methods enable us to obtain smart medicinal plants that are resistant and tolerant to climatic stress. However, the use of these techniques must be controlled by regulatory and ethical barriers. Furthermore, *in-situ* and *ex-situ* conservation of medicinal plants represents also a solid and effective way for the preservation of medicinal plants under climate stress. Indeed, *in-situ* conservation can be achieved by extending the existence of medicinal plants in suitable areas known as refugia. These areas are adapted to a wide range of environmental conditions and allow the preservation and protection of the growth and life of the plants despite the presence of unsettled weather conditions. As for the *ex-situ* conservation of medicinal plants, it can be realized by storing seeds of plant species at high risk of extinction in order to ensure their maintenance. Thus, seed banks constitute an important tool that can be used thereafter for the cultivation of threatened plants in appropriate areas (Christmas et al., 2016).

Another approach based on the use of beneficial microorganisms and plant growth-promoting rhizobacteria was reported to be a good path for enhancing stress tolerance in some medicinal plants such as basil (*Ocimum basilicum*), thyme (*Thymus vulgaris*), and summer savory (*Satureja hortensis*). Thus, performing more studies in this context will be of great interest to the protection of medicinal plants against climate change and stress (Avasiloaiei et al., 2023). Furthermore, it's important to point out the importance of the application of different *in vitro* biotechnological techniques for medicinal plant breeding. These techniques have been successfully and widely used for edible plant breeding. However, their wide application for medicinal plants is still uncommon. Among these techniques, flowering has been revealed to help the development of of breeding strategies aiming at inducing either early or late flowering cultivars of Rosaseae species such as apricot, apple, peach, and cherry, permitting therefore the control of their phenological adaptation to climate change (Zulkarnain et al., 2015).

1.5 BREEDING STRATEGIES FOR STRESS-TOLERANT PLANTS

1.5.1 Speed Breeding and Genomic Selection

The changing climate presents a significant threat to medicinal plant cultivation. Abiotic stresses such as drought, heat, and salinity are becoming more frequent and severe, impacting plant growth, yield, and the production of valuable secondary metabolites (Pandey et al., 2019). Traditional breeding programs for medicinal plants are often slow and resource-intensive, relying on field trials spanning multiple seasons. To address this challenge, advancements in breeding techniques like speed breeding and genomic selection offer promising solutions to develop stress-tolerant and climate-smart medicinal plant varieties (Cvejić et al., 2022).

Speed breeding utilizes innovative techniques to dramatically accelerate the plant breeding cycle. One approach involves manipulating photoperiod (day length) to induce multiple generations of plants per year within controlled environments (Gudi et al., 2022). This can be achieved by extending the light period or utilizing specialized chambers with precisely controlled day/night cycles. Cazzola et al. (2021) have shown that speed breeding can reduce breeding cycles by up to 60%, significantly shortening the time required to develop new varieties. Genomic selection (GS) represents another powerful tool in the medicinal plant breeder's arsenal (Collard & Mackill, 2008). GS leverages advances in DNA sequencing technology to identify genetic markers associated with desirable traits, such as stress tolerance and metabolite production (Voss-Fels et al., 2019). By analyzing these markers in young seedlings, breeders can predict the performance of mature plants without the need for lengthy field trials. This allows for the early selection of promising lines and the efficient elimination of undesirable ones, leading to faster breeding cycles and improved selection accuracy (Collard & Mackill, 2008). The integration of speed breeding and GS offers a powerful synergy for medicinal plant breeding. Speed breeding accelerates the generation of new populations, while GS facilitates the rapid identification of superior genotypes within these populations (Gudi et al., 2022). This combined approach can significantly accelerate the development of stress-tolerant medicinal plant varieties with superior metabolite profiles (Razzaq et al., 2021). For instance, Dubey et al. (2019) employed speed breeding and GS to develop drought-tolerant soybean varieties. They were able to identify and select drought-tolerant lines within a single year, a process that would traditionally take several seasons. This demonstrates the immense potential of these breeding techniques for medicinal plants, where stress tolerance is crucial for ensuring sustainable cultivation and production of valuable medicinal compounds (Wang et al., 2020).

1.5.2 Marker-Assisted Breeding

Molecular markers delineate genetic distinctions among individual organisms or species (Muhamba et al., 2013). Marker-assisted selection (MAS) involves employing morphological, biochemical, or DNA markers as indirect selection criteria for the cultivation of agriculturally significant traits in crop breeding (Ashraf et al., 2012). Typically, these markers do not directly signify the target genes themselves; rather, they serve as indicators, aiding in the integration of genes linked with economically vital traits (Collard et al., 2005). Nonetheless, these markers themselves do not influence the phenotype of the trait of interest; rather, they are situated in proximity to or are linked with genes regulating the desired traits (O'Boyle et al., 2007) (Figure 1.2). They are DNA sequences located at specific genomic sites and inherited according to standard laws of heredity from one generation to the next, the advent of molecular markers constitutes a significant advancement in overcoming a principal constraint of classical crop improvement programs, where selection is based on plant phenotypes without precise knowledge of the selected genes, often resulting in linkage drag (Ruane & Sonnino, 2007; Visscher et al., 2021). While the application of MAS for manipulating simple/qualitative traits is straightforward and extensively documented, its application for enhancing complex/polygenic traits, including plant tolerance/resistance to abiotic stresses, presents greater complexity (Ashraf et al., 2012).

Isozymes represent commonly utilized protein markers in plant breeding, these biochemical markers operate on the premise that allelic forms of enzyme variation exist among various proteins. Among the diverse strategies employed to mitigate the adverse effects of high soil or water salinity on plant growth and crop productivity, genetic approaches aimed at developing plants with enhanced salt tolerance are being extensively pursued worldwide (Ashraf et al., 2012).

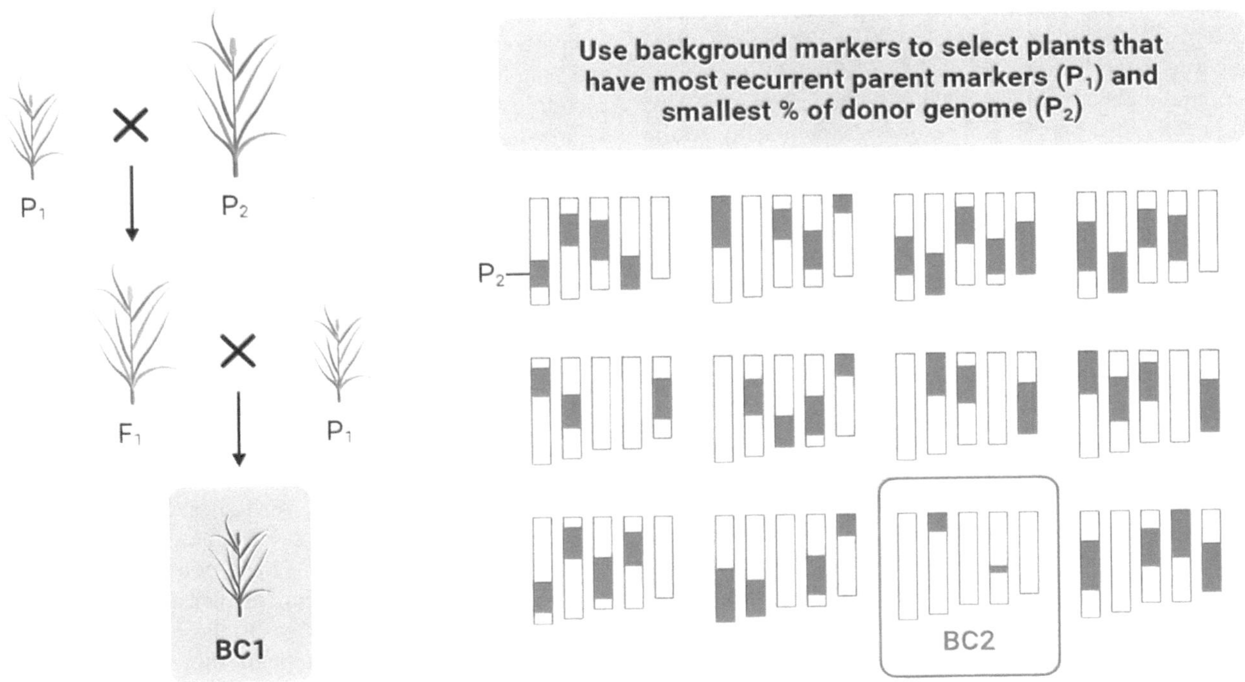

FIGURE 1.2 Marker-assisted breeding for medicinal plants.

Biochemical markers for pathogen resistance have been employed to select plants tolerant to salinity stress (Kumar et al., 2014). In stress-tolerant plants, various biochemical markers such as osmolytes, soluble sugars and proteins, polyamines, amides, quaternary ammonium compounds, amino acids, polyols, antioxidants, and ATPase levels are utilized (Ashraf & Harris, 2004). Yamada et al. (2002) highlighted the significance of an essential protein, allene oxide cyclase, in conferring salt stress tolerance in the mangrove *Bruguiera sexangular*. Ashraf and Harris (2004) demonstrated that the expression of the same protein in commercial yeast and tobacco cell lines enhances tolerance to higher salt concentrations. Proline accumulation has been utilized as a biochemical marker in salt-stressed plants, where it accumulates in higher concentrations (Abrahám et al., 2003). A population of introgression lines (ILs) was created by crossing the high-yielding cultivar 'Cheniere' with the salt-tolerant donor 'TCCP'. This IL population underwent a thorough evaluation to pinpoint quantitative trait loci (QTLs) linked to salt tolerance across various growth stages: germination, seedling, and reproductive phases. Among the identified candidate genes associated with salt tolerance were OsSDIR1 and SERF during the seedling stage, WRKY55 and OsUBC during the reproductive stage, and MYB family transcription factors throughout all three stages. Through gene ontology analysis, significant terms emerged, highlighting functions such as nucleotide binding, protein binding, protein kinase activity, antiporter activity, active transmembrane transporter activity, calcium-binding protein, and F-box protein interaction domain-containing protein. The salt-tolerant ILs display promise for potential release as varieties or as pre-breeding lines aimed at developing salt-tolerant rice varieties (Chapagain et al., 2024). The dep1 (dense and erect panicle 1) gene serves as a pleiotropic regulator influencing yield, quality, disease resistance, and stress tolerance, all highly valued traits in rice (*Oryza sativa* L.) breeding. The findings demonstrated the utility of both InDel and LAMP markers for precise genotyping, showcasing the efficacy of this approach in screening for the dep1 allele in rice breeding (Tian et al., 2024). *Fragaria* sp. (*Fragaria* x *ananassa*), commonly referred to as wild strawberry, is recognized for its medicinal attributes owing to its diverse health-enhancing properties, such as antioxidant, anti-inflammatory, digestive, and cardiovascular benefits. It stands out as one of the most sought-after and commercially significant berries within the Rosaceae family, boasting abundant reserves of vitamins and various bioactive compounds known for their health benefits. Strawberries are particularly esteemed for their remarkable antioxidant capacity, largely attributed to their high concentrations of vitamin C and polyphenolic compounds, including anthocyanins and ellagitannins (Padmanabhan et al., 2016). Razavi (2024) devised a standardized approach for evaluating drought tolerance in *Fragaria*, aiming to identify a set of linked markers associated with physiological traits involved in drought resistance. This involved a rapid screening method to assess *Fragaria* genotypes' responses to dehydration through a short-term water deficit experiment. By measuring two eco-physiological parameters leaf relative water content (RWC) and leaf water loss rate (WLR) across a total of 20 strawberry cultivars (*Fragaria* × *ananassa* Duch.) sourced from various breeding programs, Razavi observed discernible responses to drought stress. Moreover, DNA analysis, utilizing AFLP and candidate gene EST markers, provided further insights into select *Fragaria* genotypes. The findings revealed correlations among specific DNA markers, leaf WLR, and RWC (Razavi, 2024). On a separate note, cumin (*Cuminum cyminum* L.) is an aromatic herb with recognized medicinal properties, particularly through its essential oil, which finds application in antibacterial activities as well as in the formulation of creams, lotions, and perfumes. The utilization of functional markers expedites the breeding process for drought-tolerant cumin varieties. Archangi et al. (2019) investigated the correlation between plant height and different traits related to seed yield by utilizing complementary DNA-amplified fragment length polymorphism (cDNA-AFLP) markers across 15 cumin accessions exposed to both normal irrigation (NI) and drought stress (DS) conditions. The researchers employed twelve EcoRI-MseI primer combinations to amplify AFLP markers from cDNA fragments extracted from RNA samples. Following this, the identified markers were subjected to analysis to uncover marker-trait associations (MTAs) using both the generalized linear model (GLM) and mixed linear model (MLM). The findings revealed significant variability in biological yield, seed yield, and seed-related traits. These associated markers hold promise for integration into marker-assisted breeding programs aimed at enhancing seed yield, particularly under drought conditions (Archangi et al., 2019). In a separate study, Baidyussen et al. (2021) developed ASQ and Amplifluor markers for HvSAP8 and HvSAP16 genes encoding zinc finger proteins. They evaluated selected Hordeum vulgare plants and parent lines in a soil-based drought screening, revealing differential expression of HvSAP8 and HvSAP16 genes in response to stress. HvSAP8 nearly doubled its expression during early stress stages before returning to pre-stress levels or undergoing significant negative regulation in plants with Granal or Baisheshek genotypes, respectively. Similarly, HvSAP16 exhibited either a return to baseline expression levels or significant negative regulation in plants with Natali or Auksiniai-2 genotypes after high expression under drought conditions. Genotypes Granal and Natali, harboring specific HvSAP8 and HvSAP16 alleles, were associated with improved drought performance, characterized by higher grain weight and more shoots per plant (Baidyussen et al., 2021).

The study of Mollier et al. (2023) compared the characteristics of rice landraces from Northeastern India capable of withstanding climatic changes to those of a control variety. They assessed phenological, morpho-physiological, and yield-related traits under high temperatures (HTs) and elevated CO2 (eCO2) levels using molecular markers. Through SCoT-marker-assisted genotyping, they identified

77 alleles, 55 of which showed polymorphism, with polymorphism information content (PIC) values ranging from 0.22 to 0.67. The research highlighted genetic diversity among the rice lines and confirmed a close relationship between two high-performing landraces, Kohima Special and Lisem. These landraces could serve as valuable resources for future rice breeding efforts aimed at enhancing stress tolerance (Mollier et al., 2023). In the realm of plant breeding, the prioritization of cultivating pathogen-resistant crops holds significant weight. When plants are attacked by pathogens, it triggers a stress response that activates the plant's defense mechanisms, resulting in the accumulation of particular proteins that can serve as biochemical indicators. The western conifer seed bug (*Leptoglossus occidentalis* Heidemann) poses a threat to Douglas fir seeds *Pseudotsuga menziesii* [Mirbel] Franco. Utilizing antibody-based assays, certain proteins such as crystalloid storage reserve proteins, specific antibodies, and components of salivary proteins emerge as potential markers for identifying damage inflicted by insect pests (Lait et al., 2001). Sesame (*Sesamum indicum* L.) holds significant importance as an oilseed crop, serving various purposes including food, feed, medicinal, and industrial applications. Its inherently modest genetic yield potential and susceptibility to both biotic and abiotic stresses contribute to its low productivity, Enhancing the genetic potential of sesame through the development of stress-resistant varieties alongside high yields presents a promising avenue (Weldemichael & Gebremedhn, 2023) outlined that this development involves utilizing molecular marker-assisted selection and subsequently cloning functional genes associated with yield and related traits, as well as conferring tolerance to a range of biotic and abiotic stresses in *Sesamum indicum* L.

1.5.3 Applications of Genome-Editing Tools on Medicinal Plants in Response to Abiotic Stress

Abiotic stresses significantly impact the growth and metabolite production of medicinal plants (Noreen et al., 2022). Improving medicinal plants through modern selection techniques, effective agronomic practices, innovations in microbial applications, and harnessing natural variations in underutilized medicinal plants represents an excellent path forward to meet future medicinal plant needs (Dobrikova, 2023; Rahman et al., 2023). Currently, genome editing is widely more favored than conventionnel breeding and marker-aided selection programs, since its rapid generational advancement (Ahmar et al., 2021). Zinc-finger nucleases, clustered regularly interspaced short palindromic repeats and transcription activator-like effector nucleases, constitute an important tool for genome editing. These techniques allow to inclusion of desirable mutations, insertions/deletions (InDels), and precise genomic modifications through the use of specific nucleases (Bhuyan et al., 2023). While these tools have been extensively utilized in several crops, research on their application to enhance abiotic stress resistance in medicinal plants remains restricted.

1.5.3.1 Zinc Finger Nucleases (ZFNs)

ZFNs consist of zinc finger protein domains, able to bind to specific sequences linked to a nuclease domain and cause therefore DNA cleavage. Each zinc finger protein domain in a tethered array is capable of recognizing 3 base pairs of DNA (Boti et al., 2023). The bonding of target sequences is carried out by FokI nuclease enzymes (Majumdar et al., 2022). The determination of ZFNs sequence specificity is performed by zinc finger proteins (ZFPs) which are formed of tandem arrays of C2H2 zinc fingers (ZFs). These ZFs recognize a 3-base-pair DNA sequence, and 3 to 6 ZFs contribute to the generation of ZFN subunit which binds to a 9 to 18 base pair DNA sequence. Processes of DNA repair are then carried out by the double-strand breaks (DSBs) resulting from ZFN cleavage. An improvement in error-prone NHEJ came through homology-directed repair (HDR), resulting in site-specific integration or gene editing (Katayama et al., 2024; Negi et al., 2023). ZFNs facilitate the precise insertion of genetic material into specific loci in the plant genome. ZFNs have been used to add beneficial genes to enhance stress tolerance (Yadav et al., 2023). Despite its promise, this tool also faces some setbacks. One drawback of ZFN is the poor targeting density (Wang et al., 2021). It has been previously reported that although each ZF recognizes a 3-base-pair DNA sequence, no open-source collection of 64 ZFs covers all possible combinations of the triplet sites (Li et al., 2023).

1.5.3.2 Transcription Activator-Like Endonucleases (TALENs)

Over time, TALENs have constituted an alternative to ZFNs for causing genome editing and targeted double-stranded breaks. Similar to ZFNs, TALENs consist of nonspecific FokI nucleases paired with a customizable DNA-binding domain. This latter is generated from transcription activator-like effectors (TALEs), proteins secreted by *Xanthomonas* spp. bacteria (Teper et al., 2023). TALENs are favored by numerous researchers due to their capability for straightforward and swift modification using a basic protein-DNA code. This facilitates the modular adjustment of the TALE repeat DNA-binding domains to specific bases on the target site (Bhagtaney & Sundarrajan, 2023; Joung & Sander, 2013). The construction of DNA encoding TALE repeat arrays is often considered challenging due to the necessity of assembling multiple nearly identical repeat sequences (Younessi-Hamzekhanlu et al., 2022). Over time, various platforms have been developed to streamline the assembly of plasmids encoding TALE repeat arrays. Typically, these methods fall into three categories: standard restriction enzyme and ligation-based cloning; golden gate cloning; and solid-phase assembly (Huang et al., 2011). The characteristics of TALENs, like the length and sequence composition of the N- and C-terminal TALE-derived sequences, are critical factors in determining the appropriate assembly method (Joung & Sander, 2013).

TALENs can be employed to induce knockout mutations and provide plant resistance by disrupting the target sites of naturally occurring TALEs related to pathogenicity (Khan et al., 2017). On the other hand, TALENs offer the capability to manipulate genes involved in secondary metabolite biosynthesis. Through editing key genes, the production of bioactive compounds, such as alkaloids and flavonoids, has been successfully increased in medicinal plants (Alugoju & Tencomnao, 2023).

1.5.3.3 Clustered Regularly Inter Spaced Short Palindromic Repeats (CRISPR-Cas9)

To overcome limitations associated with the use of ZFNs and TALENs, the CRISPR/Cas (clustered regularly interspaced short palindromic repeats/CRISPR-associated protein) system has emerged in terms of protein design, synthesis, and validation. CRISPR was initially discovered in *E. coli* in 1987 while studying a gene associated with alkaline phosphatase conversion (Ishino et al., 2018). Compared to ZFNs and TALENs, CRISPR/Cas utilizes RNA as its DNA-binding moiety instead of protein. Moreover, it employs RNA sequences as the specificity-determining element for double-strand breaks (DSBs) and requires only a single guide RNA (sgRNA), which is relatively more cost-effective and efficient (Kozovska et al., 2021). In a laboratory setting, a single guide RNA (sgRNA) is constructed by combining a spacer, a CRISPR RNA (crRNA), and a transactivating CRISPR RNA (tracrRNA). Various CRISPR-associated proteins (Cas), initially identified as Cas9, are subsequently linked with the sgRNA, based on the specific objectives, and inserted into a vector. This assembly, either as individual units or in combination, is then introduced into plant cells through transformation processes (Ryan et al., 2023). Cas9 is a widely used multifunctional protein in genome editing, requiring a specific protospacer adjacent motif (PAM) for activation. It can induce double-strand breaks (DSBs) in plant cells, leading to DNA repair processes that modify the DNA sequence and result in desired phenotypic changes. Repair mechanisms include non-homologous end joining (NHEJ) or homology-directed repair (HDR), resulting in gene knockout modifications. The CRISPR/Cas system offers inducible expression systems for guiding RNA targeting of genes of interest. Mutation detection methods include PCR, sequencing, molecular markers, and western blot analysis (Asmamaw & Zawdie, 2021). CRISPR/Cas9 has been used successfully in various plant species but in a few medicinal plants such as opium poppy, *Salvia miltiorrhiza*, *Chrysanthemum morifolium*, *Solanum pimpinellifoilium*, *Taraxacum koksaghyz*, and orchid to respond to abiotic stress (Alok et al., 2020)

The CRISPR (Clustered Regularly Interspaced Short Palindromic Repeats) technique has proven to be an effective method in plant genetics (Semiarti et al., 2020). Various CRISPR-associated proteins (Cas) are then associated with the guide RNA according to specific objectives, and subsequently cloned into a vector, either individually or in tandem, before being introduced into plant cells (Arora & Narula, 2017) (Figure 1.3). The most commonly used Cas, Cas9, is a multifunctional protein that comprises two nuclease domains (HNH and RuvC). It also requires the presence of an adjacent motif, known as the protospacer adjacent motif (PAM) (Maikova et al., 2021). One major advantage of CRISPR over other genetic transformation methods is its ability to produce non-transgenic plants, thereby simplifying their regulation in genetic modification. CRISPR can achieve gene editing without introducing exogenous DNA, simply by reducing the expression of endogenous genes to obtain a desired trait (Erpen-Dalla Corte et al., 2019). Furthermore, it utilizes an RNA sequence as the determinant element for double-strand breaks (DSBs) specificity and requires only a single guide RNA (sgRNA), making it relatively more cost-effective and efficient (Kozovska et al., 2021) (Figure 1.3). Advancements in genome sequencing within the Orchidaceae family have influenced the development of this technique. Although the genomes of some species belonging to the genera *Phalaenopsis*, *Dendrobium*, *Cymbidium*, *Gastrodia*, *Bletilla*, *Platanthera*, *Vanilla*, and *Apostasia* have been sequenced. In *Phalaenopsis amabilis*, a successful CRISPR/Cas9 KO system was developed by targeting protocorms and PDS3 as genes of interest, resulting in an albino phenotype in the leaf tissues of transformed plants (Reyes et al., 2023). A gene editing system was developed using CRISPR/Cas9 delivered by Agrobacterium and targeting the VAR2 gene in prototypes, resulting in variegated patterns in the leaves of transformed plants. In *Phalaenopsis amabilis equestris*, CRISPR was used to create mutants in conjunction with *Agrobacterium*-mediated transformation of MADS genes, which are crucial for flower development (Nopitasari et al., 2020; Tong et al., 2020).

1.6 METABOLOMICS

Medicinal plants derive their therapeutic properties from a diverse array of secondary metabolites. These compounds, including alkaloids, terpenes, and flavonoids, play various roles in plant physiology and offer a vast array of medicinal benefits (Hussein & El-Anssary, 2019). Understanding the impact of stress on metabolite production is crucial for developing stress-tolerant medicinal plants (Ghassemi et al., 2021). This is where metabolomics, a powerful analytical technique, comes into play. Metabolomics allows for the comprehensive profiling and quantification of all small molecules within a biological sample (Muthubharathi et al., 2021). By comparing the metabolic profiles of stressed and unstressed plants, researchers can identify changes in metabolite abundance and identify stress-responsive pathways. This information is invaluable for understanding how plants adapt to stress and for pinpointing key metabolites with potential medicinal value (Xu et al., 2022). For instance, metabolomic studies on drought-stressed ginseng plants revealed significant upregulation of ginsenosides, the primary bioactive compounds responsible for ginseng's medicinal properties (Shahrajabian et al., 2023).

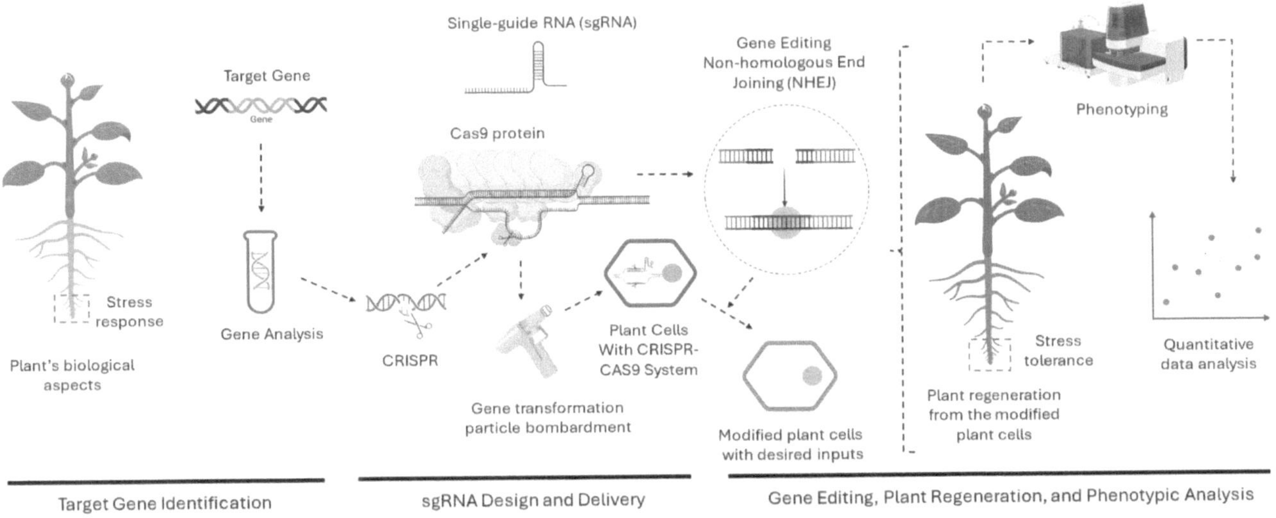

FIGURE 1.3 General workflow of CRISPR-Cas9 approach in addressing plant stress tolerance.

This finding suggests that drought stress may act as a trigger for enhanced ginsenoside production, providing valuable insights for optimizing cultivation practices or breeding for drought tolerance (Rahimi et al., 2021). Furthermore, metabolomics can be used to identify novel stress-responsive metabolites with potential medicinal applications (Kumar et al., 2021). Researchers can uncover previously unknown compounds with potential therapeutic benefits by analyzing the metabolic profiles of stress-tolerant medicinal plant genotypes. This opens doors for the discovery and development of new natural medicines based on the plant's inherent stress-adaptation mechanisms (Younessi-Hamzekhanlu et al., 2022).

1.7 HIGH-THROUGHPUT PHENOTYPING

While speed breeding and genomic selection accelerate breeding cycles and identify promising genotypes, they primarily rely on genetic information. However, the link between genes and the final expression of a trait (phenotype) can be complex and influenced by environmental factors (te Pas et al., 2017). High-throughput phenotyping (HTP) bridges this gap by providing a comprehensive picture of plant phenotypes at a large scale. HTP utilizes advanced technologies such as automated imaging systems, sensors, and robotics to collect a vast amount of phenotypic data (Li et al., 2021). The latter can encompass diverse parameters, such as plant growth rate, biomass accumulation, stress tolerance, and even metabolite production (Mir et al., 2019). The high-throughput nature of these techniques allows researchers to analyze hundreds or even thousands of plants simultaneously, significantly increasing the efficiency of phenotyping efforts (Gill et al., 2022). One example of HTP application in medicinal plant breeding involves the use of hyperspectral and multispectral imaging (Sarić et al., 2022). These techniques capture detailed spectral information from plant leaves, revealing subtle variations in pigment composition and potentially indicating stress responses or changes in metabolite abundance (Bodner et al., 2018). In that manner, stress-tolerant genotypes with specific spectral signatures can be identified by analyzing large datasets of hyperspectral images, leading to more targeted breeding efforts (Kaur et al., 2021).

For medicinal plants, a study conducted by Theerawitaya et al. (2023) endeavors to identify an optimal irrigation regimen aimed at enhancing both morphological and physiological traits, and crop productivity of *Centella asiatica* L. Urban; Apiaceae using high-throughput phenotyping. A study of (Briglia et al., 2019) investigated the potential of non-destructive RGB and NIR image-based analysis techniques to determine morphophysiological traits such as leaf area, plant water consumption, and leaf water potential in drought-stressed grapevines (*Vitis vinifera* L.). In maize research, SPAD meters, NDVI, and infrared gas analyzers (IRGA) have been employed to assess environmental impacts. Digital imaging has been utilized to evaluate early biomass changes under water stress, while thermal imaging has been used to measure leaf temperature variations during transpiration (Rorie et al., 2011).

Furthermore, HTP allows for the development of non-destructive phenotyping methods. This is crucial for medicinal plants where destructive sampling for metabolite analysis can be detrimental to plant health and limit the number of data points collected (Hamany Djande et al., 2020). Techniques such as chlorophyll fluorescence imaging can provide valuable insights into plant stress responses without harming the plant, allowing for repeated measurements throughout the growth cycle (Moustaka & Moustakas, 2023). This allows for the gathering of big genomic data afterward and performing advanced analyses, such as genomic-wide analysis (GWAS) in order to select medicinal plant varieties for stress tolerance (Figure 1.4).

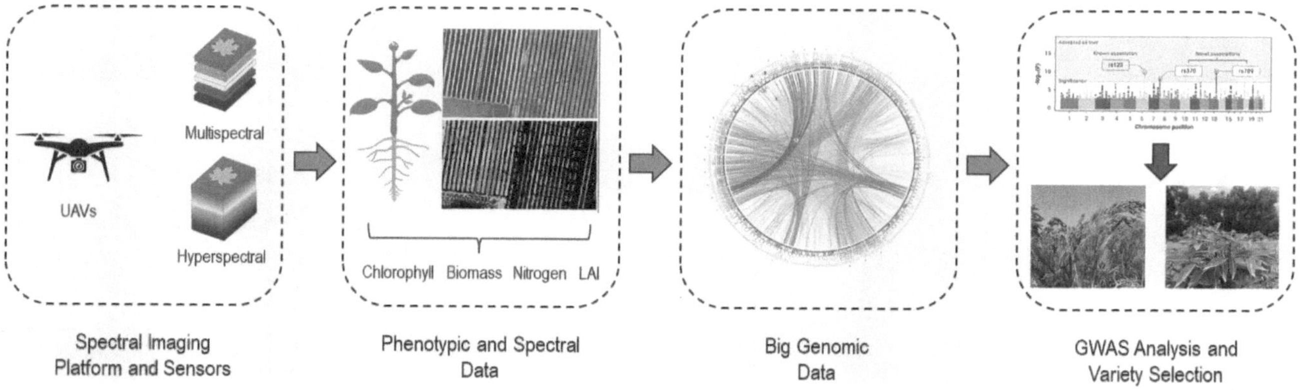

FIGURE 1.4 Workflow of the high-throughput phenotyping in medicinal plant tolerance selection for advent stresses.

1.8 ADVANTAGES AND CHALLENGES OF BREEDING STRATEGIES

The development of stress-tolerant medicinal plants presents a significant challenge in the face of climate change (Ghassemi et al., 2021). Traditional breeding approaches, while successful in the past, often struggle with the time and resource constraints associated with cultivating and analyzing large plant populations across multiple seasons (Tian et al., 2021). However, advancements in breeding techniques offer exciting possibilities for accelerating medicinal plant breeding and developing climate-resilient varieties.

1.8.1 Advantages of Novel Breeding Strategies

One of the most significant advantages of these novel breeding strategies lies in their speed and efficiency. Techniques like speed breeding, as highlighted by Watson et al. (2018), significantly reduce breeding cycles by manipulating photoperiod to induce multiple generations of plants per year. This translates to a faster development of stress-tolerant varieties compared to traditional methods that rely on field trials spanning multiple seasons (Watson et al., 2018). Furthermore, genomic selection (GS) provides breeders with a powerful tool to identify promising genotypes based on their genetic makeup (Bhat et al., 2016). Heffner et al. (2009) highlighted the potential of GS for crop improvement, and this approach can be equally valuable for medicinal plants. By analyzing genetic markers associated with desirable traits like stress tolerance and metabolite production, GS eliminates the need for lengthy field trials and allows for the early selection of plants with the potential for superior performance (Khan et al., 2017). This leads to more efficient breeding programs with a faster turnaround time for the development of new varieties.

Another advantage of these advancements lies in the deeper understanding of stress tolerance mechanisms they facilitate. Metabolomics, a powerful analytical technique, offers valuable insights into the plant's biochemical response to stress (Khan et al., 2017). for instance, employed metabolomics to reveal significant upregulation of ginsenosides, the primary bioactive compounds responsible for ginseng's medicinal properties, in drought-stressed plants. By identifying key metabolites and stress-responsive pathways, a deeper understanding of how plants adapt to stress and guide breeding efforts toward genotypes with enhanced resilience mechanisms can be gained (Raza, 2022).

Finally, high-throughput phenotyping (HTP) allows for the analysis of a vast array of phenotypic data from large plant populations (Gill et al., 2022). Zhang et al. (2019) showcased the application of hyperspectral imaging, a form of HTP, for monitoring progress in wheat breeding. This technique can be similarly valuable for medicinal plant breeding and expands the possibilities for medicinal plant development and the discovery of novel natural medicines based on the plant's inherent stress-adaptation mechanisms.

1.8.2 Challenges and Considerations

Despite the numerous advantages, implementing these novel breeding strategies for medicinal plants also presents some challenges (Nuijten et al., 2016). One major hurdle is the cost and infrastructure required. Speed breeding facilities and advanced analytical equipment for metabolomics and HTP can be expensive to establish and maintain (Bell et al., 2018). This may limit the accessibility of these techniques for smaller breeding programs or those operating in resource-constrained environments.

Another challenge lies in maintaining genetic diversity. Speed breeding combined with GS can potentially lead to genetic bottlenecks if breeding populations are not carefully managed (Samantara et al., 2022). Genetic diversity is crucial for long-term breeding success and ensuring the adaptability of future generations to new stresses. Breeders need to implement strategies to ensure sufficient genetic variation within their populations while focusing on selecting desired traits (Zhang et al., 2017).

Furthermore, while GS identifies genotypes with desirable genetic markers, environmental factors can still influence the final expression of traits. HTP helps bridge this gap by providing data on phenotypic expression, but further research is needed to fully understand the complex interplay between genes and the environment in the context of stress tolerance. Breeders need to consider incorporating field trials alongside novel breeding techniques to account for environmental influences on stress tolerance (Mir et al., 2019). Finally, regulatory considerations come into play when introducing new medicinal plant varieties developed through advanced breeding techniques (Ashraf et al., 2021). Establishing clear guidelines for the evaluation and approval of these varieties is crucial for their successful introduction into the market.

1.9 CONCLUSIONS AND FUTURE PROSPECTS

This chapter addresses the mounting challenges posed by climate change and environmental stress on the cultivation and production of medicinal plants. These threats can disrupt plant growth cycles, reduce yields, and potentially alter the quality and effectiveness of medicinal compounds. To ensure a sustainable and secure supply of these critical resources, innovative breeding strategies are urgently needed to develop stress-tolerant and climate-smart medicinal plant varieties. Various breeding approaches are explored, with conventional methods having limitations in speed and efficiency. However, integrating them with advanced techniques like marker-assisted selection (MAS) and genomic selection (GS) holds significant promise. MAS enables targeting specific stress-tolerance genes, accelerating selection, and reducing linkage drag, while GS predicts breeding values more accurately using genome-wide markers. Metabolomics, which analyzes plant traits on a large scale, is also highlighted for its potential. High-throughput phenotyping platforms allow rapid evaluation of germplasm collections under stress conditions, providing breeders with valuable insights when combined with genotypic data. Genetic diversity within existing germplasm, particularly landraces and wild relatives, is crucial. These often possess inherent stress tolerance and can be utilized through introgression breeding to enhance cultivated varieties' resilience. Advancements in gene editing, like CRISPR-Cas9, offer opportunities for targeted manipulation of stress-responsive genes. However, breeding for stress tolerance shouldn't compromise medicinal compound production, necessitating careful selection and evaluation throughout the process, aided by techniques like metabolomics. Developing climate-smart medicinal plants requires a holistic approach beyond traditional breeding. Agronomic practices promoting soil health, water conservation, and integrated pest management (IPM) can mitigate stress. Exploring alternative cultivation methods like controlled environment agriculture (CEA) offers greater control over environmental parameters. Breeding stress-tolerant and climate-smart medicinal plants demands a novel approach that combines traditional methods with advanced technologies like MAS, GS, metabolomics, high-throughput phenotyping, and gene editing. By harnessing existing genetic diversity and leveraging cutting-edge tools, researchers can develop robust varieties capable of thriving in a changing climate. Sustainable agricultural practices and alternative cultivation methods are essential for a holistic solution, ensuring the future availability of these vital resources for generations to come.

ACKNOWLEDGMENTS

The authors extend their gratitude to the Phytopathology Unit (ENA-Meknès) for their generous financial support in facilitating this research.

REFERENCES

Abrahám, E., Rigó, G., Székely, G., Nagy, R., Koncz, C., & Szabados, L. (2003). Light-dependent induction of proline biosynthesis by abscisic acid and salt stress is inhibited by brassinosteroid in Arabidopsis. *Plant Molecular Biology*, *51*(3), 363–372. https://doi.org/10.1023/a:1022043000516

Acquaah, G. (2015). Conventional plant breeding principles and techniques. In *Advances in plant breeding strategies: Breeding, biotechnology and molecular tools* (pp. 115–158). Springer.

Aftab, T., & Hakeem, K. R. (Eds.). (2021). Plant growth regulators. Springer International Publishing. https://doi.org/10.1007/978-3-030-61153-8

Ahmar, S., Ballesta, P., Ali, M., & Mora-Poblete, F. (2021). Achievements and challenges of genomics-assisted breeding in forest trees: From marker-assisted selection to genome editing. *International Journal of Molecular Sciences*, *22*(19), 10583.

Alok, A., Jain, P., Kumar, J., Yajnik, K., & Bhalothia, P. (2020). Genome engineering in medicinally important plants using CRISPR/Cas9 tool. In *Genome engineering via CRISPR-Cas9 system* (pp. 155–161). Elsevier.

Alugoju, P., & Tencomnao, T. (2023). Medicinal plants proteomics in response to abiotic stresses. In *Medicinal plants: Their response to abiotic stress* (pp. 35–107). Springer.

Applequist, W. L., Brinckmann, J. A., Cunningham, A. B., Hart, R. E., Heinrich, M., Katerere, D. R., & van Andel, T. (2020). Scientists' warning on climate change and medicinal plants. *Planta Medica*, *86*(01), 10–18. https://doi.org/10.1055/a-1041-3406

Archangi, A., Heidari, B., & Mohammadi-Nejad, G. (2019). Association between seed yield-related traits and cDNA-AFLP markers in cumin (*Cuminum cyminum*) under drought and irrigation regimes. *Industrial Crops and Products*, *133*, 276–283. https://doi.org/10.1016/j.indcrop.2019.03.038

Arora, L., & Narula, A. (2017). Gene editing and crop improvement using CRISPR-Cas9 system. *Frontiers in Plant Science*, *8*. https://doi.org/10.3389/fpls.2017.01932

Ashraf, M., & Harris, P. J. C. (2004). Potential biochemical indicators of salinity tolerance in plants. *Plant Science*, *166*(1), 3–16. https://doi.org/10.1016/j.plantsci.2003.10.024

Ashraf, M., Akram, N. A., Mehboob-ur-Rahman, & Foolad, M. R. (2012). *Marker-assisted selection in plant breeding for salinity tolerance* (pp. 305–333). https://doi.org/10.1007/978-1-61779-986-0_21

Ashraf, S. A., Siddiqui, A. J., Elkhalifa, A. E. O., Khan, M. I., Patel, M., Alreshidi, M., Moin, A., Singh, R., Snoussi, M., & Adnan, M. (2021). Innovations in nanoscience for the sustainable development of food and agriculture with implications on health and environment. *Science of the Total Environment*, *768*, 144990. https://doi.org/10.1016/j.scitotenv.2021.144990

Asmamaw, M., & Zawdie, B. (2021). Mechanism and applications of CRISPR/Cas-9-mediated genome editing. In *Biologics: Targets and therapy* (pp. 353–361). Taylor & Francis.

Avasiloaiei, D. I., Calara, M., Brezeanu, P. M., Murariu, O. C., & Brezeanu, C. (2023). On the future perspectives of some medicinal plants within Lamiaceae botanic family regarding their comprehensive properties and resistance against biotic and abiotic stresses. *Genes*, *14*(5), 955. https://doi.org/10.3390/genes14050955

Baidyussen, A., Jatayev, S., Khassanova, G., Amantayev, B., Sereda, G., Sereda, S., Gupta, N. K., Gupta, S., Schramm, C., Anderson, P., Jenkins, C. L. D., Soole, K. L., Langridge, P., & Shavrukov, Y. (2021). Expression of specific alleles of zinc-finger transcription factors, HvSAP8 and HvSAP16, and corresponding SNP markers, are associated with drought tolerance in barley populations. *International Journal of Molecular Sciences*, *22*(22), 12156. https://doi.org/10.3390/ijms222212156

Balasubramanian, V., Suresh, J., Srinivasan, R., Prabhu, M., Manikandan, E., & Nandha, V. B. G. (2020). Evaluation of soil characteristics and yield variation of coleus (*Coleus forskohlii*) in different agro climatic zones of Tamil Nadu. *International Journal of Chemical Studies*, *8*(4), 2841–2845. https://doi.org/10.22271/chemi.2020.v8.i4ah.10076

Bell, L., Calder, B., Hiller, R., Klein, A., Soares, N. C., Stoychev, S. H., Vorster, B. C., & Tabb, D. L. (2018). Challenges and opportunities for biological mass spectrometry core facilities in the developing world. *Journal of Biomolecular Techniques: JBT*, *29*(1), 4–15. https://doi.org/10.7171/jbt.18-2901-003

Bhagtaney, L., & Sundarrajan, P. (2023). An overview of tools for genome editing: ZFNs, mega nucleases, and TALENs. In *CRISPR/Cas-mediated genome editing in plants* (pp. 37–64). Apple Academic Press.

Bhat, J. A., Ali, S., Salgotra, R. K., Mir, Z. A., Dutta, S., Jadon, V., Tyagi, A., Mushtaq, M., Jain, N., Singh, P. K., Singh, G. P., & Prabhu, K. V. (2016). Genomic selection in the era of next generation sequencing for complex traits in plant breeding. *Frontiers in Genetics*, *7*. https://doi.org/10.3389/fgene.2016.00221

Bhuyan, S. J., Kumar, M., Ramrao Devde, P., Rai, A. C., Mishra, A. K., Singh, P. K., & Siddique, K. H. M. (2023). Progress in gene editing tools, implications, and success in plants: A review. *Frontiers in Genome Editing*, *5*, 1272678.

Bodner, G., Nakhforoosh, A., Arnold, T., & Leitner, D. (2018). Hyperspectral imaging: A novel approach for plant root phenotyping. *Plant Methods*, *14*(1), 84. https://doi.org/10.1186/s13007-018-0352-1

Boti, M. A., Athanasopoulou, K., Adamopoulos, P. G., Sideris, D. C., & Scorilas, A. (2023). Recent advances in genome-engineering strategies. *Genes*, *14*(1), 129.

Brandt, J. S., Haynes, M. A., Kuemmerle, T., Waller, D. M., & Radeloff, V. C. (2013). Regime shift on the roof of the world: Alpine meadows converting to shrublands in the southern Himalayas. *Biological Conservation*, *158*, 116–127. https://doi.org/10.1016/j.biocon.2012.07.026

Briglia, N., Montanaro, G., Petrozza, A., Summerer, S., Cellini, F., & Nuzzo, V. (2019). Drought phenotyping in *Vitis vinifera* using RGB and NIR imaging. *Scientia Horticulturae*, *256*, 108555. https://doi.org/10.1016/j.scienta.2019.108555

Cazzola, F., Bermejo, C. J., Gatti, I., & Cointry, E. (2021). Speed breeding in pulses: An opportunity to improve the efficiency of breeding programs. *Crop and Pasture Science*, *72*(3), 165–172. https://doi.org/10.1071/CP20462

Chapagain, S., Pruthi, R., Singh, L., & Subudhi, P. K. (2024). Comparison of the genetic basis of salt tolerance at germination, seedling, and reproductive stages in an introgression line population of rice. *Molecular Biology Reports*, *51*(1), 252. https://doi.org/10.1007/s11033-023-09049-1

Christmas, M. J., Breed, M. F., & Lowe, A. J. (2016). Constraints to and conservation implications for climate change adaptation in plants. *Conservation Genetics*, *17*(2), 305–320. https://doi.org/10.1007/s10592-015-0782-5

Collard, B. C. Y., Jahufer, M. Z. Z., Brouwer, J. B., & Pang, E. C. K. (2005). An introduction to markers, quantitative trait loci (QTL) mapping and marker-assisted selection for crop improvement: The basic concepts. *Euphytica*, *142*(1–2), 169–196. https://doi.org/10.1007/s10681-005-1681-5

Collard, B. C., & Mackill, D. J. (2008). Marker-assisted selection: An approach for precision plant breeding in the twenty-first century. *Philosophical Transactions of the Royal Society B: Biological Sciences*, *363*(1491), 557–572. https://doi.org/10.1098/rstb.2007.2170

Cvejić, S., Jocić, S., Mitrović, B., Bekavac, G., Mirosavljević, M., Jeromela, A. M., Zorić, M., Radanović, A., Kondić-Špika, A., & Miladinović, D. (2022). Innovative approaches in the breeding of climate-resilient crops. In *Climate change and agriculture* (pp. 111–156). Wiley. https://doi.org/10.1002/9781119789789.ch6

Dagar, J. C. (2014). *Greening salty and waterlogged lands through agroforestry systems for livelihood security and better environment* (pp. 273–332). https://doi.org/10.1007/978-81-322-1662-9_9

Dagar, J. C., Lal, K., Ram, J., Kumar, M., Chaudhari, S. K., Yadav, R. K., Ahamad, S., Singh, G., & Kaur, A. (2016). Eucalyptus geometry in agroforestry on waterlogged saline soils influences plant and soil traits in North-West India. *Agriculture, Ecosystems & Environment*, *233*, 33–42. https://doi.org/10.1016/j.agee.2016.08.025

Das, M., Jain, V., & Malhotra, S. K. (2016). Impact of climate change on medicinal and aromatic plants: Review. *The Indian Journal of Agricultural Sciences*, *86*(11). https://doi.org/10.56093/ijas.v86i11.62865

Devi, E. L., Devi, C. P., Kumar, S., Sharma, S. K., Beemrote, A., Chongtham, S. K., Singh, C. H., Tania, C., Singh, T. B., Ningombam, A., Akoijam, R., Singh, I. M., Singh, Y. R., Monteshori, S., Omita, Y., Prakash, N., & Ngachan, S. V. (2017). Marker assisted selection (MAS) towards generating stress tolerant crop plants. *Plant Gene*, *11*, 205–218. https://doi.org/10.1016/j.plgene.2017.05.014

Dobrikova, A. G. (2023). Abiotic stress tolerance in crop and medicinal plants. *Plants*, *12*(24), 4167. https://doi.org/10.3390/plants12244167

Dubey, P. K., Shukla, R. N., Srivastava, G., Mishra, A. A., & Pandey, A. (2019). Study on quality parameters and storage stability of mango coated with developed nanocomposite edible film. *International Journal of Current Microbiology and Applied Sciences*, *8*(4), 2899–2935.

Eisenhauer, D. E., Martin, D. L., Heeren, D. M., & Hoffman, G. J. (2021). *Irrigation systems management*. American Society of Agricultural and Biological Engineers (ASABE).

Eremina, M., Rozhon, W., & Poppenberger, B. (2016). Hormonal control of cold stress responses in plants. *Cellular and Molecular Life Sciences*, *73*(4), 797–810. https://doi.org/10.1007/s00018-015-2089-6

Erpen-Dalla Corte, L., Mahmoud, L. M., Moraes, T. S., Mou, Z., Grosser, J. W., & Dutt, M. (2019). Development of improved fruit, vegetable, and ornamental crops using the CRISPR/Cas9 genome editing technique. *Plants*, *8*(12), 601.

Field, C. B., Barros, V. R., Dokken, D. J., Mach, K. J., & Mastrandrea, M. D. (Eds.). (2014). *Climate change 2014 impacts, adaptation, and vulnerability*. Cambridge University Press. https://doi.org/10.1017/CBO9781107415379

Ghassemi, S., Delangiz, N., Asgari Lajayer, B., Saghafi, D., & Maggi, F. (2021). Review and future prospects on the mechanisms related to cold stress resistance and tolerance in medicinal plants. *Acta Ecologica Sinica*, *41*(2), 120–129. https://doi.org/10.1016/j.chnaes.2020.09.006

Gill, T., Gill, S. K., Saini, D. K., Chopra, Y., de Koff, J. P., & Sandhu, K. S. (2022). A comprehensive review of high throughput phenotyping and machine learning for plant stress phenotyping. *Phenomics*, *2*(3), 156–183.

Gholizadeh, A., Dehghani, H., Khodadadi, M., & Gulick, P. J. (2018). Genetic combining ability of coriander genotypes for agronomic and phytochemical traits in response to contrasting irrigation regimes. *Plos one*, *13*(6), e0199630.

Gudi, S., Kumar, P., Singh, S., Tanin, M. J., & Sharma, A. (2022). Strategies for accelerating genetic gains in crop plants: Special focus on speed breeding. *Physiology and Molecular Biology of Plants*, *28*(10), 1921–1938. https://doi.org/10.1007/s12298-022-01247-8

Gull, A., Lone, A. A., & Wani, N. U. I. (2019). Biotic and abiotic stresses in plants. In *Abiotic and biotic stress in plants* (pp. 1–19). IntechOpen.

Gupta, A., Singh, P. P., Singh, P., Singh, K., Singh, A. V., Singh, S. K., & Kumar, A. (2019). Medicinal plants under climate change: Impacts on pharmaceutical properties of plants. In *Climate change and agricultural ecosystems* (pp. 181–209). Elsevier. https://doi.org/10.1016/B978-0-12-816483-9.00008-6

Hamany Djande, C. Y., Pretorius, C., Tugizimana, F., Piater, L. A., & Dubery, I. A. (2020). Metabolomics: A tool for cultivar phenotyping and investigation of grain crops. *Agronomy*, *10*(6), 831. https://doi.org/10.3390/agronomy10060831

Heffner, E. L., Sorrells, M. E., & Jannink, J. (2009). Genomic selection for crop improvement. *Crop Science*, *49*(1), 1–12.

Heidari, M., & Sarani, S. (2012). Growth, biochemical components, and ion content of Chamomile (*Matricaria chamomilla* L.) under salinity stress and iron deficiency. *Journal of the Saudi Society of Agricultural Sciences*, *11*(1), 37–42. https://doi.org/10.1016/j.jssas.2011.05.002

Hokmalipour, S. (2015). Effect of salinity and temperature on seed germination and seed vigor index of chicory (*chichorium intynus* L.), cumin (*Cuminium cyminium* L.) and fennel (*Foeniculum vulgare*). *Indian Journal of Science and Technology*, *8*(35). https://doi.org/10.17485/ijst/2015/v8i35/85914

Huang, P., Xiao, A., Zhou, M., Zhu, Z., Lin, S., & Zhang, B. (2011). Heritable gene targeting in zebrafish using customized TALENs. *Nature Biotechnology*, *29*(8), 699–700.

Humphreys, A. M., Govaerts, R., Ficinski, S. Z., Nic Lughadha, E., & Vorontsova, M. S. (2019). Global dataset shows geography and life form predict modern plant extinction and rediscovery. *Nature Ecology & Evolution*, *3*(7), 1043–1047. https://doi.org/10.1038/s41559-019-0906-2

Hussein, R. A., & El-Anssary, A. A. (2019). Plants secondary metabolites: The key drivers of the pharmacological actions of medicinal plants. In *Herbal Medicine*. IntechOpen. https://doi.org/10.5772/intechopen.76139

Ishino, Y., Krupovic, M., & Forterre, P. (2018). History of CRISPR-Cas from encounter with a mysterious repeated sequence to genome editing technology. *Journal of Bacteriology*, *200*(7), 10–1128.

Jaafar, H. Z. E., Ibrahim, M. H., & Mohamad Fakri, N. F. (2012). Impact of soil field water capacity on secondary metabolites, phenylalanine ammonia-lyase (PAL), maliondialdehyde (MDA) and photosynthetic responses of Malaysian Kacip Fatimah (*Labisia pumila* Benth). *Molecules*, *17*(6), 7305–7322. https://doi.org/10.3390/molecules17067305

Joung, J. K., & Sander, J. D. (2013). TALENs: A widely applicable technology for targeted genome editing. *Nature Reviews Molecular Cell Biology*, *14*(1), 49–55. https://doi.org/10.1038/nrm3486

Katayama, S., Watanabe, M., Kato, Y., Nomura, W., & Yamamoto, T. (2024). Engineering of zinc finger nucleases through structural modeling improves genome editing efficiency in cells. *Advanced Science*, 2310255.

Kaur, B., Sandhu, K. S., Kamal, R., Kaur, K., Singh, J., Röder, M. S., & Muqaddasi, Q. H. (2021). Omics for the improvement of abiotic, biotic, and agronomic traits in major cereal crops: Applications, challenges, and prospects. *Plants*, *10*(10), 1989. https://doi.org/10.3390/plants10101989

Kaur, T., Bhat, H. A., Bhat, R., Kumar, A., Bindu, K., Koul, S., & Vyas, D. (2015). Physio-chemical and antioxidant profiling of *Salvia sclarea L.* at different climates in north-western Himalayas. *Acta Physiologiae Plantarum*, *37*(7), 132. https://doi.org/10.1007/s11738-015-1879-7

Kendall, H. (1992). *World scientists' warning to humanity (1992)*. Resource document. Union of Concerned Scientists. Retrieved August 11, 1992, from http://www.ucsusa.org/about/1992-world-scientists.html

Khan, N., Ali, S., Shahid, M. A., & Kharabian-Masouleh, A. (2017). Advances in detection of stress tolerance in plants through metabolomics approaches. *Plant Omics*, *10*(03), 153–163. https://doi.org/10.21475/poj.10.03.17.pne600

Khan, Z., Khan, S. H., Mubarik, M. S., Sadia, B., & Ahmad, A. (2017). Use of TALEs and TALEN technology for genetic improvement of plants. *Plant Molecular Biology Reporter*, *35*, 1–19.

Khanum, R., Mumtaz, A. S., & Kumar, S. (2013). Predicting impacts of climate change on medicinal asclepiads of Pakistan using Maxent modeling. *Acta Oecologica*, *49*, 23–31. https://doi.org/10.1016/j.actao.2013.02.007

Kim, N., Jayakodi, M., Lee, S., Choi, B., Jang, W., Lee, J., Kim, H. H., Waminal, N. E., Lakshmanan, M., van Nguyen, B., Lee, Y. S., Park, H., Koo, H. J., Park, J. Y., Perumal, S., Joh, H. J., Lee, H., Kim, J., Kim, I. S., . . . Yang, T. (2018). Genome and evolution of the shade-requiring medicinal herb Panax ginseng. *Plant Biotechnology Journal*, *16*(11), 1904–1917. https://doi.org/10.1111/pbi.12926

Kozovska, Z., Rajcaniova, S., Munteanu, P., Dzacovska, S., & Demkova, L. (2021). CRISPR: History and perspectives to the future. *Biomedicine & Pharmacotherapy*, *141*, 111917. https://doi.org/10.1016/j.biopha.2021.111917

Kumar, A., Sharma, M., Basu, S. K., Asif, M., Li, X. P., & Chen, X. (2014). Perspectives from plant biotechnology and marker-assisted selection. In *Omics technologies and crop improvement* (p. 153). CRC Press.

Kumar, M., Kumar Patel, M., Kumar, N., Bajpai, A. B., & Siddique, K. H. M. (2021). Metabolomics and molecular approaches reveal drought stress tolerance in plants. *International Journal of Molecular Sciences*, *22*(17), 9108.

Lait, C. G., Bates, S. L., Kermode, A. R., Morrissette, K. K., & Borden, J. H. (2001). Specific biochemical marker-based techniques for the identification of damage to Douglas-fir seed resulting from feeding by the western conifer seed bug, *Leptoglossus occidentalis* Heidemann (Hemiptera: Coreidae). *Insect Biochemistry and Molecular Biology*, *31*(6–7), 739–746. https://doi.org/10.1016/S0965-1748(00)00176-4

Lamprecht, A., Semenchuk, P. R., Steinbauer, K., Winkler, M., & Pauli, H. (2018). Climate change leads to accelerated transformation of high-elevation vegetation in the central Alps. *New Phytologist*, *220*(2), 447–459. https://doi.org/10.1111/nph.15290

Lateef, D. D. (2015). DNA marker technologies in plants and applications for crop improvements. *Journal of Biosciences and Medicines*, *03*(05), 7–18. https://doi.org/10.4236/jbm.2015.35002

Lenoir, J., & Svenning, J.-C. (2013). Latitudinal and elevational range shifts under contemporary climate change. In *Encyclopedia of biodiversity* (pp. 599–611). Elsevier. https://doi.org/10.1016/B978-0-12-384719-5.00375-0

Li, D., Quan, C., Song, Z., Li, X., Yu, G., Li, C., & Muhammad, A. (2021). High-throughput plant phenotyping platform (HT3P) as a novel tool for estimating agronomic traits from the lab to the field. *Frontiers in Bioengineering and Biotechnology*, *8*. https://doi.org/10.3389/fbioe.2020.623705

Li, D.-Y., Li, L.-Q., & Liu, J.-J. G. (2023). Nucleases in gene-editing technologies: Past and prologue. *National Science Open*, *2*(5), 20220067.

Lianopoulou, V., Bosabalidis, A. M., Patakas, A., Lazari, D., & Panteris, E. (2014). Effects of chilling stress on leaf morphology, anatomy, ultrastructure, gas exchange, and essential oils in the seasonally dimorphic plant *Teucrium polium* (Lamiaceae). *Acta Physiologiae Plantarum*, *36*(8), 2271–2281. https://doi.org/10.1007/s11738-014-1605-x

Lianopoulou, V., Patakas, A., & Bosabalidis, A. M. (2014). Seasonal dimorphism and winter chilling stress in *Thymus sibthorpii*. *Biologia Plantarum*, *58*(1), 139–146. https://doi.org/10.1007/s10535-013-0371-8

Liu, X., Liu, Y., Huang, P., Ma, Y., Qing, Z., Tang, Q., Cao, H., Cheng, P., Zheng, Y., Yuan, Z., Zhou, Y., Liu, J., Tang, Z., Zhuo, Y., Zhang, Y., Yu, L., Huang, J., Yang, P., Peng, Q., . . . Zeng, J. (2017). The genome of medicinal plant *Macleaya cordata* provides new insights into benzylisoquinoline alkaloids metabolism. *Molecular Plant*, *10*(7), 975–989. https://doi.org/10.1016/j.molp.2017.05.007

Loarie, S. R., Duffy, P. B., Hamilton, H., Asner, G. P., Field, C. B., & Ackerly, D. D. (2009). The velocity of climate change. *Nature*, *462*(7276), 1052–1055. https://doi.org/10.1038/nature08649

Mackill, D., Nguyen, H., & Zhang, J. (1999). Use of molecular markers in plant improvement programs for rainfed lowland rice. *Field Crops Research*, *64*(1–2), 177–185. https://doi.org/10.1016/S0378-4290(99)00058-1

Maikova, A., Boudry, P., Shiriaeva, A., Vasileva, A., Boutserin, A., Medvedeva, S., Semenova, E., Severinov, K., & Soutourina, O. (2021). Protospacer-adjacent motif specificity during *Clostridioides difficile* Type I-B CRISPR-Cas interference and adaptation. *MBio*, *12*(4). https://doi.org/10.1128/mBio.02136-21

Majumdar, P., Harale, G., Ganger, S., & Pardeshi, S. (2022). Zinc finger nuclease based genome editing in enhancement of antidiabetic activity of plants: Prospective applications. In *Antidiabetic potential of plants in the era of omics* (pp. 373–401). Apple Academic Press.

Mane, A. V, Karadge, B. A., & Samant, J. S. (2010). Salinity induced changes in photosynthetic pigments and polyphenols of *Cymbopogon nardus* (L.) Rendle. *Journal of Chemical and Pharmaceutical Research*, *2*(3), 338–347.

Mir, B. A., Mir, S. A., Khazir, J., Tonfack, L. B., Cowan, D. A., Vyas, D., & Koul, S. (2015). Cold stress affects antioxidative response and accumulation of medicinally important withanolides in *Withania somnifera* (L.) Dunal. *Industrial Crops and Products*, *74*, 1008–1016. https://doi.org/10.1016/j.indcrop.2015.06.012

Mir, R. R., Reynolds, M., Pinto, F., Khan, M. A., & Bhat, M. A. (2019). High-throughput phenotyping for crop improvement in the genomics era. *Plant Science*, *282*, 60–72. https://doi.org/10.1016/j.plantsci.2019.01.007

Mirmiran, S. M., Nezami, A., Kafi, M., Nabati, J., & Soureshjani, H. K. (2021). Selection of fenugreek (*Trigonella foenum-graecum* L.) landraces for fall planting and freezing tolerance. *40*, 71–82. https://doi.org/10.22099/IAR.2021.40312.1433

Mishra, M. K., Chaturvedi, P., Singh, R., Singh, G., Sharma, L. K., Pandey, V., Kumari, N., & Misra, P. (2013). Overexpression of WsSGTL1 gene of *Withania somnifera* enhances salt tolerance, heat tolerance and cold acclimation ability in transgenic arabidopsis plants. *PLoS ONE*, *8*(4), e63064. https://doi.org/10.1371/journal.pone.0063064

Mishra, T. (2016). Climate change and production of secondary metabolites in medicinal plants: A review. *International Journal of Herbal Medicine*, *4*(4), 27–30.

Mollier, M., Roychowdhury, R., Tzudir, L., Sharma, R., Barua, U., Rahman, N., Pal, S., Gogoi, B., Kalita, P., Jain, D., & Das, R. (2023). Evaluation of morpho-physiological and yield-associated traits of rice (*Oryza sativa L.*) landraces combined with marker-assisted selection under high-temperature stress and elevated atmospheric CO_2 levels. *Plants*, *12*(20), 3655. https://doi.org/10.3390/plants12203655

Moustaka, J., & Moustakas, M. (2023). Early-stage detection of biotic and abiotic stress on plants by chlorophyll fluorescence imaging analysis. *Biosensors*, *13*(8), 796. https://doi.org/10.3390/bios13080796

Muhamba, G., Amos, L., Protas, D., Mbogo, P., & Nchimbi-Msoll, S. (2013). Marker assisted selection for common bean diseases improvements in Tanzania: Prospects and future needs. In *Plant breeding from laboratories to fields*. InTech. https://doi.org/10.5772/52823

Muir, A. D., & Westcott, N. D. (Eds.). (2003). *Flax*. CRC Press. https://doi.org/10.1201/9780203437506

Mukhopadhyay, R., Sarkar, B., Jat, H. S., Sharma, P. C., & Bolan, N. S. (2021). Soil salinity under climate change: Challenges for sustainable agriculture and food security. *Journal of Environmental Management*, *280*, 111736. https://doi.org/10.1016/j.jenvman.2020.111736

Muthubharathi, B. C., Gowripriya, T., & Balamurugan, K. (2021). Metabolomics: Small molecules that matter more. *Molecular Omics*, *17*(2), 210–229. https://doi.org/10.1039/D0MO00176G

Nazari, M., Ghasemi-Soloklui, A. A., Kordrostami, M., & Abdel Latef, A. A. H. (2023). Deciphering the response of medicinal plants to abiotic stressors: A focus on drought and salinity. *Plant Stress*, *10*, 100255. https://doi.org/10.1016/j.stress.2023.100255

Negi, S., Imanishi, M., Hamori, M., Kawahara-Nakagawa, Y., Nomura, W., Kishi, K., Shibata, N., & Sugiura, Y. (2023). The past, present, and future of artificial zinc finger proteins: Design strategies and chemical and biological applications. *JBIC Journal of Biological Inorganic Chemistry*, *28*(3), 249–261.

Nogués-Bravo, D., Araújo, M. B., Errea, M. P., & Martínez-Rica, J. P. (2007). Exposure of global mountain systems to climate warming during the 21st century. *Global Environmental Change*, *17*(3–4), 420–428. https://doi.org/10.1016/j.gloenvcha.2006.11.007

Nopitasari, S., Setiawati, Y., Lawrie, M. D., Purwantoro, A., Widada, J., Sasongko, A. B., Yoshioka, Y., Matsumoto, S., Ninomiya, K., Asano, Y., & Semiarti, E. (2020). Development of an agrobacterium-delivered CRISPR/Cas9 for *Phalaenopsis amabilis* (L.) Blume genome editing system. 060014. https://doi.org/10.1063/5.0015868

Noreen, S., Aasim, M., Ummara, U., Akhter, M. S., Saleem, N., Mahmood, S., & Shah, K. H. (2022). Effect of abiotic stresses and adaptation strategies of medicinal plants. In *Environmental challenges and medicinal plants: Sustainable production solutions under adverse conditions* (pp. 203–228). Springer.

Nowak, M., Kleinwächter, M., Manderscheid, R., Weigel, H. J., & Selmar, D. (2010). Drought stress increases the accumulation of monoterpenes in sage (*Salvia officinalis*), an effect that is compensated by elevated carbon dioxide concentration. *Journal of Applied Botany and Food Quality*, *83*(2), 133–136.

Nuijten, E., Messmer, M., & Lammerts van Bueren, E. (2016). Concepts and strategies of organic plant breeding in light of novel breeding techniques. *Sustainability*, *9*(1), 18. https://doi.org/10.3390/su9010018

O'Boyle, P. D., Kelly, J. D., & Kirk, W. W. (2007). Use of marker-assisted selection to breed for resistance to common bacterial blight in common bean. *Journal of the American Society for Horticultural Science*, *132*(3), 381–386. https://doi.org/10.21273/JASHS.132.3.381

Padmanabhan, P., Correa-Betanzo, J., & Paliyath, G. (2016). Berries and related fruits. In *Encyclopedia of food and health* (pp. 364–371). Elsevier. https://doi.org/10.1016/B978-0-12-384947-2.00060-X

Pandey, V., Bhatt, I. D., & Nandi, S. K. (2019). Environmental stresses in Himalayan medicinal plants: Research needs and future priorities. *Biodiversity and Conservation*, *28*(8–9), 2431–2455. https://doi.org/10.1007/s10531-019-01776-x

Patni, B., Bhattacharyya, M., Kumari, A., & Purohit, V. K. (2022). Alarming influence of climate change and compromising quality of medicinal plants. *Plant Physiology Reports*, *27*(1), 1–10. https://doi.org/10.1007/s40502-021-00616-x

Pourmeidani, A., Ghamghami, M., Olya, H., & Ghahreman, N. (2020). Determination of suitable regions for cultivation of three medicinal plants under a changing climate. *Environmental Processes*, *7*(1), 89–108. https://doi.org/10.1007/s40710-020-00423-w

Quéro, A., Fliniaux, O., Elboutachfaiti, R., Petit, E., Guillot, X., Hawkins, S., Courtois, J., & Mesnard, F. (2015). β-Aminobutyric acid increases drought tolerance and reorganizes solute content and water homeostasis in flax (*Linum usitatissimum*). *Metabolomics*, *11*(5), 1363–1375. https://doi.org/10.1007/s11306-015-0792-9

Rahimi, S., Mohanan, P., Zhang, D., Jung, K.-H., Yang, D.-C., Mijakovic, I., & Kim, Y.-J. (2021). Metabolic dynamics and ginsenoside biosynthesis. In *The Ginseng genome* (pp. 121–141). Springer.

Rahimi-Dehgolan, R., Tahmasebi Sarvestani, Z., Rezazadeh, S. A., & Dolatabadian, A. (2012). Morphological and physiological characters of aloe vera subjected to saline water irrigation. *Journal of Herbs, Spices & Medicinal Plants*, *18*(3), 222–230. https://doi.org/10.1080/10496475.2012.671802

Rahimtoola, S. H. (2004). Digitalis therapy for patients in clinical heart failure. *Circulation*, *109*(24), 2942–2946. https://doi.org/10.1161/01.CIR.0000132477.32438.03

Rahman, S., Iqbal, M., & Husen, A. (2023). Medicinal plants and abiotic stress: An overview. In *Medicinal plants: Their response to abiotic stress* (pp. 1–34). Springer.

Raza, A. (2022). Metabolomics: A systems biology approach for enhancing heat stress tolerance in plants. *Plant Cell Reports*, *41*(3), 741–763. https://doi.org/10.1007/s00299-020-02635-8

Razavi, F. (2024). Development of drought associated AFLP and EST candidate gene markers in *Fragaria sp.* as a simple genomic model of rosaceae. *International Journal of Horticultural Science and Technology*, *11*(3), 331–352.

Razzaq, A., Kaur, P., Akhter, N., Wani, S. H., & Saleem, F. (2021). Next-generation breeding strategies for climate-ready crops. *Frontiers in Plant Science*, *12*. https://doi.org/10.3389/fpls.2021.620420

Reyes, V. P., Bulasag, A. S., Mohapatra, S. R., Samantara, K., & Wani, S. H. (2023). Genome-editing strategies for enhanced stress tolerance in medicinal plants. In *Medicinal plants* (pp. 417–442). Springer Nature Singapore. https://doi.org/10.1007/978-981-19-5611-9_16

Ripple, W. J., Wolf, C., Newsome, T. M., Galetti, M., Alamgir, M., Crist, E., Mahmoud, M. I., Laurance, W. F., & 15, 364 Scientist Signatories from 184 Countries. (2017). World scientists' warning to humanity: A second notice. *BioScience*, *67*(12), 1026–1028.

Rodríguez-Ortega, T., Olaizola, A. M., & Bernués, A. (2018). A novel management-based system of payments for ecosystem services for targeted agri-environmental policy. *Ecosystem Services*, *34*, 74–84. https://doi.org/10.1016/j.ecoser.2018.09.007

Rorie, R. L., Purcell, L. C., Karcher, D. E., & King, C. A. (2011). The assessment of leaf nitrogen in corn from digital images. *Crop Science*, *51*(5), 2174–2180. https://doi.org/10.2135/cropsci2010.12.0699

Ruane, R., & Sonnino, A. (2007). *Marker-assisted selection as a tool for genetic improvement of crops, livestock, forestry, and fish in developing countries: An overview of the issues*. FAO. ftp://ftp.fao.org/docrep/fao/010/a1120e/a1120e.pdf

Ruelland, E., Vaultier, M.-N., Zachowski, A., & Hurry, V. (2009). *Chapter 2 cold signalling and cold acclimation in plants* (pp. 35–150). https://doi.org/10.1016/S0065-2296(08)00602-2

Ryan, E. C., Huggins, L. M., & Podlevsky, J. D. (2023). The engineered single guide RNA structure as a biomarker for gene-editing reagent exposure. *Scientific Reports*, *13*(1), 10804.

Saberali, S. F., & Moradi, M. (2019). Effect of salinity on germination and seedling growth of *Trigonella foenum-graecum*, *Dracocephalum moldavica*, *Satureja hortensis* and *Anethum graveolens*. *Journal of the Saudi Society of Agricultural Sciences*, *18*(3), 316–323. https://doi.org/10.1016/j.jssas.2017.09.004

Salick, J., Ghimire, S. K., Fang, Z., Dema, S., & Konchar, K. M. (2014). Himalayan alpine vegetation, climate change and mitigation. *Journal of Ethnobiology*, *34*(3), 276. https://doi.org/10.2993/0278-0771-34.3.276

Samantara, K., Bohra, A., Mohapatra, S. R., Prihatini, R., Asibe, F., Singh, L., Reyes, V. P., Tiwari, A., Maurya, A. K., Croser, J. S., Wani, S. H., Siddique, K. H. M., & Varshney, R. K. (2022). Breeding more crops in less time: A perspective on speed breeding. *Biology*, *11*(2), 275. https://doi.org/10.3390/biology11020275

Sarić, R., Nguyen, V. D., Burge, T., Berkowitz, O., Trtílek, M., Whelan, J., Lewsey, M. G., & Čustović, E. (2022). Applications of hyperspectral imaging in plant phenotyping. *Trends in Plant Science*, *27*(3), 301–315.

Selmar, D., & Kleinwächter, M. (2013). Influencing the product quality by deliberately applying drought stress during the cultivation of medicinal plants. *Industrial Crops and Products*, *42*, 558–566. https://doi.org/10.1016/j.indcrop.2012.06.020

Semiarti, E., Nopitasari, S., Setiawati, Y., Lawrie, M. D., Purwantoro, A., Widada, J., Yoshioka, Y., Matsumoto, S., Ninomiya, K., & Asano, Y. (2020). Application of CRISPR/Cas9 genome editing system for molecular breeding of orchids.

Indonesian Journal of Biotechnology, *25*(1), 61. https://doi.org/10.22146/ijbiotech.39485

Sevillano, L., Sanchez-Ballest, M. T., Romojaro, F., & Flores, F. B. (2009). Physiological, hormonal, and molecular mechanisms regulating chilling injury in horticultural species. Postharvest technologies applied to reduce its impact. *Journal of the Science of Food and Agriculture*, *89*(4), 555–573. https://doi.org/10.1002/JSFA.3468

Shafer, S. L., Bartlein, P. J., & Thompson, R. S. (2001). Potential changes in the distributions of western North America tree and shrub taxa under future climate scenarios. *Ecosystems*, *4*(3), 200–215. https://doi.org/10.1007/s10021-001-0004-5

Shahrajabian, M. H., Kuang, Y., Cui, H., Fu, L., & Sun, W. (2023). Metabolic changes of active components of important medicinal plants on the basis of traditional Chinese medicine under different environmental stresses. *Current Organic Chemistry*, *27*(9), 782–806. https://doi.org/10.2174/1385272827666230807150910

Sharma, M., Thakur, R., Sharma, M., Kumar Sharma, A., & Kumar Sharma, A. (2020). Changing scenario of medicinal plants diversity in relation to climate change: A review. *Plant Archives*, *20*(2), 4389–4400.

Sheler, R. J. (2013). *The impact of agricultural drainage systems on hydrologic responses*. University of Iowa.

Shivaraj, S. M., Jain, A., & Singh, A. (2018). Highly preserved roles of Brassica MIR172 in polyploid Brassicas: Ectopic expression of variants of Brassica MIR172 accelerates floral transition. *Molecular Genetics and Genomics*, *293*(5), 1121–1138. https://doi.org/10.1007/s00438-018-1444-3

Sisea, C.-R. (2024). Characteristics of various types of plant breeding. In *Beyond the blueprint – decoding the elegance of gene expression [working title]*. IntechOpen. https://doi.org/10.5772/intechopen.1004008

Skaggs, R. W., & Van Schilfgaarde, J. (2015, August). Agricultural drainage. *Agricultural Drainage*, 1–1259. https://doi.org/10.1002/9780891182306

Slafer, G. A., Araus, J. L., Royo, C., & Moral, L. F. G. (2005). Promising eco-physiological traits for genetic improvement of cereal yields in Mediterranean environments. *Annals of Applied Biology*, *146*(1), 61–70. https://doi.org/10.1111/j.1744-7348.2005.04048.x

Smith, P. (2008). Land use change and soil organic carbon dynamics. *Nutrient Cycling in Agroecosystems*, *81*(2), 169–178. https://doi.org/10.1007/s10705-007-9138-y

Souther, S., & McGraw, J. B. (2014). Synergistic effects of climate change and harvest on extinction risk of American ginseng. *Ecological Applications*, *24*(6), 1463–1477. https://doi.org/10.1890/13-0653.1

Sun, X., Zhu, S., Li, N., Cheng, Y., Zhao, J., Qiao, X., Lu, L., Liu, S., Wang, Y., Liu, C., Li, B., Guo, W., Gao, S., Yang, Z., Li, F., Zeng, Z., Tang, Q., Pan, Y., Guan, M., . . . Liu, T. (2020). A chromosome-level genome assembly of garlic (*Allium sativum*) provides insights into genome evolution and allicin biosynthesis. *Molecular Plant*, *13*(9), 1328–1339. https://doi.org/10.1016/j.molp.2020.07.019

Swain, R., Sahoo, S., Behera, M., & Rout, G. R. (2023). Instigating prevalent abiotic stress resilience in crop by exogenous application of phytohormones and nutrient. *Frontiers in Plant Science*, *14*. https://doi.org/10.3389/fpls.2023.1104874

te Pas, M. F. W., Madsen, O., Calus, M., & Smits, M. (2017). The importance of endophenotypes to evaluate the relationship between genotype and external phenotype. *International Journal of Molecular Sciences*, *18*(2), 472. https://doi.org/10.3390/ijms18020472

Teper, D., White, F. F., & Wang, N. (2023). The dynamic transcription activator-like effector family of Xanthomonas. *Phytopathology®*, *113*(4), 651–666.

Theerawitaya, C., Praseartkul, P., Taota, K., Tisarum, R., Samphumphuang, T., Singh, H. P., & Cha-um, S. (2023). Investigating high throughput phenotyping based morpho-physiological and biochemical adaptations of Indian pennywort (*Centella asiatica* L. urban) in response to different irrigation regimes. *Plant Physiology and Biochemistry*, *202*, 107927. https://doi.org/10.1016/j.plaphy.2023.107927

Tian, Y., Chen, X., Xu, P., Wang, Y., Wu, X., Wu, K., Fu, X., Chin, Y., & Liao, Y. (2024). Rapid visual detection of elite erect panicle dense and erect panicle 1 allele for marker-assisted improvement in rice (*Oryza sativa* L.) using the loop-mediated isothermal amplification method. *Current Issues in Molecular Biology*, *46*(1), 498–512. https://doi.org/10.3390/cimb46010032

Tian, Z., Wang, J., Li, J., & Han, B. (2021). Designing future crops: Challenges and strategies for sustainable agriculture. *The Plant Journal*, *105*(5), 1165–1178. https://doi.org/10.1111/tpj.15107

Tong, C., Wu, F., Yuan, Y., Chen, Y., & Lin, C. (2020). High-efficiency CRISPR/Cas-based editing of *Phalaenopsis* orchid *MADS* genes. *Plant Biotechnology Journal*, *18*(4), 889–891. https://doi.org/10.1111/pbi.13264

Tyagi, S., Mazumdar, P. A., Mayee, P., Shivaraj, S. M., Anand, S., Singh, A., Madhurantakam, C., Sharma, P., Das, S., Kumar, A., & Singh, A. (2018). Natural variation in Brassica FT homeologs influences multiple agronomic traits including flowering time, silique shape, oil profile, stomatal morphology, and plant height in *B. juncea*. *Plant Science*, *277*, 251–266. https://doi.org/10.1016/j.plantsci.2018.09.018

Upchurch, R. G. (2008). Fatty acid unsaturation, mobilization, and regulation in the response of plants to stress. *Biotechnology Letters*, *30*(6), 967–977. https://doi.org/10.1007/s10529-008-9639-z

Van Noorden, R. (2010). Carbon sequestration: Buried trouble. *Nature*, *463*(7283), 871–873. https://doi.org/10.1038/463871a

Visscher, P. M., van der Beek, S., & Haley, C. S. (2021). Marker assisted selection. In *Animal breeding*. https://doi.org/10.1201/9781315137483-9

Voss-Fels, K. P., Cooper, M., & Hayes, B. J. (2019). Accelerating crop genetic gains with genomic selection. *Theoretical and Applied Genetics*, *132*(3), 669–686. https://doi.org/10.1007/s00122-018-3270-8

Wang, L., Breton, C., Warzecha, C. C., Bell, P., Yan, H., He, Z., White, J., Zhu, Y., Li, M., & Buza, E. L. (2021). Long-term stable reduction of low-density lipoprotein in nonhuman primates following in vivo genome editing of PCSK9. *Molecular Therapy*, *29*(6), 2019–2029.

Wang, W., Xu, J., Fang, H., Li, Z., & Li, M. (2020). Advances and challenges in medicinal plant breeding. *Plant Science*, *298*, 110573. https://doi.org/10.1016/j.plantsci.2020.110573

Watson, A., Ghosh, S., Williams, M. J., Cuddy, W. S., Simmonds, J., Rey, M.-D., Asyraf Md Hatta, M., Hinchliffe, A., Steed, A., Reynolds, D., Adamski, N. M., Breakspear, A., Korolev, A., Rayner, T., Dixon, L. E., Riaz, A., Martin, W., Ryan, M., Edwards, D., . . . Hickey, L. T. (2018). Speed breeding is a powerful tool to accelerate crop research and breeding. *Nature Plants*, *4*(1), 23–29. https://doi.org/10.1038/s41477-017-0083-8

Weinmann, S., Roll, S., Schwarzbach, C., Vauth, C., & Willich, S. N. (2010). Effects of Ginkgo biloba in dementia: Systematic review and meta-analysis. *BMC Geriatrics*, *10*(1), 14. https://doi.org/10.1186/1471-2318-10-14

Weldemichael, M. Y., & Gebremedhn, H. M. (2023). QTL mapping in sesame (*Sesamum indicum* L.): A review. *Journal of Biotechnology*, *376*, 11–23. https://doi.org/10.1016/j.jbiotec.2023.09.003

Xu, Y., Freund, D. M., Hegeman, A. D., & Cohen, J. D. (2022). Metabolic signatures of *Arabidopsis thaliana* abiotic stress responses elucidate patterns in stress priming, acclimation, and recovery. *Stress Biology*, *2*(1), 11. https://doi.org/10.1007/s44154-022-00034-5

Yadav, R. K., Tripathi, M. K., Tiwari, S., Tripathi, N., Asati, R., Chauhan, S., Tiwari, P. N., & Payasi, D. K. (2023). Genome editing and improvement of abiotic stress tolerance in crop plants. *Life*, *13*(7), 1456.

Yamada, A., Saitoh, T., Mimura, T., & Ozeki, Y. (2002). Expression of mangrove allene oxide cyclase enhances salt tolerance in *Escherichia coli*, yeast, and tobacco cells. *Plant and Cell Physiology*, *43*(8), 903–910. https://doi.org/10.1093/pcp/pcf108

Younessi-Hamzekhanlu, M., Ozturk, M., Jafarpour, P., & Mahna, N. (2022). Exploitation of next generation sequencing technologies for unraveling metabolic pathways in medicinal plants: A concise review. *Industrial Crops and Products*, *178*, 114669.

Yu, Y., Huang, W., Chen, H., Wu, G., Yuan, H., Song, X., Kang, Q., Zhao, D., Jiang, W., Liu, Y., Wu, J., Cheng, L., Yao, Y., & Guan, F. (2014). Identification of differentially expressed genes in flax (*Linum usitatissimum* L.) under saline–alkaline stress by digital gene expression. *Gene*, *549*(1), 113–122. https://doi.org/10.1016/j.gene.2014.07.053

Zandalinas, S. I., Balfagón, D., Gómez-Cadenas, A., & Mittler, R. (2022). Plant responses to climate change: Metabolic changes under combined abiotic stresses. *Journal of Experimental Botany*, *73*(11), 3339–3354. https://doi.org/10.1093/jxb/erac073

Zarghami Moghadam, M., Shoor, M., Nemati, H., & Nezami, A. (2021). Evaluation of 13 calendula (*Calendula officinalis*) cultivars response to drought stress. *Journal of Ornamental Plants*, *11*(2), 109–121.

Zhang, C., Pumphrey, M. O., Zhou, J., Zhang, Q., & Sankaran, S. (2019). Development of an automated high-throughput phenotyping system for wheat evaluation in a controlled environment. *Transactions of the ASABE*, *62*(1), 61–74.

Zhang, H., Mittal, N., Leamy, L. J., Barazani, O., & Song, B. (2017). Back into the wild – apply untapped genetic diversity of wild relatives for crop improvement. *Evolutionary Applications*, *10*(1), 5–24. https://doi.org/10.1111/eva.12434

Zhao, H., Shen, C., Hao, Q., Fan, M., Liu, X., & Wang, J. (2023). Metabolic profiling and gene expression analysis reveal the quality deterioration of postharvest toon buds between two different storage temperatures. *Frontiers in Plant Science*, *14*. https://doi.org/10.3389/fpls.2023.1142840

Zhao, Q., Li, R., Gao, Y., Yao, Q., Guo, X., & Wang, W. (2018). Modeling impacts of climate change on the geographic distribution of medicinal plant *Fritillaria cirrhosa* D. Don. *Plant Biosystems – An International Journal Dealing with All Aspects of Plant Biology*, *152*(3), 349–355. https://doi.org/10.1080/11263504.2017.1289273

2 Speed Breeding Technology for Enhanced Production of Secondary Metabolites in Medicinal Plants

Najwa Seddiqi Kallali, Khadija Goura, Haitam Lahmamsi, Mohammed Radi, Rachid Ezzougari, Ghizlane Echchgadda, and Rachid Lahlali

2.1 INTRODUCTION

Medicinal plants have been an essential element of traditional medicine for centuries, providing a diverse range of secondary metabolites with medicinal benefits (Cardoso et al., 2019). These natural substances have played important roles in treating diseases and stimulating well-being across cultures and civilizations. However, as modern healthcare systems increasingly rely on plant-derived therapies, the demand for these bioactive chemicals has risen to unprecedented levels (Hussein & El-Anssary, 2019). The World Health Organization estimated that 80% of the global population in developing nations relies mostly on medicinal plants for basic sanitation necessities (Vines, 2004).

Conventional cultivation methods, rooted in traditional agricultural practices, have long been employed in the production of secondary metabolites from medicinal plants. Despite their historical significance, these methods face significant challenges in meeting contemporary demands. Factors such as environmental constraints, prolonged cultivation periods, and inconsistent yields pose significant obstacles to the efficient and sustainable production of bioactive compounds (Ramawat & Arora, 2021)

The establishment of Mendelian laws was an important turning point in crop breeding. In the past 150 years, crop development has undergone a profound transformation, largely due to advancements in contemporary genomics (Collins et al., 2003). Several approaches have been used to shorten plant reproductive cycles. Recent advancements, including genomic selection, high-throughput phenotyping (HTP), and speed breeding have proven advantageous in accelerating plant breeding processes. Furthermore, genetic engineering and molecular technology have played a role in the development of crops with desirable characteristics through gene transformation (Majid et al., 2017). Other techniques, which involve large-scale sequencing, genomics, rapid gene isolation, and high-throughput molecular markers have also been proposed to improve the breeding of commercially significant crop species (Murovec et al., 2017).

Plant breeding serves as a pivotal tool for cultivating plants with desired traits. Over the past 10,000 years, humans have employed artificial selection to select and breed plants with enhanced nutritional qualities, marking a significant milestone in agricultural history (Moose & Mumm, 2008). Traditional agricultural practices have long aimed to elevate the nutritional profiles of various food crops. However, recent advancements in scientific research have unlocked a plethora of possibilities and innovations in plant breeding techniques. Meeting the escalating demand for plant-based products necessitates a doubling of the current annual yield enhancement rate in major crop species, which typically ranges from 0.8% to 1.2% (Varshney et al., 2006).

Speed breeding represents a paradigm shift in plant breeding methodologies, characterized by their ability to expedite breeding cycles and enhance the production of secondary metabolites in medicinal plants (Watson et al., 2018). In 2003, scientists at the University of Queensland invented the notion of 'speed breeding' to refer to a collection of approaches designed to accelerate wheat breeding operations (Uğur & Gören, 2023). Current efforts are focused on the development of speed breeding protocols for various crops (Ghosh et al., 2018). In contrast to the doubled haploid (DH) technique, which relies on the production of haploid embryos to generate entirely homozygous lines and often necessitates specialized equipment for in vitro culturing, speed breeding is versatile across different germplasms and does not mandate particular equipment requirements (Slama-Ayed et al., 2019).

The objective of this chapter is to assess the existing methods used in breeding medicinal plants and to critically analyze their potential for increasing the production of secondary metabolites and understanding their molecular mechanisms. Additionally, the chapter aims to explore the effectiveness of speed breeding technology within controlled environmental conditions for accelerating the growth cycle of medicinal plants. Moreover, it aims to compare the outcomes of this technology with traditional cultivation methods, particularly regarding the enhancement of key secondary metabolite production in selected medicinal plants.

DOI: 10.1201/9781003475491-2

2.2 CONVENTIONAL CULTIVATION METHODS OF MEDICINAL PLANTS AND THEIR CHALLENGES

The scientific evaluation of medicinal plant species for their potential medical applications has underscored their economic significance in the pharmaceutical industry. However, the extensive collection of these plants from their natural habitats have led to environmental degradation, endangering their survival (Ramawat & Arora, 2021). The unsustainable harvesting practices have pushed many medicinal plant species to the brink of extinction or categorized them as endangered. Being aware of the urgency of conservation challenges, companies and scientists are actively investigating alternatives for phytopharmaceutical manufacturing (Sood, 2020).

Plant primary metabolism encompasses essential life processes important for plant growth and development, including photosynthesis, respiration, and assimilation of water and nutrients (Canarini et al., 2019). These processes provide the energy and structural components necessary for plant survival and reproduction. In contrast, plant secondary metabolism operates parallel to primary metabolism, utilizing the energy and precursors generated therein to synthesize a vast array of specialized compounds. These secondary metabolites serve diverse ecological functions, particularly in enhancing plant adaptability to various environmental stresses encountered in wild habitats (Justin et al., 2014). By producing secondary metabolites, plants can defend against herbivores and pathogens, adapt to environmental changes and engage in chemical signaling with other organisms. This chemical communication facilitates complex ecological interactions, including symbiotic relationships with beneficial organisms and antagonistic interactions with potential threats. Overall, secondary metabolism significantly contributes to the ecological success and resilience of plant species by enabling them to respond effectively to biotic and abiotic challenges in their environment (Cardoso et al., 2019). On another hand, the World Health Organization's estimation suggests that over 80% of the population in developing countries primarily relies on herbal medicine for fundamental healthcare necessities (Vines, 2004). Around two-thirds of the 50,000 distinct species of medicinal plants currently in use are harvested from wild sources (Edwards, 2004). Additionally, in Europe, merely 10% of the medicinal species utilized commercially are cultivated (Vines, 2004). Moreover, there is an increasing concern regarding reduced populations, decline in genetic diversity, local extinctions, and degradation of habitats. While enhanced regulation and the use of sustainable wild harvesting methods can provide adequate protection for some species, increased domestic production of medicinal herbs is a more sustainable long-term solution (Ramawat & Arora, 2021).

The domestication of plants and animals is a centuries-long evolutionary process that involves farmer collaboration, crop adaptation, and interactions with adjacent ecosystems. This sophisticated process alters characteristics and promotes progressive co-evolution, eventually giving rise to new species or cultivated variants (Diamond, 2002). Domestication, widely recognized as one of the most significant advances in human civilization, promotes the establishment of more desirable and acceptable cultivars by a variety of selection pressures (Purugganan, 2019). The development of agricultural technologies for plant production, as well as understanding seed and floral biology, often takes several years, if not decades. Cultivation systems for many species have developed with human civilization, and agriculture currently accounts for around 38% of the total land area on Earth (Pretty & Bharucha, 2014).

In this context, horticulture holds significant potential to contribute in various ways to medicinal plant production (Paek et al., 2005). This includes domesticating wild medicinal plants, enhancing plant biomass, and increasing the yielding of plants with pharmacologically active compounds (PDMC) to encourage commercial production. Additionally, horticultural methods may use plants as natural bioreactors, thereby minimizing the irresponsible utilization of natural resources. Horticultural science offers methods to cultivate both native and exotic medicinal compounds within agroecosystems, drawing upon efficient methods commonly used in conventional horticulture for food production (Cardoso et al., 2019). These techniques encompass chemical and physical soil preparation, fertilization, and other tailored treatments (Gobbo-Neto & Lopes, 2007).

Also, cultivation provides a chance to use biotechnology to overcome inherent problems in the manufacturing of herbal medicines. These obstacles include species misidentification, genetic and phenotypic heterogeneity, extraction unpredictability and instability, and the presence of dangerous elements and pollutants (Canter et al., 2005). However, certain plants have proven difficult to cultivate due to factors such as low germination rates or particular ecological conditions (Vines, 2004). Natural selection and widespread seed dissemination are critical to crop plant adaptation and diversification. Despite having around 452,000 plant species, barely 500 of them are being farmed. This suggests that domestication has reduced the pool of genetic diversity, leading to a total loss of biodiversity (Ramawat & Arora, 2021). Consequently, to satisfy the increasing need for medicinal herbs as the population grows, there is a tendency to collect more plants from the wild, which puts species at risk and leads to habitat loss. However, this can be prevented by cultivating medicinal plants. Yet cultivating a single uniform crop could harm the biodiversity of that species (Ramawat, 2019), and if we don't practice in-situ conservation, we may lose it forever. Therefore, it's essential to practice conservation and cultivation together.

While the wild collection of medicinal plants remains popular, due to perceived potency resulting from slow growth and accumulation of secondary metabolites, cultivation offers distinct advantages in terms of quality control, uniformity, and long-term supply stability. However, not all medicinal plants could be cultivated because of biological constraints such as poor seed set, germination issues, and specific soil requirements (Schippmann et al., 2002).

Hence, the perceived qualitative superiority of wild medicinal plants compared to cultivated ones can often be attributed to cultural factors. An illustrative example is found in the case of ginseng, where wild roots are valued significantly higher than those produced through artificial propagation. This valuation stems primarily from cultural habits, particularly within the Chinese community, which constitutes the largest consumer group of wild ginsengs. In this cultural context, the resemblance of gnarled wild ginseng roots to the human body is symbolically associated with vitality and potency, enhancing the perceived efficacy of the root in traditional medicine (Robbins, 1998). Conversely, cultivated ginseng roots lack the distinctive shape and appearance of their wild counterparts, diminishing their desirability among consumers. This cultural preference underscores the intricate interplay between traditional beliefs, aesthetics, and perceived medicinal efficacy, shaping consumer attitudes toward cultivated versus wild-gathered medicinal plants (Schippmann et al., 2002).

However, studies have demonstrated that it is feasible to control the quantities of phytochemicals in cultivated medicinal plants (Palevitch, 1991). Furthermore, multiple investigations have proven that the quality of products developed from farmed Chinese medicinal plants, particularly those cultivated close to their native habitats, often surpasses that of wild populations, owing to improved cultivation techniques (Guo et al., 2009). These findings have gained acceptance among international traders and companies, highlighting the potential advantages of cultivated medicinal herbs (Laird & Pierce, 2002). Thus, cultivation emerges as a viable strategy for ensuring a consistent supply of high-quality medicinal plants while also addressing environmental concerns associated with wild harvesting practices.

Despite the significant medicinal and functional properties associated with plants like *Rosmarinus officinalis*, the majority of medicinal plant farmers continue to rely on wild types or local varieties that have not undergone targeted selection processes (Wang et al., 2020). For example, *R. officinalis* possesses crucial antibacterial and anti-inflammatory activities. However, it is noteworthy that among the 22 existing cultivars of *R. officinalis*, none have been specifically developed for medicinal applications. Instead, these cultivars are primarily utilized for culinary and ornamental uses (Begum et al., 2013). The lack of cultivar selection adapted to medicinal demands highlights an urgent need in the agricultural sector's approach to exploiting the full potential of medicinal plants for therapeutic reasons (Wei et al., 2015). Efforts to bridge this gap through targeted breeding programs could unlock valuable medicinal properties, thereby enhancing the sustainability and efficacy of medicinal plant cultivation (Wang et al., 2020).

Altogether, before approving cultivating methods, it is critical to choose medicinal plants based on demand, supply, biology, phytochemistry, and traditional use. Additionally, research is needed on aspects like collection methods, storage techniques, identifying marker compounds for quality control, and assessing microbial contamination (Chinthiya & Bhavyasree, 2019). Consequently, a balanced approach to sourcing medicinal plants is essential, integrating sustainable wild harvesting practices with cultivation efforts where feasible. This approach aims to ensure the continued availability of medicinal plant resources while mitigating the risk of overexploitation and loss of biodiversity. Conservation strategies must be implemented to protect wild species and their genetic diversity, thereby safeguarding the future of medicinal plant-based therapies (Chinthiya & Bhavyasree, 2019; Larson et al., 2014; Wei et al., 2015).

2.3 THE BASICS OF SPEED BREEDING IN THE PRODUCTION OF MEDICINAL PLANTS

2.3.1 Definition and Evolution of the Speed Breeding Concept

Speed breeding is a technology that offers great assistance to the growth and reproductive cycles of plants, and not only will it become one of the plant sciences breakthroughs. This method is growing plants under controlled lighting and temperature conditions that enhance plant growth (Wanga et al., 2021). Consequently, this enables the generation period of plants to be reduced, hence helping researchers and breeders easily accomplish goals in much reduced time (Sharma et al., 2023). Considering the production of secondary metabolites, speed breeding is of great potential to maximize the synthesis of these valuable compounds of key importance for plant defense mechanisms and which have broad applications in medicine, agriculture, and industry (Temesgen, 2022).

Secondary metabolites are those compounds synthesized in plants, and actually, they are not directly related to the process of growth, development, or reproduction, but they always maintain a certain level of interrelation with the functions of the environment (Anjali et al., 2023). Thus, such compounds are widely used for different applications of their bioactivity properties. Speed breeding could have transformative outcomes on the production of these metabolites, with the potential of rapidly generating plants under controlled conditions that would facilitate the study and manipulation of metabolic pathways involved in their biosynthesis (Isah et al., 2018).

It could therefore be grouped under one of the mainstream biotechnological tools through coupling genetic and biotechnological approaches, to achieve higher yields of secondary metabolites in this case: terpenoids. Thus, this integration offers highly promising opportunities for the intensification of the yield and diversity of secondary metabolites in much more efficient manners by capitalizing on the fast life cycle of the plant for faster and more iterated experiments (Bhatta et al., 2021).

Moreover, it is an extremely pivotal aspect to know the biosynthetic pathways of the products and the environmental control over the production of these secondary metabolites. If this power joins hands with that, the way of production of these compounds would change. The treatment with elicitors was enhanced and demonstrated to increase the production of metabolites in plants cultivated under speed breeding protocols. These approaches not only enhance metabolite production efficiency but are also in compliance with sustainability practices, as they curtail dependence on natural resources. On the other side, the environmental impact of the same is simultaneously lowered (Reshi et al., 2023).

Indeed, the development of plant breeding and pioneering activities of outstanding personalities in this field, such as Gregor Mendel and Norman E. Borlaug, have opened the way for huge improvements in agriculture, particularly in terms of cultivation and genetic improvement techniques. One such area is for secondary metabolite yield in medicinal and aromatic plants (MAPs) (Barut et al., 2022).

Mendel's pathbreaking experiments with pea plants in the mid-1800s established the fundamental laws of inheritance and laid the foundation for the modern discipline of genetics. His work observed most carefully established the inheritance of traits from parents to their descendants, a concept that eventually became important for understanding genetic selection and improvement in plant breeding (Smýkal et al., 2016). These form the basis of Mendelian genetics, and over the years, they have helped in the molecular characterization of genes. The above have contributed to breeding strategies geared to phenotypic plant trait optimization, for instance, increased yields and quality for secondary metabolites in MAPs (Sussmilch et al., 2022).

The Green Revolution, led by Norman E. Borlaug, brought to the mid-20th century renewed, progressive agricultural technologies that seriously stimulated food production worldwide. The work of Borlaug, in developing semi-dwarf, disease-resistant wheat varieties, comes out as a lead exemplar of how well-targeted plant breeding can transform the food security and agricultural productivity of the world (Sakurai, 2022). This epoch has revealed the possibility of genetic improvements in crops, including those of MAP, toward the perennial increase of demand for medicinal resources and food.

As a recent step, plant breeding largely uses biotechnological tools and genomic resources that can permit the precise manipulation of genes regulating secondary metabolite production in medicinal and aromatic plants. These advances will also further broaden the diversity and efficacy of plant-based medicinal compounds and show how breeding has evolved from Mendelian genetics to modern genomic selection and gene editing (Alseekh et al., 2023). Generally, it is considered that Mendel's work is the foundation of genetic research. At the same time, the improvements developed under the Green Revolution by Borlaug would still loom large over the whole space of plant breeding.

2.3.2 Comparison with Traditional Breeding Methods

Conventional plant breeding is the traditional method of plant breeding, which aims to reorganize plant populations into new genotypes. It strictly adheres to the natural heredity principle; in that it involves the selection of plants based on the superiority of the traits they offer for breeding (Lamichhane & Thapa, 2022). It means, in a way, mating or hybridization between closely related individuals, followed by the mixing of their genes (Anand et al., 2023). If we compare it with a long history of breeding field crops, it's relatively recent: the selection and improvement of medicinal plants using traditional methods. Research relating to the breeding of medicinal plants for these biochemicals is in its early stages (Lamichhane & Thapa, 2022).

Traditional breeding has further developed with time and has incorporated many techniques for increasing genetic diversity in plants and improving some of their characteristics. These methods included hybridization by way of multigenerational crossbreeding, using wild parent plant characteristics, mutagenesis to bring in genetic variations, and double haploid technology to reach homozygosity at a faster pace (Li, 2020). Besides, various methods of tissue culture, such as embryo rescue and egg rescue techniques, has been applied to increase plant regeneration abilities (Manchanda et al., 2023). Protoplast fusion has even been an innovative approach to overcoming barriers in traditional hybridization approaches, where genetic material from plants is not allowed to be combined by crossing naturally (Armita, 2020).

To improve the efficiency of their breeding programs, plant breeders are constantly innovating and refining techniques. Some of the most important methods that can be integrated with other approaches, include:

Selection: This basic method is used most frequently to improve crops, both by researchers and by untrained farmers. It involves making distinctions between varieties to select desirable plants for further development. In some cases, selection focuses on an individual plant; elsewhere, it involves several plants that will be chosen and selected as part of a breeding program. As part of their breeding programs, breeders employ a variety of different selection programs and breeding methods (Acquaah, 2016).

Artificial pollination: This technique has a long history and is very important in modern plant breeding. It enables genetic research, the creation of targeted variability, directed gene transfers, stock breeding, and boosting fruit and seed production, among other uses. However, challenges to successful artificial pollination is posed by problems like genetic incompatibility ('genetically induced prevention of fertilization'). Various approaches to overcoming these problems are available (Acquaah, 2012).

Hybrid breeding: This method involves merging desirable traits from two or more varieties through hybridization. The selection of parent plants is a critical step in this process, as the progeny's characteristics are directly influenced by the parents' genetics. Hybridization employs heterosis, or hybrid vigor, allowing descendants to outperform their parents (Fridman, 2015). At present, several interesting varieties have resulted from these crosses.

Embryo culture: For sterility reasons, embryos from very crosses simply don't perform properly and have to be extracted (rescued) early in the process and expanded in vitro into whole plants using tissue culture techniques (Acquaah, 2012). This procedure has been used successfully in crosses between species of the *Hylocereus* genus (Cactaceae) (Cisneros & Tel-Zur, 2010).

Bridge crossing: This technique involves indirectly creating a transient or intermediate cross with two parents of different ploidy levels (different chromosome numbers) (Yirga, 2021). This results in sterile reproduction, and it is necessary to double the chromosomes to re-establish fertility.

DNA marker-assisted breeding: DNA marker-based selection for the construction of new types of phenotypes, which include a high yield, improved quality, and more resistance of the ideal variety, DNA-based selection markers, DNA polymorphism, molecular hybridization, polymerase chain reaction, and high-speed sequencing is carried out for screening (Dong et al., 2017). Traditional breeding methods tend to rely on the visible phenotype of the plants, which might be subjected to quite a lot of fluctuation based on factors such as the genotype of the plant, management techniques used in the field, age, and environment. Generically, the process of producing medicinal varieties depends much more on the knowledge of the breeders, in general, long-drawn-out. Molecular markers are used in the identification process to speed up the selection of medicinal plants targeted. Many successful experiments in the application of molecular markers to research on medicinal plants have been noted. However, most of the researchers focused only on the subject of polymorphism and species identification, and a very small number of reports exist on the subject of marker-assisted selection of these plants (Wang et al., 2020).

Protoplast fusion: This is another technique for extending crosses. It is possible to create protoplasts (cells without a cell wall) in vitro to obtain hybridizations when crosses are made via pollination or normal fertilization are difficult or impossible, to achieve (Acquaah, 2012).

Mutagenesis: Hereditary variation is the basis of plant breeding. Hybridization makes it possible to restructure existing variations to obtain new matrices or new combinations, without creating new genes. However, to create new or non-existent genes, it is possible to use mutagens (mutant agents – chemical or physical) to create other sources of breeding variation (Acquaah, 2016). Mutations continue to be used in modern conventional plant breeding, in contrast to more precise breeding technologies offered by advanced biotechnology (Bado et al., 2013; Udage, 2021).

Currently, only a small number of species belonging to medicinal plant cultivars have been bred properly, and studies relating to the selection of medicinal plants and their genetic potential has remained poorly attended. In the world's genetic resources of medicinal plants, the study, collection, assessment, and preservation are systematized only in very rare exceptional cases. The breeding scientists of these plants frequently require basic data on medicinal plants, such as genetic parameters, physiological conditions of the characteristics, the concerned seeds, and the relation existing between the quality and efficacy of the medicinal substance with the growth factors. Medicinal plants have been in use for years, though subjected to very little selection since only a few have been chosen in such a way that a variety is derived with the mixture of varieties of medicinally active ingredients between several genotypes. According to some records, only 40% of the medicinal plants used at a larger scale have undergone genetic improvement research. Out of the 225 newly created varieties, 71.2% were derived from selection, 10.1% from hybridization, 8.6% from cloning, 4.5% from mutagenesis by chemical or radiological means, and 5.5% from other methods (Wang et al., 2020).

2.4 SPEED BREEDING KEY COMPONENTS

2.4.1 Surveillance of Environmental Conditions

While speed breeding (SB) sounds like a magnificent idea to increase the secondary metabolites production in medicinal plants, ensuring a steady yield over many growth cycles, it is essential to mention that SB strategies must be tailored for each crop type, with a meticulous examination of growth conditions. Thus this technique demands high infrastructure equipped with control facilities for temperature, photoperiod, humidity, etc. (Shanmugavel et al., 2023), which will be discussed in the following section. Generally, this technique is conducted in greenhouses where the conditions are meticulously controlled, while the approach configuration is based on individual experiments in the published literature (Watson et al., 2018).

2.4.1.1 Optimized Light Regimes

The long generation time of most plants is a major problem facing applied researchers and breeders, posing a bottleneck in the way of fast-tracking research and development of new varieties. In the case of medicinal plants, the concentration of secondary metabolites (SMs) throughout the seasons and the geographic distribution depend heavily on the environmental conditions, which represent another problem of SM instability and supply fluctuations (Zhang et al., 2021). These problems create a need for ingenious technologies to accelerate plant growth, shortening its generation turnover and ensuring a steady supply of SMs.

As described earlier, speed breeding is a new technology based on the meticulous control of environmental conditions to speed up the plant's transition from the vegetative to the reproductive phase. One of the involved procedures is the light control either by changing the light exposition period and intensity or its quality (wavelength) (Ghosh et al., 2018).

This control is done following several protocols depending on the plants; for instance, the modulation for photoperiod-sensitive plants is done by increasing the light exposition period within a range of 16 to 22 h for long-day plants like radish, turnip, wheat, and broccoli, or by decreasing it within a range of 8 to 12 h for short day plants like hemp, green and black gram, soapnut, turmeric, and ginger.

The optimization of light spectrum and photoperiod is crucial for plant growth acceleration and breeding cycle shortening. A photoperiod of 13 h at the vegetative and pod-filling stages, with a broad spectrum of 5700 K LED light for growth and an 8 h photoperiod with 735 nm far-red light for flowering, can enhance plant development. In addition, variations in photoperiod, genotype, and their interaction influence flowering time and plant height (Gangashetty et al., 2024).

To achieve faster generation times, a photoperiod of 18 to 22 hours with a brief dark period is recommended, along with attention to light quality and intensity, aiming for a photosynthetic photon flux density (PPFD) of around $450-500\ \mu mol/m2/s$ at canopy height. Consistently, light spectra covering the photosynthetically active radiation (PAR) region, especially blue, red, and far-red ranges, are suitable for plant growth and could be achieved through LEDs or supplemental lighting in glasshouses. Finally, regular monitoring of light spectrum and intensity is highly recommended to ensure optimal conditions for different plant species.

2.4.1.2 Temperature Modulation

Temperature is a key factor that significantly influences seed germination, since it directly affects the biochemical reactions involved in the germination process, thus affecting the germination rate and percentage, flowering period, plant growth, maturity, and seed set percent by regulating their metabolism. It is also an important factor that influences the plant transition from the vegetative to the flowering stage (Okello et al., 2022).

The temperature effect on the germination of different plant species encompassing medicinal plants, has been the subject of several researches. Baskin and Baskin (1998) reported that optimal germination temperatures for several plant species are confined between 10 and 20 °C, while Wanga et al. (2021) indicated that a range of temperature confined between 12 and 30°C is suitable for seed germination of a wide variety of crop plants. On the other hand $25-30°C$ is for the other developmental stages (flowering, growth and seed formation) (Wanga et al., 2021). In addition, Bannayan et al., in 2006 were reported that huge and very different impact of temperature on different medicinal plants species.

The altered temperature regime has also a significant effect on plant growth, for instance, the development of groundnut plants was faster when introduced to a temperatures regime between 17°C/32°C (Shanmugavel et al., 2023). In lentils and fava beans, the growth under a 22°C light/18°C dark temperature regime resulted in around eight generations per year (Mobini & Warkentin, 2016).

Several researchers reported the effect of temperature on the flowering time (Samineni et al., 2020) and also the concentration of secondary metabolites, suggesting that culture temperature change may affect the metabolism and physiology of the plant and subsequently affect its growth and secondary metabolite production (Ramakrishna & Ravishankar, 2011). For instance, high temperatures were proven to increase the concentration of secondary metabolites such as *Panax quinquefolius* (Jochum et al., 2007).

Moreover, seed germination is inhibited by high temperatures, which could be explained by the upregulation of abscisic acid (ABA) biosynthesis genes and downregulation of catabolic genes that increase ABA endogenous levels. In addition, high temperatures repress GA3 biosynthesis genes, decreasing GA3 content and thus inhibiting seed germination. Temperature also controls the imbibition process, which is crucial for germination, with low temperatures slowing the process and threatening cell membranes. Furthermore, the activity of some enzymes involved in germination, such as dehydrogenases, increases with temperature (Okello et al., 2022).

Since it has a huge impact on plant growth and development, temperature modulation could greatly shorten the generation time of plants. However it is a delicate parameter and its modulation has to be meticulous in order not to induce stress and affect plant productivity (Watson et al., 2018); thus, the temperature control system has to be carefully implemented within the SB system to get the desired results (Samantara et al., 2022).

2.4.1.3 Humidity Control

Another parameter to control in SB systems is humidity, which plays a critical role in influencing plant growth and breeding outcomes. Generally, a percentage of 60 to 70% is optimal to accelerate the breeding process and maximize plant growth as described by several authors (Samantara et al., 2022). Greenhouses and growth chambers are proved to be a perfect controlled environment to meticulously regulate the humidity levels, thus enhancing plant growth and prosperity, unlike field conditions where several parameters affect its growth and breeding cycle (Jähne et al., 2020).

To accelerate grain filling and maturation, some soil management strategies that include moisture level reduction after flowering were shown to be effective in crops like canola, barley, wheat, and chickpea (Ghosh et al., 2018). Also, it is important to note that humidity level management strategies must be tailored according to each plant's requirement, considering other parameters such as temperature and CO2 concentration (Iovane et al., 2022).

As with any other parameter control, whenever we are close to the two extremities the effect becomes negative to the plant's physiology, either excessively high or low humidity levels could have a detrimental effect on the reproduction process and plant physiology. In the first case, when the humidity level is high, particularly if it's conjugated with elevated temperatures, the thermotolerance and pollen viability could be significantly reduced, thus affecting the crop yield (Pacini & Dolferus, 2019); when it comes to the other case, low humidity levels are more likely to induce the pollen grain metabolic dormancy in response as a resistance strategy to environmental stress (Pacini & Dolferus, 2019). Consequently, meticulous and precise management of humidity levels is essential to maintain optimal conditions for plant reproduction and growth processes mainly in a speed breeding framework.

Finally, humidity level management coupled with other environmental parameters is crucial to ensure sustainability in the agricultural systems and guarantee global food security. By understanding the multiple interactions that took place between the humidity, temperature, light exposition, and other parameters, speed breeding programs could be optimized to mitigate the different abiotic stresses and accelerate plant growth, flowering, grain filling, and other processes.

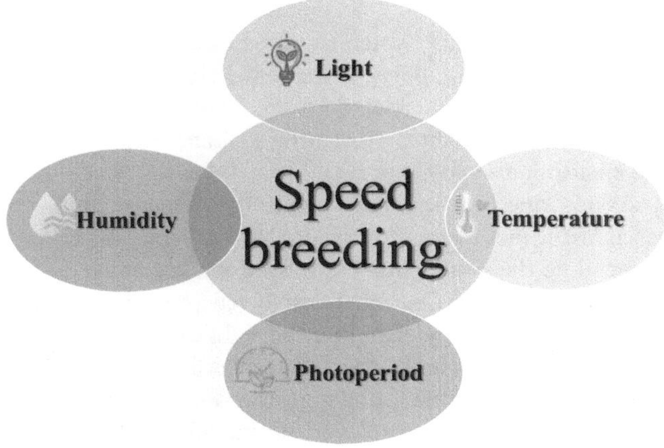

FIGURE 2.1 Primary influencing factors in speed breeding.

(The important factor is light, in the spectral range and intensity required. In the spectral range most readily increasing plant photosynthesis, it is between 400 and 700 nm, and it can be readily achieved by different sources of light, for example, light-emitting diodes (LEDs), halogen, or sodium vapor lamps (SVLs). It is important to maintain the right light intensity, with a recommended flux density of photosynthetic photons (PPFD), which ranges from about 450–500 μmol/m²/s. Besides, photoperiod is very important, and the normal cycle is 22 hours of light to 2 hours of darkness every 24 hours. Temperature is another vital element, which defines maximum and minimum temperatures adjusted to each kind of crop in the most appropriate growth. However, humidity should generally stand between 60 and 70%, but for some crops, it can be sustained at slightly lesser values because the crop can withstand drier conditions.)

2.5 MOLECULAR MECHANISMS

2.5.1 Speed-Breeding-Related Genomic and Transcriptomic Analyses

At the molecular level, the impact of speed breeding goes far beyond mere phenotypic changes to drive major overhauls at both the genomic and transcriptomic levels. Such changes comprise much more than the upregulation of growth and flowering genes, they cover a wider range of genetic changes that predispose plants to rapidly develop and increase the production of secondary metabolites (Gudi et al., 2022).

At the genomic level, speed breeding can induce epigenetic modifications. They play a vital role in the mechanisms that plants go through when breeding at high speeds. These epigenetic changes on the genomic level can readily impel, independently, the expression of several key genes in the consensus change to the DNA sequence. Such a change can either activate or deactivate genes when a response to definite environmental signals does not make a change in the underlying actual DNA sequence (Saxena et al., 2019). In other terms, the variations, such as DNA methylation or histone acetylation, may be the cause of changes in the chromatin structure by which the openness of the gene and level of expression is modulated. The accelerated flowering time may, in part, be due to the epigenetic repression of FLC, which is a major floral repressor in the context of speed breeding. The long-light conditions, which are environmental cues to the process of speed breeding can trigger histone changes in the gene, causing the silencing of the FLC, which is key to escaping from the process of precocious flowering, hence allowing the process even under non-inductive photoperiods (Chen & Penfield, 2018).

Also, transcriptomic changes brought about by speed breeding are not limited to the realm of growth and development genes but extend into the metabolism of secondary compounds, crucial for both plant defense and medicine. Upregulation of the genes that are associated with regulating this secondary metabolic pathway, as observed in the speed breeding experiments, is the cue to some complex mechanism that leads to an enhanced synthesis of natural products within the plant (Xiao et al., 2022). For example, under speed-breeding conditions, the increase in the production of certain SMs occurs due to the upregulation of gene expression encoding key enzymes within these biosynthetic pathways. This increment in the overexpression of genes involved in the mevalonate and non-mevalonate pathways could lead to an accumulation of terpenoids, which results in high levels of accumulation of terpenoid compounds widely recognized for their vast pharmaceutical applications (Kappers et al., 2005).

The combination of such speed-breeding protocols with epigenetic and transcriptomic analyses points to a multi-pronged approach to plant breeding, much above and larger in its scope than traditional methodologies, alluding to the remarkable molecular adaptability in plant systems. More specifically, such conditions can largely promote genetic system flexibility through epigenetic modifications (Ghosh et al., 2018). This adaptation shows not only the resilience of plant systems but also a valuable advantage for plant breeders for their respective breeding programs. Majorly, the plasticity of epigenetic mechanisms to modulate gene expression in response to an extended photoperiod and other environmental cues allows dynamic regulation of growth and development pathways to fasten the breeding cycle without any compromise to the genetic diversity (Uğur & Gören, 2023).

This work, in turn, suggests that the transcriptomic profiles under speed breeding conditions can be controlled, hence opening exciting new frontiers for improving the phytochemical repertoire of the plants (Xiao et al., 2022). Targeted manipulation of gene expression profiles related to secondary metabolism offers a unique opportunity that will augment the synthesis of valuable phytochemicals, hence elevating the pharmaceutical and nutraceutical potential of cultivated plants. Such an intricate control of the transcriptomic dynamics allows, within plants, not only for the fast-tracking of desired traits but also opens up the potential for a significantly increased concentration and diversity of bioactive compounds (Luo et al., 2019).

2.5.2 Upregulation of Genes in Biosynthetic Pathways

The speed breeding molecular mechanism is based on upregulating genes in biosynthetic pathways to increase the production of secondary metabolites in medicinal plants (Samanta et al., 2023). Upregulation describes that certain environmental or genetic manipulations lead to the enhancement of the expression level of certain genes with the increase in the synthesis of particular metabolites (Chaturvedi et al., 2022). The increasing gene expression is essential for amplifying the pathway synthesizing valuable phytochemicals, hence increasing the medicinal value that is contained in these plants (Chaturvedi et al., 2022; Samanta et al., 2023).

The speed of breeding changes the condition of growth, and hence, by doing this, the plant shows a faster life cycle. So it directly affects the genetic machinery of the plant. It means that the technique of speed breeding can bring about upregulation of those genes that are involved in secondary metabolite biosynthesis by creating conditions that would be ideal for gene activation (Watson et al., 2018; Xu et al., 2023). For example, the use of some light regimens, such as LEDs of distinct spectra, led to promising results in modulating gene expression linked to the synthesis of secondary metabolites in *Arabidopsis thaliana*. Targeted, such manipulation of light conditions may significantly affect plant developmental stages, including flowering, which is the most important period for metabolite accumulation (Jafari et al., 2018). This approach mirrors the possibility for speed breeding to further the production of secondary metabolites directly due to the expression of genes involved in their biosynthetic pathways, basically a fine amalgamation with plant physiology and genetic engineering for the optimization of medicinal plant yield (Jafari et al., 2018).

Further, the molecular basis that underlies the success of speed breeding is not confined to manipulating environmental cues; it includes genetic engineering and molecular breeding techniques. In other words, secondary metabolites are produced copiously and at a high rate by specifically expressed or upregulated genes in a strategic position of certain genes or modifications. This improvement in the genetic process fits into speed breeding since traditional breeding approaches are integrated with molecular biology to enhance the synthesis of necessary components (Chaturvedi et al., 2022).

2.5.3 Molecular Basis Explaining the Efficiency of Speed Breeding

Speed breeding is generally a revolutionary shift in the cultivation of medicinal plants, which involves advanced bi-rational biotechnological tools. The technique is generally exploited widely in in-vitro regeneration, cryopreservation, bioreactors, genetic transformation, and other general aspects of biotechnology to enhance production (Schulz, 2020). The use of secondary metabolites, compounds of the critical need for the medicinal value of plants. This successfully regenerated plant cells in-vitro through regeneration with precision in the process while maintaining genetic integrity for sustaining desired traits (Guo et al., 2021). In a way, cryopreservation can serve the same in case it does allow storage for an indefinitely long time without allowing any genetic drift, thus ensuring regenerations from high-quality lines even after long storage (Niazian, 2019; Tripathi & Tripathi, 2005).

A bioreactor is a facility for mass cell culture of plant cells whose growth has metabolic potential; it should be controlled in an optimized manner. The conditions are fine-tuned in a way that allows the plant cells to develop better with good metabolic profiles (Kong et al., 2014). Genetic transformation could be the most direct of them, and it includes the incorporation of specific genes within the plant genome for the overexpression of target metabolites. These also make sure that their efficient production of secondary metabolites is practiced and help in the introduction of new biosynthetic pathways that possibly could not have been naturally present within species (Kong et al., 2014; Kumar et al., 2019).

Upregulation of some strategic key genes within the respective biosynthetic pathways is responsible for the enhanced speed breeding efficacy of secondary metabolite

TABLE 2.1
Advanced Genetic and Molecular Techniques in Speed Breeding for Secondary Metabolite Enhancement

Strategy	Description	Example	Reference
Gene insertion and overexpression	This would include the insertion or overexpression of the essential enzymes of the key secondary metabolite pathways through which the production of respective targeted compounds can be increased.	*Arabidopsis* and *Petunia* have been used to investigate if the gene expression in the gibberellin biosynthesis pathways is affected by the light conditions, particularly by red and blue LED lights. This genetic manipulation reveals how the environmental factor condition can be part of the factors that, in tandem with genetic engineering, would result in optimal metabolic outputs that enhance the growth and flowering stage of the plant. Flowering is an essential stage of the plant for secondary metabolite production.	(Schrieber et al., 2019)
CRISPR/Cas9 gene editing	It employs the CRISPR/Cas9 system for precise editing of the genome, which shall include genome editing toward modification or knockout of secondary metabolite pathways regulating genes.	For the characterization of a novel gene in the GATA family of *Arabidopsis thaliana*, the application of the CRISPR/C08Cas9 system was targeted to a light signaling response GATA family gene, such as PIF4, for signal integration. The gene has an important role in the regulation of developmental and metabolic processes, and again, this work presents how accurate genetic edits can manipulate the physiology of the plant to increase the production of secondary metabolites, which might be useful in producing stronger plant varieties for use in pharmaceuticals.	(Naito et al., 2007)
Transcription factor engineering	It involves the engineering of transcription factors to activate specific genes within a secondary metabolite pathway to enhance the biosynthesis of desirable compounds.	Light-induced expression of NADPH-protochlorophyllide oxidoreductase: This research was about the expressional nature of a gene responsible for a key enzyme in chlorophyll biosynthesis under specific light conditions. Engineering transcription factors that may overexpress under controlled light environments will have important effects on the ability of the plant to synthesize chlorophyll and other secondary metabolites important to improve the medicinal properties of the plants.	(Naito et al., 2007)
Metabolic engineering	It involves reprogramming entire metabolic pathways by upregulation of the biosynthetic genes and suppression of competing pathways to channel resources towards the target metabolite production.	Schmitz et al. investigated if sugar metabolism was responsible for acclimation at high-light-intensity-related gene expression in *Arabidopsis thaliana*. In turn, the activity of all these nuclear genes might be influenced by the level of the carbohydrate as well as the genes that were involved in the process of photosynthesis and secondary metabolite pathways. This work is exemplary of how metabolic engineering can be targeted at channeling energy and precursor molecules to increase the biosynthesis of important secondary metabolites in the adaptations and productivity of plants under varying light conditions.	(Schmitz et al., 2014)

production (Luo et al., 2019). Those mentioned flexible metabolic networks of organisms become even more flexible with genetic engineering, as it allows the selective amplification of a specific metabolic pathway either through overexpression of relevant genes or the introduction of new genes from other organisms (Udage, 2021). Therefore, an increase in the content of certain metabolites, such as terpenoids and alkaloids of pharmacological interest, is expected to be indirectly enhanced. This is further potentiated by the modulation of the transcriptional factors involved in the regulation of these genes, hence making sure that the metabolic flux is pointedly redirected toward the synthesis of wanted compounds in greater quantities (Wu et al., 2020).

The adoption of speed breeding more recently to integrate genomic and transcriptomic analysis, and these have greatly enriched our knowledge of the genetic underpinning for secondary metabolite production (Singh et al., 2020). Genomic analysis will give an accurate map of the plant genome with genes identified that can correlate with high metabolites. This is very useful information in selecting for breeding the best genetically realized potential individuals of the species (Choi, 2019). It is an application to cover the gene expression patterns of transcriptomics. Meanwhile, under various environmental conditions or genetic modifications, it highlights how these factors affect the biosynthetic potentials of the plant. These will serve as powerful scaffolds not only for the identification and exploitation of the genetic factors governing metabolite production but also for the optimization of breeding for enhanced output of secondary metabolites (Wu et al., 2020).

2.6 APPLICATION TO MEDICINAL PLANTS

Speed breeding technology is an enormous leap that must be put into practice during the cultivation and research of medicinal plants (Samantara et al., 2022; Temesgen, 2022). Improvement of medicinal values of crops is a critical problem of agricultural science, like hemp (*Cannabis sativa*) and pepper (*Capsicum annuum*). The ability of this technique to bring down the time needed to go from seed to seed by many orders of magnitude is a game-changer for biologists and breeders of plants (Temesgen, 2022).

In the case of hemp (*C. sativa*), that infamous plant of diverse uses in medicine, an equally rapid-generation cycling protocol was developed for it. It could see it through a generation in less than nine weeks from seed to seed. This advance will greatly facilitate timely crosses of cultivars with defined genetic traits in hemp for the rapid development of new cultivars. It is likely to have very high value in application for medicinal cannabis either in research or agriculture. Such rapid turnover in generations is important for speeding up the process of breeding to obtain hemp varieties with improved medicinal properties and, among them, higher content of cannabidiol (CBD) its applications have been found in therapy (Schilling et al., 2023).

Another recent development of plant photoperiodic lighting dealt with the pepper (*C. annuum*) crop, which contains compounds that have potent antioxidant and anti-inflammatory properties. It used a speed breeding protocol, whereby the photoperiod and quality of light were controlled to use specifically an extended photoperiod (Epp) of 20 hours and a low red (R)-to-far-red (FR) ratio. Both these approaches helped to increase the breeding cycle, but this second one also helped in genetic analysis to understand the flowering response in the speed breeding system. Moreover, genome-wide association studies (GWAS) and gene expression analysis found a new locus of four candidate genes for flowering, which are located at the newly identified locus, further supporting the effectiveness of the protocol in dissecting genetic determinants that could be useful in modifying medicinal traits of pepper plants (Choi et al., 2023).

Further support to shed more light on the application of speed breeding to the light environment as investigated by Liu et al. (2022) showed various usages of this technology toward the promotion of phenotyping and breeding programs with accelerated growth and development features in hot peppers. Their potential to produce four generations per year opens the door for the rapid phenotypic selection and breeding of cultivars with augmented medicinal and nutritional values, underlying the efficiency gains afforded by speed breeding in the realm of agricultural research.

This has very far-reaching implications: it will allow, at least theoretically, the development of new medicinal plant varieties faster than this has ever been possible before, in addition, to increasing the precision of genetic trait selection and subsequent propagation (Cardoso et al., 2019). This confluence of speed and specificity opens up possibilities for how plant-based therapeutics and medicines could be developed far more quickly from discovery to application in a therapeutic manner. Hence, this project technology of speed breeding does not act as a tool of agricultural efficiency but becomes one of the pivotal innovations in pursuit of optimized medicinal plant varieties with the potential to be a revolutionizing technology for the field of plant-based medicine (Wanga et al., 2021).

2.7 ADVANTAGES AND CHALLENGES OF SPEED BREEDING

Speed breeding (SB) is an innovative technique that harnesses controlled environments to promote rapid vegetative growth and development, facilitating an accelerated transition to the reproductive stage within high-density plantations (Bonea, 2022). The principle of rapid selection involves utilizing optimal light intensity, temperature, and day length (22 hours of light, 22°C day/17°C night, and high light intensity) to increase photosynthesis rates, directly stimulating early flowering, associated with annual seed harvesting to shorten generation duration. By employing growth chambers or greenhouses equipped with artificial lights and other environmental regulation devices, SB enables multiple plant generations in a year, thus accelerating the plant selection and breeding process for the development of new varieties with improved traits (Chiurugwi et al., 2019; Watson et al., 2018). This approach offers considerable potential for improving and diversifying crops, including medicinal plants, by enabling faster production of varieties with increased levels of active compounds or traits suited to specific growth conditions. In contrast to the conventional pattern of two to three generations annually, SB enables numerous cycles, achieving up to nine generations per year for crops like *Triticum*, and *Hordeum vulgare* (Ghosh et al., 2018). One of the main advantages of SB lies in its ability to incorporate desirable traits into crops such as disease resistance, increased yields, enhanced nutritional value, and quality characteristics. By employing techniques such as genetic engineering or marker-assisted selection, it is possible to seamlessly integrate desirable genetic traits into crops, thus creating new crop varieties with improved characteristics (Hickey et al., 2017). This technique not only saves time but also minimizes labor costs, rendering it a cost-effective approach. Beyond its time- and resource-saving advantages, SB empowers plant breeders to quickly adapt to shifting environmental circumstances and the emergence of plant diseases, among other factors (Varshney et al., 2006). Furthermore, this method is not restricted by external environmental factors, allowing breeding activities to occur year-round. Additionally, it streamlines the incorporation of multiple favorable characteristics into a singular crop variety, enhancing its resilience. This approach also fosters informed decision-making and problem-solving across diverse research domains while promoting the optimal timing of critical growth stages necessary for seed development and maturation.

However, despite these advantages, there are several limitations and challenges associated with accelerated breeding techniques. One significant challenge is the insufficiency of qualified professionals and specialized technicians in this field, which hampers the efficient implementation of these techniques (Wanga et al., 2021). Continuous financial support for long-term research and development, as well as government policy support, are crucial for overcoming this challenge. Additionally, the adoption of major changes in the design and operations of breeding programs is necessary (Cazzola et al., 2020). The lack of appropriate infrastructure and efficient complementary plant facilities, especially in underdeveloped countries, further complicates the widespread adoption of these techniques. Ensuring consistent crop performance across diverse environments remains a challenge, requiring careful management of interactions between combined traits. Genetic variations within plant species also pose challenges in maintaining stability and consistency in crop performance. Furthermore, managing pest infestations and diseases, as well as individual monitoring for gene discovery, are critical tasks in this process (Potts et al., 2023). Lastly, the potential reduction in genetic diversity resulting from the accelerated techniques must be carefully managed to prevent similar damage and loss caused by excessive plant cultivation beyond their physiological thresholds. Novel technologies such as solar conditioning offer cost-effective solutions to efficiently facilitate this process but require further development and implementation to address these challenges comprehensively (Cazzola et al., 2020; Potts et al., 2023; Roeber et al., 2021; Thomson et al., 2022; Wanga et al., 2021; Xu et al., 2023).

Furthermore, the widespread application of SB for research and selection remains a challenge in countries with limited resources due to reduced infrastructure, lack of expertise, and limited collaboration with international organizations (Samantara et al., 2022). SB represents a promising advancement in the field of plant breeding, offering substantial benefits in terms of accelerating the plant selection and crossing process. However, to fully harness the potential of this technique, it is crucial to consider and overcome the challenges associated with preserving genetic diversity and ensuring the long-term sustainability of cropping systems.

2.8 CONCLUSIONS

The review has provided a comprehensive overview of the potential of speed breeding technology in enhancing the production of secondary metabolites in medicinal plants. The advancement of the medicinal plants sector heavily relies on breeding, which plays a pivotal role in catering to the specific requirements of stakeholders along the production chain, thereby enhancing the quality, profitability, and sustainability of production. Despite the significant potential inherent in the genetic resources of medicinal plants, their full exploitation remains at an early stage of development. Conventional breeding methods are currently predominate, benefiting from the abundant natural variability present in medicinal plant genotypes. Speed breeding is a potential strategy for modifying plants to satisfy the individual needs of consumers throughout the production process. While traditional breeding methods remain dominant, new biotechnology approaches provide potential for improvement, but at a higher cost. However, consumer preferences for natural products and worries about genetically modified plants present challenges to widespread adoption.

Nevertheless, ongoing research on genes controlling secondary metabolite formation and transmission methods indicates a promising future for biotechnological tools in plant breeding. Despite this potential, the full exploitation of medicinal herb's genetic potential through breeding remains underutilized, highlighting a significant opportunity for advancement in the phytopharmaceutical sector. Moving forward, future improvement programs should prioritize the identification and consideration of species-specific characteristics influencing postharvest processing success. By addressing these challenges and leveraging emerging technologies, speed breeding stands poised to emerge as a key driver in advancing the phytopharmaceutical industry, paving the way for the development of novel therapeutic agents and sustainable agricultural practices.

ACKNOWLEDGMENTS

This research has been generously supported financially by the Phytopathology Unit at ENA-Meknès.

REFERENCES

Acquaah, G. (2012). *Principles of plant genetics and breeding* (2nd ed.). https://doi.org/10.1002/9781118313718

Acquaah, G. (2016). Conventional plant breeding principles and techniques. In *Advances in plant breeding strategies: Breeding, biotechnology and molecular tools* (Vol. 1, pp. 115–158). https://doi.org/10.1007/978-3-319-22521-0_5

Alseekh, S., Karakas, E., Zhu, F., Wijesingha Ahchige, M., & Fernie, A. R. (2023). Plant biochemical genetics in the multiomics era. *Journal of Experimental Botany*, *74*(15), 4293–4307. https://doi.org/10.1093/jxb/erad177

Anand, A., Subramanian, M., & Kar, D. (2023). Breeding techniques to dispense higher genetic gains. *Frontiers in Plant Science*, *13*. https://doi.org/10.3389/fpls.2022.1076094

Anjali, Kumar, S., Korra, T., Thakur, R., Arutselvan, R., Kashyap, A. S., Nehela, Y., Chaplygin, V., Minkina, T., & Keswani, C. (2023). Role of plant secondary metabolites in defense and transcriptional regulation in response to biotic stress. *Plant Stress*, *8*. https://doi.org/10.1016/j.stress.2023.100154

Armita, D. (2020). Plant breeding through protoplast fusion. *Jurnal Biologi UNAND*, *8*(2), 42. https://doi.org/10.25077/jbioua.8.2.42-47.2020

Bado, S., Kozak-Stankiewicz, K., Sekander, H., Alhajaj, N., Ghanim, A., Forster, B. P., & Laimer, M. (2013). Resurgence of x-rays in plant mutation breeding. In *Plant genetics and breeding technologies, plant diseases and resistance mechanisms proceedings* (pp. 13–16). 18–20 February 2013, Vienna.

Bannayan, M., Nadjafi, F., Rastgoo, M., Tabrizi, L., Bannayan, M., Nadjafi, F., Rastgoo, M., & Tabrizi, L. (2006). Germination properties of some wild medicinal plants from Iran. *Seed Technology*, *28*(1), 80–86.

Barut, M., Nadeem, M. A., Akgür, Ö., Tansi, L. S., Aasim, M., Altaf, M. T., & Baloch, F. S. (2022). Medicinal and aromatic plants in the omics era: Application of plant breeding and-biotechnology for plant secondary metabolite production. *Turkish Journal of Agriculture and Forestry*, *46*(2), 182–203. https://doi.org/10.55730/1300-011x.2970

Baskin, C. C., & Baskin, M. (1998). *Seeds: Ecology, biogeography, and evolution of dormancy and germination* (2nd ed.). Elsevier.

Begum, A., Sandhya, S., Ali, S. S., Vinod, K. R., Reddy, S., & Banji, D. (2013). An in-depth review on the medicinal flora *Rosmarinus officinalis* (lamiaceae). *Acta Scientiarum Polonorum, Technologia Alimentaria*, *12*(1), 61–73.

Bhatta, M., Sandro, P., Smith, M. R., Delaney, O., Voss-Fels, K. P., Gutierrez, L., & Hickey, L. T. (2021). Need for speed: Manipulating plant growth to accelerate breeding cycles. *Current Opinion in Plant Biology*, *60*, 101986.

Bonea, D. (2022). Speed breeding and its importance for the improvement of agricultural crops. *Annals of the University of Craiova-Agriculture Montanology Cadastre Series*, *52*(1), 59–66.

Canarini, A., Kaiser, C., Merchant, A., Richter, A., & Wanek, W. (2019). Root exudation of primary metabolites: Mechanisms and their roles in plant responses to environmental stimuli. *Frontiers in Plant Science*, *10*, 422679.

Canter, P. H., Thomas, H., & Ernst, E. (2005). Bringing medicinal plants into cultivation: Opportunities and challenges for biotechnology. *Trends in Biotechnology*, *23*(4), 180–185. https://doi.org/10.1016/j.tibtech.2005.02.002

Cardoso, J. C., de Oliveira, M. E., & de CI Cardoso, F. (2019). Advances and challenges on the in vitro production of secondary metabolites from medicinal plants. *Horticultura Brasileira*, *37*, 124–132.

Cazzola, F., Bermejo, C. J., Guindon, M. F., & Cointry, E. (2020). Speed breeding in pea (*Pisum sativum* L.), an efficient and simple system to accelerate breeding programs. *Euphytica*, *216*(11), 178.

Chaturvedi, T., Gupta, A. K., Lal, R. K., & Tiwari, G. (2022). March of molecular breeding techniques in the genetic enhancement of herbal medicinal plants: Present and future prospects. *Nucleus (India)*, *65*(3), 413–436. https://doi.org/10.1007/s13237-022-00406-2

Chen, M., & Penfield, S. (2018). Feedback regulation of COOLAIR expression controls seed dormancy and flowering time. *Science*, *360*(6392), 1014–1017. https://doi.org/10.1126/science.aar7361

Chinthiya, A., & Bhavyasree, R. K. (2019). Domestication in plants: A key to unexplored variability. *International Journal of Current Microbiology and Applied Sciences*, *8*(1), 133–138. https://doi.org/10.20546/ijcmas.2019.801.016

Chiurugwi, T., Kemp, S., Powell, W., & Hickey, L. T. (2019). Speed breeding orphan crops. *Theoretical and Applied Genetics*, *132*(3), 607–616. https://doi.org/10.1007/s00122-018-3202-7

Choi, H., Back, S., Kim, G. W., Lee, K., Venkatesh, J., Lee, H. B., Kwon, J. K., & Kang, B. C. (2023). Development of a speed breeding protocol with flowering gene investigation in pepper (*Capsicum annuum*). *Frontiers in Plant Science*, *14*. https://doi.org/10.3389/fpls.2023.1151765

Choi, H. K. (2019). Translational genomics and multi-omics integrated approaches as a useful strategy for crop breeding. *Genes and Genomics*, *41*(2), 133–146. https://doi.org/10.1007/s13258-018-0751-8

Cisneros, A., & Tel-Zur, N. (2010). Embryo rescue and plant regeneration following interspecific crosses in the genus *Hylocereus* (Cactaceae). *Euphytica*, *174*(1), 73–82. https://doi.org/10.1007/s10681-010-0135-x

Collins, F. S., Green, E. D., Guttmacher, A. E., & Guyer, M. S. (2003). A vision for the future of genomics research. *Nature*, *422*(6934), 835–847. https://doi.org/10.1038/nature01626

Diamond, J. (2002). Evolution, consequences and future of plant and animal domestication. *Nature*, *418*(6898), 700–707.

Dong, L. L., Chen, Z. J., Wang, Y., Wei, F. G., Zhang, L. J., Xu, J., Wei, G. F., Wang, R., Yang, J., Liu, W. L., Li, X. W., Yu, Y. Q., & Chen, S. L. (2017). DNA marker-assisted selection of medicinal plants (I). Breeding research of disease-resistant cultivars of *Panax notoginseng*. *Zhongguo Zhongyao Zazhi*, *42*(1), 56–62.

Edwards, R. (2004). No remedy in sight for herbal ransack. *New Scientist*, *181*(2429), 10–11.

Fridman, E. (2015). Consequences of hybridization and heterozygosity on plant vigor and phenotypic stability. *Plant Science*, *232*, 35–40. https://doi.org/10.1016/j.plantsci.2014.11.014

Gangashetty, P. I., Belliappa, S. H., Bomma, N., Kanuganahalli, V., Sajja, S. B., Choudhary, S., Gaviyappanavar, R., Bomireddy, D., Anil Kumar, V., Pranati, J., Sharma, M., & Pandey, M. K. (2024). Optimizing speed breeding and seed/pod chip based genotyping techniques in pigeonpea: A way forward for high throughput line development. *Plant Methods*, *20*(1), 1–12. https://doi.org/10.1186/s13007-024-01155-w

Ghosh, S., Watson, A., Gonzalez-Navarro, O. E., Ramirez-Gonzalez, R. H., Yanes, L., Mendoza-Suárez, M., Simmonds, J., Wells, R., Rayner, T., Green, P., Hafeez, A., Hayta, S.,

Melton, R. E., Steed, A., Sarkar, A., Carter, J., Perkins, L., Lord, J., Tester, M., . . . Hickey, L. T. (2018). Speed breeding in growth chambers and glasshouses for crop breeding and model plant research. *Nature Protocols*, *13*(12), 2944–2963. https://doi.org/10.1038/s41596-018-0072-z

Gobbo-Neto, L., & Lopes, N. P. (2007). Plantas medicinais: Fatores de influência no conteúdo de metabólitos secundários. *Química nova*, *30*, 374–381.

Gudi, S., Kumar, P., Singh, S., Tanin, M. J., & Sharma, A. (2022). Strategies for accelerating genetic gains in crop plants: Special focus on speed breeding. *Physiology and Molecular Biology of Plants*, *28*(10), 1921–1938. https://doi.org/10.1007/s12298-022-01247-8

Guo, H. B., Song, Z. P., Liang, Z. S., & Zhang, Y. J. (2009). Domestic cultivation may abate the contradiction between sustainable utilization and genetic diversity conservation of medicinal plants. *Journal of Medicinal Plants Research*, *3*(13), 1184–1188.

Guo, J., Huang, Z., Sun, J., Cui, X., & Liu, Y. (2021). Research progress and future development trends in medicinal plant transcriptomics. *Frontiers in Plant Science*, *12*. https://doi.org/10.3389/fpls.2021.691838

Hickey, L. T., Germán, S. E., Pereyra, S. A., Diaz, J. E., Ziems, L. A., Fowler, R. A., Platz, G. J., Franckowiak, J. D., & Dieters, M. J. (2017). Speed breeding for multiple disease resistance in barley. *Euphytica*, *213*(3), 64. https://doi.org/10.1007/s10681-016-1803-2

Hussein, R. A., & El-Anssary, A. A. (2019). Plants secondary metabolites: The key drivers of the pharmacological actions of medicinal plants. *Herbal Medicine*. https://doi.org/10.5772/intechopen.76139

Iovane, M., Cirillo, A., Izzo, L. G., Di Vaio, C., & Aronne, G. (2022). High temperature and humidity affect pollen viability and longevity in olea europaea l. *Agronomy*, *12*(1), 1–12. https://doi.org/10.3390/agronomy12010001

Isah, T., Umar, S., Mujib, A., Sharma, M. P., Rajasekharan, P. E., Zafar, N., & Frukh, A. (2018). Secondary metabolism of pharmaceuticals in the plant in vitro cultures: Strategies, approaches, and limitations to achieving higher yield. *Plant Cell, Tissue and Organ Culture*, *132*(2), 239–265. https://doi.org/10.1007/s11240-017-1332-2

Jafari, T., Rahikainen, M., Puljula, E., Sinkkonen, J., & Kangasjärvi, S. (2018). The impact of light intensity on metabolomic profile of *Arabidopsis thaliana* wild type and reticulata mutant by NMR spectroscopy. *Phytochemistry Letters*, *26*, 170–178. https://doi.org/10.1016/j.phytol.2018.06.013

Jähne, F., Hahn, V., Würschum, T., & Leiser, W. L. (2020). Speed breeding short-day crops by LED-controlled light schemes. *Theoretical and Applied Genetics*, *133*(8), 2335–2342. https://doi.org/10.1007/s00122-020-03601-4

Jochum, G. M., Mudge, K. W., & Thomas, R. B. (2007). Elevated temperatures increase leaf senescence and root secondary metabolite concentrations in the understory herb *Panax quinquefolius* (Araliaceae). *American Journal of Botany*, *94*(5), 819–826. https://doi.org/10.3732/ajb.94.5.819

Justin, K., Edmond, S., Ally, M., & Xin, H. (2014, January). Plant secondary metabolites: Biosynthesis, classification, function and pharmacological properties. *Journal of Pharmacy and Pharmacology*, *2*, 377–392.

Kappers, I. F., Aharoni, A., Van Herpen, T. W. J. M., Luckerhoff, L. L. P., Dicke, M., & Bouwmeester, H. J. (2005). Plant science: Genetic engineering of terpenoid metabolism attracts bodyguards to Arabidopsis. *Science*, *309*(5743), 2070–2072. https://doi.org/10.1126/science.1116232

Kong, L., Holtz, C. T., Nairn, C. J., Houke, H., Powell, W. A., Baier, K., & Merkle, S. A. (2014). Application of airlift bioreactors to accelerate genetic transformation in American chestnut. *Plant Cell, Tissue and Organ Culture*, *117*(1), 39–50. https://doi.org/10.1007/s11240-013-0418-8

Kumar, P., Sharma, P., Kumar, V., Kumar, A., Kumar, A., Singh, R., & Sharma, A. K. (2019). Plant resources: In vitro production, challenges and prospects of secondary Metabolites from medicinal plants. *Industrial Biotechnology: Plant Systems, Resources and Products*, *2019*, 89–104. https://doi.org/10.1515/9783110563337-005

Laird, Ş. A., & Pierce, A. R. (2002). *Promoting sustainable and ethical botanicalstrategies to improve commercial raw material sourcing*. Rainforest Alliance.

Lamichhane, S., & Thapa, S. (2022). Advances from conventional to modern plant breeding methodologies. *Plant Breeding and Biotechnology*, *10*(1). https://doi.org/10.9787/PBB.2022.10.1.1

Larson, G., Piperno, D. R., Allaby, R. G., Purugganan, M. D., Andersson, L., Arroyo-Kalin, M., Barton, L., Vigueira, C. C., Denham, T., Dobney, K., Doust, A. N., Gepts, P., Gilbert, M. T. P., Gremillion, K. J., Lucas, L., Lukens, L., Marshall, F. B., Olsen, K. M., Pires, J. C., . . . Fuller, D. Q. (2014). Current perspectives and the future of domestication studies. *Proceedings of the National Academy of Sciences of the United States of America*, *111*(17), 6139–6146. https://doi.org/10.1073/pnas.1323964111

Li, C. (2020). Breeding crops by design for future agriculture. *Journal of Zhejiang University: Science B*, *21*(6), 423–425. https://doi.org/10.1631/jzus.B2010001

Liu, K., He, R., He, X., Tan, J., Chen, Y., Li, Y., Liu, R., Huang, Y., & Liu, H. (2022). Speed breeding scheme of hot pepper through light environment modification. *Sustainability (Switzerland)*, *14*(19). https://doi.org/10.3390/su141912225

Luo, T., Xian, M., Zhang, C., Zhang, C., Hu, L., & Xu, Z. (2019). Associating transcriptional regulation for rapid germination of rapeseed (*Brassica napus* L.) under low temperature stress through weighted gene co-expression network analysis. *Scientific Reports*, *9*(1). https://doi.org/10.1038/s41598-018-37099-0

Majid, A., Parray, G. A., Wani, S. H., Kordostami, M., Sofi, N. R., A. Waza, S., Shikari, A. B., & Gulzar, S. (2017). Genome editing and its necessity in agriculture. *International Journal of Current Microbiology and Applied Sciences*, *6*(11), 5435–5443. https://doi.org/10.20546/ijcmas.2017.611.520

Manchanda, P., Kaur, M., Sharma, S., & Sidhu, G. S. (2023). Biotechnological interventions for reducing the juvenility in perennials. *Horticulturae*, *9*(1). https://doi.org/10.3390/horticulturae9010033

Mobini, S. H., & Warkentin, T. D. (2016). A simple and efficient method of in vivo rapid generation technology in pea (*Pisum sativum* L.). *In Vitro Cellular and Developmental Biology – Plant*, *52*(5), 530–536. https://doi.org/10.1007/s11627-016-9772-7

Moose, S. P., & Mumm, R. H. (2008). Molecular plant breeding as the foundation for 21st century crop improvement. *Plant Physiology*, *147*(3), 969–977. https://doi.org/10.1104/pp.108.118232

Murovec, J., Pirc, Ž., & Yang, B. (2017). New variants of CRISPR RNA-guided genome editing enzymes. *Plant Biotechnology Journal*, *15*(8), 917–926. https://doi.org/10.1111/pbi.12736

Naito, T., Kiba, T., Koizumi, N., Yamashino, T., & Mizuno, T. (2007). Characterization of a unique GATA family gene that responds to both light and cytokinin in Arabidopsis thaliana.

Bioscience, Biotechnology and Biochemistry, *71*(6), 1557–1560. https://doi.org/10.1271/bbb.60692

Niazian, M. (2019). Application of genetics and biotechnology for improving medicinal plants. *Planta*, *249*(4), 953–973. https://doi.org/10.1007/s00425-019-03099-1

Okello, D., Komakech, R., Gang, R., Rahmat, E., Chung, Y., Omujal, F., & Kang, Y. (2022). Influence of various temperatures, seed priming treatments and durations on germination and growth of the medicinal plant *Aspilia africana*. *Scientific Reports*, *12*(1), 1–11. https://doi.org/10.1038/s41598-022-18236-2

Pacini, E., & Dolferus, R. (2019). Pollen developmental arrest: Maintaining pollen fertility in a world with a changing climate. *Frontiers in Plant Science*, *10*, 429534.

Paek, K. Y., Chakrabarty, D., & Hahn, E. J. (2005). Application of bioreactor systems for large scale production of horticultural and medicinal plants. *Plant Cell, Tissue and Organ Culture*, *81*(3), 287–300. https://doi.org/10.1007/s11240-004-6648-z

Palevitch, D. (1991). Agronomy applied to medicinal plant conservation. In *Conservation of Medicinal Plants* (p. 167). Cambridge University Press. https://doi.org/10.1017/cbo9780511753312.012

Potts, J., Jangra, S., Michael, V. N., & Wu, X. (2023). Speed breeding for crop improvement and food security. *Crops*, *3*(4), 276–291. https://doi.org/10.3390/crops3040025

Pretty, J., & Bharucha, Z. P. (2014). Sustainable intensification in agricultural systems. *Annals of Botany*, *114*(8), 1571–1596. https://doi.org/10.1093/aob/mcu205

Purugganan, M. D. (2019). Evolutionary insights into the nature of plant domestication. *Current Biology*, *29*(14), R705–R714.

Ramakrishna, A., & Ravishankar, G. A. (2011). Influence of abiotic stress signals on secondary metabolites in plants. *Plant Signaling and Behavior*, *6*(11), 1720–1731. https://doi.org/10.4161/psb.6.11.17613

Ramawat, K. G. (2019). An introduction to biodiversity and chemotaxonomy BT. In K. G. Ramawat (Ed.), *Biodiversity and chemotaxonomy* (pp. 1–14). Springer International Publishing. https://doi.org/10.1007/978-3-030-30746-2_1

Ramawat, K. G., & Arora, J. (2021). Medicinal plants domestication, cultivation, improvement, and alternative technologies for the production of high value therapeutics: An overview. In *Medicinal plants: Domestication, biotechnology and regional importance* (pp. 1–29). https://doi.org/10.1007/978-3-030-74779-4_1

Reshi, Z. A., Ahmad, W., Lukatkin, A. S., & Javed, S. B. (2023). From nature to lab: A review of secondary metabolite biosynthetic pathways, environmental influences, and in vitro approaches. *Metabolites*, *13*(8). https://doi.org/10.3390/metabo13080895

Robbins, C. S. (1998). American ginseng: The root of North America's medicinal herb trade. TRAFFIC North America.

Roeber, V. M., Bajaj, I., Rohde, M., Schmülling, T., & Cortleven, A. (2021). Light acts as a stressor and influences abiotic and biotic stress responses in plants. *Plant, Cell & Environment*, *44*(3), 645–664.

Sakurai, N. (2022). Recent applications of metabolomics in plant breeding. *Breeding Science*, *72*(1), 56–65. https://doi.org/10.1270/jsbbs.21065

Samanta, D., Das, D., Sinha, S., Mallick, B., Banerjee, R., Ganguli, S., & Roy, D. (2023). Transcriptome analysis reveals upregulated secondary metabolite pathways in micropropagated *Lawsonia inermis* L. *Vegetos*, *36*(3), 1130–1138. https://doi.org/10.1007/s42535-023-00613-5

Samantara, K., Bohra, A., Mohapatra, S. R., Prihatini, R., Asibe, F., Singh, L., Reyes, V. P., Tiwari, A., Maurya, A. K., Croser, J. S., Wani, S. H., Siddique, K. H. M., & Varshney, R. K. (2022). Breeding more crops in less time: A perspective on speed breeding. *Biology*, *11*(2), 275. https://doi.org/10.3390/biology11020275

Samineni, S., Sen, M., Sajja, S. B., & Gaur, P. M. (2020). Rapid generation advance (RGA) in chickpea to produce up to seven generations per year and enable speed breeding. *Crop Journal*, *8*(1), 164–169. https://doi.org/10.1016/j.cj.2019.08.003

Saxena, K. B., Saxena, R. K., Hickey, L. T., & Varshney, R. K. (2019). Can a speed breeding approach accelerate genetic gain in pigeonpea? *Euphytica*, *215*(12). https://doi.org/10.1007/s10681-019-2520-4

Schilling, S., Melzer, R., Dowling, C. A., Shi, J., Muldoon, S., & McCabe, P. F. (2023). A protocol for rapid generation cycling (speed breeding) of hemp (*Cannabis sativa*) for research and agriculture. *Plant Journal*, *113*(3), 437–445. https://doi.org/10.1111/tpj.16051

Schippmann, U., Leaman, D. J., & Cunningham, A. B. (2002, October). Impact of cultivation and gathering of medicinal plants on biodiversity: Global trends and issues. FAO. 2002. *Biodiversity and the Ecosystem Approach in Agriculture, Forestry and Fisheries. Satellite Event on the Occasion of the Ninth Regular Session of the Commission on Genetic Resources for Food and Agriculture. Rome, 12–13 October 2002. Inter-Depart*, *676*, 1–21.

Schmitz, J., Heinrichs, L., Scossa, F., Fernie, A. R., Oelze, M. L., Dietz, K. J., Rothbart, M., Grimm, B., Flügge, U. I., & Häusler, R. E. (2014). The essential role of sugar metabolism in the acclimation response of arabidopsis thaliana to high light intensities. *Journal of Experimental Botany*, *65*(6), 1619–1636. https://doi.org/10.1093/jxb/eru027

Schrieber, K., Schweiger, R., Kröner, L., & Müller, C. (2019). Inbreeding diminishes herbivore-induced metabolic responses in native and invasive plant populations. *Journal of Ecology*, *107*(2), 923–936. https://doi.org/10.1111/1365-2745.13068

Schulz, H. (2020). *Analysis of secondary metabolites in breeding research and plant breeding* (pp. 207–231). https://doi.org/10.1007/978-3-030-38792-1_2

Shanmugavel, P., Ramasamy, G., Vellingiri, G., Marimuthu, R., & Thiyagarajan, K. (2023). Speed breeding: A propitious technique for accelerated crop improvement. In *Case studies of breeding strategies in major plant species* (Vol. 11, p. 13). https://doi.org/10.5772/intechopen.105533

Sharma, S., Kumar, A., Dhakte, P., Raturi, G., Vishwakarma, G., Barbadikar, K. M., Das, B. K., Shivaraj, S. M., Sonah, H., & Deshmukh, R. (2023). Speed breeding opportunities and challenges for crop improvement. *Journal of Plant Growth Regulation*, *42*(1), 46–59. https://doi.org/10.1007/s00344-021-10551-8

Singh, R. K., Prasad, A., Muthamilarasan, M., Parida, S. K., & Prasad, M. (2020). Breeding and biotechnological interventions for trait improvement: Status and prospects. *Planta*, *252*(4). https://doi.org/10.1007/s00425-020-03465-4

Slama-Ayed, O., Bouhaouel, I., Ayed, S., De Buyser, J., Picard, E., & Amara, H. S. (2019). Efficiency of three haplomethods in durum wheat (*Triticum turgidum* subsp. durum desf.): Isolated microspore culture, gynogenesis and wheat × maize crosses. *Czech Journal of Genetics and Plant Breeding*, *55*(3), 101–109. https://doi.org/10.17221/188/2017-CJGPB

Smýkal, P., K. Varshney, R., K. Singh, V., Coyne, C. J., Domoney, C., Kejnovský, E., & Warkentin, T. (2016). From Mendel's discovery on pea to today's plant genetics and breeding: Commemorating the 150th anniversary of the reading of Mendel's discovery. *Theoretical and Applied Genetics*, *129*(12), 2267–2280. https://doi.org/10.1007/s00122-016-2803-2

Sood, H. (2020). Production of medicinal compounds from endangered and commercially important medicinal plants through cell and tissue culture technology for herbal industry. In *Bioactive compounds in nutraceutical and functional food for good human health*. IntechOpen.

Sussmilch, F. C., Ross, J. J., & Reid, J. B. (2022). Mendel: From genes to genome. *Plant Physiology*, *190*(4), 2103–2114. https://doi.org/10.1093/plphys/kiac424

Temesgen, B. (2022). Speed breeding to accelerate crop improvement. *International Journal of Agricultural Science and Food Technology*, *8*(2), 178–186. https://doi.org/10.17352/2455-815x.000161

Thomson, M. J., Biswas, S., Tsakirpaloglou, N., & Septiningsih, E. M. (2022). Functional allele validation by gene editing to leverage the wealth of genetic resources for crop improvement. *International Journal of Molecular Sciences*, *23*(12), 6565.

Tripathi, L., & Tripathi, J. N. (2005). Role of biotechnology in medicinal plants. *Tropical Journal of Pharmaceutical Research*, *2*(2). https://doi.org/10.4314/tjpr.v2i2.14607

Udage, A. C. (2021). Introduction to plant mutation breeding: Different approaches and mutageniagents. *Journal of Agricultural Sciences – Sri Lanka*, *16*(3), 466–483. https://doi.org/10.4038/jas.v16i03.9472

Uğur, N., & Gören, K. (2023). Speed breeding: An innovative approach for accelerated genetic improvement in agricultural crops and integrating with molecular-based approaches. In *Pioneer and contemporary studies in agriculture, forest and water issues* (pp. 241–264). Duvar. https://doi.org/10.59287/pcsafwi.56

Varshney, R. K., Hoisington, D. A., & Tyagi, A. K. (2006). Advances in cereal genomics and applications in crop breeding. *Trends in Biotechnology*, *24*(11), 490–499. https://doi.org/10.1016/j.tibtech.2006.08.006

Vines, G. (2004). *Herbal harvests with a future: Towards sustainable sources for medicinal plants*. Plantlife International.

Wang, W., Xu, J., Fang, H., Li, Z., & Li, M. (2020). Advances and challenges in medicinal plant breeding. *Plant Science*, *298*. https://doi.org/10.1016/j.plantsci.2020.110573

Wanga, M. A., Shimelis, H., Mashilo, J., & Laing, M. D. (2021). Opportunities and challenges of speed breeding: A review. *Plant Breeding*, *140*(2), 185–194. https://doi.org/10.1111/pbr.12909

Watson, A., Ghosh, S., Williams, M. J., Cuddy, W. S., Simmonds, J., Rey, M. D., Asyraf Md Hatta, M., Hinchliffe, A., Steed, A., Reynolds, D., Adamski, N. M., Breakspear, A., Korolev, A., Rayner, T., Dixon, L. E., Riaz, A., Martin, W., Ryan, M., Edwards, D., . . . Hickey, L. T. (2018). Speed breeding is a powerful tool to accelerate crop research and breeding. *Nature Plants*, *4*(1), 23–29. https://doi.org/10.1038/s41477-017-0083-8

Wei, J. H., Tu, P. F., Li, G., Wang, W. Q., Yang, C. M., & Sui, C. (2015). Situation and trends in development of Chinese Medicinal Agriculture in China. *Modern Chinese Medicine*, *17*, 94–100.

Wu, W. G., Dong, L. L., & Chen, S. L. (2020). Development direction of molecular breeding of medicinal plants. *Zhongguo Zhongyao Zazhi*, *45*(11), 2714–2719. https://doi.org/10.19540/j.cnki.cjcmm.20200329.105

Xiao, L., Shibuya, T., Watanabe, T., Kato, K., & Kanayama, Y. (2022). Effect of light quality on metabolomic, ionomic, and transcriptomic profiles in tomato fruit. *International Journal of Molecular Sciences*, *23*(21). https://doi.org/10.3390/ijms232113288

Xu, Y., Luo, H., Zhang, H., Yung, W.-S., Li, M.-W., Lam, H.-M., & Huang, C. (2023). Feeding the world using speed breeding technology. *Trends in Plant Science*, *28*(3), 372–373. https://doi.org/10.1016/j.tplants.2022.12.003

Yirga, M. (2021). Polyploidy and its implications in plants breeding – a review. *International Journal of Current Research in Biosciences and Plant Biology*, *8*(3), 1–9. https://doi.org/10.20546/ijcrbp.2021.803.001

Zhang, S., Zhang, L., Zou, H., Qiu, L., Zheng, Y., Yang, D., & Wang, Y. (2021, December). Effects of light on secondary metabolite biosynthesis in medicinal plants. *Frontiers in Plant Science*, *12*. https://doi.org/10.3389/fpls.2021.781236

3 Strategies for Precision Breeding of Medicinal Plants for Enhanced Secondary Metabolite Production

M. Gnanaraj, P. Samuel, T. Jebastin, M. Rajadurai, D. Rajasudhakar, and R. Manikandan

3.1 INTRODUCTION

Traditional systems worldwide have relied on medicinal plants for centuries to alleviate a diverse array of ailments. These plants are valued for their variety of phytochemicals, which contribute to their therapeutic qualities (Deepa & Sivakumar, 2020). These compounds include flavonoids, alkaloids, steroids, and saponins (Jumiati et al., 2022). According to the World Health Organization, medicinal plants are those that are used to treat and prevent illness (Nagavalli, 2022). These plants are frequently referred to in communities as 'home remedies,' highlighting their role in healthcare solutions (Matos et al., 2020).

Studies that have examined the routine use of medicinal plants within various communities have illustrated the close connection between their use and cultural traditions. (Hidayani et al., 2021). Each location possesses its distinct knowledge of medicinal plants. For instance, in the Himalayan state of Sikkim, scientists have discovered 51 species believed to have therapeutic qualities (Chettri & Kumari, 2021). According to Yang et al. (2022), studies conducted in areas such as Sumatra, China, and Vietnam have highlighted the variety of wild medicinal plants present there, as well as their usage by the native populations.

The cultivation of therapeutic plants at home has gained popularity, particularly in the context of the COVID-19 pandemic. (Siregar et al., 2021). The fascination with herbal medicine stems from the enduring belief held by numerous civilizations regarding its curative properties (Siregar et al., 2021). Therapeutic plants have garnered significant attention from both traditional medicine and contemporary science. Particular focus has been placed on researching the antibacterial properties of plant extracts and conducting phytochemical analyses (Deepa & Sivakumar, 2020). The significance of medicinal plants in healthcare and pharmaceuticals is illustrated in Figure 3.1.

3.2 SECONDARY METABOLITES

Secondary metabolites (SMs) are a specific subset of phytochemicals. Secondary metabolites are organic compounds produced by plants that are not actively involved in the plant's growth, development, or reproduction. Unlike primary metabolites, SMs are not essential for the plant's survival, but they play a crucial role in interactions with the environment, including defense mechanisms against herbivores, pathogens, and competing plant species (Hussain et al., 2012; Yeshi et al., 2022). The diversity of SMs in plants, ranging from sesquiterpene lactones to flavonoids, presents a vast array of compounds that can be utilized for pharmaceutical purposes (Jaswinder et al., 2020; Syarifah et al., 2017). Additionally, the presence of secondary metabolites like alkaloids, diterpenoid lactones, and flavonoid glycosides in medicinal plants underscores their importance in providing bioactive compounds (Ebenezer et al., 2019).

Furthermore, secondary metabolites have been identified as potential sources for drug discovery, with research focusing on their antiviral properties and their role in combating various diseases (Park et al., 2019). The biocontrol potential of endophytes in regulating secondary metabolite accumulation in medicinal plants highlights their significance not only as sources of bioactive compounds but also as alternatives to chemical control methods (Chen et al., 2021a; Szakiel et al., 2010). The influence of environmental factors on SM production, such as saponins, emphasizes the importance of understanding how external conditions can impact the bioactivity of medicinal plants (Hussein & El-Anssary, 2019). Overall, SMs in medicinal plants serve as key drivers for their therapeutic effects, offering a rich source of bioactive compounds that can be harnessed for pharmaceutical, nutraceutical, and medicinal purposes (Armarkar et al., 2021; Madariaga-Mazón et al., 2021). Researchers have found a diverse variety of SMs having pharmacological activity, including alkaloids, flavonoids, terpenoids, and phenolic compounds. These bioactive chemicals have been exploited for their therapeutic qualities, including anti-inflammatory, antioxidant, antibacterial, and anticancer activities (Cowan, 1999; Jalal et al., 2020; Escandón-Rivera et al., 2020; Mulyono et al., 2021; Bharti et al., 2023) (Figure 3.2). Table 3.1 lists several medicinal secondary metabolites along with their sources and medicinal properties. Hussein and El-Anssary (2019)

DOI: 10.1201/9781003475491-3

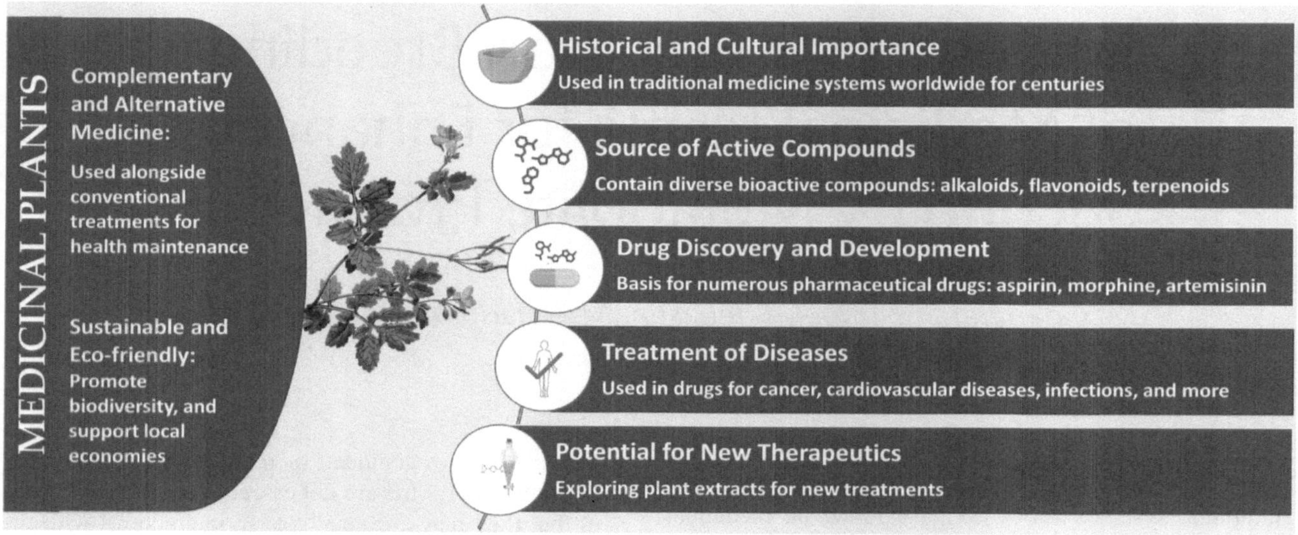

FIGURE 3.1 Significance of medicinal plants in healthcare and pharmaceuticals.

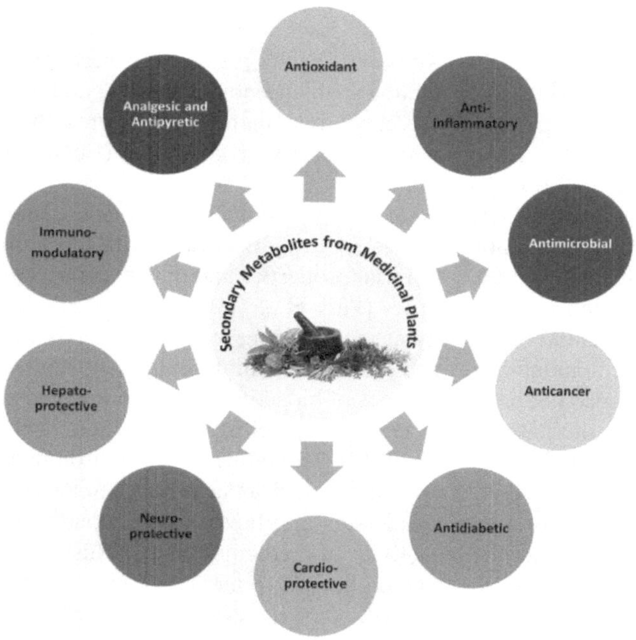

FIGURE 3.2 Medicinal plant secondary metabolite-associated activities and therapeutic potential.

have summarized various important medicinal plants and their commercially valuable SMs. The comprehensive review and characterization of these secondary metabolites, such as through databases like DiaNat-DB, provide valuable insights into their chemical composition and potential medicinal benefits (Hakim et al., 2022).

The medicinal properties of SMs have paved traditional herbal remedies as the primary healthcare system for an estimated 70–80% of the world's population (Farnsworth & Soejarto, 1991). This dependency is maintained by the broad usage of approximately 50,000 to 70,000 plant species across both traditional and modern therapeutic practices globally. As the demand for herbal medications continues to ascend, with a constant rising trend over time, it becomes evident that these natural remedies serve vital health needs. Notably, the significance of secondary metabolites extends beyond traditional medicine since roughly a quarter of pharmaceutical medications consumed worldwide originate from plant-derived chemicals (Newman & Cragg, 2016). This underlines the crucial significance of utilizing the therapeutic potential of secondary metabolites to battle a wide array of health conditions, while also stressing the lasting relevance of ancient wisdom in influencing current healthcare systems (Bharti et al., 2023).

However, the amount of SM production is influenced by several factors, including the accessibility of the plant itself, the existence of appropriate elicitors, the surrounding environment, and the method of extraction used. To increase the accumulation of certain SMs, precision breeding allows for the manipulation of elicitor pathways in plants, allowing for fast and sustained responses. Increased yields of substances of interest can be achieved by engineering plants to develop optimum responses through targeted genetic alterations. There is hope that this method can improve secondary metabolite production, which could have implications in areas as diverse as agriculture, industry, and medicine. In this chapter, we will discuss the precision breeding approaches for the development of medicinal plants with higher production of target SMs.

3.3 PRECISION BREEDING TECHNIQUES

Through methods such as gene editing, precision breeding allows for the exact modification of plant DNA. Precision breeding is a sophisticated plant breeding method that

TABLE 3.1
SMs with Their Source Plant and Medicinal Properties (Adapted from Bharti et al., 2023)

Secondary metabolites	Source	Properties
Ajmalicine	*Rauvolfia serpentine*	Antiarrhythmic, antihypertensive
Caffeine	*Coffea arabica*	Stimulant
Camptothecin	*Camptotheca acuminate*	Antineoplastic
Cocaine	*Erythroxylum coca*	Analgesic, narcotic, local anesthetic
Codeine	*Papaver somniferum*	Antitussive, analgesic
Emetine	*Carapichea ipecacuanha*	Antiamoebic, emetic, expectorant
Hyoscyamine	*Atropa belladonna*	Anticholinergic
Morphine	*Papaver somniferum*	Analgesic, narcotic
Nicotine	*Nicotiana tabacum*	Stimulant
Pilocarpine	*Pilocarpus jaborandi*	Cholinergic
Quinidine	*Cinchona* spp.	Antiarrhythmic
Quinine	*Cinchona* spp.	Antimalarial
Reserpine	*Rauwolfia serpentine*	Tranquilizer
Scopolamine	*Hyoscyamus niger*	Sedative, anticholinergic
Strychnine	*Strychnos nux-vomica*	Stimulant, poison
Taxol	*Taxus brevifolia*	Antineoplastic
Vinblastine and vincristine	*Catharanthus roseus*	Antineoplastic

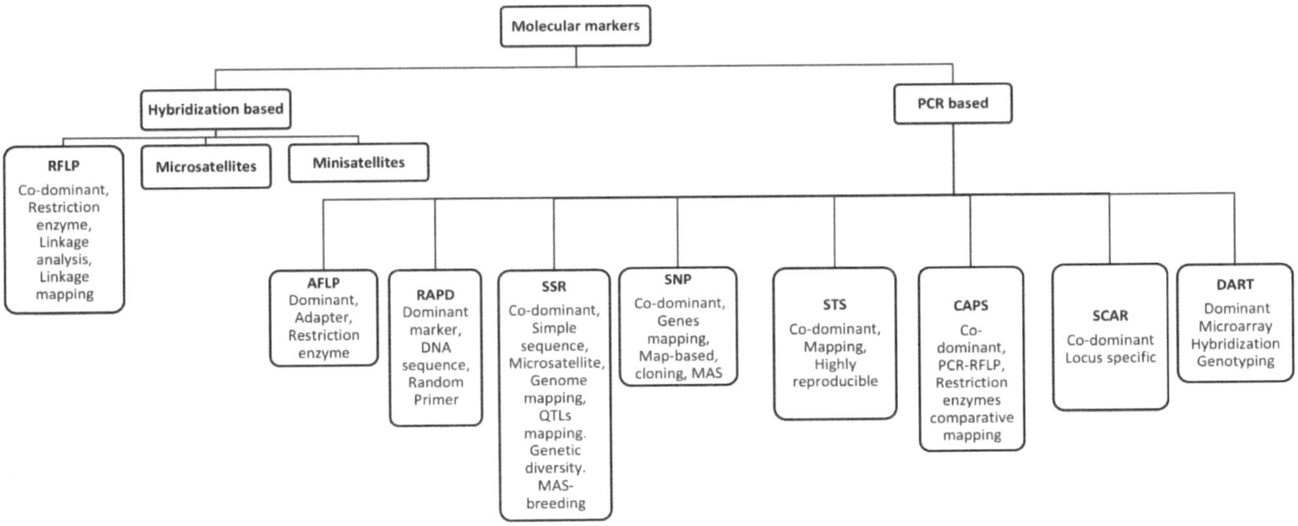

FIGURE 3.3 Molecular markers and their type used in plant breeding (Modified from Gouda et al., 2020).

efficiently and accurately produces offspring with desired features using cutting-edge genetic tools and biotechnologies (Farinati et al., 2023). Unlike conventional breeding approaches that depend on phenotypic selection and cross-breeding, precision breeding uses molecular biology tools to directly edit or select genetic features. This strategy improves both the efficiency and accuracy of breeding operations. As a result, there are now more avenues to explore in developing crops that are more nutritious, more resistant to diseases and pests, and better able to withstand environmental stresses. Techniques like molecular marker-assisted selection and gene editing tools like CRISPR-Cas9 (clustered regularly interspaced short palindromic repeats), zinc finger nucleases (ZFNs), TALENs (transcription activator-like effector nucleases), and RNA interference (RNAi) are used in precision breeding.

3.4 MARKER-ASSISTED SELECTION (MAS)

MAS is a genotype-based selection strategy where plants exhibiting desirable attributes are chosen from segregating generations employing molecular markers closely correlated to the target features (Choudhary et al., 2008; Barrett et al., 2001). Various molecular markers are used in gene mapping, marker-assisted breeding programs, and genetic diversity evaluation, including EST-SSR, ISSR, SNP, and RAPD, among others. Figure 3.3 shows the types of molecular markers used in plant breeding.

According to Babu et al. (2004), the following conditions must be met for MAS to be implemented: (1) a well-saturated genetic linkage map, (2) a tight linkage between the gene of interest and adjacent marker, (3) sufficient recombination between the marker and the rest of the genome, and (4) the capacity to screen a large number of individuals effectively and economically. The efficacy of MAS relies on the intensity of the relationship between the marker and the target gene.

There are several interdependent processes involved in marker-assisted selection (MAS) for improving plant metabolite synthesis. The process begins with the identification of the particular secondary metabolites in medicinal plants to be enhanced. Next, molecular markers linked to the desired secondary metabolite traits are identified and confirmed using genetic and Quantitative trait loci (QTL) mapping. A mapping population is formed by crossing plants with differing amounts of the target metabolite, and these plants are phenotyped to assess metabolite levels. After that, the breeding population is genotyped to find individuals with the desirable alleles, using molecular markers closely associated with the QTLs. Genotypic data is used to identify plants with the right features, followed by phenotypic and metabolite profiling to confirm their presence. To stabilize the characteristics, these plants are crossed and progressed over many generations. Field experiments are then carried out to evaluate the stability and heritability of these traits under different environmental conditions. Finally, the best-performing lines are replicated for large-scale production and introduced into breeding programs or commercial agriculture, promoting the creation of novel plant types with higher secondary metabolite production.

In *Artemisia annua*, MAS has been utilized to generate high-yielding variants of artemisinin, vital for malaria therapy (Paul et al., 2010). *Withania somnifera* has benefitted from MAS via the discovery of markers connected to withanolide production, enhancing its anti-inflammatory and anti-cancer capabilities (Singh et al., 2015). Similarly, *Catharanthus roseus* has witnessed enhanced vincristine and vinblastine synthesis, vital for cancer treatment, via MAS (Rai et al., 2022). *Glycyrrhiza glabra* has deployed MAS to enhance glycyrrhizin content, increasing its anti-inflammatory and antiviral qualities (Barut et al., 2023). Additionally, *Stevia rebaudiana* has enhanced stevioside content by MAS, enhancing its natural sweetener characteristics (Vasanthkumar et al., 2024). These examples emphasize MAS's essential function in enhancing the medical and economic utility of plant secondary metabolites.

3.5 CRISPR-CAS9

CRISPR-Cas9 is a gene-editing technology generating an enormous upheaval in biomedical research. It makes it plausible to repair defects in the genome and turn on or off genes in cells and animals rapidly, affordably, and with relative ease (Redman et al., 2016). Enhancing the production of SMs in medicinal plants using CRISPR-Cas9 takes a rigorous approach described in Figure 3.4. Researchers commence by identifying target genes associated with secondary metabolite synthesis by intensive literature analysis and database searches, complementing transcriptome and metabolomic investigations to highlight essential genes in pertinent metabolic pathways. Once discovered, guide RNAs (gRNAs) are constructed using bioinformatics methods to guarantee specificity and minimal off-target effects. The next phase entails creating CRISPR-Cas9 vectors, where gRNAs are cloned with selectable markers and promoters for Cas9 and gRNA synthesis in plant cells. Vector construction substantially determines the efficacy of future transformation. Techniques like *Agrobacterium*-mediated transformation or particle bombardment are then utilized to deliver the CRISPR-Cas9 construct into plant cells, followed by a selection of transgenic cells using selectable markers.

Regeneration of transgenic plants from changed cells is critical for developing viable organisms with desired genetic changes. This includes cultivating altered cells on regeneration media to generate plant growth, which is then transferred to soil for further development. Subsequent screening and confirmation of mutations using PCR and sequencing validates changes at target loci and assures specificity of the CRISPR-Cas9 system.

Phenotypic and metabolite analysis analyzes the influence of genetic modifications on plant growth and metabolite synthesis, employing procedures like HPLC, GC-MS, or LC-MS for metabolomic analysis. Optimization and repetition may be essential to fine-tune modifications and promote metabolite production, with recurrent adjustments assuring continuous benefits.

Field trials and stability testing assess the efficacy and stability of transformed plants under real-world situations, verifying stable and heritable features over generations. This is vital for practical implementation in agricultural and medicinal plant production systems. Once stability and enhanced metabolite production are proven, commercialization and scale-up require replicating the best-performing lines for large-scale culture and integration into agricultural systems and medicinal plant production chains. This scale-up phase is crucial for transforming laboratory triumphs into commercially viable products.

CRISPR/Cas increases plant research, allowing precise gene editing for therapeutic species like *Cannabis sativa* (Bonini et al., 2018). It shows potential for investigating the antiviral effects of latex production in plants, such as *Chelidonium majus* L. (Gracz-Bernaciak et al., 2021). Gene editing and the production of more targeted metabolites rely on efficient transformation and regeneration mechanisms. Despite the challenges, CRISPR-based mutagenesis is a well-established method for overexpressing endogenous genes. Exact control of gene expression is achieved by making use of untranslated regions (UTRs) and cis-regulatory elements. According to many studies (Shrestha et al., 2018; Wolter et al., 2019; Si et al., 2020), CRISPR/Cas is a powerful and economical method for studying medicinal plants.

Medicinal plants have made considerable use of CRISPR/Cas9 gene disruption technologies to increase levels of therapeutically beneficial metabolites or decrease levels of harmful metabolites. As an example, Zeng et al. (2021) found that plants missing anisodamine and scopolamine chemicals were produced when the hyoscyamine 6β-hydroxylase (AbH6H) gene in *Atropa belladonna* L. was changed. To improve hyoscyamine synthesis while restricting derivative formation, Zeng et al. (2021) used CRISPR/Cas9 to alter H6H. The use of CRISPR/Cas technology for targeted mutagenesis also increased *Dendrobium officinale*'s lignocellulose output (Chen et al., 2014, 2021b). Additionally, Shi et al. (2021) found that after deleting the basic leucine zipper2 (bZIP2) gene from *Salvia miltiorrhiza* using CRISPR/Cas9, the quantity of phenolic acids was increased. Further applications include targeting the farnesyl pyrophosphate synthase gene (Dzfps) in *Dioscorea zingiberensis* to modify diosgenin production (Feng et al., 2018). In *Papaver somniferum*, CRISPR/Cas9 altered the benzylisoquinoline alkaloid production pathway, diminishing alkaloid levels (Alagoz et al., 2016). Additionally, CRISPR/Cas9 targeting of the THC acid synthase gene (THCAS) in *Cannabis sativa* intended to develop plants with high cannabidiol (CBD) and low delta-9-tetrahydrocannabidiol (THC) content (Shiels et al., 2022). Despite its efficacy, CRISPR/Cas9 necessitates meticulous target selection and sgRNA design, particularly when targeting gene families (Zhou et al., 2018).

CRISPR/Cas-based technologies enable multiplex editing of gene targets, facilitating the exploration of genes with significant redundancy. To evaluate the functional relevance of genes with many isoforms or family members, simultaneous targeting of several sgRNAs and Cas proteins in a single expression vector system is employed. This method is crucial when evaluating gene families to identify the contribution of individual genes despite strong similarities with other family members. In *Salvia miltiorrhiza*, Zhou et al. (2021) used CRISPR/Cas9 for multiplex gene suppression, targeting the laccase gene family via a 'dual-locus editing' technique. Laccases, members of the multicopper-oxidase gene family, have a role in the synthesis of compounds like salvianolic acid B (SAB), significant in cardio-cerebral and vascular disease treatments (Festa et al., 2008).

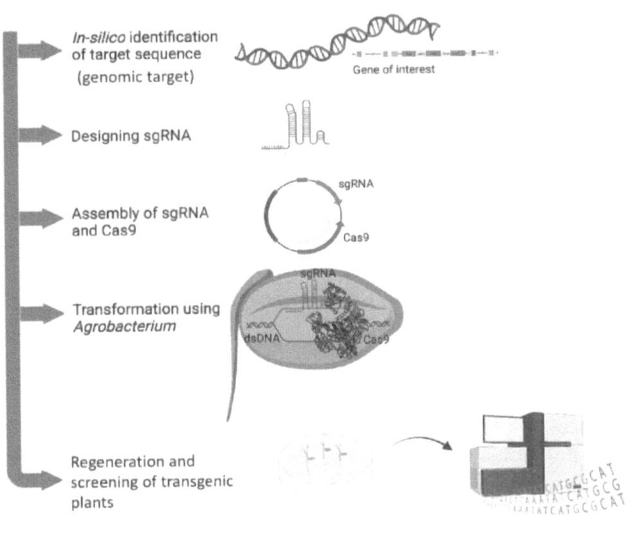

FIGURE 3.4 Steps involved in gene modification by CRISPR-Cas gene-editing system (adapted from Mipeshwaree Devi et al., 2023).

In *Nicotiana tabacum*, core fucose and xylose residues in glycoproteins induce immune responses. Strasser et al. (2008) employed RNAi to suppress FucT and XylT in *N. benthamiana*, however partial inactivation restricted efficiency. A multiplex CRISPR/Cas9 system effectively deactivated several genes, assuring comprehensive depletion of both enzymes (Li et al., 2016; Xie et al., 2015). Li et al. (2018) exploited CRISPR/Cas9 to eradicate the diterpene synthase gene (SmCPS1) in *S. miltiorrhiza*, redirecting metabolic pathways toward taxol synthesis. Multiplexed CRISPR/Cas vectors allowed simultaneous deletion of six tomato genes, exhibiting a very effective mutation method (Čermák & Curtin, 2017; Li et al., 2018). *Symphytum officinale* provides therapeutic qualities but incorporates poisonous pyrrolizidine alkaloids (PAs). CRISPR/Cas9-mediated reduction of homospermidine synthase (HSS) efficiently abolished PA production. Knockout lines demonstrated lower PA levels, perhaps permitting extraction of beneficial metabolites without toxicity (Roeder et al., 2015; Zakaria et al., 2021). Efficiency is increased using multiplex CRISPR/Cas-mediated engineering, which decreases branching routes (Alagoz et al., 2016; Sun et al., 2017). Multiplex CRISPR/Cas systems provide effective instruments for investigating gene families, diverting biosynthetic pathways, and removing objectionable characteristics in plants. These enhancements offer promise for both fundamental research and biotechnological applications in agriculture and health. Metabolites enhanced through CRISPR/Cas gene editing technology are listed by Das et al. (2024).

3.6 TRANSCRIPTION ACTIVATOR-LIKE EFFECTOR NUCLEASES (TALENs)

TALENs are a potent instrument for precisely altering the genomes of medicinal plants, hence enhancing the production of essential SMs. But, the process of designing ZFN and TALENs is challenging, time-consuming, and less specific. ZFN and TALEN were employed more extensively in the context of genome editing in model agricultural plants than in plants intended for medical applications (Das et al., 2024). To improve the target region's specificity, these genome-editing techniques use programmable nucleases (Figure 3.5). Initially, researchers identify target genes related to the creation of desired secondary metabolites by a blend of literature studies, database searches, and omics analysis. TALENs are then created, comprising customizable DNA-binding domains generated from transcription activator-like effectors (TALEs) linked to FokI nuclease domains (Figure 3.5). These TALENs are developed to detect particular DNA sequences within the plant genome, allowing for precise targeting. Upon delivery into plant

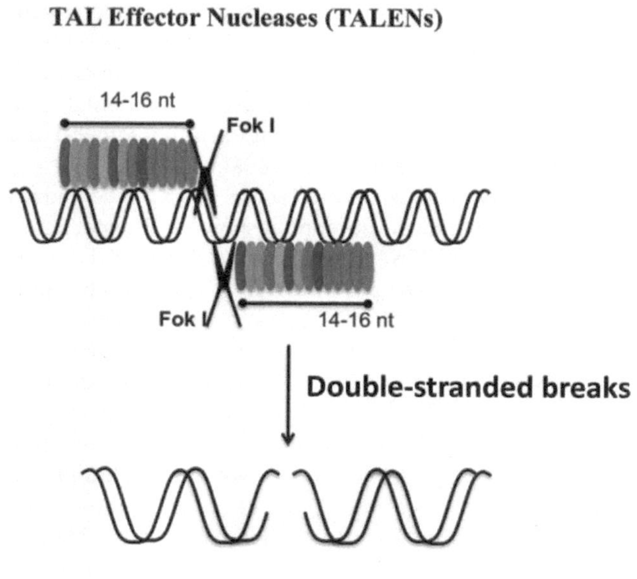

FIGURE 3.5 Transcription activator-like nucleases (TALENs): these are dimeric enzymes similar to ZFNs. Each subunit consists of a DNA binding domain (highly conserved 33–34 amino acid sequence specific for each nucleotide) and Fok1 nuclease domain (adapted from Arora & Narula, 2017).

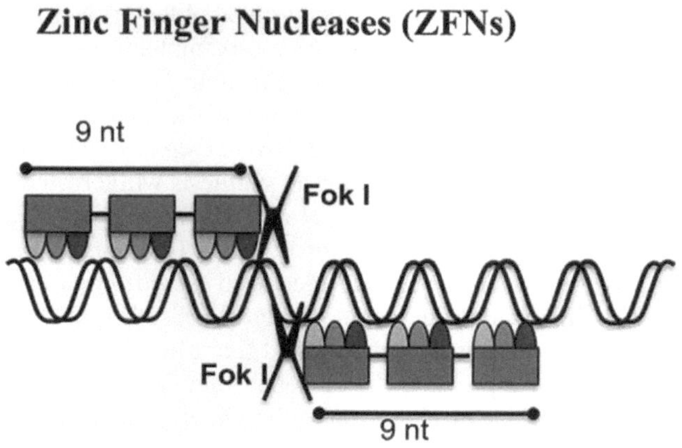

FIGURE 3.6 Zinc-finger nucleases (ZFNs) act as dimers. Each monomer consists of a DNA binding domain and a nuclease domain. Each DNA binding domain consists of an array of 3–6 zinc finger repeats that recognize 9–18 nucleotides. The nuclease domain consists of type II restriction endonuclease Fok1 (adapted from Arora & Narula, 2017).

cells, generally via mechanisms such as *Agrobacterium*-mediated transformation or particle bombardment, TALENs generate double-strand breaks (DSBs) at their target locations. The plant's natural DNA repair processes, such as non-homologous end joining (NHEJ) or homology-directed repair (HDR), are subsequently used to restore these breaks. NHEJ often leads to insertions or deletions (indels), resulting in gene inactivation, while HDR offers precise gene editing provided a repair template is supplied. By meticulously modifying genes involved in metabolic pathways, TALENs may redirect the flux of precursors toward the production of targeted metabolites. For example, in plants like *Artemisia annua* and *Catharanthus roseus*, TALENs could have been utilized to delete or downregulate genes competing for shared precursors, therefore enhancing the synthesis of compounds like artemisinin and alkaloids. TALEN technology was employed to mutate the caffeic acid O-methyltransferase (COMT) gene, which is responsible for the biosynthesis of lignin. This enhanced saccharification efficiency by 54% without altering biomass production (Jung & Altpeter, 2016). This precise alteration of plant genomes using TALENs offers tremendous promise for enhancing the production and quality of medicinal plant metabolites, hence boosting their therapeutic potential (Weeks et al., 2016).

3.7 ZINC FINGER NUCLEASES (ZFNs)

ZFNs are formidable instruments in genetic engineering that have been exploited to increase the synthesis of plant metabolites, notably in medicinal plants. The technique incorporates multiple laborious phases. (Figure 3.6). Firstly, the target gene sequence is determined, which is responsible for the desired metabolite synthesis. This gene sequence is then exploited to build modified ZFNs, which are designed proteins consisting of a DNA-binding zinc finger domain and a DNA-cleaving nuclease domain. These ZFNs are developed to identify and bind to particular sequences in the plant genome. Once within the plant cells, often supplied by *Agrobacterium*-mediated transformation or biolistic techniques, the ZFNs cause double-strand breaks at the targeted loci. The cells inherent repair mechanisms, notably non-homologous end joining (NHEJ) or homologous recombination (HR), subsequently come into action. NHEJ frequently leads in insertions or deletions (indels) that may damage the target gene, essentially blotting it out, but HR can be used to insert or replace genes with a donor template to enhance or alter metabolic pathways. Several research have proved successful gene editing using the ZFN technique, including the high-frequency alteration of tobacco genes, targeted inactivation of *Arabidopsis* genes, and insertion of a herbicide-resistant gene, as well as disruption of a target locus in maize.

For instance, in the medicinal plant *Catharanthus roseus*, ZFNs have been used to bring down the strictosidine synthase (STR) gene, resulting in an increase in the synthesis of precursors for the anti-cancer compounds vinblastine and vincristine. Similarly, in *Artemisia annua*, ZFNs have targeted genes involved in the manufacture of artemisinin, a vital anti-malarial medication, therefore enhancing its output. Moreover, these precise modifications may also be employed to upregulate genes implicated in secondary metabolite pathways. For example, ZFNs have been utilized to enhance the production of shikonin in *Lithospermum erythrorhizon* by targeting genes that suppress its biosynthetic pathway, resulting in increased accumulation of this important substance used in traditional medicine (Halder & Roy, 2023)

The flexibility and specificity of ZFNs make them an important instrument in plant biotechnology, notably for enhancing the synthesis of therapeutic metabolites. This method not only accelerates the breeding process by averting the lengthy conventional breeding cycles but also enables for the production of plant varieties with specific metabolite profiles. The advancements in ZFN technology continue to pave the way for the manufacture of high-value pharmaceutical compounds, giving a sustainable and effective alternative to conventional extraction techniques from plants.

3.8 RNA INTERFERENCE (RNAi)

RNAi is a potent and accurate approach for increasing the synthesis of essential metabolites in medicinal plants. The method begins with the identification of target genes implicated in the metabolic pathways of the targeted metabolites. Once the target gene is identified, a corresponding double-stranded RNA (dsRNA) molecule is created to be complementary to the mRNA of the target gene. This dsRNA may be injected into the plant by numerous techniques, such as *Agrobacterium*-mediated transformation, virus-induced gene silencing (VIGS), or direct injection. Upon introduction, the dsRNA is processed by the plant's cellular apparatus into minuscule interfering RNAs (siRNAs), which bind to the target mRNA and induce its destruction, thereby silencing the gene.

The silence of these genes may lead to an enhanced flow via alternate pathways that encourage the accumulation of desirable metabolites. For example, in *Artemisia annua*, RNAi has been utilized to shut down genes in competing pathways to enhance the synthesis of artemisinin, an essential antimalarial chemical. By suppressing the CYP71AV1 gene, researchers produced greater artemisinin content by shifting metabolic flux toward its manufacture (Feng et al., 2009).

Moreover, RNAi technology is not restricted to raising metabolite synthesis but also extends to fortifying plant resilience to diseases and environmental duress, thus encouraging higher yields of therapeutic compounds. For instance, RNAi-mediated silencing of pathogen susceptibility genes in tobacco has been demonstrated to enhance the plant's resistance to viral infections, providing a more constant production of therapeutic compounds (Majumdar et al., 2017).

Overall, the strategic use of RNAi in medicinal plants represents a potent instrument in metabolic engineering, enabling the exact regulation of biosynthetic pathways to enhance the synthesis of essential compounds. This strategy not only enhances the availability of these compounds for pharmaceutical application but also contributes to the sustainability of medicinal plant production by bolstering tolerance to biotic and abiotic stressors. Thus, RNAi stands out as a new strategy in the domain of plant science, bringing considerable gains in the production and quality of therapeutic plant metabolites.

3.9 ADVANCING MEDICINAL PLANT SECONDARY METABOLITE BIOSYNTHESIS: OMICS INTEGRATION, BIOINFORMATICS ANALYSIS, AND CULTIVATION OPTIMIZATION

The usage of omics technologies, including genomes, transcriptomics, and metabolomics, has changed the research of secondary metabolite production in medicinal plants. These methods enable researchers to unravel the convoluted genetic and metabolic networks underpinning the creation of bioactive substances, facilitating the finding of essential genes and pathways involved in SM biosynthesis. Genomics plays a significant role in this process by offering a complete knowledge of the plant's genetic composition. Through genome sequencing and annotation, researchers may uncover genes encoding enzymes and regulatory proteins involved in secondary metabolite production. For example, in plants like *Taxus* species, whose bark contains the anti-cancer chemical taxol, genomic investigations have uncovered genes implicated in taxol biosynthesis, driving attempts to increase taxol production via genetic engineering or breeding programs.

Transcriptomics complements genomics by analyzing gene expression patterns under various physiological situations or developmental stages. By evaluating transcriptome data, researchers may discover genes that are activated or downregulated during secondary metabolite production. This knowledge assists in comprehending the regulatory processes governing metabolic pathways. For instance, transcriptome investigations in *Catharanthus roseus* have uncovered genes involved in alkaloid biosynthesis, opening the path for targeted modification of alkaloid production by gene editing or metabolic engineering techniques.

Metabolomics focuses on the qualitative and quantitative characterization of metabolites present in plant tissues. Through metabolomic profiling, researchers may discover secondary metabolites generated by therapeutic plants and measure their abundance. This information facilitates in disclosing metabolic pathways and comprehending metabolic fluxes under varied environmental situations or genetic contexts. For example, metabolomic studies in *Hypericum perforatum* have identified the biosynthesis routes of hypericin, a crucial chemical with antidepressant characteristics, offering insights into techniques for increasing hypericin production by metabolic engineering.

Integration of bioinformatics tools is vital for assessing omics data and identifying metabolic pathways involved in SM production. Bioinformatics methods facilitate the annotation of genes, prediction of enzyme activities, and reconstruction of metabolic networks. By integrating genomes, transcriptomics, and metabolomics data, researchers can model complicated metabolic pathways and discover plausible targets for pathway optimization or metabolic engineering.

Implementation of tissue culture and cell suspension culture methods allows mass proliferation and generation of secondary metabolites in regulated conditions. Tissue culture methods, such as micropropagation and organ culture, enable the rapid reproduction of superior plant varieties with desired features. Cell suspension culture methods, on the other hand, offer a platform for the generation of SMs in vitro. By enhancing growth conditions and medium compositions, researchers may boost secondary metabolite production and surmount constraints associated with standard cultivation techniques.

Optimization of growing conditions, including light intensity, temperature, and nutrient availability, is critical for increasing secondary metabolite outputs in medicinal plants. Environmental conditions have a crucial influence in regulating plant metabolism and secondary metabolite synthesis. By fine-tuning developing conditions, researchers may modulate metabolic pathways and enhance the accumulation of specific compounds. For example, altering light quality and photoperiod has been found to impact the manufacture of flavonoids in plants like *Arabidopsis thaliana*, suggesting an approach for enhancing flavonoid outputs via controlled agricultural practices.

3.10 NAVIGATING CHALLENGES AND PROMOTING SUSTAINABILITY IN MEDICINAL PLANT PRECISION BREEDING

In the objective of leveraging precision breeding for the enhancement of medicinal plants, it is necessary to foresee and solve a variety of obstacles and limits that may occur. Regulatory difficulties comprise a considerable challenge, since the use of modern biotechnological tools like CRISPR-Cas9 and gene editing methods may fall under severe rules governing genetically modified organisms (GMOs) in numerous countries. Navigating various regulatory environments necessitates rigorous adherence to compliance standards and collaboration with regulatory authorities to ensure that research and development initiatives are performed ethically and lawfully.

Ethical issues also play a key role, particularly with the possible implications on biodiversity, environmental sustainability, and cultural practices related to traditional medicinal plant utilization. It is crucial to preserve ethical norms such as respect for indigenous knowledge, preservation of genetic variety, and equal access to the advantages derived from precision breeding technology. Transparent conversation and stakeholder participation are vital for negotiating these ethical difficulties and facilitating responsible innovation in medicinal plant breeding.

Furthermore, intellectual property rights (IPR) provide concerns regarding ownership, access, and benefit-sharing agreements, particularly when commercial interests meet with traditional knowledge and genetic resources. Developing fair and equitable IPR frameworks that balance the interests of breeders, researchers, indigenous groups, and the public is vital for stimulating innovation while maintaining socio-economic and cultural rights.

To overcome these obstacles and promote the sustainable application of precision breeding for medicinal plant enhancement, coordinated initiatives are required. Multidisciplinary research partnerships, stakeholder discussions, and policy dialogues may facilitate the creation of rigorous regulatory frameworks, ethical principles, and fair benefit-sharing systems. Additionally, capacity-building initiatives, public awareness campaigns, and knowledge-sharing platforms may enable stakeholders to negotiate the complexity of precision breeding ethically and ensure that its advantages are attained for the collective well-being of society. By proactively resolving these obstacles and adopting a collaborative approach, we may unleash the full potential of precision breeding for increasing the medical benefits of plants while respecting values of sustainability, fairness, and ethical stewardship.

3.11 CONCLUSION

In conclusion, the chapter 'Strategies for Precision Breeding of Medicinal Plants for Enhanced Secondary Metabolite Production' emphasizes the essential significance of cutting-edge approaches in altering the area of medicinal plant breeding. Through the utilization of molecular marker-assisted selection, mutagenesis using chemicals or radiation, and gene editing tools like CRISPR-Cas9, TALENs, zinc finger nucleases, and RNA interference, researchers have unlocked unprecedented opportunities for targeted enhancement of secondary metabolite production in medicinal plants. These precision breeding procedures enable exact control over genetic alterations, enabling breeders to accelerate the breeding process and precisely target desired features, including the manufacture of essential bioactive compounds. By utilizing these modern biotechnological technologies, the chapter emphasizes the potential to generate high-performance plant varieties with increased medicinal qualities, answering the rising need for natural therapeutic agents in the healthcare and pharmaceutical sectors. Moreover, the integration of omics technologies, bioinformatics analysis, and optimization of cultivation conditions further enhances the efficacy of precision breeding approaches, paving the way for sustainable and efficient production of medicinal plants with superior secondary metabolite profiles.

REFERENCES

Alagoz, Y., Gurkok, T., Zhang, B., & Unver, T. (2016). Manipulating the biosynthesis of bioactive compound alkaloids for next-generation metabolic engineering in opium poppy using CRISPR-Cas 9 genome editing technology. *Scientific Reports*, *6*(1), 30910.

Armarkar, A., Mahure, D., Pounikar, A., Bhagat, R., & Nagare, D. (2021). An overview on the biosynthetic pathways and medicinal values of secondary metabolites. *Journal of Pharmaceutical Research International*, 100–114. https://doi.org/10.9734/jpri/2021/v33i33a31777

Arora, L., & Narula, A. (2017). Gene editing and crop improvement using CRISPR-Cas9 system. *Frontiers in Plant Science*, *8*, 296675.

Babu, R., Nair, S. K., Prasanna, B. M., & Gupta, H. S. (2004). Integrating marker-assisted selection in crop breeding–prospects and challenges. *Current Science*, 607–619.

Barrett, B., Griffiths, A., Mercer, C., Ellison, N., Faville, M., Easton, S., & Woodfield, D. R. (2001, January). Marker-assisted selection to accelerate forage improvement. In *Proceedings of the New Zealand Grassland Association*, *63*, 241–245.

Barut, M., Cavdar, A. S., & Tansi, L. S. (2023). Medicinal legumes in Turkey: A gift of nature for bios. In *Legumes biofortification* (pp. 481–506). Springer International Publishing.

Bharti, P. K., Singh, S. K., & Kumari, S. (2023). Genetic manipulation in medicinal plants for enhancement of secondary metabolites. In *Genetic manipulation of secondary metabolites in medicinal plant* (pp. 129–145). Springer Nature Singapore.

Bonini, S. A., Premoli, M., Tambaro, S., Kumar, A., Maccarinelli, G., Memo, M., & Mastinu, A. (2018). *Cannabis sativa*: A comprehensive ethnopharmacological review of a medicinal plant with a long history. *Journal of Ethnopharmacology*, *227*, 300–315.

Čermák, T., & Curtin, S. J. (2017). Design and assembly of CRISPR/Cas9 reagents for gene knockout, targeted insertion, and replacement in wheat. In *Wheat biotechnology: Methods and protocols* (pp. 187–212). Humana Press.

Chen, D., Lee, J., Hou, Q., Zhao, X., & Sun, K. (2021a). Analysis of endophyte diversity of *Rheum palmatum* from different production areas in Gansu province of China and the association with secondary metabolite. *Microorganisms*, *9*(5), 978. https://doi.org/10.3390/microorganisms9050978

Chen, J., Qi, H., Li, J. B., Yi, Y. Q., Chen, D., Hu, X. H., Wang, M. L., Sun, X. L., & Wei, X. Y. (2014). Experimental study on *Dendrobium candidum* polysaccharides on promotion of hair growth. *Zhongguo Zhong yao za zhi= Zhongguo Zhongyao Zazhi= China Journal of Chinese Materia Medica*, *39*(2), 291–295.

Chen, Y., Zheng, Y. F., Lin, X. H., Zhang, J. P., Lin, F., & Shi, H. (2021b). Dendrobium mixture attenuates renal damage in rats with diabetic nephropathy by inhibiting the PI3K/Akt/mTOR pathway. *Molecular Medicine Reports*, *24*(2), 1–11.

Chettri, U., & Kumari, S. (2021). A list of medicinally important plants of Sikkim Himalayan region, India. *Journal of Medicinal Plants Studies*, *9*(4), 24–27. https://doi.org/10.22271/plants.2021.v9.i4a.1307

Choudhary, K., Choudhary, O. P., & Shekhawat, N. S. (2008). Marker assisted selection: A novel approach for crop improvement. *American-Eurasian Journal of Agronomy*, *1*(2), 26–30.

Cowan, M. (1999). Plant products as antimicrobial agents. *Clinical Microbiology Reviews*, *12*(4), 564–582. https://doi.org/10.1128/cmr.12.4.564

Das, S., Kwon, M., & Kim, J. Y. (2024). Enhancement of specialized metabolites using CRISPR/Cas gene editing technology in medicinal plants. *Frontiers in Plant Science*, *15*, 1279738.

Deepa, B., & Sivakumar, T. (2020). Phytochemical analysis and antibacterial efficacy of ethanolic extract of musa paradisiacal. *Research Biotica*, *2*(3), 126. https://doi.org/10.54083/resbio/2.3.2020.126-130

Ebenezer, K., Manivannan, R., Punniyamoorthy, A., & Tamilselvan, C. (2019). Plant secondary metabolites of antiviral properties a rich medicinal source for drug discovery: A mini review. *Journal of Drug Delivery and Therapeutics*, *9*(5), 161–167. https://doi.org/10.22270/jddt.v9i5.3471

Escandón-Rivera, S., Mata, R., & Andrade-Cetto, A. (2020). Molecules isolated from Mexican hypoglycemic plants: A review. *Molecules*, *25*(18), 4145. https://doi.org/10.3390/molecules25184145

Farinati, S., Draga, S., Betto, A., Palumbo, F., Vannozzi, A., Lucchin, M., & Barcaccia, G. (2023). Current insights and advances into plant male sterility: New precision breeding technology based on genome editing applications. *Frontiers in Plant Science*, *14*, 1223861.

Farnsworth, N. R., & Soejarto, D. D. (1991). Global importance of medicinal plants. In *The conservation of medicinal plants* (vol. 26, pp. 25–51). Cambridge University Press.

Feng, L. L., Yang, R. Y., Yang, X. Q., Zeng, X. M., Lu, W. J., & Zeng, Q. P. (2009). Synergistic re-channeling of mevalonate pathway for enhanced artemisinin production in transgenic *Artemisia annua*. *Plant Science*, *177*(1), 57–67.

Feng, S., Song, W., Fu, R., Zhang, H., Xu, A., & Li, J. (2018). Application of the CRISPR/Cas9 system in *Dioscorea zingiberensis*. *Plant Cell, Tissue and Organ Culture (PCTOC)*, *135*, 133–141.

Festa, G., Autore, F., Fraternali, F., Giardina, P., & Sannia, G. (2008). Development of new laccases by directed evolution: Functional and computational analyses. *Proteins: Structure, Function, and Bioinformatics*, *72*(1), 25–34.

Gouda, G., Gupta, M. K., Donde, R., Mohapatra, T., Vadde, R., & Behera, L. (2020). Marker-assisted selection for grain number and yield-related traits of rice (*Oryza sativa* L.). *Physiology and Molecular Biology of Plants*, *26*, 885–898.

Gracz-Bernaciak, J., Mazur, O., & Nawrot, R. (2021). Functional studies of plant latex as a rich source of bioactive compounds: Focus on proteins and alkaloids. *International Journal of Molecular Sciences*, *22*(22), 12427.

Hakim, A., Jufri, A., Jamaluddin, J., Septiani, D., Ningsih, B., & Supriadi, S. (2022). The influence of the natural product practicum on decision making skills. *Unesa Journal of Chemical Education*, *11*(2), 136–142. https://doi.org/10.26740/ujced.v11n2.p136-142

Halder, M., & Roy, S. (2023). Current status of metabolic engineering of medicinal plants for production of plant-derived secondary metabolites. In *Medicinal plants: Biodiversity, biotechnology and conservation* (pp. 819–869). Springer Nature Singapore.

Hidayani, H., Hadiprayitno, G., Kusmiyati, K., & Irawan, J. (2021). Ethnobotany of medicinal plants and traditional rituals community of Sembalun Village, East Lombok. *Jurnal Pijar Mipa*, *16*(5), 676–682. https://doi.org/10.29303/jpm.v16i5.2984

Hussain, M. S., Fareed, S., Ansari, S., Rahman, M. A., Ahmad, I. Z., & Saeed, M. (2012). Current approaches toward production of secondary plant metabolites. *Journal of Pharmacy and Bioallied Sciences*, *4*(1), 10–20.

Hussein, R. A., & El-Anssary, A. A. (2019). Plants secondary metabolites: The key drivers of the pharmacological actions of medicinal plants. *Herbal Medicine*, *1*(3), 11–30. https://doi.org/10.5772/intechopen.76139

Jalal, S., Ahmad, B., Zhang, T., Guo, L., & Huang, L. (2020). Santamarine: Mechanistic studies on multiple diseases. *Chemical Biology & Drug Design*, *95*(4), 427–434. https://doi.org/10.1111/cbdd.13666

Jaswinder, K., Vyas, M., Singh, J., Prasad, R., & Gupta, J. (2020). Therapeutic applications of naringenin, a flavanone enriched in citrus fruits, for disorders beyond diabetes. *Phyton*, *89*(4), 795–803. https://doi.org/10.32604/phyton.2020.09420

Jumiati, E., & Ismandari, T. (2022). The potency of karamunting Borneo plants from weeds into herbs. *IOP Conference Series Earth and Environmental Science*, *1083*(1), 012003. https://doi.org/10.1088/1755-1315/1083/1/012003

Jung, J. H., & Altpeter, F. (2016). TALEN mediated targeted mutagenesis of the caffeic acid O-methyltransferase in highly polyploid sugarcane improves cell wall composition for production of bioethanol. *Plant Molecular Biology*, *92*, 131–142.

Li, J., Stoddard, T. J., Demorest, Z. L., Lavoie, P. O., Luo, S., Clasen, B. M., Cedrone, F., Ray, E. E., Coffman, A. P., Daulhac, A., Yabandith, A., Retterath, A. J., Mathis, L., Voytas, D.F., D'Aoust, M. A., & Zhang, F. (2016). Multiplexed, targeted gene editing in Nicotiana benthamiana for glycoengineering and monoclonal antibody production. *Plant Biotechnology Journal*, *14*(2), 533–542.

Li, X., Wang, Y., Chen, S., Tian, H., Fu, D., Zhu, B., Luo, Y., & Zhu, H. (2018). Lycopene is enriched in tomato fruit by CRISPR/Cas9-mediated multiplex genome editing. *Frontiers in Plant Science*, *9*, 559.

Madariaga-Mazón, A., Naveja, J., Medina-Franco, J., Noriega-Colima, K., & Martinez-Mayorga, K. (2021). Dianat-DB: A molecular database of antidiabetic compounds from medicinal plants. *RSC Advances*, *11*(9), 5172–5178. https://doi.org/10.1039/d0ra10453a

Majumdar, R., Rajasekaran, K., & Cary, J. W. (2017). RNA interference (RNAi) as a potential tool for control of mycotoxin contamination in crop plants: Concepts and considerations. *Frontiers in Plant Science*, *8*, 245287.

Matos, N., Freitas, D., Ottoni, M., Caldeira, A., Souza, V., Meireles, A., Alvim Brito-Melo, G., & Freitas, B. (2020). The registration of the usage of medicinal plants on e-sus program: A case study in a city in the upper Jequitinhonha Valley. *Rodriguésia*, *71*. https://doi.org/10.1590/2175-7860202071057

Mipeshwaree Devi, A., Khedashwori Devi, K., Premi Devi, P., Lakshmipriyari Devi, M., & Das, S. (2023). Metabolic engineering of plant secondary metabolites: Prospects and its technological challenges. *Frontiers in Plant Science*, *14*, 1171154.

Mulyono, D., Pinardi, D., & Jufri, A. (2021). Development of medicinal plant culturethrough empowerment of entrepreneur. https://doi.org/10.2991/absr.k.210609.065

Nagavalli, K. (2022). Medicinal plant resource diversity of agro-ecosystem in Thiruvannamalai District, Tamil Nadu, South India. *Journal of Development Economics and Management Research Studies*, *09*(13), 16–29. https://doi.org/10.53422/jdms.2022.91303

Newman, D. J., & Cragg, G. M. (2012). Natural products as sources of new drugs over the 30 years from 1981 to 2010. *Journal of Natural Products*, *75*(3), 311–335.

Park, Y., Mishra, R., Yoon, S., Kim, H., Park, C., Seo, S., & Bae, H. (2019). Endophytic *Trichoderma citrinoviride* isolated from mountain-cultivated ginseng (*Panax ginseng*) has great potential as a biocontrol agent against ginseng pathogens. *Journal of Ginseng Research*, *43*(3), 408–420. https://doi.org/10.1016/j.jgr.2018.03.002

Paul, S., Khanuja, S. P., Shasany, A. K., Gupta, M. M., Darokar, M. P., Saikia, D., & Gupta, A. K. (2010). Enhancement of artemisinin content through four cycles of recurrent selection with relation to heritability, correlation and molecular marker in *Artemisia annua* L. *Planta Medica*, *76*(13), 1468–1472.

Rai, S. P., Rai, S. K., Apoorva, Rai, K. K., & Kumar, S. (2022). Molecular markers and descriptor's suitable for quantitative trait loci and genome mapping in *Catharanthus roseus*. In *The Catharanthus genome* (pp. 123–140). Springer International Publishing.

Redman, M., King, A., Watson, C., & King, D. (2016). What is CRISPR/Cas9? *Archives of Disease in Childhood-Education and Practice*, *101*(4), 213–215.

Roeder, E., Wiedenfeld, H., & Edgar, J. A. (2015). Pyrrolizidine alkaloids in medicinal plants from North America. *Die Pharmazie-An International Journal of Pharmaceutical Sciences*, *70*(6), 357–367.

Shi, M., Du, Z., Hua, Q., & Kai, G. (2021). CRISPR/Cas9-mediated targeted mutagenesis of bZIP2 in *Salvia miltiorrhiza* leads to promoted phenolic acid biosynthesis. *Industrial Crops and Products*, *167*, 113560.

Shiels, D., Prestwich, B. D., Koo, O., Kanchiswamy, C. N., O'Halloran, R., & Badmi, R. (2022). Hemp genome editing – challenges and opportunities. *Frontiers in Genome Editing*, *4*, 823486.

Shrestha, A., Khan, A., & Dey, N. (2018). Cis-trans engineering: Advances and perspectives on customized transcriptional regulation in plants. *Molecular Plant*, *11*, 886–898. https://doi.org/10.1016/j.molp.2018.05.008

Si, X., Zhang, H., Wang, Y., Chen, K., & Gao, C. (2020). Manipulating gene translation in plants by CRISPR–Cas9-mediated genome editing of upstream open reading frames. *Nature Protocols*, *15*(2), 338–363.

Singh, P., Guleri, R., Singh, V., Kaur, G., Kataria, H., Singh, B., Kaur, G., Kaul, S. C., Wadhwa, R., & Pati, P. K. (2015). Biotechnological interventions in *Withania somnifera* (L.) Dunal. *Biotechnology and Genetic Engineering Reviews*, *31*(1–2), 1–20.

Siregar, A., & Tanjung, A. (2021). Factors that influence the interests of households in medicinal plants farming in the COVID-19 pandemic situation. https://doi.org/10.2991/assehr.k.210909.038

Strasser, R., Stadlmann, J., Schähs, M., Stiegler, G., Quendler, H., Mach, L., Glössl, J., Weterings, K., Pabst, M., & Steinkellner, H. (2008). Generation of glyco-engineered *Nicotiana benthamiana* for the production of monoclonal antibodies with a homogeneous human-like N-glycan structure. *Plant Biotechnology Journal*, *6*(4), 392–402.

Sun, Y., Jiao, G., Liu, Z., Zhang, X., Li, J., Guo, X., Du, W., Du, J., Francis, F., Zhao, Y., & Xia, L. (2017). Generation of high-amylose rice through CRISPR/Cas9-mediated targeted mutagenesis of starch branching enzymes. *Frontiers in Plant Science*, *8*, 298.

Syarifah, V., Rafi, M., & Wahyuni, W. (2017). High perfomance liquid chromatography fingerprint analysis for quality control of brotowali (*Tinospora crispa*). *Journal of Physics Conference Series*, *835*, 012016. https://doi.org/10.1088/1742-6596/835/1/012016

Szakiel, A., Pączkowski, C., & Henry, M. (2010). Influence of environmental abiotic factors on the content of saponins in plants. *Phytochemistry Reviews*, *10*(4), 471–491. https://doi.org/10.1007/s11101-010-9177-x

Vasanthkumar, S. S., Pooja, U. K., Manjusha, M. R., & Nithiyasri, S. (2024). *Breeding of medicinal and aromatic plantsreeding of medicinal and aromatic plants* (Vol. 2). Shanlax Publications.

Weeks, D. P., Spalding, M. H., & Yang, B. (2016). Use of designer nucleases for targeted gene and genome editing in plants. *Plant Biotechnology Journal*, *14*(2), 483–495.

Wolter, F., Schindele, P., & Puchta, H. (2019). Plant breeding at the speed of light: The power of CRISPR/Cas to generate directed genetic diversity at multiple sites. *BMC Plant Biology*, *19*, 176. https://doi.org/10.1186/s12870-019-1775-1

Xie, K., Minkenberg, B., & Yang, Y. (2015). Boosting CRISPR/Cas9 multiplex editing capability with the endogenous tRNA-processing system. *Proceedings of the National Academy of Sciences*, *112*(11), 3570–3575.

Yang, J., Huang, J., Shen, H., & Xinsheng, Q. (2022). The diversity of wild medicinal plants of lufeng in Eastern Guangdong, China. https://doi.org/10.21203/rs.3.rs-1373104/v1

Yeshi, K., Crayn, D., Ritmejerytė, E., & Wangchuk, P. (2022). Plant secondary metabolites produced in response to abiotic stresses has potential application in pharmaceutical product development. *Molecules*, *27*(1), 313.

Zakaria, M. M., Schemmerling, B., & Ober, D. (2021). CRISPR/Cas9-mediated genome editing in comfrey (*Symphytum officinale*) hairy roots results in the complete eradication of pyrrolizidine alkaloids. *Molecules*, *26*(6), 1498.

Zeng, L., Zhang, Q., Jiang, C., Zheng, Y., Zuo, Y., Qin, J., Liao, Z., & Deng, H. (2021). Development of *Atropa belladonna* L. plants with high-yield hyoscyamine and without its derivatives using the CRISPR/Cas9 system. *International Journal of Molecular Sciences*, *22*(4), 1731.

Zhou, Z., Li, Q., Xiao, L., Wang, Y., Feng, J., Bu, Q., Xiao, Y., Hao, K., Guo, M., Chen, W., & Zhang, L. (2021). Multiplexed CRISPR/Cas9-mediated knockout of laccase genes in *Salvia miltiorrhiza* revealed their roles in growth, development, and metabolism. *Frontiers in Plant Science*, *12*, 647768.

Zhou, Z., Tan, H., Li, Q., Chen, J., Gao, S., Wang, Y., Chen, W., & Zhang, L. (2018). CRISPR/Cas9-mediated efficient targeted mutagenesis of RAS in Salvia miltiorrhiza. *Phytochemistry*, *148*, 63–70.

4 Advances in Medicinal Plant Metabolomics and Future Directions

Arockiya Anita Margret, Ramesh Preyenga, Tajudheen Nargis Begum, Alagarsamy Sumithra, Shanmugam Gavaskar, Gangadharan Thilagavathi, Asirvatham Alwin Robert, and Arumugam Vijaya Anand

LIST OF ABBREVIATIONS

APCI	air pressure chemical ionization
DI-MS	direct infusion–mass spectrometry
ESI	electrospray ionization
GC-MS	gas chromatography–mass spectrometry
KEGG	Kyoto Encyclopedia of Genes and Genomes
LC-MS	liquid chromatography–mass spectrometry
PMDB	Plant Metabolome Database
PPMR	Phytochemicals and Plant Metabolomics Resource
TAIR	The Arabidopsis Information Resource

4.1 INTRODUCTION

Medicinal plants and its metabolites have garnered immense interest due to their pharmacological benefits (Manoharan et al., 2008; Begum et al., 2009; Muthumani et al., 2013; Varalakshmi et al., 2014; Vijaya Anand et al., 2016, 2020; Bharathi et al., 2018; Divya et al., 2019; Akilandeswari et al., 2019; Sangeetha et al., 2022; Kaviya et al., 2022; Manikandan et al., 2022; Karthika et al., 2023; Vijayakumar et al., 2023). "Omics" is a new term that emerged to describe an immense set of biologically rich data within a cell (Karthikkeyan et al., 2020). It focuses on identifying and quantifying the biomolecules related to their molecular mechanisms that contribute to the functioning of cells. "Metabolome" plus "omics" creates "metabolomics," which includes with chemical processes within a metabolite (Buriani et al., 2012). The small molecules entangled in the process of metabolism have been noteworthy in recent progressions by contributing reflective insights into biological systems, health, and therapeutics. The furtherance of metabolomics has expanded profoundly in recent years with the advent of computational tools and its integration. Figure 4.1 illustrates the alliance of experimentation and data analysis. Various significant fields like clinical medicine, research, health sectors, and disease diagnosis benefit from metabolomics. Agricultural and environmental surveillance are being optimistically impacted by the influence of metabolomics.

4.2 FUTURE DIRECTIONS IN MEDICINAL PLANT METABOLOMICS

The complete examination of tiny molecules (metabolites) found in plants is the primary focus of the field of study known as plant metabolomics. Metabolites are essential to the growth, development, defense systems, and environmental interactions of plants (Hong et al., 2016). The goal of plant metabolomics is to comprehend the metabolic processes that take place in plants under many circumstances, including distinct growth stages, stresses from the environment, and genetic alterations (Fukusaki & Kobayashi, 2005). Plant metabolomics encompasses several key aspects that contribute to our understanding of plant metabolism, physiology, and responses to environmental factors. This field involves various aspects, such as analyzing metabolite profiles, exploring metabolic pathways, and investigating how plants react to environmental stressors like drought, salinity, extreme temperatures, and pathogen attacks (Kumar et al., 2017; Shulaev et al., 2008). In plant biology, metabolomics is also used to find biomarkers associated with specific physiological conditions or diseases, which may have benefits in precision agriculture and plant breeding. Figure 4.2 shows the promising phases of and progression of plant metabolomics. Medicinal plant metabolomics imposes advanced innovative technologies with integrative approaches to decipher their phytochemical composition and therapeutic potential to target human ailments. The impending significance of plant metabolomics categorizes novel bioactive compounds in accordance with their therapeutic usage.

4.3 PROCESSING AND ANALYZING METABOLITES IN PLANT METABOLOMICS

The multi-mechanistic and therapeutic role of medicinal plants and their derived products in the form of metabolites has to be extroverted. This requires a standard set of steps and procedures. Figure 4.3 illustrates the various facets involved in the processing of plant metabolism. Various

DOI: 10.1201/9781003475491-4

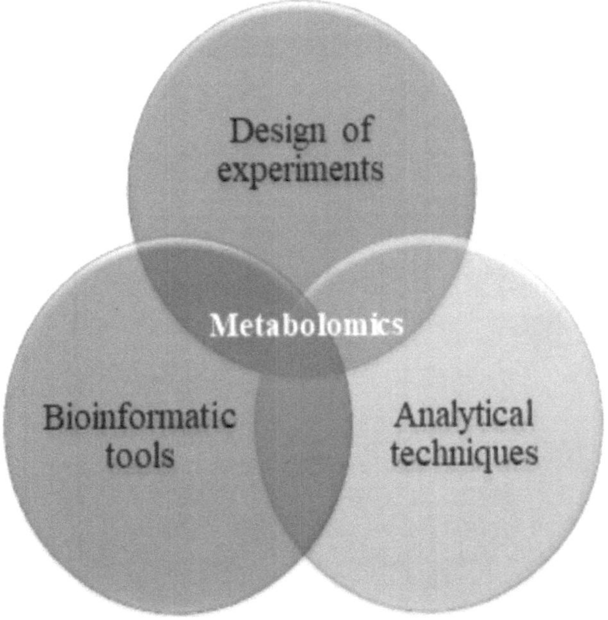

FIGURE 4.1 Metabolomics – vital tool in computational biology.

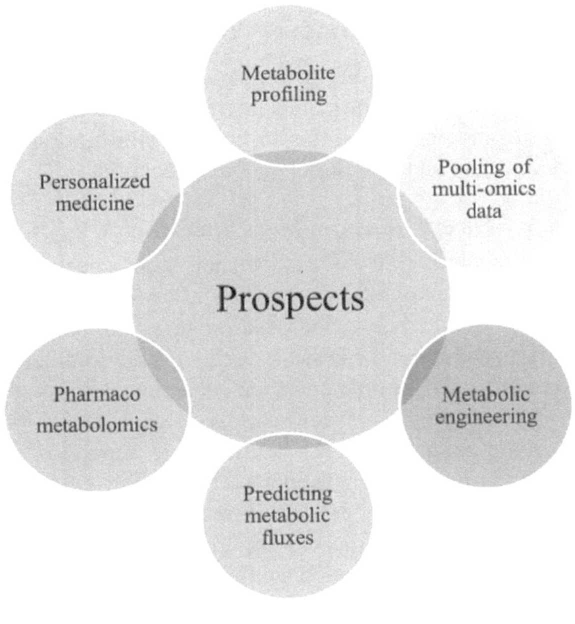

FIGURE 4.2 Progression and prospects of plant metabolomics.

techniques involved in the progression of plant metabolomics are discussed in the following sections.

4.3.1 High-Throughput Analytical Techniques in Plant Metabolomics

4.3.1.1 Mass-Spectrometry-Based Technologies

Traditionally, mass spectrometry has been used with liquid and gas chromatography primarily for two purposes:

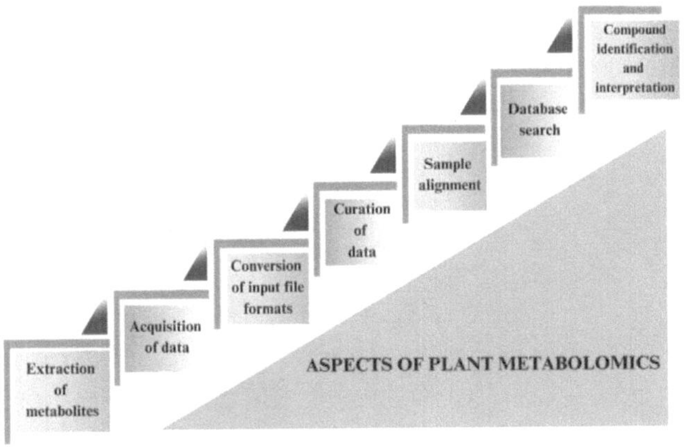

FIGURE 4.3 Aspects of plant metabolomics.

- To lessen interference between analytes and the background (i.e., the matrix)
- To compensate for the lack of separation provided by low-resolution mass spectrometers (e.g., by using quadrupoles or ion traps as mass filters).

4.3.1.1.1 Gas Chromatography–Mass Spectrometry (GC-MS)

The capacity of gas chromatography–mass spectrometry (GC-MS) to quickly and accurately evaluate a large number of metabolites has made it a well-known high-throughput analytical method in the field of plant metabolomics. It is a useful method for examining the intricate chemical makeup of plant samples because it combines the separation strength of gas chromatography with the detection and identification capabilities of mass spectrometry. Plant metabolomics uses GC-MS to examine a broad range of metabolites found in plant tissues, including secondary metabolites (such as terpenoids, phenolics and alkaloids) and primary metabolites (like sugars, organic acids and amino acids) (Lisec et al., 2006).

To isolate the desired metabolites, plant samples are extracted using the proper solvents and in order to improve the extracted samples' volatility and stability, they are subsequently derivatized for the GC-MS analysis. The samples that have been derivatized are introduced into a GC system that has a capillary column. As metabolites move through the column, the GC separates them according to physical characteristics like polarity and boiling temperatures. Metabolites fragment and become ionized using electron ionization or chemical ionization as they pass through the mass spectrometer after eluting from the GC column. The metabolites are then identified and quantified using their mass-to-charge ratios (m/z), and they serve as a basis for analysis of the resultant mass spectra (Figure 4.4). For metabolite identification, quantification, and statistical analysis, specialized software is used to process the obtained data, which includes

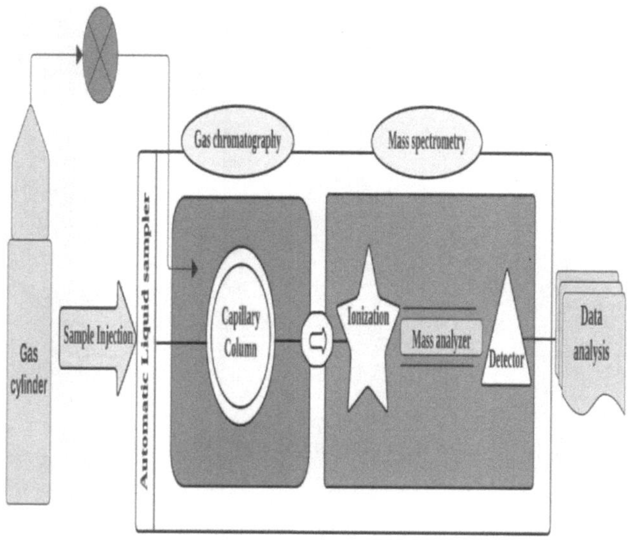

FIGURE 4.4 Gas chromatography–mass spectrometry (GC-MS).

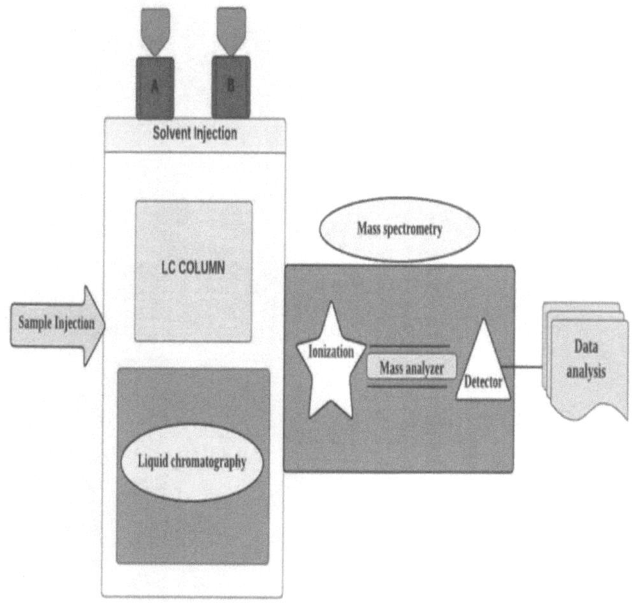

FIGURE 4.5 Liquid chromatography-mass spectrometry (LC-MS)

retention times, mass spectra, and peak intensities. Ultimately, GC-MS offers several benefits such as excellent sensitivity, comprehensive metabolite coverage, efficient high-throughput quantitative analysis, and the ability to provide structural information about metabolites (Wolfender et al., 2015).

4.3.1.1.2 Liquid Chromatography–Mass Spectrometry (LC-MS)

Comprehensive information about plant metabolism can be obtained through the analysis of a broad range of metabolites by liquid chromatography–mass spectrometry (LC-MS), including polar and nonpolar molecules, amino acids, lipids, phenolics, and more. Its ability to detect metabolites at extremely low concentrations makes it possible to analyze complicated plant samples with extreme sensitivity.

Typically, plant samples are extracted to isolate metabolites using the proper solvents. After drying and reconstitution in an appropriate solvent, these extracts are ready for LC-MS analysis. The prepared samples are introduced into a chromatographic column-equipped LC system. Individual metabolites are separated using the ionization technique according to their size, polarity, and affinity for the stationary phase. The metabolites are then identified and quantified using their mass-to-charge ratios (m/z) similar to GC-MS analysis (Figure 4.5). However, LC-MS offers benefits such as thermal stability, ease of sample preparation, and increased sensitivity of detection. It also includes a variety of ionization techniques such as air pressure chemical ionization (APCI) and electrospray ionization (ESI), offering a flexible range of ionization techniques suitable for a wide range of compound types (Mattoli et al., 2006).

4.3.2 Chromatography-Free Technologies for High-Throughput Metabolomics

A logical approach to boost sample throughput in metabolomics is to fully eliminate chromatographic separation since chromatography seems to be the main hindrance to quick analysis and seamless peak identification.

4.3.2.1 Direct Infusion–Mass Spectrometry (DI-MS)

In plant metabolomics, direct infusion–mass spectrometry (DI-MS) is an analytical method that allows one to directly evaluate the metabolite composition of a sample without any prior separation. In DI-MS, chemicals are immediately ionized from the sample and then subjected to mass spectrometry analysis. This contrasts with approaches based on chromatography, where chemicals are separated prior to MS measurements. Solvents such as methanol or acetonitrile are frequently used to extract metabolites and the solution is thereafter injected straight into the mass spectrometer. Diverse ionization methods, such as APCI and ESI, can be employed in DI-MS. Once ionized, metabolites are subjected to mass analysis in the mass spectrometer. Different mass analyzers such as quadrupole, time-of-flight, or Orbitrap can be used for accurate mass measurement and detection of metabolites (Shree et al., 2019).

4.3.2.2 Matrix-Assisted Laser Desorption/ Ionization–Mass Spectrometry (MALDI-MS)

Matrix-assisted laser desorption/ionization–mass spectrometry (MALDI-MS) involves ionizing and analyzing molecules, typically biomolecules such as proteins, peptides, lipids, and metabolites, by using a laser to desorb them from a matrix material and uses a time of flight

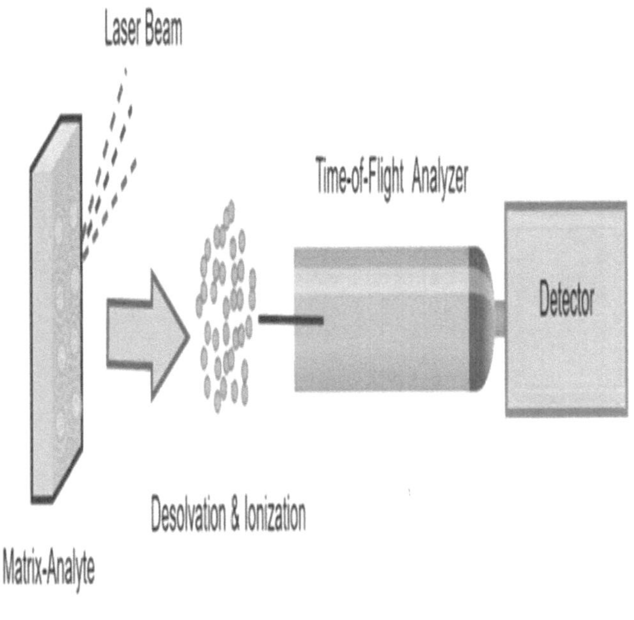

FIGURE 4.6 Matrix-assisted laser desorption/ionization–mass spectrometry (MALDI-MS)

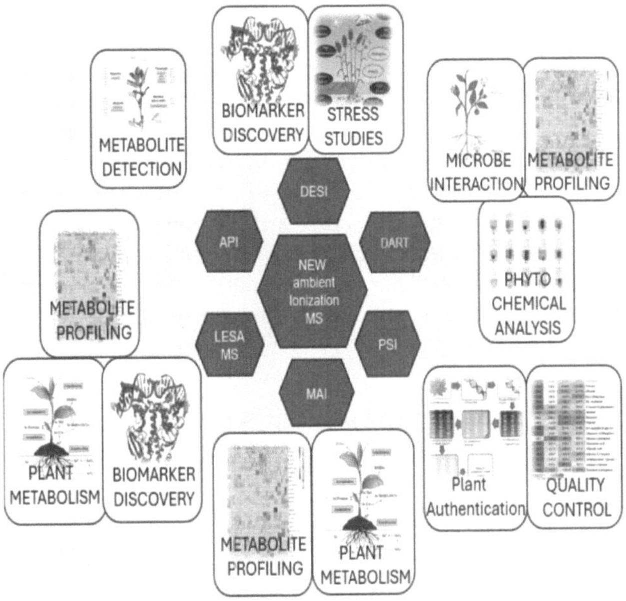

FIGURE 4.7 Techniques of new ambient mass spectrometry.

analyzer (Figure 4.6). When analyzing large numbers of plant samples or doing time-course experiments, MALDI-MS is a useful tool for high-throughput metabolomics studies because of its high sensitivity and ability to detect a wide range of metabolites present in plant samples, even at low concentrations. Plant tissues or cells can be subjected to targeted metabolite analysis using MALDI-MS. Understanding the spatial distribution of metabolites within plants and examining metabolic alterations in response to environmental stimuli or developmental phases are made possible by this technique. Since it is non-destructive to samples, it can evaluate a wide range of metabolites, including polar and nonpolar compounds (Arbona et al., 2009).

4.3.2.3 New Ambient Ionization Mass Spectrometry

The study of plant metabolomics has been greatly impacted by new ambient ionization mass spectrometry techniques that allow for direct and quick analysis of plant samples with little to no sample pretreatment (Figure 4.7). Benefits of these methods include higher throughput, shorter analysis times, and the capacity to capture a large variety of metabolites in their natural form.

4.4 DATA INTEGRATION IN PLANT METABOLOMICS

4.4.1 Multi-Omics Integration

To obtain a thorough understanding of plant biology and metabolism, multi-omics integration in plant metabolomics requires integrating data from multiple omics layers, such as proteomics, metabolomics, transcriptomics, and genomics. Knowledge regarding plant genetics, especially genes involved in metabolic processes, can be obtained through genomics. Researchers can discover potential genes for certain metabolic pathways, correlate genetic variants with metabolite profiles, and understand the genetic basis of plant metabolism by combining genomics and metabolomics data (Figure 4.8). Transcriptomics uses RNA transcript analysis to study patterns of gene expression. By combining transcriptomics and metabolomics, scientists may examine gene-metabolite interactions, discover regulatory mechanisms governing metabolic pathways, and correlate gene expression levels with metabolite abundance (Cambiaghi et al., 2017). On the other hand, the expression and roles of proteins in cells are investigated using proteomics. The study of protein-metabolite interactions that impact plant metabolism, the identification of enzymes and proteins engaged in metabolic pathways, and the clarification of post-translational changes affecting enzyme activity are all made possible by the integration of proteomics and metabolomics data. Thus, the entire collection of these tiny molecules (metabolites) present in a biological sample is examined using metabolomics. Researchers can establish thorough metabolic networks, identify important metabolites and pathways connected to plant growth, development, and responses to environmental stimuli, and find metabolic markers linked to plant traits or stress tolerance by combining metabolomics data with other omics layers (Bylesjö et al., 2007).

4.4.2 Integration of Experimental and Computational Data

The process of creating experimental data involves the measurement and identification of metabolites contained in plant samples using methods like chromatography, mass

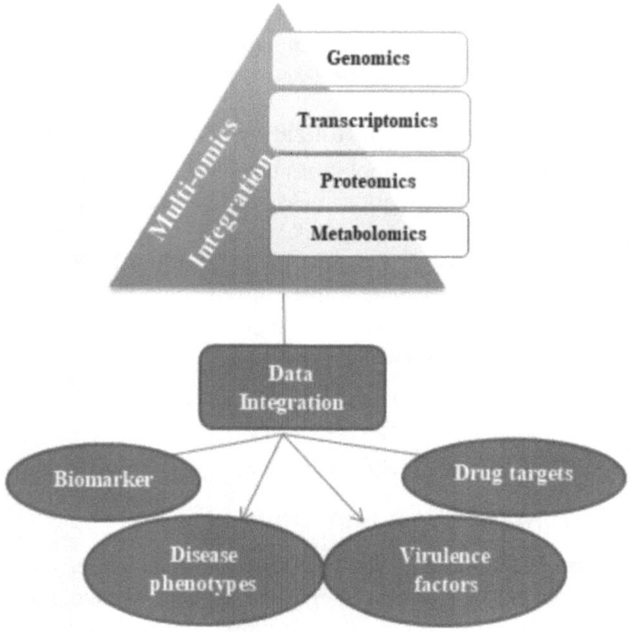

FIGURE 4.8 Multi-omics data integration.

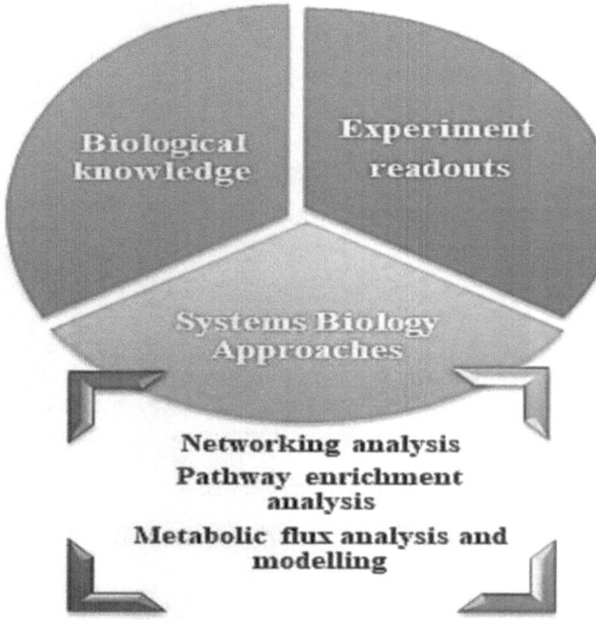

FIGURE 4.9 Systems biology approaches.

spectrometry, and nuclear magnetic resonance spectroscopy. Experiments involving isotopic labeling, like stable isotope labeling with ~13C or ~15N, can offer dynamic details regarding plant metabolic fluxes and pathways. In contrast, metabolite identification and annotation are facilitated by the use of computational tools and databases to compare experimental mass spectra or nuclear magnetic resonance (NMR) data with known metabolites (Mehrotra & Mendes, 2006). Methods such as pathway enrichment analysis, metabolic network modeling, and flux balance analysis are employed to reconstruct metabolic pathways, predict pathway activities, and simulate metabolic fluxes in plant cells. A deeper understanding of plant metabolic processes results from the integration of these experimental and computational data allowing more reliable and thorough data assessment (Sumner et al., 2007).

4.4.3 Time-Series and Spatial Data Integration

In plant metabolomics, temporal dynamics and geographical distributions of metabolites inside plant tissues or organs throughout time are captured by integrating time series and spatial data. A thorough understanding of how metabolite levels alter in response to stressors, environmental cues, and developmental phases is made possible by this integration. In time-series experiments, plant tissues or extracts are sampled at several time points, tracking changes in metabolite abundance or composition over time, and recording dynamic responses to developmental processes, diurnal rhythms, growth stages, or external stimuli. The time-resolved dataset produced by these experiments reflects metabolic fluctuations and regulatory patterns (Saito & Matsuda, 2010).

Plant metabolomics uses spatial data to shed light on the distribution of metabolites inside different plant tissues, organs, or cell types. This information can be used to understand metabolic gradients, compartmentalization, and localized metabolic activity. Integrating time series and spatial data involves correlating temporal changes in metabolite levels with their spatial distribution within plant tissues or organs. This reveals spatiotemporal patterns, metabolic gradients, and tissue-specific responses (Fukushima et al., 2009).

4.5 SYSTEMS BIOLOGY APPROACHES IN PLANT METABOLOMICS

4.5.1 Network Analysis

Plant metabolomics employs network analysis to build and examine metabolic networks in order to comprehend the connections among genes, metabolites, enzymes, and regulatory elements. Metabolite-metabolite networks depict the relationships between metabolites by means of their metabolic reactions and correlation coefficients (Weckwerth, 2003). Enzyme-metabolite networks serve as a means of connecting enzymes to the metabolites they catalyze by combining enzyme and metabolite data. Gene-metabolite networks show how genes affect metabolic pathways by linking genes to metabolites. Integrating network analysis with pathway enrichment tools helps identify enriched pathways, metabolic modules, and key metabolites/enzymes involved in biological functions or metabolic responses (Budzinski et al., 2019).

4.5.2 Machine Learning and Data Mining

Plant metabolomics is becoming more and more dependent on machine learning and data mining tools to draw insights from large datasets, identify patterns and forecast future outcomes. Using high-dimensional metabolomics datasets, machine learning algorithms may automatically identify key features (metabolites), cutting through the noise and concentrating on variables that provide useful information (Baharum & Azizan, 2018). Utilizing metabolite profiles as a basis for classification, machine learning algorithms support disease diagnosis, plant health monitoring, and the identification of biomarkers for disease susceptibility or resistance. This data can reveal regulatory networks and metabolic interactions through the use of machine learning and data mining techniques (Figure 4.9). These techniques can also identify patterns in the data, such as temporal dynamics, spatial distributions, and correlations between genes, enzymes, and metabolites (Scott et al., 2010; Okada et al., 2010).

4.6 PLANT METABOLITE DATABASES AND TOOLS

4.6.1 Plant Metabolomics Database (PMDB)

Plant Metabolome Database (PMDB) is a three-dimensional structural repository for secondary metabolites found in plants, gathered from other scientific databases and data banks. It allows users to sketch and query metabolite structures and provides an easy-to-use search interface that easily integrates with the JME Editor. With the Jmol graphics plug-in, these structures can be seen in both text and graphical versions. The database has a comprehensive help page explaining database features and a navigation link that allows users to browse metabolites by name. To improve the amount of information available on metabolites, PMDB is dynamically compiled and has internal and external links to other databases such as KEGG, PubChem, and CAS Number. It is designed to support researchers in molecular and structural biology, serving as a valuable resource freely accessible to all users 24/7 at http://www.sastra.edu/scbt/pmdb (Vinay et al., 2021).

4.6.2 The Arabidopsis Information Resource (TAIR)

For those studying plant metabolomics, the Arabidopsis Information Resource (TAIR) is an invaluable tool. TAIR offers extensive data on the model plant *Arabidopsis thaliana*, which is frequently utilized in studies because of its tiny genome, quick life cycle, and genetic tractability. For *Arabidopsis* genes, AIR offers functional data, gene annotations, and genomic sequences. For the purpose of discovering and researching genes involved in plant metabolism, this information is very essential. TAIR is responsible for curating metabolite data, which includes small molecule information, biochemical pathway information, and metabolite-gene relationships. Metabolite profiles are a valuable resource for researchers to investigate the molecular makeup and metabolic pathways of *Arabidopsis*. TAIR provides bioinformatics tools for data analysis related to metabolomics. These resources can aid in the interpretation of complicated metabolomics datasets by providing researchers with tools such as pathway enrichment analysis, metabolite identification processes, and network visualization tools (Mehrotra & Mendes, 2006).

4.6.3 Phytochemicals and Plant Metabolomics Resource (PPMR)

Plant metabolomics benefits greatly from the use of the Phytochemicals and Plant Metabolomics Resource (PPMR). For scientists interested in learning more about phytochemicals and plant metabolism, it offers an extensive resource. Plant-based phytochemicals are bioactive substances that are extensively studied by PPMR, and this contains details about the sources, chemical compositions, physical characteristics, and biological roles of phytochemicals in various plant species. Scientists can use this information to locate, describe, and investigate phytochemicals associated with plant metabolism. PPMR offers access to datasets on metabolomics generated by different kinds of plants. These databases might contain details on metabolic pathways, metabolite profiles, and quantitative measurements of metabolites in various experimental setups. These datasets can be analyzed by researchers to learn more about the metabolic and regulatory mechanisms of plants (Markley et al., 2007).

4.6.4 KEGG (Kyoto Encyclopedia of Genes and Genomes) Pathway Database

KEGG offers an extensive database of biological pathways, encompassing metabolic, genetic, and signaling routes. Plant metabolic pathways, including those involved in primary metabolism (such as glycolysis and the citric acid cycle) and secondary metabolism (such as phenylpropanoid biosynthesis and terpenoid biosynthesis), are extensively covered by KEGG. By investigating these pathways, scientists can get insight into the biochemical processes, enzyme-catalyzed events, and metabolite transformations that occur in plants. KEGG uses route maps to provide graphical representations of metabolic pathways. These maps include names of metabolites, enzymes, and related genes together with a visual representation of the sequential steps and interactions inside a pathway. Pathway maps are a useful tool for researchers to view and interpret metabolomics data in relation to biological processes. KEGG links metabolic pathways with genomic data to improve the investigation of gene-metabolite interactions. Through gene identification, scientists can explore the functions of enzymes encoding certain metabolic reactions in plant metabolism. Comprehensive details on each compound are included in its entry, including its mass, chemical structure, molecular formula, and biological characteristics (Kanehisa, 2016)

4.6.5 Metabolomics Workbench

Data processing tools and pipelines for metabolomics are offered by the Metabolomics Workbench. Preprocessing techniques for data normalization, peak identification, alignment, and feature extraction may be included in these tools. By preprocessing raw metabolomics data and preparing it for further analysis, researchers can improve the quality and dependability of metabolomics datasets used in plant studies. To find metabolic signatures linked to plant phenotypes, treatments, or genetic differences, researchers might employ statistical tests, multivariate analysis, and pattern recognition techniques to identify important metabolites and detect changes in metabolism. In addition to comparing metabolomics data from various plant species, tissues, and experimental settings, the Metabolomics Workbench functions as a centralized repository for metabolomics data. As a whole, it provides access to a wide range of metabolomics information, including those pertaining to metabolite profiles and metabolic pathways (Shulaev, 2006).

4.7 METABOLITE PROFILING

An essential component of plant metabolomics is metabolite profiling, which entails a thorough examination and measurement of the tiny compounds (metabolites) found in plant tissues or fluids. This profiling seems to describe the molecular makeup of plants, comprehend metabolic processes, and pinpoint adjustments in metabolism resulting from a range of stimuli, such as development phases, stress, environment, and genetic alterations. Metabolite profiling includes a broad spectrum of small molecules, such as secondary metabolites (e.g., phenolics, terpenoids, alkaloids) and primary metabolites (e.g., sugars, amino acids, organic acids). These metabolites are vital to a plant's ability to grow, develop, defend itself, and interact with its surroundings. This process includes comparing experimental data (e.g., mass spectra, retention times) with reference databases, spectral libraries, and authentic standards to confirm metabolite identities. It also performs quantitative analysis to determine the relative or absolute concentrations of metabolites in plant samples (Figure 4.10).

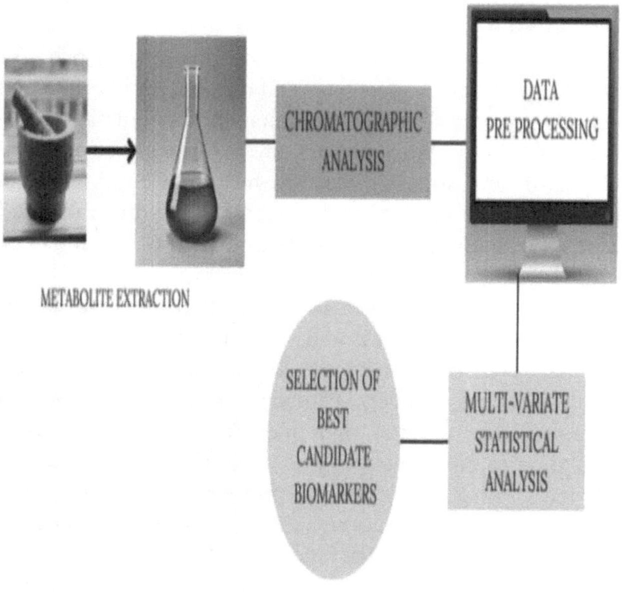

FIGURE 4.10 Metabolite fingerprinting.

4.8 CHEMICAL FINGERPRINTING AND QUALITY CONTROL

In plant metabolomics, chemical fingerprinting and quality control are crucial procedures that support the validity, robustness, and dependability of metabolomics data. By using these procedures, scientists can produce high-quality metabolomics datasets for the study of plant metabolism, bioactivity, and metabolic responses, as well as precisely define plant metabolites, monitor data quality, and identify analytical deviations. Chemical fingerprinting is the process of characterizing and contrasting intricate chemical profiles, or fingerprints, extracted from plant samples by the application of analytical methods like spectroscopy and chromatography. Chemical fingerprinting's main objective is to identify the distinct chemical makeup of plant samples, including the variety of metabolites present, their relative abundances, and changes between samples and environments (Fiehn et al., 2008).

The term "quality control" in metabolomics describes a collection of guidelines and practices used to guarantee the reliability, correctness, and consistency of metabolomics data all the way through the experimental process (Martins et al., 2018). The use of analytical and internal standards, and quality control samples are examples of quality control procedures in plant metabolomics. Internal standards support quantification and data normalization, whereas analytical standards assist in metabolite identity identification and confirmation. Data processing quality control includes stringent data preparation, filtering, and validation procedures in addition to experimental quality control approaches to reduce noise, artifacts, and analytical biases in metabolomics datasets. Pooled or reference samples are examples of quality control samples that are used to track instrument performance, evaluate data reliability, and identify analytical variances (Wang et al., 2004).

4.9 METABOLOMICS IN TRADITIONAL MEDICINE

Plant-based derivatives and extracts are profoundly used to treat various human ailments. Herbal remedies are widely incorporated into the traditional system of medicine, which comprises practices that were widely accepted during ancient times. Ayurveda is an indigenous system of Indian medicine and is recognized globally. The rise of traditional medicine faces more challenges when compared with modern medicine. Traditional medicine believes in deep-rooted knowledge, skills, and practices based on the theories and practices that are indigenous to various communities. Medicinal herbs are used in the maintenance of health and in the prevention and treatment of human ailments. Modern medicine comprises advanced technologies and provides evidence-based therapeutics. Contemporary integrative strategies in combination with traditional lead compounds can augment the potential of medicinal plants. Metabolomics is one such integrating tool that recognizes the prospects of traditional medicine and ensures the practices are effective and safe (Sharma & Yadav, 2022). Figure 4.11 illustrates common steps involved in the process of metabolomics. The vital parts of the medicinal plants are extracted to extricate significant metabolites using common solid-phase and liquid-liquid extraction methods.

Analytical techniques like chromatography (e.g., gas chromatography, liquid chromatography), mass spectrometry, and nuclear magnetic resonance (NMR) spectroscopy are used to profile the metabolites. The wide stack of raw data is analyzed and interpreted using statistical tools. Multivariate analysis techniques like principal component analysis or partial least squares discriminant analysis are used to identify significant differences in metabolite under various conditions. The metabolites are annotated by mapping them along with metabolic pathways and biochemical networks using *in silico* techniques. The metabolites are validated with the biological mechanisms and their data sets associated with gene expression. The vast reserves of secondary metabolites belonging to medicinal plants have unique therapeutic potentials that need to be explored. The

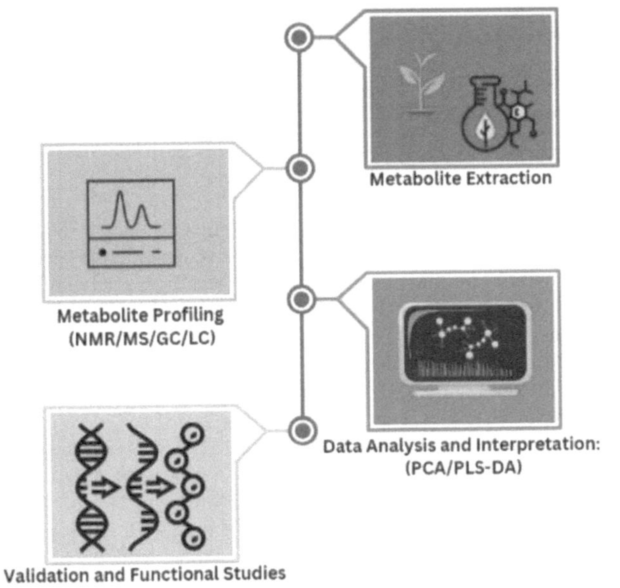

FIGURE 4.11 General phases of metabolomics in traditional medicine.

secondary metabolites of *Panax* species (*Panax ginseng*, *Panax noto ginseng*, and *Panax japonicas*) have been analyzed and reported to have pharmacological effects on the nervous system and immune disorders (Xu et al., 2016). Metabolomics approaches have explored the antitussive and expectorant potentials of *Tussilago farfara* L. (Li et al., 2013). Due to the limited metabolite information available in data of Indian medicinal plants and spices, there is a dire need to analyze the metabolite profiles of ayurvedic formulations. *Valeriana officinalis* L. (Valerianaceae), *Melissa officinalis* L. (Lamiaceae), *Hypericum perforatum* L. (Hypericaceae), and *Passiflora incarnata* L. (Passifloraceae) are used as treatments for anxiety, stress, and depression. The metabolomics studies have endorsed these herbs as brain-derived neurotrophic factor (BDNF) to modulate neurotransmitters. The Chinese medicinal materials have put forth efforts to use metabolomics and decipher several attributes of the herbs. Metabolomics plays a vital role in species identification and characterization. Three distinct species of liquorice herbs (*Glycyrrhiza uralensis*, *Glycyrrhiza inflata*, and *Glycyrrhiza glabra*) were potentials identified based on the biomarker of licochalcone A (Fukuda et al., 2010). Synchronizing molecular techniques with metabolomics has distinguished *Astragalus mongolica* and *Astragalus membranaceus* with specific DNA molecular markers. Genetic variations due to metabolite change has identified three distinct species of seabuckthorns: *Hippopahe gyantsensis*, *Hippopahe neurocarpa*, and *Hippopahe tibetana* (Liu et al., 2016). *Panax ginseng* has enormous metabolite data sets confined to its habitat and origination.

Metabolomics endorsed the origination and distribution of the plant in Korea, China, and Japan. Metabolites from dandelions (*Taraxacum mongolicum*) were identified from four different geographical regions in China (Gansu, Henan, Shanxi, and Jiangsu). The bioactive compounds of an aromatic medicinal material agarwood (*Aquilaria*), rich in 2-(2-phenylmethyl) chromones, are used in commercialized cologne and as antidepressant agents. Further, the pigmentation in two danshen (*Salvia miltiorrhiza*) species was distinguished by identifying 40 lipophilic components; 7 of them included tanshinone IIA and tanshinone I. Metabolomics plays a distinct role in heavy metal scavenging and agricultural residue detection. The metabolites of exposed the highlighted feature of *Sedum alfredii* that could tolerate super-rich heavy metal cadmium by regulating their secretion (Lu et al., 2015). Figure 4.12 shows the medicinal plants and their related metabolites that rendered significant application.

4.10 THE ALLIED SUCCESS OF METABOLOMICS WITH COMPUTATIONAL TOOLS

The groundbreaking achievement of metabolomics lies in interpreting the functional roles of metabolites through detailed annotation. Withanolides are anti-inflammatory metabolites derived from the medicinal plant *Withania somnifera* (Rahman et al., 2015). The multifaceted

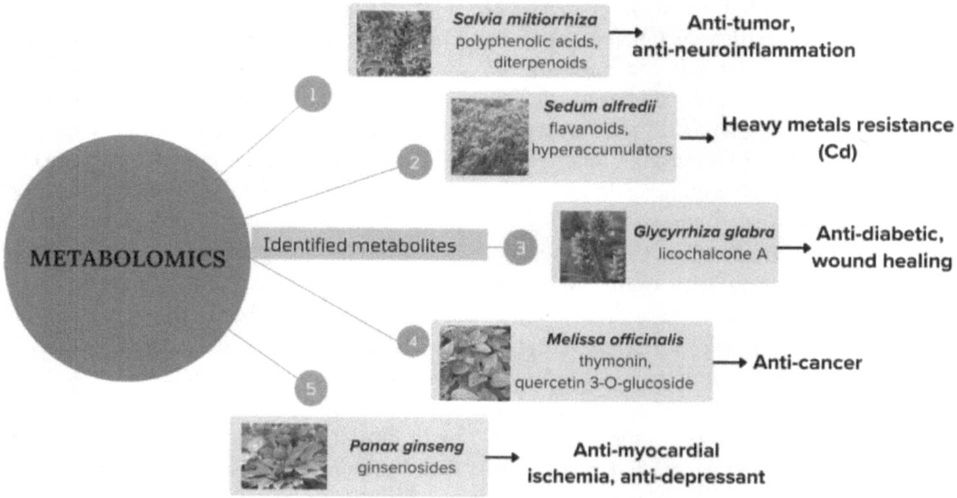

FIGURE 4.12 Prominent medicinal plants used in metabolomics.

therapeutic potential of herbal plants can be explored by mapping the relevant genes with their biological functions and classifications. This can uncover traditional medicines' multi-targeted effects and biological mechanisms using network pharmacology approaches. The medicinal benefits of *Boerhaavia diffusa* and *Tinospora cordifolia* were uncovered by associating their metabolites with several metabolic pathways. In particular, the metabolites synchronized widely with the TNF signaling pathway, RAS, and bradykinin pathway and were reported to alleviate COVID-19 symptoms (Markley et al., 2017). The pharmacological effects *Momordica charantia* revealed significant metabolites as biomarkers that interacted with several pathophysiological pathways of diabetic nephropathy oxidative stress and inflammation (Beckonert et al., 2010). The potential targets of *Physostigma venenosum* were identified using the combined techniques of metabolomics and virtual reverse pharmacology to treat vascular dementia. Metabolites like eseridine 42, physostigmine, eseramine, and physovenine were identified as effective targets to inhibit acetylcholinesterase and butyrylcholinesterase for treating dementia and Alzheimer's disease (Saleh et al., 2017). The antidiabetic potentials of *Duranta repens* were elucidated by showing interaction with the PI3K-Akt signaling route and the p53 signaling pathway to inhibit the expression of amylase in the GI tract (Vlachos et al., 2006).

4.11 CONCLUSION

Traditional medicinal plants have a multifaceted potential to treat several human ailments. The number of metabolomics-based research methods is growing in order to enhance the potential metabolites and plant-based derivatives. Therapeutic use of metabolites is explored by the advent of metabolomics techniques and mapping biological processes using network-based methodologies. Metabolomics identifies natural products as primary metabolites and explores their metabolic interactions that influence gene expression. Thus, this explicates the synergy of advanced computational and statistical tools to devise novel therapies. Metabolomics paves the way toward better clinical development of drugs through clinical trials. Because of its computational networking capabilities, it can increase the number of possible active ingredients in pre- and post-clinical formulations, hence generating interesting ways for drug assessment.

ACKNOWLEDGMENT

All the authors are thankful to their respective universities and institution for their valuable support.

DATA AVAILABILITY

The authors confirm that the data supporting the findings of this study are available in the book/chapter.

REFERENCES

Akilandeswari, G., Vijaya Anand, A., Sampathkumar, P., Vinayaga Moorthi, P., & Preethi, B. (2019). A prospective review on phyto-pharmacological aspects of *Andrographis paniculata*. *Systematic Reviews in Pharmacy*, *10*(1), 15–19.

Arbona, V., Iglesias, D. J., Talón, M., & Gómez-Cadenas, A. (2009). Plant phenotype demarcation using nontargeted LC-MS and GC-MS metabolite profiling. *Journal of Agricultural and Food Chemistry*, *57*(16), 7338–7347.

Baharum, S. N., & Azizan, K. A. (2018). Metabolomics in systems biology. In W. Aizat, H. H, Goh, & S. Baharum (Eds.), *Omics applications for systems biology*. Advances in experimental medicine and biology (Vol. 1102, pp. 51–68). Springer.

Beckonert, O., Coen, M., Keun, H. C., Wang, Y., Ebbels, T. M., Holmes, E., Lindon, J. C., & Nicholson, J. K. (2010). High-resolution magic-angle-spinning NMR spectroscopy for metabolic profiling of intact tissues. *Nature Protocols*, *5*(6), 1019–1032.

Begum, T. N., Ilyas, M. M., Kalavathy, S., Anand, A. V., Kumar, S., & Senthil, R. (2009). Effects of ethanolic leaf extract of *Azim tetracantha* Lam. on Ehrlich ascites carcinoma tumour bearing mice. *Research Journal of Medicine and Medical Sciences*, *4*(2), 351–354.

Bharathi, V., Rengarajan, R. L., Radhakrishnan, R., & Hashem, A., Abd Allah, E. F., Alqarawi, A. A., & Anand, A. V. (. (2018). Effects of a medicinal plant *Macrotyloma uniflorum* (Lam.) Verdc. Formulation (MUF) on obesity-associated oxidative stress-induced liver injury. *Saudi Journal of Biological Sciences*, *25*(6), 1115–1121.

Budzinski, I. G. F., de Moraes, F. E., Cataldi, T. R., Franceschini, L. M., & Labate, C. A. (2019). Network analyses and data integration of proteomics and metabolomics from leaves of two contrasting varieties of sugarcane in response to drought. *Frontiers in Plant Science*, *10*, 446557.

Buriani, A., Garcia-Bermejo, M. L., Bosisio, E., Xu, Q., Li, H., Dong, X., Simmonds, M. S. J., Carrara, M., Tejedor, N., & Lucio-Cazana, J. (2012). Omic techniques in systems biology approaches to Traditional Chinese Medicine Research: Present and future. *Journal of Ethnopharmacology*, *140*, 535–544.

Bylesjö, M., Eriksson, D., Kusano, M., Moritz, T., & Trygg, J. (2007). Data integration in plant biology: The O2PLS method for combined modeling of transcript and metabolite data. *The Plant Journal*, *52*(6), 1181–1191.

Cambiaghi, A., Ferrario, M., & Masseroli, M. (2017). Analysis of metabolomic data: Tools, current strategies and future challenges for omics data integration. *Briefings in Bioinformatics*, *18*(3), 498–510.

Divya, N., Rengarajan, R. L., Radhakrishnan, R., Abd Allah, E. F., Alqarawi, A. A., Hashem, A., & Anand, A. V. (2019). Phytotherapeutic efficacy of the medicinal plant *Terminalia catappa* L. *Saudi Journal of Biological Sciences*, *26*(5), 985–988.

Fiehn, O., Wohlgemuth, G., Scholz, M., Kind, T., Lee, D. Y., Lu, Y., Moon, S., & Nikolau, B. (2008). Quality control for plant metabolomics: Reporting MSI-compliant studies. *The Plant Journal*, *53*(4), 691–704.

Fukuda, E., Baba, M., Iwasaki, N., Uesawa, Y., Arifuku, K., Kamoe, O., Tsubono, K., & Okada, Y. (2010). Identification of *Glycyrrhiza* species by direct analysis in real time mass spectrometry. *Natural Product Communications*, *5*(11), 1755–1758.

Fukusaki, E., & Kobayashi, A. (2005). Plant metabolomics: Potential for practical operation. *Journal of Bioscience and Bioengineering*, *100*(4), 347–354.

Fukushima, A., Kusano, M., Redestig, H., Arita, M., & Saito, K. (2009). Integrated omics approaches in plant systems biology. *Current Opinion in Chemical Biology*, *13*(5–6), 532–538.

Hong, J., Yang, L., Zhang, D., & Shi, J. (2016). Plant metabolomics: An indispensable system biology tool for plant science. *International Journal of Molecular Sciences*, *17*(6), 767.

Kanehisa, M. (2016). KEGG bioinformatics resource for plant genomics and metabolomics. In D. Edwards (Ed.), *Plant bioinformatics*. Methods in molecular biology™ (Vol. 1374, pp. 55–70). Humana Press.

Karthika, P., Balamuralikrishnan, B., Yamini, K., Vijaya Anand, A., Durairaj, K., Maruthupandian, A., Alodaini, H. A., Hatamleh, A. A., Manikantan, P., & Arun, M. (2023). Green synthesis, characterization of silver nanoparticles using aqueous leaf extracts of *Solanum melongena* and in vitro evaluation of antibacterial, pesticidal and anticancer activity in human MDA-MB-231 breast cancer cell lines. *Journal of King Saud University – Science*, *35*(5), 102663.

Karthikkeyan, G., Pervaje, R., Subbannayya, Y., Patil, A. H., Modi, P. K., & Prasad, T. S. K. (2020). Plant omics: Metabolomics and network pharmacology of Liquorice, Indian Ayurvedic medicine Yashtimadhu. *OMICS: A Journal of Integrative Biology*, *24*, 743–755.

Kaviya, M., Balasubramanian, B., Bharathi, K., Malaisamy, A., Al-Dhabi, N. A., Mariadhas, V. A., Anand, A. V., & Liu, W. (2022). Evaluation of nutritional substances and investigation of antioxidant and antimicrobial potentials of *Boerhavia diffusa* with in silico molecular ocking. *Molecules*, *27*(4), 1280.

Kumar, R., Bohra, A., Pandey, A. K., Pandey, M. K., & Kumar, A. (2017). Metabolomics for plant improvement: Status and prospects. *Frontiers in Plant Science*, *8*, 271676.

Li, Z. Y., Zhi, H. J., Zhang, F. S., Sun, H. F., Zhang, L. Z., Jia, J. P., Xing, J., & Qin, X. M. (2013). Metabolomic profiling of the antitussive and expectorant plant *Tussilago farfara* L. by nuclear magnetic resonance spectroscopy and multivariate data analysis. *Journal of Pharmaceutical and Biomedical Analysis*, *75*, 158–164.

Lisec, J., Schauer, N., Kopka, J., Willmitzer, L., & Fernie, A. R. (2006). Gas chromatography mass spectrometry-based metabolite profiling in plants. *Nature Protocols*, *1*(1), 387–396.

Liu, Y., Liu, C., Tan, E., Fan, G., Xiang, L., Li, X. D., & Zhang, Y. (2016). Genetic and chemical discrimination of traditional Tibetan medicine seabuckthorn based on DNA barcode and ^{1}H-NMR metabolic method. *Zhongguo Zhong yao za zhi= Zhongguo Zhongyao Zazhi= China Journal of Chinese Materia Medica*, *41*(4), 578–585.

Lu, Q., Sun, L. N., & Hu, X. M. (2015). Metabonomics study on root exudates of cadmium hyperaccumulator *Sedum alfredii*. *Chinese Journal of Analytical Chemistry*, *43*(1), 7–12.

Manikandan, R., Balasubramanian, B., Punniyakotti, P., Vijaya Anand, A., Meyyazhagan, A., Velauthaprabhu, S., & Liu, W. C. (2022). Cardio-protective effects of *Terminalia catappa* leaves and *Terminalia chebula* fruit extract in doxorubicin-induced cardiomyopathy in rats. *Biomarkers*, *27*(5), 488–495.

Manoharan, N., Sampathkumar, P., Dheeba, B., Sheik Abdulla, S., Vinothprasanna, G., Vinothkannan, R., Kalavathy, S., Vijaya Anand, A., & Shanmugasundaram, A. (2008). Potential hepatoprotective effect of aqueous extract of *Gracilaria corticata* in AFB_1 induced hepatotoxicity in Wistar rats. *Journal of Biological Sciences*, *8*, 1352–1355.

Markley, J. L., Anderson, M. E., Cui, Q., Eghbalnia, H. R., Lewis, I. A., Hegeman, A. D., Li, J., Schulte, C. F., Sussman, M. R., Westler, W. M., Ulrich, E. L., & Zolnai, Z. (2007). New bioinformatics resources for metabolomics. In *Biocomputing* (pp. 157–168). World Scientific.

Markley, J. L., Bruschweiler, R., Edison, A. S., Eghbalnia, H. R., Powers, R., Raftery, D., & Wishart, D. S. (2017). The future of NMR-based metabolomics. *Current Opinion in Biotechnology*, *43*, 34–40.

Martins, M. C., Caldana, C., Wolf, L. D., & de Abreu, L. G. F. (2018). The importance of experimental design, quality assurance, and control in plant metabolomics experiments. In C. Antonio (Ed.), *Plant metabolomics*. Methods in molecular biology (Vol. 1778, pp. 3–17). Humana Press.

Mattoli, L., Cangi, F., Maidecchi, A., Ghiara, C., Ragazzi, E., Tubaro, M., Stella, L., Tisato, F., & Traldi, P. (2006). Metabolomic fingerprinting of plant extracts. *Journal of Mass Spectrometry*, *41*(12), 1534–1545.

Mehrotra, B., & Mendes, P. (2006). Bioinformatics approaches to integrate metabolomics and other systems biology data.

In *Plant metabolomics* (pp. 105–115). Springer Berlin Heidelberg.

Muthumani, G. D., Anand, A. V., & Manikandan, R. (2013). Antioxidant, antihelmintic and antimicrobial activity of *Euphorbia thymifolia* Linn whole plant. *International Journal of Current Microbiology and Applied Sciences*, *2*(1), 66–75.

Okada, T., Mochamad Afendi, F., Altaf-Ul-Amin, M., Takahashi, H., Nakamura, K., & Kanaya, S. (2010). Metabolomics of medicinal plants: The importance of multivariate analysis of analytical chemistry data. *Current Computer-Aided Drug Design*, *6*(3), 179–196.

Rahman, M. Z. F. A., Chong, H. W., & Lim, V. (2015). UV-visible metabolomics approach for the determination of selected adulterants in claimed premixed coffee. *The Malaysian Journal of Medical Sciences*, *14*, 147–152.

Saito, K., & Matsuda, F. (2010). Metabolomics for functional genomics, systems biology, and biotechnology. *Annual Review of Plant Biology*, *61*, 463–489.

Saleh, M., Siddiqui, M. J., So'Ad, S. Z. M., & Khatib, A. (2017). FT-IR-based metabolomics approach to characterize the α-glucosidase inhibitory activity of Salakfruit. *International Malayasian French Scientific Conference*.

Sangeetha, T., Balamuralikrishnan, B., Sampathkumar, P., Velayuthaprabhu, S., Senthilkumar, N., Syed Ibrahim, K., Baskaran, R., & Vijaya Anand, A. (2022). Characterization and phytoconstituents of *Petroselinum crispum* (Mill) and *Coriandrum sativum* (Linn) and their impacts on inflammation – an in vitro analysis against human adenocarcinoma cells with molecular docking. *South African Journal of Botany*, *146*, 776–788.

Scott, I. M., Vermeer, C. P., Liakata, M., Corol, D. I., Ward, J. L., Lin, W., Johnson, H. E., Whitehead, L., Kular, B., Baker, J. M., Walsh, S., Dave, A., Larson, T. R., Graham, I. A., Wang, T. L., King, R. D., Draper, J., & Beale, M. H. (2010). Enhancement of plant metabolite fingerprinting by machine learning. *Plant Physiology*, *153*(4), 1506–1520.

Sharma, B., & Yadav, D. K. (2022). Metabolomics and network pharmacology in the exploration of the multi-targeted therapeutic approach of traditional medicinal plants. *Plants*, *11*(23), 3243.

Shree, M., Lingwan, M., & Masakapalli, S. K. (2019). Metabolite profiling and metabolomics of plant systems using 1H NMR and GC-MS. *OMICS-Based Approaches Plant Biotechnol*, *13*, 129.

Shulaev, V. (2006). Metabolomics technology and bioinformatics. *Briefings in Bioinformatics*, *7*(2), 128–139.

Shulaev, V., Cortes, D., Miller, G., & Mittler, R. (2008). Metabolomics for plant stress response. *Physiologia Plantarum*, *132*(2), 199–208.

Sumner, L. W., Urbanczyk-Wochniak, E., & Broeckling, C. D. (2007). Metabolomics data analysis, visualization, and integration. In D. Edwards (Ed.), *Plant bioinformatics*. Methods in molecular biology™ (Vol. 406, pp. 409–436). Humana Press.

Varalakshmi, B., Anand, A. V., Karpagam, T., Bai, J. S., & Manikandan, R. (2014). In vitro antimicrobial and anticancer activity of *Cinnamomum zeylanicum* Linn bark extracts. *International Journal of Pharmacy and Pharmaceutical Sciences*, *6*(1), 12–18.

Vijaya Anand, A., Varalakshmi, B., Prasana, B., Sampathkumar, P., Pushpa, N., & Hedina, A. (2016). *Cinnamomum zeylanicum* Linn. The spice with multi potential. *Systematic Reviews in Pharmacy*, *7*(1), 24–29.

Vijaya Anand, A., Velayuthaprabhu, S., Rengarajan, R. L., Sampathkumar, P., & Radhakrishnan, R. (2020). Bioactive compounds of guava (*Psidium guajava* L.). In *Bioactive compounds in underutilized fruits and nuts*, Reference Series in Phytochemistry (pp. 503–527). Springer.

Vijayakumar, K., Prasanna, B., Rengarajan, R. L., Rathinam, A., Velayuthaprabhu, S., & Vijaya Anand, A. (2023). Antidiabetic and hypolipidemic effects of *Cinnamon cassia* bark extracts: An in vitro, in vivo, and in silico approach. *Archieves of Physiology and Biochemistry*, *129*(2), 338–348.

Vinay, C. M., Udayamanoharan, S. K., Prabhu Basrur, N., Paul, B., & Rai, P. S. (2021). Current analytical technologies and bioinformatic resources for plant metabolomics data. *Plant Biotechnology Reports*, *15*(5), 561–572.

Vlachos, N., Skopelitis, Y., Psaroudaki, M., Konstantinidou, V., Chatzilazarou, A., & Tegou, E. (2006). Applications of fourier transform-infrared spectroscopy to edible oils. *Analytica Chimica Acta*, 573–574, 459–465.

Wang, Y., Tang, H., Nicholson, J. K., Hylands, P. J., Sampson, J., Whitcombe, I., Stewart, C. G., Caiger, S., Our, I., & Holmes, E. (2004). Metabolomic strategy for the classification and quality control of phytomedicine: A case study of chamomile flower (*Matricaria recutita* L.). *Planta Medica*, *70*(03), 250–255.

Weckwerth, W. (2003). Metabolomics in systems biology. *Annual Review of Plant Biology*, *54*(1), 669–689.

Wolfender, J. L., Marti, G., Thomas, A., & Bertrand, S. (2015). Current approaches and challenges for the metabolite profiling of complex natural extracts. *Journal of Chromatography A*, *1382*, 136–164.

Xu, X. F., Cheng, X. L., Lin, Q. H., Li, S. S., Jia, Z., & Han, T. (2016). Identification of mountaincultivated ginseng and cultivated ginseng using UPLC/oa-TOF MSE with a multivariate statistical sample-profiling strategy. *Journal of Ginseng Research*, *40*(4), 344–350.

5 The Application of Epigenetics in Medicinal Plants

Anna Rera, Subhashree Sahu, Tuward J. Dweh, Nyashadzashaishe Shengezi, Privilege Chikove, Angeline Jurry, Jyoti Prakash Sahoo, Madhuri Pradhan, and Manaswini Mahapatra

5.1 INTRODUCTION

The use of medicinal plants, constituting about 60% of all plants, has significantly contributed to the development of therapeutic compounds for various diseases. Research in natural medicine and synthetic biology has increasingly focused on phytometabolites from secondary metabolism, showcasing their diverse polypharmacological effects in disease prevention and treatment (Qi et al., 2023). Epigenetic processes, encompassing DNA methylation, noncoding RNA, and modifications to histones, are pivotal in initiating chronic diseases by overseeing gene expression in reaction to environmental signals (Qi et al., 2023). Imbalances in the body's epigenetic equilibrium, impacted by elements such as components found in e-cigarettes, have the potential to result in diverse pathologies, including obesity and the development of type 2 diabetes mellitus (Guo et al., 2021).

Chinese herbal medicine (CHM) has gained global recognition for its preventive and therapeutic applications across diverse medical conditions. Adopting approaches that target multiple factors and consider the whole system. CHMs offer hope and are proven to be advantageous in addressing complex diseases, including cancer, diseases involving the immune system, etc. (Qi et al., 2023). Cutting-edge biomedical technologies, including genomics, proteomics, and metabolomics, have enriched CHM research, aligning with its holistic therapeutic philosophy (Qi et al., 2023). The "omics" approach, encompassing genomics, proteomics, metabolomics, and epigenomics, offers promising avenues for advancing CHM. Epigenomics, in particular, has revealed reversible and heritable epigenetic modifications capable of influencing gene expression without altering DNA sequences (Wu et al., 2023).

Traditional Chinese medicine (TCM) has gained popularity globally, especially in treating conditions such as inflammation of the nasal passage, digestive discomfort, and disease of the skeletal muscles. Among them are renowned therapies like acupuncture, herbal remedies, and qigong practices (Qi et al., 2023; Wu et al., 2023). Research in CHMs has benefited from advancements in DNA methylation techniques, shedding light on previously obscure aspects. This chapter explores the clinical utility of DNA methylation in CHM research, aiming to guide future studies. Additionally, genome editing technologies, including the CRISPR/Cas9 system, provide accurate and streamlined methods for crop domestication and enhancement, revolutionizing plant breeding practices.

5.2 REGULATION OF GENE EXPRESSION IN MEDICINAL PLANTS THROUGH DNA METHYLATION

An important epigenetic change in eukaryotes is the addition of a methyl group to a cytosine base in DNA since it regulates gene activity and maintains the stability of genetic materials. The link between DNA methylation levels, genome size, and heterochromatin concentration is complex and does not perfectly correspond to taxonomic relationships. Cytosine methylation has a profound impact on plant development, including changes in the activity of genes, the synthesis of secondary compounds, the formation of organs, and the stress reaction (Guo et al., 2021). For example, salicylic acid causes a significant reduction in VaSTS2 and VaSTS10 gene methylation in *Vitis amurensis* Rupr. cell cultures, resulting in enhanced transresveratrol synthesis. Cytosine modification was most prevalent in *Clivia miniata* Regel plantlets produced from petals, showing less variability among petal-derived plantlets compared to the differences between petals and leaves from the same plant. Autopolyploidy in *Cymbopogon* (Poaceae) enhances the formation of monoterpene essential oils, which are connected to cytosine methylation. In *Senecio* (Asteraceae), hybridization and genome duplication cause nonadditive alterations in cytosine methylation, which can result in differences in metabolic DNA regulation. The process of methyl frequencies changes between polyploid hybrid generations, and newly produced allopolyploid species exhibit substantial epigenetic diversity for natural/artificial selection (Wu et al., 2023).

Bisulfite sequencing proves to be a valuable method for analyzing the methylome, particularly when coupled with knowledge of the entire genome sequences of the plant. In situations where whole-genome sequences are unavailable, although RNA-Seq or EST data may be accessible, datasets can serve as useful references in annotating whole-genome

DOI: 10.1201/9781003475491-5

bisulfite sequencing (WGBS-Seq). Several medicinal genes, including *Cannabis*, *Citrus*, *Aristolochia*, and *Asplenium*, have been subjected to WGBS-Seq. The evolution of the methylome over extended periods is influenced by genomic changes, whereas short-term evolution is mainly driven by spontaneous epimutation events (Guo et al., 2021). The methylation-sensitively amplification polymorphism (MSAP) method utilizes isoschizomers HpaII and MspI, both of which recognize the same restriction site but differ in sensitivity to cytosine methylation at CCGG sites (Figure 5.1). In regenerates of *Codonopsis lanceolata*, 1,674 MSAP patterns were identified using 39 probe configurations, with 10.5% showing polymorphism among regenerants and/or between regenerants and the donor plant (Guo et al., 2021). Modifications in cytosine methylation trends at certain genetic sites closely resembled those of the donor plant, despite variations in the estimated total level of methylation (Guo et al., 2021), 37 different MSAP patterns were identified and transcribed, displaying significant similarity to known functional genes, chloroplast sequences, and 30 profiles with no matches in the database. The prolonged subculture of *Taxus* × *media* cells led to morphological changes, a gradual decrease in paclitaxel (PTX) yield, and reduced transcriptional levels of PTX synthesis of genes. Both MSAP and HPLC analyses indicated elevated DNA methylation levels in Taxus cells with low PTX yield after extended culture periods. The control of the chemotherapy agent PTX, along with other taxanes in Taxus cell lines, appears to be regulated by methylation of DNA (Wu et al., 2023).

The methylation of cytosines at cleavage sites evaluated by MSAP exceeded genome-wide estimates obtained through HPLC, and the percentages of methylation varied at different nucleotide contexts within and across species. Simultaneous HPLC and MSAP assessments of DNA from 200 individuals of the perennial herb *Helleborus foetidus* validated a higher frequency of methyl-cytosine in CCGG contexts compared to the overall genome. In *H. foetidus*, global methylation did not correlate with modification of the internal CG region, suggesting that genomic variability significantly contributes to intraspecific functional diversity in this species, enabling it to thrive in diverse ecological conditions despite limited genetic diversity (Wu et al., 2023). In an *in vitro* garlic culture system, induced somaclonal variation intensified both genetic and epigenetic polymorphisms. MSAP was employed to assess epigenetic transfers and genetic methylation in genes in *Cupressus sempervirens* and *Rhodiola sachalinensis* populations (Venetsky et al., 2015). A comprehensive investigation of Chinese ginseng revealed Northeast Asia has a wide range of genetic patterns and epigenetic variation (Hao et al., 2018). TCM often uses woody plants, but not much is known about the cytosine patterns of methylation in annual plants that live for a long time. Different amounts of DNA modification have been seen in different stages of tree growth and reaction to external factors. This suggests that DNA methylation plays a big role in tree development and balance. Inhibition of DNA methyltransferase in *Taxus* cultivars and *Lonicera japonica* Thunb. results in upregulation of biosynthetic pathway genes, leading to higher production of paclitaxel and flavonoids, respectively (Hao et al., 2018).

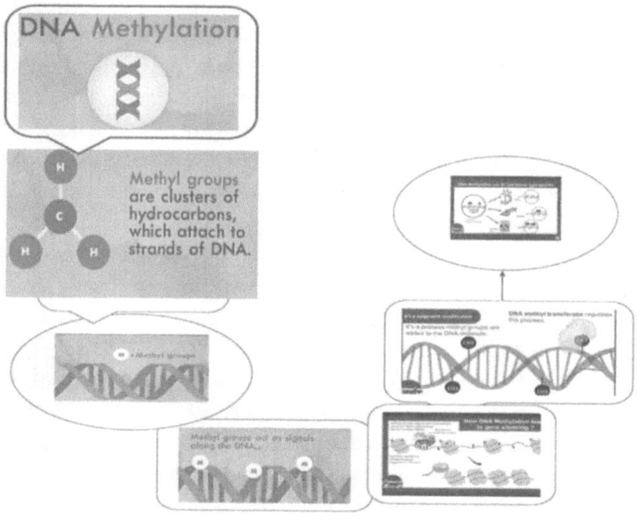

FIGURE 5.1 Epigenetic change: DNA methylation.

5.3 EPIGENETIC REGULATION IN CHINESE HERBAL MEDICINE PLANT GROWTH AND DEVELOPMENT

The process of planting and developing steps holds significance for the success of CHMs, with DNA modifications influencing various biological processes during their growth and development, including domestication and adaptation to different environments. This investigation focuses on several CHM medicinal plants:

- *Panax ginseng*: The methylation of DNA is linked with the breeding as well as quality management of *P. ginseng*. Genetic study of wild and domesticated *P. ginseng* shows unique genetic features among different landraces, showing the role of DNA methylation in the domestication process (Guo et al., 2021).
- *Panax quinquefolium*: Cold exposure during winter can lead to DNA demethylation in tender leaves, associated with increased ginsenoside collections in the roots (Hao et al., 2020).
- *Salvia miltiorrhiza*: Cytosine DNA methylation, regulated by cytosine-5 DNA methyltransferases (C5-MTases), influences secondary metabolism and stress response. Genome-wide analyses identify key genes involved in these processes (Li et al., 2018).
- *Pinellia ternata*: Shade cultivation significantly enhances its yield, possibly through increased gene DNA methylation under shaded conditions (Guo et al., 2021).

Epigenetic modifications, including DNA methylation (Figure 5.1), act as compensatory mechanisms for genetic adaptations, aiding plants in adjusting to dynamic environmental conditions. Studies on various plant populations, such as *Viola elatior* and *Helleborus foetidus*, stress the significance of DNA modification in influencing phenotypes and ecological processes, projecting both regional and local intraspecific variation in function (Guo et al., 2021). Environmental factors, like excess copper exposure, induce DNA methylation changes, impacting plant growth and adaptation. This exploration of DNA methylation in botanical studies illuminates how environmental factors shape plant development and response to natural changes, potentially affecting the quality control and clinical therapeutic outcomes of CHMs (Guo et al., 2021). Moreover, CHMs have long been viewed as a "black box" due to their complicated ingredient makeup and mysterious drug processes. However, major progress has been made in recent years with the development of science studies. DNA and other epigenetic methods have played a vital role in revealing the chemical processes underlying CHM products, plant drugs, and herbal compounds.

5.4 REMODELING OF CHROMATIN AND DNA ALTERATION IN MEDICINAL PLANTS

Guo et al. (2021) conducted a study on Chinese materia medica (CMM) medicinals, revealing that 36% of them interacted significantly with human histone-modifying enzymes, with 56% acting as chromatin condensation promoters. These CMM medicinals are likely linked with plant/microbe histone-modifying kinases, affecting the process of remodeling. Hao et al. (2018) found that 99% of TCM formulas were histone-modifying, with heterochromatic medicinals enhancing overall heterochromatinization, primarily through various mechanisms. Histone modifications play key roles in gene regulation in medicinal plants (Figure 5.2). Histone demethylases are crucial for reversible histone modification, with two main categories identified. In *Eucalyptus grandis* Hill, H3K4me3 enrichment indicates active transcription, contributing to understanding xylogenesis and lignin biosynthesis. (Guo et al., 2021) identified 23 histone methyltransferase genes in *Lonicera japonica*, showing species and organ specificity in expression, suggesting their involvement in epigenetic regulation. Treatment with hydroxamic acid-induced significant histone hyperacetylation in cassava roots, relieving salinity stress. Various epigenetic processes collectively influence secondary metabolism and defense responses (Farrow & Facchini, 2014). Transcripts associated with posttranscriptional gene silencing were upregulated in rice roots infected with giant cells, with diverse expression patterns observed in histone-modifying enzymes. A protein study on *Brassica napus* L. revealed upregulated proteins involved in histone methylation in a chlorophyll-deficient mutant, indicating the epigenetic control of expressed genes in secondary metabolite biosynthesis (Chu et al. 2017).

5.5 ROLE OF NON-CODING RNAS IN EPIGENETIC MODIFICATION

DNA and histone modifications (HM) are controlled during the process of transcription, whereas non-coding RNAs (ncRNAs) are principally responsible for post-transcriptional control. These ncRNAs, which lack protein-coding capabilities, are classified into two structural types: linear ncRNAs and circular ncRNAs. Linear ncRNAs, such as microRNAs (miRNAs) and long non-coding RNAs (lncRNAs), have an active role in plant development by influencing the processes of RNA metabolism, changes in protein levels, and chromatin remodeling epigenetically (Zhang et al., 2019). MiRNAs are tiny, evolutionarily conserved RNAs that control target mRNAs by complementary base pairing, resulting in transcriptional cleavage, degradation, or translation inhibition. They are among the most thoroughly researched ncRNAs. In contrast, long noncoding RNAs (lncRNAs) are preserved transcripts with elaborate modes of action. They interact with numerous molecules, including DNA and RNA, and function as precursors for short RNAs like miRNAs (Chen et al., 2021; Zhang et al., 2019). All the activities or processes of DNA and RNA have an impact on lncRNA transcription and processing.

Recent plant research has revealed lncRNAs' activities as cis- or trans-regulators and gene regulators (Panni et al., 2020). During post-transcriptional processing, RNA spliceosomes require an array of short nuclear RNAs (snRNAs), which are essential for mRNA precursor processing. Small nucleolar RNAs (snoRNAs) are essential in rRNA manufacturing and direct post-transcriptional changes of snRNA, tRNA, and mRNA. Furthermore, crucial siRISC complex components, such as short interfering RNAs (siRNAs), silence complementary target regions (Qi et al., 2023). Advancements in high-throughput sequencing methods continue to reveal novel types and functions of ncRNAs, with CRISPR/Cas9 expected to play a significant role in their future identification.

5.6 SMALL NON-CODING RNAs AND TRANSPOSABLE ELEMENTS (TEs)

Transcriptional gene silencing caused by small RNAs is a common process found across different species, ranging from invertebrates to humans. Small RNAs (sRNAs) carried onto AGO family proteins lead to inhibitory changes like histone alterations or DNA cytosine methylation to match genomic regions, a process known as RNA-directed DNA methylation (RdDM), which stands as a basic epigenetic mechanism in gene control (Qi et al., 2023). Nucleosomes, the building blocks of chromosomes, consist of a core octamer of histone dimers (H2A, H2B, H3, and H4) encircling DNA strands.

These histones experience different chemical changes, including methylation, acetylation, ubiquitination, phosphorylation, and glycosylation, thereby modifying gene expression by changing chromatin packing.

Post-translational changes such as histone methylation and acetylation play key roles in regulating gene transcription by modulating chromatin structure or recruiting downstream effectors (Qi et al., 2023). Histone modification in plants predominantly takes place at lysine and arginine residues of H3 and H4 histones, existing in three distinct states: monomethylated (me1), dimethylated (me2), and trimethylated (me3) forms (Qi et al., 2023).

These modifications intricately govern transcriptional activity, underscoring the significance of histone post-translational modifications in cellular processes (Qi et al., 2023). The dynamics of changes in histone entails coordinated actions among histone methyltransferases (HMTs), methylation recognition proteins, and demethylases. Primary sites of modification consist of H3K9, H3K27, H3K4, H3K36, and other lysine modification loci on H-3 histones. In general, dimethylation and trimethylation of H3K4 and H3K36 are associated with gene activation, while methylation of H3K9 and H3K27 correlates with gene repression. Notably, H3K27me3 exhibits high reversibility and is critical for the dynamic regulation of plant development (Qi et al., 2023).

5.7 HISTONE MODIFICATIONS

The changing control of gene expression includes histone changes, a key feature of epigenetic alterations. Nucleosomes, containing histone dimers surrounding DNA sequences, experience different chemical changes, including modification and acetylation, affecting chromatin shape and gene expression (Wu et al., 2023). In plants, histone methylation mainly affects specific sites on H3 and H4 histones, affecting gene expression or repression, with H3K4 and H3K36 methylation usually connected to histone changes, playing active regulatory roles in gene transcription by changing chromatin structure and recruiting downstream effectors. While methylation is linked with steady changes, acetylation, and phosphorylation serve as more dynamic regulatory sites, showing active transcription regions in chromatin. The relationship between histone modifications and DNA methylation is important for effective gene silence and transcriptional control, often involving coordinated actions between different types of histone modifications (Qi et al., 2023). In reaction to external triggers, like flooding stress plants, modulate histone methylation and acetylation to regulate gene expression, enabling rapid gene activation or repression under stress conditions, highlighting the dynamic nature of histone modifications in plant adaptation (Wu et al., 2023) (Figure 5.2).

5.8 CHINESE HERBAL MEDICINE FORMULAS

Numerous CHM formulas have demonstrated therapeutic potential through epigenetic mechanisms, particularly DNA methylation modulation, across various clinical conditions. For instance, the traditional CHM formula Liuwei Dihuang decoction (LDD) has been found to upregulate hormone receptor α (ER±) expression, thus preventing coronary artery disease (AS) by reducing ER± gene methylation and elevating ER± expression, particularly in the endothelial cells of human umbilical veins (HUVECs) (Chen et al., 2020). Similarly, San-Huang-Xie-Xin-Tang (SHXXT) has shown efficacy in AS treatment by restoring microRNA-152, reducing DNMT1, and lowering ER± methylation, thereby enhancing ER± expression (Guo et al., 2021). In the context of rheumatoid arthritis, Wutou Decoction (WTD) has demonstrated anti-inflammatory effects by downregulating DNA methylation levels and enhancing H3 acetylation, indicating its modulation of methylation of DNA and histone changes (Liu et al., 2016). Additionally, Baihu Guizhi Decoction (BGD) has displayed healing benefits on pyretic arthralgia (PA) by modulating methylation levels of characteristic genes, alleviating foot swelling, and reducing inflammatory factors (Chen et al., 2021).

In cardiovascular diseases, Xuefu Zhuyu Decoction (XZD) has shown promise in treating clopidogrel resistance (CR) by improving methylation of the P2Y12 gene, particularly in individuals with lower methylation levels of CpG island (CpGI) (Yu et al., 2023). Furthermore, Qian Yang Yu Yin Granule (QYYYG) has demonstrated renoprotective effects by inhibiting cell proliferation and downregulating NNMT expression and DNA methylation in hypertensive nephropathy (Chen et al., 2020). In hematologic disorders like myelodysplastic syndrome (MDS), Qinghuang Powder (QHP) has shown effectiveness in genome-wide DNA demethylation, suggesting its potential for hematologic improvement for MDS patients (Guo et al., 2021). Lastly, uremic clearance granule (UCG) has exhibited therapeutic effects in chronic renal failure (CRF) by restoring the methylation level of the Tgf-β1 gene, influencing its transcription and translation, indicating its role in modulating DNA methylation in CRF treatment (Guo et al., 2021).

5.9 CHINESE HERBAL MEDICINE BOTANICAL DRUGS AND COMPOUNDS

In the field of pharmacology, Ginseng Radix et Rhizoma is widely used globally, with Ginsenoside Rg3 extracted from this source demonstrating antitumor properties. Recent studies suggest a potential involvement of DNA methylation in mediating these effects (Guo et al., 2021; Wu et al., 2018). Ginsenoside Rg3 has been found to reduce overall genomic DNA methylation levels and induce changes in cytosine methylation within specific gene promoter regions. These epigenetic alterations are characterized by increased inhibition of cellular development, particularly evident in the HepG2 human hepatocarcinoma cell line (Guo et al., 2021). Furthermore, in ovarian cancer cells, Ginsenoside Rg3 boosts the production of the tumor inhibitor von Hippel-Lindau (VHL) gene by suppressing DNMT3A expression, causing a decrease of methylation in the VHL promoter (Wang et al., 2021). Additionally, Ginsenoside Rg3 can mitigate the Warburg effect, a chemical feature of malignant cells marked by elevated glycolysis and rapid energy production. This mechanism involves the downregulation of the methyltransferase (DNMT3A), resulting in reduced modification of DNA within the promoter region of the miR-519a-5p precursor gene and subsequent upregulation (Wang et al., 2021).

5.10 INHERITANCE OF EPIGENETIC MODIFICATIONS

Epigenetic modifications respond dynamically to environmental cues, potentially leading to transmissible genomic processes across generations (Saad-Hussein & Taha, 2021). These modifications, resistant to the early developmental "reset" of epigenetic features, can transmit along with their phenotypic consequences, intertwining with environmental conditions and lifestyle habits such as nutrition. The cellular epigenome reflects the impacts of environmental influences, shaping genes, genetics, epigenetics, and epigenomics. This challenges the classical notion of non-inheritance of acquired traits, highlighting the interplay between nature and nurture in shaping genes, epigenotype, and appearance. Well-being and disorders are influenced not solely by DNA as well as by interactions between genetics and the environment and external effects on hereditary genes (Saad-Hussein & Taha, 2021). Experimental studies in rodents suggest that food and treatments of the endocrine system during labor cause morphological and/or epigenetic alterations persisting across multiple filial generations (Saad-Hussein & Taha, 2021). In human studies, maternal diet supplementation with docosahexaenoic acid (DHA) during pregnancy led to increased methylation of the IGF2/H19 gene in offspring, associated with infant metabolism, growth, and development.

However, the overall methylation profile showed no effect despite sex-dependent, distinctly modified DNA areas and genes found at birth (Saad-Hussein & Taha, 2021). Rat studies showed that butyrate administration in mothers leads to children with decreased glucose tolerance and higher insulin resistance due to an increase of lipogenic genes, accompanied by increased histone acetylation in skeletal muscle, suggesting negative effects on lipid metabolism and insulin sensitivity (Saad-Hussein & Taha, 2021). Developmental undernutrition can also predispose individuals to metabolic diseases later in life. Fetuses exposed to in utero hyperglycemia due to maternal obesity are particularly susceptible to developing metabolic disorders like type 2 diabetes mellitus. The human epigenome is highly sensitive to food signals during early developmental stages, beginning from conception and continuing through the weaning phase (Saad-Hussein & Taha, 2021).

5.11 HERBAL AND DIET-DERIVED COMPOUNDS AND EPIGENETICS

Polyphenols produced from *Hibiscus sabdariffa* have been shown to control miRNA production in hyperlipidemic mice missing LDL receptors (Song et al., 2009). These changes are widely known as key processes in gene control. Dietary and external factors have the ability to affect animals from the embryo to adulthood and transgenerationally by means of epigenetic mechanisms guiding gene control. Organisms show specific tissue variations in DNA modification and chromatin change (Gluckman et al., 2009). Examples of plant-based polyphenols, such as curcumin and catechins, interact with enzymes and epigenetic regulators, including methyltransferases, acetyltransferases, kinases, miRNA, etc. (Li et al., 2016). Fatty acids play a key part in controlling gene expression through epigenetic processes, with effects on metabolism, including obesity, being both beneficial and detrimental (González-Becerra et al., 2019).

They can change the epigenome, altering genes involved in lowering resistance to insulin and obesity, as well as improving fat and glucose metabolism (Frigolet & Gutiérrez-Aguilar, 2007). Nutrients, including amino acids, minerals, vitamins, methyl donors, flavonoids, fatty acids, and other phytochemicals, hold hope for changing epigenetic parameters to control chronic diseases like obesity (Kiec-Wilk et al., 2011). Polyunsaturated fatty acids (PUFAs), such as omega-3 and omega-6, exhibit distinct effects on DNA methylation, influencing various metabolic pathways (Lee et al., 2017). Short-chain fatty acids, like butyric acid, act as histone deacetylation inhibitors, affecting gene expression related to obesity and diabetes (Tremblay et al., 2017). Omega-3 supplementation has proven to alter the methylation profiles relevant to lipid metabolism, inflammation, cardiovascular signaling, and diabetes (Tremblay et al., 2017).

Similarly, omega-3 intake influences the profile of genes involved in the metabolizing of lipids through methylation and insulin sensitivity, impacting obesity-related complications (Lee et al., 2017; Flores-Sierra et al., 2016). Oleic acid, a monounsaturated fatty acid, demonstrates notable epigenetic advantages in ways that have to do with fat, type 2 diabetes, and hypertension (Amaral et al., 2014). On the other hand, saturated fatty acids, like palmitate, can induce histone methylation alterations. Short-chain lipids, like sodium butyrate, stop the histone deacetylase enzyme from working and change gene activity that is connected to tumors, glucose control, sensitivity to insulin, and the breakdown of lipids (González-Becerra et al., 2019; Thakur et al., 2014).

5.12 EPIGENETIC MECHANISM OF HERBAL MEDICINE

DNA and histone enzymes like methyltransferases (DNMTs) and deacetylases (HDACs) play pivotal roles in the onset and advancement of human malignancies (Yuan et al., 2019). DNMTs facilitate targeted methylation transfers carried out by DNMTs, which can result in the suppression of numerous genes (Zych et al., 2019). Conversely, HDACs help get rid of acetyl chains from lysine sites in histone molecules' N-terminal tails, thereby regulating associated gene expression (Wu et al., 2023). Herbal medicine and its active constituents commonly regulate DNMT and HDAC expression or activity to combat diseases. Besides that,

TABLE 5.1
Herbal-Medicine-Induced Epigenetic Modifications on Chronic Diseases

Context/ Element	Robust design/ cell line	Epigenetic target	Attributes	Gene-edited/epigenetic modification	Outcome	References
Luteolin	MCF7-TamR cells	Chromatin proteins	10 and 20 μM luteolin	Enhances MLL3 expression; boosts overall H3K4me1 levels; reduces overall H3K4ac levels; elevates H3K4 monomethylation within the Ras gene enhancer and promoter region	Induces apoptosis in tamoxifen-resistant breast cancer cells by repressing the Ras gene, leading to the inactivation of the PI3K/AKT/mTOR pathway	Wu et al., 2023
Cotinus coggygria Scop.	MCF7-TamR cells	Deaxyribonucleic acid	40.6 μg/mL *Cotinus coggygria*, extract for 3, 24, 48, or 72 h	Suppresses DNA methyltransferase expression; inhibits deoxyribonucleic acid (DNA) methylation	Induces S phase cell cycle arrest and apoptosis; reduces colony formation; triggers DNA damage; influences cellular thermodynamic parameters	Gospodinova et al., 2021
Valeric acid	MCF-7 cells	Deaxyribonucleic acid; chromatin proteins	0–10 μM valeric acid for 48 h	Decreases HDAC activity; induces global DNA hypomethylation	Reduces proliferation of breast cancer cells	Shi et al., 2021
Luteolin	BT-20 and MDA-MB-231 cells	Chromatin proteins	10, 20, and 30 μM luteolin	In BT-20 cells: decreases H3K27ac and H3K56ac levels in MMP-9 promoter regions 2 and 3 In MDA-MB-231 cells: increases H3K27ac and H3K56ac levels on MMP-2 and MMP-9 promoter regions	Valeric acid inhibits HDAC1, HDAC2, and HDAC3 activity; causes DNA hypomethylation; inhibits the proliferation and metastasis of androgen receptor-positive triple-negative breast cancer cells	Wu et al., 2021
Silibinin	RT4, 5637, and T24 cells	Deoxyribonucleic acid	50 μM silibinin for 24 h	Silibinin induces global DNA hypomethylation	Restores p16 expression	Niu et al., 2024
Oleanolic acid	MKN-45 cells	Deoxyribonucleic acid	0–40 μM oleanolic acid for 24hrs	Oleanolic acid selectively decreases TET3 expression. This effect is observed in IL-1β-treated MKN-45 cells.	Results in hypomethylation of genetic material	Lu et al., 2021
Hesperetin	Nude mice with MKN45 cells MKN-45 and HGC-27 cells	Chromatin proteins	Mice: 50 mg/kg hesperetin for 1 week; Cell: 0–100 μM hesperetin for 48 h	Hesperetin reduces H3K79me2 and H3K79me3 levels in mice; It inhibits H3K79 methylation. Hesperetin decreases the abundance of Dot1L protein in cells.	Decreases the mobility of gastric cancer cells; inhibits the abundance of DOT1L	Tian et al., 2021
Curcumol	HepG2 and SMMC-7721 cels	Chromatin proteins	60 μM curcumol for 48 h	Curcumol increases the expression of H3K27me3 and H3K9me3 in HepG2 cells, decreases. H3K27me3 expres-sion in SMMC-7721 cells, and downregulates EZH2.	Downregulates lncRNA Hotair; Hotair downregula-tion leads to downregulation of EZH2; disrupts trimethylation of H3K9 and H3K27 catalyzed by EZH2; regulates histone modification to inhibit tumor growth and metastasis	Zain & Omar, 2022
Baicalin	BALB/c nude mouse and HepG2 cells	Deoxyribonucleic acid	Mice: 20 or 50 mg/kg baicalin for 28 days Cell: 50 mg/L baicalin for 24 h	Baicalin downregulates m6A/A, SAM/SAH, and m6A (2854) levels of HKDC1 in BALB/c mice tumor tissue; downregulates total DNA 5mC and RNA m6A levels; upregulates SAM/SAH and suppresses RNA m6A (2854) of HKDC1 in HepG2 cells.	Inhibits the progression of T2D-induced liver tumors; regulates the HKDC1/JAK2/STAT1/ caspase-3 pathway	Jiang et al., 2022
Phyllanthus debilis methano-lic extract	HT-29 cells	Deoxyribonucleic acid	0.1 mg/mL *Phyllanthus debilis* methanolic extract for 24 hrs	The extract from *Phyllanthus debilis* demonstrates anticancer properties; enhances Alu DNA methyla-tion; boosts LINE-1 methylation.	Anticancer effects	Wang et al., 2021

Zaluzanin D	PMA differentiated human monocytic THP-1 cells	Deaxyribonucleic acid	0.35 mM zaluzanin D for 24 h	Zaluzanin D decreases hypomethylation of the MMP9 gene promoter region induced by PMA activation.	Reinstates the expression of both Nur77 and NOR-1	Ghantous et al., 2012
Z-ligustilide	Acute myeloid leukemia cells	Chromatin proteins	50 μM Z-ligustilide for 1; 50 μM Z-ligustilide for 6 h	Z-ligustilide enhances the enrichment of acetylated histone H3 (K9/14) in the promoters of Nur77 and NOR-1; increases the activity of p300 acetyltransferase and reduces recruitment of HDACs (HDAC1, HDAC4/5/7) and MTA1 to the Nur77 promoter region.	Epigenetically regulates the expression of p21 and cyclin D1.	Wu et al., 2023
Naringenin and hesperetin	INS-1 cells and C57BLKS	Chromatin proteins	Cell: 100 μM naringenin or Leprdb mice hesperetin for 24 h; Mice: 50 mg/kg/day naringenin or 50 mg/kg/day hesperetin for 6 weeks	Naringenin and hesperetin suppress H3K18 and H3K27 acetylation and inhibit p300 activity; Hesperetin specifically suppresses H3K27 acetylation in the transcriptional regulatory region of the Txnip gene in INS-1 cells; Both compounds inhibit H3K18 and H3K27 acetylation in the islets of C57BLKS/Leprdb mice.	Reduces the expression of TXNIP	Park et al., 2022
Hesperetin	RAW 264.7 cells	Chromatin proteins	100 μM hesperetin overnight	Suppresses RelA/p65 acetylation by upregulating SIRT1 expression	Diminishes NF-κB activity	Goyal et al., 2020
Kaempferol	3T3-L1 cells	Chromatin proteins	100 μM kaempferol	Kaempferol reduces H3K27me3 deposition in the promoter regions of Adipoq, Fabp4, and Lpl genes	Suppresses the expression of PPARγ target genes, including Adipoq, Fabp4, and Lpl	Song et al., 2021
Sophora flavescens (SF)-F2	PBMCs	Deoxyribonucleic acid	22.5 μg/mL *Sophora flavescens* (SF)-F2 for 3 days	The combination of *Sophora flavescens* (SF)-F2 with dexamethasone decreases Foxp3 promoter methylation at CpG sites	Counters dexamethasone-induced immunosuppression	Zhang et al., 2020
Hydroxysafflor yellow A	hBMSCs	Chromatin proteins	10 μM hydroxysafflor yellow A for 48 h	Hydroxysafflor yellow A boosts KDM7A protein levels and diminishes H3K27me2 occupancy on the beta-catenin promoter.	Enhances the expression of β-catenin	Zhang et al., 2020
Qian Yang Yu Yin Granule	The renal damage model of spontaneously hypertensive rats HEK293T cells	Deoxyribonucleic; chromatin proteins	Mice: 2.1 or 8.4 g/kg/d Qian Yang Yu Yin Granule for 8 weeks Cell: 0.5 or 10 μg/mL	Qian Yang Yu Yin Granule reduces NNMT and acetylated-cortactin protein levels while enhancing H3K4me3 protein expression	Guards against hypertension-induced renal injury in spontaneously hypertensive rats and suppresses cell proliferation induced by Ang II	Zhang et al., 2020
Juglanin	High-fat-diet-fed mice	Chromatin proteins	7.5–30 mg/kg juglanin for 16 weeks	Juglanin inhibits the expression of HDAC3 at both mRNA and protein levels.	Suppresses the activation of NF-κB/HDAC3 signaling in the kidneys of HFD-challenged mice	Zhang et al., 2020

plant medicine can change the activation of linked genes by regulating various post-transcriptional activities, including methylation (Yuan et al., 2019) (Figure 5.2).

5.12.1 Brain Tumor

Gliomas represent a substantial portion of primary brain tumors and are known for their aggressiveness and poor prognosis across both pediatric and adult populations (Wu et al., 2023). Despite various treatment options like surgery, chemotherapy, and radiotherapy, survival rates for gliomas remain low (Wu et al., 2023). Research initiated in 2014 by Skala, Sitarek, and colleagues focused on exploring the effects of Chinese herbal medicine on gliomas. Their studies indicated that extracts from transgenic roots of *Leonurus sibiricus* and transformed roots of *Rhaponticum carthamoides* could induce apoptosis in glioma cells and impede their proliferation.

Mechanistically, these effects involve PARP1 cleavage and more γH2A.X histones, which are important for fixing breaks in the double strand of DNA and genomic stability. Additionally, the modulation of histone ubiquitination and DNA methylation by UHRF1 and DNMT1 contributes to the anticancer effects of these herbal extracts (Wu et al., 2023). Subsequent investigations by (Sitarek et al., 2018; Skała et al., 2018) further supported these findings, associating the anticancer effects induced by Leonurus sibiricus and Rhaponticum carthamoides extracts with alterations in γH2A.X, cleaved PARP1, UHRF1, and DNMT1 levels.

In a related study, Yan et al. (Wu et al., 2023) examined the activities of topoisomerase IIβ (TopoIIβ), an essential enzyme in neuronal development, in the context of herbal medicine. Chromatin immunoprecipitation analysis conducted by Yan et al. revealed that tetramethylpyrazine, a component of herbal medicine, makes it easier for ac-H3 and ac-H4 to join the promoter area of the TopoIIβ gene. This shows that plant medicine can help SH-SY5Y cells become neurons by boosting the production of TopoII through epigenetic regulation.

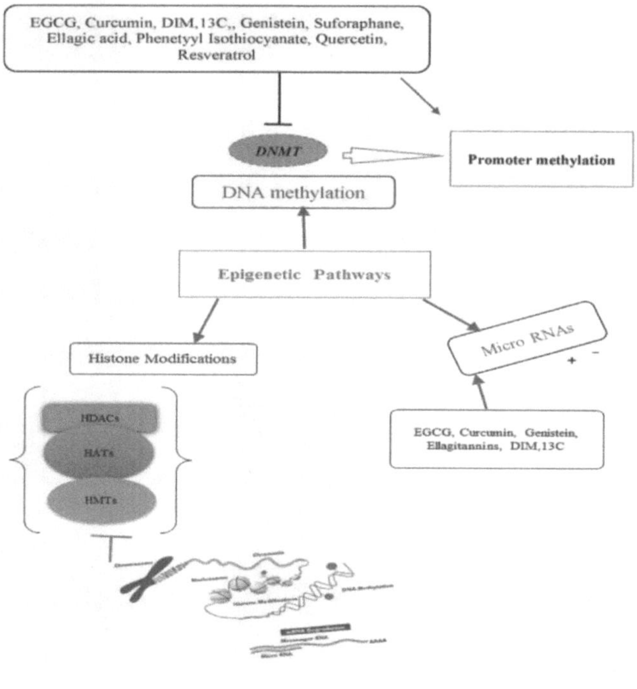

FIGURE 5.2 Epigenetic Change: histone modification.

5.12.2 Thoracic Tumor

Thoracic malignancies, which include lung and breast cancers, present significant health challenges, particularly due to the prevalence of smoking-induced lung cancer and the high incidence of female breast cancer, both contributing significantly to the escalating cancer mortality rates (Zych et al., 2019). Tanshinone I, identified in a study from 2008, demonstrated promising potential as a supplementary therapy for human breast cancer by inhibiting the proliferation of Some scientists have found a way to kill breast cancer cells called MCF-7 and MDA-MB-231 (Wu et al., 2023). Its method includes the downregulation of the Aurora A oncogene, important for mitotic development, thereby slowing breast tumor cell growth through lower acetylation of H3 in the genome promoter region of the Aurora A gene. Moreover, herbal remedies control tumor cell growth in the breast by influencing matrix metalloproteinases (MMPs), key for tumor growth and spread (Wu et al., 2023).

Luteolin targets the AKT/mTOR signaling pathway, resulting in lower amounts of p-AKT and mTOR proteins, leading to higher binding of H3K27ac and H3K56ac in the promoter sequences of MMP-2 and MMP-9, successfully blocking MMP production (Wu et al., 2023). Similarly, cucurbitacin B from *Trichosanthes cucumerina* regulates oncogenes such as c-Myc, cyclin D1, and survivin by upregulating DNMT1 and causing hypermethylation in their promoters, showing its promise as an anticancer agent (Wu et al., 2023). The epigenetic control of lung cancer development through herbal medicine is illustrated by Jinfukang (JFK), a clinical medication commonly employed in lung cancer treatment. JFK was observed to reduce H3K4me3 modification levels in various genes associated with lung cancer, indicating the therapeutic promise of herbal medicine in thoracic cancer (Wu et al., 2023).

5.12.3 Digestive System Tumor

Thoracic malignancies, such as lung and breast cancers, are significant health concerns, with smoking-induced lung cancer and female breast cancer contributing substantially to increasing cancer mortality rates (Zych et al., 2019). Tanshinone I, identified in a study from 2008, showed promise as an adjunct therapy for human breast cancer by inhibiting the proliferation of MCF-7 and MDA-MB-231 breast cancer cells and inducing apoptosis (Zych et al., 2019). Its mechanism involves downregulating the gene. Aurora A is crucial for mitotic progression, thereby inhibiting breast cancer cell growth through reduced H3 acetylation in the DNA promoter region of the Aurora A gene (Gong et al., 2012). Moreover, herbal medicine regulates breast cancer cell proliferation by epigenetically modulating matrix metalloproteinases (MMPs), crucial for cancer progression and metastasis (Ghantous et al., 2012).

Luteolin targets the AKT/mTOR signaling pathway, leading to decreased p-AKT and mTOR protein levels, thus increasing H3K27ac and H3K56ac occupancy in MMP-2 and MMP-9 promoter regions, effectively inhibiting MMP expression (Wu et al., 2020). Similarly, cucurbitacin B from Trichosanthes cucumerina modulates oncogenes like c-Myc, cyclin D1, and survivin by upregulating DNMT1 and inducing hypermethylation in their promoters, underscoring its potential as a cancer therapeutic (Wu et al., 2023). The epigenetic regulation of lung cancer development by herbal medicine is exemplified by Jinfukang (JFK), a clinical medicine commonly used for lung cancer treatment. JFK was found to decrease H3K4me3 modification levels in various lung cancer-associated genes, indicating the therapeutic potential of herbal medicine in thoracic cancer (Lu et al., 2021; Wu et al., 2023).

5.12.4 Urogenital Tumor

Despite notable advancements in cancer treatment, urogenital cancers, particularly prostate cancer, continue to present significant challenges, with prostate cancer being the second most prevalent tumor type in men (Wu et al., 2023). Since 2013, triptolide has attracted attention in prostate cancer epigenetic studies, with research conducted by Tamgue et al. demonstrating its potent ability to inhibit prostate cancer cell proliferation while concurrently reducing the expression of activator of zeste homolog 2 (EZH2). EZH2, an essential aspect of the polycomb repressor complex 2, controls the amount of H3K27me3 and plays a critical part in the initiation of tumors, growth, spread, and resistance to medication (Wu et al., 2023).

Studies have shown that even though triptolide lowered EZH2 production in PC-3 cells, it led to higher amounts of H3K27 and histone H3, suggesting additional control mechanisms. Further studies revealed large drops in mRNA and protein amounts of UTX (KDM6A) and histone demethylase JMJD3, both of which control H3K27 demethylation (Park et al., 2022). Triptolide-induced deposition of H3K9me3 at target gene promoters was proven to be SUV39H1-dependent, blocking gene expression through increased H3K9me3 deposition and promotion of heterochromatin formation (Yang et al., 2014). These results indicate how triptolide can potentially inhibit prostate tumor growth.

Additionally, components of plant therapy have been found to control NRF2, an important element of antioxidative stress defense genes in prostate tumors, thereby recovering its production (Wu et al., 2023). However, further laboratory exploration is needed to explain the specific processes involved.

5.12.5 Blood Tumor

Multiple myeloma results from the abnormal transformation of long-lived plasma cells, causing harmful lesions in patients (Moser-Katz et al., 2021). Investigating the epigenetic effects of herbal medicine on myeloma, particularly with triptolide, a study carried out in 2010, using the U266 multiple myeloma cell line, triptolide demonstrated a dose-dependent decrease in genome-wide levels of H3K4me3, H3K27me3, and H3K36me3 (Shi et al., 2021). It also suppressed the expression of SMYD3, EZH2, and nuclear receptor binding SET domain protein 1 (NSD1) (Moser-Katz et al., 2021). Notably, SMYD3, a histone methyltransferase targeting histone H3K4, is usually overexpressed in various cancer cell types, correlating with advanced stages and poor survival (Wu et al., 2023). NSD1, a dual-function transcriptional regulator that controls mono- and dimethylation of H3K36, has defined its potential role in cancer treatment (Krossa et al., 2022) (Table 5.1).

Follow-up studies in 2012 showed that triptolide dose-dependently lowered total amounts of H3K4me2 and H3K36me2 in multiple myeloma cells. It raised the levels of histone demethylase 1 of lysine (LSD1), a central demethylase of histone, while decreasing JMJD2B expression, an enzyme involved in demethylating H3K9me3 and H3K36me2 (Wu et al., 2023). Additionally, in 2015, Wen et al. showed that triptolide administered to KM3 cells decreased the expression of oncogenes c-Myc and VEGFA, an important angiogenic variable, by blocking the buildup of H3K4me3 at their promoters (Wu et al., 2023). These results show that plant medicine may have major effects against multiple myeloma via epigenetic processes. Moreover, plant medicine has been found to control the epigenetics of leukemia, preventing its growth. Z-ligustilide, studied by Wang et al., increased the levels of acetylated H3 (K9/14) in HL-60 cells, enriched acetylated H3 (K9/14) in the Nur77 and NOR-1 promoter regions, and modulated p300 acetyltransferase and HDAC recruitment, indicating the strong potential of herbal medicine in treating acute leukemia through epigenetic regulation (Wang et al., 2020) (Table 5.1).

5.13 EPIGENETIC REGULATION OF MEDICINAL-PLANT-ASSOCIATED MICROBES

Endophytes and bacteria from the rhizosphere are important for medical plant growth and the synthesis of pharmaceutical substances (Hao et al., 2021). *Cordyceps*, derived from mycoparasites or endophytes, hold significance in traditional Chinese medicine due to their diverse bioactivities (Hao et al., 2021). Cordycepin, a derivative, exhibits antitumor and antiviral properties. In cellular studies, cordycepin induces methylation of BCL7A, downregulates EBV genes, and reduces H3K4me3 histone enrichment, impacting EBV genome copy numbers (Hao et al., 2016). *Cordyceps* culture degeneration is attributed to epigenetic alterations (Zheng et al., 2023). Filamentous fungi experience during culture maintenance is often associated with genomic DNA methylation and chromosome instability (Hao et al., 2021). In non-sporulating cultures, oxidative stress is linked to fungal degeneration, with mitochondrial DNA glycation identified as a control mechanism (Zheng et al., 2023).

DNA methylation influences fungal sexual development and genome stability maintenance (Zhu et al., 2016). *Cordyceps militaris* undergoes DNA methylation reprogramming during fungal development, while mutations in the H3K4 methylation enzyme CclA activate cryptic secondary metabolite clusters (Bok et al., 2009). Exploring modifiers like HDAC inhibitors and DNMT inhibitors induces microbial secondary metabolite production (Van Der Molen et al., 2022). DNMT inhibitor 5-azacytidine and HDAC inhibitor SAHA stimulate mycotoxin production in endophytic fungi (Zheng et al., 2023). In addition, SAHA and 5-azacytidine boost secondary metabolism-related gene activity and bioactive chemical production (Qadri et al., 2017). The reduction of camptothecine formation in endophytic fungi can be corrected by 5-azacytidine (Vasanthakumari et al., 2015).

5.14 GENOME EDITING TOOLS

Advancements in sequencing technology have facilitated whole-genome sequencing in various crops, significantly enhancing our understanding of plant genetics (Chopra et al., 2021). Rapid gene editing tools have become important for improving plant study. The rise of unique-to-sequence DNA enzymes that enable the exact targeting of specific genes for change holds great hope for growth (Gaj et al., 2013). Currently, three main nucleases – zinc finger nucleases (ZFN), transcription activator-like effector nucleases (TALEN), and CRISPR/Cas – are utilized for gene editing reasons (Hussain et al., 2020). Among these, the CRISPR/Cas system-mediated genome editing stands out because of its affordability, accuracy, and speed, making it highly valuable for crop genetic enhancement and molecular breeding endeavors (Wang et al., 2017).

Genome editing technologies have revolutionized plant research, providing scientists with powerful tools to precisely modify specific genes. ZFNs, developed in the 1990s, were among the earliest genome editing tools (Qi et al., 2023). ZFNs consist of a zinc finger protein (ZFP) domain responsible for recognizing and binding to target DNA sequences, coupled with the Fok I nuclease domain for inducing double-strand DNA breaks. Despite their initial promise, ZFNs have limitations such as limited diversity in zinc finger proteins, high design costs, and restricted DNA sequence recognition, leading to reduced utilization over time. In 2009, TALENs emerged as an alternative to ZFNs, offering enhanced design flexibility (Qi et al., 2023). TALENs employ a cleavage domain derived from the Fok I restriction enzymes and a binding domain based on transcription activator-like effectors (TALEs) found in *Xanthomonas* spp. (Sekhon & Chopra, 2021).

The CRISPR/dCas9 activation (CRISPRa) system represents a significant advancement in genome editing technology, addressing traditional limitations in gene overexpression (Ding et al., 2022). This system utilizes the catalytically inactive Cas9 (dCas9) protein fused with transcription activation domains (TADs) to activate gene expression. By designing multiple guide RNAs targeting specific gene promoters, the CRISPRa method allows the concurrent gene expression of several genes within a single vector. Various modifications, such as fusing different TADs to dCas9 or utilizing multiplexed TALE-activation (mTALE-Act) systems, have further enhanced the efficiency and versatility of CRISPRa for plant genome editing applications. Many studies have proven the usefulness of CRISPRa in stimulating target genes in plants, including genes that code for proteins like anthocyanin pigment 1 (AtPAP1) and non-protein-coding genes like miR319 (Lowder et al., 2015). The development of CRISPR-Act2.0, which combines dCas9-VP64 with MS2-VP64 through gRNA2.0, has significantly increased activating transcription rate, making it a valuable tool for gene activation in crops like rice and *Arabidopsis* (Qi et al., 2023).

5.15 FUTURE PROSPECT

- *Elucidating Mechanisms:* More study is required to expand the understanding of the mechanisms of action for phytometabolites via their interactions with biological systems. Understanding how these compounds modulate cellular pathways and gene expression profiles can provide insights into their therapeutic potential and optimize their use in disease management.
- *Clinical Validation:* Undertaking thorough clinical studies to confirm the efficacy and safety of medicinal plant-derived compounds in humans is essential. Clinical studies can help validate traditional uses of herbal medicines, identify potential adverse effects, and establish guidelines based on scientific evidence for their use in healthcare settings.
- *Epigenetic Regulation:* Investigating the epigenetic regulation of gene expression by medicinal plant compounds offers a promising avenue for future research. Understanding how these compounds influence post-transcriptional modification can provide novel targets for therapeutic interventions in various diseases.
- *Personalized Medicine:* Integrating genomic and epigenomic data with clinical outcomes can pave the way for personalized medicine approaches

using medicinal plants. Tailoring treatment strategies based on individual genetic and epigenetic profiles can enhance therapeutic efficacy and minimize adverse effects.

- *Sustainable Sourcing:* Ensuring sustainable sourcing and cultivation practices for medicinal plants is crucial to preserve biodiversity and protect natural ecosystems. Research into innovative cultivation methods, conservation strategies, and alternative sources of bioactive compounds can promote environmental sustainability in herbal medicine production.
- *Regulatory Frameworks:* Developing standardized protocols, quality control measures, and regulatory frameworks for herbal medicines is essential to ensure product safety, efficacy, and consistency. Collaboration between researchers, healthcare professionals, and regulatory agencies is necessary to establish standards and scientific evidence for the usage of medicinal plants in clinical practice.
- *Education and Public Awareness:* Increasing education and public awareness about the benefits and risks of using medicinal plants is essential to promote their responsible use. Providing accurate information, training healthcare professionals, and engaging with communities can empower individuals to make informed decisions about their health and well-being.

5.16 CONCLUSION

The use of medicinal plants has been instrumental in developing therapeutic compounds for various diseases, with research in natural medicine and synthetic biology focusing on phytometabolites from secondary metabolism. Epigenetic regulations play crucial roles in chronic disease onset, emphasizing the importance of maintaining epigenetic balance for overall health. Chinese herbal medicine (CHM) has emerged as a globally recognized preventive and therapeutic approach, offering holistic strategies for complex diseases like cancer and autoimmune disorders. Advanced biomedical technologies, such as genomics and epigenomics, have enriched CHM research, aligning with its holistic therapeutic philosophy. Traditional Chinese medicine (TCM), encompassing therapies like acupuncture and herbal medicine, has gained widespread popularity, especially in treating various conditions. The exploration of DNA methylation techniques in CHM research provides valuable insights for future studies. Moreover, genome editing technologies offer precise methods for crop improvement, revolutionizing plant breeding practices and ensuring sustainable agricultural development. Overall, the integration of traditional medicine with modern scientific advancements holds immense promise for addressing global health challenges and advancing therapeutic interventions.

REFERENCES

Amaral, C. L. D., Milagro, F. I., Curi, R., & Martínez, J. A. (2014). DNA methylation pattern in overweight women under an energy-restricted diet supplemented with fish oil. *BioMed Research International*, Article ID 675021. https://doi.org/10.1155/2014/675021

Bok, J. W., Chiang, Y. M., Szewczyk, E., Reyes-Dominguez, Y., Davidson, A. D., Sanchez, J. F., Lo, H., Watanabe, K., Strauss, J., Oakley, B.R., & Wang, C. C. C., Keller, N. P. (2009). Chromatin-level regulation of biosynthetic gene clusters. *Nature Chemical Biology*, *5*(7), 462–464.

Chen, L., Hong, J., Hu, R., Yu, X., Chen, X., Zheng, S., Qin, Y., Zhou, X., Wang, Y., Zheng, L., Fang, H., Liu, P., & Huang, B. (2021). Clinical value of combined detection of serum sTim-3 and pepsinogen for gastric cancer diagnosis. *Cancer Management and Research*, 7759–7769.

Chen, X., Zhang, J., Ruan, W., Huang, M., Wang, C., Wang, H., & Lin, T. (2020). Urine DNA methylation assay enables early detection and recurrence monitoring for bladder cancer. *The Journal of Clinical Investigation*, *130*(12), 6278–6289.

Chen, Y., Cheng, C., Feng, X., Lai, R., Gao, M., Chen, W., & Wu, R. (2021). Integrated analysis of lncRNA and mRNA transcriptomes reveals the potential regulatory role of lncRNA in kiwifruit ripening and softening. *Scientific Reports*, *11*(1), 1671. https://doi.org/10.1038/s41598-021-81155-1

Chopra, S., Myers, Z., Sekhon, H., & Dufour, A. (2021). The nerves to conduct a multiple sclerosis crime investigation. *International Journal of Molecular Sciences*, *22*(5), 2498.

Chu, A. Y., Tin, A., Schlosser, P., Ko, Y. A., Qiu, C., Yao, C., Joehanes, R., Grams, M. E., Liang, L., Gluck, C. A., Liu, C., Coresh, J., Hwang, S.J., Levy, D., Boerwinkle, E., Pankow, J. S., Yang, Q., Fornage, M., Fox, C. S., Susztak, K., & Köttgen, A. (2017). Epigenome-wide association studies identify DNA methylation associated with kidney function. *Nature Communications*, *8*(1), 1286. https://doi.org/10.1038/s41467-017-01297-7

Ding, X., Yu, L., Chen, L., Li, Y., Zhang, J., Sheng, H., & Cao, J. (2022). Recent progress and future prospect of CRISPR/Cas-derived transcription activation (CRISPRa) system in plants. *Cells*, *11*(19), 3045. https://doi.org/10.3390/cells11193045

Farrow, S. C., & Facchini, P. J. (2014). Functional diversity of 2-oxoglutarate/Fe (II)-dependent dioxygenases in plant metabolism. *Frontiers in Plant Science*, *5*, 524. https://doi.org/10.3389/fpls.2014.00524

Flores-Sierra, J., Arredondo-Guerrero, M., Cervantes-Paz, B., Rodríguez-Ríos, D., Alvarado-Caudillo, Y., Nielsen, F. C., & Lund, G. (2016). The trans fatty acid elaidate affects the global DNA methylation profile of cultured cells and in vivo. *Lipids in Health and Disease*, *15*, 75. https://doi.org/10.1186/s12944-016-0243-2

Frigolet, M. E., & Gutiérrez-Aguilar, R. (2017). The role of the novel lipokine palmitoleic acid in health and disease. *Advances in Nutrition*, *8*(1), 173S–181S.

Gaj, T., Gersbach, C. A., & Barbas, C. F. (2013). ZFN, TALEN, and CRISPR/Cas-based methods for genome engineering. *Trends in Biotechnology*, *31*(7), 397–405.

Ghantous, A., Saikali, M., Rau, T., Gali-Muhtasib, H., Schneider-Stock, R., & Darwiche, N. (2012). Inhibition of tumor promotion by parthenolide: Epigenetic modulation of p21. *Cancer Prevention Research*, *5*(11), 1298–1309.

Gluckman, P. D., Hanson, M. A., Buklijas, T., Low, F. M., & Beedle, A. S. (2009). Mechanisms that underpin metabolic and cardiovascular diseases. *Nature Reviews Endocrinology*, *5*(7), 401–408.

González-Becerra, K., Ramos-López, O., Barrón-Cabrera, E., Riezu-Boj, J. I., Milagro, F. I., Martínez-López, E., & Martínez, J. A. (2019). Fatty acids, epigenetic mechanisms and chronic diseases: A systematic review. *Lipids in Health and Disease*, *18*, 178. https://doi.org/10.1186/s12944-019-1120-6

Goyal, K., Konar, A., Kumar, A., & Koul, V. (2020). Bacosides encapsulated in lactoferrin conjugated PEG-PLA-PCL-OH based polymersomes act as epigenetic modulator in chemically induced amnesia. *Neurochemical Research*, *45*, 796–808.

Guo, W., Ma, H., Wang, C. Z., Wan, J. Y., Yao, H., & Yuan, C. S. (2021). Epigenetic studies of Chinese herbal medicine: Pleiotropic role of DNA methylation. *Frontiers in Pharmacology*, *12*, 790321. https://doi.org/10.3389/fphar.2021.790321

Hao, Y., Hao, S., Andersen-Nissen, E., Mauck, W. M., Zheng, S., Butler, A., Lee, M. J., Wilk, A. J., Darby, C., Zager, M., Hoffman, P., Stoeckius, M., Papalexi, E., Mimitou, E. P., Jain, J., Srivastava, A., Stuart, T., Fleming, L. M., Yeung, B., Rogers, A. J., McElrath, J.M., Blish, C. A., Gottardo, R., Smibert, P., & Satija, R. (2021). Integrated analysis of multimodal single-cell data. *Cell*, *184*(13), 3573–3587.

Hao, Z., Hao, F., Singh, V. P., & Zhang, X. (2018). Changes in the severity of compound drought and hot extremes over global land areas. *Environmental Research Letters*, *13*(12), 124022. https://doi.org/10.1088/1748-9326/aaee96

Hao, Z., Wu, T., Cui, X., Zhu, P., Tan, C., Dou, X., Hsu, K. W., Lin, Y. T., Peng, P. H., Zhang, L. S., Gao, Y., Hu, L., Sun, H. L., Zhu, A., Liu, J., Wu, K. J., & He, C. (2020). N6-deoxyadenosine methylation in mammalian mitochondrial DNA. *Molecular Cell*, *78*(3), 382–395.

Hussain, A., Kaler, J., Tabrez, E., Tabrez, S., & Tabrez, S. S. (2020). Novel COVID-19: A comprehensive review of transmission, manifestation, and pathogenesis. *Cureus*, *12*(5), e8184. https://doi.org/10.7759/cureus.8184

Jiang, H., Yao, Q., An, Y., Fan, L., Wang, J., & Li, H. (2022). Baicalin suppresses the progression of Type 2 diabetes-induced liver tumor through regulating METTL3/m6A/HKDC1 axis and downstream p-JAK2/STAT1/clevaged Capase3 pathway. *Phytomedicine*, *94*, 153823. https://doi.org/10.1016/j.phymed.2021.153823

Kiec-Wilk, B., Sliwa, A., Mikolajczyk, M., Malecki, M. T., & Mathers, J. C. (2011). The CpG island methylation regulated expression of endothelial proangiogenic genes in response to β-carotene and arachidonic acid. *Nutrition and Cancer*, *63*(7), 1053–1063.

Krossa, I., Strub, T., Aplin, A. E., Ballotti, R., & Bertolotto, C. (2022). Lysine methyltransferase NSD1 and cancers: Any role in melanoma? *Cancers*, *14*(19), 4865.

Lee, C., Kim, B. G., Kim, J. H., Chun, J., Im, J. P., & Kim, J. S. (2017). Sodium butyrate inhibits the NF-kappa B signaling pathway and histone deacetylation, and attenuates experimental colitis in an IL-10 independent manner. *International Immunopharmacology*, *51*, 47–56.

Li, W., Guo, Y., Zhang, C., Wu, R., Yang, A. Y., Gaspar, J., & Kong, A. N. T. (2016). Dietary phytochemicals and cancer chemoprevention: A perspective on oxidative stress, inflammation, and epigenetics. *Chemical Research in Toxicology*, *29*(12), 2071–2095.

Li, Y., Zheng, H., Wang, Q., Zhou, C., Wei, L., Liu, X., Zhang, W., Zhang, Y., Du, Z., Wang, W., & Xie, W. (2018). Genome-wide analyses reveal a role of Polycomb in promoting hypomethylation of DNA methylation valleys. *Genome Biology*, *19*, 18. https://doi.org/10.1186/s13059-018-1390-8

Liu, Y. Y., Wang, Y., Walsh, T. R., Yi, L. X., Zhang, R., Spencer, J., & Shen, J. (2016). Emergence of plasmid-mediated colistin resistance mechanism MCR-1 in animals and human beings in China: A microbiological and molecular biological study. *The Lancet Infectious Diseases*, *16*(2), 161–168.

Lowder, L. G., Zhang, D., Baltes, N. J., Paul III, J. W., Tang, X., Zheng, X., & Qi, Y. (2015). A CRISPR/Cas9 toolbox for multiplexed plant genome editing and transcriptional regulation. *Plant Physiology*, *169*(2), 971–985.

Lu, X., Li, Y., Yang, W., Tao, M., Dai, Y., Xu, J., & Xu, Q. (2021). Inhibition of NF-κB is required for oleanolic acid to downregulate PD-L1 by promoting DNA demethylation in gastric cancer cells. *Journal of Biochemical and Molecular Toxicology*, *35*(1), e22621. https://doi.org/10.1002/jbt.22621

Moser-Katz, T., Joseph, N. S., Dhodapkar, M. V., Lee, K. P., & Boise, L. H. (2021). Game of bones: How myeloma manipulates its microenvironment. *Frontiers in Oncology*, *10*, 625199. https://doi.org/10.3389/fonc.2020.625199

Niu, W., Feng, Y., Peng, M., & Cai, J. (2024). A narrative review on the mechanism of natural flavonoids in improving glucolipid metabolism disorders. *Phytotherapy Research*, *38*(8), 4202–4229.

Panni, S., Lovering, R. C., Porras, P., & Orchard, S. (2020). Non-coding RNA regulatory networks. *Biochimica et Biophysica Acta (BBA)-Gene Regulatory Mechanisms*, *1863*(6), 194417.

Park, U. H., Hwang, J. T., Youn, H., Kim, E. J., & Um, S. J. (2022). Kaempferol antagonizes adipogenesis by repressing histone H3K4 methylation at PPARγ target genes. *Biochemical and Biophysical Research Communications*, *617*, 48–54.

Qadri, S. M., Bissinger, R., Solh, Z., & Oldenborg, P. A. (2017). Eryptosis in health and disease: A paradigm shift towards understanding the (patho) physiological implications of programmed cell death of erythrocytes. *Blood Reviews*, *31*(6), 349–361.

Qi, Q., Hu, B., Jiang, W., Wang, Y., Yan, J., Ma, F., & Xu, J. (2023). Advances in plant epigenome editing research and its application in plants. *International Journal of Molecular Sciences*, *24*(4), 3442. https://doi.org/10.3390/ijms24043442

Saad-Hussein, A., & Taha, M. M. (2021). Epigenetic study of global gene methylation in PON1, XRCC1 and GST s different genotypes in rural and urban pesticide exposed workers. *Journal of Complementary and Integrative Medicine*, *17*(4), 20190166. https://doi.org/10.1515/jcim-2019-0166

Shi, F., Li, Y., Han, R., Fu, A., Wang, R., Nusbaum, O., & Zhu, Y. (2021). Valerian and valeric acid inhibit growth of breast cancer cells possibly by mediating epigenetic modifications. *Scientific Reports*, *11*, 2519. https://doi.org/10.1038/s41598-021-81620-x

Sitarek, P., Kowalczyk, T., Santangelo, S., Białas, A. J., Toma, M., Wieczfinska, J., & Skała, E. (2018). The extract of *Leonurus sibiricus* transgenic roots with AtPAP1 transcriptional factor induces apoptosis via DNA damage and down regulation of selected epigenetic factors in human cancer cells. *Neurochemical Research*, *43*, 1363–1370.

Skała, E., Toma, M., Kowalczyk, T., Śliwiński, T., & Sitarek, P. (2018). Rhaponticum carthamoides transformed root extract inhibits human glioma cells viability, induces double strand DNA damage, H2A.X phosphorylation, and PARP1 cleavage. *Cytotechnology*, *70*, 1585–1594.

Song, F., Mahmood, S., Ghosh, S., Liang, P., Smiraglia, D. J., Nagase, H., & Held, W. A. (2009). Tissue specific differentially methylated regions (TDMR): Changes in DNA methylation during development. *Genomics*, *93*(2), 130–139.

Song, Y., Wang, Z. Z., Wang, L., Faybusovich, P., Srivastava, K., Liu, C., & Li, X. M. (2021). *Sophora flavescens* alkaloids and corticosteroid synergistically augment IL-10/IL-5 ratio with *Foxp3*-Gene-epigenetic modification in asthma PBMCs. *Journal of Asthma and Allergy*, 1559–1571.

Thakur, V. S., Deb, G., Babcook, M. A., & Gupta, S. (2014). Plant phytochemicals as epigenetic modulators: Role in cancer chemoprevention. *The AAPS Journal*, *16*, 151–163.

Tian, N. N., Zheng, Y. B., Li, Z. P., Zhang, F. W., & Zhang, J. F. (2021). Histone methylatic modification mediates the tumor-suppressive activity of curcumol in hepatocellular carcinoma via an Hotair/EZH2 regulatory axis. *Journal of Ethnopharmacology*, *280*, 114413. https://doi.org/10.1016/j.jep.2021.114413

Tremblay, B. L., Guénard, F., Rudkowska, I., Lemieux, S., Couture, P., & Vohl, M. C. (2017). Epigenetic changes in blood leukocytes following an omega-3 fatty acid supplementation. *Clinical Epigenetics*, *9*, 43. https://doi.org/10.1186/s13148-017-0345-3

Van Der Molen, J., Groeskamp, S., & Maas, L. R. (2022). Imminent reversal of the residual flow through the Marsdiep tidal inlet into the Dutch Wadden Sea based on multiyear ferry-borne acoustic Doppler current profiler (ADCP) observations. *Ocean Science*, *18*(6), 1805–1816.

Vasanthakumari, M. M., Jadhav, S. S., Sachin, N., Vinod, G., Shweta, S., Manjunatha, B. L., . . . Uma Shaanker, R. (2015). Restoration of camptothecine production in attenuated endophytic fungus on re-inoculation into host plant and treatment with DNA methyltransferase inhibitor. *World Journal of Microbiology and Biotechnology*, *31*, 1629–1639.

Venetsky, A., Levy-Zamir, A., Khasdan, V., Domb, K., & Kashkush, K. (2015). Structure and extent of DNA methylation-based epigenetic variation in wild emmer wheat (*T. turgidum* ssp. dicoccoides) populations. *BMC Plant Biology*, *15*, 200. https://doi.org/10.1186/s12870-015-0544-z

Wang, M., Lu, Y., Botella, J. R., Mao, Y., Hua, K., & Zhu, J. K. (2017). Gene targeting by homology-directed repair in rice using a geminivirus-based CRISPR/Cas9 system. *Molecular Plant*, *10*(7), 1007–1010.

Wang, P., Wang, M., Zhuo, T., Li, Y., Lin, W., Ding, L., & Xu, L. (2021). Hydroxysafflor yellow a promotes osteogenesis and bone development via epigenetically regulating β-catenin and prevents ovariectomy-induced bone loss. *The International Journal of Biochemistry & Cell Biology*, *137*, 106033. https://doi.org/10.1016/j.biocel.2021.106033

Wang, W., Sheng, H., Bai, Y. F., Weng, Y. Y., Fan, X. Y., & Zhang, F. (2020). Hesperetin, a SIRT1 activator, inhibits hepatic inflammation via AMPK/CREB pathway. *International Immunopharmacology*, *89*, 107036. https://doi.org/10.1016/j.intimp.2020.107036

Wu, H. T., Lin, J., Liu, Y. E., Chen, H. F., Hsu, K. W., Lin, S. H., & Chen, D. R. (2021). Luteolin suppresses androgen receptor-positive triple-negative breast cancer cell proliferation and metastasis by epigenetic regulation of MMP9 expression via the AKT/mTOR signaling pathway. *Phytomedicine*, *81*, 153437. https://doi.org/10.1016/j.phymed.2020.153437

Wu, H. T., Liu, Y. E., Hsu, K. W., Wang, Y. F., Chan, Y. C., Chen, Y., & Chen, D. R. (2020). MLL3 induced by luteolin causes apoptosis in tamoxifen-resistant breast cancer cells through H3K4 monomethylation and suppression of the PI3K/AKT/mTOR pathway. *The American Journal of Chinese Medicine*, *48*(05), 1221–1241.

Wu, K., Na, K., Chen, D., Wang, Y., Pan, H., & Wang, X. (2018). Effects of non-steroidal anti-inflammatory drug-activated gene-1 on *Ganoderma lucidum* polysaccharides-induced apoptosis of human prostate cancer PC-3 cells. *International Journal of Oncology*, *53*(6), 2356–2368.

Wu, Y. Y., Xu, Y. M., & Lau, A. T. (2023). Epigenetic effects of herbal medicine. *Clinical Epigenetics*, *15*, 85. https://doi.org/10.1186/s13148-023-01481-1

Yang, B., Yang, J., Bai, J., Pu, P., Liu, J., Wang, F., & Ruan, B. (2014). Suv39h1 protects from myocardial ischemia-reperfusion injury in diabetic rats. *Cellular Physiology and Biochemistry*, *33*(4), 1176–1185.

Yu, H., Wang, X., Bai, L., Tang, G., Carter, K. T., Cui, J., Huang, P., Liang, L., Ding, Y., Cai, M., Huang, M., Liu, H., Cao, G., Gallinger, S., Pai, R. K., Buchanan, D. D., Win, A. K., Newcomb, P. A., Wang, J., Grady, W. M., & Luo, Y. (2023). DNA methylation profile in CpG-depleted regions uncovers a high-risk subtype of early-stage colorectal cancer. *JNCI: Journal of the National Cancer Institute*, *115*(1), 52–61.

Yuan, Z., Chen, S., Gao, C., Dai, Q., Zhang, C., Sun, Q., & Jiang, Y. (2019). Development of a versatile DNMT and HDAC inhibitor C02S modulating multiple cancer hallmarks for breast cancer therapy. *Bioorganic Chemistry*, *87*, 200–208.

Zain, S. N. D. M., & Omar, W. A. W. (2022). Phyllanthus debilis methanolic extract reduces the viability of human colorectal adenocarcinoma (HT-29) cells and increases LINE-1 and alu DNA methylation. *Pertanika Journal of Tropical Agricultural Science*, *45*(1), 25--36-. -https://www.cabidigitallibrary.org/doi/full/10.5555/2022008549-8

Zhang, P., Fan, Y., Sun, X., Chen, L., Terzaghi, W., Bucher, E., & Dai, M. (2019). A large-scale circular RNA profiling reveals universal molecular mechanisms responsive to drought stress in maize and Arabidopsis. *The Plant Journal*, *98*(4), 697–713.

Zhang, S. F., Mao, X. J., Jiang, W. M., & Fang, Z. Y. (2020). Qian Yang Yu Yin Granule protects against hypertension-induced renal injury by epigenetic mechanism linked to Nicotinamide N-Methyltransferase (NNMT) expression. *Journal of Ethnopharmacology*, *255*, 112738. https://doi.org/10.1016/j.jep.2020.112738

Zheng, Y., Yu, P. P., Li, Z. K., Xiong, S. F., Zhou, L. L., Zhou, J. X., Wang, C. M., Meng, Y. M., Zhang, Y., Wang, Y. J., Xu, J., Wu, Y., Guo, L. X., & Zhao, T. (2023). Critical metals Ga, Ge and In in the global Pb-Zn deposits: Current understanding, challenges and perspectives. *Journal of Earth Science*, *34*(4), 1308–1311. https://doi.org/10.1007/s12583-023-1909-0

Zhu, Y., Tchkonia, T., Fuhrmann-Stroissnigg, H., Dai, H. M., Ling, Y. Y., Stout, M. B., Pirtskhalava, T., Giorgadze, N., Johnson, K. O., Giles, C. B., Wren, J. D., Niedernhofer, L. J., Robbins, P. D., & Kirkland, J. L. (2016). Identification of a novel senolytic agent, navitoclax, targeting the Bcl-2 family of anti-apoptotic factors. *Aging Cell*, *15*(3), 428–435. https://doi.org/10.1111/acel.12445

Zych, M., Stec, K., Bablewska, A., & Pilis, K. (2019). Importance of diet in reducing cancer incidence in Poland: A review. *Roczniki Państwowego Zakładu Higieny*, *70*(4), 337–345. https://agro.icm.edu.pl/agro/element/bwmeta1.element.agro-39fb35ce-cc84-4f80-9e3f-16ddd527b453

6 Pangenomes and Pangenomics in Medicinal Plants

S. Priyadharshini, P. Samuel, M. Gnanaraj, M. Bala Sundar, M. Rajadurai, D. Rajasudhakar, and T. Jebastin

6.1 INTRODUCTION

Plants stand out as the primary source of natural medicines, owing to their rich array of chemical compounds and diverse structures, as well as the vast biodiversity encompassing their constituents. Notable medications derived from plants include aspirin (extracted from willow tree bark) (Desborough & Keeling, 2017), digoxin (derived from the *Digitalis lanata* flower) (De Swiet, 2023), and morphine (obtained from opium) (Brook et al., 2017). Projections indicate that over 60% of pharmaceuticals in industrialized nations are either natural products or their secondary metabolites. The battle against numerous multi-resistant pathogens demands the development of advanced and highly efficient therapeutic agents. In recent years, medicinal plants (MPs) have attracted researchers for the identification of natural drugs for their pharmaceutical, economic, and social advantages (Yang et al., 2023; Zhang et al., 2023). There are '300,000–400,000' higher species in the plant kingdom. It is estimated that about 5–15% of terrestrial plant species have undergone pharmaceutical investigation. Approximately 150–200 plant species have been included in western medicine, while 10,000–15,000 plant species worldwide have been identified as medicinal. Nowadays, plants are the source of around 25% of all medications (Gurnani et al., 2014). Since they are the main source of metabolites and biologically active compounds, a genome-level understanding of MPs is required for drug development. While breeding MPs, the plant parts and active elements possessing medicinal properties, and production cycle should be considered. The application of MPs in pharmacology requires the pangenome construction to promote the analyses for the identification of their genomic diversity and distribution, cultivation period, and effect of environmental causes (Zhang et al., 2023).

A pangenome (Figure 6.1) could be defined as a set of genomes or DNA sequences of a species. The length and the presence of repetitive regions/polyploidy in the plant genomes are responsible for the complexities that occurred during the assembling and interpretation processes. High-resolution pangenomes are capable of acquiring all variants in a genome which necessitates the development of robust computational tools to assemble, visualize, and analyze them with high resolution (Lei et al., 2021). The advancements in sequencing and assembling methodologies encourage the construction of pangenomes for eukaryotes in a limited time, low cost, and with high accuracy (Golicz et al., 2020; Bayer et al., 2020; Danilevicz et al., 2020).

Pangenome assembly and analysis enhances the research on the adaptability, and evolution among/across the species. In the initial studies, the de novo assembly and comparison methodology were utilized to exert genomic level distinctions among individuals. Following this, the iterative mapping and assembly methodology constructs a pangenome by including nonredundant sequences from other individuals produced by a single reference genome. However, a single reference genome could not explain all the characteristics of a species genome (Danilevicz et al., 2020).

The pangenome analysis explores the evolutionary relationship within species, natural selection, adaptability to the environment, and breeding of species. The studies related to the identification of structural variations (SVs) and polymorphisms (like SNPs) of genomes within species or higher taxonomic groups performed in pangenome provide information about the characteristics associated with phenotype, adaptability to nature and expression data in order to define their impact on different regions of genomes ((Tranchant-Dubreuil et al., 2019; Golicz et al., 2020; Danilevicz et al., 2020; Li et al., 2022).

6.2 PLANT PANGENOME/PANGENOMICS

In plants, the mechanisms maintained by the genomic variations such as gene tandem duplications, transposable element activity, deletions, genome rearrangement and recombination across populations influence the occurrence of gene presence/absence variation (PAV)/SVs. The association between the mentioned variations and the natural events (such as phenotypic diversity, adaptability, and selection) affects genetic diversity (Danilevicz et al., 2020; Li et al., 2022). The significance of single-nucleotide polymorphism (SNP) and SVs such as presence and absence variation (PAV) and copy number variant (CNP), which are responsible for agronomical characteristics of plants, encourages the development of pangenomes (Lei et al., 2021; Li et al., 2022).

Due to the structural complexities between prokaryotes and eukaryotes, pangenome took dual description as,

DOI: 10.1201/9781003475491-6

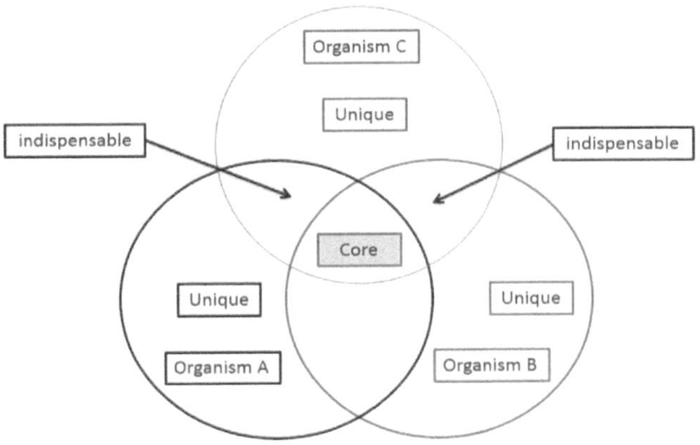

FIGURE 6.1 Illustration of pangenomes (modified from Golchha et al., 2024).

(i) function/gene-based and (ii) structure/sequence-based. The gene-based definition states that a pangenome is a whole gene of a collection of individuals and gene families; the sequence-based definition states as a compilation of nonredundant genes of a species without SNPs, InDels, and transposons (Golicz et al., 2020; Tranchant-Dubreuil et al., 2019; Lei et al., 2021). A pangenome is a composition of the entire genome including the variations within a species that enables the identification and analysis of crucial genomic segments present in it (Golicz et al., 2015; Li et al., 2022).

The pangenome concept was first applied in 2005 by Tittelin et al. for *Streptococcus agalactiae* bacteria using the de novo assembly and annotation approach. A pangenome can be divided into three regions: (i) core region, (ii) accessory region, and (iii) individual-specific genome. The core region refers to the presence of a particular region in every individual and it governs some functions, such as DNA replication, translation, cellular homeostasis, and cellular mechanisms like glycolysis. The accessory region refers to the presence of a particular region in a subgroup of individuals and this influences various functions like telomere maintenance. This region is further divided into two due to the following properties: (i) the presence of genes in two or more individuals and (ii) in only one individual. An individual-specific genome is particularly present in only one individual of a species. Hence, it is responsible for the unique characteristics of that individual. However, this region has not been taken into account during the analysis of pangenomes so far (Golicz et al., 2015, 2019; Tranchant-Dubreuil et al., 2019; Danilevicz et al., 2020; Bayer et al., 2020; Jayakodi et al., 2021; Lei et al., 2021; Yang et al., 2023). In plants, the core region of the genome is large and covers the maximum amount of genes. The accessory regions have genes that are able to adapt and survive in any kind of environment. The analogy of core region against wild species and accessory region against cultivated species unfolds the impact of domestication. A pangenome could either be open type or closed type. In open type, the insertion of individuals maximizes the size of the pangenome, and this is vice versa in close type pangenome (Golicz et al., 2015, 2019; Khan et al., 2019; Tranchant-Dubreuil et al., 2019).

The first announcement of plant pangenome was made in 2007 illustrating short accessory regions of rice and maize based on whole genome assembly and comparison (Bayer et al., 2020). From the pangenomes of *Brachypodium distachyon* (purple false brome), *Brassica oleracea* (cabbage), *Brassica napus* (rapeseed), *Capsicum* (chilli pepper), *Medicago* (medick/burclover), *Oryza sativa* (rice), *Glycine max* (soybean), *Sesamum indicum* (sesame), *Solanum lycopersicum* (tomato), *Helianthus* (sunflower), and *Triticum aestivum* (bread wheat), we could realize that a plant pangenome, in general, comprises 33% to 80% of core region (Golicz et al., 2020; Jayakodi et al., 2021).

The first high-quality pangenome was constructed by Li et al., in 2014 for the seven accessions of wild soybean that belong to a particular phylogeny and geography with the application of one of the whole genome assemblies and comparison methodology – de novo assembly, which identifies variations that are responsible for the agronomic features like biotic defiance, seed composition, flowering, and maturity time (Li et al., 2022). This study ensures the application of pangenome/pangenomics in plant genome-related analysis by proving that a single genome is not enough to recognize the diversity of a species (Golicz et al., 2015; Tranchant-Dubreuil et al., 2019; Bayer et al., 2020; Li et al., 2022; Yang et al., 2023).

Even though the earlier studies focused on transposons/SVs to determine the accessory genome, they could not acquire enough knowledge from the data due to shortcomings such as storage, time consumption, and accuracy of whole genome assembly at that time. This was rectified when three methodologies were introduced for pangenome assembly (Bayer et al., 2020; Jayakodi et al., 2021).

6.3 METHODOLOGIES FOR PANGENOME ASSEMBLY

Pangenome characterization provides in-depth knowledge of the core genome, such as the number of genes/gene families present in every individual, the genome size of the species as a whole, the genes/gene families of new individuals inserted into the pangenome and the genes absent in reference genomes (Golicz et al., 2015; Khan et al., 2019).

The whole genome assembly and comparison methodology which assembles and compares the genomes of species has evolved as three widely used methodologies (Fig.2) and they are namely, (i) *de novo* assembly, (ii) reference-based assembly and iterative mapping, (iii) graph and k-mer (Golicz et al., 2015, 2019; Bayer et al., 2020).

(i) *De novo assembly:* This methodology is applied when a reference genome is not available. Hence, the assembly of every individual genome will be performed and aligned with the related whole genome in order to find out the core and accessory regions. The annotation provides information regarding orthologous genes and families. However, this method is expensive, slow, and produces variations that occur due to errors while sequencing short reads. The introduction of growing technologies in long-read sequencing facilitates the generation ofplant genome at the chromosome level with high accuracy (Golicz et al., 2015, 2019; Khan et al., 2019; Tranchant-Dubreuil et al., 2019; Li et al., 2022).

(ii) *Reference-based assembly and iterative mapping:* As a first step, the reads will be mapped to the known (reference) genome and the unmapped contaminant-free reads will be mapped by de novo assemblers. Then the assembled reads will be appended to the pseudomolecules. These data together build the pangenome of a species and took a reference to map the reads of individuals again to find PAVs and the coverage of genes although it exerts some limitations while handling repetitive genes and multiple structural variants. On the other side, mapping and assembling of reads were performed simultaneously and the unwanted data will be removed. This method facilitates the identification of PAVs at each gene locus at a low cost (Golicz et al., 2015, 2019; Khan et al., 2019; Tranchant-Dubreuil et al., 2019; Bayer et al., 2020; Li et al., 2022).

(iii) *Graph/k-mer-based assembly:* The de Bruijn-graph and string graph algorithms-based assemblers have been used to assemble the sequence data of a genome, where the graph is an illustration of varied and conserved regions. Here, the genome will break down into several k-mers indicated as nodes of a graph, which illustrates the association between them. The nodes will be connected by an edge or numerous edges may connect one node and the origin node will be indicated by a distinctive color. Correspondingly, the whole pangenome has been produced to compare the genetic diversity and variations against the reference genomes. Like the de Bruijn graph focused on integrating genetic information, the bidirected variation graph identifies variations over the population and maps them with the reference genome. Even though this method has limitations like the other methodologies, it has grown as a promising methodology for genome assembly (Golicz et al., 2015; Khan et al., 2019; Tranchant-Dubreuil et al., 2019; Bayer et al., 2020; Li et al., 2022).

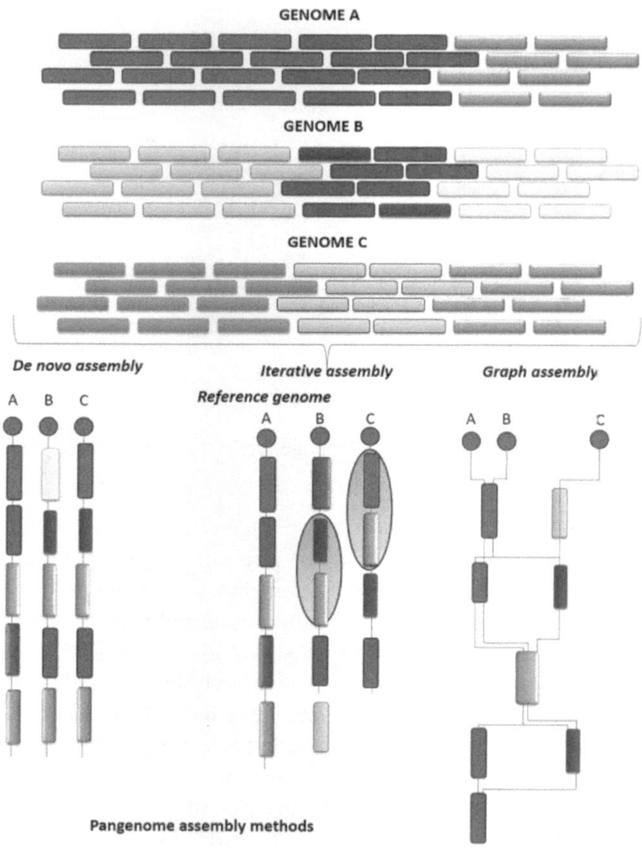

FIGURE 6.2 Illustration of different pangenome assembly methods.

6.4 PANGENOMICS IN MEDICINAL PLANTS

Research on natural products derived from medicinal plants plays a pivotal role in the quest for novel bioactive compounds, which hold immense therapeutic potential and could be instrumental in the development of new drugs to combat a wide array of diseases (Barba et al., 2014). One powerful tool in this endeavor is pangenomic analysis, which enables scientists to delve into the complete genetic makeup of medicinal plants. By doing so, they can pinpoint specific genes and pathways involved in the synthesis of bioactive compounds. This comprehensive approach not only allows for the identification and isolation of previously unknown compounds but also facilitates a deeper understanding of their biological activities (Briskin, 2000). Furthermore, pangenomics empowers researchers to compare the genomes of different medicinal plants, unveiling common gene clusters or pathways associated with the production of bioactive compounds. Armed with this knowledge, scientists can manipulate metabolic pathways to enhance the production of desired bioactive compounds or even synthesize entirely new compounds with potentially superior therapeutic properties. Moreover, pangenomic analysis aids in uncovering genetic variations responsible for the diverse bioactive compound profiles observed among different plant varieties (Hirsch et al., 2014). Understanding these variations provides insights that can be leveraged in the development of targeted breeding programs aimed at optimizing the production of specific bioactive compounds in medicinal plants (Jiménez-García et al., 2013). In addition to its scientific significance, the insights gained from pangenomic analysis contribute to the sustainable utilization of medicinal plants. By identifying the most efficient sources for particular bioactive compounds, this knowledge supports conservation efforts and ensures a stable supply of raw materials for industries such as pharmaceuticals and nutraceuticals (Dolatabadian et al., 2019). Moreover, the application of pangenomic analysis extends beyond known medicinal plants. It can also lead to the discovery of new plant species or strains harboring unique and valuable bioactive compounds (Li et al., 2023). These discoveries hold the potential to revolutionize the field of natural product drug discovery, paving the way for the development of innovative treatments for various diseases (Nasim et al., 2022).

Recent advancements in genome sequencing and synthetic biology have revived interest in natural products. Despite the availability of sequenced genomes and transcriptomes for these plants, the absence of publically available gene annotations and tabulated gene expression data has inhibited their efficient usage. To solve this difficulty, IMP (Integrated Medicinal Plantomics), a publicly available platform (https://www.bic.ac.cn/IMP), has curated a total of 8,565,672 genes from 84 high-quality genome assemblies and 2,156 transcriptome sequencing samples covering multiple organs, tissues, developmental stages, and stimuli. Featuring ten integrated analytic modules, IMP allows users to examine gene annotations, sequences, functions, distributions, and expressions in a simplified way. In order to propel advances in synthetic biology and make it easier to explore natural sources for valuable chemical constituents like drug production and discovery, IMP will be essential in improving our understanding of the molecular metabolic pathways in medicinal plants or plants with medicinal benefits. (Chen et al., 2024). As we all know, medicinal plants produce natural metabolites (e.g., pigments, condiments, insecticides, and medicines) that have been in utilization in countries like India, China, and Egypt for 5,000 years for various disease treatments. Monomers, like morphine, artemisinin, taxol, and vinblastine, and secondary metabolites, like saponins, tanshinones, alkaloids, terpenes, coumarin, and polyphenols, were obtained from plants and exert vigorous therapeutic impact in antioxidant, antibacterial, antiphlogistic, anticancer, and antidiabetic activities. The inadequate knowledge of the genome of medicinal plants restricts the massive-scale development of active compounds. As stated earlier, the high-throughput sequencing technology enhances the determination of key genes that participate in particular secondary biosynthesis pathways (Yang et al., 2023; Zhang et al., 2023).

The *Panax notoginseng* (Chinese ginseng) genomic analysis reveals that saponin, a bioactive compound exhibiting antimicrobial and anti-inflammatory properties, is a byproduct of genes involved in the biosynthesis pathway. A study on the Chinese *Papaver somniferum* (opium poppy) genome found the biosynthesis, regulation, and transportation of benzylisoquinoline alkaloid (BIA). The integrated pangenome and epigenome analysis of MP would provide information regarding phenotype and environmental flexibility, and the metabolism of bioactive compounds of (sub)species and demand the development of secondary metabolites for pharmaceutical utility.

Cajanus cajan (pigeon pea) is known for its nutritients, like vitamin B, vitamin C (ascorbic acid), carotene, and protein. Its pangenome analysis observed that the SNPs belonging to agronomic properties could impact the yield-associated phenotypes (Zhao et al., 2020). The *Catharanthus roseus* (Cape periwinkle) methylome showed that the tissue-specific covariations on context-dependent DNA methylation may affect the cellular/physiological processes. Another study on *Panax quinquefolium* (American ginseng) reveals that cold environment conditions would affect the ginsenoside accretion (Zhang et al., 2023).

The pangenomes constructed using GWAS (genome-wide association studies) for the bulk data of SNPs and PAVs reveal the genome length and key genes involved in significant traits of the medicinal plant *Vignaradiata* (mung bean) originated from India (South Asia) and spread to East Asia (China). The rich nutritional contents like carbohydrates, proteins, folic acid, iron, and flavonoids (vitexin and isovitexin) promote multi-utility (food, vegetables, and medicine) and make it one of the economic crops of China. Pangenome sequence analysis looks at the roles that core and accessory regions play, such as metabolic, catalytic, and cellular processes, as well as pathways linked to growth, stress response, and environmental adaptation. Additionally, it identifies important genes related to ribosome biogenesis, amino acid synthesis, sucrose metabolism, and GST (glycine, serine, and threonine) metabolism. Through this analysis, the functions carried out by core and accessory regions (like cellular, metabolic, and catalytic activities/pathways, stress inducers, growth-related mechanisms, adaptability to an environment), key genes of sucrose metabolism, amino acid production, biogenesis of ribosome, GST (glycine, serine, threonine) metabolism. In Dels and SVs, similar chromosomes, gene families, genetic diversity (geographical history) have been discovered (Liu et al., 2022).

6.5 PANGENOMICS AND OTHER OMICS TECHNOLOGIES

The genes maintaining the pharmaceutical properties could be exposed through the integration of pangenome with multi-omics technologies, such as metabolomics, and phenomics (Sun et al., 2022). The integrated multi-omics studies, such as taxonomics, transcriptomics, metabolomics, proteomics, genomics, pangenomics, epigenomics, phenomics, and mutagenomics with bioinformatics analyses on medicinal plants, allow us to explore and understand the biogenesis, chemical varieties, and formulation of biologically active phytocompounds that are regulated by functionally significant genes, metabolites, and enzymes (Zhang et al., 2023), molecular markers in particular species, functional genes regulating biological mechanisms, biosynthetic pathways of biologically active metabolites, and mechanisms against environmental events (Li et al., 2022; Yang et al., 2023). The plant pangenomic studies considering both genic and nongenic parts of the genome identify phenotypic variances representing agronomic features of crops through gene regulation modifications (Bayer et al., 2020).

The artificial intelligence (AI) based genomic prediction model was created using the data obtained from multi-omics analyses and complex phenotypic datasets. That predicts the genotypic and phenotypic traits, expression level, modification and regulation, and crucial functional genes/variants with accuracy. The identification of plant functional genes complementing the utility of technologies like gene editing turns our focus into the improvement of yield, quality, nutritional value, and stress resistance, leading to the molecular breeding of MPs. Hence, the integrated analysis of pangenome provides us with a descriptive understanding of MPs at the genome level (Li et al., 2022).

6.6 APPLICATIONS OF PANGENOMES

Pangenomes have the capability to represent multiple high-order taxonomic species that leads to the identification of genes present only in a particular species and the evolutionary history of the genes, which would reveal information regarding orthologous genes, characteristics of gene families, their functionalities, and mutations in genes. The patterns present in genomes, and the mapping of evolutionary divergence occurred in various lineages could studied through pangenomes (Golicz et al., 2020; Danilevicz et al., 2020; Li et al., 2022).

Genomic analyses, such as variant calling, gene expression, and PAV depiction, are performed with high quality and accuracy due to the usage of reference pangenomes that enable the identification of genes missed while breeding. To improve the production, if they exert potential agronomic traits they may be introduced into modern varieties. Likewise, the functional analysis of accessory genes allows us to identify genes involved in stress (both abiotic and biotic), and disease resistance. In account of this, the identification of accessory regions in a species helps us to determine the genomic variations, such as SNPs and PAV associated with agronomic tasks (Golicz et al., 2020; Danilevicz et al., 2020; Bayer et al., 2020).

In MPs, pangenomics facilitates the determination of the population system, unique characteristics of a population, genes associated with agronomic traits, common phenotypes, protein-coding genes, missing regions that are lost in reference, genetic variation, and evolution of the species.

6.7 TOOLS FOR PANGENOMICS

There are tools, technologies, and methodologies available especially for plants to assemble/develop, visualize, analyze, and interpret pangenome that may facilitate tasks like typical accessions selection, genome assembly and annotation, identification of SNPs, orthologous group, SVs, construction of phylogeny and pangenome, and mapping of GWAS (Li et al., 2022). With this, the inconvenience in the construction, assembly, analysis, and interpretation of genomes necessitates the development of upgraded sequencing methodologies/technologies to produce contigs with maximum accuracy at a justifiable cost. In particular, the PacBiosequencing technology satisfies all the mentioned requirements. The introduction of PacBiosequencing in DNA sequencing technology enhances the development of sequence-based plant pangenomes that consist of noncoding and repetitive sequences (Lei et al., 2021; Li et al., 2022).

6.7.1 Tools for Pangenome Assembly

Technologies such as NRGene assembly, Hi-C, 10x Genomics, PacBio, and Nanopore assemble genomes with better coverage. In addition to this, Velvet, MasuRCA, SOAPdenovo, ALLPATHS-G, and ABySS assemblers were specifically developed to produce *de novo* plant pangenome assemblies and for graph-based assembling de Bruijn, and string graphs-based assemblers and SplitMEM toolkits are available (Golicz et al., 2015; Khan et al., 2019).

6.7.2 Tools for Pangenome Visualization

Microbial pangenome has been taken as a model to develop visualization tools for plant pangenome that enable the interpretation of conserved and variable genomic characteristics by visualizing the genomic relationships. These tools are either based on reference genome structure or *de novo* assembly with variation data to show the accessory regions and pathways. The tools/databases such as RPAN for rice, GBrows (Golicz et al., 2015) for Wheat, Brachypan for *Brachypodiumdistachuon*, and EnsEMBL (Golicz et al., 2015) were developed to visualize the genomic sequences, gene annotations, PAVs, and gene expressions (Danilevicz et al., 2020; Lei et al., 2021). The cloud-based technology helps to surpass the difficulties in pangenome visualization tools development. Since the plant genome is complex, variable, and larger in size, the cloud databases fulfill the technical requirements such as storage, curation, maintenance, and increased reproducibility (Danilevicz et al., 2020).

6.7.3 Tools for Pangenome Analysis

The analysis of eukaryotic high-throughput, low-depth pangenome sequence data was performed in EUPAN (Eukaryotic PAN-genome analysis toolkit, https://cgm.sjtu.edu.cn/eupan) and the large plant genomes were analyzed by GET_HOMOLOGUES-EST (https://github.com/

eead-csic-compbio/get_homologues). Adding to this, the PanVC (https://gitlab.com/dvalenzu/PanVC) is used to analyse variant calling against pangenome reference, the CHIC aligner (https://gitlab.com/dvalenzu/CHIC) looks for repetitive references and performs mapping of individual reads of the pangenome, and theGenomeMapper (http://1001genomes.org/software/genomemapper.html) maps short reads to pangenomes simultaneously (Khan et al., 2019).

The vg (variation graph) toolkit (https://github.com/vgteam/vg), minigraph (https://github.com/lh3/minigraph), and PHG (Practical Haplotype Graph, https://www.maizegenetics.net/phg) could be used to construct, analyze, and visualize graph-based pangenome (Jayakodi et al., 2021). Especially, the vg toolkit facilitates data storage, interchange formats, alignment, genotyping, and variant calling. Plink (https://zzz.bwh.harvard.edu/plink/), EMMAX (https://genome.sph.umich.edu/wiki/EMMAX), and GEMMA (https://github.com/genetics-statistics/GEMMA) employ the linear alignment algorithms such as BLAST, BLAT, and HMMER, PLAST, and PathRacer, respectively, and used to detect SVs in graph-based pangenome and homologous sequences (Li et al., 2022).

The deep learning (DL) algorithms are also widely used in the analysis of plant pangenomes, and the DL models have the potential to identify the cause of genome variation within and across species and sequence function prediction with good accuracy, which led to the development of breeding and evolutionary analysis tools (Danilevicz et al., 2020). The recently available tools, namely, PATO toolkit (https://github.com/irycisBioinfo/) and Panache (https://github.com/SoutGreenPlatform/Panache) help to view linear pangenome. GreenPhylDB v5 is a comparative database plant pangenomes and is used to analyze gene families and homologous relationships (https://www.greenphyl.org) (Yang et al., 2023).

6.8 CURRENT PROSPECTUS, CHALLENGES, AND FUTURE DIRECTIONS

Collaborative efforts like the '10KP Genome Sequencing Project' fuel the increasing interest in researching pangenomes, which may span various species (Cheng et al., 2018). These efforts are strengthening the reservoir of sequencing data encompassing many plant species. Nonetheless, expanding research demands creative methods to expedite variation graph compilation, pangenome construction, and display.

Although pangenomes can increase the efficacy of agricultural breeding, future databases and tools will need to integrate pangenomic data with gene functional studies and evolutionary reconstructions without any voids. Understanding the convoluted connections between genomic variants (including non-coding areas) and complex agronomic variables necessitates a complete set of pangenomics techniques to integrate various omics data. There will be a growing need to expand the germplasm pool and better comprehend the genomic landscape of cereal species. This will pave the way for genome editing to enhance crops, as crop gene editing technologies become more widely used and accepted (Wang et al., 2019), in part because they are subject to less stringent regulations than genetically modified (GM) crops (Waltz et al., 2018).

Building accurate pangenomes rapidly and cheaply while integrating them with existing functional data is still beset with many obstacles. When dealing with polyploid genomes and repeated motifs, which are obstacles to assembly when using short-read sequences, long-sequence reads might be beneficial. With more precise data and decreasing manufacturing costs, this method is going to dominate plant pangenomics (Jung et al., 2019; Danilevicz et al., 2020). According to LeCun et al. (2015), deep learning might be used in pangenomics to address problems with functional characterization and assembly by discovering patterns in the data on its own.

Although many main crops currently claim accessible pangenomes, diverse in completeness, expanding these beyond species boundaries would simplify delving into wild gene sequence diversity for crop enhancement, supporting the creation of tailored crops. With pangenomes becoming available for a diversified variety of species, we may begin to comprehend how species and higher taxa are delimited at the genomic level, affording fundamental insights into plant evolution and diversity.

The growing accessibility of genome sequence data has bolstered pangenome investigations, and this tendency is only going to persist as long-read sequence data continues to see significant price and quality decreases. As our awareness of the influence of varied genes improves, singular reference assemblies may become obsolete, with the pangenome increasingly functioning as the new reference, giving comprehensive insights on evolution, selection, and especially genome functionality.

One issue of pangenome analysis arises in preserving and displaying pangenome data. Abundant long-read sequencing data facilitates the implementation of pangenome variation graphs, which record variations for whole populations, such as those implemented in variation graph (VG) (Garrison et al., 2018) or multi-genome reference (MGR) (Rabbani et al., 2020). Establishing standards for genomic structure and annotation tolerating structural genome variance is crucial. In plant breeding populations, a step in advance is the use of plausible haplotype graphs for the scalable production of pangenomes (Jensen et al., 2020). These graphs depend on a reference genome coordinate system and employ genes to anchor sequences, allowing them to surmount obstacles in aligning repetitive and highly divergent sections.

Methods for the precise and consistent functional annotation of genes and genomes significantly lag behind procedures for their assembly, and the purpose of many variable genes remains ambiguous. However, variable genes share some traits, such as being less likely to be syntenic, developing under decreased evolutionary constraints, and having lower expression levels (Golicz et al., 2016; Gordon et al., 2017; Contreras-Moreira et al., 2017). Enhancing our knowledge of the functions and interactions between core and variable genes would considerably enhance the utility of pangenome investigations. Integrative genomics methodologies, which aim to link properties of genes such as expression level, connectivity in biological networks, and sequence conservation to their function (Golicz et al., 2018; Hassani-Pak et al., 2016), offer a potential approach to gain a broader understanding of their potential function.

The preponderance of pangenome research so far has focused on the genetic portion of the genome, yet genomic areas outside of genes explain a considerable amount of phenotypic variables in cereals (Rodgers-Melnick et al., 2016). This shows that many critical agronomic properties may be influenced by changes in gene regulation rather than gene presence-absence variation (PAV). For example, areas shown to be under selection in the tomato pangenome include a promoter related to fruit taste. Combined with epigenomic functional annotations of regulators, pangenomes offer a rich resource to mine for regulatory sequence variation that may be harnessed in breeding.

Prokaryotic pangenomic investigations have proven a remarkable capacity to transcend taxonomic boundaries, as indicated by a comprehensive analysis including 7,104 genomes encompassing ten prokaryotic phyla (Maistrenko et al., 2020). The compact structure of haploid bacterial genomes permits such large-scale analysis, rendering them computationally practicable.

In contrast, plant pangenomic research has not yet reached the genus barrier, mostly owing to restrictions in computing resources and financing. However, with the continuous decline in sequencing costs and developments in computational capabilities, the horizon of plant pangenomics is growing. It is predicted that plant pangenome investigations will shortly expand beyond the species level. This development offers the prospect of joining pangenomes at higher taxonomic levels, maybe up to the genus or family level. Such developments will enable research into issues like the critical gene content of legumes.

Ultimately, this trajectory of study will enable scientists to forecast and define the gene repertoire of all plant species – a discovery with the potential to alter future genomic inquiry. The enormous span of these pangenomes will offer insights into age-old concerns, notably resolving the genetic features that identify a plant (Bayer et al., 2020).

6.9 CONCLUSION

The definition evolves as the concept of pangenome evolves. The construction of pangenome faces complexities due to the structural complexities of prokaryotes and eukaryotes. Since pangenomes provide accessibility to analyze the entire genome of a species with regard to their genetic diversity, functionalities, variations, and polymorphisms, its utility has been enlarged to plants (MPs) as well. The pangenome of a particular MP species stimulates the production of therapeutics by providing information related to metabolites, biosynthesis pathways, molecular markers, etc., as discussed. This concept could be expanded to super-pangenome (Khan et al., 2019) that contains the genome of diverse species belonging to a particular genus. The application of advanced technologies, tools, and AI (deep learning and machine learning) based models alleviate the whole genome assembly, annotation, visualization, and analysis of pangenomes. However, the pangenomes constructed so far were mainly focused on crops with agronomic importance (Li et al., 2022) rather than MPs. The development of pangenome of various MP species with the integration of multi-omics analysis provides a complete knowledge of the pharmaceutical properties of various MPs. Constructing the genome and pangenome of plants presents challenges due to widespread polyploidization and the abundance of repetitive sequences. Nonetheless, long-read sequencing technologies offer robust solutions for building pangenomes in plants with large genomes, effectively navigating complex repetitive regions to identify significant structural variations (SVs). Notably, the integration of differential gene identification and CRISPR/Cas9 techniques allows for comprehensive dissection and validation of gene functions. Pangenomic investigations in crops yield valuable insights, serving as pivotal references for constructing MP pangenomes. As pangenomics approaches progressively supersede single reference genomes, they are poised to establish a new standard for elucidating evolutionary relationships and genetic variations within plants and MPs.

REFERENCES

Barba, F. J., Esteve, M. J., & Frígola, A. (2014, January 1). Bioactive components from leaf vegetable products. https://doi.org/10.1016/b978-0-444-63294-4.00011-5

Bayer, P. E., Golicz, A. A., Scheben, A., Batley, J., & Edwards, D. (2020). Plant pan-genomes are the new reference. *Nature Plants*, *6*(8), 914–920.

Briskin, D. P. (2000, October 1). Medicinal plants and phytomedicines. Linking plant biochemistry and physiology to human health. *Plant Physiology*, *124*(2), 507–514. https://doi.org/10.1104/pp.124.2.507

Brook, K., Bennett, J., & Desai, S. P. (2017). The chemical history of morphine: An 8000-year journey, from resin to de-novo synthesis. *Journal of Anesthesia History*, *3*(2), 50–55.

Chen, T., Yang, M., Cui, G., Tang, J., Shen, Y., Liu, J., Yuan, Y., Guo, J., Huang, L. (2024). IMP: Bridging the gap for medicinal plant genomics. *Nucleic Acids Research*, 52(D1), D1347–D1354.

Cheng, S., Melkonian, M., Smith, S. A., Brockington, S., Archibald, J. M., Delaux, P. M., Li, F. W., Melkonian, B., Mavrodiev, E. V., Sun, W., Fu, Y., Yang, H., Soltis, D. E., Graham, S. W., Soltis, P. S., Liu, X., Xu, X., & Wong, G. K. S. (2018). 10KP: A phylodiverse genome sequencing plan. *Gigascience*, *7*(3), giy013.

Contreras-Moreira, B., Cantalapiedra, C. P., García-Pereira, M. J., Gordon, S. P., Vogel, J. P., Igartua, E., Casas, A. M., & Vinuesa, P. (2017). Analysis of plant pan-genomes and transcriptomes with GET_HOMOLOGUES-EST, a clustering solution for sequences of the same species. *Frontiers in Plant Science*, *8*, 184.

Danilevicz, M. F., Fernandez, C. G. T., Marsh, J. I., Bayer, P. E., & Edwards, D. (2020). Plant pangenomics: Approaches, applications and advancements. *Current Opinion in Plant Biology*, *54*, 18–25.

De Swiet, M. (2023). Digitalis purpurea The source of digitoxin: Digitalis lanata The source of digoxin. In *Modern medicines from plants* (pp. 120–128). CRC Press.

Desborough, M. J., & Keeling, D. M. (2017). The aspirin story: From willow to wonder drug. *British Journal of Haematology*, *177*(5), 674–683.

Dolatabadian, A., Bayer, P. E., Tirnaz, S., Hurgobin, B., Edwards, D., & Batley, J. (2019, October 10). Characterization of disease resistance genes in the *Brassica napus* pangenome reveals significant structural variation. https://doi.org/10.1111/pbi.13262

Garrison, E., Sirén, J., Novak, A. M., Hickey, G., Eizenga, J. M., Dawson, E. T., Jones, W., Garg, S., Markello, C., Lin, M. F., Paten, B., & Durbin, R. (2018). Variation graph toolkit improves read mapping by representing genetic variation in the reference. *Nature Biotechnology*, *36*(9), 875–879.

Golchha, N. C., Nighojkar, A., & Nighojkar, S. (2024). Bacterial pangenome: A review on the current strategies, tools and applications. *Medinformatics*, 1–13.

Golicz, A. A., Batley, J., & Edwards, D. (2015). Towards plant pangenomics. *Plant Biotechnology Journal*, 1099–1105.

Golicz, A. A., Bayer, P. E., Barker, G. C., Edger, P. P., Kim, H., Martinez, P. A., Chan, C. K. K., Ellis, A. S., McCombie, W. R., Parkin, I. A. P., Paterson, A. H., Chris Pires, J., Sharpe, A. G., Tang, H., Teakle, G. R., Town, C. D., Batley, J., & Edwards, D. (2016). The pangenome of an agronomically important crop plant Brassica oleracea. *Nature Communications*, *7*(1), 13390.

Golicz, A. A., Bayer, P. E., Bhalla, P. L., Batley, J., & Edwards, D. (2020). Pangenomics comes of age: From bacteria to plant and animal applications. *Trends in Genetics*, *36*(2), 132–145.

Golicz, A. A., Bhalla, P. L., & Singh, M. B. (2018). MCRiceRepGP: A framework for the identifcation of genes associated with sexual reproduction in rice. *The Plant Journal*, *96*, 188–202.

Gordon, S. P., Contreras-Moreira, B., Woods, D. P., Des Marais, D. L., Burgess, D., Shu, S., Stritt, C., Roulin, A. C., Schackwitz, W., Tyler, L., Martin, J., Lipzen, A., Dochy, N., Phillips, J., Barry, K., Geuten, K., Budak, H., Juenger, T. E., Amasino, R., Caicedo, A. L., Goodstein, D., Davidson, P., Mur, L. A. J., Figueroa, M., Freeling, M., Catalan, P., & Vogel, J. P. (2017). Extensive gene content variation in the Brachypodium distachyon pan-genome correlates with population structure. *Nature Communications*, *8*(1), 2184.

Gurnani, N., Mehta, D., Gupta, M., & Mehta, B. K. (2014). Natural products: Source of potential drugs. *African Journal of Basic & Applied Sciences*, *6*(6), 171–186.

Hassani-Pak, K., Castellote, M., Esch, M., Hindle, M., Lysenko, A., Taubert, J., & Rawlings, C. (2016). Developing integrated crop knowledge networks to advance candidate gene discovery. *Applied & Translational Genomics*, *11*, 18–26.

Hirsch, C. N., Foerster, J. M., Johnson, J., Sekhon, R. S., Muttoni, G., Vaillancourt, B., Peñagaricano, F., Lindquist, E., Pedraza, M. A., Barry, K., León, N. D., Kaeppler, S. M., & Buell, C. R. (2014, January 1). Insights into the Maize Pan-Genome and Pan-Transcriptome. https://doi.org/10.1105/tpc.113.119982

Jayakodi, M., Schreiber, M., Stein, N., &Mascher, M. (2021). Building pan-genome infrastructures for crop plants and their use in association genetics. *DNA Research*, *28*(1), 1–9.

Jensen, S. E., Charles, J. R., Muleta, K., Bradbury, P. J., Casstevens, T., Deshpande, S. P., Gore, M. A., Gupta, R., Ilut, D. C., Johnson, L., Lozano, R., Miller, Z., Ramu, P., Rathore, A., Romay, M. C., Upadhyaya, H. D., Varshney, R. K., Morris, G. P., Pressoir, G., Buckler, E. S., & Ramstein, G. P. (2020). A sorghum practical haplotype graph facilitates genome-wide imputation and cost-effective genomic prediction. *The Plant Genome*, *13*(1), e20009.

Jiménez-García, S. N., Vázquez-Cruz, M. A., Guevara-González, R. G., Torres-Pacheco, I., Cruz-Hernández, A., & Pérez, A. A. F. (2013, June 30). Current approaches for enhanced expression of secondary metabolites as bioactive compounds in plants for agronomic and human health purposes. *Polish Journal of Food and Nutrition Sciences*, *63*(2), 67–78. https://doi.org/10.2478/v10222-012-0072-6

Jung, H., Winefield, C., Bombarely, A., Prentis, P., & Waterhouse, P. (2019). Tools and strategies for long-read sequencing and de novo assembly of plant genomes. *Trends in Plant Science*, *24*(8), 700–724.

Khan, A. W., Garg, V., Roorkiwal, M., Golicz, A. A., Edwards, D., & Varshney, R. K. (2020). Super-pangenome by integrating the wild side of a species for accelerated crop improvement. *Trends in Plant Science*, *25*(2), 148–158.

LeCun, Y., Bengio, Y., & Hinton, G. (2015). Deep learning. *Nature*, *521*, 436.

Lei, L., Goltsman, E., Goodstein, D., Wu, G. A., Rokhsar, D. S., & Vogel, J. P. (2021). Plant pan-genomics comes of age. *Annual Review of Plant Biology*, *72*, 411–435.

Li, H., Qamar, M. T. U., Li, Y., Liang, J., You, J., & Wang, L. (2023, February 4). Current progress, applications and challenges of multi-omics approaches in sesame genetic improvement. https://doi.org/10.3390/ijms24043105

Li, W., Liu, J., Zhang, H., Liu, Z., Wang, Y., Xing, L., He, Q., & Du, H. (2022). Plant pan-genomics: Recent advances, new challenges, and roads ahead. *Journal of Genetics and Genomics*, *49*, 833–946.

Liu, C., Wang, Y., Peng, J., Fan, B., Xu, D., Wu, J., Cao, Z., Gao, Y., Wang, X., Li, S., Su, Q., Zhang, Z., Wang, S., Wu, X., Shang, Q., Shi, H., Shen, Y., Wang, B., & Tian, J. (2022). High-quality genome assembly and pan-genome studies facilitate genetic discovery in mung bean and its improvement. *Plant Communications*, *3*, 100352.

Maistrenko, O. M., Mende, D. R., Luetge, M., Hildebrand, F., Schmidt, T. S., Li, S. S., Coelho, L. P., Cepas, J. H., Sunagawa, S., & Bork, P. (2019). Disentangling the impact of environmental and phylogenetic constraints on prokaryotic strain diversity. *bioRxiv*, 735696.

Nasim, N., Sandeep, I. S., & Mohanty, S. (2022, October 18). Plant-derived natural products for drug discovery: Current approaches and prospects. *Nucleus/The Nucleus*, *65*(3), 399–411. https://doi.org/10.1007/s13237-022-00405-3

Rabbani, L., Mueller, J., & Weigel, D. (2020). An algorithm to build a multi-genome reference. *bioRxiv*. https://doi.org/10.1101/2020.04.11.036871

Rodgers-Melnick, E., Vera, D. L., Bass, H. W., & Buckler, E. S. (2016). Open chromatin reveals the functional maize genome. *Proceedings of the National Academy of Sciences of the United States of America*, *113*, E3177–E3184.

Sun, W., Xu, Z., Song, C., & Chen, S. (2022). Herbgenomics: Decipher molecular genetics of medicinal plants. *The Innovation*, *3*(6), 100322.

Tranchant-Dubreuil, C., Rouard, M., & Sabot, F. (2019). Plant pangenome: Impacts on phenotypes and evolution. *Annual Plant Reviews*, 1–8.

Waltz, E. (2018). *With a free pass, CRISPR-edited plants reach market in record time*. Nature Publishing Group.

Wang, T., Zhang, H., & Zhu, H. (2019). CRISPR technology is revolutionizing the improvement of tomato and other fruit crops. *Horticulture Research*, *6*(1), 77.

Yang, L., Yang, Y., Huang, L., Cui, X., & Liu, Y. (2023). From single- to multi-omics: Future research trends in medicinal plants. *Briefings in Bioinformatics*, *24*(1), 1–13.

Zhang, W., Zeng, Y., Jiao, M., Ye, C., Li, Y., Liu, C., & Wang, J. (2023). Integration of high-throughput omics technologies in medicinal plant research: The new era of natural drug discovery. *Frontiers in Plant Sciences*, *14*, 1073848.

Zhao, J., Bayer, P. E., Ruperao, P., Saxena, R. K., Khan, A. W., Golicz, A. A., Nguyen, H. T., Batley, J., Edwards, D., & Varshney, R. K. (2020). Trait associations in the pangenome of pigeon pea (Cajanus cajan). *Plant Biotechnology Journal*, *18*(9), 1946–1954.

7 Advances in CRISPR-Based Approaches in Medicinal Plants

Ahmed Marhri, Yousra Hammouti, Malika Abid, and Mohamed Addi

7.1 INTRODUCTION

7.1.1 Briefly Introduce the Concept of Medicinal Plants as Natural Remedies

Since time immemorial, traditional cultures have primarily been based on natural products to manufacture medicines and maintain their health. Despite the use of mineral and animal products, plants remain the primary source of natural cures (Hosseinzadeh et al., 2015). They contain substances within their tissues that can be employed for therapeutic aims or serve as precursors for the production of beneficial drugs. About 350,000 to 500,000, accounting for around 10% of vascular plants, are used as medicinal plants (Salmerón-Manzano et al., 2020). Around 80% of the global population depends on medicinal herbs rather than modern medicine (Daswani et al., 2006; Rasul, 2018).

In effect, many plants exhibit efficacy against a range of fungi, bacteria, or mycobacteria. This underscores the considerable promise of medicinal plants in creating innovative antibacterial treatments and supportive therapies (Mahady, 2005). However, the majority of medicinal plant extracts remain untested in animal trials or clinical studies (Mahady, 2005).

Due to the extraordinary biodiversity present in Amazon and Thailand, they hold the top two positions worldwide in terms of the number of plant species exhibiting antimicrobial properties. Herbal medicine offers a rich source of biologically active substances (secondary metabolites), such as phenolic compounds, nitrogen-containing molecules (alkaloids), and terpenoids, known for their therapeutic properties (Harborne, 1998). Phenolic compounds represent the main antioxidants in numerous medicinal plants. Consequently, some plants exhibit strong antioxidant capacities and could be considered a source of natural antioxidants (Gan et al., 2010).

Secondary metabolites, also called phytochemicals, are natural organic substances that play an important role in human health. Phytochemicals are found in select taxonomic groups in minimal amounts and do not contribute to the developmental or growth processes of plants. On the contrary, these compounds contribute significantly to plant protection against biotic and abiotic stressors.

Alongside utilizing medicinal plants for treating illnesses, there is an increasing focus on harnessing their potential for disease prevention. Traditional remedies sourced from these plants have been harnessed to strengthen immune responses against various pathogens (Di Pierro et al., 2012; Ramakrishna et al., 2011). Additionally, medicinal plants are also employed in managing and averting chronic illnesses (Gan et al., 2010). Moreover, the richness of medicinal plants in total phenolic contents, coupled with the highest antioxidant capacities, makes them an appropriate choice for preventing and treating diseases induced by oxidative stress (Gan et al., 2010).

In recent years, whether in developing or industrialized countries, plants are still considered important sources of medicine. In some countries like the United States, the utilization of medicinal herbs has significantly increased, with herbal sales rising from $ 2.5 billion in 1995 to $ 4.4 billion in 2005 (Ramawat & Mérillon, 2008). Moreover, traditional medicine is still widely practiced in China, boasting about 5,000 traditional remedies available for use (Li, 2000; Nestler, 2002). Furthermore, traditional herbal preparations constitute 30 to 50% of total medical consumption in China (Eddouks et al., 2012). On a global scale, the market worth of medicinal plant products surpasses $100 billion annually (Sofowora et al., 2013).

7.1.2 Highlight the Historical Significance of Medicinal Plants in Healthcare

The historical utilization of herbal medicines dates back thousands of years, in which natural remedies played a crucial role in both preventing and treating human illnesses worldwide (Newman et al., 2000). More than 12,000 herbal treatments have been used by traditional healers (Hosseinzadeh et al., 2015). The first fossil records of the rational use of herbs in antiquity date back to Mesopotamia, around 2600 BC. Substances derived from about 1,000 plants have been employed in traditional medicine practices as treatment for colds, coughs, parasitic infections and inflammation (Van Wyk & Wink, 2018). Ayurvedic medicine in India is also among the oldest of all, with some of the earliest records of plant usage as drugs found in Artharvaveda, dating back to 2000 BC (Sneader, 2005). During the era of ancient Egyptian civilization, Egyptians were renowned for their effectiveness in traditional medicine. Various treatments were recommended, including mandrake for alleviating pain and garlic for addressing heart and circulatory issues. These remedies, along with numerous others, were

DOI: 10.1201/9781003475491-7

documented in the Ebers Papyrus, which is considered to be the best pharmaceutical record dating back to 1500 BC.

The foundation of the European healing tradition can be traced back to the contributions of Hippocrates (460–377 BC) and Aristotle (384–322 BC), who were known for their attachment to ancient Egyptian and Indian beliefs. The book *De Materia Medica* is considered the first and the standard European herbal for more than 1000 years. Until now, herbal medicine in Europe has remained a rational method of treating ailments, particularly in Germany, Switzerland, France, Austria, and Great Britain. The Greeks, especially the philosopher and naturalist Theophrastus (circa 300 BC), played a substantial role in advancing the logical utilization of herbal remedies within ancient Western civilizations (around 300 BC). Remnants of this Western knowledge were disseminated and preserved by some European countries, but the Arabs are considered most responsible for preserving much of the Greco-Roman expertise in the field of herbal medicine. In addition, the Arabs integrated Western knowledge with Indian and Chinese knowledge unrecognized to the Greco-Roman world at the time. In the 8th century, the Persian pharmacist, physician, and philosopher Avicenna wrote *Canon Medicinae*, which encompassed all of Greco-Roman medicine. Next, there came the extensive gathering of medical regulations by Ibn al-Bitar, who practiced medicine in Spain (Newman et al., 2000).

7.1.3 Introduce CRISPR-Based Genome Editing as a Transformative Technology

Over the past three decades, genome-editing technology of living organisms garnered great interest. Along the way, scientists encountered several obstacles and difficulties, promoting them to investigate alternative methods, resulting in noticeable results in the field (Khalil, 2020). Consequently, several editing technologies based on the principles of DNA-protein recognition have been created. With the most commonly employed being homing endonucleases or meganucleases, transcription activator-like effector nucleases (TALENs) and zinc-finger nucleases (ZFNs). However, these traditional techniques are technically challenging, require protein engineering to target specific DNA sequences, and are expensive and require much time (Caplan et al., 2015). During the last decade, a more powerful tool has been discovered, this is about the CRISPR-Cas system. This innovative technology does not necessitate intricate expertise or costly equipment (Khalil, 2020).

Researchers in the field of biology began taking an active interest in studying CRISPR at the beginning of the 21st century. However, the description of sequencing multiple repeated DNA fragments in tandem dates back to 1987 by a Japanese group (Ishino et al., 1987). In 2002, the name of the regularly spaced repeats was changed to clustered regularly interspaced palindromic repeats (CRISPR). In 2005, it was observed that many of the spacer sequences within CRISPR arrays are derived from virus and plasmid origins. The role of CRISPR arrays in protecting the host from invasive genetic elements was discovered (Makarova et al., 2006). CRISPR-Cas is an adaptive defense mechanism utilizing antisense RNAs as memory signatures of previous invasions (Mojica et al., 2005; Pourcel et al., 2005). In 2007, it was discovered that *Streptococcus thermophilus* developed resistance to lytic phages through artificial modification of repeats within CRISPR (Barrangou et al., 2007). In 2008, it was demonstrated that mature CRISPR RNAs (crRNAs) act as guides when they form a complex with Cas proteins (Brouns et al., 2008). In 2012, research demonstrated the ability of single-guide RNA in the CRISPR-Cas system to target precise DNA segments, suggesting its potential for gene editing (Jinek et al., 2012). By 2013, the CRISPR-Cas system had been successfully used for targeted gene editing animal genomes, marking a significant breakthrough. Since then, CRISPR-Cas editing technology has gained extensive utilization across diverse domains of molecular biology (Cong et al., 2013).

Unlike the first and the second generations of gene editing systems, (CRISPR)-Cas gene editing technology allows for avoiding some defects such as limitations on multiplexed mutations, transfection inefficiency, design complexity, and other unexpected variations. The third-generation technology provides a straightforward method to accurately target, modify, regulate, and characterize genomic sites across a diverse range of cells and organisms. Its economical and easy-to-use approach resulted in its swift acceptance by laboratories worldwide.

The RNA-guided enzyme Cas9 has been transforming biology by offering a genome engineering tool grounded in the principles of Watson-Crick base pairing (Doudna & Charpentier, 2014). The CRISPR-Cas9 enzyme mechanism is characterized by utilizing guide RNA sequences, each consisting of just 20 artificially engineered nucleotides, to form complementary pairs with target DNA sequences, hence identifying specific target sites (Guo et al., 2022). CRISPR-Cas9 technology has sparked a revolution in laboratories globally and is currently being extensively applied in pioneering biological innovations. Its use in genetic manipulation now encompasses all branches of biotechnology, as well as innovative human therapeutic approaches.

7.2 THE ROLE OF MEDICINAL PLANTS IN HEALTHCARE

7.2.1 Discuss the Continued Importance of Medicinal Plants in Global Healthcare

For a long time, ancient medical knowledge was based on a cumulative body of expertise, encompassing beliefs, theories, and extensive personal experiences unique to various cultures. But presently, the advances in medical research over the last few decades have made it possible to understand the scientific rationale of traditional remedies. As a result, phytomedicines and their derivatives are frequently employed as natural remedies or to complement synthetic therapies (Rasool Hassan, 2012). In effect, contemporary medical practices are increasingly recommending the use of natural products for the healing of various illnesses all over the world (Eddouks et al., 2012). This has created enthusiasm for researching the traditional uses of medicinal plants in order to discover new drugs and incorporate them into modern pharmacies.

In addition to being highly valued as traditional remedies around the world, medicinal plants are an essential source of medicines used to address global health challenges (Howes et al., 2020). In fact, plants have remarkable chemical diversity and an impressive ability to produce highly complicated innovative compounds with 'drug-like' characteristics (Harvey et al., 2015). As well as natural products also exhibit higher success rates than synthetic compounds, with approximately 65% of drugs approved for cancer treatment derived from or inspired by natural products (Newman & Cragg, 2020).

Currently, the world witnessing a growth in the utilization of herbal remedies, and this is due to several factors. These include the high prevalence of chronic diseases, the search for treatments not available with modern therapies, and rising healthcare costs. In addition, medicines plants are readily available and affordable (Howes et al., 2020). All these factors illustrate the future importance of medicinal plants.

7.2.2 Emphasize the Need for Innovative Solutions to Enhance Productivity and Composition

Undoubtedly, modern medicine offers significant and essential benefits, and its use is widespread. Nevertheless, approximately 80% of the global population, particularly in developing countries, rely predominantly on plant remedies rather than modern medical practices (Abdin & Kamaluddin, 2006). Medicinal plants rich in bioactive compounds known for their therapeutic properties, are becoming increasingly important and popular in the pharmaceutical industry. Unfortunately, despite their acknowledged benefits and diverse uses, therapeutic herbs are positioned within the final ring of domestication. For years, the primary attention has been directed toward major crop species for food supply, leaving valuable medicinal plants neglected and overlooked (Siahsar et al., 2011). Knowing that the process of plant domestication is a constantly evolving process characterized by significant changes. Therefore, now more than ever, there is an increasing urgency to leverage modern biotechnology methods for enhancing medicinal plants.

The improvement of medicinal plants, both quantitatively and qualitatively, has become an important area of research. Various technologies have been employed to achieve this objective, including biotechnological tools, plants serving as bioreactors and genetic enhancement, which are considered the main developed technologies. Ongoing efforts seek methods to further increase pharmaceutical production and optimize yields of useful compounds, as well as to minimize costs and maintain the quality of phytochemicals (Khan et al., 2009).

7.3 CRISPR-CAS SYSTEMS IN MEDICINAL PLANT BIOTECHNOLOGY

7.3.1 Explain the Fundamentals of CRISPR-Cas Systems as Precise Genome Editing Tools

CRISPR (clustered regularly interspaced short palindromic repeats) Cas (CRISPR-associated protein) is the latest system in the field of genome editing. CRISPR-Cas technology consists of spacers (genome-targeting sequences) interspersed by the short palindromic repeat, and the Cas gene organized in operon (Bhaya et al., 2011; Tsui & Li, 2015). The CRISPR locus confers microbes such as bacteria and archaea an adaptive immunity that protects them against invasive genetic materials such as viral genomes and plasmids. The CRISPR-Cas technology incorporates small segments of foreign DNA into the CRISPR locus, which are subsequently transcribed into short RNAs. These RNAs serve as guides for targeting and dismantling foreign genetic materials. CRISPR-Cas systems differ according to the structure and sequence of Cas protein and can be applied to endogenous genes across various species and cell types. Unlike the system, CRISPR/Cas type II, the two complex systems I and III are not applied in genome editing. The system type II necessitates just one multifunctional Cas9 protein to disrupt invading genetic elements (Makarova et al., 2011). The CRISPR array is located adjacent to the CRISPR-associated system (Cas), comprising the tracrRNA (trans-activating crRNA) gene, Cas gene, and a leader sequence. The CRISPR-associated system encodes Cas proteins that contain functional domains: polymerases, endonucleases, nucleotide-binding proteins, and helicases.

CRISPR-Cas type II mediated immunity relies on functional mechanisms including spacer acquisition, CRISPR RNA (crRNA), biogenesis (expression and maturation), and interference with invading genetic elements. Concerning the acquisition phase (adaptive stage), the identification of protospacers in the viral genome is carried out through the particular identification of protospacer adjacent motifs (PAM). Subsequently, the Cas nuclease ensures the transformation of exogenous DNA into protospacers, which are small DNA fragments.

The isolated protospacers lacking the PAM site are then incorporated in a polarized manner into the host chromosome at the proximal end of the CRISPR array, separated by short palindromic repeat (Wiedenheft et al., 2012). The repeat sequences have a length ranging from 23 et 48 nucleotides while the spacer sequences have approximately 24 to 70 nucleotides in length (Hols et al., 2005; Pourcel et al., 2005).

The second step or biogenesis consists of the transcription of the CRISPR locus into the precursor CRISPR RNA (pre-crRNA). Subsequently, the RNase III cleaves the long pre-crRNA after the association of TracrRNA to the repeat sequences. As a result, mature crRNAs contain a partial repeat sequence, and only one spacer was obtained. The maturation process is performed by the Cas6 superfamily of endoribonucleases, and by the Cas5 protein in the absence of Cas6 (Haurwitz et al., 2010; Lintner et al., 2011; Nam et al., 2012).

Throughout the interference phase, the PAM sequence plays an important role in distinguishing between the foreign DNA and the CRISPR array. The CRISPR/Cas9 system encompasses Cas 9 protein and sgRNA (single-guide RNA), which consists of the pairing of trans-activating crRNA (tracrRNA) and mature crRNA. By standard Watson-Crick base pairing, the complementarity of crRNA with the invading DNA leads to the cleaving of the foreign nucleic acid. RuvC and HNH nuclease domains belonging to the Cas9 protein ensure the cleaving of the DNA and introduce double-strand breaks (DSB) in the gene of interest (Jinek et al., 2012). In which the cas9 RuvC cleaves the non-complementary strand, while the Cas9 HNH nuclease cleaves the complementary strand (Jinek et al., 2012).

In contrast to other techniques like ZFN and TALEN, which rely on artificial protein-DNA interactions, DNA site recognition in the CRISPR/Cas system relies on RNA-DNA interactions. In addition, the guide sequence within CRISPR spacers can readily be replaced with a sequence of interest to direct the Cas9 protein. Consequently, the Cas9 has the ability to be directed by crRNAs to cut specific DNA targets in vitro. In addition, Cas9 can alternatively be transformed into a nicking enzyme to assist in homology-directed repair with minimal mutagenic effects. Moreover, the introduction of two or more sgRNAs induces multiplex genome editing, enabling the simultaneous modification of multiple genes or the deletion of extensive genomic regions. (Cong et al., 2013). As a result, due to its numerous advantages, this system has been embraced in the domain of molecular biology and has emerged as a groundbreaking tool facilitating accurate and effective precise genome editing tool.

7.3.2 Describe the Impact of CRISPR Technology on Plant Biotechnology

Given the challenges posed by food shortages resulting from global population growth and climate change, leveraging the benefits of plant biotechnology has become crucial. This field focuses on enhancing the advantageous traits of crops to ensure food security. (Gan & Ling, 2022). Research on agronomic traits has revealed that many characteristics are influenced by modifications in gene bases (Li et al., 2017). Essentially, plant biotechnology encompasses a range of scientific methods and practices used to identify and modify plant genes to produce desired traits or products in plants (Kalia, 2018). The capacity to make precise alterations in plant genomes through genome editing techniques has transformed research in enhancing crops within today's molecular biology field (Azameti & Dauda, 2021). DNA-free editing strategies are considered most appropriate when aiming to bring edited plants to market. The absence of transgenes in the latest product facilitates licensing and market approval procedures, making DNA-free editing the preferred scenario (Kocsisova & Coneva, 2023). In effect, consumers widely believe that genetically modified organisms (GMOs) can potentially cause negative effects on both the environment and humans.

Advances in molecular biology have led to the development of a number of leading gene manipulation techniques, encompassing transcriptional activator-like nucleases (TALENs) and zinc finger nucleases (ZFNs). Among these, the CRISPR/Cas9 system has gained worldwide attention and found applications across numerous fields, including plant biotechnology (Zhu et al., 2020). CRISPR technology stands out as the predominant and the latest genome editing instrument, significantly expanding the horizons of plant biotechnology. It enables diverse applications such as transcriptional regulation, prime editing, gene knockout or knock-in, base editing, epigenetic modification and nucleic acid probing or detection (Pickar-Oliver & Gersbach, 2019).

CRISPR/Cas systems hold immense potential for enhancing the utilization of genome editing tools in the analysis of gene function and the enhancement of crops in both dicotyledonous and monocotyledonous plants (Sun et al., 2016). Notably, this system can effectively achieve gene modifications without transgenes. Since its initial introduction of CRISPR/Cas as a technique for editing plant

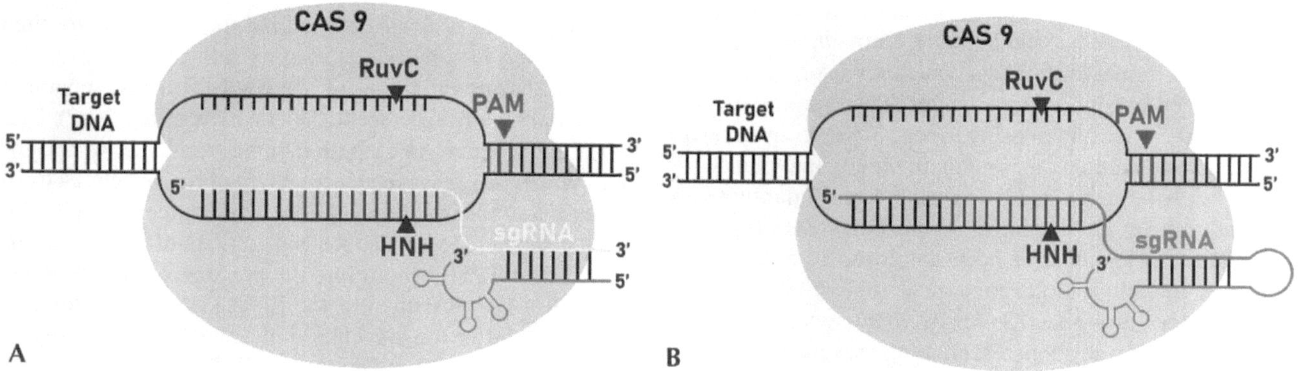

FIGURE 7.1 **(A)** CRISPR/Cas9 complex of *Streptococcus pyogenes* composed of tracrRNA and crRNA, The navy-blue triangles indicate the locations of cleavage within the target DNA. **(B)** An optimized system composed of a single-guide RNA (sgRNA) is employed for genome editing.

genomes, CRISPR/Cas has been successfully employed in nearly 120 crops and model plants (Cardi et al., 2023).

In contrast to traditional systems, the CRISPR-Cas technique allows for highly accurate and efficient targeted modifications in most crops. This system operates by altering only the sgRNA molecules, enabling editing across multiple genomic locations. As a result, it notably hastens the progress of crop enhancement compared to traditional breeding techniques (Chen et al., 2019). This efficient genome editing method has been utilized to enhance plant productivity and quality (greater nutritional value), bolster resistance against diseases and herbicides, expedite breeding across various plant species, and increase tolerance to abiotic stresses (Zhu et al., 2020). Thus, the straightforwardness, adaptability, and economical nature of this technology have sparked a significant agricultural revolution, making it the most widely used technology for crop improvement (Yu et al., 2017). In addition to the widely utilized CRISPR/SpCas9 system, other promising systems like CRISPR/Cas12a, which offer numerous advantages, have been identified, refined, and applied to plants. Moreover, in recent years, additional systems with significant potential for applications in plant biotechnology, such as the Fanzor (Fz) OMEGA-like programmable system, have been uncovered (Abdul Aziz et al., 2022; Bernabé-Orts et al., 2019).

7.4 CUSTOMIZING METABOLIC PATHWAYS IN MEDICINAL PLANTS

7.4.1 Explore the Concept of Tailoring Metabolic Pathways in Medicinal Plants Using CRISPR and the Advantages of Enhancing the Production of Secondary Metabolites

Throughout their life cycle, plants possess the ability to react to both internal and external stimuli due to the intricate metabolic pathways they have developed. In contrast to primary metabolites, essential for plant growth and development, secondary metabolites are synthesized specifically to interact with their surrounding environment (Pagare et al., 2015). These compounds allow plant adaptation and withstand various biotic (fungi, insects, bacteria, and nematodes) and abiotic (temperature fluctuations, water scarcity, light intensity, salinity, shading, and heavy metals) stress conditions (Akula & Ravishankar, 2011). Secondary metabolites, also referred to as phyto-metabolites, encompass a diverse array of biological properties applicable across pharmaceuticals, industry, and agriculture (Bennett & Wallsgrove, 1994). These bioactive compounds find utility as medications, dyes, insecticides, and toxins, among other applications. Due to the importance of benefits in research and therapeutics of bioactive compounds, medicinal plants have gained immense industrial importance (Niazian et al., 2023).

The rising need for bioactive compounds necessitates cost-effective synthesis methods to produce high concentrations, ensuring economic viability. Thus, high-biomass, short-lived, low-maintenance plants are perfect bioreactors for the synthesis of secondary metabolites (Bhojiya & Joshi, 2024). In natural habitats, medicinal plants and their derivatives are often not available in large quantities. In effect, plants often produce limited quantities of secondary metabolites and grow only in specific regions. In addition, the cultivation of certain therapeutic plants can be challenging due to environmental conditions or agricultural constraints (Halder & Roy, 2023). These challenges have prompted researchers to explore and employ innovative methods, including tissue culture techniques and transgenic approaches, to enhance the production of desired phytochemicals and meet global demand. The main objective of metabolic engineering of therapeutic herbs is always to improve the quality of the bioactive products to be extracted. The both physical and chemical indices are commonly used for evaluating medicinal plants.

Various biotechnological methods have been employed to improve the synthesis of bioactive compounds in plants, such as tissue culture, which is commonly used for plant multiplication of valuable medicinal plants (Bharti et al., 2023). Genetic engineering requires reliability, flexibility, simplicity, precision, highly effective genetic manipulation techniques and multiplexing capabilities (Shelake et al., 2023). All these valuable requirements were obtained through the revolution that occurred in the domain of genetic engineering in 2013. Owing to its straightforwardness and efficiency, the CRISPR system quickly became a focal point in molecular biology and has been effectively employed for accurate gene editing in various model plants.

In effect, the CRISPR/Cas system has found utility in plant breeding, strain enhancement, and genetic engineering, particularly in the modulation of metabolic pathways crucial for the biochemical production of commercially important bioactive compounds (Bhagwat et al., 2022). In the domain of medicinal plants, the latest gene-editing technology has predominantly concentrated on select model plants possessing comprehensive genome data and efficient genetic transformation systems. This technology has been effectively utilized for genetic modification in various medicinal plants, including *Salvia miltiorrhiza*, *Dendrobium officinale*, *Cannabis sativa*, *Monochasma savatieri*, *Cichorium intybus* L., *Papaver somniferum* L., *Atropa belladonna*, *Rehmannia glutinosa*, *Dioscorea zingiberensis*, *Monochasma savatieri*, and *Symphytum officinale* (Arun et al., 2023; Guo et al., 2022).

7.5 FUNCTIONAL GENOMICS AND TRAIT ENHANCEMENT

7.5.1 Discuss the Role of CRISPR-Cas Systems in Unraveling Complex Regulatory Networks

One of the main goals of evolutionary biology research is to properly understand how genes are expressed in different cell types (Akinci et al., 2021). In effect, the more gene regulation is understood, the more cellular identity and function can be understood and controlled, which can greatly aid in predicting disease-causing non-coding sequence changes (Zhou et al., 2020). Complex gene regulatory networks consist of genes, proteins, noncoding RNA molecules, signaling components, and metabolites. These networks can be observed within a single cell, among multiple cells, or even across different cell types (Long et al., 2008). Non-coding segments exert influence on gene expression, and alterations within these regions significantly contribute to both disease progression and evolutionary processes (Banerjee & Sherwood, 2017). Therefore, enhancing our comprehension of the regulatory genome is crucial.

The CRISPR-Cas system stands as the latest efficient tool for genome editing. CRISPR has the ability to unravel regulatory networks through highly precise perturbation experiments. CRISPR sgRNAs are employed to target regulatory regions, and the resulting changes in gene expression can then be measured to assess whether the putative regulatory regions actually cause changes in gene expression (Canver et al., 2015; Diao et al., 2016). Consequently, CRISPR-Cas9 emerges as a strong tool for elucidating and deciphering the underlying mechanisms of complex regulatory networks. Currently, CRISPR-based approaches have become indispensable in unraveling the intricacies of cellular regulation. Additionally, CRISPR technology has opened up unprecedented opportunities to study and discover complex regulatory networks governing gene expression.

7.5.2 Explain How Insights from Functional Genomics Expedite Breeding Programs and Improve Medicinal Properties

The importance of therapeutic herbs in different domains is essentially related to the bioactive substances they contain. Therefore, improving medicinal plants means improving the production of valuable secondary metabolites and their quality. For achieving this purpose, functional genomics is regarded as a crucial measure for enhancing medicinal plants and organizing effective breeding programs. As we discussed previously, CRISPR technology allows modifications to be made in the genome and also unravels complex regulatory networks, which allows the identification of underlying genes that control desired characteristics including resistance to salinity, drought, disease, and pests. Consequently, exploiting functional genomics holds significant potential for advancing the cultivation of plant varieties with optimized medicinal properties rich in valuable bioactive compounds. Consequently, ensuring high yield and quality contributes to sustainable agriculture and the development of new medicinal treatments.

7.6 THE FUTURE OF CRISPR-BASED RESEARCH IN MEDICINAL PLANTS

7.6.1 Analyze the Future Prospects of CRISPR Technology in Medicinal Plant Research

Due to its numerous characteristics, such as straightforward manipulation, high effectiveness, and broad applicability, the CRISPR system has ushered in a new era of genome exploration and engineering. This innovation has led to a growing number of medicinal plants undergoing precise gene editing. Despite ongoing efforts by the scientific community to optimize the capabilities of the CRISPR technology, its application to medicinal plants is still in its early stages (de Lange et al., 2022). In effect, despite the extensive utilization of this system in genetic engineering, transcriptional regulation, DNA tagging, and epigenome editing, the CRISPR-Cas9 system application in medicinal plants remains limited. The adoption of powerful innovative technology such as CRISPR-Cas9 could advance research in pharmaceutical botany by facilitating its application in the manipulation of biosynthetic pathways for valuable bioactive compounds. Relying on this technology, the production of bioactive compounds from medicinal plants could be optimized in a highly effective and efficient way. CRISPR is anticipated to advance scientific exploration and practical utilization of medicinal plants across fields like reverse genetics, genetic enhancement, synthetic biology, and germplasm innovation (Liu et al., 2017).

With the increasing fascination with medicinal plants and their various advantageous by-products (nutritional, therapeutic, and biomedical applications) coupled with the growing demand, it is anticipated that CRISPR technology will continue to play a pivotal function in revolutionizing biotechnology in medicinal plant research in the foreseeable future (Guo et al., 2022). Furthermore, this period is defined by significant demographic growth, highlighting the essential role that this groundbreaking technology will play in increasing agricultural productivity and expediting the creation of novel crop varieties of medicinal plants.

7.6.2 Address Ethical Considerations, Regulatory Frameworks, and Public Engagement

Since CRISPR gene editing was validated in 2013, it has undergone swift advancement and extensive utilization across various domains. It is a reliable and powerful tool that allows gene editing in any organism, CRISPR is currently employed to modify microorganisms, plants, insects, and human cells (Ledford, 2015). Given its significance and efficacy, researchers have devoted considerable attention to advancing and applying this technology across diverse fields. Nonetheless, insufficient emphasis has been placed on allocating the necessary time for experiments, which ideally span several years. This oversight has resulted in numerous ethical concerns linked to gene editing through CRISPR technology (Mulvihill et al., 2017). Apart from the discourse surrounding human germline editing, CRISPR raises a plethora of ethical issues, extending beyond human implications to include other species and environmental concerns (Zetsche et al., 2015).

Regulating and assessing the utilization of CRISPR poses a major obstacle, in which the challenge of navigating the enduring impacts of these cutting-edge technologies transcends mere technical proficiency. Therefore, the search for appropriate regulatory solutions has become an urgent need worldwide to avoid potentially significant consequences (Scheufele et al., 2021; Svingen, 2023). To accomplish this goal, it is crucial to underscore the significance of responsible and socially accountable governance concerning matters pertaining to genetic modification technology. Additionally, there needs to be political and scientific attention directed toward the institutions tasked with regulating the utilization of CRISPR across different (Caplan et al., 2015). To ensure the prudent and responsible utilization of this technology, rigorous guidelines and principles must be implemented across different domains. Establishing effective communication channels between scientists and policymakers is imperative to thoroughly examine the future implications of this technology across various levels (Mulvihill et al., 2017). In addition to engaging the public which increases the likelihood that research and innovation will be aligned with society's values, needs, and expectations, creating a more socially powerful science.

Due to its affordability, simplicity, and minimal equipment and expertise demands, CRISPR has gained global popularity as a technology. Consequently, there's a growing need for international standards concerning the testing, release, and liability of genetically edited organisms. It's imperative that modified organisms undergo thorough safety and efficacy assessments before approval (Esvelt et al., 2014). One of the advantages of CRISPR in agricultural genetic engineering is that this technique does not require the insertion of exogenous DNA into the plant genome using a vector system (virus, bacterial plasmid, etc.). Consequently, genetically modified plants using CRISPR are not considered to be transgenic organisms in the strict sense. This advantage has been exploited by the American Inspection Service for Animal and Plant Health (APHIS), which considers that the resulting products of deletion of only one gene will not be regulated. However, deregulation could lead to the widespread application of these amendments without enforceable oversight, which could harm ecosystems, biodiversity, and human health (Caplan et al., 2015).

In contrast to the United States, the European Commission upholds a notably rigorous regulatory framework for genetically modified crops used in agriculture. The European Union mandates regulation and risk assessment for all genetically modified plants, including those modified through direct genome editing using CRISPR. Nonetheless, plants edited via CRISPR or TALEN methods, devoid of foreign DNA, should not undergo the same regulatory scrutiny and risk assessment as transgenic varieties (Hirsch et al., 2019).

7.7 CONCLUSION

The revolutionary genome-editing tool CRISPR has enormous potential and has enabled the real development of biotechnology for medicinal plants. Compared to classical methods, this modern technology has enabled the development of crops with improved and desirable medicinal properties based on targeted genomic modifications. These properties are characterized by a number of desirable traits, encompassing drought resistance, improved resistance to diseases and herbicides, and resistance to various environmental stresses, which strongly support increased productivity and improved quality. In addition, CRISPR technology significantly speeds up the process of crop improvement compared to traditional breeding techniques. CRISPR has also been instrumental in boosting the production of bioactive compounds with important therapeutic properties used in the production of pharmaceuticals. Moreover, CRISPR has enabled the elimination of chemical inputs, both natural and synthetic, opening up new opportunities to improve human health and agricultural sustainability. In addition, CRISPR technology offers economic efficiency and ease of use, as well as the absence of GMO genes in the final product, streamlining licensing and regulatory processes.

REFERENCES

Abdin, M. Z., & Kamaluddin, A. (2006). Improving quality of medicinal herbs through physico-chemical and molecular approaches. In M. Z. Abdin & Y. P. Abrol (Eds.), *Traditional systems of medicine* (pp. 30–39). Narosa Publishing House Pvt. Ltd.

Abdul Aziz, M., Brini, F., Rouached, H., & Masmoudi, K. (2022). Genetically engineered crops for sustainably enhanced food production systems. *Frontiers in Plant Science*, *13*, 1027828.

Akinci, E., Hamilton, M. C., Khowpinitchai, B., & Sherwood, R. I. (2021). Using crispr to understand and manipulate gene regulation. *Development*, *148*(9), dev182667.

Akula, R., & Ravishankar, G. A. (2011). Influence of abiotic stress signals on secondary metabolites in plants. *Plant Signaling & Behavior*, *6*(11), 1720–1731.

Arun, M., Halka, J., & Kowsalya, K. (2023). CRISPR/Cas9: A novel genetic tool to manipulate plant secondary metabolite pathways. In *Genetic manipulation of secondary metabolites in medicinal plant* (pp. 45–57). Springer.

Azameti, M. K., & Dauda, W. P. (2021). Base editing in plants: Applications, challenges, and future prospects. *Frontiers in Plant Science*, *12*, 664997.

Banerjee, B., & Sherwood, R. I. (2017). A crispr view of gene regulation. *Current Opinion in Systems Biology*, *1*, 1–8.

Barrangou, R., Fremaux, C., Deveau, H., Richards, M., Boyaval, P., Moineau, S., Romero, D. A., & Horvath, P. (2007). Crispr provides acquired resistance against viruses in prokaryotes. *Science*, *315*(5819), 1709–1712.

Bennett, R. N., & Wallsgrove, R. M. (1994). Secondary metabolites in plant defence mechanisms. *New Phytologist*, *127*(4), 617–633.

Bernabé-Orts, J. M., Casas-Rodrigo, I., Minguet, E. G., Landolfi, V., Garcia-Carpintero, V., Gianoglio, S., Vázquez-Vilar, M., Granell, A., & Orzaez, D. (2019). Assessment of Cas12a-mediated gene editing efficiency in plants. *Plant Biotechnology Journal*, *17*(10), 1971–1984.

Bhagwat, A. C, Patil, A. M., & Saroj, S. D. (2022). CRISPR/Cas 9-based editing in the production of bioactive molecules. *Molecular Biotechnology*, 1–7.

Bharti, P. K., Singh, S. K., & Kumari, S. (2023). Genetic manipulation in medicinal plants for enhancement of secondary metabolites. In *Genetic manipulation of secondary metabolites in medicinal plant* (pp. 129–145). Springer.

Bhaya, D., Davison, M., & Barrangou, R. (2011). CRISPR-Cas systems in bacteria and archaea: versatile small RNAs for adaptive defense and regulation. *Annual Review of Genetics*, *45*, 273–297.

Bhojiya, A. A., & Joshi, H. (2024). Crispr gene editing for secondary metabolite production: A review. In *Gene editing in plants CRISPR-Cas and its applications* (pp. 437–475). Springer Nature.

Brouns, S. J. J., Jore, M. M., Lundgren, M., Westra, E. R., Slijkhuis, R. J. H., Snijders, A. P. L., Dickman, M. J., Makarova, K. S., Koonin, E. V., & Van Der Oost, J. (2008). Small crispr RNAs guide antiviral defense in prokaryotes. *Science*, *321*(5891), 960–964.

Canver, M. C., Smith, E. C., Sher, F., Pinello, L., Sanjana, N. E., Shalem, O., Chen, D. D., Schupp, P. G., Vinjamur, D. S., & Garcia, S. P. (2015). BCL11A enhancer dissection by Cas9-mediated in situ saturating mutagenesis. *Nature*, *527*(7577), 192–197.

Caplan, A. L., Parent, B., Shen, M., & Plunkett, C. (2015). No time to waste – the ethical challenges created by crispr: CRISPR/Cas, being an efficient, simple, and cheap technology to edit the genome of any organism, raises many ethical and regulatory issues beyond the use to manipulate human germ line cells. *EMBO Reports*, *16*(11), 1421–1426.

Cardi, T., Murovec, J., Bakhsh, A., Boniecka, J., Bruegmann, T., Bull, S. E., Eeckhaut, T., Fladung, M., Galovic, V., & Linkiewicz, A. (2023). CRISPR/Cas-mediated plant genome editing: Outstanding challenges a decade after implementation. *Trends in Plant Science*, *28*, 1144–1165.

Chen, K., Wang, Y., Zhang, R., Zhang, H., & Gao, C. (2019). CRISPR/Cas genome editing and precision plant breeding in agriculture. *Annual Review of Plant Biology*, *70*, 667–697.

Cong, L., Ran, F. A., Cox, D., Lin, S., Barretto, R., Habib, N., Hsu, P. D., Wu, X., Jiang, W., & Marraffini, L. A. (2013). Multiplex genome engineering using CRISPR/cas systems. *Science*, *339*(6121), 819–823.

Daswani, G. P., Brijesh, S., & Birdi, J. T. (2006). Preclinical testing of medicinal plants: Advantages and approaches. In *Workshop proceedings on approaches towards evaluation of medicinal plants prior to clinical trial.* Poonam G. Daswani, Brijesh S. and Tannaz J. Birdi (pp. 60–77).

de Lange, J., Nalley, L. L., Yang, W., Shew, A., & De Steur, H. (2022). The future of crispr gene editing according to plant scientists. *iScience*, *25*(9).

Di Pierro, F., Rapacioli, G., Ferrara, T., & Togni, S. (2012). Use of a standardized extract from *Echinacea angustifolia* (polinacea [R]) for the prevention of respiratory tract infections. *Alternative Medicine Review*, *17*(1), 36–42.

Diao, Y., Li, B., Meng, Z., Jung, I., Lee, A. Y., Dixon, J., Maliskova, L., Guan, K.-L., Shen, Y., & Ren, B. (2016). A new class of temporarily phenotypic enhancers identified by CRISPR/Cas9-mediated genetic screening. *Genome Research*, *26*(3), 397–405.

Doudna, J. A., & Charpentier, E. (2014). The new frontier of genome engineering with CRISPR-Cas9. *Science*, *346*(6213), 1258096.

Eddouks, M., Chattopadhyay, D., De Feo, V., & Cho, W. C. (2012). *Medicinal plants in the prevention and treatment of chronic diseases* (Vol. 2012). Hindawi.

Esvelt, K. M., Smidler, A. L., Catteruccia, F., & Church, G. M. (2014). Concerning RNA-guided gene drives for the alteration of wild populations. *Elife*, *3*, e03401.

Gan, R.-Y., Xu, X.-R., Song, F.-L., Kuang, L., & Li, H.-B. (2010). Antioxidant activity and total phenolic content of medicinal plants associated with prevention and treatment of cardiovascular and cerebrovascular diseases. *Journal of Medicinal Plants Research*, *4*(22), 2438–2444.

Gan, W. C., & Ling, A. P. K. (2022). CRISPR/Cas9 in plant biotechnology: Applications and challenges. *BioTechnologia*, *103*(1), 81–93.

Guo, M., Chen, H., Dong, S., Zhang, Z., & Luo, H. (2022). CRISPR-Cas gene editing technology and its application prospect in medicinal plants. *Chinese Medicine*, *17*(1), 33.

Halder, M., & Roy, S. (2023). Current status of metabolic engineering of medicinal plants for production of plant-derived secondary metabolites. In *Medicinal plants: Biodiversity, biotechnology and conservation* (pp. 819–69). Springer.

Harborne, A. J. (1998). *Phytochemical methods a guide to modern techniques of plant analysis*. Springer Science & Business Media.

Harvey, A. L., Edrada-Ebel, R., & Quinn, R. J. (2015). The re-emergence of natural products for drug discovery in the genomics era. *Nature Reviews Drug Discovery*, *14*(2), 111–129.

Haurwitz, R. E., Jinek, M., Wiedenheft, B., Zhou, K., & Doudna, J. A. (2010). Sequence-and structure-specific RNA processing by a CRISPR endonuclease. *Science*, *329*(5997), 1355–1358.

Hirsch, F., Iphofen, R., & Koporc, Z. (2019). Ethics assessment in research proposals adopting CRISPR technology. *Biochemia Medica*, *29*(2), 206–213.

Hols, P., Hancy, F., Fontaine, L., Grossiord, B., Prozzi, D., Leblond-Bourget, N., Decaris, B., Bolotin, A., Delorme, C., & Ehrlich, S. D. (2005). New insights in the molecular biology and physiology of streptococcus thermophilus revealed by comparative genomics. *FEMS Microbiology Reviews*, *29*(3), 435–463.

Hosseinzadeh, S., Jafarikukhdan, A., Hosseini, A., & Armand, R. (2015). The application of medicinal plants in traditional and modern medicine: A review of *Thymus vulgaris*. *International Journal of Clinical Medicine*, *6*(9), 635–642.

Howes, M.-J. R., Quave, C. L., Collemare, J., Tatsis, E. C., Twilley, D., Lulekal, E., Farlow, A., Li, L., Cazar, M.-E., & Leaman, D. J. (2020). Molecules from nature: Reconciling biodiversity conservation and global healthcare imperatives for sustainable use of medicinal plants and fungi. *Plants, People, Planet*, *2*(5), 463–481.

Ishino, Y., Shinagawa, H., Makino, K., Amemura, M., & Nakata, A. (1987). Nucleotide sequence of the iap gene, responsible for alkaline phosphatase isozyme conversion in *Escherichia coli*, and identification of the gene product. *Journal of Bacteriology*, *169*(12), 5429–5433.

Jinek, M., Chylinski, K., Fonfara, I., Hauer, M., Doudna, J. A., & Charpentier, E. (2012). A programmable dual-RNA–guided DNA endonuclease in adaptive bacterial immunity. *Science*, *337*(6096), 816–821.

Kalia, A. (2018). Nanotechnology in bioengineering: Transmogrifying plant biotechnology. In *Omics technologies and bioengineering* (pp. 211–29). Elsevier.

Khalil, A. M. (2020). The genome editing revolution. *Journal of Genetic Engineering and Biotechnology*, *18*(1), 68.

Khan, M. Y., Aliabbas, S., Kumar, V., & Rajkumar, S. (2009). Recent advances in medicinal plant biotechnology. *Indian Journal of Biotechnology*, *8*, 9–22.

Kocsisova, Z., & Coneva, V. (2023). Strategies for delivery of CRISPR/Cas-mediated genome editing to obtain edited plants directly without transgene integration. *Frontiers in Genome Editing*, *5*.

Ledford, H. (2015). CRISPR, the disruptor. *Nature*, *522*(7544), 20–25.

Li, J., Sun, Y., Du, J., Zhao, Y., & Xia, L. (2017). Generation of targeted point mutations in rice by a modified CRISPR/Cas9 system. *Molecular Plant*, *10*(3), 526–529.

Li, L. (2000). Opportunity and challenge of traditional Chinese medicine in face of the entrance to WTO (World Trade Organization). *Chinese Journal of Information on Traditional Chinese Medicine*, *7*, 7–8.

Lintner, N. G., Kerou, M., Brumfield, S. K., Graham, S., Liu, H., Naismith, J. H., Sdano, M., Peng, N., She, Q., & Copié, V. (2011). Structural and functional characterization of an archaeal clustered regularly interspaced short palindromic repeat (CRISPR)-associated complex for antiviral defense (cascade). *Journal of Biological Chemistry*, *286*(24), 21643–21656.

Liu, X., Wu, S., Xu, J., Sui, C., & Wei, J. (2017). Application of CRISPR/Cas9 in plant biology. *Acta Pharmaceutica Sinica B*, *7*(3), 292–302.

Long, T. A., Brady, S. M., & Benfey, P. N. (2008). Systems approaches to identifying gene regulatory networks in plants. *Annual Review of Cell and Developmental Biology*, *24*, 81–103.

Mahady, G. B. (2005). Medicinal plants for the prevention and treatment of bacterial infections. *Current Pharmaceutical Design*, *11*(19), 2405–2427.

Makarova, K. S., Grishin, N. V., Shabalina, S. A., Wolf, Y. I., & Koonin, E. V. (2006). A putative RNA-interference-based immune system in prokaryotes: Computational analysis of the predicted enzymatic machinery, functional analogies with eukaryotic RNAi, and hypothetical mechanisms of action. *Biology Direct*, *1*, 1–26.

Makarova, K. S., Haft, D. H., Barrangou, R., Brouns, S. J. J., Charpentier, E., Horvath, P., Moineau, S., Mojica, F. J. M., Wolf, Y. I., & Yakunin, A. F. (2011). Evolution and classification of the CRISPR–Cas systems. *Nature Reviews Microbiology*, *9*(6), 467–477.

Mojica, F. J. M., Díez-Villaseñor, C., García-Martínez, J., & Soria, E. (2005). Intervening sequences of regularly spaced prokaryotic repeats derive from foreign genetic elements. *Journal of Molecular Evolution*, *60*, 174–182.

Mulvihill, J. J., Capps, B., Joly, Y., Lysaght, T., Zwart, H. A. E., Chadwick, R., & Law International Human Genome Organisation Committee of Ethics and Society. (2017). Ethical issues of CRISPR technology and gene editing through the lens of solidarity. *British Medical Bulletin*, *122*(1), 17–29.

Nam, K. H., Haitjema, C., Liu, X., Ding, F., Wang, H., DeLisa, M. P., & Ke, A. (2012). Cas5d protein processes pre-crRNA and assembles into a cascade-like interference complex in subtype Ic/Dvulg CRISPR-cas System. *Structure*, *20*(9), 1574–1584.

Nestler, G. (2002). Traditional Chinese medicine. *Medical Clinics*, *86*(1), 63–73.

Newman, D. J., & Cragg, G. M. (2020). Plant endophytes and epiphytes: Burgeoning sources of known and "unknown" cytotoxic and antibiotic agents? *Planta Medica*, *86*(13/14), 891–905.

Newman, D. J., Cragg, G. M., & Snader, K. M. (2000). The influence of natural products upon drug discovery. *Natural Product Reports*, *17*(3), 215–234.

Niazian, M., Torkamaneh, D., & Hesami, M. (2023). Advances in biotechnology-based breeding of medicinal plants. *Frontiers in Plant Science*, *14*, 1228951.

Pagare, S., Bhatia, M., Tripathi, N., Pagare, S., & Bansal, Y. K. (2015). Secondary metabolites of plants and their role: Overview. *Current Trends in Biotechnology and Pharmacy*, *9*(3), 293–304.

Pickar-Oliver, A., & Gersbach, C. A. (2019). The next generation of CRISPR–Cas technologies and applications. *Nature Reviews Molecular Cell Biology*, *20*(8), 490–507.

Pourcel, C., Salvignol, G., & Vergnaud, G. (2005). CRISPR elements in *Yersinia pestis* acquire new repeats by preferential uptake of bacteriophage DNA, and provide additional tools for evolutionary studies. *Microbiology*, *151*(3), 653–663.

Ramakrishna, Y., Goda, H., Baliga, M. S., & Munshi, A. K. (2011). Decreasing cariogenic bacteria with a natural, alternative prevention therapy utilizing phytochemistry (plant extracts). *Journal of Clinical Pediatric Dentistry*, *36*(1), 55–64.

Ramawat, K. G., & Mérillon, J.-M. (2008). *Bioactive molecules and medicinal plants*. Springer.

Rasool Hassan, B. A. (2012). Medicinal plants (importance and uses). *Pharmaceutica Analytica Acta*, *3*(10), 2153–2435.

Rasul, M. G. (2018). Conventional extraction methods use in medicinal plants, their advantages and disadvantages. *International Journal of Basic Sciences and Applied Computing*, *2*, 10–14.

Salmerón-Manzano, E., Garrido-Cardenas, J. A., & Manzano-Agugliaro, F. (2020). Worldwide research trends on medicinal plants. *International Journal of Environmental Research and Public Health*, *17*(10), 3376.

Scheufele, D. A., Krause, N. M., Freiling, I., & Brossard, D. (2021). What we know about effective public engagement on CRISPR and beyond. *Proceedings of the National Academy of Sciences*, *118*(22), e2004835117.

Shelake, R. M., Jadhav, A. M., Bhosale, P. B., & Kim, J.-Y. (2023). Unlocking secrets of nature's chemists: Potential of CRISPR/Cas-based tools in plant metabolic engineering for customized nutraceutical and medicinal profiles. *Plant Physiology and Biochemistry*, 108070.

Siahsar, B., Rahimi, M., Tavassoli, A., & Raissi, A. (2011). Application of biotechnology in production of medicinal plants. *American-Eurasian Journal of Agricultural & Environmental Sciences*, *11*(3), 439–444.

Sneader, W. (2005). *Drug discovery: A history*. John Wiley & Sons.

Sofowora, A., Ogunbodede, E., & Onayade, A. (2013). The role and place of medicinal plants in the strategies for disease prevention. *African Journal of Traditional, Complementary and Alternative Medicines*, *10*(5), 210–229.

Sun, Y., Zhang, X., Wu, C., He, Y., Ma, Y., Hou, H., Guo, X., Du, W., Zhao, Y., & Xia, L. (2016). Engineering herbicide-resistant rice plants through CRISPR/Cas9-mediated homologous recombination of *Acetolactate synthase*. *Molecular Plant*, *9*(4), 628–631.

Svingen, M. (2023). Paving the way for a softer regulation of CRISPR in Norway: Public engagement as window dressing. *Journal of Responsible Innovation*, *10*(1), 2146863.

Tsui, T. K. M., & Li, H. (2015). Structure principles of CRISPR-Cas surveillance and effector complexes. *Annual Review of Biophysics*, *44*, 229–255.

Van Wyk, B.-E., & Wink, M. (2018). *Medicinal plants of the world*. Cabi.

Wiedenheft, B., Sternberg, S. H., & Doudna, J. A. (2012). RNA-guided genetic silencing systems in bacteria and archaea. *Nature*, *482*(7385), 331–338.

Yu, Q.-H., Wang, B., Li, N., Tang, Y., Yang, S., Yang, T., Xu, J., Guo, C., Yan, P., & Wang, Q. (2017). CRISPR/Cas9-induced targeted mutagenesis and gene replacement to generate long-shelf life tomato lines. *Scientific Reports*, *7*(1), 11874.

Zetsche, B., Gootenberg, J. S., Abudayyeh, O. O., Slaymaker, I. M., Makarova, K. S., Essletzbichler, P., Volz, S. E., Joung, J., Van Der Oost, J., & Regev, A. (2015). Cpf1 is a single RNA-guided endonuclease of a class 2 Crispr-Cas system. *Cell*, *163*(3), 759–771.

Zhou, H., Su, J., Hu, X., Zhou, C., Li, H., Chen, Z., Xiao, Q., Wang, B. O., Wu, W., & Sun, Y. (2020). Glia-to-neuron conversion by CRISPR-Casrx alleviates symptoms of neurological disease in mice. *Cell*, *181*(3), 590–603. e16.

Zhu, H., Li, C., & Gao, C. (2020). Applications of CRISPR–Cas in agriculture and plant biotechnology. *Nature Reviews Molecular Cell Biology*, *21*(11), 661–677.

8 Enhanced Secondary Metabolite Production by Genome Editing in Medicinal Plants

Rajaswini Rupali, Nandakumar Vidya, Arumugam Vijaya Anand, Muthukrishnan Anand, Karuppiah Ramalashmi, Ahamed Kabeer Umera Begam, Asirvatham Alwin Robert, and Muthukrishnan Arun

LIST OF ABBREVIATIONS

ADS	amorpha-4,11-diene synthase
BIAs	benzylisoquinoline alkaloids
C3H	coumarate 3-hydroxylase
C4H	cinnamate 4-hydroxylase
DSBs	double-strand Breaks
HDR	homology-directed repair
HSS	homospermidine synthase
HRs	hairy roots
NHEJ	non-homologous end joining
PA	pyrrolizidine alkaloid
PAL1	phenylalanine ammonia-lyase 1
PAM	protospacer-adjacent motif
RVD	repeat-variable di-residue
SgRNA	single guide RNA
STL	sesquiterpene lactones
TALENs	transcription activator-like effector nucleases
VLCFAs	very-long-chain fatty acids
ZFNs	zinc finger nucleases

8.1 INTRODUCTION

Plants contain bioactive chemicals, known as secondary metabolites, which are essential for traditional medicine and have therapeutic benefits (Manoharan et al., 2008; Begum et al., 2009; Muthumani et al., 2013; Varalakshmi et al., 2014; Vijaya Anand et al., 2016, 2020; Bharathi et al., 2018; Divya et al., 2019; Akilandeswari et al., 2019; Sangeetha et al., 2022; Kaviya et al., 2022; Manikandan et al., 2022; Karthika et al., 2023; Vijayakumar et al., 2023). These compounds encompass various chemical groups including alkaloids, terpenoids, flavonoids, saponins, phenolics, and glycosides and are essential for their unique biosynthetic pathways and bioactivities (Pagare et al., 2015). They have diverse applications in medicine and phytotherapy. The secondary metabolites found in therapeutic plants are produced through diverse pathways, such as the shikimic acid pathway for phenolic, the mevalonic acid pathway for terpenoids, and the malonic acid pathway for phenolic compounds (Verpoorte et al., 2002). The biosynthesis of these secondary metabolites is controlled by transcription factors and epigenetic modifications that include DNA methylation, histone modification, and small RNA-based mechanisms (Ferrari et al., 2019). The creation of secondary metabolites in therapeutic plants is crucial for their medicinal and commercial value. They have a range of beneficial effects, such as anti-inflammatory, anticancer, antimicrobial, antioxidant, and antidiabetic activities (Jain et al., 2019). Therefore, enhancing their creation in plants is a crucial area of research. Modern genome editing and metabolic engineering techniques offer valuable tools to optimize biosynthetic pathways.

Understanding the complex functions of secondary metabolites is crucial to grasp their importance for plants, animals, and humans in various ecosystems and industries. For plants, secondary metabolites serve as vital protectors, strengthening them against biotic and abiotic stresses aiding their survival and adaptation (Chen et al., 2022). These substances enhance plant resilience and promote interactions with the soil, neighboring plants, and animals, ultimately contributing to agroecological sustainability and resiliency. For humans, secondary metabolites obtained from plants are essential in traditional herbal medicines and modern pharmaceuticals due to their bioactive properties, including anti-inflammatory, anticancer, and antioxidant activities (Velu et al., 2018). Their application in medicinal uses benefits human health and highlights their potential as lead compounds for developing potent drugs, with over 100,000 described structures providing diverse pharmacological activities and therapeutic agents. For animals, secondary metabolites have varying impacts, from providing nutritional sources to influencing behaviors and serving as toxins or repellents against predators (Iason et al., 2006). These substances significantly impact animal health and creation, thus affecting the chemical composition of meat and dairy products consumed by humans. This underscores the interconnectedness of plant-animal interactions and emphasizes the intricate roles of secondary metabolites in shaping ecosystems and agricultural practices.

The versatility of secondary metabolites in pharmaceuticals, food processing, cosmetics, and agriculture makes

DOI: 10.1201/9781003475491-8

them highly required. They hold great importance due to their therapeutic value and ability to act as raw materials in pharmaceutical creation. These bioactive compounds are invaluable resources across different sectors, as they are crucial for synthesizing medicines, flavors, fragrances, and dyes (Pai et al., 2022). The demand for secondary metabolites is driven by the need to create new and innovative compounds that are less harmful and more efficient. This is part of the constant search for safer pharmaceuticals that can be used to treat various medical conditions. The global scientific community is increasingly focused on advancing plant biotechnology and bioprospecting research, leading to more efficient creation methods. This innovation is expected to drive further growth in demand for secondary metabolites in the predictable future. In essence, the pivotal role of secondary metabolites in various industries and research endeavor's is underscored by their diverse applications, therapeutic importance, and the continuous pursuit of safer and more efficient creation methods (Mohiuddin et al., 2019).

It is crucial to understand the difficulties in generating secondary metabolites from therapeutic plants in order to generate effective and tolerable methods for creation. These challenges arise due to low planta yield, cultivation obstacles, and seasonal variation, which eventually limit the availability of secondary metabolites (Cardoso et al., 2019). The biodiversity threats are exacerbated by the overharvesting of underground plant tissues, which pushes species toward endangerment. Drug discovery efforts are hampered due to restricted access to natural products from endangered species, while direct isolation methods contribute to overharvesting and loss of biodiversity among medicinal plant species (Moyo et al., 2015). Comprehending and optimizing creation processes are challenging due to the intricate involvement of transcription factors and epigenetic modifications in regulating secondary metabolite creation. Despite advancements in biotechnological approaches like *in vitro* root-based cultures, challenges persist due to elicitation and overexpression of biosynthetic genes, which hinder reliable commercial creation of secondary metabolites. Exploring alternative origins and approaches, such as metabolic engineering and microbial endophytes, is necessary to meet the global demand for phytochemicals (Tariq et al., 2023). Innovative solutions can be developed to ensure the sustainable and efficient formation of valuable secondary metabolites from therapeutic plants by navigating the complexities of regulatory pathways and biodiversity conservation.

Analyzing the appearance of secondary metabolites requires carefully considering the complex interplay between genomic factors and genetic regulation (Li et al., 2020). When plants are exposed to environmental stressors, specific genes are activated, producing secondary metabolites like terpenoids (Jan et al., 2021). This genetic control recreates an important part in regulating metabolite creation. Factors like drought, temperature, and salinity trigger signaling pathways that improve the synthesis of secondary metabolites (Yang et al., 2018). The plant's genotype and developmental stage also interact with these external cues and influence metabolite synthesis (Olivoto et al., 2017). Transcription factors are critical regulators that improve the creation of secondary metabolites by binding to specific gene promoters and controlling the expression level of enzyme-coding genes (Yang et al., 2012). They are practical molecular tools for metabolic engineering. Omics-based approaches, such as metabolomics and genomics, have been instrumental in discovering new secondary metabolites (Rai et al., 2017). These methods detect biosynthesis gene clusters associated with metabolite creation, providing insight into regulating metabolic pathways. The regulation of gene expression for secondary metabolite biosynthesis involves various mechanisms, including transcription, splicing, and post-translational modifications. These regulatory processes significantly impact the levels of beneficial compounds produced (Davies & Schwinn, 2003). By unravelling the complexities of gene regulation mechanisms and using omics-based techniques, researchers can develop innovative approaches to optimize the creation of valuable secondary metabolites in plants.

The optimization of medicinal plant potential relies on various methods to enhance the creation of secondary metabolites. Traditional strategies are cost-effective but yield fewer metabolites than modern methods like metabolic engineering and biotechnology (Gonçalves et al., 2018). Elicitation techniques using silver nanoparticles and salicylic acid have emerged as effective methods to stimulate the creation of several bioactive compounds in plants (Maqbool et al., 2023). The *in vitro* callus cultures and elicitation regimes have revealed favourable results in improving the creation of secondary metabolites, such as antioxidants and aervine, offering sustainable choices to conventional extraction methods (Maqbool et al., 2023). Cell culture, a biotechnological method, recreates an essential role in plant conservation and enhancing metabolite biosynthesis. These techniques provide opportunities to modify synthesis pathways for optimal secondary metabolite creation, contributing collectively to enhancing the grade and quantity of phytochemicals or secondary metabolites in therapeutic plants. Adjusting culture media composition and signaling pathways concerned in secondary metabolite creation can further enhance the potential to enrich the number of bioactive compounds present naturally in plants (Niazian et al., 2021).

8.2 GENOME EDITING TECHNOLOGIES

Optimizing secondary metabolite creation in therapeutic plants requires genome editing techniques. The CRISPR/Cas9 tool enables targeted mutagenesis, gene knockouts, and gene regulation, which facilitates modifications to genes interested in biosynthetic pathways of secondary metabolites. Its effectiveness in optimizing secondary metabolite profiles has been extensively documented, making it a valuable asset in enhancing the creation of therapeutic plants. Various genome editing systems, including mega nucleases, ODM, zinc finger nucleases (ZFNs), transcription activator-like effector nucleases (TALENs), and CRISPR/Cas9, have been widely employed to change biosynthetic pathways of secondary metabolites (Mishra et al., 2023).

These techniques provide researchers with diverse options for enhancing the creation of medicinal plants, offering flexibility and versatility. Genome editing systems have revolutionized metabolic engineering, enabling investigators to alter biosynthetic pathways of secondary metabolites. Researchers can effectively improve the creation of secondary metabolites in therapeutic plants by leveraging these techniques, exposing their full potential for medicinal and commercial applications.

8.3 OVERVIEW OF TRANSCRIPTION ACTIVATOR-LIKE EFFECTOR NUCLEASES (TALEN)

TALENs are specifically engineered nucleases utilized for precise genome editing. They consist of customizable TALE domains derived from plant pathogenic bacteria, which bind to specific DNA sequences using a series of repeat units. The FokI nuclease domain within TALENs induces double-strand breaks (DSBs) in DNA, requiring dimerization for cleavage. This technology enables targeted gene modifications by disrupting, correcting, or inserting DNA sequences. Despite being largely supplanted by CRISPR/Cas9 due to its simpler design and higher efficiency, TALENs still offer advantages in applications demanding very high targeting specificity and reduced off-target effects. Hence, ongoing research endeavors to enhance the utility of TALENs in genetic engineering and biotechnology (Gaj et al., 2013).

8.3.1 Components of TALEN

TALEN functions by creating precise breaks in the DNA at precise target sites within the genome. These breaks cause the cell to initiate a repair mechanism, including non-homologous end joining (NHEJ), and homology-directed repair (HDR) pathways, which can be utilized for targeted genetic modifications (Gaj et al., 2013). TALENs have a unique structure that includes several components: an N-terminal T3S signal, a central tandem repeat domain, a C-terminal nuclear localization signal, and an acidic transcriptional activation domain (Becker & Boch, 2021). This intricate design gives TALENs high binding specificity and imposes fewer sequence constraints when targeting the genome, setting them apart from other nucleases (Gaj et al., 2013).

8.3.2 Mechanism of TALEN

TALENs are a group of plant pathogens with unique DNA binding domains known as TALE repeats. These repeats consist of tandem collections that contain 10 to 30 repeats. Each repeat is around 33 to 35 amino acids long, and the neighbouring amino acids are called repeat-variable di-residue (RVD). These RVDs confer particularity for 1 of the 4 DNA base pairs (Becker & Boch, 2021). The RVD code has been discovered and paved the way for developing TALENs. TALENs combine TALE repeats with the FokI endonuclease domain, which enables them to generate DSBs at the

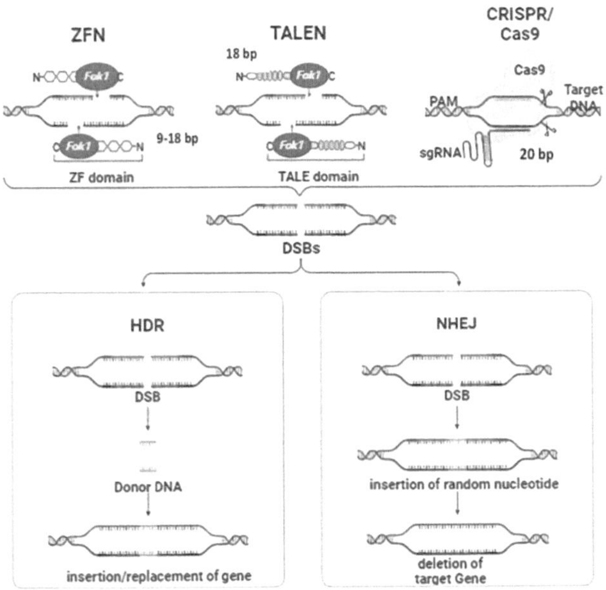

FIGURE 8.1 Mechanism of ZFN, TALEN, CRISPR/Cas.

target sites in the genome (Figure 8.1). On account of the RVD code, TALENs are more accessible to design, allowing the creation of TALE repeat collections with increased association for desired sequences of DNA. De novo engineered TALE repeat arrays have shown binding affinity rates reaching 96%. TALENs can be rapidly created and made within two days, and large libraries have successfully targeted all genes in the genome. Their flexibility in length allows them to bind longer DNA sequences, typically 18 bp or longer, although larger TALENs may compromise specificity (Becker & Boch, 2021).

Research has indicated that TALENs demonstrate a lower rate of off-target mutagenesis than ZFNs, reducing off-target mutations and decreasing cell toxicity when targeting particular genes. TALENs with obligate heterodimer FokI parts are frequently utilized to minimize off-target effects (Matsumoto et al., 2019). Nevertheless, the size of TALENs is bigger (about 3 kb for cDNA), which presents challenges in terms of delivery and expression, thereby restricting their usefulness in medicinal applications that necessitate delivery in viral vectors with little cargo size or as RNA molecules (Gaj et al., 2013). TALENs present a powerful and exact tool for genome editing, potentially improving secondary metabolite formation in therapeutic plants through targeted modifications in biosynthesis pathways. Due to their advantages in design, binding specificity, and decreased off-target effects, TALENs are promising candidates for promoting plant biotechnology and pharmacology.

8.3.3 Application of TALEN in Medicinal Plant Editing

TALEN technology can drastically improve crop traits in agriculture, including nutrient content, drought tolerance, herbicide resistance, yield potential, and post-harvest shelf life (Sharma et al., 2023). This technology can achieve this by precisely editing critical genes involved in nutrient biosynthesis pathways, water use efficiency, stress response mechanisms, herbicide sensitivity pathways, growth regulation, resource allocation, and post-harvest deterioration processes. TALENs can revolutionize crop creation, sustainability, and multinational food security by leveraging these diverse fronts of agricultural innovation (Khan et al., 2017). TALEN editing has been employed in *Streptomyces* species to enhance the yields of antibiotic biosynthetic gene clusters, showing potential for increasing drug creation. In cancer cell lines, targeting oncogenes with TALENs has inhibited cancer growth, providing a new approach to cancer therapy. TALEN editing of disease susceptibility genes in livestock has increased their resistance to specific pathogens, which could be a promising strategy for enhancing animal health. TALEN-mediated modifications of metabolic enzymes in microorganisms have led to higher creation of biofuels, demonstrating the usefulness of TALENs in metabolic engineering. These various applications highlight the transformative potential of TALEN technology in diverse fields, from pharmaceuticals to agriculture and beyond.

Artemisia annua, a medicinal plant, can potentially improve artemisinin creation, a crucial compound in malaria treatment (Shahrajabian et al., 2020), using advanced genome editing techniques such as TALEN-mediated manipulation. Researchers can increase artemisinin synthesis by precisely incorporating enhanced variants of the amorpha-4,11-diene synthase (*ADS*) gene through knock-in strategies (Arya et al., 2023). Additionally, redirecting metabolic resources toward artemisinin biosynthesis can be achieved through targeted knockouts of competing biosynthetic pathways, thus augmenting its yield. TALEN-mediated gene editing can also suppress genes responsible for artemisinin degradation pathways, ensuring the stability and accumulation of this vital compound (Arya et al., 2023). These approaches make it feasible to develop *Artemisia annua* varieties with increased artemisinin content, which can offer improvements in yield and quality, thereby effectively supporting efforts to combat malaria. TALEN technology shows the potential to enhance the medicinal benefits of *Artemisia annua* by targeting the *CYP71AV1* gene. By using precise knockout strategies, the conversion of artemisinic acid to dihydroartemisinic acid is decreased, increasing artemisinin synthesis. This increase ultimately leads to higher artemisinin levels in the plant, thus improving its effectiveness in malaria treatment (Arya et al., 2023).

Arabidopsis thaliana is utilized as an ideal crop for genetic analyses in plant biotechnology, focusing on the *PAL1* gene responsible for phenylalanine ammonia-lyase 1 (Yang et al., 2023). To enhance the creation of phenolic compounds, TALENs can be used to implement targeted gene knockout strategies that modulate *PAL1* expression. This is expected to improve the accumulation of phenolics in *Arabidopsis thaliana*, which may lead to a more useful understanding of the therapeutic benefits of these compounds. *Catharanthus roseus* shows potential for its potent alkaloid content in medicinal plant enhancement. The *TDC* gene, responsible for tryptophan decarboxylase (Bhushan et al., 2023), is targeted using precise gene editing with TALENs to elevate alkaloid creation. The expected outcome is an improvement in alkaloid content within *Catharanthus roseus*, which could expand its medicinal applications. The *SQS* gene, which is responsible for squalene synthase, is used to improve the medicinal properties of ginseng (*Panax ginseng*). Precise editing targets are established through TALEN-mediated gene knockout to increase ginsenoside creation. The application of TALEN editing on the *SQS* gene leads to higher levels of ginsenosides in ginseng, enhancing its therapeutic potential and expanding its medicinal applications (Sharma et al., 2019).

Now, researchers are focusing their efforts on enhancing the medicinal properties of opium poppy (*Papaver somniferum*) by targeting the *THAS* gene, which governs thebaine synthase (Agrataben et al., 2023). To achieve this, they are utilizing TALEN-mediated gene editing to make targeted modifications that enrich the creation of thebaine in the plant (Luyang, 2023). These modifications result in the levels of thebaine in the plant being significantly increased, which has the potential to enhance the medicinal efficacy of opium poppy and broaden its applications in the pharmaceutical industry (Tun et al., 2022). In *Artemisia annua*, the *DXS* gene can be edited to enhance artemisinin creation, increasing its antimalarial potency (Saifi et al., 2020). By targeting the *PDS* gene, tomatoes (*Solanum lycopersicum*) can develop resistance to powdery mildew, which is a promising solution for disease-resistant crops (Salava et al., 2021). In gene therapy, correcting the *CFTR* gene with TALENs in cystic fibrosis cells can restore CFTR channel activity, while editing the *LDLR* gene in hepatocytes shows potential for treating familial hypercholesterolemia (Mention et al., 2019). These discoveries represent a transformative shift in agriculture and healthcare, providing new solutions and therapeutic options.

For *Oryza sativa* (rice), TALEN-mediated gene modification focuses on the *OsSWEET14* gene to alter sugar transport mechanisms, aiming to improve sugar transport efficiency and potentially influencing the growth of the plant and stress responses in rice plants (Bi & Yang, 2017). The *OsSWEET14* gene recreates a necessary role in rice fertility and seed development, alongside other *OsSWEET* genes like *OsSWEET15*, and is susceptible to TALE TalC from the African *Xanthomonas oryzae* pv. strains (Kamhun & Rungrat, 2023). Modification of *OsSWEET14* has the potential to enhance sugar transport efficiency, which could have implications for overall plant health, stress tolerance, and yield improvement in rice cultivation. TALEN allows for precise gene modifications, including the *Xa27* gene in rice, which is linked to disease resistance pathways. By using TALEN-mediated knockout strategies, susceptibility genes in crops can be silenced, providing long-lasting resistance to diseases and reducing the need for chemical pesticides. This approach helps mitigate environmental impacts

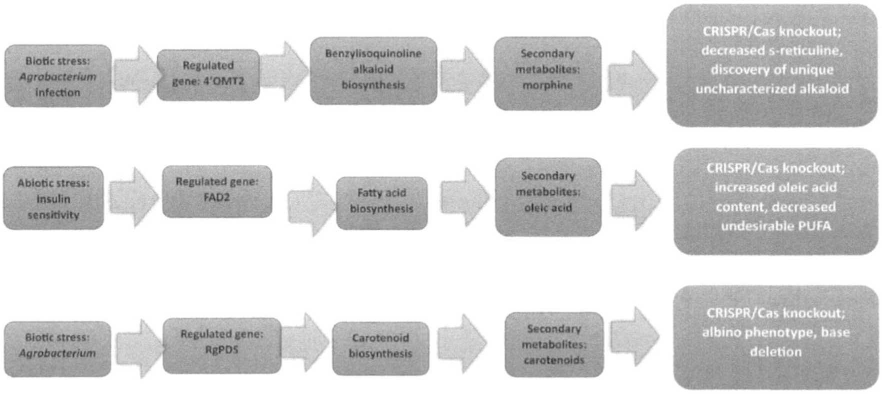

FIGURE 8.2 Integrated regulatory network of stress response and secondary metabolite biosynthesis via CRISPR-Cas.

and fosters sustainable agricultural practices by reducing chemical inputs and increasing crop resilience (Cai et al., 2017). TALENs target the GmFAD2-1A gene in Soybean to modify the fatty acid composition. The modification aims to enhance the soybean seed's oleic acid content, resulting in enriched amounts of oleic acid and improved quality of soybean oil (Fu et al., 2022).

Targeting the *MYB* gene in *Catharanthus roseus* has improved alkaloid creation, particularly enhancing the levels of vinblastine and vincristine (Bhushan et al., 2023). Similarly, manipulating the *WRKY* gene in *Withania somnifera* has enhanced pathogen resistance, boosting the creation of withanolides (Singh et al., 2017). Further studies have delved into genes like *ORCA3* in *Catharanthus roseus* (Bhushan et al., 2023), *HMGR* in *Hypericum perforatum* (Rai et al., 2021), and *PAL* in *Camellia sinensis* (Bhoi et al., 2022), showcasing the versatility of TALENs in precisely modulating gene expression to improve the biosynthesis of specific secondary metabolites. The improvements in targeted genome editing techniques have an important assurance for harnessing the therapeutic potential of medicinal plants (Figure 8.2).

8.4 OVERVIEW OF ZINC FINGER NUCLEASES (ZFN)

ZFNs are genetically engineered proteins utilized for precise genome editing. They consist of zinc finger domains that precisely bind to DNA series, along with a FokI nuclease domain that causes DSB breaks at targeted sites. This permits targeted gene disruption or modification through the cell's natural restoration mechanisms. ZFNs have been used for studies involving gene knockout, gene correction, the creation of transgenic organisms, and potential therapeutic applications. Despite their high specificity and versatility across various organisms, challenges related to design complexity and the risk of off-target effects persist. Despite the emergence of CRISPR-Cas9, ZFNs continue to be valuable for specific applications due to their unique advantages and regulatory considerations (Gaj et al., 2013)

8.4.1 Components of ZFN

These nucleases rely on the functioning properties of the Cys2-His2 zinc finger domain, which attaches to specific sequences within double-stranded DNA. Each ZFN unit typically comprises a DNA binding domain at its N-terminal end, comprising 3–6 distinct ZF repeats that can identify 9–18 base pairs. At the C-terminal end, a Fokl DNA cleaving site requires dimerization for enzymatic activity. While ZFNs offer the advantage of targeting specific DNA sites, there may be overlap in their specificities, which poses a challenge. However, advancements in modular designing, accumulating, and optimizing ZFNs against target DNA sequences have facilitated their widespread use in manipulating genomes of various organisms, for purposes such as gene inactivation, alteration, and trait stacking (Townsend et al., 2009; Osakabe et al., 2010; Ainley et al., 2013; Gaj et al., 2013; Kamburova et al., 2017).

8.4.2 Mechanism of ZFN

ZFNs operate in pairs, each detecting two adjacent sequences on the reverse and forward strands. When they bind to the specific sites, the FokI domains form a dimer and break the DNA at the location, causing a DSB with 5′ overhangs. The cells fix the DSBs through either NHEJ, which can result in incorrect repairs, or HDR, usually when a sister chromatid is present as a repair template (Figure 8.1). This repair process has several applications, including correcting genetic mutations linked to disorders such as sickle cell anemia and producing genetically modified crops with enhanced qualities like better products and resistance to pests (Shamshirgaran et al., 2022).

NHEJ can be employed to cause frameshift mutations in a gene's coding sequence, which can direct gene knockout. HDR allows the insertion of specific mutations by including a repair template having the preferred mutation flanked by homology arms. Despite its benefits, ZFNs have several potential drawbacks. Creating zinc finger domains to bind a comprehensive stretch of nucleotides with increased affinity can be challenging, making it challenging for non-professionals to utilize ZFNs. Moreover, there are restrictions in target site selection, with open-source ZFN components only capable of targeting binding sites behind every 200 bps in a random DNA sequence. Additionally, ZFNs prepared to introduce DSBs into the genome may also target off-target random sites, necessitating strategies to decrease off-target events, such as using obligate heterodimer ZFNs or introducing purified forms of ZFN proteins into cells. These considerations emphasize the complexities and difficulties of utilizing ZFNs for genome editing (Gaj et al., 2013).

8.4.3 Application of ZFN in Medicinal Plant Editing

Catharanthus roseus is a therapeutic plant that has been found to have potent anticancer properties. To enhance its potential, researchers target the *DXR* gene within its terpenoid biosynthesis pathway to enhance the formation of critical alkaloids with anticancer properties like vinblastine and vincristine. This is being done by using TALEN-mediated gene knockout to modulate the pathway. The ultimate goal is to improve the yield of valuable anticancer compounds in *Catharanthus roseus*, which could lead to new and improved cancer treatment options (Saraswat et al., 2023; Majumdar et al., 2022). *Catharanthus roseus* is known for its formation of vinblastine, an anticancer compound. Researchers used ZFNs to target and modify the tryptophan decarboxylase gene responsible for vinblastine biosynthesis. As a result, the creation of vinblastine in *Catharanthus roseus* was increased, leading to higher yields of this valuable secondary metabolite (Schluttenhofer, 2016). In *Catharanthus roseus*, manipulating the geraniol 10-hydroxylase gene has resulted in higher yields of monoterpenoid indole alkaloids, which possess significant medicinal properties (Pan et al., 2016; Ludwig-Müller et al., 2014). ZFNs have exhibited great potential in treating genetic disorders, such as HIV resistance. The researchers modified the *CCR5* gene in human T cells, which resulted in granting HIV opposition to the cells and demonstrated the effectiveness of ZFNs in correcting genetic disorders (Almeida & Matos, 2019).

In *Atropa belladonna*, enhancing the alkaloid biosynthesis gene boosted the creation of alkaloids, amplifying the plant's medicinal properties (Avsar et al., 2020). ZFN technology was employed in *Salvia miltiorrhiza* to modify genes connected to phenolic acid biosynthesis, significantly increasing phenolic compound levels and opening new avenues for medicinal applications (Mitra et al., 2023). These studies emphasize the usefulness of ZFNs in precise gene editing for optimizing the medicinal value of plants.

Although they offer several benefits, ZFNs can also pose challenges. The assembly of zinc finger domains with high affinity for specific sequences can be complex, which may make it challenging for non-specialists to access them. To address this, an open-source library and screening protocols have been developed. However, obtaining optimized ZFNs can take a significant amount of time. Additionally, target site selection is limited, and off-target effects are possible. This has prompted the improvement of strategies, such as utilizing ZFN pairs with different FokI domains to minimize accidental site binding and introducing purified ZFN proteins to enhance specificity. These efforts aim to discourse the critical challenges related to ZFN-mediated genome editing, ensuring precision and safety (Ramirez et al., 2008; Urnov et al., 2010; Maeder et al., 2008; Hockemeyer et al., 2009; Miller et al., 2007; Szczepek et al., 2007; Doyon et al., 2011; Gaj et al., 2012).

8.5 OVERVIEW OF CRISPR/CAS9

The CRISPR-Cas9 tool has altered genetics by offering a detailed and flexible method for modifying specific DNA sequences. It emanates from the adaptive immune systems of bacteria and allows scientists to make targeted DSBs in DNA, facilitating the disruption, correction, or insertion of genes. CRISPR-Cas9 is preferred over older technologies such as ZFNs and TALENs due to its simplicity, effectiveness, and ability to target nearly any genomic sequence.

8.5.1 Components of CRISPR/Cas9

The Cas9 protein and a single guide RNA (sgRNA) are the two main parts of the CRISPR-Cas9 system. Cas9 is an endonuclease that causes DSBs in target DNA. The sgRNA is a brief artificial RNA comprising a scaffold sequence necessary for Cas9 binding and a 20-nucleotide spacer sequence specified by the user to determine the target DNA site. Together, these elements create a complex that can identify and cleave specific DNA sequences in the genome.

8.5.2 Mechanism of CRISPR/Cas9

CRISPR/Cas9 combines proteins and RNAs to target the precise DNA sequences for gene modification. CRISPR/Cas9 is the natural defence system of bacteria. Bacterial vectors integrate foreign DNA sequences into their genomes to produce sgRNA (Hryhorowicz et al., 2017; Horodecka & Düchler, 2021). The process of CRISPR-Cas9 starts with the creation of a sgRNA that complements the specific DNA sequence of interest. Upon binding to the Cas9 protein, the complex searches the genome for a 20-bp nucleotide sequence that corresponds the sgRNA spacer adjoining to an NGG sequence known as the protospacer-adjacent motif (PAM). After locating the target site, the Cas9 protein induces DSBs. Afterward, the cell's natural DNA repair mechanisms, predominantly NHEJ or HDR, proceed to repair the break, resulting in gene disruption, correction, or the insertion (Figure 8.1) of new genetic material (Horodecka & Düchler, 2021). It can use multiple sgRNA simultaneously, known as multiplexing, enabling efficient mutation of multiple genes or accurate deletions within the same cell.

Despite its potential, challenges including the large size of Cas9 protein and potential off-target effect due to nucleotide mismatches need to be addressed for its widespread application. Additionally, researchers are exploring Cas9 proteins from different species with unique PAM sequence requirements to expand the system's targeting capabilities (Cong & Zhang, 2015).

8.5.3 Application of CRISPR/Cas for Enhancing Secondary Metabolite Creation in Plants

CRISPR/Cas9 is a beneficial tool in enhancing the biosynthesis of secondary metabolites in therapeutic plants. This technology allows for targeted mutagenesis and regulation of gene transcription, which in turn enables the modification of biosynthetic pathways responsible for producing specific secondary metabolites such as alkaloids, terpenoids, flavonoids, and phenolics. Through the precise editing of genes concerned with the biosynthesis of secondary metabolites, CRISPR/Cas9 makes it possible to enhance the creation of bioactive chemicals in plants, known to be rich sources of pharmaceutical compounds.

CRISPR/Cas9-based editing tools successfully produced carotenoids and gamma-aminobutyric acid (Li et al., 2018), important secondary metabolites with significant effects. Soybeans were engineered to produce isoflavones, and bananas were genetically modified to produce beta-carotene (Kaur et al., 2020). Due to the CRISPR/Cas9-established mutagenesis of the dihydroflavonol 4 reductase B, Japanese morning glories get their white colour flower from the absence of anthocyanin pigments. When the same plant is mutagenized with carotenoid-degrading dioxygenase 4, mutagenesis dehydrates carotenoids and downregulates carotenoid genes, thus reducing carotenoids in leaves (Watanabe et al., 2017).

The bZIP2 knockout lines in *Salvia miltiorrhiza* (Bunge) caused an increase in the concentration of phenolic compounds (Shi et al., 2021). CRISPR/Cas9 moderated knockout of SmbHLH60 leads to upregulating the biosynthesis of anthocyanin and phenolic acid within *Salvia miltiorrhiza* Liu et al. (2022). CRISPR/Cas9 technology targeted four genes encoding germacrene A synthase (CiGAS) in *Cichorium intybus*. This enzyme contributes to the synthesis of sesquiterpene lactones (STL). Hence, inhibition of the STL biosynthetic pathway leads to increased phenolic concentrations and squalene accumulation in the primary root (Cankar et al., 2021).

8.5.4 Utilizing CRISPR/Cas Gene Technology to Enhance Secondary Metabolites in Medicinal Plants: Case Studies

Atropa belladonna L. is an example of a plant where modifying the hyoscyamine 6b hydroxylase (*AbH6H*) gene results in plants lacking anisodamine and scopolamine, valuable anticholinergic tropane alkaloids. Researchers enhanced hyoscyamine while avoiding the creation of its derivatives, scopolamine and anisodamine, by targeting the *AbH6H* gene utilising CRISPR/Cas9 (Zeng et al., 2021). *Dendrobium officinale* is an herb used to treat conditions related to yin deficiency. It has medicinal properties, such as antioxidants and neuroprotection. CRISPR/Cas9 technique enhanced its medicinal properties by modifying gene expression, such as cinnamate 4-hydroxylase (C4H), coumarate 3-hydroxylase (C3H), and 4CL and IRX concerned in lignocellulose synthesis (Kui et al., 2017). CRISPR/Cas-based editing was utilized in *Dendrobium officinale* to modify five targeted genes responsible for lignocellulose biosynthesis. The researchers targeted enzymes C4H and C3H, which resulted in a range of alterations in the gene products (Gao et al., 2002; Chen et al., 2014).

Salvia miltiorrhiza is a therapeutic plant used for various diseases. It anticancer, antioxidant, and anti-inflammatory properties due to its components. Scientists have utilized CRISPR/Cas9 techniques to manipulate genes like SmRAS and SmCPS1 in *Salvia miltiorrhiza*, which has resulted in changes in phenolic acid content and tanshinone levels. These results indicate the possibility of enhancing desired phyto-metabolites through genetic manipulation (Fan et al., 2019). In *Salvia miltiorrhiza*, the PAL level increased in knockout mutants while decreasing in overexpression lines significantly. This indicates that the *bZIP2* gene acts as an inhibitor in the biosynthesis of phenolic acids, especially for producing rosmarinic acid and salvianolic acid B (Shi et al., 2021). *Nicotiana tabacum* is a type of plant from the Solanaceae family that contains various secondary metabolites used in medicine, pharmaceuticals, and industry. CRISPR/Cas9 technique was employed to target NtPDS and NtPDR6 in tobacco, resulting in high mutation rates (Gao et al., 2015).

In tobacco plants, 6 glycosyltransferase genes were inactivated via CRISPR/Cas9 to produce recombinant proteins (Jansing et al., 2019). CRISPR/Cas9 gene activation studies showed deletions between target sites in 7 out of 20 transformations (Mercx et al., 2016). Glycoprotein without specific glycan was generated in BY2 Cell of *Nicotiana tabacum* using multiplex CRISPR/Cas9 to target α-1,3-fucosyltransferase and β(1,2)-xylosyltransferase genes, resulting in the knockout of two *XylT* genes and four *FucT* genes. CRISPR gene editing targeted flavoproteins associated with nicotine biosynthesis in *Nicotiana tabacum* to reduce nicotine levels significantly in subsequent plant generations. The T3 generation lacked nicotine, indicating successful disruption of nicotine biosynthesis (Kajikawa et al., 2011; Schachtsiek & Stehle, 2019). Scientists used CRISPR/Cas9 to modify the ADS enzyme responsible for artemisinin creation in *Artemisia annua*. Introducing specific mutations in the *ADS* gene enhanced artemisinin synthesis (Zhang et al., 2019).

Investigators have utilized a type II CRISPR/Cas9-based mutagenesis to target 4' OMT2 in *Papaver somniferum*, a plant that can produce morphine through benzylisoquinoline alkaloids (BIAs). By manipulating the biosynthetic flux of BIAs through the delivery of synthetic binary plasmids, they observed changes in S-reticuline levels upon silencing 4'OMT2, leading to a reduction in morphine creation (Labanca et al., 2018). Moreover, the investigation highlights the possible application of CRISPR in discovering new secondary metabolites, as a previously unknown alkaloid was identified in genetically altered plants (Tong et al., 2019; Alagoz et al., 2016; Gurkok et al., 2016).

Scientists have genetically modified *Rehmannia glutinosa* (Gaertn.) DC, a traditional medicinal herb, uses CRISPR/Cas9 to target the *PDS* gene. They produced albino and chimeric albino plants in *Rehmannia glutinosa.* The study confirms the effectiveness of CRISPR/Cas9 for precise genome editing in *Rehmannia glutinosa*, delivering new possibilities for molecular breeding to enhance desirable plant traits (Li et al., 2021). The effectiveness of CRISPR/Cas9 was demonstrated by Zhou et al. (2018) in a study where decreased expression levels of phenolic acids were observed after targeting the *SmRAS* gene. Meanwhile, Shiels, Reski, et al. (2022) utilized CRISPR/Cas system to create Cannabis sativa plants with high CBD content and without THC by targeting the *THCAS* gene. Investigators have employed CRISPR/Cas9 technology to alter genes in *Panax ginseng* (Liu, 2023), *Glycyrrhiza glabra* (Chiyo et al., 2024), and *Hypericum perforatum* (Singh, 2023) to increase the formation of valuable medicinal substances. The herb *Symphytum officinale* possesses analgesic, proliferative, and anti-inflammatory qualities. Large amounts of toxic pyrrolizidine alkaloid (PA) are present throughout the plant, which limits its potential use in pharmaceuticals.

The foremost important enzyme in the biosynthesis of PA is homospermidine synthase (HSS), which can be eradicated with by using CRISPR/Cas9 technique. The hairy roots' (HRs) homospermidine and PA concentrations seemed to have dropped, based on the successfully obtained HSS-deficient HRs. This outcome demonstrated the use of gene editing and the feasibility of producing transgenic comfrey cultivars that are not harmful (Zakaria et al., 2021). Scientists utilized CRISPR/Cas9 to modify the Dzfps in *Dioscorea zingiberensis*, a plant species known for producing diosgenin with medicinal properties. The gene was edited by *Agrobacterium*-mediated transformation and promoters like 35S and OsU3 to drive Cas9 and sgRNA expression. This reduced squalene and Dzfps transcript levels, demonstrating the potential for targeted manipulation of secondary metabolites in the *Dioscorea* species (Feng et al., 2018).

In 2016, Lv and his team discovered a method to boost the amount of artemisinin in sweet wormwood, a plant with antimalarial properties. They achieved this by inhibiting specific genes, such as *BFS*, *CPS*, and *SQS*, which increased artemisinin levels in transgenic plants. This method demonstrated the potentiality of the CRISPR technique to target these genes (Lv et al., 2016). Researchers used CRISPR/Cas9 to investigate isoflavone creation in red clover. They created a mutant plant with reduced IFS1 activity, resulting in lower biochanin A, genistein, and other isoflavones expression levels. The mutant plants showed increased gene expression that produced IFS1 precursors, but no significant changes were observed in IFS1 activity or downstream isoflavone genes. This study sheds light on the genes implicated in phenylpropanoid biosynthesis in red clover.

Cichorium intybus is a kind of important industrial plant developed to extract prebiotic and very low-calorie sweeteners known as inulin. The taproot of chicory accumulates squalene and several phenolic compounds, which have a penal like and cannot be removed at the time of inulin extraction. Therefore, the CRISPR/Cas9 technology can inhibit the expression of the genes encoding the germacrene synthase (CiGAS). This blocked the STL biosynthesis reaction, resulting in a reduced STL concentration, which helps to facilitate inulin extraction without bitter-tasting chemicals, and caused a significant gain in the obtainability of farnesyl pyro-phosphate, which enhances the phenolic range in the chicory roots (Cankar et al., 2021).

Cannabis sativa is a dioecious annual plant. The need for this plant is rising significantly because of its phytocannabinoids. X. Zhang et al. (2021) used CRISPR/Cas9 to alter the phytoene desaturase, a well-linked marker gene, in *Cannabis sativa.* Additionally, they used an *Agrobacterium*-mediated transformation technique to create four transgenic plants with an albino phenotype. A perennial medicinal plant known for its wide range of uses in treating illnesses is called *Monochasma savatieri.* Using the CRISPR/Cas 9 method, Bai et al. (2023) transformed *Monochasma savatieri* hairy root mutants successfully by targeting the CSLD2/3 in a phenolic biosynthesis pathway.

8.6 FUTURE PERSPECTIVES

CRISPR/Cas9 could transform the formation of secondary metabolites in medicinal plants by enabling precise alterations to biosynthetic pathways, potentially increasing the yields of compounds. By targeting crucial genes implicated in metabolite creation, it becomes feasible to generate new compounds with improved therapeutic properties, thus aiding in the discovery of new drugs. Its multiplexing capability allows the concurrent editing of multiple genes, which can assist in optimizing metabolic networks to enhance plant resilience and efficiency. Pairing CRISPR/Cas9 with advanced omics technologies offers deeper insights into plant genetics and metabolism, guiding more effective editing strategies. As CRISPR/Cas9 progresses in applications for sustainable agriculture and pharmaceutical development, addressing ethical and regulatory considerations will be crucial due to its potential for substantial advancements in medicinal plant biotechnology.

8.7 CONCLUSION

Genome editing technology provides an advantageous strategy to improving the formation of secondary metabolites in medicinal plants. Researchers can target specific genes that are concerned with producing desired compounds with high precision, which can potentially enhance the product of secondary metabolites significantly. This innovation opens opportunities for the sustainable creation of beneficial phytochemicals and offers avenues for developing novel plant-based therapeutics with more significant bioactivity and efficacy. However, more research is required to optimize CRISPR/Cas editing strategies, ensure regulatory compliance, and address potential concerns about off-target effects and environmental impact.

ACKNOWLEDGMENT

The corresponding author expresses gratitude to the Tamil Nadu State Council for Higher Education for the financial support provided to Dr. M. Arun under the grant RGP/2019-20/BU/HECP-0018, dated April 27, 2021.

DATA AVAILABILITY

The authors confirm that the data supporting the findings of this study are available within the book/chapter.

REFERENCES

Agrataben, V., Sabreen, B., Ashiq, H. M., Madhuri, G., Deepak, K., Anil, K., Aradhana, M., Tabarak, M., & Mohan, A. (2023). Opium alkaloids, biosynthesis, pharmacology and association with cancer occurrence. *Open Biology*, *13*(5), 220355.

Ainley, W. M., Sastry-Dent, L., Welter, M. E., Murray, M. G., Zeitler, B., Amora, R., & Petolino, J. F. (2013). Trait stacking via targeted genome editing. *Plant Biotechnology Journal*, *11*(9), 1126–1134.

Akilandeswari, G., Vijaya Anand, A., Sampathkumar, P., Vinayaga Moorthi, P., & Preethi, B. (2019). A prospective review on phyto-pharmacological aspects of *Andrographis paniculata*. *Systematic Reviews in Pharmacy*, *10*(1), 15–19.

Almeida, M. J., & Matos, A. (2019). Designer nucleases: Gene-editing therapies using CCR5 as an emerging target in HIV. *Current HIV Research*, *17*(5), 306–323.

Arya, A., Gautam, S., Goel, S., Grewal, S., & Bhattacharyya, M. (2023). Metabolic engineering for high-value bioactive compounds from medicinal plants. In *Phytochemical genomics: Plant metabolomics and medicinal plant genomics* (pp. 521–544). Springer Nature Singapore.

Avsar, B., Zhao, Y., Li, W., & Lukiw, W. J. (2020). *Atropa belladonna* expresses a microRNA (aba-miRNA-9497) highly homologous to *Homo sapiens* miRNA-378 (hsa-miRNA-378); both miRNAs target the 3′-Untranslated region (3′-UTR) of the mRNA encoding the neurologically relevant, zinc-finger transcription factor ZNF-691. *Cellular and Molecular Neurobiology*, *40*, 179–188.

Bai, C., Cao, Y., Zhao, S., Wu, Z., Dai, S., Wang, H., & Fu, C. (2023). Generation of CRISPR/Cas9-mediated mutants in *Monochasma savatieri* using a hairy root system. *Industrial Crops and Products*, *191*, 116008.

Becker, S., & Boch, J. (2021). TALE and TALEN genome editing technologies. *Gene and Genome Editing*, *2*, 100007.

Begum, T. N., Ilyas, M. M., Kalavathy, S., Anand, A. V., Kumar, S., & Senthil, R. (2009). Effects of ethanolic leaf extract of *Azima tetracantha* Lam. on *Ehrlich ascites* carcinoma tumour bearing mice. *Research Journal of Medicine and Medical Sciences*, *4*(2), 351–354.

Bharathi, V., Rengarajan, R. L., Radhakrishnan, R., & Hashem, A., Abd-Allah, E. F., Alqarawi, A. A., & Anand, A. V. (2018). Effects of a medicinal plant *Macrotyloma uniflorum* (Lam.) Verdc. Formulation (MUF) on obesity-associated oxidative stress-induced liver injury. *Saudi Journal of Biological Sciences*, *25*(6), 1115–1121.

Bhoi, A., Yadu, B., Chandra, J., & Keshavkant, S. (2022). Mutagenesis: A coherent technique to develop biotic stress resistant plants. *Plant Stress*, *3*, 100053.

Bhushan, S., Sharma, D., Rakshant, Kaul, S., Dhar, M. K., & Sharma, M. (2023). Recent strategies to engineer alkaloid biosynthesis in medicinal plants. In *Medicinal plants: Their response to abiotic stress* (pp. 391–416). Springer Nature Singapore.

Bi, H., & Yang, B. (2017). Gene editing with TALEN and CRISPR/Cas in rice. *Progress in Molecular Biology and Translational Science*, *149*, 81–98.

Cai, L., Cao, Y., Xu, Z., Ma, W., Zakria, M., Zou, L., Cheng, Z., & Chen, G. (2017). A transcription activator-like effector Tal7 of Xanthomonas oryzae pv. oryzicola activates rice gene Os09g29100 to suppress rice immunity. *Scientific Reports*, *7*(1), 5089.

Cankar, K., Paul, B., Robert, S., Suvi, T. H., Johanna, C. H., Jules, B., vander Ingrid, M. M., de Both, M., & Dirk, B. (2021). Inactivation of the germacrene A synthase gene by CRISPR/Cas9 eliminates the biosynthesis of sesquiterpene lactones in *Cichorium intybus* L. *Plant Biotechnology Journal*, *19*(12), 2442–2453.

Cardoso, J. C., Oliveira, M. E., & Cardoso, F. D. C. (2019). Advances and challenges on the in vitro creation of secondary metabolites from medicinal plants. *Horticultura Brasileira*, *37*, 124–132.

Chen, D., Mubeen, B., Hasnain, A., Rizwan, M., Adrees, M., Naqvi, S. A. H., Iqbal, S., Kamran, M., El-Sabrout, A. M., Elansary, H. O., & Mahmoud, E. A. (2022). Role of promising secondary metabolites to confer resistance against environmental stresses in crop plants: Current scenario and future perspectives. *Frontiers in Plant Science*, *13*, 881032.

Chen, J., Qi, H., Li, J. B., Yi, Y. Q., Chen, D., Hu, X. H., Wang, M. L., Sun, X. L., & Wei, X. Y. (2014). Experimental study on Dendrobium candidum polysaccharides on promotion of hair growth. *Zhongguo Zhong yao za zhi= Zhongguo Zhongyao Zazhi= China Journal of Chinese Materia Medica*, *39*(2), 291–295.

Chiyo, N., Seki, H., Kanamoto, T., Ueda, H., Kojoma, M., & Muranaka, T. (2024). Glycyrrhizin creation in licorice hairy roots based on metabolic redirection of triterpenoid biosynthetic pathway by genome editing. *Plant and Cell Physiology*, *65*(2), 185–198.

Cong, L., & Zhang, F. (2015). Genome engineering using CRISPR-Cas9 system. *Chromosomal Mutagenesis*, *1239*, 197–217.

Davies, K. M., & Schwinn, K. E. (2003). Transcriptional regulation of secondary metabolism. *Functional Plant Biology*, *30*(9), 913–925.

Divya, N., Rengarajan, R. L., Radhakrishnan, R., Abd-Allah, E. F., Alqarawi, A. A., Hashem, A., & Anand, A. V. (2019). Phytotherapeutic efficacy of the medicinal plant *Terminalia catappa* L. *Saudi Journal of Biological Sciences*, *26*(5), 985–988.

Doyon, Y., Vo, T. D., Mendel, M. C., Greenberg, S. G., Wang, J., Xia, D. F., Miller, J. C., Urnov, F. D., Gregory, P. D., & Holmes, M. C. (2011). Enhancing zinc-finger-nuclease activity with improved obligate heterodimeric architectures. *Nature Methods*, *8*(1), 74–79.

Fan, M., Luo, D., Peng, L. Y., Li, X. N., Wu, X. D., Ji, X., & Zhao, Q. S. (2019). Neo-clerodane diterpenoids from aerial parts of *Salvia hispanica* L. and their cardioprotective effects. *Phytochemistry*, *166*(1), 112065.

Feng, S., Song, W., Fu, R., Zhang, H., Xu, A., & Li, J. (2018). Application of the CRISPR/Cas9 system in *Dioscorea zingiberensis*. *Plant Cell Tissue Organ Culture (PCTOC)*, *135*, 133–141.

Ferrari, A., Longo, R., Silva, R., Mitro, N., Caruso, D., De Fabiani, E., & Crestani, M. (2019). Epigenome modifiers and metabolic rewiring: New frontiers in therapeutics. *Pharmacology & Therapeutics*, *193*, 178–193.

Fu, M., Chen, L., Cai, Y., Su, Q., Chen, Y., & Hou, W. (2022). CRISPR/Cas9-mediated mutagenesis of GmFAD2-1A and/or GmFAD2-1B to create high-oleic-acid soybean. *Agronomy*, *12*(12), 3218.

Gaj, T., Gersbach, C. A., & Barbas, C. F. (2013). ZFN, TALEN, and CRISPR/Cas-based methods for genome engineering. *Trends in Biotechnology*, *31*(7), 397–405.

Gaj, T., Guo, J., Kato, Y., Sirk, S. J., & Barbas, C. F. (2012). Targeted gene knockout by direct delivery of zinc-finger nuclease proteins. *Nature Methods*, *9*(8), 805–807.

Gao, J., Jin, R., Wu, Y., Zhang, H., Zhang, D., Chang, Y., & Hu, Z. (2002). Comparative study of tissue cultured Dendrobium protocorm with natural Dendrobium candidum on immunological function. *Zhong yao cai= Zhongyaocai= Journal of Chinese Medicinal Materials*, *25*(7), 487–489.

Gao, J., Wang, G., Ma, S., Xie, X., Wu, X., Zhang, X., Wu, Y., Zhao, P., & Xia, Q. (2015). CRISPR/Cas9-mediated targeted mutagenesis in *Nicotiana tabacum*. *Plant Molecular Biology*, *87*(1), 99–110.

Gonçalves, S., & Romano, A. (2018). Creation of plant secondary metabolites by using biotechnological tools. *Secondary Metabolites-Sources and Applications*, *5*, 81–99.

Gurkok, T., Ozhuner, E., Parmaksiz, I., Özcan, S., Turktas, M., Ipek, A., Demirtas, I., Okay, S., & Unver, T. (2016). Functional characterization of 4′ OMT and 7OMT genes in BIA biosynthesis. *Frontiers in Plant Science*, *7*, 98.

Hockemeyer, D., Soldner, F., Beard, C., Gao, Q., Mitalipova, M., DeKelver, R. C., Katibah, G. E., Amora, R., Boydston, E. A., Zeitler, B., & Meng, X. (2009). Efficient targeting of expressed and silent genes in human ESCs and iPSCs using zinc-finger nucleases. *Nature Biotechnology*, *27*(9), 851–857.

Horodecka, K., & Düchler, M. (2021). CRISPR/Cas9: Principle, applications, and delivery through extracellular vesicles. *International Journal of Molecular Sciences*, *22*(11), 6072.

Hryhorowicz, M., Lipiński, D., Zeyland, J., & Słomski, R. (2017). CRISPR/Cas9 immune system as a tool for genome engineering. *Archivum Immunologiae et Therapiae Experimentalis*, *65*, 233–240.

Iason, G. R., & Villalba, J. J. (2006). Behavioral strategies of mammal herbivores against plant secondary metabolites: The avoidance-tolerance continuum. *Journal of Chemical Ecology*, *32*, 1115–1132.

Jain, C., Khatana, S., & Vijayvergia, R. (2019). Bioactivity of secondary metabolite's of various plants: A review. *Interntional Journal of Pharmaceutical Sciences & Research*, *10*(2), 494–504.

Jan, R., Asaf, S., Numan, M., Lubna, & Kim, K. M. (2021). Plant secondary metabolite biosynthesis and transcriptional regulation in response to biotic and abiotic stress conditions. *Agronomy*, *11*(5), 968.

Jansing, J., Sack, M., Augustine, S. M., Fischer, R., & Bortesi, L. (2019). CRISPR/Cas9-mediated knockout of six glycosyltransferase genes in *Nicotiana benthamiana* for the creation of recombinant proteins lacking β-1, 2-xylose and core α-1, 3-fucose. *Plant Biotechnology Journal*, *17*(2), 350–361.

Kajikawa, M., Shoji, T., Kato, A., & Hashimoto, T. (2011). Vacuole-localized berberine bridge enzyme-like proteins are required for a late step of nicotine biosynthesis in tobacco. *Plant Physiology*, *155*, 2010–2022.

Kamburova, V. S., Nikitina, E. V., Shermatov, S. E., Buriev, Z. T., Kumpatla, S. P., Emani, C., & Abdurakhmonov, I. Y. (2017). Genome editing in plants: An overview of tools and applications. *International Journal of Agronomy*, *2017*, 1–15.

Kamhun, W., & Rungrat, T. (2023). Effects of nitrogen levels on the expression of OsSWEET genes and the resistance to *Xanthomonas oryzae* pv. oryzae in bacterial blight susceptible and resistant rice cultivars (Doctoral dissertation, Naresuan University).

Karthika, P., Balamuralikrishnan, B., Yamini, K., Vijaya Anand, A., Durairaj, K., Maruthupandian, A., Alodaini, H. A., Hatamleh, A. A., Manikantan, P., & Arun, M. (2023). Green synthesis, characterization of silver nanoparticles using aqueous leaf extracts of *Solanum melongena* and in vitro evaluation of antibacterial, pesticidal and anticancer activity in human MDA-MB-231 breast cancer cell lines. *Journal of King Saud University – Science*, *35*(5), 102663.

Kaur, N., Alok, A., Shivani, P., Kumar, P., Kaur, N., Awasthi, P., Chaturvedi, S., Pandey, P., Pandey, A., Pandey, A. K., & Tiwari, S. (2020). CRISPR/Cas9 directed editing of lycopene epsilon-cyclase modulates metabolic flux for β-carotene biosynthesis in banana fruit. *Metabolic Engineering*, *59*, 76–86.

Kaviya, M., Balasubramanian, B., Bharathi, K., Malaisamy, A., Al-Dhabi, N. A., Mariadhas, V. A., Anand, A. V., & Liu, W. (2022). Evaluation of nutritional substances and investigation of antioxidant and antimicrobial potentials of *Boerhavia diffusa* with in silico molecular docking. *Molecules*, *27*(4), 1280.

Khan, Z., Khan, S. H., Mubarik, M. S., Sadia, B., & Ahmad, A. (2017). Use of TALEs and TALEN technology for genetic improvement of plants. *Plant Molecular Biology Reporter*, *35*, 1–19.

Kui, L., Chen, H., Zhang, W., He, S., Xiong, Z., Zhang, Y., Yan, L., Zhong, C., He, F., Chen, J., Zeng, P., Zhang, G., Yang, S., Dong, Y., Wang, W., & Cai, J. (2017). Building a genetic manipulation tool box for orchid biology: Identification of constitutive promoters and application of CRISPR/Cas9 in the orchid, *Dendrobium officinale*. *Frontiers in Plant Science*, *7*, 2036.

Labanca, F., Ovesnà, J., & Milella, L. (2018). *Papaver somniferum* L. taxonomy, uses and new insight in poppy alkaloid pathways. *Phytochemistry Reviews*, *17*(4), 853–871.

Li, X., Wang, Y., Chen, S., Tian, H., Fu, D., Zhu, B., Luo, Y., & Zhu, H. (2018). Lycopene is enriched in tomato fruit by CRISPR/Cas9-mediated multiplex genome editing. *Frontiers in Plant Science*, *9*, 559.

Li, X., Zuo, X., Li, M., Yang, X., Zhi, J., Sun, H., Xie, C., Zhang, Z., & Wang, F. (2021). Efficient CRISPR/Cas9-mediated genome editing in *Rehmannia glutinosa*. *Plant Cell Reports*, *40*(9), 1695–1707.

Li, Y., Kong, D., Fu, Y., Sussman, M. R., & Wu, H. (2020). The effect of developmental and environmental factors on secondary metabolites in medicinal plants. *Plant Physiology and Biochemistry*, *148*, 80–89.

Liu, C. (2023). Targeted editing of the ginsenoside biosynthesis pathway in ginseng using CRISPR/Cas9 technology and its implications. *Medicinal Plant Research*, *13*(3), 1–12.

Liu, S., Wang, Y., Shi, M., Maoz, I., Gao, X., Sun, M., Yuan, T., Li, K., Zhou, W., Guo, X., & Kai, G. (2022). SmbHLH60 and SmMYC2 antagonistically regulate phenolic acids and anthocyanins biosynthesis in Salvia miltiorrhiza. *Journal of Advanced Research*, *42*, 205–219.

Ludwig-Müller, J. (2014). Metabolic engineering of selected secondary metabolites. In *Production of biomass and bioactive compounds using bioreactor technology* (pp. 509–536). Springer Netherlands.

Luyang, H. (2023). Evaluation of biosynthetic pathway and engineered biosynthesis of morphine with CRISPR. *BIO Web of Conferences, 59*, 01022–01022.

Lv, Z., Zhang, F., Pan, Q., Fu, X., Jiang, W., Shen, Q., Yan, T., Shi, P., Lu, X., Sun, X., & Tang, K. (2016). Branch pathway blocking in Artemisia annua is a useful method for obtaining high yield artemisinin. *Plant and Cell Physiology, 57*(3), 588–602.

Maeder, M. L., Thibodeau-Beganny, S., Osiak, A., Wright, D. A., Anthony, R. M., Eichtinger, M., Jiang, T., Foley, J. E., Winfrey, R. J., Townsend, J. A., & Unger-Wallace, E. (2008). Rapid "open-source" engineering of customized zinc-finger nucleases for highly efficient gene modification. *Molecular Cell, 31*(2), 294–301.

Majumdar, P., Harale, G., Ganger, S., & Pardeshi, S. (2022). Zinc finger nuclease based genome editing in enhancement of anti-diabetic activity of plants: Prospective applications. In *Anti-diabetic potential of plants in the era of omics* (pp. 373–401). Apple Academic Press.

Manikandan, R., Balasubramanian, B., Punniyakotti, P., Vijaya Anand, A., Meyyazhagan, A., Velauthaprabhu, S., & Liu, W. C. (2022). Cardio-protective effects of *Terminalia catappa* leaves and *Terminalia chebula* fruit extract in doxorubicin-induced cardiomyopathy in rats. *Biomarkers, 27*(5), 488–495.

Manoharan, N., Sampathkumar, P., Dheeba, B., Sheik Abdulla, S., Vinothprasanna, G., Vinothkannan, R., Kalavathy, S., Vijaya Anand, A., & Shanmugasundaram, A. (2008). Potential hepatoprotective effect of aqueous extract of *Gracilaria corticata* in AFB_1 induced hepatotoxicity in Wistar rats. *Journal of Biological Sciences, 8*, 1352–1355.

Maqbool, M., Ishtiaq, M., Mazhar, M. W., Casini, R., Mahmoud, E. A., & Elansary, H. O. (2023). Enhancing bioactive metabolite creation in *Aerva sanguinolenta* callus cultures through silver nanoparticle and salicylic acid elicitation. *Sustainability, 15*(13), 10395.

Matsumoto, D., Tamamura, H., & Nomura, W. (2019). TALEN-based chemically inducible, dimerization-dependent, sequence-specific nucleases. *Biochemistry, 59*(2), 197–204.

Mention, K., Santos, L., & Harrison, P. T. (2019). Gene and base editing as a therapeutic option for cystic fibrosis-learning from other diseases. *Genes, 10*(5), 387.

Mercx, S., Tollet, J., Magy, B., Navarre, C., & Boutry, M. (2016). Gene inactivation by CRISPR-Cas9 in *Nicotiana tabacum* BY-2 suspension cells. *Frontiers in Plant Science, 7*(1), 40.

Miller, J. C., Holmes, M. C., Wang, J., Guschin, D. Y., Lee, Y. L., Rupniewski, I., Beausejour, C. M., Waite, A. J., Wang, N. S., Kim, K. A., & Gregory, P. D. (2007). An improved zinc-finger nuclease architecture for highly specific genome editing. *Nature Biotechnology, 25*(7), 778–785.

Mishra, R., Agarwal, P., & Mohanty, A. (2023). Applications of genome editing techniques for the improvement of medicinal plants. In *Phytochemical genomics: Plant metabolomics and medicinal plant genomics* (pp. 545–569). Springer Nature Singapore.

Mitra, S., Anand, U., Ghorai, M., Kant, N., Kumar, M., Radha, J. N. K., Swamy, M. K., Proćków, J., de la Lastra, J. M. P., & Dey, A. (2023). Genome editing technologies, mechanisms and improved production of therapeutic phytochemicals: Opportunities and prospects. *Biotechnology and Bioengineering, 120*(1), 82–94.

Mohiuddin, A. K. (2019). Chemistry of secondary metabolites. *Annals of Clinical Toxicology, 2*(1), 1014.

Moyo, M., Aremu, A. O., & Van Staden, J. (2015). Medicinal plants: An invaluable, dwindling resource in sub-Saharan Africa. *Journal of Ethnopharmacology, 174*, 595–606.

Muthumani, G. D., Anand, A. V., & Manikandan, R. (2013). Antioxidant, antihelmintic and antimicrobial activity of *Euphorbia thymifolia* Linn whole plant. *International Journal of Current Microbiology and Applied Sciences, 2*(1), 66–75.

Niazian, M., & Sabbatini, P. (2021). Traditional in vitro strategies for sustainable creation of bioactive compounds and manipulation of metabolomic profile in medicinal, aromatic and ornamental plants. *Planta, 254*, 1–22.

Olivoto, T., Nardino, M., Carvalho, I. R., Follmann, D. N., Szareski, V. J., Ferrari, M., de Pelegrin, A. J., & de Souza, V. Q. (2017). Plant secondary metabolites and its dynamical systems of induction in response to environmental factors: A review. *African Journal of Agricultural Research, 12*(2), 71–84.

Osakabe, K., Osakabe, Y., & Toki, S. (2010). Site-directed mutagenesis in Arabidopsis using custom-designed zinc finger nucleases. *Proceedings of the National Academy of Sciences, 10726*, 12034–12039.

Pagare, S., Bhatia, M., Tripathi, N., Pagare, S., & Bansal, Y. K. (2015). Secondary metabolites of plants and their role: Overview. *Current Trends in Biotechnology and Pharmacy, 9*(3), 293–304.

Pai, S., Hebbar, A., & Selvaraj, S. (2022). A critical look at challenges and future scopes of bioactive compounds and their incorporations in the food, energy, and pharmaceutical sector. *Environmental Science and Pollution Research, 29*(24), 35518–35541.

Pan, Q., Mustafa, N. R., Tang, K., Choi, Y. H., & Verpoorte, R. (2016). Monoterpenoid indole alkaloids biosynthesis and its regulation in *Catharanthus roseus*: A literature review from genes to metabolites. *Phytochemistry Reviews, 15*, 221–250.

Rai, A., Saito, K., & Yamazaki, M. (2017). Integrated omics analysis of specialized metabolism in medicinal plants. *The Plant Journal, 90*(4), 764–787.

Rai, N., Kumari Keshri, P., Verma, A., Kamble, S. C., Mishra, P., Barik, S., Kumar Singh, S., & Gautam, V. (2021). Plant associated fungal endophytes as a source of natural bioactive compounds. *Mycology, 12*(3), 139–159.

Ramirez, C. L., Foley, J. E., Wright, D. A., Müller-Lerch, F., Rahman, S. H., Cornu, T. I., Winfrey, R. J., Sander, J. D., Fu, F., Townsend, J. A., Cathomen, T., Voytas, D. F., & Joung, J. K. (2008). Unexpected failure rates for modular assembly of engineered zinc fingers. *Nature Methods, 5*(5), 374–375.

Saifi, M., Khan, S., Kiran, U., Fatima, S., & Abdin, M. Z. (2020). Transgenic technology to improve therapeutic efficacy of medicinal plants. In *Transgenic technology based value addition in plant biotechnology* (pp. 207–226). Academic Press.

Salava, H., Thula, S., Mohan, V., Kumar, R., & Maghuly, F. (2021). Application of genome editing in tomato breeding: Mechanisms, advances, and prospects. *International Journal of Molecular Sciences, 22*(2), 682.

Sangeetha, T., Balamuralikrishnan, B., Sampathkumar, P., Velayuthaprabhu, S., Senthilkumar, N., Syed Ibrahim, K., Baskaran, R., & Vijaya Anand, A. (2022). Characterization and phytoconstituents of *Petroselinum crispum* (Mill) and *Coriandrum sativum* (Linn) and their impacts on inflammation – an in vitro analysis against human adenocarcinoma cells with molecular docking. *South African Journal of Botany, 146*, 776–788.

Saraswat, P., Chaturvedi, A., & Ranjan, R. (2023). Zinc finger nuclease (ZFNs) and transcription activator-like effector nucleases (TALENs) based genome editing in enhancement of anticancer activity of plants. In *Plant-derived anticancer drugs in the OMICS era* (pp. 281–293). Apple Academic Press.

Schachtsiek, J., & Stehle, F. (2019). Nicotine-free, nontransgenic tobacco (*Nicotiana tabacum* L.) edited by CRISPR-Cas9. *Plant Biotechnology Journal, 17*, 2228–2230.

Schluttenhofer, C. M. (2016). *Transcriptional regulation of specialized metabolites in Arabidopsis thaliana and Catharanthus roseus*. University of Kentucky.

Shahrajabian, M. H., Wenli, S. U. N., & Cheng, Q. (2020). Exploring *Artemisia annua* L., artemisinin and its derivatives, from traditional Chinese wonder medicinal science. *Notulae Botanicae Horti Agrobotanici Cluj-Napoca, 48*(4), 1719–1741.

Shamshirgaran, Y., Liu, J., Sumer, H., Verma, P. J., & Taheri-Ghahfarokhi, A. (2022). Tools for efficient genome editing; ZFN, TALEN, and CRISPR. In *Applications of genome modulation and editing* (pp. 29–46). Springer.

Sharma, M., Koul, A., Sharma, D., Kaul, S., Swamy, M. K., & Dhar, M. K. (2019). Metabolic engineering strategies for enhancing the production of bio-active compounds from medicinal plants. In *Natural bio-active compounds: Volume 3: Biotechnology, bioengineering, and molecular approaches* (pp. 287–316). Springer Singapore.

Sharma, P., Pandey, A., Malviya, R., Dey, S., Karmakar, S., & Gayen, D. (2023). Genome editing for improving nutritional quality, post-harvest shelf life and stress tolerance of fruits, vegetables, and ornamentals. *Frontiers in Genome Editing, 5*, 1094965.

Shi, M., Du, Z., Hua, Q., & Kai, G. (2021). CRISPR/Cas9-mediated targeted mutagenesis of bZIP2 in *Salvia miltiorrhiza* leads to promoted phenolic acid biosynthesis. *Industrial Crops and Products, 167*, 113560.

Shiels, A. B., Reski, R., & Decker, E. L. (2022). CRISPR/Cas-mediated gene editing for the biosynthesis of pharmaceuticals and other chemicals in plants. *Current Opinion in Plant Biology, 65*, 102239.

Singh, A. K., Kumar, S. R., Dwivedi, V., Rai, A., Pal, S., Shasany, A. K., & Nagegowda, D. A. (2017). A WRKY transcription factor from *Withania somnifera* regulates triterpenoid withanolide accumulation and biotic stress tolerance through modulation of phytosterol and defense pathways. *New Phytologist, 215*(3), 1115–1131.

Singh, H. (2023). Enhancement of plant secondary metabolites by genetic manipulation. In *Genetic manipulation of secondary metabolites in medicinal plant* (pp. 59–90). Springer Nature Singapore.

Szczepek, M., Brondani, V., Buchel, J., Serrano, L., Segal, D. J., & Cathomen, T. (2007). Structure-based redesign of the dimerization interface reduces the toxicity of zinc-finger nucleases. *Nature Biotechnology, 25*(7), 786–793.

Tariq, H., Asif, S., Andleeb, A., Hano, C., & Abbasi, B. H. (2023). Flavonoid creation: Current trends in plant metabolic engineering and de novo microbial creation. *Metabolites, 13*(1), 124.

Tong, Y., Weber, T., & Lee, S. Y. (2019). CRISPR/Cas-based genome engineering in natural product discovery. *Natural Product Reports, 36*(9), 1262–1280.

Townsend, J. A., Wright, D. A., Winfrey, R. J., Fu, F., Maeder, M. L., Joung, J. K., & Voytas, D. F. (2009). High-frequency modification of plant genes using engineered zinc-finger nucleases. *Nature, 459*(7245), 442–445.

Tun, X., Xiaofei, Y., Yanyan, J., Zihang, L., Guangbo, T., Xiujuan, L., Bo, W., Tingjie, W., Jiadong, L., Li, G., & Kai, Y. (2022). A global survey of the transcriptome of the opium poppy (*Papaver somniferum*) based on single-molecule long-read isoform sequencing. *Plant Journal, 110*(2), 607–620.

Urnov, F. D., Rebar, E. J., Holmes, M. C., Zhang, H. S., & Gregory, P. D. (2010). Genome editing with engineered zinc finger nucleases. *Nature Reviews Genetics, 11*(9), 636–646.

Varalakshmi, B., Anand, A. V., Karpagam, T., Bai, J. S., & Manikandan, R. (2014). In vitro antimicrobial and anticancer activity of *Cinnamomum zeylanicum* Linn bark extracts. *International Journal of Pharmacy and Pharmaceutical Sciences, 6*(1), 12–18.

Velu, G., Palanichamy, V., & Rajan, A. P. (2018). Phytochemical and pharmacological importance of plant secondary metabolites in modern medicine. *Bioorganic phase in natural food: An overview* (pp. 135–156). Springer.

Verpoorte, R., Contin, A., & Memelink, J. (2002). Biotechnology for the creation of plant secondary metabolites. *Phytochemistry Reviews, 1*, 13–25.

Vijaya Anand, A., Varalakshmi, B., Prasana., Sampathkumar, P., Pushpa, N., & Hedina, A. (2016). *Cinnamomum zeylanicum* Linn. The spice with multi potential. *Systematic Reviews in Pharmacy, 7*(1), 24–29.

Vijaya Anand, A., Velayuthaprabhu, S., Rengarajan, R. L., Sampathkumar, P., & Radhakrishnan, R. (2020). Bioactive compounds of guava (*Psidium guajava* L.). In *Bioactive compounds in underutilized fruits and nuts, reference series in phytochemistry* (pp. 503–527). Springer.

Vijayakumar, K., Prasanna, B., Rengarajan, R. L., Rathinam, A., Velayuthaprabhu, S., & Vijaya Anand, A. (2023). Antidiabetic and hypolipidemic effects of *Cinnamon cassia* bark extracts: An in vitro, in vivo, and in silico approach. *Archives of Physiology and Biochemistry, 129*(2), 338–348.

Watanabe, K., Kobayashi, A., Endo, M., Sage-Ono, K., Toki, S., & Ono, M. (2017). CRISPR/Cas9-mediated mutagenesis of the *dihydroflavonol- 4-reductase-B (DFR-B)* locus in the Japanese morning glory *Ipomoea* (*Pharbitis*) nil. *Scientific Reports, 7*(1), 10028.

Yang, C. Q., Fang, X., Wu, X. M., Mao, Y. B., Wang, L. J., & Chen, X. Y. (2012). Transcriptional regulation of plant secondary metabolism. *Journal of Integrative Plant Biology, 54*(10), 703–712.

Yang, L., Wen, K. S., Ruan, X., Zhao, Y. X., Wei, F., & Wang, Q. (2018). Response of plant secondary metabolites to environmental factors. *Molecules, 23*(4), 762.

Yang, Y., & Siwen, W. (2023). *Arabidopsis thaliana*: A powerful model organism to explore histone modifications and their upstream regulations. *Epigenetics, 18*(1), 2211362.

Zakaria, M. M., Schemmerling, B., & Ober, D. (2021). CRISPR/Cas9-mediated genome editing in comfrey (*Symphytum officinale*) hairy roots results in the complete eradication of pyrrolizidine alkaloids. *Molecules, 26*(6), 1498.

Zeng, L., Zhang, Q., Jiang, C., Zheng, Y., Zuo, Y., Qin, J., Zhihua, L., Hongpeng, D. (2021). Development of *Atropa belladonna* L. plants with high-yield hyoscyamine and without its derivatives using the CRISPR/Cas9 system. *International Journal of Molecular Sciences, 22*, 1731.

Zhang, X., Gencheng, X., Cheng, C., Lei, L., Sun, J., Ying, X., Deng, C., Dai, Z., Yang, Z., Chen, X., Liu, C., Tang, Q., & Jianguang, S. (2021). Establishment of an Agrobacterium-mediated genetic transformation and CRISPR/Cas9-mediated targeted mutagenesis in Hemp (*Cannabis Sativa* L.). *Plant Biotechnology Journal, 19*(10), 1979–1987.

Zhang, Y., Malzahn, A. A., Sretenovic, S., & Qi, Y. (2019). The emerging and uncultivated potential of CRISPR technology in plant science. *Nature Plants, 5*, 778–794.

9 Mutagenesis for Enhanced Secondary Metabolites Production in Medicinal Plants

S. Tharani, V. N. Mutharaian, R. Thirugnanasampandan, Sarmad Moin, and Balasundaram Saraswathy Chithra Devi

9.1 INTRODUCTION

For centuries, medicinal plants have served as a valuable source of bioactive substances, offering a diverse array of secondary metabolites with a wide range of therapeutic benefits, including anti-inflammatory, anticancer, and antimicrobial properties (Marrelli et al., 2021). Plant secondary metabolites, with their diverse structures and functions, play crucial roles in plants, including attracting pollinators, defending against pathogens, and adapting to environmental stresses through complex interactions influenced by genetic, developmental, and environmental factors (Pierik et al., 2014). As the demand for natural remedies and pharmaceutical agents continues to rise, there is an escalating requirement to enhance the synthesis of secondary metabolites in medicinal plants. Mutagenesis is a powerful tool used in medicinal plant research to enhance the production of secondary metabolites. Mutagenesis is a technique that involves inducing genetic changes in organisms to enhance desirable traits, such as increased production of secondary metabolites, in medicinal plants (Roychowdhury & Tah, 2013).

In recent years, there has been a notable increase in interest in mutagenesis within the fields of plant biotechnology and pharmacognosy. This growing focus is primarily attributed to mutagenesis capacity to significantly boost the production of secondary metabolites in medicinal plants. Through this method, genetic modifications can be deliberately introduced into plant genomes, thereby modifying the activity of genes involved in the pathways responsible for secondary metabolite biosynthesis. Various mutagenesis techniques are available, including chemical mutagenesis, radiation-induced mutagenesis, and more recently, advanced genome editing approaches such as CRISPR-Cas9 (Alok et al., 2020). These techniques offer precise tools for manipulating plant genomes and achieving specific outcomes.

Medicinal plants serve as reservoirs for a rich assortment of plant secondary metabolites, each endowed with diverse biological activities that render them immensely valuable for various practical applications across industries. The advent of genetic engineering techniques has provided innovative solutions to the obstacles encountered in harnessing secondary metabolites from medicinal plants or resorting to chemical synthesis. By manipulating the genetic makeup of these plants, standardized protocols for genetic transformation have been instrumental in augmenting the yield of target compounds in key medicinal crops such as *Atropa belladonna, Artemisia annua, Panax ginseng, Papaver somniferum, Dioscorea sp.*, and numerous others. (Kiran & Abdin, 2020). Through meticulous experimentation and refinement, strategies aimed at optimizing secondary metabolite production encompass a spectrum of approaches, including overcoming rate-limiting steps in biosynthetic pathways, redirecting metabolic flux away from competitive pathways, minimizing catabolism of target compounds, and upregulating the expression of key regulatory genes (Yoon et al., 2013).

Moreover, the realm of genetic engineering extends beyond mere manipulation of gene expression. Metabolic engineering offers additional avenues for enhancing secondary metabolite production. By rewiring metabolic pathways and introducing novel enzymatic activities, plants can be engineered to produce higher quantities of specific compounds of interest. Furthermore, RNA interference (RNAi) technology has emerged as a powerful tool for fine-tuning gene expression, enabling precise control over metabolic processes involved in secondary metabolite biosynthesis (Wagner & Kroumova, 2008).

However, despite the significant strides made in genetic and metabolic engineering, several challenges persist. Foremost among these challenges is the scarcity of comprehensive information pertaining to the genomes of medicinal plants, the intricacies of their biosynthetic pathways, the identity of genes encoding key enzymes, and the regulatory mechanisms governing secondary metabolite production. The lack of such fundamental knowledge impedes the effective application of biotechnological tools for enhancing secondary metabolite yields.

Addressing these knowledge gaps necessitates a multidisciplinary approach that transcends traditional boundaries. In addition to genetic and metabolic engineering, leveraging insights from fields such as genomics, transcriptomics, proteomics, and metabolomics is essential. Integration of omics data offers a holistic understanding of plant

DOI: 10.1201/9781003475491-9

metabolism, enabling precise identification of potential targets for genetic manipulation and strategies for optimizing secondary metabolite production (Rai et al., 2017). Furthermore, a concerted effort to combine conventional breeding techniques with modern biotechnological approaches holds promise for accelerating progress in this field. By harnessing the natural genetic diversity present in medicinal plants, improved plant varieties with enhanced secondary metabolite profiles can be developed. These efforts, coupled with advances in genetic engineering and omics technologies, are poised to revolutionize the production of targeted metabolites for commercial applications (Saini et al., 2020).

The exploration of medicinal plants for their secondary metabolites presents a fertile ground for innovation in the realm of biotechnology. Genetic engineering, metabolic engineering, and omics-based approaches offer powerful tools for enhancing the production of valuable compounds with diverse biological activities (Alami et al., 2022). Embracing a collaborative mindset across scientific domains is imperative to bridge existing knowledge disparities and employ a multifaceted approach. This concerted effort is essential for unlocking the holistic therapeutic and commercial potential of medicinal plants, thereby catalyzing advancements not only in medicine and agriculture but also in broader domains.

The effectiveness of mutagenesis in enhancing secondary metabolite production has been demonstrated in numerous studies. For example, it has been successfully used to enhance the production of valuable compounds like artemisinin in *Artemisia annua*, taxol in *Taxus* species, and ginsenosides in *Panax ginseng*. These examples highlight the potential of mutagenesis as a powerful tool for optimizing the biosynthesis of specific secondary metabolites with pharmaceutical importance (Le et al., 2019; Al-Qurainy & Khan, 2010). Understanding the production dynamics of plant secondary metabolites is essential for comprehending plant adaptation mechanisms and ecological interactions, as these compounds reflect the intricate and adaptive nature of plant responses to their ever-changing environments (Alami et al., 2022).

The enhanced production of secondary metabolites in medicinal plants is of great interest due to the pharmacological importance of these compounds. Secondary metabolites often have medicinal properties and are used in the production of drugs. Therefore, increasing their production through mutagenesis holds significant promise for the pharmaceutical and herbal medicine industries.

9.2 SECONDARY METABOLITES IN MEDICINAL PLANTS

Medicinal plants have been valued for centuries for their therapeutic compounds, offering a diverse array of bioactive molecules known as secondary metabolites (Jain et al., 2019; Newman & Cragg, 2016) (Table 9.1). These compounds, distinct from primary metabolites essential for basic cellular functions, encompass a wide range of chemical classes, including alkaloids, flavonoids, terpenoids, and polyphenols (Alamgir, 2018; Harvey et al., 2015). Secondary metabolites are synthesized by plants primarily for ecological functions, such as defense against herbivores, pathogens, and environmental stresses (Divekar et al., 2022; Wink, 2015). However, many of these compounds also possess remarkable pharmacological properties, making them valuable resources for drug discovery and development (Jain et al., 2019).

9.2.1 Biosynthesis of Secondary Metabolites

The biosynthesis of secondary metabolites in medicinal plants is a complex and highly regulated process involving multiple enzymatic reactions and metabolic pathways. Recent advances in molecular biology and biochemistry have provided insights into the biosynthetic pathways responsible for the production of these compounds. For example, alkaloids, a class of nitrogen-containing compounds found in numerous medicinal plants, are synthesized via specialized pathways involving enzymes such as methyltransferases, cytochrome P450 monooxygenases, and alkaloid-specific biosynthetic enzymes (Satish et al.,

TABLE 9.1
Examples of Secondary Metabolites from Medicinal Plants and Their Medicinal Properties

Medicinal Plant	Secondary Metabolites	Medicinal Properties	References
Aloe vera	Aloin, anthraquinones	Treats burns, wounds, skin inflammations	Cardarelli et al., 2017.
Ginseng	Ginsenosides	Boosts immunity, improves cognitive function	Hou et al., 2021.
Turmeric	Curcumin	Anti-inflammatory, antioxidant properties	Setyawati et al., 2021.
Coneflowers	Alkamides, polysaccharides	Immune system support, cold and flu relief	Kakouri et al., 2024; Saema et al., 2023.
Garlic	Allicin, alliin	Lowers blood pressure, and cholesterol levels	Kovarovič et al., 2019.
St. John's wort	Hypericin, hyperforin	Treats depression, anxiety disorders	Kirakosyan et al., 2004.
Ginger	Gingerol	Relieves nausea, aids digestion	Arijanti & Suryaningsih, 2019.
Chamomile	Chamazulene, apigenin	Calms nerves, promotes sleep, anti-inflammatory	Mailänder et al., 2022.
Peppermint	Menthol	Relieves headaches, soothes digestive issues	Mimica-Dukic & Bozin, 2008.
Lavender	Linalool, linalyl acetate	Reduces stress and anxiety, promotes relaxation	Prusinowska & Śmigielski, 2014.

2020; Weng & Noel, 2012; Schuler & Werck-Reichhart, 2003). Similarly, flavonoids, which contribute to the color and flavor of many fruits and vegetables, are synthesized through the phenylpropanoid and flavonoid biosynthetic pathways (Santos-Buelga & González-Paramás, 2016).

Understanding the biosynthesis of secondary metabolites is crucial for manipulating their production in medicinal plants. Genetic manipulation techniques such as mutagenesis, genetic engineering, and genome editing offer promising avenues for increasing the yield and potency of desired compounds (Mishra et al., 2023). Plants have been successfully engineered to over-express key enzymes involved in secondary metabolite biosynthesis, resulting in enhanced production of specific compounds (Verpoorte et al., 2002). Additionally, metabolic engineering approaches, which involve the introduction of foreign genes or modification of endogenous pathways, have been used to redirect metabolic flux towards the biosynthesis of target secondary metabolites.

9.2.2 Pharmacological Properties of Secondary Metabolites

Secondary metabolites found in medicinal plants exhibit a wide range of pharmacological properties, including antimicrobial, anti-inflammatory, anticancer, and antioxidant activities (Bhatti et al., 2022). These bioactive compounds interact with various molecular targets in the human body, exerting therapeutic effects that may alleviate symptoms of disease or promote overall health and well-being. For example, alkaloids such as morphine and quinine have long been used as analgesic and antimalarial agents, respectively, while flavonoids and polyphenols found in fruits and vegetables exhibit antioxidant and anti-inflammatory properties (Kumar, 2014; Maleki et al., 2019; Fernie & Tohge, 2017).

The pharmacological properties of secondary metabolites have been extensively studied in pharmacognosy and natural medicine (Kaushik et al., 2021). Numerous bioactive compounds with potential applications in the treatment and management of various diseases have been identified. Resveratrol, a polyphenolic compound found in red wine and grapes, has been studied for its potential cardioprotective and anticancer effects (Bertelli & Das, 2009; Zenebe et al., 2001) Similarly, artemisinin, a sesquiterpene lactone isolated from *Artemisia annua,* has shown potent antimalarial activity and is widely used in the treatment of malaria (de Faveri Favero et al., 2014; Paddon et al., 2013).

9.2.3 Applications of Secondary Metabolites

Secondary metabolites found in medicinal plants have a wide range of applications beyond traditional medicine. Certain plant-derived compounds have been explored for their potential as functional food ingredients, dietary supplements, and nutraceuticals (Caleja et al., 2017). The antioxidant properties of polyphenols found in tea and berries, for instance, have attracted attention for their potential role in promoting cardiovascular health and preventing chronic diseases (Iqbal et al., 2023).

Furthermore, secondary metabolites from medicinal plants have been investigated for their antimicrobial properties and potential applications in agriculture and food preservation (Daraban et al., 2021). Plant-derived antimicrobial agents offer a natural alternative to synthetic chemicals for controlling pests, pathogens, and food-borne pathogens (Upadhyay et al., 2014). Essential oils containing terpenoids and phenolic compounds have been studied for their antimicrobial activity against bacteria, fungi, and viruses (Burt, 2004; Akthar et al., 2014; Grosso et al., 2015).

9.2.4 Challenges and Future Directions

Despite the immense therapeutic potential of secondary metabolites in medicinal plants, several challenges and limitations must be addressed. One major challenge is the sustainable sourcing and cultivation of medicinal plants to meet growing demand without depleting natural resources or harming ecosystems (Singh, 2006). Overharvesting, habitat destruction, and climate change pose significant threats to the biodiversity and sustainability of medicinal plant populations worldwide (Raju & Das, 2024; Saslis-Lagoudakis et al., 2012).

Another challenge is the variability in secondary metabolite content and composition among different plant species, varieties, and geographical regions (Moore et al., 2014). Factors such as genetic variation, environmental conditions, and cultivation practices can influence the biosynthesis and accumulation of secondary metabolites in medicinal plants, leading to variability in their pharmacological effects and therapeutic efficacy (Cardoso et al., 2019). Standardization of herbal extracts and traditional remedies is therefore essential to ensure consistency and quality control in medicinal plant-based products (Moussaieff & Mechoulam, 2009; Rahimi & Abdollahi, 2012).

Additionally, there is a need for further research into the safety, efficacy, and pharmacokinetics of plant-based drugs and herbal remedies (Saad et al., 2017; Fennell et al., 2004). While many traditional herbal remedies have been used for centuries with apparent safety and efficacy, rigorous scientific evaluation is required to validate their therapeutic claims and identify potential adverse effects or drug interactions (Rahimi & Abdollahi, 2012).

9.3 PRINCIPLES OF MUTAGENESIS

Mutagenesis is a powerful tool used in plant biotechnology to induce genetic variations, leading to the enhancement of desirable traits, including the production of secondary metabolites in medicinal plants (Kolakar et al., 2018). It involves the induction of heritable changes in the DNA sequence of an organism, resulting in genetic variation (Oladosu et al., 2016). This process can occur naturally through exposure to environmental factors such as radiation or chemicals, or it can be induced artificially using mutagens. In the context of medicinal plants, mutagenesis can be employed to generate genetic diversity, which may lead to the discovery of novel secondary metabolites or the enhancement of existing ones with desirable pharmacological properties (Singh, 2023).

Mutagens can be classified into two main categories: physical and chemical. Physical mutagens include ionizing radiation, such as X-rays and gamma rays, and non-ionizing radiation, such as ultraviolet (UV) light. These mutagens induce DNA damage by causing breaks in the DNA strands or by generating reactive oxygen species that modify DNA bases (Khursheed et al., 2021). Chemical mutagens, on the other hand, include alkylating agents, such as ethyl methane sulfonate (EMS), and base analogs, such as 5-bromouracil (Sawarkar & Mansoori, 2023). These chemicals alter the structure of DNA bases or interfere with DNA replication, leading to mutations.

Mutations can arise spontaneously or be induced by external agents known as mutagens (Chatterjee & Walker, 2017; Friedberg et al., 1995). Spontaneous mutations occur due to errors in DNA replication, repair, or recombination processes, leading to changes in the nucleotide sequence. In contrast, induced mutations are caused by physical or chemical agents that interact with DNA, causing alterations in its structure or stability.

Physical mutagens, such as radiation (e.g., X-rays, gamma rays, UV light), exert their effects by directly damaging DNA or generating reactive oxygen species that can induce oxidative damage (Cadet et al., 1999; Kodym & Afza, 2003; Chatterjee & Walker, 2017). Chemical mutagens, on the other hand, interact with DNA through covalent binding, base modifications, or intercalation, disrupting normal base pairing and replication fidelity (Chatterjee & Walker, 2017).

The accomplishment of mutagenesis in enhancing secondary metabolite production depends on several key principles. The mutagen must be applied at an appropriate dose and duration to induce genetic changes without causing excessive damage to the plant (Coimbra et al., 2005; Mumtaz & Tausif, 2023; Zeki & Doğan, 2022). Additionally, the mutagenized population must be carefully screened to identify individuals with desirable traits, such as increased secondary metabolite content or altered metabolic pathways. This process often involves high-throughput screening methods coupled with analytical techniques such as chromatography and mass spectrometry (Ali et al., 2021).

Mutagenesis has been successfully applied to a wide range of medicinal plants to enhance the production of secondary metabolites with therapeutic potential. For example, mutagenesis has been used to increase the levels of artemisinin, a potent antimalarial compound, in *Artemisia annua* (Al-Qurainy & Khan, 2010). By treating *A. annua* seeds with EMS, mutants with significantly higher artemisinin content compared to the wild-type plants were generated. Similarly, mutagenesis has been employed to increase the levels of ginsenosides, bioactive compounds found in *Panax ginseng*, through the induction of genetic variation (Le et al., 2019) (Table 9.2).

While mutagenesis offers a powerful approach to enhance secondary metabolite production in medicinal plants, several challenges and considerations must be taken into account. The mutagenized populations may exhibit unwanted traits or mutations unrelated to the desired phenotype, necessitating careful screening and selection (Lal et al., 2020). Additionally, the use of mutagens carries inherent risks, including potential cytotoxicity and mutagenicity, which must be carefully managed to minimize adverse effects (Serpeloni et al., 2011).

Despite these challenges, mutagenesis remains a valuable tool in the quest to enhance secondary metabolite production in medicinal plants. Advances in genome editing technologies, such as CRISPR-Cas9, have revolutionized the field of mutagenesis by enabling precise and efficient manipulation of DNA sequences (Alok et al., 2020). Future research efforts should focus on refining mutagenesis protocols, optimizing screening methods, and elucidating the molecular mechanisms underlying mutagen-induced changes in secondary metabolite biosynthesis. Furthermore, advances in genomic technologies, such as next-generation sequencing and genome editing, hold promise for accelerating the discovery and characterization of novel secondary metabolites in mutagenized plant populations (Pathak et al., 2019).

TABLE 9.2
Examples of Medicinal Plants and Their Enhanced Secondary Metabolites through Mutagenesis

S. No.	Medicinal Plant	Enhanced Secondary Metabolite	Mutagenesis Technique	Reference
1.	Ginseng *(Panax ginseng)*	Ginsenosides	Radiation mutagenesis	Le et al. (2019)
2.	Artemisia *(Artemisia annua)*	Artemisinin	Chemical mutagenesis	Al-Qurainy and Khan (2010)
3.	Tea *(Camellia sinensis)*	Catechins	EMS mutagenesis	Kuroda (1996)
4.	Ashwagandha *(Withania somnifera)*	Withanolides	EMS treatment	Sharma et al. (2023)
5.	Ginkgo tree *(Ginkgo biloba)*	Ginkgolides, bilobalide	Gamma irradiation	Pereira et al. (2015)
6.	opium poppy *(Papaver somniferum)*	Morphine	Gamma irradiation	Nalina et al. (2022)
7.	Madagascar periwinkle *(Catharanthus roseus)*	Vinblastine, vincristine	EMS mutagenesis	Mistry et al. (2022)

Mutagenesis represents a promising approach to enhance secondary metabolite production in medicinal plants, offering opportunities to discover novel bioactive compounds and optimize existing ones for therapeutic use. By understanding the principles of mutagenesis and carefully harnessing its potential, individuals can contribute to the development of new plant-based medicines and natural remedies.

9.4 INDUCING GENETIC VARIATION IN MEDICINAL PLANTS

The deliberate induction of genetic variation in medicinal plants represents a pivotal endeavor within modern plant breeding and biotechnology. This strategic approach aims to amplify both the quality and quantity of invaluable secondary metabolites inherent in these plants. Genetic variation serves as the cornerstone for plant improvement, offering a rich reservoir from which to cultivate medicinal plant varieties endowed with enhanced traits, including heightened yields of bioactive compounds (Suprasanna et al., 2015).

Mutagenesis emerges as a primary methodology for instigating genetic variation, involving the intentional introduction of alterations in the DNA sequence of plants. These alterations encompass point mutations, deletions, or insertions, which subsequently engender modified gene expression or function (Sharma et al., 2019). The spectrum of mutagenesis methodologies encompasses chemical mutagenesis, radiation-induced mutagenesis, and the cutting-edge precision genome-editing techniques exemplified by CRISPR-Cas9 (Bhojiya & Joshi, 2024).

9.4.1 Chemical Mutagenesis

Chemical mutagenesis is a powerful technique employed to induce genetic variation in medicinal plants, aiming to enhance the production of secondary metabolites with therapeutic significance (Yasmin & Nehvi, 2017; Asif, 2023). This method involves the deliberate introduction of mutations into the plant genome using mutagenic chemicals, such as ethyl methane sulfonate (EMS) or ethidium bromide. Through the induction of genetic variations, chemical mutagenesis offers a promising avenue for augmenting the biosynthesis pathways of secondary metabolites in medicinal plants (Samadi et al., 2022).

Chemical mutagenesis serves as a key strategy in plant breeding and biotechnology, facilitating the generation of genetic diversity essential for crop improvement. In the context of medicinal plants, chemical mutagenesis holds immense potential for enhancing the production of bioactive compounds with pharmaceutical applications (Rime et al., 2019). By introducing mutations into the plant genome, chemical mutagenesis provides a means to explore and exploit the genetic variability inherent in medicinal plant species.

The process of chemical mutagenesis typically involves treating plant material with mutagenic chemicals to induce DNA damage and subsequent mutations. Ethyl methane sulfonate (EMS) is a commonly used chemical mutagen that induces point mutations by alkylating DNA bases, leading to changes in the nucleotide sequence. Similarly, ethidium bromide intercalates into DNA strands, causing structural distortions and inducing mutations during DNA replication (Rime et al., 2019; Serrat et al., 2014).

Chemical mutagenesis has been widely employed to enhance the production of secondary metabolites in various medicinal plant species. Studies have utilized EMS treatment to induce mutations in genes involved in secondary metabolite biosynthesis pathways, resulting in increased yields of bioactive compounds. Additionally, ethidium bromide has been utilized to generate genetic variation in medicinal plants, leading to the discovery of novel variants with enhanced secondary metabolite profiles (Liu et al., 2024).

Numerous research studies have demonstrated the effectiveness of chemical mutagenesis in enhancing secondary metabolite production in medicinal plants. For instance, a study by Mistry et al. (2022) utilized EMS treatment to induce mutations in *Catharanthus roseus*, leading to increased production of vinblastine and vincristine, two important anticancer alkaloids. Similarly, Pandey et al. (2016) employed ethidium bromide treatment to induce mutations in *Artemisia annua*, resulting in elevated levels of artemisinin, a potent antimalarial compound. While chemical mutagenesis offers significant potential for enhancing secondary metabolite production in medicinal plants, several challenges and considerations must be addressed. These include the need for careful optimization of mutagenic treatment conditions to minimize undesirable effects such as plant lethality or reduced fertility. Additionally, thorough screening and characterization of mutant populations are essential to identify individuals with desirable secondary metabolite profiles (Singh et al., 2021).

Chemical mutagenesis holds promise as a valuable tool for inducing genetic variation and enhancing secondary metabolite production in medicinal plants. Future research efforts should focus on refining mutagenesis protocols, elucidating the underlying molecular mechanisms of mutagenesis-induced changes in secondary metabolite biosynthesis, and exploring the potential of novel mutagenic agents.

9.4.2 Radiation-Induced Mutagenesis

Radiation-induced mutagenesis is a well-established technique in plant breeding and biotechnology, leveraging the mutagenic effects of ionizing radiation to introduce genetic diversity. In the context of medicinal plants, radiation-induced mutagenesis holds significant potential for increasing the yield of bioactive compounds with pharmaceutical applications (Murthy et al., 2024).

Radiation-induced mutagenesis is a significant technique employed to induce genetic variation in medicinal plants, aimed at enhancing the production of secondary metabolites with therapeutic value. This method involves exposing plant material to ionizing radiation, such as gamma rays or X-rays, to induce DNA damage and subsequent mutations (Benslimani & Khelifi, 2014). Through the induction of genetic variations, radiation-induced mutagenesis offers a promising approach for augmenting the biosynthesis pathways of secondary metabolites in medicinal plants.

The process of radiation-induced mutagenesis involves exposing plant material to ionizing radiation, such as gamma rays or X-rays, at predetermined doses and durations. This exposure causes DNA damage, including single-strand breaks, double-strand breaks, and base modifications, leading to mutations in the plant genome. The extent and nature of mutations induced by radiation depend on factors such as the type and intensity of radiation, duration of exposure, and genetic background of the plant species (Mujib et al., 2022).

Radiation-induced mutagenesis has been widely employed to enhance the production of secondary metabolites in various medicinal plant species (Asif, 2023). Studies have demonstrated its effectiveness in inducing mutations in genes involved in secondary metabolite biosynthesis pathways, resulting in increased yields of bioactive compounds. For example, radiation mutagenesis has been used to increase the production of artemisinin in *Artemisia annua*, a potent antimalarial compound, and taxol in *Taxus* spp., a valuable anticancer agent (F. Al-Qurainy & Khan, 2010; Paddon et al., 2013).

Several research studies have highlighted the potential of radiation-induced mutagenesis in enhancing secondary metabolite production in medicinal plants. For instance, a study by D. S. Kim et al. (2013) utilized gamma radiation to induce mutations in *Panax ginseng*, resulting in increased ginsenoside content, compounds known for their pharmacological properties. Similarly, Arora et al. (2005) employed X-ray irradiation to induce mutations in *Ginkgo biloba,* leading to enhanced production of ginkgolides, bioactive compounds with neuroprotective effects.

While radiation-induced mutagenesis offers significant potential for enhancing secondary metabolite production in medicinal plants, several challenges and considerations must be addressed (Zhang & Björn, 2009). These include the need for precise control over radiation doses and exposure times to minimize detrimental effects on plant growth and development. Additionally, thorough screening and characterization of mutant populations are essential to identify individuals with desirable secondary metabolite profiles (Rifnas et al., 2020; Vardhan & Shukla, 2017).

Radiation-induced mutagenesis holds promise as a valuable tool for inducing genetic variation and enhancing secondary metabolite production in medicinal plants. Future research efforts should focus on optimizing radiation mutagenesis protocols, elucidating the molecular mechanisms underlying radiation-induced changes in secondary metabolite biosynthesis, and exploring the potential of combining radiation mutagenesis with other biotechnological approaches (El-Sayed, 2021).

9.4.3 CRISPR-Cas9 Technology

CRISPR-Cas9 technology has emerged as a revolutionary tool for precision genome editing in various organisms, including medicinal plants. This powerful molecular tool allows for the precise modification of DNA sequences, offering unprecedented opportunities for genetic manipulation and engineering (Mishra et al., 2023). In the context of medicinal plants, CRISPR-Cas9 technology holds immense potential for targeted modifications aimed at enhancing traits such as secondary metabolite production, stress tolerance, and disease resistance (Naik et al., 2023).

CRISPR-Cas9 technology is based on the natural defense mechanisms found in bacteria and archaea, which use CRISPR (clustered regularly interspaced short palindromic repeats) and Cas proteins to target and cleave foreign DNA sequences. The CRISPR-Cas9 system has been adapted for genome editing purposes to precisely target specific genomic loci and introduce modifications, such as insertions, deletions, or substitutions (Munawar & Ahmad, 2021).

The process of CRISPR-Cas9 genome editing involves several key steps. First, a guide RNA (gRNA) is designed to complement the target DNA sequence of interest. The gRNA directs the Cas9 nuclease to the target site, where it generates a double-strand break (DSB) in the DNA. The cell's natural repair mechanisms, such as non-homologous end joining (NHEJ) or homology-directed repair (HDR), can then be utilized to introduce desired modifications or insertions at the target locus (Xue & Greene, 2021; Barman et al., 2020).

CRISPR-Cas9 technology has numerous applications in medicinal plants, particularly in the context of secondary metabolite production. By precisely targeting genes involved in secondary metabolite biosynthesis pathways, one can modulate their expression to enhance the production of bioactive compounds (Shkryl et al., 2021). Additionally, CRISPR-Cas9 can be used to improve other agronomic traits, such as yield, quality, and stress tolerance, in medicinal plant species (Rao & Wang, 2021).

Several studies have demonstrated the effectiveness of CRISPR-Cas9 technology in modifying genes related to secondary metabolite production in medicinal plants. Rajyaguru et al. (2021) used CRISPR-Cas9 to knock out a gene involved in lignin biosynthesis in *Artemisia annua*, resulting in increased artemisinin content, a valuable antimalarial compound. Similarly, C. Liu (2023) employed CRISPR-Cas9 to target genes involved in triterpene biosynthesis in *Panax ginseng*, leading to enhanced ginsenoside production.

Despite its immense potential, CRISPR-Cas9 technology also presents several challenges and considerations in the context of medicinal plants. These include off-target effects, delivery methods for introducing CRISPR components into plant cells, and regulatory considerations surrounding genetically modified organisms (GMOs). Addressing these challenges will be crucial for the widespread adoption and application of CRISPR-Cas9 technology in medicinal plant research and breeding programs (Puchta, 2016).

CRISPR-Cas9 technology represents a groundbreaking approach to precision genome editing in medicinal plants, offering unprecedented opportunities for genetic manipulation and trait improvement (Mitra et al., 2023). Future research efforts should focus on refining CRISPR-Cas9 protocols, optimizing delivery methods, and addressing regulatory barriers to enable the full potential of this technology in medicinal plant research and biotechnology (Dey, 2021; Das et al., 2022).

The pursuit of inducing genetic variation in medicinal plants holds immense promise for the advancement of plant breeding and biotechnology.

The induction of genetic variation in medicinal plants represents a frontier at the nexus of plant breeding and biotechnology. Through mutagenesis methodologies encompassing chemical mutagenesis, radiation-induced mutagenesis, and precision genome editing with CRISPR-Cas9, individuals hold powerful tools to unlock the latent potential of medicinal plants (Arora et al., 2010) (**Figure 9.1**). As the journey unfolds, collaborative efforts and innovative approaches will drive transformative breakthroughs in harnessing genetic variation to optimize the production of bioactive compounds in medicinal plants (Jain & Maluszynski, 2004).

9.5 MUTAGENESIS APPLICATIONS IN MEDICINAL PLANTS

Mutagenesis applications in medicinal plants encompass a wide range of strategies aimed at inducing genetic mutations to improve the quality and quantity of secondary metabolites, which hold therapeutic value. Here are some key applications of mutagenesis in the context of medicinal plants.

9.5.1 Secondary Metabolite Enhancement

One of the primary applications of mutagenesis in medicinal plants is to amplify the production of specific secondary metabolites. Introducing targeted mutations in genes related to secondary metabolite biosynthesis provokes the overproduction of desired compounds, such as alkaloids, flavonoids, and terpenoids (Alamgir, 2018; Bharti et al., 2023). This method results in augmented yields of bioactive substances with promising pharmaceutical applications.

9.5.2 Novel Compound Development

Mutagenesis can be utilized to create entirely new secondary metabolites or novel variants of existing compounds. By inducing mutations in relevant biosynthetic pathways, it cultivates plant varieties with distinctive chemical profiles,

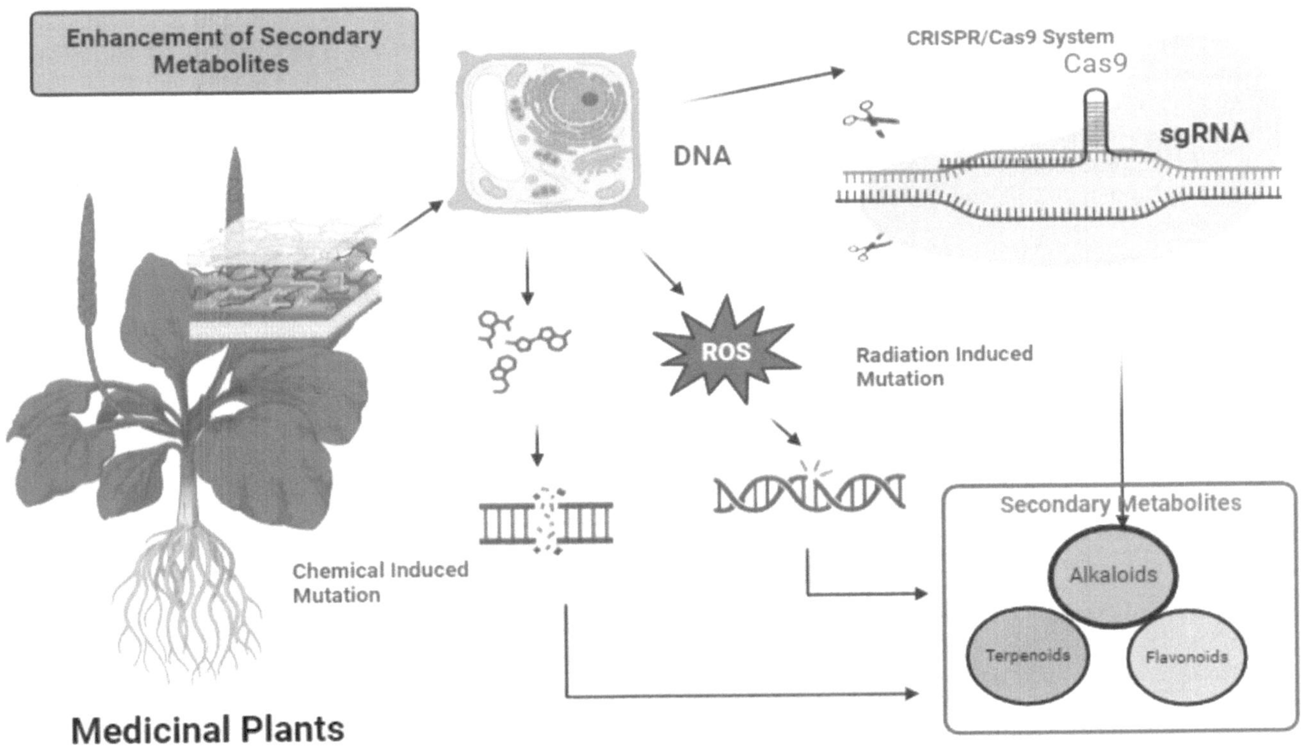

FIGURE 9.1 Enhancement of secondary metabolites through mutagenesis methodologies encompassing chemical mutagenesis, radiation-induced mutagenesis, and precision genome editing with CRISPR-Cas9 technology.

potentially yielding previously undiscovered bioactive compounds (Li et al., 2017). Various methods have been explored to manipulate the production of alkaloids from the opium poppy, *Papaver somniferum*, which are commonly used in pain relief formulations. Singh et al. (2019) employed genome editing techniques to investigate the function of the 4′OMT gene in the production of benzylisoquinoline alkaloids (BIAs) in opium poppy. Through the knockout of the 4′omt2 gene, they generated edited plants exhibiting reduced biosynthesis of several BIAs, including S-reticuline and laudanosine. This targeted editing also led to significant decreases in the levels of thebaine, codeine, noscapine, and papaverine, in a tissue-specific manner. This study demonstrated the effectiveness of genome editing in elucidating the regulatory roles of biosynthetic genes within metabolic pathways, offering opportunities for precise manipulation of metabolite flux (D'Amelia et al., 2021).

9.5.3 Stress Tolerance

Mutagenesis can be used to enhance the stress tolerance of medicinal plants. Mutants can be selected for their ability to thrive under adverse conditions, which is especially relevant for plants growing in challenging environments. These stress-tolerant mutants may exhibit altered secondary metabolite profiles in response to stress, leading to potential pharmaceutical applications (Niazian, 2019).

9.5.4 Disease Resistance

Inducing genetic mutations can also result in improved disease resistance in medicinal plants. By altering specific genes associated with defense mechanisms, one can develop plant varieties that are more resilient to pathogens. These disease-resistant plants can maintain their secondary metabolite production in the face of disease pressure (Agrawal & Kumar, 2021; Kolakar et al., 2018).

9.5.5 Accelerated Breeding

Mutagenesis expedites the traditional plant breeding process by generating genetic diversity in a shorter time frame. This can be particularly advantageous for medicinal plant breeding programs, as it allows for the rapid development of improved plant varieties with higher secondary metabolite yields (Kulkarni et al., 2016).

9.5.6 Rare and Endangered Species Conservation

Mutagenesis aids in propagating and conserving rare and endangered medicinal plant species. By inducing mutations in such species, it generates additional plants for conservation efforts, ensuring their survival while preserving their secondary metabolite content (Khan et al., 2012).

9.5.7 Genome Editing

With the advent of advanced genome-editing techniques like CRISPR-Cas9, mutagenesis has become even more precise and targeted. The capability to make specific modifications to genes involved in secondary metabolite biosynthesis leads to more controlled and predictable outcomes in terms of metabolite enhancement (Abdelrahman et al., 2018).

9.5.8 Pharmacological and Toxicological Studies

Mutagenesis allows for the generation of plants with altered secondary metabolite profiles, making it a valuable tool for studying the pharmacological and toxicological effects of these compounds. These mutant plants will be used to investigate the therapeutic potential and safety of specific bioactive compounds (Subramanian et al., 2018; Kuete, 2014). The application of mutagenesis in secondary metabolite production represents a promising avenue for improving the medicinal properties of plants and advancing the development of natural products for pharmaceutical and nutraceutical applications (Araújo et al., 2016). As research in this field continues to progress, it holds great promise for the sustainable and efficient production of bioactive compounds from medicinal plants, ultimately benefiting both the pharmaceutical industry and consumers seeking natural remedies (Hussein & El-Anssary, 2019). In conclusion, the application of mutagenesis in secondary metabolite production in medicinal plants has the potential to significantly enhance the yield and quality of bioactive compounds. The ability to manipulate the genetic makeup of medicinal plants through mutagenesis offers a powerful tool for improving the production of valuable secondary metabolites. By harnessing the inherent capacity of plants to produce bioactive compounds, industries can contribute to the development of new and effective natural products for medicinal and therapeutic purposes (Ramawat et al., 2009).

In addition to enhancing the yield and quality of known bioactive compounds, mutagenesis can also facilitate the discovery of novel metabolites with potential pharmaceutical applications. With careful selection and screening, mutated plant lines can yield a diverse array of chemical compounds, some of which may exhibit unique bioactivities that could lead to the development of new drugs or therapeutic agents (Kolakar et al., 2018).

Furthermore, the application of mutagenesis in secondary metabolite production aligns with the growing emphasis on sustainable and environmentally friendly approaches to pharmaceutical manufacturing. By optimizing the natural biosynthetic pathways within medicinal plants, it is possible to reduce reliance on synthetic production methods and promote the sustainable utilization of plant resources (Li et al., 2023).

As the field of mutagenesis continues to advance, it is essential to conduct thorough characterization and evaluation of mutated plant lines to ensure the safety and efficacy of the bioactive compounds they produce. This entails comprehensive chemical profiling, pharmacological testing, and toxicological assessments to validate the therapeutic potential and safety of the enhanced metabolites (Araújo et al., 2016).

The application of mutagenesis in secondary metabolite production in medicinal plants holds immense promise for addressing the demand for high-quality natural products with medicinal value.

9.6 SELECTION AND PROPAGATION OF HIGH-YIELDING MUTANTS

Selection and propagation of high-yielding mutants in medicinal plants involve a comprehensive approach that integrates various disciplines and methodologies. Understanding desired traits encompasses not only the yield and potency of active compounds but also other attributes, such as disease resistance, adaptability to different climates, and ecological sustainability (Kumar et al., 2019; Deng et al., 2020). For instance, in the case of *Artemisia annua*, a medicinal plant known for its antimalarial properties, traits such as artemisinin content and resistance to pests and diseases are crucial for enhancing yield and quality (Jelodar et al., 2014).

Advanced screening methods play a pivotal role in the identification and characterization of high-yielding mutants. High-throughput phenotyping platforms allow for the rapid and accurate assessment of plant traits under various environmental conditions (Shakoor et al., 2017). These platforms utilize techniques such as remote sensing, imaging, and spectroscopy to quantify plant morphology, physiology, and biochemical composition. Additionally, genomic approaches, including marker-assisted selection and next-generation sequencing, enable the identification of genetic markers associated with desirable traits (Hasan et al., 2021).

Inducing mutations is a key strategy for generating genetic variability in medicinal plants. Traditional methods such as chemical mutagenesis using agents like ethyl methane sulfonate (EMS) or physical mutagenesis through gamma irradiation have been widely employed (Singh et al., 2021). However, the advent of genome editing technologies, particularly CRISPR/Cas9, has revolutionized the field by allowing precise modifications of target genes associated with yield and quality traits (Nerkar et al., 2022). This approach offers the potential to accelerate the breeding process and overcome the limitations of traditional mutagenesis methods.

Selection criteria for high-yielding mutants must be carefully defined to ensure the success of breeding programs. These criteria may vary depending on the specific objectives of the breeding program and the intended use of the medicinal plant. For example, in the case of ginseng (*Panax ginseng*), selection criteria may include root biomass, ginsenoside content, and resistance to soil-borne pathogens (Goodwin & Best, 2023). Multi-state selection processes, combining phenotypic evaluation, molecular markers, and field trials, are often employed to identify superior mutants with desirable traits (Orf, 2008).

Propagation techniques play a critical role in the multiplication and dissemination of high-yielding mutants. Vegetative propagation methods such as cuttings, grafting, and tissue culture are commonly used to preserve the genetic integrity of elite plant lines and rapidly propagate them on a commercial scale (Hussain et al., 2012). Moreover, advancements in tissue culture techniques, including somatic embryogenesis and organogenesis, offer opportunities for the mass production of uniform and disease-free planting material (Rai et al., 2022). These techniques not only facilitate the rapid multiplication of high-yielding mutants but also enable the conservation of genetic diversity and the preservation of endangered medicinal plant species.

The selection and propagation of high-yielding mutants in medicinal plants require a multidisciplinary approach that encompasses genetics, genomics, physiology, and agronomy (Datta, 2023). By integrating advanced screening methods, mutation induction techniques, selection criteria, and propagation strategies, breeders can accelerate the development of superior cultivars with enhanced yield, quality, and resilience. Moreover, collaboration between academia, industry, and government agencies is essential to ensure the successful translation of research findings into practical applications for sustainable medicinal plant production and improved global healthcare.

9.7 EVALUATION OF MUTANTS

Evaluation of mutants for enhanced secondary metabolite production in medicinal plants is a multifaceted process that involves a comprehensive understanding of plant physiology, genetics, and biochemistry. Secondary metabolites, such as alkaloids, flavonoids, terpenoids, and phenolic compounds, play crucial roles in the therapeutic properties of medicinal plants (Alamgir, 2018; Harvey et al., 2015). Enhancing the production of these bioactive compounds can significantly increase the medicinal value and commercial potential of plant species. Various approaches, including mutation breeding, genetic engineering, and elicitation strategies, are employed to stimulate secondary metabolite biosynthesis in medicinal plants (Fazili et al., 2022). This evaluation process typically begins with the selection of promising mutant lines generated through mutagenesis techniques, such as chemical mutagenesis or radiation mutagenesis (Kodym & Afza, 2003).

Once mutant lines are obtained, screening methods are employed to identify individuals with enhanced secondary metabolite production. High-throughput screening techniques, including metabolomics and transcriptomics, allow for the rapid and simultaneous analysis of a large number of mutants to identify those with desirable phenotypes (Yang et al., 2020). For instance, metabolomic profiling can provide insights into changes in the abundance of specific secondary metabolites in mutant plants compared to wild-type controls. Transcriptomic analysis, on the other hand, can reveal alterations in gene expression patterns associated with secondary metabolite biosynthesis pathways. By correlating metabolite profiles with gene expression data, one can identify candidate genes and metabolic pathways responsible for enhanced secondary metabolite production in mutant plants (Stitt et al., 2010).

In addition to molecular and biochemical analyses, physiological parameters such as growth rate, biomass accumulation, and photosynthetic efficiency are evaluated to assess the overall performance of mutant lines (Gosa et al., 2022). While increased secondary metabolite production is desirable, it should not come at the expense of plant growth and vigor. Therefore, mutant lines exhibiting a balance between enhanced metabolite production and healthy growth characteristics are prioritized for further evaluation (Zalabák et al., 2013).

Following the initial screening phase, selected mutant lines undergo validation trials under controlled environmental conditions to confirm their stability and reproducibility (Sarsu et al., 2018). These trials involve monitoring secondary metabolite production over multiple generations to ensure that the observed enhancements are heritable and consistent across different growing conditions (Sharma et al., 2013). Additionally, field trials are conducted to evaluate the performance of mutant lines under realistic agricultural settings. Factors such as soil type, climate, and agronomic practices can influence secondary metabolite production in medicinal plants, making field trials essential for assessing the practicality and scalability of mutant lines for commercial cultivation (Pant et al., 2021).

Throughout the evaluation process, a combination of analytical techniques, including chromatography, spectroscopy, and molecular biology assays, is employed to quantify and characterize secondary metabolites in mutant plants. High-performance liquid chromatography (HPLC) and gas chromatography–mass spectrometry (GC-MS) are commonly used to separate and identify individual compounds within complex mixtures (Ni et al., 2012). Spectroscopic methods such as UV-Vis spectroscopy and nuclear magnetic resonance (NMR) spectroscopy provide insights into the chemical structures and functional groups present in secondary metabolites. Molecular biology assays, including qRT-PCR and promoter-reporter assays, can be employed to validate the upregulation of key biosynthetic genes associated with enhanced metabolite production in mutant plants (Liu et al., 2023).

The evaluation of mutants for enhanced secondary metabolite production in medicinal plants is a systematic and interdisciplinary process that involves the integration of genetics, physiology, biochemistry, and agronomy. Through a combination of mutagenesis techniques, high-throughput screening methods, and validation trials, lines with superior medicinal properties can be identified and characterized (Singh, 2023). These efforts not only contribute to the discovery of novel bioactive compounds but also facilitate the development of improved cultivars with enhanced therapeutic potential and commercial value.

9.8 ETHICAL, ENVIRONMENTAL, AND REGULATORY CONSIDERATIONS

Ethical, environmental, and regulatory considerations play a crucial role in the application of mutagenesis techniques for secondary metabolite production in medicinal plants. While mutagenesis offers promising benefits in enhancing the therapeutic potential of medicinal plants, it also raises important ethical and environmental concerns that must be addressed. Additionally, regulatory frameworks govern the use of mutagenesis in plant breeding and research to ensure safety and ethical standards are met.

9.8.1 Ethical Considerations

9.8.1.1 Biodiversity Conservation

Mutagenesis techniques should be applied with careful consideration for biodiversity conservation. Preservation of natural genetic diversity and respect for indigenous knowledge systems are essential ethical principles. The use of mutagenesis techniques for the enhancement of secondary metabolites raises important ethical considerations (Mishra et al., 2023). Preservation of natural genetic diversity and respect for indigenous knowledge systems should be central ethical principles when applying mutagenesis techniques for the enhancement of secondary metabolites. These principles are essential for maintaining the integrity of ecosystems and ensuring the long-term viability of species. Furthermore, it is crucial to consider the potential consequences and risks associated with mutagenesis techniques. This includes assessing the impact on non-target organisms, unintended genetic modifications, and the potential for loss of genetic diversity. Careful consideration and evaluation of the potential benefits and risks of mutagenesis techniques should be conducted to ensure that they align with conservation goals (S. M. Jain, 2010).

When considering the ethical implications of mutagenesis techniques for the enhancement of secondary metabolites, it is important to delve deeper into the intersection of science, ethics, and sustainable practices. Preservation of natural genetic diversity goes beyond the mere application of scientific techniques; it involves a profound understanding and respect for indigenous knowledge systems that have long coexisted with the natural world. Furthermore, assessing the potential consequences and risks associated with mutagenesis techniques requires a comprehensive evaluation (Z. Tadele, 2016). This evaluation should encompass not only the immediate impact on target species but also the potential effects on non-target organisms and the overall ecosystem. Unintended genetic modifications and the potential loss of genetic diversity must be carefully scrutinized to ensure the long-term viability of species and the integrity of ecosystems (Borsari et al., 2014).

9.8.1.2 Informed Consent

Stakeholders, including local communities and farmers, must be adequately informed and consulted about the potential risks and benefits associated with mutagenesis applications in medicinal plants (Uzochukwu et al., 2022). In order to effectively inform and consult stakeholders on mutagenesis applications in medicinal plants, it is essential to first establish a thorough understanding of the specific genetic modifications being made and their potential impact on the medicinal properties of the plants. This involves conducting comprehensive research and analysis to assess the risks and benefits associated with the use of mutagenesis techniques (George, 2011).

Furthermore, it is crucial to engage with a diverse range of stakeholders including scientists, regulatory authorities, agricultural experts, and representatives from local communities who may be impacted by the cultivation and use of these modified plants. By involving a wide array of perspectives and expertise, a more comprehensive understanding of the potential implications of mutagenesis applications can be achieved.

Additionally, fostering open and transparent communication channels is vital to ensure that stakeholders are well-informed about the process, potential outcomes, and any associated risks. This may involve organizing workshops, public forums, and informational materials that are accessible to all stakeholders (Shukla-Jones et al., 2018). As mutagenesis applications in medicinal plants can have far-reaching implications, it is important to approach the informing and consulting process with sensitivity and respect for diverse viewpoints. By incorporating diverse perspectives and engaging in open dialogue, a more informed and inclusive decision-making process can be achieved. Transparent communication of research findings, methodologies, and outcomes is essential to uphold ethical standards. Adherence to rigorous ethical guidelines and promotion of accountability in research practices are imperative (Lavery, 2007).

9.8.2 Environmental Considerations

9.8.2.1 Ecological Impact

Mutagenesis techniques may introduce genetic modifications that could have unintended ecological consequences. Assessments of the environmental impact, including potential effects on non-target organisms and ecosystems, should be conducted before widespread implementation (Wolt, 2017). Alkaloids, which are plant secondary metabolites, pose a risk to ruminants according to various studies (Adibah & Azzreena, 2019). However, strategies exist to mitigate their adverse effects. One method involves providing animals with supplements, while another entails offering a diverse array of forages containing different types of plant secondary metabolites. These supplements and varied forages can help counteract the harmful impact of alkaloids present in certain plants like endophyte-infected tall fescue and terpenes found in sagebrush. Through biochemical interactions, such as binding or other mechanisms, these approaches can minimize the negative effects of these compounds (Provenza et al., 2003).

9.8.2.2 Genetic Contamination

Cross-pollination between genetically modified plants and wild or cultivated relatives may result in genetic contamination. Preventative measures, such as isolation distances and containment strategies, should be implemented to minimize the risk of genetic spread (Abdelmigid, 2013).

Sustainable practices should be prioritized in mutagenesis research and agricultural production. This includes promoting agroecological approaches, minimizing chemical inputs, and conserving natural resources to ensure the long-term viability of medicinal plant cultivation (Patel et al., 2022).

9.8.3 Regulatory Considerations

9.8.3.1 Safety Assessment

Regulatory authorities require comprehensive safety assessments to evaluate the potential risks associated with mutagenesis-derived products. This includes assessing the toxicity, allergenicity, and nutritional composition of genetically modified plants. The assessment includes examining the potential presence of toxins or anti-nutrients that may affect consumer safety, comparing the allergenic proteins in the genetically modified plants with known allergens and conducting tests to detect allergic reactions and ensure that the nutritional quality of the food remains intact or possibly improved after the genetic modification (Booth et al., 2017).

9.8.3.2 Labeling and Traceability

Clear Labeling: Clear labeling involves providing accurate and understandable information about the mutagenesis-derived products on product packaging. This includes indicating whether the product is genetically modified, the specific traits introduced through mutagenesis, and any potential allergens present (Ribarits et al., 2014).

Traceability Systems: Traceability systems track the mutagenesis-derived products throughout the supply chain, from production to consumption. This enables the identification and recall of products if necessary and provides transparency regarding their origin, cultivation methods, and distribution (Aarts et al., 2002).

9.8.3.3 Compliance with Regulations

Breeders must comply with national and international regulations governing the use of mutagenesis techniques in plant breeding and research. This includes obtaining appropriate permits, conducting risk assessments, and adhering to ethical and environmental standards (Bhardwaj et al., 2020).

In summary, ethical, environmental, and regulatory considerations are essential aspects of applying mutagenesis techniques for secondary metabolite production in medicinal plants. Addressing these considerations thoughtfully and proactively ensures that mutagenesis applications are conducted responsibly, sustainably, and in compliance with ethical and regulatory standards. This approach is vital for maximizing the benefits of mutagenesis while minimizing potential risks to human health, biodiversity, and the environment (Mipeshwaree Devi et al., 2023).

9.9 FUTURE PROSPECTS

Prospects for mutagenesis applications in secondary metabolite production in medicinal plants are promising, with significant potential for advancements in various areas.

Continued advancements in genome editing technologies, such as CRISPR-Cas9 and base editing, will revolutionize mutagenesis approaches. These precise tools enable targeted modifications in plant genomes, allowing for more controlled enhancements in secondary metabolite production with minimal off-target effects (Das et al., 2022). The integration of genomics, transcriptomics, metabolomics, and other omics technologies will provide deeper insights into the complex regulatory networks governing secondary metabolite biosynthesis. This holistic approach will facilitate the identification of key genes, enzymes, and regulatory elements involved in secondary metabolite pathways, enabling more targeted mutagenesis strategies for enhanced metabolite production (Bindu et al., 2018).

Advances in synthetic biology and metabolic engineering will enable the design and construction of novel biosynthetic pathways for the production of desired secondary metabolites. Rational design combined with mutagenesis approaches will allow for the engineering of medicinal plants with tailored metabolic profiles optimized for specific therapeutic applications (Breitling & Takano, 2015). Harnessing the potential of plant-microbe interactions, particularly with endophytic and rhizospheric microbes, offers new avenues for enhancing secondary metabolite production in medicinal plants. Mutagenesis-based approaches can be employed to manipulate microbial communities or engineer beneficial symbiotic interactions to promote metabolite biosynthesis and plant growth (Mishra et al., 2021; Li et al., 2022).

Integration of mutagenesis with other advanced breeding techniques, such as marker-assisted selection and genomic selection, will accelerate the development of improved medicinal plant varieties with enhanced secondary metabolite yields and desirable agronomic traits. These multidisciplinary approaches will facilitate the breeding of high-performing cultivars tailored to specific environmental conditions and market demands (Barut et al., 2022).

Future research efforts will focus on developing sustainable production practices for mutagenesis-derived medicinal plants. Agro ecological approaches, organic farming methods, and precision agriculture technologies will be employed to optimize resource use efficiency, minimize environmental impacts, and ensure the long-term sustainability of medicinal plant cultivation (Alamgir & Alamgir, 2017).

With advancements in technology and research, mutagenesis-derived medicinal plant varieties are poised for commercialization. However, comprehensive regulatory frameworks will be necessary to ensure the safety, quality, and traceability of mutagenesis-derived products. Clear guidelines and standards will facilitate regulatory approval and market acceptance of these innovative plant varieties (Booth et al., 2017). Therefore, the future of mutagenesis applications in secondary metabolite production in medicinal plants holds immense potential for addressing global health challenges, advancing pharmaceutical discovery, and promoting sustainable agriculture. Through interdisciplinary collaborations, technological innovations, and responsible stewardship, mutagenesis-based approaches will continue to drive advancements in medicinal plant research and contribute to improving human health and well-being (Bahadur et al., 2007).

9.10 CONCLUSION

Enhancement of secondary metabolites holds great promise for the pharmaceutical industry as these compounds have potential therapeutic applications. The mutagenesis approach has proven to be an effective method for improving the yield of secondary metabolites in medicinal plants. Further research and development in this area could lead to the discovery of new bioactive compounds with valuable medicinal properties, thereby contributing to the advancement of plant-based medicine. The application of mutagenesis for enhanced secondary metabolite production in medicinal plants, highlights its significant potential in pharmaceutical research and agriculture, thereby opening up new possibilities for the development of novel medicines and natural products. Moreover, mutagenesis has proven to be a viable approach for optimizing plant-specific secondary metabolite production in plant cell culture. By employing mutagenesis techniques, significant increases in the production of secondary metabolites in medicinal plants have been achieved.

The increased production of secondary metabolites through mutagenesis opens up a wide array of possibilities for the pharmaceutical industry. With the potential therapeutic applications of these compounds, there is an opportunity for the development of novel medicines and natural products that could address unmet medical needs. Furthermore, the optimization of plant-specific secondary metabolite production in plant cell culture through mutagenesis techniques offers a sustainable and scalable approach for the large-scale production of these valuable compounds. Continued research and development in this area could lead to the discovery of new bioactive compounds with diverse medicinal properties, potentially impacting various therapeutic areas.

As the use of mutagenesis in medicinal plants advances, it is essential to consider regulatory frameworks and ethical practices to ensure the safe and responsible application of this technology. Collaboration with regulatory authorities and adherence to ethical guidelines will be key in navigating the development and commercialization of products derived from mutagenesis-enhanced secondary metabolites.

The findings from the mutagenesis approach for enhancing secondary metabolite production in medicinal plants present a compelling opportunity for the pharmaceutical industry and plant-based medicine. With further exploration, collaboration, and adherence to regulatory and ethical considerations, this approach has the potential to revolutionize the discovery and production of valuable bioactive compounds for therapeutic use.

REFERENCES

Aarts, H. J., Rie, J. P. P. V., & Kok, E. J. (2002). Traceability of genetically modified organisms. *Expert Review of Molecular Diagnostics*, *2*(1), 69–77.

Abdelmigid, H. M. (2013). New trends in genotoxicity testing of herbal medicinal plants. In *New insights into toxicity and drug testing* (pp. 89–120). InTech.

Abdelrahman, M., Al-Sadi, A. M., Pour-Aboughadareh, A., Burritt, D. J., & Tran, L. S. P. (2018). Genome editing using CRISPR/Cas9–targeted mutagenesis: An opportunity for yield improvements of crop plants grown under environmental stresses. *Plant Physiology and Biochemistry*, *131*, 31–36.

Adibah, K. Z. M., & Azzreena, M. A. (2019). Plant toxins: Alkaloids and their toxicities. *GSC Biological and Pharmaceutical Sciences*, *6*(2).

Agrawal, L., & Kumar, M. (2021). Improvement in ornamental, medicinal, and aromatic plants through induced mutation. *Journal of Applied Biology and Biotechnology*, *9*(4), 162–169.

Akthar, M. S., Degaga, B., & Azam, T. (2014). Antimicrobial activity of essential oils extracted from medicinal plants against the pathogenic microorganisms: A review. *Journal Issues*, *2350*, 1588.

Alamgir, A. N. M. (2018). Biotechnology, in vitro production of natural bioactive compounds, herbal preparation, and disease management (treatment and prevention). In *Therapeutic use of medicinal plants and their extracts: Vol. 2. Progress in drug research* (Vol. 74). Springer. https://doi.org/10.1007/978-3-319-92387-1_7

Alamgir, A. N. M., & Alamgir, A. N. M. (2017). Cultivation of herbal drugs, biotechnology, and in vitro production of secondary metabolites, high-value medicinal plants, herbal wealth, and herbal trade. In *Therapeutic use of medicinal plants and their extracts: Vol. 1: Pharmacognosy* (pp. 379–452). Springer.

Alami, M. M., Ouyang, Z., Zhang, Y., Shu, S., Yang, G., Mei, Z., & Wang, X. (2022). The current developments in medicinal plant genomics enabled the diversification of secondary metabolites' biosynthesis. *International Journal of Molecular Sciences*, *23*(24), 15932.

Ali, K., Ali, A., Khan, M. N., Rahman, S., Faizi, S., Ali, M. S., Khalifa., S. A. M., El-Seedi, H. R., & Musharraf, S. G. (2021). Rapid identification of common secondary metabolites of medicinal herbs using high-performance liquid chromatography with evaporative light scattering detector in extracts. *Metabolites*, *11*(8), 489.

Alok, A., Jain, P., Kumar, J., Yajnik, K., & Bhalothia, P. (2020). Genome engineering in medicinally important plants using CRISPR/Cas9 tool. In *Genome engineering via CRISPR-Cas9 system* (pp. 155–161). Academic Press.

Al-Qurainy, F., & Khan, S. (2010). Mutational approach for enhancement of artemisinin in *Artemisia annua*. *Journal of Medical Plants Research*, *4*(17), 1714–1726.

Araújo, S. D. S., Fernandes, T. C., Marin-Morales, M. A., Brasileiro-Vidal, A. C., & Benko-Iseppon, A. M. (2016). Mutagenicity, genotoxicity and cytotoxicity assays of medicinal plants: First step for drug development. In *Therapeutic medicinal plants: From lab to the market* (pp. 130–153). CRC Press.

Arijanti, S., & Suryaningsih, D. R. (2019). Biosynthesis of secondary metabolites (gingerol, shogaol, and zingerone) from callus of three ginger varieties. *Drug Invention Today*, *11*(2).

Arora, R., Gupta, D., Chawla, R., Sagar, R., Sharma, A., Kumar, R., Prasad, J., Singh, S., Samanta, N., & Sharma, R. K. (2005). Radioprotection by plant products: Present status and future prospects. *Phytotherapy Research: An International Journal Devoted to Pharmacological and Toxicological Evaluation of Natural Product Derivatives*, *19*(1), 1–22.

Arora, R., Mathur, A., & Mathur, A. K. (2010). Emerging trends in medicinal plant biotechnology. In *Medicinal plant biotechnology* (pp. 1–12). CAB International.

Asif, A. (2023). Mutagenesis in medicinal plants. In *Omics studies of medicinal plants* (pp. 211–241). CRC Press.

Bahadur, B., Reddy, K. J., & Rao, M. L. N. (2007). Medicinal plants: An overview. In *Advances in medical plants*. Univ Press Hyderabad.

Barman, A., Deb, B., & Chakraborty, S. (2020). A glance at genome editing with CRISPR–Cas9 technology. *Current Genetics*, *66*, 447–462.

Barut, M., Nadeem, M. A., Akgür, Ö., Tansi, L. S., Aasim, M., Altaf, M. T., & Baloch, F. S. (2022). Medicinal and aromatic plants in the omics era: Application of plant breeding andbiotechnology for plant secondary metabolite production. *Turkish Journal of Agriculture and Forestry*, *46*(2), 182–203.

Benslimani, N., & Khelifi, L. (2014). Use of radiomutagenesis for obtaining alkaloid-rich plant mutants. In M. Iqbal & A. Ahmad (Eds.), *Current trends in medicinal botany* (pp. 396–401). IK International Publishing House Pvt. Ltd.

Bertelli, A. A., & Das, D. K. (2009). Grapes, wines, resveratrol, and heart health. *Journal of Cardiovascular Pharmacology*, *54*(6), 468–476.

Bhardwaj, R., Sharma, K., Sharma, D. K., & Prakash, P. (2020, November 4). Conventional genetic improvement methods in medicinal and aromatic plants: A review. *International Journal of Economic Plants*, *7*, 156–161.

Bharti, P. K., Singh, S. K., & Kumari, S. (2023). Genetic manipulation in medicinal plants for enhancement of secondary metabolites. In *Genetic manipulation of secondary metabolites in medicinal plant* (pp. 129–145). Springer Nature Singapore.

Bhatti, M. Z., Ismail, H., & Kayani, W. K. (2022). Plant secondary metabolites: Therapeutic potential and pharmacological properties. In *Secondary metabolites-trends and reviews*. IntechOpen.

Bhojiya, A. A., & Joshi, H. (2024). Crispr gene editing for secondary metabolite production: A review. In *Gene editing in plants: CRISPR-Cas and its applications* (pp. 437–475). Springer.

Bindu, K. H., Mythili, J. B., & Radhika, R. M. (2018). Genetic engineering in medicinal and aromatic plants. In *Genetic engineering of horticultural crops* (pp. 249–271). Academic Press.

Booth, E. D., Rawlinson, P. J., Maria Fagundes, P., & Leiner, K. A. (2017). Regulatory requirements for genotoxicity assessment of plant protection product active ingredients, impurities, and metabolites. *Environmental and Molecular Mutagenesis*, *58*(5), 325–344.

Borsari, B., Neri, D., Ferretti, G., Tavoletti, S., & Yu, W. W. C. (2014). Critical evaluation of genetic manipulation for improved productivity: Is this a sustainable agenda? In *Sustainable food production includes human and environmental health* (pp. 43–83). Springer.

Breitling, R., & Takano, E. (2015). Synthetic biology advances for pharmaceutical production. *Current Opinion in Biotechnology, 35*, 46–51.

Burt, S. (2004). Essential oils: Their antibacterial properties and potential applications in foods – a review. *International Journal of Food Microbiology, 94*(3), 223–253.

Cadet, J., Delatour, T., Douki, T., Gasparutto, D., Pouget, J. P., & Ravanat, J. L. (1999). Hydroxyl radicals and DNA base damage. *Mutation Research/Fundamental and Molecular Mechanisms of Mutagenesis, 424*(1–2), 9–21.

Caleja, C., Ribeiro, A., Filomena Barreiro, M., & Ferreira, I. C. F. R. (2017). Phenolic compounds as nutraceuticals or functional food ingredients. *Current Pharmaceutical Design, 23*(19), 2787–2806.

Cardarelli, M., Rouphael, Y., Pellizzoni, M., Colla, G., & Lucini, L. (2017). Profile of bioactive secondary metabolites and antioxidant capacity of leaf exudates from eighteen Aloe species. *Industrial Crops and Products, 108*, 44–51.

Cardoso, J. C., Oliveira, M. E., & Cardoso, F. D. C. (2019). Advances and challenges on the in vitro production of secondary metabolites from medicinal plants. *Horticultura Brasileira, 37*, 124–132.

Chatterjee, N., & Walker, G. C. (2017). Mechanisms of DNA damage, repair, and mutagenesis. *Environmental and Molecular Mutagenesis, 58*(5), 235–263.

Coimbra, J. L. M., Carvalho, F. I. F. D., Oliveira, A. C. D., Silva, J. A. G. D., & Lorencetti, C. (2005). Comparação entre mutagênicos químico e físico em populações de aveia. *Ciência Rural, 35*, 46–55.

D'Amelia, V., Docimo, T., Crocoll, C., & Rigano, M. M. (2021). Specialized metabolites and valuable molecules in crop and medicinal plants: The evolution of their use and strategies for their production. *Genes, 12*(6), 936.

Daraban, G. M., Hlihor, R. M., Apostol, M., Bădeanu, M., & Şuteu, D. (2021, November). Secondary plant metabolites: Sources of principles for use in agriculture, food industry and for human health safety. In *2021 International Conference on e-Health and Bioengineering (EHB)* (pp. 1–4). IEEE.

Das, T., Ghorai, M., Pandey, D. K., Radha, Thakur, M., Rathour, S., Al-Tawaha, A. R., Bursal, E., Kumar, V., Nongdam, P., Shekhawat, M. S., Batiha, G. E.-S., Ghosh, A., Dwivedi, P., Kumar, V., Kumar, M., & Dey, A. (2022). CRISPR/Cas genome editing in engineering plant secondary metabolites of therapeutic benefits. In *Metabolic engineering in plants* (pp. 187–208). Springer Nature Singapore.

Datta, S. K. (2023). Technology package for induced mutagenesis. *Journal of Biology and Nature, 15*(1), 70–88.

de Faveri Favero, F., Grando, R., Nonato, F. R., Sousa, I. M., Queiroz, N. C., Longato, G. B., Galceran, H. C. T., Pupo, M. T., Gonçalves, J. E., Rodrigues, F. J., Rodrigues, T. S., França, S. C., Andreo, M. A., & Foglio, M. A. (2014). *Artemisia annua* L.: Evidence of sesquiterpene lactones' fraction antinociceptive activity. *BMC Complementary and Alternative Medicine, 14*, 1–11.

Deng, Y., Ning, Y., Yang, D. L., Zhai, K., Wang, G. L., & He, Z. (2020). Molecular basis of disease resistance and perspectives on breeding strategies for resistance improvement in crops. *Molecular Plant, 13*(10), 1402–1419.

Dey, A. (2021). CRISPR/Cas genome editing to optimize pharmacologically active plant natural products. *Pharmacological Research, 164*, 105359.

Divekar, P. A., Narayana, S., Divekar, B. A., Kumar, R., Gadratagi, B. G., Ray, A., Kumar, M., Sharma, A., Waghmare, R., Bhatt, N., Chauhan, P. S., Gupta, V., Shinde, A. K., Vennapusa, S. S. M., Jha, S., Tyagi, R., Sinha, P., & Behera, T. K. (2022). Plant secondary metabolites as defense tools against herbivores for sustainable crop protection. *International Journal of Molecular Sciences, 23*(5), 2690.

El-Sayed, E. R. (2021). Discovery of the anticancer drug vinblastine from the endophytic *Alternaria alternata* and yield improvement by gamma irradiation mutagenesis. *Journal of Applied Microbiology, 131*(6), 2886–2898.

Fazili, M. A., Bashir, I., Ahmad, M., Yaqoob, U., & Geelani, S. N. (2022). In vitro strategies for the enhancement of secondary metabolite production in plants: A review. *Bulletin of the National Research Centre, 46*(1), 35.

Fennell, C. W., Lindsey, K. L., McGaw, L. J., & Sparg, S. G. (2004). Assessing African medicinal plants for efficacy and safety: Pharmacological screening and toxicology. *Journal of Ethnopharmacology, 94*(2–3), 205–217.

Fernie, A. R., & Tohge, T. (2017). The genetics of plant metabolism. *Annual Review of Genetics, 51*, 287–310.

Friedberg, E. C., Walker, G. C., & Siede, W. (1995). *DNA repair and mutagenesis*. ASM Press.

George, P. (2011). Concerns regarding the safety and toxicity of medicinal plants-an overview. *Journal of Applied Pharmaceutical Science*, 40–44.

Goodwin, P. H., & Best, M. A. (2023). Ginsenosides and biotic stress responses of ginseng. *Plants, 12*(5), 1091.

Gosa, S. C., Koch, A., Shenhar, I., Hirschberg, J., Zamir, D., & Moshelion, M. (2022). The potential of dynamic physiological traits in young tomato plants to predict field-yield performance. *Plant Science, 315*, 111122.

Grosso, G., Stepaniak, U., Micek, A., Kozela, M., Stefler, D., Bobak, M., & Pająk, A. (2015). Dietary polyphenols are inversely associated with metabolic syndrome in Polish adults of the HAPIEE study. *European Journal of Nutrition, 54*(7), 941–950.

Harvey, A. L., Edrada-Ebel, R., & Quinn, R. J. (2015). The re-emergence of natural products for drug discovery in the genomics era. *Nature Reviews Drug Discovery, 14*(2), 111–129.

Hasan, N., Choudhary, S., Naaz, N., Sharma, N., & Laskar, R. A. (2021). Recent advancements in molecular marker-assisted selection and applications in plant breeding programmes. *Journal of Genetic Engineering and Biotechnology, 19*(1), 128.

Hou, M., Wang, R., Zhao, S., & Wang, Z. (2021). Ginsenosides in *Panax* genus and their biosynthesis. *Acta Pharmaceutica Sinica B, 11*(7), 1813–1834.

Hussain, A., Qarshi, I. A., Nazir, H., & Ullah, I. (2012). Plant tissue culture: Current status and opportunities. *Recent Advances in Plant in Vitro Culture, 6*(10), 1–28.

Hussein, R. A., & El-Anssary, A. A. (2019). Plants secondary metabolites: The key drivers of the pharmacological actions of medicinal plants. *Herbal Medicine, 1*(3), 11–30.

Iqbal, I., Wilairatana, P., Saqib, F., Nasir, B., Wahid, M., Latif, M. F., Ahmed, M., Shah, M. I., Abbas, M., Sheikh, M. A., Khan, Z. U. I., & Mubarak, M. S. (2023). Plant polyphenols and their potential benefits on cardiovascular health: A review. *Molecules, 28*(17), 6403.

Jain, C., Khatana, S., & Vijayvergia, R. (2019). Bioactivity of secondary metabolites of various plants: A review. *International Journal of Pharma Sciences and Research, 10*(2), 494–504.

Jain, S. M. (2010). Mutagenesis in crop improvement under the climate change. *Romanian Biotechnological Letters, 15*(2), 88–106.

Jain, S. M., & Maluszynski, M. (2004). Induced mutations and biotechnology in improving crops. In *In vitro application in crop improvement* (pp. 187–220). CRC Press.

Jelodar, N. B., Bhatt, A., Mohamed, K., & Keng, C. L. (2014). New cultivation approaches of *Artemisia annua* L. for a sustainable production of the antimalarial drug artemisinin. *Journal of Medicinal Plants Research, 8*(10), 441–447.

Kakouri, E., Talebi, M., & Tarantilis, P. A. (2024). *Echinacea spp.*: The cold-fighter herbal remedy? *Pharmacological Research-Modern Chinese Medicine*, 100397.

Kaushik, B., Sharma, J., Kumar, P., & Shourie, A. (2021). Phytochemical properties and pharmacological role of plants: Secondary metabolites. *Biosciences Biotechnology Research Asia: An International, Open Access, Peer-Reviewed Research Journal, 18*(1), 23.

Khan, S., Al-Qurainy, F., & Nadeem, M. (2012). Biotechnological approaches for conservation and improvement of rare and endangered plants of Saudi Arabia. *Saudi Journal of Biological Sciences, 19*(1), 1–11.

Khursheed, S., Wani, M. R., Khan, S., & Bhat, T. A. (2021). Mutagens, their types and mechanism of action with an emphasis on sodium azide and gamma radiations aamir raina, Shazia BI Ansari. In *Mutagenesis, cytotoxicity and crop improvement: Revolutionizing food science* (pp. 1–11). Cambridge Scholars Publishing.

Kim, D. S., Song, M., Kim, S. H., Jang, D. S., Kim, J. B., Ha, B. K., Lee, J. H., & Jeong, I. Y. (2013). The improvement of ginsenoside accumulation in *Panax ginseng* as a result of γ-irradiation. *Journal of Ginseng Research, 37*(3), 332.

Kirakosyan, A., Sirvent, T. M., Gibson, D. M., & Kaufman, P. B. (2004). The production of hypericins and hyperforin by in vitro cultures of St. John's wort (*Hypericum perforatum*). *Biotechnology and Applied Biochemistry, 39*(1), 71–81.

Kiran, U., & Abdin, M. Z. (Eds.). (2020). *Transgenic technology based value addition in plant biotechnology*. Academic Press.

Kodym, A., & Afza, R. (2003). Physical and chemical mutagenesis. In *Plant functional genomics* (pp. 189–203). Springer.

Kolakar, S. S., Nadukeri, S., Jakkeral, S. A., Lakshmana, D., Hanumanthappa, M., & Gangaprasad, S. (2018). Role of mutation breeding in improvement of medicinal and aromatic crops. *Journal of Pharmacognosy and Phytochemistry, 7*(3S), 425–429.

Kovarovič, J., Bystrická, J., Vollmannová, A., Tóth, T., & Brindza, J. (2019). Biologically valuable substances in garlic (*Allium sativum* L.)–a review. *Journal of Central European Agriculture, 20*(1), 292–304.

Kuete, V. (2014). *Toxicological survey of African medicinal plants*. Elsevier.

Kulkarni, R. N., Baskaran, K., & Jhang, T. (2016). Breeding medicinal plant, periwinkle [*Catharanthus roseus* (L) G. Don]: A review. *Plant Genetic Resources, 14*(4), 283–302.

Kumar, J., Choudhary, A. K., Gupta, D. S., & Kumar, S. (2019). Towards exploitation of adaptive traits for climate-resilient smart pulses. *International Journal of Molecular Sciences, 20*(12), 2971.

Kumar, S. (2014). Alkaloidal drugs-A review. *Asian Journal of Pharmaceutical Science & Technology, 4*(3), 107–119.

Kuroda, Y. (1996). Bio-antimutagenic activity of green tea catechins in cultured Chinese hamster V79 cells. *Mutation Research/Environmental Mutagenesis and Related Subjects, 361*(2–3), 179–186.

Lal, R. K., Chanotiya, C. S., & Gupta, P. (2020). Induced mutation breeding for qualitative and quantitative traits and varietal development in medicinal and aromatic crops at CSIR-CIMAP, Lucknow (India): Past and recent accomplishment. *International Journal of Radiation Biology, 96*(12), 1513–1527.

Lavery, J. V. (Ed.). (2007). *Ethical issues in international biomedical research: A casebook*. Oxford University Press.

Le, K. C., Ho, T. T., Paek, K. Y., & Park, S. Y. (2019). Low dose gamma radiation increases the biomass and ginsenoside content of callus and adventitious root cultures of wild ginseng (*Panax ginseng* Mayer). *Industrial Crops and Products, 130*, 16–24.

Li, B., Cui, G., Shen, G., Zhan, Z., Huang, L., Chen, J., & Qi, X. (2017). Targeted mutagenesis in the medicinal plant *Salvia miltiorrhiza*. *Scientific Reports, 7*(1), 43320.

Li, Z., Wang, Y., Xu, M., Liu, H., Li, L., & Xu, D. (2023). Molecular mechanism overview of metabolite biosynthesis in medicinal plants. *Plant Physiology and Biochemistry, 204*, 108125.

Li, Z., Wen, W., Qin, M., He, Y., Xu, D., & Li, L. (2022). Biosynthetic mechanisms of secondary metabolites promoted by the interaction between endophytes and plant hosts. *Frontiers in Microbiology, 13*, 928967.

Liu, A., Zhu, Y., Wang, Y., Wang, T., Zhao, S., Feng, K., Wang, Y., Yuan, Y., Xu, J., Chen, X., & Wu, P. (2023). Molecular identification of phenylalanine ammonia lyase-encoding genes EfPALs and EfPAL2-interacting transcription factors in *Euryale ferox*. *Frontiers in Plant Science, 14*, 1114345.

Liu, C. (2023). Targeted editing of the ginsenoside biosynthesis pathway in ginseng using CRISPR/Cas9 technology and its implications. *Medicinal Plant Research*, 13.

Liu, Y., Li, Y., Wang, A., Xu, Z., Li, C., Wang, Z., Wang, X., & Li, J. (2024). Enhancing cold resistance in Banana (*Musa spp.*) through EMS-induced mutagenesis, L-Hyp pressure selection: Phenotypic alterations, biomass composition, and transcriptomic insights. *BMC Plant Biology, 24*(1), 1–15.

Mailänder, L. K., Lorenz, P., Bitterling, H., Stintzing, F. C., Daniels, R., & Kammerer, D. R. (2022). Phytochemical characterization of chamomile (*Matricaria recutita* L.) roots and evaluation of their antioxidant and antibacterial potential. *Molecules, 27*(23), 8508.

Maleki, S. J., Crespo, J. F., & Cabanillas, B. (2019). Anti-inflammatory effects of flavonoids. *Food Chemistry, 299*, 125124.

Marrelli, M. (2021) Medicinal plants. *Plants, 10*(7), 1355. https://doi.org/10.3390/plants10071355

Mimica-Dukic, N., & Bozin, B. (2008). *Mentha L.* species (*Lamiaceae*) as promising sources of bioactive secondary metabolites. *Current Pharmaceutical Design, 14*(29), 3141–3150.

Mipeshwaree Devi, A., Khedashwori Devi, K., Premi Devi, P., Lakshmipriyari Devi, M., & Das, S. (2023). Metabolic engineering of plant secondary metabolites: Prospects and its technological challenges. *Frontiers in Plant Science, 14*, 1171154.

Mishra, R., Agarwal, P., & Mohanty, A. (2023). Applications of genome editing techniques for the improvement of medicinal plants. In *Phytochemical genomics: Plant metabolomics and medicinal plant genomics* (pp. 545–569). Springer Nature Singapore.

Mishra, S., Sahu, P. K., Agarwal, V., & Singh, N. (2021). Exploiting endophytic microbes as micro-factories for plant secondary metabolite production. *Applied Microbiology and Biotechnology, 105*(18), 6579–6596.

Mistry, V., Tiwari, P., Patel, P., Vishwakarma, G. S., Lee, G. J., & Sharma, A. (2022). Ethyl methane sulfonate and sodium azide-mediated chemical and X-ray-mediated physical mutagenesis positively regulate Peroxidase 1 gene activity and biosynthesis of antineoplastic vinblastine in *Catharanthus roseus*. *Plants*, *11*(21), 2885.

Mitra, S., Anand, U., Ghorai, M., Kant, N., Kumar, M., Radha, Singh, B., Bhattacharya, K., & Dey, A. (2023). Genome editing technologies, mechanisms and improved production of therapeutic phytochemicals: Opportunities and prospects. *Biotechnology and Bioengineering*, *120*(1), 82–94.

Moore, B. D., Andrew, R. L., Külheim, C., & Foley, W. J. (2014). Explaining intraspecific diversity in plant secondary metabolites in an ecological context. *New Phytologist*, *201*(3), 733–750.

Moussaieff, A., & Mechoulam, R. (2009). Boswellia resin: From religious ceremonies to medical uses; a review of in-vitro, in-vivo and clinical trials. *Journal of Pharmacy and Pharmacology*, *61*(10), 1281–1293.

Mujib, A., Fatima, S., & Malik, M. Q. (2022). Gamma ray–induced tissue responses and improved secondary metabolites accumulation in Catharanthus roseus. *Applied Microbiology and Biotechnology*, *106*(18), 6109–6123.

Mumtaz, B., & Tausif, S. (2023). Investigation of optimum dose of gamma irradiation and ems (single and combination treatments) to induce mutation in safflower (*Carthamus tinctorius*) wild cultivar. *Journal for Research in Applied Sciences and Biotechnology*, *2*(6), 135–136.

Munawar, N., & Ahmad, A. (2021). CRISPR/Cas system: An introduction. In *CRISPR crops: The future of food security* (pp. 1–35). Springer.

Murthy, H. N., Joseph, K. S., Paek, K. Y., & Park, S. Y. (2024). Production of specialized metabolites in plant cell and organo-cultures: The role of gamma radiation in eliciting secondary metabolism. *International Journal of Radiation Biology*, 1–11.

Naik, B. J., Vasamsetti, B. M. K., Kim, S. C., Gunti, M., Swamy, M. K., & Mekapogu, M. (2023). Genome editing in medicinal plants using CRISPR/Cas9 tool. In *Phytochemical genomics: Plant metabolomics and medicinal plant genomics* (pp. 571–589). Springer Nature Singapore.

Nalina, L., Rajamani, K., Shanmugasundaram, K. A., & Boomiga, M. (2022). Breeding and conservation of medicinal plants in India. In *Medicinal and aromatic plants of India* (Vol. 1, pp. 201–236). Springer International Publishing.

Nerkar, G., Devarumath, S., Purankar, M., Kumar, A., Valarmathi, R., Devarumath, R., & Appunu, C. (2022). Advances in crop breeding through precision genome editing. *Frontiers in Genetics*, *13*, 880195.

Newman, D. J., & Cragg, G. M. (2016). Natural products as sources of new drugs from 1981 to 2014. *Journal of Natural Products*, *79*(3), 629–661.

Ni, Y., Mei, M., & Kokot, S. (2012). One-and two-dimensional gas chromatography–mass spectrometry and high performance liquid chromatography–diode-array detector fingerprints of complex substances: A comparison of classification performance of similar, complex *Rhizoma curcumae* samples with the aid of chemometrics. *Analytica Chimica Acta*, *712*, 37–44.

Niazian, M. (2019). Application of genetics and biotechnology for improving medicinal plants. *Planta*, *249*, 953–973.

Oladosu, Y., Rafii, M. Y., Abdullah, N., Hussin, G., Ramli, A., Rahim, H. A., Miah, G., & Usman, M. (2016). Principle and application of plant mutagenesis in crop improvement: A review. *Biotechnology & Biotechnological Equipment*, *30*(1), 1–16.

Orf, J. H. (2008). Breeding, genetics, and production of soybeans. In *Soybeans* (pp. 33–65). AOCS Press.

Paddon, C. J., Westfall, P. J., Pitera, D. J., Benjamin, K., Fisher, K., McPhee, D., Newman, J. D., Morel, D. D., Rabinovitch-Deere, C. A., Reynolds, R. W., Tsuruta, H., Newman, K. L., Keasling, J. D., Ro, D. K., Paradise, E. M., Lee, T. S., Kirby, J., Chang, M. C., Withers, S. T., . . . Alper, H. S. (2013). High-level semi-synthetic production of the potent antimalarial artemisinin. *Nature*, *496*(7446), 528–532.

Pandey, N., Meena, R. P., Rai, S. K., & Pandey-Rai, S. (2016). In vitro generation of high artemisinin yielding salt tolerant somaclonal variant and development of SCAR marker in *Artemisia annua* L. *Plant Cell, Tissue and Organ Culture (PCTOC)*, *127*, 301–314.

Pant, P., Pandey, S., & Dall'Acqua, S. (2021). The influence of environmental conditions on secondary metabolites in medicinal plants: A literature review. *Chemistry & Biodiversity*, *18*(11), e2100345.

Patel, P. T., Desai, T. A., Virdiya, Y. A., & Kugashiya, K. G. (2022). Genetic pollution in plants: A review. *The Pharma Innovation Journal*, *11*(5), 331–334.

Pathak, S., Agarwal, A. V., Agarwal, P., & Trivedi, P. K. (2019). Secondary metabolite pathways in medicinal plants: Approaches in reconstruction and analysis. In *Molecular approaches in plant biology and environmental challenges* (pp. 339–364). Springer.

Pereira, E., Barros, L., Dueñas, M., Antonio, A. L., Santos-Buelga, C., & Ferreira, I. C. (2015). Gamma irradiation improves the extractability of phenolic compounds in *Ginkgo biloba* L. *Industrial Crops and Products*, *74*, 144–149.

Pierik, R., Ballaré, C. L., & Dicke, M. (2014). Ecology of plant volatiles: Taking a plant community perspective. *Plant, Cell & Environment*, *37*(8), 1845–1853.

Provenza, F. D., Villalba, J. J., Dziba, L. E., Atwood, S. B., & Banner, R. E. (2003). Linking herbivore experience, varied diets, and plant biochemical diversity. *Small Ruminant Research*, *49*(3), 257–274.

Prusinowska, R., & Śmigielski, K. B. (2014). Composition, biological properties and therapeutic effects of lavender L). A review. *Herba Polonica*, *60*(2), 56–66.

Puchta, H. (2016). Applying CRISPR/Cas for genome engineering in plants: The best is yet to come. *Current Opinion in Plant Biology*, *36*, 1–8. [3].

Rahimi, R., & Abdollahi, M. (2012). Herbal medicines for the management of irritable bowel syndrome: A comprehensive review. *World Journal of Gastroenterology: WJG*, *18*(7), 589.

Rai, A. C., Kumar, A., Modi, A., & Singh, M. (Eds.). (2022). *Advances in plant tissue culture: Current developments and future trends*. Academic Press.

Rai, A., Saito, K., & Yamazaki, M. (2017). Integrated omics analysis of specialized metabolism in medicinal plants. *The Plant Journal*, *90*(4), 764–787.

Raju, S., & Das, M. (2024). Medicinal plants industry in India: Challenges, opportunities and sustainability. *Medicinal Plants – International Journal of Phytomedicines and Related Industries*, *16*(1), 1–14.

Rajyaguru, R., Maheshala, N., Mootapally, C., Nathani, N., Tomar, R., Bhalani, H., & Sharma, P. (2021). CRISPR-Cas9 approaches to enhance contents of plant secondary metabolites. In *Biotechnological approaches to enhance plant secondary metabolites* (pp. 133–152). CRC Press.

Ramawat, K. G., Dass, S., & Mathur, M. (2009). The chemical diversity of bioactive molecules and therapeutic potential of

medicinal plants. In *Herbal drugs: Ethnomedicine to modern medicine* (pp. 7–32). Springer.
Rao, M. J., & Wang, L. (2021). CRISPR/Cas9 technology for improving agronomic traits and future prospective in agriculture. *Planta*, *254*, 1–16.
Ribarits, A., Brüller, W., Hartmann, J., Hochegger, R., Mechtler, K., Peterseil, V., Söllinger, J., Stepanek, W., Widhalm, I., Wögerbauer, M., & Leonhardt, C. (2014). Use of novel techniques in plant breeding and practical consequences concerning detection, traceability, labeling, and risk assessment. *AgBioForum*, *17*(2), 183–190.
Rifnas, L. M., Vidanapathirana, N. P., Silva, T. D., Dahanayake, N., Weerasinghe, S. S., & Subasinghe, S. (2020). Effect of gamma radiation on survival rate of Allamanda cathartica-an indigenous medicinal plant. *International Journal of Minor Fruits, Medicinal and Aromatic Plants*, *6*(1), 50–53.
Rime, J., Dinesh, M. R., Sankaran, M., Shivashankara, K. S., Rekha, A., & Ravishankar, K. V. (2019). Evaluation and characterization of EMS derived mutant populations in mango. *Scientia Horticulturae*, *254*, 55–60.
Roychowdhury, R., & Tah, J. (2013). Mutagenesis – a potential approach for crop improvement. In *Crop improvement: New approaches and modern techniques* (pp. 149–187). Springer.
Saad, B., Zaid, H., Shanak, S., Kadan, S. (2017). Introduction to medicinal plant safety and efficacy. In *Anti-diabetes and anti-obesity medicinal plants and phytochemicals: Safety, efficacy, and action mechanisms* (pp. 21–55). Springer.
Saema, S., Shaheen, N., & Pandey, V. (2023). Immunostimulatory properties of *Echinacea purpurea* and conservation strategy. In *Plants for immunity and conservation strategies* (pp. 153–168). Springer Nature Singapore.
Saini, P., Saini, P., Kaur, J. J., Francies, R. M., Gani, M., Rajendra, A. A., Negi, N., Jagtap, A., Kadam, A., Singh, C., & Chauhan, S. S. (2020). Molecular approaches for harvesting natural diversity for crop improvement. In *Rediscovery of genetic and genomic resources for future food security* (pp. 67–169). Springer.
Samadi, N., Naghavi, M. R., Moratalla-López, N., Alonso, G. L., & Shokrpour, M. (2022). Morphological, molecular and phytochemical variations induced by colchicine and EMS chemical mutagens in *Crocus sativus* L. *Food Chemistry: Molecular Sciences*, *4*, 100086.
Santos-Buelga, C., & González-Paramás, A. M. (2016). Flavonoids: Functions, metabolism and biotechnology. In *Industrial biotechnology of vitamins, biopigments, and antioxidants* (pp 469–495). Wiley.
Sarsu, F., Ghanim, A., Das, P., Bahuguna, R. N., Kusolwa, P. M., Ashraf, M., Jayasuriya, D. M., Usman, M., Abubakar, S., Hussein, S., Reddy, K. R., Joshi, M. V., Shah, F., Alghamdi, S. S., Girma, M., Panigrahi, R., & Ingelbrecht, I. (2018). *Pre-field screening protocols for heat-tolerant mutants in rice* (p. 39). Springer Nature.
Saslis-Lagoudakis, C. H., Savolainen, V., Williamson, E. M., Forest, F., Wagstaff, S. J., Baral, S. R., Watson, M. F., Pendry, C. A., & Hawkins, J. A. (2012). Phylogenies reveal predictive power of traditional medicine in bioprospecting. *Proceedings of the National Academy of Sciences*, *109*(39), 15835–15840.
Satish, L., Shamili, S., Yolcu, S., Lavanya, G., Alavilli, H., & Swamy, M. K. (2020). Biosynthesis of secondary metabolites in plants as influenced by different factors. In *Plant-derived bioactives: Production, properties and therapeutic applications* (pp. 61–100). Springer.
Sawarkar, H. A., & Mansoori, M. H. (2023). Methodology for physical and chemical mutagenic treatments. In *Biotechnologies and genetics in plant mutation breeding* (pp. 15–44). Apple Academic Press.
Schuler, M. A., & Werck-Reichhart, D. (2003). Functional genomics of P450s. *Annual Review of Plant Biology*, *54*(1), 629–667.
Serpeloni, J. M., Barcelos, G. R. M., Mori, M. P., Yanagui, K., Vilegas, W., Varanda, E. A., & de Syllos Colus, I. M. (2011). Cytotoxic and mutagenic evaluation of extracts from plant species of the *Miconia* genus and their influence on doxorubicin-induced mutagenicity: An in vitro analysis. *Experimental and Toxicologic Pathology*, *63*(5), 499–504.
Serrat, X., Esteban, R., Guibourt, N., Moysset, L., Nogués, S., & Lalanne, E. (2014). EMS mutagenesis in mature seed-derived rice calli as a new method for rapidly obtaining TILLING mutant populations. *Plant Methods*, *10*, 1–14.
Setyawati, A., Pujiasmanto, B., Fatawi, A., & Batubara, I. (2021, April). Secondary metabolites of turmeric and ginger on various altitudes and soil characteristics. In *IOP Conference Series: Earth and Environmental Science* (Vol. 724, No. 1, p. 012020). IOP Publishing.
Shakoor, N., Lee, S., & Mockler, T. C. (2017). High throughput phenotyping to accelerate crop breeding and monitoring of diseases in the field. *Current Opinion in Plant Biology*, *38*, 184–192.
Sharma, L., Maurya, B., & Rai, S. P. (2023). An overview of biotechnological interventions and abiotic elicitors on biomass and withanolide biosynthesis in *Withania somnifera* (L.) Dunal. *Industrial Crops and Products*, *193*, 116238.
Sharma, M., Koul, A., Sharma, D., Kaul, S., Swamy, M. K., & Dhar, M. K. (2019). Metabolic engineering strategies for enhancing the production of bio-active compounds from medicinal plants. In *Natural bio-active compounds: Vol. 3: Biotechnology, bioengineering, and molecular approaches* (pp. 287–316). Springer Nature.
Sharma, P., Padh, H., & Shrivastava, N. (2013). Hairy root cultures: A suitable biological system for studying secondary metabolic pathways in plants. *Engineering in Life Sciences*, *13*(1), 62–75.
Shkryl, Y., Yugay, Y., Avramenko, T., Grigorchuk, V., Gorpenchenko, T., Grischenko, O., & Bulgakov, V. (2021). CRISPR/Cas9-mediated knockout of HOS1 reveals its role in the regulation of secondary metabolism in Arabidopsis thaliana. *Plants*, *10*(1), 104.
Shukla-Jones, A., Friedrichs, S., & Winickoff, D. E. (2018). Gene editing in an international context: Scientific, economic and social issues across sectors. OECD Science, Technology and Industry Working Papers, 2018/04, OECD Publishing, Paris.
Singh, A., Menéndez-Perdomo, I. M., & Facchini, P. J. (2019). Benzylisoquinoline alkaloid biosynthesis in opium poppy: An update. *Phytochemistry Reviews*, *18*(6), 1457–1482.
Singh, H. (2006). Prospects and challenges for harnessing opportunities in medicinal plants sector in India. *Law, Environmemt and Development Journal*, *2*, 196.
Singh, H. (2023). Enhancement of plant secondary metabolites by genetic manipulation. In *Genetic manipulation of secondary metabolites in medicinal plant* (pp. 59–90). Springer Nature Singapore.
Singh, H., Khar, A., & Verma, P. (2021). Induced mutagenesis for genetic improvement of Allium genetic resources: A comprehensive review. *Genetic Resources and Crop Evolution*, *68*(7), 2669–2690.
Stitt, M., Sulpice, R., & Keurentjes, J. (2010). Metabolic networks: How to identify key components in the regulation of metabolism and growth. *Plant Physiology*, *152*(2), 428–444.

Subramanian, K., Sankaramourthy, D., & Gunasekaran, M. (2018). Toxicity studies related to medicinal plants. In *Natural products and drug discovery* (pp. 491–505). Elsevier.

Suprasanna, P., Mirajkar, S. J., & Bhagwat, S. G. (2015). Induced mutations and crop improvement. In *Plant biology and biotechnology: Vol. I: Plant diversity, organization, function and improvement* (pp. 593–617). Springer India.

Tadele, Z. (2016). Mutagenesis and TILLING to dissect gene function in plants. *Current Genomics*, *17*(6), 499–508.

Upadhyay, A., Upadhyaya, I., Kollanoor-Johny, A., & Venkitanarayanan, K. (2014). Combating pathogenic microorganisms using plant-derived antimicrobials: A minireview of the mechanistic basis. *BioMed Research International*, *2014*.

Uzochukwu, S., Esiobu, N. D., Okoli, A. S., Nwoba, E. G., Christpeace, E. N., Adetunji, C. O., Okonkwo, J. C., Akande, I. S., Ude, C. U., Agu, K. C., Chukwuma, S. C., Ododo, O. R., Ejike, C. E. C. C., Enweani, I. B., Adeoye, A. O., & Ubi, B. E. (Eds.). (2022). *Biosafety and bioethics in biotechnology: Policy, advocacy, and capacity building*. CRC Press.

Vardhan, P. V., & Shukla, L. I. (2017). Gamma irradiation of medicinally important plants and the enhancement of secondary metabolite production. *International Journal of Radiation Biology*, *93*(9), 967–979.

Verpoorte, R., Contin, A., & Memelink, J. (2002). Biotechnology for the production of plant secondary metabolites. *Phytochemistry Reviews*, *1*, 13–25.

Wagner, G. J., & Kroumova, A. B. (2008). The use of RNAi to elucidate and manipulate secondary metabolite synthesis in plants. In *Current perspectives in microRNAs (miRNA)* (pp. 431–459). Springer.

Weng, J. K., & Noel, J. P. (2012). Chemodiversity in Selaginella: A reference system for parallel and convergent metabolic evolution in terrestrial plants. *Frontiers in Plant Science*, *3*, 1–16.

Wink, M. (2015). Modes of action of herbal medicines and plant secondary metabolites. *Medicines*, *2*(3), 251–286.

Wolt, J. D. (2017). Safety, security, and policy considerations for plant genome editing. *Progress in Molecular Biology and Translational Science*, *149*, 215–241.

Xue, C., & Greene, E. C. (2021). DNA repair pathway choices in CRISPR-Cas9-mediated genome editing. *Trends in Genetics*, *37*(7), 639–656.

Yang, W., Feng, H., Zhang, X., Zhang, J., Doonan, J. H., Batchelor, W. D., Messina, C. D., Chapman, S. C., Zhu, J., & Yan, J. (2020). Crop phenomics and high-throughput phenotyping: Past decades, current challenges, and future perspectives. *Molecular Plant*, *13*(2), 187–214.

Yasmin, S., & Nehvi, F. A. (2017). Ethyl methane sulphonate-A potent mutagen for inducing genetic variability in saffron (*Crocus sativus* L.). *Applied Biological Research*, *19*(2), 164–171.

Yoon, J. M., Zhao, L., & Shanks, J. V. (2013). Metabolic engineering with plants for a sustainable biobased economy. *Annual Review of Chemical and Biomolecular Engineering*, *4*, 211–237.

Zalabák, D., Pospíšilová, H., Šmehilová, M., Mrízová, K., Frébort, I., & Galuszka, P. (2013). Genetic engineering of cytokinin metabolism: Prospective way to improve agricultural traits of crop plants. *Biotechnology Advances*, *31*(1), 97–117.

Zeki, K. A. R. A., & Doğan, O. (2022). Reactions of some grapevine rootstock cuttings to mutagenic applications. *Selcuk Journal of Agriculture and Food Sciences*, *36*(2), 238–246.

Zenebe, W., Pechánová, O., & Bernátová, I. (2001). Protective effects of red wine polyphenolic compounds on the cardiovascular system. *Experimental & Clinical Cardiology*, *6*(3), 153.

Zhang, W. J., & Björn, L. O. (2009). The effect of ultraviolet radiation on the accumulation of medicinal compounds in plants. *Fitoterapia*, *80*(4), 207–218.

10 GWAS Approaches in Medicinal Plants for Enhanced Secondary Metabolite Production

Kiran, Neeraj Kumar, Sudeep Sindhu, Banita Gawaria, Ashika Ansari, Aniket Dhariwal, Pankaj Sharma, and Rahul Kumar

10.1 INTRODUCTION

Metabolites are considered intermediate or end products of the physiological and developmental status of organisms. The approximate number of plant metabolites found is 20,000, which means that natural products serve diverse and extraordinary functions (Pichersky & Lewinsohn, 2011; Schwab, 2003). These metabolites perform many indispensable functions in cellular replacement, growth, complete allocation of plant resources, and environmental interaction. In addition, they are also sources of human medicines, nutrition, and energy. The metabolites are basically divided into two major categories: primary and secondary metabolites. Primary metabolites are mainly involved in the proper growth and development of plants, while secondary metabolites are involved in stress response mechanisms due to constant environmental change (Weng et al., 2012).

Metabolite diversity is an advantageous factor in taking plant models to depict metabolite biosynthesis and metabolic pathway regulation. Recently, using both forward and reverse genetic approaches, investigations have been conducted for varying plant metabolism and genetic bases for it (Cardoso et al., 2014; Meihls et al., 2013; Quadrana et al., 2014). This advancement in profiling techniques enabled the metabolite variation analysis between and within species (Chen et al., 2014). Until now, many metabolic studies have been done on the basis of population linkage mapping as recombinant inbred line sets are attained from two or more two parental accessions by using low-density markers as RFLP and SSR markers (Lisec et al., 2009; Toubiana et al., 2012; Matsuda et al., 2012). Despite some promising advances thanks to next-generation sequencing technologies, the resulting high-density maps are still not detailed enough for proper understanding and identification of mQTLs associated with particular metabolites. Until now there has been a challenge associated with analyzing this diverse and plentiful germplasm. As a method has been developed to identify mQTLs (Gong et al., 2013; Kliebenstein, 2009; Rowe et al., 2008), it has not been possible to scale it up. The constraints arise from the necessity to establish individual populations and correspondingly low-resolution stemming from both limited recombination within the linkage population and utilization of low-density markers. Consequently, this process of identifying the genes or nucleotides underlying quantitative trait loci (QTL) remains arduous and time-intensive (Alseekh et al., 2013; Huang et al., 2014).

High-throughput genotyping technologies' emergence like microarray and next-generation sequencing have prompted plant genetic research's paradigm shift, especially through the use of association mapping techniques. In its early days, association mapping was restricted to only a small subset of metabolic traits; however, it has since broadened its scope to include other plant-related metabolic characteristics. Integrating metabolomics with GWAS has proved to be an advanced and powerful forward genetics strategy for revealing complex biochemical as well as genetic mechanisms underlying plant metabolism. Recent research demonstrated this by combining GWAS and metabolomics to enhance the forward genetics approach in studying plant metabolism genetics and biochemistry complexity (Chen et al., 2014; Wen et al., 2014). The recent advances in metabolite-based GWAS in plants are highlighted in this chapter.

10.2 IMPORTANCE OF SECONDARY METABOLITES

Secondary metabolites are majorly considered essential components for the stress defense mechanism of plants. They also play a role in inducing flowering, setting roots, and the delays abscission by maintaining deciduous behavior and perineal growth of plants. They can be repellents and also act as antimicrobials. There are main five types of secondary metabolites considered: saponins, alkaloids, steroids, terpenes/terpenoids, and polyphenols.

10.3 ALKALOIDS

Alkaloids are a subgroup of secondary metabolites of plants. They have nitrogen atoms in their structure, which makes them display a variety of useful properties (e.g., anti-inflammatory, antipsychotic, and antiparasitic). Some

DOI: 10.1201/9781003475491-10

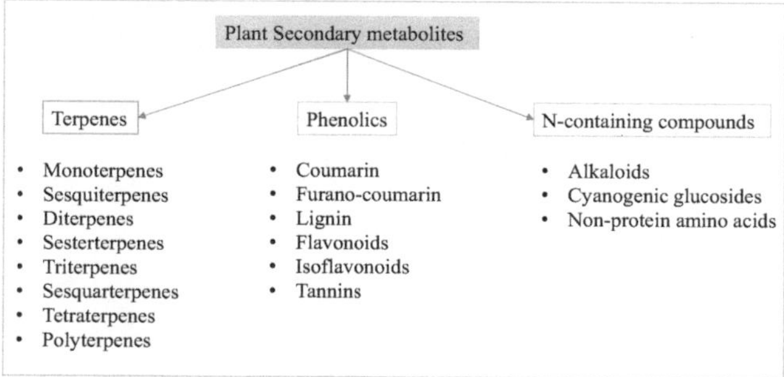

FIGURE 10.1 Classification of plant secondary metabolites based on their chemical composition.

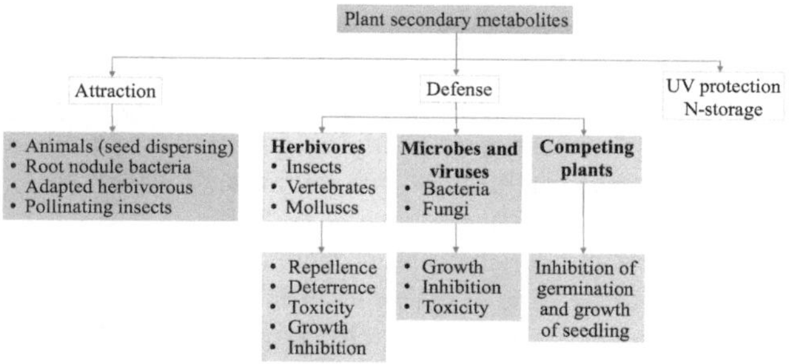

FIGURE 10.2 Functional classification of plant secondary metabolites.

alkaloids have also been shown to possess anticancer properties according to Debnath et al. (2018) and Vyas et al. (2016). According to Abdulazeem et al. (2018), I found out that the study conducted on *Opuntia polyacantha* alkaloids reveals that they may have potential for use in treating breast cancer cells (MCF-7) as opposed to the normal epithelial cell line (WRL-68).

10.4 POLYPHENOLS

Polyphenols are regarded among the broadest categories of secondary metabolites in plants. They are known for various biological activities like anti-inflammatory, cancer-fighting, and antioxidant effects, as observed by Kabera et al. (2014) and Vyas et al. (2017), with reference to these three properties. One other benefit of polyphenols beyond these is their link to decreasing chances of getting heart diseases. This was mentioned by Pastor et al. (2019). Many researches have been done on their ability to prevent aging, and one of the research projects showed that they have unique plant extracts that can be used to protect us from the harmful rays of the sun as well as prevent photo aging, one such being the polyphenols in rosemary (*Rosmarinus officinalis*) and grapefruits (*Citrus paradisi*), which has been studied by Nobile et al. (2016) in relation to protection against sun exposure and photo-inflammation. Accordingly, Ortega et al. (2019) documented the antiviral capabilities of methoxyflavones extracted from *Marcetia taxifolia* against poliovirus type-1, herpes simplex virus type-1, and hepatitis B virus.

10.5 SAPONINS

The soapwort plant gave birth to the term 'saponin', which has historical uses in soap production, mainly its roots (*sapo* means 'soap' in Latin). Most importantly, these chemicals play a vital role in plant life by acting as antimicrobials and protecting them from insects. In addition to their importance in the plant world, saponins present various biological processes, such as having hypoglycaemic effects and lowering cholesterol levels in the blood stream. They are also said to have antiviral and anti-inflammatory properties, among many others, according to Desai et al. (2009). Moreover, saponins from *Berberis vulgaris* have been shown by panels of scientists like Meliani et al. (2011) and Kaur et al. (2018) to have antidiabetic properties. In addition, they reduce the levels of both cholesterol and triglycerides in streptozotocin-induced diabetic rodents.

10.6 TERPENES/TERPENOIDS

Being parts of essential oils originating from plants, terpenes are very diverse structurally. They are widely used in manufacturing cosmetics and other health-related products, perfumery being a strong part of it. Synthetic terpenes are used as food additives, fragrances, and flavorings (Caputi & Aprea, 2011). It is worth noting that terpenes have a range of well-known health benefits. For instance, they reduce inflammation in chronic pain, and treats addiction against fungal and bacterial infections for epilepsy and cancer management. This allows us to focus more, have a better mood, deal with stress, and pay attention better. Also, it can help us fight some respiratory pathogens like MRSA. Myrcene, a prominent terpene abundant in *Cannabis*, is notably recognized for its sleep-inducing properties.

10.7 STEROIDS

Steroids are substances with low molecular weight and high fat-loving character. They originate from cholesterol, which is a very diverse group found in nature that can be obtained from land and water bodies or made artificially. Consisting of cholesterol-based molecules called sterols, steroids are widely used in medicine because they possess body-building properties that involve burning fats, resulting in reduction of weight, as well as increased muscle mass, when compared with other anabolic agents. Synthetic steroids find widespread use across various medical properties (e.g., anesthetic, anticancer, anti-asthmatic, anti-hormone, anti-inflammatory, contraceptive, cardioregulatory, antibiotic), as evidenced by studies such as those conducted by Aav et al. (2005), Jovanović-Šanta et al. (2015), and Lopez et al. (2014).

10.8 PHYTOCHEMICAL SCREENING

Initial quantitative phytochemical screening was conducted by applying the following strategies:

Steroids: By taking 1 ml of extract and dissolving it in 10 ml chloroform and adding to it equal sulphuric acid volume. Two layers are formed, upper with red and lower, yellow with green fluorescence. This indicated steroid presence (Gibbs, 1974).

Terpenoids: In 2 ml of acid anhydride with sulphuric acid (conc.), add 2 ml extract, and if blue and green rings are visible, terpenoids are present. (Ayoola et al., 2008).

Fatty Acids: The 0.5 ml from the extract was mixed with 5 ml ether which was then allowed to dry on filter paper as the final step before observing fat presence through transparently flat filter papers with fatty acids (Ayoola et al., 2008).

Tannins: A few drops of 1% lead acetate solution were mixed with 2 ml of the extract, and if there is any yellowish precipitate formed, then it means that tannins are present (Treare et al., 1985).

Saponins: A graduated cylinder was used to combine 5 ml of extract and 20 ml of distilled water, then the mixture was agitated for 15 minutes. The presence of saponins was detected by the formation of foam (Kumar et al., 2009).

Anthocyanins: The presence of anthocyanins is depicted by color changing from pink-red to blue-violet when 2 ml of an aqueous extract is mixed with 2 ml of 2N HCl and ammonia (Paris & Moyse, 1969).

Leuco-anthocyanins: The presence of leuco-anthocyanins can be noted when the combination of 5 ml of a water-based extract with 5 ml of isoamyl alcohol results in an upper red layer (Paris & Moyse, 1969).

Coumarins: A yellow color develops to indicate the presence of coumarins when a solution from many pharmaceutical investigations is combined with 3 ml of NaOH is 10% and 2 ml in volume (Rizk, 1982).

Emodins: The addition of 2 ml of NH_4OH and 3 ml of benzene treated the extract. The existence of emodins is indicated by a red color (Rizk, 1982).

10.9 NEXT-GENERATION SEQUENCING – INFORMATION ON MEDICINAL PLANTS

As Nogueira et al. (2018) noted, plants offer an enormous variety of essential chemicals supporting different industries such as food, feed, pharmaceuticals, and biomaterials hence being termed as remarkable 'green chemical factories'. Nevertheless, the excessive harvesting and overuse of herbs used in the pharmaceutical industry expose them to numerous threats such as extinction within their areas of origin (Niazian, 2019). Atanasov et al. (2015) elaborate on how human activities like environmental destruction, desertification, industrialization, wildfires, conversion of land for farming purposes, and climate change enhance the susceptibility of wild-growing medicinal plants to depletion.

Researchers must adopt biotechnological approaches when dealing with domestication deficits confronting the conservation and utilization of medicinal plants (MPs). Zhao et al. (2018) suggest that one way of achieving this is by making use of next-generation sequencing (NGS) since it allows for an in-depth examination of genomic and transcriptomic data for these plant species. Merging these genomic and transcriptomic datasets alongside proteomics as well as metabolomics data facilitates unique natural products found in herbal MPs. Furthermore, NGS gives various details on which many molecular techniques such as DNA barcoding may be built; hence, it serves as a platform that can be utilized to identify genetic diversity within different populations of MPs; in addition, this helps in accurate classifying of plants at genus and species levels, as elucidated by Techen et al. (2014).

RAD-seq is used in mediating fast and accurate location of the wild and unexplored population of MPs making possible identification of major genes and enzymes for important bio-synthetic pathways. Transcriptome analysis of medicinal plants can thus be effectively done with next-generation sequencing (NGS), which is through the sequencing of whole genomes at full length or NGS-based mapping. Following NGS analysis, the implementation of targeted genome editing methods such as ZFNs, TALENs, and CRISPR-Cas9 introduces synthetic biology for genetic and metabolic engineering in medicinal plants, paving the way for the creation of new varieties enriched with desired bio-products, as suggested by Pouvreau et al. (2018).

TABLE 10.1
List of Important Alkaloids and Polyphenols Used in Pharmacology

Secondary metabolites	Source	Type	Function
Nicotine	*Nicotiana tobaccum*	**Alkaloid**	**Stimulant, insecticide, anti-inflammatory**
Solenine	*Solanum tuberosum*	**Alkaloid**	**Anti- fungal, sedative, anticonvulsant, anti-inflammatory**
Quercetin	**Onion**	**Polyphenol**	**Reduces coronary heart disease risk, reduces arthreoscleriosis plaques**
Ferulic acid	**Vegetable and maize bran**	**Polyphenol**	**Antidiabetic**
Resveratrol	**Wine**	**Polyphenol**	**Prevents platelet aggregation, inhibits cycloxygenase I enzyme**
Codeine	*Papaver somniferum*	**Alkaloid**	**Analgesic, antidiarrheal, antidepressant, sedative, and hypnotic**

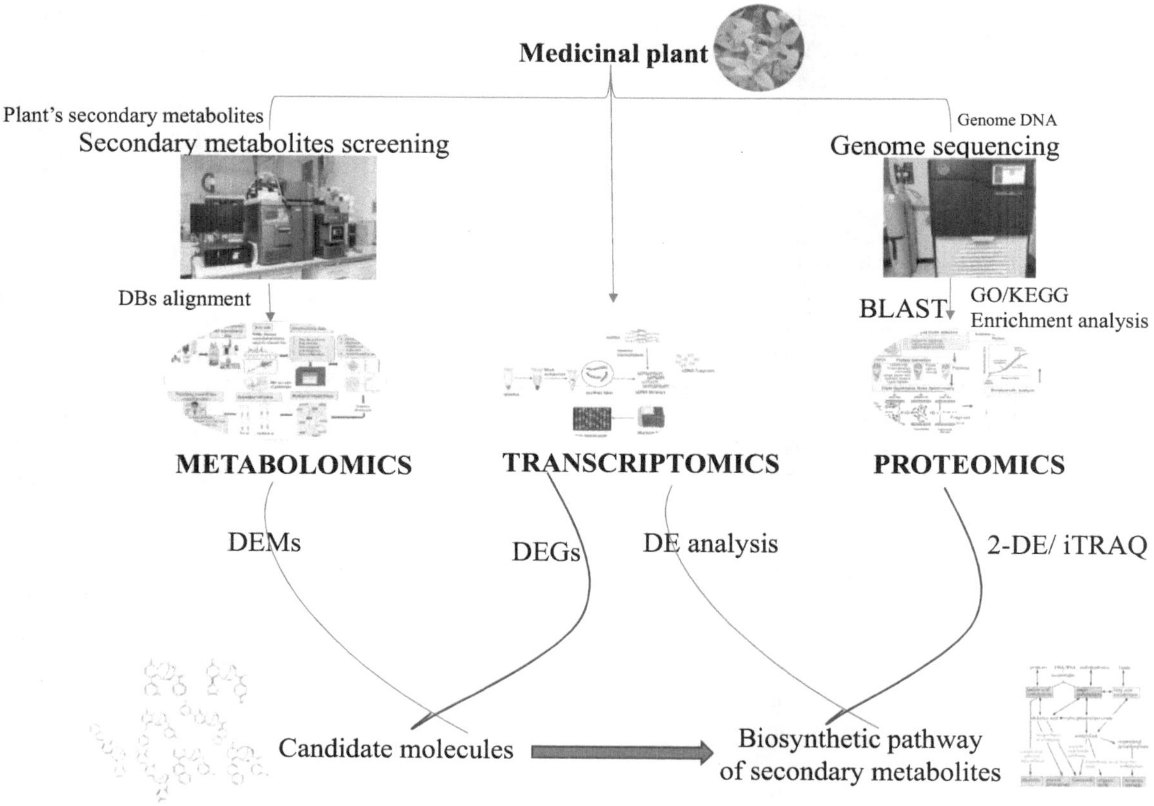

FIGURE 10.3 Roadmap for population evolution through RAD sequencing technology.

NGS techniques are known for being swift and inexpensive tools that have become very effective instruments in the identification, sequence analysis, and genotyping of several thousands of makers within a genome of interest at once as asserted by Davey et al. (2011). According to Liu et al. (2017), these strategies are crucial for studying metabolisms of non-model plants, such as MPs. The technology of NGS is used in profiling gene expression and sequencing cDNA, so-called RNA-seq, for the in-depth sequencing of target populations' transcriptomes. When compared with microarrays, NGS-based analysis reveals novel genetic information by allowing researchers to identify differing expressions of homoeologous or paralogous gene copies. Furthermore, DNA barcoding benefits from the NGS technology for the rapid determination of genetic variability in MPs at both the genus and species level, as explained by Techen et al. (2014). With no reference genome in the wild population, RAD-seq methods can be employed in genotyping utilizing reduced representation sequencing data from NGS, as described by Davey et al. (2011) and Zargar et al. (2015).

10.10 GWAS (GENOME-WIDE ASSOCIATION STUDIES) – DISCOVERY OF CANDIDATE GENES

Zhu et al. (2008) highlighted the potentiality of GWAS for finding important alleles inside plant germplasm as GWAS represents a credible way of detecting possibly important alleles in plant germplasms. For detecting associations between DNA sequence variations and phenotypic variations like single-nucleotide polymorphisms (SNPs), GWAS depends on linkage disequilibrium (LD) association mapping depends on the population's LD extent while Sullivan and Losos have been credited for deconstructing a lot of complex qualitative and quantitative features in rice using GWAS techniques (Zhang et al., 2016).

According to Mwadzingeni et al. (2017), GWAS performs a significant role as a fundamental operational method to discover the genetic bases underlying and uncover complicated genes that control economically important traits, particularly in the agriculture sector. GWAS aspires to recognize hereditary sections that encompass markers genes or quantitative trait loci (QTL) linked with principal agronomical attributes via the application of the marker-trait associations (MTAs) in heterogeneous populations such as improved cultivars and elite breeding lines. The knowledge is useful in molecular breeding methods utilizing markers, uncovering genes and putting genes from one organism into another.

To delve into the genetic and biochemical intricacies underlying plant metabolism, metabolite-based GWAS (mGWAS) emerges as a potent forward genetics technique, as highlighted in recent research. The wealth of diversity and high inheritability observed in most metabolites in plants augments the utility of mGWAS. Exploring each gene's precise role in a plant's genetic material, mGWAS supports functional genomic studies while revealing something about plant metabolism genetics (Singh et al., 2022). As an illustration, some physically and nutritionally valuable substances synthesized in plants have been suggested by mGWAS, thus opening new avenues for biomedical research and metabolic engineering processes. Besides, it is possible to integrate GWAS discoveries with other genome-wide datasets, especially the ones from transcript profiling and proteomics, making it easier to quickly detect more genes and possible metabolic routes (thereby stressing the interdisciplinary character of contemporary plant science).

mGWAS illuminate an intricate relationship between genotypes, environment, and development concerning the build-up of secondary metabolites, also providing useful perspectives on how plant metabolism varies across different tissues and maturity stages in terms of spatial and temporal regulation of natural variations in metabolites. Abhijith et al. (2022) claim that GWAS of natural genetic variation for entire genomes has proven to be a very valuable aid for speedily identifying crucial genes and alleles involved in complex traits. By combining high-throughput phenotyping with GWAS, a pioneering idea could lead to a deeper insight into the genetic structure of plants, enabling the generation of an entire inheritance map for each characteristic. In addition, the integration of high-throughput genome sequencing technology with omics technologies, including metabolomics, leads to functional genomics which is also referred to as phytochemical genomics. This interdisciplinary strategy permits pinpointing specific chemical substances produced by genes taking part in biosynthetic routes (Yamazaki et al., 2018). By linking genetic variation to specific phenotypic traits, phenotyping of diverse panels for GWAS or biparental mapping populations for quantitative trait locus (QTL) analysis aids in deciphering the genetic architecture of complex traits and understanding how specific DNA variants influence diverse phenotypic inheritance.

10.11 DETECTION OF QTLS THROUGH GWAS

Several statistical models, such as the general linear model (GLM) and mixed linear model (MLM), have been developed for GWAS computational analysis. Merrick et al. (2022) have shown that multi-locus models like Farm CPU (Fixed and random model circulating probability unification), MLMM (multi-locus mixed model), and BLINK (Bayesian-information and linkage-disequilibrium iteratively nested keyway) are superior to single-locus models. What they are emphasizing is that these multi-locus models have improved computing efficacy, more powerful statistics, and less calculation load. This indicates that they are effective in genetic analysis. Enyew et al. (2022) argue that BLINK is better than other genome-wide association studies (GWAS) methods. This is because it can easily identify important relationships between different genetic markers in a short time (superior computational power) as well as increasing the accuracy of marker-trait association detection. They also use the Bayesian information criterion (BIC) for BLINK instead of the prefix sum trick employed in GCTA. The other main difference lies with respect to how LD is incorporated: Throughout the model, Farm CPU employs a binning strategy while BLINK utilizes LD. This improvement helps it identify genetic associations more effectively (Krishnappa et al., 2023). Importantly, QTL mapping, which is used to identify the genetic architecture of complex traits in plants, is still crucial for this purpose (Holland, 2007; Devate et al., 2022).

10.12 GWAS OVER BIPARENTAL MAPPING

GWAS provides many benefits since it allows for the identification of QTLs for multiple traits with high resolution in only one analysis. It is more efficient and costs less to map QTL using different germplasm than biparental QTL mapping (Zhang et al., 2022). The resolution and power of association studies depend on the extent of linkage disequilibrium (LD) across the genome, which is influenced by factors, such as recombination frequency, population history, and mating system (Qanbari et al., 2020). Successful application of association mapping in wheat has led to the detection of marker trait associations and QTLs for various traits (Mulugeta et al., 2023). High-density SNP markers used in GWAS enable the examination of large gene pools representative of diverse breeding reservoirs. GWAS is a method that is most suitable for the identification of strong QTLs that demonstrate effects, whether under normal or stressed conditions (Sukumaran et al., 2018). Thus, GWAS has emerged as a potent and highly utilized tool for the study of complex traits.

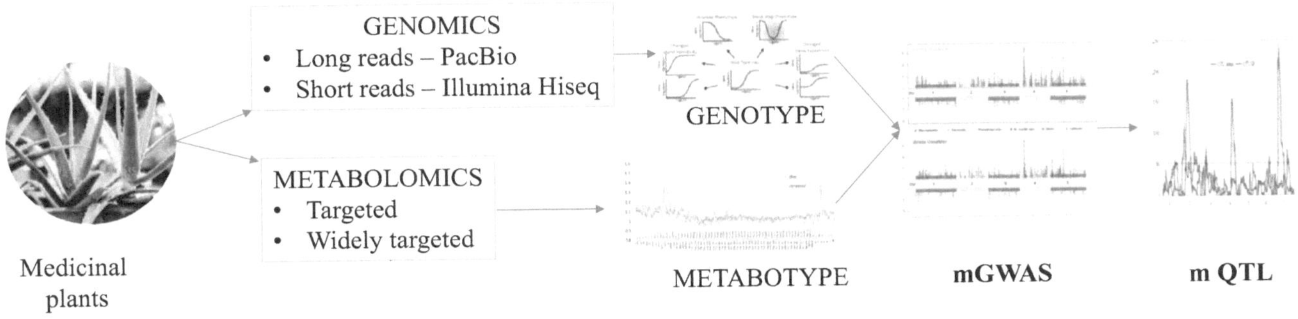

FIGURE 10.4 Discovery of QTL-mQTL conformation for secondary metabolites.

Mapping quantitative traits using platforms with high-density genotyping as SNP arrays has made GWAS stand out. It has greatly helped in pinpointing many genes for complex agronomic traits in rice, especially by drawing from unrelated, diverse germplasm pools (Wang et al., 2011). While GWAS helps in finding new alleles related to complex traits, its application in marker-assisted breeding requires additional validation in breeding populations. Therefore, a better alternate could be to conduct these association studies in breeding lines that are adapted, where results can directly be applied in contemporary breeding programs (Cobb et al., 2019). Potential candidates for GWAS have been identified by elite breeding lines that approach the end of their selection cycle keeping in mind the fact that the latter can enable breeders to design crosses that are more targeted and, consequently, retain required gene diversity in improved populations following further selection.

10.13 MARKER-ASSISTED SELECTION (MAS) IN PLANT BREEDING

The utilization of marker-assisted selection (MAS) is increasing within the plant breeding discipline. Consequently, it is increasingly preferred in breeding for plants with pathogen resistance controlled by one major gene or other such traits. To discover quantitative trait loci (QTLs) and then search for candidate genes under these QTLs, biparental recombinant inbred lines, doubled haploid sets, or genetically characterized accessions are used in population genetics (Wang et al., 2023). Molecular breeding can be improved when different phenotypes are genetically broken down for marker-assisted selection (MAS) and genomic selection (GS) purposes. For animals' genomic selection (GS) utilizes genome-wide markers for evaluating breeding value. Because in GS any gene/QTL can potentially be targeted unlike in MAS where only a subset of such genes is focused upon, it is better suited for selection on quantitative traits governed by multiple minor genes compared to MAS.

Challenges are present in both MAS and GS approaches. So that few markers can cover the whole genome, it is hard to guarantee that each major and minor effect locus for target traits is in linkage disequilibrium (LD) with at least one of the genome-wide markers (Schaid et al., 2018). However, just including high-density markers with significant effects in GS models can increase GS accuracy further. They suggested that additional markers with no effect cause disturbance in the forecasts (Cardno et al., 1999). GWAS seeks to identify relevant genotype-phenotype relationships and causative gene(s) of a given trait. Such studies use statistics to investigate genetic markers capable of predicting traits all over the human genome. GWAS looks for genes or markers that influence phenotypes by analyzing association statistics between phenotypic measurements and all markers in the genome, whereas PheWAS attempts to trace individual genetic loci and consequently genes themselves back to traits that they underlie by means of the dataset-wide spatiotemporal distribution of characteristics in organisms (Fong, 2022). PheWAS has made significant advancements toward the identification of SNP-disease associations in medical sciences. The scan may be used to detect associated variants based on GWAS that are often represented through a Manhattan plot displaying genomic locations and association strength.

10.14 FUNCTIONAL INFERENCE FROM GWAS

To elucidate the biological underpinnings of phenotypes that are heritable and act as an initial point to investigate capable therapeutic interventions the prime motive of GWAS is to employ identified associations. Broekema et al. (2020) and Uffelmann et al. (2021) describe in detail the complete mechanistic analysis process after fine-mapping a locus identified by GWAS. There is a great need to identify the immediate effects that causal variants have, find locus genes that mediate trait association, acquire downstream networks or pathways that result in modification of cellular or physiological functions, and well determine the relevant tissue/cell type/cell state where these effects take place. GWAS owes a great deal to advances in technology and methodology that have greatly simplified their execution. Several algorithms to genotype calling from SNP array data have been developed, with each new version increasing precision and call rates. Algorithms tailored for calling low-frequency and rare variants, as well as inferring haplotypes and structural variations, have also burgeoned. With a growing dependence on whole-genome sequencing (WGS), an array of variant discovery and SNP calling tools has been developed for GWAS. In addition, the statistical imputation of genotypes has gotten better. For the computational handling and analysis of whole-genome SNP array data, PLINK is a software that is purposely made. Additionally, several tools exist for the visualization of GWAS outcomes, for example, qqman as well as Locus Zoom, and also for conducting GWAS meta-analyses, such as METAL (Watanabe et al., 2017; Pruim et al., 2010).

10.15 EXPERIMENTATION

Several principles should always be kept in mind when deciding which medicinal plants to sequence for whole-genome sequencing projects. First and foremost, choose plants that have either complex medicinal molecules (CMMs) that are well-known or those chemical drugs that so many people would like to see on the market today, like *Panax ginseng* and *Artemisia annua*. Second, choose model plants that have straightforward pharmaceutical ingredients as well as standard secondary metabolism pathways, like sage. Third, think of typical plants from some great genera or families, like *Glycyrrhiza uralensis* (Chinese liquorice; Fabaceae) and *Lycium chinense* (Chinese boxthorn; Solanaceae). Additionally, when you are selecting medicinal plants, give priority to those ones that can act as model organisms and have available biological information in abundance.

Finally, select plants that have established genetic lineages, modest diploid genomes, and simple genomic structures for medicinal purposes. Combining complex plant genomes with clinical agriculture involves using genetic and physical markers. These markers facilitate the study of metabolic gene locations, enabling cloning and marker-assisted selection of beneficial traits for medical applications (Boopathi, 2013). However, developing dense genetic maps for medicinal plants remains problematic due to the lack of available parental lines and their progenies with clear genetic linkage in most cases.

10.16 GENOTYPING AND DATA PROCESSING IN GWAS

Genetic variants, when patterned in people sharing a common ancestry but having different in characteristics, are used to reveal connections between genotypes and physical traits through GWAS. A GWAS experimental workflow includes many important stages, such as gathering the cohort's DNA samples and phenotypic information, genotyping them by either arrays or NGS platforms, performing QC on data, imputing missing markers, carrying out statistical tests for the association, doing optional meta-analyses, looking for independent confirmation, etc. (Zhang et al., 2020). The results are interpreted by means of different kinds of post-GWAS analyses.

Genotyping is usually done through microarray methods for common SNPs, or NGS technologies like whole exome sequencing (WES) and whole genome sequencing (WGS) – this can be utilized when talking about common or rare variants. Microarray-based genotyping is popular because of the cost relation to NGS. The genetic information analysis file provided for GWAS comprises individual anonymized identifiers, familial relation information, demographic data, phenotype information, covariables, genotype calls, and genotyping batch characteristics (Corvin et al., 2010). Ensuring reliable results from GWAS necessitates meticulous quality control, which may involve filtering rare or monomorphic variants, variants deviating from Hardy–Weinberg equilibrium, missing SNPs, genotyping errors, and verifying phenotypic-genotypic concordance, mostly by validating self-reported sex against genetic sex. Analysis of a phenotype of interest finds association tests with genetic variation. Statistical significance and size effects are known from the P-values produced by GWAS outputs, which usually include quantile–quantile and Manhattan plots built with statistical packages like R. To understand these results, additional analysis is necessary to find possible causal variants and explore other biological pathways that might converge in processes related to possibility (Wang et al., 2021). Manhattan plots are commonly used by GWAS to show the genomic locations of identified variants and the degree to which they are connected.

10.17 ASSESSMENT OF THE NOVELTY OF IDENTIFIED MTAs

The success with GWAS for a specific trait or phenotype hinges upon several aspects, including loci number that affects this character, co-occurring distributions for effect size with allele frequency across these loci (termed genetic

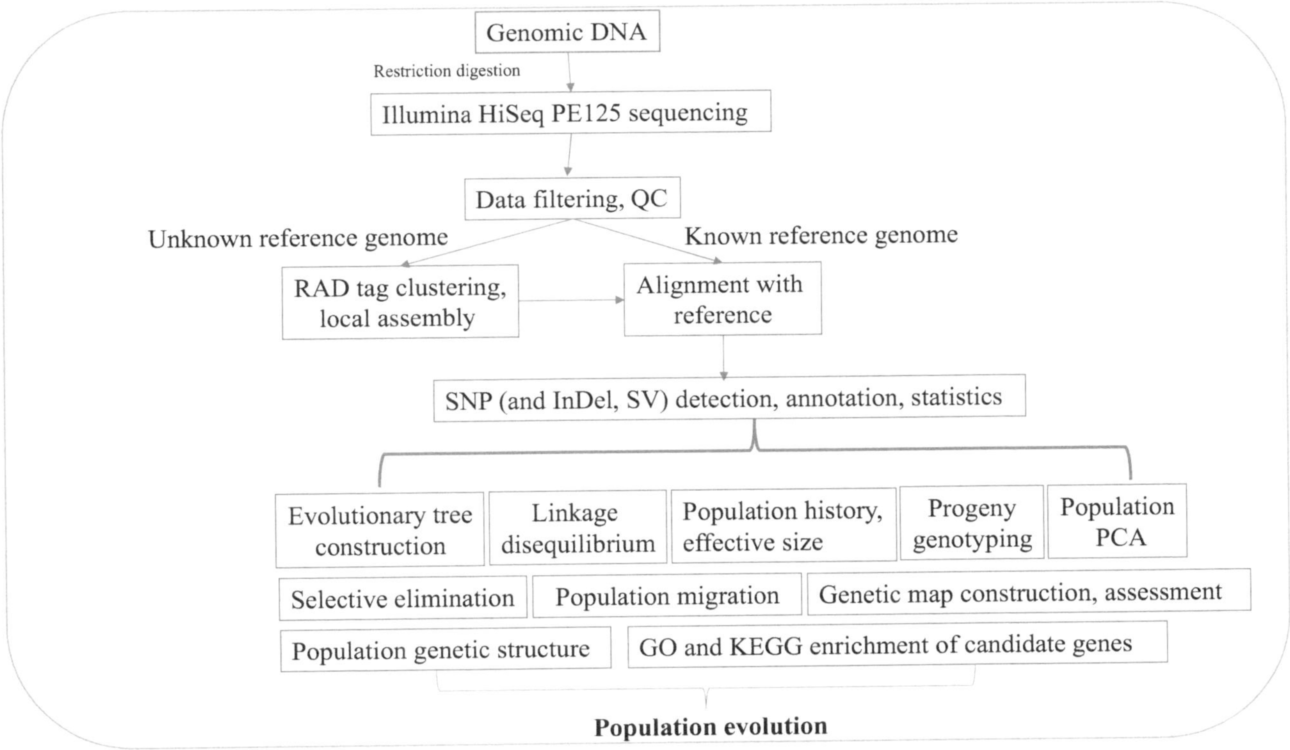

FIGURE 10.5 GWAS experimental validation and functional characterization.

architecture), experimental sample size, the genome-wide variants set that is used, and heterogeneity and precision of a trait under investigation. Considering their physical positions in relation to previously reported QTLs and candidate genes for a particular characteristic, MTAs identified in various environments are assessed for their novelty. With regard to variant discovery and SNP calling in sequencing data, multiple tools for that purpose have been created as whole-genome sequencing (WGS) is becoming more and more important in GWAS. Various tools like PLINK are good at dealing with the whole-genome SNP array data efficiently (Purcell et al., 2007). Moreover, for GWAS there are different visualization tools such as qqman and Locus Zoom and meta-analysis tools like METAL. Besides statistical imputation of genotypes has seen developments.

The way from GWAS discoveries to biological understanding is complex since the correlation between genetic variants and traits doesn't immediately clarify the underlying mechanism or candidate gene. Sequencing a lot of medicinal plants might illuminate the biosynthetic pathways as well as the regulatory mechanisms of medicinal compounds particularly secondary metabolites and further high-throughput sequencing of medicinal plants might promise molecular breeding for developing highly productive medicinal cultivars and genetic modification. New data types, molecular technologies, and analytical methods becoming available now create chances for filling the knowledge gap between sequence and effect (Zhang et al., 2022). The success with GWAS for a specific character or phenotype relies upon several aspects including loci number affecting that character; co-occurring distributions for effect size with allele frequency across these loci (termed genetic architecture), experimental sample size, the genome-wide variants set that is used, and heterogeneity and precision of a trait under investigation.

10.18 TOOLS AND PATHWAYS FOR ANALYZING MOLECULAR INFORMATION FOR SECONDARY METABOLITE PRODUCTION IN MPs

There remains a dearth of plant whole-genome sequence data, with expressed sequence tags (ESTs) contributing the major share of genomics resources. For molecular biology, the invaluable information from gene transcripts in a transcriptome is indispensable, especially where medicinal plants are concerned (Sharma et al., 2013). Transcriptome data is vital for discovering potential genes and networks linked to the formation of species-specific metabolites in medicinal plants (Choudhary et al., 2023). Furthermore, transcriptional regulation prediction by transcriptomics can help unearth transcription factors, responsive elements, and effector genes participating in the production of bioactive compounds. As an example, *Salvia militiorrhiza* had its ethylene-responsive element binding genes under analysis; in *Salvia sclarea* L., their targets and the miRNAs and transcription factors regulating pathways of secondary metabolism were defined (Sadau et al., 2024).

Following the generation and assembly of EST data, annotation becomes a crucial step. Various resources, such as KEGG genes, Swiss Prot, TAIR, and NCBI's nonredundant and nucleotide databases, offer platforms for sequence data annotation (Wang et al., 2023). EST data are also valuable for molecular marker discovery, which finds applications in linkage mapping, comparative genomics, species identification, and gene distribution on chromosomes. Simple sequence repeat (SSR) markers are advantageous among EST-based markers mainly because they are multiallelic and hence can be reproduced co-dominantly and have several alleles making them very easy to measure and are produced at higher frequency and thus can be found everywhere on the chromosome because they spanned a larger region of DNA (Hunt et al., 2023). This makes SSR Locator one such software that can be used for identifying as well as studying SSRs and other mini-satellite sequences in sets of sequences. Bioinformatics approaches further allow for the generation of co-expression networks from the transcriptome data thus providing insight into gene discovery within related plants. It is in comparative genomics that information exchange among different species is based, leading to improved comprehension of plant biology and making the detection of new genes and regulatory mechanisms easier.

10.19 STRATEGIES FOR PATHWAY ANALYSIS FROM GWAS DATA

Pathway analysis could be a helpful method in revealing genes' functional roles through Kyoto Encyclopedia of Genes and Genomes (KEGG) among other sources that provide secondary metabolite-related analyses, for different organisms. This is supported by the KEGG Drug database which shows drug-target interactions and ways of biosynthesizing natural products. There is promise in discovering drugs from plants' genetic information in the KEGG Drug database, which contains structures and components of pharmaceuticals (Romano & Tatonetti, 2019). An additional resource that can be used for pathway analysis is Path Pred, an application predicting in silico stepwise metabolic processes on query compounds based on those found within the KEGG compound database through similarity searching.

She et al. (2019) conducted a study on a gene–metabolite co-expression network within MP *Catharanthus roseus* that aimed at the detection of genes and metabolites that might be participating in the biosynthesis of terpenoid indole alkaloids. In this direction, co-expression analysis offers an area of hope when exploring metabolic regulation within non-model plants as far as medicinal uses are concerned. High-throughput technologies in biotechnology laboratories generate large volumes of data that need processing successfully for scientific exploitation. Thus,

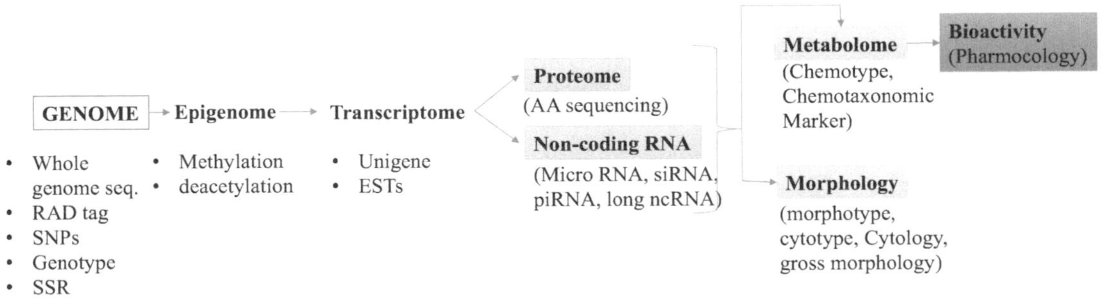

FIGURE 10.6 Muti-omics applications and strategies in recognizing secondary metabolite biosynthesis pathways.

bioinformatics serves to extract potential genes, pathways, or networks together with their functions and impacts in the production process of medicinal plant metabolites. Moreover, this field aids in merging disorderly information to create valid theories that are capable of being tested using experiments leading to proof(s).

10.20 NEW DRUG SOURCE AND BIOACTIVE MOLECULE DISCOVERY

10.20.1 DNA Barcoding (Plant Material Authenticity)

For drugs obtained from plants to work well, it is very crucial that the correct source plants are identified accurately. It is vitally important to note that there have been reports of poisoning resulting from wrong use of plant material for medicine emphasizing how crucial reliable species identification has become in this discipline. Human societies rely heavily on such herbal medicines whose efficacy largely depends on choosing appropriate plant parts, thus necessitating proper identification of plant species for safe drug production. DNA markers offer effective solutions to these issues when they are employed in species identification despite their limitation, whereby they may fail to identify some closely related species. DNA barcodes are strands of short DNA sequences that are identical in different species of the same species and can be obtained within a set part of the genome like from certain parts for instance specific areas. Other services, such as the Barcode of Life Database (BOLD), provide large libraries that contain the barcodes of plants and enable one to match DNA sequences from some unidentified organisms with their barcode (Price et al., 2020). Moreover, the Medicinal Materials DNA Barcode Database (MMDBD) is a resource that combines data about medicinal plants and their unique genetic codes called used DNA sequences, offering bioinformatics functionality for looking up and comparing these sequences (Lou et al., 2010). There are studies where a large number of genes or proteins are checked at the same time in functional genomic experiments and, in turn, used to monitor their activity in distinct conditions within high-throughput assays. Through the use of plant genomes, one can quickly identify what functions each gene performs by incorporating data obtained from both genes and orders of RNA into the proteins themselves (Holtorf et al., 2002). Grains have been using genomics so as to keep track of the genetic composition of different kinds of them, know links between different precedents, understand mutations that may contribute to the development of plants, and recognize how interactions between genotypes can be affected by situational differences. Improving the efficiency, effectiveness, and economy of cultivar improvement is the main aim of genomics research in plant breeding.

10.21 INTEGRATION OF GWAS WITH OTHER OMICS APPROACHES

The arrival of 'omics' marks a new period of collaboration between distinct modern biological methods (Deshmukh et al., 2014). It is a catch-all word that encompasses molecular, system, and computational biology tools used for studying how life-related data works in different life groups (Muthuramalingam et al., 2022). Semi-tech gives a whole perspective of the total crop mechanism permitting analysts to investigate the subtle linkages among genes, proteins, and metabolites involved in molecular mechanisms that manage the tolerance of crops to stress situations. Plants metabolic processes have been studied intensively using integrative omics methods that have enabled an understanding of how they are regulated. Different multi-omics, such as genomics, transcriptomics, proteomics, metabolomics, epigenomics, bioinformatics, proteogenomics, lipidomics, ionomics, interactomics, and phenomics, provide crucial datasets used in understanding plant physiologies (Ashraf et al., 2022). Groundbreaking methods, such as panomics, blend data from omics levels such as genomics, transcriptomics, proteomics, and metabolomics together. These composite systems that integrate information across different genomic layers could likely help in improving crop management methods (Weckwerth et al., 2020). The convergence of information across these different omics levels leads to a broader view of intricate biological processes and assists in pinpointing major molecular actors who participate in a plant's reaction under different environmental scenarios. Researchers use multi-omics data integration to understand complicated regulatory networks and identify possible genes, proteins, and metabolites, thus revealing

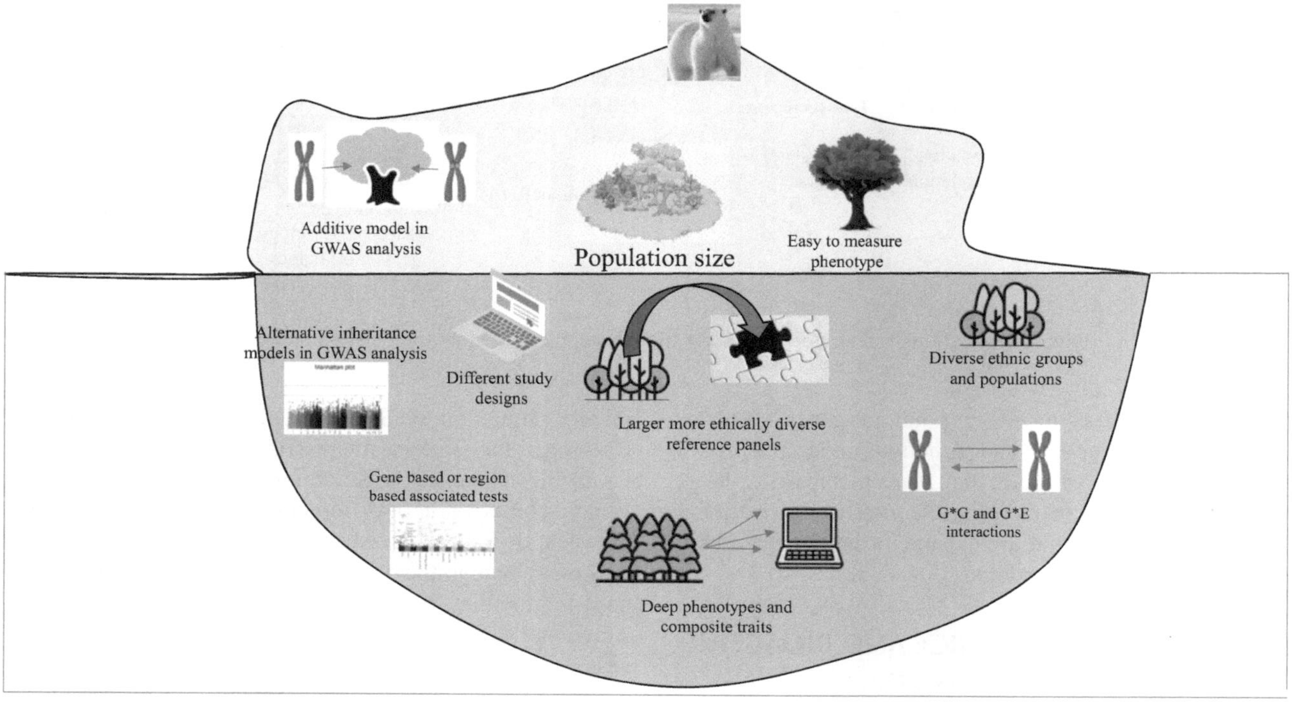

FIGURE 10.7 Genome to metabolome bioactivity.

possible biomarkers or targets for improving crop traits and resilience-enhancing agents. This unique technique supports progress in increasing crop quality.

Several omics techniques used are the following.

10.21.1 Genomics

Genomics marks the preeminent sector of omics with significance for a deep understanding of plant development (Kersey, 2019). For the investigation of gene functions and genetic variability in the future, a highly annotated genome reference provides a direction for further research on DNA sequence coding regions; it facilitates the investigation of relations between genetic constitution and phenotype as well as helps to find regulatory or other sequences engaged in definite life processes. A plant species' whole-genome DNA sequence is solely informative; it does not guarantee the ultimate phenotype. Phenotypic variations occur due to many molecular pathways, which include epigenetic changes, gene expression dynamics, posttranslational modifications on proteins, and accumulation of metabolites and fruit bodies (Yang et al., 2023). This is why not all variants in the DNA sequences have a direct effect on how the organism looks like. A single genomic analysis technique is insufficient to explain complicated molecular control networks for phenotypes of various kinds (Yang et al., 2023). Alternatively, there is a need to comprehend thoroughly by bringing together data from various omics levels which portray diverse facets of gene regulation. Groundbreaking research in plant biology has been accelerated by the publication of genomes, including that of the popularly known black cottonwood plant. These studies have focused on specific aspects of plant growth, including the annual growth cycle, how flowers are formed and the plant's reaction to unfavorable non-living external factors like drought (Ye et al., 2021; Shuai et al., 2014).

10.21.2 Transcriptomics

According to McGettigan (2013), what are some of the most important omics methodologies in plant biology today? They have been explained well in her context. One of them is transcriptomics, which is the comprehensive study of transcriptome referring to all RNA transcripts in a particular cell or tissue. Knowing transcriptome will help us know more about genome complexity and function. Transcriptome profiling methods are developed using different techniques like microarray, EST sequencing, SAGE, and RNA sequencing (Wang et al., 2009). Expression data from these methods provides a detailed measure of the abundance of messages within specific tissues. Combining genomics data with information derived from GWAS is vital for analyzing association signals, especially because variations linked with traits but not situated in coding regions are able to affect whether or not genes are turned on or off (Nicolae et al., 2010). It is within the context of examining how different factors affect each other that the value of gene expression quantitative trait locus (eQTL) studies should be perceived. Cis-EQTLs and trans-EQTLs aid in pointing out genetic variants responsible for levels of gene expression that are either near or afar respectively.

Even though GWAS successfully discovered many areas on the gene that tend to cause certain complex qualities, providing substantial evidence to show that the noncoding areas and intricate linkage disequilibrium (LD) are the major problems that hinder the identification of causal genes in many instances (Claussnitzer et al., 2015). Many of the causal genes are not known to the position of loci thus confusing them with the positional candidate genes identified in GWAS and hence the need to study other biological information. In plants, transcriptomics assists in revealing the molecular mechanisms behind complex traits accomplished by using GWAS coupled with eQTL inquiries. It is through harmonizing these tools that it becomes possible to have a clear comprehension of how genes are regulated in terms of their expressions as well as determine which specific genes cause major agricultural qualities.

10.21.3 Proteomics

Plant cells and tissues require proteins for various functions in all living organisms. Such functions are performed by almost all living things, including plants owed to proteins, which also form an integral part of cells. It encompasses a thorough examination of protein properties like their arrangement, modification, and interaction under given species' genetic sequence limits where necessary tools are those associated with a deeper understanding of their functions than could have been realized up to date (Aslam et al., 2016). This discipline deals with the investigation of genome-specific proteome localization that entails its compositional analysis. Distinctive alterations are observed in the plant proteome among different cells and in distinct developmental or environmental states according to Mergner et al. (2022). A method of candidate protein identification associated with complex traits facilitated by proteomics correlating protein expression with genetic maps in quantitative trait locus (QTL) analysis as stated in Zivy et al. (2000). Auerbach et al. (2002) stated that proteomic is highly important to the full understanding of biological process in the post-genomic period due to eukaryotic proteins undergoing some changes after being synthesized, which significantly determine the manner and levels of their expression, as well as structural organization and stability of each member or type (most eukaryotic proteins undergo post-translational modifications that profoundly impact protein expression and function). Kage et al. (2016) have been given recognition as they found protein markers that can be used to improve quality characteristics in plants, hence selecting for wanted traits. Through the use of proteomics integrated with quantitative trait loci (QTL) analysis, there has been the identification of genes encoding proteins that are functional during both normal plant growth as well as response to stresses in different plant species. Combining proteomics with genomics has helped in great insights regarding genetic variants linked to protein expression. For example, a study merged proteomic and genomic data for human plasma protein analysis. Genetic regulation and expression of proteins are intertwined due to the significant connections the authors discovered between proteins and their genetic variants. The complexities of biological systems such as plants can be unraveled through proteomics. It helps to understand plants by giving important information on the dynamics of their proteins as well as how these proteins relate to genes in the genome. This is useful for improving crops and also in understanding other basic life phenomena that are common among all living organisms.

10.21.4 Metabolomics

The technology known as metabolomics has become an important tool after genomics; its core concentration is a comprehensive qualitative and quantitative analysis of small-molecule metabolites that have molecular mass lower than 1000 and are found in cellular compounds or tissues, being the final products in the process of gene transcription and protein expression and intermediates between genotype and phenotype (Macel et al., 2010). There are numerous biological processes that are sensitive to changes in the levels of metabolic substances in an organism, and hence these substances have an effect on the appearance of genes (Fang et al., 2019). With these analyses, it's now possible to know about the main genes responsible for old and modern metabolic pathways spread across the genome as well as in an unknown dimension just by bringing metabolomics alongside transcriptomics together and then checking out the relationship among them (Kumar et al., 2017). The heritability of a majority of metabolic characteristics across different generations confirms the fact that it is essential to combine metabolomics with quantitative trait loci and genome-wide association studies. Large population metabolomic and phenotypic datasets facilitate establishing direct connections between metabolites and phenotypes (Hong et al., 2016). Identifying genes associated with metabolic traits in agricultural plants using metabolome-based GWAS (mGWAS) and metabolome QTL (mQTL) has been very successful in species like poplars and apple trees. For example, Khan et al. (2012) published a study in which they analyzed untargeted metabolic profiling data with mGWAS (metabolome QTL) technique, thereby uncovering mQTL hotspots corresponding to different types of metabolites in fruit peel and pulp, which are responsible for aroma formation. Shin et al. (2014) also identified 145 metabolic loci through a comprehensive genetic association study of plasma or serum metabolites in Europeans, showing that it was associated with other complex human characteristics and disorders similarly. When you delve into all those metabolites that vary extensively, metabolomics enhances the way we comprehend what plants are all about and helps identify genes linked to significant metabolic pathways or traits, thus revealing the hidden relationships between genes and characteristics.

10.21.5 Epigenomics

Epigenomics touches on the study of the epigenome, which involves chemical modifications of nuclear DNA, histone proteins, and ncRNA (Kumar, 2018). These modifications are not related to nucleotide sequences but are carried on from generation to generation by mitosis or imprinting. Environmental pressures or endogenous signals during plant growth usually start epigenetic changes like DNA methylation, histone modifications, and noncoding RNA regulation, leading to modification of gene expression and chromatin structure (Singh et al., 2021). Phenotypic plasticity in plants is greatly influenced by a constant interaction between genetics and environment, which also determines plant's adaptive strategies. The complex traits that occur in plants are linked to epigenomic research. The genome from plants can, in turn, be used to detect global single-nucleotide-level epigenetic differences. Base changes in plants' DNA methylation have been widely studied using such as bisulfite or long-read sequencing techniques (Gouil et al., 2019). Methylation occurs within gene bodies and transposable elements in different contexts such as CG, CHG, and CHH. It is important to note that plant species, tissues, and cells, as well as the stage of their growth and development properties, have high levels of diversity in DNA methylation. Additionally, there is a dynamic process concerning DNA methylation throughout plant growth and development, which is affected directly by methylation patterns of genes or environmental conditions (Bartels et al., 2018). These epigenetic changes also regulate growth and development processes, such as those caused by various environmental stresses on plants. Many studies have shown that DNA methylation is important in several plant processes such as flower development, drought tolerance, and wood formation, as well as an immune response against infections by pathogens in species including *Populus* (Xiao et al., 2021). It shows why epigenetic mechanisms play an important role in the determination of plant phenotypes under different environmental conditions.

10.22 GWAS FUTURE PERSPECTIVE

GWAS in plants greatly altered the way we understand some of the genetic causes that shape the metabolism process that takes place in these organisms. This kind of study enables researchers to know more about how genes interact with other genes to produce certain metabolites in their quotas as well as offer some insight into the genetic and biochemical establishment of the plant metabolomes. GWAS reveals not only the recognized biochemical pathways but also unknown pathways or interactions. These illuminate the black box of plant metabolism and offer useful aids for genetically improving essential traits in plants. Although substantial advances have been made, key metabolomic approaches continue to require further optimization: such optimization would ensure coverage of a wide range of metabolites at better spatiotemporal resolutions. Equally important is the need for creating accurate metabolite annotation databases and bioinformatics tools aimed at enhancing metabolomics research developments. As we continue to grapple with genetics that must be understood for the complexities they truly are, a possible solution to understanding

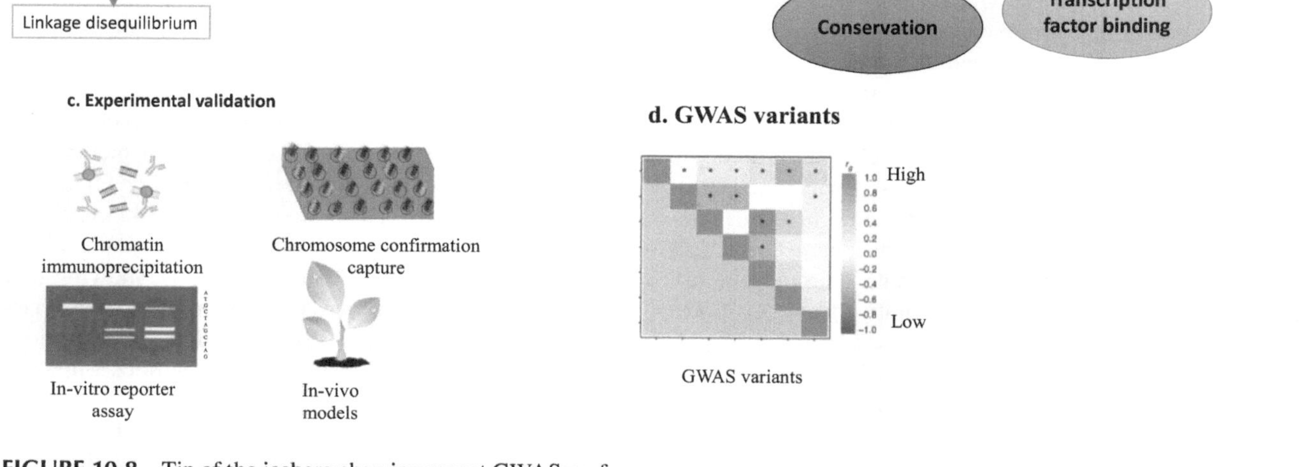

FIGURE 10.8 Tip of the iceberg showing recent GWAS performance.

the gene-phenotype connections could involve looking at metabolic traits in between the two extremes.

By correlating metabolic-genetics studies using GWAS tools against other QTL studies that have been done for disease susceptibility or resistance traits; we might be able to identify additional single-nucleotide polymorphisms (SNPs), which could offer further insight into how these work. The use of metabolite concentration ratios as features for GWAS can boost association power greatly, thereby assisting in the precise identification of genetic determinants. Through various collaborative methods, such as association panels and repeated biparental crosses, researchers can identify rare alleles or loci with small effects. This approach uncovers hidden genetic variations within germplasm supplies. GWAS found helpful in revealing hitherto unknown genetic differences in germplasm pools leading the way for any future breeding projects. Also, through locating QTLs as well as isolating some genes for further genetic analysis, it is possible to transform them by means of site-directed mutagenesis or gene-editing systems and thus discover what these genes do in terms of expression patterns that result from them being modified. GWAS is a promising methodology that can be used to uncover the intricate genetic basis of plant metabolism as well as plant trait development. One way of achieving this is through addressing challenges and taking advantage of opportunities existing in metabolomics and genetic diversity research so as to deepen our understanding of plants while enabling targeted genetic improvement strategies in agriculture.

REFERENCES

Aav, R., Kanger, T., Pehk, T., & Lopp, M. (2005). Unexpected reactivity of ethyl 2-(*Diethylphosphono*) propionate toward 2, 2-disubstituted-1, 3-cyclopentanediones. *Phosphorus, Sulfur, and Silicon, 180*(7), 1739–1748.

Abdulazeem, L., Al-Alaq, F. T., Alrubaei, H. A., Al-Mawlah, Y. H., & Alwan, W. K. (2018). Anti-cancer activity of *Opuntia polyacantha* alkaloid extract on human breast cancer cell line. *Journal of Pharmaceutical Sciences and Research, 10*(7), 1753–1754.

Abhijith, K. P., Gopala Krishnan, S., Ravikiran, K. T., Dhawan, G., Kumar, P., Vinod, K. K., Bhowmick, P. K., Nagarajan, M., Seth, R., Sharma, R., Badhran, S. K., Bollinedi, H., Ellur, R. K., & Singh, A. K. (2022). Genome-wide association study reveals novel genomic regions governing agronomic and grain quality traits and superior allelic combinations for Basmati rice improvement. *Frontiers in Plant Science, 13*, 994447.

Alseekh, S., Ofner, I., Pleban, T., Tripodi, P., Di Dato, F., Cammareri, M., & Zamir, D. (2013). Resolution by recombination: Breaking up *Solanum pennellii* introgressions. *Trends in Plant Science, 18*(10), 536–538.

Ashraf, U., Mahmood, S., Shahid, N., Imran, M., Siddique, M., & Abrar, M. (2022). Multi-omics approaches for strategic improvements of crops under changing climatic conditions. In *Principles and practices of OMICS and genome editing for crop improvement* (pp. 57–92). Springer International Publishing.

Aslam, B., Basit, M., Nisar, M. A., Khurshid, M., & Rasool, M. H. (2016). Proteomics: Technologies and their applications. *Journal of Chromatographic Science*, 1–15.

Atanasov, A. G., Waltenberger, B., Pferschy-Wenzig, E. M., Linder, T., Wawrosch, C., Uhrin, P., Temml, V., Wang, L., Schwaiger, S., Heiss, E. H., Rollinger, J. M., Schuster, D., Breuss, J. M., Bochkov, V., Mihovilovic, M. D., Kopp, B., Bauer, R., Dirsch, V. M., & Stuppner, H. (2015). Discovery and resupply of pharmacologically active plant-derived natural products: A review. *Biotechnology Advances, 33*(8), 1582–1614.

Auerbach, D., Thaminy, S., Hottiger, M. O., & Stagljar, I. (2002). The post-genomic era of interactive proteomics: Facts and perspectives. *Proteomics, 2*(6), 611–623.

Ayoola, G. A., Coker, H. A., Adesegun, S. A., Adepoju-Bello, A. A., Obaweya, K., Ezennia, E. C., & Atangbayila, T. O. (2008). Phytochemical screening and antioxidant activities of some selected medicinal plants used for malaria therapy in Southwestern Nigeria. *Tropical Journal of Pharmaceutical Research, 7*(3), 1019–1024.

Bartels, A., Han, Q., Nair, P., Stacey, L., Gaynier, H., Mosley, M., & Xiao, W. (2018). Dynamic DNA methylation in plant growth and development. *International Journal of Molecular Sciences, 19*(7), 2144.

Boopathi, N. M. (2013). *Genetic mapping and marker assisted selection*. Springer.

Broekema, R. V., Bakker, O. B., & Jonkers, I. H. (2020). A practical view of fine-mapping and gene prioritization in the post-genome-wide association era. *Open Biology, 10*(1), 190221.

Caputi, L., & Aprea, E. (2011). Use of terpenoids as natural flavouring compounds in food industry. *Recent Patents on Food, Nutrition and Agriculture, 3*(1), 9–16.

Cardno, A. G., Marshall, E. J., Coid, B., Macdonald, A. M., Ribchester, T. R., Davies, N. J., & Murray, R. M. (1999). Heritability estimates for psychotic disorders: The Maudsley twin psychosis series. *Archives of General Psychiatry, 56*(2), 162–168.

Cardoso, C., Zhang, Y., Jamil, M., Hepworth, J., Charnikhova, T., Dimkpa, S. O., & Ruyter-Spira, C. (2014). Natural variation of rice strigolactone biosynthesis is associated with the deletion of two MAX1 orthologs. *Proceedings of the National Academy of Sciences, 111*(6), 2379–2384.

Chen, W., Gao, Y., Xie, W., Gong, L., Lu, K., Wang, W., & Luo, J. (2014). Genome-wide association analyses provide genetic and biochemical insights into natural variation in rice metabolism. *Nature Genetics, 46*(7), 714–721.

Choudhary, V., Gangwar, H., & Jaiswal, V. (2023). Application of transcriptomics in exploring important genes in medicinal plants. In *Phytochemical genomics: Plant metabolomics and medicinal plant genomics* (pp. 263–284). Springer Nature Singapore.

Claussnitzer, M., Dankel, S. N., Kim, K. H., Quon, G., Meuleman, W., Haugen, C., & Kellis, M. (2015). FTO obesity variant circuitry and adipocyte browning in humans. *The New England Journal of Medicine, 373*(10), 895–907.

Cobb, J. N., Biswas, P. S., & Platten, J. D. (2019). Back to the future: Revisiting MAS as a tool for modern plant breeding. *Theoretical and Applied Genetics, 132*, 647–667.

Corvin, A., Craddock, N., & Sullivan, P. F. (2010). Genome-wide association studies: A primer. *Psychological Medicine, 40*(7), 1063–1077.

Davey, J. W., Hohenlohe, P. A., Etter, P. D., Boone, J. Q., Catchen, J. M., & Blaxter, M. L. (2011). Genome-wide genetic marker discovery and genotyping using next-generation sequencing. *Nature Reviews Genetics, 12*(7), 499–510.

Debnath, B., Singh, W. S., Das, M., Goswami, S., Singh, M. K., Maiti, D., & Manna, K. (2018). Role of plant alkaloids on human health: A review of biological activities. *Materials Today Chemistry*, *9*, 56–72.

Desai, S. D., Desai, D. G., & Kaur, H. (2009). Saponins and their biological activities. *Pharma Times*, *41*(3), 13–16.

Deshmukh, R., Sonah, H., Patil, G., Chen, W., Prince, S., Mutava, R., & Nguyen, H. T. (2014). Integrating omic approaches for abiotic stress tolerance in soybean. *Frontiers in Plant Science*, *5*, 244.

Devate, N. B., Krishna, H., Parmeshwarappa, S. K. V., Manjunath, K. K., Chauhan, D., Singh, S., Singh, J. B., Kumar, M., Patil, R., Khan, H., Jain, N., Singh, G. P., & Singh, P. K. (2022). Genome-wide association mapping for component traits of drought and heat tolerance in wheat. *Frontiers in Plant Science*, *13*, 943033.

Enyew, M., Feyissa, T., Carlsson, A. S., Tesfaye, K., Hammenhag, C., Seyoum, A., & Geleta, M. (2022). Genome-wide analyses using multi-locus models revealed marker-trait associations for major agronomic traits in Sorghum bicolor. *Frontiers in Plant Science*, *13*, 999692.

Fang, C., Fernie, A. R., & Luo, J. (2019). Exploring the diversity of plant metabolism. *Trends in Plant Science*, *24*(1), 83–98.

Fong, S. L. (2022). *Dissecting the evolution of human enhancer sequences* (Doctoral dissertation, Vanderbilt University).

Gibbs, R. D. (1974). *Chemotaxonomy of flowering plants*. McGill-Queen's University Press.

Gong, L., Chen, W., Gao, Y., Liu, X., Zhang, H., Xu, C., & Luo, J. (2013). Genetic analysis of the metabolome exemplified using a rice population. *Proceedings of the National Academy of Sciences*, *110*(50), 20320–20325.

Gouil, Q., & Keniry, A. (2019). Latest techniques to study DNA methylation. *Essays in Biochemistry*, *63*(6), 639–648.

Holland, J. B. (2007). Genetic architecture of complex traits in plants. *Current Opinion in Plant Biology*, *10*(2), 156–161.

Holtorf, H., Guitton, M. C., & Reski, R. (2002). Plant functional genomics. *Naturwissenschaften*, *89*, 235–249.

Hong, J., Yang, L., Zhang, D., & Shi, J. (2016). Plant metabolomics: An indispensable system biology tool for plant science. *International Journal of Molecular Sciences*, *17*(6), 767.

Huang, X., & Han, B. (2014). Natural variations and genome-wide association studies in crop plants. *Annual Review of Plant Biology*, *65*, 531–551.

Hunt, J. M. T., Samson, C. A., Rand, A. D., & Sheppard, H. M. (2023). Unintended CRISPR-Cas9 editing outcomes: A review of the detection and prevalence of structural variants generated by gene-editing in human cells. *Human Genetics*, *142*(6), 705–720.

Jovanović-Šanta, S. S., Petri, E. T., Klisurić, O. R., Szécsi, M., Kovačević, R., & Petrović, J. A. (2015). Antihormonal potential of selected D-homo and D-seco estratriene derivatives. *Steroids*, *97*, 45–53.

Kabera, J. N., Semana, E., Mussa, A. R., & He, X. (2014). Plant secondary metabolites: Biosynthesis, classification, function and pharmacological properties. *Journal of Pharmacy and Pharmacology*, *2*(7), 377–392.

Kage, U., Kumar, A., Dhokane, D., Karre, S., & Kushalappa, A. C. (2016). Functional molecular markers for crop improvement. *Critical Reviews in Biotechnology*, *36*(5), 917–930.162.

Kaur, P., Mittal, A., Nayak, S. K., Vyas, M., Mishra, V., & Khatik, G. L. (2018). Current strategies and drug targets in the management of type 2 diabetes mellitus. *Current Drug Targets*, *19*(15), 1738–1766.

Kersey, P. J. (2019). Plant genome sequences: Past, present, future. *Current Opinion in Plant Biology*, *48*, 1–8.

Khan, S. A., Chibon, P. Y., De Vos, R. C., Schipper, B. A., Walraven, E., Beekwilder, J., & Schouten, H. J. (2012). Genetic analysis of metabolites in apple fruits indicates an mQTL hotspot for phenolic compounds on linkage group 16. *Journal of Experimental Botany*, *63*(8), 2895–2908.

Kliebenstein, D. (2009). Advancing genetic theory and application by metabolic quantitative trait loci analysis. *The Plant Cell*, *21*(6), 1637–1646.

Krishnappa, G., Khan, H., Krishna, H., Devate, N. B., Kumar, S., Mishra, C. N., Parkash, O., Kumar, S., Kumar, M., Mamrutha, H. M., Singh, G. P., & Singh, G. (2023). Genome-Wide Association Study for grain protein, thousand kernel weight, and normalized difference vegetation index in bread wheat (*Triticum aestivum* L.). *Genes*, *14*(3), 637.

Kumar, A., Ilavarasan, R., Jayachandran, T., Decaraman, M., Aravindhan, P., Padmanabhan, N., & Krishnan, M. R. V. (2009). Phytochemicals investigation on a tropical plant, *Syzygium cumini* from Kattuppalayam, Erode dipoustrict, Tamil Nadu, South India. *Pakistan Journal of Nutrition*, *8*(1), 83–85.

Kumar, R., Bohra, A., Pandey, A. K., Pandey, M. K., & Kumar, A. (2017). Metabolomics for plant improvement: Status and prospects. *Frontiers in Plant Science*, *8*, 271676.

Kumar, S. (2018). Epigenomics of plant responses to environmental stress. *Epigenomes*, *2*(1), 6.

Lisec, J., Steinfath, M., Meyer, R. C., Selbig, J., Melchinger, A. E., Willmitzer, L., & Altmann, T. (2009). Identification of heterotic metabolite QTL in *Arabidopsis thaliana* RIL and IL populations. *The Plant Journal*, *59*(5), 777–788.

Liu, Y., Sun, G., Zhong, Z., Ji, L., Zhang, Y., Zhou, J., Zheng, X., & Deng, K. (2017). Overexpression of AtEDT1 promotes root elongation and affects medicinal secondary metabolite biosynthesis in roots of transgenic *Salvia miltiorrhiza*. *Protoplasma*, *254*, 1617–1625.

Lopez, L. M., Grimes, D. A., Schulz, K. F., Curtis, K. M., & Chen, M. (2014). Steroidal contraceptives: Effect on bone fractures in women. *Cochrane Database of Systematic Reviews*, *2014*(6), 1–65.

Lou, S. K., Wong, K. L., Li, M., But, P. P. H., Tsui, S. K. W., & Shaw, P. C. (2010). An integrated web medicinal materials DNA database: MMDBD (Medicinal Materials DNA Barcode Database). *BMC Genomics*, *11*, 1–8.

Macel, M., Van Dam, N. M., & Keurentjes, J. J. (2010). Metabolomics: The chemistry between ecology and genetics. *Molecular Ecology Resources*, *10*(4), 583–593.

Matsuda, F., Okazaki, Y., Oikawa, A., Kusano, M., Nakabayashi, R., Kikuchi, J., & Saito, K. (2012). Dissection of genotype–phenotype associations in rice grains using metabolome quantitative trait loci analysis. *The Plant Journal*, *70*(4), 624–636.

McGettigan, P. A. (2013). Transcriptomics in the RNA-seq era. *Current Opinion in Chemical Biology*, *17*(1), 4–11.

Meihls, L. N., Handrick, V., Glauser, G., Barbier, H., Kaur, H., Haribal, M. M., & Jander, G. (2013). Natural variation in maize aphid resistance is associated with 2, 4-dihydroxy-7-methoxy-1, 4-benzoxazin-3-one glucoside methyltransferase activity. *The Plant Cell*, *25*(6), 2341–2355.

Meliani, N., Dib, M. E. A., Allali, H., & Tabti, B. (2011). Hypoglycaemic effect of *Berberis vulgaris* L. in normal and streptozotocin-induced diabetic rats. *Asian Pacific Journal of Tropical Biomedicine*, *1*(6), 468–471.

Mergner, J., & Kuster, B. (2022). Plant proteome dynamics. *Annual Review of Plant Biology*, *73*, 67–92.

Merrick, L. F., Burke, A. B., Zhang, Z., & Carter, A. H. (2022). Comparison of single-trait and multi-trait genome-wide association models and inclusion of correlated traits in the dissection of the genetic architecture of a complex trait in a breeding program. *Frontiers in Plant Science*, *12*, 772907.

Mulugeta, B., Tesfaye, K., Ortiz, R., Johansson, E., Hailesilassie, T., Hammenhag, C., & Geleta, M. (2023). Marker-trait association analyses revealed major novel QTLs for grain yield and related traits in durum wheat. *Frontiers in Plant Science*, *13*, 1009244.

Muthuramalingam, P., Jeyasri, R., Rakkammal, K., Satish, L., Shamili, S., Karthikeyan, A., & Ramesh, M. (2022). Multi-Omics and integrative approach towards understanding salinity tolerance in rice: A review. *Biology*, *11*(7), 1022.

Mwadzingeni, L., Shimelis, H., Rees, D. J. G., & Tsilo, T. J. (2017). Genome-wide association analysis of agronomic traits in wheat under drought-stressed and non-stressed conditions. *PloS One*, *12*(2), e0171692.

Niazian, M. (2019). Application of genetics and biotechnology for improving medicinal plants. *Planta*, *249*, 953–973.

Nicolae, D. L., Gamazon, E., Zhang, W., Duan, S., Dolan, M. E., & Cox, N. J. (2010). Trait-associated SNPs are more likely to be eQTLs: Annotation to enhance discovery from GWAS. *PLoS Genetics*, *6*(4), e1000888.

Nobile, V., Michelotti, A., Cestone, E., Caturla, N., Castillo, J., Benavente-García, O., & Micol, V. (2016). Skin photoprotective and antiageing effects of a combination of rosemary (*Rosmarinus officinalis*) and grapefruit (*Citrus paradisi*) polyphenols. *Food and Nutrition Research*, *60*(1), 31871.

Nogueira, M., Enfissi, E. M., Almeida, J., & Fraser, P. D. (2018). Creating plant molecular factories for industrial and nutritional isoprenoid production. *Current Opinion in Biotechnology*, *49*, 80–87.

Ortega, J. T., Serrano, M. L., Suárez, A. I., Baptista, J., Pujol, F. H., Cavallaro, L. V., & Rangel, H. R. (2019). Antiviral activity of flavonoids presents in aerial parts of *Marcetia taxifolia* against Hepatitis B virus, Poliovirus, and Herpes Simplex Virus in vitro. *EXCLI Journal*, *18*, 1037.

Paris, R., & Moyse, H. (1969). *Precis de matiere medicinale*. Ed Masson Paris.

Pastor, R. F., Restani, P., Di Lorenzo, C., Orgiu, F., Teissedre, P. L., Stockley, C., & Iermoli, R. H. (2019). Resveratrol, human health and winemaking perspectives. *Critical Reviews in Food Science and Nutrition*, *59*(8), 1237–1255.

Pichersky, E., & Lewinsohn, E. (2011). Convergent evolution in plant specialized metabolism. *Annual Review of Plant Biology*, *62*, 549–566.

Pouvreau, B., Vanhercke, T., & Singh, S. (2018). From plant metabolic engineering to plant synthetic biology: The evolution of the design/build/test/learn cycle. *Plant Science*, *273*, 3–12.

Price, B. W., Briscoe, A. G., Misra, R., & Broad, G. (2020). *Evaluation of DNA barcode libraries used in the UK and developing an action plan to fill priority gaps*. Natural England.

Pruim, R. J., Welch, R. P., Sanna, S., Teslovich, T. M., Chines, P. S., Gliedt, T. P., Boehnke, M., Abecasis, G. R., & Willer, C. J. (2010). LocusZoom: Regional visualization of genome-wide association scan results. *Bioinformatics*, *26*(18), 2336–2337.

Purcell, S., Neale, B., Todd-Brown, K., Thomas, L., Ferreira, M. A., Bender, D., Maller, J., Sklar, P., I. W. de Bakker, P., J. Daly, M., & Sham, P. C. (2007). PLINK: A tool set for whole-genome association and population-based linkage analyses. *The American Journal of Human Genetics*, *81*(3), 559–575.

Qanbari, S. (2020). On the extent of linkage disequilibrium in the genome of farm animals. *Frontiers in Genetics*, *10*, 487295.

Quadrana, L., Almeida, J., Asís, R., Duffy, T., Dominguez, P. G., Bermúdez, L., & Carrari, F. (2014). Natural occurring epialleles determine vitamin E accumulation in tomato fruits. *Nature Communications*, *5*(1), 4027.

Rizk, A. M. (1982). *Constituents of plants growing in Qatar*. I. A chemical survey of sixty plants. *Fitoterapia*, *52*, 35–42.

Romano, J. D., & Tatonetti, N. P. (2019). Informatics and computational methods in natural product drug discovery: A review and perspectives. *Frontiers in Genetics*, *10*, 442506.

Rowe, H. C., Hansen, B. G., Halkier, B. A., & Kliebenstein, D. J. (2008). Biochemical networks and epistasis shape the Arabidopsis thaliana metabolome. *The Plant Cell*, *20*(5), 1199–1216.

Sadau, S. B., Liu, Z., Ninkuu, V., Guan, L., & Sun, X. (2024). DREB transcription factors are crucial regulators of abiotic stress responses in *Gossypium spp. Plant Stress*, *11*, 100350.

Schaid, D. J., Chen, W., & Larson, N. B. (2018). From genome-wide associations to candidate causal variants by statistical fine-mapping. *Nature Reviews Genetics*, *19*(8), 491–504.

Schwab, W. (2003). Metabolome diversity: Too few genes, too many metabolites? *Phytochemistry*, *62*(6), 837–849.

Sharma, V., & Sarkar, I. N. (2013). Bioinformatics opportunities for identification and study of medicinal plants. *Briefings in Bioinformatics*, *14*(2), 238–250.

She, J., Yan, H., Yang, J., Xu, W., & Su, Z. (2019). croFGD: Catharanthus roseus functional genomics database. *Frontiers in Genetics*, *10*, 431163.

Shin, S. Y., Fauman, E. B., Petersen, A. K., Krumsiek, J., Santos, R., Huang, J., & Soranzo, N. (2014). An atlas of genetic influences on human blood metabolites. *Nature Genetics*, *46*(6), 543–550.

Shuai, P., Liang, D., Tang, S., Zhang, Z., Ye, C. Y., Su, Y., & Yin, W. (2014). Genome-wide identification and functional prediction of novel and drought-responsive lincRNAs in Populus trichocarpa. *Journal of Experimental Botany*, *65*(17), 4975–4983.

Singh, D., Chaudhary, P., Taunk, J., Kumar Singh, C., Sharma, S., Singh, V. J., & Pal, M. (2021). Plant epigenomics for extenuation of abiotic stresses: Challenges and future perspectives. *Journal of Experimental Botany*, *72*(20), 6836–6855.

Singh, K. S., van der Hooft, J. J., van Wees, S. C., & Medema, M. H. (2022). Integrative omics approaches for biosynthetic pathway discovery in plants. *Natural Product Reports*, *39*(9), 1876–1896.

Sukumaran, S., Reynolds, M. P., & Sansaloni, C. (2018). Genome-wide association analyses identify QTL hotspots for yield and component traits in durum wheat grown under yield potential, drought, and heat stress environments. *Frontiers in Plant Science*, *9*, 330307.

Techen, N., Parveen, I., Pan, Z., & Khan, I. A. (2014). DNA barcoding of medicinal plant material for identification. *Current Opinion in Biotechnology*, *25*, 103–110.

Toubiana, D., Semel, Y., Tohge, T., Beleggia, R., Cattivelli, L., Rosental, L., & Fait, A. (2012). Metabolic profiling of a mapping population exposes new insights in the regulation of seed metabolism and seed, fruit, and plant relations. *PLoS Genetics*, *8*(3), e1002612.

Treare, G. E., & Evans, W. C. (1985). *Pharmacognosy* (17 ed., p. 149). Bahive Tinal.

Uffelmann, E., Huang, Q. Q., Munung, N. S., De Vries, J., Okada, Y., Martin, A. R., C. Martin, H., Lappalainen, T., & Posthuma, D. (2021). Genome-wide association studies. *Nature Reviews Methods Primers*, *1*(1), 59.

Vyas, M. (2016). A short review on anticancer investigations of *Strychnos nux-vomica*. *International Journal of Green Pharmacy (IJGP)*, *10*(3).

Vyas, M. (2017). Nutritional profile of spinach and its antioxidant and antidiabetic evaluation. *International Journal of Green Pharmacy (IJGP)*, *11*(03).

Wang, C., Ahsan, T., Ding, A., Han, D., Zang, C. Q., Huang, Y. Q., & Hussain, K. (2023). Whole genome analysis of *Bacillus amyloliquefaciens* TA-1, a promising biocontrol agent against *Cercospora arachidicola* pathogen of early leaf spot in *Arachis hypogaea* L. *BMC Plant Biology*, *23*(1), 410.

Wang, J., & Zhang, Z. (2021). GAPIT version 3: Boosting power and accuracy for genomic association and prediction. *Genomics, Proteomics and Bioinformatics*, *19*(4), 629–640.

Wang, L., Wang, A., Huang, X., Zhao, Q., Dong, G., Qian, Q., & Han, B. (2011). Mapping 49 quantitative trait loci at high resolution through sequencing-based genotyping of rice recombinant inbred lines. *Theoretical and Applied Genetics*, *122*, 327–340.

Wang, Z., Gerstein, M., & Snyder, M. (2009). RNA-Seq: A revolutionary tool for transcriptomics. *Nature Reviews Genetics*, *10*(1), 57–63, 119.

Watanabe, K., Taskesen, E., Van Bochoven, A., & Posthuma, D. (2017). Functional mapping and annotation of genetic associations with FUMA. *Nature Communications*, *8*(1), 1826.

Weckwerth, W., Ghatak, A., Bellaire, A., Chaturvedi, P., & Varshney, R. K. (2020). PANOMICS meets germplasm. *Plant Biotechnology Journal*, *18*(7), 1507–1525.

Wen, W., Li, D., Li, X., Gao, Y., Li, W., Li, H., Liu, J., Liu, H., Chen, W., Luo, J., & Yan, J. (2014). Metabolome-based genome-wide association study of maize kernel leads to novel biochemical insights. *Nature Communications*, *5*(1), 3438.

Weng, J. K., Philippe, R. N., & Noel, J. P. (2012). The rise of chemodiversity in plants. *Science*, *336*(6089), 1667–1670.

Xiao, D., Zhou, K., Yang, X., Yang, Y., Ma, Y., & Wang, Y. (2021). Crosstalk of DNA methylation triggered by pathogen in poplars with different resistances. *Frontiers in Microbiology*, *12*, 750089.

Yamazaki, M., Rai, A., Yoshimoto, N., & Saito, K. (2018). Perspective: Functional genomics towards new biotechnology in medicinal plants. *Plant Biotechnology Reports*, *12*, 69–75.

Yang, L., Yang, Y., Huang, L., Cui, X., & Liu, Y. (2023). From single-to multi-omics: Future research trends in medicinal plants. *Briefings in Bioinformatics*, *24*(1), bbac485.

Ye, Y., Xin, H., Gu, X., Ma, J., & Li, L. (2021). Genome-wide identification and functional analysis of the Basic Helix-Loop-Helix (bHLH) transcription family reveals candidate PtFBH genes involved in the flowering process of *Populus trichocarpa*. *Forests*, *12*(11), 1439.

Zargar, S. M., Raatz, B., Sonah, H., Muslima Nazir, Bhat, J. A., Dar, Z. A., & Rakwal, R. (2015). Recent advances in molecular marker techniques: Insight into QTL mapping, GWAS and genomic selection in plants. *Journal of Crop Science and Biotechnology*, *18*, 293–308.

Zhang, M., Liu, Y. H., Xu, W., Smith, C. W., Murray, S. C., & Zhang, H. B. (2020). Analysis of the genes controlling three quantitative traits in three diverse plant species reveals the molecular basis of quantitative traits. *Scientific Reports*, *10*(1), 10074.

Zhang, P., Zhong, K., Shahid, M. Q., & Tong, H. (2016). Association analysis in rice: From application to utilization. *Frontiers in Plant Science*, *7*, 1202.

Zhang, X., Ren, Z., Luo, B., Zhong, H., Ma, P., Zhang, H., & Gao, S. (2022). Genetic architecture of maize yield traits dissected by QTL mapping and GWAS in maize. *The Crop Journal*, *10*(2), 436–446.

Zhao, X., Park, S. Y., Yang, D., & Lee, S. Y. (2018). Synthetic biology for natural compounds. *Biochemistry*, *58*(11), 1454–1456.

Zhu, C., Gore, M., Buckler, E. S., & Yu, J. (2008). Status and prospects of association mapping in plants. *The Plant Genome*, *1*(1).

Zivy, M., & de Vienne, D. (2000). Proteomics: A link between genomics, genetics and physiology. *Plant Molecular Biology*, *44*, 575–580.

11 Advancements in Medicinal Plant Research

Harnessing Artificial Intelligence, Machine Learning, Deep Learning, and Bioinformatics

Indrajit Kalita, Sourav Bhattacharjee, and Manalisha Saharia

11.1 INTRODUCTION AND HISTORICAL BACKGROUND

Plants are considered to be of immense importance in traditional medicinal treatments and cures since time immemorial. The application of plants as a source of medicine happens as the various parts of plants have numerous bioactive chemical components within their phytochemical profile having wide therapeutic applications. The use of plants and plant products as indigenous medicine is a part of traditional knowledge propagated through generations, particularly among the rural population. It has been statistically shown that in developing nations, around 80% of people residing in rural areas adhere to plants as natural sources of medicine for primary healthcare (Azadnia et al., 2022). Selection, identification, and application of medicinal plants for primary healthcare happened roughly through morpho-visual and organoleptic assessment by humans for several decades until research on medicinal plants began with the formal conceptualization of laboratory-based research works in Europe during the beginning of the 17th century (Fitzgerald et al., 2020). Historically, the word 'drug' originated from 'dreug,' meaning dried plants or plant products converted to powdery form for application in primary healthcare among humans (Erhaldt, 2009). With the advent of various analytical techniques and advancements in biological research, these indigenously applied medicinal plants were taken as potential sources for drug discovery and drug development.

Drug discovery refers to the process of elucidating and characterizing specific bioactive molecules as lead candidates for targeted therapeutic action against diseases in biological systems. The timeless plant-human relationship in which plants form the main ingredient for the most common indigenous ethno-medical preparations necessary for primary healthcare makes them the most sought-after starting materials for drug discovery and drug design in a direct approach to treating diseases (Figure 11.1).

11.1.1 Concept of AI and ML in Medicinal Plant Research

The concept of AI encompasses the resemblance of human thinking capabilities via computerized programs established through a unique blend of computer science and mathematics. The concept of AI in medicine was first designated as augmented intelligence for a precise healthcare framework (Ashby, 1956). This was then followed by a unique concept and mathematical model of 'learning machines' that was put forward by Allan Turning in the year 1995 (Turing, 1995). The field gained momentum and efficacy through decades of development by various eminent workers from time to time, leading to its potential application in revolutionizing diverse areas of research by moving toward technology-based automation and formulation of algorithm-based computerized programs that normally require human professional intervention (Singh et al., 2023). One of the emerging fields of study under revolution due to the increased application of AI is medicinal plant research where the blend of artificial intelligence (AI), machine learning (ML), and bioinformatics has started a period of revolutionary advancement in the discovery of drugs for increased effectiveness and exponential progress.

The cure for various diseases lies within nature itself as humans have been applying the concoctions and indigenous preparations of plants and plant derivatives for healthcare since ancient times. The plants can synthesize many bioactive molecules for their metabolic functioning as well as for coping against biotic, abiotic, and xenobiotic stress conditions. Many of such molecules have been undiscovered compounds working against pathological conditions of human beings in indigenous medicinal preparations since time immemorial. Drug discovery and development based on medicinal plant research focuses on the use of these medicinal plants as a source of drugs for designing molecules for the effective treatment of pathological conditions. Application of AI has increased the use of a combinational approach by blending resource repositories from chemistry

DOI: 10.1201/9781003475491-11

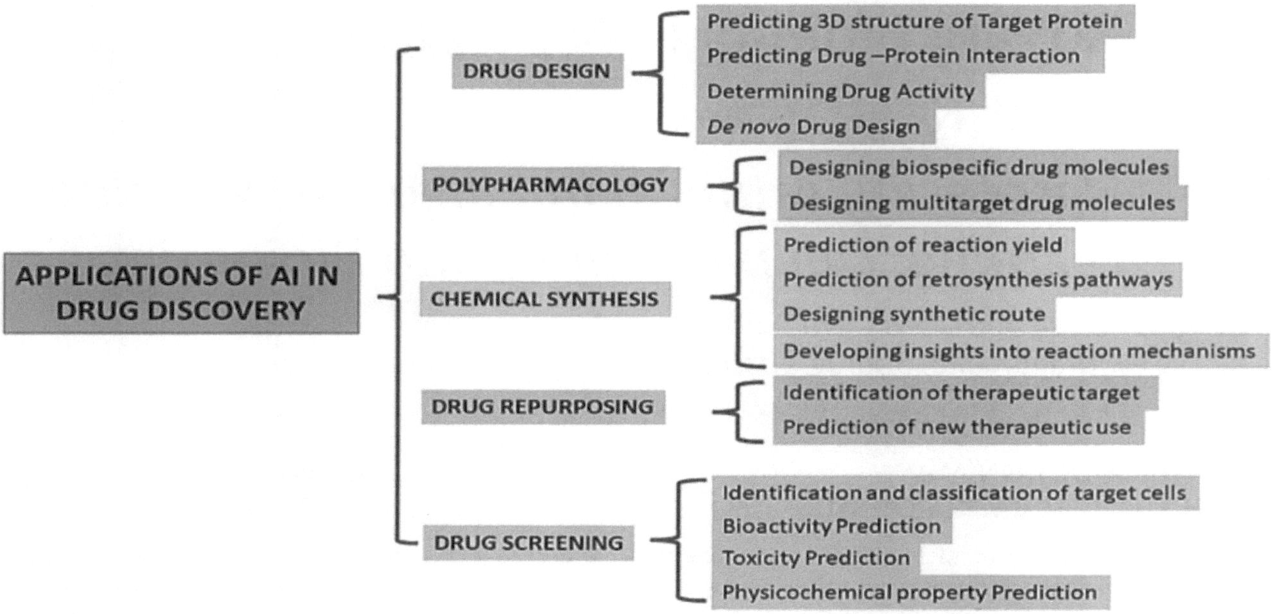

FIGURE 11.1 Flow chart depicting various applications of AI in drug discovery (Visan and Negut, 2024).

and biology to increase the accuracy of drug target elucidation, validation, lead candidate identification, and optimization to sharpen the interaction between lead candidate and drug target for better therapeutic outcomes.

11.1.2 AI-Based Approaches for the Advancement of Drug Discovery

11.1.2.1 AI-Based Advancement in Elucidation and Validation of Drug Targets

AI simulation algorithms screen the genomic, proteomic data, and regression models to suitably identify key target genes associated with specific diseases. This is done through uncovering the complex molecular pathways associated with specific pathological conditions using AI and ML (Vora et al., 2023). AI algorithms can access the molecular structure-based binding affinity and can thus predict the feasibility of selecting a potential unexplored drug target associated with a pathological condition having the most suitable characteristics for showing the desirable results (Adam et al., 2020).

11.1.2.2 AI-Based Lead Candidate Identification and Optimization

AI uses algorithms to screen molecular and chemical structure repositories to filter out lead candidates having maximum binding tendency with the selected drug target molecule based on analysis of liberation, absorption, distribution, metabolism, and excretion (LADME) parameters of the potential lead candidate compounds (Sorkun et al., 2020). The identification and screening can be on pure structure filtered through AI-based scoring and space-searching algorithms. The lead candidate identification can also be done using a more specific distinguishable molecular character-based *in-silico* quantitative structure-activity relationship (QSAR) based approach, a sub-branch of the broader computer-assisted drug design (CADD) method, which solely relies on the relatedness between the biological activity of a lead candidate compound with some specific identifiable marker features in its molecular structure termed as descriptors. Such relatedness in QSAR is established based on mathematical equations and predictive modeling. AI-based algorithms go for automated singling out of the specific distinguishable markers on molecular structure leading to the construction of relevant models such as linear discriminant analysis (LDA) in which algorithms separate classes within a relevant set of data by reducing the dimensional axes from a higher dimension to lower that reflects the classification patterns lying within the dataset (Paul & Schaefer, 2020). The dimensionality of an input molecular structural dataset is classified using a specialized low dimensional division boundary known as hyperplane after the input data is shifted using an appropriate mathematical function from a specific dimension to higher dimension space in another AI-based QSAR algorithm known as support vector machine (SVM). The geometric surface bifurcates the input data into discrete classes to improve the efficiency and precision of complex categorization, correlation, and regression problems to predict the possible links between the structure and activity of a lead candidate molecule on a quantitative level (Figure 11.2).

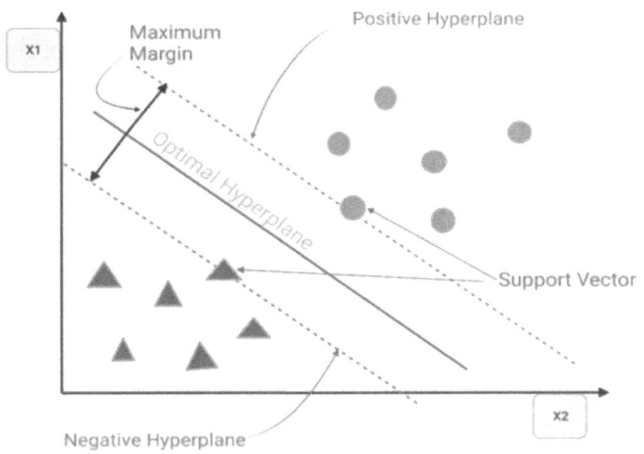

FIGURE 11.2

It is one of the most widely used AI-based compound classification techniques used in drug discovery (Maltarollo et al., 2019). A random forest (RF) algorithm is another robust and efficient AI-based approach that prepares a combinational account of multiple computer-generated flow charts elucidated from structural datasets known as decision tree outputs to produce an accurate single result for application in QSAR predictions. Such approaches open the possibilities for analyzing the various previously undiscovered lead compounds and drug targets.

AI-based algorithms allow standardization of lead compounds, causing a marked shift from wet lab-based experimental screening procedures to a muti-parametric level in which multiple parameters are optimized together to minimize the challenging countereffects of changes in one parameter on another to ensure maximum rapidity, effectiveness, specificity and safety with minimal side effects. Apart from specificity, the two most important parameters to be optimized for lead candidates are binding affinity and longevity which determines the designation and final approval of a lead compound as a suitable drug.

Recurrent neural network (RNN) is an AI-based *in-silico* approach that has revolutionized the resource-intensive conventional receptor-based *de novo* lead candidate optimization by improving the efficacy and accuracy of the drug-target interaction. It has decreased the dependence on prerequisite knowledge of chemical design rules, molecular structure distinguishing characters, etc., and works by generating some specialized molecular graphs known as 1D (one-dimensional) simplified molecular input line entry systems (SMILES) representation. SMILES is a highly simplified depiction of chemical structure based on pure symbols in a computer-readable format. RNN algorithm uses the SMILES patterns to perform a probability prediction for designing drugs similar to the SMILES profile of drugs with established target binding activities (Chen et al., 2022).

Another AI-based algorithm for drug designing and lead compound optimization is the generative adversarial network (GAN), which shows a unique blended use of two artificial neural networks that are both under the influence of training data and are under conflict or opposition. One of the networks functions as a generator to create synthesized data corresponding to the features of training data and the second network segregates the received actual input data from the network-generated data. The generative module feeds synthesized data such as a synthetically reconstructed lead candidate-like compound to the segregation module which predicts the probability of the synthesized data analyzing its closeness to the real lead candidate dataset (Lin et al., 2020).

11.1.2.3 AI-Based Toxicity Analysis

AI-based algorithms can analyze the toxicity of compounds and mark the undesirable structural characteristics that could lead to potential toxicity before clinical trials, thereby enhancing the selection of lead compounds with minimal side effects, increasing safe drug development, and safe drug administration, and enhancing inexpensive healthcare. This can lead to a marked shift from conventional lab-based toxicology studies involving *in-vitro* experimental procedures that are time-consuming with reduced accuracy. This has also resulted in a shift toward an ethical practice of using *in-silico* computerized algorithms, replacing *in-vivo* toxicity testing on experimental animal models (Pérez Santín et al., 2021). AI-based drug toxicity detection software such as PrOCTOR and TargeTox generate a lead compound score based on a combination of drug target information extracted from public information repositories and molecular structural characteristics of the compound. The score also includes an evaluation of the quantitative estimate of drug-likeness (QED), an advanced concept to estimate the efficiency of a drug compound originally put forward by Bickerton et al. (2012) that includes molecular weight, polar surface area, lipophilicity, aromatic ring count, rotatable bond count, hydrogen bond donor number, hydrogen bond acceptor number, and record of any unnecessary substructures within the molecular structure of the compound. The specific singular desirability scores for all these eight QED properties are combined by calculating their geometric mean (GM) value and then the final QED score for the lead compound is deduced for the construction of decision trees. PrOCTOR analyses 48 features corresponding to the above categories per lead compound to elucidate the probability of toxicity (Basile et al., 2019).

11.1.2.4 AI-Based Prediction of Drug Behavior

AI is used to deduce a quantitative measurement of the interaction nature of a specific lead compound with the body functioning over time after successful administration. Absorption, metabolism, distribution, and excretion (ADME) are the four vital drug behavior properties that must be analyzed to get a reliable prediction of the action of a drug compound. ADME is especially essential to know the accurate probability of a target protein or gene being 'hit' by the administered drug and the retention time of a drug compound in the blood circulation of a patient suffering from a pathological condition. AI-based examination of ADME can be a reliable shift toward a decrease in failures of clinical trials as most failure cases during clinical trials of potential drugs occur due to non-standardization of ADME properties as a part of the finalization of an efficient drug discovery procedure (Dong et al., 2023).

Conventional *in-vivo* and *in-vitro* experimental procedures for testing ADME parameters are resource-intensive, depend on animal models, and require more time for successful completion with considerable chances of error. This has driven the gradual shift toward an *in-silico* approach using AI-based algorithms that can expedite the process with high accuracy. ADMET-AI is one such reliable AI-based ADME prediction platform that also includes toxicity prediction. Chemprop-RDkitis applied by the ADMET-AI platform that uses a directed message passing neural network (DMPNN) mechanism to convey the molecular and structural information of a lead compound and its local regions of vicinity via neural network layers.

Chemprop-RDkit segregates a compound into small chemical entities such as functional groups, bonds, aromatic rings, etc. DMPNN can communicate this specific information downstream to find the matching of these chemical entities with the desirable features sought to perform the drug action after accurate administration (Walker et al., 2020).

11.1.2.5 AI-Based Prediction of Drug-Drug Interactions

Drug-drug interactions (DDI) predictions can provide a reliable estimation of the effects of combined drug compounds administered simultaneously. DDI predictions are very crucial in drug discovery in today's world as the need for polypharmacy is on the rise, especially among the older generation of people suffering from multiple associated medical conditions. Polypharmacy in treatment expedites the need for continuous monitoring of the interaction and aftermath of administering one drug with another for the well-being of a patient under multiple medications (Agarwal et al., 2021). This ensures the maintenance of safe medical practices to decelerate specific pathological conditions. Interactions between administered drugs may either cause the generation of toxic products, decrease the efficacy of a drug, or expedite the action of a particular drug. Accuracy in DDI prediction can enhance the survival rates of patients, especially under intensive care and monitoring. Lethal or undesirable combinations of potential drugs can be directly identified if DDIs are accurately predicted during drug discovery. AI-based network propagation through a link prediction approach is one of the most common methods where nodes and edges of a network topology diagram are created based on the links between two drug compounds. Nodal points of the network topology representation indicate the drug compounds and the boundary regions indicate the DDIs. This topology can then be analyzed to find out the DDIs that are unreported and need to be tested (Vo et al., 2022).

11.1.2.6 AI-Based Drug Repurposing

AI is also applied for discovering novel therapeutic applications for already approved existing drug compounds having validated safety, efficacy, and ADMET profiles. The drug repurposing (DR) process can be a rapid alternative for expediting drug discovery for specific pathological conditions by lowering developmental expenses and resource mobilization. A network-oriented *in-silico* drug repurposing system was put forward by Zhou et al. (2020) by analyzing the visible and DNA level relatedness among approved drug compounds, drug-target interactions, DDIs, and pathological condition data to filter out drug compounds for repurposing. AI approaches such as support vector machines (SVMs), network-based link prediction, and random forest (RF) are also used in the repurposing of drugs. Evaluation of electronic health records (EHR), biomedical knowledge, and literature repositories is necessary for the validation of hypotheses and potential models for repurposing approved drug compounds and designating new targets for these drugs (Zong et al., 2022).

11.1.3 Concept of Machine Learning

Machine learning involves teaching computers to extract meaningful information from data and make decisions or predictions instead of direct programming. In traditional programming, computers follow specific instructions, but in machine learning, computers are provided with examples (data) and a task, enabling them to determine how to perform the task based on the examples. Machine learning algorithms improve their accuracy and effectiveness over time by processing more data and learning from the information provided.

11.1.3.1 Application of Machine Learning in Medicinal Plant Research

Machine learning has been widely used in medicinal plant research, especially for identifying, classifying, and predicting the bioactivity of different plant species. Researchers have implemented the structured models of machine learning and deep learning for automatically identifying medicinal plant species in real-time using leaves, achieving high accuracy levels. For example, studies in Indonesia have applied techniques such as local binary patterns, SVMs, deep learning neural networks, fuzzy local binary patterns, and probabilistic neural networks to classify various medicinal plant species, with accuracies ranging from 56% to 96% (Malik et al., 2022; Zang et al., 2023; Mulugeta et al., 2024). Additionally, deep learning models like EfficientNet-B1-based models have demonstrated considerable improvements in identifying and classifying medicinal plant species. These models were trained and evaluated using both public and private datasets, reaching top-1 accuracies of 87% and 84% on test sets. Real-time testing achieved accuracies of 78.5% (top-1) and 82.6% (top-5) (Malik et al., 2022). Machine learning has contributed a significant step in forecasting the bioactivity of plant species, especially in pinpointing potential antimalarial compounds. By combining machine learning models based on distinguishing characters with ethnomedicinal insights, researchers have created innovative methods to forecast the biological activity of plants, surpassing conventional ethnobotanical selection approaches. This blend of technology and botanical expertise presents exciting possibilities for future drug discovery and emphasizes the importance of fair benefit-sharing with knowledge holders (Richard-Bollans et al., 2023). Machine learning's use in medicinal plant research has transformed the identification, categorization, and prediction of the biological activity of plants, highlighting the potential of technology to advance biodiversity studies and drug discovery. The quantitative examination of the relatedness and connecting links between chemical profiles and biological activity aids in forecasting the impacts of ethnomedicine. This is called a quantitative composition–activity relationship (QCAR) (Zhao et al., 2004). This method is akin to the quantitative structure–activity relationship (QSAR), which relies on molecular structure descriptor features to forecast activity (Brown & Fraser, 1868). Nayak et al. proposed a hybrid genetic algorithm–artificial neural network approach to create a computational QCAR model, using machine-learning techniques to develop these models (Nayak et al., 2019). The structured model examined the standard blend of components in an herbal diabetes treatment. Alternative approaches such as artificial neural networks, support vector machines, and multiple linear regression have also demonstrated efficacy in modeling the correlation between chemical composition and activity (Wang et al., 2006). Supervised learning is a subset of the machine learning method in which a programmed algorithm is automated with a distinguishable marked dataset comprising known outputs. By examining connecting links and relatedness features within the marked data, the algorithm learns to link input data with the appropriate output. This technique finds extensive applications in domains such as image recognition, natural language processing, and predictive modeling. Machine-learning-based supervised learning approach is useful for drug discovery and design, especially in the pharmaceutical and primary healthcare industry (Dara et al., 2021), predictive maintenance and quality control (Kavasidis et al., 2023), drug target identification (Bagherian et al., 2021), disease diagnosis and prognosis (Kumar et al., 2022), adverse event detection (Chapman et al., 2019), and predictive modeling for clinical trials (Elkin & Zhu, 2021). Deep learning is a subset of machine learning methods which is revolutionizing medicinal plant research by offering advanced analysis and insights that traditional methods cannot provide. Deep learning, alternatively termed deep neural networks, demonstrates proficiency in recognizing data-driven patterns and autonomously finding out character details and concealed data configurations from extensive datasets (Ma & Mei, 2021). Employing nonlinear information processing to unveil multi-level features for data investigation and efficient forecasting deep learning emerges as a potent instrument for tackling qualitative and quantitative hurdles in examining the therapeutic potential of medicinal plants, specifically during the handling of extensive sample sizes (Debus et al., 2021). Artificial neural networks (ANNs) and convolutional neural networks (CNNs) stand out as commonly employed deep learning frameworks in assessing the therapeutic potential of medicinal plants. ANN shows the presence of a simplified and concise arrangement, featuring three primary layers: input, hidden, and output. The basic step is linking several crucial constituents, and ANN constructs a sophisticated nonlinear network proficient in executing complex logical operations and establishing nonlinear associations (Yuan et al., 2023). It demonstrates effectiveness in managing both qualitative and quantitative data, and its capacity for normalization and error resistance may be uplifted through weight adjustments (Jin & Zheng, 2022). It has been reported to the contrary that CNN exhibits a more entangled architecture, comprising five principal elements: input layer, convolutional layer, pooling layer, fully connected layer, and output layer. ANN gains the character information via input mapping, but CNN extracts the same information by executing convolution operations on the output of each layer (Ma et al., 2021). The convolution layer, an essential component of the CNN framework, gains advantages from shared and sparse connections in filters, aiding in mitigating model overfitting (Zhang et al., 2021). Employing deep learning to assess the quality of medicinal plants with extensive samples presents notable benefits, converging the gap existing in conventional machine learning in handling vast amounts of data. Nonetheless, interpreting the minute character details found through deep learning is an area of difficulty that requires additional investigation.

11.1.4 Concept of Bioinformatics

In today's world, notable advancements in the integration of information technology in molecular biology have led to a proliferation of biological data generated by researchers (Vassilev et al., 2006). Bioinformatics is a growing intersection of biology, chemistry, information technology, mathematics, and statistics, emphasizing the utilization of information technology to handle biological data. This encompasses activities like gathering and organizing data, extracting insights from data, executing database queries, interpreting data, and crafting models and solutions (Degrave & Machado, 2002; Jayaram & Dhingra, 2010; Xiong, 2006). Typical tasks in bioinformatics include the mapping and examination of DNA and protein sequences, aligning diverse nucleotide and polypeptide sequences to facilitate comparison, and generating and viewing three-dimensional representations of protein structures (Luscombe et al., 2001). Bioinformatics plays a pivotal role in diagnostic pathology, facilitating a more profound comprehension of plant-pathogen biological interactions, thereby

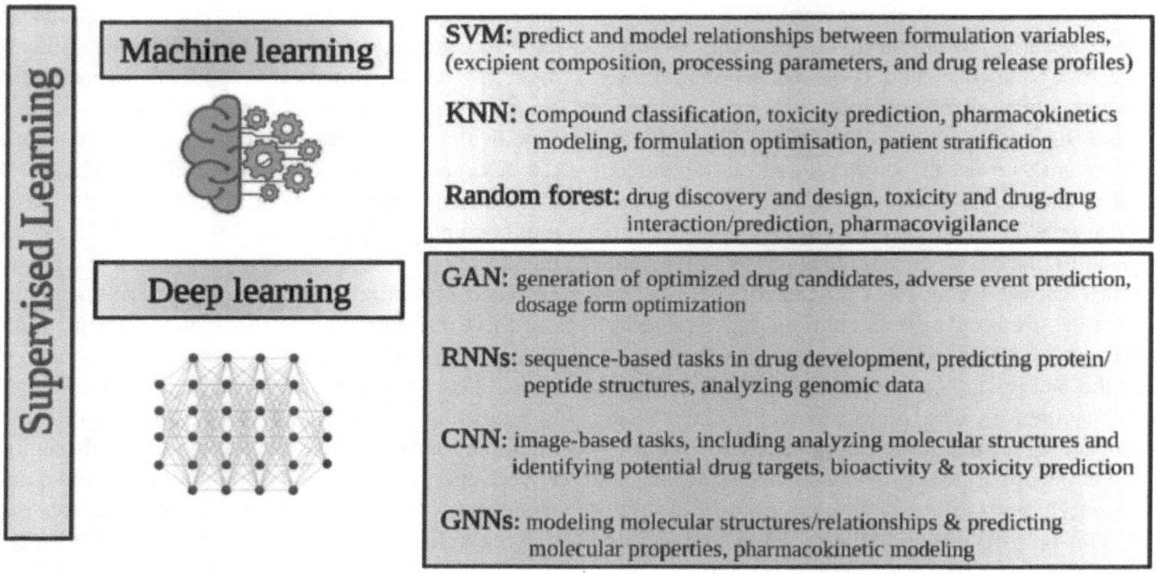

FIGURE 11.3

assisting in disease control (Koltai & Volpin, 2003). This field encompasses two primary domains: the various generations of computerized databases and assisting computational bioinformatic tools along with the use of these resources to derive biological insights and enrich our knowledge of living organisms. Tool development involves designing software for analyzing sequences, structures, and functions, as well as establishing and managing biological databases. These tools find application in three principal realms of genomic and molecular biology research: analyzing molecular sequences, examining molecular structures, and assessing molecular functions (Xiong, 2006). Sequence analysis involves tasks, such as aligning sequences, searching databases, discovering motifs and patterns, identifying genes and promoters, reconstructing evolutionary relationships, and comparing genomes. Structural analysis comprises activities like examining, comparing, classifying, and predicting the structures of proteins and nucleic acids. Functional analysis encompasses profiling gene expression, predicting protein-protein interactions, forecasting protein subcellular localization, and reconstructing and simulating metabolic pathways (Rao et al., 2008; Xiong, 2006). The primary objective of database development is to arrange data into organized records to facilitate quick and efficient extraction of meaningful information (Xiong, 2006). There are two main types of databases: primary and secondary with a less-known third type known as composite databases, an amalgamation of primary and secondary databases. Primary or archived databases consist of annotated information on DNA and protein sequences, structures, and expression profiles. Secondary or derived databases house analysis outcomes derived from primary resources, such as sequence patterns or motifs, variations and mutations, and evolutionary connections (Rao et al., 2008).

11.1.4.1 Medicinal Plant Databases

Numerous databases have been established to compile data concerning various facets of medicinal plants, encompassing ethnobotanical knowledge, bioactive compounds, pharmacological applications, genomic or transcriptomic data, and molecular targets of active constituents. Additionally, pertinent information can be accessed from broader taxonomic, chemical, and molecular repositories such as the National Center for Biotechnology Information (NCBI) (Sayers et al., 2010), the Kyoto Encyclopedia of Genes and Genomes (KEGG) (Kanehisa & Goto, 2000), and KNApSAcK (Afendi et al., 2011). The following are instances of databases providing essential insights into medicinal plants. The International Ethnobotany Database (ebDB) serves as a non-commercial archive for ethnobotanical information, offering multilingual support (http://ebdb.org). This repository boasts a wide array of features tailored for ethnobotanical investigations, including comprehensive location details and robust search functionalities facilitating data export. NAPALERT, on the other hand, is dedicated to identifying and scrutinizing experimental data of natural formulations, specifically those originating from plant-based sources (Loub et al., 1985). The United States Department of Agriculture (USDA) manages a database concerning medicinal plants, which provides concise explanations of associated bioactive components and their medicinal applications. Curated by Dr. Park, this database additionally features a combined representation of medicinally important plants correlated with six important prerequisite therapeutic properties: inflammation, heart disease, obesity, hypertension, kidney disease, and diabetes. Furthermore, Dr. Duke's Phytochemical and Ethnobotanical Database (http://www.ars-grin.gov/duke/) at the USDA encompasses comprehensive details about medicinal plants, encompassing their classification, active constituents, and therapeutic significance. Additional medicinal databases encompass TCM-ID (Traditional Chinese Medicine – Information Database) (Wang et al., 2005), which furnishes information on herbal components (HIT, Herbal Ingredient's Target) (Ye et al., 2010), molecular structures, functional attributes of active constituents, clinical applications, and the therapeutic and toxic effects associated with traditional Chinese medicine. TCM-ID also provides connections to other resources like the Therapeutic Target Database (TTD) (Liu et al., 2011), KEGG (Kanehisa & Goto, 2000), Protein Data Bank (PDB) (Bernstein et al., 1978), Uniprot (Magrane & Consortium, 2011), and Pfam (Punta et al., 2011). There exist various databases designed to cater to particular geographic regions. For example, CMKb specializes in Australia (Gaikwad et al., 2008), whereas Raintree (http://rain-tree.com/ethnic.htm) offers insights into the Amazon rainforests. In addition to compiling phytochemical data, it is advantageous to document genomic or transcriptomic information pertaining to medicinal plants. The National Library of Medicine's NCBI hosts 'Plant Genome Central,' which contains genome and expressed sequence tags (ESTs) data obtained from wide-ranging successfully accomplished sequencing projects. For transcriptomic plant therapeutic data, the EGENES database presents an automated and user-friendly platform for efficient analysis of plant ESTs via integrating whole genome-based information with higher-level functional data (Masoudi-Nejad et al., 2007). This process entails sequence cleansing, repeat and vector masking, sequence assembly, and KEGG annotation. EGENES endeavor to chalk out various biochemical reactions and metabolic pathways in plants based on EST data. A notable resource specifically designated for information on medicinally important plants is the Medicinal Plants Genomics Resource (MPGR: http://medicinalplantgenomics.msu.edu/), which aims to furnish gene expression and metabolically important biomolecule information from several geographically distributed medicinally important plant species.

11.1.4.2 Methods and Strategies for Examining the Molecular Data

Bioinformatics methods can produce coexpression networks using transcriptome data, providing potential avenues for identifying genes in plant species with ancestral relatedness. In today's world, it can be specifically stated that comparative genomics supports the exchange of information across different species. Plant-specific datasets are accessible through platforms such as PLEXdb (Wise et al., 2007), GEO (Barrett & Edgar, 2006), and EBI Array Express (Rocca-Serra et al., 2003). By studying coexpression networks, researchers may identify genes of interest and understand their functions (Mochida et al., 2011). The KEGG platform offers an avenue for examining the pathways of secondary metabolites across diverse organisms (Kanehisa et al., 2010). Additionally, the KEGG Drug database provides insights into two molecular networks: (i) the interactions between drugs and target molecules and (ii) the metabolic pathways of bioactive molecules in various organisms. KEGG Drug encompasses chemical structures or constituents of both prescription and over-the-counter (OTC) medications, along with drugs derived from traditional Chinese medicine (Kanehisa, 2009). The provided information could be beneficial for exploring drug discovery from plant genomes. Another tool for investigating pathways of secondary metabolites cataloged in KEGG is PathPred (Moriya et al., 2010). PathPred, a web-based server, forecasts the sequential steps for metabolic reactions with multiple steps for a specific lead candidate by initiating affinity extraction by matching with the information repository in the KEGG COMPOUND database. Primarily tailored for pathways linked to microbial degradation of environmental substances and the formation pathways of various therapeutically important secondary metabolite compounds in plants, PathPred also encompasses normalized reaction outcomes and processes between various compounds with structural similarity. With advancements in omics technologies, bioinformatics plays a pivotal role in facilitating a holistic comprehension of metabolic processes. The amalgamation of transcriptomic and metabolomic data, aided by data mining techniques, opens avenues for exploring metabolic pathways (Saito et al., 2008). Patterns in EST expression and categorized mass peaks via self-organizing maps have revealed associations among nutrient deprivation, primary metabolism, and glucosinolate metabolism (Hirai et al., 2004). Bioinformatics furnishes indispensable tools for dissecting the extensive data generated by high-throughput methodologies, enabling the identification of potential genes, pathways, and networks implicated in synthesizing bioactive metabolites in medicinal plants. Furthermore, bioinformatics assists in bridging disparate pieces of evidence into coherent conceptualizations offering avenues for experimental verification (Sharma & Sarkar, 2013).

The bioinformatics field holds considerable potential to leverage diverse data integration, machine learning methodologies, and network-driven strategies to uncover promising combinations of medicinal plants for enhanced therapeutic interventions.

11.2 CONCLUSION AND FUTURE PERSPECTIVES

Artificial intelligence (AI) stands poised to transform the pharmaceutical sector, accelerating the pace of drug discovery and development. Through virtual screening methods, AI can swiftly analyze extensive chemical libraries to pinpoint potential therapeutic candidates that meet specific criteria, thereby streamlining the identification of lead compounds. AI-enabled precision medicine can classify patients, predict their response to treatments, and tailor medications based on their genomes, proteomes, and clinical data. Researchers can use deep learning and generative models to design new compounds with target-binding

TABLE 11.1
AI Platforms Successfully Used in Drug Discovery (Visan & Negut, 2024)

Program/Platform	Description	Primary Use	Accession
DeepChe	Python-based AI system using MLP model	Candidate selection in drug discovery	https://github.com/deepchem/deepchem (accessed on February 1, 2024)
DeepNeuralNetQSAR	Python-based AI system	Canaid the detection of the molecular activity of compounds	https://github.com/Merck/DeepNeuralNet-QSAR (accessed on February 1, 2024)
Chemputer	Combination of Monte Carlo tree search and symbolic AI, including DNNs	Synthesize organic molecules	https://zenodo.org/record/1481731 (accessed on February 1, 2024)
DeepTox	AI system using DL	Chemical prediction toxicity	www.bioinf.jku.at/research/DeepTox (accessed on February 1, 2024)
AlphaFold	AI system using DL	Predicts the 3D structures	https://alphafold.ebi.ac.uk/ (accessed on February 1, 2024)
ORGANIC	Generative ML approaches and DNNs	Novel molecular materials	https://github.com/aspuru-guzik-group/ORGANIC (accessed on February 1, 2024)
Potential Net	Neural networks, deep attention mechanisms and descriptor embeddings	The binding affinity of ligands in protein–ligand complexes	https://www.genesistherapeutics.ai/platform.html (accessed on February 1, 2024)
Hit Dexter CNNs and ANNs	ML technique	For predicting molecules that might respond to biochemical assays	http://hitdexter2.zbh.uni-hamburg.de (accessed on February 1, 2024)
DeltaVina	ML algorithms, including XGBoo stand random forest	CNNs	https://github.com/chengwang88/deltavina (accessed on February 1, 2024)
Neural graph fingerprint		Predict properties of novel molecules	https://github.com/HIPS/neural-fingerprint (accessed on February 1, 2024)
GastroPlus predictive	AI and modelling	For pharmaceutical products (dosage form) in many animal models	https://www.simulations-plus.com/software/gastroplus/# (accessed on February 1, 2024)

features, enhancing drug efficacy and reducing side effects. The programmed algorithms based on AI, ML, and DL can also optimize medication compositions and delivery methods by accounting for patient-specific factors such as age, weight, genetics, and disease status to improve treatment outcomes. Furthermore, AI has the potential to transform safety evaluations by forecasting potential adverse effects and toxicity of drug candidates. AI technologies are also reshaping drug delivery methods, facilitating specialized, efficient, and highly targeted therapeutic practices for primary and intensive healthcare. By enhancing the range and scope of AI in optimization, streamlining and analysis of plant-based therapeutic data, pharmaceutical researchers and healthcare providers can enhance drug development, targeted drug delivery, minimize adverse effects, and improve patient outcomes. This integration of AI technologies offers substantial advantages over traditional experimental approaches, promising to expedite drug development, enhance patient care, and reshape the pharmaceutical landscape. In the realm of medicinal plants, their therapeutic efficacy stems from a combination of various compounds. The employment of machine learning algorithms has demonstrated efficacy in analyzing and managing the multi-source data associated with medicinal plants. When combined with data fusion techniques, these algorithms can amalgamate diverse data sets, thereby increasing the feasibility and applicability of categorization and forecasting models. Research findings by various workers underscored the utility of machine learning algorithms combined with data from multiple reliable sources for facilitating comprehensive quality assessments of medicinal plants. Deep learning has the potential to revolutionize medicinal plant research, contributing to drug discovery and development, conservation of plant species, and enhanced understanding of plant biology. As the field progresses, interdisciplinary collaboration will be essential to fully capitalize on the benefits of deep learning in this area. Bioinformatics methods provide essential instruments for creating efficient and focused quests for plant-derived remedies. The synergy between artificial intelligence, machine learning, deep learning, and bioinformatics has led to groundbreaking discoveries in medicinal plant research, providing new insights, accelerating drug development, and deepening our understanding of the bioactive potential of plant species. This interdisciplinary approach holds significant promise for the future of pharmaceutical research and the sustainable use of medicinal plants for human health and well-being.

REFERENCES

Adam, M., Wessel, M., & Benlian, A. (2020). AI-based chatbots in customer service and their effects on user compliance. *Electronic Markets*, *31*(2), 427–445. https://doi.org/10.1007/s12525-020-00414-7

Afendi, F. M., Okada, T., Yamazaki, M., Hirai-Morita, A., Nakamura, Y., Nakamura, K., Ikeda, S., Takahashi, H., Altaf-Ul-Amin, M., Darusman, L. K., Saito, K., & Kanaya, S. (2011). KNApSAcK family databases: Integrated metabolite–plant species databases for multifaceted plant research. *Plant and Cell Physiology*, *53*(2), e1. https://doi.org/10.1093/pcp/pcr165

Agarwal, M., Saba, L., Gupta, S. K., Carriero, A., Falaschi, Z., Paschè, A., Danna, P., El-Baz, A., Naidu, S., & Suri, J. S. (2021). A novel block imaging technique using nine artificial intelligence models for COVID-19 disease classification, characterization and severity measurement in lung computed tomography scans on an Italian cohort. *Journal of Medical Systems*, *45*(3). https://doi.org/10.1007/s10916-021-01707-w

Ashby, W. R. (1956). An introduction to cybernetics. By W. Ross Ashby. Chapman and hall, Ltd., London, 1956. *Journal of Mental Science*, *104*(435), 590592. https://doi.org/10.1192/bjp.104.435.590-b

Azadnia, R., Al-Amidi, M. M., Mohammadi, H., Cifci, M. A., Daryab, A., & Cavallo, E. (2022). An AI based approach for medicinal plant identification using deep CNN based on global average pooling. *Agronomy*, *12*(11), 2723. https://doi.org/10.3390/agronomy12112723

Bagherian, M., Sabeti, E., Wang, K., Sartor, M. A., Nikolovska-Coleska, Z., & Najarian, K. (2021). Machine learning approaches and databases for prediction of drug–target interaction: A survey paper. *Briefings in Bioinformatics*, *22*(1), 247–269. https://doi.org/10.1093/bib/bbz157

Barrett, T., & Edgar, R. (2006). Gene expression omnibus: Microarray data storage, submission, retrieval, and analysis. *Methods in Enzymology*, *411*, 352–369. https://doi.org/10.1016/S0076-6879(06)11019-8

Basile, A. O., Yahi, A., & Tatonetti, N. P. (2019). Artificial intelligence for drug toxicity and safety. *Trends in Pharmacological Sciences*, *40*(9), 624–635. https://doi.org/10.1016/j.tips.2019.07.005

Bernstein, F. C., Koetzle, T. F., Williams, G. J., Meyer, Jr., E. F., Brice, M. D., Rodgers, J. R., & Tasumi, M. (1978). The protein data bank: A computer-based archival file for macromolecular structures. *Archives of Biochemistry and Biophysics*, *185*(2), 584–591. https://doi.org/10.1016/0003-9861(78)90204-7

Bickerton, G. R., Paolini, G. V., Besnard, J., Muresan, S., & Hopkins, A. L. (2012). Quantifying the chemical beauty of drugs. *Nature Chemistry*, *4*(2), 90–98. https://doi.org/10.1038/nchem.1243

Brown, A. C., & Fraser, T. R. (1868). On the connection between chemical constitution and physiological Action. Part. I. – On the physiological action of the salts of the ammonium bases, derived from strychnia, Brucia, Thebaia, Codeia, morphia, and Nicotia. *Transactions of the Royal Society of Edinburgh*, *25*(1), 151–203. https://doi.org/10.1017/s0080456800028155

Chapman, A. B., Peterson, K. S., Alba, P. R., DuVall, S. L., & Patterson, O. V. (2019). Detecting adverse drug events with rapidly trained classification models. *Drug Safety*, *42*(1), 147–156. https://doi.org/10.1007/s40264-018-0763-y

Chen, Y., Moreira, P., Liu, W., Monachino, M., Nguyen, T. L., & Wang, A. (2022). Is there a gap between artificial intelligence applications and priorities in health care and nursing management? *Journal of Nursing Management*, *30*(8), 3736–3742. https://doi.org/10.1111/jonm.13851

Dara, S., Dhamercherla, S., Jadav, S. S., Babu, C. M., & Ahsan, M. J. (2021). Machine learning in drug discovery: A review. *Artificial Intelligence Review*, *55*(3), 1947–1999. https://doi.org/10.1007/s10462-021-10058-4

Debus, B., Parastar, H., Harrington, P., & Kirsanov, D. (2021). Deep learning in analytical chemistry. *TrAC Trends in Analytical Chemistry*, *145*, 116459. https://doi.org/10.1016/j.trac.2021.116459

Degrave, W., & Machado, C. J. (2012). Bioinformatics and health: An overview. *RECIIS*, *1*(2). https://doi.org/10.3395/reciis.v1i2.sup.97en

Dong, J., Wu, Z., Xu, H., & Ouyang, D. (2023). FormulationAI: A novel web-based platform for drug formulation design driven by artificial intelligence. *Briefings in Bioinformatics*, *25*(1). https://doi.org/10.1093/bib/bbad419

Elkin, M. E., & Zhu, X. (2021). Predictive modeling of clinical trial terminations using feature engineering and embedding learning. *Scientific Reports*, *11*(1). https://doi.org/10.1038/s41598-021-82840-x

Erhaldt, L. (2009). Cigarette smoking: An undertreated risk factor for cardiovascular disease. *Atherosclerosis*, *205*(1), 23–32. https://doi.org/10.1016/j.atherosclerosis.2009.01.007

Fitzgerald, M., Heinrich, M., & Booker, A. (2020). Medicinal plant analysis: A historical and regional discussion of emergent complex techniques. *Frontiers in Pharmacology*, *10*. https://doi.org/10.3389/fphar.2019.01480

Gaikwad, J., Khanna, V., Vemulpad, S., Jamie, J., Kohen, J., & Ranganathan, S. (2008). CMKb: A web-based prototype for integrating Australian Aboriginal customary medicinal plant knowledge. *BMC Bioinformatics*, *9*(S12). https://doi.org/10.1186/1471-2105-9-s12-s25

Hirai, M. Y., Yano, M., Goodenowe, D. B., Kanaya, S., Kimura, T., Awazuhara, M., Arita, M., Fujiwara, T., & Saito, K. (2004). Integration of transcriptomics and metabolomics for understanding of global responses to nutritional stresses in *Arabidopsis thaliana*. *Proceedings of the National Academy of Sciences*, *101*(27), 10205–10210. https://doi.org/10.1073/pnas.0403218101

Jayaram, B., & Dhingra, P. (2010). *Bioinformatics for a better tomorrow*. Indian Institute of Technology.

Jin, X., & Zheng, S. (2022). Measurement and calibration of optical instruments based on metrological calibration method and artificial neural network. *Optik*, 170479. https://doi.org/10.1016/j.ijleo.2022.170479

Kanehisa, M., & Goto, S. (2000). KEGG: Kyoto encyclopedia of genes and genomes. *Nucleic Acids Research*, *28*(1), 27–30. https://doi.org/10.1093/nar/28.1.27

Kanehisa, M. (2009). Representation and analysis of molecular networks involving diseases and drugs. *Genome Informatics*, *2009*. https://doi.org/10.1142/9781848165632_0020

Kanehisa, M., Goto, S., Furumichi, M., Tanabe, M., & Hirakawa, M. (2010). KEGG for representation and analysis of molecular networks involving diseases and drugs. *Nucleic Acids Research*, *38*(Suppl_1), D355–D360. https://doi.org/10.1093/nar/gkp896

Kavasidis, I., Lallas, E., Gerogiannis, V. C., Charitou, T., & Karageorgos, A. (2023). Predictive maintenance in pharmaceutical manufacturing lines using deep transformers. *Procedia Computer Science*, *220*, 576–583. https://doi.org/10.1016/j.procs.2023.03.073

Koltai, H., & Volpin, H. (2003). Agricultural genomics: An approach to plant protection. *European Journal of Plant Pathology*, *109*, 101–108.

Kumar, Y., Koul, A., Singla, R., & Ijaz, M. F. (2022). Artificial intelligence in disease diagnosis: A systematic literature review, synthesizing framework and future research agenda. *Journal of Ambient Intelligence and Humanized Computing*, *14*(7), 8459–8486. https://doi.org/10.1007/s12652-021-03612-z

Lin, E., Lin, C., & Lane, H. (2020). Relevant applications of generative adversarial networks in drug design and discovery: Molecular de Novo design, dimensionality reduction, and de Novo peptide and protein design. *Molecules*, *25*(14), 3250. https://doi.org/10.3390/molecules25143250

Liu, X., Zhu, F., Ma, X., Tao, L., Zhang, J., Yang, S., Wei, Y., & Chen, Y. Z. (2011). Undefined. *Expert Opinion on Therapeutic Targets*, *15*(8), 903–912. https://doi.org/10.1517/14728222.2011.586635

Loub, W. D., Farnsworth, N. R., Soejarto, D. D., & Quinn, M. L. (1985). Napralert: Computer handling of natural product research data. *Journal of Chemical Information and Computer Sciences*, *25*(2), 99–103. https://doi.org/10.1021/ci00046a009

Luscombe, N. M., Greenbaum, D., & Gerstein, M. (2001). What is bioinformatics? A proposed definition and overview of the Field. *Methods of Information in Medicine*, *40*(04), 346–358. https://doi.org/10.1055/s-0038-1634431

Ma, W., Liu, Z., Kudyshev, Z. A., Boltasseva, A., Cai, W., & Liu, Y. (2021). Deep learning for the design of photonic structures. *Nature Photonics*, *15*(2), 77–90. https://doi.org/10.1038/s41566-020-0685-y

Ma, Z., & Mei, G. (2021). Deep learning for geological hazards analysis: Data, models, applications, and opportunities. *Earth-Science Reviews*, *223*, 103858. https://doi.org/10.1016/j.earscirev.2021.103858

Magrane, M., & Consortium, U. (2011). UniProt knowledgebase: A hub of integrated protein data. *Database*, *2011*(bar009), bar009–bar009. https://doi.org/10.1093/database/bar009

Malik, O. A., Ismail, N., Hussein, B. R., & Yahya, U. (2022). Automated real-time identification of medicinal plants species in natural environment using deep learning models – a case study from Borneo region. *Plants*, *11*(15), 1952. https://doi.org/10.3390/plants11151952

Maltarollo, V. G., Kronenberger, T., Espinoza, G. Z., Oliveira, P. R., & Honorio, K. M. (2019). Advances with support vector machines for novel drug discovery. *Expert Opinion on Drug Discovery*, *14*(1), 23–33. https://doi.org/10.1080/17460441.2019.1549033

Masoudi-Nejad, A., Goto, S., Jauregui, R., Ito, M., Kawashima, S., Moriya, Y., Endo, T. R., &Kanehisa, M. (2007). EGENES: Transcriptome-based plant database of genes with metabolic pathway information and expressed sequence tag indices in KEGG. *Plant Physiology*, *144*(2), 857–866. https://doi.org/10.1104/pp.106.095059

Mochida, K., Uehara-Yamaguchi, Y., Yoshida, T., Sakurai, T., & Shinozaki, K. (2011). Global landscape of a Co-expressed gene network in Barley and its application to gene discovery in Triticeae crops. *Plant and Cell Physiology*, *52*(5), 785–803. https://doi.org/10.1093/pcp/pcr035

Moriya, Y., Shigemizu, D., Hattori, M., Tokimatsu, T., Kotera, M., Goto, S., &Kanehisa, M. (2010). PathPred: An enzyme-catalyzed metabolic pathway prediction server. *Nucleic Acids Research*, *38*(Web Server), W138–W143. https://doi.org/10.1093/nar/gkq318

Mulugeta, A. K., Sharma, D. P., & Mesfin, A. H. (2024). Deep learning for medicinal plant species classification and recognition: A systematic review. *Frontiers in Plant Science*, *14*. https://doi.org/10.3389/fpls.2023.1286088

Nayak, B., Padhi, S. K., & Pattnaik, P. K. (2019). Optimization of cloud datacenter using heuristic strategic approach. In *Advances in intelligent systems and computing* (pp. 91–100). https://doi.org/10.1007/978-981-13-3600-3_9

Paul, A. K., & Schaefer, M. (2020). Safeguards for the use of artificial intelligence and machine learning in global health. *Bulletin of the World Health Organization*, *98*(4). https://doi.org/10.2471/blt.19.237099

Pérez Santín, E., Rodríguez Solana, R., González García, M., García Suárez, M. D., Blanco Díaz, G. D., Cima Cabal, M. D., Moreno Rojas, J. M., & López Sánchez, J. I. (2021). Toxicity prediction based on artificial intelligence: A multidisciplinary overview. *WIREs Computational Molecular Science*, *11*(5). https://doi.org/10.1002/wcms.1516

Punta, M., Coggill, P. C., Eberhardt, R. Y., Mistry, J., Tate, J., Boursnell, C., Pang, N., Forslund, K., Ceric, G., Clements, J., Heger, A., Holm, L., Sonnhammer, E. L., Eddy, S. R., Bateman, A., & Finn, R. D. (2011). The Pfam protein families database. *Nucleic Acids Research*, *40*(D1), D290–D301. https://doi.org/10.1093/nar/gkr1065

Rao, V. S., Das, S. K., Rao, V. J., & Srinubabu, G. (2008). Recent developments in life sciences research: Role of Bioinformatics. *African Journal of Biotechnology*, *7*(5), 495–503.

Richard-Bollans, A., Aitken, C., Antonelli, A., Bitencourt, C., Goyder, D., Lucas, E., Ondo, I., Pérez-Escobar, O. A., Pironon, S., Richardson, J. E., Russell, D., Silvestro, D., Wright, C. W., & Howes, M. R. (2023). Machine learning enhances prediction of plants as potential sources of antimalarials. *Frontiers in Plant Science*, *14*. https://doi.org/10.3389/fpls.2023.1173328

Rocca-Serra, P., Brazma, A., Parkinson, H., Sarkans, U., Shojatalab, M., Contrino, S., Vilo, J., Abeygunawardena, N., Mukherjee, G., Holloway, E., Kapushesky, M., Kemmeren, P., Lara, G. G., Oezcimen, A., & Sansone, S. (2003). ArrayExpress: A public database of gene expression data at EBI. *Comptes Rendus. Biologies*, *326*(10–11), 1075–1078. https://doi.org/10.1016/j.crvi.2003.09.026

Saito, K., Hirai, M. Y., & Yonekura-Sakakibara, K. (2008). Decoding genes with coexpression networks and metabolomics – 'majority report by precogs'. *Trends in Plant Science*, *13*(1), 36–43. https://doi.org/10.1016/j.tplants.2007.10.006

Sayers, E. W., Barrett, T., Benson, D. A., Bolton, E., Bryant, S. H., Canese, K., Chetvernin, V., Church, D. M., DiCuccio, M., Federhen, S., Feolo, M., Fingerman, I. M., Geer, L. Y., Helmberg, W., Kapustin, Y., Landsman, D., Lipman, D. J., Lu, Z., Madden, T. L., Madej, T., Maglott, D. R., Bauer, A. M., Miller, V., Mizarchi, I. K., Ostell, J., Panchenko, A., Phan, L., Pruitt, K. D., Schuler, G. D., Sequeira, E., Sherry, S. T., Shumway, M., Sirotkin, K., Slotta, D., Souvorov, A., Starcheenko, G., Tatusova, G. A., Wagner, A., Wang, Y., Wilbur, W. J., Yaschenko, E., & Ye, J. (2010). Database resources of the national center for biotechnology information. *Nucleic Acids Research*, *39*(Database), D38–D51. https://doi.org/10.1093/nar/gkq1172

Sharma, V., & Sarkar, I. N. (2013). Bioinformatics opportunities for identification and study of medicinal plants. *Briefings in Bioinformatics*, *14*(2), 238–250. https://doi.org/10.1093/bib/bbs021

Singh, A., Kanaujia, A., Singh, V. K., & Vinuesa, R. (2023). Artificial intelligence for Sustainable development Goals: Bibliometric patterns and concept evolution trajectories. *Sustainable Development*, *32*(1), 724–754. https://doi.org/10.1002/sd.2706

Sorkun, M. C., Astruc, S., Koelman, J. M., & Er, S. (2020). An artificial intelligence-aided virtual screening recipe for two-dimensional materials discovery. *NPJ Computational Materials*, *6*(1). https://doi.org/10.1038/s41524-020-00375-7

Turing, A. M. (1995). Computing machinery and intelligence. In *Mind design II* (pp. 29–56). https://doi.org/10.7551/mitpress/4626.003.0002

Vassilev, D., Leunissen, J., Atanassov, A., Nenov, A., & Dimov, G. (2006). Application of bioinformatics in plant breeding. *Biotechnology & Biotechnological Equipment*, *19*(Suppl. 3), 139–152. https://doi.org/10.1080/13102818.2005.10817293

Vo, T. H., Nguyen, N. T., Kha, Q. H., & Le, N. Q. (2022). On the road to explainable AI in drug-drug interactions prediction: A systematic review. *Computational and Structural Biotechnology Journal*, *20*, 2112–2123. https://doi.org/10.1016/j.csbj.2022.04.021

Vora, L. K., Gholap, A. D., Jetha, K., Thakur, R. R., Solanki, H. K., & Chavda, V. P. (2023). Artificial intelligence in pharmaceutical technology and drug delivery design. *Pharmaceutics*, *15*(7), 1916. https://doi.org/10.3390/pharmaceutics15071916

Walker, P. A., Ryder, S., Lavado, A., Dilworth, C., & Riley, R. J. (2020). The evolution of strategies to minimise the risk of human drug-induced liver injury (DILI) in drug discovery and development. *Archives of Toxicology*, *94*(8), 2559–2585. https://doi.org/10.1007/s00204-020-02763-w

Wang, J., Zhou, H., Han, L., Chen, X., Chen, Y., & Cao, Z. (2005). Traditional Chinese medicine information database. *Clinical Pharmacology & Therapeutics*, *78*(1), 92–93. https://doi.org/10.1016/j.clpt.2005.03.010

Wang, Y., Wang, X., & Cheng, Y. (2006). A computational approach to botanical drug design by modeling quantitative composition–activity relationship. *Chemical Biology & Drug Design*, *68*(3), 166–172. https://doi.org/10.1111/j.1747-0285.2006.00431.x

Wise, R. P., Caldo, R. A., Hong, L., Shen, L., Cannon, E., & Dickerson, J. A. (2007). BarleyBase/PLEXdb. In *Plant bioinformatics* (pp. 347–363). https://doi.org/10.1007/978-1-59745-535-0_17

Xiong, J. (2006). *Essential bioinformatics*. Cambridge University Press.

Ye, H., Ye, L., Kang, H., Zhang, D., Tao, L., Tang, K., Liu, X., Zhu, R., Liu, Q., Chen, Y. Z., Li, Y., & Cao, Z. (2010). HIT: Linking herbal active ingredients to targets. *Nucleic Acids Research*, *39*(Database), D1055–D1059. https://doi.org/10.1093/nar/gkq1165

Yuan, Z., Niu, M., Ma, H., Gao, T., Zang, J., Zhang, Y., & Chen, L. (2023). Predicting mechanical behaviors of rubber materials with artificial neural networks. *International Journal of Mechanical Sciences*, *249*, 108265. https://doi.org/10.1016/j.ijmecsci.2023.108265

Zhang, X., Yang, J., Lin, T., & Ying, Y. (2021). Food and agro-product quality evaluation based on spectroscopy and deep learning: A review. *Trends in Food Science & Technology*, *112*, 431–441. https://doi.org/10.1016/j.tifs.2021.04.008

Zhao, X. P., Fan, X. H., Yu, J., & Cheng, Y. Y. (2004). A method for predicting activity of traditional Chinese medicine based on quantitative composition-activity relationship of neural network model. *Zhongguo Zhong Yao Za Zhi= ZhongguoZhongyaoZazhi= China Journal of Chinese Materia Medica*, *29*(11), 1082–1085.

Zhou, Y., Wang, F., Tang, J., Nussinov, R., & Cheng, F. (2020). Artificial intelligence in COVID-19 drug repurposing. *The Lancet Digital Health*, *2*(12), e667–e676. https://doi.org/10.1016/s2589-7500(20)30192-8

Zong, N., Wen, A., Moon, S., Fu, S., Wang, L., Zhao, Y., Yu, Y., Huang, M., Wang, Y., Zheng, G., Mielke, M. M., Cerhan, J. R., & Liu, H. (2022). Computational drug repurposing based on electronic health records: A scoping review. *NPJ Digital Medicine*, *5*(1). https://doi.org/10.1038/s41746-022-00617-6

12 Genetic Engineering of Medicinal Plants for Enhanced Secondary Metabolite Production

Jayachandran Halka, Kumaresan Kowsalya, Arumugam Vijaya Anand, Muthukrishnan Anand, Karuppiah Ramalashmi, Ahamed Kabeer Umera Begam, Asirvatham Alwin Robert, and Muthukrishnan Arun

LIST OF ABBREVIATIONS

CRISPR	clustered regularly interspaced short palindromic repeats
gRNA	guide RNA
PAM	protospacer adjacent motif
RNAi	RNA interference
SgRNA	single guide RNA
TALENs	transcription activator-like effector nucleases

12.1 INTRODUCTION

Medicinal plants provide essential bioactive compounds that effectively treat various health issues such as asthma, cancer, heart disease, wound healing, diabetes, and hypertension (Manoharan et al., 2008; Begum et al., 2009; Muthumani et al., 2013; Varalakshmi et al., 2014; Vijaya Anand et al., 2016, 2020; Bharathi et al., 2018; Divya et al., 2019; Akilandeswari et al., 2019; Sangeetha et al., 2022; Kaviya et al., 2022; Manikandan et al., 2022; Karthika et al., 2023; Vijayakumar et al., 2023; Halka et al., 2024). These bioactive compounds, known as secondary metabolites, are produced in small quantities and even hold a major part during plant growth and development (Kim et al., 2002). They are vital for enhancing a wide array of processes that include biological and physiological, helping plants to adapt to and withstand both biological and environmental stresses (Hussain et al., 2012; Yeshi et al., 2022).

It is stated that there are approximately 100,000 types of metabolites found in various plant species. Over 4000 unique phytocompounds are discovered and analysed from plant species. Since antiquity, humans have been utilizing these valuable secondary metabolites to combat numerous life-threatening diseases. Further, secondary metabolites are characterized by their diverse chemical compositions and also have a distinct role in various fields, such as therapeutic, agrochemical, nutritional, aromatic, coloring, insecticidal, and pesticide agents (Balandrin & Klocke, 1988; Bharti et al., 2023). Nearly 80% of people use traditional medicine, particularly herbal treatments, to meet their basic healthcare desires (Farnsworth & Soejarto, 1991). Globally the call for pharmaceutical drugs derived from natural sources, such as herbal medicine, is steadily growing with 25% of the total usage (Newman & Cragg, 2012).

Manipulating the genetic pathways responsible for secondary metabolism in medicinal plants are offering numerous opportunities to boost secondary metabolite production. However, the genetic engineering approach has technical expertise and a deep insight into plant secondary metabolite pathways. Progress in molecular biology has facilitated the discovery and replication of numerous genes associated with secondary metabolite pathways, setting the stage for molecular-level genetic manipulation (Li et al., 2023b). The two steps involved in enhancing metabolite production are, initially, individual genes are altered, including those encoding key enzymes, transcription factors, single-chain antibodies, and transporters (Huang et al., 2015; Hu et al., 2023; Chang et al., 2023).

Multiple genes are manipulated using techniques such as sexual hybridization, co-transformation, recombinant construction, polycistron formation, and retransformation (Chen et al., 2021; Chen et al., 2022; Xu et al., 2022). Additionally, methods like cell suspension culture, micro propagation, hairy root culture, shoot culture, elicitation, callus culture, and genetic engineering are being investigated to satisfy the growing demand for medically significant phytocompounds (Niazian & Sabbatini, 2021). This chapter offers an in-depth overview of the diverse metabolic engineering strategies employed to boost lead metabolite production in various medicinal plants.

12.2 CLASSIFICATION OF METABOLITES

Secondary metabolites are classified based on factors such as their structure (including features like rings or sugars), their composition (such as the presence of nitrogen), their solubility in various solvents, or the biosynthetic pathways that produce them (Tiwari & Rana, 2015). Typically, these compounds are grouped by their biosynthetic origins like phenolic, terpenes, flavonoids, alkaloids, tannins, glycosides, and essential oils.

DOI: 10.1201/9781003475491-12

12.2.1 Polyphenolics

Polyphenols are natural antioxidants found abundantly in whole grains. They come in various forms, including freely soluble, insoluble-conjugated, soluble-conjugated, and esterified. Different types of polyphenols encompass phenolic acids, tannins, flavonoids, lignans, and coumarins (Ma et al., 2022). Polyphenols perform a range of functions, such as neutralizing free radicals (Abd El-Hack et al., 2023), reducing inflammation (Li et al., 2023a), enhancing endothelial function (Iqbal et al., 2023), inhibiting the proliferation of cancer cells (Farghadani & Naidu, 2023), protecting neurons from damage, and boosting immune response (Gairola et al., 2023). For example, the main phenolic acids found in rice are sinapic acid, ferulic acid, vanillic acid, and p-coumaric acid (Zhang et al., 2020; Goufo et al., 2014; Acosta-Estrada et al., 2014; Wu et al., 2009; Zaupa et al., 2015; Tanaka et al., 2019).

12.2.2 Terpenes

Terpenes are known for their vast chemical diversity and are widely distributed natural compounds in plants. They possess unique hydrocarbon-based structures typically derived from isoprene and are categorized based on five-carbon units. Terpenes have varied roles as pharmacological agents (Wróblewska-Łuczka et al., 2023), the modulating endocannabinoid system (Wang et al., 2023), therapeutic efficacy (Trepa et al., 2024), aromatherapy, and flavor and fragrance (Jiang et al., 2023). Even essential oils of peppermint, *i.e.*, menthol, a monoterpene, has a specific role to act as a central nervous system stimulant (Singh et al., 2023). Compounds like curcumin, from turmeric, and taxol, from the Pacific yew tree, have demonstrated its efficacy in combating inflammation and cancer (Bharti et al., 2023).

12.2.3 Flavonoids

The chemical compounds produced by plants in the form of secondary metabolites are flavonoids that play an indispensable role in imparting color to fruits and flowers. The major flavonoids that are found to treat various diseases are quercetin, rutin, and kaempferol. They are found in an extensive range, offer health benefits, and are indispensable components in various nutraceuticals, pharmaceuticals, and cosmetic applications. Because of their potent antioxidant properties and capacity to regulate crucial cellular enzyme functions, they play an irreplaceable part in protecting plants from oxidative stress (Karak, 2019; Carsono et al., 2022). Further *in vivo* and *in vitro* studies disclosed the diverse therapeutic potential of flavonoids, such as their anti-inflammatory, antioxidant, anti-allergic, antipyretic, anticancer, antiviral, antiprotozoal, antiplatelet, anti-ulcer, antibacterial, and anti-atherogenic activities (Lešnik & Bren, 2021; Mumtaz et al., 2021; Demir et al., 2022; Puangpraphant et al., 2022; Vo et al., 2022).

12.2.4 Alkaloids

Alkaloids are essential for plants' defense mechanisms owing to their possible role in treating cancer and other medical conditions (Srivastava & Tiwari, 2022). These alkaloids are naturally occurring substances recognized for their alkyl-substituted peptide rings that have the capability to chemically interact with various molecules. Remarkably, alkaloids possess antibacterial, antimicrobial properties and anti-inflammatory (Yan et al., 2021). They hinder the advancement of cancer by imparting signaling pathways associated with cell division, proliferation, and metastasis. Consequently, alkaloids are pivotal in numerous clinical trials exploring cancer treatments (Bello-Martínez et al., 2022). The various key alkaloid-based compounds, such as vinblastine, vincristine, vitexin, and paclitaxel are extensively employed in clinical research as anticancer agents (Desam & Al-Rajab, 2022).

12.2.5 Tannins

Tannins are a broad category of phenolic compounds consisting of various oligomers and polymers that can precipitate proteins. They play a unique role in forming complexes with starch, proteins, minerals, and cellulose. Additionally, tannins are active ingredients used for their astringent, diuretic, and anti-inflammatory properties and in the treatment of diarrhea (Kabera et al., 2014).

12.2.6 Glycosides

Glycosides are compounds that can be alcohols, phenols, or sulfur-containing substances, characterized by having a sugar moiety attached to a non-sugar component. The glycosides are diverse in nature, featuring glycosidic bonds linked to monosaccharides, oligosaccharides, or polysaccharides. They exist in various forms, including flavonoid glycosides, saponin glycosides, plant-derived cardiac glycosides, anthraquinone glycosides, coumarin glycosides, cyanogenic glycosides, glucosinolates, steroidal glycosides, and phenolic glycosides (Bartnik & Facey, 2024).

12.2.7 Essential Oils

Essential oils are derived naturally from various plant parts. They play valuable roles in aromatherapy, food products, pharmaceuticals, cosmetics, and beverages. Their fat solubility makes them effective in relieving congestion and promoting relaxation. Well-known medicinal plants that produce essential oils include lavender, peppermint, eucalyptus, rosemary, and lemon. Additionally, essential oils hold a varied range of beneficial properties, including anticancer, antiviral, antimutagenic, antioxidant, antidiabetic, and anti-inflammatory effects (Raut & Karuppayil, 2014).

12.3 TECHNIQUES TO IMPROVE SECONDARY METABOLITES IN MEDICINAL PLANTS

The molecular breeding approach is favored due to its effectiveness in addressing agronomic and environmental challenges compared to natural breeding programs. Genetic engineering and omics technology have fastened up promising avenues for boosting the yield of nutraceuticals, pharmaceuticals, and other metabolites in plants (Bindu et al., 2018; Wink, 2018). Since various compounds exhibit therapeutic and medicinal properties, it is essential to increase their production in medicinal plants, which can be achieved through a variety of methods.

12.3.1 Breeding

Classical breeding has factually been employed to progress crop varieties capable of withstanding various climatic conditions. These methods entail identifying and crossing plants with desired characteristics to procreate offspring with enhanced characteristics in coping with climatic conditions. It includes selection, hybridization, and mutations to enhance stress endurance in various plants (Dwivedi et al., 2016; Zafar et al., 2020). Selection involves identifying and promoting plants with exceptional traits, like drought tolerance or disease resistance, in specific environments. Repeated selection for these desirable traits over generations has encouraged breeders to develop crop varieties with enhanced stress tolerance. On the other hand, hybridization encompasses crossing genetically diverse plants to combine favorable traits from different parental lines, leading to hybrid offspring with improved stress tolerance and performance.

12.3.2 Biotechnological Approaches

Genetic engineering can alter the genetic composition of plants, either by interposing genes from other plants or organisms to increase the formation of lead metabolites. The various techniques involved in improving secondary metabolites are overexpression, chloroplast transformation, RNAi (ribonucleic acid inference), genome editing, and ploidy engineering (Bindu et al., 2018).

12.3.3 Elicitors and Plant Growth Regulators

They are utilized to boost the production of lead metabolites by activating various genes involved in the pathways (Vidya et al., 2021). Elicitors refer to substances capable of enriching metabolites in plants. The use of elicitors enhance secondary metabolites in different plants through foliar spraying or soil-drenching methods (Karuppusamy, 2009).

Plant breeding using biotechnological methods is the only approach among these that manipulates the genetic makeup of plants, resulting in transmissible changes over generations (Kuiper et al., 2001; Narula et al., 2005). Additionally, new techniques are developed to increase secondary metabolite production by optimizing environmental conditions for these metabolic processes.

12.4 GENETIC ALTERATION IN MEDICINAL PLANTS

Genetic alteration involves intentionally altering an organism's genetic makeup through techniques like gene transfer between organisms of the same or different species or even across kingdoms. This manipulation aims to enhance characteristics in resulting organisms, such as medicinal plants, either by improving their medicinal properties or increasing their productivity (**Figure 12.1**). The application of genetic alteration has been extensively researched and developed in medicinal plants over many years. The genes responsible for producing bioactive compounds are regulated through both transcriptional and post-transcriptional mechanisms. Normally, the genetic activity is low, but it can be enhanced by external factors called elicitors. Various methods are employed to enrich the levels of phytocompounds in specific plants (Yuan et al., 2020).

12.4.1 Techniques Involved in Genetic Alteration

12.4.1.1 Genetic Engineering

Genetic engineering involves the modification of plants to enhance their medicinal properties by incorporating genetic material from other sources into their genome. This process aims to boost the production of key metabolites by introducing foreign genes that encode enzymes responsible for production. These genes are usually obtained from a range of microorganisms or different plant species. Various transformation methods are employed to integrate these genes into the plant genome with these techniques (Niazian, 2019; Nielsen et al., 2019). The process of genetic engineering includes targeted modifications to gene sequences using recombinant DNA techniques (Sinha et al., 2019). This precise genetic alteration can be attained directly through methods such as chloroplast transformation, particle bombardment, and electroporation or indirectly via microbial vectors like *agrobacterium*-mediated gene transfer. The applications of genetic alteration techniques to improve secondary metabolites are summarized in Table 12.1.

12.4.1.2 *Agrobacterium*-Mediated Gene Transfer

The primary method for inserting desired genes into specific plants involves using *Agrobacterium tumefaciens*, a gram-negative soil-borne bacterium that induces crown gall disease in plants. This process can target undifferentiated callus cells or specialized plant cells, leveraging *Agrobacterium*'s natural infection mechanism to deliver genes. Advances in regeneration techniques utilizing *Agrobacterium* have greatly advanced biotechnology (Chilton et al., 1977; De La Riva et al., 1998). The process of transferring

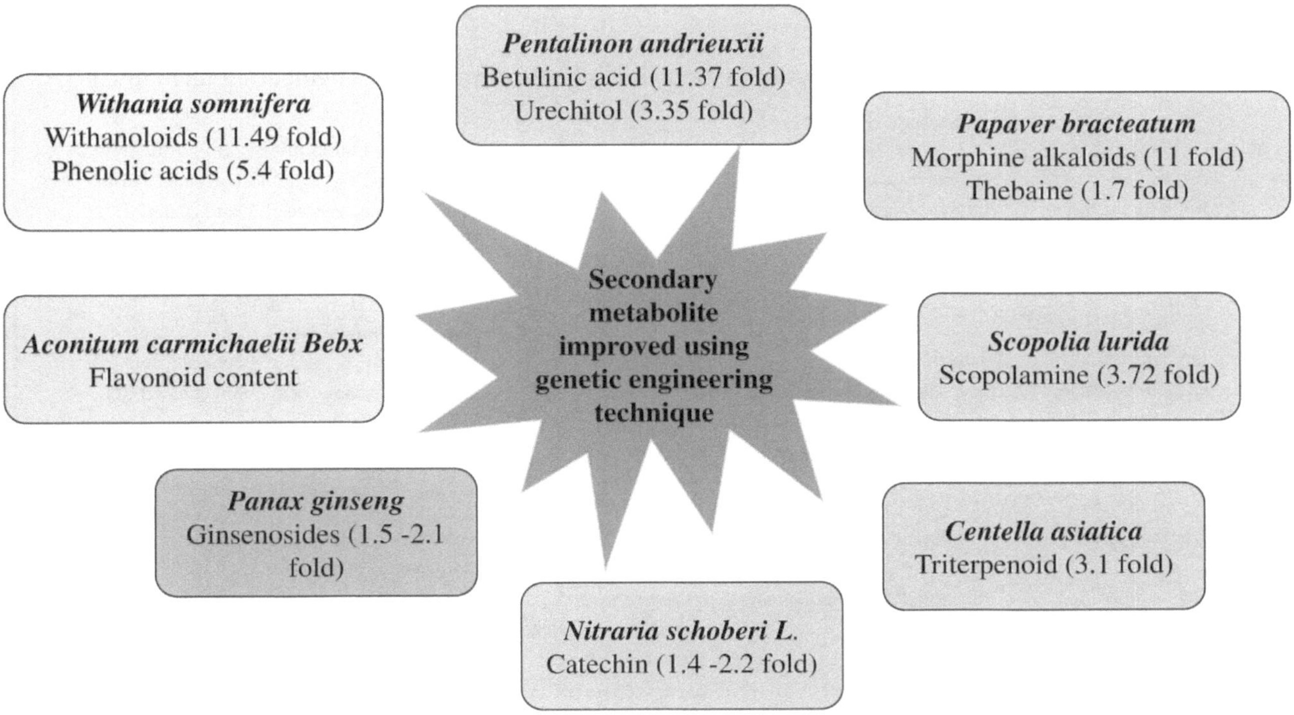

FIGURE 12.1 Secondary metabolites improved using genetic engineering technique.

TABLE 12.1
Summary of Secondary Metabolites Improved in medicinal plants using genetic engineering techniques

S. no.	Plant species	Method of delivery/ technique	Fold improvement	Secondary metabolite	References
			Overexpression		
1	*Panax ginseng*	*Agrobacterium rhizogenes by transformation*	1.5-to-2.1-fold	Ginsenosides	Van Binh et al. (2024)
2	*Salvia castanea Diels*	*Agrobacterium rhizogenes by electroporation*	–	Salvianolic acid B and rosmarinic acid	Zhang et al. (2024)
3	*Withania somnifera* L.	*Agrobacterium rhizogenes*	11.49-fold, 5.34-fold	Withanolides, phenolic acids	Halder and Ghosh (2024)
4	*Nitraria schoberi* L.	*Agrobacterium rhizogenes*	1.4-to-2.2-fold	Catechin	Zheleznichenko et al. (2023)
5	*Thymus daenensis*	*Agrobacterium rhizogenes*	–	Phenolics and flavonoid	Alamholo and soltani (2023)
6	*Ocimum basilicum*	*Agrobacterium rhizogenes*	–	Rosmaric acid and phenolic content	Sathasivam et al. (2022)
7	*Pentalinon andrieuxii*	*Agrobacterium rhizogenes*	11.37-fold, 3.35-fold	Betulinic acid and urechitol A	Hiebert-Giesbrecht et al. (2021)
8	*Aconitum carmichaelii Debx*	*Agrobacterium*-mediated transformation	39.13–63.63%	Flavonoid content	Nguyen et al. (2022)
9	*Veratrum dahuricum*	*Agrobacterium*-mediated transformation	–	Steroid alkaloids	Ma et al. (2022)
10	*Panax ginseng*	*Agrobacterium*-mediated transformation	–	Ginsenosides	Degtyarenko et al. (2021)
11	*Papaver bracteatum*	*Agrobacterium*-mediated transformation	1.7-fold	Thebaine	Qaderi et al. (2021)
12	*Centella asiatica*	*Agrobacterium rhizogenes*	3.1-fold	Triterpenoid	Ganie et al. (2022)
13	*Trachyspermum ammi* L.	*Agrobacterium*-mediated transformation	55.07%	Thymol	Niazian et al. (2019)

(*Continued*)

TABLE 12.1 ***(Continued)***
Summary of Secondary Metabolites Improved in medicinal plants using genetic engineering techniques

S. no.	Plant species	Method of delivery/ technique	Fold improvement	Secondary metabolite	References
14	*Scopolia lurida*	*Agrobacterium rhizogenesis*	3.72-fold	Scopolamine	Lan et al. (2018)
15	*Catharanthus roseus*	*Agrobacterium rhizogenes*	-	Enhanced tabersonince	Paul et al. (2017)
16	*Papaver bracteatum*	*Agrobacterium rhizogenes*	11-fold	Morphinan alkaloids (codeine)	Sharafi et al. (2013)

CRISPR/Cas 9

S. no.	Plant species	Gene of interest	Method of delivery	Secondary metabolite improved	References
1	*Salvia miltiorrhiza*	*Basic leucine zipper* (*bZIP*)2	*Agrobacterium rhizogenes*	Enhanced phenolic acid content	Shi et al. (2021)
2	*Atropa belladonna*	*Hyoscyamine 6 β-hydroxylase* (*AbH6H*)	*Agrobacterium tumifaciens*–mediated gene transfer	Hyoscyamine	Zeng et al. (2021)
3	*Cichorium intybus* L.	Germacrene A	Protoplast transformation	Phenolic content and squalene	Cankar et al. (2021)
4	*Salvia miltorrhiza*	*Committed diterpene synthase* (*SmCPS1*)	*Agrobacterium tumifaciens*–mediated gene transfer	Tanshinone biosynthesis	Li et al. (2017)
5	*Papaver bracteatum*	*Codeinone resuctase*	*Agrobacterium rhizogenes*	Codeine and morphine 0.28% and 0.04%	Sharafi et al. (2013)
6	*Rauwolfia serpentina*	*Catharanthus tryptophan decarboxylase* (*Crtdc*)	*Agrobacterium rhizogenes*–mediated	Reserpine and ajmalicine (2 fold)	Mehrotra et al. (2013)
7	*Salvia miltiorrhiza*	*Cinnamic acid 4-hydroxylase* (c4h), *tyrosine aminotransferase* (tat), *4-hydroxyphenylpyruvate reductase* (hppr) and *4-hydroxyphenylpyruvatedioxygenase* (hppd)	*Agrobacterium rhizogenes*–mediated	Rosmaric acid (RA) and lithospermic acid (LAB) RA: 4.3 LAB: 3.2	Xiao et al. (2011)
8	*Papaver somniferum*	*Salutaridinol 7-o-acetyl transferase*	*Agrobacterium*-mediated transformation	Morphine Codeine Thebaine	Allen et al. (2008)

Ploidy engineering

S. no.	Plant species	Method of delivery/ technique	Level of improvement	Secondary metabolite	References
1	*Papaver bracteatum Lindl*	Ploidy engineering	5.56-fold, 30.55-fold	Enhanced thebaine and noscapine	Madani et al. (2019)
2	*Linum album*	Ploidy engineering	1.39-fold	Increased podophyllotoxin	Javadian et al. (2017)
3	*Artemisia cina*	Ploidy engineering	–	Enhanced flavonoid content	Kasmiyati et al. (2020)

RNA inference

S. No	Plant species	Gene of interest	Method of delivery	Secondary metabolite improved	References
1	*Salvia miltiorrhiza*	Chalcone synthase	RNAi	Enhanced phenolic acid	Zhang et al. (2015)
2	*Mentha spicata*	MsYABBY5 (transcription factor)	RNAi	Enhanced terpene content	Wang et al. (2016)

genetic material into plant cells begins with the induction of wounds in plant tissue, which triggers the release of chemical signals. Subsequently, the genetic material attaches to the plant cells. Once attached, virulence genes become activated. This activates the production of the T-complex, facilitating the transport of foreign genes into the nucleus. Finally, the transfer DNA integrates into the plant genome, leading to the expression of the transferred genes. The process culminates in the selection of transgenic plants (Tiwari et al., 2022).

The various factors that influence the efficacy of *Agrobacterium*-mediated transformation include plant genotypes, types of *Agrobacterium* strains used, types of explants, environmental conditions, use of antibiotics, co-cultivation duration, and incubation periods (Hao et al., 2024). Niazian et al. (2019) found that thymol content in *Trachyspermum ammi* (L.) Sprague was increased to 55.07% compared to control plants at 39.20% through *Agrobacterium*-mediated transformation. Furthermore, Amani et al. (2020) stated that in *Ficus carica* L., six phenolic acids (caffeic acid, gallic acid, rosmarinic acid, chlorogenic acid, cinnamic acid) and flavonoids (quercetin, kaempferol, rutin, apigenin) were enhanced using this technique. Basharat et al. (2021) stated that in *Stevia rebaudiana*, steviol glycoside content was improved using both conventional and biotechnological approaches. Frick et al. (2007) observed that increasing the production of *cyp80b3* led to a 45% rise in alkaloid levels in *Papaver somniferum*. Takemura et al. (2010) found that introducing (S)-scoulerine 9-O-methyltransferase resulted in *Eschscholzia californica* resulted in a heightened accumulation of benzylisoquinoline alkaloids. Ghimire et al. (2017) demonstrated that overexpressing γ-tocopherol methyltransferase in *Perilla frutescens* using *Agrobacterium*-mediated transformation increased phenolic compound levels rather than non-transformed plants. Sharan et al. (2019) genetically modified *Ocimum sanctum* with *Agrobacterium rhizogenes* to enhance eugenol and ursolic acid production, utilizing biotic (yeast extract) and abiotic (methyl jasmonate and salicylic acid) elicitors. Zheleznichenko et al. (2023) reported a 2.2-fold enhancement of catechin content in *Nitraria schoberi* through the use of a hairy root strain. In *Panax ginseng*, ginsenoside production was enhanced 2.1-fold using *Agrobacterium rhizogenes* as a delivery method. Alamholo and Soltani (2023) showed that thymol and carvacrol content increased in *Thymus daenensis* following *Agrobacterium rhizogenesis*–mediated transformation.

12.4.1.3 Particle Bombardment

Particle bombardment is another approach for plant transformation where plant cells are bombarded using tungsten or gold (1–4 μm in diameter) particles coated with DNA, where naked DNA can be effectively moved into the cells (Klein et al., 1987). This method presents an efficient way to deliver DNA in species that do not naturally undergo transformation by *Agrobacterium* (Dai et al., 2001). One of the key advantages of this technique is it produces DNA-free gene-edited plants. It also helps in delivering RNAs, ribonucleoproteins and proteins into the plants directly (Ozyigit et al., 2020). In this approach, the target DNA is coated with gold or tungsten that acts as a DNA-coated micro carrier. When a high helium pulse is created, it moves toward the DNA-coated micro carrier using an electric discharge. These particles have the capacity to move into the plant cell and get integrated into the plant chromosome to produce transgenic plants. Various explants chosen for the study, including callus, cell-suspended cells, embryos, and meristems, have been utilized as targets for transformation. When compared with *Agrobacterium*-mediated transformation, the technique mentioned by Su et al. (2023) has several drawbacks. These include high cost, low transformation efficiency, and a limitation to transferring fragments smaller than 10 kb. For example, Zeef et al. (2000) utilized cell suspension cultures from gentian leaf explants (*Gentiana triflora*) to generate transformed plants through this technique. He also observed an enrichment of tropane alkaloids in *Hyoscyamus muticus* through the particle bombardment method of delivery. Furthermore, Sawahel (2002) demonstrated the successful integration of a DNA fragment into the chromosomal DNA of garlic plants using this method. The potential of this technique for future applications in the genetic modification of organisms is significant.

12.4.1.4 Electroporation

The process of electroporation employs electrical pulses to assist plant protoplasts in absorbing large molecules from their surrounding fluid. The cells without protective cell walls; i.e., protoplasts are typically cultured in a medium. Furthermore, introducing known DNA into the protoplast culture medium enables its penetration into the cells. Subsequently, transformed cells develop new cell walls and mature into complete, fertile transgenic plants with improved secondary metabolite. However, organ development from protoplasts of woody medicinal plants poses challenges. This process can be expedited by subjecting them to three consecutive voltage pulses ranging from 250 to 1250 V per centimeter for durations of 10 to 50 microseconds (Chand et al., 1988). Tissues derived from treated protoplasts exhibit accelerated morphogenesis within a shorter culture period compared to untreated tissues. Furthermore, regenerated shoots demonstrated stronger root systems compared to shoots derived from untreated protoplasts. Poompavai et al. (2023) claimed that phenolic content (12.5%) was increased when electroporation techniques were used as a genetic engineering tool to improve metabolites in *Azadirachta indica*. On the other hand, electrostimulation enhanced flavonoid content up to 2-fold compared to the control in *Scutellaria baicalensis georgi* (Grzelka et al., 2023).

12.4.1.5 Chloroplast Transformation

In chloroplast transformation, vectors are designed specifically or using heterologous methods to introduce foreign genes into the plastome via homologous recombination (Khan, 2012). When the foreign gene is delivered into the plasmid, only a few plastomes undergo transformation. By sub-culturing the explants, those with transformed plastomes containing the transgene are developed into plants. For example, chloroplast transformation involves the insertion of external genes directly into the genome of a green alga, *Chlamydomonas reinhardtii* (Boynton et al., 1988). Chloroplast transformation is a technique aimed at achieving elevated levels of valuable compounds within a controlled and stable environment. This method has been rapidly adopted for producing various medicinal compounds in *Nicotiana tabacum* (tobacco) plants and sporadically in *Arabidopsis thaliana*. According to Adem et al. (2017), the diverse traits targeted for chloroplast transformation encompass improving agronomic traits, phytoremediation, producing antigens, industrial enzymes, and biofuels. Daniell et al. (2005) reported that over 40 genes encoding vaccines, antigens, antibodies, and pharmaceutical compounds have been successfully integrated and expressed in the chloroplast genome, ensuring stable integration and expression. Chloroplast transformation shows promising potential for enhancing compound production in medicinal plants, facilitating the establishment of plant-based systems for pharmaceutics (Kumar et al., 2021). Chloroplast transformation is a technique that offers several advantages, one of which is the potential to enhance the yield of crucial drugs such as artemisinin in *Artemisia annua* and ginsenosides in *Panax ginseng*. Another advantage is the ability to achieve higher expression levels of transgenes compared to nuclear transformation. Additionally, chloroplast transformation allows for the simultaneous expression of several genes from a single transformation to occur. Furthermore, this technique provides the ground study to silence different genes indirectly distinguishing it from nuclear transformation as a whole (Adem et al., 2017).

12.4.1.6 Genome Editing

The traditional approach to genetic engineering encounters several obstacles, including challenging insertion procedures, concerns about public acceptance, and issues related to reporter genes (Naqvi et al., 2010). Despite these challenges, it remains essential to incorporate genome editing tools, which facilitate precise insertion *via* customized constructs tailored to specific genomic regions. This targeted insertion method allows PCR-based detection, unlike detection techniques that rely on markers (Liu et al., 2013). Genome editing like transcription activator-like effector nucleases (TALENs), zinc finger nucleases, clustered regularly interspaced short palindromic repeats (CRISPR), and CRISPR-associated proteins (CRISPR/Cas9) involves precisely altering various genes within the plant genome. Among these techniques, CRISPR/Cas has significantly contributed to enhancing secondary metabolites in medicinal plants. This method enables genes to be deleted, inserted, or modified to produce secondary metabolites. In this method, the CRISPR components sgRNA (single guide RNA) and Cas form a sgRNA-Cas complex. This complex binds to the target DNA, cleaving, and identifies using the Protospacer Adjacent Motif (PAM) near the target sequence. The gRNA (guide RNA) matches and binds to the target sequence. The Cas enzyme acts as an endonuclease, recognizing smaller gRNA. This molecule contains RNAase, which processes crRNA to create staggered DNA for detecting T-rich PAM sites (Das et al., 2024). Further, the customized constructs designed against the target genes enable nucleases to cleave DNA double strands and are later joined by non-homologous end joining and homology-directed repair (Gaj et al., 2013). Alagoz et al. (2016) demonstrated that 4- *OMT2* was knocked out using CRISPR/Cas9 to enhance benzylisoquinoline alkaloids, such as noscapine, thebaine, morphine, codeine, s-reticuline, papaverine, and laudanosine in opium poppy. In 2021, Zeng et al. pioneered the modification of *Atropa belladonna* plants by targeting *hyoscyamine 6-hydroxylase* (*AbH6H*) using CRISPR/Cas9 technology, thereby eliminating anisodamine and scopolamine to increase secondary metabolites (**Figure 12.2**). Genome editing offers several advantages, including precise modifications to the plant genome, which helps reduce the pleiotropic effects associated with modifying or inserting target genes.

12.4.1.7 RNA Inference (RNAi)

RNA interference (RNAi) technology is a modern and potent technique utilized for gene suppression. It offers distinct advantages over transgenic methods by preventing the synthesis of transgene proteins within RNAi-modified plants (Rajam, 2020). Currently, this technique has the ability to enrich plant architecture, plant blooming time, plant nutritional composition, disease resistance, pest management (biotic stress), and adaptability to unfavorable environmental conditions (abiotic stress) (Das et al., 2020). Unlike other methods, CRISPR and TALENs that achieve complete gene knockout, RNAi enables targeted gene knockdown, thereby facilitating precise modulation of gene expression levels in plants.

The RNAi strategy serves as an alternative method for gene modification, facilitating the heightened buildup of enzymes essential for secondary metabolite production. It has small RNA molecules, known as short-interfering RNAs, which are crafted to target specific genes. Upon binding to the target gene, these small RNAs initiate its degradation through mechanisms involving both post-transcriptional and transcriptional gene silencing (Jain & Khurana, 2009). This pioneering approach has reshaped the landscape of molecular biology research and sparked the emergence of biotechnology enterprises specializing in RNAi technology. Despite its potential to significantly influence the production of secondary metabolites in plants, RNAi, a potent gene-silencing method, remains largely underutilized (Ossowski et al., 2008). This technique has shown its effectiveness in enhancing secondary metabolites in several plant species such as *Panax ginseng*, *Papaver somniferum*, *Allium cepa*, *Withania somnifera*, and *Artemisia annua*, all of which hold considerable pharmaceutical promise (Hussain et al., 2012).

Researchers have demonstrated the potential of RNAi in various plant species to modulate key metabolic pathways and enhance desirable traits. For instance, Ogita et al. (2003) effectively reduced caffeine content in coffee plants by 70% through RNAi-mediated suppression of three caffeine biosynthetic genes (*CaXMT1*, *CaMXMT1*, and *CaDXMT1*), enabling the development of decaffeinated coffee beans. Similarly, Eady et al. (2008) utilized RNAi to downregulate lachrymatory factor synthase in *Allium cepa*,

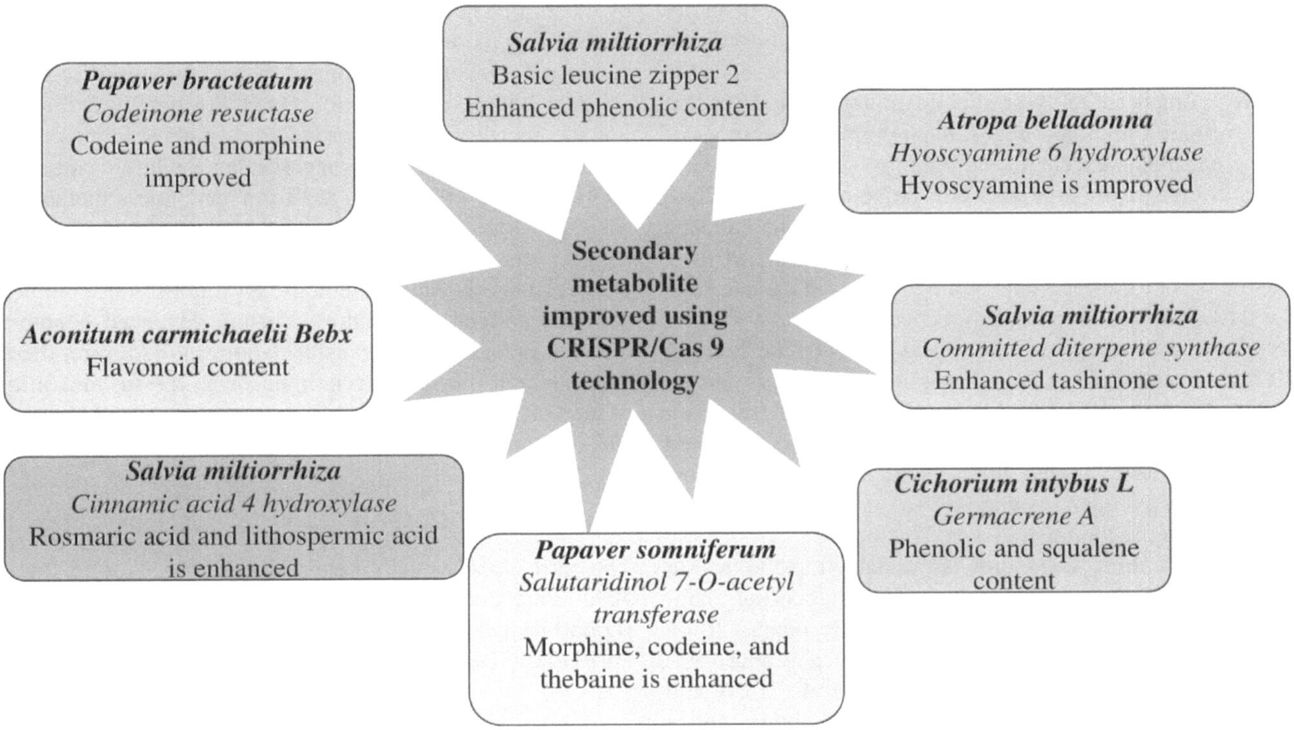

FIGURE 12.2 Secondary metabolites enhanced using CRISPR/Cas 9 technique.

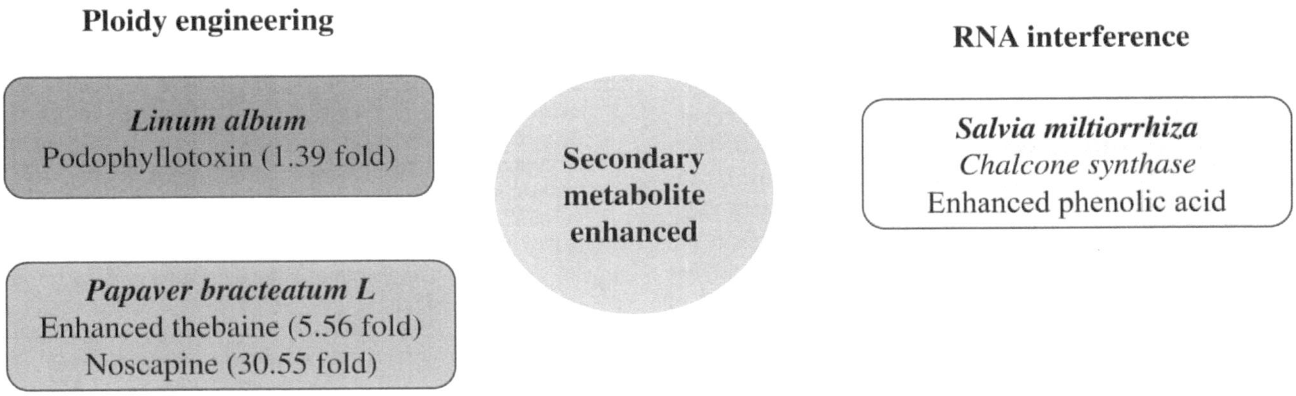

FIGURE 12.3 Secondary metabolites improved using ploidy engineering and RNA interference technique.

resulting in decreased levels of this compound in onions. Another investigation by Mohanpuria et al. (2011) showed that RNAi lowers *caffeine synthase* expression in *Camellia sinensis*, achieving a substantial reduction (44–61%) in caffeine content in tea leaves. Furthermore, researchers have harnessed RNAi to manipulate alkaloid biosynthesis in medicinal plants. For example, Wijekoon and Facchini (2012) inhibited codeinone reductase, a pivotal enzyme in morphine biosynthesis in opium poppies, thereby promoting the synthesis of reticuline, a non-narcotic alkaloid with superior medicinal properties. Additionally, Zhang et al. (2015) reported increased phenolic content in *Salvia miltiorrhiza* through RNAi-based strategies rather than untreated plants. Wang et al. (2016) investigated the effects of RNAi on MsYABBY5 in *Mentha spicata*, showing that silencing this gene increased terpene levels, while overexpression reduced them (**Figure 12.3**). These studies collectively highlight RNAi as a powerful weapon for metabolic engineering in plants, enabling targeted modifications to improve crop yield.

12.4.1.8 Ploidy Engineering

In the realm of medicinal plant research, manipulating chromosome numbers through ploidy engineering is employed to enrich both the quantity and quality of secondary metabolites. This method aims to boost the genetic composition of organisms, thereby increasing the diversity and effectiveness of products. Studies consistently indicate that polyploid variants exhibit greater vigor and superior quality compared to their diploid counterparts. Specifically, research findings suggest that polyploid genotypes of medicinal plants excel in accumulating secondary metabolites compared to diploids. The polyploids produced in various medicinal plants are being explored to enhance their breeding efficiency and improve the quality of their chemical composition. This strategy seeks to elevate both the quantitative and qualitative aspects of the plant's performance, according to Salma et al. (2017). The level of 20-hydroxyecdysone in Brazilian ginseng (*Pfaffia glomerata*) was improved by 31% compared to diploid plants when polyploidy was introduced. Polyploidization was induced using anti-mitotic agents, with colchicine being the most commonly used agent as opposed to oryzalin and trifluralin, according to Salma et al. (2017).

In the field of enhancing secondary metabolites in medicinal plants through polyploidy engineering, haploids are considered a crucial factor (Germana, 2011). Various methods based on tissue culture, such as androgenesis (Kasha, 2005), gynogenesis (Piosik et al., 2016), and hybridization combined with chromosome abnormalities (Forster et al., 2007), are employed to produce haploids in different plant species. Notably, another culture and microspore culture are prominent techniques for inducing haploids via the androgenic pathway in many medicinal plants (Sharma et al., 2018). A highly effective method for generating doubled haploid plants in borage (*Borago officinalis* L.) involved treating anthers with 0.2% n-butanol for 5 hours followed by 200 mg/L colchicine for 4 days (Hoveida et al., 2017). This approach successfully increased the ploidy level in borage. Similarly, in *Nigella sativa* (black cumin), haploid production was enhanced by treating ovules with varying concentrations of plant hormones (El-Mahrouk et al., 2018). According to Kasmiyati et al. (2020), employing ploidy engineering in *Artemisia cina* enhanced the levels of flavonoids and chlorophyll (Figure 12.3).

12.5 CONCLUSION AND FUTURE PROSPECTIVE

Medicinal plants can produce secondary metabolites of commercial importance. Further, the resurgence in public fascination with natural remedies stems from concerns surrounding the adverse effects and prolonged health risks often tied to synthetic medications. This heightened demand motivates the pharmaceutical industry to redouble its efforts in uncovering and utilizing fresh bioactive compounds sourced from nature, to develop effective treatments for various ailments. The emergence of cutting-edge genetic engineering methodologies further enhances our comprehension of traditional and natural therapeutics, fostering novel perspectives on their therapeutic potential. Improved metabolite production efficiency can be attained by combining new technologies with metabolic and genetic engineering methodology. Identifying additional functional genes and enzymes that control the production of secondary metabolites is crucial for advancing rapidly and making the biotechnological production of these compounds economically viable. This necessitates a multidisciplinary approach that integrates data from genomics, metabolomics, proteomics, and synthetic biology. Bindu et al. (2018) suggest that enhancing materials for breeders involves efficient gene isolation methods, and modifying complex secondary metabolite pathways using advanced genomic techniques and genetic transformation. In the future, these methods will play a key role in ensuring the sustainability of medicinal plants as reliable sources of medically beneficial substances (Bharti et al., 2023).

ACKNOWLEDGMENT

The corresponding author expresses gratitude to the Tamil Nadu State Council for Higher Education for the financial support provided to Dr. M. Arun under the grant RGP/2019-20/BU/HECP-0018, dated April 27, 2021.

DATA AVAILABILITY

The authors confirm that the data supporting the findings of this study are available within the book/chapter.

REFERENCES

Abd El-Hack, M. E., de Oliveira, M. C., Attia, Y. A., Kamal, M., Almohmadi, N. H., Youssef, I. M., Khalifa, N. E., Moustafa, M., Shehri, M. A., & Taha, A. E. (2023). The efficacy of polyphenols as an antioxidant agent: An updated review. *International Journal of Biological Macromolecules*, 126525.

Acosta-Estrada, B. A., Gutiérrez-Uribe, J. A., & Serna-Saldívar, S. O. (2014). Bound phenolics in foods, a review. *Food Chemistry, 152*, 46–55.

Adem, M., Beyene, D., & Feyissa, T. (2017). Recent achievements obtained by chloroplast transformation. *Plant Methods, 13*, 1–11.

Akilandeswari, G., Vijaya Anand, A., Sampathkumar, P., Vinayaga Moorthi, P., & Preethi, B. (2019). A prospective review on phyto-pharmacological aspects of Andrographis paniculata. *Systematic Reviews in Pharmacy, 10*(1), 15–19.

Alagoz, Y., Gurkok, T., Zhang, B., & Unver, T. (2016). Manipulating the biosynthesis of bioactive compound alkaloids for next-generation metabolic engineering in opium poppy using CRISPR-Cas 9 genome editing technology. *Scientific Reports, 6*(1), 30910.

Alamholo, M., & Soltani, J. (2023). *Agrobacterium rhizogenes*-mediated transformation as a means for induction of hairy roots and secondary metabolites in *Thymus daenensis*. *Biology Bulletin, 50*(4), 621–627.

Allen, R. S., Miller, J. A., Chitty, J. A., Fist, A. J., Gerlach, W. L., & Larkin, P. J. (2008). Metabolic engineering of morphinan alkaloids by over-expression and RNAi suppression of salutaridinol 7-O-acetyltransferase in opium poppy. *Plant Biotechnology Journal, 6*(1), 22–30.

Amani, S., Mohebodini, M., Khademvatan, S., & Jafari, M. (2020). *Agrobacterium rhizogenes* mediated transformation of *Ficus carica* L. for the efficient production of secondary metabolites. *Journal of the Science of Food and Agriculture, 100*(5), 2185–2197.

Balandrin, M. F., & Klocke, J. A. (1988). Medicinal, aromatic, and industrial materials from plants. In *Medicinal and aromatic plants I* (pp. 3–36). Springer Berlin Heidelberg.

Bartnik, M., & Facey, P. (2024). Glycosides. In *Pharmacognosy* (pp. 103–165). Academic Press.

Basharat, S., Huang, Z., Gong, M., Lv, X., Ahmed, A., Hussain, I., Li, J., Du, G., & Liu, L. (2021). A review on current conventional and biotechnical approaches to enhance biosynthesis of steviol glycosides in *Stevia rebaudiana. Chinese Journal of Chemical Engineering, 30*, 92–104.

Begum, T. N., Ilyas, M. M., Kalavathy, S., Anand, A. V., Kumar, S., & Senthil, R. (2009). Effects of ethanolic leaf extract of *Azim tetracantha* Lam. on *Ehrlich ascites* carcinoma tumour bearing mice. *Research Journal of Medicine and Medical Sciences, 4*(2), 351–354.

Bello-Martínez, G., García-Ramírez, G., Olea-Flores, M., Navarro-Tito, N., Hernández-Moreno, A., Avila-Caballero, L. P., Torres-Moreno, H., & Bello-Martínez, J. (2022). Biological activity of *Haematoxylum brasiletto* in MCF7 and MDA-MB-231 breast cancer cell lines. *South African Journal of Botany, 146*, 528–537.

Bharathi, V., Rengarajan, R. L., Radhakrishnan, R., & Hashem, A., Abd-Allah, E. F., Alqarawi, A. A., & Anand, A. V. (2018). Effects of a medicinal plant *Macrotyloma uniflorum* (Lam.) Verdc. Formulation (MUF) on obesity-associated oxidative stress-induced liver injury. *Saudi Journal of Biological Sciences, 25*(6), 1115–1121.

Bharti, P. K., Singh, S. K., & Kumari, S. (2023). Genetic manipulation in medicinal plants for enhancement of secondary metabolites. In *Genetic manipulation of secondary metabolites in medicinal plant* (pp. 129–145). Springer Nature Singapore.

Bindu, K. H., Mythili, J. B., & Radhika, R. M. (2018). Genetic engineering in medicinal and aromatic plants. In *Genetic engineering of horticultural crops* (pp. 249–271). Academic Press.

Boynton, J. E., Gillham, N. W., Harris, E. H., Hosler, J. P., Johnson, A. M., Jones, A. R., Randolph-Anderson, B. L., Robertson, D., Klein, T. M., Shark, K. B., & Sanford, J. C. (1988). Chloroplast transformation in Chlamydomonas with high velocity microprojectiles. *Science, 240*(4858), 1534–1538.

Cankar, K., Bundock, P., Sevenier, R., Häkkinen, S. T., Hakkert, J. C., Beekwilder, J., van der Meer, I. M., de Both, M., & Bosch, D. (2021). Inactivation of the germacrene A synthase genes by CRISPR/Cas9 eliminates the biosynthesis of sesquiterpene lactones in *Cichorium intybus* L. *Plant Biotechnology Journal, 19*(12), 2442–2453.

Carsono, N., Tumilaar, S. G., Kurnia, D., Latipudin, D., & Satari, M. H. (2022). A review of bioactive compounds and antioxidant activity properties of piper species. *Molecules, 27*(19), 6774.

Chand, P. K., Ochatt, S. J., Rech, E. L., Power, J. B., & Davey, M. R. (1988). Electroporation stimulates plant regeneration from protoplasts of the woody medicinal species *Solarium dulcamara* L. *Journal of Experimental Botany, 39*(9), 1267–1274.

Chang, Y. L., Huang, L. M., Kuo, X. Z., Chen, Y. Y., Lin, S. T., Jeng, M. F., Yeh, H. H., Tsai, W. C., & Chen, H. H. (2023). PbABCG1 and PbABCG2 transporters are required for the emission of floral monoterpenes in *Phalaenopsis bellina. The Plant Journal, 114*(2), 279–292.

Chen, J., Tomes, S., Gleave, A. P., Hall, W., Luo, Z., Xu, J., & Yao, J. L. (2022). Significant improvement of apple (*Malus domestica* Borkh.) transgenic plant production by pre-transformation with a Baby boom transcription factor. *Horticulture Research, 9*, uhab014.

Chen, X., Lai, H., Li, R., Yao, Y., Liu, J., Yuan, S., Fu, S., Hu, X., & Guo, J. (2021). Character changes and Transcriptomic analysis of a cassava sexual Tetraploid. *BMC Plant Biology, 21*, 1–17.

Chilton, M. D., Drummond, M. H., Merlo, D. J., Sciaky, D., Montoya, A. L., Gordon, M. P., & Nester, E. W. (1977). Stable incorporation of plasmid DNA into higher plant cells: The molecular basis of crown gall tumorigenesis. *Cell, 11*(2), 263–271.

Dai, S., Zheng, P., Marmey, P., Zhang, S., Tian, W., Chen, S., Beachy, R. N., & Fauquet, C. (2001). Comparative analysis of transgenic rice plants obtained by Agrobacterium-mediated transformation and particle bombardment. *Molecular Breeding, 7*, 25–33.

Daniell, H., Kumar, S., & Dufourmantel, N. (2005). Breakthrough in chloroplast genetic engineering of agronomically important crops. *TRENDS in Biotechnology, 23*(5), 238–245.

Das, P. R., & Sherif, S. M. (2020). Application of exogenous dsRNAs-induced RNAi in agriculture: Challenges and triumphs. *Frontiers in Plant Science, 11*, 946.

Das, S., Kwon, M., & Kim, J. Y. (2024). Enhancement of specialized metabolites using CRISPR/Cas gene editing technology in medicinal plants. *Frontiers in Plant Science, 15*, 1279738.

De la Riva, G. A., González-Cabrera, J., Vázquez-Padrón, R., & Ayra-Pardo, C. (1998). Agrobacterium tumefaciens: A natural tool for plant transformation. *Electronic Journal of Biotechnology, 1*(3), 24–25.

Degtyarenko, A. I., Gorpenchenko, T. Y., Grigorchuk, V. P., Bulgakov, V. P., & Shkryl, Y. N. (2021). Optimization of the transient Agrobacterium-mediated transformation of Panax ginseng shoots and its use to change the profile of ginsenoside production. *Plant Cell, Tissue and Organ Culture (PCTOC), 146*, 357–373.

Demir, T., Akpınar, Ö., Kara, H., & Güngör, H. (2022). Phenolic profile and investigation of biological activities of *Allium scorodoprasum* L. subsp. rotundum. *Food Bioscience, 46*, 101548.

Desam, N. R., & Al-Rajab, A. J. (2022). Herbal biomolecules: Anticancer agents. In *Herbal biomolecules in healthcare applications* (pp. 435–474). Academic Press.

Divya, N., Rengarajan, R. L., Radhakrishnan, R., Abd-Allah, E. F., Alqarawi, A. A., Hashem, A., & Anand, A. V. (2019). Phytotherapeutic efficacy of the medicinal plant *Terminalia catappa* L. *Saudi Journal of Biological Sciences, 26*(5), 985–988.

Dwivedi, S. L., Ceccarelli, S., Blair, M. W., Upadhyaya, H. D., Are, A. K., & Ortiz, R. (2016). Landrace germplasm for improving yield and abiotic stress adaptation. *Trends in Plant Science, 21*(1), 31–42.

Eady, C. C., Kamoi, T., Kato, M., Porter, N. G., Davis, S., Shaw, M., Kamoi, A., & Imai, S. (2008). Silencing onion lachrymatory factor synthase causes a significant change in the sulfur secondary metabolite profile. *Plant Physiology, 147*(4), 2096–2106.

El-Mahrouk, M. E., Maamoun, M. K., El-Banna, A. N., Omran, S. A., Dewir, Y. H., & El-Hendawy, S. (2018). In vitro gynogenesis and flow cytometry analysis of the regenerated haploids of black cumin (*Nigella sativa*). *HortScience*, *53*(5), 681–686.

Farghadani, R., & Naidu, R. (2023). The anticancer mechanism of action of selected polyphenols in triple-negative breast cancer (TNBC). *Biomedicine & Pharmacotherapy*, *165*, 115170.

Farnsworth, N. R., & Soejarto, D. D. (1991). Global importance of medicinal plants. *The Conservation of Medicinal Plants*, *26*(26), 25–51.

Forster, B. P., Heberle-Bors, E., Kasha, K. J., & Touraev, A. (2007). The resurgence of haploids in higher plants. *Trends in Plant Science*, *12*(8), 368–375.

Frick, S., Kramell, R., & Kutchan, T. M. (2007). Metabolic engineering with a morphine biosynthetic P450 in opium poppy surpasses breeding. *Metabolic Engineering*, *9*(2), 169–176.

Gairola, K., Gururani, S., & Dubey, S. K. (2023). Polyphenols and its effect on the immune system. In *Nutraceuticals and functional foods in immunomodulators* (pp. 121–140). Springer Nature Singapore.

Gaj, T., Gersbach, C. A., & Barbas, C. F. (2013). ZFN, TALEN, and CRISPR/Cas-based methods for genome engineering. *Trends in Biotechnology*, *31*(7), 397–405.

Ganie, I. B., Ahmad, Z., Shahzad, A., Zaushintsena, A., Neverova, O., Ivanova, S., Wasi, A., & Tahseen, S. (2022). Biotechnological intervention and secondary metabolite production in *Centella asiatica* L. *Plants*, *11*(21), 2928.

Germana, M. A. (2011). Gametic embryogenesis and haploid technology as valuable support to plant breeding. *Plant Cell Reports*, *30*, 839–857.

Ghimire, B. K., Yu, C. Y., & Chung, I. M. (2017). Assessment of the phenolic profile, antimicrobial activity and oxidative stability of transgenic *Perilla frutescens* L. overexpressing tocopherol methyltransferase (γ-tmt) gene. *Plant Physiology and Biochemistry*, *118*, 77–87.

Goufo, P., Pereira, J., Figueiredo, N., Oliveira, M. B. P., Carranca, C., Rosa, E. A., & Trindade, H. (2014). Effect of elevated carbon dioxide (CO2) on phenolic acids, flavonoids, tocopherols, tocotrienols, γ-oryzanol and antioxidant capacities of rice (*Oryza sativa* L.). *Journal of Cereal Science*, *59*(1), 15–24.

Grzelka, K., Matkowski, A., & Ślusarczyk, S. (2023). Electrostimulation improves plant growth and modulates the flavonoid profile in aeroponic culture of *Scutellaria baicalensis* Georgi. *Frontiers in Plant Science*, *14*, 1142624.

Halder, T., & Ghosh, B. (2024). *Withania somnifera* (L.) Dunal: Enhanced production of withanolides and phenolic acids from hairy root culture after application of elicitors. *Journal of Biotechnology*, *388*, 59–71.

Halka, J., Vidya, N., Gurusaravanan, P., Sivaranjini, A., Anand, A. V., & Arun, M. (2024). CRISPR/Cas genome editing in fruit crops: Recent advances, challenges, and future prospects. In *Plant genome editing technologies: Speed breeding, crop improvement and sustainable agriculture* (pp. 261–278). Springer Nature.

Hao, S., Zhang, Y., Li, R., Qu, P., & Cheng, C. (2024). Agrobacterium-mediated in planta transformation of horticultural plants: Current status and future prospects. *Scientia Horticulturae*, *325*, 112693.

Hiebert-Giesbrecht, M. R., Avilés-Berzunza, E., Godoy-Hernández, G., & Peña-Rodríguez, L. M. (2021). Genetic transformation of the tropical vine *Pentalinon andrieuxii* (Apocynaceae) via *Agrobacterium rhizogenes* produces plants with an increased capacity of terpenoid production. *In Vitro Cellular & Developmental Biology-Plant*, *57*, 21–29.

Hoveida, Z. S., Abdollahi, M. R., Mirzaie-Asl, A., Moosavi, S. S., & Seguí-Simarro, J. M. (2017). Production of doubled haploid plants from anther cultures of borage (*Borago officinalis* L.) by the application of chemical and physical stress. *Plant Cell, Tissue and Organ Culture (PCTOC)*, *130*, 369–378.

Hu, B., Zhao, X., Wang, E., Zhou, J., Li, J., Chen, J., & Du, G. (2023). Efficient heterologous expression of cytochrome P450 enzymes in microorganisms for the biosynthesis of natural products. *Critical Reviews in Biotechnology*, *43*(2), 227–241.

Huang, Z., Lin, J., Cheng, Z., Xu, M., Huang, X., Yang, Z., & Zheng, J. (2015). Production of dammarane-type sapogenins in rice by expressing the dammarenediol-II synthase gene from *Panax ginseng* CA Mey. *Plant Science*, *239*, 106–114.

Hussain, M. S., Fareed, S., Ansari, S., Rahman, M. A., Ahmad, I. Z., & Saeed, M. (2012). Current approaches toward production of secondary plant metabolites. *Journal of Pharmacy and Bioallied Sciences*, *4*(1), 10–20.

Iqbal, I., Wilairatana, P., Saqib, F., Nasir, B., Wahid, M., Latif, M. F., Iqbal, A., Naz, R., & Mubarak, M. S. (2023). Plant polyphenols and their potential benefits on cardiovascular health: A review. *Molecules*, *28*(17), 6403.

Jain, M., & Khurana, J. P. (2009). Transcript profiling reveals diverse roles of auxin-responsive genes during reproductive development and abiotic stress in rice. *The FEBS Journal*, *276*(11), 3148–3162.

Javadian, N., Karimzadeh, G., Sharifi, M., Moieni, A., & Behmanesh, M. (2017). In vitro polyploidy induction: Changes in morphology, podophyllotoxin biosynthesis, and expression of the related genes in *Linum album* (Linaceae). *Planta*, *245*, 1165–1178.

Jiang, H., & Wang, X. (2023). Biosynthesis of monoterpenoid and sesquiterpenoid as natural flavors and fragrances. *Biotechnology Advances*, *65*, 108151.

Kabera, J. N., Semana, E., Mussa, A. R., & He, X. (2014). Plant secondary metabolites: Biosynthesis, classification, function and pharmacological properties. *Journal of Pharmacy and Pharmacololgy*, *2*(7), 377–392.

Karak, P. (2019). Biological activities of flavonoids: An overview. *International Journal of Pharma Sciences and Research*, *10*(4), 1567–1574.

Karthika, P., Balamuralikrishnan, B., Yamini, K., Vijaya Anand, A., Durairaj, K., Maruthupandian, A., Alodaini, H. A., Hatamleh, A. A., Manikantan, P., & Arun, M. (2023). Green synthesis, characterization of silver nanoparticles using aqueous leaf extracts of *Solanum melongena* and in vitro evaluation of antibacterial, pesticidal and anticancer activity in human MDA-MB-231 breast cancer cell lines. *Journal of King Saud University – Science*, *35*(5), 102663.

Karuppusamy, S. (2009). A review on trends in production of secondary metabolites from higher plants by in vitro tissue, organ and cell cultures. *Journal of Medicinal Plants Research*, *3*(13), 1222–1239.

Kasha, K. J. (2005). Chromosome doubling and recovery of doubled haploid plants. In *Haploids in crop improvement II* (pp. 123–152). Springer Berlin Heidelberg.

Kasmiyati, S., Kristiani, E. B. E., & Herawati, M. M. (2020). Effect of induced polyploidy on plant growth, chlorophyll and flavonoid content of *Artemisia cina*. *Biosaintifika: Journal of Biology & Biology Education*, *12*(1), 90–96.

Kaviya, M., Balasubramanian, B., Bharathi, K., Malaisamy, A., Al-Dhabi, N. A., Mariadhas, V. A., Anand, A. V., & Liu, W. (2022). Evaluation of nutritional substances and

investigation of antioxidant and antimicrobial potentials of *Boerhavia diffusa* with in silico molecular docking. *Molecules*, *27*(4), 1280.
Khan, M. S. (2012). Plastid genome engineering in plants: Present status and future trends. *Molecular Plant Breeding*, *3*(9).
Kim, Y. J., Weathers, P. J., & Wyslouzil, B. E. (2002). Growth of *Artemisia annua* hairy roots in liquid-and gas-phase reactors. *Biotechnology and Bioengineering*, *80*(4), 454–464.
Klein, T. M., Wolf, E. D., Wu, R., & Sanford, J. C. (1987). High-velocity microprojectiles for delivering nucleic acids into living cells. *Nature*, *327*(6117), 70–73.
Kuiper, H. A., Kleter, G. A., Noteborn, H. P., & Kok, E. J. (2001). Assessment of the food safety issues related to genetically modified foods. *The Plant Journal*, *27*(6), 503–528.
Kumar, A. U., & Ling, A. P. K. (2021). Gene introduction approaches in chloroplast transformation and its applications. *Journal of Genetic Engineering and Biotechnology*, *19*, 1–10.
Lan, X., Zeng, J., Liu, K., Zhang, F., Bai, G., Chen, M., Iqbal, A., Naz, R., & Huang, L. (2018). Comparison of two hyoscyamine 6β-hydroxylases in engineering scopolamine biosynthesis in root cultures of *Scopolia lurida*. *Biochemical and Biophysical Research Communications*, *497*(1), 25–31.
Lešnik, S., & Bren, U. (2021). Mechanistic insights into biological activities of polyphenolic compounds from rosemary obtained by inverse molecular docking. *Foods*, *11*(1), 67.
Li, B., Cui, G., Shen, G., Zhan, Z., Huang, L., Chen, J., & Qi, X. (2017). Targeted mutagenesis in the medicinal plant *Salvia miltiorrhiza*. *Scientific Reports*, *7*(1), 43320.
Li, L., Peng, P., Ding, N., Jia, W., Huang, C., & Tang, Y. (2023a). Oxidative stress, inflammation, gut dysbiosis: What can polyphenols do in inflammatory bowel disease? *Antioxidants*, *12*(4), 967.
Li, Z., Wang, Y., Xu, M., Liu, H., Li, L., & Xu, D. (2023b). Molecular mechanism overview of metabolite biosynthesis in medicinal plants. *Plant Physiology and Biochemistry*, *204*, 108125.
Liu, W., Yuan, J. S., & Stewart, Jr., C. N. (2013). Advanced genetic tools for plant biotechnology. *Nature Reviews Genetics*, *14*(11), 781–793.
Ma, Y., Zhang, S., Rong, L., Wu, Z., & Sun, W. (2022). Polyphenol composition and antioxidant activity of *Japonica* rice cultivars and intake status. *Foods*, *11*(23), 3788.
Madani, H., Hosseini, B., Karimzadeh, G., & Rahimi, A. (2019). Enhanced thebaine and noscapine production and modulated gene expression of tyrosine/dopa decarboxylase and salutaridinol 7-O-acetyltransferase genes in induced autotetraploid seedlings of *Papaver bracteatum* Lindl. *Acta Physiologiae Plantarum*, *41*(12), 194.
Manikandan, R., Balasubramanian, B., Punniyakotti, P., Vijaya Anand, A., Meyyazhagan, A., Velauthaprabhu, S., & Liu, W. C. (2022). Cardio-protective effects of *Terminalia catappa* leaves and *Terminalia chebula* fruit extract in doxorubicin-induced cardiomyopathy in rats. *Biomarkers*, *27*(5), 488–495.
Manoharan, N., Sampathkumar, P., Dheeba, B., Sheik Abdulla, S., Vinothprasanna, G., Vinothkannan, R., Kalavathy, S., Vijaya Anand, A., & Shanmugasundaram, A. (2008). Potential hepatoprotective effect of aqueous extract of *Gracilaria corticata* in AFB_1 induced hepatotoxicity in Wistar rats. *Journal of Biological Sciences*, *8*, 1352–1355.
Mehrotra, S., Srivastava, V., Rahman, L. U., & Kukreja, A. K. (2013). Overexpression of a *Catharanthus* tryptophan decarboxylase (tdc) gene leads to enhanced terpenoid indole alkaloid (TIA) production in transgenic hairy root lines of *Rauwolfia serpentina*. *Plant Cell, Tissue and Organ Culture (PCTOC)*, *115*, 377–384.
Mohanpuria, P., Kumar, V., Ahuja, P. S., & Yadav, S. K. (2011). Producing low-caffeine tea through post-transcriptional silencing of caffeine synthase mRNA. *Plant Molecular Biology*, *76*, 523–534.
Mumtaz, M. Z., Kausar, F., Hassan, M., Javaid, S., & Malik, A. (2021). Anticancer activities of phenolic compounds from *Moringa oleifera* leaves: In vitro and in silico mechanistic study. *Beni-Suef University Journal of Basic and Applied Sciences*, *10*, 1–11.
Muthumani, G. D., Anand, A. V., & Manikandan, R. (2013). Antioxidant, antihelmintic and antimicrobial activity of *Euphorbia thymifolia* Linn whole plant. *International Journal of Current Microbiology and Applied Sciences*, *2*(1), 66–75.
Naqvi, S., Farré, G., Sanahuja, G., Capell, T., Zhu, C., & Christou, P. (2010). When more is better: Multigene engineering in plants. *Trends in Plant Science*, *15*(1), 48–56.
Narula, A., Kumar, S., Bansal, K. C., & Srivastava, P. S. (2005). Biotechnological approaches towards improvement of medicinal plants. In *Plant biotechnology and molecular markers* (pp. 78–116). Springer.
Newman, D. J., & Cragg, G. M. (2012). Natural products as sources of new drugs over the 30 years from 1981 to 2010. *Journal of Natural Products*, *75*(3), 311–335.
Nguyen, T. N. L., Hoang, T. T. H., Nguyen, H. Q., Tu, Q. T., Tran, T. H., Lo, T. M. T., Vu, T. T. T., & Chu, H. M. (2022). Agrobacterium tumefaciens–mediated genetic transformation and overexpression of the flavonoid 3′ 5′-hydroxylase gene increases the flavonoid content of the transgenic *Aconitum carmichaelii* Debx. plant. *In Vitro Cellular & Developmental Biology-Plant*, *58*(1), 93–102.
Niazian, M. (2019). Application of genetics and biotechnology for improving medicinal plants. *Planta*, *249*, 953–973.
Niazian, M., & Sabbatini, P. (2021). Traditional in vitro strategies for sustainable production of bioactive compounds and manipulation of metabolomic profile in medicinal, aromatic and ornamental plants. *Planta*, *254*, 1–22.
Niazian, M., Sadat-Noori, S. A., Tohidfar, M., Galuszka, P., & Mortazavian, S. M. M. (2019). Agrobacterium-mediated genetic transformation of ajowan (*Trachyspermum ammi* (L.) Sprague): An important industrial medicinal plant. *Industrial Crops and Products*, *132*, 29–40.
Nielsen, E., Temporiti, M. E. E., & Cella, R. (2019). Improvement of phytochemical production by plant cells and organ culture and by genetic engineering. *Plant Cell Reports*, *38*(10), 1199–1215.
Ogita, S., Uefuji, H., Yamaguchi, Y., Koizumi, N., & Sano, H. (2003). Producing decaffeinated coffee plants. *Nature*, *423*(6942), 823–823.
Ossowski, S., Schwab, R., & Weigel, D. (2008). Gene silencing in plants using artificial microRNAs and other small RNAs. *The Plant Journal*, *53*(4), 674–690.
Ozyigit, I. I., & Yucebilgili Kurtoglu, K. (2020). Particle bombardment technology and its applications in plants. *Molecular Biology Reports*, *47*(12), 9831–9847.
Paul, P., Singh, S. K., Patra, B., Sui, X., Pattanaik, S., & Yuan, L. (2017). A differentially regulated AP 2/ERF transcription factor gene cluster acts downstream of a MAP kinase cascade to modulate terpenoid indole alkaloid biosynthesis in *Catharanthus roseus*. *New Phytologist*, *213*(3), 1107–1123.
Piosik, Ł., Zenkteler, E., & Zenkteler, M. (2016). Development of haploid embryos and plants of *Lactuca sativa* induced by distant pollination with *Helianthus annuus* and *H. tuberosus*. *Euphytica*, *208*(3), 439–451.

Poompavai, S., Gowri Sree, V., & RamSugitha, B. (2023). Enhanced extraction and improved anticancer activity of neem (*Azadirachta indica*) leaves by electroporation technique. *IETE Journal of Research*, *69*(8), 4961–4971.

Puangpraphant, S., Cuevas-Rodríguez, E. O., & Oseguera-Toledo, M. (2022). Anti-inflammatory and antioxidant phenolic compounds. In *Current advances for development of functional foods modulating inflammation and oxidative stress* (pp. 165–180). Academic Press.

Qaderi, A., Qavami, N., Ghezelbash, S., & Zarinpanjeh, N. (2021). Enhanced production of thebaine in cell suspension cultures of *Papaver bracteatum* by overexpression of SalAT gene. *Journal of Medicinal Plants and By-Product*, *10*(Special), 69–75.

Rajam, M. V. (2020). RNA silencing technology: A boon for crop improvement. *Journal of Biosciences*, *45*(1).

Raut, J. S., & Karuppayil, S. M. (2014). A status review on the medicinal properties of essential oils. *Industrial Crops and Products*, *62*, 250–264.

Salma, U., Kundu, S., & Mandal, N. (2017). Artificial polyploidy in medicinal plants: Advancement in the last two decades and impending prospects. *Journal of Crop Science and Biotechnology*, *20*, 9–19.

Sangeetha, T., Balamuralikrishnan, B., Sampathkumar, P., Velayuthaprabhu, S., Senthilkumar, N., Syed Ibrahim, K., Baskaran, R., & Vijaya Anand, A. (2022). Characterization and phytoconstituents of *Petroselinum crispum* (Mill) and *Coriandrum sativum* (Linn) and their impacts on inflammation – an in vitro analysis against human adenocarcinoma cells with molecular docking. *South African Journal of Botany*, *146*, 776–788.

Sathasivam, R., Choi, M., Radhakrishnan, R., Kwon, H., Yoon, J., Yang, S. H., Kim, J. K., Chung, Y. S., & Park, S. U. (2022). Effects of various *Agrobacterium rhizogenes* strains on hairy root induction and analyses of primary and secondary metabolites in *Ocimum basilicum*. *Frontiers in Plant Science*, *13*, 983776.

Sawahel, W. A. (2002). Stable genetic transformation of garlic plants using particle bombardment. *Cellular and Molecular Biology Letters*, *7*(1), 49–60.

Sharafi, A., Sohi, H. H., Mousavi, A., Azadi, P., Razavi, K., & Ntui, V. O. (2013). A reliable and efficient protocol for inducing hairy roots in *Papaver bracteatum*. *Plant Cell, Tissue and Organ Culture (PCTOC)*, *113*, 1–9.

Sharan, S., Sarin, N. B., & Mukhopadhyay, K. (2019). Elicitor-mediated enhanced accumulation of ursolic acid and eugenol in hairy root cultures of *Ocimum tenuiflorum* L. is age, dose, and duration dependent. *South African Journal of Botany*, *124*, 199–210.

Sharma, S., Satardekar, K. V., & Barve, S. S. (2018). Genetic improvement of medicinal and aromatic plants through haploid and double haploid development. In *Biotechnological approaches for medicinal and aromatic plants: Conservation, genetic improvement and utilization* (pp. 523–556). Springer Nature.

Shi, M., Du, Z., Hua, Q., & Kai, G. (2021). CRISPR/Cas9-mediated targeted mutagenesis of bZIP2 in Salvia miltiorrhiza leads to promoted phenolic acid biosynthesis. *Industrial Crops and Products*, *167*, 113560.

Singh, L. J., Challam, D. A., & Senjam, B. D. (2023). Medicinal plants as a source of terpenoids and their impact on central nervous system disorders: A review. *The Journal of Phytopharmacology*, *12*(2), 104–110.

Sinha, S., Sandhu, K., Bisht, N., Naliwal, T., Saini, I., & Kaushik, P. (2019). Ascertaining the paradigm of secondary metabolism enhancement through gene level modification in therapeutic plants. *Journal of Young Pharmacists*, *11*(4), 337.

Srivastava, R., & Tiwari, P. (2022). Medicinal plant used against cancer: A review. *Asian Journal of Pharmaceutical Research and Development*, *10*(4), 76–85.

Su, W., Xu, M., Radani, Y., & Yang, L. (2023). Technological development and application of plant genetic transformation. *International Journal of Molecular Sciences*, *24*(13), 10646.

Takemura, T., Ikezawa, N., Iwasa, K., & Sato, F. (2010). Metabolic diversification of benzylisoquinoline alkaloid biosynthesis through the introduction of a branch pathway in *Eschscholzia californica*. *Plant and Cell Physiology*, *51*(6), 949–959.

Tanaka, T., Onuma, H., Shigihara, T., Kimura, E., Fukuta, Y., Shirasaka, N., Moriyama, T., & Homma, Y. (2019). Anti-osteoporotic effects of syringic acid and vanilic acid in the extracts of waste beds after mushroom cultivation. *Journal of Bioscience and Bioengineering*, *128*(5), 622–629.

Tiwari, M., Mishra, A. K., & Chakrabarty, D. (2022). Agrobacterium-mediated gene transfer: Recent advancements and layered immunity in plants. *Planta*, *256*(2), 37.

Tiwari, R., & Rana, C. S. (2015). Plant secondary metabolites: A review. *International Journal of Engineering Research and General Science*, *3*(5), 661–670.

Trepa, M., Sułkowska-Ziaja, K., Kała, K., & Muszyńska, B. (2024). Therapeutic potential of fungal terpenes and terpenoids: Application in skin diseases. *Molecules*, *29*(5), 1183.

Van Binh, N., Kim, M. J., Giang, V. N. L., Lee, Y. S., Jayakodi, M., Park, H. S., Mohanan, P., Kang, K. B., Ryu, B., Park, E. J., & Park, T. K. (2024). Improved biomass and metabolite production in hairy root culture in various genotypes of *Panax ginseng* through genetic transformation. *Plant Cell, Tissue and Organ Culture (PCTOC)*, *156*(2), 43.

Varalakshmi, B., Anand, A. V., Karpagam, T., Bai, J. S., & Manikandan, R. (2014). In vitro antimicrobial and anticancer activity of *Cinnamomum zeylanicum* Linn bark extracts. *International Journal of Pharmacy and Pharmaceutical Sciences*, *6*(1), 12–18.

Vidya, N., Saravanan, K., Halka, J., Kowsalya, K., Preetha, J. S. Y., Gurusaravanan, P., Radhakrishnan, R., Nanthini, A. U. R., & Arun, M. (2021). An insight into in vitro strategies for bioproduction of isoflavones. *Plant Biotechnology Reports*, 1–24.

Vijaya Anand, A., Varalakshmi, B., Prasana, Sampathkumar, P., Pushpa, N., & Hedina, A. (2016). *Cinnamomum zeylanicum* Linn. The spice with multi potential. *Systematic Reviews in Pharmacy*, *7*(1), 24–29.

Vijaya Anand, A., Velayuthaprabhu, S., Rengarajan, R. L., Sampathkumar, P., & Radhakrishnan, R. (2020). Bioactive compounds of guava (*Psidium guajava* L.). *Bioactive Compounds in Underutilized Fruits and Nuts*, Reference Series in Phytochemistry (pp. 503–527). Springer.

Vijayakumar, K., Prasanna, B., Rengarajan, R. L., Rathinam, A., Velayuthaprabhu, S., & Vijaya Anand, A. (2023). Antidiabetic and hypolipidemic effects of *Cinnamon cassia* bark extracts: An in vitro, in vivo, and in silico approach. *Archieves of Physiology and Biochemistry*, *129*(2), 338–348.

Vo, G. T., Liu, Z., Chou, O., Zhong, B., Barrow, C. J., Dunshea, F. R., & Suleria, H. A. (2022). Screening of phenolic compounds in Australian grown grapes and their potential antioxidant activities. *Food Bioscience*, *47*, 101644.

Wang, Q., Reddy, V. A., Panicker, D., Mao, H. Z., Kumar, N., Rajan, C., Venkatesh, P. N., Chua, N. H., & Sarojam, R. (2016). Metabolic engineering of terpene biosynthesis in plants using a trichome-specific transcription factor

Ms YABBY 5 from spearmint (*Mentha spicata*). *Plant Biotechnology Journal*, *14*(7), 1619–1632.

Wang, Q., Zhao, X., Jiang, Y., Jin, B., & Wang, L. (2023). Functions of representative terpenoids and their biosynthesis mechanisms in medicinal plants. *Biomolecules*, *13*(12), 1725.

Wijekoon, C. P., & Facchini, P. J. (2012). Systematic knockdown of morphine pathway enzymes in opium poppy using virus-induced gene silencing. *The Plant Journal*, *69*(6), 1052–1063.

Wink, M. (2018). Plant secondary metabolites modulate insect behavior-steps toward addiction? *Frontiers in Physiology*, *9*, 360017.

Wróblewska-Łuczka, P., Cabaj, J., Bargieł, J., & Łuszczki, J. J. (2023). Anticancer effect of terpenes: Focus on malignant melanoma. *Pharmacological Reports*, *75*(5), 1115–1125.

Wu, H. S., Wang, Y., Bao, W., Liu, D. Y., Raza, W., Huang, Q. W., Mao, Z. S., & Shen, Q. R. (2009). Responses of *Fusarium oxysporum* f. sp. niveum to exogenously added sinapic acid in vitro. *Biology and Fertility of Soils*, *45*, 443–447.

Xiao, Y., Zhang, L., Gao, S., Saechao, S., Di, P., Chen, J., & Chen, W. (2011). The c4h, tat, hppr and hppd genes prompted engineering of rosmarinic acid biosynthetic pathway in *Salvia miltiorrhiza* hairy root cultures. *PLoS One*, *6*(12), e29713.

Xu, Y., Jia, H., Tan, C., Wu, X., Deng, X., & Xu, Q. (2022). Apomixis: Genetic basis and controlling genes. *Horticulture Research*, *9*, 1–10.

Yan, Y., Li, X., Zhang, C., Lv, L., Gao, B., & Li, M. (2021). Research progress on antibacterial activities and mechanisms of natural alkaloids: A review. *Antibiotics*, *10*(3), 318.

Yeshi, K., Crayn, D., Ritmejerytė, E., & Wangchuk, P. (2022). Plant secondary metabolites produced in response to abiotic stresses has potential application in pharmaceutical product development. *Molecules*, *27*(1), 313.

Yuan Y, Chung, J. D., & Fu, X (2020). Strategies for increasing the accumulation of valuable compounds in plant cells and tissues. *Plant Cell Reports*, *39*(3), 261–277.

Zafar, S. A., Zaidi, S. S. E. A., Gaba, Y., Singla-Pareek, S. L., Dhankher, O. P., Li, X., Mansoor, S., & Pareek, A. (2020). Engineering abiotic stress tolerance via CRISPR/Cas-mediated genome editing. *Journal of Experimental Botany*, *71*(2), 470–479.

Zaupa, M., Calani, L., Del Rio, D., Brighenti, F., & Pellegrini, N. (2015). Characterization of total antioxidant capacity and (poly) phenolic compounds of differently pigmented rice varieties and their changes during domestic cooking. *Food Chemistry*, *187*, 338–347.

Zeef, L. A., Christou, P., & Leech, M. J. (2000). Transformation of the tropane alkaloid-producing medicinal plant *Hyoscyamus muticus* by particle bombardment. *Transgenic Research*, *9*, 163–168.

Zeng, L., Zhang, Q., Jiang, C., Zheng, Y., Zuo, Y., Qin, J., Liao, Z., & Deng, H. (2021). Development of *Atropa belladonna* L. plants with high-yield hyoscyamine and without its derivatives using the CRISPR/Cas9 system. *International Journal of Molecular Sciences*, *22*(4), 1731.

Zhang, G., Sun, Y., Uallah, N., Kasote, D., Zhu, L., Liu, H., & Xu, L. (2024). Changes in secondary metabolites contents and stress responses in *Salvia miltiorrhiza* via Scwrky35 overexpression: Insights from a wild relative *Salvia castanea*. *Plant Physiology and Biochemisty*, *211*, 1–13.

Zhang, H., Yang, X., Mei, X., Wang, H., Huang, H., Yang, M., He, X., Zhu, S., & Liu, Y. (2020). Phenolic acids released in maize rhizosphere during maize-soybean intercropping inhibit Phytophthora blight of soybean. *Frontiers in Plant Science*, *11*, 532738.

Zhang, S., Li, H., Liang, X., Yan, Y., Xia, P., Jia, Y., & Liang, Z. (2015). Enhanced production of phenolic acids in *Salvia miltiorrhiza* hairy root cultures by combing the RNAi-mediated silencing of chalcone synthase gene with salicylic acid treatment. *Biochemical Engineering Journal*, *103*, 185–192.

Zheleznichenko, T., Voronkova, M., Asbaganov, S., Kukushkina, T., Filippova, E., Protsenko, M., Mazurkova, N., & Novikova, T. (2023). Impact of different *Agrobacterium rhizogenes* strains on secondary metabolites accumulation in *Nitraria schoberi* L. hairy roots and antiviral activity of their extracts against influenza virus of subtypes A (H5N1) and A (H3N2). *In Vitro Cellular & Developmental Biology-Plant*, *59*(3), 378–392.

13 Molecular Marker-Assisted Breeding in Medicinal Plants for Enhanced Secondary Metabolites Production

Kumaresan Kowsalya, Jayachandran Halka, Karuppiah Ramalashmi, Arumugam Vijaya Anand, Muthuswamy Razia, Muthukrishnan Anand, Ahame Kabeer Umera Begam, and Muthukrishnan Arun

LIST OF ABBREVIATIONS

RFLP restriction fragment length polymorphism
RAPD random amplified polymorphic DNA
AFLP amplified fragment length polymorphism
ISSR inter simple sequence repeat
SCAR sequence characterized amplified region
SNP single-nucleotide polymorphisms
DNA deoxyribonucleic acid
SSR simple sequence repeat

13.1 INTRODUCTION

Plants produce secondary metabolites, and natural products through secondary metabolism. It plays a crucial role in adaptation to environmental challenges and morphological differentiation, influencing growth and also offering potential benefits for human nutrition and antioxidant defense against oxidative damage induced by various environmental stressors. Enhancing the production of these compounds through breeding is a key objective to meet increasing pharmaceutical demands. Molecular marker-assisted breeding (MAB) has arisen as a powerful tool to accelerate and refine the breeding process for this valuable trait (Ozyigit et al., 2023). For centuries, these secondary metabolites from traditional medicinal plants have been consumed to treat various ailments and chronic diseases (Manoharan et al., 2008; Begum et al., 2009; Varalakshmi et al., 2014; Vijaya Anand et al., 2016; Bharathi et al., 2018; Divya et al., 2019; Akilandeswari et al., 2019; Vijaya Anand et al., 2020; Sangeetha et al., 2022; Kaviya et al., 2022; Manikandan et al., 2022; Karthika et al., 2023; Vijayakumar et al., 2023). For this purpose, two-thirds of 50,000 wild medicinal plants have been utilized, and 10% of medicinal plants are commercially cultivated. Hence, authenticating medicinal plants using DNA-based molecular markers is quite essential for eliminating other species' adulterants and producing large-scale secondary metabolites (Pourmohammad, 2013). Molecular markers are recognized as short DNA fragments present in a specific position of the chromosome. These DNA-based molecular markers are widely applied in food crops and horticultural plants for various purposes, such as genetic variation studies, phylogenetic studies, cultivar authentication, and disease-resistant specific gene detection (Srivastava & Mishra, 2009). The detectable differences in the genomic DNA sequences of particular plants are utilized as molecular markers. These specific markers are closely related to the gene of interest that can be used to identify the desirable trait or allele. Hence, this method is called marker-assisted selection (MAS) (Jonah et al., 2011).

Molecular markers precisely identify the closely interrelated species in medicinal plants. This method is completely reliable and reproducible as DNA is stable and devoid of seasonal variation, environmental factors, and the age of plants. These marker systems have also found the overall population and genetic variation of significant medicinal plants which is essential for the conservation strategy of endangered therapeutic plants (Arumugam et al., 2019). These DNA molecular marker systems are categorized as codominant or dominant markers. Dominant markers such as amplified fragment length polymorphism (AFLP) and random amplified polymorphic DNA (RAPD). Codominant markers involve simple sequence repeat (SSR) and restriction fragment length polymorphism (RFLP) (Negi et al., 2000). Another type of classification represents the emergence of hybridization-based markers such as RFLP marker, other PCR-based markers like amplified fragment length (AFLP), random amplified polymorphism DNA (RAPD), and inter simple sequence repeat (ISSR), followed by sequence-based markers, such as ISSRs, sequence characterized amplified region (SCAR), single-nucleotide polymorphism (SNP) have been generated.

Over two decades of the 20th century, these markers were addressing the various needs in breeding programs and secondary metabolite profiling of medicinal plants. Even though all the molecular markers are different, the overall process of developing PCR-based molecular markers shares a few common steps, namely, DNA isolation from medicinal plants, PCR-based amplification, and restriction digestion using specific restriction enzymes, followed by gel electrophoresis and visualization of bands through autoradiography or imaging techniques (Figure 13.1). Therefore,

DOI: 10.1201/9781003475491-13

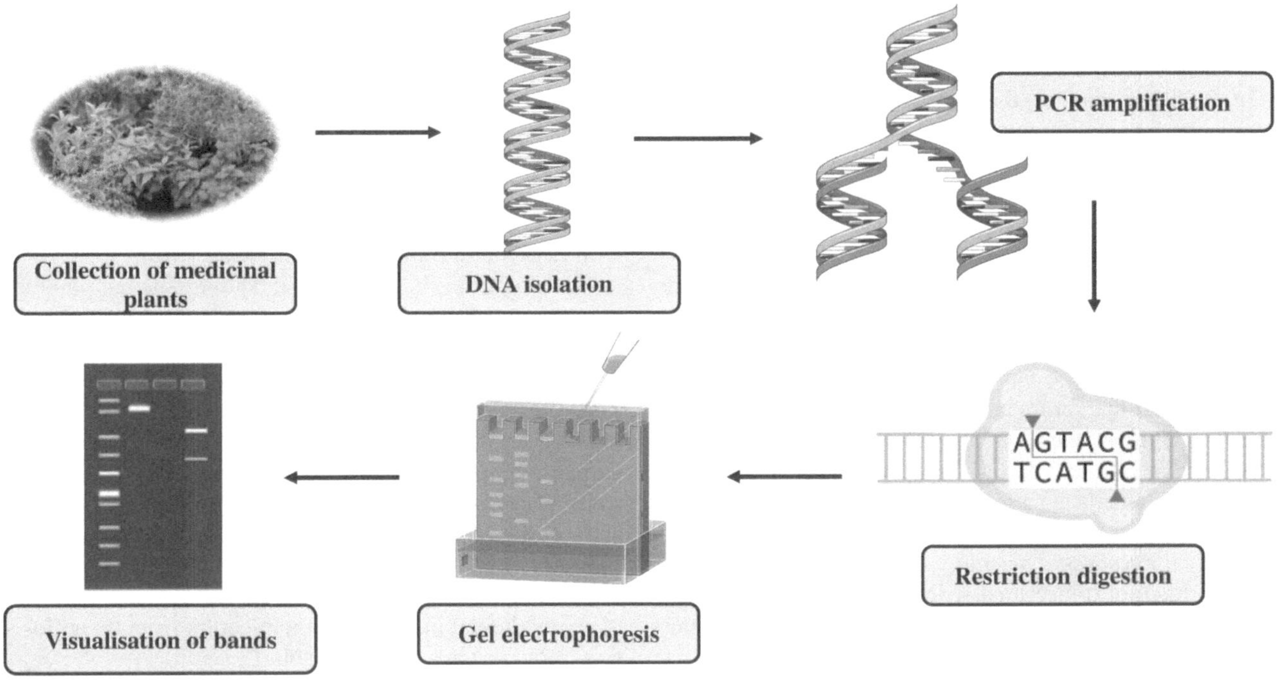

FIGURE 13.1

these markers have been continuously refined to improve their utility in gene tagging, genome analysis, the selection of desirable genotypes and phylogenetic analysis. Molecular markers, unlike morphological markers, offer numerous advantages as they provide objective data, revolutionizing breeding practices and significantly reducing the time needed to create new crop cultivar varieties in medicinal plants (Figure 13.2). Furthermore, the molecular marker approach has several applications in disciplines like quantitative genetics and plant taxonomy, for advanced plant breeding programs (Jonah et al., 2011). Hence, this chapter emphasizes the importance and application of molecular marker-assisted breeding in several therapeutically significant medicinal plants for secondary metabolite production.

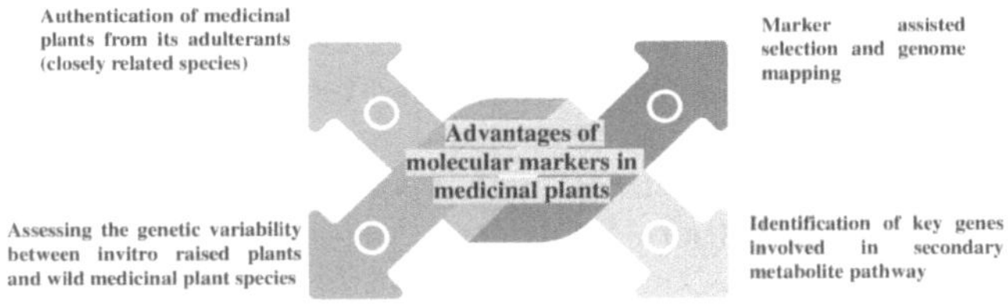

FIGURE 13.2

13.1.1 Perspectives, Classification of Molecular Markers Used for Plant Breeding

Molecular markers play crucial roles across various domains, including genetics, plant breeding, forensics, and medicine, offering insights into genetic diversity, evolutionary relationships, and disease predisposition. These markers can be categorized in diverse ways. Firstly, they vary in genetic alteration, such as SNPs (single-nucleotide polymorphisms), SSRs (simple sequence repeats), and INDELs (insertions/deletions), representing distinct genomic changes. It has several applications such as genetic mapping, phylogenetics, diagnostics, and forensic identification. Detection methods further diversify these markers, including PCR-based (RFLP, AFLP, SCAR), sequencing-based, and hybridization-based approaches, each offering unique advantages. Additionally, markers differ in genome coverage, with some spanning the entire genome (genome-wide markers) while others target specific genes or regions (gene-specific markers) (Nadeem et al., 2018).

13.2 RESTRICTION FRAGMENT LENGTH POLYMORPHISM (RFLP)

RFLP is the most commonly used molecular technique to check the specific variations present in the double-stranded DNA. This method produces unique RFLP patterns based on the cleaved DNA fragments in gel electrophoresis. It is mainly based on the restriction endonucleases, that recognize the restriction sites to cut the DNA at those sites. This method can be paired with PCR to amplify the fragments (Enayat, 1996). Two different organisms will have different restriction sites, which is indicated by the distance between specific restriction endonuclease cleavage sites. When the genomic DNA sequence is digested with a specific restriction enzyme, the resulting DNA fragments will have different lengths This method is commonly called PCR-RFLP and is extensively utilized in DNA fingerprinting, genotyping, and detecting genetic disorders (Chaudhary & Maurya, 2020).

In the herbal market, the authentication of herbal medicines is the first and foremost step which plays a central role in producing effective medicinal products. The quality of herbal products is majorly hindered by adulterant species of plants that are morphologically similar to medicinal plants. Hence, there is a constant need for a reliable method to verify the original medicinal plants from their adulterants. One such chinese medicinal plant, *Portulaca oleracea*, has been identified using ITS1, ITS2, *trn*L, *and rbc*L as molecular markers. Among these markers, ITS2 was the most prominent molecular marker for authenticating *Portulaca oleracea* from its adulterant plant, *Bacopa monnieri* (Xu et al., 2022).

Kunchang Wu et al. (2022) developed a PCR-RFLP method to authenticate the medicinal plant *Adenostemma lavenia* from its adulterants *Wedelia biflora*, *Wedelia chinensis*, and *Sigesbeckia orientalis*. They have amplified and used the internal transcribed spacer 2 (ITS2) regions of *A. lavenia* as a specific DNA marker. Subsequently, the blasticidin *BsrI* restriction enzyme was used to cleave the particular restriction sites in all four plants, which resulted in varying DNA sequence lengths of 253bp for *Adenostemma lavenia,* 473bp for *Wedelia biflora*, 472bp for *Sigesbeckia orientalis*, and 483bp for *Wedelia chinensis* (Wu et al., 2022).

Another study in *Dasamula* species shows the RFLP patterns of the internal transcriber region (ITS), which ranges from 650bp to 850bp. Among different restriction enzymes (ApaI, AluI, BamHI, EcoRV, EcoRI, Hind III, HinfI, MspI, MboI), EcoRI and MspI generated unique patterns, emphasizing that these enzymes are more suitable than other enzymes for these six selected *Dasamula* species (*Oroxylum indicum*, *Aegle marmelos*, *Desmodium giganiticum*, *Solanum indicum*, *Solanum xanthocarpum*, and *Tribulus terresteris*) (Biswas & Biswas, 2013).

In a study, three morphologically similar varieties of *Phyllanthus*, such as *P. amarus*, *P. fraternus*, and *P. urinaria*, have been clearly distinguished using the RFLP technique with amplifying the conserved regions of chloroplast (cp) genome, followed by restriction with *HinfI* and *TaqI* enzymes. This study claims that using these RFLP methods combined with universal cpDNA can also be potentially applied to other medicinal plants with similar indistinguishable morphological issues (Sarin et al., 2013).

In 50 Indian citrus genotypes, the phylogeny and taxonomy were identified using polymorphic patterns of PCR-RFLP with cpDNA containing two intergenic spacer regions (trnD-trnT and rbcL). Four restriction enzymes (Msp I, Hinf I, Hae III, Alu I) out of 13, have produced 47 fragments, in which polymorphic bands were 24, which constitutes 51% of polymorphism. The indigenous species, including *C. karna*, *C. megaloxycarpa*, *C. jambhiri*, and *C. limettoides*, and wild species, like *C. latipes*, *C. indica*, and *C. macroptera* var. *annamensis*, have been well differentiated systematically through this RFLP approach (Jena et al., 2009).

The phylogenetic relationship between Asteraceae and Anthemideae has been evaluated using RAPD, RFLP, and SSR markers in Egypt. Twenty-six RFLP bands were identified in which *Artemisia arborescens*, *Achillea fragrantissima*, *Artemisia judaica*, *Glebionis coronaria*, *Matricaria aurea*, *Cotula barbata*, and *Cotula cinerea* samples possessed, whereas *Anacyclus monanthos*, *Matricaria recutita*, and *Achillea santolina* detected no bands, after cutting with *Eco*RI and *Bam*HI (Abd El-Twad, 2010).

Vitex glabrata is a traditional Thai medicinal plant that has a higher 20-hydroxyecdysone content used for reproductive health. Among different *Vitex* species – *V. negundo*, *V. trifolia*, *V. rotundifolia*, *V. glabrata*, and *V. limonifolia* – *Vitex glabrata* was authenticated using the leaf-derived *mat*K gene, and the *psb*A-*trn*H intergenic spacer region, which is done by Phoolcharoen and Sukrong (2013). This study claims that the RFLP technique is a rapid and reliable method for discriminating closely related species of *Vitex* that lack distinguishable morphology (Phoolcharoen & Sukrong, 2013). Hence, the RFLP technique is utilized in many medicinal plants for authentication from their closely associated species listed in Table 13.1.

13.3 RANDOM AMPLIFIED POLYMORPHIC DNA (RAPD)

RAPD technique produces DNA fragments resulting from random fragments amplification by PCR using a single primer sequence. This technique operates differently from traditional PCR analysis as it doesn't rely on the genomic DNA sequence of the target organism. Based on DNA sequence complementarity, it utilizes 10-mer primers which are identical and will be either amplified or not amplified. For instance, if the 3' ends of the DNA sequence are not aligned or if the primers are too distantly placed, there will be no fragment production. Consequently, if a mutation in the template DNA at a site previously complementary to the primer, it will prevent PCR product formation, leading to amplified DNA segments with a distinct pattern (Williams et al., 1990).

The basis of RAPD lies in the genomic DNA PCR amplification, highlighting DNA polymorphisms resulting from deletions or rearrangements between the primer binding locations. Short random oligonucleotide sequences, typically ten bases long, are employed to infer these polymorphisms. RAPD is applicable across species using universal primers, as it doesn't require prior knowledge of the genomic DNA sequences being analyzed. The main advantage of RAPD is its simplicity and cost-effectiveness, requiring minimal prior sequence information and allowing for rapid screening of genetic variation. The drawback of this technique is its dependency on reaction conditions, that can vary between laboratories. Additionally, each primer amplifies multiple loci in the genome, so the profiles cannot be effectively distinguished between heterozygous and homozygous individuals (Gantait et al., 2014).

TABLE 13.1
List of Medicinal Plant Species Authenticated Using Different Molecular Markers

Plant name	Family	Technique	References
Matricaria recutita L.	Asteraceae	RAPD, ISSR, RFLP	Abd El-Twad et al., 2010
Achillea fragrantissima		RAPD, ISSR, RFLP	Abd El-Twad et al., 2010
Artesimia arborences		RAPD, ISSR, RFLP	Abd El-Twad et al., 2010
Achillea santolina		RAPD, ISSR, RFLP	Abd El-Twad et al., 2010
Artemisia judaica		RAPD, ISSR, RFLP	Abd El-Twad et al., 2010
Glebionis coronaria		RAPD, ISSR, RFLP	Abd El-Twad et al., 2010
Cotula cinerea		RAPD, ISSR, RFLP	Abd El-Twad et al., 2010
Cotula barbata		RAPD, ISSR, RFLP	Abd El-Twad et al., 2010
Matricaria recutita		RAPD, ISSR, RFLP	Abd El-Twad et al., 2010
Matricaria aurea		RAPD, ISSR, RFLP	Abd El-Twad et al., 2010
Citrus	Rutaceae	PCR-RFLP	Jena et al., 2009
Pulsatilla chinensis	Ranunculaceae	PCR-RFLP	Shi et al., 2017
Calotropis gigantea	Asclepiadaceae	PCR-RFLP	Biswas, Garg, Biswas, 2017
Maytenus aquifolia *Maytenus ilicifolia*	Celastraceae	PCR-RFLP	Nakamura et al., 2013
Ascocenda, Phalaenopsis	Orchidaceae	PCR-RFLP	Liu et al., 2016
Murraya paniculata	Rutaceae	PCR-RFLP	Wang et al., 2024
Actinidia macrosperma	Actinidiaceae	PCR-RFLP	Zhao, Qiu, Gong et al., 2007
Hypericum	Hypericaceae	PCR-RFLP	Hazler et al., 2010
Pyropia species	Bangiaceae	PCR-RFLP	Kim et al., 2017
Vitex glabrata	Lamiaceae	PCR-RFLP	Phoolcharoen & Sukrong, 2013
Paris polyphylla Smith var. *yunnanensis*	Melanthiaceae	PCR-RFLP	Liu & Ji, 2012
Boerhavia diffusa L.	Nyctaginaceae	PCR-RFLP	Biswas, 2013
Glehnia littoralis	Umbelliferae	PCR-RFLP	Mizukami, Ohbayashi, Umetsu et al., 1993
Desmodium giganicum, *Aegle marmelos*, *Solanum xanthocarpum*, *Solanum indicum*, *Tribulus terresteris*, *Oroxylum indicum*	Fabaceae, Rutaceae, Solanaceae, Solanaceae, Zygophyllaceae, Bignoniaceae	PCR-RFLP	Biswas, Biswas, 2013
Phyllanthus amarus, *Phyllanthus fraternus*, *Phyllanthus urinaria*	Phyllanthaceae	PCR-RFLP	Sarin et al., 2013
Duboisia leicchardtii, *Duboisia myoporodes*, *Duboisia hopwoodii*	Solanaceae	PCR-RFLP	Mizukami, Ohbayashi, Kitamura et al., 1993
Adenostemma lavenia	Asteraceae	PCR-RFLP	Wu et al., 2022
Adathoda beddomei, *Adathoda zeylanica*, *Barleria prionitis*, *Justicia gendarussa*, *Nigirianthus* sp., and *Hemigraphis colorata*	Acathaceae	RAPD	Prasad, 2014
Ocimum sanctum, *O. basilicum*, and *O. gratissimum*	Lamiaceae	RAPD, ISSR	Sarwat, Srivastava, Khan, 2016
Rauwolfia tetraphylla L.	Apocynaceae	SCoT, ISSR, and RAPD	Rohela, Jogam, Bylla et al., 2018
Curcuma longa	Zingiberaceae	RAPD	Shivashankar, 2014

(Continued)

TABLE 13.1 (*Continued*)
List of Medicinal Plant Species Authenticated Using Different Molecular Markers

Plant name	Family	Technique	References
Pittosporum eriocarpum	Pittosporaceae	SCoT, ISSR, and RAPD	Thakur, Dwivedi, Sourabh et al., 2016
Hemigraphis Colorata	Acanthaceae,	RAPD	Paramanik & Chikkaswamy, 2014
Cuscuta reflexa	Convolvulaceae	RAPD	Khan, Mirsa, Abdin, 2010
Senna angustifolia	Fabaceae	RAPD	Khan, Mirza, Al-qurainey, 2011
Gymnema sylvestre	Asclepiadaceae	AFLP	Osman et al., 2011
Embelia ribes Burm. F and *Embelia tsjeriam cottama*	Myrsinaceae	AFLP	Gowda et al., 2010
Rosa damascena	Rosaceae	AFLP	Baydar et al., 2004
Plectranthus grandis, P. barbatus, P. ornatus	Lamiaceae	AFLP	Passinho-Soares et al., 2006
Ocimum basilicum	Lamiaceae	AFLP	Labra et al., 2004
Plumbago indigo	Plumbaginaceae	ISSR	Li et al., 2023
Stevia rebaudiana	Asteraceae	ISSR	Luz et al., 2020
Eleusine coracana L.	Poaceae	ISSR	Venkatesan et al., 2021
Papaver bracteatum	Papaveraceae	ISSR, AFLP	Hadipour et al., 2020
Silybum marianum	Asteraceae	ISSR	Hamouda, 2019
Malva sp.	Malvaceae	ISSR	Jedrzejczyk, 2020
Hypericum L.	Hypericaceae	ISSR	Ma et al., 2021
Thymus sp.	Lamiaceae	ISSR	Yousefi et al., 2015
Ocimum sp.	Lamiaceae	ISSR	Patel et al., 2015
Ocimum sanctum, O. basilicum, and *O. gratissimum*	Lamiaceae	ISSR	Sarwat, Srivastava, Khan, 2016
Cassia auriculata, Cassia surattensis	Fabaceae	ISSR	Panapitiya & Welikala, 2018
Rauwolfia species	Apocynaceae	ISSR	Rohela, Bylla, Pendli et al., 2018
Hedera helix	Araliaceae	ISSR	Shakoor et al., 2022
Kaempferia galanga L.	Zingiberaceae	ISSR	Subositi et al., 2020
Oroxylum indicum	Bignoniaceae	ISSR	Rajasekharan et al., 2017
Passiflora	Passifloraceae	ISSR	Dos Santos et al., 2011
Juniperus procera	Cupressaceae	ISSR	Al-Yasi et al., 2024
Dioscorea sp.	Dioscoreaceae	ISSR	Velasco-Ramírez et al., 2014
Uritica dioica	Urticaceae	ISSR	Haghpanah et al., 2016
Zhumeria majdea	Lamiaceae	ISSR	Daryaii et al., 2020
Ziziphus spina-christi L.	Rhamnaceae	ISSR	Alansi et al., 2016
Commiphora wightii and *C. myrrha*	Burseraceae	SCAR	Sairkar et al., 2016
Lonicera japonica	Caprifoliaceae	SCAR	Yang et al., 2014
Angelica sinensis, Angelica acutiloba	Apiaceae	SCAR	Zhang et al., 2015
Phyllanthus emblica	Phyllanthaceae	SCAR	Dnyaneshwar et al., 2006
P. amarus, P. debilis, and *P. urinaria*	Phyllanthaceae	SCAR	Theerakulpisut et al., 2008
Litchi chinensis	Sapindaceae	SCAR	Cheng et al., 2015
Podophyllum hexandrum	Berberidaceae	SCAR	Al-shaqha et al., 2014
Liriope platyphylla, Ophiopogon japonicus	Asparagaceae	SCAR	Li & Park, 2012
Cinnamomum verum and *Cinnamomum cassia*	Lauraceae	SCAR	Gangwar et al., 2023
Iris ensata	Iridaceae	SCAR	Mughees et al., 2017
Ipomoea mauritiana	Convolvulaceae	SCAR	Devaiah et al., 2011
Viola serpens	Violaceae	SCAR	Jha et al., 2020
Lycium L.	Solanaceae	SNP	Zhao, Xu, Li et al., 2019
Matricaria recutita	Asteraceae	SNP	Otto et al., 2017
Panax ginseng	Araliaceae	SNP	Chen et al., 2013

TABLE 13.1 (*Continued*)
List of Medicinal Plant Species Authenticated Using Different Molecular Markers

Plant name	Family	Technique	References
Ananas comosus	Bromeliaceae	EST	Sahu et al., 2014
Catharanthus roseus	Apocynaceae	EST	Sahu et al., 2014
Dioscorea alata	Dioscoreaceae	EST	Sahu et al., 2014
Mangifera indica	Anacardiaceae	EST	Sahu et al., 2014
Nelumbo nucifera	Nelumbonaceae	EST	Sahu et al., 2014
Epimedium sagittatum	Berberidaceae	EST	Zeng et al., 2010
Plantago major	Plantaginaceae	EST	Pommerrenig et al., 2006
Panax ginseng	Araliaceae	EST	Choi et al., 2011
Scutellaria baicalensis	Lamiaceae	ScoT	Gawroński & Dyduch-Siemińska, 2022
Crepedium acuminatum	Orchidaceae	ScoT	Thakur, Dwivedi, Singh et al., 2021
Tylophora indica	Asclepiadaceae	ScoT	Mamgain et al., 2022
Echinacea	Asteraceae	ScoT	Jedrzejczyk & Rewers, 2020
Stellaria media	Caryophyllaceae	ScoT	Ma et al., 2022
Anabasis setifera	Amaranthaceae		Alotaibi & Abd-Elgawad, 2022
Calotropus procera	Apocynaceae	ScoT	Alotaibi & Abd-Elgawad, 2022
Cyperus conglomeratus	Cyperaceae	ScoT	Alotaibi & Abd-Elgawad, 2022
Fagonia bruguieri	Zygophyllaceae	ScoT	Alotaibi & Abd-Elgawad, 2022
Halopeplis perfoliata	Amaranthaceae	ScoT	Alotaibi & Abd-Elgawad, 2022
Heliotropium bacciferum	Boraginaceae	ScoT	Alotaibi & Abd-Elgawad, 2022
Haloxylon salicornicum	Amaranthaceae	ScoT	Alotaibi & Abd-Elgawad, 2022
Juncus rigidus	Juncaceae	ScoT	Alotaibi & Abd-Elgawad, 2022
Salsola imbricate	Amaranthaceae	ScoT	Alotaibi & Abd-Elgawad, 2022
Seidlitzia rosmarinus	Amaranthaceae	ScoT	Alotaibi & Abd-Elgawad, 2022
Zygophyllum qatarens	Zygophyllaceae	ScoT	Alotaibi & Abd-Elgawad, 2022

This RAPD marker was used to distinguish between *Senna acutifolia, Senna angustifolia, Senna sophera, and Senna tora*. Among 32 decamer primers tested, six primers were found to generate species-specific bands. Further, the analysis revealed 24 polymorphic, 2 monomorphic, and 16 unique loci, enabling differentiation and identification of these morphologically similar *Senna* species, facilitating their distinction in local herbal markets (Khan et al., 2011).

Another study utilized the RAPD technique to authenticate *Cuscuta chinensis, and Cuscuta reflexa*. Thirty-two decamer oligonucleotide primers were involved in the amplication of the genomic DNA of both species. Among them, seven primers produced species-specific reproducible unique bands. Hence, this process is more suitable for determining *C. reflexa* and its adulterant *C. chinensis* depicting RAPD as a valuable supplementary tool for the quality management of herbals (Khan et al., 2010).

Rabin Chandra Paramanik and Chikkaswamy (2014) analyzed the genetic associations among 18 medicinal plants (*Marjorana hortensis, Hemigraphis colorata, Artemisia vulgaris, Artemisia pallens, Ocimum sanctum, Ocimum basilicum, Mentha piparita, Ocimum hratissimum, Acorus calamus, Bacopa moninierii, Mentha spicata, Mentha citrate, Clitoria ternatea, Centella asiatica, Piper nigrum, Piper longum, Stevia rebaudiana, Aloe vera*) using RAPD markers, revealing individual distinctions and underlying variations. Thirteen decamer RAPD primers were employed, generating over 77.23 reproducible bands, with 77.23 identified as polymorphic, indicating 97.6% polymorphism. All primers exhibited consistent banding across medicinal plants, affirming the test's suitability for plant identification. Most individuals displayed unique molecular genotypes (Paramanik & Chikkaswamy, 2014).

These molecular markers, such as RAPD, validate the genetic homogeneity of micropropagated plants by comparing their DNA fingerprints with the donor mother plant. These molecular markers help assess the similarity between micro-propagated plants and the donor mother plant, providing valuable information on the genetic stability and fidelity of the propagation process. A protocol for micropropagation of *Pittosporum eriocarpum*, an endangered therapeutic plant in the Himalayas, has been developed to address the challenges in seed-based propagation. In vitro-raised plantlets and the mother plant showed genetic homogeneity in RAPD, SCoT, and ISSR markers, showing monomorphic bands identical to the mother plant. UPGMA analysis represented 97% similarity between the micropropagated plant and the donor mother plant, confirming the genetic homogeneity.

Thus, the RAPD technique is indirectly used for the verification of micropropagated plants enables large-scale production for medicinal purposes and facilitates the restoration of populations in their natural habitats (Thakur et al., 2016). Therefore, many medicinal plants were successfully authenticated using RAPD markers, which are presented in Table 13.1.

13.4 AMPLIFIED FRAGMENT LENGTH POLYMORPHISM (AFLP)

The AFLP method integrates both RFLP and PCR techniques, starting with DNA digestion, adapter attachment, and selective PCR amplification. Restriction enzymes ensure the fragment reproducibility, reflecting genetic diversity in electrophoretic patterns. After DNA isolation, adapters are ligated, and fragments are amplified using selective PCR primers. AFLPs serve as the dominant marker in which the gel electrophoresis reveals genetic polymorphisms through the presence or absence of bands (Amiteye, 2021). AFLP is utilized in genetic mapping, population genetics, and plant breeding for high-resolution genotyping. AFLP offers high reproducibility, and multiplexing capability, and can detect a large number of polymorphisms across the genome. However, AFLP requires specialized equipment, and optimization and can be labor-intensive, making it less suitable for high-throughput analyses compared to some other marker techniques (Savelkoul et al., 1999).

Another investigation assessed the polymorphism in 12 *Gymnema sylvestre* progenies using AFLP markers and multiplex PCR with 64 primer combinations. Fourteen primers were identified as optimal. Analysis of the progenies using these primers resulted in 1689 fragments, of which 972 were polymorphic (57.5%), and 485 were unique to specific genotypes (28.7%). Fragment numbers varied from 55 to 225 per primer, with polymorphic fragments ranging from 34 to 157. Certain combinations of primers, particularly E-AGG/CTG, E-AGG/M-CAC, E-ACA/CAT, and E-AGG/CAG, were most effective in revealing polymorphism. Similitude coefficients varied from 0.212 to 0.731, concluding with the greatest divergence between S3 and S11 and the highest similarity followed between progenies S8 and S9 (73%) (Osman et al., 2011).

A linkage map for ryegrass (*Lolium perenne* L.) was successfully created utilizing the AFLP markers. The study involved 95 species from a testcross with a doubled-haploid tester. The genetic map spanned 930 cm across seven linkage groups and utilized 463 AFLP markers from 17 primer pairs, along with 3 isozymes and 5 EST markers. This work lays the groundwork for genome mapping in *Lolium*, paving the way for identifying quantitative trait loci (QTL) associated with various agronomic traits in ryegrass (Bert et al., 1999).

The *Ocimum* genus comprises over 150 species, yet precise attributions are challenging due to human intervention through selection, cultivation, and hybridization, alongside significant morphological diversity among species. A standardized descriptor system, primarily reliant on volatile oils, has been suggested but faces limitations due to various environmental factors affecting plant chemical composition. In a study, the researchers explore the efficacy of DNA polymorphism-based molecular markers, specifically AFLP analysis, in resolving these disputed attributions. Their findings suggested a comprehensive approach combining morphological traits, molecular markers, and volatile oil composition, which offers the most effective means to validate and relate the taxonomy and the agronomic traits (Labra et al., 2004).

Baydar et al. (2004) explored the genetic associations among *Rosa damascena* cultivated in Turkey utilizing microsatellite and AFLP markers. In Turkey, despite originating from the same wild populations, *R. damascena* species exhibit slight variations in morphology. However, analysis revealed no detectable polymorphism, indicating identical patterns on markers across different specimens. Consequently, all *R. damascena* plants likely stem from the same actual genotype through vegetative propagation. Nevertheless, the observed differences in morphology are attributed to point mutations, undetectable by molecular markers, commonly noted in cut and garden rose types (Baydar et al., 2004).

A study in *Aloe barbadensis* (aloe vera) utilizing AFLP was conducted to evaluate the diversity among 12 *Aloe barbadensis* elite lines sourced in Madhya Pradesh, India. AFLP primer combinations yielded a total of 410 fragments, averaging 51.2 fragments per primer combination. Different marker attributes, like percent polymorphism, polymorphism information content (PIC), and unique alleles, were computed. Genotyping data from polymorphic fragments were utilized for clustering the analyzed samples through principal component analysis (PCA) and UPGMA-phenogram that revealed significant variability among the collected *Aloe vera* genotypes (Tripathi et al., 2011).

The growing interest in medicinal plants like *Echinacea* sp. has highlighted the need for species-specific identification due to their diverse significant therapeutic properties. While the morphological features of the three *Echinacea* species were quite identical and difficult to differentiate, from the other two species. Using AFLP analysis on cultivated *Echinacea*, 994 fragments were detected, with 565 being polymorphic. Specific primer combinations provided distinct markers for each species. A validation trial on diverse accessions supported these findings, showcasing the potential applications in medicine and adulterant detection within the Echinacea genus (Russil et al., 2009). Hence, the AFLP marker has been introduced in numerous therapeutically significant medicinal plants that are summarized in Table 131.

13.5 INTER SIMPLE SEQUENCE REPEAT (ISSR)

ISSR is a technique that targets the microsatellite regions in the genome. ISSR techniques closely resemble RAPD techniques but differ in ISSR primers are developed from microsatellite parts and require higher annealing temperatures.

This technique primarily addresses the RAPD's limitations. ISSR markers, based on various studies, show promising potential for natural population studies. ISSRs work by evenly distributing primer-annealing sites throughout the genome. If these sites align appropriately, the region between them amplifies via PCR. Structural changes or divergence at binding sites can prevent this type of amplification. ISSRs are now used in natural populations, notably addressing hybridization. Studies across various plant families (Asteraceae, Brassicaceae, etc.) have showcased its versatile applications (Kurane et al., 2009). Corresponded to other molecular markers like RAPD, AFLP, and SSR, ISSR markers offer distinct advantages: (1) They don't require prior sequence information. (2) They're simple and fast to use. (3) They're adaptable to lab settings. (4) They exhibit high stability. (5) They provide abundant genomic information. (6) They don't need radioactivity. (7) Finally, they demonstrate high polymorphism. ISSR markers tap into microsatellite regions spread across genomes, especially in the nuclear genome, sidestepping the need to characterize particular loci. One limitation of ISSR markers is their potential for poor reproducibility due to sensitivity to variations in PCR conditions. Microsatellites, short DNA stretches with variable repeat patterns, particularly di- and trinucleotide repeats targeted in ISSRs, are highly variable within and among plant populations and species (Kurane et al., 2009).

ISSR markers are valuable for precise botanical identification, effectively discerning variations within and between species. Multiple studies employ these markers for species or cultivar distinction. In a study on the medicinal plant nettle (*Urtica dioica* L.) in Mazandaran province, Iran, genetic polymorphism was examined using AFLP and ISSR markers. AFLP revealed 90.21% polymorphism with 830 scorable bands, while ISSR showed 77.3% polymorphism with 234 bands. The average heterozygosity was 0.25 for AFLP and 0.23 for ISSR, with marker index values of 22.25 and 15.57, respectively. Both markers identified the same number of clusters, indicating AFLP's potential for detecting genetic diversity in *U. dioica* (Haghpanah et al., 2016).

Another study evaluates the *Dioscorea* sp. genetic diversity in Mexico, crucial for conservation efforts due to their declining populations. RAPD, AFLP, ISSR, and Inverse Sequence-Tagged Repeat (ISTR) markers were noted and compared, showing similar polymorphic information content (~0.48). While ISTR exhibited lower average heterozygosity (0.36), RAPD, ISSR, and AFLP markers demonstrated higher values (0.43, 0.45, and 0.47, respectively). Despite asexual propagation, significant genetic differences were observed. Although any marker can be recommended for population genetics studies, the low correlations among markers emphasize the value of using multiple markers for comprehensive genetic analyses (Velasco- Ramírez et al., 2014).

Fingerprinting and molecular marker techniques like ISSR were utilized to investigate genetic variety and associations among five Solanaceae species. The Solanaceae family, encompassing over 3000 plant species, holds particular importance due to its bioactive components. The HB15 primer exhibited the highest polymorphism (58.33%) with 12 bands, while the 49A primer showed the lowest polymorphism (16.66%) with 6 bands among the plant populations. These findings underscore the effectiveness of ISSR as a potent tool for assessing genetic diversity in inter-specific species (Salah et al., 2018).

Al- Yasi and Al-Qthanin (2024) has focused on assessing genetic diversity among 23 juniper genotypes in Saudi Arabia, using six ISSR and six SCoT primers. Out of 103 ISSR bands and 105 SCoT bands, 29 and 26 were polymorphic, respectively. The discovery of 120 genotype-specific markers, such as 60 each for ISSR and SCoT, revealed diverse juniper varieties. The findings underscore the significance of conserving *Juniperus procera*'s genetic diversity for medicinal advancements and provide insights into western Saudi Arabian juniper populations for future conservation efforts. These profiling techniques offer valuable long-term fingerprinting and diagnostic markers (Al- Yasi & Al-Qthanin, 2024).

Further, a study in Iran investigates the genetic diversity of 13 geographically isolated *Zhumeria majdea* genotypes, an endangered medicinal plant exclusive to Hormozgan Province, Iran. Leaf samples from Haji Abad, Geno, and Bastak regions underwent DNA extraction and PCR analysis using ISSR markers. Twelve markers identified 121 polymorphic bands, delineating distinctions within and among *Z. majdea* genotypes. Molecular variance analysis (AMOVA) revealed 86% within-population and 14% among-population variations. Cluster analysis categorized genotypes into four main groups, with principal coordinate analysis explaining 28.81% and 15.71% of the deviations. The observed intra-population variation in *Z. majdea* may stem from inherent differences among individuals and variations in endophyte presence, type, and genetic patterns (Daryaii et al., 2020).

Genetic variation within different sweet, yellow, and purple passion fruit accessions was evaluated utilizing ISSR markers, employing 18 primers among 45 accessions. Cluster analysis identified 11 major clusters, irrespective of geographic origin, suggesting a lack of structure in the gene bank. In addition to that, the Shannon–Weaver diversity index averaged 0.32, representing the robust genetic diversity resolution. ISSR proves valuable for genetic diversity studies, aiding parental selection, breeding, and conservation strategies (Dos Santos et al., 2011). Consequently, the ISSR marker has been implemented in numerous medically important plant species, as outlined in Table 13.1.

13.6 SEQUENCE CHARACTERIZED AMPLIFIED REGION (SCAR)

SCAR markers are PCR-based codominant markers utilizing two specific primers derived from nucleotide sequences within cloned RAPD fragments associated with a trait of curiosity. SCAR markers, derived from RAPD fragments, offer robust and reliable genotyping. They utilize specific primers designed from cloned sequences, ensuring accuracy and reproducibility. Their versatility allows for screening various samples simultaneously, enhancing cost efficiency. SCARs, though not highly polymorphic, enable rapid marker development and can serve as allele-specific primers for specific trait identification. Utilizing ASAPs, they generate allele-specific DNA fragments, eliminating the need for electrophoresis in genotyping (Kiran et al., 2010). SCAR markers are utilized in genetic mapping, marker-assisted selection, and molecular diagnostics. They offer high specificity, reproducibility, and easy detection through gel electrophoresis or sequencing. Initial development may require sequence information, primer design can be labor-intensive, and they are limited to detecting variations in targeted sequences (Nadeem et al., 2018).

SCAR markers exhibit both dominant and codominant characteristics. In asparagus, they were treated as dominant markers. At a 60°C annealing temperature, locus M was amplified in both parental lines, but at 67°C, only males showed amplification. However, digesting the SCAR fragments with various endonucleases at 60°C didn't yield small fragment-length polymorphisms. These results were contrary to previous findings where codominant markers (SCAR) generated monomorphic bands produced by digesting with restriction enzymes. These markers have been utilized in authenticating various therapeutic plants prone to adulteration, such as *Panax japonicas*, *Phyllanthus*, *Artemisia*, *Atractylodes*, *Panax*, *Echinacea*, *Pueraria tuberose* Roxb, *Peucedanum praeruptorum*, *Angelica decursiva*, *Jatropha curcas*, *Anthricus sylvestris*, and *Curcuma alismatifolia* (Kiran et al., 2010).

Li and Park (2012) developed species-specific molecular markers to ensure the accurate identification of *Ophiopogon* and *Liriope* species, crucial for the quality of medicinal products derived from these plants. Over 40 primers were screened for polymorphism, and species-specific RAPD bands were isolated, cloned, and sequenced. Subsequently, SCAR primers were developed based on these sequences. Two SCAR markers (SA06 and SB05), specific to the plant *Ophiopogon japonicus,* produced 460 and 553 base pairs DNA fragments, correspondingly, while another marker SA12 intensified a 485 bp fragment precise to *Liriope platyphylla*. These results represent the instance of a species-specific SCAR marker for the group of plants, facilitating the quick authentication of closely related *Ophiopogon* and *Liriope* species (Li & Park, 2012).

Researchers engineered a SCAR molecular marker to swiftly identify *Podophyllum hexandrum* Royle, also known as the Indian May apple, a globally esteemed medicinal plant. This marker stems from the analysis of RAPD, where three RAPD outcomes common to all acquisitions were isolated, cloned, and sequenced. Using these sequences, SCAR primers were synthesized. They acquired several species-specific RAPD amplicons by conducting RAPD analysis on three distinct populations of *P. hexandrum*. From the nucleotide sequences of these amplicons, they developed SCAR markers for the precise marker of *P. hexandrum*. These SCAR markers facilitate swift and reliable authentication of closely related *P. hexandrum* species, aiding in the prevention of adulteration during distribution in herbal markets (Al-shaqha et al., 2014).

Another study created stable SCAR markers from RAPD fragments of *L. chinensis*, a traditional Chinese medicinal plant. Sequencing revealed three distinct clones, each with specific nucleotide compositions. PCR amplification demonstrated marker specificity for *L. chinensis* samples. These markers offer a reliable method for *L. chinensis* identification and genetic analysis (Cheng et al., 2015).

Molecular cloning was employed to enhance the RAPD amplification of *Levisticum officinale*, *Angelica acutiloba*, and *Angelica sinensis*, yielding unexplored SCAR markers (A13, A1-34, A23, and A1-0). Their sequence information was recorded under accession numbers (KP641315-KP641318) in GenBank. Through optional PCR amplification, A1-0 serves as specific markers for *Angelica sinensis*, A23 and A13 for *Levisticum officinale,* while A1-34 is specific to *Angelica acutiloba*. Hence, these SCAR markers offer valuable tools for genetic authentication, aiding in ecological conservation and detecting adulterants within these medicinal plants (Zhang et al., 2015).

Phyllanthus amarus Schum. & Thonn. holds a significant place in Thai traditional medicine, renowned for its antipyretic, diuretic, and liver-disease-treating properties, along with its efficacy against viral infections. However, confusion arises due to the morphological similarities between *P. amarus* and its less potent counterparts, *P. urinaria* Klein ex Wild. and *P. debilis* L., often leading to inadvertent collection of the latter two species for medicinal purposes. To address this, DNA fingerprints were generated using RAPD-PCR methods, unveiling specific RAPD markers for each species. From these markers, highly targeted SCAR primers were developed, yielding SCAR markers of 408 bp for *P. amarus*, 501 bp for *P. debilis*, and 319 bp for *P. urinaria*. This identification method proved high specificity when tested against various closely related species and even in mixed DNA samples (Theerakulpisut et al., 2008). In addition to that, another study aimed to establish a SCAR marker for identifying *Phyllanthus emblica* Linn. A putative marker specific to *P. emblica* was detected through the RAPD technique. Subsequently, a SCAR marker was successfully generated from the RAPD amplicon. SCAR marker proved effective in identifying *P. emblica* in both its commercial samples and in the Ayurvedic formulation Triphalachurna (Dnyaneshwar et al., 2006). Therefore, medicinal plants within species have been authenticated using SCAR markers which are listed in Table 13.1.

13.7 SINGLE-NUCLEOTIDE POLYMORPHISMS (SNP)

The fundamental concept of single-nucleotide polymorphism (SNP) detection involves a non-sequencing method distinguishing between discovering a new genetic variation that hasn't been characterized before and identifying a known genetic variation. The principle of SNP detection relies on identifying single-nucleotide variations within DNA sequences. It involves comparing sequences from different individuals or populations to pinpoint differences at specific nucleotide positions. A wide array of in vitro methods, like resequencing, non-sequencing, and sequencing techniques, and in silico methods exists for SNP detection, offering diverse options. These technologies have advanced over time through the introduction of new techniques involving reporter systems, fluorescent probes, enzymatic assays, highly sensitive instruments, and notably, the rapid progress of high-throughput sequencing technology along with accompanying bioinformatic tools (Morgil et al., 2016). SNPs offer high abundance and widespread distribution across genomes, providing abundant genetic markers for diverse applications. SNP detection can be expensive and requires complex infrastructure (Canturk et al., 2014).

Chen et al. (2013) examined 377 sequences from four species within the *Panax* genus, a well-renowned traditional Chinese therapeutic plant, which belongs to the Araliaceae family. The aim was to address issues of adulteration in commercial *Panax* products due to their rarity and high demand. By analyzing SNPs in the internal transcribed spacer 2 (ITS2) region of these species, this work also discusses the intragenomic variations. Additionally, intraspecific variations were studied using three DNA barcoding methods (ITS2, psbA-trnH, and matK). In conclusion, results indicated the minimal genetic distances within *Panax quinquefolius* and *Panax ginseng*, suggesting that ITS2 could function effectively as a single locus in their genomes. They also identified five SNPs in the ITS2 region for distinguishing *Panax* medicinal species. Notably, when analyzing mixed powders of *P. quinquefolius,* and *P. ginseng*, distinct double peaks were observed at SNP positions, allowing for an estimation of the mixed ratio. These findings propose the development of SNP-based molecular barcodes as a standard technique for accurately identifying *Panax* species, particularly those closely related, and for detecting adulteration in mixed *Panax* products (Chen et al., 2013).

Chamomile (*Matricaria recutita* L.), genotyping-by-sequencing (GBS) was utilized to evaluate genetic diversity and conduct a genome-wide association study (GWAS), targeting flowering time responsible genes and alpha-bisabolol. From 33 origins, analysis revealed 6495 high-quality SNP markers across 91 *M. recutita* plants, distinguishing them from the outgroup *M. discoidea.* Tetraploid genotypes showed lower diversity compared to diploids, likely due to artificial polyploidization. Diploids have shorter flowering times than tetraploids, and several tetraploid genotypes exhibited high alpha-bisabolol content. GWAS identified significant SNPs associated with alpha-bisabolol range, and flowering period with one sequence linked to *Arabidopsis thaliana* self-pollination and other four sequences linked with stress responses in different plants (Otto et al., 2017). Further, SNPs identified in other therapeutic plants are listed in Table 13.1, and the comparison of different characteristics present in molecular markers used for medicinal plants has been outlined in Table 13.2.

13.8 EXPRESSED SEQUENCE TAG (EST)

Expressed sequence tag (EST) is a marker with short DNA sequences representing expressed genes, aiding in gene discovery and expression analysis. This method involves the principle of capturing short sequences from cDNA (complementary DNA) libraries, which represent actively transcribed genes. This process begins with the isolation of mRNA molecules from cells or tissues of interest. The mRNA is then reverse transcribed into cDNA utilizing reverse transcriptase enzyme. The resultant cDNA library contains fragments representing genes that were actively transcribed at the time of sample collection. Subsequently, these cDNA fragments are sequenced, typically yielding relatively short sequences, which provide insights into gene expression patterns under specific conditions. ESTs serve as a beneficial resource for gene discovery, functional genomics, and molecular marker development. EST markers are applied in gene discovery, gene expression profiling, marker development, functional genomics,

TABLE 13.2
Comparison of Different Characteristics Present in Molecular Markers Used for Medicinal Plants

Techniques	PCR based	Availability in genome	Polymorphism level	Reproducibility	Inheritance	Radiolabeling
RFLP	No	High	Moderate	High	Co-dominance	Used
RAPD	Yes	High	High	Low	Dominance	Not used
AFLP	Yes	High	High	Moderate	Dominance	Both
ISSR	Yes	Moderate to high	High	High	Co-dominance	Not used
SNP	Yes	Very high	High	High	Co-dominance	Not used
SCAR	Yes	Moderate to high	High	High	Co-dominance	Used

and comparative genomics studies. The advantage of EST markers is they are derived from transcribed regions, offering insights into gene function and regulation, and aiding in identifying candidate genes for traits of interest. Limited representation of the genome is the major limitation compared to other markers like SNPs, and potential difficulty in designing primers due to variable transcript sequences (Kalia et al., 2011).

A comprehensive EST dataset for *Epimedium sagittatum* was generated, revealing 76,459 consensus sequences and 2,810 EST-SSRs. Transferability tests across 52 *Epimedium* species show promising results, with 85.7% successful primer transfer and high genetic diversity. These findings underscore the significance of ESTs in elucidating secondary metabolism and advancing research in the taxonomy, molecular breeding, and genomics of *Epimedium* species (Zeng et al., 2010).

Another study developed 70 SSR markers from EST for *Panax ginseng*, showing reproducible polymorphism among nineteen *P. ginseng* cultivars and four *Panax* species. These markers exhibited Mendelian inheritance in an F2 population, indicating their reliability for genetic mapping. Phylogenetic analysis revealed distinct groups, with *P. ginseng* and *P. japonicus* clustering closely. *P. ginseng* cultivars formed minor subgroups. These results suggest the existence of genetic diversity within cultivated varieties, and a highly replicated paleopolyploid genome structure in ginseng (Choi et al., 2011).

The transcriptome analysis of *Plantago major* vascular tissue revealed 5,900 ESTs and 3,247 mRNAs, validated by phloem and xylem marker genes. This specific study identified novel vascular-specific genes and pathways, confirmed via various analyses. Plantain transformation validated the vascular explicitness of a *Plantago* promoter-GUS construct, with implications for functional genomics in other species like *Arabidopsis* (Pommerrenig et al., 2006). For many medicinal plants, EST databases were developed, which are listed in Table 13.1, and the advantages and limitations of EST are given in Table 13.3.

TABLE 13.3
Advantages and Disadvantages of Different Molecular Markers in Medicinal Plants

Techniques	Advantages	Disadvantages
RFLP	• Co-dominant • High genomic abundance • Highly reproducible	• Laborious • High quality of DNA is necessary • Radioactive labelling
RAPD	• Can opt for automation • Radioactive labelling is not needed • Less amount of DNA is sufficient	• Not reproducible • Alleles are not detected • Dominant inheritance
AFLP	• Polymorphism rate is higher • Small fragments can be used • Genomic abundance is high	• Not reproducible • High quality DNA and primers are needed • Alleles are not detected • Dominant inheritance
SCAR	• Simple and quick method • Better reproducibility • Locus specific	• Radioactive isotopes is required • Gene sequence information is needed
SNP	• Genome wide distribution • Co-dominant inheritance • Prior sequence information is not needed	• Higher cost • Interference of DNA secondary structure
ISSR	• DNA sequence information is not needed. • Low cost • Multiple alleles	• Alleles are not detected • No homology • Non reproducible
EST	• Detection of variation at expressed regions of the genome • No direct cost is involved	• Higher rate of sequencing error
ScoT	• Effective detection of polymorphic loci • Higher capacity for detecting genetic diversity	• Poor transferability

13.9 START CODON TARGETED MARKER (SCoT)

SCoT (start codon targeted) markers utilize conserved regions flanking start codons in protein-coding genes for PCR amplification. The principle behind SCoT markers involves targeting conserved regions flanking the start codons (ATG) of protein-coding genes. PCR primers are developed to amplify these areas, resulting in the generation of DNA fragments that serve as molecular markers for genetic analysis. SCoT markers are mainly based on the regions surrounding start codons that are relatively conserved across species. It allows for the amplification of homologous sequences from different genomes. This approach facilitates the detection of genetic variation and polymorphisms within and among populations or species, contributing to genetic diversity studies, population genetics, and evolutionary analyses. SCoT markers are utilized in genetic diversity analysis, population genetics, molecular breeding, and evolutionary studies due to their ability to detect polymorphisms across diverse genomes. SCoT markers offer high reproducibility, multilocus amplification, and codominant inheritance, providing valuable genetic information for various genomic analyses. Potential limitations include the requirement for prior sequence information for primer design and the possibility of generating nonspecific amplicons with careful optimization (Rai, 2023).

A study in Al Jubail, Saudi Arabia, examined hereditary differences in 29 wild plants using ISSR and SCoT

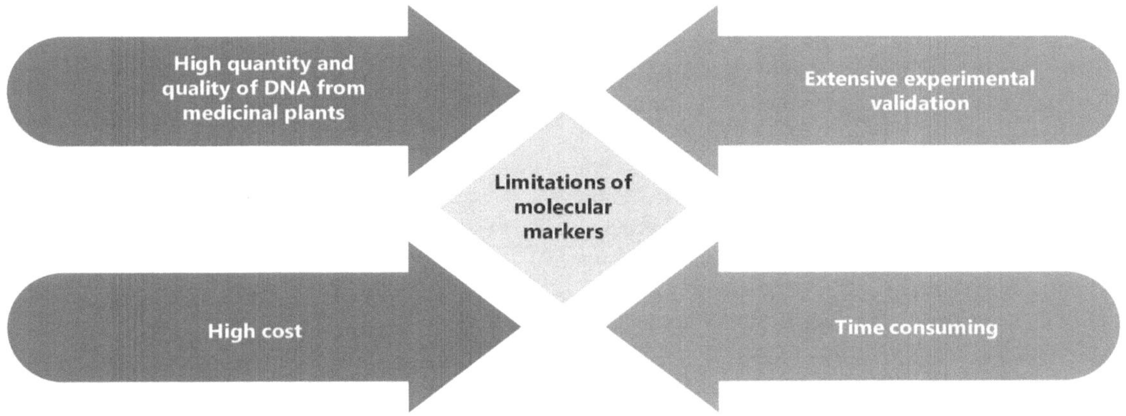

FIGURE 13.3

molecular markers. Results showed 87% and 84% polymorphism, including an average of 12.4 and 13.7 polymorphic bands per primer combination, respectively. The species with the highest genetic similarity were *Juncus rigidus*-23 and *Zygophyllum qatarense*-22. Conversely, the lowest genetic similarity was observed between *Salsola imbricate*-18 and *Zygophyllum qatarense*-6 for ISSR (Alotaibi & Abd-Elgawad, 2022).

Ma et al. (2021) examined the genetic variety in 65 *Stellaria media* individuals from 13 groups employing ten SCoT markers, revealing high polymorphism (96.38%), polymorphic data range (0.34), and allele number (1.38). This investigation mainly sought to identify the genetic variation within and among *S. media* populations. It also identifies genetic groupings within 13 chickweed groups, revealing significant genetic heterogeneity within these populations (Ma et al., 2022).

A micropropagation protocol along with ScoT marker was successfully developed for *Pittosporum eriocarpum* Royle, an endangered therapeutic plant endemic to the Uttarakhand area of the Himalayas. High shoot regeneration (95%) was achieved using specific hormone combinations, with subsequent successful rooting (90%) and acclimatization (73%) of the regenerated shoots in the field. Genetic homogeneity among the micropropagated plants was documented utilizing SCoT, ISSR, and RAPD markers, with similarity values ranging from 0.89 to 1.00, 0.91 to 1.00, and 0.95 to 1.00, respectively. The outcome of this study highlights the potential for protection and large-scale production of *Pittosporum eriocarpum* through ScoT development, aiding in habitat restoration efforts (Thakur et al., 2016). Hence, SCoT markers, with their simplicity and versatility, revolutionize crop breeding, enabling precise genetic assessments and facilitating QTL mapping and cultivar identification across various commercially important species. Table 13.1 lists the developed SCoT markers for various medicinal plants, while Table 13.3 outlines the advantages and disadvantages of SCoT markers.

13.10 LIMITATIONS OF MOLECULAR MARKERS IN MEDICINAL PLANTS

In medicinal plants, secondary metabolite production is governed by intricate biochemical pathways influenced by multiple genes and environmental factors. So molecular markers may not capture this complexity comprehensively. On the other hand, the molecular markers availability, specific to genes concerned with secondary metabolite pathways may be limited, particularly in non-model medicinal plants (Zhao et al., 2023). Also, environmental factors like light, temperature, soil composition, and different geographical places significantly influence secondary metabolite production. However, molecular markers may capture these environmental influences, but not at the exact levels. When the alleles associate non-randomly at different loci, then it is called linkage disequilibrium, which can affect the accuracy of marker-trait associations, particularly in diverse and outbred populations (Ali et al., 2013).

Validating the association between molecular markers and enhanced secondary metabolite production requires extensive experimental validation, which can be resource-intensive and time-consuming (Fang et al., 2021). The major limitation of medicinal herbs is the presence of secondary metabolites like polysaccharides, polyphenols, alkaloids, flavonoids, tannins, and terpenes, which pose challenges in obtaining high-quality genomic DNA. These compounds can disrupt DNA isolation procedures, yielding low-quality genomic DNA (Figure 13.3). Additionally, expensive commercial kits may sometimes prove ineffective in producing high-quality genomic DNA (Nazarian- Firouzabadi, 2014). The benefits and drawbacks of different molecular markers have been summarized in Table 13.3.

13.11 FUTURE PERSPECTIVES

The genetic profile of medicinal plant species holds immense potential for ensuring the quality control of drugs by both pharmaceutical industries and governmental agencies.

However, while DNA markers offer valuable insights into plant identification, they must be complemented by chemical analysis to authenticate drugs effectively (Ganie et al., 2015). While using DNA markers, challenges arise due to their lack of tissue specificity, particularly when different parts of the same plant are sold as medicinal products. In such cases, markers like RAPD, ISSR, and AFLP should only be utilized alongside techniques like loop-mediated isothermal amplification (LAMP) or SCAR conversion for precise authentication (Le & Vu, 2017). Ultimately, SCAR, LAMP, and DNA barcoding, alongside morphological, anatomical, and chemical markers, offer robust authentication methods (Amiteye, 2021). Furthermore, while traditional molecular markers do not directly capture epigenetic modifications, integrating epigenetic data into genetic studies can nourish useful understandings of the regulatory mechanisms influencing trait variation in medicinal plants (Ferrari et al., 2019). In the future, DNA markers can aid in building reference libraries for traditional medicines, combating adulteration, and enhanced breeding ensuring the integrity of medicinal plant products.

ACKNOWLEDGMENT

The corresponding author expresses his gratitude toward Bharathiar Cancer Theranostics Research Centre (BU/RUSA2.0/BCTRC/2020/BCTRC-CT08/14.12.2020), Bharathiar University, and Tamil Nadu State Council for Higher Education (RGP/2019-20/BU/HECP-0018, Dt.27.04.2021), for generously extending the financial support.

DATA AVAILABILITY

The authors confirm that the data supporting the findings of this study are available within the book/chapter.

REFERENCES

Abd El-Twad, M. H., & Zahran, F. A. (2010). RAPD, ISSR and RFLP analysis of phylogenetic relationships among congeneric species (Anthemideae, Astraceae) in Egypt. *International Journal of Botany*, *6*, 1–10.

Akilandeswari, G., Vijaya Anand, A., Sampathkumar, P., Vinayaga Moorthi, P., & Preethi, B. (2019). A prospective review on phyto-pharmacological aspects of *Andrographis paniculata*. *Systematic Reviews in Pharmacy*, *10*(1), 15–19.

Alansi, S., Tarroum, M., Al-Qurainy, F., Khan, S., & Nadeem, M. (2016). Use of ISSR markers to assess the genetic diversity in wild medicinal *Ziziphus spina-christi* (L.) Willd. collected from different regions of Saudi Arabia. *Biotechnology & Biotechnological Equipment*, *30*(5), 942–947. https://doi.org/10.1080/13102818.2016.1199287

Ali, Z., Ganie, S. H., Narula, A., Sharma, M. P., & Srivastava, P. S. (2013). Intra-specific genetic diversity and chemical profiling of different accessions of *Clitoria ternatea* L. *Industrial Crops and Products*, *43*(1), 768–773. https://doi.org/10.1016/j.indcrop.2012.07.070

Alotaibi, M. O., & Abd-Elgawad, M. E. (2022). ISSR and SCoT for evaluation of hereditary differences of 29 wild plants in Al Jubail Saudi Arabian. *Saudi Journal of Biological Sciences*, *29*(5), 3223–3231. https://doi.org/10.1016/j.sjbs.2022.01.053

Al-shaqha, W. M., Khan, M., & Chaudhary, A. A. (2014). SCAR molecular markers for identification and authentication of medicinal plants *Podophyllum hexandrum* Royle. *Asian Journal of Biochemical and Pharmaceutical Research*, *4*(4), 66–75.

Al-Yasi, H. M., & Al-Qthanin, R. (2024, March). Comparing genetic differentiation and variation using ISSR and SCoT among Juniper plant markers in Saudi Arabia. *Frontiers in Plant Science*, *15*, 1–9. https://doi.org/10.3389/fpls.2024.1356917

Amiteye, S. (2021). Basic concepts and methodologies of DNA marker systems in plant molecular breeding. *Heliyon*, *7*(10), e08093. https://doi.org/10.1016/j.heliyon.2021.e08093

Arumugam, T., Jayapriya, G., & Sekar, T. (2019). Molecular fingerprinting of the Indian medicinal plant *Strychnos minor* Dennst. *Biotechnology Reports*, *21*(2018), e00318. https://doi.org/10.1016/j.btre.2019.e00318

Baydar, N. G., Baydar, H., & Debener, T. (2004). Analysis of genetic relationships among *Rosa damascena* plants grown in Turkey by using AFLP and microsatellite markers. *Journal of Biotechnology*, *111*(3), 263–267. https://doi.org/10.1016/j.jbiotec.2004.04.014

Begum, T. N., Ilyas, M. M., Kalavathy, S., Anadh, A. V., Kumar, S., & Senthil, R. (2009). Effects of ethanolic leaf extract of *Azim tetracantha* Lam. on *Ehrlich ascites* carcinoma tumour bearing mice. *Research Journal of Medicine and Medical Sciences*, *4*(2), 351–354.

Bert, P. F., Charmet, G., Sourdille, P., Hayward, M. D., & Balfourier, F. (1999). A high-density molecular map for ryegrass (*Lolium perenne*) using AFLP markers. *Theoretical and Applied Genetics*, *99*(3–4), 445–452. https://doi.org/10.1007/s001220051256

Bharathi, V., Rengarajan, R. L., Radhakrishnan, R., & Hashem, A., Abd-Allah, E. F., Alqarawi, A. A., & Anand, A. V. (2018). Effects of a medicinal plant *Macrotyloma uniflorum* (Lam.) Verdc. Formulation (MUF) on obesity-associated oxidative stress-induced liver injury. *Saudi Journal of Biological Sciences*, *25*(6), 1115–1121.

Biswas, K., & Biswas, R. (2013). Identification of medicinal plants using PCR- RFLP in Dasamula – an ayurvedic drug. *Journal of Pharmaceutical and Biosciences*, *3*, 94–99.

Biswas, K., Garg, N., & Biswas, R. (2017). PCR-RFLP of *Calotropis gigantea* (L): A tool for forensic application. *International Journal of Applied Agricultural Research*, *12*(1), 61–67.

Canturk, K. M., Emre, R., Kínoglu, K., Başpínar, B., Sahin, F., & Ozen, M. (2014). Current status of the use of single-nucleotide polymorphisms in forensic practices. *Genetic Testing and Molecular Biomarkers*, *18*(7), 455–460. https://doi.org/10.1089/gtmb.2013.0466

Chaudhary, R., & Maurya, G. (2020, April). Restriction fragment length polymorphism. In *Encyclopedia of animal cognition and behavior*. https://doi.org/10.1007/978-3-319-47829-6

Chen, X., Liao, B., Song, J., Pang, X., Han, J., & Chen, S. (2013). A fast SNP identification and analysis of intraspecific variation in the medicinal Panax species based on DNA barcoding. *Gene*, *530*(1), 39–43. https://doi.org/10.1016/j.gene.2013.07.097

Cheng, J., Long, Y., Khan, M. A., Wei, C., Fu, S., & Fu, J. (2015). Development and significance of RAPD-SCAR markers for the identification of *Litchi chinensis* Sonn. by improved RAPD amplification and molecular cloning.

Electronic Journal of Biotechnology, *18*(1), 35–39. https://doi.org/10.1016/j.ejbt.2014.11.004

Choi, H. Il, Kim, N. H., Kim, J. H., Choi, B. S., Ahn, I. O., Lee, J. S., & Yang, T. J. (2011). Development of reproducible EST-derived SSR markers and assessment of genetic diversity in *Panax ginseng* cultivars and related species. *Journal of Ginseng Research*, *35*(4), 399–412. https://doi.org/10.5142/jgr.2011.35.4.399

Daryaii, L. B., Samsampour, D., Bagheri, A., Seyahooei, M. A., & Raam, M. (2020). Molecular characterization of *Zhumeria majdea* Iranian germplasms using ISSR markers. *International Journal of Horticultural Science and Technology*, *7*(4), 343–352. https://doi.org/10.22059/ijhst.2020.296060.335

Devaiah, K., Balasubramani, S. P., & Venkatasubramanian, P. (2011). Development of randomly amplified polymorphic DNA based SCAR marker for identification of *Ipomoea mauritiana* Jacq (Convolvulaceae). *Evidence-Based Complementary and Alternative Medicine*, *2011*(1), 868720.

Divya, N., Rengarajan, R. L., Radhakrishnan, R., Abd-Allah, E. F., Alqarawi, A. A., Hashem, A., & Anand, A. V. (2019). Phytotherapeutic efficacy of the medicinal plant *Terminalia catappa* L. *Saudi Journal of Biological Sciences*, *26*(5), 985–988.

Dnyaneshwar, W., Preeti, C., Kalpana, J., & Bhushan, P. (2006). Development and application of RAPD-SCAR marker for identification of *Phyllanthus emblica* Linn. *Biological and Pharmaceutical Bulletin*, *29*(11), 2313–2316. https://doi.org/10.1248/bpb.29.2313

Dos Santos, L. F., De Oliveira, E. J., Dos Santos Silva, A., De Carvalho, F. M., Costa, J. L., & Pádua, J. G. (2011). ISSR markers as a tool for the assessment of genetic diversity in Passiflora. *Biochemical Genetics*, *49*(7–8), 540–554. https://doi.org/10.1007/s10528-011-9429-5

Enayat, M. (1996). Restriction fragment length polymorphism. *Acta Horticulturae*, *374*(1), 29–35. https://doi.org/10.17660/actahortic.1996.374.7

Fang, K., Xia, Z., Li, H., Jiang, X., Qin, D., Wang, Q., Wang, Q., Pan, C., Li, B., & Wu, H. (2021). Genome-wide association analysis identified molecular markers associated with important tea flavor-related metabolites. *Horticulture Research*, *8*(1). https://doi.org/10.1038/s41438-021-00477-3

Ferrari, L., Pavanello, S., & Bollati, V. (2019). Molecular and epigenetic markers as promising tools to quantify the effect of occupational exposures and the risk of developing non-communicable diseases. *La Medicina Del Lavoro*, *110*(3), 168–190. https://doi.org/10.23749/mdl.v110i3.8538

Gangwar, H., Gahlaut, V., Chauhan, R., Singh, S., & Jaiswal, V. (2023). Development of species-specific ISSR-derived SCAR marker for early discrimination between *Cinnamomum verum* and *Cinnamomum cassia*. *Molecular Biology Reports*, *50*(8), 6311–6321.

Ganie, S. H., Upadhyay, P., Das, S., & Prasad Sharma, M. (2015). Authentication of medicinal plants by DNA markers. *Plant Gene*, *4*, 83–99. https://doi.org/10.1016/j.plgene.2015.10.002

Gantait, S., Debnath, S., & Nasim Ali, M. (2014). Genomic profile of the plants with pharmaceutical value. *3 Biotech*, *4*(6), 563–578. https://doi.org/10.1007/s13205-014-0218-9

Gawroński, J., & Dyduch-Siemińska, M. (2022). Potential of *in vitro* culture of *Scutellaria baicalensis* in the formation of genetic variation confirmed by ScoT markers. *Genes*, *13*(11), 2114.

Gowda, B., Chandrika, K., Prasanna, K. T., & Kirana, V. C. (2010). AFLP authentication of *Embelia ribes* Burm. f. and *Embelia tsjeriam*-cottam A. DC. *International Journal of Science and Nature*, *1*(1), 58–60.

Hadipour, M., Kazemitabar, S. K., Yaghini, H., & Dayani, S. (2020). Genetic diversity and species differentiation of medicinal plant Persian Poppy (*Papaver bracteatum* L.) using AFLP and ISSR markers. *Ecological Genetics and Genomics*, *16*, 100058.

Haghpanah, M., Kazemitabar, S. K., Hashemi, S. H., & Alavi, S. M. (2016). Comparison of ISSR and AFLP markers in assessing genetic diversity among Nettle (*Uritica dioica* L.) populations. *Journal of Plant Molecular Breeding*, *4*(1), 10–16.

Hamouda, M. (2019). Molecular analysis of genetic diversity in population of *Silybum marianum* (L.) Gaertn in Egypt. *Journal of Genetic Engineering and Biotechnology*, *17*(1), 12.

Hazler Pilepić, K., Morović, M., Orač, F., Šantor, M., & Vejnović, V. (2010). RFLP analysis of cpDNA in the genus Hypericum. *Biologia*, *65*(5), 805–812.

Jedrzejczyk, I. (2020). Genome size and SCoT markers as tools for identification and genetic diversity assessment in *Echinacea* genus. *Industrial Crops and Products*, *144*, 112055.

Jedrzejczyk, I., & Rewers, M. (2020). Identification and genetic diversity analysis of edible and medicinal Malva species using flow cytometry and ISSR molecular markers. *Agronomy*, *10*(5), 650.

Jena, S. N., Kumar, S., & Nair, N. K. (2009). Molecular phylogeny in Indian Citrus L. (*Rutaceae*) inferred through PCR-RFLP and trnL-trnF sequence data of chloroplast DNA. *Scientia Horticulturae*, *119*(4), 403–416. https://doi.org/10.1016/j.scienta.2008.08.030

Jha, S. R., Naz, R., Asif, A., Okla, M. K., Soufan, W., Al-Ghamdi, A. A., & Ahmad, A. (2020). Development of an in vitro propagation protocol and a sequence characterized amplified region (SCAR) marker of *Viola serpens* Wall. ex Ging. *Plants*, *9*(2), 246.

Jonah, P. M., Bello, L. L., Lucky, O., Midau, A., & Moruppa, S. M. (2011). Review: The importance of molecular markers in plant breeding programmes. *Global Journal of Science Frontier Research*, *11*(5), 5–12.

Kalia, R. K., Rai, M. K., Kalia, S., Singh, R., & Dhawan, A. K. (2011). Microsatellite markers: An overview of the recent progress in plants. *Euphytica*, *177*(3), 309–334. https://doi.org/10.1007/s10681-010-0286-9

Karthika, P., Balamuralikrishnan, B., Yamini, K., Vijaya Anand, A., Durairaj, K., Maruthupandian, A., Alodaini, H. A., Hatamleh, A. A., Manikantan, P., & Arun, M. (2023). Green synthesis, characterization of silver nanoparticles using aqueous leaf extracts of *Solanum melongena* and in vitro evaluation of antibacterial, pesticidal and anticancer activity in human MDA-MB-231 breast cancer cell lines. *Journal of King Saud University – Science*, *35*(5), 102663.

Kaviya, M., Balasubramanian, B., Bharathi, K., Malaisamy, A., Al-Dhabi, N. A., Mariadhas, V. A., Anand, A. V., & Liu, W. (2022). Evaluation of nutritional substances and investigation of antioxidant and antimicrobial potentials of *Boerhavia diffusa* with in silico molecular docking. *Molecules*, *27*(4), 1280.

Khan, S., Mirza, K. J., & Abdin, M. Z. (2010). Development of RAPD markers for authentication of medicinal plant *Cuscuta reflexa*. *EurAsian Journal of BioSciences*, (4), 1–7. https://doi.org/10.5053/ejobios.2010.4.0.1

Khan, S., Mirza, K. J., Al-Qurainy, F., & Abdin, M. Z. (2011). Authentication of the medicinal plant *Senna angustifolia* by RAPD profiling. *Saudi Journal of Biological Sciences*, *18*(3), 287–292. https://doi.org/10.1016/j.sjbs.2011.03.001

Kim, Y., Choi, S. J., & Choi, C. (2017). An efficient PCR-RFLP method for the rapid identification of Korean Pyropia species. *Molecules*, *22*(12), 2182.

Kiran, U., Khan, S., Mirza, K. J., Ram, M., & Abdin, M. Z. (2010). SCAR markers: A potential tool for authentication of herbal drugs. *Fitoterapia*, *81*(8), 969–976. https://doi.org/10.1016/j.fitote.2010.08.002

Kurane, J., Shinde, V., & Harsulkar, A. (2009). PHCOG REV: Review article application of ISSR marker in pharmacognosy: Current update. *Pharmocognosy Reviews*, *3*(6), 216–228. www.phcog.net

Labra, M., Miele, M., Ledda, B., Grassi, F., Mazzei, M., & Sala, F. (2004). Morphological characterization, essential oil composition and DNA genotyping of *Ocimum basilicum* L. cultivars. *Plant Science*, *167*(4), 725–731. https://doi.org/10.1016/j.plantsci.2004.04.026

Le, D. T., & Vu, N. T. (2017). Progress of loop-mediated isothermal amplification technique in molecular diagnosis of plant diseases. *Applied Biological Chemistry*, *60*(2), 169–180. https://doi.org/10.1007/s13765-017-0267-y

Li, G., & Park, Y. J. (2012). SCAR markers for discriminating species of two genera of medicinal plants, Liriope and Ophiopogon. *Genetics and Molecular Research: GMR*, *11*(3), 2987–2996. https://doi.org/10.4238/2012.May.18.14

Li, Y., Cheng, X., Lai, J., Zhou, Y., Lei, T., Yang, L., Li, J., Yu, X., & Gao, S. (2023). ISSR molecular markers and anatomical structures can assist in rapid and directional screening of cold-tolerant seedling mutants of medicinal and ornamental plant in *Plumbago indica* L. *Frontiers in Plant Science*, *14*, 1149669.

Liu, T., & Ji, Y. (2012). Molecular authentication of the medicinal plant *Paris polyphylla* Smith var. *yunnanensis* (Melanthiaceae) and its related species by polymerase chain reaction–restriction fragment length polymorphism (PCR-RFLP). *Journal of Medicinal Plant Research*, *6*(7), 1181–1186.

Liu, W. L., Shih, H. C., Weng, I. S., Ko, Y. Z., Tsai, C. C., Chou, C. H., & Chiang, Y. C. (2016). Characterization of genomic inheritance of intergeneric hybrids between Ascocenda and Phalaenopsis cultivars by GISH, PCR-RFLP and RFLP. *PloS One*, *11*(4), e0153512.

Luz, G. C., Strioto, D. K., Mangolin, C. A., & Maria, D. F. (2020). ISSR markers to assess genetic diversity of cultivated populations from artificial selection of *Stevia rebaudiana* (Bert.) Bertoni. *Breeding Science*, *70*(4), 508–514.

Ma, S., Khayatnezhad, M., & Minaeifar, A. A. (2021). Genetic diversity and relationships among *Hypericum* L. species by ISSR markers: A high value medicinal plant from Northern of Iran. *Caryologia*, *74*(1), 97–107.

Ma, X., Tian, H., & Minaeifar, A. A. (2022). SCoT molecular markers and genetic diversity of *Stellaria media* (L.) Vill. *Cytologia*, *87*(1), 23–28. https://doi.org/10.1508/CYTOLOGIA.87.23

Mamgain, J., Mujib, A., Ejaz, B., Gulzar, B., Malik, M. Q., & Syeed, R. (2022). Flow cytometry and start codon targeted (SCoT) genetic fidelity assessment of regenerated plantlets in Tylophora indica (Burm. f.) Merrill. *Plant Cell, Tissue and Organ Culture*, *150*(1), 129–140.

Manikandan, R., Balasubramanian, B., Punniyakotti, P., Vijaya Anand, A., Meyyazhagan, A., Velauthaprabhu, S., & Liu, W. C. (2022). Cardio-protective effects of *Terminalia catappa* leaves and *Terminalia chebula* fruit exstract in doxorubicin-inuced cardiomyopathy in rats. *Biomarkers*, *27*(5), 488–495.

Manoharan, N., Sampathkumar, P., Dheeba, B., Sheik Abdulla, S., Vinothprasanna, G., Vinothkannan, R., Kalavathy, S., Vijaya Anand, A., & Shanmugasundaram, A. (2008). Potential hepatoprotective effect of aqueous exract of *Gracilaria corticata* in AFB_1 induced hepatotoxicity in WIstar rats. *Journal of Biological Sciences*, *8*, 1352–1355.

Mizukami, H., Ohbayashi, K., Kitamura, Y., & Ikenaga, T. (1993a). Restriction fragment length polymorphisms (RFLPs) of medicinal plants and crude drugs. I. RFLP probes allow clear identification of *Duboisia* interspecific hybrid genotypes in both fresh and dried tissues. *Biological and Pharmaceutical Bulletin*, *16*(4), 388–390.

Mizukami, H., Ohbayashi, K., Umetsu, K., & Hiraoka, N. (1993b). Restriction fragment length polymorphisms of medicinal plants and crude drugs. II. Analysis of *Glehnia littoralis* of different geographical origin. *Biological and Pharmaceutical Bulletin*, *16*(6), 611–612.

Morgil, H., Gercek, Y. C., & Tulum, I. (2016). *Single nucleotide polymorphisms (SNPs) in plant genetics and breeding*. Intech. https://www.intechopen.com/books/advanced-biometric-technologies/liveness-detection-in-biometrics

Mughees, M., Jhab, S. R., Ahmad, J., & Ahmad, A. (2017). SCAR marker development for the correct identification of *Iris ensata*. *International Journal of Pharmacy and Pharmaceutical Sciences*, *9*(12), 201–205.

Nadeem, M. A., Nawaz, M. A., Shahid, M. Q., Doğan, Y., Comertpay, G., Yıldız, M., Hatipoğlu, R., Ahmad, F., Alsaleh, A., Labhane, N., & Özkan, H. (2018). DNA molecular markers in plant breeding: Current status and recent advancements in genomic selection and genome editing. *Biotechnology and Biotechnological Equipment*, *32*(2), 261–285. https://doi.org/10.1080/13102818.2017.1400401

Nakamura, S. S., do Valle, J. S., Jacomassi, E., Linde, G. A., & Colauto, N. B. (2013). Molecular authentication of Maytenus sp. by PCR-RFLP. *Semina: Ciências Agrárias*, *34*(2), 627–633.

Nazarian-Firouzabadi, F., Ismaili, A., & Zabeti, S. M. (2014). Phenol-stacked carbon nanotubes: A new approach to genomic DNA isolation from plants. *Molecular Biology Research Communications*, *3*(3), 205–213.

Negi, M. S., Singh, A., & Lakshmikumaran, M. (2000). Genetic variation and relationship among and within Withania species as revealed by AFLP markers. *Genome*, *43*(6), 975–980. https://doi.org/10.1139/g00-073

Osman, M. A., Dhawan, S. S., Bahl, J. R., Darokar, M. P., & Khanuja, S. P. S. (2011). AFLP marking and polymorphism among progenies of *Gymnema sylvestre*: An important medicinal plant of India. *Natural Product Communications*, *6*(11), 1679–1682. https://doi.org/10.1177/1934578x1100601129

Otto, L. G., Mondal, P., Brassac, J., Preiss, S., Degenhardt, J., He, S., Reif, J. C., & Sharbel, T. F. (2017). Use of genotyping-by-sequencing to determine the genetic structure in the medicinal plant chamomile, and to identify flowering time and alpha-bisabolol associated SNP-loci by genome-wide association mapping. *BMC Genomics*, *18*(1), 1–18. https://doi.org/10.1186/s12864-017-3991-0

Ozyigit, I. I., Dogan, I., Hocaoglu-Ozyigit, A., Yalcin, B., Erdogan, A., Yalcin, I. E., Cabi, E., & Kaya, Y. (2023). Production of secondary metabolites using tissue culture-based biotechnological applications. *Frontiers in Plant Science*, *14*, 1–28. https://doi.org/10.3389/fpls.2023.1132555

Panapitiya, P. D. D. M., & Welikala, N. (2018). Development of RAPD, DAMD and ISSR markers for authentication of medicinal plant *Cassia auriculata* and its adulterant *Cassia surattensis*. *International Journal of Environment, Agriculture and Biotechnology*, *3*(5), 266206. http://dx.doi.org/10.22161/ijeab/3.5.43

Paramanik, R. C., & Chikkaswamy, B. K. (2014). Study on genetic diversity relationship some medicinal plants using RAPD molecular marker. *International Journal of Current Microbiology and Applied Sciences*, *3*(6), 10041015.

Passinho-Soares, H., Felix, D., Kaplan, M. A., Margis-Pinheiro, M., & Margis, R. (2006). Authentication of medicinal plant botanical identity by amplified fragmented length polymorphism dominant DNA marker: Inferences from the Plectranthus genus. *Planta Medica*, *72*(10), 929–931.

Patel, H. K., Fougat, R. S., Kumar, S., Mistry, J. G., & Kumar, M. (2015). Detection of genetic variation in Ocimum species using RAPD and ISSR markers. *3 Biotech*, *5*, 697–707.

Phoolcharoen, W., & Sukrong, S. (2013). Molecular analysis of Vitex species using candidate DNA barcoding and PCR-RFLP of the matK gene for authentication of *Vitex glabrata*. *Natural Product Communications*, *8*(1), 125–128. https://doi.org/10.1177/1934578x1300800130

Pommerrenig, B., Barth, I., Niedermeier, M., Kapp, S., Schmid, J., Dwyer, R. A., & Sauer, N. (2006). Common plantain. A collection of expressed sequence tags from vascular tissue and a simple and efficient transformation method. *Plant Physiology*, *142*(4), 1427–1441. https://doi.org/10.1104/pp.106.089169

Pourmohammad, A. (2013). Application of molecular markers in medicinal plant studies. *Acta Universitatis Sapientiae, Agriculture and Environment*, *5*(1), 80–90. https://doi.org/10.2478/ausae-2014-0006

Prasad, M. P. (2014). Molecular characterization of medicinal plant species of Acathaceae family using rapd molecular markers. *International Journal of Scientific Research*, *3*(9), 505–507.

Rai, M. K. (2023). Start codon targeted (SCoT) polymorphism marker in plant genome analysis: Current status and prospects. *Planta*, *257*, 34.

Rajasekharan, P. E., Kareem, V. A., Ravish, B. S., & Mini, S. (2017). Genetic diversity in *Oroxylum indicum* (L.) Vent., a threatened medicinal plants from India by ISSR analysis. *Indian Journal of Biotechnology*, *16*(1), 357–363.

Rohela, G. K., Bylla, P., Pendli, S., Rajender, K., Dharavath, B., & Thammidala, C. R. (2018). ISSR marker based DNA profiling studies in Rauwolfia species. *Annals of Plant Sciences*, *7*(5), 2289. http://dx.doi.org/10.21746/aps.2018.7.5.13

Russil, L., Moretti, C., Raggi, L., Albertini, E., & Falistocco, E. (2009). Identifying commercially relevant Echinacea species by AFLP molecular markers. *Genome*, *52*(11), 912–918.

Sahu, J., Sen, P., Choudhury, M. D., Dehury, B., Barooah, M., Modi, M. K., & Talukdar, A. D. (2014). Rediscovering medicinal plants' potential with OMICS: Microsatellite survey in expressed sequence tags of eleven traditional plants with potent antidiabetic properties. *OMICS: A Journal of Integrative Biology*, *18*(5), 298–309.

Sairkar, P. K., Sharma, A., & Shukla, N. P. (2016). SCAR marker for identification and discrimination of *Commiphora wightii* and *C. myrrha*. *Molecular Biology International*, *2016*(1), 1482796.

Salah, S. A., Masoad, E. M., El-bakry, A., & Ibrahim, S. A. (2018). Genetic assessment in some medicinal plants utilizing molecular markers. *Arab Universities Journal of Agricultural Sciences*, *26*(5), 2003–2009. https://doi.org/10.21608/ajs.2018.31667

Sangeetha, T., Balamuralikrishnan, B., Sampathkumar, P., Velayuthaprabhu, S., Senthilkumar, N., Syed Ibrahim, K., Baskaran, R., & Vijaya Anand, A. (2022). Characterization and phytoconstituents of *Petroselinum crispum* (Mill) and *Coriandrum sativum* (Linn) and their impacts on inflammation – an in vitro analysis against human adenocarcinoma cells with molecular docking. *South African Journal of Botany*, *146*, 776–788.

Sarin, B., Clemente, J. P. M., & Mohanty, A. (2013). PCR-RFLp to distinguish three *phyllanthus sp.*, commonly used in herbal medicines. *South African Journal of Botany*, *88*, 455–458. https://doi.org/10.1016/j.sajb.2013.09.011

Sarwat, M., Srivastava, S., & Khan, T. H. (2016a). RAPD and ISSR polymorphism in the medicinal plants: *Ocimum sanctum*, *O.basilicum* and *O.gratissimum*. *International Journal of Pharmacognosy and Phytochemical Research*, *8*(8), 1417–1424.

Sarwat, M., Srivastava, S., & Naved, T. (2016b). Diversity analyses in Ocimum species: Why and how. *International Journal of Pharmacognosy and Phytochemical Research*, *8*(2), 252–262.

Savelkoul, P. H. M., Aarts, H. J. M., De Haas, J., Dijkshoorn, L., Duim, B., Otsen, M., Rademaker, J. L. W., Schouls, L., &. Lenstra, J. A. (1999). Amplified-fragment length polymorphism analysis: The state of an art. *Journal of Clinical Microbiology*, *37*(10), 3083–3091. https://doi.org/10.1128/jcm.37.10.3083-3091.1999

Shakoor, A., Zaib, G., Zhao, F., Li, W. Y., Lan, X. C., & Esfandani-Bozchaloyi, S. (2022). ISSR markers and morphometry determine genetic diversity and population structure in *Hedera helix* L. *Czech Journal of Genetics and Plant Breeding*, 58(2), 73–82.

Shi, Y., Zhao, M., Yao, H., Yang, P., Xin, T., Li, B., Sun, W., & Chen, S. (2017). Rapidly discriminate commercial medicinal *Pulsatilla chinensis* (Bge.) Regel from its adulterants using ITS2 barcoding and specific PCR-RFLP assay. *Scientific Reports*, *7*(1), 40000.

Shivashankar, M. (2014). Random amplified polymorphic DNA (RAPD) markers in anticancer drug plants. *International Journal of Current Microbiology and Applied Sciences*, *3*(7), 1091–1101.

Srivastava, S., & Mishra, N. (2009). Genetic markers – a cutting-edge technology in herbal drug research. *Journal of Chemical and Pharmaceutical Research*, *1*(1), 1–18.

Subositi, D., Kurnianingrum, N., Mujahid, R., & Widiyastuti, Y. (2020). *Kaempferia galanga* L. A medicinal plant used by Indonesian ethnic groups: Genetic diversity based on inter-simple sequence repeats (ISSR). *AGRIVITA Journal of Agricultural Science*, *42*(1), 45–52.

Thakur, J., Dwivedi, M. D., Singh, N., Uniyal, P. L., Goel, S., & Pandey, A. K. (2021). Applicability of Start Codon Targeted (SCoT) and Inter Simple Sequence Repeat (ISSR) markers in assessing genetic diversity in *Crepidium acuminatum* (D. Don) Szlach. *Journal of Applied Research on Medicinal and Aromatic Plants*, *23*, 100310.

Thakur, J., Dwivedi, M. D., Sourabh, P., Uniyal, P. L., & Pandey, A. K. (2016). Genetic homogeneity revealed using SCoT, ISSR and RAPD markers in micropropagated *Pittosporum eriocarpum* Royle – an endemic and endangered medicinal plant. *PLoS ONE*, *11*(7), 1–17. https://doi.org/10.1371/journal.pone.0159050

Theerakulpisut, P., Kanawapee, N., Maensiri, D., Bunnag, S., & Chantaranothai, P. (2008). Development of species-specific SCAR markers for identification of three medicinal species of Phyllanthus. *Journal of Systematics and Evolution*, *46*(4), 614–621. https://doi.org/10.3724/SP.J.1002.2008.07123

Tripathi, N., Saini, N., & Tiwari, S. (2011). Assessment of genetic diversity among Aloe vera accessions using amplified fragment length polymorphism. *International Journal of Medicinal and Aromatic Plants*, *1*, 115–121.

Varalakshmi, B., Anand, A. V., Karpagam, T., Bai, J. S., & Manikandan, R. (2014). In vitro antimicrobial and anticancer activity of *Cinnamomum zeylanicum* Linn bark extracts.

International Journal of Pharmacy and Pharmaceutical Sciences, *6*(1), 12–18.

Velasco-Ramírez, A. P., Torres-Morán, M. I., Molina-Moret, S., de Jesús Sánchez-González, J., & Santacruz-Ruvalcaba, F. (2014). Efficiency of RAPD, ISSR, AFLP and ISTR markers for the detection of polymorphisms and genetic relationships in camote de cerro (*Dioscorea spp.*). *Electronic Journal of Biotechnology*, *17*(2), 65–71. https://doi.org/10.1016/j.ejbt.2014.01.002

Venkatesan, J., Ramu, V., Sethuraman, T., Sivagnanam, C., & Doss, G. (2021). Molecular marker for characterization of traditional and hybrid derivatives of *Eleusine coracana* (L.) using ISSR marker. *Journal of Genetic Engineering and Biotechnology*, *19*(1), 178.

Vijaya Anand, A., Varalakshmi, B., Prasana, Sampathkumar, P., Pushpa, N., & Hedina, A. (2016). *Cinnamomum zeylanicum* Linn. The spice with multi potential. *Systematic Reviews in Pharmacy*, *7*(1), 24–29.

Vijaya Anand, A., Velayuthaprabhu, S., Rengarajan, R. L., Sampathkumar, P., & Radhakrishnan, R. (2020). Bioactive compounds of guava (*Psidium guajava* L.). In *Bioactive compounds in underutilized fruits and nuts*, Reference Series in Phytochemistry (pp. 503–527). Springer.

Vijayakumar, K., Prasanna, B., Rengarajan, R. L., Rathinam, A., Velayuthaprabhu, S., & Vijaya Anand, A. (2023). Antidiabetic and hypolipidemic effects of *Cinnamon cassia* bark extracts: An in vitro, in vivo, and in silico approach. *Archieves of Physiology and Biochemistry*, *129*(2), 338–348.

Wang, B., Chen, Z., Hua, Z., Tian, H., Xie, W., & Yuan, Y. (2024). A Rapid PCR-RFLP method for assessing heterozygosity of *Murraya paniculata* germplasm. *Chinese Journal of Experimental Traditional Medical Formulae*, *24*, 29–34.

Williams, J. G. K., Kubelik, A. R., Livak, K. J., Rafalski, J. A., & Tingey, S. V. (1990). DNA polymorphisms amplified by arbitrary primers are useful as genetic markers. *Nucleic Acids Research*, *18*(22), 6531–6535. https://doi.org/10.1093/nar/18.22.6531

Wu, K., Liu, Y., Yang, B., Kung, Y., Chang, K., & Lee, M. (2022). Rapid discrimination of the native medicinal plant *Adenostemma lavenia* from its adulterants using PCR-RFLP. *PeerJ*, *10*. https://doi.org/10.7717/peerj.13924

Xu, M. R., Yang, B. C., Chang, H. C., Kuo, C. L., Lin, C. H., Chen, H. J., Cheng J. H., & Lee, M. S. (2022). Molecular authentication of the medicinal crop *Portulaca oleracea* and discrimination from its adulterants in herbal markets using PCR-restriction fragment length polymorphism (PCR-RFLP) analysis. *Industrial Crops and Products*, *183*, 114934. https://doi.org/10.1016/j.indcrop.2022.114934

Yang, L., Khan, M. A., Mei, Z., Yang, M., Zhang, T., Wei, C., Yang, W., Zhu, L., Long, Y., & Fu, J. (2014). Development of RAPD-SCAR markers for *Lonicera japonica* (Caprifoliaceae) variety authentication by improved RAPD and DNA cloning. *Revista de Biología Tropical*, *62*(4), 1649–1657.

Yousefi, V., Najaphy, A., Zebarjadi, A., & Safari, H. (2015). Molecular characterization of Thymus species using ISSR markers. *JAPS: Journal of Animal & Plant Sciences*, *25*(4).

Zeng, S., Xiao, G., Guo, J., Fei, Z., Xu, Y., Roe, B. A., & Wang, Y. (2010). Development of a EST dataset and characterization of EST-SSRs in a traditional Chinese medicinal plant, *Epimedium sagittatum* (Sieb. Et Zucc.) Maxim. *BMC Genomics*, *11*(1). https://doi.org/10.1186/1471-2164-11-94

Zhang, C., Mei, Z., Cheng, J., He, Y., Khan, M. A., Luo, P., Imani, S., & Fu, J. (2015). Development of SCAR markers based on improved RAPD amplification fragments and molecular cloning for authentication of herbal medicines *Angelica sinensis*, *Angelica acutiloba* and *Levisticum officinale*. *Natural Product Communications*, *10*(10), 1743–1747. https://doi.org/10.1177/1934578x1501001027

Zhao, J., Xu, Y., Li, H., Yin, Y., An, W., Li, Y., Wang, Y., Fan, Y., Wan, R., Guo, X., & Cao, Y. (2019). A SNP-based high-density genetic map of leaf and fruit related quantitative trait loci in wolfberry (Lycium Linn.). *Frontiers in Plant Science*, *10*, 977.

Zhao, Y., Liu, G., Yang, F., Liang, Y., Gao, Q., Xiang, C., Li, X., Yang, R., Zhang, G., Jiang, H., & Yu, L. (2023). Multilayered regulation of secondary metabolism in medicinal plants. *Molecular Horticulture*, *3*(11).

Zhao, Y. P., Qiu, Y. X., Gong, W., Li, J. H., & Fu, C. X. (2007). Authentication of *Actinidia macrosperma* using PCR-RFLP based on trnK sequences. *Botanical Studies*, *48*(3), 239–242.

14 Bioinformatics Tools for Research on Plant Secondary Metabolites

Arokiaraj Sherlin Rosita, Jambulingam Shivani, Aarsha Saji, Muthukumar Bala Sundar, Tajudheen Nargis Begum, Gangadharan Thilagavathi, Alagarsamy Sumithra, and Arumugam Vijaya Anand

LIST OF ABBREVIATIONS

ADMET	absorption, distribution, metabolism, excretion, and toxicity
AI	artificial intelligence
antiSMASH	antibiotics and secondary metabolite analysis shell
BFE	binding free energy
BGCs	biosynthetic gene clusters
BLAST	Basic Local Alignment Search Tool
CDK 2	cyclin-dependent kinase-2
DL	deep learning
DNNs	deep neural networks
FLEX	flexibility
GANs	generative adversarial networks
GC-MS	gas chromatography–mass spectrometry
INSATU	insaturation
INSOLU	insolubility
KEGG	Kyoto Encyclopaedia of Genes and Genomes
LIPO	lipophilicity
ML	machine learning
MPA	metabolic pathway analysis
NLP	natural language processing
PDB	Protein Data Bank
POLAR	polarity
PRISM	Prediction Informatics for Secondary Metabolomes
PyRx	Python Prescription
QSAR	quantitative structure-activity relationship
RNA-seq	RNA sequencing
SAR	structure-activity relationship
SIB	Swiss Institute of Bioinformatics
TCMSP	Traditional Chinese Medicine Systems Pharmacology Database and Analysis Platform,
VAEs	variational autoencoders
WGCNA	weighted gene co-expression network analysis

14.1 INTRODUCTION

Plants are not passive organisms. They actively respond to their environment, defending themselves against predators, pathogens, and environmental stresses while communicating with other organisms and attracting pollinators. One of the fascinating ways plants achieve these functions is by producing secondary metabolites (Crozier et al., 2006). Unlike primary metabolites, which are necessary for basic life processes, secondary metabolites are produced by plants primarily for ecological interactions.

Secondary metabolites are a diverse array of organic compounds plants synthesize through complex biochemical pathways (Crozier et al., 2006). They are not directly concerned with growth, development, or reproduction but serve specialized functions, such as defense, attraction, and signaling. These compounds have garnered immense interest due to their pharmacological, agricultural, and ecological significance (Manoharan et al., 2008; Begum et al., 2009; Muthumani et al., 2013; Varalakshmi et al., 2014; Vijaya Anand et al., 2016, 2020; Bharathi et al., 2018; Divya et al., 2019; Akilandeswari et al., 2019; Sangeetha et al., 2022; Kaviya et al., 2022; Manikandan et al., 2022; Karthika et al., 2023; Vijayakumar et al., 2023). The formation of secondary metabolites permits plants to adapt and thrive in diverse ecological niches. In response to herbivory, for instance, plants can synthesize toxins or deterrents to deter feeding by herbivores. Secondary metabolites also play crucial roles in defense against pathogens, protecting plants from microbial infections (Crozier et al., 2006). Additionally, these compounds can act as allelochemicals, influencing the growth and behavior of neighboring plants and organisms.

Beyond defense, secondary metabolites contribute to the rich tapestry of colors, scents, and flavors observed

DOI: 10.1201/9781003475491-14

in the plant kingdom (Anulika et al., 2016). They are reliable for the vibrant colors of flowers, fruits, and leaves, attracting pollinators and seed dispersers. Furthermore, many secondary metabolites possess medicinal properties and have been utilized by humans for centuries in traditional medicine systems. The chemical diversity of plant secondary metabolites is staggering, with thousands of compounds identified to date. They encompass various chemical classes, including phenolics, alkaloids, terpenoids, glycosides, and many others, each with unique structures and bioactivities (Anulika et al., 2016). Advances in analytical techniques and genomic studies have facilitated the discovery and elucidation of biosynthetic pathways for these compounds.

Understanding plant secondary metabolites is not only crucial for unraveling the intricacies of plant biology but also holds promise for applications in agriculture, medicine, and industry. Harnessing the potential of these compounds could lead to the development of novel drugs, eco-friendly pesticides, and flavorsome food additives. Moreover, studying plant secondary metabolites provides insights into ecological interactions, biodiversity, and evolutionary adaptations in the natural world. In this chapter, we provide an overview of recent aspects of tools and databases used for plant secondary metabolites prediction for plant metabolomics analysis and limitations of current tools and report results based on the tools described toward plant secondary metabolite investigation. We have highlighted the emergence of artificial intelligence (AI) in secondary metabolite prediction.

14.1.1 Significance of Secondary Metabolites Isolated from Plants

The importance of plant secondary metabolites spans a broad spectrum of ecological, agricultural, medicinal, and industrial applications. Here are some key significances of the plant's secondary metabolites:

Firstly, secondary metabolites serve as a frontline defense for plants against herbivores, pathogens, and competing plants. These compounds deter feeding by herbivores through their toxicity or repellent properties or by interfering with digestive processes (Bennett & Wallsgrove, 1994). They also exhibit antimicrobial properties, protecting plants from fungal, bacterial, and viral infections.

Secondary metabolites play pivotal roles in mediating ecological interactions within plant communities and between plants and other organisms (Crozier et al., 2006). They can influence the growth, behavior, and interactions of neighboring plants, herbivores, pollinators, and seed dispersers. Allelopathic compounds released by plants can inhibit the germination or growth of nearby competitors, shaping plant community composition.

Many secondary metabolites, such as floral pigments and fragrances, attract pollinators like bees, butterflies, and birds (Anulika et al., 2016). These compounds enhance the reproductive success of plants by facilitating pollination and seed dispersal, ultimately contributing to ecosystem biodiversity and stability (Wink, 1988).

Plant secondary metabolites have been a profitable source of medicinal compounds for human health and well-being. Numerous drugs and therapeutics, including antibiotics, anticancer agents, pain relievers, and cardiovascular medications, have been derived from plant secondary metabolites (Badyal et al., 2020). These compounds often possess potent biological activities and therapeutic potentials. These metabolites also contribute to the sensory qualities, nutritional value, and culinary appeal of fruits, vegetables, herbs, and spices. They impart distinctive flavors, colors, and aromas to foods, enhancing their palatability and consumer appeal (Anulika et al., 2016; Badyal et al., 2020). Additionally, some secondary metabolites have antioxidant properties and may confer health benefits when ingested as part of a balanced diet.

They have diverse industrial applications, including the production of natural dyes, fragrances, flavors, and bioactive compounds. They are utilized in cosmetics, perfumes, food additives, and herbal supplements (Anulika et al., 2016). Additionally, secondary metabolites can be used as biopesticides and bioherbicides, offering eco-friendly alternatives to synthetic chemicals in agriculture. Finally, advances in biotechnology have enabled the manipulation and production of plant secondary metabolites through metabolic engineering, synthetic biology, and bioreactor technology (Sabitha Rani et al., 2018). These techniques hold promise for the sustainable production of valuable compounds with pharmaceutical, agricultural, and industrial applications.

Therefore, the main roles of plant secondary metabolites include the defense against pathogens and herbivores. These act as allelochemicals and have antioxidant properties. These protect the plants from UV radiation and thus act as natural sunscreens for plants (Korkina et al., 2018). They attract pollinators and seed dispersers. Secondary metabolites regulate the growth and development of plants and the survival of plants in extreme conditions. Plants use secondary metabolites as signaling molecules to communicate with other organisms in their environment (Reichling, 2010). For example, volatile organic compounds emitted by damaged plants can alert neighboring plants to impending threats, triggering defense responses and priming their immune systems.

Overall, secondary metabolites are integral to plant biology, providing plants with essential tools for adaptation, defense, communication, and reproduction in their ever-changing environments. Their diverse chemical structures and biological activities underscore their importance in shaping plant ecology, evolution, and interactions with other organisms in the ecosystem. This chapter follows with the work in regard to the plant's secondary metabolites. The identification and description of the plant's secondary metabolites is done using various databases such as PubChem. The target protein is selected using the available protein databases, such as Protein Data Bank (PDB). The finalized compounds and the target protein are docked using the AutoDock software.

14.2 STUDY OF PLANTS SECONDARY METABOLOME USING EMERGING TOOLS AND TECHNIQUES *IN SILICO*

14.2.1 Multi-Omics-Based Approaches to Study Plant Secondary Metabolites

Omics techniques refer to high-throughput approaches that analyze large sets of biological molecules or data (Dai & Shen, 2022). When studying secondary metabolites, which are organic compounds not directly concerned with the normal growth and development (Palazzotto & Weber, 2018) or reproduction of the plant, various omics techniques can be employed to understand their biosynthesis, regulation, and function (Rai et al., 2017). Here are some commonly used omics tools in the study of plant secondary metabolites:

14.2.1.1 Basic Local Alignment Search Tool (BLAST)

BLAST (https://blast.ncbi.nlm.nih.gov/Blast.cgi) is used for sequence similarity searching, which helps in identifying homologous sequences in databases (Hautbergue et al., 2018). It is useful for comparing plant secondary metabolite biosynthetic genes with known sequences.

14.2.1.2 Kyoto Encyclopaedia of Genes and Genomes (KEGG)

KEGG (https://www.genome.jp/kegg/) provides a complete resource for understanding metabolic pathways, including those involved in the biosynthesis of secondary metabolites. It allows for the analysis and visualization of metabolic pathways (Kanehisa, 2016).

14.2.1.3 PlantCyc

PlantCyc (https://www.plantcyc.org/) is a collection of curated databases containing metabolic pathways and enzymes from various plant species (Zhang et al., 2010). It can be used to explore the biosynthesis of secondary metabolites and their associated genes.

14.2.1.4 Pathway Tools or Metabolic Pathway Analysis (MPA)

It's a software that enables the reconstruction and analysis of metabolic pathways involved in secondary metabolite biosynthesis (Jaiswal & Usadel, 2016). They facilitate the prediction of gene functions and metabolic fluxes.

14.2.1.5 Gene Ontology Enrichment Analysis

This helps in identifying characterized biological and molecular functions (Zheng & Wang, 2008) as well as cellular components related to genes involved in secondary metabolite biosynthesis.

Tools like **MAKER** and **Augustus** are used for genome annotation, which involves identifying genes, regulatory elements, and other genomic features relevant to secondary metabolite biosynthesis (Inglis et al., 2013). Phylogenetic analysis tools such as **Phylogenetic Tree Viewer or MEGA** (https://www.megasoftware.net/) (Ziemert et al., 2012) enable the reconstruction of evolutionary relationships among genes involved in secondary metabolite biosynthesis, aiding in understanding their functional diversification. These tools, when used in combination, enable researchers to gain a thorough knowledge of the biosynthesis, regulation, and ecological functions of plant secondary metabolites.

14.2.2 Toxicological Evaluation of Metabolites Using ADMET (Absorption, Distribution, Metabolism, Excretion, and Toxicity) Profile Analysis

ADMET (absorption, distribution, metabolism, excretion, and toxicity) (Anza et al., 2021) profile analysis is a crucial aspect of drug finding and development, but it's also increasingly applied to the study of plant secondary metabolites. Here's how ADMET profile analysis can be utilized in this context: Understanding how secondary metabolites are absorbed in the human body or in other organisms is crucial for determining their bioavailability and potential efficacy (Acamovic & Brooker, 2005). Computational methods can predict parameters such as intestinal permeability and blood-brain barrier permeability, providing insights into the likelihood of a secondary metabolite reaching its target site. Secondary metabolites must be distributed to their target tissues or organs to exert their biological effects (Pagare et al., 2015). ADMET analysis can predict parameters, such as plasma protein binding and tissue distribution, helping researchers understand how secondary metabolites are distributed within the body and their potential for accumulation in specific tissues. Secondary metabolites are often metabolized by enzymes in the body, which can affect their bioactivity, toxicity, and duration of action (Jain et al., 2019).

Predicting metabolic pathways and identifying potential metabolites through computational methods can give a useful insight into the biotransformation of secondary metabolites and help anticipate potential drug-drug interactions or adverse effects. Understanding the routes of excretion for secondary metabolites is important for determining their elimination half-life and potential for accumulation in the body. ADMET analysis can predict parameters such as renal clearance and biliary excretion, aiding in the assessment of secondary metabolite pharmacokinetics and safety (Moroy et al., 2012). Assessing the potential toxicity of secondary metabolites is critical for ensuring their safety in pharmaceutical or agricultural applications (Gola et al., 2006). Computational toxicity prediction models can evaluate parameters such as mutagenicity, hepatotoxicity, and cardiotoxicity, providing early indications of potential safety concerns and guiding the selection of secondary metabolites with favorable ADMET profiles, several tools and databases are used in ADMET (Moroy et al., 2012) profile analysis of plant secondary metabolites. These tools aid in predicting various pharmacokinetic and pharmacodynamics characteristics of these compounds. The commonly used tools are SwissADME, ADMETlab, PharmaDB, ADMET Predictor, Toxtree, and TCMSP.

14.2.2.1 SwissADME

SwissADME (https://www.swissadme.ch/) is a web tool designed by the Swiss Institute of Bioinformatics (SIB) (Kamble & Mitkar, 2023) that predicts various pharmacokinetic characteristics ADMET of small molecules including plant secondary metabolites. It provides predictions for physicochemical characteristics, drug-likeness, and medicinal chemistry parameters.

14.2.2.2 ADMETlab

ADMETlab (https://admetmesh.scbdd.com/) is an integrated platform that offers multiple tools for predicting ADMET properties of small molecules (James et al., 2023). It includes modules for ADMET and various other pharmacokinetic and pharmacodynamics properties. ADMETlab provides both standalone software and web-based services.

14.2.2.3 PharmDB-K

PharmDB-K (http://pharmdb-k.org/) is a database containing curated information on the pharmacokinetic and pharmacodynamics properties of small molecules (Lee et al., 2015). It includes data on ADMET, and other related parameters. PharmDB-K can be utilized for data mining and analysis of ADMET profiles of plant secondary metabolites.

14.2.2.4 ADMET Predictor

ADMET Predictor (https://www.simulations-plus.com/software/admetpredictor/) is a software tool developed by Simulations Plus for predicting ADMET characteristics of small molecules. It integrates various algorithms and models to predict ADMET properties (Abdul-Hammed et al., 2022). ADMET Predictor can be used for virtual screening and profiling of plant secondary metabolites.

14.2.2.5 Toxtree

Toxtree (https://toxtree.sourceforge.net/) is an open-source software tool for the prediction of toxicological characteristics of chemical substances. It provides a platform for structure-activity relationship (SAR) investigation and prediction of toxicity endpoints (Nisa et al., 2023). Toxtree can be used to assess the potential toxicity of plant secondary metabolites based on their chemical structures.

14.2.2.6 Traditional Chinese Medicine Systems Pharmacology Database and Analysis Platform

TCMSP (https://www.tcmsp-e.com/#/home) is a resource that integrates data on traditional Chinese medicines and their pharmacological properties. It provides information on the ADMET of natural compounds including plant secondary metabolites (Liu et al., 2024). TCMSP can be utilized for exploring the pharmacological effects and ADMET profiles of plant-derived compounds (Ru et al., 2014). These tools and databases offer valuable resources for researchers to predict and analyze the ADMET profiles of plant secondary metabolites, aiding in drug discovery and development from natural sources.

This type of analysis involves data collection and curation, absorption analysis, distribution analysis, metabolism analysis, excretion analysis, toxicity analysis, integration and interpretation, and finally, validation. By incorporating ADMET profile analysis into the study of plant secondary metabolites, researchers can identify compounds with optimal pharmacokinetic properties and safety profiles for further development as pharmaceuticals, nutraceuticals, or agrichemicals. This approach can streamline the process of secondary metabolite discovery and optimization, ultimately accelerating the translation of plant-derived compounds into clinically or commercially valuable products.

14.2.3 Interaction Analysis of Secondary Metabolites with the Target Protein through Molecular Docking Analysis

Molecular docking is an effective computational approach used in bioinformatics and computational biology to study the interaction between small molecules, such as plant secondary metabolites, and target proteins (Illian et al., 2022). In the context of studying secondary metabolites of plants, molecular docking can provide insights into their potential binding affinity and way of interchange with target proteins involved in various biological processes, including ADMET (Doraiswamy et al., 2014) profile analysis. Some of the commonly used bioinformatics tools for molecular docking studies of plant secondary metabolites are AutoDock, DOCK, PyRx, and SwissDock.

14.2.3.1 AutoDock

AutoDock (https://autodock.scripps.edu/) is a widely used software for molecular docking studies. It allows for the prediction of ligand-receptor interactions and the estimation of binding affinities (Park et al., 2006) AutoDock Vina, an improved version of AutoDock, offers enhanced efficiency and accuracy in docking calculations.

14.2.3.2 DOCK

DOCK (https://dock.compbio.ucsf.edu/) is a suite of programs designed for molecular docking and virtual screening (Moustakas et al., 2006). It includes various algorithms and scoring functions for predicting ligand-protein interactions. DOCK provides flexibility in the selection of quest algorithms and scoring functions, allowing researchers to tailor docking studies to their specific needs.

14.2.3.3 PyRx

Python Prescription (https://pyrx.sourceforge.io/) is a user-friendly virtual screening software based on AutoDock, AutoDock Vina, and other popular docking tools. It provides a graphical interface for setting up, running, and analyzing docking experiments (Yuliana et al., 2013). PyRx simplifies the process of performing molecular docking studies and allows researchers to visualize and interpret docking results efficiently.

14.2.3.4 SwissDock

SwissDock (https://www.swissdock.ch/) is an online molecular docking service provided by the SIB. It offers a user-friendly web interface for conducting docking simulations with small molecules and protein targets (Lohning et al., 2017). SwissDock incorporates advanced algorithms for ligand docking and scoring, making it suitable for studying plant secondary metabolites and their relations with target proteins.

These tools can be used to investigate the interactions between plant secondary metabolites and target proteins involved in various biological processes, supplying useful insights into their possible pharmacological activities and medicinal applications. Additionally, combining molecular docking with other computational techniques, like molecular dynamics simulations and structure-based drug design, can further enhance our comprehension of the SAR of plant secondary metabolites.

14.3 THE EMERGENCE OF ARTIFICIAL INTELLIGENCE (AI) IN SECONDARY METABOLITE PREDICTION

AI techniques are increasingly being employed in the study of plant secondary metabolites due to their capability to analyse large datasets, predict molecular properties, and specify potential bioactive compounds. Machine learning (ML) algorithms, including random forests, and neural networks are applied to analyze molecular descriptors and biological data related to plant secondary metabolites (Moore et al., 2019). ML models can be introduced to predict various properties of secondary metabolites, including bioactivity, toxicity, and ADMET properties, based on their chemical structures and other features. Deep Learning (DL) (Vasseur et al., 2022) techniques, particularly deep neural networks (DNNs), have revealed assurance in learning complex patterns from molecular data. DL models can be used for tasks such as compound classification, virtual screening, and *de novo* drug design, leveraging their ability to capture intricate associations between chemical structures and biological actions (Yoo et al., 2020).

Quantitative Structure-Activity Relationship (QSAR) models utilize AI algorithms to show quantitative relationships between chemical structures and biological activities. By training QSAR models on datasets of plant secondary metabolites and their bioactivity profiles, researchers can predict the biological effects of new compounds and prioritize candidates for experimental validation (Rastija et al., 2021). AI-driven virtual screening methods involve the computational screening of large compound libraries to determine the possible lead compounds with desired bioactivities. Techniques such as ligand and structure -based virtual screening can be enhanced using AI algorithms to improve hit identification and reduce false positives (Rathore & Shakya, 2024). Natural Language Processing (NLP) techniques are employed to drag useful data from the scientific literature, patents, and databases related to plant secondary metabolites. By analyzing textual data using NLP algorithms, researchers can identify novel compounds, elucidate their biological activities, and discover new relationships between metabolites and biological targets (Liu et al., 2015).

Generative models, like generative adversarial networks (GANs) and variational autoencoders (VAEs), can be utilized for *de novo* molecular design (Samanta et al., 2020). These models learn the underlying distribution of chemical structures from training data and generate novel compounds with desired properties, offering a potential avenue for drug discovery and optimization of plant secondary metabolites. AI techniques are increasingly employed to integrate multi-omics data, including genomics, transcriptomics, metabolomics, and proteomics, to gain a holistic experience of plant metabolism and secondary metabolite biosynthesis pathways.

AI and bioinformatics tools can be employed in various steps to study plant secondary metabolites. AI algorithms can be used to automate the collection and curation of data from diverse sources such as scientific literature, databases (e.g., PubChem, ChEMBL, KEGG), and experimental datasets. NLP techniques can assist in extracting relevant information from text-based sources. AI techniques like ML (Cuperlovic-Culf et al., 2016) and DL can be applied to data pre-processing tasks, including normalization, feature extraction, and dimensionality reduction. For instance, DL models such as autoencoders can learn latent representations of chemical structures from molecular fingerprints or descriptors. AI-based SAR modeling techniques, such as QSAR models (Kar & Roy, 2012), can be utilized to predict the relationship between chemical structure and biological action of plant secondary metabolites.

ML algorithms such as random forests, support vector machines, and neural networks are commonly utilized for QSAR modeling. AI algorithms are used to perform molecular docking simulations and predict the binding affinity and way of interchange between plant secondary metabolites and target proteins. Docking algorithms may be augmented with ML approaches for scoring and ranking ligand-protein complexes more accurately. AI models, such as ML classifiers and regression algorithms, are employed to predict the ADMET (Siramshetty et al., 2023) properties of plant secondary metabolites. These models can predict various pharmacokinetic and toxicological properties based on molecular descriptors and chemical features. AI-driven virtual screening approaches utilize ML models to prioritize candidate compounds from large chemical libraries for experimental validation (Gentile et al., 2022).

Techniques like similarity-based screening, pharmacophore modeling, and ligand-based virtual screening can be enhanced using AI algorithms for more accurate prediction of bioactive compounds. AI techniques, particularly network-based analysis and pathway modeling, can help in elucidating the biological pathways and mechanisms associated with plant secondary metabolites. Integration of omics data (e.g., transcriptomics, metabolomics) (Flores et al., 2023) with AI algorithms enables the identification of key metabolic pathways and regulatory networks influenced by secondary metabolites. AI plays a crucial role in predictive modeling and rational drug design by generating hypotheses for designing novel plant secondary metabolites or optimizing existing ones (Singh et al., 2023). Generative models, reinforcement learning, and evolutionary algorithms can be employed to explore chemical space and design molecules with desired biological activities and ADMET profiles.

By integrating AI into these steps, bioinformatics tools can accelerate the discovery and characterization of secondary metabolites of plants, facilitate the understanding of their natural activities, and aid in the development of potential pharmaceuticals or agricultural products.

14.4 ADVERSE EFFECTS OF PLANT-DERIVED FUNCTIONAL METABOLITES

Secondary plant metabolites, also known as phytochemicals, have significant impacts on human health due to their diverse biological activities (Compean & Ynalvez, 2014). These compounds are synthesized by plants to defend against environmental stresses, attract pollinators, and deter herbivores. When consumed by humans through dietary sources, secondary plant metabolites can exert various beneficial effects on health. Some of the key impacts of secondary plant metabolites on human health are as follows:

14.4.1 Antioxidant Activity

Many secondary plant metabolites, like flavonoids, carotenoids, and phenolic acids, possess antioxidant properties (Lavecchia et al., 2013). They scavenge dangerous free radicals in the body, thereby reducing oxidative stress and safeguarding cells from injury. Antioxidant-rich diets have been related to the lower risk of chronic diseases.

14.4.2 Anti-inflammatory Effects

Some secondary plant metabolites exhibit anti-inflammatory action by interfering with pro-inflammatory enzymes and cytokines (Pushpangadan et al., 2015). By reducing inflammation, these compounds may help relieve symptoms associated with inflammatory diseases like asthma, arthritis, and inflammatory bowel disease.

14.4.3 Cardioprotective Effects

Certain secondary plant metabolites, such as flavonoids found in plants, such as tea, have been linked to a reduced risk of cardiovascular diseases (Shah et al., 2019). These compounds help improve vascular function, lower blood pressure, inhibit platelet aggregation, and reduce cholesterol levels, thereby promoting heart health.

14.4.4 Cancer Prevention

Several secondary plant metabolites have been shown to possess chemo-preventive properties by impeding the development and proliferation of cancer cells, inducing apoptosis, and suppressing tumor formation and progression (Korkina & Kostyuk, 2012). Examples include polyphenols, glucosinolates, and terpenoids found in various plant foods.

14.4.5 Immune Modulation

Certain secondary plant metabolites can modulate immune function by enhancing immune response or exerting immunosuppressive effects (Ali Reza et al., 2023). For example, polysaccharides from medicinal mushrooms have been shown to stimulate immune cells, while alkaloids from plants like goldenseal possess immunosuppressive activity and may be used to treat autoimmune diseases.

14.4.6 Neuroprotective Effects

Some secondary plant metabolites exhibit neuroprotective properties by promoting neuronal survival, enhancing cognitive function, and decreasing the threat of neurodegenerative diseases (Kumar & Khanum, 2012). Examples include flavonoids, polyphenols, and omega-3 fatty acids seen in specific plant-based foods.

14.4.7 Antimicrobial Activity

Secondary plant metabolites possess antimicrobial properties, making them effective against bacteria, fungi, viruses, and parasites (Compean & Ynalvez, 2014; Wallace, 2004). These compounds have been traditionally utilized in herbal medication to treat various infections and promote wound healing.

14.4.8 Antidiabetic Effects

Some secondary plant metabolites, particularly polyphenols and flavonoids, have been shown to improve insulin sensitivity, control blood glucose, and reduce the risk of diabetes (Sukhikh et al., 2023). These compounds may exert their effects by enhancing pancreatic function, inhibiting carbohydrate digestion, and promoting glucose uptake by cells.

Overall, secondary plant metabolites play a vital role in promoting human health and controlling chronic diseases. Incorporating a diverse range of phytochemical-rich foods into the diet is essential for reaping the health benefits associated with these bioactive compounds. Nevertheless, it's essential to note that the effects of secondary plant metabolites can vary depending on factors like dose, bioavailability, and individual susceptibility.

14.5 CURRENT APPLICATIONS OF BIOINFORMATICS TOOLS IN PLANT SECONDARY METABOLITE RESEARCH

Bioinformatics tools have been extensively employed in recent research to study plant secondary metabolites, offering insights into their biosynthesis, regulation, bioactivity, and potential applications. Here are some key applications of bioinformatics tools in this field:

14.5.1 Genome Mining and Gene Discovery

Bioinformatics tools enable the identification and annotation of genes concerned with the biosynthesis of secondary metabolites. Through genome mining and comparative genomics approaches, researchers can predict biosynthetic gene clusters (BGCs) (Guo & Wang, 2014) reliable for the formation of specific secondary metabolites in plants. Tools like antiSMASH (Antibiotics and Secondary Metabolite Analysis Shell) and PRISM (Prediction Informatics for Secondary Metabolomes) (Blin et al., 2019) aid in the identification and characterization of BGCs, facilitating the discovery of unexplored secondary metabolites and biosynthetic pathways.

14.5.2 Transcriptomics Analysis

Transcriptomic data analysis allows researchers to study gene expression patterns associated with secondary metabolite biosynthesis in plants (Guo et al., 2021). By using bioinformatics tools such as RNA sequencing (RNA-seq) and differential gene expression study, researchers can identify genes and transcription factors concerned with the regulation of secondary metabolism pathways. This approach provides insights into the dynamic changes in gene expression under different developmental stages, environmental conditions, or elicitor treatments.

14.5.3 Metabolomics Profiling

Metabolomics analysis involves the comprehensive profiling of small molecules, including secondary metabolites, in plant tissues (Cubero-Leon et al., 2018). Bioinformatics tools, such as XCMS, MetaboAnalyst, and MZmine, facilitate the processing, alignment, and statistical analysis of metabolomic data generated from methods including, liquid chromatography–mass spectrometry and gas chromatography–mass spectrometry (GC-MS) (Kivilompolo et al., 2007) Metabolomics profiling allows for the identification and quantification of secondary metabolites, as well as the elucidation of metabolic pathways and metabolic networks associated with plant secondary metabolism.

14.5.4 Network Analysis and Systems Biology

Network analysis approaches, such as pathway enrichment analysis and co-expression network analysis, are used to explore the complex interactions and regulatory networks underlying plant secondary metabolism (Hochberg et al., 2013). Bioinformatics tools such as Cytoscape and WGCNA (Weighted Gene Co-Expression Network Analysis) enable the visualization and analysis of gene regulatory networks, metabolic networks, and protein-protein interaction networks associated with secondary metabolite biosynthesis pathways (Saito & Matsuda, 2010). Systems biology approaches integrate multi-omics data to model and simulate the dynamic behavior of biological systems, delivering a holistic knowledge of plant secondary metabolism and its regulatory mechanisms.

Overall, the application of bioinformatics tools in studying plant secondary metabolites has facilitated the discovery of unknown compounds, the explanation of biosynthetic pathways, the prediction of bioactivities, and the understanding of their roles in plant physiology and human health (Raskin et al., 2002). These tools resume playing an important role in advancing our knowledge of plant secondary metabolism and harnessing the potential of plant-derived compounds for various applications in medicine, agriculture, and biotechnology.

14.6 BIOINFORMATICS TOOLS IMPLEMENTED FOR ANALYSIS OF PLANT SECONDARY METABOLIC PREDICTION

Bioinformatics tools are essential for analyzing, interpreting, and visualizing the complex data generated from research on plant secondary metabolites. Some of the common bioinformatics tools used for research in this field are sequence databases, MPA software, structural prediction tools, *in vitro* screening, molecular dynamics simulation analysis, omics techniques, binding free energy (BFE) analysis, ADMET profile analysis, and ML models. It mainly involves molecular docking, QSAR examination, and AI. The study here is based on finding a suitable complex that can effectively help cure colorectal cancer. The secondary metabolites are selected from the plant *Catharanthus roseus*. The pharmacological effects of the plant's metabolites have been found to make an effective novel anticancer drug. The compounds are then used in identifying their unique pharmacological properties and further evolving to molecular docking studies. The detailed study of the analysis carried out for plant secondary metabolite prediction is presented in **Figure 14.1**.

14.6.1 Data Acquisition

The analysis of plant secondary metabolites requires the usage of various bioinformatics tools and databases. In this chapter, the data required for the analysis are the target protein and the ligands which are further used for the identification of the pharmacological properties and their molecular docking process.

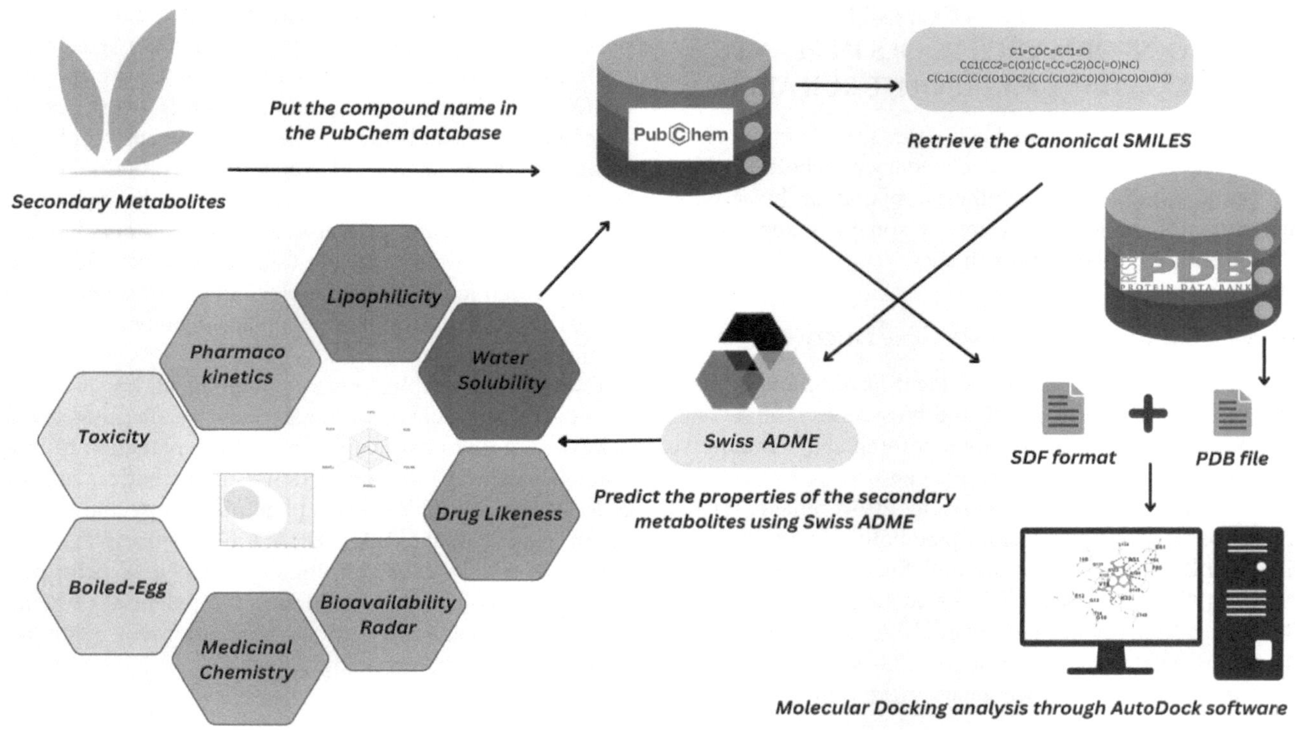

FIGURE 14.1 Workflow for the bioinformatics tools required in the analysis of plant secondary metabolites.

14.6.2 Target Protein Selection

The protein was selected using varied research through literature. The research articles regarding colorectal cancer were selected and studied to identify the protein that could possibly cause the disease. The protein was identified to be cyclin-dependent kinase -2 (PDB ID: 1DI8), which was obtained from the PDB (https://www.rcsb.org/) database (Berman et al., 2000).

Cyclin-dependent kinase-2 (CDK-2) is known to play controversial parts in various cancers. It has been found to be overexpressed in cancer patients, thus inducing tumorigenesis (Shi et al., 2015). It forms complexes with cyclins and phosphorylates key substrates, driving cell cycle passage from the G1 and S phases. This transition is necessary for the process of replication of DNA and cell division. The protein expression of CDK 2 was found to be more elevated in primary carcinoma tissues than the adjacent normal tissues (Kim et al., 1999).

14.6.3 Ligand Preparation

The compounds are based on the study by Rani and Kapoor (2019) who have performed GC-MS analysis of the selected plant *Catharanthus roseus* and have evaluated the compounds with antioxidant activity (Rani & Kapoor, 2019). The compounds listed for this chapter are selected from the above-mentioned study and the compounds only with antioxidant properties are selected for further analysis with the plant secondary metabolites. The metabolites are searched through the PubChem database (https://pubchem.ncbi.nlm.nih.gov/) (Kim et al., 2023) to require their canonical smiles, their structure and other required information. The structure is also downloaded in their SDF formats for the docking analysis. Table 14.1 provides detailed information and characterization of all of the selected 15 compounds of the literature from the PubChem Database.

14.6.4 ADMET Analysis

ADMET analysis is done with the help of the SwissADME tool (Kamble et al., 2023) in this analysis. It is an accessible web server helpful in estimating the individual ADME behaviors of the compounds. The compiled list of compounds SMILES is used to retrieve results from the web server. The physicochemical properties are tabulated in Table 14.2.

14.6.5 Lipophilicity

Lipophilicity is scientifically defined as partition coefficients or as distribution coefficients (Ghose et al., 1999). SwissADME provides many freely available modes, such as XLOGP3, WLOGP, MLOGP, SILOCS-IT, and iLOGP, which are provided to be atomic corrective factors and knowledge-based libraries (Hossain et al., 2016). The results are provided in Table 14.2.

TABLE 14.1
The Compounds Selected from the GC-MS Analysis of the Selected Plant *Catharanthus roseus*

S. No.	Compound	Molecular Formula	CID No.	Canonical SMILES	2D Structure
1	4H-Pyran-4-one	C5H4O	7698	C1=COC=CC1=O	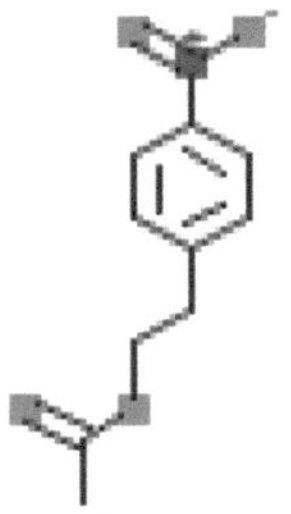 A two-dimensional chemical structure of 4H-pyran-4-one shows the bonding of a benzene ring to a linear chain.
2	Benzofuran,2,3-dihydro	C11H13NO3	15279	CC1(CC2=C(O1)C(=CC=C2)OC(=O)NC)	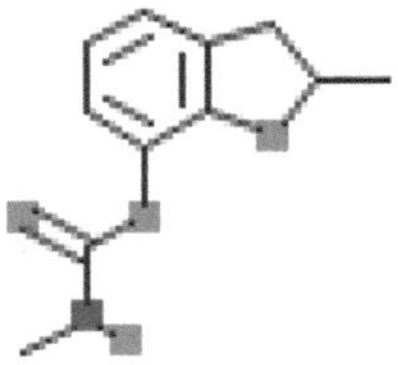 A two-dimensional chemical structure of benzofuran,2,3-dihydro shows the bonding of two rings to a linear chain.
3	Sucrose	C12H22O10	13371686	C(C1C(C(C(C(O1)OC2(C(C(C(O2)CO)O)O)CO)O)O)O)O)	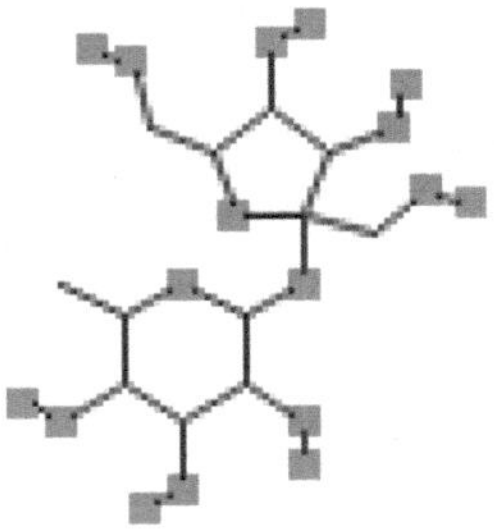 The two-dimensional chemical structure of sucrose shows the bonding of carbon, hydrogen, and oxygen.
4	Desulfosinigrin	C10H17NO6S	548248	C=CCC(=NO)SC1OC(CO)C(C(C1O)O)O	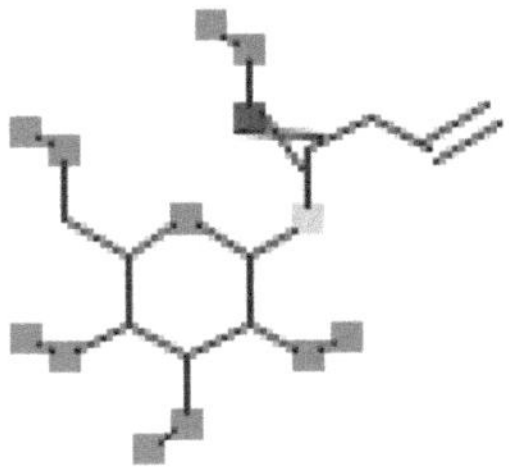 A two-dimensional chemical structure of desulfosinigrin shows the bonding of carbon, hydrogen, nitrogen, Sulphur, and oxygen.

(*Continued*)

TABLE 14.1 (*Continued*)
The Compounds Selected from the GC-MS Analysis of the Selected Plant *Catharanthus roseus*

S. No.	Compound	Molecular Formula	CID No.	Canonical SMILES	2D Structure
5	3',5'-Dimethoxy-acetophenone	C9H10O4	19830	CC(=O)C1=CC(=C(C(=C1)OC)O)O	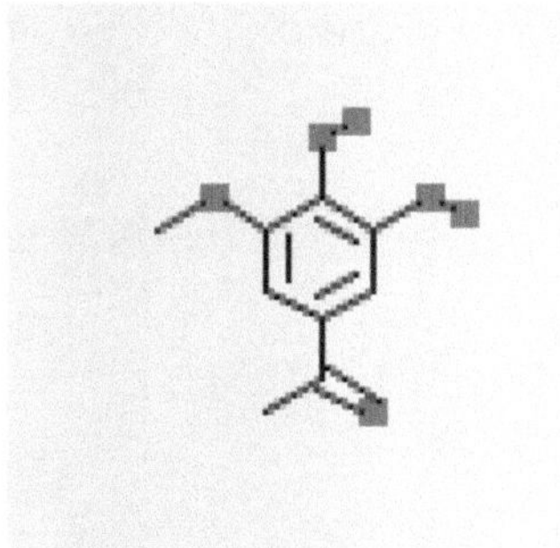 A two-dimensional chemical structure of 3',5'-dimethoxyacetophenone shows the bonding of carbon, hydrogen, and oxygen.
6	1,2,3,5-Cyclo-hexanetetrol	C6H12O3	222962	C1C(CC(C(C1O)O)O)	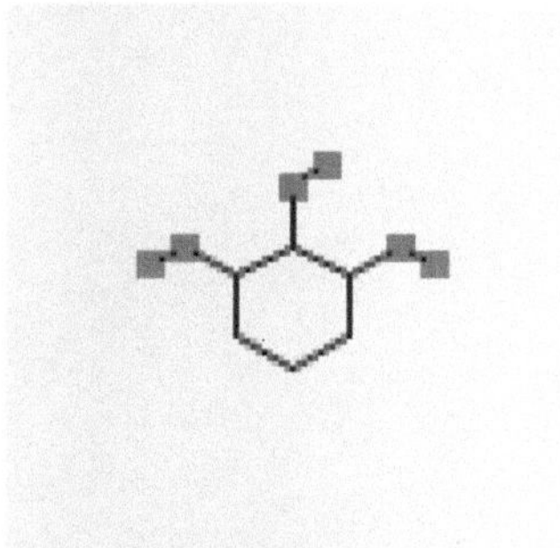 A two-dimensional chemical structure of 1,2,3,5-cyclohexanetetrol shows the bonding of carbon, hydrogen, and oxygen.
7	Myo-Inositol,4-C-methyl	C7H14O5	15119096	CC1(C(C(C(C(C1O)O)O)O)O)	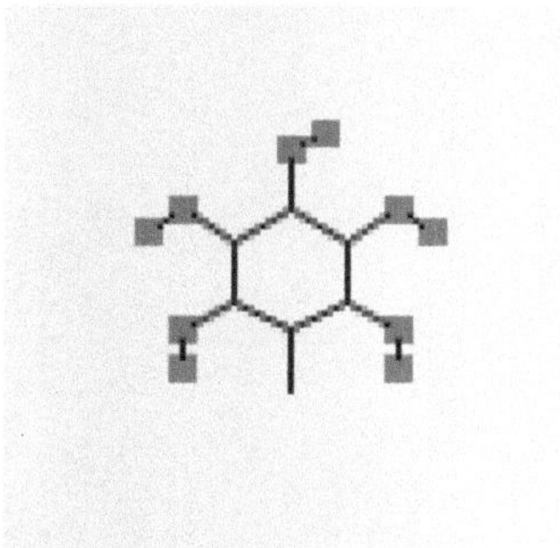 A two-dimensional chemical structure of myo-inositol,4-C-methyl shows the bonding of a ring, carbon, hydrogen, and oxygen atoms.
8	Psicofuranine	C11H15N5O	237254	C1=NC(=C2C(=N1)N(C=N2)C3(C(C(C(O3)CO)O)O)CO)N	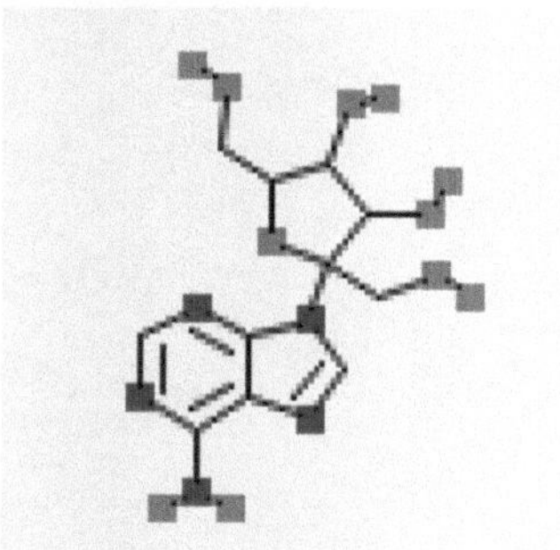 A two-dimensional chemical structure of psicofuranine shows the bonding of rings, chloride, carbon, hydrogen, and oxygen atoms.

TABLE 14.1 (*Continued*)
The Compounds Selected from the GC-MS Analysis of the Selected Plant *Catharanthus roseus*

S. No.	Compound	Molecular Formula	CID No.	Canonical SMILES	2D Structure
9	Hexadecanoic acid, methyl ester	C16H32O2	985	CCCCCCCCCCCCCCCC(=O)O	 A two-dimensional chemical structure of hexadecanoic acid, methyl ester shows the bonding of carbon, hydrogen, and oxygen atoms.
10	Pentadecanoic acid	C15H30O	17697	CCCCCCCCCCCCCCC(=O)	 A two-dimensional chemical structure of pentadecanoic acid shows the bonding of carbon and oxygen atoms.
11	9,12,15-Octadecatrienoic acid, methyl ester	C18H30O2	860	CCC=CCC=CCC=CCCCCCCCC(=O)O	 A two-dimensional chemical structure of 9,12,15-octadecatrienoic acid, methyl ester shows the double bonding of carbon and oxygen atoms.
12	Phytol	C20H40O	145386	CC(C)CCCC(C)CCCC(C)CCCC(=CCO)C	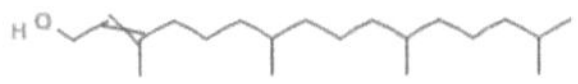 A two-dimensional chemical structure of phytol shows the double bonding of carbon and oxygen atoms.

(*Continued*)

TABLE 14.1 ***(Continued)***
The Compounds Selected from the GC-MS Analysis of the Selected Plant *Catharanthus roseus*

S. No.	Compound	Molecular Formula	CID No.	Canonical SMILES	2D Structure
13	Methyl8,11,14-heptadecatrie-noate	C18H30O2	85978449	CCC=CCC=CCC=CCCCCCCC(=O)O	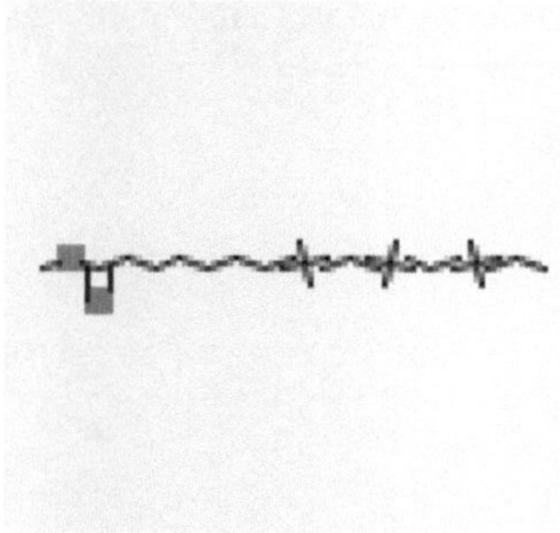 A two-dimensional chemical structure of methyl 8,11,14-heptadecatrienoate shows the double bonding of carbon, hydrogen, and oxygen atoms.
14	Aspidospermi-dine-3-carbox-ylic acid	C45H54N4O	5672	CCC1=CC2CC(C3=C(CN(C2)C1)C4=CC=CC=C4N3)(C5=C(C=C6C(=C5)C78CCN9C7C(=CC9)(C(C(C8N6C)(C(=O)OC)O)OC(=O)C)CC)OC)C(=O)OC	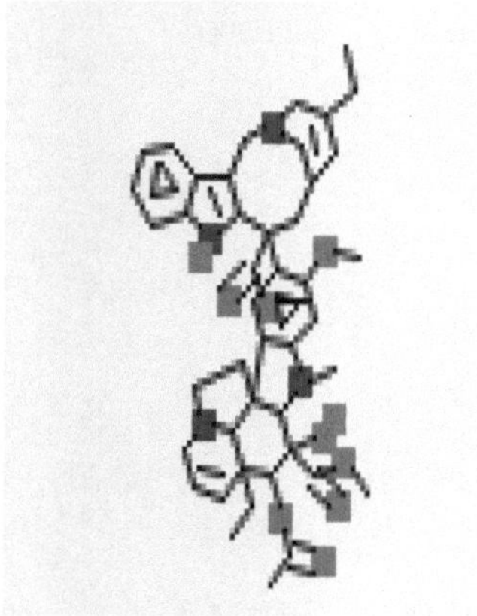 A two-dimensional chemical structure of aspidospermidine-3-carboxylic acid shows the double bonding of rings, carbon, nitrogen, hydrogen, and oxygen atoms.
15	20-methyl-8,16-diaza-hexacyclo	C20H22N2O	169197421	CC1C23CC(C14C5(C2N(CC5)CC=C3)C6=CC=CC=C6N4)C(=O)O	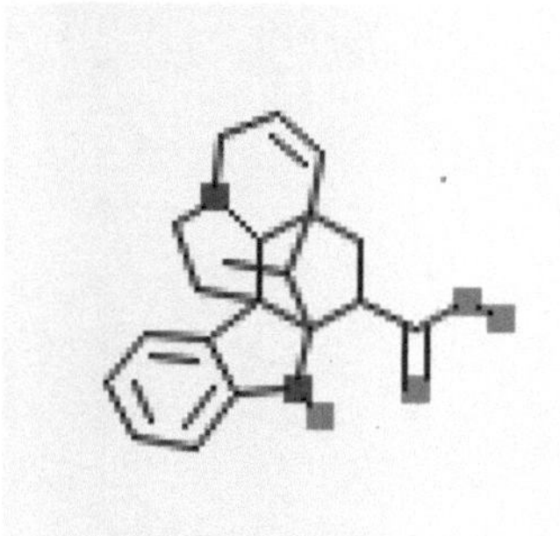 A two-dimensional chemical structure of 20-methyl-8,16-diazahexacyclo shows the double bonding of rings, carbon, hydrogen, nitrogen, and oxygen atoms.

14.6.6 Boiled-EGG Plot

Boiled-EGG model is used in evaluating gastrointestinal absorption and blood-brain barrier permeant P-gp substrate. The white part refers to the high possibility of passive GI absorption, and the yellow part (yolk) represents the high likelihood of brain penetration (Hsiao et al., 2021). The red-colored compound representation is anticipated to not be P-gp substrate (P-GP-) and colored blue is found to be energetically refluxed by P-gp (P-GP+). The selected 15 compounds have been represented as a Boiled-EGG plot and are shown in **Figure 14.2**.

TABLE 14.2
Phytochemical Properties of the Phytoconsituents of *Cantharanthas roseus*

S. no.	Compound	Molecular weight (g/mol)	TPSA	HBA	HBD
1	4H-Pyran-4-one	96.08	30.21	2	0
2	Benzofuran, 2,3-dihydro	207.23	47.56	3	1
3	Sucrose	326.30	169.30	10	7
4	Desulfosinigrin	279.31	148.04	7	5
5	3',5'-Dimethoxyacetophenone	182.17	66.76	4	2
6	1,2,3,5-Cyclohexanetetrol	132.16	60.69	3	3
7	Myo-Inositol, 4-C-methyl	178.18	101.15	5	5
8	Psicofuranine	297.27	159.77	8	5
9	Hexadecanoic acid, methyl ester	256.42	37.30	2	1
10	Pentadecanoic acid	226.40	17.07	1	0
11	9,12,15-Octadecatrienoic acid, methyl ester	278.43	37.30	2	1
12	Phytol	296.53	20.23	1	1
13	Methyl8,11,14-heptadecatrienoate	278.43	26.30	2	0
14	Aspidospermidine-3-carboxylic acid	778.93	133.87	10	2
15	20-methyl-8,16-diazahexacyclo	322.40	52.57	3	2

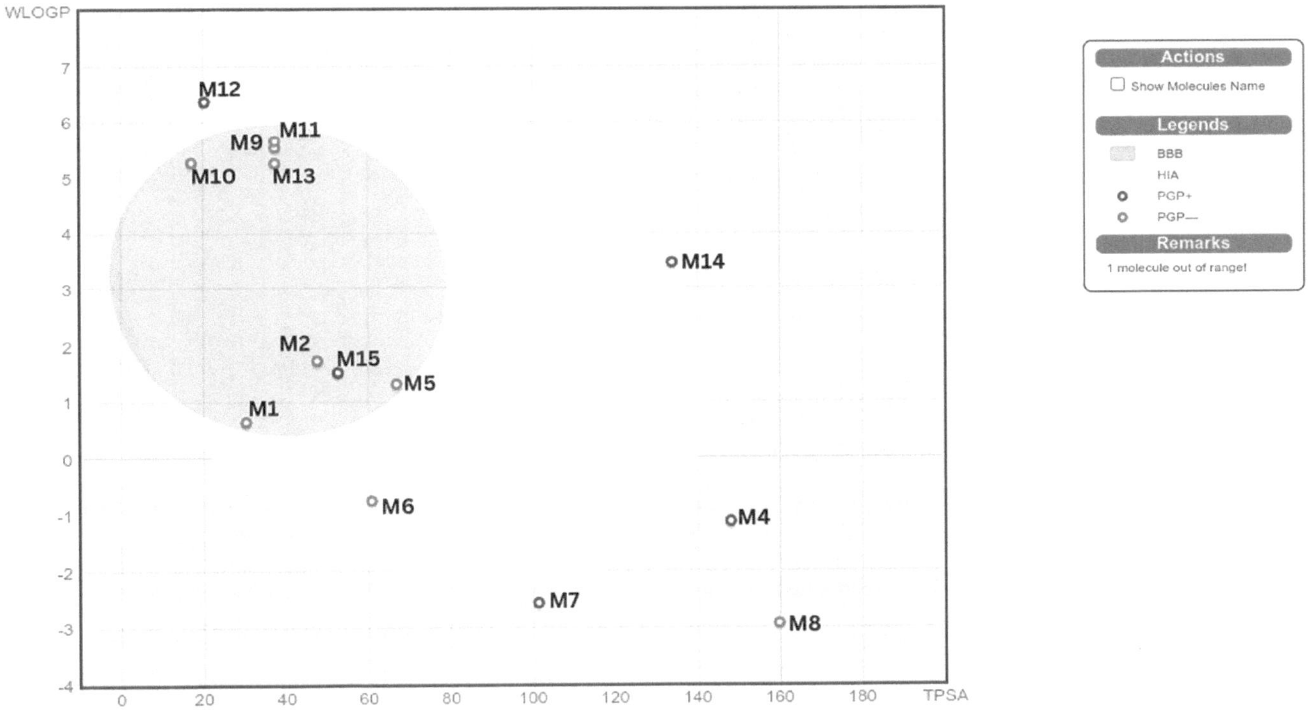

FIGURE 14.2 Boiled-EGG model for the pharmacokinetic constituents of the plant.

14.6.6.1 *Catharanthus roseus*

The structure and bioavailability radar is a unique result provided in the SwissADME web server. It aids in retrieving a preliminary glimpse at the drug likeliness of the molecule and considers six other physiochemical properties such as LIPO (lipophilicity), SIZE, POLAR (polarity), INSOLU (insolubility), INSATU (insaturation), and FLEX (flexibility). The results for all the 15 compounds are provided in **Figure 14.3a** and **Figure 14.3b**.

The compounds benzofuran, 2,3-dihyro, hexadecanoic acid, and pentadecanoic acid are found to be orally bioavailable which are highlighted in Figure 14.3 and can

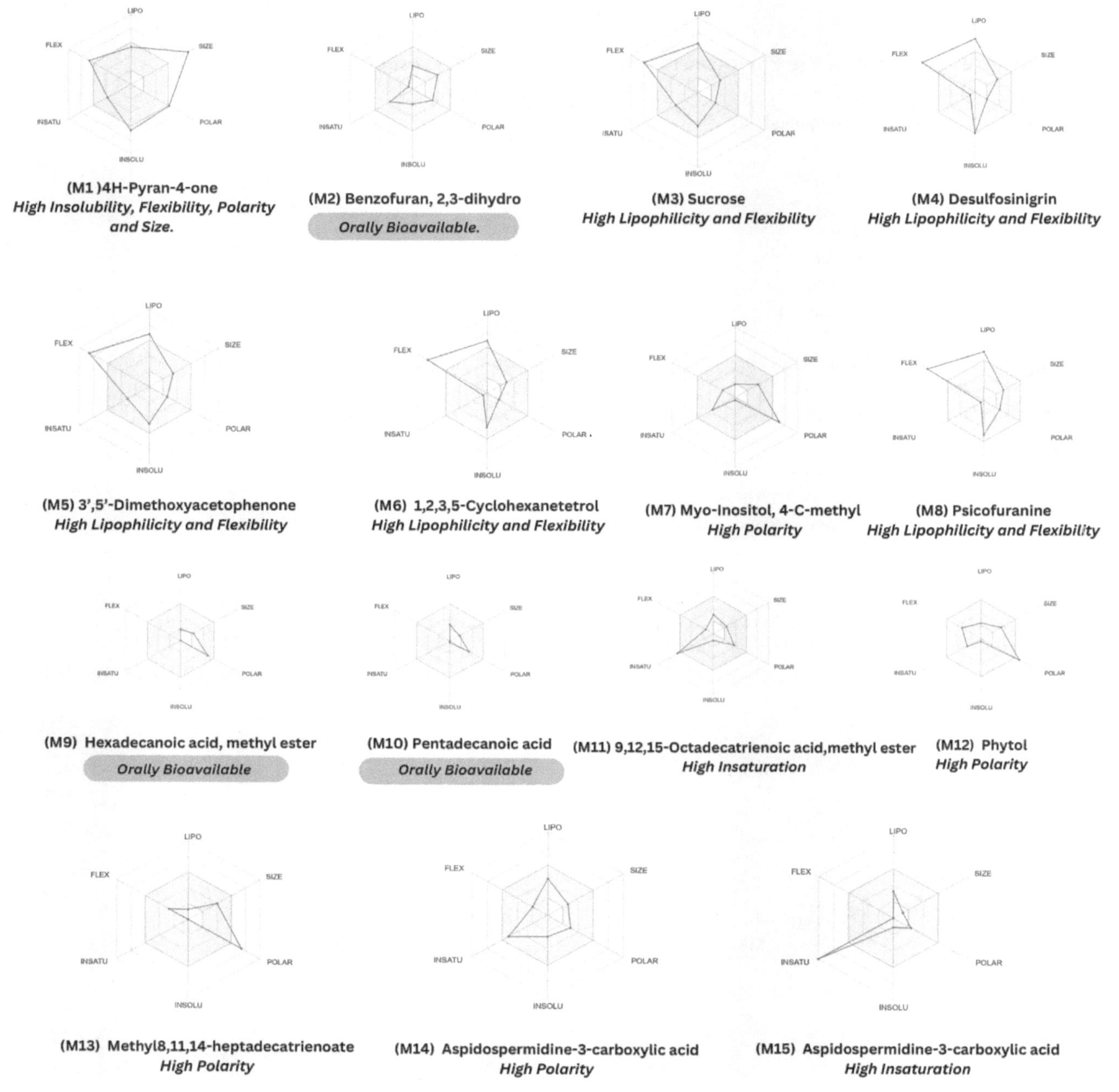

FIGURE 14.3A AND FIGURE 14.3B Structure and bioavailability radar for the selected 15 metabolites.

be considered for the docking studies. All of the three compounds were found to be within the allocated limits (lipophilicity: XLOGP3 is between −0.7 and +0.5 size; molecular weight is within the range of 150 to 500 g/mol; polarity: TPSA between 20 and 130^{-} ; solubility is said to be not higher than 6; the sp3 hybridization of fraction of carbons is not less than 0.25; the flexibility of the rotatable bonds is not higher than 9 (Delaney, 2004). Table 14.3 represents the lipophilicity of the various phytoconstituents and provides insights into the ADMET properties.

14.6.7 Molecular Docking Studies

The docking studies are conducted using the AutoDock software. Computational docking is employed in studying protein-ligand relations. The AutoDock suite is helpful in identifying selective receptor flexibility, active site prophecy, and the entire process of docking (Forli et al., 2016). The process involves using the target protein acquired from the PDB with the PDB id: 1DI8 and the orally bioactive compounds SDF formats are acquired for the docking analysis. The target protein and the ligands are docked

TABLE 14.3
Lipophilicity of the Phytoconstituents of *Catharanthas roseus*

S. no.	Compound	iLOGP	XLOGP3	WLOGP	MLOGP	SILICOS-IT	Consensus Log Po/w
1	4H-Pyran-4-one	1.31	0.13	0.64	−0.60	1.52	0.60
2	Benzofuran, 2,3-dihydro	2.31	2.14	1.73	1.40	1.59	1.83
3	Sucrose	1.09	−3.20	−4.37	−3.60	−3.28	−2.67
4	Desulfosinigrin	0.85	−1.28	−1.12	−2.09	−0.95	−0.92
5	3', 5' - Dimethoxyacetophenone	1.36	−0.12	1.31	0.24	1.21	0.80
6	1,2,3,5-Cyclohexanetetrol	0.98	−0.74	−0.75	−0.67	-0.30	−0.30
7	Myo-Inositol, 4-C-methyl	0.81	−2.30	−2.56	−1.99	-2.07	−1.62
8	Psicofuranine	1.63	−2.22	−2.93	−3.20	-2.53	−1.85
9	Hexadecanoic acid, methyl ester	3.85	7.17	5.55	4.19	5.25	5.20
10	Pentadecanoic acid	3.84	6.51	5.28	4.06	5.52	5.04
11	9,12,15-Octadecatrienoic acid, methyl ester	3.36	6.46	5.66	4.38	5.59	5.09
12	Phytol	4.71	8.19	6.36	5.25	6.57	6.22
13	Methyl 8,11,14-heptadecatrienoate	3.91	5.92	5.27	4.14	5.15	4.88
14	Aspidospermidine-3-carboxylic acid	5.53	3.77	3.46	2.87	4.96	4.12
15	20-methyl-8,16-diazahexacyclo	2.20	0.30	1.52	2.90	1.85	1.75

TABLE 14.4
Molecular Docking Results of the Orally Bioactive Compounds and the Protein Cyclin-Dependent Kinase-2

Protein	Ligand	Dock Score	Interaction Residues
Cyclin-dependent kinase -2	Benzofuran, 2,3-dihydro	−7.1	ILE10, GLY11, GLU12 GLY13, THR14, GLY16, VAL18, ALA31, LYS33, VAL64, PHE80, GLU81, PHE82, LEU83, HIS84, GLN85, ASP86, LYS89, LYS129, GLN131, ASN132, LEU133, LEU134, ALA144, ASP145, LEU148
	Hexadecanoic acid, methyl ester	−6.4	ILE10, GLY11, GLU12, GLY13, THR14, GLY16, VAL18, ALA31, LYS33, VAL64, PHE80, GLU81, PHE82, LEU83, HIS84, GLN85, ASP86, LYS89, LYS129, GLN131, ASN132, LEU134, LEU143, ALA144, ASP145, PHE146, LEU148
	Pentadecadonic acid	−6.2	ILE10, GLY11, GLU12, GLY13, THR14, GLY16, VAL18, ALA31, LYS33, VAL64, PHE80, GLU81, PHE82, LEU83, HIS84, GLN85, ASP86, LYS89, LYS129, GLN131, ASN132, LEU134, ALA144, ASP145, LEU148

individually and the dock scores are recorded. All of the compounds are blindly docked in the AutoDock suite and the interaction residues are also provided in Table 14.4. From the accounted docking scores the metabolite benzofuran, 2,3-dihydro is found to have the lowest binding energy required for binding with the CDK 2 protein, and there are about 26 interaction residues found in the complex. The complexes are shown in **Figure 14.4**, which represents the 2D interaction between the target protein and benzofuran, 2,3-dihydro, **Figure 14.5** representing the hexadecadonic acid and the target protein, and **Figure 14.6** representing the interchange between pentadecadonic acid and the target proteins.

The study thus provides the usage of employing molecular docking studies using AutoDock suite and *in silico* ADMET evaluations to predict the pharmacokinetic properties of the compound isolated from the plant against colorectal cancer. Molecular docking was evolved to identify the potential interaction modes between the target protein and the compound of interest, thus providing valuable insights into the designation of the leads and then the rational design of medicinal procedures for treating colorectal cancer.

14.7 CONCLUSION

The use of bioinformatics technologies has greatly advanced our experience of plant secondary metabolites. Computational analysis and the integration of diverse chemical structures derived from secondary metabolites in plants can be facilitated by bioinformatics technologies. These would aid in the knowledge of biosynthetic processes, the identification of important genes and enzymes, the prediction of bioactive chemicals, and the identification of binding interactions between the drug target mechanism and secondary metabolism. This discovery sheds

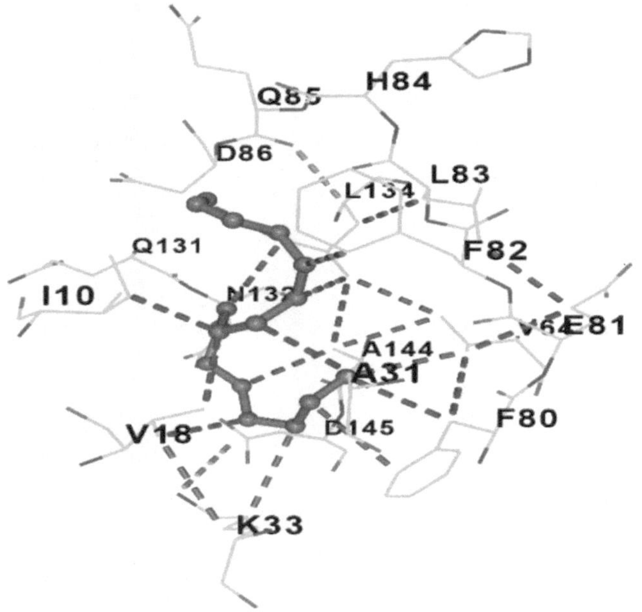

FIGURE 14.4 Molecular docking interactions between the target protein CDK 2 and the ligand benzofuran, 2,3-dihydro.

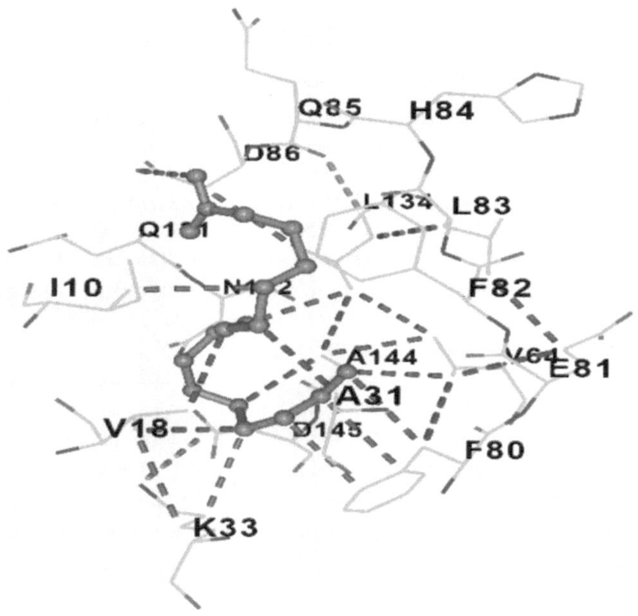

FIGURE 14.5 Molecular docking interaction between the target protein CDK 2 and the ligand hexadecanoic acid, methyl ester.

light on certain aspects of the evolution of plant secondary metabolites, and it also speeds up the identification of new druggable compounds from plant sources. It goes without saying that as technology advances the usage of various bioinformatics tools and databases unlocks the potential of plant-derived bioactive substances in many areas, including the identification of new medications and significant compounds.

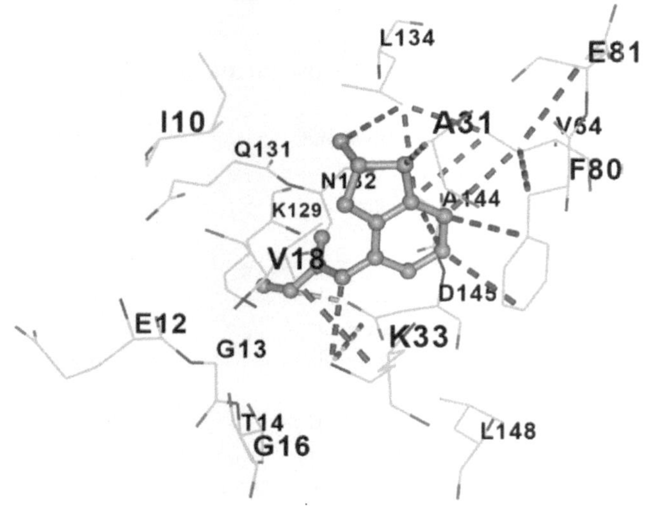

FIGURE 14.6 Molecular docking interaction between the target protein CDK 2 and the ligand pentadecanoic acid.

ACKNOWLEDGMENT

All the authors are thankful to their respective universities and institutions for their valuable supports.

DATA AVAILABILITY

The authors confirm that the data supporting the findings of this study are available within the book/chapter.

REFERENCES

Abdul-Hammed, M., Adedotun, I. O., Olajide, M., Irabor, C. O., Afolabi, T. I., Gbadebo, I. O., Rhyman, L., & Ramasami, P. (2022). Virtual screening, ADMET profiling, PASS prediction, and bioactivity studies of potential inhibitory roles of alkaloids, phytosterols, and flavonoids against COVID-19 main protease (M^{pro}). *Natural Product Research*, *36*(12), 3110–3116.

Acamovic, T., & Brooker, J. D. (2005). Biochemistry of plant secondary metabolites and their effects in animals. *Proceedings of the Nutrition Society*, *64*(3), 403–412.

Akilandeswari, G., Vijaya Anand, A., Sampathkumar, P., Vinayaga Moorthi, P., & Preethi, B. (2019). A prospective review on phyto-pharmacological aspects of *Andrographis paniculata*. *Systematic Reviews in Pharmacy*, *10*(1), 15–19.

Ali Reza, A. S. M., Nasrin, M. S., Hossen, M. A., Rahman, M. A., Jantan, I., Haque, M. A., & Sobarzo-Sánchez, E. (2023). Mechanistic insight into immunomodulatory effects of food-functioned plant secondary metabolites. *Critical Reviews in Food Science and Nutrition*, *63*(22), 5546–5576.

Anulika, N. P., Ignatius, E. O., Raymond, E. S., Osasere, O. I., & Abiola, A. H. (2016). The chemistry of natural product: Plant secondary metabolites. *International Journal of Emerging Engineering Research and Technology*, *4*(8), 1–9.

Anza, M., Endale, M., Cardona, L., Cortes, D., Eswaramoorthy, R., Zueco, J., Rico, H., Trelis, M., & Abarca, B. (2021). Antimicrobial activity, *in silico* molecular docking, ADMET and DFT analysis of secondary metabolites from roots of three

Ethiopian medicinal plants. In *Advances and applications in bioinformatics and chemistry* (pp. 117–132). Taylor & Francis.

Badyal, S., Singh, H., Yadav, A. K., Sharma, S., & Bhushan, I. (2020). Plant secondary metabolites and their uses. *Plant Archives*, *20*(2), 3336–3340.

Begum, T. N., Ilyas, M. M., Kalavathy, S., Anand, A. V., Kumar, S., & Senthil, R. (2009). Effects of ethanolic leaf extract of *Azim tetracantha* Lam. on *Ehrlich ascites* carcinoma tumour bearing mice. *Research Journal of Medicine and Medical Sciences*, *4*(2), 351–354.

Bennett, R. N., & Wallsgrove, R. M. (1994). Secondary metabolites in plant defence mechanisms. *New Phytologist*, *127*(4), 617–633.

Berman, H. M., Westbrook, J., Feng, Z., Gilliland, G., Bhat, T. N., Weissig, H., Shindyalov, I. N., & Bourne, P. E. (2000). The protein data bank. *Nucleic Acids Research*, *28*(1), 235–242.

Bharathi, V., Rengarajan, R. L., Radhakrishnan, R., & Hashem, A., Abd-Allah, E. F., Alqarawi, A. A., & Anand, A. V. (2018). Effects of a medicinal plant *Macrotyloma uniflorum* (Lam.) Verdc. Formulation (MUF) on obesity-associated oxidative stress-induced liver injury. *Saudi Journal of Biological Sciences*, *25*(6), 1115–1121.

Blin, K., Shaw, S., Steinke, K., Villebro, R., Ziemert, N., Lee, S. Y., Medema, M. H., & Weber, T. (2019). antiSMASH 5.0: Updates to the secondary metabolite genome mining pipeline. *Nucleic Acids Research*, *47*(W1), W81–W87.

Compean, K. L., & Ynalvez, R. A. (2014). Antimicrobial activity of plant secondary metabolites: A review. *Research Journal of Medicinal Plant*, *8*, 204–213.

Crozier, A., Clifford, M. N., & Ashihara, H. (2006). *Plant secondary metabolites: Occurrence, structure and role in the human diet*. Blackwell-Publishers.

Cubero-Leon, E., De Rudder, O., & Maquet, A. (2018). Metabolomics for organic food authentication: Results from a long-term field study in carrots. *Food Chemistry*, *239*, 760–770.

Cuperlovic-Culf, M., Rajagopalan, N., Tulpan, D., & Loewen, M. C. (2016). Metabolomics and cheminformatics analysis of antifungal function of plant metabolites. *Metabolites*, *6*(4), 31.

Dai, X., & Shen, L. (2022). Advances and trends in omics technology development. *Frontiers in Medicine*, *9*, 911861.

Delaney, J. S. (2004). ESOL: Estimating aqueous solubility directly from molecular structure. *Journal of Chemical Information and Computer Sciences*, *44*(3), 1000–1005.

Divya, N., Rengarajan, R. L., Radhakrishnan, R., Abd-Allah, E. F., Alqarawi, A. A., Hashem, A., & Anand, A. V. (2019). Phytotherapeutic efficacy of the medicinal plant *Terminalia catappa* L. *Saudi Journal of Biological Sciences*, *26*(5), 985–988.

Doraiswamy, H., Kathavarayan, N., Sharma, R. C., & Krishnamurthy, V. (2014). Molecular docking analysis of secondary metabolites of *Trigonella foenum* graecum and *Carica papaya* with FTO: An *in silico* approach. *International Journal of Pharmaceutical Sciences Review and Research*, *27*(1), 105–110.

Lohning, A. E., Levonis, S. M., Williams-Noonan, B., & Schweiker, S. S. (2017). A practical guide to molecular docking and homology modelling for medicinal chemists. *Current Topics in Medicinal Chemistry*, *17*(18), 2023–2040.

Flores, J. E., Claborne, D. M., Weller, Z. D., Webb-Robertson, B. J. M., Waters, K. M., & Bramer, L. M. (2023). Missing data in multi-omics integration: Recent advances through artificial intelligence. *Frontiers in Artificial Intelligence*, *6*, 1098308.

Forli, S., Huey, R., Pique, M. E., Sanner, M. F., Goodsell, D. S., & Olson, A. J. (2016). Computational protein-ligand docking and virtual drug screening with the AutoDock suite. *Nature Protocols*, *11*(5), 905–919.

Gentile, F., Yaacoub, J. C., Gleave, J., Fernandez, M., Ton, A. T., Ban, F., Stern, A., & Cherkasov, A. (2022). Artificial intelligence-enabled virtual screening of ultra-large chemical libraries with deep docking. *Nature Protocols*, *17*(3), 672–697.

Ghose, A. K., Viswanadhan, V. N., & Wendoloski, J. J. (1999). A knowledge-based approach in designing combinatorial or medicinal chemistry libraries for drug discovery. 1. A qualitative and quantitative characterization of known drug databases. *Journal of Combinatorial Chemistry*, *1*(1), 55–68.

Gola, J., Obrezanova, O., Champness, E., & Segall, M. (2006). ADMET property prediction: The state of the art and current challenges. *QSAR & Combinatorial Science*, *25*(12), 1172–1180.

Guo, C. J., & Wang, C. C. (2014). Recent advances in genome mining of secondary metabolites in *Aspergillus terreus*. *Frontiers in Microbiology*, *5*, 124539.

Guo, J., Huang, Z., Sun, J., Cui, X., & Liu, Y. (2021). Research progress and future development trends in medicinal plant transcriptomics. *Frontiers in Plant Science*, *12*, 691838.

Hautbergue, T., Jamin, E. L., Debrauwer, L., Puel, O., & Oswald, I. P. (2018). From genomics to metabolomics, moving toward an integrated strategy for the discovery of fungal secondary metabolites. *Natural Product Reports*, *35*(2), 147–173.

Hochberg, U., Degu, A., Toubiana, D., Gendler, T., Nikoloski, Z., Rachmilevitch, S., & Fait, A. (2013). Metabolite profiling and network analysis reveal coordinated changes in grapevine water stress response. *BMC Plant Biology*, *13*, 1–16.

Hossain, M. U., Khan, M. A., Rakib-Uz-Zaman, S. M., Ali, M. T., Islam, M. S., Keya, C. A., & Salimullah, M. (2016). Treating diabetes mellitus: Pharmacophore based designing of potential drugs from *Gymnema sylvestre* against insulin receptor protein. *BioMed Research International*, 3187647.

Hsiao, Y., Su, B. H., & Tseng, Y. J. (2021). Current development of integrated web servers for preclinical safety and pharmacokinetics assessments in drug development. *Briefings in Bioinformatics*, *22*(3), bbaa160.

Illian, D. N., Widiyana, A. P., Hasana, A. R., Maysarah, H., Al Mustaniroh, S. S., & Basyuni, M. (2022). *In silico* approach: Prediction of ADMET, molecular docking, and QSPR of secondary metabolites in Mangroves. *Journal of Applied Pharmaceutical Science*, *12*(11), 021–029.

Inglis, D. O., Binkley, J., Skrzypek, M. S., Arnaud, M. B., Cerqueira, G. C., Shah, P., Wymore, F., Wortman, J. R., & Sherlock, G. (2013). Comprehensive annotation of secondary metabolite biosynthetic genes and gene clusters of *Aspergillus nidulans*, *A. fumigatus*, *A. niger* and *A. oryzae*. *BMC Microbiology*, *13*, 1–23.

Jain, C., Khatana, S., & Vijayvergia, R. (2019). Bioactivity of secondary metabolites of various plants: A review. *International Journal of Pharmaceutical Sciences and Research*, *10*(2), 494–504.

Jaiswal, P., & Usadel, B. (2016). Plant pathway databases. In *Plant bioinformatics: Methods and protocols* (pp. 71–87). Humana Press.

James, J. P., Ail, P. D., Crasta, L., Kamath, R. S., Shura, M. H., & Sindhu, T. J. (2023). *In silico* ADMET and molecular interaction profiles of phytochemicals from medicinal plants in Dakshina Kannada. *Journal of Health and Allied Sciences NU*, 190–201.

Kamble, A. N. S., & Mitkar, A. A. (2023). Swiss ADME predictions of pharmacokinetics and drug-likeness properties of secondary metabolites present in *Trigonella foenum-graecum*.

Journal of Pharmacognosy and Phytochemistry, *12*(5), 341–349.

Kanehisa, M. (2016). KEGG bioinformatics resource for plant genomics and metabolomics. In *Plant bioinformatics: Methods and protocols* (pp. 55–70). Humana Press.

Kar, S., & Roy, K. (2012). QSAR of phytochemicals for the design of better drugs. *Expert Opinion on Drug Discovery*, *7*(10), 877–902.

Karthika, P., Balamuralikrishnan, B., Yamini, K., Vijaya Anand, A., Durairaj, K., Maruthupandian, A., Alodaini, H. A., Hatamleh, A. A., Manikantan, P., & Arun, M. (2023). Green synthesis, characterization of silver nanoparticles using aqueous leaf extracts of Solanum melongena and *in vitro* evaluation of antibacterial, pesticidal and anticancer activity in human MDA-MB-231 breast cancer cell lines. *Journal of King Saud University – Science*, *35*(5), 102663.

Kaviya, M., Balasubramanian, B., Bharathi, K., Malaisamy, A., Al-Dhabi, N. A., Mariadhas, V. A., Anand, A. V., & Liu, W. (2022). Evaluation of nutritional substances and investigation of antioxidant and antimicrobial potentials of *Boerhavia diffusa* with in silico molecular ocking. *Molecules*, *27*(4), 1280.

Kim, J. H., Kang, M. J., Park, C. U., Kwak, H. J., Hwang, Y., & Koh, G. Y. (1999). Amplified CDK2 and cdc2 activities in primary colorectal carcinoma. Cancer: *Interdisciplinary International Journal of the American Cancer Society*, *85*(3), 546–553.

Kim, S., Chen, J., Cheng, T., Gindulyte, A., He, J., He, S., Li, Q., Shoemaker, B. A., Thiessen, P. A., Yu, B., & Zaslavsky, L. (2023). PubChem 2023 update. *Nucleic Acids Research*, *51*(D1), D1373–D1380.

Kivilompolo, M., Obůrka, V., & Hyötyläinen, T. (2007). Comparison of GC-MS and LC-MS methods for the analysis of antioxidant phenolic acids in herbs. *Analytical and Bioanalytical Chemistry*, *388*, 881–887.

Korkina, L., & Kostyuk, V. (2012). Biotechnologically produced secondary plant metabolites for cancer treatment and prevention. *Current Pharmaceutical Biotechnology*, *13*(1), 265–275.

Korkina, L., Kostyuk, V., Potapovich, A., Mayer, W., Talib, N., & De Luca, C. (2018). Secondary plant metabolites for sun protective cosmetics: From pre-selection to product formulation. *Cosmetics*, *5*(2), 32.

Kumar, G. P., & Khanum, F. (2012). Neuroprotective potential of phytochemicals. *Pharmacognosy Reviews*, *6*(12), 81.

Lavecchia, T., Rea, G., Antonacci, A., & Giardi, M. T. (2013). Healthy and adverse effects of plant-derived functional metabolites: The need of revealing their content and bioactivity in a complex food matrix. *Critical Reviews in Food Science and Nutrition*, *53*(2), 198–213.

Lee, J. H., Park, K. M., Han, D. J., Bang, N. Y., Kim, D. H., Na, H., Lim, S., Kim, T. B., Kim, D. G., Kim, H. J., & Chung, Y. (2015). PharmDB-K: Integrated bio-pharmacological network database for traditional Korean medicine. *PLoS One*, *10*(11), e0142624.

Liu, Y., Liang, Y., & Wishart, D. (2015). PolySearch2: A significantly improved text-mining system for discovering associations between human diseases, genes, drugs, metabolites, toxins and more. *Nucleic Acids Research*, *43*(W1), W535–W542.

Liu, Z., Ran, Q., Luo, J., Shen, Q., Zhang, T., Fang, S., Pan, K., & Long, L. (2024). Correlation analysis of secondary metabolites and disease resistance activity of different varieties of Congou black tea based on LC-MS/MS and TCMSP. *Food Chemistry, X*, 101331.

Manikandan, R., Balasubramanian, B., Punniyakotti, P., Vijaya Anand, A., Meyyazhagan, A., Velauthaprabhu, S., & Liu, W. C. (2022). Cardio-protective effects of *Terminalia catappa* leaves and *Terminalia chebula* fruit extract in doxorubicin-induced cardiomyopathy in rats. *Biomarkers*, *27*(5), 488–495.

Manoharan, N., Sampathkumar, P., Dheeba, B., Sheik Abdulla, S., Vinothprasanna, G., Vinothkannan, R., Kalavathy, S., Vijaya Anand, A., & Shanmugasundaram, A. (2008). Potential hepatoprotective effect of aqueous extract of *Gracilaria corticata* in AFB_1 induced hepatotoxicity in Wistar rats. *Journal of Biological Sciences*, *8*, 1352–1355.

Moore, B. M., Wang, P., Fan, P., Leong, B., Schenck, C. A., Lloyd, J. P., Lehti-Shiu, M. D., Last, R. L., Pichersky, E., & Shiu, S. H. (2019). Robust predictions of specialized metabolism genes through machine learning. *Proceedings of the National Academy of Sciences*, *116*(6), 2344–2353.

Moroy, G., Martiny, V. Y., Vayer, P., Villoutreix, B. O., & Miteva, M. A. (2012). Toward *in silico* structure-based ADMET prediction in drug discovery. *Drug Discovery Today*, *17*(1–2), 44–55.

Moustakas, D. T., Lang, P. T., Pegg, S., Pettersen, E., Kuntz, I. D., Brooijmans, N., & Rizzo, R. C. (2006). Development and validation of a modular, extensible docking program: DOCK 5. *Journal of Computer-aided Molecular Design*, *20*, 601–619.

Muthumani, G. D., Anand, A. V., & Manikandan, R. (2013). Antioxidant, antihelmintic and antimicrobial activity of *Euphorbia thymifolia* Linn whole plant. *International Journal of Current Microbiology and Applied Sciences*, *2*(1), 66–75.

Nisa, N., Dinata, R., Arati, C., Abdelgani-Baraka, G. A., Bhanushree, B., Bidanchi, R. M., Manikandan, B., Saeed, A. L., Abinash, G., Pori, B., & Lalrinzuali, S. (2023). Computational toxicology and food safety assessment of *Parkia timoriana* phytoconstituents using quantitative structure-activity relationship (QSAR) modeling approaches. *Indian Journal of Biochemistry and Biophysics (IJBB)*, *60*(12), 896–918.

Pagare, S., Bhatia, M., Tripathi, N., Pagare, S., & Bansal, Y. K. (2015). Secondary metabolites of plants and their role: Overview. *Current Trends in Biotechnology and Pharmacy*, *9*(3), 293–304.

Palazzotto, E., & Weber, T. (2018). Omics and multi-omics approaches to study the biosynthesis of secondary metabolites in microorganisms. *Current Opinion in Microbiology*, *45*, 109–116.

Park, H., Lee, J., & Lee, S. (2006). Critical assessment of the automated AutoDock as a new docking tool for virtual screening. *Proteins: Structure, Function, and Bioinformatics*, *65*(3), 549–554.

Pushpangadan, P., Ijinu, T. P., & George, V. (2015). Plant based anti-inflammatory secondary metabolites. *Annals of Phytomedicine*, *4*(1), 17–36.

Rai, A., Saito, K., & Yamazaki, M. (2017). Integrated omics analysis of specialized metabolism in medicinal plants. *The Plant Journal*, *90*(4), 764–787.

Rani, J., & Kapoor, M. (2019). Gas chromatography-mass spectrometric analysis and identification of bioactive constituents of *Catharanthus roseus* and its antioxidant activity. *GAS*, *12*(3).

Raskin, I., Ribnicky, D. M., Komarnytsky, S., Ilic, N., Poulev, A., Borisjuk, N., Brinker, A., Moreno, D. A., Ripoll, C., Yakoby, N., & O'Neal, J. M. (2002). Plants and human health in the twenty-first century. *Trends in Biotechnology*, *20*(12), 522–531.

Rastija, V., Vrandečić, K., Ćosić, J., Majić, I., Šarić, G. K., Agić, D., Karnaš, M., Lončarić, M., & Molnar, M. (2021). Biological activities related to plant protection and environmental effects of coumarin derivatives: QSAR and molecular docking studies. *International Journal of Molecular Sciences*, *22*(14), 7283.

Rathore, P., & Shakya, R. (2024). Next-generation techniques in ecophysiology: Metabolomics, proteomics, SAR/QSAR. In *Ecophysiology of tropical plants* (pp. 284–297). CRC Press.

Reichling, J. (2010). Plant-microbe interactions and secondary metabolites with antibacterial, antifungal and antiviral properties. In *Annual plant reviews volume 39: Functions and biotechnology of plant secondary metabolites* (pp. 214–347). Wiley Online Library.

Ru, J., Li, P., Wang, J., Zhou, W., Li, B., Huang, C., Li, P., Guo, Z., Tao, W., Yang, Y., & Xu, X. (2014). TCMSP: A database of systems pharmacology for drug discovery from herbal medicines. *Journal of Cheminformatics*, *6*, 1–6.

Sabitha Rani, A., Neelima, G., Mukhopadhyay, R., Jyothi, K. S. N., & Sulakshana, G. (2018). *In silico* characterization of plant secondary metabolites. In *In silico approach for sustainable agriculture* (pp. 251–273). Springer.

Saito, K., & Matsuda, F. (2010). Metabolomics for functional genomics, systems biology, and biotechnology. *Annual Review of Plant Biology*, *61*, 463–489.

Samanta, S., O'Hagan, S., Swainston, N., Roberts, T. J., & Kell, D. B. (2020). VAE-Sim: A novel molecular similarity measure based on a variational autoencoder. *Molecules*, *25*(15), 3446.

Sangeetha, T., Balamuralikrishnan, B., Sampathkumar, P., Velayuthaprabhu, S., Senthilkumar, N., Syed Ibrahim, K., Baskaran, R., & Vijaya Anand, A. (2022). Characterization and phytoconstituents of *Petroselinum crispum* (Mill) and *Coriandrum sativum* (Linn) and their impacts on inflammation – an *in vitro* analysis against human adenocarcinoma cells with molecular docking. *South African Journal of Botany*, *146*, 776–788.

Shah, S. M. A., Akram, M., Riaz, M., Munir, N., & Rasool, G. (2019). Cardioprotective potential of plant-derived molecules: A scientific and medicinal approach. *Dose-Response*, *17*(2), 1559325819852243.

Shi, X. N., Li, H., Yao, H., Liu, X., Li, L., Leung, K. S., Kung, H. F., & Lin, M. C. M. (2015). Adapalene inhibits the activity of cyclin-dependent kinase 2 in colorectal carcinoma. *Molecular Medicine Reports*, *12*(5), 6501–6508.

Singh, A. V., Chandrasekar, V., Paudel, N., Laux, P., Luch, A., Gemmati, D., Tisato, V., Prabhu, K. S., Uddin, S., & Dakua, S. P. (2023). Integrative toxicogenomics: Advancing precision medicine and toxicology through artificial intelligence and OMICs technology. *Biomedicine & Pharmacotherapy*, *163*, 114784.

Siramshetty, V. B., Xu, X., & Shah, P. (2023). Artificial intelligence in ADME property prediction. In *Computational drug discovery and design* (pp. 307–327). Springer US.

Sukhikh, S., Babich, O., Prosekov, A., Kalashnikova, O., Noskova, S., Bakhtiyarova, A., Krol, O., Tsvetkova, E., & Ivanova, S. (2023). Antidiabetic properties of plant secondary metabolites. *Metabolites*, *13*(4), 513.

Varalakshmi, B., Anand, A. V., Karpagam, T., Bai, J. S., & Manikandan, R. (2014). *In vitro* antimicrobial and anticancer activity of *Cinnamomum zeylanicum* Linn bark extracts. *International Journal of Pharmacy and Pharmaceutical Sciences*, *6*(1), 12–18.

Vasseur, F., Cornet, D., Beurier, G., Messier, J., Rouan, L., Bresson, J., Ecarnot, M., Stahl, M., Heumos, S., Gérard, M., & Reijnen, H. (2022). A perspective on plant phenomics: Coupling deep learning and near-infrared spectroscopy. *Frontiers in Plant Science*, *13*, 836488.

Vijaya Anand, A., Varalakshmi, B., Prasana, B., Sampathkumar, P., Pushpa, N., & Agaath Hedina (2016). *Cinnamomum zeylanicum* Linn. The spice with multi potential. *Systematic Reviews in Pharmacy*, *7*(1), 24–29.

Vijaya Anand, A., Velayuthaprabhu, S., Rengarajan, R. L., Sampathkumar, P., & Radhakrishnan, R. (2020). Bioactive compounds of guava (*Psidium guajava* L.). In *Bioactive compounds in underutilized fruits and nuts*, Reference Series in Phytochemistry (pp. 503–527). Springer.

Vijayakumar, K., Prasanna, B., Rengarajan, R. L., Rathinam, A., Velayuthaprabhu, S., & Vijaya Anand, A. (2023). Antidiabetic and hypolipidemic effects of *Cinnamon cassia* bark extracts: An *in vitro*, *in vivo*, and *in silico* approach. *Archieves of Physiology and Biochemistry*, *129*(2), 338–348.

Wallace, R. J. (2004). Antimicrobial properties of plant secondary metabolites. *Proceedings of the Nutrition Society*, *63*(4), 621–629.

Wink, M. (1988). Plant breeding: Importance of plant secondary metabolites for protection against pathogens and herbivores. *Theoretical and Applied Genetics*, *75*, 225–233.

Yoo, S., Yang, H. C., Lee, S., Shin, J., Min, S., Lee, E., Song, M., & Lee, D. (2020). A deep learning-based approach for identifying the medicinal uses of plant-derived natural compounds. *Frontiers in Pharmacology*, *11*, 584875.

Yuliana, D., Bahtiar, F. I., & Najib, A. (2013). *In silico* screening of chemical compounds from roselle (*Hibiscus sabdariffa*) as angiotensin-I converting enzyme inhibitor used PyRx program. *ARPN Journal of Science and Technology*, *3*, 1158–1160.

Zhang, P., Dreher, K., Karthikeyan, A., Chi, A., Pujar, A., Caspi, R., Karp, P., Kirkup, V., Latendresse, M., Lee, C., & Mueller, L. A. (2010). Creation of a genome-wide metabolic pathway database for Populus trichocarpa using a new approach for reconstruction and curation of metabolic pathways for plants. *Plant Physiology*, *153*(4), 1479–1491.

Zheng, Q., & Wang, X. J. (2008). GOEAST: A web-based software toolkit for Gene Ontology enrichment analysis. *Nucleic Acids Research*, 36(Suppl_2), W358–W363.

Ziemert, N., Podell, S., Penn, K., Badger, J. H., Allen, E., & Jensen, P. R. (2012). The natural product domain seeker NaPDoS: A phylogeny based bioinformatic tool to classify secondary metabolite gene diversity. *PloS One*, *7*(3), e34064.

15 Network Pharmacology for Exploring Key Genes and Signaling Networks in Plant Secondary Metabolites

Arokiaraj Sherlin Rosita, Jambulingam Shivani, Mathivanan Jay Shree, Tajudheen Nargis Begum, Asirvatham Alwin Robert, Shanmugam Gavaskar, Alagarsamy Sumithra, and Arumugam Vijaya Anand

LIST OF ABBREVIATIONS

DAVID	Database for Annotation, Visualization, and Integrated Discovery
GO	Gene Ontology
KEGG	Kyoto Encyclopaedia of Genes and Genome
NP	network pharmacology
PPI	protein-protein interaction
PSMs	plant specialized metabolites

15.1 INTRODUCTION

Plants deliver a vast array of small organic molecules known as 'secondary metabolites' which are unrelated to the growth, development, or reproduction of plants and an important sources of therapeutic properties (Manoharan et al., 2008; Begum et al., 2009; Muthumani et al., 2013; Varalakshmi et al., 2014; Vijaya Anand et al., 2016, 2020; Bharathi et al., 2018; Divya et al., 2019; Akilandeswari et al., 2019; Sangeetha et al., 2022; Kaviya et al., 2022; Manikandan et al., 2022; Karthika et al., 2023; Vijayakumar et al., 2023). They do, however, mediate some processes that improve the fecundity and survival of the plant. A vast array of complex secondary metabolites, derived from various abiotic stresses, are synthesized by plants to accomplish vital physiological functions such as establishing symbiosis, attracting pollinators, and supplying structural elements to lignified cell walls of vascular tissue (Ncube & Van Staden, 2015). Terpenoids, phenolic compounds, sulfur-containing compounds, and terpenoids are the four main groups of secondary metabolites found in plants (Guerriero et al., 2018). The synthesis of phytochemicals in plants known as secondary metabolites is called the physical and biochemical pathways (Delgoda & Murray, 2017).

Network pharmacology (NP) is a discipline grounded in systems biology and multidirectional pharmacology. The advancements that have been revealed in the use of the NP-based approach have pivoted the archetype from a 'one target, one drug' mode to a 'network-target, multiple-component therapeutics' mode (Noor et al., 2022). This has emerged to help understand the complex associations between medicines and natural systems. This approach has been used in the relevance of the study of plant secondary metabolites, which are known for their diverse pharmacological properties (Batool et al., 2022).

The discovery and description of key genes associated with the production of secondary metabolites could offer valuable information and be beneficial for enhancing the outcome of a few significant compounds in plants (Bolhassani et al., 2021). Systems biology is one of the comprehensive approaches used in ascertaining target genes involved in target biological functions via the implementation of complex studies llike co-expression and NP (Liu et al., 2019)

Protein signaling networks are concerned with the regulation of plant secondary metabolism, thus validating the focus back to secondary metabolites. Protein interactions are those between two proteins or extend to include more complicated networks within local protein modules. However, signal propagation in protein networks typically involves multiple intermediates. Thus, studying characteristic subsets of relations can obscure the general signaling mechanism. Although analyzing protein networks has enabled robust bioinformatics forecasts, a gap persists between theoretical investigations of extensive regulatory networks and the experimental claims of experimentalists (Bulgakov & Tsitsiashvili, 2013; Potapov et al., 2008). The hub genes found through many known protein-protein interactions are the main characters of signaling pathways. Some of the mediated signaling discussed in the review article by Bulgakov et al. (2017) have presented their critical review on varied subnetworks of the anthocyanin such as light-anthocyanin network: HY5-mediated signaling, calcium signaling, etc.

Several information sources, such as pathway databases like the Kyoto Encyclopaedia of Genes and Genome (KEGG) pathway (Kanehisa et al., 2023), Reactome (Milacic et al., 2024), and BioCyc (Karp et al., 2019), incorporated details sources like ONDEX (Taubert et al., 2014) and Biomine (Eronen & Toivonen, 2012), and the scientific publications itself, may be utilized to build the signaling network topology. When constructing the signaling network architecture,

 DOI: 10.1201/9781003475491-15

information must be retrieved from the biological literature because the majority of human biological understanding is still exclusively kept in the silos of biological literature (Miljkovic et al., 2012).

Pathway enrichment analysis has emerged as a leading method for interpreting biological data by condensing high-dimensional information into a few key biological processes that underlie specific phenotypes (Mubeen et al., 2022). It is here considered for elucidating biological functions of key genes and pathways identified through NP. Many sources are now developed for the elucidation and interpretation of the metabolomics data such as RaMP 2.0 (Relational database of Metabolomics Pathways) has enhanced the interpretation of pathway enrichment study by calculating pathway overrepresentation using three databases (WikiPathways, KEGG, and Reactome) (Zhang et al., 2018).

The case studies provided in the chapter help to provide a demonstration of the applications of NP in predicting targets for plant secondary metabolites, thus being utilized in drug discovery and development. The intricacies involved in network modeling and visualization are also discussed in this chapter. The potential directions for the area of NP for secondary metabolites in plants are gone over, emphasizing new developments in trends and technology that might further the knowledge of the pharmacological effects of these substances. The potential of this technique is highlighted to advance drug discovery and development in the realm of natural products by summarizing the main conclusions and insights from using NP to the study of plant secondary metabolites.

15.2 ROLE OF NETWORK PHARMACOLOGY IN EXPLORING KEY GENES IN PLANTS SECONDARY METABOLITES

While secondary metabolites in plants exhibit a range of functions, including growth regulation (Pang et al., 2021), environmental interaction (Yang et al., 2018), and stress alleviation (Yang et al., 2018; Kosová et al., 2018), their structural and metabolic pathways remain poorly understood. Identifying the pivotal genes in these pathways is crucial for enhancing large-scale production and fully exploiting the potential of these metabolites in biotechnology, horticulture, and pharmacy. The biosynthesis of the secondary metabolites in plants often involves genes and the closely related ones can be identified using co-expression analysis (Moghadam et al., 2023) and NP.

The varied plant-related target information can be collected from many databases and a network map can be constructed in response to the targets selected (Zhong et al., 2023). Evidence suggests that the introduction of secondary metabolite pathways into plants may have occurred through horizontal gene transfer from secondary-metabolite-producing bacteria, as indicated by the distribution of specific key genes or proteins. Additionally, it has been proposed that protobacteria transformed mitochondria, while cyanobacteria evolved into plastids (Wink, 2008).

The compounds identified from the plants of a medicinal nature were determined through varied analyses such as gas chromatography – mass spectrometry, high-resolution liquid chromatography – mass spectrometry, etc., and are subjected to NP investigation in response to the identification of key genes (Mir et al., 2023). NP offers a systemic methodology for comprehending disease pathogenesis, and it can be applied to lead finding, target identification, and prediction of indications. This aligns with the holistic and systematic principles of traditional Chinese medicine theory (Hopkins, 2008; Tang & Aittokallio, 2014).

The potentials of a particular drug with disease-associated or target-associated genes are identified using NP techniques. Even with the provided advantages, the application of NP to evaluate the active components and potential activities of plant extracts has certain limitations (Kalaimathi et al., 2021). In this context, a study in 2019 by Piao et al. highlighted that the interaction relationships among nodes in network construction strategies remain uncertain. They also noted that unexpected inconsistencies in the pathways and functions explored could influence the predicted outcomes. Piao et al. (2019) suggested that additional investigations into active compounds are necessary to validate these predictions made using NP. NP recreates a crucial part in elucidating the role and functions of key genes in plant secondary metabolites, furnishing deeper insights into their pharmacological effects and potential therapeutic applications.

15.3 CORRELATING NETWORK PHARMACOLOGY TO GENES INVOLVED IN SIGNAL TRANSDUCTION

Signal transduction serves as the fundamental mechanism through which cellular activities are regulated in response to environmental cues and regulatory signals. This process is especially crucial for plants, as their survival depends on the appropriate physiological and developmental responses to environmental changes (Wang & Yang, 2012). In order to appropriately respond to specific environmental stimuli such as pathogen attack, drought, high soil salinity, oxidative stress and so on, plants employ signal transduction pathways, intricate networks of interactions concerning signaling elements that propagate via the plant cell. Indeed, cell signaling impacts nearly all aspects of plant cell structure and function (Chen & Heidari, 2020).

In accordance with Chen and Heidari (2020) the construction of a sophisticated and interconnected regulating network involves components, such as receptors, secondary messengers, transcription factors, protein kinases, and plant hormones that control or facilitate other components. Signaling pathways in plants employ intricate networks of interactions to coordinate biochemical and physiological responses (Mulligan et al., 1997).

Compared with the traditional 'one gene, one drug, one disease' mode, NP concentrates on 'multi-targets, multi-components' remedy for diseases. Signal network analysis is a highly attractive tool to explore complex associations. NP is now being used as a holistic tool in describing complex interactions between medicines and biological systems from a network perspective (Zhang et al., 2016). It is now effectively employed combining with pharmacology and pharmacodynamics, to elucidate the molecular network-level mechanisms of TCM (Yue et al., 2017).

The genes involved in signal transduction can be correlated with NP, and this can be achieved through several approaches. KEGG pathway enrichment investigation can be conducted to comprehend the correlation between KEGG pathways and targets that are selected for the analysis (Wu et al., 2020). Regarding the comprehension of key enzymes engaged in secondary metabolite biosynthesis, Zhang et al. (Zhan & Shen, 2022) have outlined the fundamental biosynthetic pathways of usual PSMs. They have endeavored to enhance understanding of the possible targets for stress-mediated management. NP provides a valuable framework for correlating networks with genes involved in signal transduction. So, integrating it with omics data offers a wider, comprehensive approach to studying molecular mechanisms of drug actions and novel drug targets within signaling pathways.

15.4 UTILIZING NETWORK PHARMACOLOGY IN EXPLORING KEY GENES

The following steps exemplify the utilization of NP in examining plant secondary metabolites. **Figure 15.1** depicts the workflow of NP construction and employing it for exploring the key genes and signaling networks in plant secondary metabolites. These instances underscore NP's diverse applications in predicting targets, modeling networks, and annotating functions, thereby providing valuable insights into the pharmacological effects of these compounds.

15.4.1 Target Prediction in Plant Secondary Metabolites

Target prediction is a critical step in comprehending the activity of plant secondary metabolites. This process entails identifying the molecular targets, such as proteins or enzymes, that interact with these compounds to produce their pharmacological effects. By elucidating the targets of plant secondary metabolites, researchers can gain insights into the biological pathways and processes influenced by these compounds. This understanding can direct the evolution of more precise and effective therapies.

NP is a trendy technique for determining active components and potential molecular targets in a range of herbal

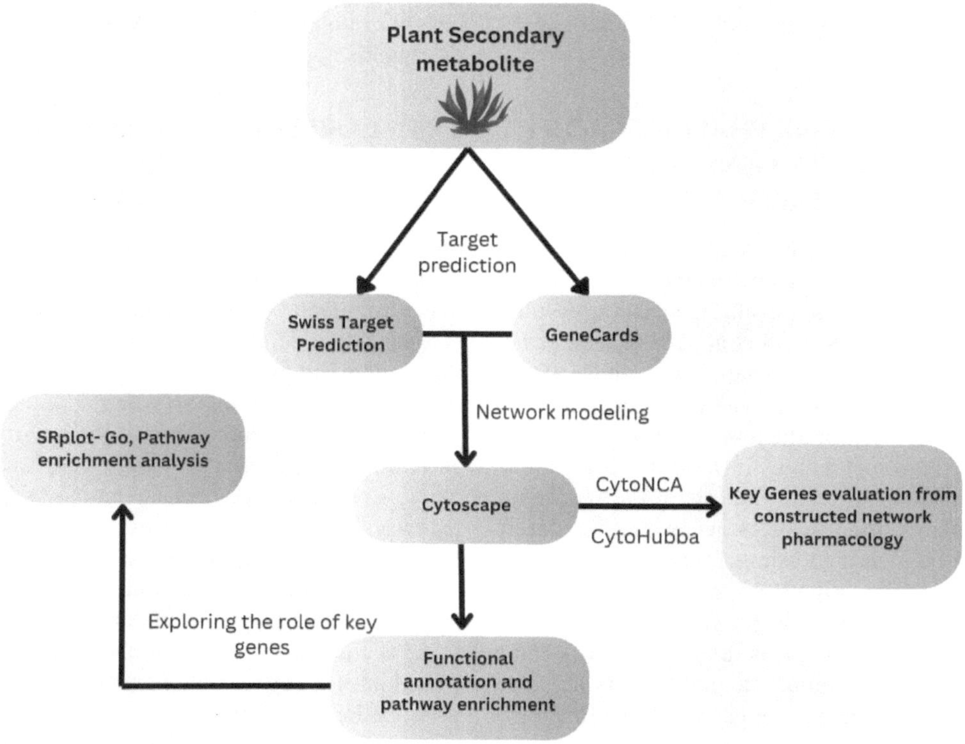

FIGURE 15.1 Flowchart of network pharmacology construction from target prediction to exploring key genes in plant secondary metabolite

mixtures or basic plants and their metabolites (Hao & Xiao, 2014; Yang et al., 2021). It is useful in exploring therapeutic targets of the plants and developing a multitarget drug. It has been said to enhance the safety and effectiveness of existing drugs (Sharma & Yadav, 2022). The targets are compiled in response to an *in silico* evaluation and are selected based on both their retail availability and the likelihood of occurrence based on the prevalence of the plants from which these metabolites originate (Nanusha et al., 2020).

NP is now being used to analyze the active compounds and secondary metabolites present in the plant that target any diseases (Thakur et al., 2023). This technique enables the elucidation of how and where a specific target within the illness network can impact the phenotype of the disease. Consequently, it helps minimize side effects and reduce the risk of drug resistance (Hopkins, 2007; Zhang et al., 2013). A study by Garcia et al. (2023) examined the cytotoxic impacts of extracts from different components of *Perilla frutescens* (leaves, stems, and seeds) on prostate cancer cells (DU-145) *in vitro*. Additionally, it explored potential interactions of putative metabolites with relevant targets associated with prostate cancer using *in silico* methods. NP has been used in revealing relations from multiple affected targets and metabolic pathways (Garcia et al., 2023). The target prediction may usually involve various high throughput analyses like gas chromatography–mass spectrometry and *in silico* analysis using the carefully constructed databases that manage the already curated data of the plants and in this case on the plant secondary metabolites responsible for a particular function in plants.

Herein, the pharmacologically important plant secondary metabolites atropine (PubChem CID: 174174) and codeine (PubChem CID: 5284371) from the plant source *Atropa belladonna* and *Papaver somniferum* (Badyal et al., 2020), respectively, were taken for the construction of their NP. This ensures a way to find the key genes that have a prospective role in the signaling pathways. The targets of the selected plant secondary metabolites were identified using the SwissTargetPrediction software tool (Gfeller et al., 2014) and GeneCards database (Safran et al., 2010), to explore the corresponding molecular targets that potentially interfere with the selected compounds. The predicted targets utilizing this tool and the database revealed 398 genes from the GeneCards database and 73 genes from the SwissTargetPrediction for atropine. In the case of codeine, the identified genes were 204 and 90 from GeneCards and SwissTargetPrediction. The identified genes from the GeneCards database and SwissTargetPrediction were represented in the Venn diagram using the FunRich tool (Pathan et al., 2015) (**Figures 15.2A and 15.2B**). This forms the preliminary step in the construction of the network that depicts the pharmacological interaction of the targets with compounds of interest.

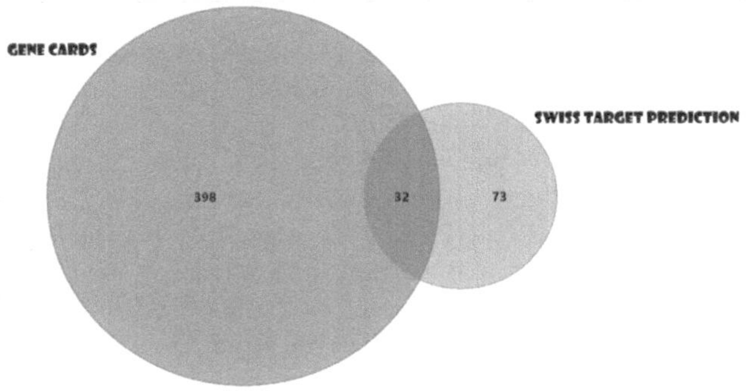

FIGURE 15.2A Venn diagram of the targets predicted from the GeneCards and SwissTarget prediction for the secondary metabolite atropine.

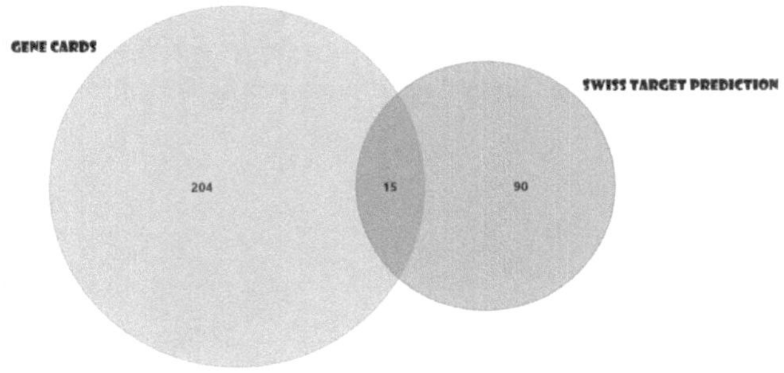

FIGURE 15.2B: Venn diagram of the targets predicted from the GeneCards and SwissTarget prediction for the secondary metabolite codeine.

15.4.2 Network Modeling, Visualization, and Interpretation

NP involves network modeling and visualization as an essential component as it enables researchers to construct and analyze complex biological networks to unravel the pharmacological effects of plant secondary metabolites. It has been shown that a precise depiction of biological reactions – which include metabolites and enzymes, the definition of substrate–product relationships, mass balance, and thermodynamic constraints – is essential for comprehending cellular function, in order for network-based methods to readily understand plant metabolism (Lewis et al., 2012).

Dynamic network modeling seeks to explain how well-known interactions between well-defined components impact the evolution of the elements' states over time as well as the state of the system as a whole under various circumstances. Researchers can monitor the behavioral changes in the system as a result of disturbances by using a passionate model that accurately represents experimentally noted normal behavior. The complementary methodologies of network analysis and dynamic network modeling are best suited for various network sizes. Tens of thousands of nodes and boundaries can be easily analyzed using network analysis, but the material and quantitative features of the procedures associated with the network's edges cannot be explicitly included (Albert, 2007).

The structure of a metabolic network involves the knowledge collected from various sources including the KEGG (Kanehisa et al., 2023), BioCyc (Karp et al., 2019), the Plant Reactome database, and the Gramene database (Tello-Ruiz et al., 2022), even then the information may be insufficient and so the gene-protein, as well as the protein-reaction information, are retrieved from other database sources. Metabolic networks are often depicted as homogeneous networks where the nodes and boundaries reflect the metabolites and responses respectively (Desmet et al., 2021).

Visualization is the most important part of understanding the complex relationships within the biological or metabolic networks. Network graphs and heat maps are some of the visualization techniques used to represent the connection between nodes and edges of a network. This allows for easier identification of clusters and central nodes within the network, aiming to further improve the understanding of plant secondary metabolism. Software such as Cytoscape (Shannon et al., 2003) can be used in the integration of network databases and thus also encompass visualization options in them. It is used to identify the key nodes and modules within the network thus helping in further analysis.

Figures 15.3a and 15.3b depict the network of atropine and codeine with the nodes illustrating the selected compounds and the edges showing the targets interacting with these compounds. This type of visualization helps us to comprehend the interconnection between the compounds and their targets using which the closely interacting targets could be inferred later. Further targets that are closely interacting with the compounds could be identified using a plugin such as CytoNCA (Tang et al., 2015) that is available within the Cytoscape software. From this, the top interacting targets with the selected compounds could be spotted. The top 10 interacting targets for the compound's atropine and

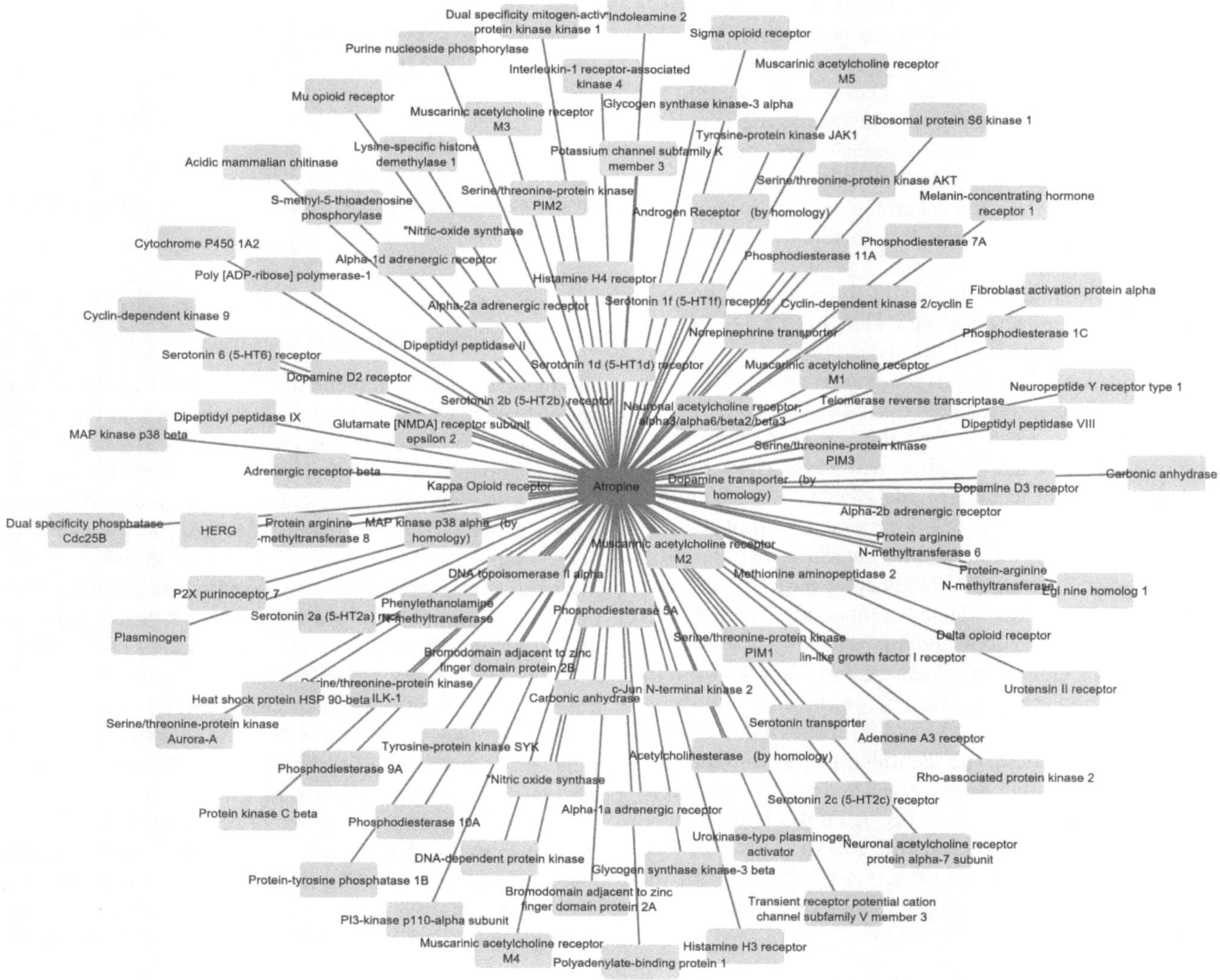

FIGURE 15.3A Network of the secondary metabolite atropine with the edges representing its predicted targets.

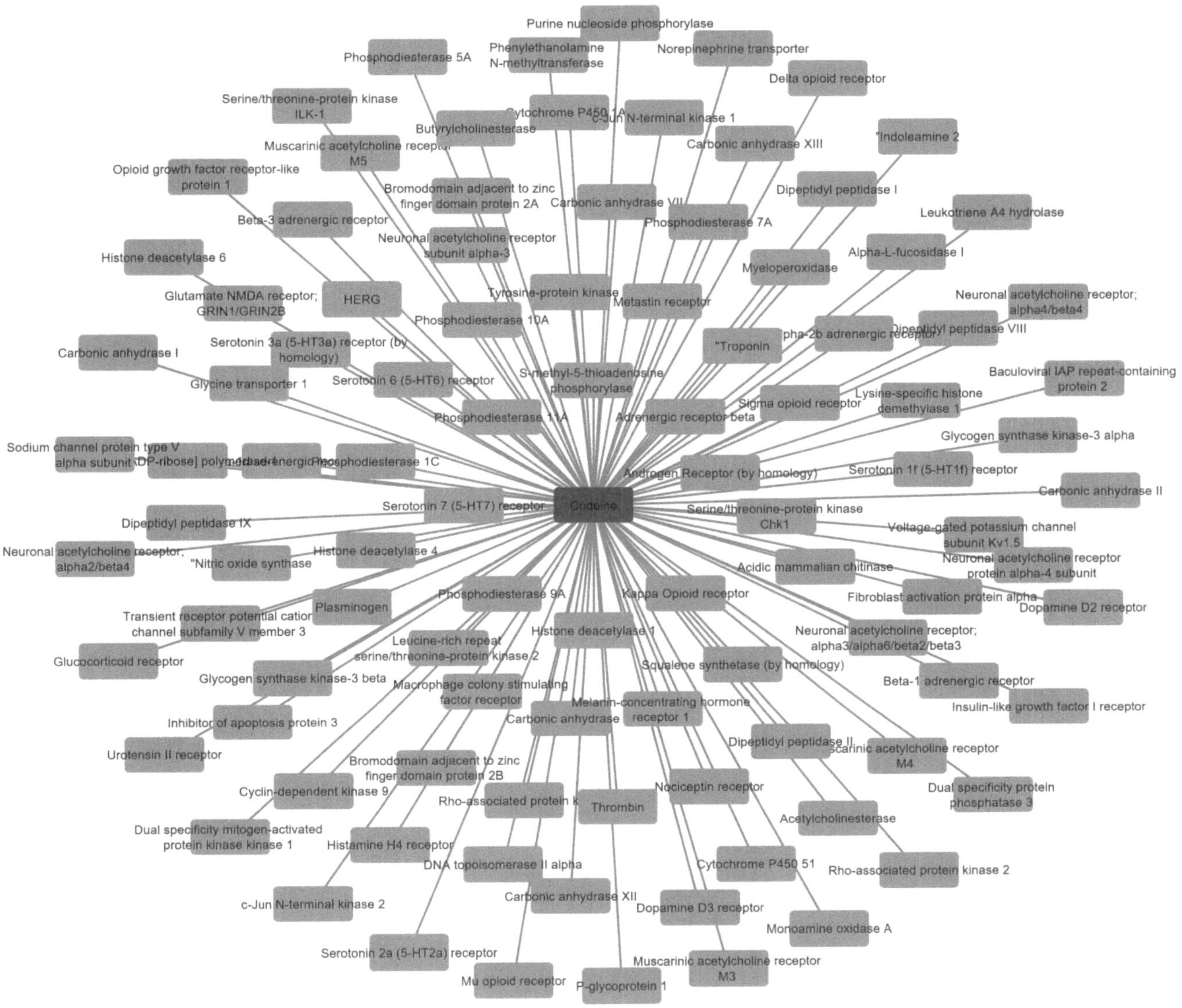

FIGURE 15.3B Network of the secondary metabolite codeine with the edges representing its predicted targets.

codeine from the list of interacting targets were presented in **Figures 15.4a and 15.4b**. With this type of analysis, the targets that are closely associated with the compounds could be identified and can be used for further studies.

In computational biology and bioinformatics, networks and network analysis techniques are fundamental tools that are being utilized more and more to examine biological and clinical data holistically. A group of methods with a common methodological viewpoint, known as network analysis, enables the representation of relationships between items and the examination of the structures that result from these relationships' repetition. The underlying presumption is that studying the relationships between things will lead to more insightful explanations of various events. Networks constructed from omics data can be subjected to network analysis in order to obtain the graph's topological features. Then, knowledge is inferred using these attributes (Milano & Cannataro, 2023).

15.4.3 Evaluation of the Key Genes from the Network Pharmacology

Key genes, which influence the entire network and enhance the production pathways and quantities of secondary metabolites (Bolhassani et al., 2021) can be identified through the application of systems biology methodologies (Pazhamala et al., 2021; Rani et al., 2021). Consequently, the integrative omics approach emerges as a potent analytical tool for biological systems (Pazhamala et al., 2021; Rani et al., 2021), facilitating a deeper comprehension of the interplay between transcriptomes and metabolomes, along with the elucidation of valuable gene networks related to secondary metabolites (Bolhassani et al., 2021; Tahmasebi et al., 2018; Higashi & Saito, 2013). According to Amiri et al. (2023) weighted gene co-expression network analysis, an R package, was utilized for the co-expression analysis of the

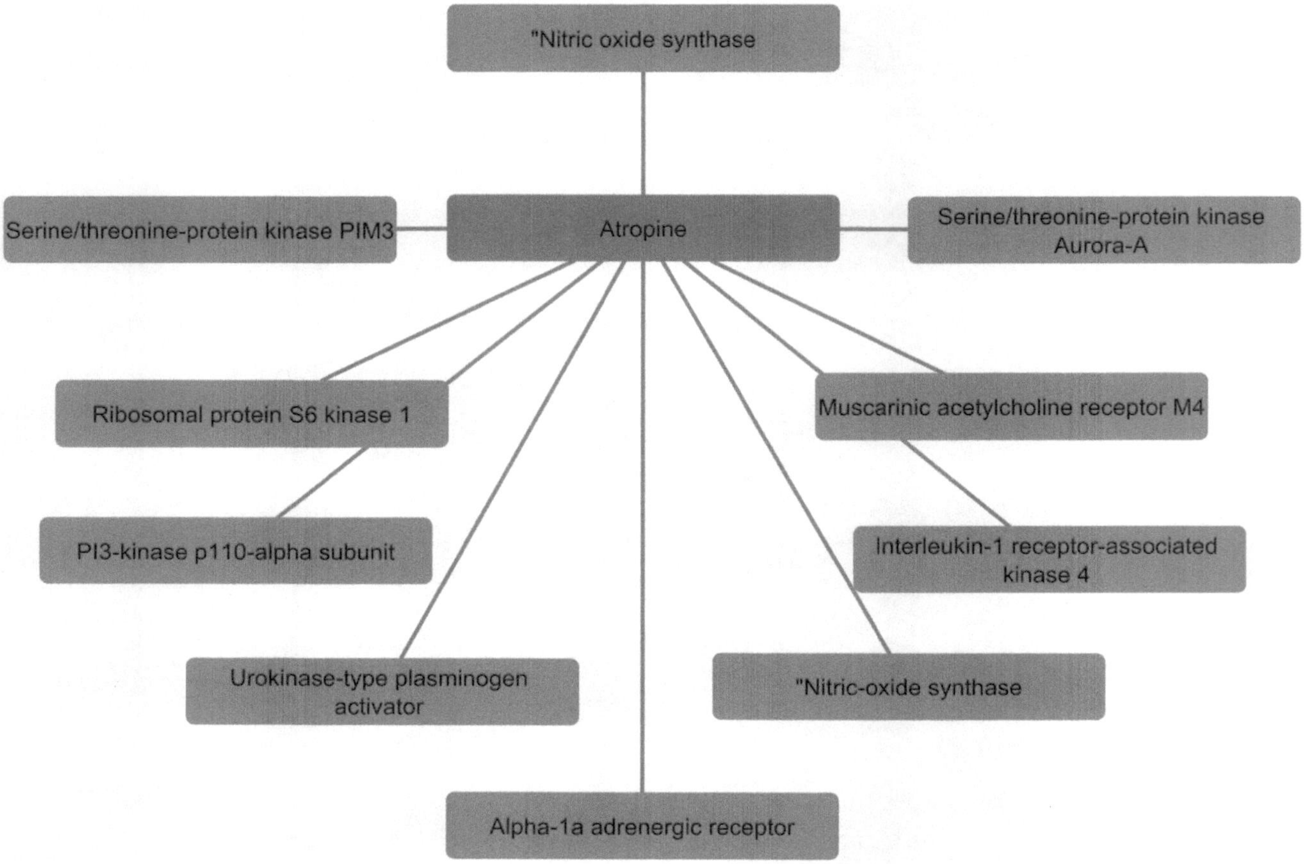

FIGURE 15.4A Top 10 targets that are closely interacting with the compound atropine.

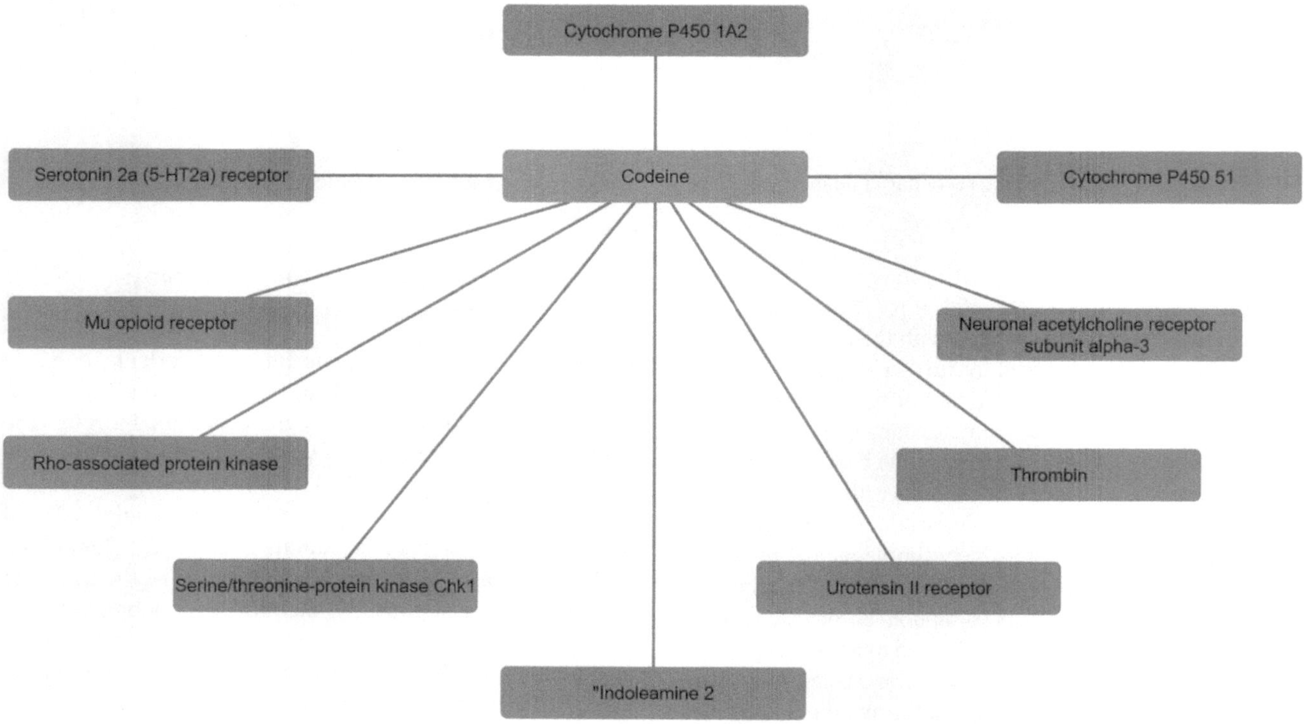

FIGURE 15.4B Top 10 targets that are closely interacting with the compound codeine.

transcriptome and metabolome informatoion. This analysis led to the identification of protein kinases, transcriptional factors, transcriptional regulators, transporters, and hub genes that play crucial roles in the biosynthesis of secondary metabolites in *Digitalis purpurea*.

The identification of key genes is generally followed by the functional evaluation of the genes and an understanding of their roles in controlling the effects of the plant's secondary metabolites. The functional analysis is conducted using the Gene Ontology (GO) database (Aleksander et al., 2023) and the Database for Annotation, Visualization, and Integrated Discovery (DAVID) database (Sherman et al., 2022) where the genes are categorized into three different types – biological processes, molecular functions, and cellular components – to predict their functions. (Amiri et al., 2023)

Experimental validation of key genes in plants involves techniques that confirm the role of genes in biological processes. Several functional genomics research, including quantitative trait locus, genome-wide association research, forward and reverse mutagenesis screens, targeted mutagenesis procedures, and omics techniques, might identify the gene networks' essential genes (Rani et al., 2021; Garagounis et al., 2021). Mechanistic insights into the actions of plant secondary metabolites are provided by experimental validation, which also serves to corroborate the results of NP research.

The key genes that are associated with the plant secondary metabolites can be established by the building of the protein-protein interaction (PPI) network. Here the common genes found within SwissTargetPrediction and GeneCards for the compounds atropine and codeine are shortlisted and used for the construction of the network. The constructed network illustrates the functional relationship of the genes with the selected compounds. Using this the functionally important key gene for the particular plant's secondary metabolite could be highlighted and later on, their specific functions could be annotated using the pathway enrichment analysis. The genes selected for the network construction are shown in **Table 15.1**. Then the PPI network was created using the Cytoscape software. Furthermore, the hub genes were discovered using the cytohubba (Chin et al., 2014) plugin within the Cytoscape established on the closeness of the genes. **Figure 15.5** and **Figure 15.6** show the constructed protein-protein network of the genes identified

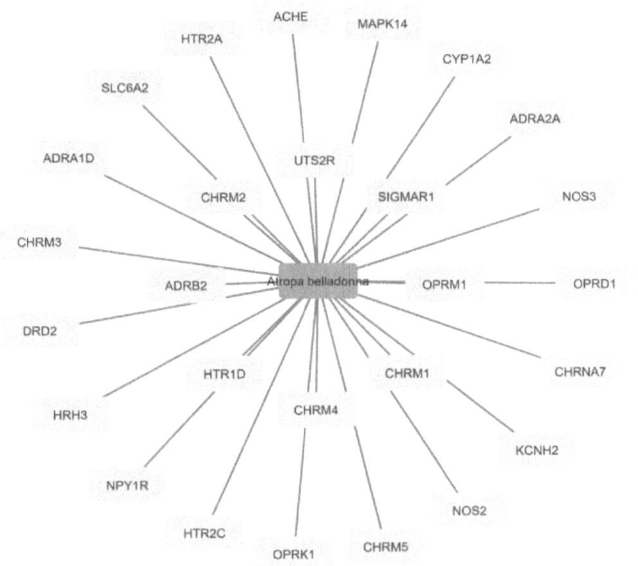

FIGURE 15.5 Functional interaction network of genes from the plant *Atropa belladonna*'s secondary metabolite atropine.

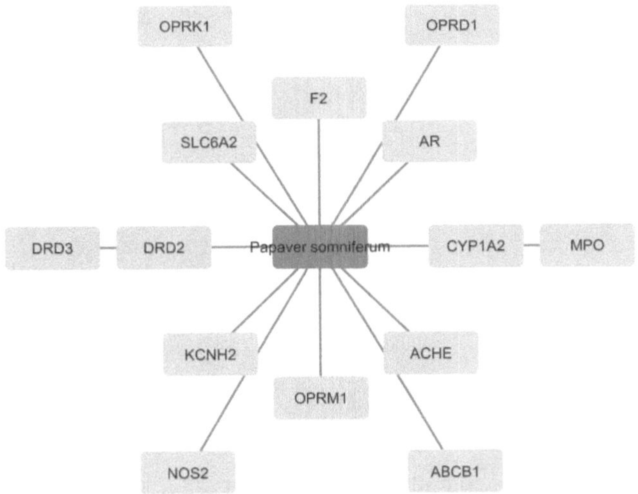

FIGURE 15.6 Functional interaction network of genes from the plant *Papaver somniferum*'s secondary metabolite codeine.

TABLE 15.1
List of Genes Shortlisted from GeneCards and SwissTargetPrediction

S. No.	Plant Source	Secondary Metabolite	Genes
1.	*Atropa belladonna*	Atropine	ACHE, CHRM3, CHRM1, CHRM2, CHRM4, CHRM5, CHRNA7, ADRA2A, HTR2A, DRD2, OPRM1, ADRB2, HTR2C, NOS2, OPRD1, CYP1A2, SLC6A2, HRH3, SIGMAR1, HTR1D, OPRK1, KCNH2, NOS3, MAPK14, ADRA1D, NPY1R, UTS2R.
2.	*Papaver somniferum*	Codeine	OPRM1, OPRD1, OPRK1, ABCB1, CYP1A2, AR, F2, SLC6A2, DRD2, KCNH2, MPO, DRD3, NOS2, ACHE

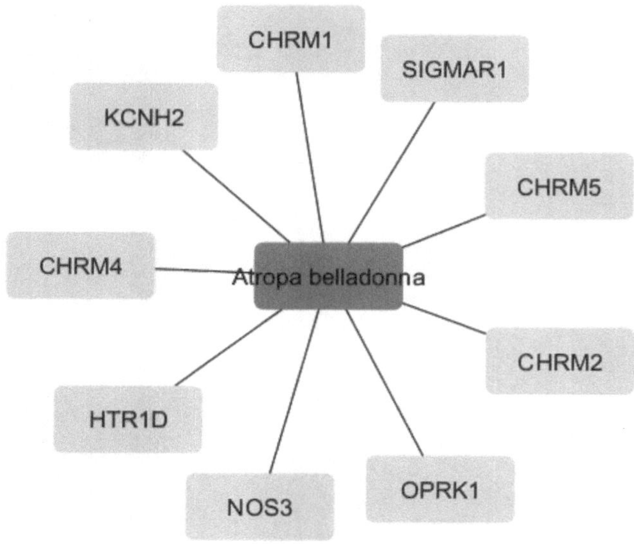

FIGURE 15.7 Top 10 hub genes of the plant secondary metabolite atropine from the plant source *Atropa belladonna*.

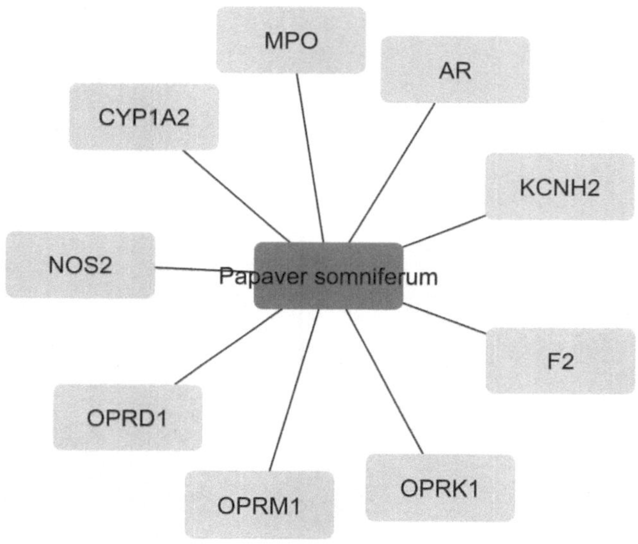

FIGURE 15.8 Top 10 hub genes of the plant secondary metabolite codeine from the plant source *Papaver somniferum*.

from the plant source *Atropa belladonna* and *Papaver somniferum*. **Figure 15.7** demonstrates the key genes from the *Atropa belladonna*'s secondary metabolite atropine and **Figure 15.8** illustrates the key genes from the *Papaver somniferum*'s secondary metabolite codeine.

15.4.4 Functional Annotation and Pathway Enrichment Analysis

Functional annotation is a systematic process that involves gathering and documenting comprehensive information about a gene's biological properties. This includes identifying its various aliases, elucidating its specific molecular functions, defining its biological roles or activities, determining its subcellular localization within the plant cell, and characterizing its expression patterns across different plant tissues or developmental stages. This process provides a detailed understanding of the gene's biological identity and its potential roles in plant physiology, development, and reactions to environmental triggers (Berardini et al., 2004). It is also indentured to validate the gene variation, here in this case in plants.

There are several databases and tools that can be used in functional annotation. There is manual annotation which is difficult to scale, while automatic annotation can easily be used and the prediction can be achieved using local alignment tools such as BLAST (Altschul et al., 1990), where a protein database is searched for high-scoring alignments. The database specifically used for functional annotation is GO (Aleksander et al., 2023), which covers three aspects of gene functions such as the molecular function, the cellular component and the biological process.

Pathways are subjected to statistical tests to determine if they are over-represented in the observed gene index compared to what would be expected by chance. These tests utilize various common statistical methods that take into account factors such as the number of genes detected in the investigation, their comparative ranking, and the number of genes annotated to a pathway of interest. The study on pathways majorly involves covering pathway enrichment examination of a large gene list generally obtained from genome-scale omics techniques (Reimand et al., 2019).

Pathway enrichment investigation also known as functional enrichment study or over-representation examination is a statistical approach utilized to discern pathways that exhibit a higher-than-expected number of genes from the gene list resulting from omics experiments, like gene expression microarrays or germline and somatic genome sequencing., relative to what is predicted by possibility (Reimand et al., 2019) Many freely available tools are known for their user-friendly GUI, extensive documentation and free access for pathway enrichment, such as g:Profiler (Reimand et al., 2016), Gene Set Enrichment Analysis (Subramanian et al., 2005), Cytoscape (Shannon et al., 2003), and EnrichmentMap (Merico et al., 2010).

The biological significance of the selected gene sets was explored using the GO examination. The role of GO investigation is to organize the genes of interest into hierarchical categories established on their general biological functions, cellular locations, and molecular processes. Here the GO research was carried out employing the SR-plot GO, pathway enrichment analysis (Tang et al., 2023). **Figures 15.9a, 15.9b**, and **15.9c** represent the biological functions, cellular locations and molecular processes of the genes from the plant *Atropa belladonna*'s secondary metabolite atropine and their enrichment score was presented in **Figure 15.10**. Similarly, the biological functions, cellular locations and molecular processes of the genes from the plant *Papaver somniferous* secondary metabolite codeine were represented in **Figures 15.11a, 15.11b**, and **15.11c**, and their enrichment analysis was given in **Figure 15.12**.

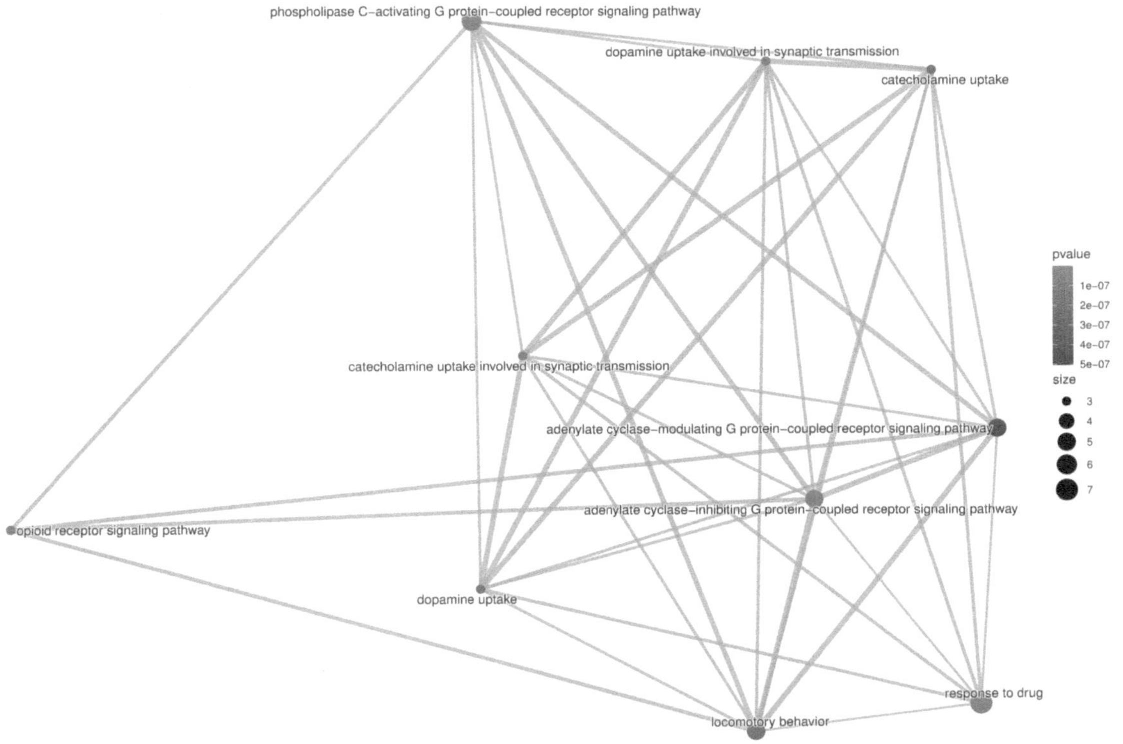

FIGURE 15.9A Emap plot representing the biological functions of the shortlisted genes from GeneCards and SwissTargetPrediction for the compound atropine generated using SRplot.

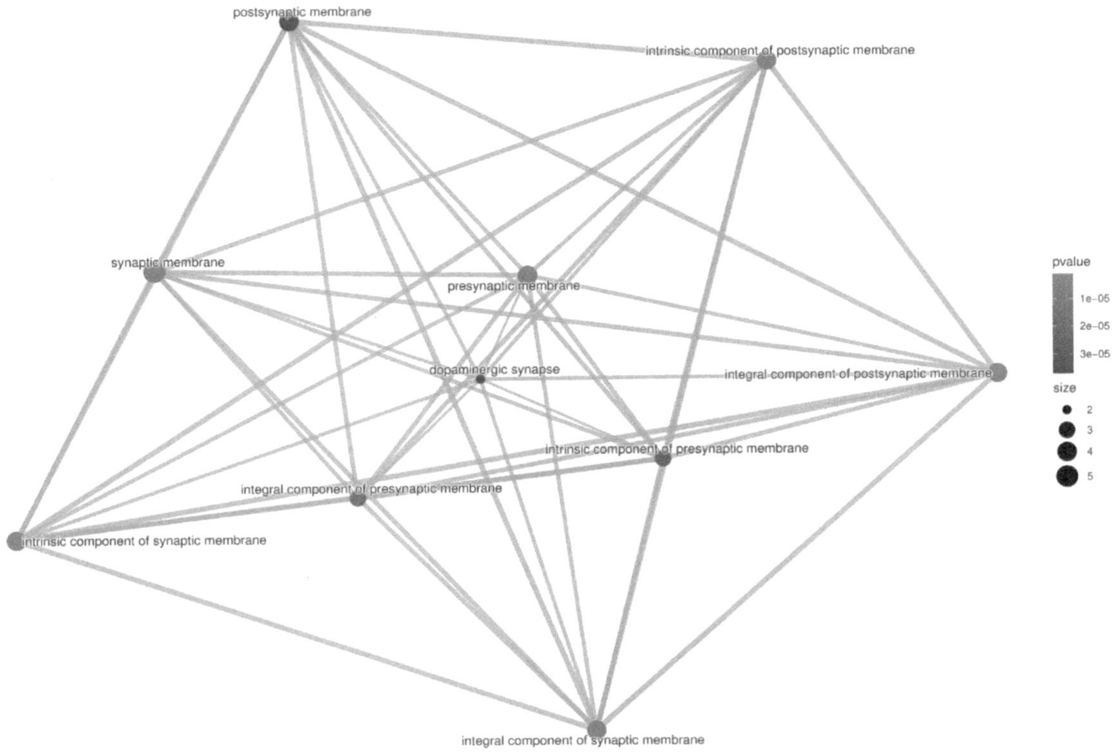

FIGURE 15.9B Emap plot representing the cellular locations of the shortlisted genes from GeneCards and SwissTargetPrediction for the compound atropine generated using SRplot.

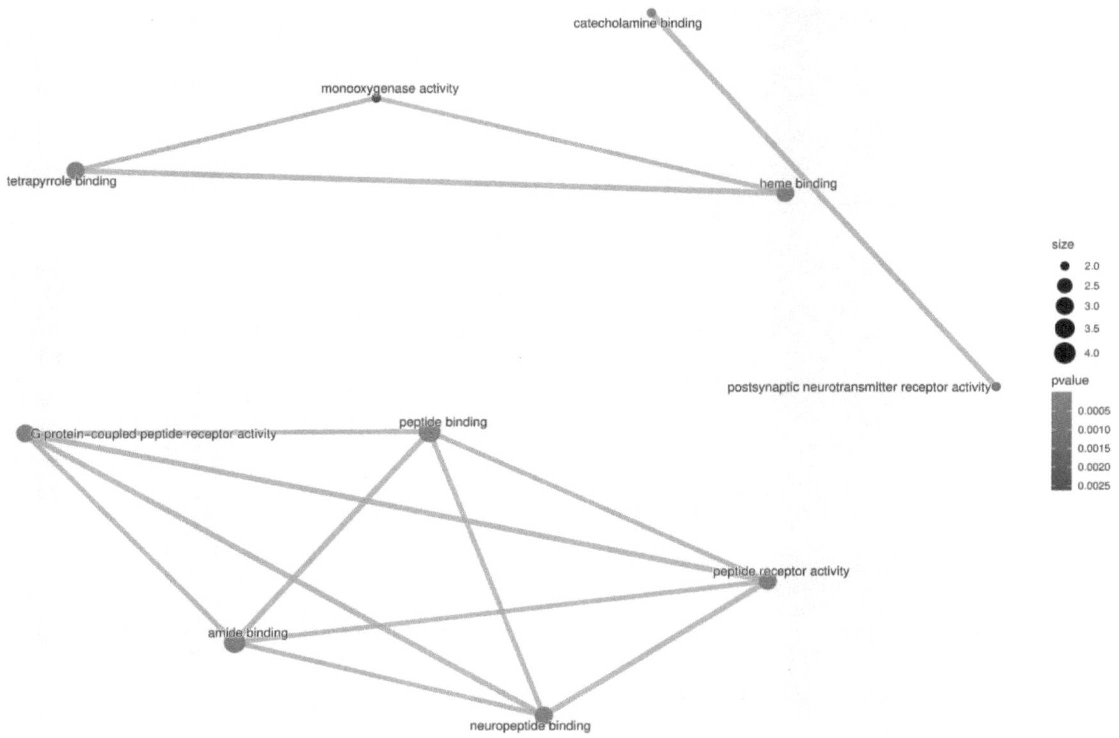

FIGURE 15.9C Emap plot representing the molecular process of the shortlisted genes from the GeneCards and SwissTargetPrediction for the compound atropine generated using SRplot.

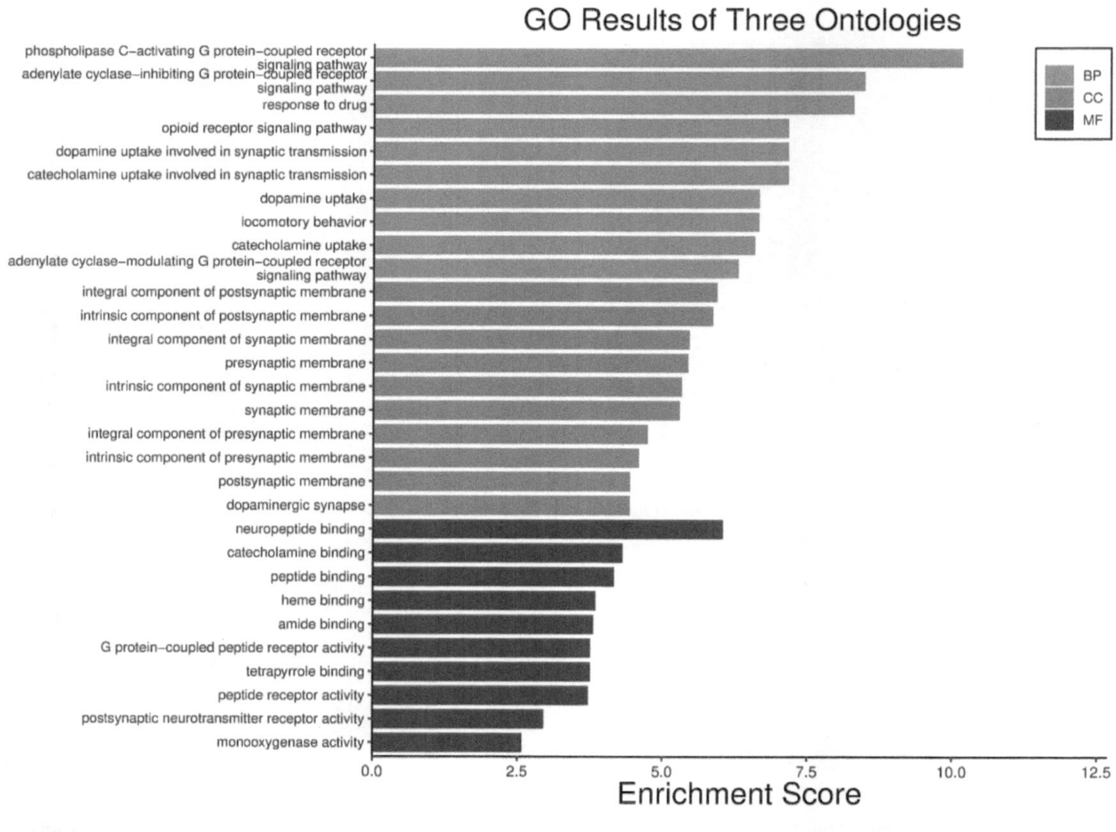

FIGURE 15.10 Barplot representing the enrichment score for all the three GO terms of the shortlisted genes from the plant secondary metabolite atropine.

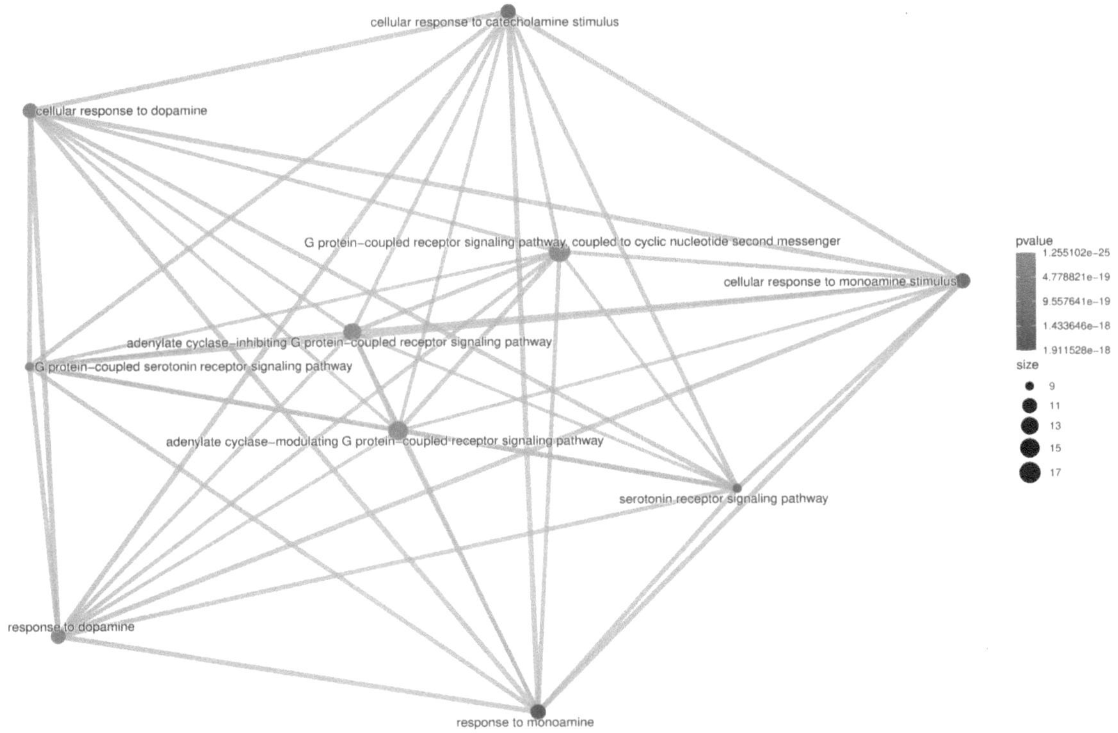

FIGURE 15.11A Emap plot representing the biological functions of the shortlisted genes from the GeneCards and SwissTargetPrediction for the compound codeine generated using SRplot.

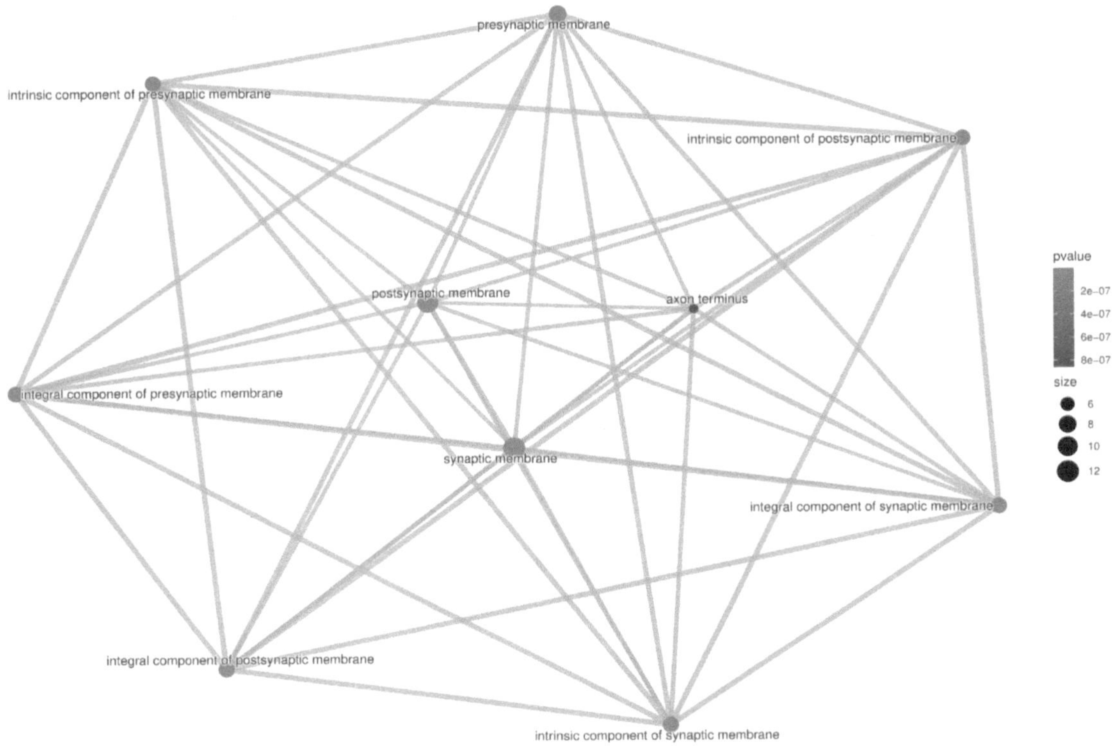

FIGURE 15.11B Emap plot representing the cellular location of the shortlisted genes from the GeneCards and SwissTargetPrediction for the compound codeine generated using SRplot.

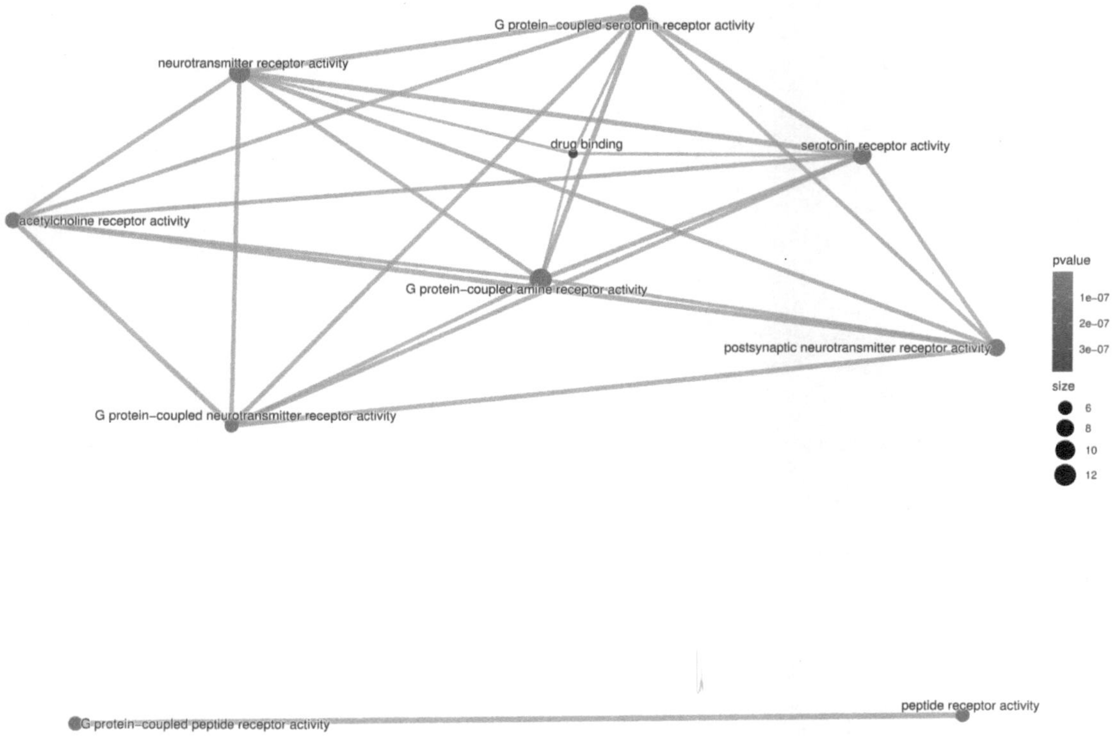

FIGURE 15.11C Emap plot representing the molecular processes of the shortlisted genes from the GeneCards and SwissTargetPrediction for the compound codeine generated using SRplot.

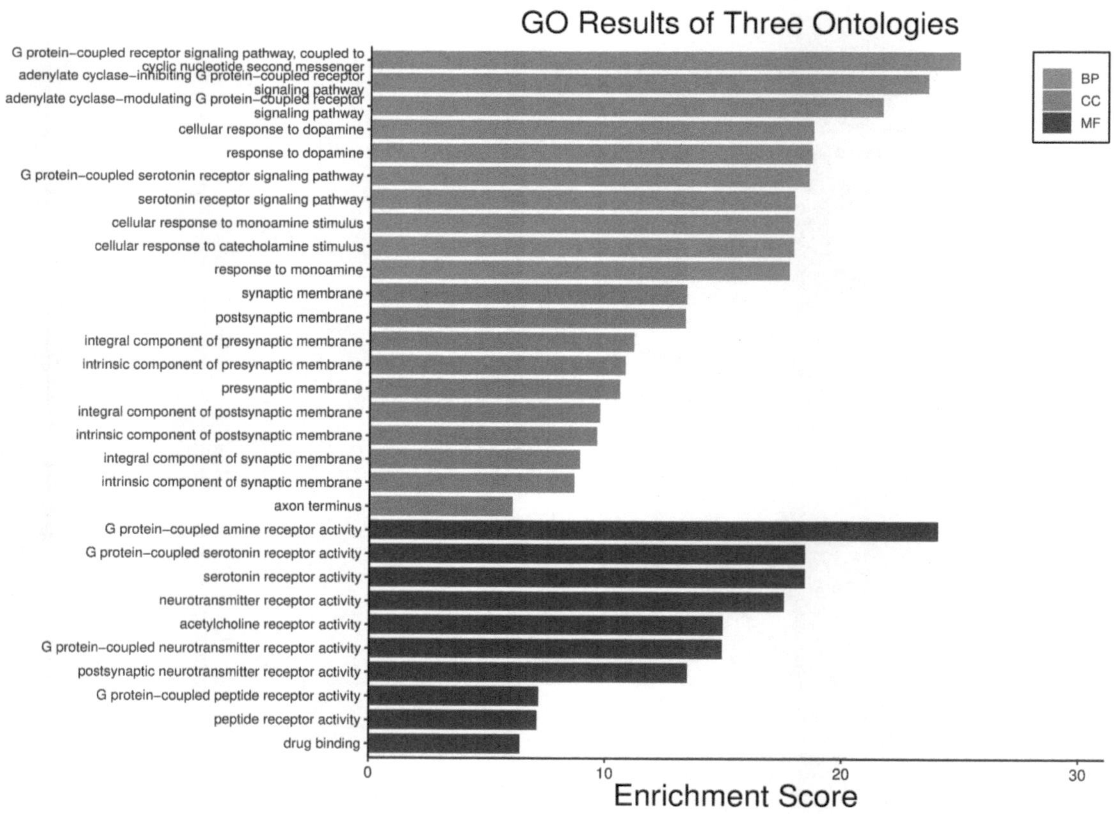

FIGURE 15.12 Barplot representing the enrichment score for all the three Go terms of the shortlisted genes from the plant secondary metabolite codeine

15.4.5 Exploring the Roles of Key Genes in the Signaling Networks

Key genes are integral in signaling networks as they are useful in transmitting signals and modulating various cellular processes. According to Hu et al. (Hu et al., 2019), a thorough grasp of the regulation of nutrient uptake is necessary to guarantee high crop yields sustainably. The signaling networks that regulate the coordinated uptake of nitrogen and phosphorus, the two most sought-after mineral nutrients, are especially crucial. The molecular targets that were identified by NP analysis are substantially connected with common signaling pathways. Numerous activities, such as angiogenesis, invasion, metastasis, and cell proliferation, are regulated by these pathways (Guo et al., 2019).

Signaling or metabolic pathways consist of clusters of genes that interact in a coordinated way to carry out specific biological functions or processes. For example, in a specific signaling pathway, KEGG employs nodes to symbolize genes or gene outcomes and edges to denote signals, like activation or inhibition, transmitted from one gene to another. Similarly, a specific metabolic path is represented by nodes indicating biochemical compounds and edges indicating reactions that convert one or more compounds into others (Chicco & Agapito, 2022).

Certain plant-specialized metabolites (PSMs) function as pivotal elements within intricate signaling pathways. They stimulate the comprehensive expression of genes associated with defense mechanisms and facilitate the assembly of other metabolites (Maag et al., 2015). Signaling pathways were elucidated through the analysis of PPI, GO, and pathway enrichment analysis (Jiang et al., 2021). Functional enrichment investigation helps in finding the target proteins that could be enriched significantly in signaling pathways.

Two of the most extensively studied molecular pathways in breast cancer are the cell cycle protein kinases and the PI3K/AKT signaling path (Olawale et al., 2021). Olawale et al. (2021) conducted a study in which they recognized new inhibitors by computationally filtering a library of therapeutic plant compounds against cyclin-dependent kinase 2 and phosphoinositide-3-kinase A. Exploring the roles of key genes in signaling networks is crucial for understanding the mechanisms of plant secondary metabolites.

The role of the key genes discovered from the NP was subjected to the pathway enrichment analysis using SR plot – Go, pathway enrichment analysis (Tang et al., 2023), to find their role in signaling pathways. **Figure 15.13** illustrates the pathway enrichment of genes identified for the compound atropine and **Figure 15.14** depicts the pathway enrichment of genes identified for the compound codeine. The key genes, such as CHRM1, SIGMAR1, CHRM5, CHRM2, OPRK1, NOS3, HTR1D, CHRM4, and KCNH2, identified from the plant secondary atropine are revealed to be playing a role in pathways, namely, neuroactive ligand-receptor interaction, dopaminergic synapse, biosynthesis of arginine, and chemical carcinogenesis-receptor activation. Likewise, the genes from the compound codeine such as MPO, AR, KCNH2, F2, OPRK1, OPRM1, OPRD1, NOS2, and CYP1A2 are found to have a role in Neuroactive ligand-receptor interaction, calcium signaling path, cholinergic pathway, cAMP signaling path, cGMP-PKG signaling path, pathways of neurodegenerative -multiple diseases, regulation of actin cytoskeleton, Arginine biosynthesis, gap junction and salivary secretion.

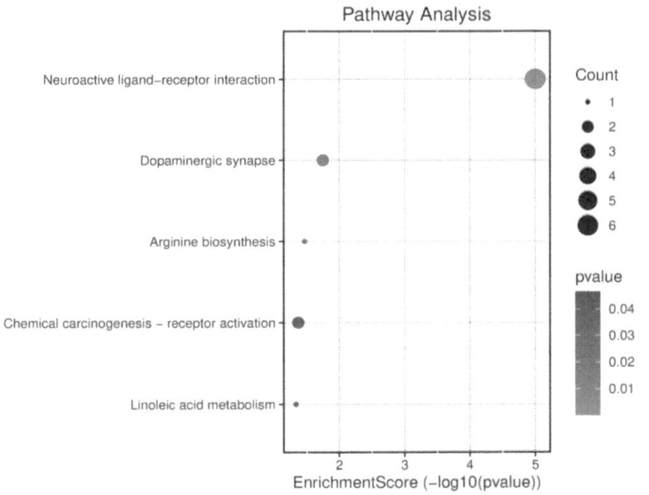

FIGURE 15.13 Pathway enrichment analysis of the genes from the compound atropine.

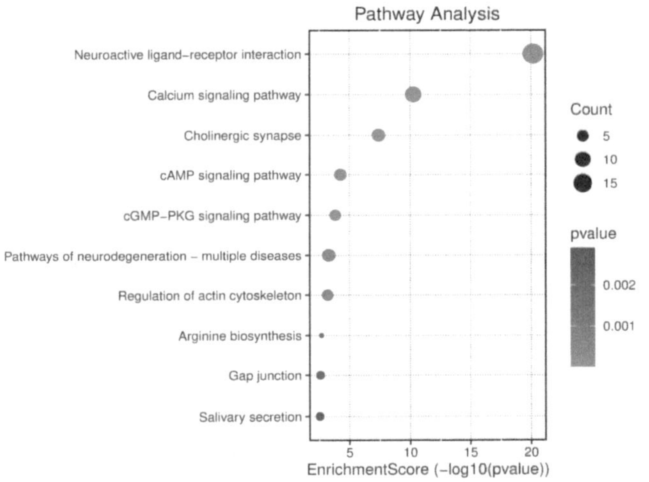

FIGURE 15.14 Pathway enrichment analysis of the genes from the compound codeine.

15.5 FUTURE PERSPECTIVES

The present and the future of NP are promising as they aid in understanding the omics data and of course in understanding the biological workings of a particular plant metabolite or an organism. By collaborating computational and experimental approaches, NP provides comprehensive knowledge of the molecular tools underlying the plant's secondary metabolites and thus helps in the development of novel therapeutics from the source. Signal transduction has been playing an important part in mediating the effects of drugs and other bioactive compounds. Advances in omics technologies like single-cell sequencing and spatial transcriptomic further enhance the understanding of drug action and signal transduction. According to Wani et al. (2021), WRKY transcription factors can help in the future development of crop-improving strategies that is there is a wide scope for further exploration of WRKY transcription factors regarding their functional roles, evolutionary patterns, and interactions with other transcription factors, both cooperatively and antagonistically. Investigating their crosstalk with noncoding RNAs and epigenetic modifications may reveal a common regulatory mechanism within complex signaling pathways, leading to enhancements in agricultural crop yield and quality. The functional annotation and pathway enrichment study of the plant secondary metabolites' future is headed toward the development of more advanced and comprehensive computational tools and databases that could improve the efficiency and accuracy of the analysis. The use of predictive models can help enhance the understanding of the biological pathways in plants.

15.6 CONCLUSION

NP is a computational approach that uses network analysis to analyze the interactions between the compounds and the biological systems. When this method is involved in investigating the plant's secondary metabolites, NP can shed light on the important genes and signaling networks implicated in their regulation, biosynthesis, and interaction with the biological pathways. The predicted key genes and the signaling networks identified through NP can be validated using experimental studies. Overall, NP provides a holistic approach to exploring the interactions between the plant's secondary metabolites, genes, and the target, leading to a better understanding of the compound's biological functions and its potential therapeutic applications.

ACKNOWLEDGEMENT

All the authors are thankful to their respective universities and institutions for their valuable support.

DATA AVAILABILITY

The authors confirm that the data supporting the findings of this study are available within the book/chapter.

REFERENCES

Akilandeswari, G., Vijaya Anand, A., Sampathkumar, P., Vinayaga Moorthi, P., & Preethi, B. (2019). A prospective review on phyto-pharmacological aspects of *Andrographis paniculata*. *Systematic Reviews in Pharmacy*, *10*(1), 15–19.

Albert, R. (2007). Network inference, analysis, and modeling in systems biology. *The Plant Cell*, *19*(11), 3327–3338.

Aleksander, S. A., Balhoff, J., Carbon, S., Cherry, J. M., Drabkin, H. J., Ebert, D., Feuermann, M., Gaudet, P., Harris, N. L., & Hill, D. P. (2023). The gene ontology knowledgebase in 2023. *Genetics*, *224*(1), iyad031.

Altschul, S. F., Gish, W., Miller, W., Myers, E. W., & Lipman, D. J. (1990). Basic local alignment search tool. *Journal of Molecular Biology*, *215*(3), 403–410.

Amiri, F., Moghadam, A., Tahmasebi, A., & Niazi, A. (2023). Identification of key genes involved in secondary metabolite biosynthesis in *Digitalis purpurea*. *Plos One*, *18*(3), e0277293.

Badyal, S., Singh, H., Yadav, A. K., Sharma, S., & Bhushan, I. (2020). Plant secondary metabolites and their uses. *Plant Archives*, *20*(2), 3336–3340.

Batool, S., Javed, M. R., Aslam, S., Noor, F., Javed, H. M. F., Seemab, R., Rehman, A., Aslam, M. F., Paray, B. A., & Gulnaz, A. (2022). Network pharmacology and bioinformatics approach reveals the multi-target pharmacological mechanism of *Fumaria indica* in the treatment of liver cancer. *Pharmaceuticals*, *15*(6), 654.

Begum, T. N., Ilyas, M. M., Kalavathy, S., Anand, A. V., Kumar, S., & Senthil, R. (2009). Effects of ethanolic leaf extract of *Azim tetracantha* Lam. on Ehrlich ascites carcinoma tumour bearing mice. *Research Journal of Medicine and Medical Sciences*, *4*(2), 351–354.

Berardini, T. Z., Mundodi, S., Reiser, L., Huala, E., Garcia-Hernandez, M., Zhang, P., Mueller, L. A., Yoon, J., Doyle, A., Lander, G., & Moseyko, N. (2004). Functional annotation of the Arabidopsis genome using controlled vocabularies. *Plant Physiology*, *135*(2), 745–755.

Bharathi, V., Rengarajan, R. L., Radhakrishnan, R., & Hashem, A., Abd-Allah, E. F., Alqarawi, A. A., & Anand, A. V. (2018). Effects of a medicinal plant *Macrotyloma uniflorum* (Lam.) Verdc. Formulation (MUF) on obesity-associated oxidative stress-induced liver injury. *Saudi Journal of Biological Sciences*, *25*(6), 1115–1121.

Bolhassani, M., Niazi, A., Tahmasebi, A., & Moghadam, A. (2021). Identification of key genes associated with secondary metabolites biosynthesis by system network analysis in *Valeriana officinalis*. *Journal of Plant Research*, *134*(3), 625–639.

Bulgakov, V. P., & Tsitsiashvili, G. S. (2013). Bioinformatics analysis of protein interaction networks: Statistics, topologies, and meeting the standards of experimental biologists. *Biochemistry (Moscow)*, *78*, 1098–1103.

Bulgakov, V. P., Avramenko, T. V., & Tsitsiashvili, G. S. (2017). Critical analysis of protein signaling networks involved in the regulation of plant secondary metabolism: Focus on anthocyanins. *Critical Reviews in Biotechnology*, *37*(6), 685–700.

Chen, J. T., & Heidari, P. (2020). Cell signaling in model plants. *International Journal of Molecular Sciences*, *21*(17), 6062.

Chicco, D., & Agapito, G. (2022). Nine quick tips for pathway enrichment analysis. *PLoS Computational Biology*, *18*(8), e1010348.

Chin, C. H., Chen, S. H., Wu, H. H., Ho, C. W., Ko, M. T., & Lin, C. Y. (2014). cytoHubba: Identifying hub objects and subnetworks from complex interactome. *BMC Systems Biology*, *8*, 1–7.

Delgoda, R., & Murray, J. E. (2017). Evolutionary perspectives on the role of plant secondary metabolites. In *Pharmacognosy* (pp. 93–100). Academic Press.

Desmet, S., Brouckaert, M., Boerjan, W., & Morreel, K. (2021). Seeing the forest for the trees: Retrieving plant secondary biochemical pathways from metabolome networks. *Computational and Structural Biotechnology Journal*, *19*, 72–85.

Divya, N., Rengarajan, R. L., Radhakrishnan, R., Abd-Allah, E. F., Alqarawi, A. A., Hashem, A., & Anand, A. V. (2019). Phytotherapeutic efficacy of the medicinal plant *Terminalia catappa* L. *Saudi Journal of Biological Sciences*, *26*(5), 985–988.

Eronen, L., & Toivonen, H. (2012). Biomine: Predicting links between biological entities using network models of heterogeneous databases. *BMC Bioinformatics*, *13*, 1–21.

Garagounis, C., Delkis, N., & Papadopoulou, K. K. (2021). Unraveling the roles of plant specialized metabolites: Using synthetic biology to design molecular biosensors. *New Phytologist*, *231*(4), 1338–1352.

Garcia, P. J. B., Huang, S. K. H., De Castro-Cruz, K. A., Leron, R. B., & Tsai, P. W. (2023). An in vitro evaluation and network pharmacology analysis of prospective anti-prostate cancer activity from *Perilla frutescens*. *Plants*, *12*(16), 3006.

Gfeller, D., Grosdidier, A., Wirth, M., Daina, A., Michielin, O., & Zoete, V. (2014). SwissTargetPrediction: A web server for target prediction of bioactive small molecules. *Nucleic acids Research*, *42*(W1), W32–W38.

Guerriero, G., Berni, R., Muñoz-Sanchez, J. A., Apone, F., Abdel-Salam, E. M., Qahtan, A. A., Alatar, A. A., Cantini, C., Cai, G., Hausman, J. F., & Siddiqui, K. S., Faisal, M. (2018). Production of plant secondary metabolites: Examples, tips and suggestions for biotechnologists. *Genes*, *9*(6), 309.

Guo, W., Huang, J., Wang, N., Tan, H. Y., Cheung, F., Chen, F., & Feng, Y. (2019). Integrating network pharmacology and pharmacological evaluation for deciphering the action mechanism of herbal formula zuojin pill in suppressing *Hepatocellular carcinoma*. *Frontiers in Pharmacology*, *10*, 466724.

Hao, D. C., & Xiao, P. G. (2014). Network pharmacology: A Rosetta stone for traditional Chinese medicine. *Drug Development Research*, *75*(5), 299–312.

Higashi, Y., & Saito, K. (2013). Network analysis for gene discovery in plant-specialized metabolism. *Plant, Cell & Environment*, *36*(9), 1597–1606.

Hopkins, A. L. (2007). Network pharmacology. *Nature Biotechnology*, *25*(10), 1110–1111.

Hopkins, A. L. (2008). Network pharmacology: The next paradigm in drug discovery. *Nature Chemical Biology*, *4*(11), 682–690.

Hu, B., Jiang, Z., Wang, W., Qiu, Y., Zhang, Z., Liu, Y., Li, A., Gao, X., Liu, L., Qian, Y., & Huang, X. (2019). Nitrate–NRT1.1B–SPX4 cascade integrates nitrogen and phosphorus signalling networks in plants. *Nature Plants*, *5*(4), 401–413.

Jiang, N., Li, H., Sun, Y., Zeng, J., Yang, F., Kantawong, F., & Wu, J. (2021). Network pharmacology and pharmacological evaluation reveals the mechanism of the *Sanguisorba officinalis* in suppressing *Hepatocellular carcinoma*. *Frontiers in Pharmacology*, *12*, 618522.

Kalaimathi, K., Thiyagarajan, G., Vijayakumar, S., Bhavani, K., Karthikeyan, K., Rani, J. M. J., Dass, K., Sureshkumar, J., & Prabhu, S. (2021). Molecular docking and network pharmacology-based approaches to explore the potential of terpenoids for *Mycobacterium tuberculosis*. *Pharmacological Research-Modern Chinese Medicine*, *1*, 100002.

Kanehisa, M., Furumichi, M., Sato, Y., Kawashima, M., & Ishiguro-Watanabe, M. (2023). KEGG for taxonomy-based analysis of pathways and genomes. *Nucleic Acids Research*, *51*, D587–D592.

Karp, P. D., Billington, R., Caspi, R., Fulcher, C. A., Latendresse, M., Kothari, A., Keseler, I. M., Krummenacker, M., Midford, P. E., Ong, Q., & Ong, W. K. (2019). The BioCyc collection of microbial genomes and metabolic pathways. *Briefings in Bioinformatics*, *20*(4), 1085–1093.

Karthika, P., Balamuralikrishnan, B., Yamini, K., Vijaya Anand, A., Durairaj, K., Maruthupandian, A., Alodaini, H. A., Hatamleh, A. A., Manikantan, P., & Arun, M. (2023). Green synthesis, characterization of silver nanoparticles using aqueous leaf extracts of *Solanum melongena* and in vitro evaluation of antibacterial, pesticidal and anticancer activity in human MDA-MB-231 breast cancer cell lines. *Journal of King Saud University – Science*, *35*(5), 102663.

Kaviya, M., Balasubramanian, B., Bharathi, K., Malaisamy, A., Al-Dhabi, N. A., Mariadhas, V. A., Anand, A. V., & Liu, W. (2022). Evaluation of nutritional substances and investigation of antioxidant and antimicrobial potentials of *Boerhavia diffusa* with in silico molecular docking. *Molecules*, *27*(4), 1280.

Kosová, K., Vítámvás, P., Urban, M. O., Prášil, I. T., & Renaut, J. (2018). Plant abiotic stress proteomics: The major factors determining alterations in cellular proteome. *Frontiers in Plant Science*, *9*, 325823.

Lewis, N. E., Nagarajan, H., & Palsson, B. O. (2012). Constraining the metabolic genotype–phenotype relationship using a phylogeny of in silico methods. *Nature Reviews Microbiology*, *10*(4), 291–305.

Liu, W., Lin, L., Zhang, Z., Liu, S., Gao, K., Lv, Y., Tao, H., & He, H. (2019). Gene co-expression network analysis identifies trait-related modules in *Arabidopsis thaliana*. *Planta*, *249*, 1487–1501.

Maag, D., Erb, M., Köllner, T. G., & Gershenzon, J. (2015). Defensive weapons and defense signals in plants: Some metabolites serve both roles. *BioEssays*, *37*(2), 167–174.

Manikandan, R., Balasubramanian, B., Punniyakotti, P., Vijaya Anand, A., Meyyazhagan, A., Velauthaprabhu, S., & Liu, W. C. (2022). Cardio-protective effects of *Terminalia catappa* leaves and *Terminalia chebula* fruit extract in doxorubicin-induced cardiomyopathy in rats. *Biomarkers*, *27*(5), 488–495.

Manoharan, N., Sampathkumar, P., Dheeba, B., Sheik Abdulla, S., Vinothprasanna, G., Vinothkannan, R., Kalavathy, S., Vijaya Anand, A., & Shanmugasundaram, A. (2008). Potential hepatoprotective effect of aqueous extract of *Gracilaria corticata* in AFB_1 induced hepatotoxicity in Wistar rats. *Journal of Biological Sciences*, *8*, 1352–1355.

Merico, D., Isserlin, R., Stueker, O., Emili, A., & Bader, G. D. (2010). Enrichment map: A network-based method for gene-set enrichment visualization and interpretation. *PloS One*, *5*(11), e13984.

Milacic, M., Beavers, D., Conley, P., Gong, C., Gillespie, M., Griss, J., Haw, R., Jassal, B., Matthews, L., May, B., & Petryszak, R. (2024). The reactome pathway knowledgebase 2024. *Nucleic Acids Research*, *52*(D1), D672–D678.

Milano, M., & Cannataro, M. (2023). Network models in bioinformatics: Modeling and analysis for complex diseases. *Briefings in Bioinformatics*, *24*(2), bbad016.

Miljkovic, D., Stare, T., Mozetič, I., Podpečan, V., Petek, M., Witek, K., Dermastia, M., Lavrač, N., & Gruden, K. (2012). Signalling network construction for modelling plant defence response. *PLoS One*, *7*(12), e51822.

Mir, W. R., Bhat, B. A., Kumar, A., Dhiman, R., Alkhanani, M., Almilaibary, A., Dar, M. Y., Ganie, S. A., & Mir, M. A. (2023). Network pharmacology combined with molecular docking and in vitro verification reveals the therapeutic potential of *Delphinium roylei* munz constituents on breast carcinoma. *Frontiers in Pharmacology*, *14*, 1135898.

Moghadam, A., Foroozan, E., Tahmasebi, A., Taghizadeh, M. S., Bolhassani, M., & Jafari, M. (2023). System network analysis of *Rosmarinus officinalis* transcriptome and metabolome – key genes in biosynthesis of secondary metabolites. *Plos One*, *18*(3), e0282316.

Mubeen, S., Tom Kodamullil, A., Hofmann-Apitius, M., & Domingo-Fernández, D. (2022). On the influence of several factors on pathway enrichment analysis. *Briefings in Bioinformatics*, *23*(3), bbac143.

Mulligan, R. M., Chory, J., & Ecker, J. R. (1997). Signaling in plants. *Proceedings of the National Academy of Sciences*, *94*(7), 2793–2795.

Muthumani, G. D., Anand, A. V., & Manikandan, R. (2013). Antioxidant, antihelmintic and antimicrobial activity of *Euphorbia thymifolia* Linn whole plant. *International Journal of Current Microbiology and Applied Sciences*, *2*(1), 66–75.

Nanusha, M. Y., Krauss, M., Schönsee, C. D., Günthardt, B. F., Bucheli, T. D., & Brack, W. (2020). Target screening of plant secondary metabolites in river waters by liquid chromatography coupled to high-resolution mass spectrometry (LC–HRMS). *Environmental Sciences Europe*, *32*, 1–11.

Ncube, B., & Van Staden, J. (2015). Tilting plant metabolism for improved metabolite biosynthesis and enhanced human benefit. *Molecules*, *20*(7), 12698–12731.

Noor, F., Rehman, A., Ashfaq, U. A., Saleem, M. H., Okla, M. K., Al-Hashimi, A., AbdElgawad, H., & Aslam, S. (2022). Integrating network pharmacology and molecular docking approaches to decipher the multi-target pharmacological mechanism of *Abrus precatorius* L. acting on diabetes. *Pharmaceuticals*, *15*(4), 414.

Olawale, F., Iwaloye, O., Elekofehinti, O. O., Kikiowo, B., Oluwarotimi, E. A., Ilesanmi, K. M., Akinropo, I. D., Akinlosotu, O. B., Adegboyega, A. E., & Ologuntere, T. E. (2021). A multi-target approach for the discovery of anti-breast cancer agents from plants secondary metabolites. *Letters in Drug Design & Discovery*, *18*(10), 1009–1023.

Pang, Z., Chen, J., Wang, T., Gao, C., Li, Z., Guo, L., Xu, J., & Cheng, Y. (2021). Linking plant secondary metabolites and plant microbiomes: A review. *Frontiers in Plant Science*, *12*, 621276.

Pathan, M., Keerthikumar, S., Ang, C. S., Gangoda, L., Quek, C. Y., Williamson, N. A., Mouradov, D., Sieber, O. M., Simpson, R. J., Salim, A., & Bacic, A. (2015). FunRich: An open access standalone functional enrichment and interaction network analysis tool. *Proteomics*, *15*(15), 2597–2601.

Pazhamala, L. T., Kudapa, H., Weckwerth, W., Millar, A. H., & Varshney, R. K. (2021). Systems biology for crop improvement. *The Plant Genome*, *14*(2), e20098.

Piao, C. L., Luo, J. L., Jin, D., Tang, C., Wang, L., Lian, F. M., & Tong, X. L. (2019). Utilizing network pharmacology to explore the underlying mechanism of *Radix salviae* in diabetic retinopathy. *Chinese Medicine*, *14*, 1–12.

Potapov, A. P., Goemann, B., & Wingender, E. (2008). The pairwise disconnectivity index as a new metric for the topological analysis of regulatory networks. *BMC Bioinformatics*, *9*, 1–15.

Rani, S. G., Kumari, H., Pati, P., Naaz, S., Prasad, T., Kumari, R., Qayyum, H., Naaz, N., & Nitin, M. (2021). Recent approaches of systems biology and omics in plant research. *Journal of Current Opinion in Crop Science*, *2*(2), 288–305.

Reimand, J., Arak, T., Adler, P., Kolberg, L., Reisberg, S., Peterson, H., & Vilo, J. (2016). g:Profiler – a web server for functional interpretation of gene lists (2016 update). *Nucleic Acids Research*, *44*(W1), W83–W89.

Reimand, J., Isserlin, R., Voisin, V., Kucera, M., Tannus-Lopes, C., Rostamianfar, A., Wadi, L., Meyer, M., Wong, J., Xu, C., & Merico, D. (2019). Pathway enrichment analysis and visualization of omics data using g:Profiler, GSEA, Cytoscape and EnrichmentMap. *Nature Protocols*, *14*(2), 482–517.

Safran, M., Dalah, I., Alexander, J., Rosen, N., Iny Stein, T., Shmoish, M., Nativ, N., Bahir, I., Doniger, T., Krug, H., & Sirota-Madi, A. (2010). GeneCards Version 3: The human gene integrator. *Database*, *2010*.

Sangeetha, T., Balamuralikrishnan, B., Sampathkumar, P., Velayuthaprabhu, S., Senthilkumar, N., Syed Ibrahim, K., Baskaran, R., & Vijaya Anand, A. (2022). Characterization and phytoconstituents of *Petroselinum crispum* (Mill) and *Coriandrum sativum* (Linn) and their impacts on inflammation – an in vitro analysis against human adenocarcinoma cells with molecular docking. *South African Journal of Botany*, *146*, 776–788.

Shannon, P., Markiel, A., Ozier, O., Baliga, N. S., Wang, J. T., Ramage, D., Amin, N., Schwikowski, B., & Ideker, T. (2003). Cytoscape: A software environment for integrated models of biomolecular interaction networks. *Genome Research*, *13*(11), 2498–2504.

Sharma, B., & Yadav, D. K. (2022). Metabolomics and network pharmacology in the exploration of the multi-targeted therapeutic approach of traditional medicinal plants. *Plants*, *11*(23), 3243.

Sherman, B. T., Hao, M., Qiu, J., Jiao, X., Baseler, M. W., Lane, H. C., Imamichi, T., & Chang, W. (2022). DAVID: A web server for functional enrichment analysis and functional annotation of gene lists (2021 update). *Nucleic Acids Research*, *50*(W1), W216–W221.

Subramanian, A., Tamayo, P., Mootha, V. K., Mukherjee, S., Ebert, B. L., Gillette, M. A., Paulovich, A., Pomeroy, S. L., Golub, T. R., Lander, E. S., & Mesirov, J. P. (2005). Gene set enrichment analysis: A knowledge-based approach for interpreting genome-wide expression profiles. *Proceedings of the National Academy of Sciences*, *102*(43), 15545–15550.

Tahmasebi, A., Aram, F., Pakniyat, H., Niazi, A., Tavakol, E., & Ebrahimie, E. (2018). Global analysis of gene expression and identification of modules in *Echinacea purpurea* using systems biology approach. *Journal of Cell and Molecular Research*, *10*(1), 18–26.

Tang, D., Chen, M., Huang, X., Zhang, G., Zeng, L., Zhang, G., Wu, S., & Wang, Y. (2023). SRplot: A free online platform for data visualization and graphing. *PLoS One*, *18*(11), e0294236.

Tang, J., & Aittokallio, T. (2014). Network pharmacology strategies toward multi-target anticancer therapies: From computational models to experimental design principles. *Current Pharmaceutical Design*, *20*(1), 23–36.

Tang, Y., Li, M., Wang, J., Pan, Y., & Wu, F. X. (2015). CytoNCA: A cytoscape plugin for centrality analysis and evaluation of protein interaction networks. *Biosystems*, *127*, 67–72.

Taubert, J., Hassani-Pak, K., Castells-Brooke, N., & Rawlings, C. J. (2014). Ondex Web: Web-based visualization and exploration of heterogeneous biological networks. *Bioinformatics*, *30*(7), 1034–1035.

Tello-Ruiz, M. K., Jaiswal, P., & Ware, D. (2022). Gramene: A resource for comparative analysis of plants genomes and pathways. In *Plant bioinformatics: Methods and protocols* (pp. 101–131). Springer US.

Thakur, P., Kumar, R., Choudhary, N., Sharma, R., & Chaudhary, A. (2023). Network pharmacology on mechanistic role of *Thymus linearis* Benth. against gastrointestinal and neurological diseases. *Phytomedicine*, *121*, 155098.

Varalakshmi, B., Anand, A. V., Karpagam, T., Bai, J. S., & Manikandan, R. (2014). In vitro antimicrobial and anticancer activity of *Cinnamomum zeylanicum* Linn bark extracts.

International Journal of Pharmacy and Pharmaceutical Sciences, *6*(1), 12–18.

Vijaya Anand, A., Varalakshmi, B., Prasana., Sampathkumar, P., Pushpa, N., & Hedina, A. (2016). *Cinnamomum zeylanicum* Linn. The spice with multi potential. *Systematic Reviews in Pharmacy*, *7*(1), 24–29.

Vijaya Anand, A., Velayuthaprabhu, S., Rengarajan, R. L., Sampathkumar, P., & Radhakrishnan, R. (2020). Bioactive compounds of guava (*Psidium guajava* L.). *Bioactive Compounds in Underutilized Fruits and Nuts*, Reference Series in Phytochemistry (pp. 503–527). Springer.

Vijayakumar, K., Prasanna, B., Rengarajan, R. L., Rathinam, A., Velayuthaprabhu, S., & Vijaya Anand, A. (2023). Anti-diabetic and hypolipidemic effects of *Cinnamon cassia* bark extracts: An in vitro, in vivo, and in silico approach. *Archieves of Physiology and Biochemistry*, *129*(2), 338–348.

Wang, Z. Y., & Yang, Z. (Eds.). (2012). *Plant signalling networks: Methods and protocols*. Humana Press.

Wani, S. H., Anand, S., Singh, B., Bohra, A., & Joshi, R. (2021). WRKY transcription factors and plant defense responses: Latest discoveries and future prospects. *Plant Cell Reports*, *40*, 1071–1085.

Wink, M. (2008). Plant secondary metabolism: Diversity, function and its evolution. *Natural Product Communications*, *3*(8), 1934578X0800300801.

Wu, J., Luo, D., & Li, S. (2020). Network pharmacology-oriented identification of key proteins and signaling pathways targeted by Xihuang pill in the treatment of breast cancer. *Breast Cancer: Targets and Therapy*, 267–277.

Yang, H., Cheung, M. K., Yue, G. G. L., Leung, P. C., Wong, C. K., & Lau, C. B. S. (2021). Integrated network pharmacology analysis and in vitro validation revealed the potential active components and underlying mechanistic pathways of herba patriniae in colorectal cancer. *Molecules*, *26*(19), 6032.

Yang, L., Wen, K. S., Ruan, X., Zhao, Y. X., Wei, F., & Wang, Q. (2018). Response of plant secondary metabolites to environmental factors. *Molecules*, *23*(4), 762.

Yue, S. J., Xin, L. T., Fan, Y. C., Li, S. J., Tang, Y. P., Duan, J. A., Guan, H. S., & Wang, C. Y. (2017). Herb pair Danggui-Honghua: Mechanisms underlying blood stasis syndrome by system pharmacology approach. *Scientific Reports*, *7*(1), 40318.

Zhan, X., & Shen, C. (2022). Environmental and genetic factors involved in plant protection-associated secondary metabolite biosynthesis pathways. *Frontiers in Plant Science*, *13*, 877304.

Zhang, B., Hu, S., Baskin, E., Patt, A., Siddiqui, J. K., & Mathé, E. A. (2018). RaMP: A comprehensive relational database of metabolomics pathways for pathway enrichment analysis of genes and metabolites. *Metabolites*, *8*(1), 16.

Zhang, G. B., Li, Q. Y., Chen, Q. L., & Su, S. B. (2013). Network pharmacology: A new approach for Chinese herbal medicine research. *Evidence-based Complementary and Alternative Medicine*, *2013*.

Zhang, Y. Q., Mao, X., Guo, Q. Y., Lin, N., & Li, S. (2016). Network pharmacology-based approaches capture essence of Chinese herbal medicines. *Chinese Herbal Medicines*, *8*(2), 107–116.

Zhong, H., Duan, B. H., Du, F. M., Wang, W. M., & Qiao, H. (2023). Identification of key genes, biological functions, and pathways of empagliflozin by network pharmacology and its significance in the treatment of type 2 diabetes mellitus. *Annals of Translational Medicine*, *11*(2), 123.

16 Nanomaterials for Enhanced Plant Secondary Metabolite Production

Adalberto Benavides-Mendoza, Gonzalo Tortella, Paola Fincheira, Julia Medrano-Macías, Susana González-Morales, and Laura Olivia Fuentes-Lara

16.1 INTRODUCTION

The secondary metabolites of medicinal plants, including, among others, alkaloids, terpenoids, flavonoids, glycosides, saponins, and phenolics, play a pivotal role in plant adaptive and defense mechanisms against herbivores, pathogens, and environmental stresses. Unlike primary metabolites, which are crucial for vital life functions such as cell structure formation and energy supply, plant secondary metabolites serve as sources of information perceived through cellular receptors that activate or regulate metabolic pathways associated with plant acclimatization, resilience, and survival (Khare et al., 2020; Divekar et al., 2022).

The evolutionary genesis of secondary metabolites can be traced back to the natural selection pressures exerted by herbivores, pathogens, and environmental stressors. Plants that evolved the capability to synthesize and accumulate bioactive secondary metabolites gained a competitive advantage through enhanced protective mechanisms against biotic and abiotic challenges, thereby ensuring their survival and reproductive success (Hadacek, 2002). From an ecological perspective, secondary metabolites serve multiple functions, including deterring herbivory through toxic or repellent properties, protecting against microbial infections via antimicrobial activities, and mitigating abiotic stress through antioxidant or membrane or protein-protection properties (Srivastava et al., 2021).

The metabolic utility of secondary metabolites in medicinal plants is multiple for humans. It is known that diverse metabolites and enzyme complexes are present in all living organisms, playing roles in the synthesis, detoxification, and metabolism of various compounds. Likewise, a complex network of interconnected, ancient signaling molecules and pathways is maintained in both plants and animals. Moreover, numerous cellular and redox signaling elements are preserved across different species, featuring comparable gene expression in reaction to cellular stressors controlled by shared transcription factors (Sen & Samanta, 2015). For instance, hydrogen sulfide (H_2S) is crucial for cellular signaling in plants and animals (Powell et al., 2018; Medrano-Macías et al., 2022).

As a result of the foregoing, the secondary and signaling plant compounds interact with diverse molecular unspecific and specific receptor targets within human cellular systems, exhibiting a wide range of pharmacological activities. Therefore, secondary metabolites of medicinal plants have been extensively used in pharmaceuticals, nutraceuticals, and cosmetic industries for their health-promoting properties and their roles as anticancer, antimicrobial, and anti-inflammatory agents. Medicinal plants are crucial to human health, the World Health Organization reports that around 60% of the global population and 80% of people in developing countries rely on traditional medicine, particularly plant-based drugs, for their primary healthcare (Srivastava et al., 2021).

Environmental factors profoundly influence the biosynthesis and accumulation of secondary metabolites in medicinal plants (Khare et al., 2020; Srivastava et al., 2021). For example, light significantly affects the production of secondary metabolites through photoreceptor-mediated pathways. Light quality (wavelength), quantity (irradiance), and photoperiod (duration) can modulate the biosynthesis of these compounds (Pech et al., 2022). For instance, ultraviolet (UV) radiation acting as an elicitor can increase the production of flavonoids, which absorb UV light and quench reactive oxygen and nitrogen species, protecting plant tissues from damage (Alyas et al., 2023).

Temperature is another important environmental inductor of secondary metabolites. Temperature extremes can induce plant stress responses, leading to altered metabolite profiles (Yeshi et al., 2022). For instance, mild heat stress can enhance the accumulation of alkaloids and terpenoids, which serve as protective agents against thermal damage. Conversely, cold stress can trigger the accumulation of flavonoids and other phenolic antioxidant compounds, contributing to the plant's cold tolerance mechanisms (Mahajan et al., 2020).

In the same way, drought stress induces a water deficit that can lead to oxidative stress, cellular damage, and impaired plant metabolic functions. In response, the medicinal plants can accumulate osmoprotectants and antioxidants that mitigate damage and ensure survival. Terpenoids and phenolic compounds, known for their antioxidant activity, are upregulated under drought conditions (Mahajan et al., 2020).

Despite some secondary metabolites showing constitutive expression, producing many secondary metabolites in medicinal plants is normally limited due to the characteristics of the environmental-inducible or environmental-activating nature of the metabolic pathways or precursors that

DOI: 10.1201/9781003475491-16

produce them (Srivastava et al., 2021). This means that the production of secondary metabolites in medicinal plants can be highly variable, leading to significant challenges in meeting the growing demand for these compounds. This demand is often unfulfilled owing to adverse weather conditions and the lack of appropriate agricultural technologies for their domestication and cultivation (Bernela et al., 2023).

On the other hand, given medicinal plants' critical role in addressing numerous health conditions, unregulated harvesting has resulted in the reduction and sometimes extinction of various species within their native environments (Bernela et al., 2023). Furthermore, the complexity of these compounds, coupled with the specificity of their biosynthetic pathways, makes chemical synthesis a viable option only for certain specific cases (Gecibesler et al., 2021), thereby necessitating the development of novel and sustainable agricultural approaches for their enhanced production.

Oxygen, nitrogen, and sulfur radicals are common factors in abiotic and biotic stresses. These reactive molecules can induce plant defense responses against stress and the induction of secondary metabolite production in response to molecular damage (Medrano-Macías et al., 2022). Consequently, from a pharmacological perspective, the enhanced production of secondary metabolites under oxidative stress conditions can increase the nutraceutical or medicinal value of plants (Yeshi et al., 2022).

Physical, chemical, or biological environmental stimuli capable of inducing a mild stress response (which commonly depends on controlled oxidative or nitrosative stress) that trigger the induction of secondary metabolism and an increase in the accumulation of bioactive biomolecules are known as biostimulants. Many kinds of biostimulants are known, such as UV radiation, beneficial elements (such as Se and Si), compost, humic substances, beneficial bacteria, and beneficial fungi, among others. Recently, it has been recognized that NMs are a class of biostimulants, and their application in agricultural practice is beginning to spread (Garza-Alonso et al., 2022).

Among the new approaches to increasing the content of secondary metabolites in medicinal plants is using nanomaterials (NMs) as biostimulant chemical agents and promotors of controlled cellular stress. NMs, through the induction of a controlled level of oxidative or nitrosative stress, can mimic the biological context or the impact of environmental stresses without the disadvantage that plants suffer due to real extremes of stress, like temperature or water deficit. NMs are potent agents that can modulate plant metabolic pathways, potentially increasing secondary metabolite yield *in vitro*, in organ cultures, in greenhouse, and open field (Juárez-Maldonado et al., 2019; Kralova & Jampilek, 2021; Bernela et al., 2023). Consequently, nanotechnology can provide emerging methodologies and technological innovations with the potential to optimize the biosynthesis, extraction, and purification of valuable phytochemicals (Kralova & Jampilek, 2021; McClements & Öztürk, 2022).

The recent scientific literature has many studies in which medicinal plant extracts are used for the green synthesis of NMs (see, e.g., Ettadili et al., 2022; Antonio-Pérez et al., 2023; Antunes Filho et al., 2023; Chaudhary et al., 2023; Godeto et al., 2023; Ai et al., 2024), but a much smaller number of works in which NMs are applied to induce metabolic adjustments and quality amelioration of medicinal plants (see, e.g., Kralova & Jampilek, 2021; Bernela et al., 2023; Velázquez-Juárez et al., 2023). NMs are applied on plants or plant substrates or culture media to increase the concentration of active metabolites, biomass, and productivity. This chapter explores the last approach: the impact of different NMs on various medicinal plant species' physiological and metabolic behavior. In all cases, the goal of the studies is to improve the production and pharmaceutical quality of the plant organs used to obtain diverse products containing active ingredients with therapeutic action.

16.2 SUMMARY OF THE METABOLISM ASSOCIATED WITH THE SYNTHESIS OF ACTIVE COMPOUNDS OF MEDICINAL PLANTS

Secondary metabolism produces a wide range of bioactive compounds that contribute to the medicinal properties of plants. The most studied compounds include alkaloids, terpenoids, and phenolics. Each class of compounds is synthesized through distinct metabolic pathways specifically regulated by the external and internal environment.

Alkaloid biosynthesis. Alkaloids are nitrogen-containing compounds with diverse pharmacological activities. They are synthesized from amino acids such as lysine, ornithine, and tyrosine. For example, the biosynthesis of morphine in *Papaver somniferum* begins with the condensation of tyrosine-derived dopamine and the phenylalanine-derived benzylisoquinoline pathway (Li et al., 2020). It has been shown that exposure of *Papaver orientale* cell cultures to 20 mg of Cu NPs for 48 h induces oxidative stress associated with increased gene expression related to the biosynthesis of thebaine, codeine, and morphine (Khaldari et al., 2024).

Terpene biosynthesis. Terpenes are derived from isoprene units and exhibit a wide range of biological activities. The mevalonate (MVA) pathway and the methylerythritol phosphate (MEP) pathway are two primary routes for terpene biosynthesis. Monoterpenes, sesquiterpenes, diterpenes, and terpenoids are common terpenes with significant medicinal properties. It has been found that foliar spraying of Fe NMs in *Ocimum basilicum* increases the synthesis of sesquiterpenes, substantially increasing the plants' antioxidant and antimicrobial activity (Tavallali et al., 2019).

Phenolic compound biosynthesis. Phenolic compounds, including flavonoids, phenolic acids, and tannins, are synthesized via the shikimate and phenylpropanoid pathways. These compounds possess antioxidant, anti-inflammatory, and anticancer activities. In the manuscript by (Khan et al., 2021), different techniques for applying NMs to induce the biosynthesis of flavonoids *in vitro* cultures are described.

16.2.1 Regulation of Metabolism of Active Compounds in Medicinal Plants

The regulation of secondary metabolism in medicinal plants is a complex and finely tuned process influenced by endogenous and exogenous factors (Figure 16.1). Endogenous regulation primarily involves genetic and biochemical controls that govern the synthesis of secondary metabolites. Transcription factors play a crucial role in this regulation by binding to specific promoter regions of genes encoding enzymes involved in secondary metabolic pathways (Mahajan et al., 2020; Bernela et al., 2023). For instance, MYB, bHLH, and WD40 transcription factors form a regulatory complex that modulates the expression of flavonoid biosynthesis genes (Li, 2014). Additionally, post-translational modifications of enzymes, such as phosphorylation and glycosylation, can alter their activity and stability, further fine-tuning the production of secondary metabolites (Kokkinidis et al., 2020).

Moreover, biotic stresses such as pathogen attack or herbivory trigger the activation of secondary metabolic pathways. Plants perceive these biotic stresses through receptors that recognize specific pathogen-associated molecular patterns (PAMPs) or herbivore-associated molecular patterns (HAMPs). This recognition initiates a cascade of signaling events, including producing reactive oxygen species (ROS) and phytohormones like jasmonic acid and salicylic acid, activating gene expression in secondary metabolism (Abdul Malik et al., 2020). Nutrient availability is another critical factor, with deficiencies or excesses in nitrogen, phosphorus, and other elements, altering the metabolic flux toward specific secondary metabolites in medicinal plants (Mahajan et al., 2020).

An interesting topic concerns the impact of the exogenous and endogenous microbiome on secondary metabolism and the ability of medicinal plants to accumulate active compounds. Diverse studies have revealed that plants can influence their microbiome by producing various metabolites in root exudates, and, in turn, the microbiome may also impact the metabolome of the partner plant (Pang et al., 2021). However, it is also known that NMs modify the species composition and abundance in soil microbiomes (González-Morales et al., 2020). Still, as far as we know, the indirect impact of NMs on the secondary metabolism of medicinal plants via modification in the metabolism or specific composition of the microbiome has not yet been explored.

In the case of NMs, plants probably perceive the interfacial interactions between the surface of NMs and the cell walls and membranes as equivalent to damage-associated molecular patterns (DAMPs), activating cellular defensive responses against stress (Juárez-Maldonado et al., 2019). Consequently, applying NMs in plants triggers a sequence of events of perception, transduction, signaling, signal amplification, and response in the form of changes in metabolism and gene expression that induce the activation

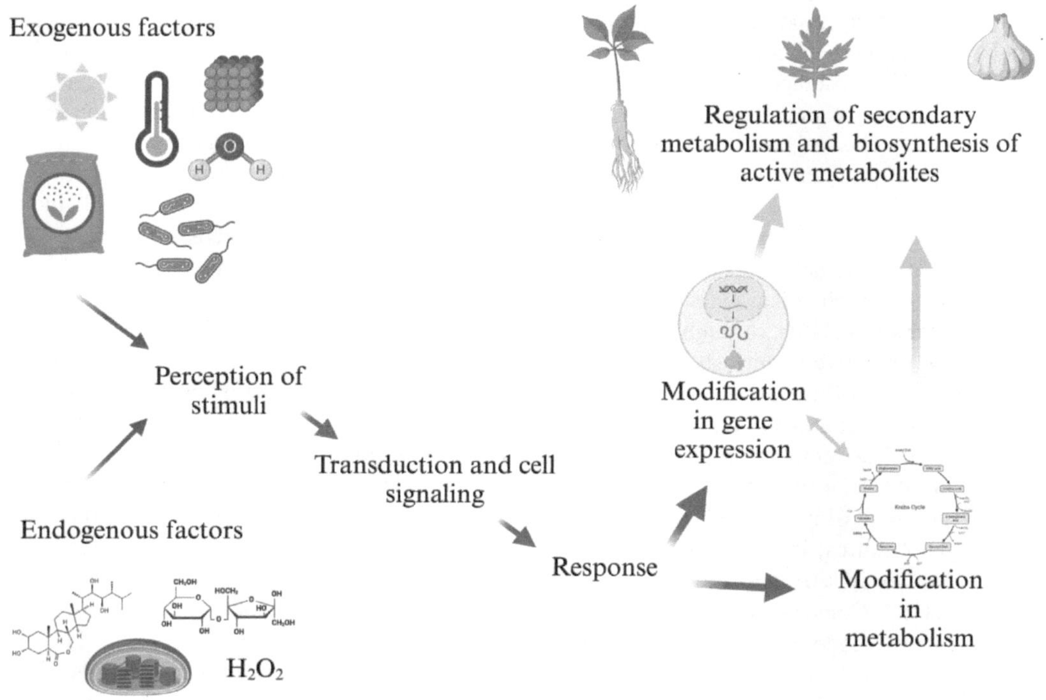

FIGURE 16.1 The regulation of secondary metabolism in medicinal plants is a complex and finely tuned process influenced by endogenous and exogenous factors.

of metabolic pathways as complex as the one shown below. In other words, the response to an environmental stimulus associated with light, temperature, the presence of microorganisms, or the availability of nutrients can be emulated with the application of NMs (Juárez-Maldonado et al., 2019, 2021).

The following is an example of the complexity of the biosynthetic pathway of an active biomolecule (Figure 16.2). The example refers to taxol, and as explained in the previous paragraph, it is possible to induce the metabolic pathways related to its biosynthesis with the application of NMs or some other stress-inducing agent at controlled levels.

Taxus spp. plants synthesize taxanes, a class of diterpenoids with a potent anticancer activity. Taxanes biosynthesis involves the MEP pathway and several cytochrome P450-mediated hydroxylation steps. Although it is possible to obtain synthetic taxol, the complexity of the synthesis makes it economically unviable. Therefore, the action of living cells is necessary to obtain the taxol used in therapeutics (Wang et al., 2021). The efforts to increase taxane production include biostimulation *in vitro* and metabolic engineering. It is worth mentioning that the synthesis of taxanes in *Taxus baccata* cell cultures is increased by biostimulating the cell cultures with gold NPs at a concentration of 2 mg L^{-1} (Golinejad et al., 2023).

In taxanes, particularly taxol, biosynthesis is a highly intricate process involving numerous enzymatic steps, beginning with forming its diterpenoid core and proceeding through a series of modifications to yield the final active compound. The pathway can be broadly divided into several key stages: formation of the core taxane structure, side chain attachment, and post-synthetic modifications (Shao et al., 2021; Wang et al., 2021).

Formation of the core taxane structure. Taxol biosynthesis starts with the universal diterpene precursor, geranylgeranyl diphosphate (GGPP). GGPP is synthesized through the methylerythritol phosphate (MEP) pathway, which takes place in the plastids of plant cells. The key steps in this initial phase include:

Isopentenyl diphosphate formation: The MEP pathway synthesizes isopentenyl diphosphate (IPP) and dimethylallyl diphosphate (DMAPP) from glyceraldehyde-3-phosphate and pyruvate.

GGPP synthesis: IPP and DMAPP are condensed to form GGPP by the enzyme geranylgeranyl diphosphate synthase.

Taxadiene formation: GGPP undergoes cyclization by the enzyme taxadiene synthase to produce taxa-4(5),11(12)-diene, the core taxane structure.

Side chain attachment. The attachment of the taxane side chain is a critical step in the biosynthesis of taxol, involving several intermediate compounds and enzymes:

Hydroxylation: The taxane core undergoes multiple hydroxylation reactions catalyzed by cytochrome P450 monooxygenases, introducing hydroxyl groups at specific positions.

Oxidation: Further oxidation steps are required to modify the hydroxyl groups, creating various oxo- and keto-derivatives.

Oxetane ring formation: The formation of the oxetane ring, a distinctive feature of taxol, involves a series of oxygenation and cyclization reactions.

Side chain attachment: The C13 side chain, essential for the biological activity of taxol, is attached via a series of esterification reactions. This involves the enzyme 10-deacetylbaccatin III-10-O-acetyl transferase (DBAT), which converts 10-deacetylbaccatin III to baccatin III, and subsequently, phenylpropanoyl transferase (PPT), which attaches the C13 side chain.

Post-synthetic modifications. The final steps in the biosynthesis of taxol involve a series of acylations and epoxidations that lead to the completion of the taxol molecule:

Esterification: Several acyltransferases catalyze the addition of acyl groups at specific hydroxyl positions.

Epoxidation: The formation of the epoxide ring involves an epoxidase enzyme that converts a double bond into an epoxide.

Final assembly: The final taxol molecule is assembled through these modifications, resulting in the complete, biologically active compound.

Regulation of taxol biosynthesis. A network of transcription factors and signaling pathways tightly regulates taxol biosynthesis. Environmental factors such as light, temperature, and nutrient availability can influence the expression of genes involved in the taxol biosynthetic pathway. Additionally, biostimulants such as NMs have been shown to enhance taxol production by modifying the cellular redoxtasis, inducing posttranslational modifications of enzymes in the biochemical pathways related to the taxol biosynthesis, and upregulating genes of key enzymes in the pathways (Golinejad et al., 2023).

This chapter's second section showed that NMs application can induce active compound synthesis in medicinal plants. Although the technique may still be incipient in the commercial field, it represents a promising technique in plant biotechnology. NMs can induce a controlled stress stimulus and enhance the bioavailability and uptake of nutrients, thereby stimulating metabolic pathways involved in the biosynthesis of valuable secondary metabolites. The following section presents the mechanisms through which NMs give rise to the above-mentioned responses.

16.3 NANOMATERIALS AS BIOSTIMULANTS AND PROMOTERS OF ACTIVE METABOLITES

Biostimulation denotes the biological adaptive process through which physical stimuli, microorganisms, or endogenous or exogenous substances, including NMs, can boost plant growth, development, or stress tolerance. It involves modifying the metabolome, ionome, and transcriptome, allowing for more efficient resource utilization, increased growth, quality, and yield, and enhanced resilience to

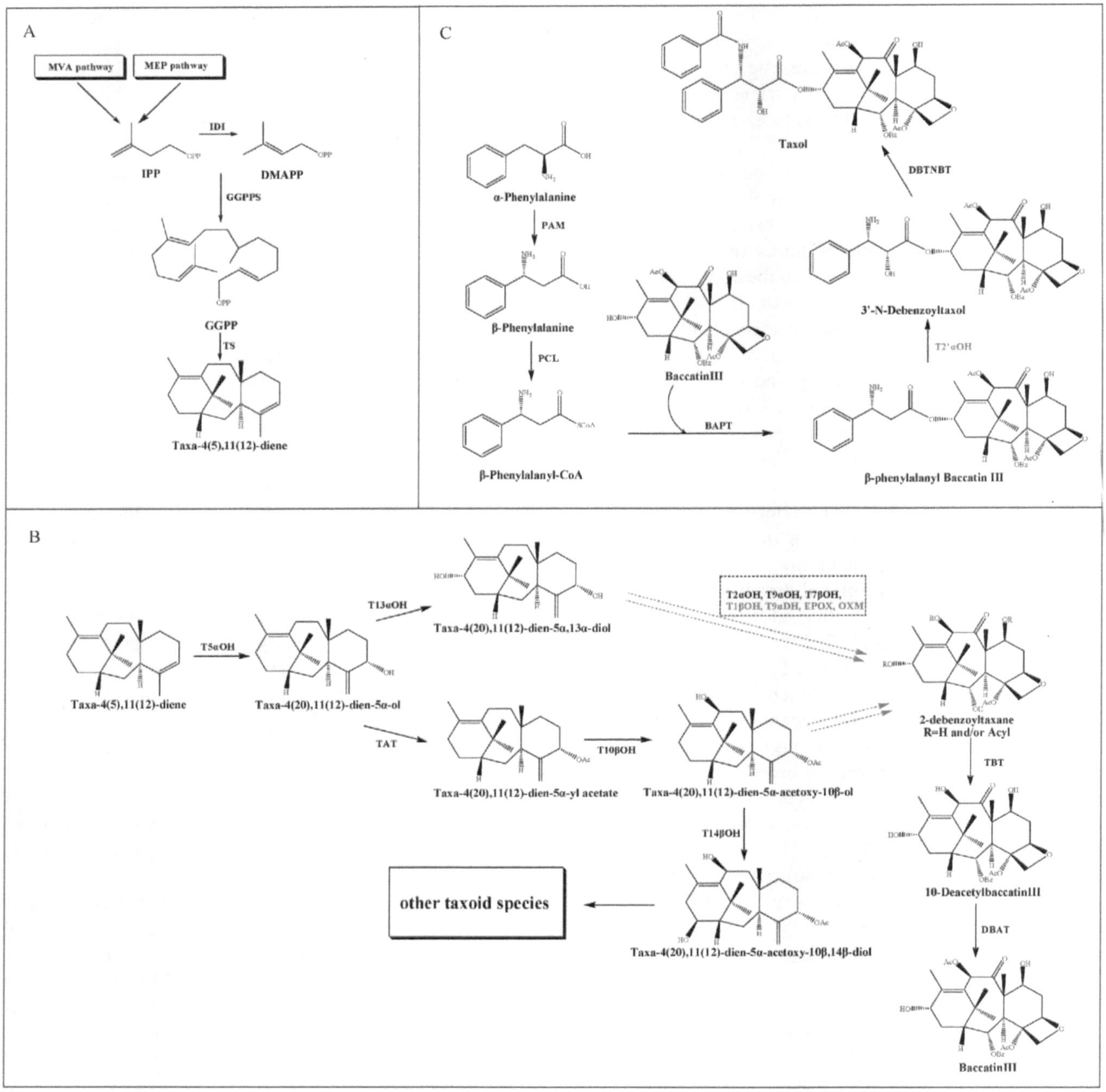

FIGURE 16.2 An example of a complex biosynthetic pathway is the synthesis of taxol in *Taxus brevifolia*. Figure from Wang et al. (2021). The explanation of the process is found in the main text.

adverse environmental conditions (Juárez-Maldonado et al., 2019).

In medicinal plants, biostimulation with NMs modulates plant physiological processes and can yield better growth and production of active compounds. Biologically, the main difference between bulk materials (like copper or iron) and NMs (such as nanocopper or nanoiron) is the high reactivity of NMs with cell surfaces, inducing responses like those seen under mild stress or exposure to other biostimulants like humic substances, chitosan, or polysaccharides from fungal or bacterial pathogens (Juárez-Maldonado et al., 2021).

The increased chemical and biological reactivity of NMs is due to their surface energy, which is inversely proportional to particle size; smaller particles possess higher surface energy. Higher surface energy enhances biostimulant capacity, strengthening interactions with cell walls, membranes, proteins, and nucleic acids (Abbasi et al., 2023).

The impact of NMs on the behavior of plant cells starts with surface interactions between plant surfaces and the

NMs corona or functionalized surface of the material. The corona comprises unfolded proteins and metabolites perceived as DAMPs and can trigger a defensive response in plant cells. This constitutes the initial phase of biostimulation. A subsequent phase, occurring almost simultaneously, involves releasing the corona or the NM core components, which can happen in the apoplast or symplast. These components (e.g., unfolded proteins, peptides, amino acids, or ions in the case of metal nanoparticles [NPs]) then interact with protein receptors of plant surfaces or with the membranes, metabolites, and proteins inside the cells and their organelles. This interaction drives the changes in the metabolic and physiological activities of plants exposed to NMs (González-Morales et al., 2022, 2024) (Figure 16.3).

Interactions between NPs and plant cells alter cell wall proteins and ion transporters, affecting metabolism and gene expression. Like other biostimulants, the ability of NMs to induce specific responses depends on concentration and dosage over time. Typically, NMs exhibit a biphasic or hormetic response, with low concentrations eliciting positive effects and higher concentrations causing negative effects (Agathokleous et al., 2019). Plant-favorable responses to NMs are observed within a 5 to 250 mg L^{-1} concentration range, with beneficial effects noted even at higher concentrations for certain combinations (Afzal et al., 2022; Agrawal et al., 2022).

Responses to NMs at concentrations below 5 mg L^{-1} are less studied in medicinal plants and, in general, in all plant species. Still, they are expected to be biologically significant, especially with long-term exposure, as the accumulated dose over time (threshold of biological activity) is crucial (González-Morales et al., 2020). The consideration of using low doses of NMs to biostimulate the secondary metabolism of medicinal plants may be especially important for commercializing the bulk biomass of medicinal plants since high doses could likely cause distrust in the consumer.

The potential for diverse biological responses in medicinal plants to the different NMs and their combinations is enormous. Variations in NM composition, size, shape, surface charges, and hydrophobic/hydrophilic sites lead to different biological responses, making each variation unique

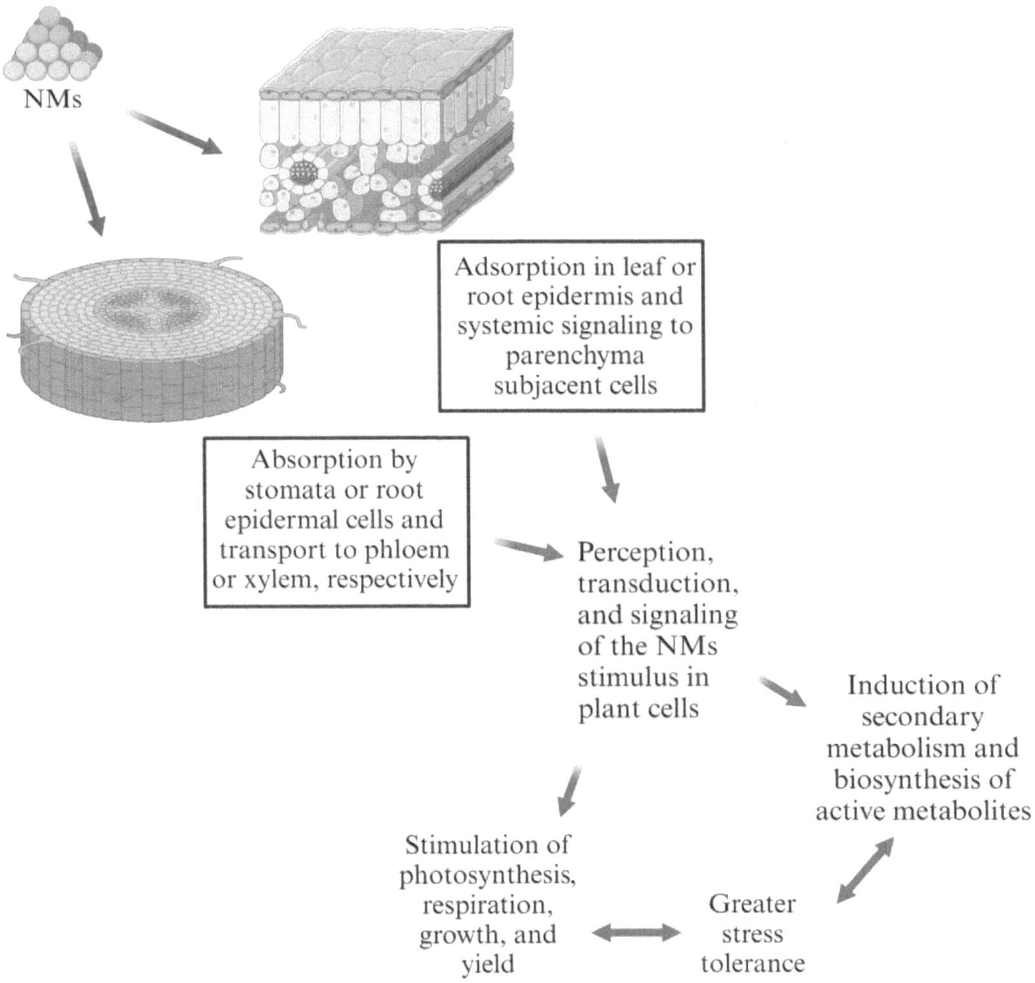

FIGURE 16.3 Scheme of the biostimulation mechanism of NMs on medicinal plants.

biologically (Barciela et al., 2023). Research thus far has explored only a small fraction of NMs' potential as agricultural biostimulants.

Following the last paragraph, in a non-systematic exercise of searching using the Google browser, starting with a base of 60 common medicinal plant species, the following 20 species were found to be the most reported or commonly studied globally: ginseng (*Panax ginseng*), echinacea (*Echinacea* spp.), *Ginkgo biloba*, turmeric (*Curcuma longa*), garlic (*Allium sativum*), St. John's wort (*Hypericum perforatum*), aloe vera (*Aloe barbadensis*), lavender (*Lavandula* spp.), peppermint (*Mentha piperita*), chamomile (*Matricaria chamomilla*), valerian (*Valeriana officinalis*), licorice root (*Glycyrrhiza glabra*), ginger (*Zingiber officinale*), cinnamon (*Cinnamomum* spp.), saw palmetto (*Serenoa repens*), bilberry (*Vaccinium myrtillus*), hawthorn (*Crataegus* spp.), milk thistle (*Silybum marianum*), feverfew (*Tanacetum parthenium*), and black cohosh (*Actaea racemosa*). These 20 species names were searched in Scopus in conjunction with 'application of nanomaterials or nanoparticles'. From the reported works, those referring to the application of NMs to enhance active biomolecules in the plants were selected. Studies in which plant extracts were used for NMs biofabrication, NMs used as carriers or encapsulating agents of active metabolites, or studies using nanoformulations of active biomolecules were excluded. With the selected studies from 2010 to 2024, Table 16.1 was constructed. It tells us if the species in question has been recently studied from the point of view of biostimulation with NMs and – if this was positive – it contains the selected works, information about the study conditions and NMs application, and references. For cases without recent reports, this indicates an opportunity to conduct research.

Table 16.1 includes only a part of the medicinal plant species used globally. However, after its examination, we can tentatively conclude that a valuable amount of work has been carried out that has generated knowledge on the induction of active biomolecules in cultivated medicinal plants. However, the number of studies is small compared to the number of reports about using other approaches to harness the biostimulant potential of NMs, such as using plant extracts for the green synthesis of NMs or the therapeutic use of NMs in combination with bioactive compounds or plant extracts.

16.4 TECHNIQUES FOR APPLYING NANOMATERIALS IN CULTIVATING MEDICINAL PLANTS *IN VITRO*, GREENHOUSES, AND OPEN FIELDS

Given the ecological challenges of producing and collecting wild medicinal plants, innovative methods are needed to enhance their yield and production (Theodoridis et al., 2023). In recent years, nanotechnology has emerged as an avenue to partially resolve this requirement. Nanotechnology's application in medicinal plant production can be broadly categorized into *in vitro*, organ cultures, greenhouse, and open field cultures.

16.4.1 *In Vitro* Cultures

In vitro plant culture techniques are fundamental to producing plant-derived active metabolites, which are valuable in the pharmaceutical, cosmetic, and food industries. These techniques offer several advantages over traditional field agriculture, such as producing metabolites irrespective of climatic conditions and enhancing metabolite yield through controlled conditions. Using NMs to induce the production of active metabolites *in vitro* cultures or organ cultures is a kind of biostimulation termed commonly *elicitation* (Rivero-Montejo et al., 2021).

According to the literature, practically any kind of NM can be used as an elicitor since the surface properties of NMs largely explain their eliciting capacity (Juárez-Maldonado et al., 2019; Javed et al., 2022), as explained in the previous section of this chapter (Figure 16.3). However, metal NMs are commonly used for *in vitro* cultures (Inam et al., 2023) (Table 16.2). The use of NMs for medicinal plants cultivated *in vitro* is a practical method of increasing usage since NMs are added directly to the culture media in the appropriate concentrations, which allows good control of doses and exposure times, and, when carried out under laboratory or industrial infrastructure conditions, it reduces the possible dispersion of NMs toward ecosystems (Bernela et al., 2023).

The following are some techniques used to produce active metabolites with medicinal plants. Each technique includes examples of applying NMs as elicitors to promote the active metabolite yield.

16.4.1.1 Tissue Culture

Tissue culture involves growing plant cells, tissues, or organs under sterile conditions on a known composition nutrient culture medium. Tissue culture can be used to mass-produce plant material from a single source plant, allowing the production of metabolites once the cells or tissues are treated with an appropriate elicitor or environmental stimulus (Chandran et al., 2020). An example of the augmentation of the production of medicinal metabolites in tissue culture applying nanoelicitors is that reported by (Ahmad et al., 2020), who examined the impact of ZnO and CuO NPs on the root formation, antioxidant activities, and steviol glycosides production in *Stevia rebaudiana in vitro* cultures. They found that low concentrations of 2 mg L^{-1} ZnO and 20 mg L-1 CuO significantly enhanced rooting in plant shoots, with 91% and 94% success rates. Compared to the control, the above concentrations yielded the highest levels of steviol glycosides, with rebaudioside A and stevioside peaking from 2.96% to 4.42% and stevioside from 1.01% to 1.28, both metabolites with ZnO 2 mg L^{-1}. Furthermore, the highest total flavonoid, phenolic content, and DPPH-free radical scavenging activity were recorded at the above NPs concentrations. In a second example, (Ghorbanpour &

TABLE 16.1
Studies Reported between 2010 and 2024 on Using NMs as Biostimulants to Induce a More Significant Accumulation of Active Compounds in Medicinal Plants

Plant species	NMs applied	Conditions of the study	Reference
Allium sativum	CeO_2 NPs (1-10 mg L^{-1}, 18 days)	Hydroponically grown plants	(Xu et al., 2024)
Echinacea purpurea	ZnO NPs (1-3 g L^{-1})	Foliar spraying in greenhouse	(Mohit Rabary et al., 2022)
Hypericum perforatum	Se NPs (6-12 mg L^{-1}, in the rosette stage)	Greenhouse	(Rezaei Nazari et al., 2022)
Lavandula spp.	Ag NPs (20-80 mg L^{-1})	*In vitro* micropropagation	(Amnuaykan et al., 2023)
Lavandula spp.	Si NPs (50 mg L^{-1}) and Ag NPs (20 mg L^{-1})	*In vitro* micropropagation	(Khattab et al., 2022)
Matricaria chamomilla	Silica NPs (2-6 g L^{-1})	*In vitro* culture	(Torabzadeh et al., 2019)
Matricaria chamomilla	Chitosan NPs (100-400 mg L^{-1})	Foliar spray	(Mazrou et al., 2021)
Matricaria chamomilla	Fe3O4 NPs (25-100 mg L^{-1})	*In vitro* culture	(Rastegaran et al., 2022)
Mentha piperita	NFe (0.5-1.5 g L^{-1})	Leaf spray in open field	(Lafmejani et al., 2018a)
Mentha piperita	Se NPs (2-20 mg L^{-1})	Leaf spray in open field	(Nazerieh et al., 2018)
Mentha piperita	Cu NPs	Leaf spray in open field	(Lafmejani et al., 2018b)
Mentha piperita	NZn (0.5-1.5 g L^{-1})	Leaf spray in open field	(Nemati Lafmejani et al., 2021)
Mentha piperita	Chitosan-coated Fe_2O_3 NPs (5-10 μM, applied in soil)	Soil, in open field	(Giglou et al., 2022)
Mentha piperita	TiO_2 NPs (150 mg L^{-1})	Leaf spray in open field	(Shenavaie Zare et al., 2022)
Mentha piperita	TiO_2 NPs (100 mg L^{-1}, applied in soil at sowing)	Soil, in open field	(Ostadi et al., 2023)
Silybum marianum	Graphene oxide (0, 0.01, 0.05, 0.1%) in soil	Open field	(Safikhan et al., 2018)
Silybum marianum	ZnO NPs (0.15 mg L^{-1})	Callus culture	(Shehzad et al., 2021)
Silybum marianum	Cyclodextrins coated with magnetic Fe_3O_4 NPs (2.5-15 g L^{-1})	Cell cultures	(Corchete et al., 2022)
Silybum marianum	Chitosan and TiO_2 NPs (50 and 100 mg L^{-1})	Foliar spraying in open field	(Jafari et al., 2022)
Silybum marianum	TiO_2 NPs (0-100 mg L^{-1})	Foliar spraying in open field	(Jafari et al., 2023)
Silybum marianum	Ag NPs (1 μg L^{-1})	Hydroponically grown plants	(Mubeen et al., 2023)
Silybum marianum	SiO_2 NPs (10-100 mg L^{-1})	Foliar spraying in greenhouse	(Almulhim et al., 2024)
Tanacetum parthenium	Spherical nano-carbon (125-1000 mg L^{-1})	Leaf spraying in greenhouse	(Ahmadi et al., 2018)
Tanacetum parthenium	TiO_2 and SiO_2 NPs (25 and 50 mg L^{-1})	Greenhouse	(Khajavi et al., 2019)
Tanacetum parthenium	TiO_2 NPs (1000-2000 mg L^{-1})	Foliar spraying in greenhouse	(Shahhoseini & Daneshvar, 2023)
Tanacetum parthenium	Ca NPs (0.1-0.4 %)	Foliar spraying in greenhouse	(Jahromi et al., 2024)
Tanacetum parthenium	Ag NPs (200-600 mg L^{-1})	Foliar spraying in greenhouse	(Shahhoseini, 2024)
Valeriana officinalis	ZnO NPs (1-3 g L^{-1})	Foliar spraying in greenhouse	(Mohit Rabary et al., 2022)
Zingiber officinale	Si NPs (50-150 mg L^{-1})	Spraying of rhizomes in postharvest	(Peng et al., 2022)
Actaea racemosa	No reports were found about this species.		
Aloe barbadensis	No reports were found about this species.		
Cinnamomum spp.	No reports were found about this species.		
Crataegus spp.	No reports were found about this species.		
Curcuma longa	No reports were found about this species.		
Ginkgo biloba	No reports were found about this species.		
Glycyrrhiza glabra	No reports were found about this species.		
Panax ginseng	No reports were found about this species.		
Serenoa repens	No reports were found about this species.		
Vaccinium myrtillus	No reports were found about this species.		

Hadian, 2015) reported the impact of multi-walled carbon nanotubes on callus induction and secondary metabolism in *Satureja khuzestanica*. The authors noticed that including the carbon NMs in the culture medium (at concentrations of 25 and 50 mg L^{-1}) led to optimal callus growth and biomass accumulation. In contrast, the most significant levels of secondary metabolites, rosmarinic acid, and caffeic acid, were achieved at 100 and 250 mg L^{-1} concentrations.

16.4.1.2 Cell Suspension Cultures

Plant cells are grown in liquid culture media, facilitating cells' rapid growth and division, allowing for the efficient extraction of desired metabolites (Chandran et al., 2020). Regarding this technique, (Chung et al., 2019) carried out a study with *Gymnema sylvestre* to investigate the production of gymnemic acid II (GA II) and phenolic compounds in cell suspension cultures treated with CuO NPs at concentrations of 1, 3, and 5 mg L^{-1}. The treatment with 3 mg L^{-1} CuO NPs for 48 hours yielded nine-fold levels of gymnemic acid, total phenolics, and flavonoids. On the other hand, (Chung, Rekha et al., 2018) investigated the effects of Ag NPs at concentrations of 0, 1, 5, and 10 mg L^{-1} on phytochemical yield in cell suspension cultures of bitter gourd (*Momordica charantia*). Silver, malondialdehyde, and hydrogen peroxide levels were significantly higher with 10 mg/L Ag NPs compared to untreated culture. The treatment with 5 mg L^{-1} Ag NPs exhibited a substantial increase in total phenolic (3.5 mg g^{-1}) and flavonoid (2.5 mg g^{-1}) levels compared to the control showing 2.5 and 1.6 mg g^{-1}, respectively. Additionally, the exposition to 5 mg L^{-1} Ag NPs demonstrated elevated levels of flavonols (1822.37 µg g^{-1}), hydroxybenzoic acid (1713.40 µg g^{-1}), and hydroxycinnamic acid (1080.10 µg g^{-1}), surpassing those in the control (1199, 1394.42, and 944.52 µg g^{-1}, respectively).

16.4.1.3 Callus Cultures

A callus is a mass of unorganized plant cells induced from plant tissue explants on an appropriate nutrient medium under specific light and temperature conditions. Callus cultures can be maintained indefinitely and used to extract valuable secondary metabolites (Efferth, 2019). An interesting example of the application of elicitation with NMs in callus cultures was described by (Ali et al., 2019); in their study, different concentrations of Ag NPs (0, 30, 60, and 90 µg L^{-1}) were tested in callus cultures of *Caralluma tuberculata* for enhanced biomass and production of antioxidant secondary metabolites. Better biomass production was noted with 60 µg L^{-1} Ag NPs, yielding the highest recorded fresh (0.78 g L^{-1}) and dry (0.051 g L^{-1}) biomass. The biochemical assessments revealed increased levels of phenolics (3.0 mg g-1), flavonoids (1.8 mg g-1), phenylalanine ammonia-lyase activity (5.8 U mg^{-1}), and antioxidant activity (90%) in callus cultures treated with 90 µg L^{-1} Ag NPs. Furthermore, significant increases in antioxidant enzyme activities, including superoxide dismutase (4.8 U mg^{-1}), peroxidase (3.3 U mg^{-1}), catalase (2.5 U mg^{-1}), and ascorbate peroxidase (1.9 U mg^{-1}), were observed at 90 µg L-1 Ag NPs. In their study with *Artemisia annua*, Fatima et al. (2020) induced greater antioxidant capacity and accumulation of rutin, gallic acid, and caffeic acid by applying ZnO, CuO, and CoO NPs (0.01, 0.05, 0.1, 0.2, and 0.3 mg L^{-1}). The best results were obtained with CuO NPs at 0.05 y 0.1 mg L^{-1}.

16.4.1.4 Hairy Root Cultures

Initiated by infection with the soil bacterium *Agrobacterium rhizogenes*, hairy root cultures are used to produce root-specific metabolites. These cultures grow rapidly and stably, often producing higher levels of secondary metabolites than their natural root counterparts. Hairy roots have the advantage of genetic stability and high biochemical productivity (Gutierrez-Valdes et al., 2020). In the study reported by Shakeran et al. (2015), which aimed to boost atropine production by *Datura metel*, various elicitors, including biotic (*Bacillus cereus* and *Staphylococcus aureus*) and abiotic ($AgNO_3$ and Ag NPs), were introduced into the hairy root cultures. Each elicitor contributed to increased biomass and atropine yields, but notably, Ag NPs proved to be the most effective at enhancing the atropine content in the hairy roots. On the other hand, in their study, Zhang et al. (2013) explored the elicitation effects of Ag NPs on artemisinin concentration within hairy root cultures of *Artemisia annua*. The research demonstrated that artemisinin levels rose from 1.67 mg g^{-1} dry weight to 2.86 mg g^{-1} dry weight following the application of Ag NPs at a concentration of 900 mg L^{-1} over three days. The Ag NP treatment prompted oxidative stress, evidenced by increased hydrogen peroxide production. This oxidative stress led to lipid peroxidation, the subsequent accumulation of malondialdehyde, and increased catalase activity. As a result of the elicitation, there was a significant increase in the artemisinin content within the hairy roots. The application of Ag NPs enhanced artemisinin production in 20-day-old hairy root cultures to 13.3 mg L^{-1}, marking a 3.9-fold increase compared to the untreated control.

These described techniques of *in vitro* culture can be escalated to the level of industrial bioreactors, which provide a controlled environment for optimized growth and metabolite production. Parameters such as pH, temperature, oxygen levels, and nutrient feed are carefully controlled to maximize yield.

To conclude this section, Table 16.2 summarizes some examples of using NMs as elicitors in medicinal plants grown *in vitro*.

16.4.2 Greenhouse and Open Field Cultures

As occurs in many plant species, NMs are helpful in medicinal plants to improve growth and development, increase physiological activities associated with productivity, and increase tolerance to biotic and abiotic stress (Bernela et al., 2023). However, considering the focus of this chapter, the following section mainly refers to using NMs as elicitors for higher productivity of bioactive secondary compounds.

In greenhouse and open-field cultures, medicinal plants can be grown under relatively well-controlled environmental conditions, which can be manipulated to some extent to induce the production of biomass and secondary metabolites. Environmental manipulation is more straightforward in greenhouse production compared to open-field crops. Greenhouse cultivation provides a more technical and controlled environment that can be manipulated to optimize conditions for the growth of specific medicinal plants. Temperature, humidity, irradiance, light quality, and substrate characteristics can be adjusted to boost plant biomass production and the composition of desired phytochemicals. Moreover, the controlled environment of a greenhouse reduces the risks of pests and diseases, leading to higher-quality plant material. This method is particularly advantageous for delicate or rare species that require specific climatic conditions (Zobayed et al., 2005).

TABLE 16.2
Summary of Some Studies in Which NMs Were Used as Elicitors to Induce Active Compounds in Medicinal Plants Grown *In Vitro*

Nanomaterial and plant species	Concentration	Effects	Reference
Ag *Corylus avellana*	2.5, 5, 10 mg L^{-1} Added one time to the culture media	Increased taxanes content (taxol and baccatin III), decreased the content of total soluble phenolics and total flavonoids.	(Jamshidi & Ghanati, 2017)
Ag *Cucumis anguria*	0.5, 1.0, 2.0 mg L^{-1} Added one time to the culture media	More biomass accumulation, total phenolics content and biological activity.	(Chung, Rajakumar et al., 2018)
Ag *Isatis constricta*	0.25, 0.5, 1, 1.5, 2.0 mg L^{-1} Added one time to the culture media	Risen of indigo and tryptanthrin but declined on 10 and 15 days post-treatment; the indirubin decreased.	(Karakaş, 2020)
Ag and Au *Prunella vulgaris*	30 µg L^{-1} Added one time to the culture media	Increased biomass, total phenolics and flavonoids, and DPPH-radical scavenging activity	(Fazal et al., 2019)
Ag and Cu alone or in combination (1:2, 2:1, 1:3, and 3:1) *Artemisia absinthium*	30 µg L^{-1}	Ag and Cu NPs (1:2, 2:1) and 4 mg L^{-1} TDZ showed 100% biomass accumulation. The Cu NPs enhanced the superoxide dismutase activity and peroxidase and peroxidase activity. The antioxidant activity was highest in the Ag and Cu NPs (3:1) media.	(Hayat et al., 2021)
Carbon nanotubes *Salvia nemorosa*	25–100 mg L^{-1}	The content of rosmarinic acid, salvianolic acid B, ferulic acid, and cinnamic acid was 13-fold, 14.2-fold, 20-fold, and 3-fold higher than wild *S. nemorosa* plants at the flowering stage, respectively. These levels were recorded in the callus cultures treated with 100 mg L^{-1} carbon nanotubes.	(Heydari et al., 2020)
Chitosan, chitin, and cellulose nano fibers (NFs) *Betula pendula*	0.5, 0.75, and 1 mg L^{-1}	The heaviest fresh and dry weights (1.5-fold) were obtained with 1% chitosan NFs. The highest amount of betulin (0.7 mg g^{-1}) was obtained with 0.5% of cellulose NFs, and the highest amount of betulinic acid (0.98 mg g^{-1}) was obtained with 0.5% chitin NFs. The maximum antioxidant activity was achieved utilizing 0.5% chitosan NFs.	(Vahide et al., 2021)
Co *Artemisia annua*	0.25, 2.5, and 5 mg L^{-1} Added one time to the cell suspension culture	The highest artemisinin content 2.25-fold (113.35 mg g^{-1} dw) was observed 24 h after 5 mg L^{-1} Co NPs	(Ghasemi et al., 2015)
Fe *Hyoscyamus reticulatus*	0, 450, 900, 1800, and 3600 mg L^{-1} at 24, 48, and 72 h exposure time	Fe NPs at 450 and 900 mg L^{-1} (24 and 48 h) increased five-fold the level of hyoscyamine and scopolamine.	(Moharrami et al., 2017)
Fe_2O_3 *Dracocephalum kotschyi*	0, 25, 50,75,100 mg L^{-1} at exposures of 24 and 48 h	The biomass accumulation in hairy roots was enhanced by Fe_2O_3 NPs. Rosmarinic acid content increased 9.7-fold (1194 µg g^{-1} FW) with 24 h of exposure to 75 mg L^{-1} Fe NPs.	(Nourozi et al., 2019)
Mesoporous silica functionalized with amines and TiO_2 *Solidago nemoralis*	1, 10, 100 µg L^{-1} 1, 2.5 mg L^{-1} Added one time to the culture media	More antioxidant activity and flavonoids.	(Khan et al., 2020)
Mn_2O_3 *Atropa belladonna*	25, 50, 100 and 200 mg L^{-1} Added one time to the culture media	At 25 mg L^{-1}, more production of antioxidants and alcaloids.	(Tian et al., 2018)
Se (biogenic NPs) *Caralluma tuberculata*	50, 100, 200, and 400 µg L^{-1}	Se NPs at 100 µg L^{-1} produced the best response for producing bioactive antidiabetic compounds: coumarins, gallic acid, caffeic acid, ferulic acid, catechin, quercetin, and rutin.	(Ali et al., 2023)
SiO_2 *Dracocephalum kotschyi*	(0, 25, 50, 100, and 200 mg L^{-1}) at 24 and 48 h exposure time	Rosmarinic acid content increased 8.26-fold (1016.6 µg g^{-1} FW) in the hairy roots culture with SiO_2 100 mg L^{-1}.	(Nourozi et al., 2019)

(Continued)

TABLE 16.2 ***(Continued)***
Summary of Some Studies in Which NMs Were Used as Elicitors to Induce Active Compounds in Medicinal Plants Grown *In Vitro*

Nanomaterial and plant species	Concentration	Effects	Reference
SiO_2 and graphene oxide-SiO_2 (GO-SiO_2) *Ammi visnaga*	0, 15, 25 mg L^{-1} in callus culture	15 mg L^{-1} GO-SiO_2 NPs was the most appropriate dosage for enhancing the enzymatic antioxidant activities.	(Golkar et al., 2023)
TiO_2 and ZnO *Linum usitatissimum*	TiO_2 (0, 50, 100, and 150 mg L^{-1}) and ZnO (0, 30, 60, and 120 mg L^{-1}) NPs at 0, 24, 48, and 72 h exposition time in cell suspension cultures	TiO_2 increased total lignan at all concentrations and phenolic levels at 150 mg L^{-1}. The highest amount of total phenolics and lignans was observed with 30 and 60 mg L^{-1} ZnO.	(Karimzadeh et al., 2019)
ZnO *Thymus* spp. and *Zataria multiflora*	100, 150 mg L^{-1} Added one time to the culture media	More thymol and carvacrol content.	(Mosavat et al., 2019)

However, greenhouse production is not always practical or advisable for all species of medicinal plants, especially those grown in large areas. In the case of medicinal plants that grow under wild conditions, such as many medicinal plants used in traditional medicine by many communities, the application of NMs is very complicated, both from an economic, ecological, and even social point of view. To the best of the authors' knowledge, this kind of application of NMs in wild medicinal plants has not been carried out.

Some examples of medicinal species that can be produced under greenhouses are echinacea (*Echinacea* spp.), ginseng (*Panax ginseng*), St. John's wort (*Hypericum perforatum*), lavender (*Lavandula* spp.), and aloe vera (*Aloe barbadensis*). However, aloe vera and lavender are versatile species produced in large areas in open fields (Zobayed et al., 2005; Jang et al., 2015; Mambrí et al., 2018; Karimi et al., 2022; Tavali & Ok, 2022).

On the other hand, growing medicinal plants in open fields is chosen for its economic efficiency and scalability. Growing in open fields takes advantage of natural sunlight and the nutrients and microbiome of the soil, which can enhance photosynthesis, plant growth, and the synthesis of phytochemicals. However, open-field cultivation exposes plants to variable weather conditions, pests, and diseases, adversely affecting growth and medicinal quality. Even variable environmental factors can lead to fluctuations in the biochemical composition of plants. Consequently, the open-field production system is used for medicinal crops grown on a larger scale or less sensitive to environmental conditions (Alamgir, 2017).

Some examples of medicinal species that can be produced in open field are chamomile (*Matricaria chamomilla*), peppermint (*Mentha piperita*), calendula (*Calendula officinalis*), valerian (*Valeriana officinalis*), and turmeric (*Curcuma longa*) (Alamgir, 2017; Jampílek & Kráľová, 2023).

Both greenhouse and open-field cultivation methods have implications for the sustainability and quality of medicinal plant production. Using crop rotation and organic farming practices, open-field cultivation can be more sustainable with ecological sensitivity. Conversely, while more resource-intensive, greenhouse cultivation offers precision in producing high-quality medicinal compounds and can extend the growing season (Alamgir, 2017).

The mechanisms of elicitation of the synthesis of active biomolecules in medicinal plants grown in a greenhouse or the open field are practically the same as those *in vitro* cultivation (Figure 16.3). The difference lies in the scale of the cropping, which can cover a few dozen square meters up to several hectares in greenhouses and fractions of a hectare up to dozens of hectares in open field crops. As expected, the larger-scale limits the precision of management and the control over metabolic processes. Still, on the other hand, the infrastructure cost and access to the knowledge necessary to operate an *in vitro* culture unit explains that the more significant part of the volume of medicinal plants used globally is produced in open fields or greenhouses or collected from wild sites (Theodoridis et al., 2023).

16.4.2.1 Techniques for Nanomaterial Application in Medicinal Plant Cultivation in Greenhouses and Open Field

The application of NMs for medicinal plant cultivation in the greenhouse and the open field encompasses a variety of techniques that can be integrated with traditional agronomic practices (Figure 16.4). These techniques can provide targeted solutions to common challenges in open-field medicinal plant production, enhancing the yield, quality, and concentration consistency of the pharmacologically active compounds. The following are some of the techniques that can be used.

Soil amendment. One of the primary methods of applying NMs in agricultural settings is soil amendment. Due to their minuscule size and high surface area, NMs can alter the physicochemical properties of the soil,

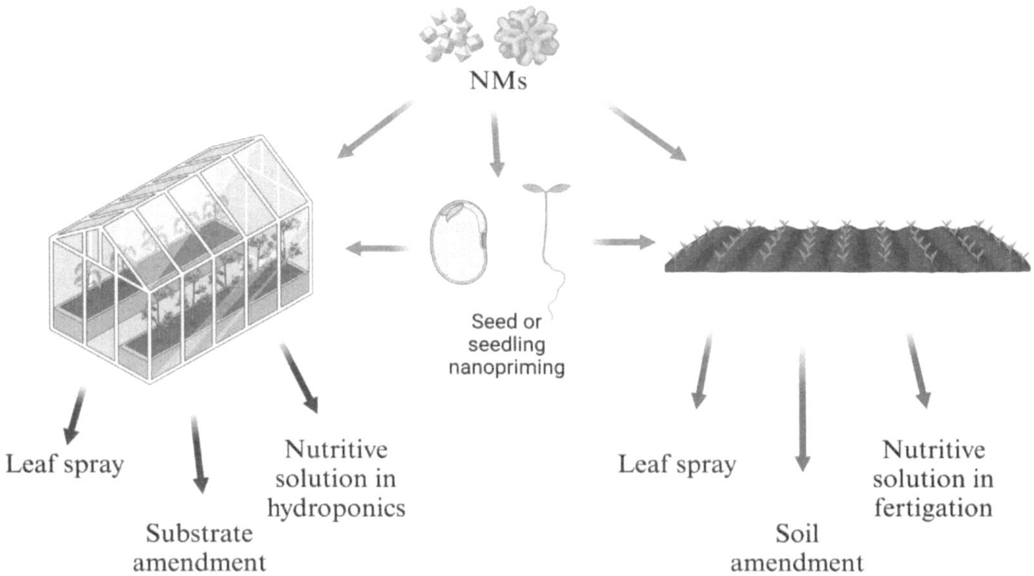

FIGURE 16.4 Techniques for NM Application in Medicinal Plant Cultivation in Greenhouses and Open Fields

improving nutrient availability and uptake by plants. In the context of medicinal plants, soil amendment with NPs like ZnO and CuO have been shown to enhance growth parameters and secondary metabolite content by facilitating nutrient assimilation and inducing beneficial stress responses (Zafar et al., 2020). However, this method poses challenges in terms of controlling the bioavailability and the soil matrix interactions of NMs, which can affect their efficiency and potentially lead to environmental accumulation. In addition, the potential negative impact of NMs on the soil microbiome must be considered (González-Morales et al., 2020).

Foliar spraying. The foliar application involves directly spraying nanoparticle solutions onto the plant leaves, allowing for rapid absorption and utilization of the NMs. This technique is particularly effective for delivering targeted nutrients or elicitors resulting from the fast absorption through stomata and cuticles, leading to quicker plant responses than root uptake mechanisms. Another advantage of this method is that it bypasses the complexities of soil dynamics and microbiome-root interactions, providing a simple pathway to enter NMs in plant tissues. The foliar application also minimizes the quantities of NMs used, reducing environmental risks and costs (Fageria et al., 2009; Ojeda-Barrios et al., 2020).

For example, foliar spraying at 160 mg L^{-1} with ZnO and CuO NPs in *Dracocephalum moldavica* has been reported to increase essential oil production and quality, total chlorophyll, aboveground biomass, flavonoid, and anthocyanin, which are key contributors to the therapeutic properties of the plants (Nekoukhou et al., 2024).

Although foliar applications of NMs are considered more effective than other application techniques, the alternative of application to the substrate or nutrient solution could be preferred if the root tissues produce active compounds. However, even in this case, a foliar application could be effective since the signaling triggered by the interactions between the cells of the leaf tissues is carried systemically to the rest of the plant, including the roots.

Nutrient solution. This method incorporates NMs into the water or nutrient solutions used in hydroponic systems or soil irrigations. While this method ensures that NMs are delivered directly to the roots, it requires ensuring adequate dispersion of the NMs in the volume of nutrient solution that will be applied, in addition to precise control over concentrations to avoid phytotoxicity (Kralova & Jampilek, 2021). For example, Velázquez-Gamboa et al. (2020) describe using NMs in a hydroponic system in *Stevia rebaudiana*. Upon receiving 75 mg L^{-1} of ZnO, the plants increased the contents of Zn, stevioside, rebaudioside A, phenolics, flavonoids, and antioxidant capacity.

Seed priming. Seed priming with nanoparticles treats seeds before planting, enhancing germination rates, seedling vigor, and stress tolerance. This method prepares the plant for improved metabolic activity and secondary metabolite production from the outset of its growth cycle (Rhaman et al., 2022). Priming *Moringa oleifera* seeds with ZnO NPs has been shown to promote early plant development and increase the accumulation of secondary metabolites in this species (Garza-Alonso et al., 2021).

16.5 CHALLENGES AND OPPORTUNITIES FOR APPLYING NANOMATERIALS IN MEDICINAL PLANT PRODUCTION

Biostimulation of medicinal plants with NMs offers a promising approach to their sustainable production. However, several issues must be resolved to ensure the widespread adoption of NMs to produce medicinal plants.

Developing tailored solutions for the different scenarios of medicinal plant production is essential. Future advancements could lead to customized NMs or mixtures of NMs designed for specific medicinal plants or production systems. Integrating NMs with advanced precision agriculture and real-time monitoring systems could also enable more targeted and efficient biostimulant applications.

Another limitation is the lack of comprehensive understanding and research data on how NMs interact with medicinal plants, their microbiomes, and soil and water over the long term. Moreover, concerns about the health and safety of ecosystems, food chains, and humans cannot be ignored. Limited information exists on NMs' ecosystem dynamics and long-term impact, especially in soil and various organisms' microbiomes. Comprehensive toxicity studies and environmental risk assessments are essential for regulatory compliance and commercializing medicinal plants obtained using NMs as biostimulants.

While nanotechnology has advanced significantly, the cost of producing NMs at a significant scale enough to be practical for agricultural use remains a major barrier. This includes challenges related to manufacturing processes, quality control, and the feasibility of delivering and using these materials on farms efficiently and effectively. Demonstrating the tangible benefits of NMs in agriculture is vital for market acceptance, and effectively communicating these benefits to potential users is essential for driving adoption.

These points highlight areas requiring investment of talent, resources, and time to resolve. New research opportunities in nanocomposites are likely to emerge in the coming years.

16.6 CONCLUSION

The techniques for *in vitro* applications of NMs or their use in the field or greenhouse cultivation of medicinal plants represent a frontier in enhancing the profile and content of secondary metabolites. Through cell or tissue biostimulation, soil amendment, foliar spraying, hydroponic systems, and seed priming, nanotechnology provides versatile tools for optimizing the therapeutic potential of medicinal plants. As research in this field advances, it holds the promise of revolutionizing the cultivation of medicinal crops, offering pathways to sustainable industrial and agricultural production of bioactive compounds for pharmaceutical use.

REFERENCES

Abbasi, R., Shineh, G., Mobaraki, M., Doughty, S., & Tayebi, L. (2023). Structural parameters of nanoparticles affecting their toxicity for biomedical applications: A review. *Journal of Nanoparticle Research, 25*(3), 43.

Abdul Malik, N. A., Kumar, I. S., & Nadarajah, K. (2020). Elicitor and receptor molecules: Orchestrators of plant defense and immunity. *International Journal of Molecular Sciences, 21*(3), 963.

Afzal, S., Aftab, T., & Singh, N. K. (2022). Impact of zinc oxide and iron oxide nanoparticles on uptake, translocation, and physiological effects in *Oryza sativa* L. *Journal of Plant Growth Regulation, 41*, 1445–1461.

Agathokleous, E., Feng, Z., Iavicoli, I., & Calabrese, E. J. (2019). The two faces of nanomaterials: A quantification of hormesis in algae and plants. *Environment International, 131*, 105044.

Agrawal, S., Kumar, V., Kumar, S., & Shahi, S. K. (2022). Plant development and crop protection using phytonanotechnology: A new window for sustainable agriculture. *Chemosphere, 299*, 134465.

Ahmad, M. A., Javed, R., Adeel, M., Rizwan, M., Ao, Q., & Yang, Y. (2020). Engineered ZnO and CuO nanoparticles ameliorate morphological and biochemical response in tissue culture regenerants of candyleaf (*Stevia rebaudiana*). *Molecules, 25*(6), 1356.

Ahmadi, S. Z., Ghorbanpour, M., Hadian, J., & Salehi-Arjmand, H. (2018). Impact of foliar spray of spherical nano-carbon and salicylic acid on physiological traits and parthenolide content in two feverfew cultivars (*Tanacetum parthenium* Linn. Cv. Pharmasaat and Jelitto). *Journal of Medicinal Plants, 17*(68), 82–98.

Ai, S., Li, Y., Zheng, H., Zhang, M., Tao, J., Liu, W., Peng, L., Wang, Z., & Wang, Y. (2024). Collision of herbal medicine and nanotechnology: A bibliometric analysis of herbal nanoparticles from 2004 to 2023. *Journal of Nanobiotechnology, 22*(1), 140.

Alamgir, A. N. M. (2017). Cultivation of herbal drugs, biotechnology, and in vitro production of secondary metabolites, high-value medicinal plants, herbal wealth, and herbal trade. In A. N. M. Alamgir (Ed.), *Therapeutic use of medicinal plants and their extracts: Vol. 1: Pharmacognosy* (pp. 379–452). Springer International Publishing.

Ali, A., Mashwani, Z.-R., Raja, N. I., Mohammad, S., Luna-Arias, J. P., Ahmad, A., & Kaushik, P. (2023). Phytomediated selenium nanoparticles and light regimes elicited in vitro callus cultures for biomass accumulation and secondary metabolite production in *Caralluma tuberculata*. *Frontiers in Plant Science*, 14.

Ali, A., Mohammad, S., Khan, M. A., Raja, N. I., Arif, M., Kamil, A., & Mashwani, Z.-R. (2019). Silver nanoparticles elicited in vitro callus cultures for accumulation of biomass and secondary metabolites in *Caralluma tuberculata*. *Artificial Cells, Nanomedicine, and Biotechnology, 47*(1), 715–724.

Almulhim, B. F., El Sherif, F., & Khattab, S. (2024). Silicon dioxide nanoparticles, elicitation of silybin concentration in *Silybum marianum* l. through chalcone synthase gene expression modification. *Applied Ecology and Environmental Research, 22*(1), 681–693.

Alyas, J., Khalid, N., Ishaque, S., Fatima, H., Hashim, M., Hassan, S., Bukhari, S., Hano, C., Abbasi, B. H., & Anjum, S. (2023). Light (high light/UV radiation) modulates adaptation mechanisms and secondary metabolite production in medicinal plants. In A. Husen & M. Iqbal (Eds.), *Medicinal plants: Their response to abiotic stress* (pp. 363–390). Springer Nature.

Amnuaykan, P., Nathewet, P., & Thepsukhon, A. (2023). Silver nanoparticles eliminate hyperhydricity in micropropagated Lavender. *Ornamental Horticulture, 29*, 354–364.

Antonio-Pérez, A., Durán-Armenta, L. F., Pérez-Loredo, M. G., & Torres-Huerta, A. L. (2023). Biosynthesis of copper nanoparticles with medicinal plants extracts: From extraction methods to applications. *Micromachines, 14*(10), 1882.

Antunes Filho, S., dos Santos, M. S., dos Santos, O. A. L., Backx, B. P., Soran, M.-L., Opriş, O., Lung, I., Stegarescu, A., & Bououdina, M. (2023). Biosynthesis of nanoparticles using plant extracts and essential oils. *Molecules*, *28*(7), 3060.

Barciela, P., Carpena, M., Li, N.-Y., Liu, C., Jafari, S. M., Simal-Gandara, J., & Prieto, M. A. (2023). Macroalgae as biofactories of metal nanoparticles; biosynthesis and food applications. *Advances in Colloid and Interface Science*, *311*, 102829.

Bernela, M., Seth, M., Kaur, N., Sharma, S., & Pati, P. K. (2023). Harnessing the potential of nanobiotechnology in medicinal plants. *Industrial Crops and Products*, *194*, 116266.

Chandran, H., Meena, M., Barupal, T., & Sharma, K. (2020). Plant tissue culture as a perpetual source for production of industrially important bioactive compounds. *Biotechnology Reports*, *26*, e00450.

Chaudhary, J., Tailor, G., Yadav, M., & Mehta, C. (2023). Green route synthesis of metallic nanoparticles using various herbal extracts: A review. *Biocatalysis and Agricultural Biotechnology*, *50*, 102692.

Chung, I.-M., Rajakumar, G., & Thiruvengadam, M. (2018). Effect of silver nanoparticles on phenolic compounds production and biological activities in hairy root cultures of *Cucumis anguria*. *Acta Biologica Hungarica*, *69*(1), 97–109.

Chung, I.-M., Rajakumar, G., Subramanian, U., Venkidasamy, B., & Thiruvengadam, M. (2019). Impact of copper oxide nanoparticles on enhancement of bioactive compounds using cell suspension cultures of *Gymnema sylvestre* (Retz.) R. Br. *Applied Sciences*, *9*(10), 2165.

Chung, I.-M., Rekha, K., Rajakumar, G., & Thiruvengadam, M. (2018). Elicitation of silver nanoparticles enhanced the secondary metabolites and pharmacological activities in cell suspension cultures of bitter gourd. *3 Biotech*, *8*(10), 412.

Corchete, P., Almagro, L., Gabaldón, J. A., Pedreño, M. A., & Palazón, J. (2022). Phenylpropanoids in *Silybum marianum* cultures treated with cyclodextrins coated with magnetic nanoparticles. *Applied Microbiology and Biotechnology*, *106*(7), 2393–2401.

Divekar, P. A., Narayana, S., Divekar, B. A., Kumar, R., Gadratagi, B. G., Ray, A., Singh, A. K., Rani, V., Singh, V., Singh, A. K., Kumar, A., Singh, R. P., Meena, R. S., & Behera, T. K. (2022). Plant secondary metabolites as defense tools against herbivores for sustainable crop protection. *International Journal of Molecular Sciences*, *23*(5), 2690.

Efferth, T. (2019). Biotechnology applications of plant callus cultures. *Engineering*, *5*(1), 50–59.

Ettadili, F. E., Aghris, S., Laghrib, F., Farahi, A., Saqrane, S., Bakasse, M., Lahrich, S., & El Mhammedi, M. A. (2022). Recent advances in the nanoparticles synthesis using plant extract: Applications and future recommendations. *Journal of Molecular Structure*, *1248*, 131538.

Fageria, N. K., Filho, M. P. B., Moreira, A., & Guimarães, C. M. (2009). Foliar fertilization of crop plants. *Journal of Plant Nutrition*, *32*(6), 1044–1064.

Fatima, K., Abbas, S. R., Zia, M., Sabir, S. M., Khan, R. T., Khan, A. A., Hassan, Z., & Zaman, R. (2020). Induction of secondary metabolites on nanoparticles stress in callus culture of *Artemisia annua* L. *Brazilian Journal of Biology*, *81*, 474–483.

Fazal, H., Abbasi, B. H., Ahmad, N., Ali, M., Shujait Ali, S., Khan, A., & Wei, D.-Q. (2019). Sustainable production of biomass and industrially important secondary metabolites in cell cultures of selfheal (*Prunella vulgaris* L.) elicited by silver and gold nanoparticles. *Artificial Cells, Nanomedicine, and Biotechnology*, *47*(1), 2553–2561.

Garza-Alonso, C. A., González-García, Y., Cadenas-Pliego, G., Olivares-Sáenz, E., Trejo-Téllez, L. I., & Benavides-Mendoza, A. (2021). Seed priming with ZnO nanoparticles promotes early growth and bioactive compounds of *Moringa oleifera*. *Notulae Botanicae Horti Agrobotanici Cluj-Napoca*, *49*(4), 12546–12546.

Garza-Alonso, C. A., Olivares-Sáenz, E., González-Morales, S., Cabrera-De la Fuente, M., Juárez-Maldonado, A., González-Fuentes, J. A., Tortella, G., Valdés-Caballero, M. V., & Benavides-Mendoza, A. (2022). Strawberry biostimulation: From mechanisms of action to plant growth and fruit quality. *Plants*, *11*(24), 3463.

Gecibesler, I. H., Disli, F., Bayindir, S., Toprak, M., Tufekci, A. R., Sahin Yaglıoglu, A., Altun, M., Kocak, A., Demirtas, I., & Adem, S. (2021). The isolation of secondary metabolites from *Rheum ribes* L. and the synthesis of new semi-synthetic anthraquinones: Isolation, synthesis and biological activity. *Food Chemistry*, *342*, 128378.

Ghasemi, B., Hosseini, R., & Nayeri, F. (2015). Effects of cobalt nanoparticles on artemisinin production and gene expression in *Artemisia annua*. *Turkish Journal of Botany*, *39*(5), 769–777.

Ghorbanpour, M., & Hadian, J. (2015). Multi-walled carbon nanotubes stimulate callus induction, secondary metabolites biosynthesis and antioxidant capacity in medicinal plant Satureja khuzestanica grown in vitro. *Carbon*, *94*, 749–759.

Giglou, M. T., Giglou, R. H., Esmaeilpour, B., Azarmi, R., Padash, A., Falakian, M., Śliwka, J., Gohari, G., & Lajayer, H. M. (2022). A new method in mitigation of drought stress by chitosan-coated iron oxide nanoparticles and growth stimulant in peppermint. *Industrial Crops and Products*, *187*, 115286.

Godeto, Y. G., Ayele, A., Ahmed, I. N., Husen, A., & Bachheti, R. K. (2023). Medicinal plant-based metabolites in nanoparticles synthesis and their cutting-edge applications: An overview. In R. K. Bachheti & A. Bachheti (Eds.), *Secondary metabolites from medicinal plants* (pp. 1–34). CRC Press.

Golinejad, S., Mirjalili, M. H., Rezadoost, H., & Ghorbanpour, M. (2023). Molecular, biochemical, and metabolic changes induced by gold nanoparticles in *Taxus baccata* L. cell culture. *Industrial Crops and Products*, *192*, 115988.

Golkar, P., Akbari, R., Bazarganipour, M., & Javed, R. (2023). Biochemical and phytochemical responses of *Ammi visnaga* L. (Apiaceae) callus culture elicited by SiO_2 and graphene Oxide-SiO_2 nanoparticles. *Plant Physiology and Biochemistry*, *200*, 107741.

González-Morales, S., Cárdenas-Atayde, P. A., Garza-Alonso, C. A., Robledo-Olivo, A., & Benavides-Mendoza, A. (2022). Plant biostimulation with nanomaterials: A physiological and molecular standpoint. In L. Fernandes Fraceto, H. W. Pereira de Carvalho, R. de Lima, S. Ghoshal, & C. Santaella (Eds.), *Inorganic nanopesticides and nanofertilizers: A view from the mechanisms of action to field applications* (pp. 153–185). Springer International Publishing.

González-Morales, S., Parera, C. A., Juárez-Maldonado, A., la Fuente, M. C. D., & Benavides-Mendoza, A. (2020). Chapter 13 – The ecology of nanomaterials in agroecosystems. In A. Husen & M. Jawaid (Eds.), *Nanomaterials for agriculture and forestry applications* (pp. 313–355). Elsevier.

Gutierrez-Valdes, N., Häkkinen, S. T., Lemasson, C., Guillet, M., Oksman-Caldentey, K.-M., Ritala, A., & Cardon, F. (2020). Hairy root cultures – a versatile tool with multiple applications. *Frontiers in Plant Science*, *11*, 33.

Guzmán-Mendoza, J. J., Sánchez-Ramírez, B., & Talamás-Rohana, P. (2024). Chapter 4 – Biological interactions and fate of nanomaterials in living systems. In R. Vijayamma, N.

Kalarikkal, & S. Thomas (Eds.), *Nanostructured materials for biomedical applications* (pp. 89–159). Elsevier.

Hadacek, F. (2002). Secondary metabolites as plant traits: Current assessment and future perspectives. *Critical Reviews in Plant Sciences*, *21*(4), 273–322.

Hayat, K., Ali, S., Ullah, S., Fu, Y., & Hussain, M. (2021). Green synthesized silver and copper nanoparticles induced changes in biomass parameters, secondary metabolites production, and antioxidant activity in callus cultures of *Artemisia absinthium* L. *Green Processing and Synthesis*, *10*(1), 61–72.

Heydari, H. R., Chamani, E., & Esmaielpour, B. (2020). Cell line selection through gamma irradiation combined with multi-walled carbon nanotubes elicitation enhanced phenolic compounds accumulation in *Salvia nemorosa* cell culture. *Plant Cell, Tissue and Organ Culture (PCTOC)*, *142*(2), 353–367.

Inam, M., Attique, I., Zahra, M., Khan, A. K., Hahim, M., Hano, C., & Anjum, S. (2023). Metal oxide nanoparticles and plant secondary metabolism: Unraveling the game-changer nano-elicitors. *Plant Cell, Tissue and Organ Culture (PCTOC)*, *155*, 327–344.

Jafari, S., Mousavi-Fard, S., Rezaei Nejad, A., Mumivand, H., Sorkheh, K., & Nikoloudakis, N. (2023). Foliar application of chitosan and titanium dioxide enhances silybin content by orchestrating miRNA and gene targets transcription in Milk thistle (*Silybum marianum* L.). *Current Plant Biology*, *35–36*, 100302.

Jafari, S., Mousavi-Fard, S., Rezaei Nejad, A., Mumivand, H., Sorkheh, K., Nikoloudakis, N., & Fanourakis, D. (2022). Chitosan and titanium dioxide are more effective in improving seed yield and quality in nanoparticle compared to non-structured form: A case study in five milk thistle ecotypes (*Silybum marianum* (L.) Gaertn. *Agronomy*, *12*(8), 1827.

Jahromi, M. H. G., Tavallali, H., Tavallali, V., & Karimi, M. A. (2024). Differential phytochemical responses of Feverfew to green synthesized calcium nanostructures formed on pectin and *Aloe vera* extract. *Food Chemistry Advances*, *4*, 100676.

Jampílek, J., & Kráľová, K. (2023). Impact of abiotic stresses on production of secondary metabolites in medicinal and aromatic plants. In T. Aftab (Ed.), *New frontiers in plant-environment interactions: Innovative technologies and developments* (pp. 169–252). Springer Nature Switzerland.

Jamshidi, M., & Ghanati, F. (2017). Taxanes content and cytotoxicity of hazel cells extract after elicitation with silver nanoparticles. *Plant Physiology and Biochemistry*, *110*, 178–184.

Jang, I.-B., Lee, D.-Y., Yu, J., Park, H.-W., Mo, H.-S., Park, K.-C., Hyun, D.-Y., Lee, E.-H., Kim, K.-H., & Oh, C.-S. (2015). Photosynthesis rates, growth, and ginsenoside contents of 2-yr-old *Panax ginseng* grown at different light transmission rates in a greenhouse. *Journal of Ginseng Research*, *39*(4), 345–353.

Javed, R., Yucesan, B., Zia, M., & Gurel, E. (2022). Nanoelicitation: A promising and emerging technology for triggering the sustainable in vitro production of secondary metabolites in medicinal plants. In J.-T. Chen (Ed.), *Plant and nanoparticles* (pp. 265–280). Springer Nature.

Juárez-Maldonado, A., Ortega-Ortíz, H., Morales-Díaz, A. B., González-Morales, S., Morelos-Moreno, Á., Cabrera-De la Fuente, M., Sandoval-Rangel, A., Cadenas-Pliego, G., & Benavides-Mendoza, A. (2019). Nanoparticles and nanomaterials as plant biostimulants. *International Journal of Molecular Sciences*, *20*(1), 162.

Juárez-Maldonado, A., Tortella, G., Rubilar, O., Fincheira, P., & Benavides-Mendoza, A. (2021). Biostimulation and toxicity: The magnitude of the impact of nanomaterials in microorganisms and plants. *Journal of Advanced Research*, *31*, 113–126.

Karakaş, Ö. (2020). Effect of silver nanoparticles on production of indole alkaloids in *Isatis constricta*. *Iranian Journal of Science and Technology, Transactions A: Science*, *44*(3), 621–627.

Karimi, M., Ahmadi, N., & Ebrahimi, M. (2022). Red LED light promotes biomass, flowering and secondary metabolites accumulation in hydroponically grown *Hypericum perforatum* L. (cv. Topas). *Industrial Crops and Products*, *175*, 114239.

Karimzadeh, F., Haddad, R., Garoosi, G., & Khademian, R. (2019). Effects of nanoparticles on activity of lignan biosynthesis enzymes in cell suspension culture of *Linum usitatissimum* L. *Russian Journal of Plant Physiology*, *66*(5), 756–762.

Khajavi, M., Rahaie, M., & Ebrahimi, A. (2019). The effect of TiO_2 and SiO_2 nanoparticles and salinity stress on expression of genes involved in parthenolide biosynthesis in Feverfew (*Tanacetum parthenium* L.). *Caryologia*, *72*(1), 3–14.

Khaldari, I., Naghavi, M. R., Motamedi, E., & Zargar, M. (2024). The effects of green and chemically-synthesized copper oxide nanoparticles on the production and gene expression of morphinan alkaloids in Oriental poppy. *Scientific Reports*, *14*(1), 6000.

Khan, A. K., Kousar, S., Tungmunnithum, D., Hano, C., Abbasi, B. H., & Anjum, S. (2021). Nano-elicitation as an effective and emerging strategy for in vitro production of industrially important flavonoids. *Applied Sciences*, *11*(4), 1694.

Khan, M. A., Wallace, W. T., Sambi, J., Rogers, D. T., Littleton, J. M., Rankin, S. E., & Knutson, B. L. (2020). Nanoharvesting of bioactive materials from living plant cultures using engineered silica nanoparticles. *Materials Science and Engineering: C*, *106*, 110190.

Khare, S., Singh, N. B., Singh, A., Hussain, I., Niharika, K., Yadav, V., Bano, C., Yadav, R. K., & Amist, N. (2020). Plant secondary metabolites synthesis and their regulations under biotic and abiotic constraints. *Journal of Plant Biology*, *63*(3), 203–216.

Khattab, S., El Sherif, F., AlDayel, M., Yap, Y.-K., Meligy, A., & Ibrahim, H. I. M. (2022). Silicon dioxide and silver nanoparticles elicit antimicrobial secondary metabolites while enhancing growth and multiplication of *Lavandula officinalis* in-vitro plantlets. *Plant Cell, Tissue and Organ Culture (PCTOC)*, *149*, 411–421.

Kokkinidis, M., Glykos, N. M., & Fadouloglou, V. E. (2020). Chapter four – Catalytic activity regulation through post-translational modification: The expanding universe of protein diversity. In T. Karabencheva-Christova & C. Christov (Eds.), *Advances in protein chemistry and structural biology* (pp. 97–125). Academic Press.

Kralova, K., & Jampilek, J. (2021). Responses of medicinal and aromatic plants to engineered nanoparticles. *Applied Sciences*, *11*(4), 1813.

Lafmejani, Z. N., Jafari, A. A., Moradi, P., & Moghadam, A. L. (2018a). Impact of foliar application of iron-chelate and iron nano particles on some morpho-physiological traits and rssential oil composition of peppermint (*Mentha piperita* L.). *Journal of Essential Oil Bearing Plants*, *21*(5), 1374–1384.

Lafmejani, Z. N., Jafari, A. A., Moradi, P., & Moghadam, A. L. (2018b). Impact of foliar application of copper sulphate and copper nanoparticles on some morpho-physiological traits and essential oil composition of peppermint (*Mentha piperita* L.). *Herba Polonica*, *64*(2), 13–24.

Li, S. (2014). Transcriptional control of flavonoid biosynthesis: Fine-tuning of the MYB-bHLH-WD40 (MBW) complex. *Plant Signaling & Behavior*, *9*(1), e27522.

Li, Y., Winzer, T., He, Z., & Graham, I. A. (2020). Over 100 million years of enzyme evolution underpinning the production of morphine in the papaveraceae family of flowering plants. *Plant Communications*, *1*(2).

Mahajan, M., Kuiry, R., & Pal, P. K. (2020). Understanding the consequence of environmental stress for accumulation of secondary metabolites in medicinal and aromatic plants. *Journal of Applied Research on Medicinal and Aromatic Plants*, *18*, 100255.

Mambrí, A. P., Andriolo, J. L., Manfron, M. P., Pinheiro, S. M., Cardoso, F. L., & Neves, M. G. (2018). Yield and composition of lavender essential oil grown in substrate. *Horticultura Brasileira*, *36*, 259–264.

Mazrou, R., Ali, E. F., Hassan, S., & Hassan, F. A. S. (2021). A pivotal role of chitosan nanoparticles in enhancing the essential oil productivity and antioxidant capacity in *Matricaria chamomilla* L. *Horticulturae*, *7*(12), 574.

McClements, D. J., & Öztürk, B. (2022). Utilization of nanotechnology to improve the application and bioavailability of phytochemicals derived from waste streams. *Journal of Agricultural and Food Chemistry*, *70*(23), 6884–6900.

Medrano-Macías, J., Flores-Gallegos, A. C., Nava-Reyna, E., Morales, I., Tortella, G., Solís-Gaona, S., & Benavides-Mendoza, A. (2022). Reactive oxygen, nitrogen, and sulfur species (RONSS) as a metabolic cluster for signaling and biostimulation of plants: An overview. *Plants*, *11*(23), 3203.

Moharrami, F., Hosseini, B., Sharafi, A., & Farjaminezhad, M. (2017). Enhanced production of hyoscyamine and scopolamine from genetically transformed root culture of *Hyoscyamus reticulatus* L. elicited by iron oxide nanoparticles. *In Vitro Cellular & Developmental Biology – Plant*, *53*(2), 104–111.

Mohit Rabary, P., Movahedi, Z., Ghabooli, M., & Rostami, M. (2022). Effects of foliar application of zinc oxide nanoparticles on traits of several medicinal plants under aeroponic system conditions. *International Journal of Horticultural Science and Technology*, *9*(4), 445–452.

Mosavat, N., Golkar, P., Yousefifard, M., & Javed, R. (2019). Modulation of callus growth and secondary metabolites in different Thymus species and *Zataria multiflora* micropropagated under ZnO nanoparticles stress. *Biotechnology and Applied Biochemistry*, *66*(3), 316–322.

Mubeen, B., Hasnain, A., Jie, W., Zheng, H., Peijnenburg, W. J. G. M., Rozali, S. E., Rasool, R., Naqvi, S. A. H., Rao, M. J., Sohail, M. A., Moustafa, M., Al-Shehri, M., & Negm, S. (2023). Enhanced production of active photosynthetic and biochemical molecules in *Silybum marianum* L. using biotic and abiotic elicitors in hydroponic culture. *Molecules*, *28*(4), 1716.

Nazerieh, H., Ardebili, Z. O., & Iranbakhsh, A. (2018). Potential benefits and toxicity of nanoselenium and nitric oxide in peppermint. *Acta Agriculturae Slovenica*, *111*(2), 357–368.

Nekoukhou, M., Fallah, S., Pokhrel, L. R., Abbasi-Surki, A., & Rostamnejadi, A. (2024). Foliar co-application of zinc oxide and copper oxide nanoparticles promotes phytochemicals and essential oil production in dragonhead (*Dracocephalum moldavica*). *Science of the Total Environment*, *906*, 167519.

Nemati Lafmejani, Z., Jafari, A. A., Moradi, P., & Ladan Moghadam, A. (2021). Application of chelate and nano-chelate zinc micronutrient onmorpho-physiological traits and essential oil compounds of peppermint (*Mentha piperita* L.). *Journal of Medicinal Plants and By-Products*, *10*(Special), 21–28.

Nourozi, E., Hosseini, B., Maleki, R., & Abdollahi Mandoulakani, B. (2019). Iron oxide nanoparticles: A novel elicitor to enhance anticancer flavonoid production and gene expression in Dracocephalum kotschyi hairy-root cultures. *Journal of the Science of Food and Agriculture*, *99*(14), 6418–6430.

Nourozi, E., Hosseini, B., Maleki, R., & Mandoulakani, B. A. (2019). Pharmaceutical important phenolic compounds overproduction and gene expression analysis in *Dracocephalum kotschyi* hairy roots elicited by SiO_2 nanoparticles. *Industrial Crops and Products*, *133*, 435–446.

Ojeda-Barrios, D. L., Morales, I., Juárez-Maldonado, A., Sandoval-Rangel, A., Fuentes-Lara, L. O., & Benavides-Mendoza, A. (2020). Importance of nanofertilizers in fruit nutrition. In A. K. Srivastava & C. Hu (Eds.), *Fruit crops* (pp. 497–508). Elsevier.

Ostadi, A., Javanmard, A., Amani Machiani, M., & Kakaei, K. (2023). Optimizing antioxidant activity and phytochemical properties of peppermint (*Mentha piperita* L.) by integrative application of biofertilizer and stress-modulating nanoparticles under drought stress conditions. *Plants*, *12*(1), 151.

Pang, Z., Chen, J., Wang, T., Gao, C., Li, Z., Guo, L., Xu, J., & Cheng, Y. (2021). Linking plant secondary metabolites and plant microbiomes: A review. *Frontiers in Plant Science*, *12*.

Pech, R., Volná, A., Hunt, L., Bartas, M., Červeň, J., Pečinka, P., Špunda, V., & Nezval, J. (2022). Regulation of phenolic compound production by light varying in spectral quality and total irradiance. *International Journal of Molecular Sciences*, *23*(12), 6533.

Peng, H., Hu, H., Xi, K., Zhu, X., Zhou, J., Yin, J., Guo, F., Liu, Y., & Zhu, Y. (2022). Silicon nanoparticles enhance ginger rhizomes tolerance to postharvest deterioration and resistance to *Fusarium solani*. *Frontiers in Plant Science*, *13*.

Powell, C. R., Dillon, K. M., & Matson, J. B. (2018). A review of hydrogen sulfide (H_2S) donors: Chemistry and potential therapeutic applications. *Biochemical Pharmacology*, *149*, 110–123.

Rastegaran, Z., Hassanpour, H., & Ziyadi, H. (2022). Synthesized Fe_3O_4 nanoparticles induce antioxidant activity and total phenolic and flavonoid contents in *Matricaria chamomilla* seedlings. *Iranian Journal of Plant Physiology*, *12*(1), 4003–4012.

Rezaei Nazari, M., Abdossi, V., Zamani Hargalani, F., & Larijani, K. (2022). The effect of nano selenium foliar application on some secondary metabolites of *Hypericum perforatum* L. *Journal of Medicinal Plants*, *21*(81), 67–78.

Rhaman, M. S., Tania, S. S., Imran, S., Rauf, F., Kibria, M. G., Ye, W., Hasanuzzaman, M., & Murata, Y. (2022). Seed priming with nanoparticles: An emerging technique for improving plant growth, development, and abiotic stress tolerance. *Journal of Soil Science and Plant Nutrition*, *22*, 4047–4062.

Rivero-Montejo, S. de J., Vargas-Hernandez, M., & Torres-Pacheco, I. (2021). Nanoparticles as novel elicitors to improve bioactive compounds in plants. *Agriculture*, *11*(2), 134.

Safikhan, S., Chaichi, M. R., Khoshbakht, K., Amini, A., & Motesharezadeh, B. (2018). Application of nanomaterial graphene oxide on biochemical traits of Milk thistle (*Silybum marianum* L.) under salinity stress. *Australian Journal of Crop Science*, *12*(6), 931–936.

Sen, T., & Samanta, S. K. (2015). Medicinal plants, human health and biodiversity: A broad review. In J. Mukherjee (Ed.), *Biotechnological applications of biodiversity* (pp. 59–110). Springer.

Shahhoseini, R. (2024). Silver nanoparticles: Elicitor of anticancer metabolite biosynthesis and growth-metabolites parameters in *Tanacetum parthenium* L. *Environmental Technology & Innovation*, *35*, 103725.

Shahhoseini, R., & Daneshvar, H. (2023). Phytochemical and physiological reactions of feverfew (*Tanacetum parthenium* (L.) Schultz Bip to TiO_2 nanoparticles. *Plant Physiology and Biochemistry*, *194*, 674–684.

Shakeran, Z., Keyhanfar, M., Asghari, G., & Ghanadian, M. (2015). Improvement of atropine production by different biotic and abiotic elicitors in hairy root cultures of Datura metel. *Turkish Journal of Biology*, *39*(1), 111–118.

Shao, F., Wilson, I. W., & Qiu, D. (2021). The research progress of taxol in taxus. *Current Pharmaceutical Biotechnology*, *22*(3), 360–366.

Shehzad, M. aamir, Khan, M. A., Ali, A., Mohammad, S., Noureldeen, A., Darwish, H., Ali, A., Ahmad, A., Khan, T., & Khan, R. S. (2021). Interactive effects of zinc oxide nano particles and different light regimes on growth and silymarin biosynthesis in callus cultures of *Silybum marianum* L. *Artificial Cells, Nanomedicine, and Biotechnology*, *49*(1), 523–535.

Shenavaie Zare, A., Ganjeali, A., Vaezi Kakhki, M. R., Cheniany, M., & Mashreghi, M. (2022). Plant elicitation and TiO_2 nanoparticles application as an effective strategy for improving the growth, biochemical properties, and essential oil of peppermint. *Physiology and Molecular Biology of Plants*, *28*(7), 1391–1406.

Srivastava, A. K., Mishra, P., & Mishra, A. K. (2021). 3 – Effect of climate change on plant secondary metabolism: An ecological perspective. In A. K. Srivastava, V. K. Kannaujiya, R. K. Singh, & D. Singh (Eds.), *Evolutionary diversity as a source for anticancer molecules* (pp. 47–76). Academic Press.

Tavali, I. E., & Ok, H. (2022). Comparison of heat-treated and unheated vermicompost on biological properties of calcareous soil and aloe vera growth under greenhouse conditions in a mediterranean climate. *Agronomy*, *12*(11), 2649.

Tavallali, V., Kiani, M., & Hojati, S. (2019). Iron nano-complexes and iron chelate improve biological activities of sweet basil (*Ocimum basilicum* L.). *Plant Physiology and Biochemistry*, *144*, 445–454.

Theodoridis, S., Drakou, E. G., Hickler, T., Thines, M., & Nogues-Bravo, D. (2023). Evaluating natural medicinal resources and their exposure to global change. *The Lancet Planetary Health*, *7*(2), e155–e163.

Tian, H., Ghorbanpour, M., & Kariman, K. (2018). Manganese oxide nanoparticle-induced changes in growth, redox reactions and elicitation of antioxidant metabolites in deadly nightshade (*Atropa belladonna* L.). *Industrial Crops and Products*, *126*, 403–414.

Torabzadeh, D., Hassanpour, H., Asgarpanah, J., & Rezayian, M. (2019). Nanoparticles induced antioxidative compounds in *Matricaria chamomilla*. *Iranian Journal of Plant Physiology*, *9*(4), 2955–2961.

Vahide, P., Negar, K., & Hajati Razieh, J. (2021). Bio nanoparticles as elicitors increase accumulation of betulin and betulinic acid in callus cultures. *South African Journal of Botany*, *141*, 431–439.

Velázquez-Gamboa, M. C., Rodríguez-Hernández, L., Abud-Archila, M., Gutiérrez-Miceli, F. A., González-Mendoza, D., Valdez-Salas, B., González-Terreros, E., & Luján-Hidalgo, M. C. (2020). *Agronomic biofortification of Stevia rebaudiana with zinc oxide (ZnO) phytonanoparticles and antioxidant compounds*. Sugar Tech.

Velázquez-Juárez, G., León-Morales, J. M., Castañeda-Nava, J. J., & García-Morales, S. (2023). In vitro production of plant bioactive compounds enhanced by nanoparticle elicitation. In J.-T. Chen (Ed.), *Advanced nanotechnology in plants* (p. 16). CRC Press.

Wang, T., Li, L., Zhuang, W., Zhang, F., Shu, X., Wang, N., & Wang, Z. (2021). Recent research progress in taxol biosynthetic pathway and acylation reactions mediated by *Taxus acyltransferases*. *Molecules*, *26*(10), 2855.

Xu, Y., Tao, M., Xu, W., Xu, L., Yue, L., Cao, X., Chen, F., & Wang, Z. (2024). Nano-CeO_2 activates physical and chemical defenses of garlic (*Allium sativum* L.) for reducing antibiotic resistance genes in plant endosphere. *Ecotoxicology and Environmental Safety*, *276*, 116289.

Yeshi, K., Crayn, D., Ritmejerytė, E., & Wangchuk, P. (2022). Plant secondary metabolites produced in response to abiotic stresses has potential application in pharmaceutical product development. *Molecules*, *27*(1), 313.

Zafar, H., Aziz, T., Khan, B., Mannan, A., ur Rehman, R., & Zia, M. (2020). CuO and ZnO nanoparticle application in synthetic soil modulates morphology, nutritional contents, and metal analysis of *Brassica nigra*. *ACS Omega*, *5*(23), 13566–13577.

Zhang, B., Zheng, L. P., Li, W. Y., & Wang, J. W. (2013). Stimulation of artemisinin production in *Artemisia annua* hairy roots by Ag-SiO_2 core-shell nanoparticles. *Current Nanoscience*, *9*(3), 363–370.

Zobayed, S. M. A., Afreen, F., & Kozai, T. (2005). Necessity and production of medicinal plants under controlled environments. *Environment Control in Biology*, *43*(4), 243–252.

17 Integrated Molecular Strategies for Improving Traditional Chinese Medicinal Plants

Arumugam Sembiyaa, Jayakumar Dharshini, Balasubramaniam Rashmika, Mani Manoj, Shanmugam Velayuthaprabhu, Asirvatham Alwin Robert, Jen-Tsung Chen, and Arumugam Vijaya Anand

LIST OF ABBREVIATIONS

2DGE	2D gel electrophoresis
ABA	abscisic acid
bZIP2	bZIP domain of the basic leucine zipper2
DNMT	DNA methyltransferases
GC-MS	gas chromatography–mass spectrometry
HMT	histone methyltransferases
HP-TLC	high-performance thin-layer chromatography
ICP-MS	inductively coupled plasma–mass spectrometry
LC-MS	liquid chromatography coupled with mass spectrometry
SFC	supercritical fluid chromatography
TAs	tropane alkaloids
TCM	traditional Chinese medicine
TFs	transcription factors
TLC	thin-layer chromatography

17.1 INTRODUCTION

Bioactive secondary metabolites can be noted in therapeutic plants, which provide an abundance of chemicals with potential applications in medicine. There is an increasing need to improve the formation of these secondary metabolites due to the increased demand for drugs and nature generated from these plants. The pharmacological qualities of their secondary metabolites effectively treat various disorders, from ordinary ailments to life-threatening conditions. However, a variety of circumstances may impact the unexpected formation of these secondary bioactive compounds, resulting in variations in both product and quality (Manoharan et al., 2008; Begum et al., 2009; Muthumani et al., 2013; Varalakshmi et al., 2014; Vijaya Anand et al., 2016; Anand et al., 2020; Bharathi et al., 2018; Divya et al., 2019; Akilandeswari et al., 2019; Thangavelu et al., 2022; Kaviya et al., 2022; Manikandan et al., 2022; Karthika et al., 2023; Vijayakumar et al., 2023; Halka et al., 2024).

17.1.1 Overview of Traditional Chinese Medicinal Plants

Traditional Chinese medicine (TCM) is known for its broad clinical applications, numerous prescriptions, multiple targets, and diverse ingredients, akin to the strategy of combination therapy. These attributes make TCM a significant source for new drug discovery (Li et al., 2022). The increasing benefit of medicinal plants can be attributed to the consequences associated with synthetic drugs, the inadequacies of current treatments for chronic diseases, microbial resistance, and the high initial acquisition in pharmaceutical investigation and development (Pan et al., 2010). The foundational text *Shen Nong's Materia Medica*, first documented during the Eastern Han Dynasty, laid the groundwork for TCM, which has since been extensively studied (Guo et al., 2020).

Over the centuries, TCM has gained global recognition for its effectiveness in treating various complex diseases by utilizing a systematic approach that involves principal herbs, delegate herbs, associate herbs, and diplomat herbs (Han et al., 2018). President Obama first proposed 'precision medicine' in 2015, which targets inaccurate medical plans in consideration of the basics of the genome of each, clinical symptoms, and health status, with the major persistence of modifying the most suitable medicine for each patient (Kasper et al., 2015). However, due to the chemical complications of TCM, traditional TCM procedures are inefficient in discovering biologically active compounds and accompanying medicinal benefits (Buriani et al., 2012).

17.1.2 Need for Molecular Strategies

Global pharmaceutical companies are increasingly turning to herbs as a conceivable source of new drugs. This trend highlights a continued interest in natural products for medical development and finding (Seidl, 2002). Various pharmacopoeias employ both microscopic and macroscopic evaluations (such as morphology and histology) along with chemical profiling techniques to ensure the quality and consistency of both processed and raw herbs (Heubl, 2010). However, the inherent chemical variability in plant materials poses challenges for verifying botanical identity, as it is influenced by factors like growth conditions, storage, and harvesting practices (Chan, 2003). Sustainable sourcing of natural derivatives from TCM can be achieved through cultivating high-yield varieties or using biotechnological

DOI: 10.1201/9781003475491-17

methods. Over time, these natural products from TCM could play a significant role in developing molecular target-guided treatments and personalized therapeutic approaches (Efferth et al., 2007).

17.2 GENETIC CHARACTERIZATION AND MARKER-ASSISTED SELECTION

17.2.1 DNA Barcoding and Molecular Markers

As the TCM market grows, the incidence of low-quality or counterfeit herbal products has increased (Zhu et al., 2022). Research by Han et al. on 1260 genuine samples from 295 species with medicinal properties across seven TCM needs in China noted that approximately 4.2% were adulterated (Han et al., 2016). Additionally, a separate study on 400 seeds intended for TCM derivatives revealed that 7.5% were mislabeled (Xiong et al., 2018). As the global consumer market continues to expand, the importance of authenticating and accurately identifying plant products has become more critical for the benefit and commercialization of therapeutic plants and supplements (Yu et al., 2021). Numerous molecular markers are available, utilizing techniques like restriction fragment length polymorphisms, amplified fragment length polymorphisms, randomly amplified polymorphic DNA, simple sequence repeats, inter simple sequence repeats, sequence characterized amplified regions, loop-mediated isothermal amplification and single nucleotide polymorphisms. Recent advancements include DNA barcoding, microarray-based markers, and next-generation sequencing-based markers (Ganie et al., 2015).

In 2003, Hebert and colleagues introduced the DNA barcoding method to accurately specify species. This technique relies on a standardized, short DNA sequence called a DNA barcode, facilitating quick, precise, and automated species identification (Hebert et al., 2003; Hebert & Gregory, 2005). Considering its development, DNA barcoding has been widely employed in various fields and is regarded as a rejuvenation of taxonomy (Chen et al., 2014). Scientists have utilized multiple or single markers for DNA barcoding, including the whole genome of chloroplast, to detect species across diverse genera and families of medicinal property-containing plants (He et al., 2017).

In DNA barcoding, considerations of cost, efficiency, and convenience are paramount. Single-locus DNA barcode identifications are cost-effective, whereas utilizing multiple loci can significantly enhance marker efficiency. However, for taxa requiring high resolution at the intraspecies level, comprehensive datasets like the whole genome of chloroplast are necessary as super barcodes to improve resolution (Zhou et al., 2018). Following rigorous experimentation and validation, Chen and colleagues recommended the ITS2 part as the preliminary DNA barcode and trnH-psbA as a supplementary sequence for detecting species of therapeutic plants (Chen et al., 2010). Due to the limitations of single-locus marker sequences in providing sufficient information for lower-level identification, scientists often combine multiple markers to identify medicinal plants. The most frequently used varieties of DNA markers include trnH-psbA, rbcL, matK, and ITS sequences (Newmaster et al., 2013). The publication of chloroplast genomes for 3452 plants on NCBI marks a significant advancement in using chloroplast genomes as DNA barcodes for plant identification and differentiation (Govindaraghavan et al., 2012).

17.2.2 Marker-Assisted Breeding

Advances in molecular biology techniques now provide extensive insights into genetic structures. Molecular markers are widely employed to construct genetic maps, offering comprehensive data on phylogeny and evolution (Nadeem et al., 2018). Genetic markers are valuable tools for evaluating genetic assortment and categorizing genetic materials (Wang et al., 2015a). DArT and SNP markers, in particular, are commonly utilized to evaluate genetic assortment in various crop species (Baloch et al., 2017). Introgression involves the transfer of specific genes from plant genetic resources to crop types. This process involves selecting preferred characteristics from exotic germplasm and incorporating them into plant crops through backcrossing (Simmonds, 1993).

Marker-assisted selection has greatly advanced the integration of wild genes into cultivated crops. This technique has enabled the successful transfer of numerous beneficial genes from wild species into nearly all major crops. Simple sequence repeat (SSR) markers are frequently utilized in these processes. For example, two SSR markers were satisfactorily exploited to incorporate the Lgc-1 locus, which is linked to low gluten levels in japonica rice, achieving selection efficiencies between 93% and 97% (Wang et al., 2005). In barley, genes such as rym4 and rym5 provide resistance against the barley yellow mosaic virus, with a variety of characteristics conceived to aid in their selection (You-Xin et al., 2012).

17.3 GENETIC ENGINEERING AND CRISPR/CAS9 APPLICATIONS

17.3.1 Gene Editing for Enhanced Metabolite Production

Enhancing specialized metabolites in crop plants has been accomplished by methodically altering key biosynthetic or metabolic pathways within their genomes (Das et al., 2024). The CRISPR-Cas9 method facilitates different precise genetic modifications, like the induction of frameshift mutations, premature termination codons, gene substitutions, targeted single-gene knockouts, fragment deletions, and the insertion of specific target genes (Dong et al., 2020).

17.3.1.1 CRISPR/Cas9-Mediated Targeted Mutagenesis/Knockout in Medicinal Plants

CRISPR/Cas9-based gene disturbance approaches have been widely used in plants with medicinal value to either enhance beneficial specialized metabolites or reduce harmful metabolite levels (Lossen et al., 1864). For instance, editing the hyoscyamine 6β-hydroxylase (AbH6H) gene in *Atropa belladonna* results in plants that are deficient in anisodamine and scopolamine (Zeng et al., 2021). *Atropa belladonna* is notable for its production of anticholinergic tropane alkaloids (TAs), like hyoscyamine, scopolamine, and anisodamine, which are commercially valuable in the pharmaceutical sector (Gieger & Hesse, 1833). Hyoscyamine is used therapeutically to treat arrhythmias and organophosphate poisoning, while anisodamine shows potential for gastrointestinal colic and vascular spasms, and scopolamine is used for motion-associated illness. Although hyoscyamine is a principal component of TAs, anisodamine and scopolamine are considered minor derivatives (Yun et al., 1992; Poupko et al., 2007).

CRISPR/Cas9 system of knocking out of genes often results in indel mutations, fragment deletions, or large deletions through the repair pathway of NHEJ, leading to a loss of the functions of the gene. This genetics reverse approach is instrumental for investigations of functional genes or transcription factors (TFs) that regulate various gene regulatory pathways. Researchers can identify whether a TFs serves as a positive or negative controller in a synthetic biological pathway by creating knockout lines that eliminate its critical domains (Feng et al., 2018). For instance, in *Salvia miltiorrhiza,* knockout lines were created targeting the bZIP domain of the basic leucine zipper2 (bZIP2) gene. This study found that bZIP2 functions as a suppressor of phenolic acid production. Furthermore, investigations into the unexplored bZIP component bZIP2 in *Salvia miltiorrhiza* showed a notable response to abscisic acid (ABA) induction (Shi et al., 2021). While earlier research has provided some evidence of ABA's role in regulating the phenolic acid production pathway in the hairy roots of *Salvia miltiorrhiza* (Cui et al., 2012), the applicable importance of the bZIP gene family in enhancing phenolic acid content through ABA-mediated regulation remains largely unstudied (Bhatnagar et al., 2017).

17.3.1.2 Multiplex Gene Editing with CRISPR/Cas9 in Medicinal Plants

The use of CRISPR/Cas9 multiplex editing techniques improves the study of genes with substantial monotony. A more specific method for examining the active roles of genes with considerable isoforms or family members in reverse genetics concerns targeting multiple sgRNAs and Cas proteins within a unique expression vector method (Strasser et al., 2008). A recent study applied CRISPR/Cas9 for multiplex gene knockout to focus on *Salvia miltiorrhiza* using a 'dual-locus editing' approach. *Salvia miltiorrhiza*, a plant generally utilized in TCM, is renowned for its bioactive compounds, including tanshinones, phenolics like salvianic acid, salvianolic acid B, and rosmarinic acid. These compounds are effective in treating various cardiovascular diseases (Zhou et al., 2021).

The *Salvia miltiorrhiza* database has identified approximately twenty-nine laccases, termed SmLACs. Specifically, SmLAC7 and SmLAC20 have been studied for their impact on salvianolic acid B production (Li et al., 2019). Due to the extensive monotony among laccase genes, determining the effect of a single SmLAC gene on plant traits is challenging. Further research into phenolic acid metabolism has shown that SmLACs significantly affect lignin formation, which is vital for root development and phenolic acid metabolism (Simões et al., 2020). A marked decrease in the expression of key genes involved in phenolic acid synthesis suggests that SmLAC genes may play a crucial role in this biosynthetic pathway (Zhou et al., 2021).

17.3.2 Disease and Pest Resistance

When the natural immune responses of our body are not able to fight off a disease, chemotherapeutic agents like antibiotics are most commonly used for treatment (Koh et al., 2011). While antibiotics have significantly lowered the mortality rate from transmittable diseases, their misuse has also led to the sudden development of drug resistance in important bacterial species (Patrick et al., 2004). Scientific investigations on TCM have predominantly concentrated on evaluating its efficacy in treating many diseases, like hypertension, diabetes, hepatitis, and pneumonia, and also exploring its antibacterial properties through biochemical, laboratory, and microbiological studies (Tang & Wong, 1998).

TCM has long depended on plant-based materials within its well-established medical frameworks, playing a significant role in treating diseases and enhancing health. Chinese herbal medicines are extensively documented in numerous texts, ranging from ancient classics like *Shen Nong's Materia Medica* and the *Compendium of Materia* Medica to contemporary sources, reflecting their widespread use both in China and among various Chinese communities (Zuo et al., 2012). The usage of pesticides may significantly decrease labor power and raise the output of crops (Xie et al., 2019). By 2015, the usage of fertilizer in China reached 506.12 kg per hectare, well above the global average for developed countries, and the usage of pesticides in China has been steadily increasing since 2000 (Cao et al., 2021). An investigation of 300 different Chinese herbal materials revealed that most of the samples had residues of organochlorine pesticides, which may cause hepatomegaly, liver cell degeneration, bone marrow issues, and CNS damage (Li et al., 2015).

The usage of synthetic fertilizers and pesticides poses significant risks to medication safety, soil health, and ecological systems. Studies have found residues of at least one of 38 types of pesticides, including organophosphates, organochlorines, pyrethroids, carbamates, triazoles, and amides, in every sample tested (Li et al., 2017). In China, six pesticides – carbofuran, phorate, methamidophos, fipronil, profenofos, and aldicarb – are banned. Additionally, excessive use of nitrogen and phosphorus fertilizers can cause soil acidification and decreased productivity (Butnariu, 2018). Only 10–30% of the pesticides applied are adequate for crops, with 50–60% staying in the soil, compromising soil reprocess and posing risks to life (Hua et al., 2000). Moreover, the overuse of chemicals is harmful to various forms of life on Earth (Hill et al., 2008).

17.4 OMICS TECHNOLOGIES

17.4.1 Genomics: Sequencing and Annotating TCM Plant Genomes

The TCM Genome Project strives to uncover the genetic and regulatory mechanisms underlying herbal treatments, and to understand their molecular mechanisms at a granular level for gene discovery, synthetic biology, drug development, and disease management (Mochida et al., 2017). Recent advances in drug discovery involve extracting active compounds from TCMs and employing omics technologies to analyze their mechanisms of action, providing new strategies for disease treatment (Harvey et al., 2015). The TCM Basic Local Alignment Search Tool offers a web-based platform for corresponding protein and DNA sequences from TCMs (Zhang et al., 2016), encompassing about 40 GB of genomic data from 36 TCMs of notable medicinal importance (Chen et al., 2021).

- Plant Material and Sequencing

In a specific study, plants were grown, and fresh leaves from one-year-old specimens were gathered for DNA extraction. A total of 5 µg of DNA was utilized to create sequencing libraries employing the Illumina TruSeq DNA Preparation Kit. Four libraries with insert sizes of 550 bp, 2 to 3 kb, 5 to 7 kb, and 10 to 15 kb were sequenced on the Illumina HiSeq platform, and 250-bp paired-end reads were also developed using the Illumina HiSeq 2500 platform (Zhang et al., 2019).

- Genome Assembly

Using Illumina PE150 sequencing data, genome assembly was performed with Jellyfish software using the guidelines provided on the GenomeScope website (Luo et al., 2012). The assembly process utilized two primary tools: SOAP*denovo* for managing short-read data and Platanus for genomes with high heterozygosity. Subsequently, GapFiller was employed to address gaps in the scaffolds (Nadalin et al., 2012). To minimize redundancy in the Platanus assembly, scaffold filtering was executed with Redundans, and BLAST analysis was conducted on scaffolds smaller than 1 kb (Pryszcz et al., 2016).

- Gene Prediction and Annotation

Gene prediction was achieved by utilizing three methods: de novo, homolog-based, and transcriptome-based. Augustus software was used for de novo prediction with a transposable element (TE)-masked genome (Stanke et al., 2006). The homolog-based strategy utilized GeMoMa for gene prediction (Keilwagen et al., 2016). For transcriptome-based prediction, HISAT (version 2.0.4) and StringTie were employed for assembling transcripts (Kim et al., 2015), and TransDecoder was employed for predicting gene models. The gene models from these methods were incorporated using EVM software (Haas et al., 2008).

- Gene Family Identification and Analysis

Gene families were identified through comparisons with other plants using OrthoMCL analysis, which involved BLASTp study (E-value < 1e-5) and subsequent grouping into gene families utilizing OrthoMCL (Li et al., 2003). To study gene family contraction and expansion and to infer lineage-specific gene gains and failures, the genomes of five species were analyzed with OrthoFinder and CAFE with default settings (Emms et al., 2015; De Bie et al., 2006). A dated species tree from the TimeTree database was employed as a guide, and the birth-death parameter (lambda) was assessed by utilizing orthogroups with no more than 100 genes per genome (Kumar et al., 2017).

17.4.2 Transcriptomics: Gene Expression Profiling in Response to Various Conditions

The transcriptome encompasses the whole set of RNAs transcribed by a particular cell or tissue in a specific functioning condition, like mRNA and noncoding RNA. Transcriptomics, the study of gene transcription and its regulation, explores the types, structures, functions, and regulatory mechanisms of these transcripts, as well as the functions of noncoding regions, the structure of transcripts, and the discovery of new transcriptional regions (Lian et al., 2015). This field enables quantitative analysis of plant gene expression changes at specific times and conditions, helps elucidate complex regulatory networks and genome-wide expression responses under stress, and aids in identifying new genes associated with plant resistance (Anderson & Mitchell-Olds, 2011).

Transcriptome sequencing has proven highly valuable for discovering key markers in human cancers and for identifying conceivable drug targets when combined with epigenetic modifications including methylation of DNA and modifications in histone (Pathania et al., 2015). For decades, gene expression studies have relied on the assumption that transcript levels reflect active protein quantities, despite the genome's relative stability in contrast to the rapidly changing transcriptome in reaction to environmental constraints and growth (Edgar et al., 2002). This dynamic nature of cells and tissues can be assessed by measuring the transcriptome, which signifies the expressed portion in the whole genome (Kumar et al., 2016).

17.4.3 Proteomics and Metabolomics: Identifying Proteins and Metabolites Involved in Medicinal Properties

Comparative proteomic and metabolomic approaches have been utilized to explore the differences in proteins and metabolites in various medicinal plants under stress conditions and among different cultivars (Kim et al., 2016). Metabolite formation is a multi-step process beginning with

DNA, involving several intermediaries (Verpoorte et al., 2005). Changes in the metabolome often reflect changes in the transcriptome, which in turn influence enzyme levels or activities, making metabolomics a valuable tool for studying gene functions (Sumner et al., 2003). Proteomics, which involves the comprehensive investigation of all proteins encoded by an organism, is crucial for understanding cellular processes at a molecular level and complements both genomics and metabolomics (Baginsky, 2009).

Recently, proteomics has emerged as a significant method for examining plant reactions to environmental strains and for elucidating the pharmacological mechanisms of herbal remedies (Hashimoto et al., 2004). Databases such as the Human Metabolome Database, METLIN, and MetaboLights collect data on metabolites identified through techniques like chromatography, NMR, and MS, along with relevant metadata (Wishart et al., 2013). These databases enable the development of precise diagnostics and the exploration of molecular pathways associated with specific conditions or traits by tracking changes in metabolite levels (Alonso et al., 2015). Among various proteomic techniques, 2D gel electrophoresis (2DGE) with high resolution is extensively used for separating complex mixtures such as blood plasma and tissue samples before protein identification via mass spectrometry (Ferber, 2002). 2DGE can separate thousands of proteins with high resolution, and mass spectrometry methods are widely applied for the accurate identification of these proteins (Riaz, 2015). Protein and metabolite identification is depicted in **Figure 17.1**.

17.4.4 Integrative Omics: Combining Data from Transcriptomics, Proteomics, and Metabolomics to Understand Complex Pathways

Combining experimental results from various omics platforms is an emerging strategy designed to uncover hidden

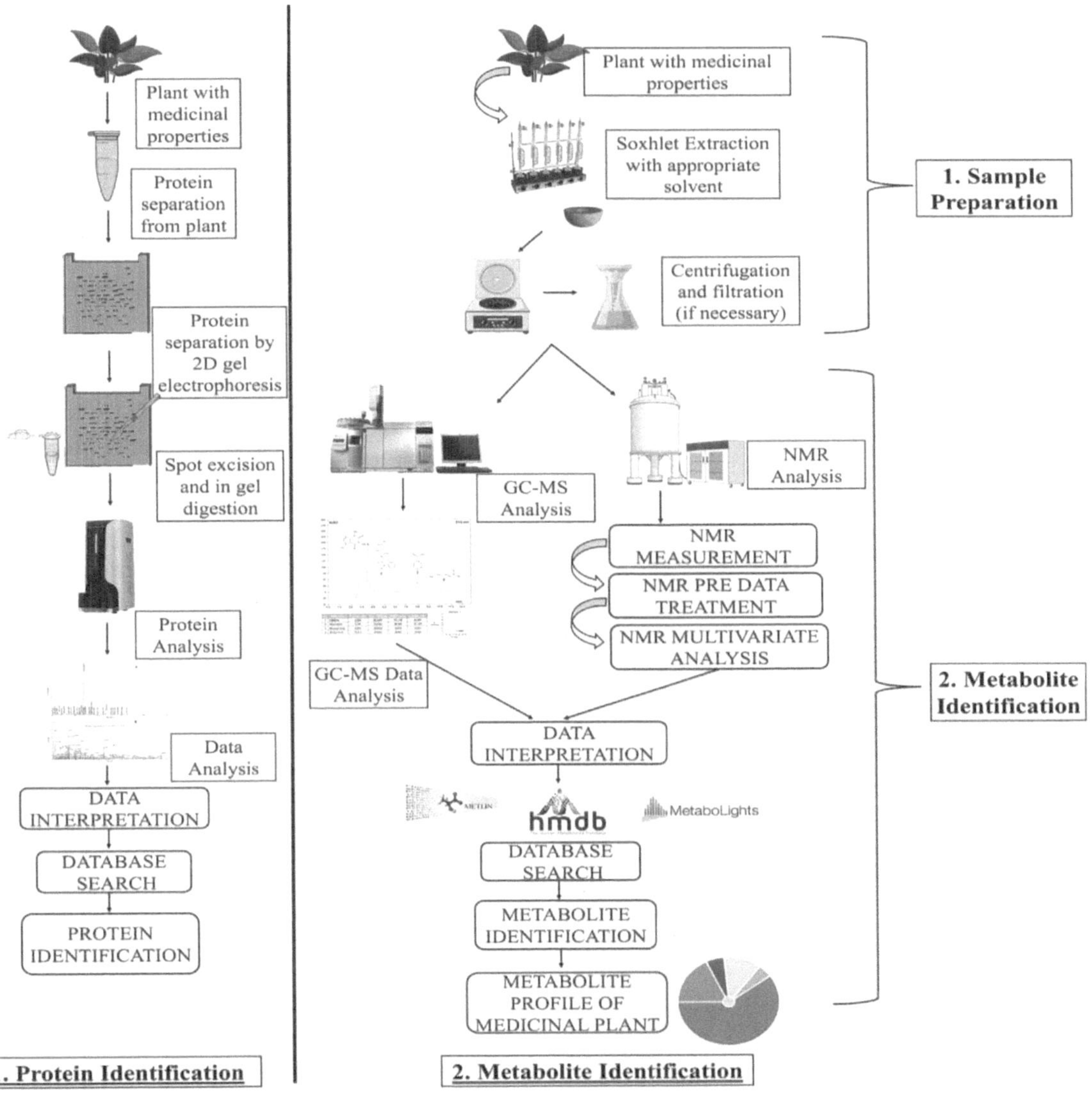

FIGURE 17.1 Protein and metabolite identification.

biological relationships that might only become apparent through comprehensive analyses that integrate data across various biochemical domains (Wanichthanarak et al., 2015). Integration involves merging different categories of omics data such as genomic mutations, mRNA levels from transcriptomics, protein quantity and types from proteomics, and metabolite statuses from metabolomics into a combined framework. This approach provides a more informative and holistic view compared to isolated omics analyses (Kristensen et al., 2014).

17.5 METABOLIC ENGINEERING TO INFLUENCE THE MEDICINAL PLANT PRODUCTION

Secondary metabolites, unlike primary metabolites crucial for plant growth, are chemical compounds that, while not essential for plant growth, play a significant part in defence against various abiotic and biotic strains (Fu et al., 2018). Historically, humans have utilized plant extracts rich in these specialized metabolites for medicinal purposes. Ancient records from around 2600 BC, such as those from Mesopotamia, document the use of around 1000 herbs and their applications (Cragg & Newman, 2013).

Traditional Chinese medicine, as detailed in historical texts like the *Compendium of Materia Medica*, the *Shennong Herbal*, and the *Tang Herbal*, has used numerous plants to treat various ailments (Tu, 2016). In recent times, advancements in chemical technologies have facilitated the extraction and identification of specialized metabolites from herbal plants (Corsello & Garg, 2015). According to Stephanopoulos (1998), metabolic engineering involves 'the direct improvement of product formation through the modification of specific biochemical reactions or the introduction of new ones'.

17.5.1 Cordycepin Biosynthesis Pathway Strategies

Cordycepin is a highly valued traditional Chinese medicine known for its diverse bioactivities, including hypoglycaemic, anti-inflammatory, anticancer, antihyperlipidemic, anti-aging, antiplatelet aggregation, antimicrobial, analgesic, and immunomodulatory effects (Sharma et al., 2022). The fungus *Cordyceps militaris* is a member of the Ascomycota phylum, Cordyceps class, and Clavicipitaceae family, and has been utilized medicinally in China for a long period. Additionally, it shows promise in treating COVID-19 due to its inhibitory effect on the COVID-19 polymerases (Verma, 2022).

Cordycepin was obtained from the species *Cordyceps militaris,* which has a concentration of 2–3 mg/kg. To satisfy the market demand for cordycepin, a mutation breeding method was introduced to induce the genetic mutation to create a new trait of *Cordyceps militaris* for increasing cordycepin production (Yang et al., 2020). Adenosine is a precursor for the biosynthesis of cordycepin in *Cordyceps militaris* was first proposed by Kredich and Guarino in 1961 (Kredich & Guarino, 1961). According to the hypothesis by Lennon and Suhadolnik in 1976, by transforming the adenosine to 3'-deoxyadenosine (dA) without hydrolyzing the N-riboside bond cordycepin can be produced which is identical to 2'dA synthesis (Lin et al., 2016; Peng et al., 2024). Advances in genetic engineering offer significant potential for enhancing cordycepin production. Techniques such as the split-marker approach, homologous recombination, protoplast-polyethylene glycol-mediated transformation, *Agrobacterium tumefaciens*–mediated transformation, and CRISPR (Chen et al., 2018; Liu et al., 2023) are being investigated to optimize this process.

17.6 MICROBES ROLE IN PRODUCTION OF MEDICINAL PLANTS

The rhizosphere environment of therapeutic plants is inhabited by approximately 10^{11} microbial cells per gram of root, including over 30,000 prokaryotes (Yu et al., 2023). Before medicinal plants reach herbal markets, they must undergo a specialized processing system to ensure long-term production. Recent research highlights that bacteria and fungi play an important part in determining the quality of the plants (Liu et al., 2017). The main aim is to minimize toxicity or enhance clinical efficacy. Xu et al. (2013) examined the microbial communities during the fermentation process of Massa Medicata Fermentata, noting distinct phases in the process. Their study identified that pre-fermentation is predominantly influenced by bacterial communities (such as *Enterobacter* sp., *Pediococcus* sp., and *Pseudomonas* sp.), while post-fermentation is dominated by fungal species (such as *Saccharomyces* sp. and *Mucor* sp.) (Xu et al., 2013; Guo et al., 2016). Additionally, differences in digestive enzyme activities during fermentation align with these microbial shifts, occurring in two separate stages (Zhang et al., 2021).

The cultivation of therapeutic plants has increasingly focused on developing beneficial microorganisms that are environmentally friendly, non-toxic to plants, and cost-effective (Mastan et al., 2019). These microorganisms not only promote plant development but also serve as alternatives to conventional fertilizers and pesticides. Advances in culturomics and metagenomics have improved the isolation, identification, and cultivation of functional strains (Hassan, 2017). Using a metagenomics-culturomics approach, synthetic microbial communities composed of functional strains have been demonstrated to enhance plant development, improve resistance to biotic stress, and improve the formation of natural products (Khalil et al., 2021). Specific medicinal plants like *Astragalus mongholicus*, *Glycyrrhiza uralensis*, and *Schisandra sphenanthera* illustrate the benefits of these synthetic microbial communities (Hassan, 2017).

17.6.1 Rhizosphere Engineering: Manipulating Soil Microbiomes to Support Traditional Chinese Medicine Plants

A certain number of bacteria contain a phyto-therapeutic compound that is related to the rhizosphere of plants which accelerates the medicinal plant's growth. The information about medicinal plants-rhizosphere associated is applicable in developing a biofertilizer for commercial purpose (Solaiman & Anawar, 2015). *Angelica sinensis* (medicinal plant) isolated from rhizosphere soil contains a strain DRP-35(T) and a DRP-9(T) are non-motile, gram-negative, short oxidase-negative rods, catalase-positive, and exopolysaccharide synthesizing bacterium (Whang et al., 2014). Phylogenetic tree analyses of the sequences of gene 16S rRNA signified that the DRP-35(T) strain comes under the genus *Terriglobus* in the phylum Acidobacteria was a resemblance to *Terriglobus roseus* KBS63(T) and *Terriglobus saanensis* SP1PR4(T). Whereas the DR-9(T) strain accord to the genus *Mucilaginibacter* was the similarity to *Mucilaginibacter myungsuensis* HMD1056(T), *Mucilaginibacter boryungensis* BDR-9(T), and *Mucilaginibacter ximonensis* XM-003(T) (Lee et al., 2013).

Hyoscyamus muticus L., an Egyptian medicinal plant belonging to the family Solanaceae mainly present in desert regions produces important secondary metabolites, also colonized by endophytic fungi. Medicinal plants such as *Centella asiatica* and *Ocimum sanctum* in rhizosphere soil contain 16–17 fungal species (El-Zayat et al., 2008). *Arbuscular mycorrhizal* (AM) can change the volatile organic carbon released from sorghum plants of morphological roots for adapting to the soil environment. Rhizospheric soil contains 3 genera of a total of 31 AM fungal species as *Glomus* genus contains 21 fungal species, 7 of *Acaulospora,* and the remaining 3 of *Scutellospora* (Shi et al., 2013).

17.7 EPIGENETIC REGULATION FOR IMPROVING CHINESE TRADITIONAL MEDICINE

In 1942, Conrad H. Waddington, a developmental biologist, introduced the term 'epigenetics', which merges the concepts of 'epigenesist' and 'genetics'. Epigenetics refers to the study of how gene expression and function can be altered without changing the underlying DNA sequence. This involves reversible modifications that affect phenotypes, such as DNA methylation, histone modifications, and miRNA interference (Hu & Su, 2017). Various proteins play a role in regulating these epigenetic changes, including histone methyltransferases (HMT), polycomb group proteins, histone acetyltransferases, histone deacetylases, DNA methyltransferases (DNMT), and methyl CpG-binding proteins (Ballestar & Wolffe, 2001). Key mechanisms involve altering chromatin structure to either suppress or activate gene expression. For instance, hypermethylation of gene promoters or hypoacetylation of histone tails can silence transcription, while demethylation or acetylation can enhance it (Ho et al., 2011; Kabesch, 2014).

Chinese medicine (CM) shares several principles with epigenetics, including holism, reversibility, and balance, as well as the dynamic nature of yin and yang and the importance of visceral health. The concept of holism in CM, rooted in ancient Chinese philosophy, emphasizes the interconnectedness of nature, the environment, and the human body (Hu & Su, 2017). According to CM, the physical and mental states of the body are interconnected, and external factors can influence physical health. Imbalances in histone acetylation and deacetylation, DNA methylation and demethylation, or histone modifications can lead to diseases (Calvanese et al., 2009; Chase et al., 2013). Research indicates that during development, DNA methylation levels increase, and the removal of cytosine methyltransferases, which generate DNMTs through cell divisions, can result in embryonic and neonatal death in mice (Li et al., 1992).

17.7.1 Epigenetic Regulation and Plant Development

17.7.1.1 Mechanism of DNA Methylation

Methylation of DNA is mediated by the enzyme DNA methyltransferase (DNMT), which specifically targets the 5' cytosine residue within CpG dinucleotides. Key DNMT enzymes include DNMT1, DNMT3a, DNMT3b, and DNMT3L (Subramaniam et al., 2014). During embryogenesis, DNMT3a and DNMT3b are the reason for de novo methylation, while DNMT1 maintains methylation patterns during somatic cell division (Zhou et al., 2017). Although DNMT3L does not possess catalytic methyltransferase action, it is essential for de novo methylation in germ cells through its interaction with unmethylated lysine4 of histone H3 (Groth et al., 2007). The ten-eleven translocation (Tet) enzymes Tet1, Tet2, and Tet3 are central to the activation of DNA demethylation processes (Chase et al., 2013).

DNA methylation encompasses both hypermethylation and hypomethylation, each affecting gene expression differently. Hypermethylation of CpG sites often leads to transcriptional silencing of tumor suppressor genes (Cao et al., 2017). In contrast, genome-wide hypomethylation can activate oncogenes and transposable elements, contributing to genetic instability. Therefore, DNA methylation is crucial in the regulation of genes with DNMT and Tet enzymes playing pivotal roles in anticancer strategies (Shen & Laird, 2013).

17.7.1.2 Mechanism of Histone Modification

Histone protein modifications are crucial for regulating gene expression through epigenetic mechanisms. In eukaryotes, nucleosomes are structures consisting of DNA bound around a histone octamer, which includes two copies of each of the H2A, H2B, H3, and H4 histones. This arrangement forms chromatin fibers, about 147 base pairs of DNA bound around the histone core (Strahl & Allis, 2000). Modifications to the histone tails, particularly at their N-terminal regions, are involved in various chromatin-related processes like DNA repair, transcription, and the replication of DNA (Peterson & Laniel, 2004).

Posttranslational histone modification integrates methylation, phosphorylation, acetylation, ubiquitination, proline isomerization, deamination, ADP ribosylation, and sumoylation, these modifications make different activities based on the residues that are modified (Kouzarides, 2007). Based on the site methylated lysine has a distinctive activity on lysine residues in histone methylation. For instance, gene expression activations usually on H3K3, H3K36, and H3K79 methylation. Enzymes such as HMTs on lysine and histone demethylases have been used to catalyze the methylation of lysine residue (Black et al., 2012; Zhang & Reinberg, 2001).

17.7.1.3 Mechanism of miRNA

MiRNA has a role in controlling the stability and translation of RNA and has 21–25 nucleotides of non-coding RNAs. When miRNA binds to the targeted RNA of 3' untranslated regions (3'-UTR), then it initiates degradation of mRNA and translational repression (Jansson & Lund, 2012). miRNAs are connected to pathological and biological functions like metabolism, apoptosis, and proliferation, and also play a necessary part in imprinted loci of de novo methylation (Lan et al., 2011).

Even more, epigenetic mechanism regulates the miRNA expressions including histone modification and DNA methylation. miRNA shows explicit evidence that is linked to cancer, which acts tumor suppressors or potential oncogenes and biomarkers in carcinogenesis (Bader, 2012). For instance, overexpressed miR-155 has been seen in different cancers such as colon, breast, hematopoietic, and lung cancer. Analyzing the role of miRNA helped in cancer diagnosis, treatment, and prevention (Bouchie, 2013).

17.7.2 Role of Epigenetic Mechanisms in Response to Plant Stress

Methylation of DNA involves adding a methyl group to cytosine to form 5-methylcytosine, which influences several biological processes, including transcriptional suppression, developmental stress responses, genome stability, and reactions to environmental stress (Abdulraheem et al., 2024). This mechanism can impact crop yield by either enhancing it or causing environmental stress, and it also plays a role in disease resistance. Research has shown that in salt-sensitive cultivars, there is an upregulation of the DNA demethylase DRM2, which is not observed in salt-tolerant cultivars. This upregulation in salt-sensitive varieties is linked to a reduction in DNA methylation levels (Ferreira et al., 2015; Wang et al., 2015b). According to Pathak et al., salt stress induces both hypomethylation and hypermethylation at specific genomic regions in rice (Pathak et al., 2021).

To mitigate agricultural failures, it is important to develop stress-tolerant crops capable of thriving in challenging conditions, and understanding how histone modifications influence gene expression is essential (Zhang et al., 2018). Post-translational modifications (PTMs) play an important role in processes like flowering, stress response, and seed formation (Abdulraheem et al., 2024). Recent studies have shown that decreasing H3K27me3 deposition within the gene body of drought stress-responsive TFs improves drought toleration in *Arabidopsis*. Further, inoculation with *Botrytis cinerea* has been found to considerably upregulate the expression of genes *DOX1*, *DES*, and *LoxD*, which are vital for the oxylipin pathway, and WRKY75, a stress-responsive TFs in tomato (*Lycopersicon esculentum*) (Crespo-Salvador et al., 2020; Zhang et al., 2020).

Small, non-coding RNAs play a necessary part in either the degradation of mRNA or the inhibition of translational expression. Plants can adjust to abiotic stress including heat and drought due to the presence of miRNA (Ferdous et al., 2015). For example, *Arabidopsis thaliana* has a miR398 that will activate in the presence of oxidative stress by targeting the copper superoxide dismutase (CSD) transcription. Small RNA molecules control both stages of the post-transcriptional and transcriptional processes (Begum, 2022). These small RNA molecules also influence the gene expression and histone modifiers by the interaction of chromatin remodelling complexes. Small RNAs can guide the histone modifiers at specific genomic regions, which leads to histone modifications and varying the gene expression. Studies state that miR156 targets squamosa promoter-binding protein-like TFs to regulate flowering time in *Arabidopsis* (Hu et al., 2015).

17.8 BIOTECHNOLOGY TOOLS FOR CULTIVATION IMPROVEMENT

17.8.1 Tissue Culture and *In Vitro* Propagation

In China, therapeutic plants play a crucial position in the country's extensive healthcare scheme. However, due to excessive harvesting of periodic and endangered plants, these natural resources are rapidly depleting. To protect these valuable medicinal plants, the Chinese government has introduced good agricultural practice programs aimed at cultivating key medicinal plant species within the country (Gao et al., 2005). Plant tissue culture strategies have been utilized in plant applications, micropropagation, conservation, and the formation of specific pharmaceutical compounds from therapeutic plants. Additionally, developed biotechnological methods, like precise genome editing, offer the potential to develop therapeutic plants with tailored secondary metabolite profiles (Niazian, 2019).

Medicinal plants produce numerous expensive drugs known as secondary metabolites or organic substances. While a few therapeutic plants are cultivated, the majority are still harvested from the wild (Ramawat & Arora, 2021). Tissue culture methods are currently utilized for the formation of secondary metabolites *in vitro*. A notable example of tissue culture technology applied in commercial contexts is micropropagation illustrated in **Figure 17.2** (Marthe, 2018). *In vitro* culture proposes several benefits, including the establishment of germplasm banks with living, actively growing therapeutic plant tissues, as well as mass propagation and long-term depository under consistent physiological conditions (Murch et al., 2004). Micropropagation is a technique used to rapidly multiply high-quality plants through advanced tissue culture techniques. It is widely recognized for its use in agriculture, horticulture, and forestry (Moraes et al., 2021). Plants produced via micropropagation often show improved branching, flowering, and overall vigor, as well as increased yields, mainly due to the potential removal of diseases (Lone et al., 2020).

17.8.2 Bioreactors: Scale-Up Production of Medicinal Compounds in a Controlled Environment

The increasing demand for plant secondary metabolites in many industries, including pharmaceuticals, food, flavorings, beverages, and cosmetics, has driven the need for efficient and large-scale production methods. Recent advancements in plant biotechnology have highlighted bioreactor cultivation of adventitious roots as a viable alternative to traditional whole plant, or cell cultures for generating bioactive compounds and biomass (Rohini & Rajasekharan, 2022; Baque et al., 2012). However, scaling up bioreactors for the production of tissue-active molecules presents challenges due to the variability in production kinetics. This variability can be influenced by factors such as the gaseous environment, oxygen levels, CO2 exchange, pH, mineral content, carbohydrate concentrations, liquid medium properties, and cell density, all of which affect growth and

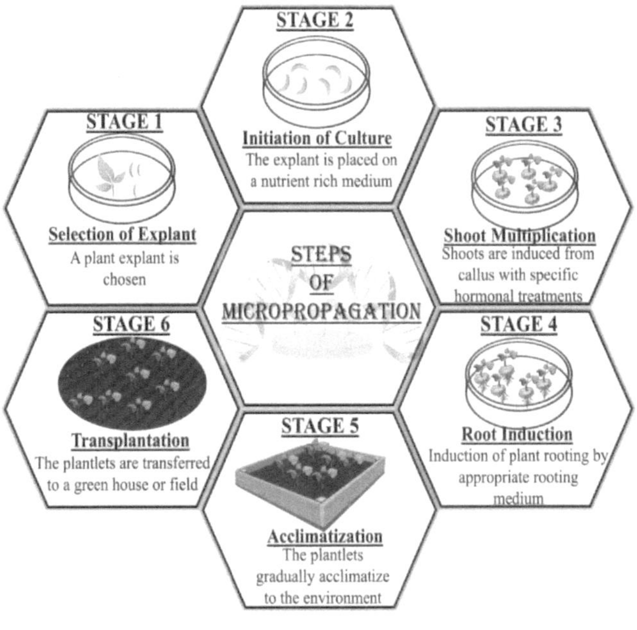

FIGURE 17.2 Steps of micropropagation.

metabolite production in bioreactors (Verma et al., 2021). Currently, bioreactor cultures, which integrate process control, scalability, automation, and cost reduction, are the primary approaches to addressing these challenges (Abraham et al., 2017). Recent efforts have focused on improving the efficiency and cost-effectiveness of bioreactor systems for biomass and secondary metabolite formation in therapeutic plants. Researchers are continuously working to enhance these systems by exploring new bioreactor designs, optimizing cultivation conditions, and developing innovative methods to boost the formation of beneficial compounds derived from medicinal plants (Park & Paek, 2014; Xu & Xu, 2024).

17.9 PHYTOCHEMICAL ANALYSIS AND QUALITY CONTROL

17.9.1 Advanced Analytical Techniques

Herbal medicines have long been employed to treat many ailments, and despite their historical effectiveness, considerations about their safeness and effectiveness persist. Ensuring the grade of botanical products, like plant parts, extracts, and herbal therapies, remains a significant challenge (Klein-Junior et al., 2021). Phytochemicals in herbal medicines show potential for addressing conditions like cancer, cardiovascular diseases, diabetes, and hypertension, offering considerable benefits for disease prevention and personalized therapy (Cao et al., 2014). Various analytical techniques, like UV and IR spectroscopy, are employed for qualitative investigation, while techniques like high-performance thin-layer chromatography (HP-TLC), high-performance layer chromatography (HP-LC), supercritical fluid chromatography (SFC), thermal analysis, inductively coupled plasma–mass spectrometry (ICP-MS), liquid chromatography coupled with mass spectrometry (LC-MS), and gas chromatography–mass spectrometry (GC-MS) are employed for quantitative evaluation to provide the quality of herbal by-products (Balekundri & Mannur, 2020).

HP-TLC enhances the capabilities of traditional thin-layer chromatography (TLC) (Srivastava, 2010). HP-TLC is known for being cost-effective, rapid, and reliable, making it a popular choice for quality control (Kumar & Sharma, 2018). It is extensively used in the pharmaceutical industry for process development, contamination detection in herbal products, and identifying pesticide residues and mycotoxins, thereby ensuring the quality assurance of herbs and healthy foods (Beressa et al., 2021). TLC, and more recently HP-TLC, has traditionally been utilized to substantiate the individuality of herbal medications and detect potential adulterations or falsifications, which is crucial for maintaining purity (Cañigueral et al., 2018). HP-LC is widely utilized for analyzing a broad range of compounds in herbal medicines due to its comprehensive analytical capabilities over recent decades (Farooqui et al., 2014). HP-LC is favored for analyzing herbal medicines due to its flexibility and ease of use, as it can assess nearly all compounds present regardless of their volatility or stability (Liang et al., 2004). In HP-LC, the sample solution interacts with a stationary phase, leading to the separation of various components established on their interactions with the stationary phase (Kupiec, 2004). Optimal separation in HP-LC depends on several factors, including pH adjustments, mobile phase composition, and pump pressures (Shulammithi et al., 2016). HP-LC is recognized as an effective method for both quantitative and qualitative assessment of active chemical constituents and biological components in herbal preparations (Hossain et al., 2022).

Recently, SFC has gained prominence as an alternative to traditional methods. With the development of next-generation SFC instruments and diverse column chemistries (He et al., 2019), SFC offers distinct advantages over HPLC due to the benefit of supercritical fluids like carbon dioxide. SFC is capable of analyzing lipids, flavonoids, phenolics, alkaloids, saponins, carbohydrates, and other analytes, and is involved in a broad range of materials like natural products, medicines, and food (Choudhary & Sekhon, 2011; Tsao & Deng, 2004). ICP-MS is a widely used technique for the quantitative analysis of trace elements, especially toxic ones (Rawat et al., 2024). It is valuable for detecting trace elements in biological fluids and ensuring the safety of herbal products by testing for heavy metals. ICP-MS offers advancements like high elemental specificity, multi-isotope detection, and the ability to detect trace and ultra-trace elements at very low concentrations (Balekundri & Mannur, 2020).

LC-MS has seen increased use in natural product investigation due to its superior separation and detection capabilities. LC-MS, especially with advancements in ionization techniques such as electrospray, thermospray, and ionspray, provides high sensitivity and specificity and is preferred in various stages of drug development (Patil & Shettigar, 2010; Wu et al., 2013). GC-MS is another widely employed technique for both quantitative and qualitative analysis of phytochemicals. This method involves converting samples into a gaseous state for analysis established on the mass-to-charge ratio (Beressa et al., 2021).

17.9.2 Standardization Protocols Ensuring Consistent Quality and Efficacy of Traditional Chinese Medicine Products

Standardizing herbal medicines involves setting specific standards and parameters that ensure consistency in quality, efficacy, safety, and reproducibility. This process is vital for quality control (Kumari & Kotecha, 2016). TCM, widely practiced across Asia and gaining global recognition, uses Chinese herbal medicine (CHM) as a core component, which relies on herbs with complex chemical profiles (Zhou et al., 2019). The quality of TCM is increasingly scrutinized due to its complex, multi-component nature, necessitating a unique evaluation system distinct from Western medicine. TCM is typically examined as a natural product or herbal medicine (Yang et al., 2017). Phytotherapeutic agents or phytomedicines, which are standardized herbal mixtures from various plants, are commonly used worldwide for treating a range of diseases. The World Health Organization defines herbal drugs as containing active ingredients from plant materials or parts, whether raw or processed, along with specific excipients (Calixto, 2000).

Effective quality control involves standardized procedures for processing, sourcing, and preparing herbal ingredients. This ensures that herbal remedies are effective, safe, and compliant with regulations, supporting consumer well-being and the advancement of the herbal medicine field (Wang, 2023). The concept of quality markers (Q-markers) is based on the 'property-effect-component' theory. Q-markers are active components derived from primary or secondary metabolites in plants, serving as indicators of TCM quality, safety, and efficacy. Their development is crucial for assessing and controlling TCM product quality. Q-markers have both qualitative and quantitative properties (Chen et al., 2017). Despite growing interest in TCM, inconsistent clinical trial quality remains a barrier to its wider acceptance (Dai et al., 2019).

17.10 CLINICAL IMPLICATIONS AND PHARMACOLOGICAL STUDIES

Translational medicine has emerged as a key facilitator for moving discoveries from basic science into clinical practice, focusing primarily on advancing drugs, medical devices, and diagnostics from preclinical research to human applications (Wehling, 2006). Biomarkers play a crucial role in defining translational medicine and act as the universal language of translational sciences (Wagner & Kroetz, 2016). Historically, pharmacology has been perceived as a cohesive, hierarchically structured field that remains neutral toward technology. In addition to its pivotal role in developing new medications, pharmacological research has enhanced our comprehension of disease pathogenesis and progression (Enna & Williams, 2009). Translational research effectiveness often hinges on the potential for exploratory or developmental studies to produce viable therapies. Pharmaceutical manufacturing has transformed into a technical sector deeply involved in the translational investigation, mirroring similar engagement within the academic medical research community, largely driven by the NIH's Roadmap initiative (Barrett, 2008). Translational medicine, introduced in recent decades, aims at bridging basic medical research with clinical application (Wang et al., 2015c).

17.10.1 Personalized Medicine Approaches Based on Genetic Information

Personalized medicine aspires to provide the right medicine to the right individual in the proper quantity. Pharmacogenomics seeks to identify genetic variants that can affect medicine's usefulness and toxicity (Mirsadeghi &

Larijani, 2017). Pharmacogenetics and pharmacogenomics focus on the heritable factors that influence how individual patients respond to drugs. The emergence of new technologies has resulted in the development of pharmacogenomics, which aims to pinpoint genetic variations affecting drug responses in individuals (Mancinelli et al., 2000).

Pharmacogenomics is a crucial element of personalized medicine, focusing on the patient and their genome to determine the most suitable medication and dosage (Whirl-Carrillo et al., 2021). So far, pharmacogenomic diversity has been recognized for its impact on drug effectiveness and safety, prompting international scientific consortia to develop treatment guidelines that guide the clinical use of pharmacogenomics (Cecchin & Stocco, 2020). Pharmacogenomics seeks to explore the genetic factors behind individual variations in drug responses, including efficacy, dosage needs, and adverse events (Wei et al., 2012).

Recent developments in pharmacogenomics have enabled the understanding of the reasons behind different drug responses. Identifying genetic variants and their association with varying drug responses forms the foundation for recommending specific drugs and dosages for individual patients (Singh, 2020). Pharmacogenomics can improve patient care through the optimization of medication selection and dosage, minimizing the likelihood of adverse events and enhancing satisfaction among both patients and providers through the implementation of personalized medicine (Moc, 2020).

17.11 CONCLUSION

The chapter explores integrated molecular strategies for enhancing TCM plants. The integration of genomics and evolution into medicinal plant research is becoming increasingly evident. Hence, it is essential to review the current advancements in these fields to fully leverage evolutionary biology and genomics, thereby transforming the approach to medicinal plant studies. Integrated molecular strategies offer a promising avenue for improving medicinal plants, ensuring an endurable supply of high-quality raw materials for the pharmaceutical industry. By leveraging the ability of biotechnology, genomics, and molecular biology, investigators can enhance the production, consistency, and efficacy of medicinal compounds, ultimately benefiting global healthcare and the economy.

TCM is widely used and plays a crucial role in treating numerous diseases, but its production and use often generate a substantial number of herbal residues. The prospects of emerging technologies in the improvement of TCM plants are vast and promising. By leveraging advancements in genomic editing, omics technologies, metabolic engineering, synthetic biology, and precision agriculture, researchers and practitioners can enhance the quality, consistency, and sustainability of medicinal plants used in TCM. These improvements will not only benefit traditional practices but also support the integration of TCM into modern healthcare systems, providing reliable and effective treatment options for a wide range of diseases.

ACKNOWLEDGMENT

All the authors are thankful to their respective universities and institutions for their valuable support.

DATA AVAILABILITY

The authors confirm that the data supporting the findings of this study are available within the book/chapter.

REFERENCES

Abdulraheem, M. I., Xiong, Y., Moshood, A. Y., Cadenas-Pliego, G., Zhang, H., & Hu, J. (2024). Mechanisms of plant epigenetic regulation in response to plant stress: Recent discoveries and implications. *Plants*, *13*(2), 163.

Abraham, E., Gupta, S., Jung, S., & McAfee, E. (2017). Bioreactor for scale-up: Process control. In *Mesenchymal stromal cells* (pp. 139–178). Academic Press.

Akilandeswari, G., Anand, A. V., Sampathkumar, P., Moorthi, P. V., & Preethi, B. (2019). A prospective review on phytopharmacological aspects of *Andrographis paniculata*. *Systematic Reviews in Pharmacy*, *10*(1). 15–19.

Alonso, A., Marsal, S., & Julià, A. (2015). Analytical methods in untargeted metabolomics: State of the art in 2015. *Frontiers in Bioengineering and Biotechnology*, *3*, 23.

Anand, A. V., Velayuthaprabhu, S., Rengarajan, R. L., Sampathkumar, P., & Radhakrishnan, R. (2020). Bioactive compounds of guava (*Psidium guajava* L.). In *Part VIII, Underutilized fruits and nuts rich in volatile compounds, Ch. 30. Bioactive compounds in underutilized fruits and nuts*. Reference series in phytochemistry (pp. 503–527). Springer.

Anderson, J. T., & Mitchell-Olds, T. (2011). Ecological genetics and genomics of plant defences: Evidence and approaches. *Functional Ecology*, *25*(2), 312–324.

Bader, A. G. (2012). miR-34–a microRNA replacement therapy is headed to the clinic. *Frontiers in Genetics*, *3*, 120.

Baginsky, S. (2009). Plant proteomics: Concepts, applications, and novel strategies for data interpretation. *Mass Spectrometry Reviews*, *28*(1), 93–120.

Balekundri, A., & Mannur, V. (2020). Quality control of the traditional herbs and herbal products: A review. *Future Journal of Pharmaceutical Sciences*, *6*, 1–9.

Ballestar, E., & Wolffe, A. P. (2001). Methyl-CpG-binding proteins: Targeting specific gene repression. *European Journal of Biochemistry*, *268*(1), 1–6.

Baloch, F. S., Alsaleh, A., Shahid, M. Q., Çiftçi, V., E. Sáenz de Miera, L., Aasim, M., & Hatipoğlu, R. (2017). A whole genome DArTseq and SNP analysis for genetic diversity assessment in durum wheat from central Fertile Crescent. *Plos One*, *12*(1), e0167821.

Baque, M. A., Moh, S. H., Lee, E. J., Zhong, J. J., & Paek, K. Y. (2012). Production of biomass and useful compounds from adventitious roots of high-value added medicinal plants using bioreactor. *Biotechnology Advances*, *30*(6), 1255–1267.

Barrett, J. S. (2008). The role of quantitative pharmacology in an academic translational research environment. *The AAPS Journal*, *10*, 9–14.

Begum, T. N., Ilyas, M. M., Kalavathy, S., Anadh, A. V., Kumar, S., & Senthil, R. (2009). Effects of ethanolic leaf extract of *Azim tetracantha* Lam. on *Ehrlich ascites* carcinoma tumour bearing mice. *Research Journal of Medicine and Medical Sciences*, *4*(2), 351–354.

Begum, Y. (2022). Regulatory role of microRNAs (miRNAs) in the recent development of abiotic stress tolerance of plants. *Gene*, *821*, 146283.

Beressa, A., Wariyo, A., Chala, G., & Tefera, Y. (2021). Analytical method development for quality control and standardization of medicinal plants: A critical review. *Journal of Herbal Science*, *10*(1), 9–20.

Bharathi, V., Rengarajan, R. L., Radhakrishnan, R., Hashem, A., Abd-Allah, E. F., Alqarawi, A. A., & Anand, A. V. (2018). Effects of a medicinal plant *Macrotyloma uniflorum* (Lam.) Verdc. formulation (MUF) on obesity-associated oxidative stress-induced liver injury. *Saudi Journal of Biological Sciences*, *25*(6), 1115–1121.

Bhatnagar, N., Min, M. K., Choi, E. H., Kim, N., Moon, S. J., Yoon, I., & Kim, B. G. (2017). The protein phosphatase 2C clade A protein OsPP2C51 positively regulates seed germination by directly inactivating OsbZIP10. *Plant Molecular Biology*, *93*, 389–401.

Black, J. C., Van Rechem, C., & Whetstine, J. R. (2012). Histone lysine methylation dynamics: Establishment, regulation, and biological impact. *Molecular Cell*, *48*(4), 491–507.

Bouchie, A. (2013). First microRNA mimic enters clinic. *Nature Biotechnology*, *31*(7), 577.

Buriani, A., Garcia-Bermejo, M. L., Bosisio, E., Xu, Q., Li, H., Dong, X., & Hylands, P. J. (2012). Omic techniques in systems biology approaches to traditional Chinese medicine research: Present and future. *Journal of Ethnopharmacology*, *140*(3), 535–544.

Butnariu, M. (2018). The nitrogen cycle in the soil, impact on the environment. *Crimson Publishers*, *4*(4), 1–3.

Calixto, J. B. (2000). Efficacy, safety, quality control, marketing and regulatory guidelines for herbal medicines (phytotherapeutic agents). *Brazilian Journal of Medical and Biological Research*, *33*, 179–189.

Calvanese, V., Lara, E., Kahn, A., & Fraga, M. F. (2009). The role of epigenetics in aging and age-related diseases. *Ageing Research Reviews*, *8*(4), 268–276.

Cañigueral i Folcarà, S., Frommenwiler, D., Reich, E., & Vila Casanovas, R. (2018). High performance thin-layer chromatography (HPTLC) in the quality control of herbal products. In D. Muñoz-Torrero, Y. Cajal, & J. M. Llobet (Eds.), *Recent advances in pharmaceutical sciences VIII* (Chap. 7, pp. 119–136). Research Signpost. ISBN: 978-81-308-0579-5.

Cao, P., Wang, G., Wei, X. M., Chen, S. L., & Han, J. P. (2021). How to improve CHMs quality: Enlighten from CHMs ecological cultivation. *Chinese Herbal Medicines*, *13*(3), 301–312.

Cao, W., Yi, L., Cao, J., Hu, S. S., & Li, P. (2014). Recent advances in the methodology and application for the metabolism of phytochemical compounds-an update covering the period of 2009–2014. *Current Drug Metabolism*, *15*(10), 966–987.

Cao, Y., Chen, Y., Huang, Y., Liu, Z., & Li, G. (2017). In vitro study of human mutL homolog 1 hypermethylation in inducing drug resistance of esophageal carcinoma. *Irish Journal of Medical Science*, *186*, 257–263.

Cecchin, E., & Stocco, G. (2020). Pharmacogenomics and personalized medicine. *Genes*, *11*(6), 679.

Chan, K. (2003). Some aspects of toxic contaminants in herbal medicines. *Chemosphere*, *52*(9), 1361–1371.

Chase, K. A., Gavin, D. P., Guidotti, A., & Sharma, R. P. (2013). Histone methylation at H3K9: Evidence for a restrictive epigenome in schizophrenia. *Schizophrenia Research*, *149*(1–3), 15–20.

Chen, A., Sun, L., Yuan, H., Wu, A., Lu, J., & Ma, S. (2017). A holistic strategy for quality and safety control of traditional Chinese medicines by the "iVarious" standard system. *Journal of Pharmaceutical Analysis*, *7*(5), 271–279.

Chen, B. X., Wei, T., Ye, Z. W., Yun, F., Kang, L. Z., Tang, H. B., & Lin, J. F. (2018). Efficient CRISPR-Cas9 gene disruption system in edible-medicinal mushroom *Cordyceps militaris*. *Frontiers in Microbiology*, *9*, 1157.

Chen, S., Pang, X., Song, J., Shi, L., Yao, H., Han, J., & Leon, C. (2014). A renaissance in herbal medicine identification: From morphology to DNA. *Biotechnology Advances*, *32*(7), 1237–1244.

Chen, S., Yao, H., Han, J., Liu, C., Song, J., Shi, L., & Leon, C. (2010). Validation of the ITS2 region as a novel DNA barcode for identifying medicinal plant species. *PloS One*, *5*(1), e8613.

Chen, Z., Li, J., Hou, N., Zhang, Y., & Qiao, Y. (2021). TCM-Blast for traditional Chinese medicine genome alignment with integrated resources. *BMC Plant Biology*, *21*, 1–7.

Choudhary, N., & Sekhon, B. S. (2011). An overview of advances in the standardization of herbal drugs. *Journal of Pharmaceutical Education and Research*, *2*(2), 55.

Corsello, M. A., & Garg, N. K. (2015). Synthetic chemistry fuels interdisciplinary approaches to the production of artemisinin. *Natural Product Reports*, *32*(3), 359–366.

Cragg, G. M., & Newman, D. J. (2013). Natural products: A continuing source of novel drug leads. *Biochimica et Biophysica Acta (BBA)-General Subjects*, *1830*(6), 3670–3695.

Crespo-Salvador, Ó., Sánchez-Giménez, L., López-Galiano, M. J., Fernández-Crespo, E., Scalschi, L., García-Robles, I., & González-Bosch, C. (2020). The histone marks signature in exonic and intronic regions is relevant in early response of tomato genes to Botrytis cinerea and in miRNA regulation. *Plants*, *9*(3), 300.

Cui, B., Liang, Z., Liu, Y., Liu, F., & Zhu, J. (2012). Effects of ABA and its biosynthetic inhibitor fluridone on accumulation of penolic acids and activity of PAL and TAT in hairy root of *Salvia miltiorrhiza*. *China Journal of Chinese Materia Medica*, *37*(6), 754–759.

Dai, L., Cheng, C. W., Tian, R., Zhong, L. L., Li, Y. P., Lyu, A. P., & Bian, Z. X. (2019). Standard protocol items for clinical trials with traditional Chinese medicine 2018: Recommendations, explanation and elaboration (SPIRIT-TCM Extension 2018). *Chinese Journal of Integrative Medicine*, *25*, 71–79.

Das, S., Kwon, M., & Kim, J. Y. (2024). Enhancement of specialized metabolites using CRISPR/Cas gene editing technology in medicinal plants. *Frontiers in Plant Science*, *15*, 1279738.

De Bie, T., Cristianini, N., Demuth, J. P., & Hahn, M. W. (2006). CAFE: A computational tool for the study of gene family evolution. *Bioinformatics*, *22*(10), 1269–1271.

Divya, N., Rengarajan, R. L., Radhakrishnan, R., Abd-Allah, E. F., Alqarawi, A. A., Hashem, A., & Anand, A. V. (2019). Phytotherapeutic efficacy of the medicinal plant *Terminalia catappa* L. *Saudi Journal of Biological Sciences*, *26*(5), 985–988.

Dong, O. X., Yu, S., Jain, R., Zhang, N., Duong, P. Q., Butler, C., & Ronald, P. C. (2020). Marker-free carotenoid-enriched rice generated through targeted gene insertion using CRISPR-Cas9. *Nature Communications*, *11*(1), 1178.

Edgar, R., Domrachev, M., & Lash, A. E. (2002). Gene Expression Omnibus: NCBI gene expression and hybridization array data repository. *Nucleic Acids Research*, *30*(1), 207–210.

Efferth, T., Fu, Y. J., Zu, Y. G., Schwarz, G., Konkimalla, V. S. B., & Wink, M. (2007). Molecular target-guided tumor therapy with natural products derived from traditional

Chinese medicine. *Current Medicinal Chemistry*, *14*(19), 2024–2032.

El-Zayat, S. A., Nassar, M. S., El-Hissy, F. T., Abdel-Motaal, F. F., & Ito, S. I. (2008). Mycoflora associated with *Hyoscyamus muticus* growing under an extremely arid desert environment (Aswan region, Egypt). *Journal of Basic Microbiology*, *48*(2), 82–92.

Emms, D. M., & Kelly, S. (2015). OrthoFinder: Solving fundamental biases in whole genome comparisons dramatically improves orthogroup inference accuracy. *Genome Biology*, *16*, 1–14.

Enna, S. J., & Williams, M. (2009). Defining the role of pharmacology in the emerging world of translational research. In *Advances in pharmacology* (Vol. 57, pp. 1–30). Academic Press.

Farooqui, N. A., Dey, A., Singh, G. N., Easwari, T. S., & Pandey, M. K. (2014). Analytical techniques in quality evaluation of herbal drugs. *Asian Journal of Pharmaceutical Research*, *4*(3), 112–117.

Feng, S., Song, W., Fu, R., Zhang, H., Xu, A., & Li, J. (2018). Application of the CRISPR/Cas9 system in *Dioscorea zingiberensis*. *Plant Cell, Tissue and Organ Culture (PCTOC)*, *135*, 133–141.

Ferber, G. (2002). Biomarkers and proof of concept. *Methods and Findings in Experimental and Clinical Pharmacology*, *24*, 35–40.

Ferdous, J., Hussain, S. S., & Shi, B. J. (2015). Role of micro RNA s in plant drought tolerance. *Plant Biotechnology Journal*, *13*(3), 293–305.

Ferreira, L. J., Azevedo, V., Maroco, J., Oliveira, M. M., & Santos, A. P. (2015). Salt tolerant and sensitive rice varieties display differential methylome flexibility under salt stress. *PLoS One*, *10*(5), e0124060.

Fu, R., Martin, C., & Zhang, Y. (2018). Next-generation plant metabolic engineering, inspired by an ancient Chinese irrigation system. *Molecular Plant*, *11*(1), 47–57.

Ganie, S. H., Upadhyay, P., Das, S., & Sharma, M. P. (2015). Authentication of medicinal plants by DNA markers. *Plant Gene*, *4*, 83–99.

Gao, W., Jia, W., Gao, X., Wang, R., & Xiao, P. (2005). In vitro culture and cultivation of Chinese medicinal plants for industrial utilization and genetic resource conservation. *Plant Genetic Resources*, *3*(2), 116–126.

Geiger, P. L., & Hesse, H. (1833). Darstellung des atropins. *Annalen der Pharmacie*, *5*(1), 43–81.

Govindaraghavan, S., Hennell, J. R., & Sucher, N. J. (2012). From classical taxonomy to genome and metabolome: Towards comprehensive quality standards for medicinal herb raw materials and extracts. *Fitoterapia*, *83*(6), 979–988.

Groth, A., Rocha, W., Verreault, A., & Almouzni, G. (2007). Chromatin challenges during DNA replication and repair. *Cell*, *128*(4), 721–733.

Guo, J. J., Su, M. S., Wang, L. Y., Ren, J. X., Yang, M., & Xie, X. M. (2016). Identification of dominant microbes from traditional Chinese medicine Pinelliae Rhizoma Fermentata. *China Journal of Chinese Materia Medica*, *41*(16), 3027–3031.

Guo, R., Luo, X., Liu, J., Liu, L., Wang, X., & Lu, H. (2020). Omics strategies decipher therapeutic discoveries of traditional Chinese medicine against different diseases at multiple layers molecular-level. *Pharmacological Research*, *152*, 104627.

Haas, B. J., Salzberg, S. L., Zhu, W., Pertea, M., Allen, J. E., Orvis, J., & Wortman, J. R. (2008). Automated eukaryotic gene structure annotation using EVidenceModeler and the program to assemble spliced alignments. *Genome Biology*, *9*, 1–22.

Halka, J., Saravanan, K., Vidya, N., Kowsalya, K., Senthilvelan, T., Gurusaravanan, Packiaraj Gurusaravanan, Vijaya anand Arumugam. (2024). Current elicitation strategies for improving secondary metabolites in medicinal plants with antiallergy properties. In *Biotechnology of medicinal plants with antiallergy properties: Research trends and prospects* (pp. 409–432). Springer Nature Singapore.

Han, J., Pang, X., Liao, B., Yao, H., Song, J., & Chen, S. (2016). An authenticity survey of herbal medicines from markets in China using DNA barcoding. *Scientific Reports*, *6*(1), 18723.

Han, P., Huang, Y., Xie, Y., Yang, W., Xiang, W., Hylands, P. J., & Legido-Quigley, C. (2018). Metabolomics reveals immunomodulation as a possible mechanism for the antibiotic effect of Persicaria capitata (Buch.-Ham. ex D. Don) H. Gross. *Metabolomics*, *14*, 1–9.

Harvey, A. L., Edrada-Ebel, R., & Quinn, R. J. (2015). The re-emergence of natural products for drug discovery in the genomics era. *Nature Reviews Drug Discovery*, *14*(2), 111–129.

Hashimoto, K., Shimizu, E., & Iyo, M. (2004). Critical role of brain-derived neurotrophic factor in mood disorders. *Brain Research Reviews*, *45*(2), 104–114.

Hassan, S. E. D. (2017). Plant growth-promoting activities for bacterial and fungal endophytes isolated from medicinal plant of *Teucrium polium* L. *Journal of Advanced Research*, *8*(6), 687–695.

He, L., Qian, J., Li, X., Sun, Z., Xu, X., & Chen, S. (2017). Complete chloroplast genome of medicinal plant Lonicera japonica: Genome rearrangement, intron gain and loss, and implications for phylogenetic studies. *Molecules*, *22*(2), 249.

He, P. X., Zhang, Y., Zhou, Y., Li, G. H., Zhang, J. W., & Feng, X. S. (2019). Supercritical fluid chromatography–a technical overview and its applications in medicinal plant analysis: An update covering 2012–2018. *Analyst*, *144*(18), 5324–5352.

Hebert, P. D., & Gregory, T. R. (2005). The promise of DNA barcoding for taxonomy. *Systematic Biology*, *54*(5), 852–859.

Hebert, P. D., Cywinska, A., Ball, S. L., & DeWaard, J. R. (2003). Biological identifications through DNA barcodes. *Proceedings of the Royal Society of London. Series B: Biological Sciences*, *270*(1512), 313–321.

Heubl, G. (2010). New aspects of DNA-based authentication of Chinese medicinal plants by molecular biological techniques. *Planta Medica*, *76*(17), 1963–1974.

Hill, S. N., & Hausbeck, M. K. (2008). Virulence and fungicide sensitivity of Phytophthora cactorum isolated from American ginseng gardens in Wisconsin and Michigan. *Plant Disease*, *92*(8), 1183–1189.

Ho, E., Beaver, L. M., Williams, D. E., & Dashwood, R. H. (2011). Dietary factors and epigenetic regulation for prostate cancer prevention. *Advances in Nutrition*, *2*(6), 497–510.

Hossain, C. M., Gera, M. E. E. T. A., & Ali, K. A. (2022). Current status and challenges of herbal drug development and regulatory aspect: A global perspective. *Asian Journal of Pharmaceutical and Clinical Research*, *15*, 31–41.

Hu, X. Q., & Su, S. B. (2017). An overview of epigenetics in Chinese medicine researches. *Chinese Journal of Integrative Medicine*, *23*(9), 714–720.

Hu, Z., Song, N., Zheng, M., Liu, X., Liu, Z., Xing, J., & Sun, Q. (2015). Histone acetyltransferase GCN 5 is essential for heat stress-responsive gene activation and thermotolerance in Arabidopsis. *The Plant Journal*, *84*(6), 1178–1191.

Hua, X. M., & Jiang, X. L. (2000). Characteristics and control countermeasures of pesticide pollution and its damage on environment in China. *Research of Environmental Sciences*, *13*(3), 40–43.

Jansson, M. D., & Lund, A. H. (2012). MicroRNA and cancer. *Molecular Oncology*, *6*(6), 590–610.

Kabesch, M. (2014). Epigenetics in asthma and allergy. *Current Opinion in Allergy and Clinical Immunology*, *14*(1), 62–68.

Karthika, P., Balamuralikrishnan, B., Yamini, K., Vijaya Anand, A., Durairaj, K., Maruthupandian, A., Alodaini, H. A., Hatamleh, A. A., Manikantan, P., & Arun, M. (2023). Green synthesis, characterization of silver nanoparticles using aqueous leaf extracts of *Solanum melongena* and in vitro evaluation of antibacterial, pesticidal and anticancer activity in human MDA-MB-231 breast cancer cell lines. *Journal of King Saud University – Science*, *35*(5), 102663.

Kasper, D. L., Fauci, A. S., Hauser, S. L., Longo, D. L., Jameson, J. L., & Loscalzo, J. (Eds.). (2015). *Harrison's principles of internal medicine (2 vols)*. McGraw-Hill Education.

Kaviya, M., Balasubramanian, B., Bharathi, K., Malaisamy, A., Al-Dhabi, N. A., Mariadhas, V. A., & Liu, W. (2022). Evaluation of nutritional substances and investigation of antioxidant and antimicrobial potentials of *Boerhavia diffusa* with in silico molecular docking. *Molecules*, *27*(4), 1280.

Keilwagen, J., Wenk, M., Erickson, J. L., Schattat, M. H., Grau, J., & Hartung, F. (2016). Using intron position conservation for homology-based gene prediction. *Nucleic Acids Research*, *44*(9), e89–e89.

Khalil, A. M. A., Hassan, S. E. D., Alsharif, S. M., Eid, A. M., Ewais, E. E. D., Azab, E., & Fouda, A. (2021). Isolation and characterization of fungal endophytes isolated from medicinal plant *Ephedra pachyclada* as plant growth-promoting. *Biomolecules*, *11*(2), 140.

Kim, D., Langmead, B., & Salzberg, S. L. (2015). HISAT: A fast spliced aligner with low memory requirements. *Nature Methods*, *12*(4), 357–360.

Kim, S. W., Gupta, R., Lee, S. H., Min, C. W., Agrawal, G. K., Rakwal, R., & Kim, S. T. (2016). An integrated biochemical, proteomics, and metabolomics approach for supporting medicinal value of *Panax ginseng* fruits. *Frontiers in Plant Science*, *7*, 994.

Klein-Junior, L. C., de Souza, M. R., Viaene, J., Bresolin, T. M., de Gasper, A. L., Henriques, A. T., & Vander Heyden, Y. (2021). Quality control of herbal medicines: From traditional techniques to state-of-the-art approaches. *Planta Medica*, *87*(12/13), 964–988.

Koh, K. H., & Tham, F. Y. (2011). Screening of traditional Chinese medicinal plants for quorum-sensing inhibitors activity. *Journal of Microbiology, Immunology and Infection*, *44*(2), 144–148.

Kouzarides, T. (2007). Chromatin modifications and their function. *Cell*, *128*(4), 693–705.

Kredich, N. M., & Guarino, A. J. (1961). Studies on the biosynthesis of cordycepin. *Biochimica et Biophysica Acta*, *47*(3), 529–534.

Kristensen, V. N., Lingjærde, O. C., Russnes, H. G., Vollan, H. K. M., Frigessi, A., & Børresen-Dale, A. L. (2014). Principles and methods of integrative genomic analyses in cancer. *Nature Reviews Cancer*, *14*(5), 299–313.

Kumar, D., & Sharma, U. (2018). High-performance thin-layer chromatography: An economical alternative for the quality control of medicinal plants and derived products. *Separation Science Plus*, *1*(2), 100–134.

Kumar, D., Bansal, G., Narang, A., Basak, T., Abbas, T., & Dash, D. (2016). Integrating transcriptome and proteome profiling: Strategies and applications. *Proteomics*, *16*(19), 2533–2544.

Kumar, S., Stecher, G., Suleski, M., & Hedges, S. B. (2017). Time tree: A resource for timelines, timetrees, and divergence times. *Molecular Biology and Evolution*, *34*(7), 1812–1819.

Kumari, R., & Kotecha, M. (2016). A review on the standardization of herbal medicines. *International Journal of Pharma Sciences and Research*, *7*(2), 97–106.

Kupiec, T. (2004). Quality-control analytical methods: High-performance liquid chromatography. *International Journal of Pharmaceutical Compounding*, *8*, 223–227.

Lan, F. F., Wang, H., Chen, Y. C., Chan, C. Y., Ng, S. S., Li, K., & Kung, H. F. (2011). Hsa-let-7g inhibits proliferation of hepatocellular carcinoma cells by downregulation of c-Myc and upregulation of p16INK4A. *International Journal of Cancer*, *128*(2), 319–331.

Lee, H. R., Han, S. I., Rhee, K. H., & Whang, K. S. (2013). Mucilaginibacter *herbaticus sp*. nov., isolated from the rhizosphere of the medicinal plant *Angelica sinensis*. International *Journal of Systematic and Evolutionary Microbiology*, *63*(Pt_8), 2787–2793.

Li, E., Bestor, T. H., & Jaenisch, R. (1992). Targeted mutation of the DNA methyltransferase gene results in embryonic lethality. *Cell*, *69*(6), 915–926.

Li, J., Gu, Y., Xue, J., Jin, H. Y., & Ma, S. (2017). Analysis and risk assessment of pesticide residues in a Chinese herbal medicine, *Lonicera japonica* Thunb. *Chromatographia*, *80*, 503–512.

Li, L., Stoeckert, C. J., & Roos, D. S. (2003). OrthoMCL: Identification of ortholog groups for eukaryotic genomes. *Genome Research*, *13*(9), 2178–2189.

Li, Q., Feng, J., Chen, L., Xu, Z., Zhu, Y., Wang, Y., & Chen, W. (2019). Genome-wide identification and characterization of *Salvia miltiorrhiza* laccases reveal potential targets for salvianolic acid B biosynthesis. *Frontiers in Plant Science*, *10*, 435.

Li, X., Chen, Y., Lai, Y., Yang, Q., Hu, H., & Wang, Y. (2015). Sustainable utilization of traditional Chinese medicine resources: Systematic evaluation on different production modes. *Evidence-Based Complementary and Alternative Medicine*, *2015*(1), 218901.

Li, X., Ren, J., Zhang, W., Zhang, Z., Yu, J., Wu, J., & Wang, W. (2022). LTM-TCM: A comprehensive database for the linking of traditional Chinese medicine with modern medicine at molecular and phenotypic levels. *Pharmacological Research*, *178*, 106185.

Lian, J., Zhang, X. C., & Gu, J. T. (2015). Advance in transcriptomics and its application in olericulture research. *Zhongguo Nongxue Tongbao*, *31*(08), 118–122.

Liang, Y. Z., Xie, P., & Chan, K. (2004). Quality control of herbal medicines. *Journal of Chromatography B*, *812*(1–2), 53–70.

Lin, S., Liu, Z. Q., Xue, Y. P., Baker, P. J., Wu, H., Xu, F., & Zheng, Y. G. (2016). Biosynthetic pathway analysis for improving the cordycepin and cordycepic acid production in Hirsutella sinensis. *Applied Biochemistry and Biotechnology*, *179*, 633–649.

Liu, Q., Meng, G., Wang, M., Li, X., Liu, M., Wang, F., & Dong, C. (2023). Safe-harbor-targeted CRISPR/Cas9 system and Cmhyd1 overexpression enhances disease resistance in *Cordyceps militaris*. *Journal of Agricultural and Food Chemistry*, *71*(41), 15249–15260.

Liu, T., Jia, T., Chen, J., Liu, X., Zhao, M., & Liu, P. (2017). Analysis of microbial diversity in Shenqu with different fermentation times by PCR-DGGE. *Brazilian Journal of Microbiology*, *48*(2), 246–250.

Lone, S. M., Hussain, K., Malik, A., Magray, M., Hussain, S. M., Rashid, M., & Farwah, S. (2020). Plant propagation through tissue culture-a biotechnological intervention. *International Journal of Current Microbiology and Applied Sciences*, *9*(7), 2176–2190.

Lossen, W., Lossen, W., Lossen, W., & Lossen, W. (1864). Ueber das Atropin. *Annalen der Chemie und Pharmacie*, *131*(1), 43–49.

Luo, R., Liu, B., Xie, Y., Li, Z., Huang, W., Yuan, J., & Wang, J. (2012). SOAPdenovo2: An empirically improved memory-efficient short-read de novo assembler. *Gigascience*, *1*(1), 2047–217X.

Mancinelli, L., Cronin, M., & Sadée, W. (2000). Pharmacogenomics: The promise of personalized medicine. *Aaps Pharmsci*, *2*, 29–41.

Manikandan, R., Balasubramanian, B., Punniyakotti, P., Vijaya Anand, A., Meyyazhagan, A., Velayuthaprabhu, S., & Liu, W. C. (2022). Cardio-protective effects of *Terminalia catappa* leaves and *Terminalia chebula* fruit extract in doxorubicin-induced cardiomyopathy in rats. *Biomarkers*, *27*(5), 488–495.

Manoharan, N., Sampathkumar, P., Dheeba, B., Sheik Abdulla, S., Vinothprasanna, G., Vinothkannan, R., Kalavathy, S., Vijaya Anand, A., & Shanmugasundaram, A. (2008). Potential hepatoprotective effect of aqueous extract of *Gracilaria corticata* in AFB_1 induced hepatotoxicity in Wistar rats. *Journal of Biological Sciences*, *8*, 1352–1355.

Marthe, F. (2018). Tissue culture approaches in relation to medicinal plant improvement. *Biotechnologies of Crop Improvement*, *1*, 487–497.

Mastan, A., Bharadwaj, R. K. B., Kushwaha, R. K., & Vivek Babu, C. S. (2019). Functional fungal endophytes in *Coleus forskohlii* regulate labdane diterpene biosynthesis for elevated forskolin accumulation in roots. *Microbial Ecology*, *78*(4), 914–926.

Mirsadeghi, S., & Larijani, B. (2017). Personalized medicine: Pharmacogenomics and drug development. *Acta Medica Iranica*, *55*(3), 150–165.

Moc, C. (2020). Pharmacogenomics: An evolving clinical tool for precision medicine. *Cleveland Clinic Journal of Medicine*, *87*(2), 91.

Mochida, K., Sakurai, T., Seki, H., Yoshida, T., Takahagi, K., Sawai, S., & Saito, K. (2017). Draft genome assembly and annotation of *Glycyrrhiza uralensis*, a medicinal legume. *The Plant Journal*, *89*(2), 181–194.

Moraes, R. M., Cerdeira, A. L., & Lourenço, M. V. (2021). Using micropropagation to develop medicinal plants into crops. *Molecules*, *26*(6), 1752.

Murch, S. J., Peiris, S. E., Liu, C. Z., & Saxena, P. K. (2004). In vitro conservation and propagation of medicinal plants. *Biodiversity*, *5*(2), 19–24.

Muthumani, G. D., Anand, A. V., & Manikandan, R. (2013). Original research article antioxidant, antihelmintic and antimicrobial activity of *Euphorbia thymifolia* Linn whole plant. *International Journal of Current Microbiology and Applied Sciences*, *2*(1), 66–79.

Nadalin, F., Vezzi, F., & Policriti, A. (2012). GapFiller: A de novo assembly approach to fill the gap within paired reads. *BMC Bioinformatics*, *13*, 1–16.

Nadeem, M. A., Nawaz, M. A., Shahid, M. Q., Doğan, Y., Comertpay, G., Yıldız, M., & Baloch, F. S. (2018). DNA molecular markers in plant breeding: Current status and recent advancements in genomic selection and genome editing. *Biotechnology and Biotechnological Equipment*, *32*(2), 261–285.

Newmaster, S. G., Grguric, M., Shanmughanandhan, D., Ramalingam, S., & Ragupathy, S. (2013). DNA barcoding detects contamination and substitution in North American herbal products. *BMC Medicine*, *11*, 1–13.

Niazian, M. (2019). Application of genetics and biotechnology for improving medicinal plants. *Planta*, *249*, 953–973.

Pan, S. Y., Pan, S., Yu, Z. L., Ma, D. L., Chen, S. B., Fong, W. F., & Ko, K. M. (2010). New perspectives on innovative drug discovery: An overview. *Journal of Pharmacy and Pharmaceutical Sciences*, *13*(3), 450–471.

Park, S. Y., & Paek, K. Y. (2014). Bioreactor culture of shoots and somatic embryos of medicinal plants for production of bioactive compounds. In *Production of biomass and bioactive compounds using bioreactor technology* (pp. 337–368). Springer.

Pathak, H., Kumar, M., Molla, K. A., & Chakraborty, K. (2021). Abiotic stresses in rice production: Impacts and management. *Oryza*, *58*(4), 103–125.

Pathania, R., Ramachandran, S., Elangovan, S., Padia, R., Yang, P., Cinghu, S., & Thangaraju, M. (2015). DNMT1 is essential for mammary and cancer stem cell maintenance and tumorigenesis. *Nature Communications*, *6*(1), 6910.

Patil, P. S., & Shettigar, R. (2010). An advancement of analytical techniques in herbal research. *Journal of Advanced Scientific Research*, *1*(01), 8–14.

Patrick, D. M., Marra, F., Hutchinson, J., Monnet, D. L., Ng, H., & Bowie, W. R. (2004). Per capita antibiotic consumption: How does a North American jurisdiction compare with Europe? *Clinical Infectious Diseases*, *39*(1), 11–17.

Peng, T., Guo, J., & Tong, X. (2024). Advances in biosynthesis and metabolic engineering strategies of cordycepin. *Frontiers in Microbiology*, *15*, 1386855.

Peterson, C. L., & Laniel, M. A. (2004). Histones and histone modifications. *Current Biology*, *14*(14), R546–R551.

Poupko, J. M., Baskin, S. I., & Moore, E. (2007). The pharmacological properties of anisodamine. *Journal of Applied Toxicology: An International Journal*, *27*(2), 116–121.

Pryszcz, L. P., & Gabaldón, T. (2016). Redundans: An assembly pipeline for highly heterozygous genomes. *Nucleic Acids Research*, *44*(12), e113–e113.

Ramawat, K. G., & Arora, J. (2021). Medicinal plants domestication, cultivation, improvement, and alternative technologies for the production of high value therapeutics: An overview. In *Medicinal plants: Domestication, biotechnology and regional importance* (pp. 1–29). Springer Nature Switzerland.

Rawat, H., Bhat, S. A., Dhanjal, D. S., Singh, R., Gandhi, Y., Mishra, S. K., & Acharya, R. (2024). Emerging techniques for the trace elemental analysis of plants and food-based extracts: A comprehensive review. *Talanta Open*, *10*, 100341.

Riaz, S. (2015). Study of protein biomarkers of diabetes mellitus type 2 and therapy with vitamin B1. *Journal of Diabetes Research*, *2015*(1), 150176.

Rohini, M. R., & Rajasekharan, P. E. (2022). Scale-up production of bioactive compounds using bioreactors. In *Nutraceuticals production from plant cell factory* (pp. 69–81). Springer Nature Singapore.

Seidl, P. R. (2002). Pharmaceuticals from natural products: Current trends. *Anais da Academia Brasileira de Ciências*, *74*, 145–150.

Sharma, S., Madaan, K., & Kaur, R. (2022). Cordycepin as a metabolite with pharmacological potential: A review. *International Journal of Medicinal Mushrooms*, *24*(8), 1–20.

Shen, H., & Laird, P. W. (2013). Interplay between the cancer genome and epigenome. *Cell*, *153*(1), 38–55.

Shi, M., Du, Z., Hua, Q., & Kai, G. (2021). CRISPR/Cas9-mediated targeted mutagenesis of bZIP2 in *Salvia miltiorrhiza* leads to promoted phenolic acid biosynthesis. *Industrial Crops and Products*, *167*, 113560.

Shi, Z., Chen, Y., Hou, X., Gao, S., & Wang, F. (2013). *Arbuscular mycorrhizal* fungi associated with tree peony in 3 geographic locations in China. *Turkish Journal of Agriculture and Forestry*, *37*(6), 726–733.

Shulammithi, R., Sharanya, M., Tejaswini, R., & Kiranmai, M. (2016). Standardization and quality evaluation of herbal drugs. *IOSR Journal of Pharmacy and Biological Sciences*, *11*(i5–v1), 89–100.

Simmonds, N. W. (1993). Introgression and incorporation. Strategies for the use of crop genetic resources. *Biological Reviews*, *68*(4), 539–562.

Simões, M. S., Carvalho, G. G., Ferreira, S. S., Hernandes-Lopes, J., de Setta, N., & Cesarino, I. (2020). Genome-wide characterization of the laccase gene family in *Setaria viridis* reveals members potentially involved in lignification. *Planta*, *251*, 1–18.

Singh, D. B. (2020). The impact of pharmacogenomics in personalized medicine. *Current Applications of Pharmaceutical Biotechnology*, *171*, 369–394.

Solaiman, Z. M., & Anawar, H. M. (2015). Rhizosphere microbes interactions in medicinal plants. Plant-growth-promoting rhizobacteria (PGPR) and medicinal plants, *Soil Biology*, *42*, 19–41.

Srivastava, M. M. (2010). An overview of HPTLC: A modern analytical technique with excellent potential for automation, optimization, hyphenation, and multidimensional applications. In *High-performance thin-layer chromatography (HPTLC)* (pp. 3–24). Springer Science & Business Media.

Stanke, M., Keller, O., Gunduz, I., Hayes, A., Waack, S., & Morgenstern, B. (2006). AUGUSTUS: Ab initio prediction of alternative transcripts. *Nucleic Acids Research*, *34*(Suppl_2), W435–W439.

Stephanopoulos, G., Aristidou, A. A., & Nielsen, J. (1998). *Metabolic engineering: Principles and methodologies*. ScienceDirect.

Strahl, B. D., & Allis, C. D. (2000). The language of covalent histone modifications. *Nature*, *403*(6765), 41–45.

Strasser, R., Stadlmann, J., Schähs, M., Stiegler, G., Quendler, H., Mach, L., & Steinkellner, H. (2008). Generation of glyco-engineered *Nicotiana benthamiana* for the production of monoclonal antibodies with a homogeneous human-like N-glycan structure. *Plant Biotechnology Journal*, *6*(4), 392–402.

Subramaniam, D., Thombre, R., Dhar, A., & Anant, S. (2014). DNA methyltransferases: A novel target for prevention and therapy. *Frontiers in Oncology*, *4*, 80.

Sumner, L. W., Mendes, P., & Dixon, R. A. (2003). Plant metabolomics: Large-scale phytochemistry in the functional genomics era. *Phytochemistry*, *62*(6), 817–836.

Tang, J. L., & Wong, T. W. (1998). The need to evaluate the clinical effectiveness of traditional Chinese medicine. *Hong Kong Medical Journal*, *4*, 208–210.

Thangavelu, S., Balasubramanian, B., Palanisamy, S., Shanmugam, V., Natchiappan, S., Kalibulla, S. I., & Arumugam, V. A. (2022). Characterization and phytoconstituents of *Petroselinum crispum* (Mill) and *Coriandrum sativum* (Linn) and their impacts on inflammation – an in vitro analysis against human adenocarcinoma cells with molecular docking. *South African Journal of Botany*, *146*, 776–788.

Tsao, R., & Deng, Z. (2004). Separation procedures for naturally occurring antioxidant phytochemicals. *Journal of Chromatography B*, *812*(1–2), 85–99.

Tu, Y. (2016). Artemisinin – a gift from traditional Chinese medicine to the world (nobel lecture). *Angewandte Chemie International Edition (England)*, *55*(35), 102100–102126.

Varalakshmi, B., Anand, A. V., Karpagam, T., Bai, J. S., & Manikandan, R. (2014). In vitro antimicrobial and anticancer activity of *Cinnamomum zeylanicum* Linn bark extracts. *International Journal of Pharmaceutical Sciences*, *6*(1), 12–18.

Verma, A. K. (2022). Cordycepin: A bioactive metabolite of *Cordyceps militaris* and polyadenylation inhibitor with therapeutic potential against COVID-19. *Journal of Biomolecular Structure and Dynamics*, *40*(8), 3745–3752.

Verma, P., Khan, S. A., Alhandhali, A. J. A., & Parasharami, V. A. (2021). Bioreactor upscaling of different tissue of medicinal herbs for extraction of active phytomolecules: A step towards industrialization and enhanced production of phytochemicals. In *Plant growth regulators: Signalling under stress conditions* (pp. 455–481). Springer.

Verpoorte, R., Choi, Y. H., & Kim, H. K. (2005). Ethnopharmacology and systems biology: A perfect holistic match. *Journal of Ethnopharmacology*, *100*(1–2), 53–56.

Vijaya Anand, A., Varalakshmi, B., Prasana, Sampathkumar, P., Pushpa, N., & Hedina, A. (2016). *Cinnamomum zeylanicum* Linn. The spice with multi potential. *Systematic Reviews in Pharmacy*, *7*(1), 24–29.

Vijayakumar, K., Prasanna, B., Rengarajan, R. L., Rathinam, A., Velayuthaprabhu, S., & Vijaya Anand, A. (2023). Antidiabetic and hypolipidemic effects of *Cinnamon cassia* bark extracts: An in vitro, in vivo, and in silico approach. *Archives of Physiology and Biochemistry*, *129*(2), 338–348.

Wagner, J. A., & Kroetz, D. L. (2016). Transforming translation: Impact of clinical and translational science. *Clinical and Translational Science*, *9*(1), 3.

Wang, H., Chen, Y., Wang, L., Liu, Q., Yang, S., & Wang, C. (2023). Advancing herbal medicine: Enhancing product quality and safety through robust quality control practices. *Frontiers in Pharmacology*, *14*, 1265178.

Wang, W., Huang, F., Qin, Q., Zhao, X., Li, Z., & Fu, B. (2015b). Comparative analysis of DNA methylation changes in two rice genotypes under salt stress and subsequent recovery. *Biochemical and Biophysical Research Communications*, *465*(4), 790–796.

Wang, J., Wang, T., Li, Y. S., Zheng, Y. F., Li, L. R., & Wang, Q. (2015c). Research on constitution of Chinese medicine and implementation of translational medicine. *Chinese Journal of Integrative Medicine*, *21*, 389–393.

Wang, X., Zhang, J., He, S., Gao, Y., Ma, X., Gao, Y., . . . Dong, Y. (2018). HMOD: An omics database for herbal medicine plants. *Molecular Plant*, *11*(5), 757–759.

Wang, Y. H., Liu, S. J., Ji, S. L., Zhang, W. W., Wang, C. M., Jiang, L., & Wan, J. M. (2005). Fine mapping and marker-assisted selection (MAS) of a low glutelin content gene in rice. *Cell Research*, *15*(8), 622–630.

Wang Yunsheng, W. Y., Shahid, M. Q., Ghouri, F., Baloch, F. S., Wang Ying, W. Y., & Huang HongWen, H. H. (2015a). Evaluation of the geographical pattern of genetic diversity of *Glycine soja* and *Glycine max* based on four single copy nuclear gene loci: For conservation of soybean germplasm. *Biochemical Systematics and Ecology*, *62*(2), 229–235.

Wanichthanarak, K., Fahrmann, J. F., & Grapov, D. (2015). Genomic, proteomic, and metabolomic data integration strategies. *Biomarker Insights*, *10*, BMI-S29511.

Wehling, M. (2006). Translational science in medicine: Implications for the pharmaceutical industry. *International Journal of Pharmaceutical Medicine*, *20*, 303–310.

Wei, C. Y., Michael Lee, M. T., & Chen, Y. T. (2012). Pharmacogenomics of adverse drug reactions: Implementing personalized medicine. *Human Molecular Genetics*, *21*(R1), R58–R65.

Whang, K. S., Lee, J. C., Lee, H. R., Han, S. I., & Chung, S. H. (2014). *Terriglobus tenax* sp. nov., an exopolysaccharide-producing acidobacterium isolated from rhizosphere soil of a medicinal plant. *International Journal of Systematic and Evolutionary Microbiology*, *64*(Pt_2), 431–437.

Whirl-Carrillo, M., Huddart, R., Gong, L., Sangkuhl, K., Thorn, C. F., Whaley, R., & Klein, T. E. (2021). An evidence-based framework for evaluating pharmacogenomics knowledge for personalized medicine. *Clinical Pharmacology and Therapeutics*, *110*(3), 563–572.

Wishart, D. S., Jewison, T., Guo, A. C., Wilson, M., Knox, C., Liu, Y., & Scalbert, A. (2012). HMDB 3.0-the human metabolome database in 2013. *Nucleic Acids Research*, *41*(D1), D801–D807.

Wu, H., Guo, J., Chen, S., Liu, X., Zhou, Y., Zhang, X., & Xu, X. (2013). Recent developments in qualitative and quantitative analysis of phytochemical constituents and their metabolites using liquid chromatography–mass spectrometry. *Journal of Pharmaceutical and Biomedical Analysis*, *72*, 267–291.

Xie, S., Yang, F., Feng, H., & Wei, C. (2019). General characteristics of the chemical fertilizers and pesticides use and the analysis of use reduction effect in China. *Environmental Pollution and Control*, *41*(4), 490–495.

Xiong, C., Sun, W., Li, J., Yao, H., Shi, Y., Wang, P., & Chen, S. (2018). Identifying the species of seeds in traditional Chinese medicine using DNA barcoding. *Frontiers in Pharmacology*, *9*, 701.

Xu, M., & Xu, D. (2024). Advanced systems and bioreactors for large-scale secondary metabolite production in medicinal plants using suspension cultured cells. In *Biosynthesis of natural products in plants: Bioengineering in post-genomics era* (pp. 293–313). Springer Nature Singapore.

Xu, Y., Xie, Y. B., Zhang, X. R., Chen, C., Xiang, H., & Xie, Q. (2013). Monitoring of the bacterial and fungal biodiversity and dynamics during Massa Medicata Fermentata fermentation. *Applied Microbiology and Biotechnology*, *97*, 9647–9655.

Yang, L., Li, G., Chai, Z., Gong, Q., & Guo, J. (2020). Synthesis of cordycepin: Current scenario and future perspectives. *Fungal Genetics and Biology*, *143*, 103431.

Yang, W., Zhang, Y., Wu, W., Huang, L., Guo, D., & Liu, C. (2017). Approaches to establish Q-markers for the quality standards of traditional Chinese medicines. *Acta Pharmaceutica Sinica B*, *7*(4), 439–446.

You-Xin, Y., Yan-Hong, L., Jing-Fei, T., Qasim, S. M., Zhi-Xiong, C., Lan, W., & Yong-Gen, L. (2012). Wide-compatibility gene s5n exploited by functional molecular markers and its effect on fertility of intersubspecific rice hybrids. *Crop Science*, *52*(2), 669–675.

Yu, J., Wu, X. I., Liu, C., Newmaster, S., Ragupathy, S., & Kress, W. J. (2021). Progress in the use of DNA barcodes in the identification and classification of medicinal plants. *Ecotoxicology and Environmental Safety*, *208*, 111691.

Yu, J., Zheng, Y., Song, C., & Chen, S. (2023). New insights into the roles of fungi and bacteria in the development of medicinal plant. *Journal of Advanced Research*, S2090-1232(23)00394-6.

Yun, D. J., Hashimoto, T., & Yamada, Y. (1992). Metabolic engineering of medicinal plants: Transgenic Atropa belladonna with an improved alkaloid composition. *Proceedings of the National Academy of Sciences*, *89*(24), 11799–11803.

Zeng, L., Zhang, Q., Jiang, C., Zheng, Y., Zuo, Y., Qin, J., & Deng, H. (2021). Development of *Atropa belladonna* L. plants with high-yield hyoscyamine and without its derivatives using the CRISPR/Cas9 system. *International Journal of Molecular Sciences*, *22*(4), 1731.

Zhang, H., Li, Y., & Zhu, J. K. (2018). Developing naturally stress-resistant crops for a sustainable agriculture. *Nature Plants*, *4*(12), 989–996.

Zhang, J., Tian, Y., Yan, L., Zhang, G., Wang, X., Zeng, Y., & Sheng, J. (2016). Genome of plant maca (*Lepidium meyenii*) illuminates genomic basis for high-altitude adaptation in the central Andes. *Molecular Plant*, *9*(7), 1066–1077.

Zhang, J. B., He, S. P., Luo, J. W., Wang, X. P., Li, D. D., & Li, X. B. (2020). A histone deacetylase, GhHDT4D, is positively involved in cotton response to drought stress. *Plant Molecular Biology*, *104*, 67–79.

Zhang, Y., & Reinberg, D. (2001). Transcription regulation by histone methylation: Interplay between different covalent modifications of the core histone tails. *Genes and Development*, *15*(18), 2343–2360.

Zhang, Y., Zheng, L., Zheng, Y., Zhou, C., Huang, P., Xiao, X., & Li, C. (2019). Assembly and annotation of a draft genome of the medicinal plant *Polygonum cuspidatum*. *Frontiers in Plant Science*, *10*, 1274.

Zhang, Z., Li, H., Xu, T., Xu, H., He, S., Li, Z., & Zhang, Z. (2021). Jianqu fermentation with the isolated fungi significantly improves the immune response in immunosuppressed mice. *Journal of Ethnopharmacology*, *267*, 113512.

Zhou, J., Cui, Y., Chen, X., Li, Y., Xu, Z., Duan, B., & Yao, H. (2018). Complete chloroplast genomes of *Papaver rhoeas* and *Papaver orientale*: Molecular structures, comparative analysis, and phylogenetic analysis. *Molecules*, *23*(2), 437.

Zhou, X., Li, C. G., Chang, D., & Bensoussan, A. (2019). Current status and major challenges to the safety and efficacy presented by Chinese herbal medicine. *Medicines*, *6*(1), 14.

Zhou, Z., Li, Q., Xiao, L., Wang, Y., Feng, J., Bu, Q., & Zhang, L. (2021). Multiplexed CRISPR/Cas9-mediated knockout of laccase genes in *Salvia miltiorrhiza* revealed their roles in growth, development, and metabolism. *Frontiers in Plant Science*, *12*, 647768.

Zhou, Z. H., Yang, J., & Kong, A. N. (2017). Phytochemicals in traditional Chinese herbal medicine: Cancer prevention and epigenetics mechanisms. *Current Pharmacology Reports*, *3*, 77–91.

Zhu, S., Liu, Q., Qiu, S., Dai, J., & Gao, X. (2022). DNA barcoding: An efficient technology to authenticate plant species of traditional Chinese medicine and recent advances. *Chinese Medicine*, *17*(1), 112.

Zuo, G. Y., Zhang, X. J., Yang, C. X., Han, J., Wang, G. C., & Bian, Z. Q. (2012). Evaluation of traditional Chinese medicinal plants for anti-MRSA activity with reference to the treatment record of infectious diseases. *Molecules*, *17*(3), 2955–2967.

18 Integration of Omics and High-Throughput Technologies for Enhancing Plant Secondary Metabolites

A Comprehensive Overview

Sujaan Aehsas, Sampurna Banerjee, Jyoti Prakash Sahoo, Madhuri Pradhan, Manaswini Mahapatra, and Prayasi Nayak

18.1 INTRODUCTION

Plant cells are made up primarily of metabolites. These substances are essential for human health providing carbohydrates, fatty acids, vitamins, minerals, and other necessary bioactive secondary metabolites. They also play important functions in plants as energy providers, signaling regulators, and cofactors for enzymes (Seeland et al., 2019). In addition, additional metabolites such as alkaloids, terpenoids, and polyphenols serve as significant regulators. These plant metabolites are known as secondary metabolites because they not only support healthy plant growth and development processes but also protect against biotic and abiotic stresses (Shao et al., 2016). They also exhibit high bioactivities, which enable them to enhance defense against human inflammation, cardiovascular disease, and cancer. Recently, there has been an emphasis on expanding our knowledge of the metabolic processes involved in biosynthesis and catabolism. As a result, significant progress has been made, including extensive transcriptome and genome analysis (Song et al., 2018).

Secondary metabolites can be sorted into various categories according to their biological roles and molecular structures. For example, terpenes, which are volatile metabolites of terpenoids, are crucial for signal transduction between different plants and pollinator plants (Strohl et al., 2000). Moreover, major chemically diverse polyphenols found in nature include chalcones, xanthones, and flavonoids. These compounds have a broad spectrum of biological tasks encompassing the inhibition of ±-glucosidase (Kimotho & Maina, 2024). Research on natural and synthetic polyphenol derivatives has produced several promising alpha-glucosidase inhibitors – *e.g.*, galloyl, caffeoyl, and hex hydroxy diphenyl esters of dihydrochalcone glucosides, sulfonamide chalcones, planar catechin analogues, macluraxanthone, mangostin xanthones, and flavanone and acetophenone glucosides – that may be developed into anti-diabetic medications (Chao et al., 2024). Moreover, dietary polyphenols and their subgroups used in omics are indexed in Figure 1. Furthermore, several polyphenols, such isoeugenol and naphthyl chalcones, show encouraging in vivo antihyperglycemic action by lowering blood sugar, promoting insulin secretion, and blocking ± - glucosidase at the nanomolar range. Moreover, polyphenols are good candidates for antidiabetic medication discovery due to their promising ± -glucosidase inhibitory action (Zhang et al., 2023).

Omics technologies, which encompasses genomics, transcriptomics, proteomics, metabolomics, and other high-throughput methodologies, have revolutionized our understanding of plant biology and have provided powerful tools for unraveling complex biochemical pathways involved in secondary metabolite biosynthesis (Zhang et al., 2023). Integrating these omics approaches with high-throughput technologies offers a comprehensive and systematic way to explore the intricate regulatory networks governing secondary metabolism in plants (Zhang et al., 2019).

In recent years, there has been a surge in research focused on integrating omics and high-throughput technologies to enhance the production of plant secondary metabolites (Zhang et al., 2017). This chapter aims to summarize the latest advancements in this field, highlighting the synergistic approaches and strategies employed to decipher the molecular mechanisms underlying secondary metabolite biosynthesis and to engineer plants for improved production of valuable compounds. By providing a comprehensive overview of the integration of omics and high-throughput technologies for enhancing plant secondary metabolites, this chapter also aims to not only consolidate the current state of knowledge in the field but also inspire future research directions and applications aimed at harnessing the full potential of plants for sustainable and value-added production of secondary metabolites.

DOI: 10.1201/9781003475491-18

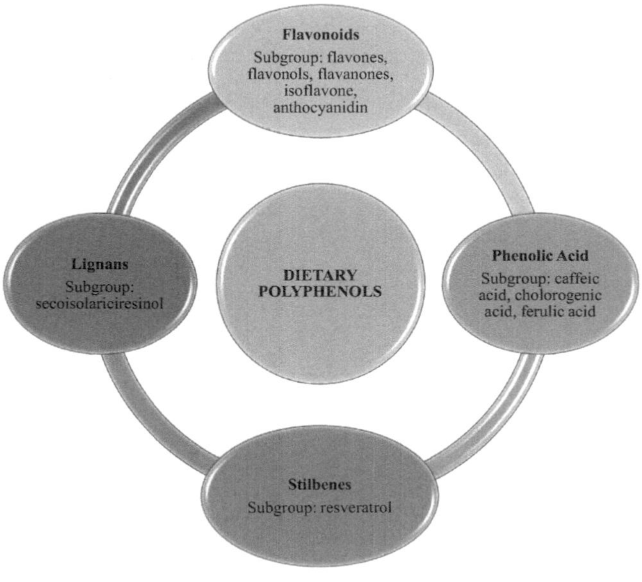

FIGURE 18.1 Dietary polyphenols and their subgroups used in omics.

18.2 THE MECHANICAL PATHWAY OF PLANT SECONDARY METABOLITE PRODUCTION

It has been widely considered that plants are the primary source of nourishment for all life on Earth. Because of adverse climatic variations, plants in the natural environment are subject to a variety of stresses, the most common ones being heat stress (HS), chilling stress (CS), and freezing stress (FS). These stresses are thought to pose a serious threat to sustainable agriculture because they impair plant growth and development, inflict damage, and ultimately result in yield loss, globally, undermining the Food and Agricultural Organization (FAO) of the United Nations' goal of 'zero hunger' (Strohl et al., 2000). The availability of data generated by many omics techniques provides the chance to address complex questions during in-plant research. However, gaps in the data generated can be a drawback. To collect more comprehensive datasets pertaining to precise biological development, *i.e.,* plant systems' reactions to external stressors, corresponding technologies are essential. RNA-seq has revolutionized the transcriptomics field with the advent of high throughput next-generation sequencing (htNGS) technology (Sun et al., 2017). It has overcome many limitations posed by the commonly used microarray techniques and does not require prior information about genome analysis or sequencing inquiry. This allows for the unbiased detection of new and known targeted datasets across the whole genome (Tardieu et al., 2017).

Since the inception of the *one-KP project*, which aims to sequence 1000 plant transcriptomes, transcriptome-base investigations of many plant species have extensively employed RNA-seq. Furthermore, the integration of RNA-seq with additional omics instruments, such as miRNA-seq, Chip-seq, Ribo-seq, and GRO-seq, has enabled the comprehensive exploration of various facets of plant omics (transcriptome) (Urasaki et al., 2017). Because of its improved noise proportion and understanding of the genomic sequence, ChIP-seq is currently the best technique for examining the interaction between DNA and particular proteins (Watson et al., 2022).

The massive amounts of data generated by multiple high-throughput techniques are often sharp and contain a lot of unwanted changes and technical items. It is an exam for the accurate analysis of exceptional datasets in order to identify reliable signals, syndicate data kinds that can be changed, and understand the biological relationships between them. RNA-seq can be performed before ChIP-seq and Ribo-seq when designing integrated omics (RNA-seq, proteomics, and metabolomics) experiments. As a result, ChIP-seq analysis measures the most enhanced transcription factors identified in differentially expressed genes revealed by RNA-seq (Wong et al., 2018). Consequently, a richer understanding of potential future paths in systems biology and molecular biology, among other fields, can be obtained by the integration of multi-omics data.

Utilizing high-performance mass spectrometers, the division specializes mostly in metabolomic studies. The division seeks to identify critical elements governing metabolic controls that support plant growth and metabolite production for human health through the merging of omic analysis and bioinformatics. The division is identifying helpful metabolites that contribute to the flavors and tastes of foods and utility plants as applied science. Additionally, the section creates analytical and separation procedures for chemical substances as needed. Thus, the merging of biology and technology is known as multi-omics. This is a novel biological strategy that involves combining and analyzing various sets of omic groupings. Genomics, proteomics, epigenomics, microbiomics, meta-transcriptomics, metabolomics, and transcriptomics are a few examples of the various omics techniques. The method of merging genome sequencing data, automated gene cluster prediction, and MS-based metabolomics was dubbed 'metabologenomics' (Xu et al., 2016). It has been demonstrated that combined omics techniques are an effective way to find novel chemicals in bacteria, especially in actinomycetes (Ashraf et al., 2014). Moreover, the mechanical pathway of plant secondary metabolite production is indexed in Figure 18.2.

18.3 INTEGRATED ANALYSIS OF OMICS

While certain plants adjust by altering their morphological structure, others adapt by changing their physiology or exhibiting changes in gene expression, which modifies their growth activities and allows them to withstand and tolerate such environments (Bassel et al., 2012). Plants therefore possess sophisticated mechanisms that allow them to adapt their physiological and developmental processes to play in unfavorable stressful conditions. On the other hand,

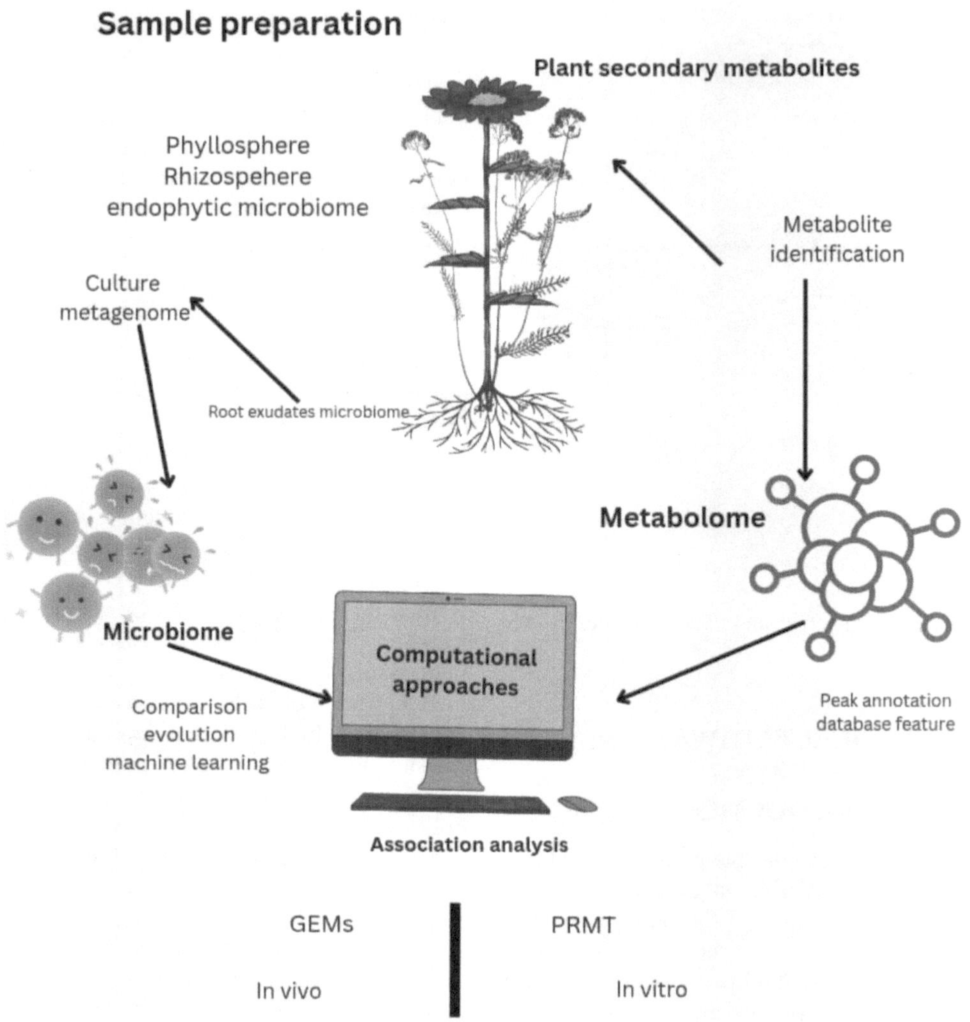

FIGURE 18.2 The mechanical pathway of plant secondary metabolite production.

environmental stressors can alter and interfere with their fundamental processes of operation, which include changes in gene expression, the creation of unique proteins and secondary metabolites, adjustments to hormonal signaling, and the actions of antioxidant enzymes, among others. However, increasing plant productivity in the face of changing climate circumstances requires an understanding of how plants respond, adapt, and withstand temperature swings.

Examining how plants have evolved stress tolerance and survival strategies will help to lessen the negative effects of temperature stress, and inspire new ideas for breeding climate-resilient crops (Bhattacharyya et al., 2013). To put it briefly, these computational techniques will assist the researchers link disparate bits of information to form coherent ideas that will be tested further. Lastly, we suggest an integrated analysis workflow of the omics data using plants of the *Dendrobium* genus as an example. Given that the phenotypic characteristics were influenced by environmental stimuli, it was suggested that chemical and genomic markers be used as essential standards for plant authentication. Furthermore, an integrated analysis of data from the fields of genomes, transcriptomics, proteomics, and metabolomics was required to identify the regulatory mechanisms underpinning the production pathways of *Dendrobium* and polysaccharides. Notably, particular bioinformatics toolkits and resources were added to this study framework, demonstrating their capacity for effective omics data handling (Bhawna et al., 2015). An integrated analysis of omics is indexed in Figure 18.3.

18.4 STUDY OF PHONEMICS AND TAXONOMICS

Plant phenotypes, which include morphological and physiological characteristics exhibited by a plant genome under particular growth conditions, are the outcome of the interaction between genes and the environment (Zhang et al., 2018). Phenotyping, the act of recording phenotypic attributes, is an important stage of recognition and taxonomy

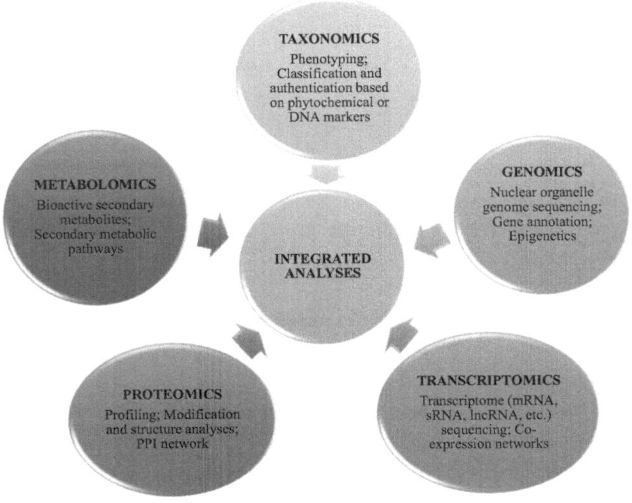

FIGURE 18.3 Integrated analysis of omics.

of plant species as well as the study of the effects of genes and environment on plant phenotypes (Li et al., 2018). The subfield of phenomics, which uses imaging technologies, image processing tools, and controlled experiments, was born out of the importance of phenotyping in plant biology (da Silva et al., 2016). A typical plant phenomics study entails designing the experiment, taking pictures with different cameras, managing the image data, and quantifying the morphology, growth dynamics, and physiological status of the plants (ElManawy et al., 2022).

High-resolution imaging systems and computational tools have been established for efficient phenotyping, which was previously dependent on manual labor (Fahlgren et al., 2015). These instruments, designed for specific uses such as root or stress phenotyping, have shown to be extremely useful for research on medicinal plants as well as model plant analysis and crop development. Differentiating between medicinal plant species is made easier by the extraction of morphological features, and investigating true quality production requires an awareness of ideal growing circumstances. Medicinal plant quality control is facilitated by tracking associated data and keeping track of important characteristics (Fu et al., 2017). The sections that follow present databases and analytical toolkits that may be useful for taxonomic and phenomics research on medicinal plants (Tables 18.1 and 18.2).

The PlantCLEF database (2019) has been updated recently and contains an extensive collection of plant photos taken from different organs, sizes, and visual viewpoints (Fukushima et al., 2009). PlantCLEF used raw pictures and a convolutional neural network-based machine learning algorithm to mimic a realistic automated plant identification system. The method's goal was to extract common features unique to a certain plant species or family. PlantCLEF is now concentrated on 1000 plant species that are representative of the flora of Western Europe. With community-based image collecting, PlantCLEF has the potential to expand globally and greatly facilitate the identification and categorization of new therapeutic plants. The most extensive archive for phenomics investigations on medicinal plants is the Hong Kong Baptist University–hosted MPID (Medicinal Plant Images Database), which is accessible to the public. For more than a thousand species of medicinal plants, MPID not only supplies representative photos but also their scientific names, morphological characteristics, environmental conditions, and therapeutic properties.

The MPDB (Medicinal Plant Database of Bangladesh) 1.0 is an additional specialized resource that focuses on Bangladeshi medicinal plants. This database includes information on 406 medicinal plants, including their scientific and common names, and information on their medicinal parts for taxonomy and standard nomenclature (Graw et al., 2021). In addition to phenotypic databases for medicinal plants, computer-assisted picture processing and analysis toolkits exist. For high-throughput plant phenotyping, the Python-based platform PlantCV (Plant Computer Vision) provides a selection of open-source software packages and algorithms developed by the community (Griffiths-Jones et al., 2006). With the use of distance-based watershed transformation, this integrated platform can analyze photos including many plants, execute leaf segmentation, and evaluate plant forms using landmark recognition. On GitHub (https://github.com/danforthcenter/plantcv), PlantCV is available without charge. ImageJ is another Java-based framework for image processing and analysis.

Currently, ImageJ supports more than 500 plugins that offer different functionalities including graphic segmentation, normalization, and background correction. Plant growth status and fitness can be ascertained by a high-throughput color picture analysis pipeline presented by HTPheno, which is a plug-in for ImageJ (Hill et al., 2017). Image retrieval, defining target regions, segmenting plants, extraction of plants, constructing morphology and presenting results are the six main processes in the HTPheno pipeline. Combinational analysis is used to extract phenotypic data, such as height, width, and predicted shoot area for growing plants, from images of each plant taken at various angles. To improve the calibre of taxonomic research, a uniform nomenclature system is essential in addition to an optimal phenotyping approach. Inconsistent nomenclature and misspelled species names make it difficult to compile and integrate data from various ethnic pharmacopoeias, which might result in inconsistent records and erroneous counts of medicinal plant species.

An online tool called TNRS (Taxonomic Name Resolution Service) is used to automatically standardize scientific names for plants (Kim et al., 2009). Tropicos (http://www.tropicos.org), the Global Compositae Checklist (http://compositae.landcareresearch.co.nz/), USDA Plants (http://plants.usda.gov/java/), and NCBI Taxonomy (http://www.ncbi.nlm.nih.gov/Taxonomy/taxonomyhome.html/) are the four optional taxonomic sources that TNRS offers its users. Compared to previous applications, TNRS stands out for a

TABLE 18.1
A Collection of Databases Offering Useful Omics Resources for Research on Medicinal Plants

Classification	Database name	Brief description
Phenomics + taxonomics	MPDB 1.0[a]	Database of Medicinal Plants in Bangladesh
	MPID[a]	Database of Botanical Imagery for Plants with Medical Values
	EbDB[a]	The International Ethnobotany Database is a compact commercial repository designed to store ethnobotanical information with robust support for multilingual capabilities.
	PlantCLEF 2019[a]	Identification of plant species through image analysis
Genomics	CmMDb[a]	Flexible repository for microsatellite markers in *Cucumis melo*, emphasizing their economic and medicinal significance, along with applications in broader horticultural crop research.
	HopBase[a]	Centralized platform for the genomics of Humulus.
	MMDBD[a]	Database for DNA Barcoding of Plant Materials of Medical values
	MGH[a]	Centralized repository for the genetic information of the medicinal crop maca *Lepidium meyenii*.
Transcriptomics	DsTRD[a]	Danshen transcriptomic data repository
	GarlicESTdb[a]	Web-based repository and exploration instrument designed for analyzing garlic EST (Expressed Sequence Tag) data.
	miRBase[b]	Supplying fundamental details about microRNA (miRNA) genes across various species, encompassing certain medicinal plants.
	MepmiRDB[b]	Database of MiRNA for Medicinal Plants
	ArrayExpress[a]	An EBI-hosted non-private repository housing microarray gene expression data.
	GEO[a]	The Gene Expression Omnibus (GEO) serves as a database for excellent data sequencing curated by the National Center for Biotechnology Information (NCBI).
	SRA[a]	The Sequence Read Archive (SRA) serves as a database for highly advanced sequencing of data curated by the NCBI.
	croFGD[b]	*Catharanthus roseus* genomics database for functional analysis
Proteomics	UniProt[b]	Furnishing the scientific community with an extensive, top-tier, and readily available repository containing protein sequence data and functional insights.
	Pfam[b]	Extensive array of protein families
	BioGRID[b]	A specialized database designed for the meticulous curation and archival preservation of interactions among proteins, genes, and chemicals across various model organism species and human subjects.
	STRING[b]	Interactions among proteins within a network of connections
	InAct[b]	Offering an openly accessible, open-source database system along with analytical tools designed for the examination of molecular interaction data.
	PAIR[b]	Anticipated protein interactome database for *Arabidopsis thaliana*.
	PRIN[b]	Anticipated interactome dataset for proteins in *Oryza sativa*
Metabolomics	CathaCyc[b]	A metabolic pathway database constructed using RNA-Seq data from *Catharanthus roseus*.
	KEGG Pathway Database[b]	Compilation of hand-drawn pathway diagrams illustrating the understanding of molecular interactions, reactions, and relational networks, particularly in the context of metabolic processes.
Integrated omics	MPGR[a]	The Medicinal Plants Genomics Resource offers comprehensive genomic, transcriptomic, and metabolomic data for a curated selection of 14 medicinal plant species.
	HMOD[a]	Database for herbal medicine plant omics in a professional context
Biotarget discovery and medicinal value exploration	AromaDb[b]	Database containing aroma molecules, phytochemical information, and therapeutic potentials of medicinal and aromatic plants.
	Phytochemica[b]	A platform designed for the investigation of phytochemicals found in medicinal plants.
	SerpentinaDB[b]	Repository containing information on bioactive compounds sourced from *Rauvolfia serpentina*
	TCM Database@ Taiwan[b]	Database housing a complimentary collection of 3D molecular structures of traditional Chinese medicine (TCM), accessible for virtual screening and molecular simulation purposes.
	IMPPAT[b]	A carefully compiled database containing information on 27,074 connections between plants and phytochemicals, as well as 11,514 associations between plants and therapeutic uses in the context of Indian medicinal flora.
	InDiaMed[b]	Extensive repository of Indian botanicals beneficial for managing diabetes.
	MAPS[b]	Database of medicinal plant functions, phytochemicals, and structures.
	NeMedPlant[b]	Collection of therapeutic uses and chemical components of medicinal plants found in the northeastern region of India.

TABLE 18.1 (*Continued*)
A Collection of Databases Offering Useful Omics Resources for Research on Medicinal Plants

Classification	Database name	Brief description
	NPASS[b]	Database for researching, discovering, and developing tools related to the activity of natural products and their respective species sources.
	CMAUP[b]	Database containing information on the combined molecular actions of beneficial plants, furnishing details about the active components of the plants.
	DIACAN[b]	Consolidated database encompassing medicinal plants with both antidiabetic and anticancer properties.
	SACPD[b]	Database of anti-cancer plants in Saudi Arabia.
	TarNet[b]	A platform and database created through manual curation, showcasing traditional medicinal plants along with their natural compounds and relevant bio-target information.
	KNApSAcK[b]	A database of 9584 triplet relationships, encompassing metabolites, biological activity, and target species, is available.
	TCMGeneDIT[b]	Creating a database that integrates information on Traditional Chinese Medicine (TCM), genes, and diseases through the application of text-mining techniques.
	HIT[b]	Thorough and meticulously organized repository of targets for herbal ingredients in a database.

NB: [a]Omics Repository Databases [b]Annotation Database

number of advanced features, including the ability to process data in batches and the use of a fuzzy matching algorithm, which improves the system's ability to identify bogus names and rectify spelling problems. There are still few bioinformatics resources specifically designed for taxonomic and phenomic research on medicinal plants. To overcome the bottleneck in systematic phenotyping and accurate classification of medicinal plants, comprehensive data storage, software packages for parsing high-throughput data, and the establishment of a standardized pipeline for recording phenotypic data are necessary (Kim et al., 2009).

18.5 A WORKFLOW ON INTEGRATED OMICS STUDIES: A DENDROBIUM CASE STUDY

There have been various advances in omics research on medicinal plants, including phenomics, taxonomics, genomics, transcriptomics, proteomics, and metabolomics. However, integrated analysis of numerous omics data types are required for many research activities, including plant authentication and mechanistic studies on plant metabolism. The plants of the *Dendrobium* genus, which are both attractive and therapeutic, have previously been the subject of numerous omics research (Li et al., 2018). In this case study, we suggest a workflow supported by bioinformatics for integrated omics investigations on medicinal plants, using *Dendrobium* as an example. In this sense, every species of *Dendrobium* should have it's own morphological and physiological characteristics documented and kept in public databases like MPID.

Certain phenotypic traits may be responsive to environmental stimuli. Furthermore, the sole raw medication used in TCM is the dried stems of certain *Dendrobium* species. Plant material authentication based on phenomics data will encounter a barrier in both the scenarios. Thankfully, a variety of DNA markers have been created and have shown to be an effective tool for *Dendrobium* species categorization and authentication. The DNA markers for a total of 39 *Dendrobium* species are stored in MMDBD, a DNA barcode database of medicinal materials (Lin et al., 2021).

Additionally, MMDBD may accept fresh data contributions. A recent study summarizes the many R packages that are currently available for integrated analysis of several omics data types. Here, the R package is offered a helpful pipeline for biological model prediction and omics data integration. MixOmics is superior to the other programs in many ways. First, it works with studies of single and multiple omics datasets. A single omics data collection might be supervisedly analyzed using sPLS-DA (sparse partial least square-discriminant analysis). N-integration analysis of various omics data (transcriptomics, proteomics, and metabolomics data) from specific biological samples is made possible by DIABLO. On the other hand, MINT allows P integration of omics datasets from several separate research projects (Meng et al., 2010). Second, mixOmics gives consumers access to unique sparse variations, allowing feature selection. Third, users may display their chosen features with informative graphical outputs using several essential functions provided by mixOmics, including plotIndiv, plotArrow, network, and circosPlot (Lux et al., 2011).

In the past few years, there has been a surge in the quantity of comparative genomic studies aimed at identifying the genes of medicinal plants connected to biosynthesis. For instance, by doing genomic comparisons, it was shown that genes encoding oligopeptide transporters (OPT) mediated the movement of several bioactive chemical substances in *Panax ginseng* and other flowering plants. Researchers can improve strain selection, adjust growth conditions, and create microbial systems for increased production and quality of medicinal compounds by studying the genetic

TABLE 18.2
A Non-exhaustive List of Bioinformatics Resources Helpful for Studying Medicinal Plants

Classification	Software name	Description
Phenomics and taxonomics	TNRS	The Taxonomic Name Resolution Service is an internet-based tool that automates and allows users to oversee the standardization of scientific names for plants.
	ImageJ	This is a Java-based framework for processing and analyzing images.
	HTPheno	A computational framework for analyzing images to study plant characteristics in a fast and efficient manner.
	PlantCV	A compilation of image analysis software for efficient plant phenotyping, derived from many source packages and methods.
Basic tools for genomics and/or transcriptomic studies	PLACE	Database of cis-acting regulatory DNA elements in plants.
	WebLogo	Creating a visual depiction of the pattern of sequence conservation in a multiple sequence alignment.
	PlantCARE	A collection of cis-acting regulatory elements for plants and a gateway to tools for conducting in silico analysis on promoter sequences.
	Vienna RNA web server	Supplying software bundles for the analysis of RNA secondary structures, such as the RNAfold tool.
	RNAshapes	An RNA secondary structure prediction program that is installed on the local system.
	Bowtie	A short read aligner that is exceptionally fast and utilizes memory efficiently.
	Bowtie 2	A high-speed and memory-efficient solution designed for aligning sequencing reads to reference sequences, with a particular emphasis on efficiently mapping long reads to extensive reference sequences.
	Trinity	To do the assembly of whole transcripts using RNA-seq data.
	SOAPdenovo-Trans	An additional instrument for creating a transcriptome assembly directly from RNA-seq data.
	TopHat+ Cufflinks and HISAT+ StringTie+ Ballgown	In the context of aligning RNA-seq reads using a reference genome, conducting assembly for full-length transcriptomes, and analyzing gene expression comparatively; and a recently enhanced software package for aligning RNA-seq reads to a reference genome, assembling full-length transcriptomes, and doing comparative gene expression analysis.
	PEA	A comprehensive R toolbox designed for the analysis of the epitranscriptome in plants.
	SAM	Significance analysis of microarray.
Non-coding RNA idenntification and design	miRPlant	A comprehensive instrument designed for the detection of plant miRNA using RNA sequencing data.
	miRDee-P	A computational method for studying the miRNA transcriptome in plants.
	PmiRDiscVali	A comprehensive system for the identification and confirmation of plant miRNAs.
	P-SAMS	A website for creation of artificial miRNA for plants and the design of synthetic trans-acting small interfering RNA on the web.
	NATpipe	An integrated approach for systematical identification of naturally occurring antisense transcripts and phase-distributed nat-siRNAs from freshly constructed transcriptomes.
	PLncPRO	A bioinformatics application designed to forecast long non-coding RNAs in plant organisms.
	PcircRNA_finder	A program for non-coding circular RNA estimation in crops.
Small RNA target prediction and validation	psRNATarget	A server for analyzing small RNA targets in plants.
	TAPIR	A server designed to predict targets of plant miRNAs, encompassing the identification of target mimics.
	CleaveLand	A methodology for degradome-seq data-based recognition of fragmented tiny RNA target.
	sPARTA	A methodology for degradome-seq data-based recognition of fragmented tiny RNA target.
	IIKmTA	MiRNA-target analyzer for both inter- and intra-kingdom interactions.
Functional analyses (gene annotations, networks, and pathways)	PlanTFcat	An online tool for classifying and examining plant transcription factors and transcriptional regulators.
	Cytoscape	A program environment designed for cohesive models of networks involving interactions among biomolecules.
	agriGO v2.0	An online platform designed for conducting GO term enrichment analysis.
	WGCNA	Package designed for weighted correlation network analysis in the R programming environment.
	INFORM	An R program allowing non-expert users to find, analyze and pick gene clusters that have significant statistical and biological importance.
	PathPred	An enzyme-catalyzed process of metabolism prediction service.
Integrated analyses	mixOmics	A comprehensive R package that provides diverse multivariate techniques for the combined analysis of various omics data types.
	iArray	Integrated Array Analyzer is a software solution designed to analyze microarray data across different platforms and species.
	BRB-ArrayTools	An integrated R tool for the viewing and statistical evaluation on microarray expression, gene copy number, methylation, and RNA-Seq data.

and epigenetic mechanisms driving metabolite synthesis. Precision in bioproduction processes is made easier by this multidisciplinary approach, which eventually helps the pharmaceutical sector (Meng et al., 2016).

18.6 DIFFICULTIES IN CELL-SPECIFIC AND MULTI-OMICS STRATEGIES FOR TRANSLATIONAL PLANT RESEARCH

There is no denying the success of using individual omics technologies in plant research, particularly in the past ten years (Mochida et al., 2019). However, each technique has unique data analysis and technological difficulties, which get even more severe at the single-cell level. For example, every plant cell has a polysaccharide cell wall that needs to be broken down in order to make the cell available for scRNA-seq. Protoplast isolation is the method by which this happens (Pinu et al., 2019). However, the thickness and composition of plant cell walls vary among species, tissues, developmental phases, and abiotic stressors (Rai et al., 2018). For this reason, protoplast isolation is a difficult task anytime a novel experimental system is investigated. Additionally, the cells may experience unintended transcriptional or metabolic alterations as a result of protoplast isolation (Zhang et al., 2023). This, along with the fact that plant cells are typically larger (100 μm) than mammalian cells (10–30 μm), as well as the fact that plant species, organs, and tissues all have different cell sizes, could make it difficult for the most widely used single-cell platforms to capture and profile some cell types, which would bias experiments (Ashraf et al., 2014). The development of optimized techniques for tissue and cellular separation that are applicable to many plant cell types is necessary for increasing the accessibility and adoption of scRNA-seq analysis among researchers. These challenges may be addressed by future efforts to use fix or isolate nuclei prior to profiling.

In genomics studies, the use of multi-omics methods is becoming standard practice. Prior research frequently conducted separate analyses of 'omics datasets' before combining the findings to offer a more comprehensive understanding of the biological system under study (Chao et al., 2024). True integrative analysis of these individual modalities, however, can more properly depict these processes. This is difficult on many levels and calls for careful evaluation of a number of aspects, including the lack of heterogeneity across omics technologies and the computing load associated with storing and analyzing different forms of data. The prevalence of missing values and noise across modalities are two further analysis issues brought on by the growing interest in single-cell studies and the enormous potential that single-cell multi-omics offers. The performance of specialized software and processes with plant single-cell datasets and their ability to handle the analysis of new modalities for plant single-cell research remain unknown, even though they are constantly being developed and improved to meet these difficulties. Utilizing high-performance mass spectrometers, the division specializes mostly in metabolomic studies (ElManawy et al., 2022).

18.7 OTHER OMICS TECHNIQUES

As sequencing technology advances, an increasing amount of data is produced, leading to the exploration of new and complex issues. For example, epigenomics characterizes modifications in gene regulation that function apart from variations in gene sequences. Stated differently, epigenomics is the study of modifications to histones and chromatin that do not affect the sequence of DNA. In Interactomics, the physical interactions between proteins and other biological components are investigated (Bhawna et al., 2015). These interactions might be between proteins, between various types of proteins, or between proteins and DNA. Gaining knowledge about the omics of secondary plant metabolites will help you to better understand the intricate networks controlling their production and regulation. This multidisciplinary strategy, which includes transcriptomics, proteomics, metabolomics, and genomes, improves our comprehension of the complex processes influencing the diversity of plant metabolites. Researchers may be able to enhance crop resilience, maximize plant features, and investigate uses in businesses and medicines by deciphering these molecular processes. To gain a thorough understanding of secondary metabolite regulation in plants, however, technological and bioinformatics developments must continue due to obstacles including data integration and functional annotation (da Sliva et al., 2016).

18.8 CONCLUSION

The integration of omics and high-throughput technologies represents a powerful strategy for enhancing the production of plant secondary metabolites. By combining genomics, transcriptomics, proteomics, metabolomics, and advanced screening techniques, researchers can gain deeper insights into the biosynthetic pathways and regulatory mechanisms underlying secondary metabolite production in plants. This comprehensive understanding enables targeted manipulation of key genes, enzymes, and regulatory elements to enhance metabolite yields. Moreover, high-throughput screening methods facilitate the rapid identification and selection of plant varieties or engineered lines with desired metabolic traits, accelerating the breeding and biotechnological approaches for trait improvement. Computational tools play a crucial role in integrating multi-omics data, deciphering complex metabolic networks, and predicting metabolic engineering targets, thereby guiding rational design strategies for enhancing secondary metabolite production. The enhanced production of plant secondary metabolites holds great promise for various applications, including the development of novel pharmaceuticals, nutraceuticals, and agrochemicals.

REFERENCES

Ashraf, M. A., Khatun, A., Sharmin, T., Mobin, F., Tanu, A. R., Morshed, T., & Nabi, A. N. (2014). MPDB 1.0: A medicinal plant database of Bangladesh. *Bioinformation*, *10*(6), 384–386. https://doi.org/10.6026/97320630010384

Bassel, G. W., Gaudinier, A., Brady, S. M., Hennig, L., Rhee, S. Y., & De Smet, I. (2012). Systems analysis of plant functional, transcriptional, physical interaction, and metabolic networks. *The Plant Cell*, *24*(10), 3859–3875. https://doi.org/10.1105/tpc.112.100776

Bhattacharyya, P., Kumaria, S., Kumar, S., & Tandon, P. (2013). Start Codon Targeted (SCoT) marker reveals genetic diversity of *Dendrobium nobile* Lindl., an endangered medicinal orchid species. *Gene*, *529*(1), 21–26. https://doi.org/10.1016/j.gene.2013.07.096

Bhawna, Chaduvula, P. K., Bonthala, V. S., Manjusha, V., Siddiq, E. A., Polumetla, A. K., & Prasad, G. M. (2015). CmMDb: A versatile database for *Cucumis melo* microsatellite markers and other horticulture crop research. *PloSone*, *10*(4), e0118630. https://doi.org/10.1371/journal.pone.0118630

Chao, H., Zhang, S., Hu, Y., Ni, Q., Xin, S., Zhao, L., & Chen, M. (2024). Integrating omics databases for enhanced crop breeding. *Journal of Integrative Bioinformatics*, *20*(4), 20230012. https://doi.org/10.1515/jib-2023-0012

da Silva, J. A. T., Jin, X., Dobránszki, J., Lu, J., Wang, H., Zotz, G., & Zeng, S. (2016). Advances in Dendrobium molecular research: Applications in genetic variation, identification and breeding. *Molecular Phylogenetics and Evolution*, *95*, 196–216. https://doi.org/10.1016/j.ympev.2015.10.012

ElManawy, A. I., Sun, D., Abdalla, A., Zhu, Y., & Cen, H. (2022). HSI-PP: A flexible open-source software for hyperspectral imaging-based plant phenotyping. *Computers and Electronics in Agriculture*, *200*, 107248. https://doi.org/10.1016/j.compag.2022.107248

Fahlgren, N., Feldman, M., Gehan, M. A., Wilson, M. S., Shyu, C., Bryant, D. W., & Baxter, I. (2015). A versatile phenotyping system and analytics platform reveals diverse temporal responses to water availability in Setaria. *Molecular Plant*, *8*(10), 1520–1535. http://dx.doi.org/10.1016/j.molp.2015.06.005

Fu, Y., Li, L., Hao, S., Guan, R., Fan, G., Shi, C., & Lee, S. M. Y. (2017). Draft genome sequence of the Tibetan medicinal herb *Rhodiola crenulata*. *Gigascience*, *6*(6), gix033. https://doi.org/10.1093/gigascience/gix033

Fukushima, A., Kusano, M., Redestig, H., Arita, M., & Saito, K. (2009). Integrated omics approaches in plant systems biology. *Current Opinion in Chemical Biology*, *13*(5–6), 532–538. https://doi.org/10.1016/j.cbpa.2009.09.022

Graw, S., Chappell, K., Washam, C. L., Gies, A., Bird, J., Robeson, M. S., & Byrum, S. D. (2021). Multi-omics data integration considerations and study design for biological systems and disease. *Molecular Omics*, *17*(2), 170–185. https://doi.org/10.1039/D0MO00041H

Griffiths-Jones, S., Grocock, R. J., Van Dongen, S., Bateman, A., & Enright, A. J. (2006). miRBase: microRNA sequences, targets and gene nomenclature. *Nucleic Acids Research*, *34*(Suppl_1), D140–D144. https://doi.org/10.1093/nar/gkj112

Hill, S. T., Sudarsanam, R., Henning, J., & Hendrix, D. (2017). HopBase: A unified resource for Humulus genomics. *Database*, bax009. https://doi.org/10.1093/database/bax009

Kim, D. W., Jung, T. S., Nam, S. H., Kwon, H. R., Kim, A., Chae, S. H., & Park, H. S. (2009). GarlicESTdb: An online database and mining tool for garlic EST sequences. *BMC Plant Biology*, *9*, 61. https://doi.org/10.1186/1471-2229-9-61

Kimotho, R. N., & Maina, S. (2024). Unraveling plant–microbe interactions: Can integrated omics approaches offer concrete answers? *Journal of Experimental Botany*, *75*(5), 1289–1313. https://doi.org/10.1093/jxb/erad448

Li, M., Frank, M. H., Coneva, V., Mio, W., Chitwood, D. H., & Topp, C. N. (2018). The persistent homology mathematical framework provides enhanced genotype-to-phenotype associations for plant morphology. *Plant Physiology*, *177*(4), 1382–1395. https://doi.org/10.1104/pp.18.00104

Lin, X. L., Mo, L., Bu, W. J., & Wang, X. H. (2021). The first comprehensive DNA barcode reference library of Chinese Tanytarsus (Diptera: Chironomidae) for environmental DNA metabarcoding. *Diversity and Distributions*, *27*(10), 1932–1941.

Lux, A., Martinka, M., Vaculík, M., & White, P. J. (2011). Root responses to cadmium in the rhizosphere: A review. *Journal of Experimental Botany*, *62*(1), 21–37.

Meng, Y., Gou, L., Chen, D., Mao, C., Jin, Y., Wu, P., & Chen, M. (2010). PmiRKB: A plant microRNA knowledge base. *Nucleic Acids Research*, *39*(Suppl_1), D181–D187. https://doi.org/10.1093/nar/gkq721

Meng, Y., Yu, D., Xue, J., Lu, J., Feng, S., Shen, C., & Wang, H. (2016). A transcriptome-wide, organ-specific regulatory map of *Dendrobium officinale*, an important traditional Chinese orchid herb. *Scientific Reports*, *6*(1), 18864. https://doi.org/10.1038/srep18864

Mochida, K., Koda, S., Inoue, K., Hirayama, T., Tanaka, S., Nishii, R., & Melgani, F. (2019). Computer vision-based phenotyping for improvement of plant productivity: A machine learning perspective. *GigaScience*, *8*(1), giy153. https://doi.org/10.1093/gigascience/giy153

Pinu, F. R., Beale, D. J., Paten, A. M., Kouremenos, K., Swarup, S., Schirra, H. J., & Wishart, D. (2019). Systems biology and multi-omics integration: Viewpoints from the metabolomics research community. *Metabolites*, *9*(4), 76. https://doi.org/10.3390/metabo9040076

Rai, A., Nakaya, T., Shimizu, Y., Rai, M., Nakamura, M., Suzuki, H., & Yamazaki, M. (2018). De novo transcriptome assembly and characterization of *Lithospermum officinale* to discover putative genes involved in specialized metabolites biosynthesis. *Planta Medica*, *84*(12/13), 920–934. https://doi.org/10.1055/a-0630-5925

Seeland, M., Rzanny, M., Boho, D., Wäldchen, J., & Mäder, P. (2019). Image-based classification of plant genus and family for trained and untrained plant species. *BMC Bioinformatics*, *20*, 4. https://doi.org/10.1186/s12859-018-2474-x

Shao, Y., Wei, J., Wu, F., Zhang, H., Yang, D., Liang, Z., & Jin, W. (2016). DsTRD: Danshen transcriptional resource database. *PloS One*, *11*(2), e0149747. https://doi.org/10.1371/journal.pone.0149747

Song, C., Liu, Y., Song, A., Dong, G., Zhao, H., Sun, W., & Chen, S. (2018). The Chrysanthemum nankingense genome provides insights into the evolution and diversification of chrysanthemum flowers and medicinal traits. *Molecular Plant*, *11*(12), 1482–1491. https://doi.org/10.1016/j.molp.2018.10.003

Strohl, W. R. (2000). The role of natural products in a modern drug discovery program. *Drug Discovery Today*, *5*(2), 39–41. https://doi.org/10.1016/s1359-6446(99)01443-9

Sun, W., Wang, B., Yang, J., Wang, W., Liu, A., Leng, L., & Chen, S. (2017). Weighted gene co-expression network analysis of the dioscin rich medicinal plant *Dioscorea nipponica*. *Frontiers in Plant Science*, *8*, 789. https://doi.org/10.3389/fpls.2017.00789

Tardieu, F., Cabrera-Bosquet, L., Pridmore, T., & Bennett, M. (2017). Plant phenomics, from sensors to knowledge. *Current*

Biology, *27*(15), R770–R783. http://dx.doi.org/10.1016/j.cub.2017.05.055

Urasaki, N., Takagi, H., Natsume, S., Uemura, A., Taniai, N., Miyagi, N., & Matsumura, H. (2017). Draft genome sequence of bitter gourd (*Momordica charantia*), a vegetable and medicinal plant in tropical and subtropical regions. *DNA Research*, *24*(1), 51–58. https://doi.org/10.1093/dnares/dsw047

Watson, E. R., Taherian Fard, A., & Mar, J. C. (2022). Computational methods for single-cell imaging and omics data integration. *Frontiers in Molecular Biosciences*, *8*, 768106. https://doi.org/10.3389/fmolb.2021.768106

Wong, T. H., But, G. W. C., Wu, H. Y., Tsang, S. S. K., Lau, D. T. W., & Shaw, P. C. (2018). Medicinal materials DNA barcode database (MMDBD) version 1.5 – one-stop solution for storage, BLAST, alignment and primer design. *Database*, *2018*, bay112. https://doi.org/10.1093/database/bay112

Xu, H., Song, J., Luo, H., Zhang, Y., Li, Q., Zhu, Y., & Chen, S. (2016). Analysis of the genome sequence of the medicinal plant *Salvia miltiorrhiza*. *Molecular Plant*, *9*(6), 949–952. http://dx.doi.org/10.1016/j.molp.2016.03.010

Zhang, D., Jiang, C., Huang, C., Wen, D., Lu, J., Chen, S., & Chen, S. (2019). The light-induced transcription factor FtMYB116 promotes accumulation of rutin in *Fagopyrum tataricum*. *Plant, Cell & Environment*, *42*(4), 1340–1351. https://doi.org/10.1111/pce.13470

Zhang, D., Li, W., Xia, E. H., Zhang, Q. J., Liu, Y., Zhang, Y., & Gao, L. Z. (2017). The medicinal herb Panax notoginseng genome provides insights into ginsenoside biosynthesis and genome evolution. *Molecular Plant*, *10*(6), 903–907. http://dx.doi.org/10.1016/j.molp.2017.02.011

Zhang, W., Zeng, Y., Jiao, M., Ye, C., Li, Y., Liu, C., & Wang, J. (2023). Integration of high-throughput omics technologies in medicinal plant research: The new era of natural drug discovery. *Frontiers in Plant Science*, *14*, 1073848. https://doi.org/10.3389/fpls.2023.1073848

19 Advances in Medicinal Plant Biotechnology and Future Directions

Devendra Singh, Stuti Srivastava, Sunil Kumar Verma, and Veena Singh

19.1 INTRODUCTION

There is a lengthy history – spanning centuries – of using medicinal plants as a kind of conventional medicine to treat various illnesses. With the advancement in biotechnology, there has been a significant focus on the use of medicinal plant biotechnology to improve the production and efficacy of medicinal compounds (Khan et al., 2009). This chapter aims to provide an overview of the recent advances in medicinal plant biotechnology and its potential applications in the pharmaceutical industry.

19.2 SCIENTIFIC ROOTS OF PLANT BIOTECHNOLOGY

Mary-Dell Chilton, PhD, has revolutionized plant science. The father of plant biotechnology is Haberlandt. He described the mesenchymal cell culture in *Lamium purpureum* and *Eichhornia crassipes*. Using fragments of mature potatoes, he observed this cell division almost without exception when the specimen contained a vascular strand (Vasil, 2008). One of the outstanding triumphs of plant biology in the 20th century has been the genetic transformation of somatic cells in plants, whereby whole plants can be stably transformed and transmit inserted genetic material to the next generation. It is now possible to play. This result grew out of his three independent research areas that began in the early 20th century: PTC (plant tissue culture), plant regeneration from individual somatic cells, and research into clubroot diseases (Alfermann & Petersen, 1995). These discoveries were made through a mix of basic scientific study and technology breakthroughs, and they resulted in the creation of genetically modified crop species that express features that conventional breeding cannot.

19.3 PLANT TISSUE CULTURE

In the 1930s, Philip White conducted experiments that allowed him to study the metabolism of completely undifferentiated tissues in which all cells were identical and, therefore, influenced each other's movements in similar ways. Developing the system, he defined PTC as a system where cells meet two major requirements: they are 'undifferentiated and yet capable of unlimited growth'. Previous attempts, beginning in the late 19th century, to cultivate plant parts isolated from living organisms did not meet these requirements and were subject to microbial contamination, inappropriate culture media, as well as poor selection of tissues to culture (Singh et al., 2023; Sussex, 2008).

The cultivation of root tips from corn, peas, and cotton was one of the earliest successful attempts. These root tips could be removed from the plants with minimal damage and grown in a sterile growth medium for a few weeks. However, their growth eventually stopped. On the other hand, the researcher achieved continuous cultivation of tomato root tips by placing them in a liquid medium containing mineral salts, 0.01% yeast extract along 2% sucrose. He cut the tips from the roots of the seedlings and subcultured them periodically. Even after 52 passages over 400 days, the cultured roots exhibit no reduction in growth rate. White determined that at this stage, the concentration of nutrients and regulators in the root tip had become diluted to approximately 10–40. As a result, he concluded that the nutrient medium provided unrestricted growth for the root tip. However, this finding contradicted the second part of his definition of tissue culture, as the roots were clearly not undifferentiated (Akiyoshi et al., 1984; Gutierrez-Valdes et al., 2020). Hybrid plants, on the other hand, developed tumor-like calluses and galls on their stems and leaves. To cultivate these tissues, aseptically collected samples from young stems were cultured in the same nutrient medium used for tomato roots. The growing mass from the initial explant was then divided and subcultured into 40 subcultures on a weekly basis (White, 1939).

The undifferentiated nature of the cells in the culture was still uncertain, but upon histological examination, it was found that there were mature parenchymal cells, areas of meristematic cells, and occasional isolated xylem cells (Haque et al., 2022). Despite this cellular heterogeneity, White concluded that these tissues were almost undifferentiated and had the potential to proliferate indefinitely, making them true tissue cultures. These studies demonstrated the ability of these cultures to proliferate undifferentiated cells and generate roots, shoots, and whole plants. However, due to the inability to determine the origin of the regenerated cells and their clones, these findings could have contributed further to the questions being investigated. Nonetheless, research on plant tissue and suspension culture continues to explore various avenues, such as the commercial production of secondary products and the cultivation of trees, crops, and horticultural plants, particularly orchid species (Krishna et al., 2022; Kumar, Rai et al., 2022) (Table 1).

DOI: 10.1201/9781003475491-19

TABLE 19.1
Bioactive Metabolites (Secondary Metabolites) Produced from Tissue Culture

S. No.	Plant	Common Name	Culture type	Active compounds
1.	*M. arvensis*	Corn mint	Shoot	Terpenoid
2.	*W. somnifera*	Ashwagandha	Shoot	Withaferin
3.	*S. fruticosa*	Glossy buckthorn	Callus	Rosmarinic acid
4.	*R. javanica*	Chinese sumac	Root	Gallotannins
5.	*S. marianum*	Milk thistle	Root	Flavonolignan

19.4 TRADITIONAL PLANT BIOTECHNOLOGY TECHNIQUES

19.4.1 Pedigree Breeding: Interacting with the Heritage of Parents

The bulk of self-pollinating species' released cultivars were created via the pedigree approach. The process of pedigree breeding includes crossing parents to generate segregating populations. These populations are then bred through generations of self-pollination as well as selection to produce a group of derived lines that combine the best traits of both parents. This technique, which is beneficial for breeding qualitative features like disease resistance or easily categorized traits like plant architecture, color, or form of plant components, is based on the complementarity of traits. Breeders find the pedigree approach appealing because it makes it possible to combine desirable traits from many elements into a single plant, resulting in improved variations (Putri et al., 2023). Since every crossing is monitored, it is feasible to ascertain each cultivar's lineage. The primary flaw in the pedigree method is that yield is only effectively assessed on inbred lines at the very end of the process when the seed is available for repeated experiments. But at this point, there is only space for an increase in yield potential once a lot of lines have progressed. Pedigree-based techniques never result in an annual rise of more than 1% and frequently lead to a moderate pace of yield growth (Chaturvedi et al., 2004).

19.5 IDEOTYPE BREEDING (IB)

The IB approach is a strategic method aimed at enhancing the effectiveness of the pedigree method in achieving improvements for quantitative traits, particularly yield. This approach is based on the belief that by modifying simpler traits that are positively associated with complex traits, one can enhance the overall quality of these traits (Ibáñez et al., 2009). The key advantage of this approach lies in the potential to achieve significant yield gains, even with a smaller breeding program, by adopting a more intelligent approach rather than relying solely on large-scale breeding programs. Moreover, this method holds great scientific appeal for breeders as it offers the opportunity to revolutionize their preferred crop. However, it is crucial to acknowledge that adverse genetic correlations can counterbalance the benefits derived from the traits that constitute the ideotype (Pattanaprateeb et al., 2005).

Finding new variations in crop ideotypes often involves exploring beyond the existing elite germplasm, which typically represents the current type OF plant. However, incorporating landraces as breeding parents often leads to a significant decrease in yield. Hence, it becomes essential to backcross with elite materials to regain a competitive offspring (Kumar, Verma et al., 2022; Sahoo & Rout, 2014). Multiple generations may face ongoing challenges when a wild relative is the donor of a new trait, including issues like flower sterility, sexual compatibility, seed dispersal, and grain quality (such as grain shattering).

To prevent negative consequences resulting from the direct introduction of foreign substances into high-quality breeding populations, these substances are typically initially utilized in a pre-breeding phase. During this phase, breeders aim to separate the advantageous traits from the undesirable ones. Once the desired trait is successfully incorporated into a superior genetic background, these vector lines can then be integrated into the elite breeding program (Welehaweria & Sbhatu, 2023).

19.6 POPULATION BREEDING (PB)

Population breeding is a breeding technique that aims to enhance the overall performance of a population by amplifying the occurrence of advantageous alleles that regulate desired traits. This method is commonly used in cross-pollinated plant species, where the improved population is directly utilized as a cultivar. Over time, more advanced schemes of PB have been developed, contributing to the development of quantitative genetics theory. In present-day PB, the focus is elevating the population's significance as a repository of elite lineages (Hailu et al., 2020). By improving the overall quality of the population while maintaining its genetic variation, superior individuals are produced, surpassing the performance of existing lines. These remarkable lines may be utilized as parental contributors for hybrids in cross-pollinating plant species or as cultivars in self-pollinating species.

PB, also referred to as recurrent selection breeding, is a continuous process involving multiple rounds of selection and recombination. However, this approach can pose challenges for self-pollinating species due to the need for an efficient crossing system. When population breeding is repeatedly applied to small populations, it can result in a reduction of genetic variation and a decrease in genetic gains over time. Therefore, it is crucial to maintain a large, effective population size to prevent genetic drift from overpowering other factors that contribute to genetic improvement (Cantabella et al., 2022). The population breeding scheme offers great flexibility, enabling customization to meet the specific needs and objectives of various species and breeding programs. (Rathore et al., 2015).

19.7 MODERN PLANT BIOTECHNOLOGY: FUTURE ASPECTS

19.7.1 Gene Mapping

When a group of plants that originated from a single cross is analyzed, the recombination frequency between the molecular markers can provide valuable information about their order and relative distance on the chromosomes. This information can then be used to construct a genetic map. If these plants, or their offspring, are also assessed for a quantitative trait, a statistical model can be developed (Kidasi et al., 2023). This model can help determine the extent to which certain markers explain the phenotypic variance, indicating their linkage to the genes responsible for the trait. This approach ultimately leads to the creation of QTL (quantitative trait loci) maps, which serve as an initial step in unraveling the genetic control of a quantitative trait.

A QTL map provides valuable information to plant breeders regarding the number of loci that control a specific trait, their relative significance, and their approximate location in the genome. When dealing with QTLs that have a significant impact, breeders can utilize marker-assisted selection by employing the closest markers available on the map (Tkachenko et al., 2023). Nevertheless, for the majority of cases, a more extensive fine-mapping approach is needed, which includes a larger population and increased marker density in the specific genomic region. This step is crucial for the development of effective selection tools. Ultimately, this strategy leads to the recognition of the gene and the precise genetic variation in its nucleotide sequence that is accountable for the observed variations in phenotype.

CSSL libraries consist of a group of lines that are derived from a superior variety. The entire genome of a wild species is included in the CSSL library and is arranged into bins. Each line in the library represents a single chromosome segment that has been replaced with the appropriate segment from the wild species of interest. The evaluation of the gene set can be carried out by comparing the phenotypic effects of the genes in each bin with the original elite line, which serves as a reference (Romadanova et al., 2022).

19.8 MARKER-ASSISTED SELECTION

Resistance gene pyramiding is a common application of marker-assisted selection. In cases where diseases caused by highly variable pathogens are involved, a single resistance gene might provide complete resistance for a limited number of plant generations before being overcome by pathogen strains. However, by accumulating multiple resistance genes in a variety, durable resistance can be achieved (Phour & Sindhu, 2022). The reason for this is that the pathogen would need to overcome all the genes at once, which significantly decreases the likelihood of resistance breakdown.

Classical breeding methods make resistance gene pyramiding nearly impossible, as the gene effect may only be observable in the absence of other resistance genes. In contrast, the combination of marker-tagged resistance genes can be achieved effortlessly without the requirement of phenotypic screening. This allows for the effective breeding of resistance to pathogens that are currently absent from a region, as phenotypic screening would not be possible in the absence of the pathogens (Jat et al., 2021).

Molecular markers of interest can be combined by breeders of a particular species to create optimized sets for simultaneous genotyping, which are commonly referred to as 'multiplexed' sets. These sets are particularly useful in applied breeding programs that focus on a wider genetic base. However, the successful implementation of such marker sets relies on extensive prior research. Each trait of interest requires several research steps, typically starting with the QTL, then by fine mapping, and finally positional cloning (Etesami & Maheshwari, 2018; Singh et al., 2023). Once the causative gene is recognized, depending upon the pair of contrasting alleles, breeders can explore diverse germplasm to discover novel alleles, thereby expanding the range of trait variation or enhancing the toolbox to address specific stresses (Singh, Dhiman et al., 2022; Yadav et al., 2023).

19.9 GENOMIC SELECTION (GS)

Genomic selection serves as a strategy to optimize the outcomes of the phenotypic assessment stage within the breeding cycle. By utilizing the data obtained from a thorough phenotypic evaluation, this knowledge can be passed down to future generations. The performance of an individual in different environments can be predicted by using their unique marker profile once models are developed for each environment. This greatly enhances the accuracy of evaluating individual performance and significantly reduces evaluation costs. Additionally, this approach shows promise in selecting parents for new breeding populations. However, as generations progress, the ability to predict the phenotype from the marker profile diminishes, necessitating the input of new phenotypic information to update the statistical model (Rai et al., 2022). This approach proves particularly valuable in scenarios where phenotypic evaluation is challenging, costly, or dependent on uncertain factors such as pest outbreaks or specific weather conditions.

GS has the potential to expedite the breeding process for yield potential, which is often regarded as the most challenging trait for marker-assisted selection. While in extremely competitive seed industries like maize and soybean, even a slight increase in the rate of genetic improvement can have a significant impact on market share. Consequently, major seed companies are making substantial investments in high-throughput analysis (DNA) as well as bioinformatics to flawlessly integrate genomic selection into their breeding plans (Das et al., 2022). By genotyping seeds before planting, expensive field plots can be reserved exclusively for preselected individuals. The complete implementation of genomic selection is expected to be more feasible for big private firms as compared to small-scale companies, public universities, and institutions. This is due to the substantial investment required to establish and maintain a recent genotyping pipeline, the need for quick turnaround times in the analysis of DNA and data processing, as well as the necessity of a highly-trained team.

19.10 ADVANCES IN PLANT TISSUE CULTURE

Plant tissue culture has revolutionized the production of medicinal compounds from plants. The development of efficient protocols for callus induction, organogenesis, and somatic embryogenesis has enabled the mass production of plant-derived pharmaceuticals. Furthermore, the use of bioreactors and suspension cultures has enhanced the scalability and productivity of medicinal plant cell cultures (Chaudhary et al., 2023).

Plant tissue culture (PTC) has evolved from its early stages to become a crucial tool in modern plant science. By cultivating plant cells, organs, or tissues in controlled and sterile conditions, researchers and industry professionals can unlock a wide range of applications that have the potential to transform agriculture, conservation, and biomanufacturing. Let us delve deeper into this dynamic field, exploring its recent advancements and future possibilities. One of the significant advancements in plant tissue culture is genome editing (Ma et al., 2023). Techniques such as CRISPR-Cas9 and other targeted editing methods empower researchers to modify plant genomes precisely, introducing desirable traits such as disease resistance, improved yields, and enhanced nutritional value. Bioreactors play a vital role in large-scale cultivation of plant cells. These sophisticated vessels provide optimized environments, enabling efficient production of valuable SM (secondary metabolites) like pharmaceuticals, cosmetics, and biofuels (Gandhi et al., 2023).

Cryopreservation is another powerful technique in plant tissue culture (PTC). It allows for the long-term storage of germplasm at ultra-low temperatures, ensuring the preservation of genetic diversity in valuable crop species and endangered plants for future generations (Yang et al., 2023). Microfluidics, on the other hand, offers precise control over tissue culture environments through miniaturized platforms. This facilitates rapid experimentation, protocol optimization, and single-cell analysis, leading to deeper insights into plant development. Artificial intelligence (AI) has also made its mark in PTC. Machine learning algorithms are utilized to analyze vast datasets from tissue culture experiments, enabling researchers to predict outcomes, optimize protocols, and accelerate discoveries (Babalola & Adedayo, 2023).

The benefits of advanced PTC extend to humanity in various ways. Firstly, it contributes to global food security by facilitating the rapid propagation of high-yielding, disease-resistant crops, ensuring a steady food supply for a growing population. Additionally, PTC allows for the enrichment of crops with essential vitamins and minerals, addressing malnutrition and promoting public health. Lastly, PTC plays a crucial role in preserving biodiversity by safeguarding the genetic diversity of valuable crop species and endangered plants (Jia et al., 2023).

In conclusion, plant tissue culture has emerged as a pivotal tool in modern plant science, with its applications ranging from genome editing to bioreactors, cryopreservation, microfluidics, and artificial intelligence. The advancements in PTC have the potential to address global challenges, such as food security, nutritional enhancement, and biodiversity conservation (Li et al., 2023).

19.11 METABOLIC ENGINEERING (ME)

The field of ME focuses on altering the metabolic pathways of plants to increase the synthesis of targeted compounds. Through genetic manipulation and gene editing techniques such as CRISPR/Cas9, researchers have been able to increase the yield of bioactive molecules such as alkaloids, flavonoids, and terpenoids in medicinal plants. This approach holds great promise for developing high-yielding plant varieties with improved therapeutic properties. Metabolic engineering is an ever-evolving discipline that seeks to manipulate and optimize the metabolic processes within cells to achieve desired outcomes (Ramatsitsi et al., 2023). These outcomes can range from increasing the production of specific compounds, such as biofuels or pharmaceuticals, to modifying an organism's metabolism for various purposes, such as enhancing its ability to withstand stress. Metabolic engineering is a branch of biotechnology that combines principles from engineering, biochemistry, and genetics to design and implement targeted modifications in an organism's metabolic pathways. It involves the identification and manipulation of metabolic pathways, which are intricate networks of chemical reactions that take place within cells. Various tools and techniques, including genetic engineering, enzyme manipulation, and computational modeling, are utilized in this field (Loyola-Vargas & Ochoa-Alejo, 2018).

Metabolic engineering holds great potential in addressing a wide range of challenges across diverse fields, including the development of sustainable biofuels and renewable chemicals to reduce reliance on fossil fuels, the engineering of microbes to produce valuable bioproducts, such as pharmaceuticals, nutraceuticals, and industrial enzymes, improving the efficiency and yield of crops by optimizing their metabolic pathways, designing microorganisms for environmental remediation to eliminate pollutants and toxins, and developing innovative therapeutic strategies by engineering cells to treat diseases like cancer and metabolic disorders (Dionisio et al., 2023). The working principles of metabolic engineering are shown in Figure 19.1.

19.12 PHYTOCHEMICAL ANALYSIS

Advances in analytical methods, e.g., HPLC, nuclear magnetic resonance, and mass spectrometry, have facilitated the identification and quantification of bioactive compounds present in medicinal plants. This has led to a better understanding of the chemical composition and pharmacological

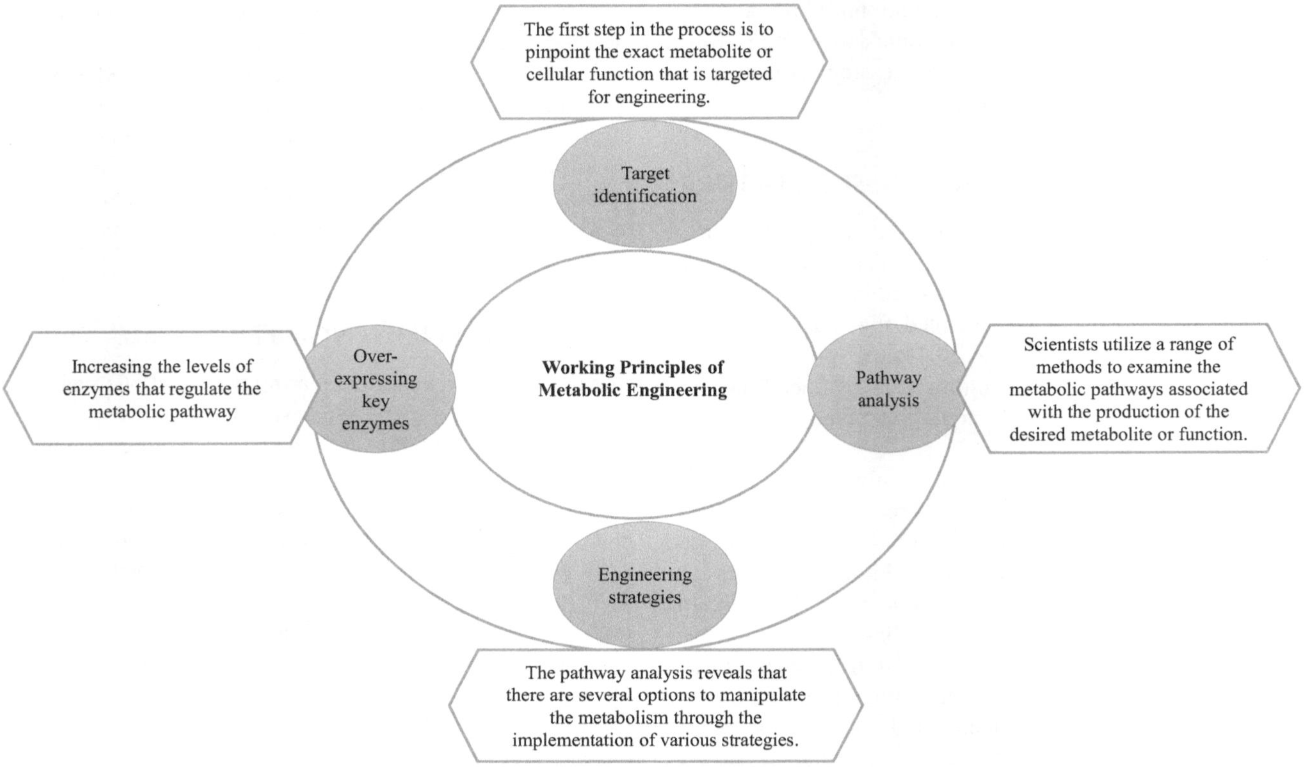

FIGURE 19.1 Showing the diagrammatic working principles of metabolic engineering

properties of medicinal plants, enabling researchers to optimize extraction methods and develop standardized herbal products (Wilson & Roberts, 2014; Kalia et al., 2020; Devendra & Vishnu, 2021; Singh et al., 2021; Singh & Agarwal, 2021, 2022; Singh, Singh et al., 2022).

Phytochemical analysis delves into the fascinating world of identifying, quantifying, and characterizing the diverse chemical compounds found in plants. These natural treasures, known as phytochemicals, play a crucial role in plant growth, defense, and interaction with the environment. Studying them unlocks a wealth of information about the plant's properties, potential uses, and relationship to the ecosystem (Mora-Vásquez et al., 2022; Das et al., 2022; Singh, Padey, Dhiman et al., 2022; Singh, Pandey, Singh et al., 2022; Singh & Agarwal, 2023).

19.13 DIFFERENT TYPES OF PHYTOCHEMICAL ANALYSIS

- *Qualitative analysis* focuses on identifying the types of phytochemicals present in a sample.
- *Quantitative analysis* determines the precise amount of each identified phytochemical.
- *Bioassays* assess the biological activity of the extracted compounds to explore their potential therapeutic or other applications.
- *Targeted analysis* focuses on specific phytochemicals of interest based on prior knowledge or research questions.
- *Untargeted analysis* explores the entire spectrum of phytochemicals present in a sample to discover novel compounds or understand the plant's overall chemical profile (Yamada et al., 2021).

Phytochemical analysis encompasses a wider range of techniques and methods employed to study the chemical composition of plant materials. It involves extraction, separation, purification, identification, and quantification of various phytochemicals like alkaloids, terpenes, phenolics, flavonoids, and many more. This analysis can be conducted on different plant parts, including leaves, roots, flowers, fruits, and seeds (Babar et al., 2017). The working principles, as well as the importance of phytochemical analysis, were explained in Figure 19.2.

19.14 CHALLENGES AND FUTURE DIRECTIONS

- Phytochemical analysis can be quite intricate due to the vast diversity of plant compounds and their complex interactions.
- Selecting the appropriate techniques and interpreting the data requires specialized knowledge and expertise.
- Advancements in technology, such as high-throughput screening and sophisticated analytical methods, are continuously pushing the boundaries of this field.

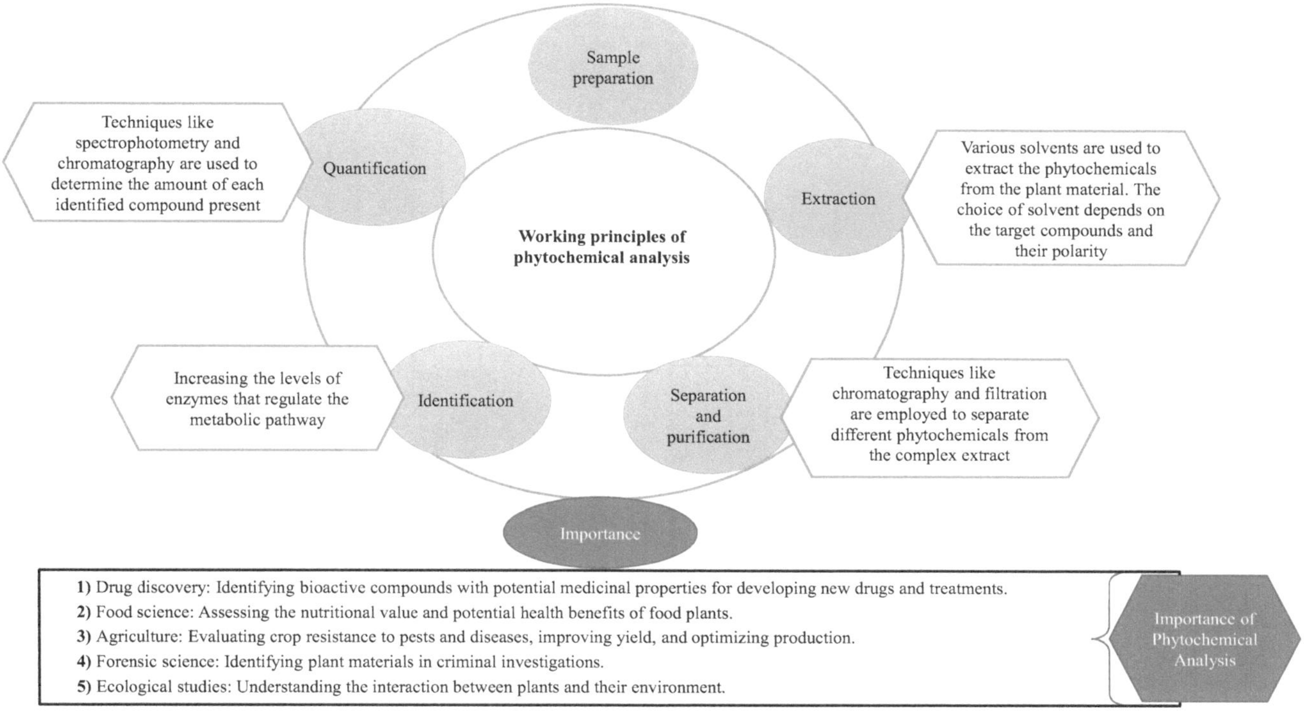

FIGURE 19.2 Showing the diagrammatic working principles and importance of phytochemical analysis.

Overall, phytochemical analysis offers a powerful tool for unlocking the secrets hidden within plants. By harnessing this knowledge, we can unlock numerous new possibilities in medicine, agriculture, food science, and environmental conservation (Ma et al., 2020).

19.15 BIOPHARMING

Biopharming involves using transgenic plants as biofactories for producing valuable pharmaceutical proteins and vaccines. Several studies have demonstrated the successful expression of therapeutic proteins in plants, including antibodies, antigens, and enzymes (Sharma et al., 2022). This approach offers a cost-effective and sustainable alternative to traditional protein production systems, with potential applications in personalized medicine and immunotherapy (Chauhan et al., 2023; Sharma & Sarkar, 2013).

Biopharming, also referred to as plant molecular farming, is an emerging field that involves the use of genetically modified plants as small-scale factories to produce pharmaceutical proteins, vaccines, and other bioproducts. This innovative approach offers a promising alternative to traditional methods of biopharmaceutical production, bringing potential cost-effectiveness, scalability, and unique advantages to the forefront (Lee, 2015; Singh & Verma, 2023; Singh & Singh, 2022). The biopharming process along with the benefits are shown in Figure 19.3.

Challenges and considerations faced by this modern technique of biopharming are regulatory hurdles; i.e., biopharming is a relatively new field, and regulatory frameworks are still being developed, which adds complexity to the process and field optimization; i.e., achieving sufficient protein yields is an ongoing challenge that requires continuous improvement and optimization.

19.16 BIOINFORMATICS AND SYSTEMS BIOLOGY

The integration of bioinformatics tools with systems biology approaches has facilitated comprehensive analysis of medicinal plant genomes, transcriptomes, and metabolomes. This has led to the identification of key regulatory genes and pathways involved in the biosynthesis of bioactive compounds (Verma et al., 2024; Vikram et al., 2014). Furthermore, computational modeling has enabled the prediction of metabolic fluxes and optimizing metabolic engineering strategies for enhancing pharmaceutical compound production. The power duo of bioinformatics and system biology is transforming the study of medicinal plants, offering an exciting new lens to unlock their therapeutic potential (Singh et al., 2020). Let's explore how these tools work together to unlock the secrets of these natural treasure troves:

Bioinformatics:

- *Data management and analysis*: This field excels at handling the massive datasets generated by modern techniques like genomics, transcriptomics, and metabolomics. It helps researchers make sense of the complex chemical language within plants.

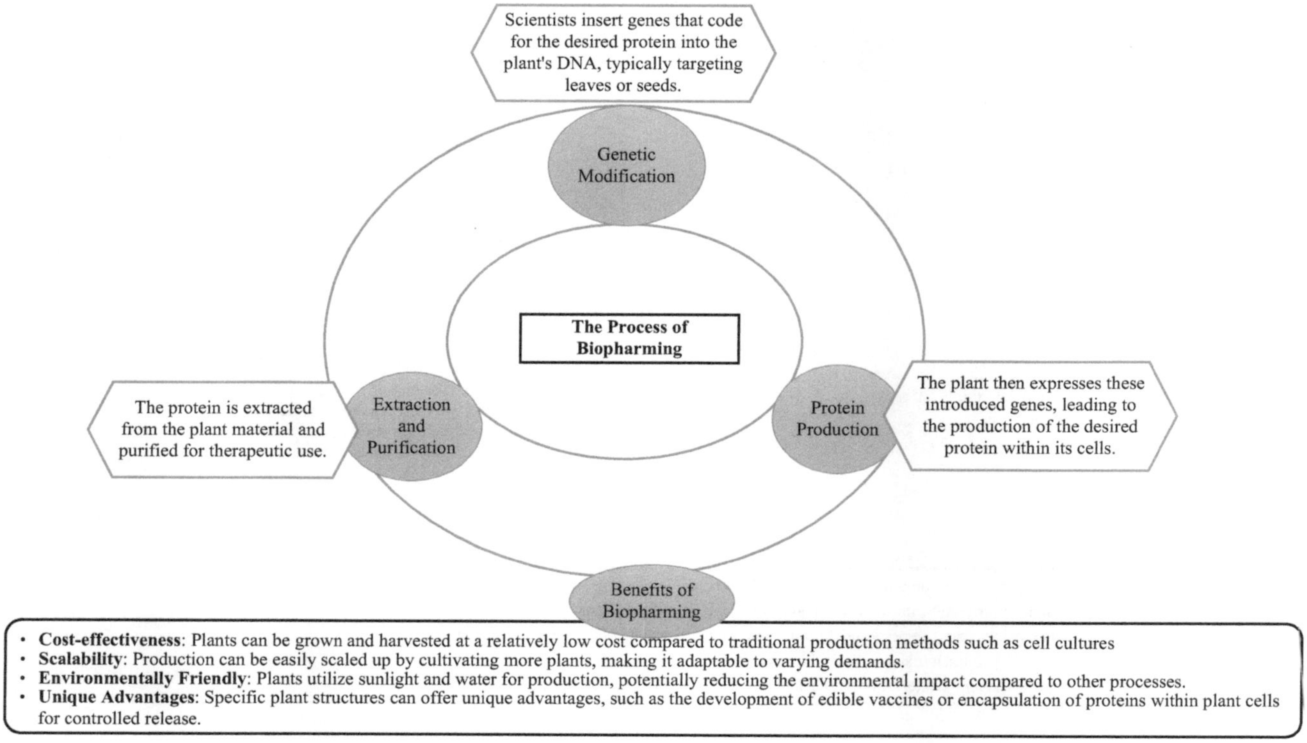

FIGURE 19.3 The diagrammatic representation shows the process and benefit of biopharming.

- *Gene discovery and characterization*: By analyzing DNA sequences, bioinformatics tools identify genes encoding potentially useful compounds. It unravels gene functions, paving the way for targeted manipulation.
- *Network analysis*: Bioinformatics helps build maps of interacting genes, proteins, and metabolites, revealing how different components work together within the plant's metabolic machinery. This understanding is crucial for optimizing the production of desired compounds (Courdavault et al., 2021).

Systems Biology:

- *Holistic approach*: This field goes beyond individual genes and takes a system-wide view, considering how all cellular components interact and influence each other. It paints a complete picture of plant metabolism.
- *Modeling and simulation*: System biology builds dynamic models that simulate plant response to environmental factors or genetic modifications. This allows researchers to virtually test scenarios and optimize strategies for improved production.
- *Predictive power*: By integrating various datasets and models, system biology helps predict the effects of genetic engineering or changing environmental conditions on the plant's production of medicinal compounds (Arshid Shabir, 2021).

Synergy for Success:

The combined power of these two fields offers several advantages:

- *Identification of novel therapeutic targets*: By integrating gene and pathway information, researchers can identify previously unknown genes or pathways with potential medicinal value.
- *Improved breeding and genetic engineering*: Understanding the complex interplay of genes and pathways allows targeted breeding or genetic engineering to enhance the production of specific compounds.
- *Personalized medicine*: By examining a person's genetic makeup, it might be possible to predict their response to precise medicinal plant treatments, paving the way for personalized medicine approaches (Desgagné-Penix, 2021).

Challenges and the Future:

Despite their potential, several challenges remain:

- *Data integration*: Seamlessly integrating vast amounts of diverse data from different sources can be complex.
- *Model complexity*: Building accurate and comprehensive models of plant metabolism remains a challenge.

- *Regulatory considerations*: Utilizing genetically modified plants for medicinal purposes requires careful consideration of regulatory frameworks (Dugé De Bernonville et al., 2020).

Despite these challenges, the future of bioinformatics, as well as system biology in medicinal plant research, is bright. Advancements in technology, data analysis methods, and model building are continuously pushing the boundaries. This research holds immense potential to discover new and more effective natural medicines, develop sustainable and efficient production methods for existing treatments, and personalized medicine based on individual needs (Hao et al., 2021).

19.17 CHALLENGES AND FUTURE PERSPECTIVES

Despite significant progress in medicinal plant biotechnology, several challenges remain to be addressed. These include regulatory hurdles associated with genetically modified organisms (GMOs), intellectual property rights issues related to bioprospecting, and ethical considerations regarding the conservation of biodiversity. Additionally, there is a need for further research on understanding plant-microbe interactions for improving secondary metabolite biosynthesis (Kowalczyk et al., 2020).

Medicinal plant biotechnology, harnessing the power of modern tools to unlock the therapeutic potential of plants, presents immense promise for the pharmaceutical industry. However, like any developing field, it faces unique challenges and exciting prospects. Challenges faced by medicinal plant biotechnology are:

- *Standardization and quality control*: Variability in plant growth conditions and extraction methods can lead to inconsistencies in active ingredient content and quality. Developing standardized protocols and robust quality control measures is crucial (Shih & Morgan, 2020).
- *Scalability and cost-effectiveness*: Scaling up production from controlled lab settings to large-scale cultivation can be challenging and expensive. Optimizing production methods and exploring cost-efficient alternatives are essential.
- *Regulatory hurdles*: Regulations for using genetically modified plants or novel natural products in medications are still evolving. Navigating these hurdles requires collaboration between researchers, industry, and regulatory bodies.
- *Sustainability and ethical considerations*: Sustainable harvesting practices and fair trade approaches are critical to ensure the long-term availability and equitable distribution of plant-based resources.
- *Understanding complex mechanisms*: Fully comprehending the intricate pathways responsible for therapeutic compounds' biosynthesis and interactions within plants remains challenging. Advanced research in plant biology and metabolomics is needed (Kumar et al., 2020).

Future Aspects and Applications:

- *Biomarker discovery and personalized medicine*: Identifying specific plant-derived biomarkers could enable personalized medicine approaches, tailoring treatments based on individual needs (Negi et al., 2021; Tyagi et al., 2024).
- *Precision breeding and metabolic engineering*: Advanced tools can optimize plant varieties for improved yield and quality of desired compounds, leading to more efficient production (Pandey et al., 2021).
- *Development of novel drug delivery systems*: Plant-based carriers or encapsulation methods can offer targeted delivery, improved bioavailability, and potentially reduced side effects.
- *Harnessing the microbiome*: Exploring the complex interactions between plants and their associated microbial communities could unlock new therapeutic potential.
- *Integration with AI and computational tools*: Leveraging artificial intelligence and advanced computational tools can accelerate the discovery, design, and optimization of plant-based medicines (Li et al., 2020).

Overall, medicinal plant biotechnology holds immense potential to revolutionize the pharmaceutical industry by providing sustainable and renewable sources of therapeutic compounds, secondly by offering new treatment options for various diseases with fewer side effects, and lastly by contributing to personalized medicine approaches for more effective treatment strategies (Pan et al., 2021; Rather et al., 2020). Addressing the challenges and actively exploring the future aspects will be crucial to unlocking the full potential of this field and shaping a future where plant-based medicines play a significant role in global healthcare.

19.18 CONCLUSIONS

The antimicrobial activity of medicinal plants offers a promising solution to combat the growing threat of antimicrobial resistance. As a result, it is critical to discover and extract novel bioactive compounds from these plants, which are mostly untapped. The wide variety of these compounds has shown promising therapeutic capabilities as antimicrobials and antimicrobial resistance modifiers. However, the utilization of these new bioactive compounds presents challenges. It is crucial to highlight the importance of conducting extensive in vitro and in vivo tests to ensure the selection of effective and non-toxic antimicrobial compounds derived from plants. Furthermore, determining the potential synergistic or antagonistic effects of chemicals found in medicinal plant extracts is quite difficult.

As biotechnology continues to advance, it becomes evident that we will have the ability to delve deeper into medicinal plants' chemical composition and develop or design more sophisticated techniques for extracting, fractionating, and identifying bioactive compounds. Standardizing extraction methods and in vitro testing would be beneficial to make the search more systematic and facilitate the interpretation of results. Additionally, the use of reference models in studying plant extract mixtures has yet to be explored, and future studies will shed light on their applicability in this approach. It is critical to prioritize research into the mechanisms of action, as well as interactions with antibiotics, medicinal herbs, and other substances. Additionally, understanding the extracts' pharmacokinetic and pharmacodynamic profiles is critical. In summary, advances in medicinal plant biotechnology offer tremendous potential for developing novel therapeutics with improved efficacy and safety profiles. Continued research efforts aimed at harnessing the power of biotechnology will undoubtedly lead to breakthroughs in drug discovery and development from natural sources.

REFERENCES

Akiyoshi, D. E., Klee, H., Amasino, R. M., Nester, E. W., & Gordon, M. P. (1984). T-DNA of *Agrobacterium tumefaciens* encodes an enzyme of cytokinin biosynthesis. *Proceedings of the National Academy of Sciences of the United States of America*, *81*(19), 5994–5998.

Alfermann, A. W., & Petersen, M. (1995). Natural product formation by plant cell biotechnology: Results and perspectives. *Plant Cell, Tissue and Organ Culture*, *43*(2), 199–205. https://doi.org/10.1007/BF00052176

Arshid Shabir, P. (2021). Chapter 9 – CRISPR/Cas9-mediated genome editing in medicinal and aromatic plants: Developments and applications. In T. Aftab & K. R. Hakeem (Eds.), *Medicinal and aromatic plants* (pp. 209–221). Academic Press. https://doi.org/10.1016/B978-0-12-819590-1.00009-4

Babalola, O. O., & Adedayo, A. A. (2023). Endosphere microbial communities and plant nutrient acquisition toward sustainable agriculture. *Emerging Topics in Life Sciences*, *7*(2), 207–217. https://doi.org/10.1042/ETLS20230069

Babar, M. M., Zaidi, N.-S. S., Pothineni, V. R., Ali, Z., Faisal, S., Hakeem, K. R., & Gul, A. (2017). Application of bioinformatics and system biology in medicinal plant studies. In K. R. Hakeem, A. Malik, F. Vardar-Sukan, & M. Ozturk (Eds.), *Plant bioinformatics* (pp. 375–393). Springer International Publishing. https://doi.org/10.1007/978-3-319-67156-7_15

Cantabella, D., Dolcet-Sanjuan, R., & Teixidó, N. (2022). Using plant growth-promoting microorganisms (PGPMs) to improve plant development under in vitro culture conditions. *Planta*, *255*(6), 117. https://doi.org/10.1007/s00425-022-03897-0

Chaturvedi, R., Razdan, M. K., & Bhojwani, S. S. (2004). In vitro clonal propagation of an adult tree of neem (*Azadirachta indica* A. Juss.) by forced axillary branching. *Plant Science*, *166*(2), 501–506. https://doi.org/10.1016/j.plantsci.2003.10.021

Chaudhary, S., Sindhu, S. S., Dhanker, R., & Kumari, A. (2023). Microbes-mediated sulphur cycling in soil: Impact on soil fertility, crop production and environmental sustainability. *Microbiological Research*, *271*, 127340. https://doi.org/10.1016/j.micres.2023.127340

Chauhan, D., Singh, D., Pandey, H., Khan, Z., Srivastava, R., Dhiman, V. K., & Dhiman, V. K. (2023). Impact of transcription factors in plant abiotic stress: A recent advancement for crop improvement. In *Plant transcription factors* (pp. 271–286). Elsevier. https://doi.org/10.1016/B978-0-323-90613-5.00005-4

Courdavault, V., O'Connor, S. E., Jensen, M. K., & Papon, N. (2021). Metabolic engineering for plant natural products biosynthesis: New procedures, concrete achievements and remaining limits. *Natural Product Reports*, *38*(12), 2145–2153. https://doi.org/10.1039/D0NP00092B

Das, R., Deb, P., Pandey, H., Shyam, P., & Singh, D. (2022). Botanical synthesis of silver nanoparticles (Agnps) and its antifungal effect against *Alternaria porri* causing purple blotch of onion: An in vitro and natural epiphytic study. *Journal of Agriculture and Food Research*, *10*, 100390.

Desgagné-Penix, I. (2021). Biosynthesis of alkaloids in Amaryllidaceae plants: A review. *Phytochemistry Reviews*, *20*(2), 409–431. https://doi.org/10.1007/s11101-020-09678-5

Devendra, S., & Vishnu, A. (2021). A review on the COVID-19, its history, diagnostic approaches, role of herbs and current world scenario. *Biomedicine*, *41*(2), 328–332.

Dionisio, G., Thakur, J. K., & Kaul, T. (2023). Editorial: Biopharming, volume II: New plant breeding technologies for metabolic engineering or recombinant proteins production. *Frontiers in Plant Science*, *14*, 1231422. https://doi.org/10.3389/fpls.2023.1231422

Dugé De Bernonville, T., Maury, S., Delaunay, A., Daviaud, C., Chaparro, C., Tost, J., O'Connor, S. E., & Courdavault, V. (2020). Developmental methylome of the medicinal plant *Catharanthus roseus* Unravels the tissue-specific control of the monoterpene indole alkaloid pathway by DNA methylation. *International Journal of Molecular Sciences*, *21*(17), 6028. https://doi.org/10.3390/ijms21176028

Etesami, H., & Maheshwari, D. K. (2018). Use of plant growth promoting rhizobacteria (PGPRs) with multiple plant growth promoting traits in stress agriculture: Action mechanisms and future prospects. *Ecotoxicology and Environmental Safety*, *156*, 225–246. https://doi.org/10.1016/j.ecoenv.2018.03.013

Gandhi, S., Saha, M. R., & Dey, P. (2023). Improved antioxidant activities of spice require enrichment of distinct yet closely-related metabolic pathways. *Heliyon*, *9*(11), e21392. https://doi.org/10.1016/j.heliyon.2023.e21392

Gutierrez-Valdes, N., Häkkinen, S. T., Lemasson, C., Guillet, M., Oksman-Caldentey, K.-M., Ritala, A., & Cardon, F. (2020). Hairy root cultures – a versatile tool with multiple applications. *Frontiers in Plant Science*, *11*. https://www.frontiersin.org/journals/plant-science/articles/10.3389/fpls.2020.00033

Hailu, A., Sbhatu, D. B., & Abraha, H. B. (2020). In vitro micropropagation of industrially and medicinally useful plant *Aloe trichosantha* Berger using offshoot cuttings. *The Scientific World Journal*, *2020*, 3947162. https://doi.org/10.1155/2020/3947162

Hao, X., Xie, C., Ruan, Q., Zhang, X., Wu, C., Han, B., Qian, J., Zhou, W., Nützmann, H.-W., & Kai, G. (2021). The transcription factor OpWRKY2 positively regulates the biosynthesis of the anticancer drug camptothecin in *Ophiorrhiza pumila*. *Horticulture Research*, *8*(1), 7. https://doi.org/10.1038/s41438-020-00437-3

Haque, M. I., Singh, P. K., Ghuge, S., Kumar, A., Chandra Rai, A., Kumar, A., & Modi, A. (2022). Chapter 1 – A general introduction to and background of plant tissue culture: Past, current, and future aspects. In A. Chandra Rai, A. Kumar, A. Modi, & M. Singh (Eds.), *Advances in plant tissue Culture* (pp. 1–30). Academic Press. https://doi.org/10.1016/B978-0-323-90795-8.00019-9

Ibáñez, C. M., Camus, P. A., & Rocha, F. J. (2009). Diversity and distribution of cephalopod species off the coast of Chile. *Marine Biology Research*, *5*(4), 374–384. https://doi.org/10.1080/17451000802534873

Jat, S. L., Suby, S. B., Parihar, C. M., Gambhir, G., Kumar, N., & Rakshit, S. (2021). Microbiome for sustainable agriculture: A review with special reference to the corn production system. *Archives of Microbiology*, *203*(6), 2771–2793. https://doi.org/10.1007/s00203-021-02320-8

Jia, J., Chen, L., Yu, W., Cai, S., Su, S., Xiao, X., Tang, X., Jiang, X., Chen, D., Fang, Y., Wang, J., Luo, X., Li, J., Huang, Y., & Su, J. (2023). The novel nematicide chiricanine A suppresses *Bursaphelenchus xylophilus* pathogenicity in *Pinus massoniana* by inhibiting Aspergillus and its secondary metabolite, sterigmatocystin. *Frontiers in Plant Science*, *14*, 1257744. https://doi.org/10.3389/fpls.2023.1257744

Kalia, M., Singh, D., Sharma, D., Narvi, S. S., & Agarwal, V. (2020). *Senna alexandriana* mill as a potential inhibitor for quorum sensing-controlled virulence factors and biofilm formation in *Pseudomonas aeruginosa* PAO1. *Pharmacognosy Magazine*, *16*(72), 797–802.

Khan, M. Y., Aliabbas, S., Kumar, V., & Rajkumar, S. (2009, January). Recent advances in medicinal plant biotechnology. *IJBT*, *8*(1). http://nopr.niscpr.res.in/handle/123456789/2963

Kidasi, P. C., Kilalo, D. C., & Mwang'ombe, A. W. (2023). Effect of sterilants and plant growth regulators in regenerating commonly used cassava cultivars at the Kenyan coast. *Heliyon*, *9*(6), e17263. https://doi.org/10.1016/j.heliyon.2023.e17263

Kowalczyk, T., Wieczfinska, J., Skała, E., Śliwiński, T., & Sitarek, P. (2020). Transgenesis as a tool for the efficient production of selected secondary metabolites from plant in vitro cultures. *Plants*, *9*(2), 132. https://doi.org/10.3390/plants9020132

Krishna, R., Khandagale, K., Benke, A. P., Soumia, P. S., Manjunathagowda, D. C., Ansari, W. A., Mokat, D. N., Gawande, S. J., Ade, A. B., & Singh, M. (2022). Chapter 11 – Embryo rescue: A potential tool for improvement of economically important crops. In A. Chandra Rai, A. Kumar, A. Modi, & M. Singh (Eds.), *Advances in plant tissue culture* (pp. 259–282). Academic Press. https://doi.org/10.1016/B978-0-323-90795-8.00009-6

Kumar, P., Rai, S., Verma, S. K., Prakash, P. S., & Chitara, D. (2022). Classification, mode of action and uses of various immunomodulators. In R. K. Kesharwani, R. K. Keservani, & A. K. Sharma (Eds.), *Immunomodulators and human health* (pp. 3–38). Springer Nature Singapore. https://doi.org/10.1007/978-981-16-6379-6_1

Kumar, P., Verma, S. K., Rai, S., Prakash, P. S., & Chitara, D. (2022). Role of homocysteine metabolism in cardiovascular diseases. In G. P. Dubey, K. Misra, R. K. Kesharwani, & R. P. Ojha (Eds.), *Homocysteine metabolism in health and disease* (pp. 257–276). Springer Nature Singapore. https://doi.org/10.1007/978-981-16-6867-8_14

Kumar, S. R., Rai, A., Bomzan, D. P., Kumar, K., Hemmerlin, A., Dwivedi, V., Godbole, R. C., Barvkar, V., Shanker, K., Shilpashree, H. B., Bhattacharya, A., Smitha, A. R., Hegde, N., & Nagegowda, D. A. (2020). A plastid-localized bona fide geranylgeranyl diphosphate synthase plays a necessary role in monoterpene indole alkaloid biosynthesis in *Catharanthus roseus*. *The Plant Journal*, *103*(1), 248–265. https://doi.org/10.1111/tpj.14725

Lee, S. (2015). Systems biology – a pivotal research methodology for understanding the mechanisms of traditional medicine. *Journal of Pharmacopuncture*, *18*(3), 11–18. https://doi.org/10.3831/KPI.2015.18.020

Li, L., Hu, Z., Tan, G., Fan, J., Chen, Y., Xiao, Y., Wu, S., Zhi, Q., Liu, T., Yin, H., & Tang, Q. (2023). Enhancing plant growth in biofertilizer-amended soil through nitrogen-transforming microbial communities. *Frontiers in Plant Science*, *14*, 1259853. https://doi.org/10.3389/fpls.2023.1259853

Li, Q., Zhu, T., Zhang, R., Bu, Q., Yin, J., Zhang, L., & Chen, W. (2020). Molecular cloning and functional analysis of hyoscyamine 6β-hydroxylase (H6H) in the poisonous and medicinal plant *Datura innoxia* mill. *Plant Physiology and Biochemistry*, *153*, 11–19. https://doi.org/10.1016/j.plaphy.2020.04.021

Loyola-Vargas, V. M., & Ochoa-Alejo, N. (2018). An introduction to plant tissue culture: Advances and perspectives. In V. M. Loyola-Vargas & N. Ochoa-Alejo (Eds.), *Plant cell culture protocols* (Vol. 1815, pp. 3–13). Springer New York. https://doi.org/10.1007/978-1-4939-8594-4_1

Ma, X., Meng, Y., Wang, P., Tang, Z., Wang, H., & Xie, T. (2020). Bioinformatics-assisted, integrated omics studies on medicinal plants. *Briefings in Bioinformatics*, *21*(6), 1857–1874. https://doi.org/10.1093/bib/bbz132

Ma, Y., Li, Y., Yang, S., Li, Y., & Zhu, Z. (2023). Biocontrol potential of *Trichoderma asperellum* Strain 576 against *Exserohilum turcicum* in *Zea mays*. *Journal of Fungi* (Basel, Switzerland), *9*(9), 936. https://doi.org/10.3390/jof9090936

Mora-Vásquez, S., Wells-Abascal, G. G., Espinosa-Leal, C., Cardineau, G. A., & García-Lara, S. (2022). Application of metabolic engineering to enhance the content of alkaloids in medicinal plants. *Metabolic Engineering Communications*, *14*, e00194. https://doi.org/10.1016/j.mec.2022.e00194

Negi, S., Singh, D., Pandey, H., & Singhal, R. K. (2021). Biofortification for high Fe and Zn in various Poaceae crops by using different molecular breeding and biotechnological approaches. *Plant Physiology Reports*, 1–11.

Pan, C., Sretenovic, S., & Qi, Y. (2021). CRISPR/dCas-mediated transcriptional and epigenetic regulation in plants. *Current Opinion in Plant Biology*, *60*, 101980. https://doi.org/10.1016/j.pbi.2020.101980

Pandey, H., Singh, D., Das, R., & Pandey, D. (2021). Precision farming and its application. In A. Choudhury, A. Biswas, T. P. Singh, & S. K. Ghosh (Eds.), *Smart agriculture automation using advanced technologies. Transactions on computer systems and networks*. Springer. https://doi.org/10.1007/978-981-16-6124-2_2

Pattanaprateeb, P., Ruangrungsi, N., & Cordell, G. A. (2005). Cytotoxic constituents from *Cratoxylum arborescens*. *Planta Medica*, *71*(2), 181–183. https://doi.org/10.1055/s-2005-837788

Phour, M., & Sindhu, S. S. (2022). Mitigating abiotic stress: Microbiome engineering for improving agricultural production and environmental sustainability. *Planta*, *256*(5), 85. https://doi.org/10.1007/s00425-022-03997-x

Putri, A. I., Kartikawati, N. K., Nirsatmanto, A., Sunarti, S., Haryjanto, L., Herawan, T., Santosa, P. B., Wahyuningtyas, R. S., Lestari, F., & Rimbawanto, A. (2023). Tissue culture of gerunggang (*Cratoxylum arborescens* (Vahl) Blume): Multipurpose native species of Indonesian peatland. *Forest Science and Technology*, *19*(3), 171–178. https://doi.org/10.1080/21580103.2023.2220010

Rai, S., Omar, A. F., Rehan, M., Al-Turki, A., Sagar, A., Ilyas, N., Sayyed, R. Z., & Hasanuzzaman, M. (2022). Crop microbiome: Their role and advances in molecular and omic techniques for the sustenance of agriculture. *Planta*, *257*(2), 27. https://doi.org/10.1007/s00425-022-04052-5

Ramatsitsi, M. N., Khosa, M. C., Mashamaite, C. V., & Ramachela, K. (2023). In vitro assessment of eight selected indigenous fungal isolates tolerance to various abiotic stresses and their effects on seed germination. *Current Microbiology*, *80*(12), 386. https://doi.org/10.1007/s00284-023-03507-6

Rather, G. A., Sharma, A., Misra, P., Kumar, A., Kaul, V., & Lattoo, S. K. (2020). Molecular characterization and overexpression analyses of secologanin synthase to understand the regulation of camptothecin biosynthesis in *Nothapodytes nimmoniana* (Graham.) Mabb. *Protoplasma*, *257*(2), 391–405. https://doi.org/10.1007/s00709-019-01440-9

Rathore, J. S., Phulwaria, M., Rai, M. K., Shekhawat, S., & Shekhawat, N. S. (2015). Use of liquid culture medium and ex vitro rooting for micropropagation of *Acacia nilotica* (L.) Del. Ssp. Cupressiformis. *Indian Journal of Plant Physiology*, *20*(2), 172–176. https://doi.org/10.1007/s40502-015-0149-4

Romadanova, N. V., Tolegen, A. B., Kushnarenko, S. V., Zholdybayeva, E. V., & Bettoni, J. C. (2022). Effect of plant preservative mixtureTM on endophytic bacteria eradication from in vitro-grown apple shoots. *Plants* (Basel, Switzerland), *11*(19), 2624. https://doi.org/10.3390/plants11192624

Sahoo, S., & Rout, G. R. (2014). Plant regeneration from leaf explants of *Aloe barbadensis* Mill. And genetic fidelity assessment through DNA markers. *Physiology and Molecular Biology of Plants: An International Journal of Functional Plant Biology*, *20*(2), 235–240. https://doi.org/10.1007/s12298-014-0226-6

Sharma, S. K., Singh, D., Pandey, H., Jatav, R. B., Singh, V., & Pandey, D. (2022). An overview of roles of enzymatic and nonenzymatic antioxidants in plant. In *Antioxidant defense in plants: Molecular basis of regulation* (pp. 1–13). Springer. https://doi.org/10.1007/978-981-16-7981-0_1

Sharma, V., & Sarkar, I. N. (2013). Bioinformatics opportunities for identification and study of medicinal plants. *Briefings in Bioinformatics*, *14*(2), 238–250. https://doi.org/10.1093/bib/bbs021

Shih, M.-L., & Morgan, J. A. (2020). Metabolic flux analysis of secondary metabolism in plants. *Metabolic Engineering Communications*, *10*, e00123. https://doi.org/10.1016/j.mec.2020.e00123

Singh, D., & Agarwal, V. (2021). Screening of antimicrobial, anti-quorum sensing activity and cytotoxicity of origanum oil against Gram-positive and Gram-negative bacteria. *Biomedicine*, *41*(3), 599–603.

Singh, D., & Agarwal, V. (2022). Herbal antibacterial remedy against upper respiratory infection causing bacteria and in vivo safety analysis. *Vegetos*, *35*(1), 264–268.

Singh, D., & Agarwal, V. (2023). An in vivo acute toxicity and anti-shigellosis effect of designed formulation on rat. *Journal of Ayurveda and Integrative Medicine*, *14*(1), 100536. https://doi.org/10.1016/j.jaim.2021.100536

Singh, D., & Verma, S. K. (2023). Impacts of particulate matter pollution on plants. In T. Aftab (Ed.), *New frontiers in plant-environment interactions* (pp. 483–500). Springer Nature Switzerland. https://doi.org/10.1007/978-3-031-43729-8_17

Singh, D., Dhiman, V. K., Pandey, H., Dhiman, V. K., & Pandey, D. (2022). Crosstalk between salicylic acid and auxins, cytokinins and gibberellins under biotic stress. In *Auxins, cytokinins and gibberellins signaling in plants* (pp. 249–262). Springer International Publishing.

Singh, D., Pandey, H., & Singh, V. (2022). Natural products that target cancer stem cells. In A. Pandurangan, S. Anandasadagopan, & F. Alhumaydhi (Eds.), *Handbook of research on natural products and their bioactive compounds as cancer therapeutics* (pp. 169–186). IGI Global. https://doi.org/10.4018/978-1-7998-9258-8.ch008

Singh, D., Pandey, H., Dhiman, V. K., Bang, N. T. H., Dhiman, V. K., Singh, V., & Pandey, D. (2022). Secondary metabolite engineering for plant immunity against various pathogens. In *Metabolic engineering in plants* (pp. 123–143). Springer Nature Singapore.

Singh, D., Sharma, D., & Agarwal, V. (2021). Screening of antimicrobial, anti-biofilm activity, and cytotoxicity analysis of a designed polyherbal formulation against shigellosis. *Journal of Ayurveda and Integrative Medicine*, *12*(4), 601–606. https://doi.org/10.1016/j.jaim.2021.06.007

Singh, D., Singh, V., Bhushan Mishra, S., Sharma, D., & Agarwal, V. (2022). Evaluation of anti-biofilm, anti-quorum, anti-dysenteric potential of designed polyherbal formulation: In vitro and in vivo study. *Journal of Applied Biomedicine*, *20*(1), 7–14. https://doi.org/10.32725/jab.2022.005

Singh, D., Verma, S. K., Singh, V., & Shyam, P. (2023). Green functional nanomaterials: Synthesis and application. In J. A. Malik & M. J. Sadiq Mohamed (Eds.), *Modern nanotechnology* (pp. 45–65). Springer Nature Switzerland. https://doi.org/10.1007/978-3-031-31104-8_3

Singh, S. K., Patra, B., Paul, P., Liu, Y., Pattanaik, S., & Yuan, L. (2020). Revisiting the ORCA gene cluster that regulates terpenoid indole alkaloid biosynthesis in *Catharanthus roseus*. *Plant Science*, *293*, 110408. https://doi.org/10.1016/j.plantsci.2020.110408

Singh, V., & Singh, D. (2022). Biocompatible herbal polymeric nano-formulation of[6]-gingerol: Development, optimisation, andcharacterization. https://www.envirobiotechjournals.com/article_abstract.php?aid=12941&iid=362&jid=3

Sussex, I. M. (2008). The scientific roots of modern plant biotechnology. *The Plant Cell*, *20*(5), 1189–1198. https://doi.org/10.1105/tpc.108.058735

Tkachenko, O. V., Evseeva, N. V., Kargapolova, K. Y., Denisova, A. Y., Pozdnyakova, N. N., Kulikov, A. A., & Burygin, G. L. (2023). Rhizobacteria increase the adaptation potential of potato microclones under aeroponic conditions. *Microorganisms*, *11*(7), 1866. https://doi.org/10.3390/microorganisms11071866

Tyagi, S., Dhiman, V. K., Dhiman, V. K., Pandey, H., Singh, D., Sharma, A., Sharma, P., Kumar, R., Lee, K. J., & Saharan, B. S. (2024). Plant defense strategies and biomarkers against heavy metal-induced stress: A comprehensive review. *ACS Agricultural Science & Technology*. *4*(2), 129–143. https://doi.org/10.1021/acsagscitech.3c00271

Vasil, I. K. (2008). A history of plant biotechnology: From the cell theory of Schleiden and Schwann to biotech crops. *Plant Cell Reports*, *27*(9), 1423–1440. https://doi.org/10.1007/s00299-008-0571-4

Verma, S. K., Pandey, M., Sharma, A., & Singh, D. (2024). Exploring ayurveda: Principles and their application in modern medicine. *Bulletin of the National Research Centre*, *48*(1), 77. https://doi.org/10.1186/s42269-024-01231-0

Vikram, P., Chiruvella, K. K., Ripain, I. H. A., & Arifullah, M. (2014). A recent review on phytochemical constituents and medicinal properties of kesum (*Polygonum minus* Huds.). *Asian Pacific Journal of Tropical Biomedicine*, *4*(6), 430–435. https://doi.org/10.12980/APJTB.4.2014C1255

Welehaweria, M., & Sbhatu, D. B. (2023). In vitro micropropagation of Aloe elegans Tod. Using offshoot cuttings. *BMC Research Notes*, *16*(1), 215. https://doi.org/10.1186/s13104-023-06490-0

White, P. R. (1939). Potentially unlimited growth of excised plant callus in an artificial nutrient. *American Journal of Botany*, *26*(2), 59–64. https://doi.org/10.2307/2436709

Wilson, S. A., & Roberts, S. C. (2014). Metabolic engineering approaches for production of biochemicals in food and medicinal plants. *Current Opinion in Biotechnology*, *26*, 174–182. https://doi.org/10.1016/j.copbio.2014.01.006

Yadav, G., Prajapati, P., Singh, D., Hora, S., Singh, S., Vishwakarma, K., & Ahmad, I. Z. (2023). Emerging trends in plant metabolomics and hormonomics to study abiotic stress tolerance associated with rhizospheric probiotics. In *Plant hormones in crop improvement* (pp. 283–306). Academic Press. https://doi.org/10.1016/B978-0-323-91886-2.00012-4

Yamada, Y., Urui, M., Oki, H., Inoue, K., Matsui, H., Ikeda, Y., Nakagawa, A., Sato, F., Minami, H., & Shitan, N. (2021). Transport engineering for improving the production and secretion of valuable alkaloids in *Escherichia coli*. *Metabolic Engineering Communications*, *13*, e00184. https://doi.org/10.1016/j.mec.2021.e00184

Yang, Z., Qiao, Y., Konakalla, N. C., Strøbech, E., Harris, P., Peschel, G., Agler-Rosenbaum, M., Weber, T., Andreasson, E., & Ding, L. (2023). Streptomyces alleviate abiotic stress in plant by producing pteridic acids. *Nature Communications*, *14*(1), 7398. https://doi.org/10.1038/s41467-023-43177-3

20 Integrated Molecular Strategies for Enhanced Production of Anti-Cancer Compounds from Medicinal Plants

Ankan Das, Ritashree Pramanik, Dyuti Ballav Chongder, Piyush Bania, and Sandip Debnath

20.1 INTRODUCTION

The battle against cancer remains an enduring global health challenge, exacting a significant toll on human lives each year (Mao et al., 2021). Despite strides made in conventional therapies, the imperative for novel, potent, and readily available anti-cancer agents persists. Medicinal plants, repositories of bioactive molecules with profound therapeutic potential, stand as a beacon of hope in this relentless pursuit. Traditionally revered for their healing properties, medicinal plants have been integral to humanity's pharmacopeia for centuries, with historical records dating back to ancient Egyptian and Chinese civilizations, circa 1500 BC.

The 20th century witnessed a significant leap forward with the discovery of vinca alkaloids from *Catharanthus roseus* (rosy periwinkle), leading to the development of vincristine and vinblastine, two cornerstones of modern cancer chemotherapy. This breakthrough spurred further research, culminating in the isolation of numerous other potent anti-cancer compounds from diverse plant sources, including paclitaxel (Taxol) from *Taxus brevifolia* (Pacific yew) and camptothecin from *Camptotheca acuminata*.

However, traditional methods of harvesting whole plants for drug extraction faced significant challenges. Reliance on wild populations led to unsustainable resource depletion and potential environmental damage. Additionally, the low abundance of the target compounds within the plant often resulted in insufficient yields, hindering large-scale production and clinical development.

The latter half of the 20th century witnessed a paradigm shift with the advent of powerful molecular tools. The advent of recombinant DNA technology in the 1970s enabled the introduction of genes encoding enzymes involved in the biosynthesis pathways of anti-cancer compounds into other organisms (heterologous expression). This paved the way for the production of these compounds in more readily available and controllable systems, such as microbial hosts.

The turn of the 21st century ushered in an era of even greater precision with the development of CRISPR/Cas9 (clustered regularly interspaced short palindromic repeats/Cas9) in 2012. This revolutionary gene-editing tool allows for highly targeted manipulation of plant genomes, enabling researchers to overexpress genes (responsible for the biosynthesis of anti-cancer compounds, thereby significantly increasing their yield), knock out or downregulate genes (compete for resources or interfere with the desired pathway, further optimizing production), and introduce specific mutations (to create novel variants equipped with enhanced anti-cancer compounds accumulation). These capabilities offer immense potential for revolutionizing the production of plant-based anti-cancer drugs.

Metabolic engineering, complementary to gene editing, focuses on optimizing the metabolic pathways within the plant itself. By understanding and manipulating the intricate network of biochemical reactions involved in the biosynthesis of the desired compound, researchers can significantly enhance its production. A prime example of metabolic engineering success is the 10-fold increase in vinblastine production achieved in Catharanthus roseus in 2009 (Zhou et al., 2009). This breakthrough not only improved the availability of this life-saving drug but also highlighted the immense potential of metabolic engineering in this field.

The field of omics encompasses a suite of powerful technologies that provide a comprehensive understanding of the biological machinery within plants. These technologies include – genomics (identification of potential genes involved in the biosynthesis of anti-cancer compounds), transcriptomics (analyzes gene expression patterns, revealing which genes are actively producing proteins required for the desired pathway), proteomics (identifies and quantifies the proteins involved in the anti-cancer compound biosynthesis pathway), metabolomics (profiles the entire spectrum of metabolites within a plant, allowing researchers to track the flow of precursors and identify potential bottlenecks in the pathway). By integrating these omics technologies, researchers can gain a holistic understanding of the plant's metabolic network. This knowledge facilitates the development of targeted strategies for manipulating the biosynthesis pathway to maximize the production of the desired anti-cancer compound. These advancements, coupled with continued research in gene editing, metabolic engineering, and omics technologies, hold immense promise for the future of plant-based anti-cancer drug discovery.

DOI: 10.1201/9781003475491-20

20.2 ANTI-CANCER COMPOUNDS IN MEDICINAL PLANTS

TABLE 20.1
Various Plant-Based Anti-cancer Compounds and Their Sources

Compounds	Source	Genes producing the primary enzymes	Action mechanism	Works against	References
Curcumin	Present in the rhizomes of *Curcuma longa*	CURS1, CURS2 & CURS3	It induces apoptosis, suppresses NF-κB activation, and stops the cell cycle.	Breast, cervical, prostate, and uterine cancers	(Das & Roychoudhury, 2023)
Vincristine	Present in leaves and roots of *Catharanthus roseus*	G10H TDC STR	It interacts with tubulin and suppresses mitosis.	Glioma, neuroblastoma, sarcoma, lymphoma, and nephroblastoma	(PubChem, 2004a)
Vinblastine	Present in the leaves and roots of *Catharanthus roseus*	TDC STR SGD VINS	It suppresses mitosis.	Breast, choriocarcinoma, Hodgkin's lymphoma, lung cancer, lymphosarcoma, and lymphocytic leukemia	(PubChem, 2004b)
Resveratrol	Present in the roots of *Veratrum grandiflorum*	STS	It induces apoptosis and suppresses NF-κB activation.	Cervical carcinoma, colorectal, skin, pancreatic, breast, prostate, and lung cancers	(Gambini et al., 2015)
Cyclopamine	Present in *Veratrum californicum*		It stops hedgehog signaling.	Lung, pancreatic, ovarian, prostate, and basal cell carcinoma	(NCBI, 2024d)
Glucoraphanin	Present in *Brassica* sp.	Myrosinase MAM CYP450 GSMT	It causes apoptosis, cell cycle arrest, and angiogenesis inhibition.	Gastric, colon, breast, prostate, and cervical cancers	(NCBI, 2024e)
Paclitaxel	Present in the barks of *Texas brevifolia*	TS DBTNBT DBTNAT	It stops mitosis.	Ovarian, breast, gastric, neck, bladder, and lung cancers	(NCBI, 2024b)
Ingenol mebutate	Present in the latex of *Euphorbia peplus*	CYP450s diterpene synthetase	It encourages apoptosis, necrosis, and antibody-dependent cellular cytotoxicity (ADCC).	Squamous cell carcinoma	(Aditya & Gupta, 2013)
Betulinic acid	Present in the bark of *Betula pubescens*	OSCs CYP450s UGTs	It modulates NF-κB activity and promotes apoptosis.	Melanoma, breast, pancreatic, ovarian, hepatoma cellular, and bladder cancers	(NCBI, 2024c)
Camptothecin	Present in the bark and stem of *Camptotheca acuminata*	TRPS GGPPS CPT STR	It attaches to topoisomerase I-DNA complex and triggers apoptosis.	Ovarian, small lung, cervical and colorectal cancers	(NCBI, 2024a)

20.3 UNDERSTANDING THE BIOSYNTHESIS OF ANTI-CANCER COMPOUNDS

Biosynthetic pathways play a major role in the quantitative and qualitative production of secondary metabolites. Having a detailed knowledge and comprehending the biosynthetic mechanism of production of anti-cancer compounds is essential for better medication development and various other research initiatives. Researchers can study the checkpoints and the limiting factors and alter them accordingly to have better bioavailability of the metabolites. The careful examination of these biosynthetic pathways will also help in reducing the side effects of the metabolites, resulting in higher efficacy of the therapies. An understanding of the enzymes involved will also aid in providing targets for genetic engineering and targeted drug development. In addition to that, there are possibilities for the discovery of new and unidentified intermediaries and compounds that

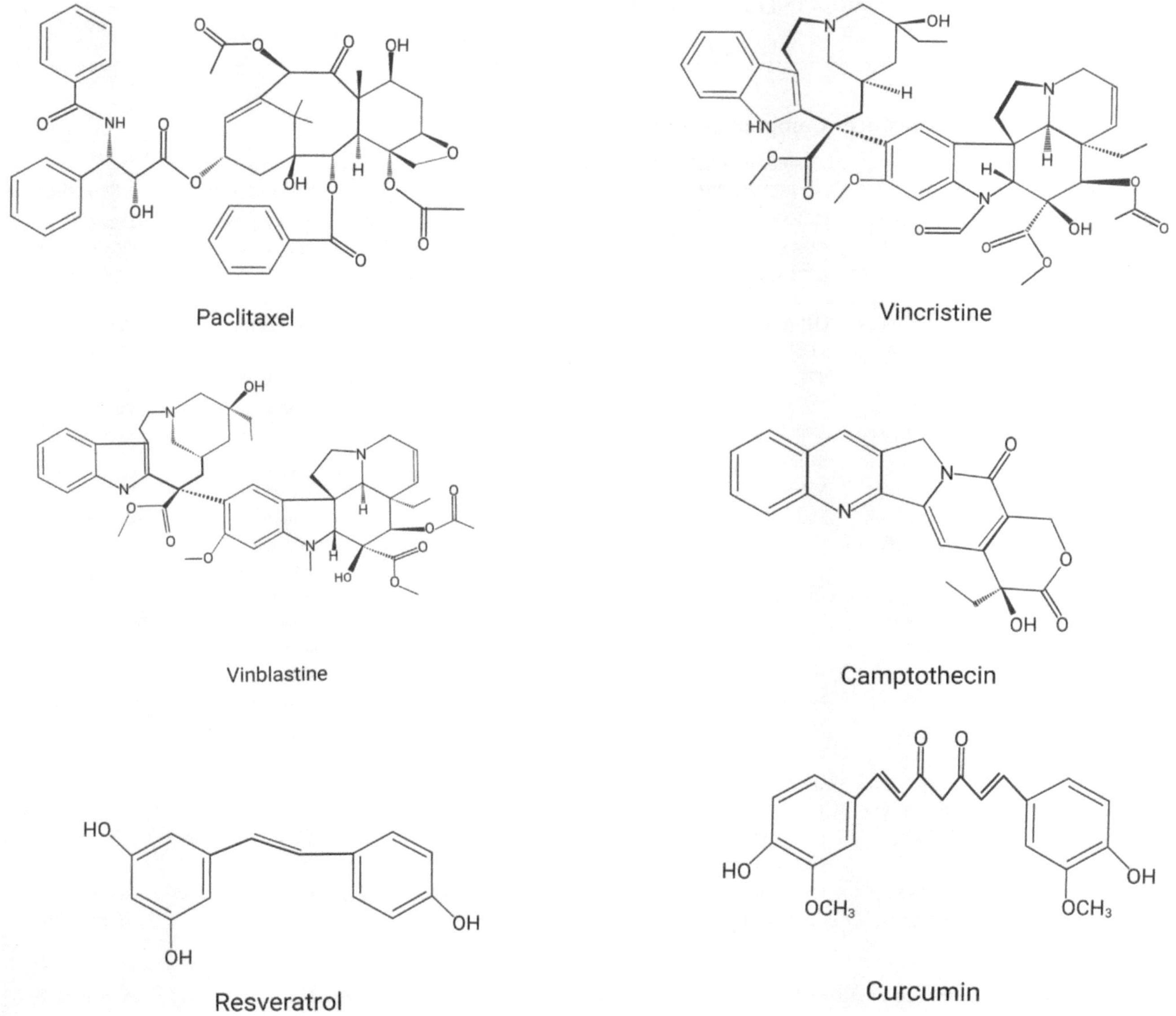

FIGURE 20.1 Some of the important plant-derived anti-cancer compounds with their chemical structures.

may have an anti-cancer effect. Therefore, a thorough grasp of biosynthetic pathways not only enables researchers to fully use the potential of chemicals generated from plants but also boosts ongoing innovation and advancement in the steadfast discovery of efficient cancer therapies.

20.3.1 Polyphenols

The synthesis of phenolic compounds in plants involves two primary metabolic pathways: the shikimate pathway and the aceto-malonate pathway. The shikimate pathway, named after shikimic acid, utilizes substances from primary metabolism to produce aromatic amino acids and phenolic compounds. Shikimic acid acts as a precursor for various hydroxybenzoic acids and plays a crucial role in the synthesis of aromatic amino acids like phenylalanine and tyrosine. Enzymes such as phenylalanine ammonia-lyase (PAL) catalyze the conversion of phenylalanine to trans-cinnamic acid, which initiates the phenylpropanoid pathway. Tyrosine can also be converted to p-coumaric acid, although with less efficiency. Hydroxylation of trans-cinnamic acid leads to the formation of other hydroxycinnamic acids, which further transform into different phenolic compounds.

The aceto-malonate pathway complements the phenylpropanoid pathway and is primarily involved in the biosynthesis of flavonoids. Through condensation reactions catalyzed by chalcone synthase (CHS), p-coumaroyl-CoA, and malonyl-CoA combine to form chalcone, a precursor for flavonoids. Subsequent transformations result in various classes of flavonoids, including flavanones and anthocyanins. Most monomeric phenolic compounds do not accumulate as end products but instead contribute to the

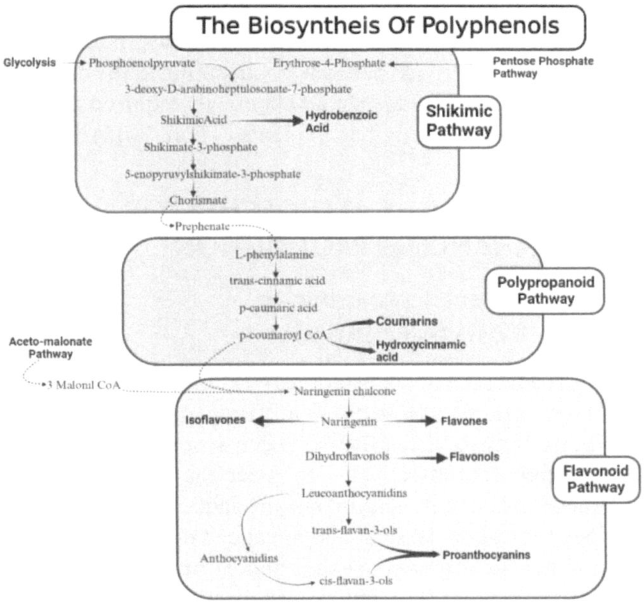

FIGURE 20.2 A schematic diagram showing the biosynthesis of polyphenols.

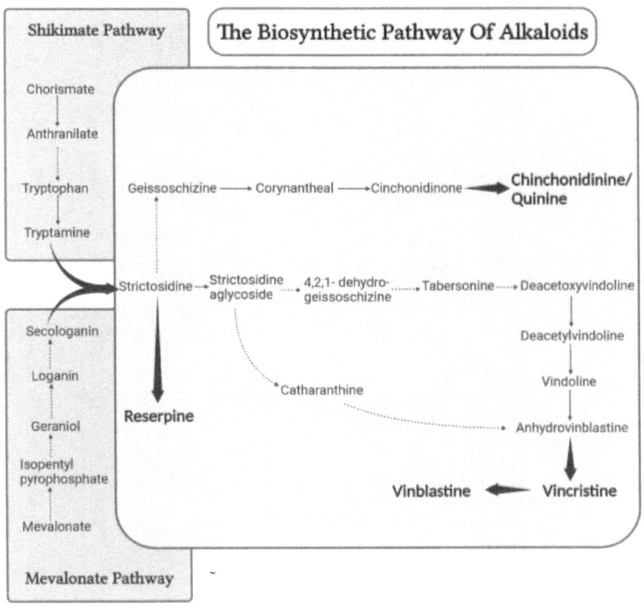

FIGURE 20.3 A schematic diagram showing the biosynthesis of alkaloids.

formation of complex oligomeric and polymeric structures, such as proanthocyanidins.

Advancements in molecular biology and functional genomics have facilitated the study of genes responsible for phenolic metabolism and their regulatory mechanisms. Transcription factors (TFs) play a crucial role in regulating gene expression in the phenylpropanoid and flavonoid pathways. The MYB TF family, in particular, is highly significant in this context, with its members acting as either activators or repressors of enzyme genes. TFs often form complexes, such as the MBW complex, which regulates flavonoid biosynthesis by activating structural genes.

Exploring the gene regulation mechanisms in phenolic metabolism shows great potential in improving the yield of medically important secondary metabolites. Analyzing transcriptomic data and studying TFs offer key information on how phenolic compounds are produced, leading to the enhancement of plant phenolic profiles and the creation of innovative therapeutic drugs. (Zagoskina et al., 2023)

20.3.2 Alkaloids

The biosynthesis of monoindole alkaloids begins with the creation of the key intermediate strictosidine via a condensation process involving tryptamine and secologanin, which is mediated by strictosidine synthase. *Catharanthus roseus* has at least seven isoforms of STR, which are most likely the consequence of posttranslational changes such as glycosylation. Strictosidine is subsequently deglucosylated by strictosidine-b-D-glucosidase (SGD), resulting in cathenamine, which acts as a junction point for three distinct routes. Cathenamine can be reduced to ajmalicine, tetrahydroalstonine, and subsequently alstonine, or reversibly changed into 4,21-dehydrogeissoschizine, resulting in the synthesis of catharanthine and tabersonine/vindoline. The catharanthine biosynthesis pathway's enzymes and genes have yet to be discovered. However, the vindoline biosynthesis route is well-established, with six enzyme-catalyzed stages. Different enzyme isoforms, such as T16H1 and T16H2, contribute to vindoline production and have unique expression patterns in diverse plant tissues.

Alternative paths in roots from tabersonine lead to the synthesis of other kinds of alkaloids, such as lochnericine and ho¨rhammericine. Epidermis cells in leaves, stems, and flower buds have STR mRNA, but idioblast and laticifer cells embedded in palisade tissue contain only D4H and DAT mRNA transcripts. T16H2 and SGD mRNA are mostly detected in the epidermis. Subcellular localization studies indicate the locations of enzymes involved in the vindoline process. T16H binds to the endoplasmic reticulum (ER), whereas 16OMT produces homodimers in the cytoplasm. NMT is found in the thylakoid membrane of chloroplasts, whereas D4H and DAT are monomers acting in the nucleocytoplasmic compartment (Pan et al., 2016).

20.3.3 Terpenoids

The mevalonic acid (MVA) pathway and the 2-C-methyl-D-erythritol 4-phosphate (MEP) pathway are two important pathways for terpenoid production in plants. The MVA route has six stages, beginning with the condensation of acetyl-CoA molecules to generate acetoacetyl-CoA, which is mediated by acetoacetyl-CoA thiolase (AACT). Subsequent aldol condensation events generate S-3-hydroxy-3-methylglutaryl-CoA (HMG-CoA), which is then transformed to mevalonate (MVA) by HMG-CoA reductase

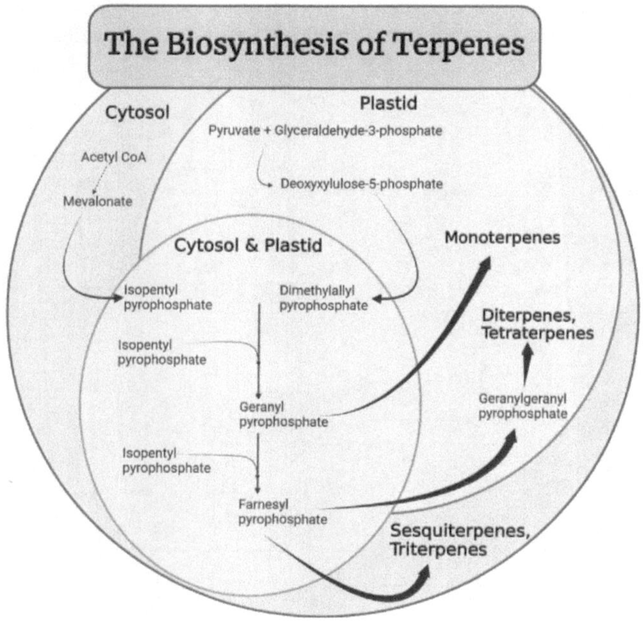

FIGURE 20.4 A schematic diagram showing the biosynthesis of terpenes.

(HMGR). HMGR regulates the manufacture of many terpenoids, including sterols and phytoalexins. MVA is then transformed into isopentenyl diphosphate (IPP) through three enzymatic processes catalyzed by mevalonate kinase (MK), phosphomevalonate kinase (PMK), and mevalonate diphosphate decarboxylase.

In contrast, the MEP route, which has seven enzymatic steps, is found in photosynthetic eukaryotes, cyanobacteria, apicomplexan protozoa, and the majority of eubacteria. 1-deoxy-D-xylulose 5-phosphate synthase (DXS) catalyzes the first step in the MEP pathway, converting pyruvate and glyceraldehyde-3-phosphate into 1-deoxy-D-xylulose 5-phosphate (DXP). DXS has an important regulatory and rate-limiting role in plastidial terpene biosynthesis. The second phase is carried out by 1-deoxy-D-xylulose 5-phosphate reductoisomerase (DXR), which transforms DXP to 2-C-methyl-D-erythritol 4-phosphate (MEP). MEP is then transformed into 4-diphosphocytidyl-2-C-methyl-D-erythritol (CDP-ME), which is phosphorylated to generate 4-diphosphocytidyl-2-C-methyl-D-erythritol 2-phosphate (CDP-ME2P). Following subsequent cyclization reactions, 2-C-methyl-D-erythritol 2,4-cyclodiphosphate (MEcPP) is formed, which is then converted to isopentenyl diphosphate (IPP) and dimethylallyl diphosphate (DMAPP) by 4-hydroxy-3-methylbut-2-enyl diphosphate synthase (HDS) and 4-hydroxy-3-methylbut-2-enyl diphosphate reductase (HDR).

Mutants of enzymes implicated in the MEP route show albino phenotypes and deficits in chloroplast formation, emphasizing the process's importance in plant development. Furthermore, several mutants exhibit increased resistance to biotrophic diseases, indicating a connection between the MEP system and plant defensive mechanisms. Overall, both the MVA and MEP routes play key roles in terpenoid biosynthesis, with each pathway contributing to the creation of a varied variety of terpenoid chemicals required for distinct physiological activities in plants (Tholl, 2015).

20.4 MOLECULAR APPROACHES FOR ENHANCING PRODUCTION

20.4.1 Genetic Engineering of Biosynthetic Pathways

The pharmaceutical industry heavily relies on secondary metabolites derived from plants and fungi due to their inherent bioactivity, which synthetic derivatives often struggle to replicate cost-effectively. To meet the global demand for these industrially significant products, there's a pressing need for both classical and advanced engineering techniques that not only enhance the quantity but also the quality of these bioactives. Recognizing the limitations of conventional breeding methods, researchers have increasingly turned to genetic-editing tools, with the advent of CRISPR/Cas9 marking a significant breakthrough in genome editing (Bhagwat et al., 2022). Derived from *Streptococcus pyogenes*, CRISPR/Cas9 is a third-generation genome editing tool characterized by its reliance on RNA–DNA base pairing, simplifying its construction and implementation. Its versatility extends across various organisms, including plants, mammals, flies, yeast, and filamentous fungi, making it a potent tool in industrial biotechnology for engineering pathways involved in producing crucial metabolites or bioactives. CRISPR/Cas9 primarily functions through the introduction of double-strand breaks (DSBs) via the Cas9 endonuclease guided by single-guide RNA (sgRNA). The repair of these DSBs through non-homologous end joining (NHEJ) often results in gene knockouts or silencing, which can be strategically employed to redirect metabolic flux toward desirable bioactives while eliminating undesired effects. Furthermore, multiplexed CRISPR Cas allows for the simultaneous editing of multiple genes, enabling intricate modifications in various industrially important organisms. By harnessing the cell's homology-directed repair, CRISPR/Cas9 can also facilitate gene knock-ins, altering the coding region by inserting donor DNA at specified sites (Balasubramani et al., 2023). Notably, CRISPR can be employed for DNA-free editing using RNA or protein components, mitigating the risks associated with vector-mediated genetic changes (Bhagwat et al., 2022). Despite its immense potential, the widespread implementation of CRISPR/Cas9 in plants has been hampered by the lack of adequate gene sequences (Balasubramani et al., 2023). Nonetheless, its efficiency, ease of use, and adaptability make it a promising tool for enhancing pharmaceutical output in heterologous medicinal plants through the precise manipulation of biosynthetic pathways. The third generation of CRISPR, emphasizing DNA-RNA recognition, offers advantages over protein-DNA interaction systems,

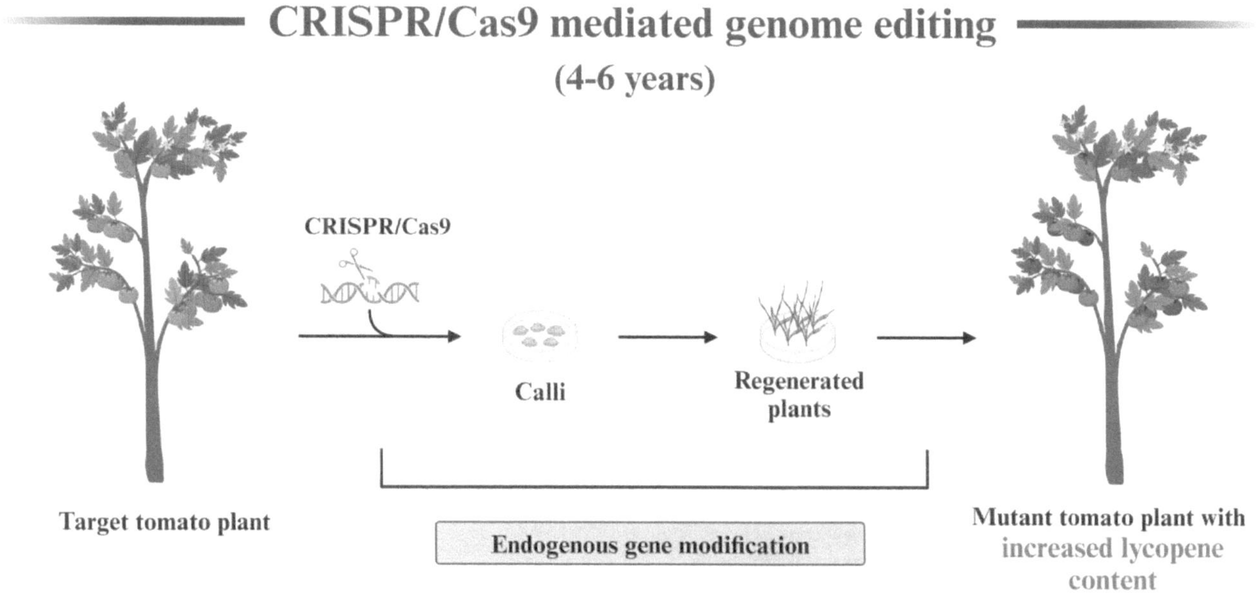

FIGURE 20.5 A general pathway for genome editing in medicinal crops.

enabling targeted DSB induction with minimal effort and expense. The resulting DNA modifications induced by CRISPR/Cas systems, including deletions, insertions, or substitutions, hold promise for revolutionizing various fields, particularly in the production of plant-derived pharmaceuticals (Das et al., 2024).

Mechanistically, the CRISPR/Cas9 system mimics the immune defense mechanism of bacteria against foreign DNA invasion, with Cas9 and guide RNA playing pivotal roles in DNA cleavage and repair. This natural defense system has been harnessed and optimized to achieve precise genome editing across diverse organisms, with CRISPR/Cas9 classified into different types based on its operation mechanism. The sgRNA, a fusion of transactivating RNA and CRISPR RNA, guides the Cas complex to the desired DNA sequence, facilitating DNA cleavage and repair. Fundamentally, the CRISPR/Cas tool aims to induce DSBs at specific genomic loci, initiating DNA modifications through repair pathways like homology-directed repair (HDR) or non-homologous end joining (NHEJ). In conclusion, the advent of CRISPR/Cas9 represents a paradigm shift in genome editing technology, offering unparalleled precision and efficiency in modifying genetic material across various organisms, with profound applications for pharmaceutical production and beyond (Bhagwat et al., 2022).

20.4.2 Use of Cell, Tissue, and Organ Culture as Tools to Enhance the Anti-cancer Phytochemicals Production

Several plants in our plant kingdom contain such secondary metabolites or anti-cancer phytochemicals that are usually used as natural solutions for several ailments. These phytochemicals are found in only some amount of plant tissue, example like only *Crocus* flowers contain the spice compound of Saffron (Cardone et al., 2020), due to the less amount of availability of the active compounds the cost of production is too high, so to increase the production of these secondary metabolites the biotechnological measures should be adopted rather than chemical method.

By this biotechnological remedy method, several targets should be fulfilled like firstly the uniform and economical production of the metabolites and moreover the rapid production and product recovery.

In production strategies, two major approaches can be applied for the required targeted phyto biochemicals. These are organ/in vitro tissue culture and genetic engineering.

20.4.2.1 Production Technology by Using Cell, Organ, and Tissue Culture

As the name suggests, tissues and cells of the plant are grown in the form of suspension cells for both semi-differentiated and undifferentiated states. This is the strategy to go around the boundaries imposed by the cell suspension stage and calli. In a certain era, about 30–40 years ago, only the root parts were cultured to gain anti-cancer drugs like hyoscyamine (Sinha et al., 2023). Also nowadays, hairy roots are rarely used. This is a very rapid process and more stable genetically. There are some restrictions to using this technique; for example, this approach can't be implicated to the type of chemicals that are produced in the organism's aerial part. But sometimes this method should not be used because many of the plants contain inner toxic substances that create problems in extracting the target compound (e.g., *Planta*)

There are some general techniques for enhancing the production of desired phytochemicals given by Zenk in 1991(Nielsen et al., 2019).

TABLE 20.2
Utilization of CRISPR/Cas9 in Medicinal Plants for Enhancing the Production of Scarce Anti-cancer Secondary Metabolites

Crop	Target anti-cancer compound	Target genes and their roles	Gene editing strategy	Results	References
Salvia mittiorrhiza	Rosmarinic acid (RA)	*SmRAS* gene	Site-specific mutagenesis leading to overexpression	Alleviated levels of 3,4-dihydroxy phenyl lactic acid (a precursor for RA)	(Bhagwat et al., 2022)
Glycine max	Isoflavonoids	*GmFNSII-1*, *GmF3H2*, and *GmF3H1* (associated with isoflavone synthase expression)	Triple mutation leading to overexpression	2-fold increase in isoflavone content	(Bhagwat et al., 2022)
Solanum lycopersicum	Lycopene	*BLC*, *SGR1*, *LCY-E*, *LCY-B1*, *LCY-B2*, *and PSYI* (responsible for conversion of lycopene to carotene on ripening)	CRISPR/Cas9-mediated gene knockout	5 times increment in lycopene content	(Bhagwat et al., 2022)
Solanum lycopersicum	GABA	*GABA-TP1*, *GABA-TP2*, *GABA-TP3*, *CAT9*, and *SSADH* (critical for GABA metabolism in tomato)	Multiplex pYLCRISPR/Cas9 system–based gene knockout	19-fold increment in GABA accumulation in the mutants	(Nielsen et al., 2019)
Medicago truncatula	Hemolytic sapogenins (e.g., medicagenic acid, bayogenin, hederagenin)	*CYP93E2* (responsible for the conversion of β-amyrin to soyasapogenin, non-hemolytic sapogenins)	CRISPR/Cas9-mediated gene knockout of *CYP93E2* channelizes the metabolic pathway from β-amyrin toward hemolytic sapogenins	Increase in the hemolytic sapogenin content as compensation for the loss of soyasapogenins	(Confalonieri et al., 2021)
S. miltiorrhiza	Phenolic acids like rosmarinic acid and salvianolic acid B and PAL (phenylalanine ammonia-lyase)	*bZIP2* (suppresses of phenolic acid pathway)	CRISPR/Cas9-mediated gene knockout of *bZIP2*, thereby facilitating the phenolic acid pathway	Up to 25% increased production of phenolic acids and PAL	(Das et al., 2024)
Cannabis sativa	Cannabidiol (CBD)	*THCAS* (involved in the synthesis of tetrahydrocannabinolic acid synthase [THCAS], which acts as a catalyst for tetrahydrocannabinolic acid)	CRISPR/Cas9-mediated gene knockout of THCAS, thereby channelizing the metabolic flux toward beneficial cannabinoids like cannabidiol.	Increased production of cannabinoids	(Das et al., 2024)

The first method is the identification of the initial material that contains the vast amount of the desired secondary metabolite. If the plant experimental material has been recognized, then follows the selection of the medium of culture that favors the perfect growth and the synthesis of the metabolite. The maximum medium factors like temperature, and sunlight directly affect the growth and the production of metabolite. The phosphate ion concentration plays a major role in producing the major secondary metabolites in the cell culture system. For example, *Solanum laciniatum*'s high amount of phosphorus can enhance cell growth but diminish the production of the secondary metabolite (Biswas et al., 2023). There is also a contradiction of a lower phosphate level in the production of phenolics and ajmalicine in *Catharantus roseus* (Barrales-Cureño et al., 2021).

To add in vitro cultures of the precursors, the technique of precursor feeding is adopted. This method is used in producing vanillin, which is an aromatic molecule used in several other economic interests.

As with artemisinin, there is an indirect way to increase the production of phytochemicals.

Elicitor molecules play a crucial role in the enhancement of the production of the secondary metabolite. These are like the molecular signals that are received by the plants to synthesize the secondary metabolite that reduces stress and also possesses pharmaceutical properties (Chamkhi et al., 2021).

The addition of the elicitor molecules to the culture medium of the origin of fungi and bacteria of metal salts enhances the production of many types of secondary metabolites; for example, the addition of *Botrytis* to the *Papaver somniferum*'s cell suspension increases the amount of the sanguinarine to 3–4% of the dry cell weight (Bhatia et al., 2015).

Many mutation strategies have been adopted to gain the high productive lines of cells. The selective agent's use is also another approach to select the lines that contain the higher content of the secondary metabolites.

Traditional techniques like organ culture, tissue culture, and cell culture have many future scopes for the enhancement of the production of secondary metabolites and of the anti-cancer properties in medicinal plants. People should be more conscious about the conservation of this type of medicinal plants as they are readily to be extinct. These natural sources of secondary metabolites contain a large number of useful metabolites. Various methods like using stem cells bypass the limitations of undifferentiated thalii, and the hairy roots help in the production of secondary metabolites. Many techniques like genetic engineering, metabolic engineering, and OMICS technologies are collectively working for the development of anti-cancer properties as future aspects.

20.5 INTEGRATION OF OMICS TECHNOLOGIES

A plentiful supply of pharmaceutical metabolites is found in plants. In addition to their inherent ability to treat wounds, medicinal plants are an excellent source of molecules that fight cancer. Examples of these compounds are vinblastine and vincristine, which are derived from periwinkle, and curcumin, which is found in turmeric. We can produce more anti-cancer chemicals originating from plants thanks to contemporary technology like molecular and genetic engineering. However, it's not so easy since, before using any technology, we must thoroughly examine the plants. This includes studying their ecology, evolution, and environmental aspects – but most crucially, their genetic constitutions. Every compound has a unique biosynthesis pathway that is controlled by one gene, multiple genes, or

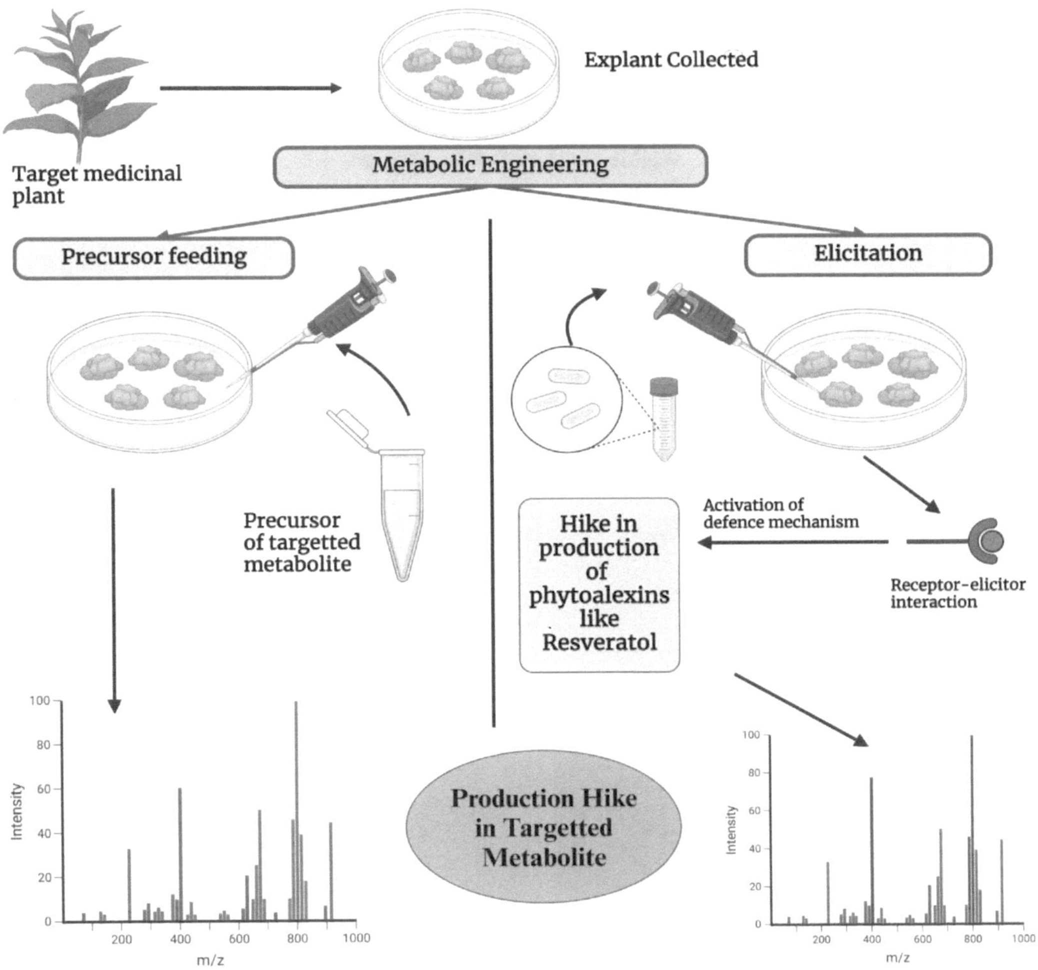

FIGURE 20.6 Metabolic engineering for enhanced anti-cancer compound production.

gene clusters – a collection of two or more genes found in the DNA of an organism that together encodes for proteins or polypeptides with similar functions. These gene clusters also form extremely complex systems with interconnecting elements. Studying the biosynthesis of anti-cancer chemicals presents several challenges, such as the complexity of plant genomes, huge genomic sizes, genetic redundancy, and others. Finding the correct gene that codes for a certain anti-cancer qualitative chemical is difficult. Plant biologists employ omics-based methods to pinpoint the specific gene that produces a given chemical in order to overcome these issues. By utilizing omics datasets, we may obtain a comprehensive understanding of the static or dynamic status of biological systems. Their integration can be a useful tool in validating genuine findings and minimizing false positives or negatives. By using strong statistical testing, omics datasets provide comprehensive systems knowledge that can be utilized to create metabolic models that forecast the behavior of the system. Metabolic models serve as a foundation for forecasting a system's behavior under particular conditions by representing biological components and their intricate interactions as mathematical functions and equations (Rai et al., 2017). It is greatly wanted to create metabolic models for medicinal plants in order to forecast the switch or strategy for prioritizing the synthesis of specialized metabolites. For techniques like synthetic biology, tissue engineering, and genetic engineering to produce a sustainable supply of these bio-products, it is crucial to comprehend the biosynthetic pathways and their control for pharmacologically significant metabolites.

20.6 COMPONENTS OF OMICS

The metabolic pathways are intricate, highly regulated, and interconnected systems. Any disturbance introduced into a biological system, whether by genetic engineering or biotic or abiotic systemic therapy, modifies the system's metabolic state. In order to establish homeostasis, this starts a process of readjustment that involves the activation of important regulators, which in turn activates particular metabolic network components (Rai et al., 2017).

High-volume comprehensive measurements of cellular constituents, including transcript, protein, and metabolite levels, among others, enable the identification of all relevant components, as well as the relationships and interactions between them (Rai et al., 2017). Many omics areas, such as phenomics,

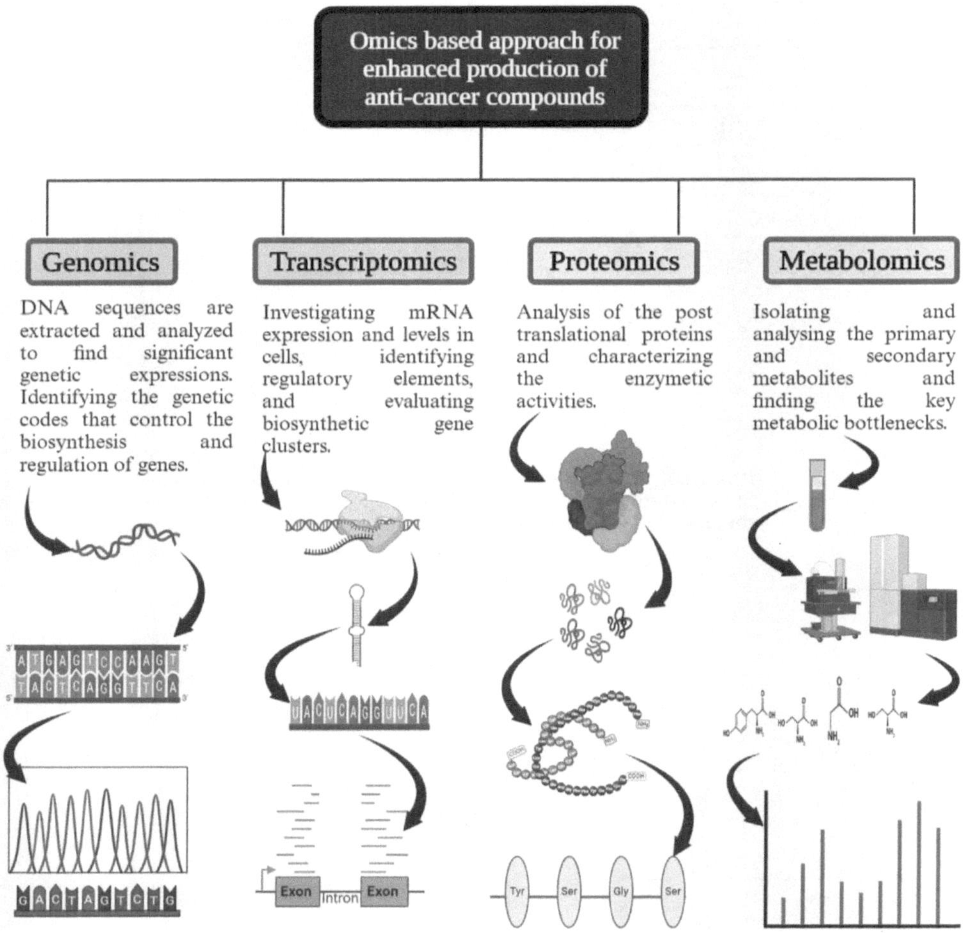

FIGURE 20.7 Omics databases used as components for new drug discovery.

glycomics, food omics, interact omics, orfeomics, physiomics, microbiomics, and cell omics, are being suggested by researchers as a result of technological advancements and our growing capacity to quantify many elements of a biological system. The four basic omics components – genomics, transcriptomics, proteomics, and metabolomics, which center on certain facets of a biological system – were developed and are briefly discussed in the subsections that follow.

20.7 GENOMICS

We require the complete genomic sequence of a plant in order to fully comprehend it. In addition to providing information about the biosynthetic pathway, these sequences also aid in our understanding of the origin, evolution, inheritable traits, developmental and physiological schematics, epigenomic regulations, and metabolic potentials. These concepts form the foundation for our understanding of genome diversity and chemo-diversity at the molecular level, as well as aid in the selection of the most advantageous trait for the production of desired metabolites. With 23 plant species being used primarily for medicinal purposes or described in various traditional medicinal practice(s), whole-genome sequences of 183 land plants have so far been released in the NCBI Genome database. Of these, 58 were completed and submitted in 2016 alone (for a total of 68 genome sequences released in 2016).

The creation of robust and easily available functional genomic resources, like as complete sets of complementary DNA clones, tagged mutant lines, and a quick and easy transformation technique, is made possible by genomic sequences. Additionally, genomic sequences offer essential details about several homologous genes from certain metabolic pathways, enhancing our capacity to delete the genes and determine their functionalities. Gene clusters implicated in genome sequences are linked to many kinds of metabolites, including polyketides, benzylisoquinoline alkaloids, cyanogenic glycosides, diterpenes, and triterpenes. Although it is improbable that these gene clusters have arisen only as a result of horizontal gene transfer, the mechanisms behind cluster formation remain unclear (Rai et al., 2017).

While describing the marneral gene cluster in *Arabidopsis*, it was suggested that another gene cluster (the thalianol gene cluster) may have evolved from a similar OSC/CYP705 gene pair. A comparative investigation of these two gene clusters revealed distinct evolutionary processes, which were postulated to be the outcome of gene relocation because both gene clusters are located in dynamic chromosomal areas with genes prone to ectopic translocation. The authors found that the formation and management of plant gene clusters is the consequence of a similar process shared by other eukaryotes such as filamentous fungi, whose gene clusters linked with secondary metabolism are commonly subtelomeric. Yu et al. (2016) showed that metabolic gene clusters are distinguished by different chromatin signatures linked to cluster suppression and activation, namely, histone 3 lysine trimethylation (H3K27me3) and histone 2 variant H2AZ, respectively (Yu et al., 2016). The scientists proposed that these unique chromatin signatures indicate discrete windows of coregulation in the genome that help coordinate expression. Because a plant's chemical-based defense tactics rely on evolution/polymorphism to provide them with the skills they need to overcome quickly changing ecological challenges, dynamic chromosomal regions may be a significant hotspot for biosynthetic genes involved in the production of specialized molecules. The existence of gene clusters in highly transposable areas of chromosomes shows that nature is still attempting to discover the best candidate genes linked with the manufacture of specialized metabolites in the processes of improving the pathways (both biosynthesis and regulation (Rai et al., 2017).

Understanding gene clusters and their roles in various biosynthetic pathways can help us understand coregulation and coinheritance in plants.

20.7.1 New Findings

Genomics-based approaches, such as genome-wide association studies (GWAS) and expression quantitative trait loci (eQTLs), have been successful in identifying metabolite pathways in plants like Arabidopsis, rice, maize, and tomatoes. The GWAS and eQTL methodologies are sophisticated genomic analytic methods that have proven useful for the functional characterization of target genes and the identification of genes responsible for desired phenotypes. Gene mapping, cloning, and studying the composition and functions of a genome are all based on the creation of a genetic linkage map. But still, genetic mapping is difficult because of the long development cycle, restricted culture history, complex genetic background, and highly heterozygous genes found in most therapeutic plants. Using conventional molecular markers, including SSR, AFLP, RFLP, RAPD, and EST-SSR, genetic maps of medicinal plants, including artichokes (*Cynara scolymus*), passion fruit (*Passiflora edulis*), bladder campion (*Silene vulgaris*), and *Trifolium pratense* were created (Rai et al., 2017). Highly saturated and flexible molecular linkage maps of medicinal plants have been produced recently as a result of the rapid generation of genome-wide molecular markers. Recently, the rapid generation of genome-wide molecular markers has led to the development of highly saturated and flexible molecular linkage maps of therapeutic plants. The increasing quantity and quality of genomes of medicinal plants will likely lead to a boom in the creation of genetic maps of phenotypes linked to bioactive compounds in medicinal plants.

20.8 TRANSCRIPTOMICS

Transcriptomics, or the thorough examination of the transcriptome (the totality of RNA molecules), also known as expression profiling, investigates mRNA expression levels in cells. The introduction of microarray-based transcriptome profiling transformed gene functional characterization, illuminating their roles within metabolic pathways and accelerating the proliferation of diverse bioinformatics tools, databases, and resources for co-expression and comparative transcriptome analysis.

Developed to facilitate co-expression analysis using extensive microarray datasets from Arabidopsis and other model plants, some of the most notable databases are ATTED-II, Genevestigator, AtGenExpress, OryzaExpress, BAR, At-TAX, PRIMe, CORNET, PathoPlant, PLEXdb, and ePlant (Rai et al., 2017).

RNA sequencing (RNA-seq) based on next-generation sequencing (NGS) represents a critical breakthrough in transcriptome research. Transcriptome profiling and functional characterization of transcripts are made easier by RNA-seq-based methods, which do not require previous genomic information like microarray analysis requires. This is especially true for non-model plants such as medicinal ones. Simultaneous identification of exons, introns, splicing sites, transcribed regions, and repetitive sequences is possible using RNA sequencing. Understanding co-expression patterns and doing comparative analysis across different medicinal plant species are made easier with the use of RNA-seq-based transcriptome profiling. For such investigations, a plethora of data is accessible, spanning over 2400 plant species versus 183 known plant genome sequences.

Co-expression analysis and transcriptome profiling based on RNA-seq are essential for locating biosynthetic enzymes required to synthesize important anti-cancer drugs. 29 potential genes in *P. hexandrum* were found by transcriptome mining and co-expression analysis, and six genes from the podophyllotoxin production pathways were found by combinatorial expression of these genes in *Nicotiana benthamiana*.

Specialized metabolite biosynthesis pathways have been identified thanks in large part to RNA-seq-based transcriptome research, which has also sparked several de novo transcriptome assembly and characterizations of various plant species throughout the world. Transcriptome study across species has been facilitated by the abundance of transcriptome datasets and assembly sequences that have been made available in public databases (such as NCBI-SRA, NCBI-GEO, and NCBI-TSA). Characterizing essential pathway enzymes may be possible by comparative transcriptome analysis of species that produce comparable classes of specialized metabolites. For example, 28 CYP450-encoding transcripts with significant expression in saponin-accumulating tissues were found by a comparative transcriptome analysis of de novo transcriptome assemblies of four important saponin-producing plants: *Panax japonicus*, *Panax ginseng*, *Panax quinquefolium*, and *Panax notoginseng*. This method verified a number of transcripts that had previously been identified as playing roles in saponin biosynthesis, suggesting possible candidates for further functional characterization.

Additionally, Pathak et al. (2013) used high-papaverine mutants of *P. somniferum* to identify potential genes linked to papaverine production using a comparative transcriptome analysis technique. The expanding transcriptomics databases hold the potential to accelerate our comprehension of the evolutionary dynamics of medicinal plants and the biosynthetic pathways connected to diverse anti-cancer compounds.

20.9 PROTEOMICS

Proteins are the most significant molecules in a plant's body since they are the ones that control and operate every biological activity. They also form the foundation of the biological system's strict homeostasis. Although proteomics has not yet been extensively used to investigate the production of anti-cancer compounds and their control in non-model medical plants, it has played a significant role in our knowledge of plant stress responses under various contexts.

20.10 METABOLOMICS

The great biochemical diversity of plant metabolites is being explored by researchers in an effort to find new anti-cancer treatments. The thorough examination of tiny molecules inside biological systems, known as metabolomics, has shown to be an effective method for deciphering the metabolic processes that underlie the creation of anti-cancer substances in plants. This study examines the potential use of metabolomics in targeted breeding, genetic engineering, culture optimization, and metabolic engineering techniques to enhance plant production of anti-cancer chemicals. We go over the use of metabolic engineering, pathway elucidation, and metabolite profiling to pinpoint important enzymes, metabolites, and regulatory genes involved in the manufacture of anti-cancer compounds. Moreover, we emphasize how metabolomics may direct metabolic engineering efforts to improve metabolic flux along desired pathways and optimize culture conditions to increase the output of target chemicals. We demonstrate the efficacy of metabolomics-guided strategies in boosting the synthesis of anti-cancer chemicals in plants via thorough case studies and experimental data, opening the door for the creation of innovative plant-based cancer therapy agents (Shafi & Zahoor, 2021).

20.11 METABOLOMICS IN ANTI-CANCER COMPOUND DISCOVERY

Metabolomics allows for the systematic profiling of plant metabolites, enabling the identification of specific compounds having anti-cancer properties. Advanced analytical methods such as the use of nuclear magnetic resonance spectroscopy and mass spectrometry have revolutionized metabolite profiling by allowing for the simultaneous identification and measurement of several metabolites inside the tissues of plants. Using metabolomics-driven investigations, researchers have identified interesting candidate compounds for further inquiry based on their structural features and bioactivity (Johnson et al., 2016).

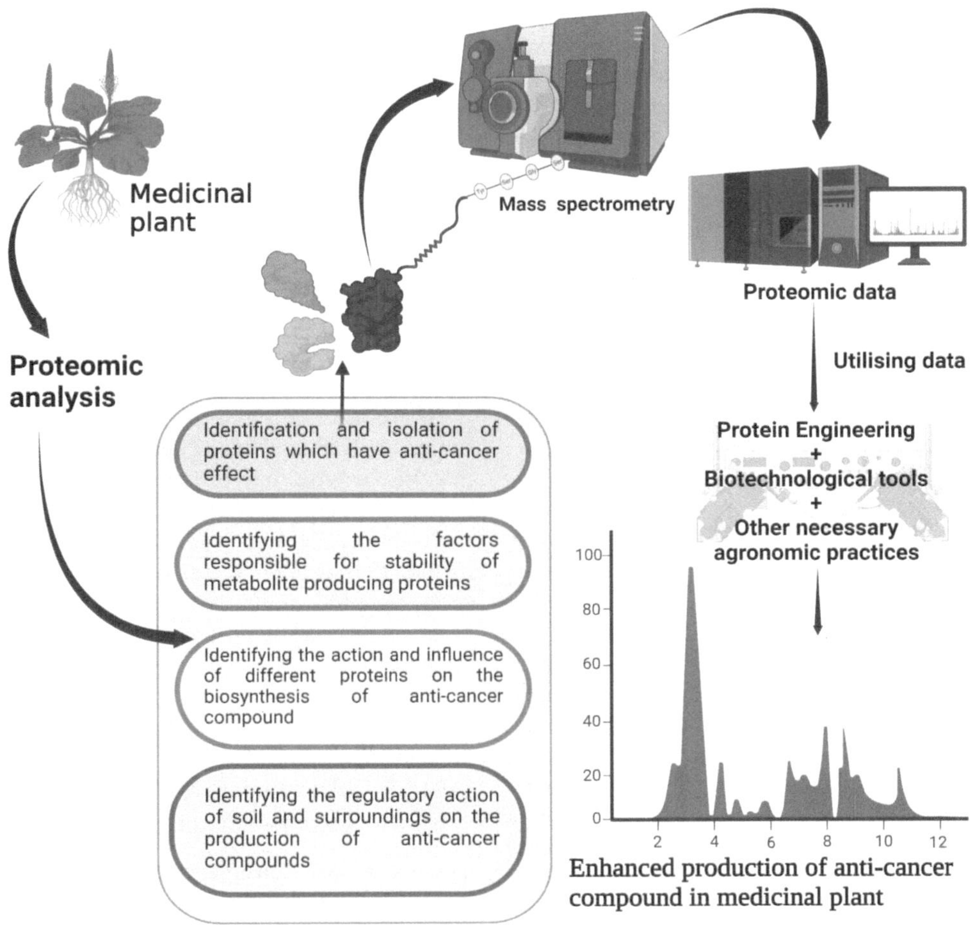

FIGURE 20.8 Environmental influence in protein productions (Zhang et al., 2023).

20.12 ELUCIDATION OF METABOLIC PATHWAYS

Understanding the metabolic processes involved in the manufacture of anti-cancer chemicals in plants is made possible through the application of metabolomics. Through the integration of genetic and enzymatic data with changes in metabolite levels, researchers are able to identify regulatory genes and important enzymes within these pathways. Analyzing metabolic pathways with a focus on metabolomics provides important information on the biosynthesis of important anti-cancer substances like resveratrol in grapes and taxol in *Taxus* species. (Liu et al., 2017) Additionally, metabolomics was used in different research to point out the metabolic pathways that *Camellia sinensis* uses to synthesize bioactive polyphenols. Their research not only clarified the biosynthetic pathways but also suggested possible targets for metabolic engineering to increase the synthesis of polyphenols (Deng et al., 2018; Zhang et al., 2020). Similar methods could be applied to increase the production of secondary metabolites in different plant varieties, including *Curcuma longa* and *Catharanthus roseus*.

20.13 TARGETED BREEDING AND GENETIC ENGINEERING

Equipped with an understanding of crucial metabolic routes and regulatory genes, scientists may utilize focused breeding and genetic modification techniques to enhance the synthesis of anti-cancer substances in plants. The production of desired molecules can be increased by either introducing genes encoding rate-limiting enzymes into plant genomes or selectively selecting plants with naturally raised quantities of target metabolites. Measuring target metabolites in modified plant lines, metabolomics is a vital technique for assessing the effectiveness of various tactics (Xiao et al., 2013).

20.14 OPTIMIZATION OF CULTIVATION CONDITIONS

The adjustment of growing settings to optimize the accumulation of anti-cancer chemicals in plants is made easier by metabolomics. By examining the impact of various environmental parameters, like temperature, light, water availability, and nutrient levels, on the production of metabolites, scientists may create customized growth environments that enhance the output of desired molecules. It has been demonstrated that utilizing cultivation optimization tactics guided by metabolomics can increase the production of anti-cancer metabolites such as flavonoids and phenolic compounds in some varieties of plant species. Furthermore, a different study showed that altering the growth medium's phosphorus and nitrogen concentrations greatly boosted the accumulation of anti-cancer alkaloids in *Catharanthus roseus* plants. This demonstrates how useful metabolomics may be in directing cultivation methods to increase plant production of anti-cancer chemicals (Chang et al., 2014).

20.15 METABOLIC ENGINEERING TARGETS

The discovery of metabolic engineering targets to increase the synthesis of particular anti-cancer drugs is guided by metabolomics. Researchers can identify possible bottlenecks in metabolic pathways and come up with workarounds by analyzing metabolic fluxes and enzyme activity. This might entail creating novel enzyme variations, modifying the expression of important enzymes, or using pathway engineering to reroute metabolic fluxes. Metabolomics tracks alterations in metabolite profiles and yields to offer insightful feedback on the effectiveness of these engineering efforts. It proved how well metabolomics data might be used with genetic engineering methods to increase *Catharanthus roseus*'s ability to produce anti-cancer alkaloids. They successfully increased the yield of important alkaloids, including vincristine and vinblastine, in *Catharanthus roseus* hairy roots by using targeted metabolic engineering techniques, demonstrating the critical role that metabolomics plays in directing metabolic engineering strategies for the production of anti-cancer compounds (Acharjee et al., 2022).

20.16 INTEGRATION OF MULTI-OMICS DATABASES

As discussed, genomics involves a detailed study of genes, transcriptomics involves detailed studies of DNA and proteomics involves a detailed study of proteins, these studies help us to understand different aspects of plants – like expressions of genes and their involvement in the biosynthesis of medicinal compounds, the whole plant regulatory system of biosynthesis in an integrative way. This provides us with previous data about the locations of genes through this we can apply different methods of genetic engineering like the application of CRISPR-Cas9 to increase the production of secondary metabolites.

Proteomics deals with the study of post-transitional modifications and protein expressions of organisms. As already discussed, proteomics can aid in boosting the production of the secondary metabolites, by providing detailed data about the enzyme and proteins associations related to biosynthesis of compounds. Proteomics has also paved the pathway for researchers to exploit medicinal plants using technologies like mass spectrometers and quantify the protein synthesis and identify the potential targets of these in biosynthesis and other pathways. In 2005, nearly 58 proteins were isolated from *C. roseus* cell culture that was involved in the biosynthesis of mono-indole alkaloids. Proteomics also helps us to get reliable data in the absence of genomic and transcriptomic data of medicinal plants (Rai et al., 2017). Through metabolomic profiling, we get an assured set of data that we can be applied to enable the production of bioactive compounds under various physiological conditions.

There are two types of omics data approaches single-omics analysis and multi-omics analysis. In single-omics analysis, we use only one type of database like transcriptomics or metabolomics, or maybe something elsewhere, multi-omics analysis is based on integrative analysis of different types of databases like proteomics-metabolomics, transcriptomics-proteomics, etc. Multi-omics database helps us to understand the relations between the layers of epigenomics.

20.17 METHODS OF DATA INTEGRATION

20.17.1 Multivariate Methods for Data Integration

The most basic unsupervised multi-omics data integration technique is joint non-negative matrix factorization (NMF), which works by splitting a non-negative matrix into non-negative loadings and non-negative factors. The approach projects different data types into a uniform coordinate system, generating modules out of heterogeneous variables projecting in the same direction. Analyzing gene expressions is made easier when you combine methylation data from NMF with expression data from mRNA and microRNA. Mo-Cluster applies a multiblock multivariate analysis and clustering to find common trends across multiple omics data. The gene clusters in biosynthetic pathways were identified through Mo-Cluster by the combination of mRNA, protein, and methylation data. To sum up, each of the methods mentioned above has one thing in common: they all place multi-omics datasets into a stacked matrix that acts as input for the subsequent cluster analysis (Menyhárt & Győrffy, 2021).

20.18 STATISTICAL METHODS OF DATA INTEGRATION

Using a Bayesian framework permits assumptions about diverse data kinds with a range of distributions and connections across data sets. Incorporating multiple genomic characteristics without nonnegative constraints, robust clustering strategies like iCluster use a Gaussian latent variable model, where the latent variables form a set of principle coordinates that collectively capture the correlative

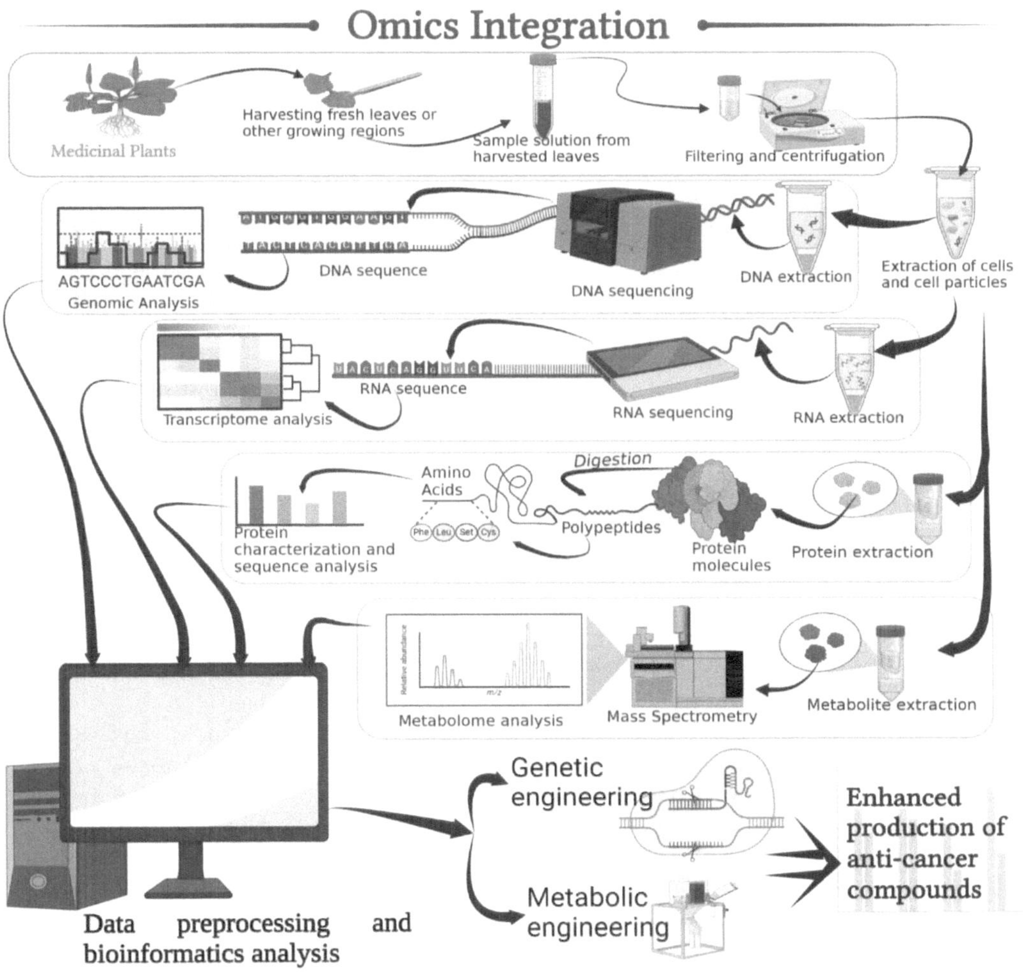

FIGURE 20.9 Use of omics profiling in the production of anti-cancer components.

structure of multi-omics data (Menyhárt & Győrffy, 2021). The iCluster+ program breaks down each form of omics data into loading factors (gene characteristics) and component factors (latent cancer subtype) based on presumptions related to various omics data types. For simulated datasets, where sample labels were unknown prior to integration, iCluster+ yielded the best accurate classification when compared to the other three approaches

Gene-to-gene pathway interactions are incorporated into the widely used probabilistic graphical model (PGM) framework known as pathway representation and analysis by direct reference to graphical models (PARADIGM). Data interpretation may benefit from the use of databases like KEGG, which feature interaction topology among genes. The method depends on factor graphs, which show genes as nodes and relationships between genes as edges, to characterize gene expression and activity as a collection of interrelated variables. A wide range of omics data may be included in the model, including DNA copy number changes, mRNA and miRNA expression, promoter methylation, and mutations. However, correct data regarding the magnitudes of biochemical interactions are required for PARADIGM to function (Menyhárt & Győrffy, 2021).

20.19 NETWORK-BASED INTEGRATION

By calculating biological linkage scores for each molecular interaction in the network to facilitate predictive subnetwork identification, iOmicsPASS carries out a controlled approach to integrating transcriptomics, proteomics, and DNA copy number data. By calculating biological association counts for each molecular interaction in the network to aid in predictive subnetwork identification, iOmicsPASS carries out a supervised integration of transcriptomics, proteomics, and DNA copy number data (Menyhárt & Győrffy, 2021).

20.20 SINGLE-OMICS ANALYSIS

Understanding how chemical changes to DNA, nuclear RNA, histone and nonhistone chromatin proteins can impact gene expression without changing the base sequence is made easier with the use of this kind of study. Epigenetic markers serve as possible indicators for customized treatment since they are very tissue-specific and very much dependent on environmental influence and disease-related modifiers. Genomic analysis can predict the efficacy of immune-checkpoint inhibitors by evaluating mutational load and signatures in addition to specific targetable changes.

20.21 PROMISING APPLICATIONS OF MULTI-OMICS APPROACHES

The information flow from one layer to the next must be taken into account while integrating various data types. It is not always clear how mRNA expression and protein levels relate to one another, though. Because of alternative splicing and the proteome's post-translational modifications, the complexity of the information transferred from the genome to the transcriptome grows dramatically. The rates of transcription, translation, mRNA decay, and protein degradation are the four primary processes that ascertain the level of expression of proteins in a cell. The conclusion reached in an ongoing discussion over the degree of correlation between mRNA and protein levels is that there is moderate to poor correlation (correlation coefficients of 0.4). Predictability was increased using a tailored proteomic approach to internal standards rather than earlier label-free absolute protein quantifications of protein transcriptome RNA levels.

While some research shows that translation rate is a significant factor in determining protein abundance, other findings imply that gene expression is regulated at the mRNA level. In summary, some protein abundances may be accurately predicted by RNA levels. Generally speaking, the first step in creating frameworks for studies unique to a certain disease is gene identification.

In addition, researchers and scientists can identify, locate, and establish different epigenetic modifications in plants with the aid of data obtained from various omics approaches. This allows the trait of increased production of bioactive compounds from the plant to be passed on to its progeny and reversed as needed.

Epigenetic alterations influence and govern a plant's growth and psyche. DNA methylation is one type of epigenetic change that affects gene expression, transposon activity, and chromosomal interaction. Using methods like bisulfite sequencing and methylation-sensitive restriction enzyme-based assays, it is possible to do DNA methylation profiling, which offers insights into the epigenetic control of metabolic biosynthesis (Zhao et al., 2023).

Histone proteins are found in DNA and are essential for the activation and inactivation of important regulatory gene systems. They may also be modified by acetylation, methylation, phosphorylation, and ubiquitination, which have a significant impact on the structure of the chromatin and the expression of genes.

When combined with multi-omics analysis of the secondary metabolite, the epigenetic profiling of such histone modifications can assist in identifying the location of these modifications, which can lead to improvements in the generation of anti-cancer compounds in medicinal plants in terms of both quantity and quality (Zhao et al., 2023).

20.22 CHALLENGES AND FUTURE OPPORTUNITIES

The study of transcriptome protein and metabolomics of cells and metabolites requires high-end advanced software and tools. Also, the number of genes, proteins, and metabolites in plants is very large due to which their detailed study and profiling require a lot of time. The data sets received from the omics study are also large and require very sophisticated and advanced bioinformatic tools and computational algorithms. Moreover, the key regulatory genes and enzymes identified required experimental validation and functional characterization in order to confirm their roles in metabolic pathways and their further exploitation up to their full potential. These are a few major challenges that lie behind the proper and smooth integration of omics technology with increasing the production of anti-cancer compounds in medicine plants.

Almost 30% of drugs used nowadays are natural products. Medicinal plants like *C. roseus* are the major producers of pharmacologically important anti-cancer drugs like vinblastine and vincristine. Similarly, in nature, many other domesticated and wild plants have inert potential; thus, the metabolites produced in them and their anti-cancer properties and therapeutic efficacy must be properly identified and studied.

Omics technology is at a budding stage in the present era. They have not only revolutionized our understanding of metabolites and their biosynthesis but also provided us with ways to increase their production at multiple points in the biosynthetic and metabolic pathway and also paved the way for multiple studies based on genetic and metabolomic engineering.

As we are progressing and focusing on the increasing number of cancer patients every year, there is an urgent need to develop rapidly and develop more informative knowledge and comprehensive analytical approaches for profiling and characterizing natural compounds, which have fewer side effects and greater efficacy, and their production is sustainable in nature (Scossa et al., 2018).

Moreover, due to the limited amount of production of anti-cancer compounds in medicinal plants, their pharmacological value is very high and makes them inaccessible to poor and marginalized people. Through an omics-based approach novelizing the production of these anti-cancer compounds in medicinal plants can be exploited up to their maximum potential and this will also lead to more affordable anti-cancer drugs in therapeutics. Finding ways for in vitro synthesis of these can also work in multiplying the production of anti-cancer compounds and will also help in preserving plant genetic resources and preventing their over-exploitation (Scossa et al., 2018).

20.23 CONCLUSION

The chapter presented herein provides a comprehensive elucidation of innovative molecular methodologies deployed to augment the yield of anti-cancer agents sourced from medicinal flora. It accentuates the centrality of sophisticated genetic engineering modalities, notably CRISPR-Cas9, and metabolic engineering in refining biosynthetic pathways governing the synthesis of these pharmacologically potent entities. Furthermore, the amalgamation of omics technologies encompassing genomics, transcriptomics, proteomics, and metabolomics furnishes a holistic comprehension of the intricate molecular mechanisms underpinning the biosynthesis of secondary metabolites. This integrated framework not only facilitates the delineation of pivotal genes, enzymes, and regulatory elements but also facilitates precision interventions aimed at maximizing the output of therapeutically relevant compounds. The discourse underscores the transformative capacity of these molecular strategies in reshaping the landscape of cancer therapeutics, thereby unleashing nature's repository of anti-cancer agents and forging novel trajectories for the exploration and synthesis of life-saving pharmaceuticals. Ultimately, the chapter contends that through the concerted application of genetic engineering, metabolic engineering, and omics technologies, scholars are poised to catalyze substantial advancements in the realm of plant-derived anti-cancer drug discovery.

REFERENCES

Acharjee, S., Kumar, R., & Kumar, N. (2022). Role of plant biotechnology in enhancement of alkaloid production from cell culture system of *Catharanthus roseus*: A medicinal plant with potent anti-tumor properties. *Industrial Crops and Products*, *176*, 114298. ISSN: 0926-6690. https://doi.org/10.1016/j.indcrop.2021.114298.

Aditya, S., & Gupta, S. (2013, July). Ingenol mebutate: A novel topical drug for actinic keratosis. *Indian Dermatology Online Journal*, *4*(3), 246–249. https://doi.org/10.4103/2229-5178.115538. PMID: 23984250; PMCID: PMC3752492.

Balasubramani, S., Chen, Q., Zhou, Z., Moola, A. K., Duraisamy, S. M., Prakash, P., Gayathiri, E., Satish, L., & Swamy, M. K. (2023). Transgenic medicinal plants for improved plant metabolites production. In *Phytochemical genomics: Plant metabolomics and medicinal plant genomics* (pp. 403–415). Springer Nature. https://doi.org/10.1007/978-981-19-5779-6_16

Barrales-Cureño, H. J., Montiel-Montoya, J., Espinoza-Pérez, J., Cortez-Ruiz, J. A., Lucho-Constantino, G. G., Zaragoza-Martínez, F., . . . López-Valdez, L. G. (2021). Metabolomics and fluxomics studies in the medicinal plant Catharanthus roseus. In *Medicinal and aromatic plants* (pp. 61–86). Academic Press.

Bhagwat, A. C., Patil, A. M., & Saroj, S. D. (2022). CRISPR/Cas 9-based editing in the production of bioactive molecules. *Molecular Biotechnology*, *64*(3), 245–251. Humana Press Inc. https://doi.org/10.1007/s12033-021-00418-4

Bhatia, S., Bera, T., Dahiya, R., Bera, T., Bhatia, S., & Bera, T. (2015). Classical and nonclassical techniques for secondary metabolite production in plant cell culture. In *Modern applications of plant biotechnology in pharmaceutical sciences* (pp. 231–291). Academic Press.

Biswas, D., Chakraborty, A., Mukherjee, S., & Ghosh, B. (2023). Hairy root culture: A potent method for improved secondary metabolite production of Solanaceous plants. *Frontiers in Plant Science*, *14*, 1197555.

Cardone, L., Castronuovo, D., Perniola, M., Cicco, N., & Candido, V. (2020). Saffron (Crocus sativus L.), the king of spices: An overview. *Scientia Horticulturae*, *272*, 109560.

Chamkhi, I., Benali, T., Aanniz, T., El Menyiy, N., Guaouguaou, F. E., El Omari, N., El-Shazly, M., Zengin, G., & Bouyahya, A. (2021). Plant-microbial interaction: The mechanism and the application of microbial elicitor induced secondary metabolites biosynthesis in medicinal plants. *Plant Physiology and Biochemistry*, *167*, 269–295.

Chang, B. W., Cong, W. W., Chen, Q., Zu, Y.-G., & Tang, Z. H. (2014). The influence of different forms and concentrations of potassium nutrition on growth and alkaloid metabolism in *Catharanthus roseus* seedlings. *Journal of Plant Interactions*, *9*(1), 370–377. https://doi.org/10.1080/17429145.2013.838312

Confalonieri, M., Carelli, M., Gianoglio, S., Moglia, A., Biazzi, E., & Tava, A. (2021). CRISPR/Cas9-mediated targeted mutagenesis of *CYP93E2* modulates the Triterpene Saponin Biosynthesis in *Medicago truncatula*. *Frontiers in Plant Science*, *12*. https://doi.org/10.3389/fpls.2021.690231

Das, N., & Roychoudhury, A. (2023, September 5). Genetic engineering of plants for enhancing secondary metabolites with anticancerous properties. In *Plant-derived anticancer drugs in the OMICS era* (pp. 127–150). https://doi.org/10.1201/9781003377412-6

Das, S., Kwon, M., & Kim, J. Y. (2024). Enhancement of specialized metabolites using CRISPR/Cas gene editing technology in medicinal plants. In *Frontiers in plant science* (Vol. 15). Frontiers Media SA. https://doi.org/10.3389/fpls.2024.1279738

Deng, W., Han, J., Fan, Y., Tai, Y., Zhu, B., Lu, M., Wang, R., Wan, X., & Zhang, Z. (2018). Uncovering tea-specific secondary metabolism using transcriptomic and metabolomic analyses in grafts of *Camellia sinensis* and *C. oleifera*. *Tree Genetics & Genomes*, *14*(2). https://doi.org/10.1007/s11295-018-1237-1

Gambini, J., Inglés, M., Olaso, G., Lopez-Grueso, R., Bonet-Costa, V., Gimeno-Mallench, L., Mas-Bargues, C., Abdelaziz, K. M., Gomez-Cabrera, M. C., Vina, J., & Borras, C. (2015). Properties of resveratrol: In vitro and in vivo studies about metabolism, bioavailability, and biological effects in animal models and humans. *Oxidative Medicine and Cellular Longevity*, *2015*, 13, Article ID 837042.

Johnson, C. H., Ivanisevic, J., & Siuzdak, G. (2016). Metabolomics: Beyond biomarkers and towards mechanisms. *Nature Reviews Molecular Cell Biology*, *17*(7), 451–459. https://doi.org/10.1038/nrm.2016.25

Liu, X., Ding, W., & Jiang, H. (2017). Engineering microbial cell factories for the production of plant natural products: From design principles to industrial-scale production. *Microbial Cell Factories*, *16*, 125. https://doi.org/10.1186/s12934-017-0732-7

Mao, J. J., Pillai, G. G., Andrade, C. J., Ligibel, J. A., Basu, P., Cohen, L., Khan, I. A., Mustian, K. M., Puthiyedath, R., Dhiman, K. S., Lao, L., Ghelman, R., Cáceres Guido, P., Lopez, G., Gallego-Perez, D. F., & Salicrup, L. A. (2021). Integrative oncology: Addressing the global challenges of cancer prevention and treatment. *CA: A Cancer Journal for Clinicians*, *72*(2), 144–164. https://doi.org/10.3322/caac.21706

Menyhárt, O., & Győrffy, B. (2021). Multi-omics approaches in cancer research with applications in tumor subtyping, prognosis, and diagnosis. *Computational and Structural Biotechnology Journal*, *19*, 949–960. https://doi.org/10.1016/j.csbj.2021.01.009

National Center for Biotechnology Information. (2024a). PubChem compound summary for CID 24360, camptothecin. https://pubchem.ncbi.nlm.nih.gov/compound/Camptothecin

National Center for Biotechnology Information. (2024b). PubChem compound summary for CID 36314, paclitaxel. https://pubchem.ncbi.nlm.nih.gov/compound/Paclitaxel

National Center for Biotechnology Information. (2024c). PubChem compound summary for CID 64971, betulinic acid. https://pubchem.ncbi.nlm.nih.gov/compound/Betulinic-Acid

National Center for Biotechnology Information. (2024d). PubChem compound summary for CID 442972, cyclopamine. https://pubchem.ncbi.nlm.nih.gov/compound/Cyclopamine

National Center for Biotechnology Information. (2024e). PubChem compound summary for CID 6602383, glucoraphanin. https://pubchem.ncbi.nlm.nih.gov/compound/Glucoraphanin

Nielsen, E., Temporiti, M. E. E., & Cella, R. (2019). Improvement of phytochemical production by plant cells and organ culture and by genetic engineering. *Plant Cell Reports*, *38*(10), 1199–1215. Springer Verlag. https://doi.org/10.1007/s00299-019-02415-z

Pan, Q., Mustafa, N. R., Tang, K., Choi, H. Y., & Verpoorte, R. (2016). Monoterpenoid indole alkaloids biosynthesis and its regulation in Catharanthus roseus: A literature review from genes to metabolites. *Phytochemistry Reviews*, *15*, 221–250. https://doi.org/10.1007/s11101-015-9406-4

Pathak, S., Lakhwani, D., Gupta, P., Mishra, B. K., Shukla, S., Asif, M. H., & Trivedi, P. K. (2013). Comparative transcriptome analysis using high papaverine mutant of papaver somniferum reveals pathway and uncharacterized steps of papaverine biosynthesis. *PLoS ONE*, *8*(5), e65622. https://doi.org/10.1371/journal.pone.0065622

PubChem [Internet]. Bethesda (MD): National Library of Medicine (US), National Center for Biotechnology Information. (2004a). PubChem compound summary for CID 5978, vincristine. https://pubchem.ncbi.nlm.nih.gov/compound/Vincristine

PubChem [Internet]. Bethesda (MD): National Library of Medicine (US), National Center for Biotechnology Information. (2004b). PubChem compound summary for CID 13342, vinblastine. https://pubchem.ncbi.nlm.nih.gov/compound/Vinblastine

Rai, A., Saito, K., & Yamazaki, M. (2017, March 30). Integrated omics analysis of specialized metabolism in medicinal plants. *The Plant Journal*, *90*(4), 764–787. https://doi.org/10.1111/tpj.13485

Scossa, F., Benina, M., Alseekh, S., Zhang, Y., & Fernie, A. (2018, May 29). The integration of metabolomics and next-generation sequencing data to elucidate the pathways of natural product metabolism in medicinal plants. *Planta Medica*, *84*(12/13), 855–873. https://doi.org/10.1055/a-0630-1899

Shafi, A., & Zahoor, I. (2021). Metabolomics of medicinal and aromatic plants: Goldmines of secondary metabolites for herbal medicine research. In *Medicinal and aromatic plants* (pp. 261–287). https://doi.org/10.1016/b978-0-12-819590-1.00012-4

Sinha, D., Odoh, U. E., Ganguly, S., Muhammad, M., Chatterjee, M., Chikeokwu, I., & Egbuna, C. (2023). Phytochemistry, history, and progress in drug discovery. In *Phytochemistry, computational tools and databases in drug discovery* (pp. 1–26). Elsevier.

Tholl, D. (2015). Biosynthesis and biological functions of terpenoids in plants. In J. Schrader & J. Bohlmann (Eds.), *Biotechnology of isoprenoids. Advances in biochemical engineering/biotechnology* (Vol. 148). Springer. https://doi.org/10.1007/10_2014_295

Xiao, M., Zhang, Y., Chen, X., Lee, E., Barber, C. J., Chakrabarty, R., Desgagné-Penix, I., Haslam, T. M., Kim, Y., Liu, E., MacNevin, G., Masada-Atsumi, S., Reed, D. W., Stout, J. M., Zerbe, P., Zhang, Y., Bohlmann, J., Covello, P. S., De Luca, V., . . . Sensen, C. W. (2013). Transcriptome analysis based on next-generation sequencing of non-model plants producing specialized metabolites of biotechnological interest. *Journal of Biotechnology*, *166*(3), 122–134. https://doi.org/10.1016/j.jbiotec.2013.04.004

Yu, N., Nützmann, H., MacDonald, J. T., Moore, B., Field, B., Berriri, S., Trick, M., Rosser, S. J., Kumar, S. V., Freemont, P. S., & Osbourn, A. (2016). Delineation of metabolic gene clusters in plant genomes by chromatin signatures. *Nucleic Acids Research*, *44*(5), 2255–2265. https://doi.org/10.1093/nar/gkw100

Zagoskina, N. V., Zubova, M. Y., Nechaeva, T. L., Kazantseva, V. V., Goncharuk, E. A., Katanskaya, V. M., Baranova, E. N., & Aksenova, M. A. (2023). Polyphenols in plants: Structure, biosynthesis, abiotic stress regulation, and practical applications (review). *International Journal of Molecular Sciences*, *24*(18), 13874. https://doi.org/10.3390/ijms241813874

Zhang, S., Chen, Y., He, X., Du, J., Zhang, R., Ma, Y., Hu, X., Zhang, Z., Chen, Q., & Wan, X. (2020, January 3). Identification of MYB transcription factors regulating theanine biosynthesis in tea plant using Omics-based gene coexpression analysis. *Journal of Agricultural and Food Chemistry*, *68*(3), 918–926. https://doi.org/10.1021/acs.jafc.9b06730

Zhang, W., Zeng, Y., Jiao, M., Ye, C., Li, Y., Liu, C., & Wang, J. (2023). Integration of high-throughput omics technologies in medicinal plant research: The new era of natural drug discovery. *Frontiers in Plant Science*, *14*, 1073848. https://doi.org/10.3389/fpls.2023.1073848

Zhao, Y., Liu, G., Yang, F., Liang, Y., Gao, Q., Xiang, C., Li, X., Yang, R., Zhang, G., Jiang, H., Yu, L., & Yang, S. (2023). Multilayered regulation of secondary metabolism in medicinal plants. *Molecular Horticulture*, *3*(1). https://doi.org/10.1186/s43897-023-00059-y

Zhou, M. L., Shao, J. R., & Tang, Y. X. (2009). Production and metabolic engineering of terpenoid indole alkaloids in cell cultures of the medicinal plant *Catharanthus roseus* (L.) G. Don (*Madagascar periwinkle*). *Biotechnology and Applied Biochemistry*, *52*(4), 313–323.

21 Improved Production of Anti-Cancer Compounds From Medicinal Plants Through Molecular and Biotechnological Mutagenesis Strategies

Shital Rani, Shweta Sharma, Dilbag Singh, Reena Sheoran, and Rahul Kumar

21.1 INTRODUCTION

Cancer is right now the prominent contributor to global mortality, after cardiovascular diseases. By 2025, estimates suggest that the number of cancer cases annually will increase to 19.3 million. The death rate from cancer is greater in countries that are developing than in developed nations because of socioeconomic factors. In 2020, approximately 19.3 million new cancer cases and 10 million deaths related to cancer were reported globally (Debela et al., 2021). Cancer remains a serious concern for healthcare systems amid years of medical investigations. In addition to significant economic costs to society on a larger scale, cancer has severe biological and behavioral impacts on patients and their families (Yabroff et al., 2011). Cancer shows some properties like growth without signals, insensitivity to anti-growth responses, apoptosis escape, infinite replicative capacity, increased angiogenesis, and potential for tissue invasion and metastasis. The key characteristics common in all types of cancer is the acquisition of a subpopulation of cells that possess the capacity for accelerated and unregulated proliferation. Usually, this phenomenon is initiated with the development of localized primary tumors inside the affected tissue (Marco, 2013). Genetic factors as well as environmental factors, viral, and some bacterial infections, can result in cancer. These elements can cause mutations and epigenetic modifications in deoxyribose nucleic acid (DNA) structure. Epigenetic modifications can be of microRNAs, DNA methylation, chromatin remodelers, histone readers and modifiers, and other chromatin elements. Mutation acts in two ways and can either activate progenes or inactivate tumor suppressor genes (Baylin & Jones, 2016). The primary cause for mutations is mutagens which can be external chemicals or physical substances. Furthermore, abnormalities that occur during the replication of DNA, repair, and recombination can result in spontaneous mutations. Mutations can be classified as positive, fatal, sub-lethal, and negative (Słoczyńska et al., 2014). Certain substances, such as alkylating agents, nucleoside analogs, or intercalating compounds, have the direct ability to induce point mutations. Others, including persistent inflammation or infections that encourage cell division, have an indirect effect (Weber, 2015). Anticancer therapies like surgery, chemotherapy, and radiation therapy have shown restricted efficacy in treating cancer, and consequently, it is important to explore alternative methods for combating this chronic illness. The most widely used traditional method for cancer treatment involves surgical removal of tumors, often followed by the application of radiotherapy or chemotherapy utilizing X-rays in subsequent stages. During the early stages of the disease, surgical intervention emerges as the most effective among the aforementioned methods (Debela et al., 2021). Different forms of chemotherapy may have varied effects on the target cells. The stage of cancer is the primary factor that influence the choice of chemomedications (Anand et al., 2023). Organs, tissues, and healthy cells can all be harmed by radiation treatment. Even though chemotherapy effectively reduces morbidity and mortality, the majority of chemotherapeutic drugs adversely affect healthy cells, especially those having a high rate of division and growth (Debela et al., 2021). In radiotherapy, tumor tissue is selectively placed under X-rays or gamma rays to cause irreversible DNA damage in cancer cells that divide quickly, leading to the cells' halt during DNA replication and subsequent apoptosis. Novel chemicals/metabolites generated from plants have gained a lot of attention recently as alternatives to traditional therapy because of their anticancer potential and very few side effects. These metabolites have low availability and are very tough in extraction. Biotechnological strategies help us to boost the production of secondary metabolites of plants (Swamy et al., 2018). This chapter will look at plant's secondary metabolites as natural sources of anti-cancer chemicals and the strategies to increase the production of these metabolites.

DOI: 10.1201/9781003475491-21

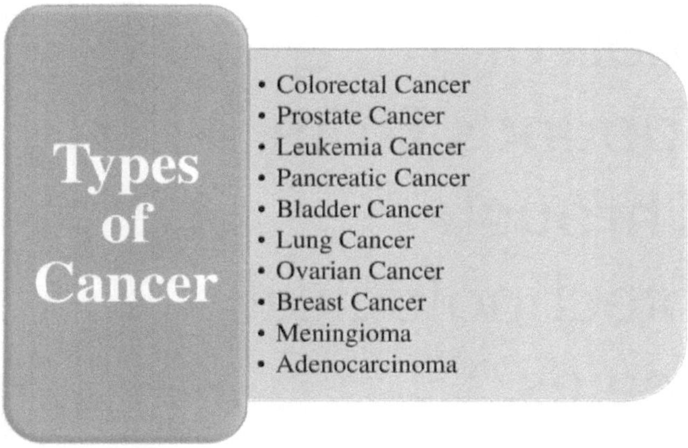

FIGURE 21.1 Different types of cancer.

21.1.1 Important Plants Used for Secondary Metabolites for Cancer

Echinacea, often known as coneflower, has been utilized in cancer treatment due to its capacity to stimulate the immune system and raise the number of natural killer cells, because of its sulfur-containing component and antioxidant qualities. Extracts from *Echinacea purpurea* have immunomodulatory properties that rely on phagocytosis mechanism and immune cell activation. Plant tissue culture is used for the synthesis of secondary metabolite flavonoid with anti-cancer potential compared to direct extraction from plant biomass (Karimi et al., 2018). Ginger, or *Zingiber officinale*, is a popular cancer therapy. Ginger extracts exhibit considerable antiproliferative effects against TNBC, inhibiting the growth and colony formation of MDA-MB-231 and MCF-7 cell lines (Chen et al., 2023). Previous studies have demonstrated that ginger types, Fulbaria and Syedpuri, have demonstrated remarkable anti-cancer efficacy in MCF-7 and MDA-MB-231 cell lines (Rahman et al., 2011). *Allium sativum*, widely known as garlic, exhibits anti-cancer effects. Garlic has various phytoconstituents with anti-cancer properties, including allyl mercaptan (AM), diallyl disulfide (DADS), diallyl trisulfide (DATS), allicin, and diallyl sulfide (DAS). Numerous garlic-derived compounds, phytochemicals, and nanoformulations have been investigated for their effects on various types of cancers. *Curcuma longa*, also known as turmeric, exhibits anti-cancer properties because of phenolic compounds. Curcumin-loaded nanoparticles have anti-inflammatory action and very little toxicity. It acts on many transcription factors and controls cell proliferation (Wilken et al., 2011). *Arctium lappa L.*, often known as Burdock, has anti-cancer chemicals such as Lappaol F (LAF) and Arctigenin. *A lappa* L. often poses lignin, fatty acids, terpenoids, and volatile oil. In traditional Chinese medicine, burdock is frequently used as an anti-cancer. Research has demonstrated that LAF has a potent cytotoxic effect on various cancer cell lines by inducing cell cycle arrest while causing negligible damage to healthy cell lines (Li et al., 2021). Wedelolactone, a chemical found in

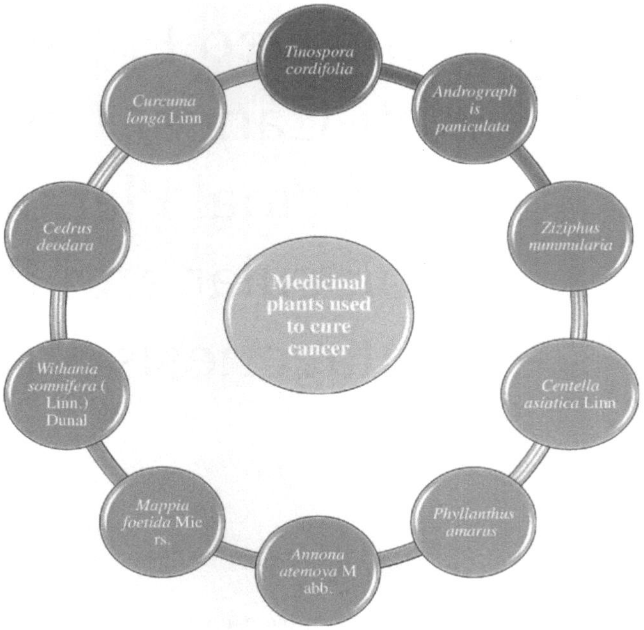

FIGURE 21.2 Different medicinal plants used to cure cancer.

Eclipta alba, downregulates the anti-apoptotic proteins and upregulates the apoptotic pathway in cancer cell lines. *Synadeniumcupulare* leaf water extract has also been shown to be beneficial because of its anti-cancer potential. The purified molecules as well as the extract of *Cimicifuga foetida*, popularly known as black cohosh, show a clear cellular proliferation inhibition. It also reduces TNF-induced IKK and IKK phosphorylation. Citral is found in *Cymbopogon citratus*, a citrus-based plant that shows anti-cancer properties in MDA-MB-231 breast cancer cells. Sumac, or *Rhus coriaria*, includes phenolic acids, flavonoids, and tannins that decrease breast cancer metastasis by inhibiting STAT3, NKB, and NO pathways. *Ricinus communis* L. fruit extract inhibits growth in MDA-MB-231 and MCF-7 cells. The sundew plant, *Droserabormannii*, exhibits anti-inflammatory and anti-cancer effects. It causes apoptosis in MCF-7 cells by cell cycle arrest in the G2/M phase.

21.2 SECONDARY METABOLITES FROM PLANTS FOR CANCER

Plants may supplement current healthcare systems with natural and recombinant anti-cancer medicines. Plants produce a wide range of organic molecules, which are primarily categorized as primary and secondary metabolites. Primary metabolites are necessary for fundamental functions such as respiration, photosynthesis, development, and growth. Secondary metabolites are phytochemicals that are specifically accumulated rather than being present as byproducts of chemical activities (Seca& Pinto, 2018). Plant secondary metabolites are often generated in lower amounts; they frequently accumulate in specific plant organs. Humans initially employed herbal cures because of these pharmacologically active secondary metabolites. Recently, the utilization of medicinal plants as an alternative medicine for preventing and treating cancer among patients has seen a notable rise. The use of such plants was prompted by their significant anticarcinogenic and chemoprotective properties. Furthermore, it is noteworthy that these plants possess a lower toxicity profile in comparison to conventional therapeutic drugs, hence exhibiting promising potential as anti-cancer, anti-tumor, and anti-proliferation agents. The anti-tumor compounds function through multiple mechanisms, including mitochondrial permeabilization, inhibition of essential enzymes involved in the transmission of signals or metabolism in cells, inducing apoptosis through DNA cleavage, and blocking tumor-induced vascular development. Currently, the majority of anticarcinogenic chemotherapeutics used in clinical settings are vinblastine and vincristine, paclitaxel, and camptothecin.

Catharanthus roseus is the source of vinca monoterpene indole alkaloids including vinblastine and vincristine. Vinblastine and vincristine have extremely low plant productivity (0.001–0.0003%), which accounts for their very expensive cost. The methyl group of vinblastine is oxidized to produce vincristine. Svoboda et al. first isolated and described Vincristine in 1961, with the US Food and Drug Administration (FDA) initially approving it for leukemia in 1963. At present, vincristine is mostly used in conjunction with other chemotherapy drugs. Hodgkin's lymphoma and other malignancies are frequently treated with vinblastine (Banyal et al., 2023).

Paclitaxel (PTX), a mitostatic medication, inhibits cell division and treats cancer. PTX is present in the Pacific tew tree (*Taxus brevifolia*) bark extract. In addition to its role in inducing cell-cycle arrest by targeting tubulin, PTX seems to enhance signaling factors in cancer therapy (Nawara et al., 2021). Because of its efficacy, PTX is potentially the most significant anti-cancer drug in history. In 2021, the global market for PTX reached a value of $4.51 billion and is predicted to exceed more than $11.16 billion by 2030 (Perez-Matas et al., 2022).

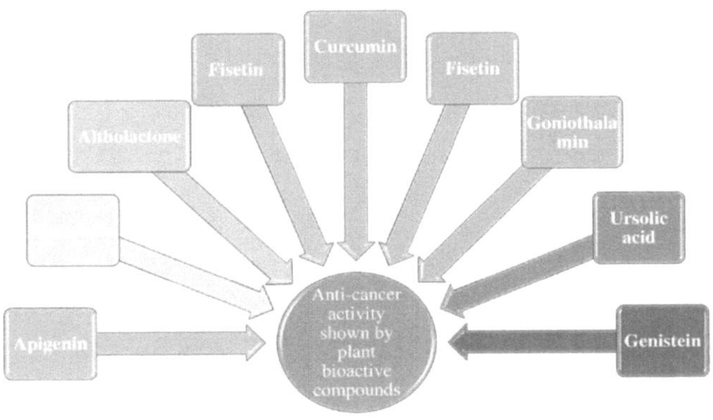

FIGURE 21.3 Anti-cancer activity shown by plant bioactive compounds.

Camptothecin, a modified monoterpene indole alkaloid, is naturally synthesized by specific angiosperm plants. Topotecan and irinotecan, both water-soluble derivatives of camptothecin, are also employed in clinical trials. Camptothecin's biological target is the type I DNA topoisomerase. Camptothecin is a treatment for recurrent colon cancer that also exhibits extraordinary action against lung, ovarian, and uterine cancer (Behera & Padhi, 2020).

21.3 BIOTECHNOLOGICAL APPROACHES TO ENHANCE SECONDARY METABOLITE PRODUCTION

The field of biotechnology offers a potential solution for addressing the limited availability or challenges associated with obtaining plant secondary metabolites from their natural sources. Mutagenesis strategies have been used for many years to improve crops, livestock, and other organisms and are now being increasingly used to develop new and improved medical treatments and industrial products (Sharma et al., 2023). Overall, molecular and biotechnological mutagenesis methods have the potential to transform how anti-cancer drugs are developed. These approaches may lead to the development of novel, more potent anticancer therapies in addition to a decrease in the cost and increase in accessibility of anti-cancer medications. These strategies can be used to increase the expression of genes associated with the biosynthesis of anti-cancer compounds, overexpress transcription factors that regulate their biosynthesis, and create plants with mutations that disrupt genes responsible for the degradation of anti-cancer compounds, this can be done by inserting additional copies of the gene into the plant genome, or by altering the regulation of the gene to increase its expression. However, it is important to note that these strategies are still under development, and more research is needed to optimize them and ensure the safety and efficacy of the plants that are produced. The utilization of biotechnological approaches and *in vitro* culture offers a valuable, sustainable, and eco-friendly solution for producing bioactive compounds (Sharma et al., 2024). This helps to minimize the need for chemically synthesized compounds while reducing the overexploitation of natural resources.

21.3.1 Metabolic Engineering

Plant metabolic engineering involves studying enzymatic reactions and biosynthetic processes at genomic, transcriptomic, and proteomic levels and manipulating genes encoding critical enzymes, to enhance the synthesis of a desired metabolite while decreasing the yield of undesired products, particularly concerning bioactive secondary compounds. Manipulation of biosynthetic pathways offers a promising approach for enhancing high productivity in plant cell cultures. This aim can be accomplished through the overexpression of enzymes, enhancing the efficiency of

TABLE 21.1
Structure of Secondary Metabolites Produced by Different Plant Species and Their Role

Sr. No.	Plant Species	Secondary Metabolites	Role	Structure	Reference
1.	*Camptotheca acuminate*	Camptothecin	Anti-tumor activity	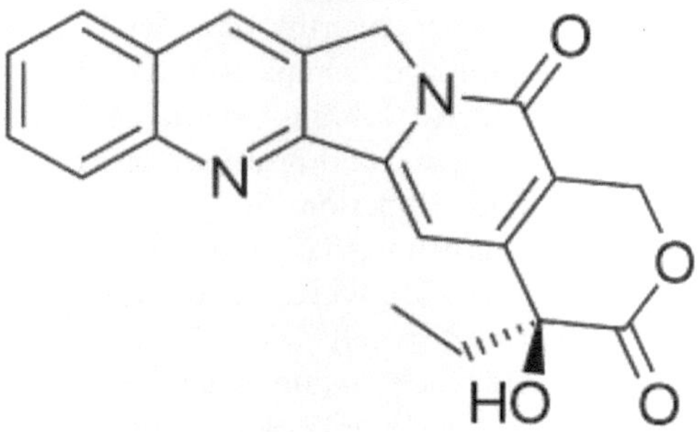	Lorence et al., 2004
2.	*Curcuma longa*	Curcumin	Anti-cancer, Anti-oxidant	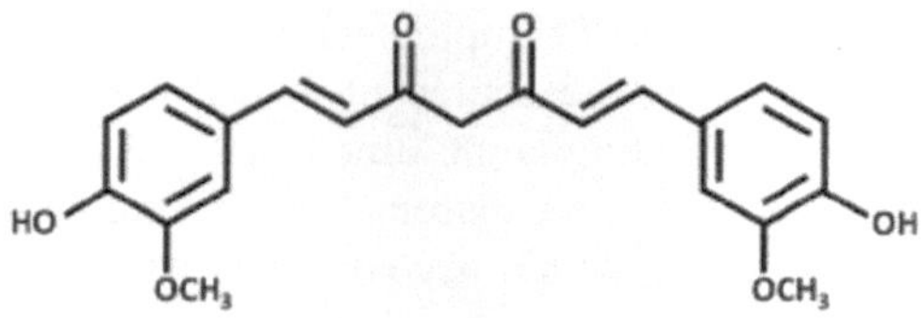	Rahmani et al., 2018
3.	*Podophyllum emodi, Podophylumpeltatum*	Epipodophyllotoxins	Anti-tumor agent	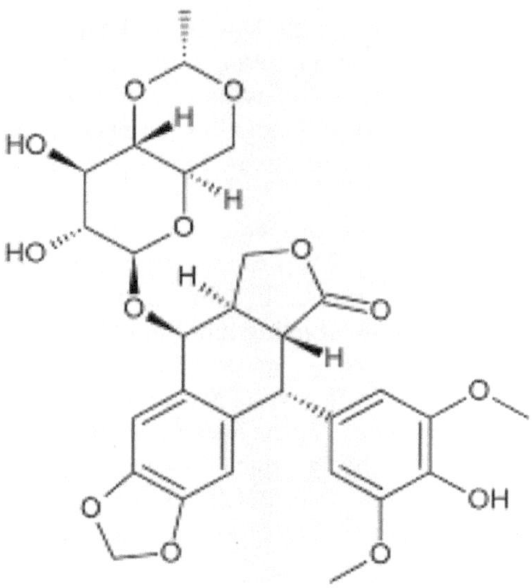	Giri & Lakshmi Narasu, 2000
4.	*Arctium lappa*	Arctigenin	Anti-cancer	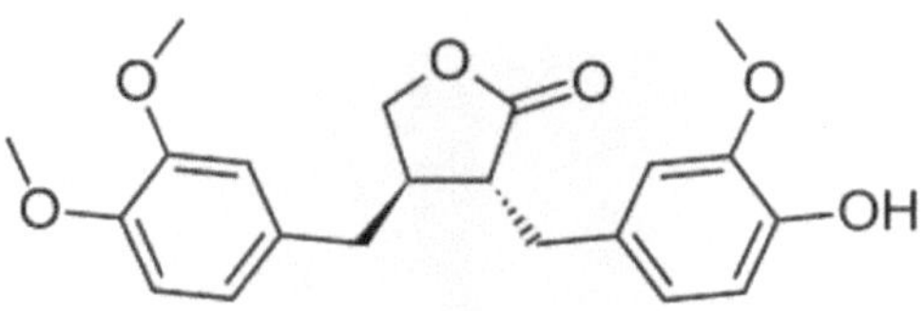	Li et al., 2021

TABLE 21.1 (*Continued*)
Structure of Secondary Metabolites Produced by Different Plant Species and Their Role

Sr. No.	Plant Species	Secondary Metabolites	Role	Structure	Reference
5.	*Taxus brevifolia*	Paclitaxel	Anti-ovarian, breast and non-small cell lung cancer	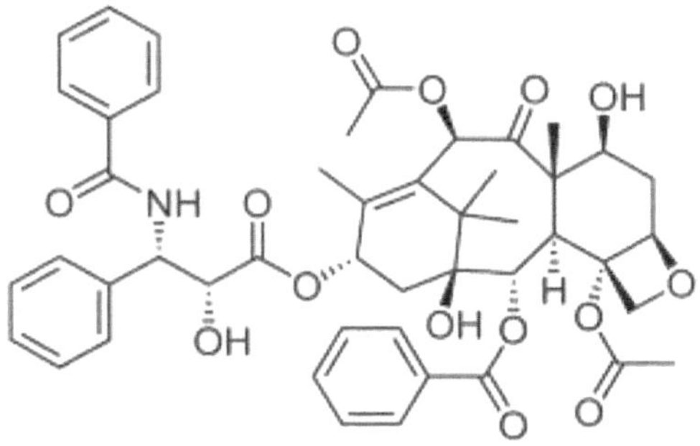	Wahab et al., 2020
6.	*Tabebuia avellanedae*	Beta-lapachone	Anti-inflammatory and anti-neoplastic activity	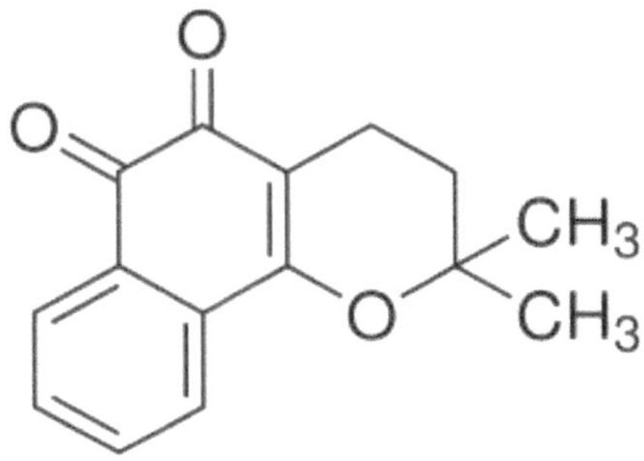	Lee ta al., 2018
7.	*Colchicum autumnale*	Colchicine	Acute Pericarditis therapy	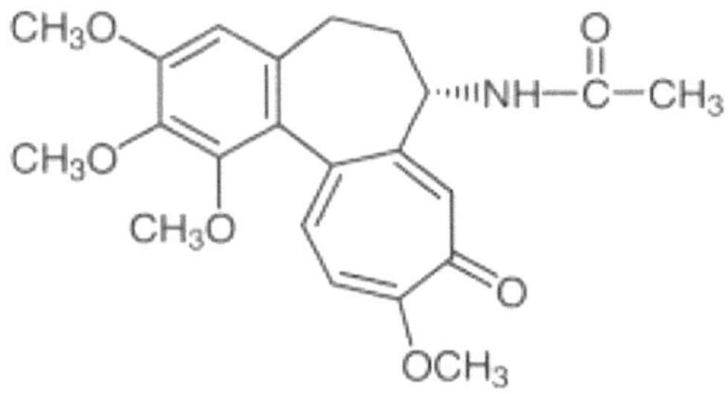	Imazio, 2015
8.	*Eclipta alba*	Wedelolactone	Anti-cancer, causes apoptosis of cancerous cells	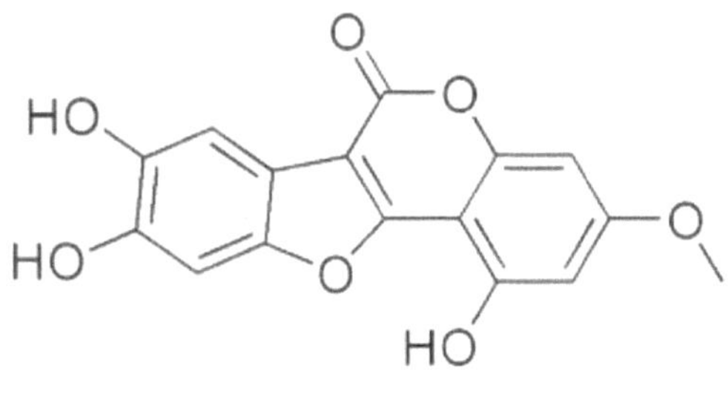	Sarveswaran et al., 2012

(*Continued*)

TABLE 21.1 ***(Continued)***
Structure of Secondary Metabolites Produced by Different Plant Species and Their Role

Sr. No.	Plant Species	Secondary Metabolites	Role	Structure	Reference
9.	*Zingiber officinale*	Gingerol	Anti-inflammatory, anti-cancer, and anti-oxidant	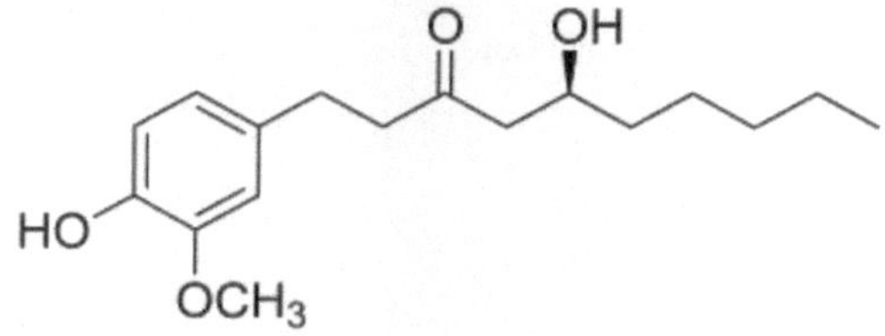	Mohd Yusof, 2016
10.	*Capsicum annuum*	Luteolin	Cardioprotective, anti-inflammatory, anti-cancer	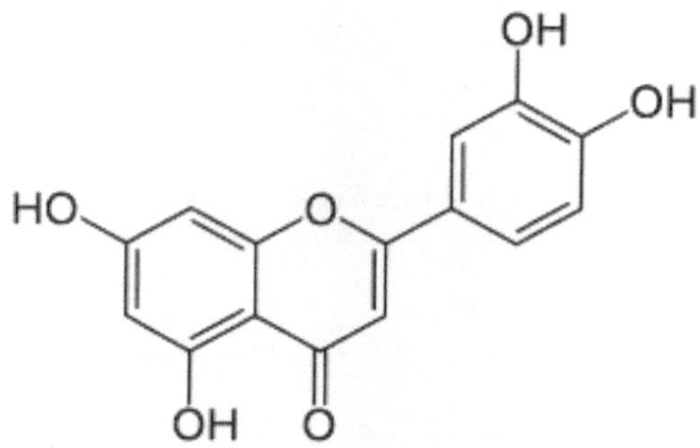	Materska et al., 2003
11.	*Gossypium hirsutum*	Gossypol	Anti-neoplastic agent	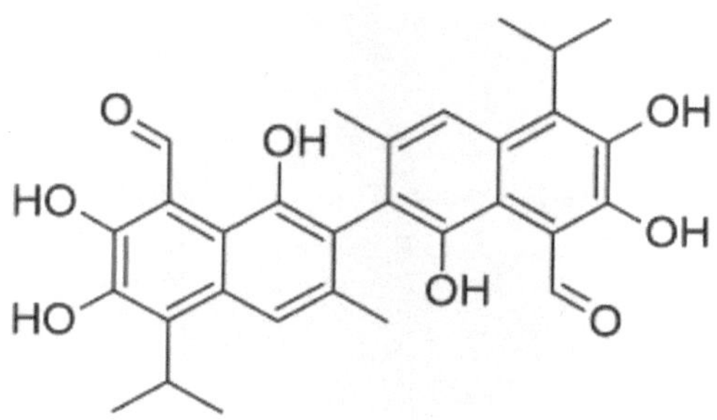	Stipanovic et al., 2005
12.	*Alore vera*	Aloin B	Anti-cancer activity	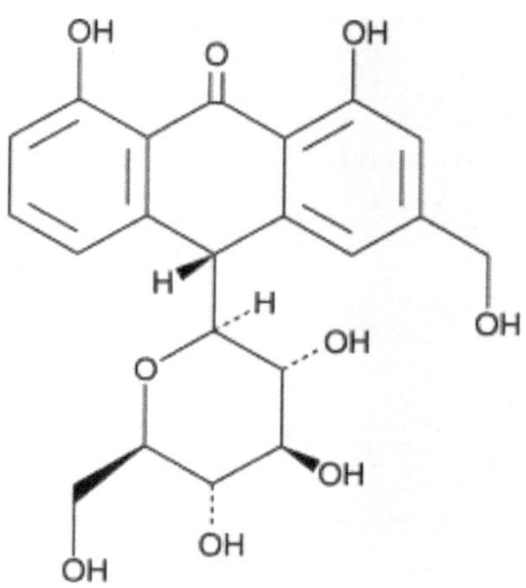	Bayati and Moradi, 2014

TABLE 21.1 (*Continued*)
Structure of Secondary Metabolites Produced by Different Plant Species and Their Role

Sr. No.	Plant Species	Secondary Metabolites	Role	Structure	Reference
13.	*Allium sativum*	Allicin	Anti-cancer activity	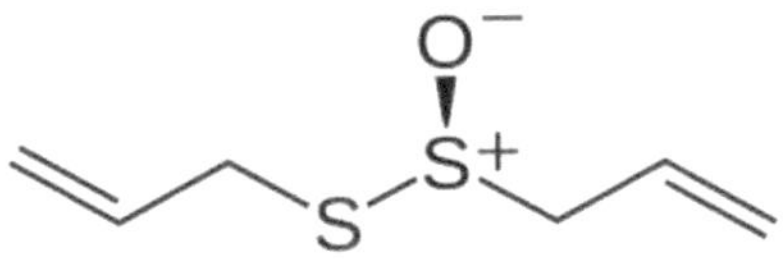	Li et al., 2018
14.	*Cymbopogon citratus*	Citral	Anti-cancer activity	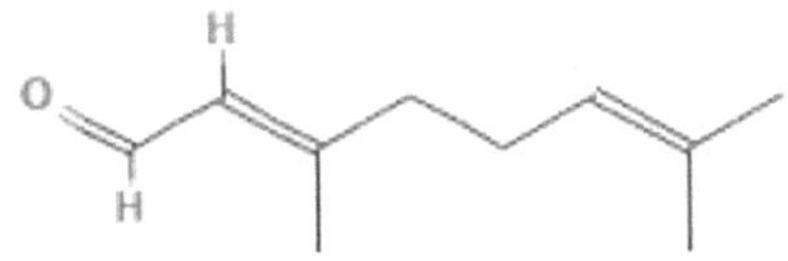	Rojas-Armas et al., 2020
15.	*Withaniasomnifera*	Withaferin A	Anti-cancer activity	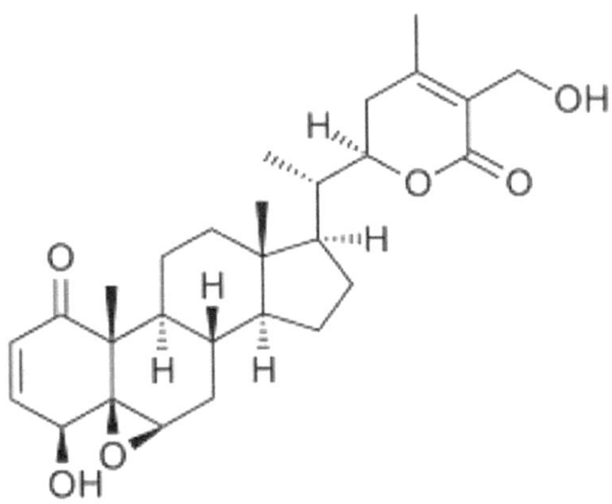	Atteeq, 2022

rate-limiting enzymes, inhibiting catabolic processes, and establishing specialized compartments to store the desired metabolite. However, enhancing the expression of specific genes might not necessarily result in improved production. Metabolic engineering also includes the suppression of competitive pathways to optimize the metabolic flux of targeted intermediates of biosynthetic pathways to increase production levels through various implemented strategies. Hence, by altering metabolic pathways, enhancing precursor availability, ensuring balanced metabolic fluxes, and employing synthetic biology techniques, metabolic engineers can maximize the efficacy of biological systems for producing valuable secondary compounds with anti-cancer properties. There have been notable developments in unraveling the biosynthetic pathways of various anti-cancer compounds, particularly indole alkaloids from *Catharanthus roseus* and taxol. In relation to this, Sharma et al. found a twofold increase in total alkaloid content in comparison to the untransformed control, by inserting extra copies of strictosidine synthase and tryptophan

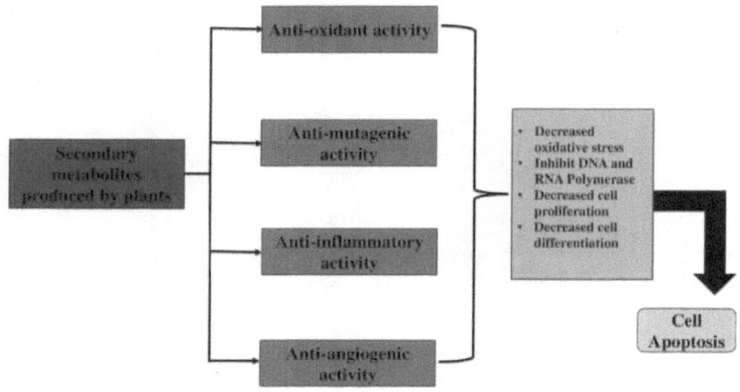

FIGURE 21.4 Different activities are shown by secondary metabolites produced by plants and their role in apoptosis.

decarboxylase genes into *Catharanthus roseus* callus (Sharma et al., 2018). Similarly, Sun and Peebles achieved a 47% rise in the content of six-root specific terpenoid indole alkaloids by introducing *Catharanthus roseus* hairy roots with enhanced expression of the transcriptional regulator ORCA3 (octadecanoid-responsive *Catharanthus* AP2-domain protein) gene and the strictosidine glucosidase gene (Sun & Peebles, 2016). Using a different approach, Wang et al. observed that overexpressing of the CrTPT2 gene, encoding an ATP-binding cassette transporter for catharanthine, in *Catharanthus roseus* hairy roots led to the accumulation of catharanthine, a fivefold increase compared to wild-type control roots (Wang et al., 2019). Onrubia et al. identified taximin, a novel regulator of taxol biosynthesis, by utilizing jasmonate-elicited *Taxus baccata* cell cultures and complementary DNA-amplified fragment length polymorphism (Onrubia et al., 2014). Several researchers have investigated the impact of elicitors like methyl jasmonate, fungal filtrate, and yeast extract, on the biosynthesis of indole alkaloids. In this context, Guo et al. examined how various elicitors affected the biosynthesis of vinblastine, vindoline, and tabersonine. Their findings revealed that tryptophan and acetyl-CoA emerged as crucial factors in vinblastine biosynthesis (Guo et al., 2013).

However, there are few studies related to the production of camptothecin and podophyllotoxin. Research conducted by Cui et al. showed a remarkable increase of over 50% in camptothecin production in the transgenic hairy root culture of *Ophiorrhiza pumila* (Cui et al., 2015). Yousefzadi et al. demonstrated that salicylic acid boosts the synthesis of podophyllotoxin in *Linum album* cultures by potentially modulating the expression of genes associated with the early stages of the phenylpropanoid pathway without affecting the genes involved in later steps of the biosynthetic process (Yousefzadi et al., 2010). The aforementioned studies emphasize the significance of understanding complex pathways essential for the efficient metabolic engineering of plants to increase the synthesis of medicinally valuable compounds. However, the lack of in-depth understanding of the biosynthetic pathways still poses a barrier to the practical implementation of this strategy to augment production.

21.3.2 Plant Tissue Culture

Plant tissue culture is at the core of medicinal biotechnology, enabling increased metabolite production even in medicinal plants that exhibit low productivity and susceptibility to biotic and abiotic stresses. The successful application of several other approaches, including *in situ* and *ex situ* conservation, micropropagation, polyploidy induction, bioengineering, and genetic transformation crucial for the improvement of medicinal plants are intricately linked to plant tissue culture either directly or indirectly. Furthermore, the effect of different experimental factors and elicitors on the synthesis of secondary metabolites in medically valuable plants, along with elucidating signaling, transport mechanisms, and hormone metabolism, can be achieved by cultivating the plants under *in vitro* conditions.

This approach not only facilitates the conservation of endangered medicinal plants and the enhancement of bioactive compounds but also allows for the production of engineered molecules (Espinosa-Leal et al., 2018). Thus, plant tissue culture opens up the possibility of cultivating genetically modified plants capable of synthesizing novel secondary metabolites, such as anti-cancer compounds, especially in cases where natural supply is limited. Key biotechnological applications in synthesizing these metabolites rely on advanced tissue culture techniques, including hairy root culture, callus culture, protoplast culture, and micropropagation. Various secondary compounds synthesized by applying tissue culture techniques sourced from a variety of explants are monoterpene-derived indole alkaloid camptothecin from the radicle-derived roots of *Pyrenacanthavolubilis*, alkaloids (vincristine, and vinblastine) from *Catharanthus roseus*, rosmarinic acid from the leaves and shoots of *Mentha spicata*, and taxol in *Taxus baccata* subsp. *wallichiana* (Hima et al., 2020; Mishra et al., 2019; Yousefian et al., 2020). Therefore, by applying plant cell and tissue culture techniques, it is feasible to rapidly multiply and recover true-to-type plants which is especially crucial for endangered species such as several medicinal plants with anti-cancer properties.

21.4 USE OF THE COMMON MOLECULAR AND BIOTECHNOLOGICAL MUTAGENESIS

Mutagenesis strategies can be used to enhance the production of anti-cancer compounds from medicinal plants in a number of ways. For example, mutations can be introduced into genes responsible for the biosynthesis of anti-cancer compounds to boost their production. Mutations can also be deliberately introduced into genes that compete with the biosynthesis of anti-cancer compounds to divert more resources to their production. Additionally, mutations can be introduced to create new or modified anti-cancer compounds.

- Chemical mutagenesis: This process involves the exposure of cells or tissues to chemicals that have the potential to harm DNA. Some of the

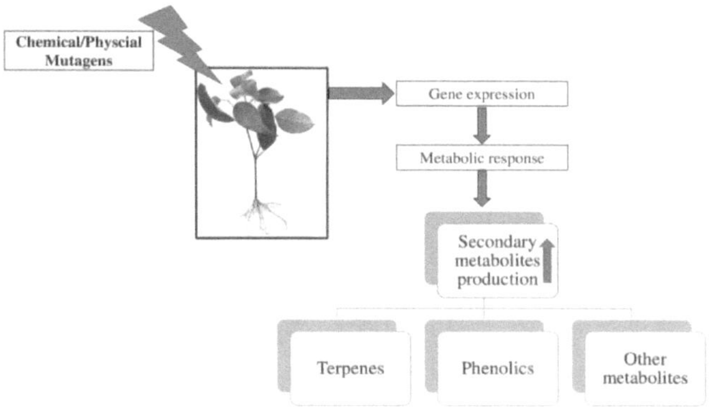

FIGURE 21.5 Use of mutagens to enhance secondary metabolites production.

most commonly used chemical mutagens include N-methyl-N-nitrosourea (MNU), ethyl methane sulfonate (EMS), sodium azide, and nitrous acid. Chemical mutagenesis is a relatively simple and inexpensive technique, but it can also be very random, resulting in a large number of mutations, many of which may be harmful. Elicitors are chemicals that can trigger the plant's defense response. Eliciting *Artemisia annua* plants with methyl jasmonate led to a 50-fold increase in the production of artemisinin, an anti-cancer compound that is also used to treat malaria.

- UV mutagenesis: This involves exposing cells or tissues to ultraviolet (UV) light. UV light can damage DNA by causing cross-linking between adjacent thymine bases. UV mutagenesis is more specific than chemical mutagenesis and is often used to generate targeted mutations in specific genes.
- Transposable element mutagenesis: Transposable elements are mobile pieces of DNA that can insert themselves into different locations in the genome. This can disrupt genes and cause mutations. Transposable element mutagenesis is a powerful technique for generating random mutations, but it can be difficult to control. Overexpression of the taxane biosynthetic pathway in *Catharanthus roseus* was achieved using transposable element mutagenesis. The resulting plants produced 10 times more taxol, an important anti-cancer drug.
- Genome editing: Genome editing is a relatively new technology, but it is rapidly becoming the most widely used mutagenesis strategy. Conventional methods of genetic engineering face challenges in inducing significant alterations in plant genomes due to variations in copy numbers and the random insertion of DNA constructs into single or multiple genetic sites on chromosomes. However, the emergence of diverse technologies such as genome editing tools, 'tunable' transcription factors, site-specific recombinase, and synthetic promoters has facilitated a more precise and rapid improvement in plants. Sequence-specific nucleases like CRISPR/Cas9, TALENs, and ZFNs have found extensive application in generating precise and site-specific double-stranded breaks at predetermined locations within the gene of interest. This makes it possible to generate targeted mutations with a high degree of specificity. Knockout of the shikonin reductase gene in *Lithospermum erythrorhizon* was done by using CRISPR/Cas9 genome editing. This results in 20 times more shikonin production, an anti-cancer compound renowned for its broad spectrum of biological activities. The introduction of the CYP71D20 gene into *Nicotiana tabacum*done by *Agrobacterium*-mediated transformation increased the paclitaxel, another important anti-cancer drug. MYB12 plays a crucial role as a transcription factor in the biosynthesis of terpenoid indole alkaloids, such as vincristine and vinblastine, which are two important anti-cancer drugs. Overexpressing MYB12 in *Catharanthus roseus* led to a notable rise in the production of vincristine and vinblastine.

21.5 CONCLUSION

Although significant advancements have been achieved in cancer research regarding rapid diagnosis and chemotherapy, it persists as a formidable threat, contributing significantly to global mortality rates. Therefore, it is crucial to implement strategies to combat and avert cancer. The use of chemically synthesized drugs is restricted due to their tendency to induce toxicity in healthy tissues, leading to numerous health complications. In this context, there is a growing appeal for alternative treatments utilizing natural anti-cancer compounds extracted from plants. Plant-based compounds have shown promising potential in fighting against cancer due to their effectiveness, minimal side effects, and multiple mechanisms of action. Plants offer a diverse array of secondary metabolites with various biological activities, making them valuable for agrochemicals and pharmaceutical use. While many plant species contain cytotoxic compounds, only a few have been proven to have clinical anti-cancer activity, such as podophyllotoxin, camptothecin derivatives, indole monoterpene alkaloids, and taxanes. Biotechnological advancements in key medicinal plant species should focus on boosting regeneration rates and elevating the accumulation of specific pharmaceutically active metabolites. Furthermore, by utilizing plant tissue culture techniques, medicinal plants with anti-cancer properties can be efficiently propagated and preserved in a relatively brief timeframe, ensuring the conservation and availability of these valuable species. Additionally, metabolic engineering plays a crucial role in augmenting the production of secondary metabolites by amplifying the expression of targeted genes. This process can potentially increase the activity of enzymes engaged in metabolism, resulting in increased accumulation of desired compounds.

However, the limited understanding of various biosynthetic pathways for bioactive compounds poses a challenge to effectively implementing this strategy to achieve significant yields of metabolites. Further research is needed to better understand the biosynthetic pathways and their regulation which will help in the creation of affordable technologies for producing anti-cancer compounds on a greater scale.

REFERENCES

Anand, U., Dey, A., Chandel, A. K. S., Sanyal, R., Mishra, A., Pandey, D. K., De Falco, V., Upadhyay, A., Kandimalla, R., Chaudhary, A., & Dhanjal, J. K. (2023). Cancer chemotherapy and beyond: Current status, drug candidates, associated risks and progress in targeted therapeutics. *Genes & Diseases*, *10*(4), 1367–1401.

Atteeq, M. (2022, October 19). Evaluating anticancer properties of Withaferin A – A potent phytochemical. *Frontiers in Pharmacology*, *13*, 975320.

Banyal, A., Tiwari, S., Sharma, A., Chanana, I., Patel, S. K. S., Kulshrestha, S., & Kumar, P. (2023). Vinca alkaloids as a potential cancer therapeutics: Recent update and future challenges. *3 Biotech*, *13*(6).

Bayati, Z. J., & Moradi, K. N. (2014). Component and application Aloe vera plant in medicine. *International Journal of Advanced Biological and Biomedical Research*, *2*(5), 1876–1882.

Baylin, S. B., & Jones, P. A. (2016). Epigenetic determinants of cancer. *Cold Spring Harbor Perspectives in Biology*, *8*(9), 1–35.

Behera, A., & Padhi, S. (2020). Passive and active targeting strategies for the delivery of the camptothecin anticancer drug: A review. *Environmental Chemistry Letters*, *18*(5), 1557–67.

Chen, C. Y., Chen, Y. N., Shiau, J. P., Tang, J. Y., Hou, M. F., & Chang, H. W. (2023). Ginger-derived 3HDT exerts antiproliferative effects on breast cancer cells by apoptosis and DNA damage. *International Journal of Molecular Sciences*, *24*(6).

Cui, L., Ni, X., Ji, Q., Teng, X., Yang, Y., Wu, C., Zekria, D., Zhang, D., & Kai, G. (2015). Co-overexpression of geraniol-10-hydroxylase and strictosidine synthase improves anti-cancer drug camptothecin accumulation in *Ophiorrhiza pumila*. *Scientific Reports*, *5*, 1–9.

Debela, D. T., Muzazu, S. G. Y., Heraro, K. D., Ndalama, M. T., Mesele, B. W., Haile, D. C., Kitui, S. K., & Manyazewal, T. (2021). New approaches and procedures for cancer treatment: Current perspectives. *SAGE Open Medicine*, *9*.

Espinosa-Leal, C. A., Puente-Garza, C. A., García-Lara, S. (2018). In vitro plant tissue culture: Means for production of biological active compounds. *Planta*, *248*(1), 1–18.

Giri, A., & Lakshmi Narasu, M. (2000, October). Production of podophyllotoxin from *Podophyllum hexandrum*: A potential natural product for clinically useful anticancer drugs. *Cytotechnology*, *34*, 17–26.

Guo, Z. G., Liu, Y., Gong, M. Z., Chen, W., & Li, W. Y. (2013). Regulation of vinblastine biosynthesis in cell suspension cultures of *Catharanthus roseus*. *Plant Cell, Tissue and Organ Culture*, *112*(1), 43–54.

Hima, S., Midhu, C. K., Krishnakumar, G., & Satheeshkumar, K. (2020). In vitro seedlings as dynamic explants for establishment of root cultures of *Pyrenacantha volubilis* Hook. for camptothecin production. *Proceedings of the National Academy of Sciences, India Section B*, *90*(2), 405–13.

Imazio, M. (2015, February 1). Colchicine for pericarditis. *Trends in Cardiovascular Medicine*, *25*(2), 129–36.

Karimi, N., Behbahani, M., Dini, G., & Razmjou, A. (2018). Enhancing the secondary metabolite and anticancer activity of *Echinacea purpurea* callus extracts by treatment with biosynthesized ZnO nanoparticles. *Advances in Natural Sciences: Nanoscience and Nanotechnology*, *9*(4).

Lee, M., Ban, J. J., Chung, J. Y., Im, W., & Kim, M. (2018, May 9). Amelioration of Huntington's disease phenotypes by Beta-Lapachone is associated with increases in Sirt1 expression, CREB phosphorylation and PGC-1α deacetylation. *PloS One*, *13*(5), e0195968.

Li, C., Jing, H., Ma, G., & Liang, P. (2018, April 1). Allicin induces apoptosis through activation of both intrinsic and extrinsic pathways in glioma cells. *Molecular Medicine Reports*, *17*(4), 5976–5981.

Li, X., Lin, Y. Y., Tan, J. Y., Liu, K. L., Shen, X. L., Hu, Y. J., Yang, R. Y., & Lappaol, F. (2021). Lappaol F, an anticancer agent, inhibits YAP via transcriptional and post-translational regulation. *Pharmaceutical Biology*, *59*(1), 619–628.

Lorence, A., Medina-Bolivar, F., & Nessler, C. L. (2004, January). Camptothecin and 10-hydroxycamptothecin from *Camptotheca acuminata* hairy roots. *Plant Cell Reports*, *22*, 437–441.

Marco, E. (2013). Semblanza de Ada Raquel Aular Alfonzo. *An Venez Nutr*, *26*(1), 62.

Materska, M., Piacente, S., Stochmal, A., Pizza, C., Oleszek, W., & Perucka, I. (2003, August 1). Isolation and structure elucidation of flavonoid and phenolic acid glycosides from pericarp of hot pepper fruit *Capsicum annuum* L. *Phytochemistry*, *63*(8), 893–898.

Mishra, M. R., Srivastava, R. K., & Akhtar, N. (2019). Effect of nitrogen, phosphorus and medium pH to enhance alkaloid production from *Catharanthus roseus* cell suspension culture. *International Journal of Secondary Metabolite*,*6*(2), 137–153.

Mohd Yusof, Y. A. (2016). Gingerol and its role in chronic diseases. In *Drug discovery from mother nature* (pp. 177–207). Springer International Publishing.

Nawara, H. M., Afify, S. M., Hassan, G., Zahra, M. H., Seno, A., & Seno, M. (2021). Paclitaxel-based chemotherapy targeting cancer stem cells from mono- to combination therapy. *Biomedicines*, *9*(5), 1–17.

Onrubia, M., Pollier, J., Vanden Bossche, R., Goethals, M., Gevaert, K., Moyano, E., Vidal-Limon, H., Cusido, R. M., Palazo, J., & Goossens, A. (2014). Taximin, a conserved plant-specific peptide is involved in the modulation of plant-specialized metabolism. *Plant Biotechnology Journal*, *12*(7), 971–983.

Perez-Matas, E., Hanano, A., Moyano, E., Bonfill, M., Cusido, R. M., & Palazon, J. (2022, July). Insights into the control of taxane metabolism: Molecular, cellular, and metabolic changes induced by elicitation in *Taxus baccata* cell suspensions. *Frontiers in Plant Science*, *13*, 1–19.

Rahman, S., Salehin, F., & Iqbal, A. (2011). Retracted article: In vitro antioxidant and anticancer activity of young *Zingiber officinale* against human breast carcinoma cell lines. BMC *Complementary and Alternative Medicine*, *11*, 1–7.

Rahmani, A. H., Alsahli, M. A., Aly, S. M., Khan, M. A., &Aldebasi, Y. H. (2018, January 1). Role of curcumin in disease prevention and treatment. *Advanced Biomedical Research*, *7*(1), 38.

Rojas-Armas, J. P., Arroyo-Acevedo, J. L., Palomino-Pacheco, M., Herrera-Calderón, O., Ortiz-Sánchez, J. M., Rojas-Armas, A., Calva, J., Castro-Luna, A., & Hilario-Vargas, J. (2020,

July 20). The essential oil of *Cymbopogon citratus*stapt and carvacrol: An approach of the antitumor effect on 7, 12-dimethylbenz-[α]-anthracene (DMBA)-induced breast cancer in female rats. *Molecules*, *25*(14), 3284.

Sarveswaran, S., Gautam, S. C., & Ghosh, J. (2012, December 1). Wedelolactone, a medicinal plant-derived coumestan, induces caspase-dependent apoptosis in prostate cancer cells via downregulation of PKCε without inhibiting Akt. *International Journal of Oncology*, *41*(6), 2191–2199.

Seca, A. M. L., & Pinto, D. C. G. A. (2018). Plant secondary metabolites as anticancer agents: Successes in clinical trials and therapeutic application. *International Journal of Molecular Sciences*, *19*(1).

Sharma, A., Verma, P., Mathur, A., & Mathur, A. K. (2018). Genetic engineering approach using early Vinca alkaloid biosynthesis genes led to increased tryptamine and terpenoid indole alkaloids biosynthesis in differentiating cultures of *Catharanthus roseus*. *Protoplasma*, *255*(1), 425–35.

Sharma, M. M. M., Kapoor, D., Loyal, A., Kumar, R., Sharma, P., & Husen, A. (2024). Environmental toxicity of engineered carbon nanoparticles. In *Carbon-based nanomaterials: Synthesis, agricultural, biomedical, and environmental interventions* (pp. 337–353). Springer Nature Singapore.

Sharma, M. M. M., Kapoor, D., Rohilla, R., & Sharma, P. (2023). Nanomaterials and their toxicity to beneficial soil microbiota and fungi associated plants rhizosphere. In *Nanomaterials and nanocomposites exposures to plants: Response, interaction, phytotoxicity and defense mechanisms* (pp. 353–380). Springer Nature Singapore.

Słoczyńska, K., Powroźnik, B., Pekala, E., & Waszkielewicz, A. M. (2014). Antimutagenic compounds and their possible mechanisms of action. *Journal of Applied Genetics*, *55*(2), 273–285.

Stipanovic, R. D., Puckhaber, L. S., Bell, A. A., Percival, A. E., & Jacobs, J. (2005, August 10). Occurrence of (+)-and (−)-gossypol in wild species of cotton and in *Gossypium hirsutum* var. marie-galante (Watt) Hutchinson. *Journal of Agricultural and Food Chemistry*, *53*(16), 6266–6271.

Sun, J., & Peebles, C. A. M. (2016). Engineering overexpression of ORCA3 and strictosidine glucosidase in *Catharanthus roseus* hairy roots increases alkaloid production. *Protoplasma*, *253*(5), 1255–1264.

Swamy, M. K., Sinniah, U. R., & Ghasemzadeh, A. (2018). Anticancer potential of rosmarinic acid and its improved production through biotechnological interventions and functional genomics. *Applied Microbiology and Biotechnology*, *102*(18), 7775–7793.

Wahab, N. A., Othman, Z., Nasri, N. W., Mokhtar, M. H., Ibrahim, S. F., Hamid, A. A., Raja Ali, R. A., & Mokhtar, N. M. (2020, April). Inhibition of miR-141 and miR-200a increase DLC-1 and ZEB2 expression, enhance migration and invasion in metastatic serous ovarian cancer. *International Journal of Environmental Research and Public Health*, *17*(8), 2766.

Wang, Y., Yang, B., Zhang, M., Jia, S., & Yu, F. (2019). Application of transport engineering to promote catharanthine production in *Catharanthus roseus* hairy roots. *Plant Cell, Tissue and Organ Culture*, *139*(3), 523–530.

Weber, G. F. (2015). *Molecular therapies of cancer* (pp. 1–455). Springer.

Wilken, R., Veena, M. S., Wang, M. B., & Srivatsan, E. S. (2011). Curcumin: A review of anti-cancer properties and therapeutic activity in head and neck squamous cell carcinoma. *Molecular Cancer*, *10*(1), 12.

Yabroff, K. R., Lund, J., Kepka, D., & Mariotto, A. (2011). Economic burden of cancer in the United States: Estimates, projections, and future research. *Cancer Epidemiology, Biomarkers & Prevention*, *20*(10), 2006–2014.

Yousefian, S., Lohrasebi, T., Farhadpour, M., & Haghbeen, K. (2020). Production of phenolic acids in hairy root cultures of medicinal plant *Mentha spicata* L. in response to elicitors. *Molecular Biology Research Communications*, *9*(1), 23–34.

Yousefzadi, M., Sharifi, M., Behmanesh, M., Ghasempour, A., Moyano, E., & Palazon, J. (2010). Salicylic acid improves podophyllotoxin production in cell cultures of Linum album by increasing the expression of genes related with its biosynthesis. *Biotechnology Letters*, *32*(11), 1739–1743.

22 Integrated Molecular Strategies for Enhanced Production of Medicinal Plant-Derived Alkaloids

Balasundaram Saraswathy Chithra Devi, M. Krishnasreya, S. Tharani, and M. R. Elakkiya

22.1 INTRODUCTION

Alkaloids constitute a vast category of naturally existing organic compounds, comprising the majority of the extensive group of phytochemicals (Rosales et al., 2020). Nitrogen atoms play a crucial role in differentiating alkaloids, leading to their alkaline characteristics and therapeutic effects (Kurek, 2019). Plant species with alkaloid content exceeding 0.001% are known as alkaloid sources. The majority of the plant families are rich in alkaloid content and play a significant role in pharmaceutical applications (Yang & Stöckigt, 2010).

Alkaloids are primarily categorized into two groups, isoquinoline alkaloids and indole alkaloids, each consisting of approximately four thousand molecules. These alkaloids are typically categorized based on their chemical structure. Additionally, there are other notable classes of these phytochemicals, such as pyrrolizidine, tropane, pyridine, and steroidal alkaloids. Alkaloids can also be categorized based on their plant origin. For example, plants from the Papaveraceae family, like papaver plants, contain alkaloids, including opium alkaloids (Dey et al., 2020).

For centuries, alkaloids have been employed in human medicine to tackle a wide range of health concerns such as neurological issues, tumors, metabolic disorders, and infectious diseases (Faisal et al., 2023). Many significant biotechnological breakthroughs have been made to improve the availability and quality of alkaloids, which hold immense value in the medical field, consequently amplifying their production (Guaadaoui et al., 2014; Espinosa-Leal et al., 2018).

Metabolic engineering is a notable advancement that boosts the natural cellular functions of a particular organism and amplifies the output of targeted products (García-Granados et al., 2019). The technique holds substantial importance in the business sector, particularly in the manufacturing of medications like paclitaxel and artemisinin, in the synthesis of various amino acids, and in the excessive production of L-valine (Kulkarni, 2016).

In recent years, there has been notable advancement in the field of developing tools and understanding of plant biosynthetic pathways. However, it is important to note that this field is still actively being researched, and there is still much to uncover in terms of discovering new enzyme functions, identifying their localization, and understanding the regulation of metabolic pathways (Shih & Morgan, 2020). In this chapter, a comprehensive compilation was made, summarizing the bioactive potential of alkaloids as well as the efficacious approaches to raising the alkaloid content in medicinal plants.

22.2 BIOACTIVITY POTENTIAL OF ALKALOIDS

Extensive studies have been carried out on organic substances to discover novel antimicrobial compounds to address the growing issue of microbial resistance. Indeed, the challenge with existing antiviral medications lies in the microorganism's ability to develop resistance.

According to the heterocyclic ring structure and precursor involved in biosynthesis, alkaloids are categorized into various classes. These classes include tropanes, isoquinoline, pyrrolizidines, purines, piperidines, imidazoles, quinolizidines, indoles, pyrrolidines, and purines. The chemical makeup and biosynthetic precursors of alkaloids have attracted significant interest due to their biological and therapeutic activities. Commonly used indole alkaloids, such as serotonin and related compounds, are extensively studied. Approximately 2000 compounds fall under the category of indole alkaloids, including ajmaline, vincamine, vinblastine, ajmalicine, strychnine, and vincristine, which are among the most widely researched due to their pharmacological effects (refer to Table 22.1). For instance, vinblastine and vincristine are commonly used as antitumor agents. Quinine, extracted from *Calendula officinalis* L. and *Cinchona ledgeriana*, has been demonstrated to be toxic to *P. vivax* and protozoans or single-celled organisms responsible for malaria. Additional quinine alkaloids consist of camptothecin, homocamptothecin, cinchonidin, chinidin, folipdine, echinopsine and dihydroquinine (Kainsa et al., 2012). These chemical compounds have shown notable pharmaceutical effects, including anticonvulsant, pain-relieving, antifungal, anthelmintic, anti-inflammatory, antimalarial, antibacterial, and cardiotonic effects (Marella et al., 2013).

Isoquinoline is the source of several noteworthy alkaloids. Alkaloids are classified into distinct categories according to the inclusion of particular groups: basic isoquinolines, benzylisoquinolines, and phthalide isoquinolines. Several

DOI: 10.1201/9781003475491-22

notable alkaloids fall under this classification, such as thebaine, narcotines, morphines, protopines, and codeine. Moreover, this cluster of alkaloids exhibits a diverse array of pharmaceutical effects, including antimicrobial, antidiabetic, antitumor, etc. (Ziegler & Facchini, 2008).

Theophylline, aminophylline, and caffeine are three significant alkaloids belonging to the purine class (xanthines). These alkaloids display a variety of biological properties, including inflammatory-reducing, antioxidant, antidiabetic, anti-obesity, and antihyperlipidemic effects (Nassiri, 2013).

Imidazole alkaloids are chemical compounds derived from L-histidine that feature an imidazole ring within their molecular structure. Pilocarpine, derived from *Pilocarpus cearensis*, is the most renowned constituent of this category. It serves as a medicinal agent in ophthalmic formulations for the treatment of glaucoma (Machado et al., 2012).

22.3 ANTIBACTERIAL ACTIVITY OF ALKALOIDS

Bacterial infections pose a significant global health issue, with a rising trend attributed to multidrug resistance leading to increased mortality and morbidity. As a result, there is a growing demand for new antibacterial treatments, with plants offering a diverse array of potential natural products. Three alkaloids – solasodine, B-solamarine, and solanine – have been isolated from the plant *Solanum dulcamara* belonging to the Solanaceae family and commonly referred to as bittersweet plant. These alkaloids have exhibited noteworthy antibacterial properties against *Staphylococcus aureus* (Kumar et al., 2009).

Marine invertebrates have been found to possess bisindole alkaloids that display antibacterial properties against *S. aureus*, including methicillin-resistant *Staphylococcus aureus* (Zoraghi et al., 2011). Hydrastine and berberine alkaloids were isolated from *Hydrastis canadensis* and have shown strong antimicrobial properties, specifically against *S. pyogenes* and *S. aureus* (Villinski et al., 2003). *Epinetrum villosum* is a plant that contains cocsoline alkaloid, which exhibits a broad spectrum of antibacterial properties. It effectively hinders the growth of *Shigella* strains, as well as *Campylobacter jejuni* and *Campylobacter coli* (Otshudi et al., 2005).

Pyrrolidine alkaloids consist of aza five-membered rings and are formed from lysine and ornithine. Within this class, there are several members, such as hygrine, cuscohygrine, and putrescine. Extensive biological studies have demonstrated noteworthy antifungal, antitubercular, and antibacterial activities in numerous compounds belonging to this class (Parmar et al., 2012).

The antimicrobial activity of the I-mahanine indole alkaloid derived from *Micromelum minutum* has been proven against *B. cereus* and *S. aureus* (Nakahara et al., 2002).

Hapalindole alkaloids, specifically hapalindole X, deschlorohapalindole I, and 13-hydroxy dechlorofontonamide, as well as hapalindoles A, C, G, H, I, J, and U, fischerindole -L, anhydrohapaloxindole A, and hapalonamide H, were isolated from cyanobacteria sp. The findings of the study revealed that hapalindole X and hapalonamide exhibited strong antimicrobial activity against *C. albicans* and *M. tuberculosis* showing minimum inhibitory concentration values ranging from 0.6 to 2.5 μM (Kim et al., 2012).

2, 3-dihydro-1H-indolizinium chloride, derived from *Prosopis glandulosa*, has been identified as a potent antifungal and antibacterial agent. It exhibits strong activity against *C. neoformans* and *A. fumigatus*, with IC50 values of 0.4 μg/mL and 3.0 μg/mL, respectively. Additionally, it demonstrates antibacterial activity against methicillin-resistant *S. aureus* and *Mycobacterium*, with IC50 values of 0.35 μg/mL and 0.9 μg/mL, respectively (Samoylenko et al., 2009).

Dehydrocavidine, coptisine, dehydroapocavidine, and tetradehydroscoulerine are the key components of Yanhuanglian, derived from *Corydalis saxicola*, utilized in conventional medicine in China. These compounds have demonstrated antimicrobial and antitumor properties in various animal models (Li et al., 2006).

Five recently discovered quinolone alkaloids, euocarpines A–E, were extracted from *Evodia rutaecarpa* fruit extract. These alkaloids demonstrated moderate antibacterial properties, with minimum inhibitory concentration values ranging from 4 to 128 μg/mL . Among the compounds, those with a side chain containing 13 carbons displayed the most potent antibacterial effects, while maintaining low cytotoxicity levels (Wang et al., 2013).

22.4 ANTIFUNGAL ACTIVITY OF ALKALOIDS

Over the last couple of years, there has been a significant rise in the proliferation of antifungal resistance. This has led to an increased percentage of morbidity and mortality, as the effectiveness of these agents has diminished. The similarity in molecular processes between humans and fungi poses a constant risk, as the cytotoxic substances used to combat fungal infections can also harm the host cells. As a result, patients with weakened immune systems, including transplant recipients, cancer patients and diabetic patients who exhibit a poor response to treatment with existing medicines, require novel antifungal treatments (Wang et al., 2014).

β-Carboline, a tryptamine and two phenylethylamine-derived alkaloids, as well as N-methylN-formyl-4-hydroxy-beta-phenylethylamine from *Cyathobasis fruticulosa*, and haloxylines A and B, a new piperidine from *Haloxylon schmittianum* Pomel, have demonstrated promising antifungal activity (Singh et al., 2003).

Jatrorrhizine, found in *Mahonia aquifolium*, exhibited the strongest antifungal inhibitory effect among all fungi studied, with MIC ranging from 62.5 to 125 μg/mL. On the other hand, the natural extract, palmatines and berberine displayed lower inhibitory activity, with MIC values of 500 greater than or equal to 1000 μg/mL (Jung et al., 2006).

Four alkaloids – glaucine, dicentrine, alpha-allocryptopin, and protopine – were extracted from *Glaucium oxylobum* and have shown promising efficacy against

Microsporum canis, *Microsporum gypseum*, *Epidermophyton floccosum*, and *T. mentagrophytes* (Singh et al., 2000).

Alkaloids derived from *Ruta graveolens* L. leaves exhibit fungicidal properties, indicating potential as alternative antifungal agents due to the serious side effects and resistance issues associated with current treatments for fungal infections (Oliva et al., 2003). A glycoalkaloid, tomatine, formed from tomatoes, has been demonstrated to possess antifungal properties (Simons et al., 2006). Quinoline alkaloids and flavonoids derived from *Waltheria indica* root extract exhibited antifungal activity against *C. albicans* (Cretton et al., 2016).

Moreover, anodic alkali aninolinate has exhibited inhibitory effects against many fungal strains that were examined. Notably, this compound displayed a remarkable sensitivity, resulting in germination levels below 10% (Cantrell et al., 2005). Two antifungal fructoxin alkaloids were extracted from *Dictamnus dasycarpus* roots. The compound 3-methoxisampangin from *Cleistopholis patens* exhibited significant inhibitory effects on *C. albicans*, *A. fumigatus*, and *C. neoformans* (Salehi et al., 2019).

22.5 ANTICANCER ACTIVITY OF ALKALOIDS

Cancer ranks second among the leading causes of death globally. Alkaloids, which are present in natural herbs, are known for their high pharmaceutical activity and are responsible for the production of several significant drugs available in the market. Among these drugs are anticancer agents like camptothecin (CPT) and vinblastine. The derived alkaloids, including vasicinone, matrine, evodiamine, piperine, sanguinarine, tetrandrine, piplartine, aporphine, harmalacidine, harmine, harmaline, and berberine, exhibit cytotoxicity and operate through various mechanisms of action (Amin et al., 2019).

Phytochemicals offer a valuable resource for developing anticancer agents, owing to their high efficacy and minimal toxicity. Among the major alkaloids discovered in the *Uncaria* genus, the hersutin alkaloid has demonstrated its ability to induce apoptosis in breast cancer cells that are HER2-positive and carry p53 mutations (Lou et al., 2015). Oxymatrine, a naturally occurring alkaloid derived from the roots of *Sophora chrysophylla* (Salisb.), has demonstrated anticancer properties in human cervical cancer Hela cells by inducing cytotoxicity and apoptosis (Li et al., 2014).

Berberine, derived from *Coptis chinensis* Franch, Berberidaceae, is an isoquinoline derivative with a range of biological activities such as antibacterial, antidiabetic, anti-inflammatory, and antiulcer properties. Moreover, it has also been discovered to have positive impacts on the circulatory system (Cretton et al., 2016). Studies have shown that *Coptis chinensis* Franch, exhibits inhibitory effects on the growth of breast cancer and liver cancer cells. The antitumor properties of berberine were investigated in human pharyngeal squamous carcinoma cells revealing a concentration-dependent cytotoxic effect that reduced cell viability (Seo et al., 2015).

A quinoline alkaloid called evodiamine, isolated from *Evodia rutaecarpa*, a Chinese medicinal plant, exhibits selectivity toward tumor cells while being less harmful to normal cells, including peripheral blood mononuclear cells. Research has indicated its ability to suppress the growth of adriamycin-resistant human breast cancer cells, both in laboratory settings and in vivo, as well as shown effectiveness when taken orally (Fei et al., 2003).

Piperlongumine, also called piplartine, is derived from *Piper longum*. It has been found to exhibit tumor growth inhibition in mice with transplanted Sarcoma 180 cells. This antitumor activity is attributed to its ability to suppress cell proliferation (Wang et al., 2014).

Eight aporphine alkaloids are derived from the aerial portion of *Pseuduvaria setosa*. These alkaloids have displayed a moderate level of antitumor efficacy against breast cancer and lung cancer cells. Furthermore, they exhibit remarkable antitumor activity against breast cancer and epidermoid carcinoma (Wirasathien et al., 2006).

Bisbenzylisoquinoline alkaloid, tetrandrine is extracted from *Stephania tetrandra*. This compound serves various purposes, such as immune modulation, anti-arrhythmic, anti-inflammatory, anti–portal hypertension, anti-hepatofibrogenetic, anticancer, and anti-neurodegenerative activities (Li et al., 2001). Tetrandrine inhibits the production of vascular endothelial growth factor in glioma cells, demonstrating antitumor properties in human umbilical vein endothelial cells, and possesses anti-angiogenic activities (Chen et al., 2009).

Peganum harmala has a history of traditional use in cancer therapy, with harmine and harmaline being the primary alkaloids present in the plant at concentrations of 4.2% and 5.7%, respectively. Four alkaloids – harmaline, harmine, vasicinone, and harmalacidine – were extracted from plant seeds and exhibited notable cytotoxic effects (Kittakoop et al., 2014). The process of action of *P. harmala* seeds reveals that the alkaloids found in these seeds inhibit topoisomerase 1, leading to the suppression of cancer cell proliferation (Sobhani et al., 2002).

The pyrrolizidine alkaloids can be found exclusively in plants in Boraginaceae, Compositae, Leguminosae, Convulvulaceae, Orchidaceae, and Poaceae. Some of the well-known members of this group include clivorine, senecionine, echinatine, and heliotrine, which plants produce for defense against herbivores. These compounds are hepatotoxic and can lead to various diseases, including liver cancer. They are also utilized for their glycosidase inhibition properties to cure diabetes and tumors (Cronemberger et al., 2012).

22.6 ANTIVIRAL ACTIVITY OF ALKALOIDS

Viruses contain DNA and RNA as their genetic material, along with protein envelopes in certain instances. In order to reproduce and flourish, they depend on their host's metabolism and the surrounding environment. By hijacking the cellular machinery of the host, they spread across the entire organism (Badshah et al., 2020). Alkaloid phytochemicals possess the ability to inhibit and influence viruses using diverse mechanisms. They have the capability to hinder the fusion or attachment of viruses with the receptors on the surface of host cells, thereby preventing their entry. Additionally, they can disrupt the synthesis of viral DNA or RNA (Fikatas et al., 2021).

In vitro studies have examined the impact of the berberine alkaloid on influenza A viruses (IAV) infections. The results indicate that berberine has the ability to suppress IAV type A/PR/8/34 in RAW 264.7 cells at concentrations exceeding 1.5 μM, with an IC50 value of 0.01 μM. Additionally, this alkaloid has demonstrated the ability to inhibit the growth of different H1N1 influenza A strain viruses in vitro, giving an IC50 value of 0.43 μM. The mechanism of action of berberine suggests that it influences the development and transportation of virus proteins, thereby slowing down virus development. Furthermore, berberine has been found to suppress the production of tumor necrosis factor a and prostaglandin E2 in cells infected with H1N1 (PR8) virus (Cecil et al., 2011).

Another group of researchers discovered homonojirimycin, an alkaloid found in *Commelina communis,* and evaluated its effectiveness against the strain IAV/PR/8/34 (H1N1). This exhibited significant antiviral activity against influenza A virus, demonstrating an EC50 inhibitory concentration of 10.3 μg/mL (Zhang et al., 2013).

A study was conducted to investigate the antiviral activity of various phytoconstituents, including phenolics, flavonoids, and alkaloids on herpes simplex virus. The results revealed that several types of alkaloids exhibited strong antiviral activity against herpes simplex virus in the Madin-Darby bovine kidney cell line. The examined alkaloids effectively inhibited and reduced the pathogenic effect caused by the virus in the bovine kidney cell culture. Yohimbine, vincamine, atropine, trigonelline, and capsaicin demonstrated inhibitory concentrations of 0.7 μg/mL and 0.5 μg/mL, respectively. These alkaloids were found to be non-toxic to the culture cells at concentrations below 3.2 μg/mL. Additionally, other alkaloids such as allantoin, scopolamine, synephrine, octopamine, trigonelline, and colchicines exhibited significant inhibitory activity against herpes simplex viral strain at a concentration of 1.6 μg/mL (Ozcelik et al., 2011).

Manzamine A, an alkaloid, was examined for its suppressive impact on herpes simplex virus -1 in rabbit corneal cells. It was discovered that Manzamine-A exhibits potent inhibitory activity on the replication process of simplex virus-I at a concentration of 1.1 μM in the corneal cells of rabbits. Acyclovir, the currently used antiviral medicine for simplex virus-I, was also tested and showed identical inhibitory activity at a concentration of 50 μM. Furthermore, manzamine A significantly reduced the release of pathogenic viruses. Treatment with manzamine A also decreased the host shutoff function of herpes simplex virion and its ICP0 transcription process, as confirmed by RTPCR tests. The IC50 of manzamine A against this virus was reported to be 5.5 μM. These results hold that manzamine exhibits potential as a viable remedy for diminishing viral infection in corneal cells and averting eye infections, such as keratitis triggered by simplex virus I (Palem et al., 2011).

During a chemical screening of metabolites against chikungunya virus, it was observed that berberine exhibited potent activity against chikungunya virus in Baby Hamster Kidney Fibroblast Cells (BHK-21) cells while maintaining the cell line viability. The experiment utilized a chikungunya virus replicon containing the *Renilla* luciferase gene. Berberine has been shown to target multiple stages of chikungunya infection, inhibiting viral protein synthesis and suppressing both genomic and anti-genomic RNA synthesis. Additionally, it was noted that berberine interfered with the process of replication in chikungunya virus infection. The EC50 range of berberine against chikungunya virus was determined to be 1.8 μM (Varghese et al., 2016).

Tomatadine alkaloid underwent testing in various cell lines to assess its effectiveness against different strains of chikungunya virus. In hepatoma tissue (Huh7) cells, it successfully suppressed infection from three distinct genotypes of chikungunya virus. The EC50 value of tomatadine was measured at 1.3 μM, indicating its potency, and it exhibited a selectivity index (SI) of 120. Notably, tomatadine demonstrated repression and inhibition of the chikungunya virus La Reunion strain. The African and Asian chikungunya virus genotypes displayed EC50 values of 3.7 μM and 2 μM, respectively. Findings from investigations unveiled that the antiviral action of tomatadine takes place after being infected and remains effective for up to six hours after the drug is administered (Troost et al., 2020).

Berberine chloride alkaloid has demonstrated significant anti-HCMV hHuman cytomegalovirus) activity, comparable to the standard antiviral drug ganciclovir. The plaque assay revealed potent inhibition of HCMV by berberine, with an IC50 of 0.69 μM. In comparison, ganciclovir exhibited an IC50 of 0.91 μM against human cytomegalovirus. Both berberine and ganciclovir effectively inhibited human cytomegalovirus proliferation. Interestingly, berberine did not affect the entry of human cytomegalovirus into host cells but interfered with viral replication post-entry and pre-DNA replication stages (Hayashi et al., 2007).

The anti–hepatitis B virus activities of alkaloids derived from the roots of *Z. nitidum* were assessed in HepG2 cells. Initially, the cultured cells underwent transfection with hepatitis B virus DNA to generate infectious particles. It was observed that the examined alkaloids exhibited strong inhibitory effects on hepatitis B virus. Specifically, the isolated alkaloids such as skimmianine, 5,6-dihydro-6-methoxynitidine and 5-methoxydictamine, demonstrated substantial repression of hepatitis B surface antigen and hepatitis B e antigen secretions. The inhibitory concentrations for the first two alkaloids were 0.3 μmol/mL 0.3 μmol/mL, resulting in a 43.3% and 49.3% reduction in hepatitis B surface antigen secretions, respectively. Skimmianine alkaloid suppressed the hepatitis B e antigen secretion at a concentration of 0.3 μmol/mL, and 5-methoxydictamine alkaloid targeted hepatitis B antigen secretion and exhibited a greater repressive effect against the hepatitis B virus. The results indicate that these secondary metabolites have the potential to effectively control hepatitis B virus infection (Yang & Chen, 2008).

A computational study utilizing molecular docking and MD simulation techniques has successfully determined several alkaloids that exhibit strong affinities with the protein, SARS-CoV-2 Nsp-15. Among these alkaloids, aranotin, ajmalicine, and piperine, have emerged as promising inhibitors of Nsp-15 in the context of SARS-CoV-2. This in silico study suggests that these alkaloids could potentially serve as valuable leads in combating the virus by targeting its replication process. By interfering with these alkaloid-based compounds, it may be possible to prevent viral replication, considering the significance of Nsp-15 in the SARS-CoV-2 protein structure (Kumar et al., 2021).

A study conducted on alkaloid compounds through in vitro screening, particularly focusing on antimalarial inhibitors, revealed that these compounds possess significant inhibitory effects. Drugs based on hydroxychloroquine, chloroquine, and pyronaridine alkaloids exhibited notable antiviral properties against SARS-CoV-2, with EC50 concentrations of 1.5 μM, 2.1 μM, and 1.8 μM. Additionally, other compounds, such as mefloquine, desethylamodiaquine and quinine, demonstrated effective inhibitory activities, with EC50 values of 1.8 μM, 0.52 μM, and 10.7 μM. These findings indicate the potential effectiveness of these compounds in combating this rapidly evolving infectious virus (Gendrot et al., 2020).

Another bioinformatics investigation demonstrated that camptothecin from *Camptotheca acuminata* demonstrated beneficial interactions with the primary enzymes Mpro and RdRp of SARS-CoV-2. Through molecular docking analysis, it was discovered that camptothecin and its derivatives, irinotecan and topotecan displayed a stronger binding potential to RdRp and Mpro. In silico investigations further indicated that camptothecin specifically interacted with the amino acids of the viral S protein of SARS-CoV-2, in contrast to an antiviral medication, lopinavir, which solely engaged in Van Der Waal's force of interactions with the S protein and lacked hydrogen bond interactions. The findings suggest that camptothecin has the potential to impede and significantly obstruct the binding of the S protein and the ACE 2 receptor, indicating its efficacy in managing SARS-CoV-2 infection (Devasia et al., 2021).

Additional alkaloids have been identified to exhibit high binding energies with key enzymes of SARS-CoV-2 in in silico research (Faisal et al., 2022). Through computational methods, researchers discovered that when compared to the antiviral drug Lopinavir, Schizanthine Z, an alkaloid, exhibits a higher affinity for the protease enzyme of SARS-CoV-2 (Alfaro et al., 2020). In addition, two additional alkaloids, namely thalimonine and sophaline D, have been discovered to exhibit favorable binding affinities toward the main-protease enzyme. The discovery suggests that these alkaloids have great potential as inhibitors for the virus (Garg & Roy, 2020).

22.7 ALKALOIDS AGAINST AGRICULTURAL PESTS

Glycoalkaloids derived from the leaves of potatoes have been proven to have detrimental impacts on both the successful hatching of eggs from *Spodoptera exigua* and the contractile activity of the heart in three species of beetles – *Tenebrio molitor, Leptinotarsa decemlineata*, and *Zophobas atratus* (Marciniak et al., 2010).

Comparable outcomes were observed in *Zophobas atratus* and *Tenebrio molitor* when exposed to water extracts from *Solanum tuberosum*, *Lycopersicon esculentum*, and *Solanum nigrum* (Marciniak et al., 2019).

The impact of leaf extracts from *S. tuberosum*, *L. esculentum*, and *S. nigrum*, as well as individual pure glycoalkaloids, on the reproduction and development of *Drosophila melanogaster* has been proven (Ventrella et al., 2016; Spochacz et al., 2018).

22.8 BOTTLENECKS IN THE ISOLATION AND QUANTIFICATION OF ALKALOIDS FROM MEDICINAL PLANTS

Plant extracts usually possess various bioactive agents or secondary metabolites such as terpenoids, alkaloids, flavonoids, tannins, saponins, phenolics, etc. The separation of alkaloids from this mixture of compounds is still a big challenge. The conventional method of separation is not only difficult but also tedious and inconvenient. For the isolation of pure compounds, column chromatography, thin-layer chromatography, high-performance liquid chromatography, flash chromatography, and Sephadex chromatography are utilized. The processing of crude material and the choice of solvent play an important role in the successful isolation of compounds (Sasidharan et al., 2011).

Since alkaloids possess a wide range of medicinal properties, their isolation and quantification aid in the development of new drugs. Rather than being the end product, alkaloids are by-products or components of several metabolic pathways. These chemicals can be converted into other alkaloids through synthesis or degradation mechanisms. The ultimate result is determined by the physiological demands and signaling of each plant. Hence, predicting the concentration or existence of every single alkaloid is challenging (Aniszewski, 2007). Isolating alkaloids in their pure form enables extensive pharmacological investigations, evaluation of structure-activity relationship (SAR), and the development of novel medications to address multiple ailments.

The general techniques of isolation of alkaloids rely on a number of vital factors, such as alkaline nature, the ability to form alkaloidal salts with acids, and the relative solubilities of alkaloidal salts in polar solvents.

Extraction is the first crucial step in the isolation of alkaloids. This involves the separation of secondary metabolites from the plant matrix. Various classical and new extraction methods have been developed to obtain plant metabolites

TABLE 22.1
The Medicinal Activities and Plant Sources of Alkaloids Discussed Here Are Summarized Below

Alkaloids	Biological activities	Source plant	Reference
Ajmaline, vinblastin, vincristine, strychnine, quinine, cinchonidin, dihydroquinine, chinidin	Anticonvulsant, analgesic, antifungal, antihemintic, anti inflammatory, antimalarial, antibacterial, cardiotonic effects	*Rauwolfia serpentine*, *Calendula officinalis*, *Cinchona ledgeriana*, *Cinchona officinalis*, *Strychnos nux-vomica*, *Vinca rosea*	Kainsa et al., 2012
Narcotines, protopines, thebaine	Antitumor, Antihyperglycemic, antibacterial	*Papaver bracteatum*, Papavaraceae plants	Ziegler and Facchini, 2008
Aminophylline, theophylline, caffeine	Anti-inflammatory, antioxidant, antidiabetic, anti-obesity, antihyperlipidemic	*Thea sinensis*, *Coffea arabica*, *Theobroma cacao*	Nassiri, 2013
Solanine, solasodine, B-solamarine	Antibacterial activity against *S. aureus*	*Solanum dulcamara*	Kumar et al., 2009
Berberine, hydrastine	Antibacterial against *Streptococcus pyogenes* and *Staphylococcus aureus*	*Coptis chinensis*, *Hydrastis canadensis*	Villinski et al., 2003
Pilocarpine	Treatment of eye irritations, medicinal agent in ophthalmic formulations	*Pilocarpus cearensis*	Machado et al., 2012
Cocsoline	Antibacterial, hinders the growth of *Shigella* strains, *Campylobacter jejuni*, *Campylobacter coli*	*Epinetrum villosum*	Otshudi et al., 2005
Hygrine, cuscohygrine, putrescine	Anti-tubercular, antibacterial, antifungal	Coca leaves	Parmar et al., 2012
I-Mahanine	Antibacterial against *B. cereus*, *S. aureus*	*Micromelum miutum*	Nakahara et al., 2002
Hapalindole, fischerindole, hapalonamide, anhydrohapaloxindo-le	Antibacterial and antifungal against *Micobacterium tuberculosis*, *C. albicans*	Cyanobacteria	Kim et al., 2012
2,3-dihydro-1H-indolizinium	Antifungal, antibacterial against *Cryptococcus neoformans*, *A. fumigatus*, *S. aureus*, *Micobacterium*	*Prosopis glandulosa*	Samoylenko et al., 2009
Dehydrocavidine, coptisine, dehydroapocavidine, tetradehydroscoulerine	Antibacterial, antiviral, anticancer	*Corydalis saxicola*	Li et al., 2006
Euocarpines A, B, C, D, E	Antibacterial, cytotoxic	*Evodia rutaecarpa*	Wang et al., 2013
B-carboline, A-tryptamine	Antifungal	*Cyathobasis fruticulosa*	Singh et al., 2003
Haloxylines A, B	Antifungal	*Haloxylon schmittianum*	Singh et al., 2003
Jatrorrhizine	Antifungal	*Mahonia aquifolium*	Jung et al., 2006
Dicentrine, Glaucine, Protopine, Alpha-allocryptopine	Antifungal against *M. gypseum*, *M. canis*, *T. mentagrophytes*, *Epidermophyton flocossum*	*Glaucium oxylobum*	Singh et al., 2000
Tomatine – glycoalkaloids	Antifungal	Tomatoes	Simons et al., 2006
Fructoxin	Antifungal	*Dictamnus dasycarpus*	Salehi et al., 2019
3-Methoxisampangin	Antifungal against *C. albicans*, *A. fumigatus*, *C. neoformans*	*Cleistopholis patens*	Salehi et al., 2019
Oxymatrine	Anticancer against cervical cancer, cytotoxic	*Sophora chrysophylla*	Li et al., 2014
Hirsutine	Apotosis induction, anticancer	*Uncaria rhynchophylla*	Lou et al., 2015
Evodiamine	Antitumor against human breast cancer	*Evodia rutaecarpa*	Fei et al., 2003
Piplartine	Antitumor, suppresses cell proliferation	*Piper longum*	Wang et al., 2014
Aporphine	Antitumor against lung cancer and breast cancer	*Pseuduvaria setose*	Wirasathein et al., 2006
Tetrandrine	Immune modulation, anti-hepatofibrogenetic, antiarrhythmic, anti-inflammatory, antihypertension, anticancer, anti-angiogeneic	*Stephania tetrandra*	Chen et al., 2009
Harmine, harmaline, harmalacidine, vasicinone	Anticancer, cytotoxic	*Peganum Harmala*	Kittakoop et al., 2014, Sobhani et al., 2002

(*Continued*)

TABLE 22.1 (*Continued*)
The Medicinal Activities and Plant Sources of Alkaloids Discussed Here Are Summarized Below

Alkaloids	Biological activities	Source plant	Reference
Homonojirimycin	Antiviral activity against influenza A virus	*Commelina communis*	Zhang et al., 2013
Yohimbine, capsaicin	Antiviral activity against herpes simplex virus	*Rauwolfia root*, *Pausinystalia*, *Capsicum* sp.	Ozcelik et al., 2011
Heliotrine, echinatine, senecionine, clivorine	Anticancer against liver cancer, antidiabetic	*Heliotropium arbainense*, *Heliotropium lasiocarpum*, *Critonia portoricensis*, *Eupatorium fortune*, *Senecio jacobaea*, *Ligularia dentate*, *Ligularia hodgsonii*	Cronemberger et al., 2012
Allantoin, octopamine, synephrine, colchicines, trigonelline, manzamine-A	Antiviral against herpes simplex virus	*Comfrey* plant, *Citus aurantium*, *Colchicum autumnale*, marine algal sp. (manzamine A)	Palem et al., 2011, Ozcelik et al., 2011
Methoxynitidine, skimmianine, methoxydictamine,	Antiviral against hepatitis B virus	*Z. nitidum*	Yang and Chen, 2008
Topotecan, irinotecan	Anti-SARS-CoV	*Camptotheca acuminate*	Mamkulathil Devasia et al., 2021
Schizanthine Z	Antiviral	*Schizanthus pinnatus*	Garg and Roy, 2020

(Abizov et al., 2012). A wide range of techniques is used for alkaloid extraction, including maceration, decoction, percolation, infusion, ultrasound-assisted extraction, and supercritical fluid extraction. Excessive heat, enzymatic reactions within the cells, and the presence of certain chemicals can potentially destroy the compounds during extraction and isolation. It is important to understand the buffer system when working with water-soluble alkaloids. The formation of foams is another challenge that needs to be addressed. This can be overcome by using appropriate antifoaming agents (Bhargavi et al., 2018; Fabricant & Farnsworth, 2001; Huie, 2002).

Alkaloids can be extracted under neutral or basic conditions after the plant materials have been basified to form free bases using organic solvents. It is important to identify the nature of the alkaloid before extraction. For example, tryptamine can only be extracted at higher pH levels, while alkaloids containing a phenolic group like morphine can become dissociated at higher pH levels (Verpoorte & Schripsema, 1994). The extraction procedure must be carefully controlled to prevent degradation or alterations.

Quantification of secondary metabolites is another important step. For the identification and quantification of alkaloids, gas chromatography and high-performance liquid chromatography methods with mass spectroscopic detection are widely used. However, insufficient sensitivity, mass accuracy and separation efficiency are some of the bottlenecks that have to be considered in these methods. Moreover, certain unstable alkaloids are not amenable to analysis in this way. Hence, identification and quantification of isolated alkaloids can be easily accomplished with the help of advanced techniques like HPLC with UV detection coupled with electrospray ionization MS/MS. Other techniques are ultra-HPLC coupled with quadrupole time-of-flight mass spectrometry (UHPLC–Q-TOF-MS), high-speed counter-current chromatography (HSCCC), or high-performance thin-layer chromatography (Zhao et al., 2012; Takla et al., 2018; Sai et al., 2020).

Temperature has a dubious impact on alkaloid extraction. Alkaloid yield reaches its peak and then starts decreasing again when the temperature is increased since high temperature leads to degradation and artifact formation (Abizov et al., 2012).

Another factor that usually affects the extraction performance of alkaloids is the solid-to-liquid ratio. A lower solid-to-liquid ratio causes wastage of the content. Hence, in order to avoid this bottleneck a higher solid-to-liquid ratio is required during the extraction of alkaloids (Fakhari et al., 2010).

Besides temperature and solid-to-liquid ratio, an important bottleneck to be considered in the quantification of alkaloids is the particle size. Smaller particles give higher yields. However, if the powder is too fine during the maceration or percolation, it may limit the contact between the solvent and the plant material and thus decrease the solvent flow. Moreover, small particles can interfere with the filtration process (Handa et al., 2008).

In addition to all these factors, solvent nature is an important parameter that can affect the quantity of alkaloids. Among polar, nonpolar, and aqueous solvents, methanol, which is a highly polar solvent, gives the better yield. However, considering the toxicity of methanol, hydroethanolic extraction can be carried out since ethanol is less toxic compared to methanol (Kaiser et al., 2013).

In conclusion, isolating plant alkaloids is critical for improving our understanding of their pharmacological characteristics, toxicological profiles, chemical structures, ecological functions, and uses in drug development and quality control. Proper isolation techniques and analytical procedures are required for obtaining pure alkaloid samples and conducting extensive research that adds to scientific understanding as well as practical applications in a variety of sectors.

22.9 MOLECULAR APPROACHES TO INCREASE THE PRODUCTION OF ALKALOIDS

22.9.1 Genetic Transformation

Alkaloids are a broad group of nitrogen-containing compounds with low molecular weight and play an important role in the mechanism of plant defense. The potent biological activity of alkaloids has led to their wide use as medications, stimulants, drugs, and poisons (Peter, 2001). Due to the growing demand for plant secondary metabolites, a lot of interest has been evolved on secondary metabolism by means of genetic engineering. Genetic transformation by particle bombardment or *Agrobacterium*-mediated transformation has demonstrated considerable potential to increase the biosynthetic capacity of plants and plant cells (Zhang et al., 2005). Since alkaloids are highly beneficial as medications, various biotechnological developments have been made to enhance these compounds for commercial purposes (Guaadaoui et al., 2014; Espinosa-Leal et al., 2018).

The two primary methods for introducing foreign DNA into the host plant cells are mediated by *A. tumefaciens* and *A. rhizogenes* resulting in the genetically transformed plants and genetically modified hairy roots. These two methods are mostly used in the overexpression of alkaloids in plants (Ricigliano et al., 2016; Ma et al., 2017). The extent of transformation is influenced by various factors such as plant species, type of explants used, specific genotype, growth conditions, regeneration frequency, etc.

Hairy root culture provides a different approach for growing plants for the production and retrieval of secondary compounds. This creates new opportunities for the synthesis of a number of naturally occurring compounds with biological activity, including paclitaxel, scopolamine, artemisinin, etc. (Sheludko, 2010). In this case, the explants are subjected to infection by Agrobacterium, which results in the limitless expansion of hairy roots. Transgenic hairy roots take less than a month to generate, unlike transgenic plants that take 2–4 months to develop. The main advantages of this system are the absence of plant growth regulators, fast-growing hairy roots, genetic stability, and can be effectively executed (Kowalczyk et al., 2020). Hairy root cultures allow regulating metabolism across multiple pathways, including tropane and indole alkaloid biosynthesis (Soledad et al., 2022).

The hairy root system of *Rauwolfia serpentina*, transformed with *Agrobacterium* rhizogenes, has been found to produce 20 indole alkaloids, with being newly discovered six compounds. Root cultures are genetically and biochemically more stable than cell cultures, with greater differentiation and potential for genetic transformation (Yang & Stöckigt, 2010).

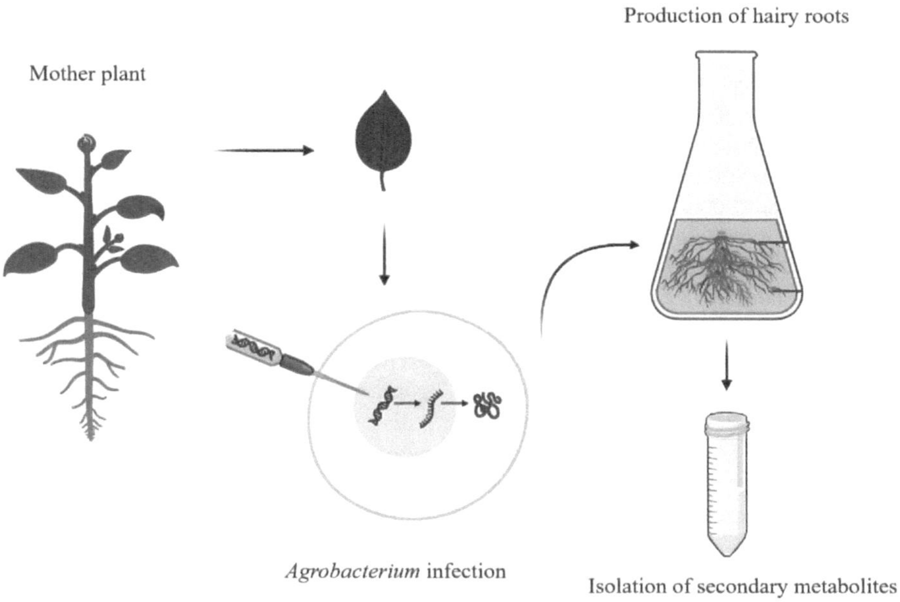

FIGURE 22.1 The production of hairy roots.

In order to integrate foreign genes into medicinal plants, a binary vector system conjugated with Ri plasmid has been widely employed. When compared to control cultures, which are untransformed, the shooty teratomas of *C. roseus* epicotyls and stem nodal explants produced ten times more vincristine (Begum et al., 2009).

22.10 METABOLIC ENGINEERING

Metabolic engineering utilizes recombinant DNA technology to enhance industrial fermentation processes by making targeted genetic modifications. To optimize the production of natural products, biosynthetic pathways are often reconstructed in different microorganisms or plant cells, resulting in engineered and genetically manipulable strains. The success of metabolic engineering relies on the development of improved strains through genetic engineering, which requires a comprehensive analysis of cellular function. By implementing metabolic engineering, it is possible to greatly enhance the yields of natural products on a large scale in microorganisms or plants. (Miaoxian et al., 2024).

22.11 OVEREXPRESSION OF ENZYMES

Once the enzyme-level biosynthesis of secondary metabolites is established, the building up of distinct phytoconstituents occurs due to the increased expression of crucial rate-limiting enzymes. These enzymes facilitate a diverse array of complex processes, including glycosylation, condensation, methylation, acylation, and isomerization (Staniek et al., 2013).

C. roseus is known for producing an extensive variety of terpenoid indole alkaloids. These alkaloids are categorized as monomeric indole alkaloids, which are strictosidine, ajmalicine, serpentine, vindoline, catharanthine, tabersonine, and more, as well as dimeric indole alkaloids, such as vincristine and vinblastine (Dug´e de Bernonville et al., 2020). The synthesis of strictosidine, a fundamental component of all terpenoid indole alkaloids, involves the condensation of an indole moiety, tryptamine, and a monoterpenoid, secologanin, facilitated by strictosidine synthase enzyme (Peebles et al., 2011). Tryptophan decarboxylase is responsible for the biosynthesis of tryptamine from tryptophan. On the other hand, the biosynthesis of secologanin initiates with geraniol (Kumar et al., 2015).

The biosynthetic pathway complexity encompasses over twenty enzymes and a significant amount of gene regulators. Stress and methyl jasmonate therapy have also been shown to trigger the indole alkaloid pathway. Thus, increased biosynthesis of these alkaloids has also been caused by the overexpression of enzymes like MAP kinases and apoplastic peroxidases, which are recognized as being activated by a range of abiotic and biotic stresses (Jaggi et al., 2011; Raina et al., 2012).

Putrescine is a necessary precursor that allows for the production of tropane alkaloids from both ornithine and arginine. Ornithine can be decarboxylated to produce putrescine. In *Atropa belladonna* hairy roots, overexpression of the ornithine decarboxylase gene led to a substantial rise in anisodamine, N-methylputrescine, hyoscyamine, and putrescine (Zhao et al., 2020). The primary enzyme responsible for converting hyoscyamine into scopolamine is 6β -hydroxylase. Significantly higher levels of scopolamine were observed in *Datura innoxia* as a result of hyoscyamine 6β -hydroxylase overexpression (Li et al., 2020).

A crucial enzyme in the biosynthesis of benzylisoquinoline is codeinone reductase. This enzyme has an important function in converting baine into morphine. Overexpressing this enzyme in *Papaver bracteatum* led to a remarkable boost in codeine and morphine levels in transgenic hairy root lines. *Papaver bracteatum* is known for its high baine content but low levels of codeine and morphine (Sharafi et al., 2013).

In order to increase a precursor, one gene overexpression may have unforeseen effects on the system. Feedback inhibition is one such issue. Anthranilate synthase, the enzyme involved in the indole alkaloid pathway, helps to catalyze the first step in the synthesis of tryptophan. But tryptophan inhibits in response (Peebles et al., 2011). Since feedback-insensitive anthranilate synthase is present in a variety of higher plants that are able to accumulate tryptophan, this inhibition issue can be resolved by either engineering an endogenous enzyme or by overexpressing feedback-insensitive homolog from different plant species (Tozawa et al., 2001).

When using metabolic engineering to increase the biosynthesis of a particular phytoconstituent, it's important that the relevant metabolic processes, related enzymes, ideal subcellular location, networks regulating gene expression, and epigenetic control should be taken into account. Functional genomics and proteomics are valuable tools that have provided regulatory enzyme-coding biosynthetic genes (Labanca et al., 2018; Huang et al., 2019; Lichman, 2021; Desgagn´e-Penix, 2021).

The expression of an alien enzyme is a helpful tactic to boost the build-up of alkaloids in plants. When converting a particular metabolite, the rate-limiting foreign enzymes from different plant species could exhibit superior catalytic efficiency. In a comparative study, the hyoscyamine 6β -hydroxylase genes from *Hyoscyamus niger* and *Scopolia lurida* were overexpressed in the hairy root system of *Scopolia lurida*. In this case, the accumulation of scopolamine content was found to be much higher in the hyoscyamine 6β -hydroxylase genes of *Hyoscyamus niger* than in the overexpressing lines of *Scopolia lurida*. This indicates that even though the amino acid sequences of both foreign and native enzymes are similar, the foreign enzyme is much more efficient in the conversion of hyoscyamine into scopolamine than the native enzyme (Lan et al., 2018).

The ability to manipulate the metabolic flux in favor of a particular metabolite has also been made possible by the overexpression of heterologous genes. Singh et al., in 2011, found that increasing the expression of the mouse (ODC) ornithine decarboxylase gene, which is a key gene in the tropane biosynthesis pathway of *Datura innoxia*, significantly increased the content of scopolamine. Ornithine decarboxylase is vital in maintaining the presence of putrescine, a key alkaloid precursor necessary for cell growth and survival. Additionally, it aids in regulating primary and secondary phytochemicals in diverse species of plants (Mohapatra et al., 2009).

The scopolamine content in the hairy root culture of *Hyoscyamus niger*, was increased by the expression of *Vitreoscilla* hemoglobin. This improved oxygen usage, speeding up metabolic processes that require oxygen, like the synthesis of tropane alkaloids (Guo et al., 2018).

Modular proteins called transcription factors (TFs) are necessary for regulating the transcription of genes. They engage with cis-elements through a DNA-binding domain to activate or suppress target genes (Samad et al., 2017). Numerous genes involved in the synthesis of phytoconstituents may have their expression changed by the overexpression of a single transcription factor overexpression, either native or alien gene (Decendit et al., 2009). Tissue-specific abiotic and biotic stressors trigger the production of terpenoid indole alkaloids, suggesting intricate transcriptional regulation. It has been demonstrated that a gene cluster consisting of ORCA2 to ORCA6 transcription factors in *Catharanthus roseus* regulates the production of terpenoid indole alkaloids (TIAs) in the presence of ethylene and methyl-jasmonate (Singh et al., 2020).

Important genes in the indole and secoiridoid pathways and strictosidine synthase expression were markedly upregulated in *C. roseus* hairy root system by the ORCA4 transcription factor overexpression. Consequently, there was a significant accumulation of tabersonine, which plays a crucial role in the biosynthesis of vinblastine (Paul et al., 2017).

22.12 GENE SILENCING

Gene silencing is a viable strategy for diminishing the production of unwanted phytochemicals or enhancing the biosynthesis of specific phytochemicals by suppressing competing pathways. To augment alkaloid levels, gene silencing can be accomplished through various RNA silencing techniques, such as overexpression of RNAi of miRNA and virus-induced gene silencing. This method has been used to enhance the alkaloid content such as indole alkaloids and tropane alkaloids (McCue et al., 2018; Singh et al., 2018; Li et al., 2015; Liu et al., 2017).

QPT (Quinolinic acid phosphoribosyl transferase) gene activity was successfully suppressed in *Duboisia leichhardtii* hairy roots system. This manipulation led to increased production of scopolamine in the transgenic hairy roots by redirecting the tropane alkaloid pathway and nicotine toward scopolamine biosynthesis. The QPT enzyme is responsible for catalyzing the biosynthesis of nicotinate mononucleotide, which is a precursor for nicotine production. By silencing the QPT gene, the methylpyrrolinium cation was channeled toward scopolamine biosynthesis instead of nicotine. Despite the fact that the QPT enzyme has a part in the primary metabolism of plants, the silencing of this gene did not have any negative impact on the development of the root hairs (Singh et al., 2018).

In regulating gene expression at the post-transcriptional level, the miRNAs play a significant role. MiRNAs achieve this by downregulating their targets, which primarily consist of transcription factors responsible for activating or repressing their self-transcriptional target genes (Samad et al., 2017). While the research on how miRNA overexpression influences alkaloid biosynthesis is limited, the overexpression or silencing of one or more miRNAs shows promise in manipulating key genes and enhancing the production of phytoconstituents in plants (Gupta et al., 2017).

Three new miRNAs – pso miR408, pso miR13, and pso miR2161 – have been discovered as potential candidates associated with the production of benzylisoquinoline in *Papaver somniferum* through thorough sequencing. The target identification analysis result indicated that pso-miR13 and pso-miR2161 were responsible for cleaving the transcripts of S-adenosyl-L-methionine 30-hydroxy-N-methyl-coclaurine 40-O-methyltransferase 2 and 7-O-methyltransferase, respectively, which play crucial roles as enzymes in benzylisoquinoline alkaloids pathway. Additionally, pso miR408 was found to potentially target the mRNA produced by a gene that encodes BBE domain-containing protein and FAD-binding (Boke et al., 2015).

Gene silencing works by decreasing the post-transcriptional level of gene expression. Consequently, a noticeable reduction in gene expression without eliminating gene function can be recognized. It has been proposed that utilizing gene editing tools for a gene knock-out strategy would be more beneficial in comprehending gene function and enhancing the yield of particular metabolites (Alagoz et al., 2016).

22.13 GENE EDITING

Despite the success of the metabolic engineering techniques discussed, drug shortages are becoming a significant issue in public health. There is still a high demand for alkaloids used in cancer treatment and anesthesia. This is caused not only by widespread pandemics, calamities, and the eradication of natural routines but also by the structural complexity, difficulty in synthesis, and extremely low content of these molecules in plants (Courdavault et al., 2021).

Among a range of plant species, the CRISPR/Cas9 technology is widely regarded as an optimal method for gene editing and targeted mutagenesis (Zakaria et al., 2021). The RNA-guided endonuclease Cas9 in this system uses a one-guide RNA to identify the target site and initiate a break in the double-strand (Westermann et al., 2021). Two methods of DNA modification occur after the double-strand breaks are created: one is the joining of nonhomologous ends which causes nucleotide insertion, deletion, and substitution, and the other one is the recombination of homologous ends, which makes it easier to insert DNA fragments if homologous donor templates are available (Leitao et al., 2017).

This has made it possible to identify the regulatory mechanisms or genes that govern secondary metabolite pathways like transcription factors, to disrupt several genes, and to characterize the gene families (Tang et al., 2017). Additional benefits of CRISPR-Cas include stable genomic changes that are passed down to the progeny and the total knockout of a target gene (Dey, 2021).

Although *Symphytum officinale*, which is well-known as a comfrey plant, is rich in analgesic and antiinflammatory qualities, it also possesses toxic alkaloids like pyrrolizidines. The homospermidine synthase gene in the hairy roots of *S. officinale* was subjected to some mutational changes using a CRISPR/Cas9-based method. The biosynthesis of pyrrolizidine alkaloid precursor, homospermidine is catalyzed by HSS. This HSS-encoding gene was successfully knocked out, and pyrrolizidine alkaloids were eliminated as a result (Zakaria et al., 2021).

In a different instance, for the biosynthesis of benzylisoquinoline in *Papaver somniferum*, the regulatory gene, 3′ hydroxy-N-methylcoclaurine 4′ -O-methyltransferase was eliminated using the CRISPR-Cas9 system. Consequently, the production of these alkaloids was greatly decreased, and a new, unidentified alkaloid was produced through biosynthesis (Alagoz et al., 2016).

Over the last few years, notable progress has been made in the development of tools and understanding of plant biosynthetic pathways. However, it is important to note that this field remains an active area of research, with discoveries yet to be made regarding the regulation of metabolic pathways, function, as well as the localization of enzymes (Shih & Morgan, 2020).

22.14 CONCLUSION

Advancements in CRISPR suggest that shortly, using effector proteins such as epigenetic regulators, repressors and transcriptional activators fused with dead Cas, which is an inactive protein variant, it will be feasible to manipulate the regulation of one or more genes in medicinal plants. This innovative approach holds great promise in enhancing the levels of plant alkaloids and various other secondary compounds. Alkaloids, which are utilized in cancer treatment and anaesthesia, are in great demand despite being scarce in plants. The application of metabolic engineering techniques described in this chapter can effectively boost alkaloid biosynthesis, thereby decreasing the reliance on excessive exploitation of limited natural resources.

REFERENCES

Abizov, E. A., & Tolkachev, O. N. (2012). Effects of extractant on the practical yield of β-carbonyl alkaloids from the bark of various species of the genus *Elaeagnus L. Pharmaceutical Chemistry Journal*, *45*, 671–674. https://doi.org/10.1007/s110940120701-8.

Alagoz, Y., Gurkok, T., Zhang, B., & Unver, T. (2016). Manipulating the biosynthesis of bioactive compound alkaloids for next-generation metabolic engineering in *Opium poppy* Using CRISPR-Cas 9 genome editing technology. *Scientific Reports*, *6*, 1–9. https://doi.org/10.1038/srep30910

Alfaro, M., Alfaro, I., & Angel, C. (2020). Identification of potential inhibitors of SARS- CoV-2 papain-like protease from tropane alkaloids from *Schizanthus porrigens*: A molecular docking study. *Chemical Physics Letters*, *761*, 138068. https://doi.org/10.1016/j.cplett.2020.138068

Amin, T., Salma, J., Mariam, B., Donia, K., Laura, S., Bufo, S. A., & Rafik, K. (2019). The biological activity of natural alkaloids against herbivores, cancerous cells and pathogens. *Toxins*, *11*, 656. https://doi.org/10.3390/toxins11110656

Aniszewski, T. (2007). Alkaloids – secrets of life. In *Alkaloid chemistry, biological significance, application and ecological role* (p. 334). Elsevier.

Badshah, S. L., Ullah, A., & Syed, S. (2020). The role of zinc-finger antiviral proteins in immunity against viruses. *Molecular Genetics, Microbiology and Virology*, *35*, 78–84. https://doi.org/10.3103/S0891416820020020.

Begum, F., Nageswara Rao, S. S., Rao, K., Prameela Devi, Y., Giri, A., & Giri, C. C. (2009). Increased vincristine production from *Agrobacterium tumefaciens* C58 induced shooty teratomas of Catharanthus roseus G. Don. *Natural Product Research*, *23*, 973–981. https://doi.org/10.1080/14786410802131153

Bhargavi, G., Rao, P. N., & Renganathan, S. (2018). Review on the extraction methods of crude oil from all generation biofuels in last few decades. *IOP Conference Series: Materials Science and Engineerin*, 012024. https://doi.org/10.1088/1757-899X/330/1/012024

Boke, H., Ozhuner, E., Turktas, M., Parmaksiz, I., Ozcan, S., & Unver, T. (2015). Regulation of the alkaloid biosynthesis by miRNA in *Opium poppy*. *Plant Biotechnology Journal*, *13*, 409–420. https://doi.org/10.1111/pbi.12346

Cantrell, C., Schrader, K., Mamonov, L., Sitpaeva, G., Kustova, T., Dunbar, C., & Wedge, D. (2005). Isolation and identification of antifungal and antialgal alkaloids from *Haplophyllum sieversii*. *Journal of Agricultural and Food Chemistry*, *53*, 7741–7748. https://doi.org/10.1021/jf051478v.

Cecil, C. E., Davis, J. M., Cech, N. B., & Laster, S. M. (2011). Inhibition of H1N1 influenza A virus growth and induction of inflammatory mediators by the isoquino- line alkaloid berberine and extracts of goldenseal (*Hydrastis canadensis*). *International Immunopharmacology*, *11*, 1706–1714. https://doi.org/10.1016/j.intimp.2011.06.002.

Chen, Y., Chen, J. C., & Tseng, S. H. (2009). Tetrandrine suppresses tumor growth and angiogenesis of gliomas in rats. *International Journal of Cancer*, *124*, 2260–2269. https://doi.org/10.1002/ijc.24208

Courdavault, V., O'Connor, S. E., Jensen, M. K., & Papon, N. (2021). Metabolic engineering for plant natural products biosynthesis: New procedures, concrete achievements and remaining limits. *Natural Product Reports*. https://doi.org/10.1039/d0np00092b

Cretton, S., Dorsaz, S., Azzollini, A., Favre Godal, Q., Marcourt, L., Ebrahimi, S. N., Voinesco, F., Michellod, E., Sanglard, D., & Gindro, K. (2016). *Antifungal quinoline* alkaloids fromWaltheria indica. *Journal of Natural Products*, *79*, 300–307. https://doi.org/10.1021/acs.jnatprod.5b00896.

Cronemberger, S., Calixto, N., Moraes, M. N., Castro, I. D., Lana, P. C., & Loredo, A. F. (2012). Efficacy of one drop of 2% pilocarpine to reverse the intra ocular pressure peak at 6: 00 am in early glaucoma. *The Pan American Journal of Ophthalmology*, *11*, 14–16.

Decendit, A., Mars, M., & Delrot, S. (2009). Overexpression of a grapevine R2R3-MYB factor in tomato affects vegetative development, flower morphology and flavonoid and

terpenoid metabolism. *Plant Physiology and Biochemistry, 47*, 551–561. https://doi.org/10.1016/j.plaphy.2009.02.015
Desgagn´e-Penix, I. (2021). Biosynthesis of alkaloids in Amaryllidaceae plants: A review. *Phytochemistry Reviews, 20*, 409–431. https://doi.org/10.1007/s11101-020-09678-5
Dey, A. (2021). CRISPR/Cas genome editing to optimize pharmacologically active plant natural products. *Pharmacological Research, 164*, 105359. https://doi.org/10.1016/j.phrs.2020.105359
Dey, P., Kundu, A., Kumar, A., Gupta, M., Lee, B. M., Bhakta, T., Dash, S., & Kim, H. S. (2020). Analysis of alkaloids (indole alkaloids, isoquinoline alkaloids, tropane alkaloids). In *Recent advances in natural products analysis* (pp. 505–567). https://doi.org/10.1016/B978-0-12-816455-6.00015-9.
Dug´e de Bernonville, T., Maury, S., Delaunay, A., Daviaud, C., Chaparro, C., Tost, J., O'connor, S. E., & Courdavault, V. (2020). Developmental methylome of the medicinal plant catharanthus roseus unravels the tissue-specific control of the monoterpene indole alkaloid pathway by DNA methylation. *International Journal of Molecular Sciences, 21*, 1–26. https://doi.org/10.3390/ijms21176028
Espinosa-Leal, C. A., Puente-Garza, C. A., & García-Lara, S. (2018). In vitro plant tissue culture: Means for production of biological active compounds. *Planta*, 248. https://doi.org/10.1007/s00425-018-2910-1.
Fabricant, D. S., & Farnsworth, N. R. (2001). The value of plants used in traditional medicine for drug discovery. *Environmental Health Perspectives, 109*, 69. https://doi.org/10.1289/ehp.01109s169
Faisal, S., Badshah, S. L., Emwas, A. H., & Jaremko, M. (2023). Alkaloids as potential antivirals. A comprehensive review. *Natural Products and Bioprospecting, 13*, 4. https://doi.org/10.1007/s13659-022-00366-9.
Faisal, S., Badshah, S. L., Kubra, B., Sharaf, M., Emwas, A. H., Jaremko, M., & Abdalla, M. (2022). Identification and inhibition of the druggable allosteric site of SARS-CoV-2 NSP10/NSP16 methyltransferase through computational approaches. *Molecules*. https://doi.org/10.3390/molecules27165241.
Fakhari, A. R., Nojavan, S., Ebrahimi, S. N., & Evenhuis, C. J. (2010). Optimized ultrasound-assisted extraction procedure for the analysis of opium alkaloids in papaver plants by cyclodextrin-modified capillary electrophoresis. *Journal of Separtion Science, 33*, 14, 2153–2159. https://doi.org/10.1002/jssc.201000135
Fei, X. F., Wang, B. X., Li, T. J., Tashiro, S.-i., Minami, M., Xing, D. J., & Ikejima, T. (2003). Evodiamine, a constituent of Evodiae Fructus, induces anti-proliferating effects in tumor cells. *Cancer Science, 94*, 92–98. https://doi.org/10.1111/j.13497006.2003.tb01358.x
Fikatas, A., Vervaeke, P., Meyen, E., Llor, N., Ordeix, S., Boonen, I., Bletsa, M., Kafetzopoulou, L. E., Lemey, P., Amat, M., Pannecouque, C., & Schols, D. (2021). A novel series of indole alkaloid derivatives inhibit dengue and zika virus infection by interference with the viral replication complex. *Antimicrobial Agents and Chemotherapy*. https://doi.org/10.1128/AAC.02349-20
García-Granados, R., Lerma-Escalera, J. A., & Morones-Ramírez, J. R. (2019). Metabolic engineering and synthetic biology: Synergies, future, and challenges. *Frontiers in Bioengineering and Biotechnology, 7*, 1–4. https://doi.org/10.3389/fbioe.2019.00036.
Garg, S., & Roy, A. (2020). In silico analysis of selected alkaloids against main pro- tease (Mpro) of SARS-CoV-2. *Chemico Biological Interactions*. https://doi.org/10.1016/j.cbi.2020.109309
Gendrot, M., Andreani, J., Boxberger, M., Jardot, P., Fonta, I., Le Bideau, M., Duflot, I., Mosnier, J., Rolland, C., Bogreau, H., Vermerie, M., & Pradines, B. (2020). Antimalarial drugs inhibit the replication of SARS-CoV-2: An in vitro evaluation. *Travel Medicine and Infectious Disease, 37*, 101873. https://doi.org/10.1016/j.tmaid.2020.101873
Guaadaoui, A., Benaicha, S., Elmajdoub, N., Bellaoui, M., & Hamal, A. (2014). What is a bioactive compound? A combined definition for a preliminary consensus. *International Journal of Food Science and Nutrition, 3*, 17–179. https://doi.org/10.11648/j.ijnfs.20140303.16.
Guo, Z., Tan, H., Lv, Z., Ji, Q., Huang, Y., Liu, J., Chen, D., Diao, Y., Si, J., & Zhang, L. (2018). Targeted expression of *Vitreoscilla hemoglobin* improves the production of tropane alkaloids in Hyoscyamus Niger hairy roots. *Scientific Reports, 8*, 1–12. https://doi.org/10.1038/s41598-018-36156-y
Gupta, O. P., Karkute, S. G., Banerjee, S., Meena, N. L., & Dahuja, A. (2017). Contemporary understanding of miRNA-based regulation of secondary metabolites biosynthesis in plants. *Frontiers in Plant Science, 8*. https://doi.org/10.3389/fpls.2017.00374
Handa, S. S., Khanuja, S. P. S., Longo, G., & Rakesh, D. D. (2008). Extraction technologies for medicinal and aromatic plants. In *Earth, environmental and marine sciences and technologies*. Springer.
Hayashi, K., Minoda, K., Nagaoka, Y., Hayashi, T., & Uesato, S. (2007). Antiviral activity of berberine and related compounds against human cytomegalovirus. *Bioorganic & Medicinal Chemistry Letters, 17*(6), 1562–1564.
Huang, J. P., Fang, C., Ma, X., Wang, L., Yang, J., Luo, J., Yan, Y., Zhang, Y., & Huang, S. X. (2019). Tropane alkaloids biosynthesis involves an unusual type III polyketide synthase and non-enzymatic condensation. *Nature Communications, 10*, 1–8. https://doi.org/10.1038/s41467-019-11987-z
Huie, C. W. (2002). A review of modern sample preparation techniques for the extraction and analysis of medicinal plants. *Analytical and Bioanalytical Chemistry, 373*, 23–30. https://doi.org/10.1007/s00216-002-1265-3
Jaggi, M., Kumar, S., & Sinha, A. K. (2011). Overexpression of an apoplastic peroxidase gene CrPrx in transgenic hairy root lines of Catharanthus roseus. *Applied Microbiology and Biotechnology, 90*, 1005–1016. https://doi.org/10.1007/s00253-011-3131-8
Jung, H. J., Sung, W. S., Yeo, S. H., Kim, H. S., Lee, I. S., Woo, E. R., & Lee, D. G. (2006). Antifungal effect of a mentoflavone derived from *Selaginella tamariscina*. *Archives of Pharmacal Research, 29*, 746. https://doi.org/10.1007/BF02974074.
Kainsa, S., Kumar, P., & Rani, P. (2012). Medicinal plants of Asian origin having anticancer potential: Short review. *Asian Journal of Biomedical and Pharmaceutical Sciences, 2*, 1–7.
Kaiser, S., Verza, S. G., Moraes, R. C., Pittol, V., Peñaloza, E. M. C., Pavei, C., & Ortega, G. G. (2013). Extraction optimization of polyphenols, oxindole alkaloids and quinovic acid glycosides from cat's claw bark by Box-Behnken design. *Industrial Crops and Products, 47*, 153–161. https://doi.org/10.1016/j.indcrop.2013.04.026
Kim, H., Lantvit, D., Hwang, C. H., Kroll, D. J., Swanson, S. M., Franzblau, S. G., & Orjala, J. (2012). Indole alkaloids from two cultured cyanobacteria, *Westiellopsis sp.* and *Fischerella muscicola*. *Bioorganic and Medicinal Chemistry, 20*, 5290–5295. https://doi.org/10.1016/j.bmc.2012.06.030.

Kittakoop, P., Mahidol, C., & Ruchirawat, S. (2014). Alkaloids as important scaffolds in therapeutic drugs for the treatments of cancer, tuberculosis, and smoking cessation. *Current Topics in Medicinal Chemistry*, *14*, 239–252.

Kowalczyk, T., Wieczfinska, J., Skała, E., Sliwinski,´ T., & Sitarek, P. (2020). Transgenesis as a tool for the efficient production of selected secondary metabolites from in vitro plant cultures. *Plants*, *9*. https://doi.org/10.3390/plants9020132

Kulkarni, R. (2016). Metabolic engineering: Biological art of producing useful chemicals. *Resonance*, *21*, 233–237. https://doi.org/10.1007/s12045-016-0318-4.

Kumar, K., Kumar, S. R., Dwivedi, V., Rai, A., Shukla, A. K., Shanker, K., & Nagegowda, D. A. (2015). Precursor feeding studies and molecular characterization of geraniol synthase establish the limiting role of geraniol in monoterpene indole alkaloid biosynthesis in *Catharanthus roseus* leaves. *Plant Science*, *239*, 56–66. https://doi.org/10.1016/j.plantsci.2015.07.007

Kumar, P., Sharma, B., & Bakshi, N. (2009). Biological activity of alkaloids from *Solanum dulcamara* L. *Natural Product Research*, *23*, 719–723. https://doi.org/10.1080/14786410802267692.

Kumar, S., Kashyap, P., Chowdhury, S., Kumar, S., Panwar, A., & Kumar, A. (2021). Identification of phytochemicals as potential therapeutic agents that binds to Nsp15 protein target of coronavirus (SARS-CoV-2) that are capable of inhibiting virus replication. *Phytomedicine*, *85*, 153317. https://doi.org/10.1016/j.phymed.2020.153317

Kurek, J. (2019). Introductory chapter: Alkaloids – their importance in nature and for human life. In *Alkaloids – their importance in nature and human life*. InTechOpen.

Labanca, F., Ovesna, J., & Milella, L. (2018). *Papaver somniferum* L. taxonomy, uses and new insight in poppy alkaloid pathways. *Phytochemistry Review*, *17*, 853–871. https://doi.org/10.1007/s11101-018-9563-3

Lan, X., Zeng, J., Liu, K., Zhang, F., Bai, G., Chen, M., Liao, Z., & Huang, L. (2018). Comparison of two hyoscyamine 6β-hydroxylases in engineering scopolamine biosynthesis in root cultures of Scopolia lurida. *Biochemical and Biophysical Research Communications*, *497*, 25–31. https://doi.org/10.1016/j.bbrc.2018.01.173

Leitao, A. L., Costa, M. C., & Enguita, F. J. (2017). Applications of genome editing by programmable nucleases to the metabolic engineering of secondary metabolites. *Journal of Biotechnology*, *241*, 50–60. https://doi.org/10.1016/j.jbiotec.2016.11.009

Li, D. G., Wang, Z. R., & Lu, H. M. (2001). Pharmacology of tetrandrine and its therapeutic use in digestive diseases. *World Journal of Gastroenterology*, *7*, 627. https://doi.org/10.3748/wjg.v7.i5.627.

Li, F., Wang, W., Zhao, N., Xiao, B., Cao, P., Wu, X., Ye, C., Shen, E., Qiu, J., Zhu, Q. H., Xie, J., Zhou, X., & Fan, L. (2015). Regulation of nicotine biosynthesis by an endogenous target mimicry of microRNA in tobacco. *Plant Physiology*, *169*, 1062–1071. https://doi.org/10.1104/pp.15.00649

Li, H. L., Zhang, W. D., Zhang, C., Liu, R. H., Wang, X. W., Wang, X. L., Zhu, J. B., & Chen, C. L. (2006). Bioavailabilty and pharmacokinetics of four active alkaloids of traditional Chinese medicine Yanhuanglian in rats following intravenous and oral administration. *Journal of Pharmaceutical and Biomedical Analysis*, *41*, 1342–1346. https://doi.org/10.1016/j.jpba.2006.03.029.

Li, M., Su, B. S., Chang, L. H., Gao, Q., Chen, K. L., An, P., Huang, C., Yang, J., & Li, Z. F. (2014). Oxymatrine induces apoptosis in human cervical cancer cells through guanine nucleotide depletion. *Anti-Cancer Drugs*, *25*, 161–173. https://doi.org/10.1097/CAD.0000000000000012

Li, Q., Zhu, T., Zhang, R., Bu, Q., Yin, J., Zhang, L., & Chen, W. (2020). Molecular cloning and functional analysis of hyoscyamine 6β-hydroxylase (H6H) in the poisonous and medicinal plant *Datura innoxia* mill. *Plant Physiology and Biochemstry*, *153*, 11. https://doi.org/10.1016/j.plaphy.2020.04.021

Lichman, B. R. (2021). The scaffold-forming steps of plant alkaloid biosynthesis. *Natural Products Report*, *38*, 103–129. https://doi.org/10.1039/d0np00031k

Liu, J., Gao, F., Ren, J., Lu, X., Ren, G., & Wang, R. (2017). A novel AP2/ERF transcription factor CR1 regulates the accumulation of vindoline and serpentine in *Catharanthus roseus*. *Frontiers in Plant Science*, *8*, 1–11. https://doi.org/10.3389/fpls.2017.02082

Lou, C., Yokoyama, S., Saiki, I., & Hayakawa, Y. (2015). Selective anticancer activity of hirsutine against HER2- positive breast cancer cells by inducing DNA damage. *Oncology Reports*, *33*, 2072–2076. https://doi.org/10.3892/or.2015.3796.

Ma, D., Li, G., Zhu, Y., & Xie, D. Y. (2017). Overexpression and suppression of Artemisia annua 4-hydroxy-3-methylbut-2-enyl diphosphate reductase 1 gene (AaHDR1) differentially regulate artemisinin and terpenoid biosynthesis. *Frontier of Plant Science*, *8*, 1–13. https://doi.org/10.3389/fpls.2017.00077

Machado, P. A., Hilário, F. F., Carvalho, L. O., Silveira, M. L., Alves, R. B., Freitas, R. P., & Coimbra, E. S. (2012). Effect of 3-alkylpyridine marine alkaloid analogues in Leishmania species related to American *Cutaneous leishmaniasis*. *Chemical Biology and Drug Design*, *80*, 745–751. https://doi.org/10.1111/cbdd.12017

Mamkulathil Devasia, R., Altaf, M., Fahad Alrefaei, A., & Manoharadas, S. (2021). Enhanced production of camptothecin by immobilized callus of *Ophiorrhiza mungos* and a bioinformatic insight into its potential antiviral effect against SARS-CoV-2. *Journal of King Squad University*, *33*, 101344. https://doi.org/10.1016/j. jksus.2021.101344

Marciniak, P., Adamski, Z., Bednarz, P., Slocinska, M., Ziemnicki, K., Lelario, F., Scrano, L., & Bufo, S. A. (2010). Cardioinhibitory properties of potato glycoalkaloids in beetles. *Bulletin of Environmental Contamination and Toxicology*, *84*, 153–156. https://doi.org/10.1007/s00128-009-9921-3.

Marciniak, P., Kolińska, A., Spochacz, M., Chowański, S., Adamski, Z., Scrano, L., Falabella, P., Bufo, S. A., & Rosiński, G. (2019). Differentiated effects of secondary metabolites from solanaceae and brassicaceae plant families on the heartbeat of *Tenebrio molitor* Pupae. *Toxins*, *11*, 287. https://doi.org/10.3390/toxins11050287

Marella, A., Tanwar, O. P., Saha, R., Ali, M. R., Srivastava, S., Akhter, M., Shaquiquzzaman, M., & Alam, M. M. (2013). Quinoline: A versatile heterocyclic. *Saudi Pharmaceutical Journal*, *21*, 1–12. https://doi.org/10.1016/j.jsps.2012.03.002

McCue, K. F., Breksa, A., Vilches, A., & Belknap, W. R. (2018). Modification of potato steroidal glycoalkaloids with silencing RNA constructs. *American Journal of Potato Research*, *95*, 9–14. https://doi.org/10.1007/s12230-017-9609-x

Miaoxian, G., Haizhou, Lv., Hongyu, C., Shuting, D., Jianhong, Z., Wanjing, L., Liu, H., Yimian, M., Hua, Y., Shilin, C., & Hongmei, L. (2024). Strategies on biosynthesis and production of bioactive compounds in medicinal plants. *Chemistry*, *16*, 13–26. https://doi.org/10.1016/j.chmed.2023.01.007

Mohapatra, S., Minocha, R., Long, S., & Minocha, S. C. (2009). Putrescine overproduction negatively impacts the oxidative state of poplar cells in culture. *Plant Physiology and Biochemisrty*, *47*, 262–271. https://doi.org/10.1016/j.plaphy.2008.12.007

Nakahara, K., Trakoontivakorn, G., Alzoreky, N. S., Ono, H., Onishi-Kameyama, M., & Yoshida, M. (2002). Antimutagenicity of some edible Thai plants, and a bioactive carbazole alkaloid, mahanine, isolated from *Micromelum minutum*. *Journal of Agricultural and Food Chemistry*, *50*, 4796–4802. https://doi.org/10.1021/jf025564w.

Nassiri, M. (2013). Simple, one-pot, and three-component coupling reactions of azaarenes (phenanthridine, isoquinoline, and quinoline), with acetylenic esters involving methyl propiolate or ethyl propiolate in the presence of nh-heterocyclicor1,3-dicarbonylcompounds. *Synthetic Communications*, *43*, 157–168. https://doi.org/10.1080/00397911.2011.589559.

Oliva, A., Meepagala, K. M., Wedge, D. E., Harries, D., Hale, A. L., Aliotta, G., & Duke, S. O. (2003). Natural fungicides from *Ruta graveolens* L. leaves, including a new quinolone alkaloid. *Journal of Agricultural and Food Chemistry*, *51*, 890–896. https://doi.org/10.1021/jf0259361

Otshudi, A. L., Apers, S., Pieters, L., Claeys, M., Pannecouque, C., De Clercq, E., Van Zeebroeck, A., Lauwers, S., Frederich, M., & Foriers, A. (2005). Biologically active bisbenzylisoquinoline alkaloids from the root bark of *Epinetrum villosum*. *Journal of Ethnopharmacology*, *102*, 89–94. https://doi.org/10.1016/j.jep.2005.05.021.

Ozcelik, B., Kartal, M., & Orhan, I. (2011). Cytotoxicity, antiviral and antimicrobial activities of alkaloids, flavonoids, and phenolic acids. *Pharmaceutical Biology*, *49*, 396–402. https://doi.org/10.3109/13880209.2010.519390.

Palem, J. R., Bedadala, G., El Sayed, K. A., & Hsia, S. C. V. (2011). Manzamine A as a novel inhibitor of herpes simplex virus type-1 replication in cultured corneal cells. *Planta Medica*, *77*, 46–51. https://doi.org/10.1055/s-0030-1250093.

Parmar, N. J., Pansuriya, B. R., Barad, H. A., Kant, R., & Gupta, V. K. (2012). An improved microwave assisted one-pot synthesis, and biological investigations of some novel aryldiazenylchromen of used pyrrolidines. *Bioorganic and Medicinal Chemistry Letters*, *22*, 4075–4079. https://doi.org/10.1016/j.bmcl.2012.04.070.

Paul, P., Singh, S. K., Patra, B., Sui, X., Pattanaik, S., & Yuan, L. (2017). A differentially regulated AP2/ERF transcription factor gene cluster acts downstream of a MAP kinase cascade to modulate terpenoid indole alkaloid biosynthesis in *Catharanthus roseus*. *New Phytologist*, *213*, 1107–1123. https://doi.org/10.1111/nph.14252

Peebles, C. A. M., Sander, G. W., Hughes, E. H., Peacock, R., Shanks, J. V., & San, K. Y. (2011). The expression of 1-deoxy-d-xylulose synthase and geraniol-10-hydroxylase or anthranilate synthase increases terpenoid indole alkaloid accumulation in *Catharanthus roseus* hairy roots. *Metabolic Engineering*, *13*, 234–240. https://doi.org/10.1016/j.ymben.2010.11.005

Peter, J. F. (2001). Alkaloid biosynthesis in plants: Biochemistry, cell biology, molecular regulation, and metabolic engineering applications. *Annual Review of Plant Physiology and Plant Molecular Biology*, *52*, 29–66. https://doi.org/10.1146/annurev.arplant.52.1.29

Raina, S. K., Wankhede, D. P., Jaggi, M., Singh, P., Jalmi, S. K., Raghuram, B., Sheikh, A. H., & Sinha, A. K. (2012). CrMPK3, a mitogen activated protein kinase from *Catharanthus roseus* and its possible role in stress induced biosynthesis of monoterpenoid indole alkaloids. *BMC Plant Biology*, *12*. https://doi.org/10.1186/14712229-12-134

Ricigliano, V., Kumar, S., Kinison, S., Brooks, C., Nybo, S. E., Chappell, J., & Howarth, D. G. (2016). Regulation of sesquiterpenoid metabolism in recombinant and elicited *Valeriana officinalis* hairy roots. *Phytochemistry*, *125*, 43–53. https://doi.org/10.1016/j.phytochem.2016.02.011

Rosales, P. F., Bordin, G. S., Gower, A. E., & Moura, S. (2020). Indole alkaloids: 2012 until now, highlighting the new chemical structures and biological activities. *Fitoterapia*, *143*, 104558. https://doi.org/10.1016/j.fitote.2020.104558.

Sai, C., Wang, J., Li, B., Ding, L., Wang, H., Wang, Q., Zhang, F., & Ren, Q. (2020). Isolation and identification of alkaloids from macleaya microcarpa by UHPLC–Q-TOF-MS and their cytotoxic activity in vitro, antiangiogenic activity in vivo. *BMC Chemistry*, *14*, 5. https://doi.org/10.1186/s13065-020-0660-1

Salehi, B., Sharopov, F., Boyunegmez Tumer, T., Ozleyen, A., Rodríguez-Pérez, C., Ezzat, S. M., Azzini, E., Hosseinabadi, T., Butnariu, M., & Sarac, I. (2019). Symphytum species: A comprehensive review on chemical composition, food applications and phytopharmacology. *Molecules*, *24*, 2272. https://doi.org/10.3390/molecules24122272

Samad, A. F. A., Sajad, M., Nazaruddin, N., Fauzi, I. A., Murad, A. M. A., Zainal, Z., & Ismail, I. (2017). MicroRNA and transcription factor: Key players in plant regulatory network. *Frontiers in Plant Science*, *8*, 1–18. https://doi.org/10.3389/fpls.2017.00565

Samoylenko, V., Ashfaq, M. K., Jacob, M. R., Tekwani, B. L., Khan, S. I., Manly, S. P., Joshi, V. C., Walker, L. A., & Muhammad, I. (2009). Indolizidine, antiinfective and antiparasitic compounds from *Prosopis glandulosa* Torr. var. glandulosa. *Planta Medica*, *75*, 48. https://doi.org/10.1055/s-2009-1216486.

Sasidharan, Y., Chen1, D., Saravanan, K. M., Sundram, L., & Yoga, L. (2011). Extraction, isolation and characterization of bioactive compounds from plants' extracts. *African Journal of Traditional Complementary and Alternative Medicines*, *8*(1), 1–101. https://doi.org/10.4314/ajtcam.v8i1.60483

Seo, Y. S., Yim, M. J., Kim, B. H., Kang, K. R., Lee, S. Y., Oh, J. S., You, J. S., Kim, S. G., Yu, S. J., & Lee, G. J. (2015). Berberine-induced anticancer activities in FaDu head and neck squamous cell carcinoma cells. *Oncolology Reports*, *34*, 3025–3034. https://doi.org/10.3892/or.2015.4312

Sharafi, A., Sohi, H. H., Mousavi, A., Azadi, P., Khalifani, B. H., & Razavi, K. (2013). Metabolic engineering of morphinan alkaloids by over-expression of codeinone reductase in transgenic hairy roots of *Papaver bracteatum*, the Iranian poppy. *Biotechnology Letters*, *35*, 445–453. https://doi.org/10.1007/s10529-012-1080-7

Sheludko, Y. V. (2010). Recent advances in plant biotechnology and genetic engineering for production of secondary metabolites. *Cytology and Genetics*, *44*, 65–75. https://doi.org/10.3103/S009545271001010X

Shih, M. L., & Morgan, J. A. (2020). Metabolic flux analysis of secondary metabolism in plants. *Metabolic Engineering Communications*, *10*, e00123 https://doi.org/10.1016/j.mec.2020.e00123

Simons, V., Morrissey, J. P., Latijnhouwers, M., Csukai, M., Cleaver, A., Yarrow, C., & Osbourn, A. (2006). Dual effects of plant steroidal alkaloids on *Saccharomyces cerevisiae*. *Antimicrobial Agents and Chemotherapy*, *50*, 2732–2740. https://doi.org/10.1128/aac.00289

Singh, A., Nirala, N. K., Das, S., Narula, A., Rajam, M. V., & Srivastava, P. S. (2011). Overexpression of odc (*Ornithine decarboxylase*) in *Datura innoxia* enhances the yield of scopolamine. *Acta Physiologiae Plantarum*, *33*, 2453–2459. https://doi.org/10.1007/s11738-011-0787-8

Singh, N., Azmi, S., Maurya, S., Singh, U., Jha, R., & Pandey, V. (2003). Two plant alkaloids isolated from *Corydalis longipes* as potential antifungal agents. *Folia Microbiologica*, *48*, 605–609. https://doi.org/10.1007/BF02993466

Singh, P., Prasad, R., Tewari, R., Jaidi, M., Kumar, S., Rout, P. K., & Rahman, L. (2018). Silencing of quinolinic acid phosphoribosyl transferase (QPT) gene for enhanced production of scopolamine in hairy root culture of *Duboisia leichhardtii*. *Scientific Reports*, *8*, 13939. https://doi.org/10.1038/s41598-018-32396-0

Singh, S. K., Patra, B., Paul, P., Liu, Y., Pattanaik, S., & Yuan, L. (2020). Revisiting the ORCA gene cluster that regulates terpenoid indole alkaloid biosynthesis in *Catharanthus roseus*. *Plant Science*, *293*, 110408. https://doi.org/10.1016/j.plantsci.2020.110408

Singh, U., Sarma, B., Mishra, P., & Ray, A. (2000). Antifungal activity of venenatine, an indole alkaloid isolated from *Alstonia venenata*. *Folia Microbiologica*, *45*, 173. https://doi.org/10.1007/BF02817419.

Sobhani, A. M., Ebrahimi, S. A., & Mahmoudian, M. (2002). An in vitro evaluation of human DNA topoisomerase I inhibition by *Peganum harmala* L. seeds extract and its beta-carboline alkaloids. *Journal of Pharmaceutical Sciences*, *5*, 19–23.

Soledad, M. V., Guillermo Gael, W. A., Epinosa Leal, C., Guy, A. C., & Silverio, G. L. (2022). Application of metabolic engineering to enhance the content of alkaloids in medicinal plants. *Metabolic Engineering Communications*, *14*, e00194. https://doi.org/10.1016/j.mec.2022.e00194

Spochacz, M., Chowanski, S., Szymczak, M., Lelario, F., Bufo, S., & Adamski, Z. (2018). Sublethal effects of *Solanum nigrum* fruit extract and its pure glycoalkaloids on the physiology of *Tenebrio molitor* (Mealworm). *Toxins*, *10*, 504. https://doi.org/10.3390/toxins10120504.

Staniek, A., Bouwmeester, H., Fraser, P. D., Kayser, O., Martens, S., Tissier, A., van der Krol, S., Wessjohann, L., & Warzecha, H. (2013). Natural products – modifying metabolite pathways in plants. *Biotechnology Journal*, *8*, 1159–1171. https://doi.org/10.1002/biot.201300224

Takla, S. S., Shawky, E., Hammoda, H. M., & Darwish, F. A. (2018). Green techniques in comparison to conventional ones in the extraction of Amaryllidaceae alkaloids: Best solvents selection and parameters optimization. *Journal of Chromatography A*, *14*, 99–110. https://doi.org/10.1016/j.chroma.2018.07.009

Tang, X., Lowder, L. G., Zhang, T., Malzahn, A. A., Zheng, X., Voytas, D. F., Zhong, Z., Chen, Y., Ren, Q., Li, Q., Kirkland, E. R., Zhang, Y., & Qi, Y. (2017). A CRISPR-Cpf1 system for efficient genome editing and transcriptional repression in plants. *Native Plants*, *3*, 1–5. https://doi.org/10.1038/nplants.2017.18

Tozawa, Y., Hasegawa, H., Terakawa, T., & Wakasa, K. (2001). Characterization of rice anthranilate synthase α-subunit genes OASA1 and OASA2. Tryptophan accumulation in transgenic rice expressing a feedback-insensitive mutant of OASA1. *Plant Physiology*, 126, 1493–1506. https://doi.org/10.1104/pp.126.4.1493

Troost, B., Mulder, L. M., Diosa-Toro, M., van de Pol, D., Rodenhuis-Zybert, I. A., & Smit, J. M. (2020). Tomatidine, a natural steroidal alkaloid shows antiviral activity towards chikungunya virus in vitro. *Scientific Reports*. https://doi.org/10.1038/s41598-020-63397-7

Varghese, F. S., Kaukinen, P., Glasker, S., Bespalov, M., Hanski, L., Wennerberg, K., Kummerer, B. M., & Ahola, T. (2016). Discovery of berberine, abamectin and iver- mectin as antivirals against chikungunya and other alpha viruses. *Antiviral Research*, *126*, 117–124. https://doi.org/10.1016/j.antiviral.2015.12.012

Ventrella, E., Adamski, Z., Chudzińska, E., Miądowicz-Kobielska, M., Marciniak, P., Büyükgüzel, E., Büyükgüzel, K., Erdem, M., Falabella, P., & Scrano, L. (2016). *Solanum tuberosum* and *Lycopersicon esculentum* leaf extracts and single metabolites affect development and reproduction of *Drosophila melanogaster*. *Plos One*, *11*, e0155958. https://doi.org/10.1371/journal.pone.0155958.

Verpoorte, R., & Schripsema, J. (1994). Isolation, identification, and structure elucidation of alkaloids a general overview. In *Modern methods of plant analysis* (p. 15). Springer.

Villinski, J., Dumas, E., Chai, H. B., Pezzuto, J., Angerhofer, C., & Gafner, S. (2003). Antibacterial activity and alkaloid content of *Berberis thunbergii*, *Berberis vulgaris* and *Hydrastis canadensis*. *Pharmaceutical Biology*, *41*, 551–557. https://doi.org/10.1080/13880200390500768.

Wang, X. X., Zan, K., Shi, S. P., Zeng, K. W., Jiang, Y., Guan, Y., Xiao, C. L., Gao, H. Y., Wu, L. J., & Tu, P. F. (2013). Quinolone alkaloids with antibacterial and cytotoxic activities from the fruits of *Evodia rutaecarpa*. *Fitoterapia*, *89*, 1–7. https://doi.org/10.1016/j.fitote.2013.04.007.

Wang, Y. H., Morris-Natschke, S. L., Yang, J., Niu, H. M., Long, C. L., & Lee, K. H. (2014). Anticancer principles from medicinal Piper plants. *Journal of Traditional and Complementary Medicine*, *4*, 8–16. https://doi.org/10.4103/2225-4110.124811.

Westermann, L., Neubauer, B., & Köttgen, M. (2021). Nobel prize 2020 in chemistry honors CRISPR: A tool for rewriting the code of life. *Pflugers Archiv-European Journal of Physiology*, *473*, 1–2. https://doi.org/10.1007/s00424-020-02497-9

Wirasathien, L., Boonarkart, C., Pengsuparp, T., & Suttisri, R. (2006). Biological activities of alkaloids from *Pseuduvaria setosa*. *Pharmaceutical Biology*, *44*, 274–278. https://doi.org/10.1080/13880200600714111

Yang, G., & Chen, D. (2008). Alkaloids from the roots of *Zanthoxylum nitidum* and their antiviral and antifungal effects. *Chemistry and Biodiversity*, *5*, 1718–1722. https://doi.org/10.1002/cbdv.200890160

Yang, L., & Stöckigt, J. (2010). Trends for diverse production strategies of plant medicinal alkaloids. *Natural Product Reports*, *27*, 1469–1479. https://doi.org/10.1039/C005378C

Zakaria, M. M., Schemmerling, B., & Ober, D. (2021). CRISPR/Cas9-Mediated genome editing in comfrey (*Symphytum officinale*) hairy roots results in the complete eradication of pyrrolizidine alkaloids. *Molecules*, *26*, 14–98. https://doi.org/10.3390/molecules26061498

Zhang, G. B., Zhang, B., Zhang, X. X., & Bing, F. H. (2013). Homonojirimycin, an alkaloid from dayflower inhibits the growth of influenza A virus in vitro. *Acta Virologica*, *57*, 85–86.

Zhang, L., Guo, Y. K., Bei-Bei, L. U., Han-Ming, Z., Ke-Xuan, T., Ji-Hong, J., & Wan-Sheng, C. (2005). Metabolic engineering of tropane alkaloid biosynthesis in plants. *Journal of Integrative Plant Biology*, *47*(2), 136–143. https://doi.org/10.1111/j.1744-7909.2005.00024.x

Zhao, T., Li, S., Wang, J., Zhou, Q., Yang, C., Bai, F., Lan, X., Chen, M., & Liao, Z. (2020). Engineering tropane alkaloid production based on metabolic characterization of ornithine decarboxylase in atropa belladonna. *ACS Synthetic Biology*, *9*, 437–448. https://doi.org/10.1021/acssynbio.9b00461

Zhao, W., He, J., Zhang, Y., Ito, Y., Su, Q., & Sun, W. (2012). Preparative isolation and purification of alkaloids from *Picrasma quassiodes* (D. Don) Benn. by high-speed countercurrent chromatography. *Journal of Liquid Chromatography and Related Techniques*, *5*, 11. https://doi.org/10.1080/10826076.2011.621150

Ziegler, J., & Facchini, P. J. (2008). Alkaloid biosynthesis: Metabolism and trafficking. *Annual Review of Plant Biology*, *59*, 735–769. https://doi.org/10.1146/annurev.arplant.59.032607.092730.

Zoraghi, R., Worrall, L., See, R. H., Strangman, W., Popplewell, W. L., Gong, H., Samaai, T., Swayze, R. D., Kaur, S., & Vuckovic, M. (2011). Methicillin-resistant *Staphylococcus aureus* (MRSA) pyruvate kinase as a target for bis-indole alkaloids with antibacterial activities. *Journal of Biological Chemistry*, *286*, 44716–44725. https://doi.org/10.1074/jbc.M111.289033

23 Integrated Molecular Strategies for Enhanced Production of Medicinal Plant-Derived Flavonoids

Dwaipayan Sinha, Zoofishan Kazi, Rashmi Muherjee, Arindam Ganguly, Swastika Banerjee, Arun Kumar Maurya, Abhra Abir Chowdhury, Aqsa Majgaonkar, and Shreyasi Dutta

23.1 INTRODUCTION

Flavonoids, a significant group of organic compounds, function as secondary metabolites in plants and are characterized by their polyphenolic architecture. They are typically present in fruits, vegetables, and beverages (Dias et al., 2021). During the early 1900s, scientists identified and analyzed the chemical composition of flavonoids and related compounds found in many plants. They also successfully created these molecules in a laboratory setting. The primary focus of the investigations was on the role of anthocyanins as pigments. Over 9,000 derivatives of flavonoids have been reported since that time (Roy et al., 2022).

The flavone skeleton, or C6-C3-C6, is the 15-carbon core structure of flavonoids. A pair of benzene rings are joined by a three-carbon pyran ring to form this structure. The orientation of the catechol (1,2-dihydroxybenzene) B-ring on the pyran C-ring, as well as the quantity and location of hydroxy groups on the B-ring, regulate the antioxidant qualities of flavonoids (D'Amelia et al., 2018). The flavonoid hydroxy groups can transfer electrons through resonance, enhancing antioxidant defense and neutralizing free radicals (Šamec et al., 2021). Based on their structural properties, flavonoids are classified into six primary groups: anthocyanins, flavan-3-ols, flavonols, flavanones, flavones, and isoflavones (Šamec et al., 2021).

Flavonoids exist in the forms of glycosides, methylated derivatives, and aglycones. The fundamental framework of flavonoids is aglycone (Figure 23.1). A six-member ring fused with a benzene ring can be classified as either an alpha-pyrone or its dihydro derivative (Narayana et al., 2001). On numerous occasions, flavonoids are hydroxylated in different places, and methyl ethers and acetyl esters also occur. Glycosides typically have bonds at positions 3 or 7 with carbohydrates such as D-(+)-glucose, L-rhamnose, glucorhamnose, arabinose, or galactose (Middleton, 1984).

Flavonoids possess strong antioxidant qualities, which enable plants to manage oxidative stressors by neutralizing free radicals. As a result, plants are shielded against cellular peroxidation. Flavonoids suppress the generation of reactive oxygen species (ROS) through four pathways: (i) limiting the production of $^1O_2$, (ii) inhibiting enzymes that produce ROS (prostaglandin-endoperoxide synthase, lipoxygenase, monooxygenase, and xanthine oxidase), (iii) binding to transition metal ions, and (iv) regenerating other antioxidants (Shomali et al., 2022). Additionally, they enhance the pigmentation of flowers, leaves, and fruits, thus attracting pollinators (Davies, 2003; Iwashina, 2015; Khoo et al., 2017). Flavonoids also have multiple functions, including acting as UV filters to shield plants from UV radiation, functioning as signalling molecules, serving as defense substances against microbes, and acting as chelators for harmful heavy metals (Vargas et al., 2013; Iranshahi et al., 2015; Cherrak et al., 2016; Ferreyra et al., 2021; Bag et al., 2022; Ghitti et al., 2022). Flavonoids in the human body have several wellness advantages, such as being anticancer (Li et al., 2022), antioxidant (Alghamdi et al., 2022), anti-inflammatory (Wadhwa et al., 2021), and antiviral (Rowaiye et al., 2022). In addition, they possess neuroprotective (Calderaro et al., 2022; Bellavite, 2023) and cardioprotective properties (Agrawal et al., 2014; Ahmad et al., 2016).

The present techniques for synthesizing flavonoids heavily rely on plant extraction, which has many restrictions and disadvantages. First and foremost, numerous plants abundant in flavonoids also serve as a source of nourishment. Therefore, using flavonoids in their manufacturing may lead to a reduction in the overall availability of food (Mark et al., 2019; Tian et al., 2021). Furthermore, low concentrations or losses during extraction may limit the extraction yields. In addition, variables, including long growing seasons, unfavorable weather patterns, small plantation dimensions, challenges in cultivation, and species differences, might affect yield (Shah et al., 2019; Mark et al., 2019; Du et al., 2020a).

Plant extraction of flavonoids has several additional drawbacks, such as the high cost of production, significant energy consumption due to low flavonoid concentration, and the susceptibility of plant cultivation to climate variations. Conversely, the chemical formation of flavonoids often requires multiple complex reaction stages, involves severe reaction conditions, and utilizes hazardous chemical reagents. These factors render the process unsuitable for meeting sustainable

DOI: 10.1201/9781003475491-23

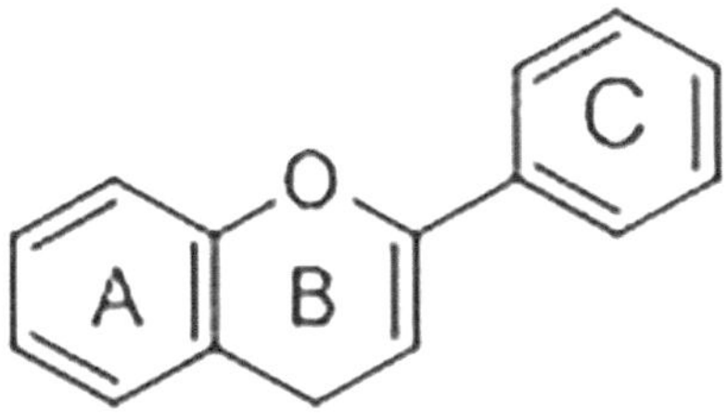

FIGURE 23.1 Basic structure of a flavonoid molecule.

growth demands and plant-extracted flavonoids' market availability (Sun et al., 2022). Reconstructing biosynthetic gene fragments in concerned plants and microbes has excellent potential for exploring flavonoids and achieving large-scale production (Marsafari et al., 2020; Sajid et al., 2021). Enhancing the production of particular beneficial flavonoids by plant metabolic engineering might lead to a greater variety of newly found flavonoids.

However, it remains arduous and technically demanding to manipulate the genetic makeup of a plant. Editing a plant genome to improve metabolite synthesis frequently presents obstacles and constraints related to variations in quantity, sensitivity to aggregation, growth rate, stress, and cultural heterogeneity (Tariq et al., 2023a). Recently, there has been significant focus on the microbial synthesis of flavonoids because of its minimal energy demands, exceptional product purity, and minimal waste generation. Certain microorganisms, including *E. coli*, yeast, and *Yarrowiali polytica*, exhibit rapid growth rates. Extensive research has been conducted on these species, exploring genetic manipulations and metabolic modifications. As a result, these microorganisms have emerged as feasible hosts for flavonoid production (Rodriguez et al., 2017; Levisson et al., 2018; Shrestha et al., 2018; Lv et al., 2019). Over the past decade, synthetic biology has made significant progress, enhancing our comprehension of life's systems and creating new opportunities for microbial engineering.

Many microbiological and computational techniques have been devised to fabricate artificial systems. Advancements in synthetic biology have been extensively utilized in metabolic engineering, which focuses on modifying microorganisms to generate desired compounds (Lee et al., 2018). Synthetic biology offers the necessary components and knowledge about various biological processes. At the same time, metabolic engineering aims to utilize this information to enhance the production of a specific molecule through biological synthesis (García-Granados et al., 2019). This strategy has numerous advantages, including a fast rate of development, safety, economical substrates, ease of cultivation, and an extensive array of genetic techniques, resulting in elevated concentrations of desired metabolites compared to the native host (Muhammad et al., 2020). The constraints of the conventional monoculture approach have given rise to a novel technique known as modular co-culture engineering. Several strains of genetically modified microbes are used in this procedure to recreate the necessary metabolic pathway. Consequently, each unique microbial strain experiences a large reduction in both the biosynthetic workload and the metabolic burden (Thuan et al., 2018; Maglangit et al., 2020; Wang et al., 2020a; Kapoore et al., 2022).

23.2 FLAVONOID BIOSYNTHETIC PATHWAY IN PLANTS

Plants have two separate biosynthetic routes for producing flavonoid-based chemicals. The shikimic acid pathway (Figure 23.2) produces the phenyl propanoids (C6-C3) skeleton, while the acetate pathway creates the basic units of two-carbon polymers (Nabavi et al., 2020). One molecule of 4-coumaroyl-CoA and three molecules of malonyl-CoA are combined to produce different flavonoid compounds to create naringenin chalcone (Chen et al., 2023). Chalcone synthase (CHS) is the enzyme that catalyzes this reaction. The two primary precursors, which generate rings A and B, respectively, and chain connections resulting in ring C, originate from two different cellular metabolic pathways: the acetate and shikimate. Ring A is produced using malonyl-CoA by using the acetate route to carboxylate acetyl-CoA. On the other hand, coumaroyl-CoA is converted into ring B and the connecting chain (ring C) via the shikimate pathway. Coumaroyl-CoA is generated through three enzymatic steps of the phenylpropanoid pathway, originating from the amino acid phenylalanine (Shah & Smith, 2020).

These aromatic rings combine via the acetate and shikimate routes to generate chalcone, which is subsequently transformed into flavanone by an isomerase-catalyzed cyclization process. The resultant molecules then go through further changes like methylation, glycosylation, or hydroxylation, which add to the wide range of flavonoid structures and hues seen in modern times (Shah & Smith, 2020). The phenylpropanoid route in plants requires the shikimate pathway to produce phenylpropanoids. The shikimate route produces phenylalanine, which is used as a starting amino acid in the phenylpropanoid biosynthesis route to synthesize flavonoids. This metabolic route is exclusive to plants as it synthesizes flavonoids and yields lignin, coumarins, hydroxycinnamates, and esters, among other secondary metabolites distinct to plants (Venkateswara Rao et al., 2017).

The first step in the phenylpropanoid pathway is the enzyme phenylalanine ammonia-lyase (PAL)-mediated deamination of the amino acid phenylalanine to cinnamic acid (Gho et al., 2020). Subsequently, the hydroxylation of cinnamic acid into p-coumaric acid is catalyzed by cinnamate-4-hydroxylase (C4H) (Kong et al., 2014). Then, para-coumarate-CoA ligase (PCL) activates p-coumaric acid to generate p-coumaroyl-CoA. Then, three acetate units from malonyl-CoA are combined with p-coumaroyl-CoA by the enzyme chalcone synthase (CHS), forming naringenin chalcone (Liu et al., 2022). Chalcone isomerase (CHI) converts naringenin chalcone into either flavanone or naringenin during the last step of flavonoid biosynthesis (Ni et al., 2020). Naringenin undergoes structural modifications to produce

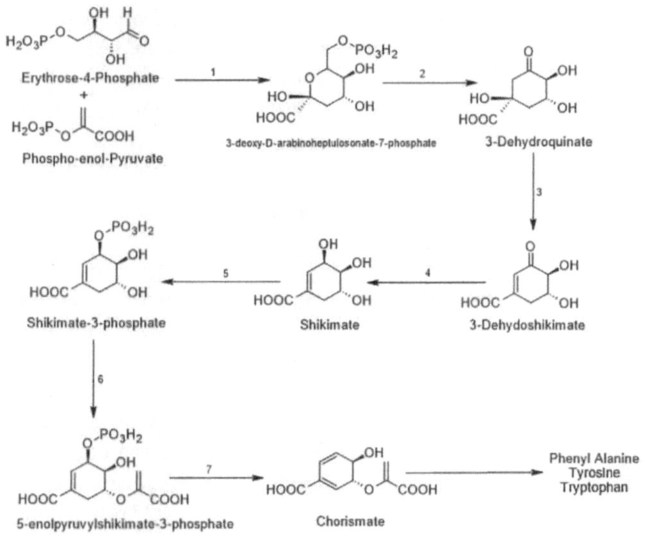

FIGURE 23.2 The shikimic acid pathway. **1- 3-deoxy-d-arabino-heptulosonate-7-phosphate synthase; 2- 3-dehydroquinate synthase; 3- 3-dehydroquinate dehydratase; 4-shikimate 5-dehydrogenase; 5-shikimate kinase; 6- 5-enolpyruvylshikimate-3-phosphate synthase; 7- chorismate synthase.**

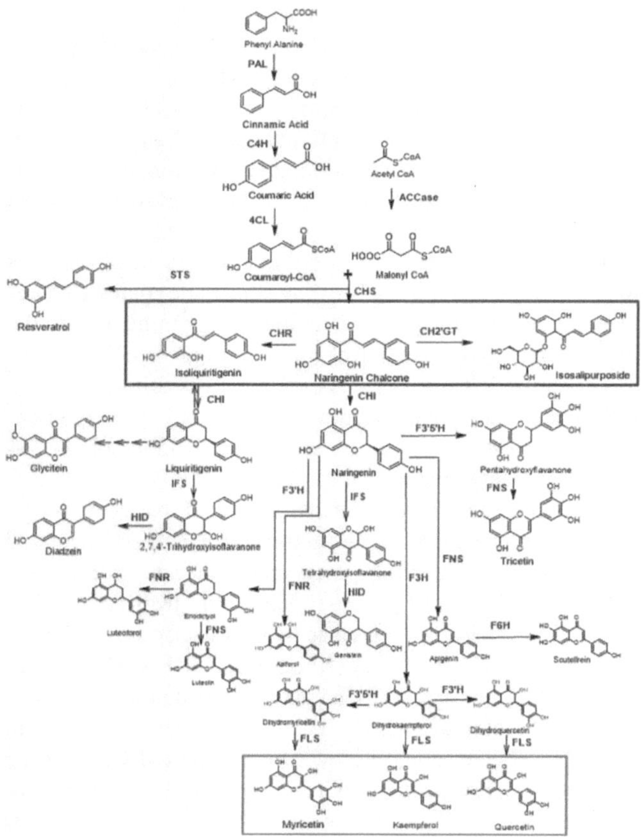

FIGURE 23.3 The phenyl propanoid pathway of flavonoid biosynthesis.

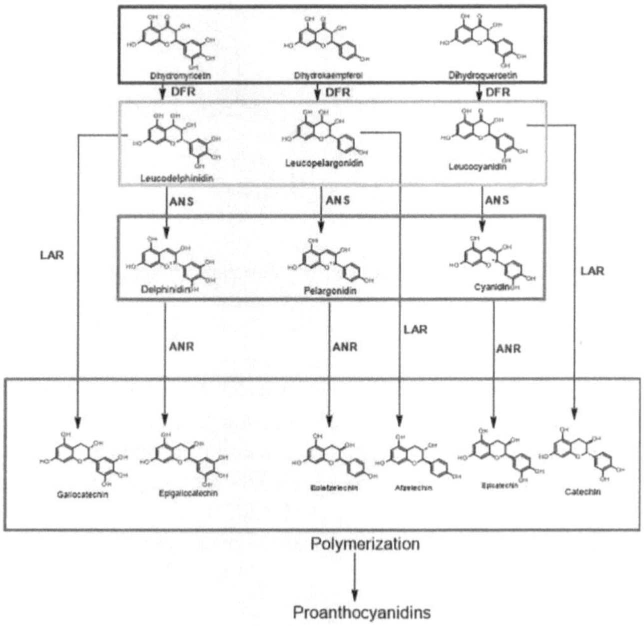

FIGURE 23.4 The phenyl propanoid pathway of flavonoid biosynthesis (continued). **PAL**: phenylalanine ammonia lyase; **C4H**: cinnamic acid 4-hydroxylase; **4CL**: 4-coumarate: CoA ligase; ACCase: acetyl-CoA carboxylase; **STS**: stilbene synthase; **CHS**: chalcone synthase; **CHR**: chalcone reductase; **CH2′GT**: chalcone 2′-glucosyltransferase; **CHI**: chalcone isomerase; **FNS**: flavone synthase; **F6H**: flavonoid 6-hydroxylase; **IFS**: isoflavone synthase; **HID**: 2-hydroxyisoflavanone dehydratase; **FNR**: flavanone-4-reductase; **F3H**: flavanone-3-hydroxylase; **F3′5′H**: flavanone-3′,5′-hydroxylase; **FLS**: flavonol synthase; **DFR**: dihydroflavonol 4-reductase; **ANS**: anthocyanidin synthase; **ANR**: anthocyanidin reductase; **LAR**: leucoanthocyanidin reductase.

various flavonoids, including flavones, isoflavones, and anthocyanidins (Tohge et al., 2017) (Figures 23.3 and 23.4).

The catalytic activity of CHS results in the production of both the A and B phenyl rings, which make up the flavonoid structure's core. This enzyme creates a chalcone by combining three molecules of malonyl-CoA with p-coumaroyl-CoA (Yonekura-Sakakibara et al., 2019). Flavanone (naringenin) is produced when CHI contributes to the biosynthesis of the essential C ring of the flavonoid backbone (Teles et al., 2018). The enzyme isoflavone synthase (IFS) assists in converting flavanone into isoflavone. Flavonoid-3-hydroxylase (F3H) may hydroxylate naringenin at the C ring's 3-position, making it a substantial precursor (Rencoret et al., 2022). Three different forms of dihydro flavonols,), DHQ (dihydroquercetin), DHK (dihydrokaempferol), and DHM (dihydromyricetin) may be produced by hydroxylating it using flavonoid-3′-hydroxylase (F3′H) or flavonoid-3′5′-hydroxylase (F3′5′H) (Tanaka & Brugliera, 2013). Dihydroflavonol 4-reductase [DFR] is an enzyme catalyst that converts dihydroflavonols to leucoanthocyanidins which act as precursors during the synthesis of anthocyanidins, flavan-3-ols, and Pas (Hua et al., 2013).

23.3 BIOTECHNOLOGICAL PRODUCTION OF FLAVONOIDS IN PLANTS

Plants of various species include a variety of essential enzymes involved in controlling flavonoid production. Researchers have worked together for several decades to perform substantial genetic investigations in this sector, leading to the identification of these enzymes (Liu et al., 2021; Cao et al., 2024). The enzymatic pathways are potential targets for metabolic engineering, including genetic modifications to boost yields, resistance to drought and insects, and the development of hybrid plants with better nutritional value and desirable features (Bai et al., 2018). Many studies have been conducted to alter plant flavonoid pathways in the last few years with four main motives: (i) modify model plants to study critical factors in flavonoid biosynthetic pathways, (ii) modify ornamental plants to produce vibrant pigmented flavonoids that enhance their decorative appeal, (iii) modify plants to increase their tolerance to both biotic and abiotic stress, and (iv) modify crop plants to boost flavonoid accumulation (Nabavi et al., 2020). The latest approaches to improve the biotechnological utilization of flavonoids in plants are outlined in this section.

23.3.1 Forward Approach Involving Overexpression of Transcriptional Factors

Both structural and regulatory genes are involved in the flavonoid biosynthesis pathway (Jiang et al., 2021). The synchronized expression of early biosynthetic genes (EBGs) and late biosynthetic genes (LBGs), which are affected by developmental and environmental signals, substantially controls the synthesis of flavonoids. Colorlessdihydroflavonol molecules are first synthesized in part by EBGs like flavanone 3-hydroxylase (F3H), chalcone synthase (CHS), and chalcone isomerase (CHI). On the other hand, the latter steps that result in the synthesis of colorful anthocyanins are caused mainly by LBGs, such as dihydroflavonol 4-reductase (DFR), UDP-glucose: flavonoid 3-O-glucosyltransferase (UFGT), and anthocyanidin synthase (ANS), (Baudry et al., 2004).

Regulatory genes regulate structural genes' activity. MBW complexes can be formed by the WD-repeat protein TRANSPARENT TESTA GLABRA1 (TTG1), MYB, and bHLH transcription factors (TFs). Regarding the anthocyanin route, these complexes control the encoding of structural genes (Hichri et al., 2011; Petroni & Tonelli, 2011). The two imperfect repeats (R2 and R3) that are conserved motifs found at the N-terminus of the R2R3 MYB proteins allow them to bind to both DNA and the homologous bHLH proteins. The C-terminal region of these MYB proteins is responsible for their transcription-regulatory function. By interacting with both R2R3 MYB and WD40 proteins, bHLH proteins generate transcriptional complexes that directly control the functioning of genes involved in the manufacture of anthocyanins.

The WD40 proteins enhance the contact stability between the MYB and bHLH transcription factors through a docking platform. Nevertheless, they do not possess the ability to regulate transcription (Lim et al., 2022). Numerous MYBs, essential elements of the MBW complex, are known to control anthocyanin accumulation in fruit crops such as grapes (Kobayashi et al., 2004), pear (Yao et al., 2017), apple (Hu et al., 2017), bayberry (Niu et al., 2010), and citrus (Butelli et al., 2012), have been well-documented. For instance, it has been demonstrated that MdMYB1 and MdMYBA regulate the anthocyanin production in apple peel. MdMYB1 can attach itself to the regulatory regions of the MdDFR and MdUFGT genes, while MdMYBA can attach itself to the regulatory region of the MdANS gene (Takos et al., 2006; Ban et al., 2007). By interacting with MdbHLH3/33, MdMYB10, another MYB protein, improves the red coloring of apple flesh. In red-fleshed apple types, this interaction boosts the MdDFR promoter's activity, which causes the generation of red pigments (Espley et al., 2009). Other transcription factors (TFs) have also been demonstrated to impact anthocyanin biosynthesis by binding to the promoters of MYB TFs, including Bzip (Wang et al., 2020c), WRKY (Wang et al., 2018), NAC (Sun et al., 2019), MADS (Nesi et al., 2002), and BBX (Bai et al., 2019).

The MdBBX20 zinc finger protein, which binds to the MdMYB1 promoter to control anthocyanin accumulation, is one example of this (Fang et al., 2019; Zhang et al., 2021). Furthermore, to regulate the production of anthocyanins in apples, MdERF1B and MdNAC52 interact with the promoters of MdMYB9 and MdMYB11, respectively (Zhang et al., 2018; Sun et al., 2019). Additionally, by binding to the G-box region in the promoter of MdMYB10, the bZIP transcription factor MdHY5 encourages the accumulation of anthocyanins (An et al., 2017). Additionally, it has been found that the analogous transcription factor CsAN1, which is comparable to AtMYB75, positively controls the expression of genes that encode the enzymes needed for the manufacture of anthocyanins in *Arabidopsis* (Huang et al., 2022). Purple tea's unusual anthocyanin accumulation results from the particular upregulation of bHLH transcription factors (CsGL3 and LBGs) caused by CsAN1 activation. The fundamental helix-loop-helix transcription factors CsGL3 and CsEGL3 combine with CsAN1 to form a dimer, and this dimer then joins forces with CsTTG1 (WD40) to create the MBW complex. This complex is essential for increasing the expression of the gene CsLDOX, which is involved in the biosynthesis of anthocyanins and promoting the accumulation of anthocyanins (Sun et al., 2016).

Concurrently, CsAN1 expression is controlled by epigenetic processes; a reduction in promoter methylation positively correlates with CsAN1 expression levels. CsMYB75, also called CsAN1, has a great deal in common with AtPAP1 in *Arabidopsis*, according to a transcriptome study of genetic populations of tea plants, with an emphasis on leaf color variation (Wei et al., 2019). Empirical data has validated that CsMYB75 modulates the CsGSTF1 gene promoter, which governs anthocyanin transport, amplifying anthocyanin accumulation within vacuoles. Transactivation experiments have demonstrated that CsMYB75 stimulates the temporary expression of CsDFR and enhances the CsCHS and CsA3T genes (He et al., 2018). According to genetic research, CsMYB75 has been discovered to be in the same genomic area as QTLs, which control the color of the shoots in tea plants. A comparative investigation was conducted between purple shoot varieties (Zijuan, Ziyan, and their natural hybrid ZZ) and the usual green-leaf Fudingdabai (FD). The analysis found that the expression levels of CsMYB75 in the purple shoot types were roughly 15 times greater than in FD (Tan et al., 2023).

According to recent reports (Naing et al., 2017, 2018; Fatihah et al., 2019), the simultaneous expression of the snapdragon transcription factors DELILA (DEL, a bHLH-type) and ROSEA1 (ROS1, a MYB-type) enhances anthocyanin accumulation (specifically, delphinidin and cyanidin). It also improves the tolerance of tobacco flowers and *Lilium* tepals to abiotic stress. According to different research, tomatoes that had their MYB transcription factor anthocyanin2 (LeAN2) overexpressed accumulated more anthocyanin and had better resilience to oxidative stress and freezing (Meng et al., 2014). Furthermore, it has been shown that elevated expression of the MYB-type transcription factor SlMYB75 promotes the accumulation of anthocyanins and increases the generation of volatile smells in tomato fruits (Jian et al., 2019).

A separate study discovered that the overexpression of EsMYB90, a MYB transcription factor derived from *Eutrema salsugineum*, in tobacco and *Arabidopsis* plants resulted in elevated pigmentation and accumulation of anthocyanin in several organs. The transcriptome study showed that EsMYB90 increased 42 genes in 35S:EsMYB90 tobacco transgenic plants essential for anthocyanin production. In addition, our qRT-PCR findings demonstrated that EsMYB90 enhanced the expression of both early (PAL, CHS, and CHI) and late (DFR, ANS, and UFGT) genes involved in the production of anthocyanin in theflowers, leaves, and stems of tobacco transgenic plants carrying the 35S:EsMYB90 construct (Qi et al., 2020). In a different study, it was shown that *Prunus salicina* PsMYB10.2 stimulates the anthocyanin pathway in tobacco leaves by triggering the promoters of the genes PsUFGT and PsGST that synthesize anthocyanin (Fang et al., 2021). Hop plants were genetically transformed using *Agrobacterium* to introduce the *Arabidopsis thaliana* regulatory factor PAP1/AtMYB75. This resulted in increased levels of anthocyanins, α-acids, isoquercitin, rutin, kaempferol-glucoside, desmethylxanthohumol, kaempferol-glucoside-malonate, xanthohumol, and β-acids. As a result, unlike wild-type plants, transgenic plants' female flowers and cones changed from reddish to pink in hue (Gatica-Arias et al., 2012). Therefore, by controlling and modifying key transcription factors involved in the production of flavonoids, it is possible to eliminate the processes that limit the production rate in different plants. This presents an excellent opportunity to reveal the regulatory mechanisms that control the biosynthesis of flavonoids. Furthermore, the studies that employed MYB transcription factors to generate plants with heightened anthocyanin levels illustrate the economic benefits of utilizing transgenic plants to produce specialized flavonoids (Marsafari et al., 2020).

23.3.2 Reverse Genetic Engineering and Accumulation of Flavonoids and Anthocyanins

One method for understanding a gene's function is reverse genetics, which looks at the phenotypic effects of altered gene sequences. Unlike traditional forward genetic screens, which begin with identifying a mutant phenotype and then identifying the corresponding gene, reverse genetics starts with the gene itself. Enabled by recombinant DNA technology, reverse genetics involves creating a mutant gene or protein and observing the resulting mutant phenotype. Reverse genetics relies heavily on techniques like in vitro mutagenesis and gene disruption, frequently called gene knockout. Adding a selective marker to the cloned gene and employing this altered construct to reshape a recipient organism that carries the wild-type gene is a successful tactic. By selecting the marker, certain transformants are obtained where the mutant gene replaces the original wild-type allele in its original location (Griffiths et al., 2000).

The goal of reverse genetics is to determine a gene's function by examining the effects of purposefully made alterations to that gene. Gene silencing and homologous recombination are examples of targeted alteration processes. Non-targeted random disruptions like chemical mutagenesis or transposon-mediated mutagenesis can also be involved in this alteration process. Next, a library screening program is used to find lesions at certain spots on people. Reverse genetics has two main approaches: (1) RNA interference, which is targeted gene silencing, and (2) TILLING (targeting induced local lesions in genomes), an additional recent approach (Aklilu, 2021). In eukaryotes, RNA interference (RNAi) is a highly conserved process that takes place after transcription. Destroying the target mRNA causes the selective silence of genes.

Double-stranded RNA (dsRNA), which has a similar sequence to the silenced gene, aids in the process. The dicer enzyme converts the dsRNA into small interfering RNA (siRNA), subsequently included in an RNA-induced silencing complex comprising many components. The target mRNA is then located and cleaved by this complex (Stanisławska & Olszewski, 2005). One use of reverse genetics is TILLING, which implements chemical or physical mutagenesis to generate new genetic variations to discover superior or novel variants (Singh et al., 2023).

A study was conducted where RNA interference was utilized to silence chalcone synthase (CHS) to create seedless tomatoes. This led to alterations in the auxin distribution and reduced the quantity of flavonoids (Schijlen et al., 2007). In a different research, the tropical actinorhizal tree *Casuarina glauca* was treated to RNA interference to suppress the activity of chalcone synthase, an enzyme essential to the first stage of the flavonoid production pathway. The researchers noted that when the expression of chalcone synthase was suppressed, there was a decrease in the concentration of specific flavonoids, and this led to a significant decline in nodulation (Abdel-Lateif et al., 2013). The FLS silenced lines exhibited a 20–80% reduction in FLS expressing gene expression and a 25–93% reduction in flavonoid (quercetin) content. It's interesting to note that these tobacco lines silenced by FLS also showed lower anthocyanidin levels. On the other hand, it was discovered that the FLS silenced lines had higher levels of flavan-3-ols, specifically catechin, epi-gallocatechin, and epi-catechin (Mahajan et al., 2011).

In a separate investigation, Transgenic tomatoes that express a hpRNA molecule that suppresses DET1, a gene that regulates photo-morphogenesis and inhibits many light-controlled signalling pathways, exhibited a substantial elevation in the concentrations of flavonoid and carotenoid compounds (Davuluri et al., 2004). This study has verified that altering plant regulatory pathways can simultaneously effectively modify many separate phytonutrient biosynthesis pathways. A separate investigation involving tobacco suggested that the suppression of two DFR genes in the flavonoid biosynthesis route resulted in a modification of the blossom color. Furthermore, the suppression of DFR genes has impacted the gene expression in the biological synthesis of flavonoids in flowers. Additionally, the amount of short RNA produced in DFR RNAi tobacco is in line with the alteration in the color of the bloom and seed (Lim et al., 2016).

The TILLING technique was employed in *A. thaliana* to discover the genes responsible for synthesizingflavonols, specifically in pollen. Examination of flavonoid buildups in various parts of plants, such as structural and reproductive organs calyx, corolla, androecium, and gynoecium of normal plants and plants with male sterility mutations, has verified that kaempferol/quercetin 3-O- β -d-glucopyranosyl-(1→2) β -d-glucopyranosides are flavonols specific to pollen. The microarray analysis comparisonofthe wild type and male sterility indicated that UDP-glycosyltransferases 79B6 (UGT79B6) is a regulatory enzyme that determines the presence of pollen-specific flavonols (Yonekura-Sakakibara et al., 2014).

TaCYP71F53 is a component of a biosynthetic gene cluster responsible for producing a specific type of isoflavone called triticein, which is produced from naringenin and undergoes O-methylation and hydroxylation at particular positions. In a study, two distinct TILLING lines, Cad1793 and Cad0227, were discovered in wheat. The coding area of TaCYP71F53 in these lines was found to have mutations resulting in an early stop codon. Triticein and 2'-O-demethyl-triticein were detected in the parent 'Cadenza' plants by LC-MS analysis of blade extracts from KASP-genotyped, MeJA-treated plants. However, these compounds were not detectable from the Cad0227 and Cad1793 mutant lines. The lack of these compounds coincided with a noted rise in the proposed substrate of TaCYP71F53, artocarpanone A, and a decrease in the levels of naringenin 4',7-dimethyl ether relative to the original Cadenza line. The TaOMT8 (Cad1684) and TaOMT3 (Cad1682) mutant lines with premature stop codons were also examined. These lines showed an increase of the suggested substrates (naringenin and isosakuranetin) and an absence of TaOMT3/TaOMT8 products as well as downstream metabolites, just like the TaCYP71F53 mutant lines (Polturak et al., 2023).

A TILLING line contained an insertion of a termination codon in the R2R3-MYB transcription factor gene BrMYB4, which causes the corresponding protein's C-terminal to be lost in *Brassica rapa* subsp. *rapa* cv. Tsuda. The mutant line showed a buildup of anthocyanin in the underground part, which led scientists to look into the expression of genes involved in anthocyanin production in the outer layer of preservation roots. These genes include three homologs of CHS, ANS1, DFRANS2, and C4H. This analysis was done concerning UVB therapy and exposure to sunshine. When compared to the natural type, the mutant lines showed a marked upregulation of the genes responsible for producing anthocyanins when exposed to dark conditions. Nonetheless, it was found that BrMYB4 expression was comparable in both the wild-type and mutant lines (Zhang et al., 2014).

23.3.3 Plant Cell Suspension Culture

Plant cell suspension culture is a modern method for efficiently manufacturing various secondary plant metabolites in large quantities within a shorter timeframe. In this specific culture system, the nutritional elements within the basal medium play an essential role in initiating and enhancing the production of secondary metabolites. This is achieved through the regulation of enzyme activity and gene expression (Haida et al., 2019). A study observed that methyl jasmonate significantly impacted the development of cells and the generation of flavonoids in *Hypericum perforatum.* The most effective method of obtaining information was treating the cell cultures with a concentration of 100 μmol/L MeJA on the 15th day. This treatment led to the maximum production of flavonoids, with a concentration of 280 mg per liter, which was 2.7 times more than that of the control cultures (Wang et al., 2015). Another study elicited the synthesis of phenolics and flavonoids in the cell suspension cultures of *Cascabelathevetia* (Mendoza et al., 2018). Suspension cultures from the leaf explant of *Ficus deltoidea* were also undertaken to generate flavonoids. The findings indicated that the most significant amount of biomass (0.65 grams per 25 mL of media) was achieved when the initial inoculum strength was 2.0 grams per 25 mL of media. Similarly, the utmost concentration of flavonoid (3.3 mg RE/g DW) was observed in 0.5 grams per 25 mL of media. The cell suspension fractions are categorized according to their sizes into three groups: 500 – 750 μm , 250 – 500 μm , and < 250 μm . The maximum cell biomass (0.28 gram per 25 mL medium) and flavonoid content (3.3 mg RE/g DW) were seen when large cell aggregates with a size of 500 – 750 μm were cultivated at pH 5.75 (Haida et al., 2019). In another study, a refined cell suspension of *Saussurea medusa* is created from the light-yellow calluses using a liquid medium containing 30 g/L sucrose, 0.5 mg/L BA, and 2.0 mg/L NAA. This suspension is used for the production of flavonoids. After 15 days, the dry weight and flavonoid production reached their maximum values of 17.2 g/L and 607.8 mg/L , respectively. The cell suspension cultures of S. medusa exhibit a notable increase in antioxidant activity and the accumulation of flavonoids (Liu & Saxena, 2009). A study examined the use of metabolically modified Silybum marianum cultures to produce t-resveratrol. Cell suspensions were genetically modified with *Vitis vinifera* stilbene synthase 3, producing t-resveratrol outside the cells at a concentration of milligrams per liter when stimulated. The synthesis of resveratrol resulted in the depletion of coniferyl alcohol. The generation of Mariadistel was relatively unaffected in the transgenic cultures as the synthesis of this compound is limited by the flavonoid pathway, which is favored over the monolignol branch in terms of precursor supply (Hidalgo et al., 2017) (Figure 23.5).

23.4 METABOLIC ENGINEERING: AN EFFICIENT APPROACH TO ENHANCE FLAVONOID SYNTHESIS IN PLANTS

23.4.1 Metabolic Engineering in Plants

Flavonoid engineering solutions must tackle four primary challenges: structural, regulatory, metabolic flow, and storage. Structural problems entail the incorporation of genes from the metabolic pathway, which can be accomplished by amplifying the expression of enzymes that control the rate of the route. Regulation problems entail influencing the transcriptional and translational machinery, which can be achieved by finding transcription factors specific to metabolites and increasing their expression. Metabolic flux issues can be resolved by employing genome-scale modelling and analysis, which enables the redirection of metabolic fluxes and enhances the production yield of products. Efficient management of storage difficulties necessitates strict control, wherein designated carriers of secondary metabolites assume pivotal functions in systematic metabolic engineering (Wang et al., 2011). Plant metabolic engineering, a process that has existed for 50 years, seeks to boost the manufacture of flavonoids by manipulating plant pathways. It identifies the components involved in

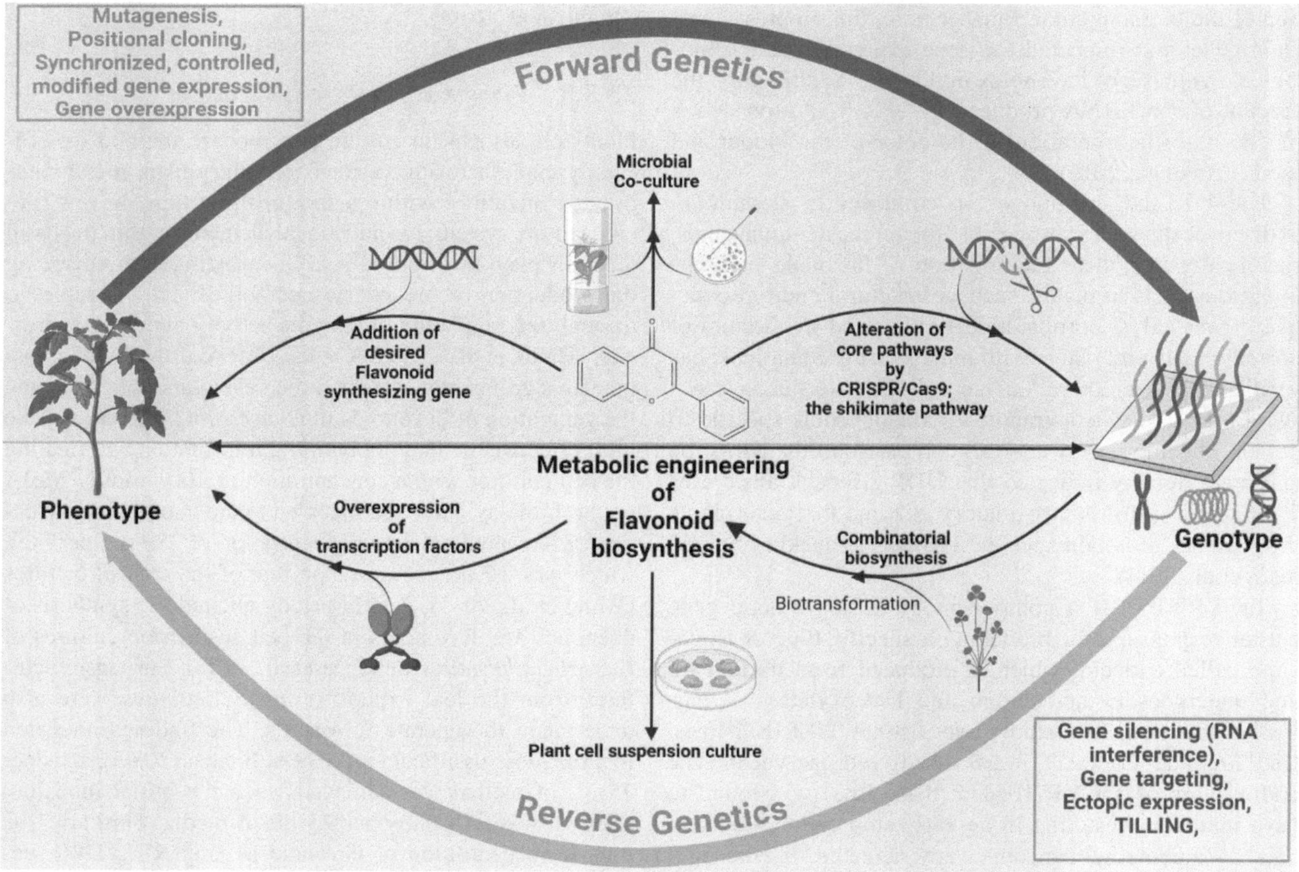

FIGURE 23.5 Biotechnological and metabolic engineering strategies for the Flavanoid biosynthesis.

flavonoid biosynthesis, promotes the production of pigments, increases plant tolerance, and improves the accumulation of flavonoids (Marsafari et al., 2020).

In a breakthrough for metabolic engineering, the first documented case occurred in 1987. The color of *Petunia* flowers was altered due to the addition of a specific gene called dihydro quercetin 4-reductase (A1) from maize, which led to this transformation by producing pelargonidin (Meyer et al., 1987). The anthocyanin formation has been improved by reducing the expression of flavonol synthase (FLS) (Davies, 2003). Likewise, in tobacco plants, the excessive production of stilbene synthase resulted in a decrease in blossom coloration and the inability of male reproduction, most likely because stilbene synthase and the natural chalcone synthase are vying for the same substances: 4 coumaroyl CoA and malonyl-CoA (Parage et al., 2012).

The MYB family, found in plants and mammals, is an essential class of transcription factors involved in regulatory processes (Sun et al., 2016). Increased yield and output and improved anthocyanin accumulation have been linked to the overexpression of MYB transcription factors (Junping et al., 2018).

Agrobacterium-mediated genetic transformation (AbMGT) was used to introduce MYB gene transcription factors from *A. thaliana* into hop plants, significantly enhancing flavonoid production (Xie et al., 2017). Table 23.1 represents selected plants that are subjected to the metabolic engineering of flavonoids

Reverse genetics is a valuable technique for evaluating gene activity and pinpointing areas to improve plant production of flavonoids. This method links genetic sequences to an organism's phenotype and is essential to functional genomics. The techniques used are gene targeting, screening mutant populations, gene silencing, and transgenics-based ectopic expression (Ben-Amar et al., 2016). RNA interference has been used to lower the activity of chalcone synthase (CHS), which has led to tomato plants lacking seeds, changes in the distribution of auxin, and a decrease in the amounts of flavonoids (Schijlen et al., 2007). Similarly, decreased coloration and modifications in flavonoid composition have been associated with blocking the chalcone isomerase gene (CHI) in genetically engineered tobacco plants (Nishihara et al., 2005).

The TILLING strategy comprises a series of fundamental procedures, including the induction of the mutation, DNA preparation, PCR amplification, production of heteroduplexes, denaturing HPLC, screening of mutant individuals, and sequencing of the mutant PCR output, all without the need for genetic transformation tools (McCallum et al., 2000) The TILLING technique was employed in *A. thaliana* to discover genes that produce flavonols, specifically in pollen. The study of flavonoid buildup in different sections of flowers from normal plants and plants with male sterility mutations verified that kaempferol/quercetin 3-O-β-d-glucopyranosyl-(1→2)-β-d-glucopyranosides are flavonols specific to pollen (Yonekura-Sakakibara et al., 2019).

Plant cell suspension cultures help increase the synthesis of flavonoids and polyphenols. Methyl jasmonate (MeJA) has been shown in studies on suspension cultures of *Hypericum perforatum* to affect cell proliferation substantially and boost flavonoid synthesis 2.7 times compared to untreated cultures. MeJA increases the synthesis of flavonoids via stimulating phenylalanine ammonia-lyase (PAL) and suppressing catalase (CAT) (Wang et al., 2011). Using leaf, fruit, and root explants of the Indian mulberry (*Morinda citrifolia*), the effects of subculturing on the synthesis of anthraquinones, flavonoids, and phenolics were examined (Tiwari et al., 2016). The effective synthesis of flavonoids in plant cell cultures is advised to use bioreactor-based systems; nevertheless, their economic viability is still questionable because of the engineering difficulties involved in large-scale cultivation.

Combinatorial biosynthesis is an in vivo approach that generates novel chemicals by utilizing specific enzyme-substrate combinations. The techniques encompass biotransformation, nourishing altered biocatalysts with non-natural precursors, mutasynthesis, combining metabolism in hybrids, activating dormant metabolism through transcription factors, and creating genes in other heterologous hosts (Bhan et al., 2015). Two methods create flavonoids in heterologous hosts. Microbial cultures replenish the pathway with precursors, which is expensive and limits commercial use. Glycerol and glucose are favoured for pharmaceutical, industrial, and protein manufacturing because they are renewable, easy, and affordable. Genetically modifying the host strain to create aromatic amino acids is the second method to produce a low-energy fermentation product. The most prevalent heterologous flavonoid production platforms are *Escherichia coli* and *S. cerevisiae*. Genetic flexibility speeds up and predicts experimentation and data collection, reducing deficiencies (Trantas et al., 2015). For example, in a study, the 4-coumarate: coenzyme ligase 2 gene was amplified and cloned into a vector from a young damaged *Nicotiana tabacum* cv. Samsun leaf. The stilbene synthase gene from *Vitis vinifera* was inserted into a vector after amplification from a plasmid. The plasmid was introduced to *E. coli*. Extract and resveratrol analysis were done following 24 hours of growth. Resveratrol increased to 12–20 mg/liter during cell growth, but p-coumaric acid declined to 0.2 mM (Beekwilder et al., 2006) (Figure 23.1).

23.4.2 Metabolic Engineering in Microbes

There has been an ever-increasing demand for flavonoids since they have widespread applications in pharmaceutics and nutraceuticals. Traditional methods of production of flavonoids, such as extraction and chemical synthesis, frequently encounter challenges related to supply shortages, low yields, and environmental impact. Plant extraction suffers from high costs, energy consumption, climate, and geographical dependency. Chemical synthesis poses challenges of complexity like specific reactions, temperature, pH requirements, etc. Furthermore, it poses environmental harm

TABLE 23.1
Selected Plants Subjected to Metabolic Engineering for Enhanced Production of Flavonoid

Plant name	Flavonoid enhanced	Target gene	Mechanism	Reference
Arabidopsis thaliana	Anthocyanin	CsCYT75B1	The gene was amplified and cloned in a vector and transferred to *A. thaliana*.	(Rao et al., 2020)
Camellia sinensis	Anthocyanin	CsAN1	The gene was amplified and inserted in a vector and transformed into *C. sinensis* L.	(Sun et al., 2016)
Dracaena cambodiana	Anthocyanin	DcCHI1 or DcCHI4	Amplified genes were put into the overexpression vector under CaMV 35S promoter and injected into *A. tumefaciens* strain and infiltrated into *D. cambodiana* leaves.	(Zhu et al., 2021)
Malus domestica	Anthocyanin	MdMyb10	The *M. domestica* leaf was inoculated with the apple transcription factor MdMyb10 gene controlled by the CaMV 35S promoter and the NOS terminator by *A. tumefaciens*.	(Rihani et al., 2017)
Marchantia polymorpha	Anthocyanin	R2REMB	Both MpMyb14 overexpression constructs and CRISPR/Cas9 constructs in the pMpU6pro/pENTR vector using oligonucleotides create pMpGE010; binary vector. *Marchantia* sporelings were transformed using binary vector-inoculated *A. tumefaciens*	(Albert et al., 2018)
Nicotiana tabacum	Anthocyanin	AtPAP1	The binary vector of AtPAP1-pK2GW7 introduced into competent *A. tumefaciens* cells and transformed into *N. tabacum* leaves.	(He et al., 2017)
Petunia hybrida	Anthocyanin and flavonol	Fh3GT1	Binary vectors were introduced into competent *A. tumefaciens* EHA105 cells. The transformation used mature leaves from a white-blossomed petunia cultivar.	(Sun et al., 2017)
Glycyrrhiza uralensis	Flavanones	CHS	Chalcone synthase cloned from G. uralensis produces expression vector pCA-CHS. Electrotransformation inserted vector into *A. rhizogenes* which then infected G uralensis cotyledons and hypocotyls.	(Yin et al., 2020)
Salvia miltiorrhiza	Anthocyanin	SmANS	SmANS was fused to pA7-GFP. Recombinant plasmid p0390A7-GFP-SmANS was created by linking pA7-GFP-SmANS and pCAMBIA0390. It is then transformed in *A. tumefaciens* and then to *S. miltiorrhiza* and *S. miltiorrhiza*	(Li et al., 2019)
Astragalus trigonus	Apigenin	chiA	*A. tumefaciens* strain LBA4404 was suspended in media with petunia chiA gene-containing plant vector pFGCS-chiA. *A. trigonuscalli* explants were submerged in Agrobacterium suspension.	(Elarabi et al., 2021)

because it uses toxic chemicals, heavy metals, etc. (Trantas et al., 2015; Sheng et al., 2020). Since 2003, synthetic biology and metabolic engineering have been successfully used to create microbial hosts for efficient flavonoid synthesis, with pinocembrin and naringenin first being synthesized in *E. coli* (Hwang et al., 2003). Since then, this venue has considerably advanced. This technique overcomes the issues of plant extraction or chemical synthesis, like lengthy growth seasons and expensive requirements, and dramatically reduces environmental hazards. Microbial Production offers advantages like shorter production cycles, reduced waste, and scalability. Hence, this technique is energy efficient and emerges as a sustainable solution to meet the increasing demand for flavonoids. Microbial hosts promise efficient and scalable production by integrating synthetic biology and metabolic engineering, offering a viable alternative to conventional methods (Tariq et al., 2023).

Flavonoids, including resveratrol, silybin, naringenin, pinocembrin, anthocyanins, kaempferol, quercetin, and scutellarin have been produced as a result of the restoration of flavonoid pathways in several microbial species over the past few years, including the yeast *Saccharomyces cerevisiae* and the bacteria *Escherichia coli* (Milke et al., 2019; Marsafari et al., 2020; Isogai et al., 2022)

With the help of metabolic engineering, one can induce changes in the compounds produced in microorganisms by altering their genetic material by inserting desired foreign DNA sequences. As a result, the flavonoid biosynthetic pathways are reconstructed in the hosts, utilizing precursors already present in the organism. This way, a variety of compounds like stilbenes, anthocyanins, and flavonoids, among others, are produced (Wang et al., 2011).

Several important aspects are considered to get the most efficient flavonoid production. Foremost is the choice of the host, its mutability, growth pattern, and metabolic potential. Microorganisms like yeast or *E. coli* bacteria are the most common choices (Pandey et al., 2016). Moreover, *S. cerevisiae* is reliable, robust, and suitable for complex pathway reconstructions (Jiang et al., 2005). *Escherichia coli,* on the other hand, has quick growth cycles and is genetically well-established. Furthermore, *E. coli* is a great host for the production of flavonoids because of several benefits, including (a) well-understood genetics, (b) easy access to rich media, and (c) a brief doubling time that shortens the amount of time needed for extraction of the product (Hwang et al., 2003).

Another significant factor is the necessity of advanced knowledge of the plant biosynthesis pathways, its enzymes, the intermediates, and the regulatory components, which is imperative to identify appropriate target genes and reconstruction of the same pathways in the host (Wang et al., 2011; Sheng et al., 2020). One also has to ensure the optimization of gene components, the presence of precursor molecules in the host, balanced expression levels, and optimum culture conditions.

23.4.3 Alterations to the Core Pathways, Such As the Shikimate Pathway, Generate Precursors

Escherichia coli strains like WSH-Z06, HD-A2, Xllp21, etc., producing phenylalanine have been developed by inserting, overexpression, and inactivating specific genes (Liu et al., 2019). Similarly, tyrosine has also been successfully produced in strains of *E. coli* (Ping et al., 2023) and yeast (Guo et al., 2023), among others. Production of phosphoenolpyruvate along with erythrose-4-phosphate, the immediate precursors of the shikimic acid pathway, have also been increased mainly for enhanced availability and production of flavonoids. PEP and E4P accumulation was improved in *E. coli* by the selective expression and activation of specific genes (Fordjour et al., 2019). Overexpression of the PEP synthase gene induces high production of PEP, while overexpression of tatB and tktA genes (encoding transketolase and transaldolase) increases E4P productivity (Jiang & Zhang, 2016; Liu et al., 2018b).

A different approach for increased flavonoid production uses exogenous precursors, intermediates, and other cofactors. For instance, about 107 mg/ml of the eriodyctyol output was achieved in an *E. coli* strain with an exogenous supply of l-tyrosine and overexpression of ACC genes (Zhu et al., 2014). Similarly, feeding p-coumaric acid exogenously to *S. cerevisiae* with the SmCHS2 gene from *Silybum marianum* yielded 648.63 mg/L of naringenin (Gao et al., 2020).

Managing toxicity and regulatory constraints is critical for maximizing flavonoid synthesis in microbial hosts, assuring product quality and host cell survival. This problem stems from the possibility of hazardous intermediates building up during flavonoid production, which can stunt cell division and lower total yield. Metabolic engineering techniques are used to find and reduce harmful intermediates to remedy this (Tariq et al., 2023). Chalcone synthase (CHS) and cytochrome P450 reductase usually impede the production of hydroxylated flavonoids (CPR). But following the removal of precursor pathway restrictions and additional optimization, a modified strain of *Yarrowia lipolytica* produced noticeably higher yields from glucose in flask cultures: 15.8 times more naringenin (252.4 mg/L), 6.9 times more eriodyctiol (134.2 mg/L), and 8.8 times more taxifolin (110.5 mg/L) (Lv et al., 2019). Several optimization approaches successfully developed a high kaempferol-yielding (86mg/ml) strain of *S. cerevisiae* YL-4, including deleting the phenyl ethanol biosynthesis pathway (Lyu et al., 2019).

Several other strategies are employed to enhance the yield of the desired flavonoids. One of the most common is to remove feedback inhibition of genes or enzymes and other regulatory elements in the biosynthesis pathway. The inhibition of chorismate mutase (CM), which converts chorismate into L-tyrosine, and 3-deoxy-D-arabinoheptulosonate 7-phosphate (DAHP) synthase (DS), which is involved in the formation of chorismate, the first precursor of the shikimate pathway, are two examples of this (Braus, 1991; Isogai et al., 2022).

A strategy involving strains of *E. coli* with feedback-resistant DS and CM, with various other optimizations of the polyphenol pathway, was employed, which resulted in naringenin production up to 80–100 mg/mL from glucose (Santos et al., 2011; Wu et al., 2014; Milke et al., 2019). Introduction of aroG and tyrA feedback insensitive mutants, knocking out of tyR, and deletingpheA in *E. coli* resulted in 484 mg/mL of naringenin production. In *E. coli*, AroG is an isoenzyme for DS, while tyrA and pheA encode the enzyme CS (Dunstan et al., 2020; Isogai et al., 2022).

Problems are also associated with low production of target flavonoid compounds because of competing intermediates or pathway flux. Techniques are employed to manipulate the hosts to eradicate, obstruct, or disable competition. In *S. cerevisiae*, an increase in the production of resveratrol (201mg/mL) was observed when the competing pathways for the precursor phenylalanine were eliminated with several other modifications (Li et al., 2016). The competing factors responsible for the degradation of the precursor molecules in phenylpropanoid pathways were eliminated in *Corynebacterium glutamicum* before they were appropriately aimed at polyphenol synthesis (Kallscheuer et al.,

2016; Milke et al., 2019). Similar to this, Zhu et al. (2014) deleted the acetate kinase gene (ackA) to stop the acetate competition pathway and overexpressed the genes for native ACS (acs) and heterologous ACC (accBC along with dtsR1 of *C. glutamicum*) to synthesize 107 mg/L of eriodyctiol from tyrosine in *E. coli* (Zhu et al., 2014). In *C. glutamicum*, naringenin production was increased tenfold to 19 mg/L with 10% residual CS activity and 70% of the usual development rate. Achieving this involved suppressing citrate synthase (CS) via promoter exchange, consequently diminishing malonyl-CoA influx into the Krebs cycle.

Furthermore, a 20% drop in growth rate and a 50% decrease in final biomass resulted from deleting the sdhCAB operon, which encodes the succinate dehydrogenase complex (SDH), in *C. glutamicum*. This led to a twofold rise in naringenin accumulation to 4 mg/L (Milke et al., 2019).

Microbial hosts can also produce flavonoids through de novo synthesis, demonstrating their capacity to create these compounds from basic building blocks. For the practical synthesis of flavanones, which are essential for the de novo synthesis of complex flavonoids, 4-coumaroyl-CoA ligase (4CL) must CoA-esterify cinnamates like p-coumaric acid and cinnamic acid. The condensation of three malonyl-CoA molecules is subsequently catalyzed by chalcone synthase (CHS), and ring closure is aided by chalcone isomerase (CHI). A wide range of flavonoid molecules are produced by subsequent chemical changes in flavanones, such as acylations, glycosylations, methoxylations, hydroxylations, and methylations (Dudnik et al., 2018).

Pterostilbene with a yield of 33.6 $\pm$ 4.1 mg/L was produced in *E. coli* strain grown in L-methionine-containing medium. The strain was engineered to contain the resveratrol O-methyltransferase (ROMT) from *Arabidopsis*, which converts resveratrol to pterostilbene (He et al., 2017). Compounds de novo eriodictyol (134.2 mg/L), naringenin (252.4 mg/L), and taxifolin (110.5 mg/L) were produced from glucose by genetically modifying *Y. lipolytica* (Lv et al., 2019).

In a study, *E. coli* strains we developed by engineering design and process optimization, yielding the highest amounts of naringenin (484 mg/l) and improved eriodictyol (55 mg/l from caffeic acid) and pinocembrin (198 mg/l). The initial example of homoeriodictyol production in vivo was obtained through glycerol (17 mg/l) (Dunstan et al., 2020). The flavonoid-7-O-glucuronides, like apigenin, scutellarin, and baicalin, were produced de novo in naringenin (Liu et al., 2018a) and pinocembrin (Liu et al., 2020), producing strains of *S. cerevisiae* (Isogai et al., 2022).

New metabolic engineering methods like synthetic biology and CRISPR/Cas9 could significantly advance flavonoid production. Researchers can build microbial hosts with specific alterations to optimize flavonoid production pathways thanks to the precision genome editing made possible by such tools (Chu et al., 2024). UsingCRISPRi, the TCA cycle, ethanol synthesis, and fatty acid synthesis – three processes involved in malonyl-CoA consumption – were downregulated, increasing the intracellular supply of malonyl-CoA in *E. coli*. As a result, 1,073 mg/L of naringenin was produced from tyrosine. (Wu et al., 2015). Furthermore, the polyphenol titer of naringenin (422 mg/L) increased sevenfold, and the stilbene pinosylvin (47 mg/L) increased twice as a result of CRISPRi-mediated inhibition of genes involved in fatty acid production (Cress et al., 2015, 2017). Recently, *S. cerevisiae* strain containing CRISPR/Cas9 inserted P450-reductase enzyme complex (SmF3′H/ScCPR) was able to 435.7 $\pm$ 7.6 mg/L of eriodietyol from naringenin (Yu et al., 2022).

Conclusively, there is great potential for augmenting flavonoid synthesis at the nexus of metabolic engineering, synthetic biology, and evolving genomic technologies. Thanks to systems metabolic engineering and genome-scale engineering, researchers can now create, change, and optimize metabolic pathways with unprecedented accuracy, facilitated by an enhanced understanding of microbial genomes and their regulatory mechanisms. Using data-driven methods, such as omics-based analysis and in silico predictions, has sped up the creation of microbial strains for commercial flavonoid production.

In addition, developments in synthetic biology methods, such as CRISPR interference, can potentially improve strain formation and substrate utilization by employing genome-wide analyses and high-throughput screening. Accepting these improvements will be essential to overcoming current obstacles and achieving the globalization of microbial-based flavonoid production to its full potential (Marsafari et al., 2020; Tariq et al., 2023).

23.4.4 Microbial Co-culture Strategy for Flavonoid Biosynthesis

Microbial consortia, or co-cultures, are becoming a viable method of improving the efficiency of metabolic pathways. Co-cultures have a number of benefits over monocultures, including labor division, response compartmentalization, and heightened resilience to environmental disturbances. Microbial cells can be engineered to generate valuable bioactive substances. Advances in protein designing, recombinant DNA technology, genome editing, and metabolic remodelling have paved the way to inculcate desired characteristics into microorganisms.

The microbial culture methodology provides a reliable and effective way around the drawbacks of conventional methods. Microbial consortiums have emerged as a viable method for expressing complex flavonoid biosynthesis processes (Lou et al., 2021). Microorganisms, including *E. coli*, *Streptomyces albidoflavus*, *Saccharomyces cerevisiae*, and *Yarrowia lipolytica* show a faster growth rate and can effectively produce desired metabolites (Madzak, 2021; Pérez-Valero et al., 2023). However, the monoculture system with metabolically engineered strains faces significant challenges, such as severe metabolic imbalances, unsuitable circumstances for enzymes specific to a route, increased metabolic stress due to complex pathways, and unnecessary accumulation of by-products (Naseema

Rasheed et al., 2023; Tariq et al., 2023). These obstacles have complicated achieving higher yields for the finished product. The co-culture technique has emerged over the monoculture technique nowadays by utilizing the combined capabilities of two or more engineered microbial strains to reconstruct the desired metabolic pathway (Thuan, Tatipamula, Canh et al., 2022). The method has also reduced the metabolic burden from the organism compared to that of a monoculture system (Lou et al., 2021). In addition, it has proved to be a low-cost, high-yield microbial flavonoid manufacturing technique (Sun et al., 2022). Microbial cocultures thus offer several benefits over monocultures in optimizing metabolic productivity and eliminating obstacles in pathways. These advantages stem from metabolic workload distribution, shared gene expression responsibilities, and facilitation of cross-metabolite exchange to enhance metabolic performance and resilience. The whole metabolic pathway is customized to optimize each module for optimum production and metabolite harvest. Furthermore, microbial consortia can handle environmental variations and complex activities due to cellular specialization and compartmentalization.

The microbial co-culture technique was first utilized for flavonoid production in 2007. Cross-species co-cultures, which combine the biosynthetic capacities of various microorganisms to fulfill the demands of metabolism, have also been attempted. Genistein, a naturally occurring isoflavone, was produced by co-cultivation of *E. coli* and yeast. *E. coli* directly converts tyrosine into (S)-naringenin, the direct precursor of genistein (Katsuyama et al., 2007). This co-culture technique generated 6 mg/L of genistein, 17.6 times more than what could be obtained from an *S. cerevisiae* monoculture. Furthermore, a unique *E. coli–E. coli* co-culture method was created, which resulted in a 970-fold increase in flavan-3-ol production over monoculture yields (Jones et al., 2016).

A study developed a mutualistic co-culture system for the naringenin synthesis from carbohydrates (D-xylose) by modifying the microorganisms *E. coli* and *Saccharomyces cerevisiae* (Zhang et al., 2017). In another research, two independent *E. coli* strains were used in a co-culture for heterogeneous naringenin production (Ganesan et al., 2017). The co-cultures have functionally expressed separate portions of the metabolic pathway. Another study employed four distinct *E. coli* strains in the co-culture system to overexpress 15 different enzymes and transcriptional factors, causing the pelargonidin-3-O-glucoside synthesis pathway to split into four modules with glycerol, xylose, and glucose as substrates, the first strain generated phenylpropionic acid (Jones et al., 2017). The third strain transformed the flavanones produced by the second strain – which employed phenylpropionic acid and externally generated malonate – into flavan-3-ols. To convert flavan-3-ols into anthocyanidins, the fourth strain overexpressed the genes for anthocyanin synthase and 3-O-glycosyltransferase (3GT). A study employed a co-culture approach to manufacture apigetrinsing a modified *E. coli* strain (Thuan et al., 2018). In this approach, the upstream module of the culture system was designed to include enzymes (chalcone synthase, chalcone isomerase, 4-coumarate: CoA ligase, and flavone synthase I), responsible for synthesizing apigenin from p-coumaric acid. Conversely, the downstream system included an expressed glycosyltransferase to change apigenin into apigetrin and a metabolic module to increase UDP-glucose synthesis. Through this co-culture approach, a yield of 16.6 mg/L apigetrin was achieved. A study used recombinant strains of *E. coli* in a co-culture technique to increase the production of pyranoanthocyanins (Akdemir & Isidro-Sánchez, 2019). They designed the modules to produce 4-hydroxystyrene and 3,4-dihydroxy styrene. Both modified strains were co-cultured with recombinant cells that produced cyanidin-3-O-glucoside (C3G). Pyranocyanidin-3-O-glucoside-phenol, as well as pyranocyanidin-3-O-glucoside-catechol, were formed as a result of this co-cultivation method.

Wang et al. (2020a) extensively studied and optimized two *E. coli* strains to synthesize the flavonoid sakuranetin from the simple carbon source glucose. They produced sakuranetin (29.7 mg/L) from glucose (5 g/L) in 48 h by stepwise optimization of the co-culture system. Sakuranetin was created by the sequential action of upstream and downstream *E. coli* strains via the intermediate p-coumaric acid synthesis process (Wang et al., 2020b). The co-culture biosynthesis was further expanded in a fed-batch bioreactor, resulting in a substantially higher yield than the conventional monoculture-based technique. A study conducted a comprehensive investigation on the capacity of *S. cerevisiae* to produce flavonoids via both monoculture and coculture techniques (Du et al., 2020). The whole flavonoid production route was divided into two independent pathways for this study, each housed inside a different *S. cerevisiae* cell. Genetically modifying the first cell type, sNAR5, to express the naringenin biosynthetic pathway resulted in the production of 144.1 mg/L of naringenin. The second cell type received genetic alterations that incorporated the heterologous naringenin-to-delphinidin mechanism.

Gram-negative *E. coli* has been employed in co-cultures, with each strain carrying a portion of the biosynthetic pathway, enabling the effective production of acetin from uncomplicated carbon sources (Wang et al., 2020b). This approach resulted in acacetin titers reaching 20.3 mg/L . A co-culture method was also used to produce genistein (Kim, 2020).

The *E. coli* DH5± strain was employed for plasmid amplification and gene cloning, while the E. coli BL21 (DE3) strain acted as the host for the biotransformation process. Four genes (Os4CL, PeCHS, RcIFS, and OsCPR) were utilized in this study. Genistein was produced using naringenin and p-coumaric acid as substrates, achieving yields of up to 35 mg per L and 18.6 mg per L under optimal cultivation conditions. An icaritin level of 19.7 mg per L was obtained by utilizing a co-culture technique that divided the synthesis route between *S. cerevisiae* and *E. coli* (Wang et al., 2021). They co-cultured an *S. cerevisiae* strain that produced 8-prenyl kaempferol with an *E. coli* strain that expressed GmOMT2 to further increase the synthesis of icaritin. A highly effective biocatalytic cascade was created to synthesize orientin and vitexin (Qiu et al., 2021). Through the use of *E. coli* co-culture engineering, the catalytic efficiency of TcCGT and GmSUS was enhanced. Optimization efforts aimed to enhance enzyme compatibilities and perfect reaction conditions.

Thuan, Tatipamula, Viet et al. (2022) utilized an *E. coli*–based co-culture system to produce eriodictyol from D-(+)-glucose. Genes that produce p-coumaric acid from D-(+)-glucose were found in the first strain of *E. coli*, but the genes in the second strain produced eriodictiol from 4-hydroxycinnamic acid. Compared to the monoculture output of 21.3 mg/L , the eriodictiol yield improved dramatically to 51.5 mg/L under stepwise optimum growing conditions (Thuan, Tatipamula, Viet et al., 2022). They are especially interesting because methylation flavonoids are more stable and absorb better through the digestive tract than their non-methylated cousins. Because of their increased oral accessibility, methylated flavonoids are particularly desirable. Several strains of *Streptomyces albidoflavus* were created using CRISPR-Cas9 technology (Pérez-Valero et al., 2023). In this work, NL333 medium was used to produce *S. albidoflavus* UO-FLAV-003-NAR and *S. albidoflavus* UO-FLAV-004-NAR in triplicate. The main focus was on methylated flavonoids, highly sought-after bioactive substances for pharmaceutical and nutraceutical applications. Examples of these include (S)-(-)-4',5-dihydroxy-7-methoxyflavanone, acacetin, and puddumetin. Compared to traditional methods, this manufacturing produced a notable 4.4-fold rise in Puddumetin (also called as gengkwanin) titer, with a value of 3.5 mg/L . A co-culture method employing recombinant *E. coli* was developed in 2023 to synthesize genkwanin from D-(+)-glucose (Thuan et al., 2023). The modified gengkwanin biosynthetic pathway consisted of two modules: an upstream strain with genes for manufacturing p-coumaric acid from D-(+)-glucose and a downstream module with a gene cluster that produced both the end product, genkwanin, and its precursor, apigenin (Thuan et al., 2023).

The capacity of the host organisms to degrade feedstock, precursor selectivity, cofactor needs, side-chain diversification chemistry and dynamic population interactions are essential factors for creating a well-controlled and stable consortium. Genetically encoded biosensors are promising tools for tuning these population dynamics and redistributing carbon flux among interacting species. The combination of recombinant protein technology, gene editing, metabolic engineering, and fermentation technology is creating novel strategies for the biosynthesis of valuable compounds through co-culture systems. Additionally, scientists are concentrating their efforts on finding novel microbial species that may effectively produce a variety of advantageous chemicals. Overall, microbial co-cultures show great potential to enhance biosynthetic capabilities and produce complex flavonoids more cost-effectively and efficiently.

23.5 CONCLUSION

Flavonoids, polyphenolic compounds originating in plants, show varied health benefits and are gradually being recognized for their prospective role in disease treatment and diverse industrial applications, with an expected market value of $1.5 billion by 2025. Nevertheless, sustainable harvesting practices must be implemented to avoid biodiversity loss and ensure the continual availability of these essential compounds. On the other hand, biotechnological procedures and metabolic engineering provide substantial benefits over traditional ways of producing plant flavonoids. Flavonoid production can be augmented by manipulating plant metabolic pathways using technologies such as genetic engineering and synthetic biology, enabling fine control over biosynthesis, resulting in greater yields, increased purity, and the development of new molecules with tailored features. Furthermore, biotechnological technologies are often more sustainable, resource-efficient, and ecologically friendly than conventional extraction methods, making them a possible route for long-term flavonoid production.

Overexpression of transcription factors in plant-derived flavonoid synthesis shows potential for increasing productivity and quality by modifying critical regulatory genes. This technique allows for the exact regulation of biosynthetic pathways, which results in greater synthesis of selected chemicals. Furthermore, this technique can create novel plant types with higher flavonoid content, allowing for a more sustainable source of important bioactive chemicals for pharmacological, nutritional, and commercial applications. Reverse genetic engineering appears to be a powerful approach for controlling flavonoid and anthocyanin accumulation in plants by finding and altering genes implicated in certain biosynthetic processes, thereby significantly increasing their synthesis. This strategy not only promotes the production of plant varieties with enhanced nutritional and therapeutic capabilities but also offers ecologically friendly farming and cultivation of high-value crops suitable for a wide range of industrial specifications.

Similarly, plant suspension culture is a potential way to extract flavonoids, facilitating the regulated development of plant cells in a sterile environment leading to better yield of target substances. Furthermore, plant suspension culture may be stepped up for manufacturing, resulting in a sustainable and economical source of flavonoids for therapeutic, nutritional supplement, and cosmetic applications. Plant-microbial metabolic technology represents an innovative approach to increasing plant-derived flavonoid production by using genetic engineering to enhance metabolic pathways, boosting the production and purity of specific flavonoids.

Microbial coculture offers a promising approach to improving flavonoid biosynthesis through synergistic interactions between organisms. Prospective omics technologies may have an exciting scope in comprehending and optimizing the biology of plant flavonoids. The researchers may be able to uncover the complex regulatory relationships and metabolic pathways involved in the production of flavonoids. This knowledge can improve flavonoid yield, quality, and diversity, paving the way for developing new plant varieties with flavonoid markers targeted for various applications.

REFERENCES

Abdel-Lateif, K., Vaissayre, V., Gherbi, H., Verries, C., Meudec, E., Perrine-Walker, F., Cheynier, V., Svistoonoff, S., Franche, C., Bogusz, D., & Hocher, V. (2013). Silencing of the chalcone synthase gene in *Casuarina glauca* highlights the important role of flavonoids during nodulation. *New Phytologist*, *199*, 1012–1021.

Agrawal, Y. O., Sharma, P. K., Shrivastava, B., Ojha, S., Upadhya, H. M., Arya, D. S., & Goyal, S. (2014). Hesperidin produces cardioprotective activity via PPAR-γ pathway in ischemic heart disease model in diabetic rats. *PLoS One*, *9*, e111212.

Ahmad, S., Alam, K., Hossain, M. M., Fatima, M., Firdaus, F., Zafeer, M. F., Arif, Z., Ahmed, M., & Nafees, K. A. (2016). Anti-arthritogenic and cardioprotective action of hesperidin and daidzein in collagen-induced rheumatoid arthritis. *Molecular and Cellular Biochemistry*, *423*, 115–127.

Akdemir, D., & Isidro-Sánchez, J. (2019). Design of training populations for selective phenotyping in genomic prediction. *Scientific Reports*, *9*, 1446.

Aklilu, E. (2021). Review on forward and reverse genetics in plant breeding. *All Life*, *14*, 127–135.

Albert, N. W., Thrimawithana, A. H., McGhie, T. K., Clayton, W. A., Deroles, S. C., Schwinn, K. E., Bowman, J. L., Jordan, B. R., & Davies, K. M. (2018). Genetic analysis of the liverwort *Marchantia polymorpha* reveals that R2R3 MYB activation of flavonoid production in response to abiotic stress is an ancient character in land plants. *New Phytologist*, *218*, 554–566.

Alghamdi, A., Almuqbil, M., Alrofaidi, M. A., Burzangi, A. S., Alshamrani, A. A., Alzahrani, A. R., Kamal, M., Imran, M., Alshehri, S., Mannasaheb, B. A., Alomar, N. F., & Asdaq, S. M. B. (2022). Potential antioxidant activity of apigenin in the obviating stress-mediated depressive symptoms of experimental mice. *Molecules*, *27*, 9055.

An, J.-P., Qu, F.-J., Yao, J.-F., Wang, X.-N., You, C.-X., Wang, X.-F., & Hao, Y.-J. (2017). The bZIP transcription factor MdHY5 regulates anthocyanin accumulation and nitrate assimilation in apple. *Horticulture Research*, *4*, 17023.

Bag, S., Mondal, A., Majumder, A., Mondal, S. K., & Banik, A. (2022). Flavonoid mediated selective cross-talk between plants and beneficial soil microbiome. *Phytochemistry Reviews*, *21*, 1739–1760.

Bai, C., Xu, J., Cao, B., Li, X., & Li, G. (2018). Transcriptomic analysis and dynamic expression of genes reveal flavonoid synthesis in *Scutellaria viscidula*. *Acta Physiologiae Plantarum*, *40*, 161.

Bai, S., Tao, R., Tang, Y., Yin, L., Ma, Y., Ni, J., Yan, X., Yang, Q., Wu, Z., Zeng, Y., & Teng, Y. (2019). BBX16, a B-box protein, positively regulates light-induced anthocyanin accumulation by activating *MYB10* in red pear. *Plant Biotechnology Journal*, *17*, 1985–1997.

Ban, Y., Honda, C., Hatsuyama, Y., Igarashi, M., Bessho, H., & Moriguchi, T. (2007). Isolation and functional analysis of a MYB transcription factor gene that is a key regulator for the development of red coloration in apple skin. *Plant and Cell Physiology*, *48*, 958–970.

Baudry, A., Heim, M. A., Dubreucq, B., Caboche, M., Weisshaar, B., & Lepiniec, L. (2004). TT2, TT8, and TTG1 synergistically specify the expression of *BANYULS* and proanthocyanidin biosynthesis in *Arabidopsis thaliana*. *The Plant Journal*, *39*, 366–380.

Beekwilder, J., Wolswinkel, R., Jonker, H., Hall, R., de Vos, C. H. R., & Bovy, A. (2006). Production of resveratrol in recombinant microorganisms. *Applied and Environmental Microbiology*, *72*, 5670–5672.

Bellavite, P. (2023). Neuroprotective potentials of flavonoids: Experimental studies and mechanisms of action. *Antioxidants*, *12*, 280.

Ben-Amar, A., Daldoul, S., Reustle, G. M., Krczal, G., & Mliki, A. (2016). Reverse genetics and high throughput sequencing methodologies for plant functional Genomics. *Current Genomics*, *17*, 460–475.

Bhan, N., Li, L., Cai, C., Xu, P., Linhardt, R. J., & Koffas, M. A. G. (2015). Enzymatic formation of a resorcylic acid by creating a structure-guided single-point mutation in stilbene synthase. *Protein Science*, *24*, 167–173.

Braus, G. H. (1991). Aromatic amino acid biosynthesis in the yeast *Saccharomyces cerevisiae*: A model system for the regulation of a eukaryotic biosynthetic pathway. *Microbiological Reviews*, *55*, 349–370.

Butelli, E., Licciardello, C., Zhang, Y., Liu, J., Mackay, S., Bailey, P., Reforgiato-Recupero, G., & Martin, C. (2012). Retrotransposons control fruit-specific, cold-dependent accumulation of anthocyanins in blood oranges. *Plant Cell*, *24*, 1242–1255.

Calderaro, A., Patanè, G. T., Tellone, E., Barreca, D., Ficarra, S., Misiti, F., & Laganà, G. (2022). The neuroprotective potentiality of flavonoids on alzheimer's disease. *International Journal of Molecular Sciences*, *23*, 14835.

Cao, Y., Mei, Y., Zhang, R., Zhong, Z., Yang, X., Xu, C., Chen, K., & Li, X. (2024). Transcriptional regulation of flavonol biosynthesis in plants. *Horticulture Research*, *11*.

Chen, S., Wang, X., Cheng, Y., Gao, H., & Chen, X. (2023). A review of classification, biosynthesis, biological activities and potential applications of flavonoids. *Molecules*, *28*, 4982.

Cherrak, S. A., Mokhtari-Soulimane, N., Berroukeche, F., Bensenane, B., Cherbonnel, A., Merzouk, H., & Elhabiri, M. (2016). In vitro antioxidant versus metal ion chelating properties of flavonoids: A structure-activity investigation. *PLoS One*, *11*, e0165575.

Chu, L. L., Tran, C. T. B., Pham, D. T. K., Nguyen, H. T. A., Nguyen, M. H., Pham, N. M., Nguyen, A. T. V., Phan, D. T., Do, H. M., & Nguyen, Q. H. (2024). Metabolic engineering of *Corynebacterium glutamicum* for the production of flavonoids and stilbenoids. *Molecules*, *29*, 2252.

Cress, B. F., Leitz, Q. D., Kim, D. C., Amore, T. D., Suzuki, J. Y., Linhardt, R. J., & Koffas, M. A. (2017). CRISPRi-mediated metabolic engineering of *E. coli* for O-methylated anthocyanin production. *Microbial Cell Factories*, *16*, 10.

Cress, B. F., Toparlak, Ö. D., Guleria, S., Lebovich, M., Stieglitz, J. T., Englaender, J. A., Jones, J. A., Linhardt, R. J., & Koffas, M. A. (2015). CRISPathBrick: Modular combinatorial assembly of type II-A CRISPR arrays for dCas9-mediated multiplex transcriptional repression in *E. coli*. *ACS Synthetic Biology*, *4*, 987–1000.

D'Amelia, V., Aversano, R., Chiaiese, P., & Carputo, D. (2018). The antioxidant properties of plant flavonoids: Their exploitation by molecular plant breeding. *Phytochemistry Reviews*, *17*, 611–625.

Davies, K. M. (2003). Enhancing anthocyanin production by altering competition for substrate between flavonol synthase and dihydroflavonol, 4-reductase. *Euphytica*, *131*, 259–268.

Davuluri, G. R., van Tuinen, A., Mustilli, A. C., Manfredonia, A., Newman, R., Burgess, D., Brummell, D. A., King, S. R., Palys, J., Uhlig, J., Pennings, H. M., & Bowler, C. (2004). Manipulation of *DET1* expression in tomato results in photomorphogenic phenotypes caused by post-transcriptional gene silencing. *The Plant Journal*, *40*, 344–354.

Dias, M. C., Pinto, D. C. G. A., & Silva, A. M. S. (2021). Plant flavonoids: Chemical characteristics and biological activity. *Molecules*, *26*, 5377.

Du, Y., Yang, B., Yi, Z., Hu, L., & Li, M. (2020). Engineering *Saccharomyces cerevisiae* coculture platform for the production of flavonoids. *Journal of Agricultural and Food Chemistry*, *68*, 2146–2154.

Dudnik, A., Gaspar, P., Neves, A. R., & Forster, J. (2018). Engineering of microbial cell factories for the production of plant polyphenols with health-beneficial properties. *Current Pharmaceutical Design*, *24*, 2208–2225.

Dunstan, M. S., Robinson, C. J., Jervis, A. J., Yan, C., Carbonell, P., Hollywood, K. A., Currin, A., Swainston, N., Feuvre, R. L., Micklefield, J., Faulon, J. L., Breitling, R., Turner, N., Takano, E., & Scrutton, N. S. (2020). Engineering *Escherichia coli* towards de novo production of gatekeeper (2S)-flavanones: Naringenin, pinocembrin, eriodictyol and homoeriodictyol. *Synthetic Biology*, *5*.

Elarabi, N. I., Abdelhadi, A. A., Sief-Eldein, A. G. M., Ismail, I. A., & Abdallah, N. A. (2021). Overexpression of chalcone isomerase A gene in Astragalus trigonus for stimulating apigenin. *Scientific Reports*, *11*, 24176.

Espley, R. V., Brendolise, C., Chagné, D., Kutty-Amma, S., Green, S., Volz, R., Putterill, J., Schouten, H. J., Gardiner, S. E., Hellens, R. P., & Allan, A. C. (2009). Multiple repeats of a promoter segment causes transcription factor autoregulation in red apples. *Plant Cell*, *21*, 168–183.

Fang, H., Dong, Y., Yue, X., Hu, J., Jiang, S., Xu, H., Wang, Y., Su, M., Zhang, J., Zhang, Z., Wang, N., & Chen, X. (2019). The B-box zinc finger protein MdBBX20 integrates anthocyanin accumulation in response to ultraviolet radiation and low temperature. *Plant, Cell & Environment*, *42*, 2090–2104.

Fang, Z.-Z., Lin-Wang, K., Zhou, D.-R., Lin, Y.-J., Jiang, C.-C., Pan, S.-L., Espley, R. V., Andre, C. M., & Ye, X. F. (2021). Activation of PsMYB10.2 transcription causes anthocyanin accumulation in flesh of the red-fleshed mutant of 'sanyueli' (*Prunus salicina* Lindl.). *Frontiers in Plant Science*, *12*.

Fatihah, H. N. N., Moñino López, D., van Arkel, G., Schaart, J. G., Visser, R. G. F., & Krens, F. A. (2019). The ROSEA1 and DELILA transcription factors control anthocyanin biosynthesis in *Nicotiana benthamiana* and Lilium flowers. *Scientia Horticulturea*, *243*, 327–337.

Ferreyra, M. L. F., Serra, P., & Casati, P. (2021). Recent advances on the roles of flavonoids as plant protective molecules after UV and high light exposure. *Physiologia Plantarum*, *173*, 736–749.

Fordjour, E., Adipah, F. K., Zhou, S., Du, G., & Zhou, J. (2019). Metabolic engineering of *Escherichia coli* BL21 (DE3) for de novo production of l-DOPA from d-glucose. *Microbial Cell Factories*, *18*, 74.

Ganesan, V., Li, Z., Wang, X., & Zhang, H. (2017). Heterologous biosynthesis of natural product naringenin by co-culture engineering. *Synthetic and Systems Biotechnology*, *2*, 236–242.

Gao, S., Lyu, Y., Zeng, W., Du, G., Zhou, J., & Chen, J. (2020). Efficient biosynthesis of (2S)-naringenin from p-coumaric acid in *Saccharomyces cerevisiae*. *Journal of Agricultural and Food Chemistry*, *68*, 1015–1021.

García-Granados, R., Lerma-Escalera, J. A., & Morones-Ramírez, J. R. (2019). Metabolic engineering and synthetic biology: Synergies, future, and challenges. *Frontiers in Bioengineering and Biotechnology*, *7*.

Gatica-Arias, A., Farag, M. A., Stanke, M., Matoušek, J., Wessjohann, L., & Weber, G. (2012). Flavonoid production in transgenic hop (*Humulus lupulus* L.) altered by PAP1/MYB75 from *Arabidopsis thaliana* L. *Plant Cell Reports*, *31*, 111–119.

Ghitti, E., Rolli, E., Crotti, E., & Borin, S. (2022). Flavonoids are intra- and inter-kingdom modulator signals. *Microorganisms*, *10*, 2479.

Gho, Y.-S., Kim, S., & Jung, K.-H. (2020). Phenylalanine ammonia-lyase family is closely associated with response to phosphate deficiency in rice. *Genes Genomics*, *42*, 67–76.

Griffiths, A., Miller, J., Suzuki, D., Lewontin, R., & Gelbart, W. (2000). *An introduction to genetic analysis*. W.H. Freeman and Company.

Guo, X., Wu, X., Ma, H., Liu, H., & Luo, Y. (2023). Yeast: A platform for the production of L-tyrosine derivatives. *Yeast*, *40*, 214–230.

Haida, Z., Syahida, A., Ariff, S. M., Maziah, M., & Hakiman, M. (2019). Factors affecting cell biomass and flavonoid production of *Ficus deltoidea* var. kunstleri in cell suspension culture system. *Scientific Reports*, *9*, 9533.

He, X., Li, Y., Lawson, D., & Xie, D. (2017). Metabolic engineering of anthocyanins in dark tobacco varieties. *Physiologia Plantarum*, *159*, 2–12.

He, X., Zhao, X., Gao, L., Shi, X., Dai, X., Liu, Y., Xia, T., & Wang, Y. (2018). Isolation and characterization of key genes that promote flavonoid accumulation in purple-leaf tea (*Camellia sinensis* L.). *Scientific Reports*, *8*, 130.

Hichri, I., Barrieu, F., Bogs, J., Kappel, C., Delrot, S., & Lauvergeat, V. (2011). Recent advances in the transcriptional regulation of the flavonoid biosynthetic pathway. *Journal of Experimental Botany*, *62*, 2465–2483.

Hidalgo, D., Martínez-Márquez, A., Cusidó, R., Bru-Martínez, R., Palazón, J., & Corchete, P. (2017). *Silybum marianum* cell cultures stably transformed with *Vitis vinifera* stilbene synthase accumulate t-resveratrol in the extracellular medium after elicitation with methyl jasmonate or methylated β-cyclodextrins. *Engineering in Life Sciences*, *17*, 686–694.

Hu, D. G., Li, Y. Y., Zhang, Q. Y., Li, M., Sun, C. H., Yu, J. Q., & Hao, Y. J. (2017). The R2R3-MYB transcription factor Md MYB 73 is involved in malate accumulation and vacuolar acidification in apple. *The Plant Journal*, *91*, 443–454.

Hua, C., Linling, L., Shuiyuan, C., Fuliang, C., Feng, X., Honghui, Y., & Conghua, W. (2013). Molecular cloning and characterization of three genes encoding dihydroflavonol-4-reductase from *Ginkgo biloba* in anthocyanin biosynthetic pathway. *PLoS One*, *8*, e72017.

Huang, F., Duan, J., Lei, Y., Kang, Y., Luo, Y., Chen, Y., Ding, D., & Li, S. (2022). Metabolomic and transcriptomic analyses reveal a MYB gene, CsAN1, involved in anthocyanins accumulation separation in F1 between 'Zijuan' (*Camellia sinensis* var. assamica) and 'Fudingdabaicha' (*C. sinensis* var. sinensis) tea plants. *Frontiers in Plant Science*, *13*.

Hwang, E. Il, Kaneko, M., Ohnishi, Y., & Horinouchi, S. (2003). Production of plant-specific flavanones by *Escherichia coli* Containing an artificial gene cluster. *Applied and Environmental Microbiology*, *69*, 2699–2706.

Iranshahi, M., Rezaee, R., Parhiz, H., Roohbakhsh, A., & Soltani, F. (2015). Protective effects of flavonoids against microbes and toxins: The cases of hesperidin and hesperetin. *Life Sciences*, *137*, 125–132.

Isogai, S., Tominaga, M., Kondo, A., & Ishii, J. (2022). Plant flavonoid production in bacteria and yeasts. *Frontiers in Chemical Engineering*, *4*.

Iwashina, T. (2015). Contribution to flower colors of flavonoids including anthocyanins: A review. *Natural Product Communications, 10*, 529–44.

Jian, W., Cao, H., Yuan, S., Liu, Y., Lu, J., Lu, W., Li, N., Wang, J., Zou, J., Tang, N., Xu, C., Cheng, Y., Gao, Y., Xi, W., Bouzayen, M., & Li, Z. (2019). SlMYB75, an MYB-type transcription factor, promotes anthocyanin accumulation and enhances volatile aroma production in tomato fruits. *Horticulture Research, 6*, 22.

Jiang, H., Wood, K. V., & Morgan, J. A. (2005). Metabolic engineering of the phenylpropanoid pathway in *Saccharomyces cerevisiae. Applied and Environmental Microbiology, 71*, 2962–2969.

Jiang, M., & Zhang, H. (2016). Engineering the shikimate pathway for biosynthesis of molecules with pharmaceutical activities in *E. coli. Current Opinion in Biotechnology, 42*, 1–6.

Jiang, S., Sun, Q., Zhang, T., Liu, W., Wang, N., & Chen, X. (2021). MdMYB114 regulates anthocyanin biosynthesis and functions downstream of MdbZIP4-like in apple fruit. *Journal of Plant Physiology, 257*, 153353.

Jones, J. A., Vernacchio, V. R., Collins, S. M., Shirke, A. N., Xiu, Y., Englaender, J. A., Cress, B. F., McCutcheon, C. C., Linhardt, R. J., Gross, R. A., & Koffas, M. A. G. (2017). Complete biosynthesis of anthocyanins using *E. coli* polycultures. *mBio, 8*.

Jones, J. A., Vernacchio, V. R., Sinkoe, A. L., Collins, S. M., Ibrahim, M. H. A., Lachance, D. M., Hahn, J., & Koffas, M. A. G. (2016). Experimental and computational optimization of an *Escherichia coli* co-culture for the efficient production of flavonoids. *Metabolic Engineering, 35*, 55–63.

Junping, X., Naing, A. H., & Kim, C. K. (2018). Transcriptional activation of anthocyanin structural genes in Torenia 'Kauai Rose' via overexpression of anthocyanin regulatory transcription factors. *3 Biotech, 8*, 476.

Kallscheuer, N., Vogt, M., Kappelmann, J., Krumbach, K., Noack, S., Bott, M., & Marienhagen, J. (2016). Identification of the phd gene cluster responsible for phenylpropanoid utilization in *Corynebacterium glutamicum. Applied Microbiology and Biotechnology, 100*, 1871–1881.

Kapoore, R. V., Padmaperuma, G., Maneein, S., & Vaidyanathan, S. (2022). Co-culturing microbial consortia: Approaches for applications in biomanufacturing and bioprocessing. *Critical Reviews in Biotechnology, 42*, 46–72.

Katsuyama, Y., Miyahisa, I., Funa, N., & Horinouchi, S. (2007). One-pot synthesis of genistein from tyrosine by coincubation of genetically engineered *Escherichia coli* and *Saccharomyces cerevisiae* cells. *Applied Microbiology and Biotechnology, 73*, 1143–1149.

Khoo, H. E., Azlan, A., Tang, S. T., & Lim, S. M. (2017). Anthocyanidins and anthocyanins: Colored pigments as food, pharmaceutical ingredients, and the potential health benefits. *Food & Nutrition Research, 61*, 1361779.

Kim, B.-G. (2020). Biological synthesis of genistein in *Escherichia coli. Journal of Microbiology and Biotechnology, 30*, 770–776.

Kobayashi, S., Goto-Yamamoto, N., & Hirochika, H. (2004). Retrotransposon-induced mutations in grape skin color. *Science, 1979*(304), 982–982.

Kong, J.-Q., Lu, D., & Wang, Z.-B. (2014). Molecular cloning and yeast expression of cinnamate, 4-hydroxylase from *Ornithogalum saundersiae* Baker. *Molecules, 19*, 1608–1621.

Lee, H.-M., Vo, P., & Na, D. (2018). Advancement of metabolic engineering assisted by synthetic biology. *Catalysts, 8*, 619.

Levisson, M., Patinios, C., Hein, S., de Groot, P. A., Daran, J.-M., Hall, R. D., Martens, S., & Beekwilder, J. (2018). Engineering de novo anthocyanin production in *Saccharomyces cerevisiae. Microbial Cell Factories, 17*, 103.

Li, H., Liu, J., Pei, T., Bai, Z., Han, R., & Liang, Z. (2019). Overexpression of SmANS enhances anthocyanin accumulation and alters phenolic acids content in *Salvia miltiorrhiza* and *Salvia miltiorrhiza* Bge f. alba plantlets. *International Journal of Molecular Sciences, 20*, 2225.

Li, M., Schneider, K., Kristensen, M., Borodina, I., & Nielsen, J. (2016). Engineering yeast for high-level production of stilbenoid antioxidants. *Scientific Reports, 6*, 36827.

Li, X.-X., Liu, C., Dong, S.-L., Ou, C.-S., Lu, J.-L., Ye, J.-H, Liang, Y. R., & Zheng, X. Q. (2022). Anticarcinogenic potentials of tea catechins. *Frontiers in Nutrition, 9*.

Lim, S.-H., You, M.-K., Kim, D.-H., Kim, J. K., Lee, J.-Y., & Ha, S.-H. (2016). RNAi-mediated suppression of dihydroflavonol, 4-reductase in tobacco allows fine-tuning of flower color and flux through the flavonoid biosynthetic pathway. *Plant Physiology and Biochemistry, 109*, 482–490.

Lim, S.-H., Kim, D.-H., & Lee, J.-Y. (2022). RsTTG1, a WD40 Protein, Interacts with the bHLH transcription factor RsTT8 to regulate anthocyanin and proanthocyanidin biosynthesis in *Raphanus sativus. International Journal of Molecular Sciences, 23*, 11973.

Liu, C.-Z., & Saxena, P. K. (2009). Saussurea medusa cell suspension cultures for flavonoid production. *Methods in Molecular Biology, 547*, 53–59.

Liu, D., Sica, M. S., Mao, J., Chao, L.F.-I., & Siewers, V. (2022). A *p* -Coumaroyl-CoA biosensor for dynamic regulation of naringenin biosynthesis in *Saccharomyces cerevisiae. ACS Synthetic Biology, 11*, 3228–3238.

Liu, W., Feng, Y., Yu, S., Fan, Z., Li, X., Li, J., & Yin, H. (2021). The flavonoid biosynthesis network in plants. *International Journal of Molecular Sciences, 22*, 12824.

Liu, X., Cheng, J., Zhang, G., Ding, W., Duan, L., Yang, J., Kui, L., Cheng, X., Ruan, J., Fan, W., Chen, J., Long, G., Zhao, Y., Cai, J., Wang, W., Ma, Y., Dong, Y., Yang, S., & Jiang, H. (2018). Engineering yeast for the production of breviscapine by genomic analysis and synthetic biology approaches. *Nature Communications, 9*, 448.

Liu, X., Cheng, J., Zhu, X., Zhang, G., Yang, S., Guo, X., Jiang, H., & Ma, Y. (2020). De novo biosynthesis of multiple pinocembrin derivatives in *Saccharomyces cerevisiae. ACS Synthetic Biology, 9*, 3042–3051.

Liu, X., Niu, H., Li, Q., & Gu, P. (2019). Metabolic engineering for the production of l-phenylalanine in *Escherichia coli. 3 Biotech, 9*, 85.

Liu, Y., Xu, Y., Ding, D., Wen, J., Zhu, B., & Zhang, D. (2018). Genetic engineering of *Escherichia coli* to improve L-phenylalanine production. *BMC Biotechnology, 18*, 5.

Lou, H., Hu, L., Lu, H., Wei, T., & Chen, Q. (2021). Metabolic engineering of microbial cell factories for biosynthesis of flavonoids: A review. *Molecules, 26*.

Lv, Y., Marsafari, M., Koffas, M., Zhou, J., & Xu, P. (2019). Optimizing oleaginous yeast cell factories for flavonoids and hydroxylated flavonoids biosynthesis. *ACS Synthetic Biology, 8*, 2514–2523.

Lyu, X., Zhao, G., Ng, K. R., Mark, R., & Chen, W. N. (2019). Metabolic engineering of *Saccharomyces cerevisiae* for de novo production of kaempferol. *Journal of Agricultural and Food Chemistry, 67*, 5596–5606.

Madzak, C. (2021). *Yarrowia lipolytica* strains and their biotechnological applications: How natural biodiversity and metabolic engineering could contribute to cell factories improvement. *Journal of Fungi*, *7*, 548.

Maglangit, F., Fang, Q., Kyeremeh, K., Sternberg, J. M., Ebel, R., & Deng, H. (2020). A co-culturing approach enables discovery and biosynthesis of a bioactive indole alkaloid metabolite. *Molecules*, *25*, 256.

Mahajan, M., Ahuja, P. S., & Yadav, S. K. (2011). Post-transcriptional silencing of flavonol synthase mRNA in tobacco leads to fruits with arrested seed set. *PLoS One*, *6*, e28315.

Mark, R., Lyu, X., Ng, K. R., & Chen, W. N. (2019). Gene source screening as a tool for naringenin production in engineered *Saccharomyces cerevisiae*. *ACS Omega*, *4*, 12872–12879.

Marsafari, M., Samizadeh, H., Rabiei, B., Mehrabi, A., Koffas, M., & Xu, P. (2020). Biotechnological production of flavonoids: An update on plant metabolic engineering, microbial host selection, and genetically encoded biosensors. *Biotechnology Jourtnal*, *15*.

McCallum, C. M., Comai, L., Greene, E. A., & Henikoff, S. (2000). Targeting induced local lesions IN genomes (TILLING) for plant functional genomics. *Plant Physiology*, *123*, 439–442.

Mendoza, D., Cuaspud, O., Arias, J. P., Ruiz, O., & Arias, M. (2018). Effect of salicylic acid and methyl jasmonate in the production of phenolic compounds in plant cell suspension cultures of *Thevetia peruviana*. *Biotechnology Reports*, *19*, e00273.

Meng, X., Yin, B., Feng, H.-L., Zhang, S., Liang, X.-Q., & Meng, Q.-W. (2014). Overexpression of R2R3-MYB gene leads to accumulation of anthocyanin and enhanced resistance to chilling and oxidative stress. *Biologica Plantarum*, *58*, 121–130.

Meyer, P., Heidmann, I., Forkmann, G., & Saedler, H. (1987). A new petunia flower colour generated by transformation of a mutant with a maize gene. *Nature*, *330*, 677–678.

Middleton, E. (1984). The flavonoids. *Trends in Pharmacological Sciences*, *5*, 335–338.

Milke, L., Ferreira, P., Kallscheuer, N., Braga, A., Vogt, M., Kappelmann, J., Oliveira, J., Silva, A. R., Rocha, I., Bott, M., Noack, S., Faria, N., & Marienhagen, J. (2019). Modulation of the central carbon metabolism of *Corynebacterium glutamicum* improves malonyl-CoA availability and increases plant polyphenol synthesis. *Biotechnology and Bioengineering*, *116*, 1380–1391.

Muhammad, A., Feng, X., Rasool, A., Sun, W., & Li, C. (2020). Production of plant natural products through engineered *Yarrowia lipolytica*. *Biotechnology Advances*, *43*, 107555.

Nabavi, S. M., Šamec, D., Tomczyk, M., Milella, L., Russo, D., Habtemariam, S., Suntar, I., Rastrelli, L., Daglia, M., Xiao, J., Giampieri, F., Battino, M., Sobarzo-Sanchez, E., Nabavi, S. F., Yousefi, B., Jeandet, P., Xu, S., & Shirooie, S. (2020). Flavonoid biosynthetic pathways in plants: Versatile targets for metabolic engineering. *Biotechnology Advances*, *38*, 107316.

Naing, A. H., Ai, T. N., Lim, K. B., Lee, I. J., & Kim, C. K. (2018). Overexpression of rosea1 from snapdragon enhances anthocyanin accumulation and abiotic stress tolerance in transgenic tobacco. *Frontiers in Plant Science*, *9*.

Naing, A. H., Park, K. Il, Ai, T. N., Chung, M. Y., Han, J. S., Kang, Y.-W., Lim, K. B., & Kim, C. K. (2017). Overexpression of *snapdragon* Delila (Del) gene in tobacco enhances anthocyanin accumulation and abiotic stress tolerance. *BMC Plant Biology*, *17*, 65.

Narayana, K., Reddy, M., Chaluvadi, M., & Krishna, D. (2001). Bioflavonoids classification, pharmacological, biochemical effects and therapeutic potential. *Indian Journal of Pharmacology*, *33*, 2–16.

Naseema Rasheed, R., Pourbakhtiar, A., Mehdizadeh Allaf, M., Baharlooeian, M., Rafiei, N., Alishah Aratboni, H., Morones-Ramirez, J. R., & Winck, F. V. (2023). Microalgal co-cultivation -recent methods, trends in omic-studies, applications, and future challenges. *Frontiers in Bioengineering and Biotechnology*, *11*.

Nesi, N., Debeaujon, I., Jond, C., Stewart, A. J., Jenkins, G. I., Caboche, M., & Lepiniec, L. (2002). The *TRANSPARENT TESTA16* locus encodes the ARABIDOPSIS BSISTER MADS domain protein and is required for proper development and pigmentation of the seed coat. *The Plant Cell*, *14*, 2463–2479.

Ni, R., Zhu, T.-T., Zhang, X.-S., Wang, P.-Y., Sun, C.-J., Qiao, Y.-N., Lou, H. X., & Cheng, A. X. (2020). Identification and evolutionary analysis of chalcone isomerase-fold proteins in ferns. *Journal of Experimental Botany*, *71*, 290–304.

Nishihara, M., Nakatsuka, T., & Yamamura, S. (2005). Flavonoid components and flower color change in transgenic tobacco plants by suppression of chalcone isomerase gene. *FEBS Letters*, *579*, 6074–6078.

Niu, S.-S., Xu, C.-J., Zhang, W.-S., Zhang, B., Li, X., Lin-Wang, K., Ferguson, I. B., Allan, A. C., & Chen, K. S. (2010). Coordinated regulation of anthocyanin biosynthesis in Chinese bayberry (*Myrica rubra*) fruit by a R2R3 MYB transcription factor. *Planta*, *231*, 887–899.

Pandey, R. P., Parajuli, P., Koffas, M. A. G., & Sohng, J. K. (2016). Microbial production of natural and non-natural flavonoids: Pathway engineering, directed evolution and systems/synthetic biology. *Biotechnology Advances*, *34*, 634–662.

Parage, C., Tavares, R., Réty, S., Baltenweck-Guyot, R., Poutaraud, A., Renault, L., Heintz, D., Lugan, R., Marais, G. A., Aubourg, S., & Hugueney, P. (2012). Structural, functional, and evolutionary analysis of the unusually large stilbene synthase gene family in grapevine. *Plant Physiology*, *160*, 1407–1419.

Pérez-Valero, Á., Ye, S., Magadán-Corpas, P., Villar, C. J., & Lombó, F. (2023). Metabolic engineering in *Streptomyces albidoflavus* for the biosynthesis of the methylated flavonoids sakuranetin, acacetin, and genkwanin. *Microb Cell Factories*, *22*, 234.

Petroni, K., & Tonelli, C. (2011). Recent advances on the regulation of anthocyanin synthesis in reproductive organs. *Plant Science*, *181*, 219–229.

Ping, J., Wang, L., Qin, Z., Zhou, Z., & Zhou, J. (2023). Synergetic engineering of *Escherichia coli* for efficient production of l-tyrosine. *Synthetic and Systems Biotechnology*, *8*, 724–731.

Polturak, G., Misra, R. C., El-Demerdash, A., Owen, C., Steed, A., McDonald, H. P., Wang, J., Saalbach, G., Martins, C., Chartrain, L., Wilkinson, B., Nicholson, P., & Osbourn, A. (2023). Discovery of isoflavone phytoalexins in wheat reveals an alternative route to isoflavonoid biosynthesis. *Nature Communications*, *14*, 6977.

Qi, Y., Gu, C., Wang, X., Gao, S., Li, C., Zhao, C., Li, C., Ma, C., & Zhang, Q. (2020). Identification of the *Eutrema salsugineum* EsMYB90 gene important for anthocyanin biosynthesis. *BMC Plant Biology*, *20*, 186.

Qiu, C., Wang, H., Zhao, L., & Pei, J. (2021). Orientin and vitexin production by a one-pot enzymatic cascade of a glycosyltransferase and sucrose synthase. *Bioorganic Chemistry*, *112*, 104926.

Rao, M. J., Xu, Y., Tang, X., Huang, Y., Liu, J., Deng, X., & Xu, Q. (2020). CsCYT75B1, a citrus CYTOCHROME P450 gene, is involved in accumulation of antioxidant flavonoids and induces drought tolerance in transgenic arabidopsis. *Antioxidants*, *9*, 161.

Rencoret, J., Rosado, M. J., Kim, H., Timokhin, V. I., Gutiérrez, A., Bausch, F., Rosenau, T., Potthast, A., Ralph, J., & Del Río, J. C. (2022). Flavonoids naringenin chalcone, naringenin, dihydrotricin, and tricin are lignin monomers in papyrus. *Plant Physiology*, *188*, 208–219.

Rihani, K. A. L., Jacobsen, H.-J., Hofmann, T., Schwab, W., & Hassan, F. (2017). Metabolic engineering of apple by overexpression of the MdMyb10 gene. *Journal of Genetic Engineering and Biotechnology*, *15*, 263–273.

Rodriguez, A., Strucko, T., Stahlhut, S. G., Kristensen, M., Svenssen, D. K., Forster, J., Nielsen, J., & Borodina, I. (2017). Metabolic engineering of yeast for fermentative production of flavonoids. *Bioresource Technology*, *245*, 1645–1654.

Rowaiye, A B., Nnamdi Oli, A., Onuh, O. A., Emeter, N. W., Bur, D., Obideyi, O. A., Dayisi, O. C., Akpa, J. N., Birah, L., Omaka, E. E., Iseghohi, F. O., Otitoju, A. P., Uzor, P. F., & Okoyeh, J. N. (2022). Rhamnetin is a better inhibitor of SARS-COV-2 2'-O-methyltransferase than dolutegravir: A computational prediction. *African Journal of Infectious Diseases*, *16*, 80–96.

Roy, A., Khan, A., Ahmad, I., Alghamdi, S., Rajab, B. S., Babalghith, A. O., Alshahrani, M. Y., Islam, S., & Islam, M. R. (2022). Flavonoids a bioactive compound from medicinal plants and its therapeutic applications. *BioMed Research International*, *2022*, 1–9.

Sajid, M., Channakesavula, C. N., Stone, S. R., & Kaur, P. (2021). Synthetic biology towards improved flavonoid pharmacokinetics. *Biomolecules*, *11*, 754.

Šamec, D., Karalija, E., Šola, I., Vujčić Bok, V., & Salopek-Sondi, B. (2021). The role of polyphenols in abiotic stress response: The influence of molecular structure. *Plants*, *10*, 118.

Santos, C. N. S., Koffas, M., & Stephanopoulos, G. (2011). Optimization of a heterologous pathway for the production of flavonoids from glucose. *Metabolic Engineering*, *13*, 392–400.

Schijlen, E. G. W. M., de Vos, C. H. R., Martens, S., Jonker, H. H., Rosin, F. M., Molthoff, J. W., Tikunov, Y. M., Angenent, G. C., van Tunen, A. J., & Bovy, A. G. (2007). RNA interference silencing of chalcone synthase, the first step in the flavonoid biosynthesis pathway, leads to parthenocarpic tomato fruits. *Plant Physiology*, *144*, 1520–1530.

Shah, A., & Smith, D. L. (2020). Flavonoids in agriculture: Chemistry and roles in, biotic and abiotic stress responses, and microbial associations. *Agronomy*, *10*, 1209.

Shah, F. L. A., Ramzi, A. B., Baharum, S. N., Noor, N. M., Goh, H.-H., Leow, T. C., Oslan, S. N., & Sabri, S. (2019). Recent advancement of engineering microbial hosts for the biotechnological production of flavonoids. *Molecular Biology Reports*, *46*, 6647–6659.

Sheng, H., Sun, X., Yan, Y., Yuan, Q., Wang, J., & Shen, X. (2020). Metabolic engineering of microorganisms for the production of flavonoids. *Frontiers in Bioengineering and Biotechnology*, *8*.

Shomali, A., Das, S., Arif, N., Sarraf, M., Zahra, N., Yadav, V., Aliniaeifard, S., Chauhan, D. K., & Hasanuzzaman, M. (2022). Diverse physiological roles of flavonoids in plant environmental stress responses and tolerance. *Plants*, *11*, 3158.

Shrestha, A., Pandey, R. P., Dhakal, D., Parajuli, P., & Sohng, J. K. (2018). Biosynthesis of flavone C-glucosides in engineered *Escherichia coli*. *Applied Microbiology and Biotechnology*, *102*, 1251–1267.

Singh, D., Chaudhary, P., Taunk, J., Singh, C. K., Chinnusamy, V., Sevanthi, A. M., Singh, V. J., & Pal, M. (2023). Targeting induced local lesions in genomes (TILLING): Advances and opportunities for fast tracking crop breeding. *Critical Reviews in Biotechnology*, *1*, 20.

Stanisławska, J., & Olszewski, W. L. (2005). RNA interference – significance and applications. *Archivum Immunologiae et Therapiae Experimentalis (Warsz)*, *53*, 39–46.

Sun, B., Zhu, Z., Cao, P., Chen, H., Chen, C., Zhou, X., Mao, Y., Lei, J., Jiang, Y., Meng, W., Wang, Y., & Liu, S. (2016). Purple foliage coloration in tea (*Camellia sinensis* L.) arises from activation of the R2R3-MYB transcription factor CsAN1. *Scientific Reports*, *6*, 32534.

Sun, J., Sun, W., Zhang, G., Lv, B., & Li, C. (2022). High efficient production of plant flavonoids by microbial cell factories: Challenges and opportunities. *Metabolic Engineering*, *70*, 143–154.

Sun, Q., Jiang, S., Zhang, T., Xu, H., Fang, H., Zhang, J., Su, M., Wang, Y., Zhang, Z., Wang, N., & Chen, X. (2019). Apple NAC transcription factor MdNAC52 regulates biosynthesis of anthocyanin and proanthocyanidin through MdMYB9 and MdMYB11. *Plant Science*, *289*, 110286.

Sun, W., Meng, X., Liang, L., Li, Y., Zhou, T., Cai, X., Wang, L., & Gao, X. (2017). Overexpression of a Freesia hybrida flavonoid, 3-O-glycosyltransferase gene, Fh3GT1, enhances transcription of key anthocyanin genes and accumulation of anthocyanin and flavonol in transgenic petunia (*Petunia hybrida*). *In Vitro Cellular & Developmental Biology – Plant*, *53*, 478–488.

Takos, A. M., Jaffé, F. W., Jacob, S. R., Bogs, J., Robinson, S. P., & Walker, A. R. (2006). Light-induced expression of a *MYB* gene regulates anthocyanin biosynthesis in red apples. *Plant Physiology*, *142*, 1216–1232.

Tan, L., Zhang, P., Cui, D., Yang, X., Zhang, D., Yang, Y., Chen, W., Tang, D., Tang, Q., & Li, P. (2023). Multi-omics analysis revealed anthocyanin accumulation differences in purple tea plants 'Ziyan', 'Zijuan' and their dark-purple hybrid. *Scientica Horticulturae*, *321*, 112275.

Tanaka, Y., & Brugliera, F. (2013). Flower colour and cytochromes P450. *Philosophical Transactions of the Royal Society B: Biological Sciences*, *368*, 20120432.

Tariq, H., Asif, S., Andleeb, A., Hano, C., & Abbasi, B. H. (2023). Flavonoid production: Current trends in plant metabolic engineering and de novo microbial production. *Metabolites*, *13*, 124.

Teles, Y., Souza, M., & Souza, M. (2018). Sulphated flavonoids: Biosynthesis, structures, and biological activities. *Molecules*, *23*, 480.

Thuan, N. H., Chaudhary, A. K., Van Cuong, D., & Cuong, N. X. (2018). Engineering co-culture system for production of apigetrin in *Escherichia coli*. *Journal of Industrial Microbiology and Biotechnology*, *45*, 175–185.

Thuan, N. H., Tatipamula, V. B., Canh, N. X., & Van Giang, N. (2022). Recent advances in microbial co-culture for production of value-added compounds. *3 Biotech*, *12*, 115.

Thuan, N. H., Tatipamula, V. B., Trung, N. T., & Van Giang, N. (2023). Metabolic engineering and optimization of *Escherichia coli* co-culture for the *de novo* synthesis of genkwanin. *Journal of Industrial Microbiology and Biotechnology*, *50*.

Thuan, N. H., Tatipamula, V. B., Viet, T. T., Tien, N. Q. D., & Loc, N. H. (2022). Bioproduction of eriodictyol by *Escherichia coli* engineered co-culture. *World Journal of Microbiology and Biotechnology*, *38*, 112.

Tian, B., Pei, Y., Huang, W., Ding, J., & Siemann, E. (2021). Increasing flavonoid concentrations in root exudates enhance associations between arbuscular mycorrhizal fungi and an invasive plant. *The ISME Journal*, *15*, 1919–1930.

Tiwari, S., Tripathi, N., Sapre, S., Gontia-Mishra, I., & Prakash, V. (2016). Bioactive natural products from plants and biotechnological approaches for their production. *International Journal of Biotechnology for Wellness Industries*, *5*, 91–110.

Tohge, T., de Souza, L. P., & Fernie, A. R. (2017). Current understanding of the pathways of flavonoid biosynthesis in model and crop plants. *Journal of Experimental Botany*, *68*, 4013–4028.

Trantas, E. A., Koffas, M. A. G., Xu, P., & Ververidis, F. (2015). When plants produce not enough or at all: Metabolic engineering of flavonoids in microbial hosts. *Frontiers in Plant Science*, *6*.

Vargas, P., Farias, G. A., Nogales, J., Prada, H., Carvajal, V., Barón, M., Rivilla, R., Martín, M., Olmedilla, A., & Gallegos, M. T. (2013). Plant flavonoids target *Pseudomonas syringae* pv. tomato DC 3000 flagella and type III secretion system. *Environmental Microbiology Reports*, *5*, 841–850.

Venkateswara Rao, P., Kiran, S., Rohini, P., & Bhagyasree, P. (2017). Flavonoid: A review on Naringenin. *Journal of Pharmacognosy and Phytochemistry*, *6*, 2778–2783.

Wadhwa, R., Paudel, K. R., Chin, L. H., Hon, C. M., Madheswaran, T., Gupta, G., Panneerselvam, J., Lakshmi, T., Singh, S. K., Gulati, M., Dureja, H., Hsu, A., Mehta, M., Anand, K., Devkota, H. P., Chellian, J., Chellappan, D. K., Hansbro, P. M., & Dua, K. (2021). Anti-inflammatory and anticancer activities of Naringenin-loaded liquid crystalline nanoparticles in vitro. *Journal of Food Biochemistry*, *45*.

Wang, J., Qian, J., Yao, L., & Lu, Y. (2015). Enhanced production of flavonoids by methyl jasmonate elicitation in cell suspension culture of *Hypericum perforatum*. *Bioresour Bioprocess*, *2*, 5.

Wang, N., Liu, W., Zhang, T., Jiang, S., Xu, H., Wang, Y., Zhang, Z., Wang, C., & Chen, X. (2018). Transcriptomic analysis of red-fleshed apples reveals the novel role of MdWRKY11 in flavonoid and anthocyanin biosynthesis. *Journal of Agricultural and Food Chemistry*, *66*, 7076–7086.

Wang, P., Li, C., Li, X., Huang, W., Wang, Y., Wang, J., Zhang, Y., Yang, X., Yan, X., Wang, Y., & Zhou, Z. (2021). Complete biosynthesis of the potential medicine icaritin by engineered *Saccharomyces cerevisiae* and *Escherichia coli*. *Science Bulletin (Beijing)*, *66*, 1906–1916.

Wang, X., Ding, G., Liu, B., & Wang, Q. (2020). Flavonoids and antioxidant activity of rare and endangered fern: Isoetes sinensis. *PLoS One*, *15*, e0232185.

Wang, X., Li, Z., Policarpio, L., Koffas, M. A. G., & Zhang, H. (2020). De novo biosynthesis of complex natural product sakuranetin using modular co-culture engineering. *Applied Microbiology and Biotechnology*, *104*, 4849–4861.

Wang, Y., Chen, S., & Yu, O. (2011). Metabolic engineering of flavonoids in plants and microorganisms. *Applied Microbiology and Biotechnology*, *91*, 949–956.

Wang, Y., Zhang, X., Zhao, Y., Yang, J., He, Y., Li, G., Ma, W., Huang, X., & Su, J. (2020). Transcription factor PyHY5 binds to the promoters of PyWD40 and PyMYB10 and regulates its expression in red pear 'Yunhongli No. 1.' *Plant Physiology and Biochemistry*, *154*, 665–674.

Wei, K., Wang, L., Zhang, Y., Ruan, L., Li, H., Wu, L., Xu, L., Zhang, C., Zhou, X., Cheng, H. & Edwards, R. (2019). A coupled role for *CsMYB75* and *CsGSTF1* in anthocyanin hyperaccumulation in purple tea. *The Plant Journal*, *97*, 825–840.

Wu, J., Du, G., Chen, J., & Zhou, J. (2015). Enhancing flavonoid production by systematically tuning the central metabolic pathways based on a CRISPR interference system in Escherichia coli. *Scientific Reports*, *5*, 13477.

Wu, J., Zhou, T., Du, G., Zhou, J., & Chen, J. (2014). Modular optimization of heterologous pathways for de novo synthesis of (2S)-naringenin in *Escherichia coli*. *PLoS One*, *9*, e101492.

Xie, Y., Zhao, Y., Tan, H., Chen, Q., & Huang, J. (2017). An ubiquitin-like protein SDE2 negatively affects sucrose-induced anthocyanin biosynthesis in Arabidopsis. *Science Bulletin (Beijing)*, *62*, 1585–1592.

Yao, G., Ming, M., Allan, A. C., Gu, C., Li, L., Wu, X., Wang, R., Chang, Y., Qi, K., Zhang, S., & Wu, J. (2017). Map-based cloning of the pear gene *MYB 114* identifies an interaction with other transcription factors to coordinately regulate fruit anthocyanin biosynthesis. *The Plant Journal*, *92*, 437–451.

Yin, Y.-C., Hou, J.-M., Tian, S.-K, Yang, L., Zhang, Z.-X, Li, W.-D., & Liu, Y. (2020). Overexpressing chalcone synthase (CHS) gene enhanced flavonoids accumulation in *Glycyrrhiza uralensis* hairy roots. *Botony Letters*, *167*, 219–231.

Yonekura-Sakakibara, K., Higashi, Y., & Nakabayashi, R. (2019). The origin and evolution of plant flavonoid metabolism. *Frontiers in Plant Science*, *10*.

Yonekura-Sakakibara, K., Nakabayashi, R., Sugawara, S., Tohge, T., Ito, T., Koyanagi, M., Kitajima, M., Takayama, H., & Saito, K. (2014). A flavonoid, 3- *O* -glucoside:2″- *O* -glucosyltransferase responsible for terminal modification of pollen-specific flavonols in *Arabidopsis thaliana*. *The Plant Journal*, *79*, 769–782.

Yu, S., Li, M., Gao, S., & Zhou, J. (2022). Engineering *Saccharomyces cerevisiae* for the production of dihydroquercetin from naringenin. *Microbial Cell Factories*, *21*, 213.

Zhang, B., Zhu, Z.-Z., Qu, D., Wang, B.-C., Hao, N.-N., Yang, Y.-Z., Yang, H. J., & Zhao, Z. Y. (2021). MdBBX21, a B-box protein, positively regulates light-induced anthocyanin accumulation in apple peel. *Frontiers in Plant Science*, *12*.

Zhang, J., Xu, H., Wang, N., Jiang, S., Fang, H., Zhang, Z., Yang, G., Wang, Y., Su, M., Xu, L., & Chen, X. (2018). The ethylene response factor MdERF1B regulates anthocyanin and proanthocyanidin biosynthesis in apple. *Plant Molecular Biology*, *98*, 205–218.

Zhang, L., Wang, Y., Sun, M., Wang, J., Kawabata, S., & Li, Y. (2014). BrMYB4, a suppressor of genes for phenylpropanoid and anthocyanin biosynthesis, is down-regulated by UV-B but not by pigment-inducing sunlight in *Turnip cv.* Tsuda. *Plant and Cell Physiology*, *55*, 2092–2101.

Zhang, W., Liu, H., Li, X., Liu, D., Dong, X. T., Li, F. F., Wang, E. X., Li, B. Z., & Yuan, Y. J. (2017). Production of naringenin from D-xylose with co-culture of *E. coli* and *S. cerevisiae*. *Engineering in Life Sciences*, *17*, 1021–1029.

Zhu, J., Zhao, W., Li, R., Guo, D., Li, H., Wang, Y., Mei, W., & Peng, S. (2021). Identification and characterization of chalcone isomerlase genes involved in flavonoid production in *Dracaena cambodiana*. *Frontiers in Plant Science*, *12*.

Zhu, S., Wu, J., Du, G., Zhou, J., & Chen, J. (2014). Efficient synthesis of eriodictyol from L-tyrosine in *Escherichia coli*. *Applied and Environmental Microbiology*, *80*, 3072–3080.

24 Integrated Molecular Strategies for Enhanced Production of Terpenes, Terpenoids, and Essential Oils from Medicinal Plants

Yousra Hammouti, Ahmed Marhri, and Mohamed Addi

24.1 INTRODUCTION

For millennia, humanity has maintained a symbiotic relationship with nature, exploiting its resources to meet a multitude of needs, ranging from the most fundamental to the most sophisticated. At the heart of this interaction lie medicinal plants, true treasures of terrestrial biodiversity. Among the jewels of these plants are essential oils, complex concentrates of bioactive compounds, which confer upon each species its unique and coveted properties. Within this array of molecules, terpenes emerge as major players, capturing the attention of both researchers and industrial stakeholders due to their functional diversity and invaluable worth [1].

Traditionally, the extraction of these precious compounds has relied on ancestral techniques such as steam distillation, cold expression, and solvent extraction. However, these methods, while effective, have their limitations, particularly in terms of environmental impact. They generate harmful solvent waste, consume significant energy, and limit extraction yields. Contemporary challenges of sustainability and efficiency have prompted researchers to explore new molecular avenues to exploit these resources more responsibly [1].

In this quest for innovative solutions, the marriage between biotechnology and omics sciences has opened promising horizons. Genetic engineering offers the possibility of manipulating plants to increase the production of bioactive compounds, while omics technologies, such as genomics, transcriptomics, and metabolomics, enable a comprehensive understanding of the metabolic pathways involved in the synthesis of these precious constituents [2].

Thus, this chapter aims to explore in detail the importance of essential oils and bioactive compounds derived from aromatic medicinal plants, with a particular emphasis on terpenes and terpenoids. We will delve into the fascinating perspectives offered by genetic engineering and omics technologies in the aromatic medicinal plant industry, highlighting opportunities for more sustainable and efficient exploitation of these precious natural resources, while simultaneously promoting both human health and environmental preservation.

24.2 TERPENES, TERPENOIDS, AND ESSENTIAL OILS IN MEDICINAL PLANTS

24.2.1 Roles and Applications in Medicine, Cosmetics, and Industry

Terpenes and terpenoids are organic compounds widely distributed in the plant kingdom, playing a fundamental role in plant chemistry and their interactions with the environment. Terpenes, composed of isoprenic units, are classified according to the number of units, ranging from monoterpenes to polyterpenes. Mainly present in essential oils, they give plants their distinctive aromas and medicinal properties. Terpenoids, derived from terpenes, undergo chemical modifications such as oxidation or cyclization, giving them greater structural and functional complexity and diversity. These compounds are of particular interest in fields such as pharmacology, biotechnology and plant chemistry due to their numerous properties and therapeutic potential [3].

In addition to their widespread presence in the plant kingdom, terpenes and terpenoids are of crucial importance in many aspects of plant and human life. Beyond their contribution to the pleasant aromas of plants, they are at the heart of interactions between plants and their environment. Terpenes act as chemical mediators, participating in inter-plant communication, defense against herbivores and pathogens, and attraction of pollinators. For example, volatile monoterpenes act as chemical signals to attract insect pollinators, while sesquiterpenes act as defense mechanisms by repelling herbivores. In addition, terpenoids play a crucial role in the biosynthesis of plant hormones and pigments, contributing to plant growth, development, and stress response. Outside the plant kingdom, terpenes and terpenoids are widely used in various industries, including pharmaceuticals, cosmetics, and food, for their bioactive properties [3].

In the healthcare field, terpenes and terpenoids are commonly used for their therapeutic benefits, a practice that dates back to traditional medicine. At the same time, terpenoids are emerging as key players in the medical field due to their diverse therapeutic actions. Their beneficial

DOI: 10.1201/9781003475491-24

effects extend to a multitude of diseases and health conditions, making them promising candidates for various medical interventions.

More specifically, certain monoterpenes such as D-limonene and perillyl alcohol have been shown to be effective in the prevention and treatment of various types of cancer, including those of the breast, liver, skin, lung, colon, stomach, prostate, and pancreas [4]. Monoterpenes, such as espintanol, piquerol A, thymol, and menthol derivatives, have also shown some efficacy against protozoan parasites. Plant triterpenes, including betulinic acid, also exhibit anticancer activities, inducing apoptosis of several tumor cell types. In addition, diterpenes such as paclitaxel have been approved by the FDA as effective anticancer agents, targeting breast, ovarian, and Kaposi's sarcoma cancers [5]. Diterpenes, such as dehydroabietinol isolated from *Hyptis suaveolens*, have shown antimalarial and leishmanicidal activities. Artemisinin, a widely used sesquiterpenoid lactone extracted from *Artemisia annua*, is a leading example of this class of compounds. Its efficacy against malaria, attributed to its unique peroxide bridge, is well documented, even against drug-resistant strains. Other sesquiterpenes such as yingzhaosu A and yingzhaosu C have also shown antiparasitic activity, notably against *Plasmodium berghei* [6].

In addition to their anticancer potential, terpenoids are also exploited for their anti-inflammatory properties, with examples such as 1,8-cineole and pseudopterosins offering promising solutions for chronic inflammatory conditions such as bronchitis and asthma [7]. Despite their multiple beneficial applications, the limited availability of certain terpenoids from natural sources remains a challenge to their effective use in the treatment of human diseases. Squalamine, in addition to its role as an anticancer agent, has also been reported to have antibacterial, antifungal, and antiprotozoal properties [8].

In the field of cosmetics, essential oils and their terpene components play an essential role. Their antioxidant and antimicrobial properties make them attractive preservatives, extending the life of products while avoiding the use of potentially harmful synthetic preservatives. Terpenes are proving to be key players in the cosmetics industry, thanks to their intrinsic properties and benefits for the skin. Their use as skin-penetration enhancers facilitates the transdermal delivery of drugs, increasing the efficacy of dermatological products and cosmetics [9].

The exploitation of terpenes, such as menthol and 1,8-cineole, which contain polar groups, opens the way to improved penetration of hydrophobic compounds through the skin barrier. Among them, monoterpenoid cyclic ethers, such as 1,8-cineole, stand out for their enhanced efficacy in this function, significantly outperforming hydrocarbon terpenes. In addition, monoterpenes such as linalool, carvone and thymol have demonstrated their ability to increase the permeability of drugs such as 5-fluorouracil through skin and mucous membranes. Traditionally used as aromas and fragrances in cosmetic formulations for centuries, terpenes offer a pleasing sensory dimension to products [10].

Finally, in industry, terpenes offer sustainable and versatile alternatives to many synthetic chemicals. Their properties as solvents, flavoring agents and fragrances make them valuable in a variety of industrial applications, from cleaning to chemical production. In addition, their use in the food industry as natural preservatives testifies to their potential in preserving food and combating the growth of undesirable microorganisms [11]. In short, the terpenes and terpenoids present in essential oils offer a panoply of promising applications in fields as varied as medicine, cosmetics, and industry, paving the way for new advances in these fields.

24.2.2 Pharmacological Properties and Biodiversity

Terpenes, as major components of essential oils, have been widely studied for their therapeutic potential. Their antioxidant properties are particularly important, as they can neutralize the free radicals responsible for oxidative damage in the body, helping to prevent various degenerative diseases such as cancer, diabetes and neurodegenerative diseases. In addition, terpenes have shown potential anticancer activities by inhibiting the growth of cancer cells and inducing their programmed cell death, or apoptosis. Their antimicrobial, antifungal, and antiviral properties make them valuable agents in the fight against bacterial, fungal, and viral infections, offering potential alternatives to conventional treatments and contributing to the fight against antibiotic resistance. In addition, terpenes have important anti-inflammatory properties, which can help reduce inflammation and relieve symptoms associated with various inflammatory conditions such as arthritis and inflammatory bowel disease. Their ability to modulate the immune system also makes them of interest in the treatment of autoimmune disorders [12].

24.2.3 Environmental Impact of Conventional Extraction Methods

The growing demand for essential oils in recent years has highlighted the environmental and social challenges associated with their extraction. Conventional methods such as steam distillation and chemical solvent extraction, while effective, have significant drawbacks in terms of environmental footprint and sustainability.

Steam distillation, while widely used to produce high-quality essential oils, requires a significant amount of energy to heat water and produce steam. This high energy consumption contributes to greenhouse gas emissions, exacerbating the problem of climate change. In addition, steam distillation often requires a large amount of plant material to produce a small quantity of essential oil, which can lead to an overuse of natural resources.

Chemical solvent extraction, while effective in extracting aromatic compounds from plants, poses significant risks to the environment and human health. The chemical solvents used, such as hexane, can contaminate soil and water, causing long-term environmental damage. In addition, the disposal of chemical solvent residues is a major concern, as inadequate disposal can lead to serious environmental pollution [1].

Faced with these challenges, biotechnology offers promising solutions for more sustainable extraction of bioactive compounds. Molecular biology and omics approaches provide a deeper understanding of the biochemical mechanisms involved in the production of aromatic compounds. This opens the way to more efficient and environmentally friendly extraction methods [13].

By adopting these sustainable extraction practices, the essential oils industry can not only reduce its environmental impact but also improve the quality and purity of the final products. This helps to ensure a sustainable supply of essential oils while preserving natural ecosystems and supporting the local communities that depend on these resources.

24.3 GENETIC ENGINEERING STRATEGIES TO ENHANCE BIOSYNTHESIS

24.3.1 Using Genetics to Improve Terpene Biosynthetic Pathways

Using genetics to improve biosynthetic pathways is a revolutionary approach in plant biotechnology. Indeed, genetic engineering techniques make it possible to isolate specific genes and introduce them into the genome of other species, thus offering the possibility of modifying and optimizing plant biosynthetic pathways to produce compounds of interest. For example, by inserting a gene encoding a key enzyme into a biosynthetic pathway, it is possible to increase the production of a particular compound, such as secondary metabolites with pharmaceutical properties or aromatic compounds in plants.

However, the implementation of these techniques is complex and requires painstaking work in the laboratory to construct gene constructs. These constructs include sequences enabling them to be integrated into the host cell genome, as well as markers to identify them and regulatory sequences to control the expression of the gene of interest. As a rule, a gene construct combines sequences from several different organisms.

Once the gene constructs have been randomly inserted into the plant genome, a complex selection process is used to identify transgenic plants containing the new construct, with no visible change in appearance. These transgenic plants can then be used in breeding programs to improve biosynthetic pathways and produce plants with desired characteristics, such as increased production of specific compounds such as terpenes and terpenoids [14].

In short, the use of genetics to improve biosynthetic pathways offers innovative possibilities for the production of plants with enhanced characteristics, opening up new perspectives in fields such as medicine, agriculture, and industry.

24.3.2 Synthetic Biology Approaches to Optimize Yields

Synthetic biology is a discipline that applies engineering principles to the manipulation of biological systems, with the aim of designing and creating artificial biological systems. It relies on concepts such as standardization, modularity, predictability, reliability, and modeling, borrowed from engineering, to manipulate cellular components and design tailor-made biological systems [15].

In the field of synthetic biology applied to plants, these principles are used to increase agricultural yields by genetically manipulating plants to express desired characteristics, such as disease resistance, tolerance to harsh environmental conditions, or improved yields of bioactive compounds.

DNA assembly techniques play a crucial role in this approach, enabling the construction of new biological systems using well-characterized genetic components. This involves the assembly of DNA fragments to form a target assembly, which is then integrated into the host plant genome using specific vectors. These advances make it possible to rapidly construct multi-gene cassettes and genetically modify plants with greater precision, helping to improve the viability and transformability of recombinant plants [16].

In another context, biologists design biological devices that regulate the flow of information, perform metabolic functions and interact with the environment. These devices are composed of biochemical reactions, such as transcription, translation, and protein phosphorylation. Synthetic biologists can use unmodified devices from nature or build devices from modified biochemical reagents to create these artificial biological systems.

For example, the first synthetic biological devices were created to control transcription by modifying promoter sequences to bind novel transcriptional activators and repressors. More recently, efforts have been made to extend control to non-transcriptional levels, using, for example, non-coding RNAs to regulate translational events [17].

These advances testify to the diversity and richness of synthetic biology applications in the design of artificial biological systems, offering promising prospects for manipulating and improving the yields of bioactive molecules for agricultural, medical, and industrial purposes.

24.3.3 Metabolic Engineering Methods for Precise Production Control

Whole-plant metabolic engineering offers considerable advantages, as the complete metabolic pathway, or a significant part of it, for the production of specialized terpenoids is often pre-existing in the plant. Thus, targeted modification of the limiting, committed, or branching steps of this pathway can lead to a significant increase in the production of the desired metabolites. Several studies have confirmed this approach, demonstrating enhanced production of various specialized terpenoids in plants, including mono-, sesqui-, di-, tri- and tetra-terpenoids [18, 19].

A notable illustration of this approach is the engineering of the menthol biosynthetic pathway in peppermint (*Mentha piperita* L.), which represents one of the first successes in the homologous engineering of specialized terpenoid biosynthesis. By deleting the menthofuran synthase gene, it was possible to obtain an essential oil containing less than half of the undesirable menthofuran (+) in MFS7A transgenic lines. In a subsequent study, cosuppression of the limonene hydroxylase gene resulted in a change in oil composition, with up to 80% limonene in the essential oil, compared with only 2% in wild-type plants, without affecting oil yield [20]. The additional introduction of DXR gene overexpression in MFS7A lines resulted in an increase of over 50% in essential oil yield. Field trials of transgenic peppermint lines showed similar results in terms of oil yield and composition to greenhouse-grown lines, demonstrating the feasibility and stability of these traits [21].

Another striking example of successful homologous engineering of specialized terpenoids is *Artemisia annua*, where combined expression of the HMGR, FPPS, and DBR2 genes resulted in enhanced artemisinin accumulation (>3 times dry weight, or 3.2%). These transgenic plants have already been tested in the field with a view to commercialization, offering a potentially cheaper source of antimalarial drugs.

In the case of *Catharanthus roseus*, overexpression of the bifunctional G(G)PPS and co-expression of G(G)PPS and GES led to a significant improvement in iridoid secologanin, of around 2-fold and 3.5-fold respectively, without affecting plant growth [22].

Similarly, manipulating precursor levels is an effective strategy for increasing terpenoid production. For example, overexpression of the FaNES1 gene in strawberries promotes the conversion of GDP and FDP precursors into linalool and nerolidol, thereby increasing their production. The use of powerful promoters, such as the CaMV 35S promoter, ensures high and widespread expression of the transgene, guaranteeing significant production of the targeted terpenoids [23].

In conclusion, homologous engineering of specialized terpenoid biosynthesis using transgenic approaches has succeeded in obtaining the desired end products in significant quantities in just a few plant species, and for certain metabolic pathways. These successes demonstrate the promising potential of this approach for improving the production of specific metabolites in plants.

24.4 GENOME EDITING WITH CRISPR-CAS9 FOR TARGETED PRODUCTION

24.4.1 Introduction to CRISPR-Cas9 Technology

In recent decades, spectacular advances in DNA manipulation have opened up exciting new avenues for biotechnology research. Emerging techniques such as DNA sequencing and genome editing have revolutionized our ability to understand and modify genetic codes with unprecedented precision. Among the various approaches, zinc finger nucleases (ZFNs) and transcription activator-like nucleases (TALENs) have pioneered targeted DNA manipulation, ushering in an era of promising possibilities. However, the advent of the CRISPR system combined with the Cas endonuclease (CRISPR-Cas) to specifically edit any DNA sequence has truly revolutionized the field of genome editing [24].

This breakthrough has encouraged the adoption of CRISPR-Cas genome editing for translational research, particularly in medicine and agriculture. This technology makes it possible to precisely modify the genome of plants, opening up a wide range of possibilities, particularly in crop improvement to address challenges such as food insecurity and climate change.

With CRISPR-Cas, it is now possible to rapidly modify plant traits, such as yield or drought tolerance, in a much shorter time than traditional methods. This means that new plant varieties can be developed more rapidly, which is crucial to meeting the growing food needs of the world's population and the challenges posed by changing environmental conditions.

The CRISPR-Cas9 system is widely used for genome editing due to its simplicity and efficiency. It consists of a guide RNA molecule and the Cas9 endonuclease, which form a complex that binds to the target sequence. When this complex binds to the desired DNA site, Cas9 induces a double-strand break, enabling precise modifications of the genomic sequence [25].

Compared with other genome-editing tools such as ZFNs and TALENs, protein design for CRISPR-Cas9 is simpler, less laborious and less costly. This technology uses guide RNAs to specifically target DNA sequences, avoiding the need for complex proteins. What's more, CRISPR-Cas9 offers greater flexibility by recognizing a PAM (protospacer adjacent motif) sequence of just three base pairs downstream of the target site, reducing constraints linked to the orientation and length of target sequences [26].

Since its first application in plants, this technology has been widely used in many plant species, including agriculturally important crops such as corn, wheat, and rice, paving the way for significant improvements in their productivity and resilience. In addition, it offers promising potential for specifically targeting key metabolic pathways, such as terpenoid biosynthesis in medicinal plants. This opens up exciting prospects for the discovery of therapeutic compounds and the development of innovative treatments. As a result, CRISPR-Cas9 has become a flagship technology with considerable impact on biotechnology and medicine, thanks to its ease of use and unprecedented targeting efficiency [24].

24.4.2 Applications for Optimizing Specific Terpenoids

To optimize the production of specific terpenoids, the researchers employed the CRISPR/Cas9 tool to perform targeted genetic modifications on plants. This technique involves introducing CRISPR/Cas9 into plant cells using the bacterium *Agrobacterium tumefaciens* as a delivery vector. Two promoters, OsU3 and 35S, were used to respectively regulate the expression of the specific guide RNA (sgRNA) and the Cas9 protein, the essential components of the CRISPR/Cas9 system. By specifically targeting the gene responsible for farnesyl pyrophosphate synthase (Dzfps), a key element in the biosynthesis of squalene, an important secondary metabolite, the researchers created mutants of *D. zingiberensis*. These mutants showed significantly different levels of Dzfps expression and squalene production compared with non-modified plants [27]. This achievement demonstrates the effectiveness of genetic manipulation to control the production of secondary metabolites, paving the way for future research aimed at improving the production of specific metabolites in plants, with potential implications in various fields such as medicine and industry.

In another study, researchers used CRISPR/Cas9 to regulate the production of specific terpenoids in *Salvia miltiorrhiza*, a traditional Chinese medicinal plant. Terpenoids, such as diterpenoids, are important bioactive compounds found in this plant. By targeting a gene called SmCPS1, involved in the biosynthesis of tanshinone, a diterpenoid valuable for its medicinal properties, the researchers created *S. miltiorrhiza* mutants with altered production of specific tanshinones [28]. This approach demonstrates the ability of gene editing to selectively control terpenoid production in plants, opening up new prospects for improving the quality and quantity of bioactive compounds in medicinal plants.

24.4.3 Benefits of Genome Editing in Precision Agriculture

Genome editing represents a major revolution in precision agriculture. The CRISPR-Cas system offers exceptional precision and specificity, enabling researchers to target specific DNA sequences in the plant genome with great precision. For example, scientists have used CRISPR-Cas to develop rice resistant to bacterial blight by precisely targeting and modifying genes involved in the plant's immune response to this disease [29].

The speed and efficiency of CRISPR-Cas are also major assets in agriculture. For example, whereas traditional methods of crossing and selecting plants can take years, CRISPR-Cas enables precise genetic modifications to be made in just a few months. For example, wheat varieties resistant to powdery mildew have been developed rapidly thanks to CRISPR-Cas, by targeting the genes responsible for resistance to this disease.

In the field of food safety, CRISPR-Cas has been used to develop corn resistant to insect pests, thus reducing the need to use chemical pesticides that are potentially harmful to the environment.

The flexibility of the CRISPR-Cas system makes it possible to adapt crops to specific environmental and agricultural challenges. For example, researchers have used CRISPR-Cas to develop corn capable of tolerating extreme drought conditions by modifying the genes involved in the response to water stress. These corn varieties can thrive in water-scarce environments, contributing to food security in drought-prone regions [30].

In addition, by reducing dependence on chemical inputs, CRISPR-Cas promotes more sustainable and environmentally friendly agriculture. For example, researchers have used CRISPR-Cas to develop disease-resistant potato varieties, reducing the need for pesticides. These potato varieties require fewer chemicals to grow, reducing the environmental impact of agricultural production.

Finally, greater access to CRISPR-Cas technology encourages international collaboration and knowledge exchange, accelerating the development and dissemination of new and improved crop varieties. For example, researchers around the world are collaborating to use CRISPR-Cas to develop disease-resistant tomato varieties capable of growing in difficult climatic conditions, thereby contributing to global food security [31]. These advantages make CRISPR-Cas a promising tool for meeting the challenges of modern agriculture and helping to ensure global food security in a sustainable way.

24.5 OMIC ANALYSIS FOR UNDERSTANDING CONTROL NETWORKS

24.5.1 Genomic Approaches to Identify Key Genes

To identify the key genes involved in terpene biosynthesis and regulation, several genomic approaches are being used, offering crucial insights into these complex biological processes. Messenger RNA sequencing (RNA-seq) is an essential first step. By isolating the essential oil-producing glandular trichomes of *Mentha* x *piperita* and sequencing their RNA, we can identify genes specifically expressed in these structures, providing potential candidates for terpene biosynthesis. For example, RNA-seq could reveal the genes encoding key enzymes in the terpene biosynthetic pathway, as well as the transcription factors regulating their expression.

In addition, differential expression analysis makes it possible to compare gene expression levels between different experimental conditions. By studying the transcriptomic responses of *Mentha* x *piperita* plants subjected to biotic or abiotic stresses, for example, we can identify the genes whose expression is altered in response to these stimuli, thus revealing the mechanisms by which terpenes are regulated as an adaptive response [32].

Whole genome sequencing (WGS) offers a global view of the genetic variation associated with phenotypic traits in essential oil-producing plants. By identifying genetic loci associated with important traits such as terpene content or essential oil composition, we can map genome regions containing candidate genes for terpene biosynthesis and regulation [33].

Genome-wide association studies (GWAS) can also be used to identify genetic variants associated with complex traits. GWAS also helps to understand the biological processes underlying molecule biosynthesis. By analyzing associations between genetic variants and molecular phenotypes, GWAS provides valuable information on the metabolic pathways and genetic regulations involved in the production of molecules of interest [33].

By combining these genomic approaches, it is possible to draw up an exhaustive map of the key genes involved in terpene biosynthesis and regulation, paving the way for genetic engineering applications to improve the production, quality and resilience of essential oil-producing plants in an agricultural and industrial context.

24.5.2 Transcriptomics to Understand the Regulation of Biosynthesis

Transcriptomics, firmly rooted in the central dogma of molecular biology, is proving to be a pivotal discipline in the quest to understand the mechanisms governing plant biosynthesis. Scrutinizing gene expression through RNA, offers us a panoramic view of genomic structure and function, shedding light on the fundamental molecular pathways governing various biological processes, of which plant oil biosynthesis presents itself as a privileged field of study.

At the crossroads of genomics, proteomics and metabolomics, transcriptomics is positioned as an essential key to interactomics, weaving crucial links between genome, proteome and cellular phenotype. At the heart of this discipline, the concept of expression profiling, initiated by Charles Auffray, deploys an exhaustive vision of transcripts, exploring the multiple facets of RNAs and their roles in gene regulation [34].

Thanks to the evolution of next-generation sequencing technologies, transcriptomics is deploying its high-throughput potential, revealing with increased resolution the intricacies of the gene regulatory network in plants. Transcriptomic studies on organisms such as tobacco, peppermint, and lavender have highlighted specific regulation of the genes involved in terpene biosynthesis, offering valuable insights for the flavor, fragrance, and natural medicine industries [35].

By scrutinizing the different types of RNA, from mRNAs to microRNAs and long non-coding RNAs, transcriptomics offers a holistic perspective on plant gene dynamics. Methodological advances, such as RNA-seq sequencing and microarrays, open up new avenues for characterizing cellular states, exploring the molecular mechanisms underlying oil biosynthesis and identifying biomarkers for improving plant oil production [36].

In short, transcriptomics is proving to be an invaluable tool for deciphering the regulation of oil biosynthesis in plants, offering crucial insights into the gene dynamics involved in this vital process. With its comprehensive approach and constant technological advances, it opens up promising prospects for optimizing agricultural production and plant biotechnology.

24.5.3 Metabolomics to Optimize Essential Oil Composition

Metabolomics is a scientific approach that studies all the metabolites present in an organism. Unlike genomics, which studies DNA, and transcriptomics, which studies RNA, metabolomics offers a direct perspective on the biochemical activity and current physiological state of cells or tissues.

In the context of optimizing the composition of plant oils, metabolomics offers an in-depth understanding of the metabolites present in plants, including the primary and secondary metabolites involved in oil biosynthesis. By identifying and quantifying these metabolites, metabolomics makes it possible to map the metabolic pathways involved in plant oil production.

This approach provides valuable information for understanding how metabolic processes are modulated in response to environmental, genetic, or cultural factors. Using metabolomics, it is possible to identify biomarkers associated with specific characteristics of plant oils, such as their fatty acid profile or heat stability [37].

In short, metabolomics offers a powerful method for studying and optimizing the composition of plant oils, providing detailed information on the biochemical mechanisms underlying their production.

24.6 CONCLUSIONS AND PERSPECTIVES

In conclusion, advances in molecular biology, notably genetic engineering, CRISPR-Cas9 genome editing and omics technologies, mark a crucial step in optimizing the production of bioactive molecules. These advances offer precise and efficient solutions to meet the growing worldwide demand for these compounds. In addition to their effectiveness, these approaches help promote more sustainable agricultural practices by minimizing the use of natural resources, while contributing to the conservation of precious medicinal plant resources.

The optimization of bioactive molecules by molecular biotechnologies reduces dependence on plant harvesting, thus preserving biodiversity and ensuring more responsible use of ecosystems. This approach not only benefits human well-being by providing more effective and better-adapted therapies but also contributes to environmental health, in particular by reducing pressure on natural habitats.

Looking to the future, these technologies promise to open up exciting new perspectives in the discovery and production of natural products. Thanks to molecular biology and biotechnology, it is possible to develop innovative and sustainable production methods, guaranteeing a continuous supply of bioactive molecules while preserving ecosystems. Ultimately, molecular approaches represent a powerful engine for progress, both for industrial demand and for the preservation of our environment, shaping a healthier, more sustainable future for us all.

REFERENCES

[1] Aziz, Z. A., Ahmad, A., Setapar, S. H. M., Karakucuk, A., Azim, M. M., Lokhat, D., Rafatullah, M., Ganash, M., Kamal, M. A., & Ashraf, G. M. (2018). Essential oils: Extraction techniques, pharmaceutical and therapeutic potential–a review. *Current Drug Metabolism*, *19*(13), 1100–1110.

[2] Musiol-Kroll, E. M., Tocchetti, A., Sosio, M., & Stegmann, E. (2019). Challenges and advances in genetic manipulation of filamentous actinomycetes–the remarkable producers of specialized metabolites. *Natural Product Reports*, *36*(9), 1351–1369.

[3] Gershenzon, J., & Dudareva, N. (2007). La fonction des produits naturels terpéniques dans le monde naturel. *Biologie chimique de la nature*, *3*(7), 408–414.

[4] Okabe, H., Takayama, K., Ogura, A., & Nagai, T. (1989). Effect of limonene and related compounds on the percutaneous absorption of indomethacin. *Drug Design and Delivery*, *4*(4), 313–321.

[5] Cipak, L., Grausova, L., Miadokova, E., Novotny, L., & Rauko, P. (2006). Dual activity of triterpenoids: Apoptotic versus antidifferentiation effects. *Archives of Toxicology*, *80*, 429–435.

[6] Kayser, O., Kiderlen, A. F., & Croft, S. L. (2003). Produits naturels comme médicaments antiparasitaires. *Recherches en parasitologie*, *90*, S55–S62.

[7] Look, S. A., Fenical, W., Jacobs, R. S., & Clardy, J. (1986). The pseudopterosins: Anti-inflammatory and analgesic natural products from the sea whip *Pseudopterogorgia elisabethae*. *Proceedings of the National Academy of Sciences*, *83*(17), 6238–6240.

[8] Moore, K. S., Wehrli, S., Roder, H., Rogers, M., Forrest, Jr., J. N., McCrimmon, D., & Zasloff, M. (1993). Squalamine: An aminosterol antibiotic from the shark. *Proceedings of the National Academy of Sciences*, *90*(4), 1354–1358.

[9] Williams, A. C., & Barry, B. W. (2012). Penetration enhancers. *Advanced Drug Delivery Reviews*, *64*, 128–137.

[10] Narishetty, S. T., & Panchagnula, R. (2005). Effet du L-menthol et du 1, 8-cinéole sur le comportement de phase et l'organisation moléculaire des lipides SC et la perméation cutanée de la zidovudine. *Journal de libération contrôlée*, *102*(1), 59–70.

[11] Brown, L. M., Springer, J., & Bower, M. (1992). Chemical substitution for 1, 1, 1-trichloroethane and methanol in an industrial cleaning operation. *Journal of Hazardous Materials*, *29*(2), 179–188.

[12] Ajikumar, P. K., Tyo, K., Carlsen, S., Mucha, O., Phon, T. H., & Stephanopoulos, G. (2008). Terpenoids: Opportunities for biosynthesis of natural product drugs using engineered microorganisms. *Molecular Pharmaceutics*, *5*(2), 167–190.

[13] Qi, S., Wang, J., Zhang, Y., Naz, M., Afzal, M. R., Du, S., & Dai, Z. (2023). Approches omiques en biologie des invasions: compréhension des mécanismes et des impacts sur la santé écologique. *Plantes*, *12*(9), 1860. https://doi.org/10.3390/plants12091860

[14] Nygren, P.-Å., Stefan, S., & Uhlén, M. (1994). Engineering proteins to facilitate bioprocessing. *Trends in Biotechnology*, *12*(5), 184–188. https://doi.org/10.1016/0167-7799(94)90080-9

[15] Andrianantoandro, E., Basu, S., Karig, D. K., & Weiss, R. (2006). Synthetic biology: New engineering rules for an emerging discipline. *Molecular Systems Biology*, *2*(1), 1–14.

[16] Guesdon, G. (2022). *Développement de méthodes de clonage de génomes entiers chez la levure pour la construction de souches châssis semi-synthétiques de Bacillus subtilis* (Doctoral dissertation, Université de Bordeaux).

[17] Celińska, E., Ledesma-Amaro, R., Larroude, M., Rossignol, T., Pauthenier, C., & Nicaud, J. M. (2017). Système Golden Gate Assembly dédié à la manipulation de voies complexes chez Yarrowia lipolytica. *Biotechnologie microbienne*, *10*(2), 450–455.

[18] Chen, F., Tholl, D., Bohlmann, J., & Pichersky, E. (2011). The family of terpene synthases in plants: A mid-size family of genes for specialized metabolism that is highly diversified throughout the kingdom. *The Plant Journal*, *66*(1), 212–229.

[19] Dudareva, N., Negre, F., Nagegowda, D. A., & Orlova, I. (2006). Plant volatiles: Recent advances and future perspectives. *Critical Reviews in Plant Sciences*, *25*(5), 417–440.

[20] Mahmoud, S. S., & Croteau, R. B. (2001). Metabolic engineering of essential oil yield and composition in mint by altering expression of deoxyxylulose phosphate reductoisomerase and menthofuran synthase. *Proceedings of the National Academy of Sciences*, *98*(15), 8915–8920.

[21] Lange, B. M., Mahmoud, S. S., Wildung, M. R., Turner, G. W., Davis, E. M., Lange, I., Baker, R. C., Boydston, R. A., & Croteau, R. B. (2011). Improving peppermint essential oil yield and composition by metabolic engineering. *Proceedings of the National Academy of Sciences*, *108*(41), 16944–16949.

[22] Kumar, S. R., Shilpashree, H. B., & Nagegowda, D. A. (2018). Terpene moiety enhancement by overexpression of geranyl (geranyl) diphosphate synthase and geraniol synthase elevates monomeric and dimeric monoterpene indole alkaloids in transgenic *Catharanthus roseus*. *Frontiers in Plant Science*, *9*, 383826.

[23] Aharoni, A., Giri, A. P., Deuerlein, S., Griepink, F., de Kogel, W. J., Verstappen, F. W. A., Verhoeven, H. A., Jongsma, M. A., Schwab, W., & Bouwmeester, H. J. (2003). Terpenoid metabolism in wild-type and transgenic Arabidopsis plants. *The Plant Cell*, *15*(12), 2866–2884.

[24] Dey, A. (2021). CRISPR/Cas genome editing to optimize pharmacologically active plant natural products. *Pharmacological Research*, *164*, 105359.

[25] Shalem, O., Sanjana, N. E., & Zhang, F. (2015). High-throughput functional genomics using CRISPR–Cas9. *Nature Reviews Genetics*, *16*(5), 299–311.

[26] Cho, S. W., Kim, S., Kim, J. M., & Kim, J. S. (2013). Targeted genome engineering in human cells with the Cas9 RNA-guided endonuclease. *Nature Biotechnology*, *31*(3), 230–232.

[27] Feng, S., Song, W., Fu, R., Zhang, H., Xu, A., & Li, J. (2018). Application du système CRISPR/Cas9 chez *Dioscorea zingiberensis*. *Culture de cellules, de tissus et d'organes végétaux (PCTOC)*, *135*, 133–141.

[28] Li, G., Zhang, X., Zhong, C., Mo, J., Quan, R., Yang, J., Liu, D., Li, Z., Yang, H., & Wu, Z. (2017). Small molecules enhance CRISPR/Cas9-mediated homology-directed genome editing in primary cells. *Scientific Reports*, *7*, 1–11.

[29] Ghoshal, B. (2024). *Une vue d'ensemble du système CRISPR-Cas dans l'agriculture* (pp. 1–8). Le noyau.

[30] Zhu, H., Li, C., & Gao, C. (2020). Applications of CRISPR–Cas in agriculture and plant biotechnology. *Nature Reviews Molecular Cell Biology*, *21*(11), 661–677.

[31] Ortigosa, A., Gimenez-Ibanez, S., Leonhardt, N., & Solano, R. (2019). Design of a bacterial speck resistant tomato by

CRISPR/Cas9-mediated editing of Sl JAZ 2. *Plant Biotechnology Journal*, *17*(3), 665–673.

[32] Mahmoud, S. S., Williams, M., & Croteau, R. (2004). La cosuppression de la limonène-3-hydroxylase dans la menthe poivrée favorise l'accumulation de limonène dans l'huile essentielle. *Phytochimie*, *65*(5), 547–554.

[33] Hébert, F., Emily, M., & Causeur, D. (2020). Combinaison de tests dépendants en études d'association pangénomiques. *Genetic Epidemiology*, *5*(2), 1–6.

[34] Zheng, P. Z., Wang, K. K., Zhang, Q. Y., Huang, Q. H., Du, Y. Z., Zhang, Q. H., Xiao, D. K., Shen, S. H., Imbeaud, S., Éveno, É., Zhao, C. J., Chen, Y. L., Yong, F. H., Waxman, S., Auffray, C., Jin, G., Chen, S. J., Chen, Z., & Zhang, J. (2005). Analyse des systèmes du transcriptome et du protéome dans la différenciation cellulaire/apoptose induite par l'acide rétinoïque/trioxyde d'arsenic de la leucémie promyélocytaire. *Actes de l'Académie nationale des sciences*, *102*(21), 7653–7658.

[35] Chang, W., Zhao, H., Yu, S., Yu, J., Cai, K., Sun, W., Liu, X., Li, X., Yu, M., Ali, S., Zhang, K., Qu, C., Lei, B. et Lu, K. (2020). Le transcriptome comparatif et le profilage métabolomique révèlent les mécanismes complexes qui sous-tendent la dynamique de développement des feuilles de tabac. *Génomique*, *112* (6), 4009–4022.

[36] Patro, R., Mount, S. M., & Kingsford, C. (2014). Sailfish enables alignment-free isoform quantification from RNA-seq reads using lightweight algorithms. *Nature Biotechnology*, *32*(5), 462–464.

[37] Tohge, T., & Fernie, A. R. (2010). Combining genetic diversity, informatics and metabolomics to facilitate annotation of plant gene function. *Nature Protocols*, *5*(6), 1210–1227.

25 Advanced Systems and Bioreactors for Large-Scale Secondary Metabolite Production in Medicinal Plants

Samrat Das, Debdut Manna, Tanmoy Mondal, and Partha Mondal

25.1 INTRODUCTION

Secondary metabolites are petite organic compounds that are produced from primary metabolites during plant metabolism, typically possessing molecular masses under 3000 daltons (Twaij et al., 2022). Metabolites in plants exhibit diverse chemical characteristics and compositions among different species. Numerous bioactive secondary metabolites found in plant organs like leaves, stems, roots, and flowers are derived from plants. While they often offer benefits to plants, these secondary metabolites are not crucial for processes related to growth, development, and reproduction. Their structural diversity and potential as drug candidates or antioxidants have made them interesting. The chemical diversity of plant metabolites is limited, making them complex compounds that cannot be easily synthesized by the industry (Sanchez & Demain, 2000; Bourgaud et al., 2001).

25.2 ROLE OF SECONDARY METABOLITES

In the early days, plant extracts were perceived as waste or detoxification byproducts (Jarvis & Williams, 1989). Until then, research endeavour focused on secondary metabolites had not commenced. It wasn't until technology-enabled exploration into the potential biological functions of secondary metabolites in nature that research efforts began in earnest. While the functions of the majority of secondary metabolites remain undisclosed, they are primarily utilized in various industries such as dye production, food processing, cosmetics, agriculture, and pharmaceuticals, serving as essential raw materials. Due to high demand, there is a need to acquire secondary metabolites in large quantities. This can be produced through large-scale production using both traditional and biotechnological methods.

They typically result from alterations in biosynthesis, such as glycosylation, hydroxylation, and methylation. These molecules exhibit greater structural complexity and in side chains than primary metabolites (Figueredo et al., 2008; Ehrlich & Holdren, 1971). Plant metabolites are categorized into three primary classes according to their biosynthetic pathway: (i) phenolic compounds, consisting of basic sugars and benzene rings; (ii) steroids and terpenes, primarily composed of carbon and hydrogen; and (iii) nitrogen-containing compounds (Saxena et al., 2013).

25.3 WAY TO PRODUCE SECONDARY METABOLITES

In the commercial utilization of plant cell cultures to produce valuable secondary metabolites, ensuring high yields and consistent output is crucial. Since the production of these metabolites in plants varies depending on genetics, the initial step involves selecting parent plants with elevated levels of the desired secondary product for initiating cell or organ cultures. Subsequently, high-producing cells or organ lines are chosen to facilitate optimal production. Selection is conducted through an analysis of the growth of a cell or organ, and quantification of the chromatographic and spectroscopic quantification of products (Murthy et al., 2014). Despite selecting highly productive lines, achieving satisfactory production yields isn't always guaranteed, and over prolonged cultivation periods, they may experience a decline in efficiency. Consequently, various alternative strategies, such as traditional and metabolic engineering approaches, can be employed to enhance the production of secondary metabolites and ensure efficient yields (Isah et al., 2018; Khani et al., 2012).

25.3.1 Traditional Strategies

Secondary metabolites are formed through primary metabolism and are dependent on factors like temperature, humidity, light intensity growth and physiology. Metabolite productivity in *in vitro* cultures is also influenced by media pH, composition, inoculum density, and temperature, agitation, light density, and aeration (Ochoa-Villarreal et al., 2016; Isah et al., 2018). Optimizing growth and metabolite productivity requires optimizing factors like culture media, which include macro- and micronutrients, amino acids, vitamins, carbohydrates, and plant growth regulators like cytokinins, auxins, gibberellins, jasmonates, and salicylates (Cardoso et al., 2019). Biotic or abiotic sources of elicitors, with abiotic elicitors being nonbiological compounds like heavy metal salts (Gorelick & Bernstein, 2014; Ramirez-Estrada et al., 2016). Physical stresses like cold shock, UV,

DOI: 10.1201/9781003475491-25

osmotic, and moisture stress induce enzymatic activity and secondary metabolism (Akul & Ravishankar, 2011).

Microbes trigger secondary metabolic pathways, influencing secondary metabolites. Bacterial, fungal, and yeast extracts improve yield and quality, with fungal elicitors being the most effective method for increasing PTC productivity (Takeuchi et al., 2013). Plant disease resistance is induced by the expression of definite genes, triggering secondary metabolic pathways and enhancing secondary metabolites (Zhang et al., 2013). In order to induce secondary metabolites in medicinal plants, fungi such as *Fusarium*, *Pythium*, yeast, *Aspergillus*, *Penicillium*, and *Trichoderma*, are commonly extracted as fungal elicitors (Namdeo et al., 2002; Liang et al., 2018). Cell-free extracts from *Aspergillus niger* and *Saccharomyces cerevisiae* were used as fungal induction to enhance the production of gymnemic acid from cell suspension cultures of *Gymnema sylvestre* (Chodisetti et al., 2013). Similarly, *Aspergillus flavus* was found to boost the synthesis of terpenoid, indole alkaloids in *Catharanthus* spp. (Liang et al., 2018). Bacterial elicitors, from both gram-positive and gram-negative strains, were utilized to stimulate the biosynthesis of tropane alkaloids in adventitious hairy root cultures of *Scopolia parviflora* (Jung et al., 2003). Kang et al. (2009) observed application of *Staphylococcus aureus* extract increased the synthesis of bilobalide and ginkgolide in *Gingko biloba* cell suspension cultures. *Rhizobium leguminosarum* and *Agrobacterium tumefaciens* bacterial elicitors were found to improve the synthesis of glycyrrhizic acid in root cultures of *Taverniera cuneifolia*.

25.3.2 Metabolic Engineering

Metabolic engineering involves modifying the metabolic pathways within an organism by utilizing advanced biological tools such as genomics, proteomics, metabolomics, and others, with the aim of generating commercially valuable metabolites (DellaPenna, 2001). Recombinant DNA technology enables the manipulation of endogenous biochemical pathways in plants, allowing overexpression or downregulation of common precursors, enzymes, and regulatory proteins. Plants have numerous metabolic pathways for complex metabolite biosynthesis, with the shikimate pathway being the main source (Herrmann & Weaver, 1999). With regard to carbon flow, it is thought that plants require this metabolic pathway in order to fix more than 30% of fixed carbon (Maeda & Dudareva, 2012).

25.3.2.1 Upregulating Pathways

Transcription factors are proteins that interact with specific regions of DNA, such as promoters or enhancers, to regulate the transcription or expression of particular genes. They interpret genetic information encoded in the genome and regulate the rate at which transcription occurs (Latchman, 2010; Broun, 2004). Transcription factors typically consist of two distinct regions: the DNA binding domain and the activator domain. Specific amino acids of the DNA binding domain recognize and bind to particular DNA sequences within regulatory elements. Meanwhile, the activator domain contains binding sites known as activation functions and a transactivation domain, which interacts with other proteins (transcription coregulators) to regulate gene expression.

25.3.2.2 Redirecting Common Precursors

Enzymes in a biosynthetic pathway compete for precursors. Boosting metabolite production involves modifying these precursors to favour the synthesis of desired metabolites, achieved by either impeding competing pathways or enhancing gene expression in the precursor pathway (Verpoorte & Alfermann, 2000). The expression of the gene that decodes taxadiene synthase, a precursor in the carotenoid pathway, can be boosted to increase the production of tomato mutant taxadiene. Jadaun et al. (2017) demonstrated that increasing the expression of 1-deoxy-D-xylulose-5-phosphate synthase (GrDXS) resulted in higher levels of terpenoidal secondary metabolites in *Pelargonium* spp. (essential oil) and *Withania somnifera* (withanolides). The overexpression of genes in precursor pathways caused the monoterpene content of essential oils to rise in peppermint and lavender (Mahmoud & Croteau, 2001). Direct overexpression of genes involved in the alkaloid pathway proved efficient in enhancing alkaloid accumulation in *C. roseus* (Canel et al., 1998).

25.3.2.3 Targeting Metabolites to Specific Cell Compartments

Secondary metabolic pathways' enzymes and other metabolites are restricted to the cytoplasm and other cell organelles such as peroxisomes, vesicles, and vacuoles. Enzymes involved in secondary metabolic pathways are organized into multifunctional complexes that facilitate substrate channeling, where the product of one enzyme serves as the substrate for the next enzyme in the pathway (Winkel, 2004). It has been found that particular amino acid sequences target and keep proteins in specific organelles (Lessard et al., 2002). In *C. roseus*, enzymes involved in the synthesis of monoterpenoid indole alkaloids can be found in roots, leaves, flowers, buds, and so on, while the metabolite accumulates in specific cells known as idioblasts and laticifers. The enzymes and metabolites created must be delivered not to a cell but between cells belonging to different types of tissues (St-Pierre et al., 1999). Cells compartmentalize the processes that produce toxic compounds in order to shield themselves from self-toxicity. Nicotine in tobacco and the alkaloid berberine produced by *Coptis japonica* are bound in vacuoles (Saunders, 1979). In *Arabidopsis thaliana*, dihydroflavonol 4-reductase, chalcone isomerase and chalcone synthase work together to synthesize flavonoids (Burbulis & Winkel-Shirley, 1999). According to Zhang et al., anthocyanins in *Eustoma grandiflorum* are transported from petals to the central vacuole by vesicles or prevacuolar compartments (Zhang et al., 2006).

25.3.2.4 Downregulating Pathways (Silencing)

The synthesis of certain metabolites can be decreased by reducing enzyme levels, boosting catabolism, and carbon

(Verpoorte & Alfermann, 2000). Unwanted metabolites can be inhibited by masking genes responsible for upregulation of the metabolic process or catabolism enhancer (Gómez-Galera et al., 2007). To eliminate unwanted metabolites, advanced recombinant DNA technologies such as antisense RNA, RNA interference (RNAi), and co-suppression can be employed. RNA interference has been applied effectively to modulate plant secondary metabolites such as phenylpropanoids, alkaloids, and terpenoids. Specifically targeting the phenylpropanoid pathway through RNAi has resulted in various outcomes, including vivid coloration in flowers and reduced pigmentation in seed coats. Furthermore, RNA interference has been instrumental in increasing the production of codeine and morphine in opium poppy, reducing cyanogenic glucoside levels in cassava, augmenting flavonoids and carotenoids in tomatoes, and regulating fragrance in petunia flowers, among other applications (Wagner & Kroumova, 2008).

25.4 FIELD CULTURE OF SECONDARY METABOLITES PRODUCTION

Farming aromatic and medicinal species may be one way to improve the quality and quantity of secondary metabolites. One intensive technique for producing secondary metabolites is field culture, which involves both production and processing. Appropriate field management is crucial because it is an agronomic process. Timing of sowing, plant density, nutrient management, crop harvesting, and intercultural activities are some of the essential components of field management. Other essential components include site selection, the use of high-quality, genetically stable planting materials, the application of required harvesting procedures, etc. (Nguyen et al., 2013; Cardoso et al., 2019).

25.4.1 Factors Affecting in Field Culture of Secondary Metabolites Production

Under natural conditions, various factors, including internal ones like genotype and external ones such as photoperiod, light intensity, temperature, water availability, cropping system, soil type, and atmospheric composition, can independently or together influence both the makeup and the levels of secondary metabolites in plants (Cardoso et al., 2019).

25.4.1.1 Geography

Geographic site selection is essential to ensuring that a particular botanical or collection of botanicals meets the ecological equality criteria. For example, latitude has a major impact on biomass composition, as evidenced by *Atropa belladonna*. Agroclimatic conditions that are ideal for growth and development vary depending on important parameters like slope face and elevation. Profitable species that thrive in the low-fertility soils of warmer climates, like *Psyllium*, *Cassia*, *Catharanthus*, *Withania*, *Rauvolfia*, and *Cymbopogon*, can be found on the plantation. Profitable cultivation is possible for *Pyrethrum*, *Solanum* spp., *Jasminium*, and *Ocimum* spp. on soils with moderate fertility. Plantations with *Papaver*, *Dioscorea* spp., *Mentha* spp., and *Cymbopogon* spp. need abundant irrigation and high-fertility soil to thrive. Except for *Withania*, *Cassia*, *Vetiveria*, and *Eucalyptus citriodora*, rain-fed cultivation is permitted in the case of other medicinal plants (Chatterjee, 2002).

25.4.1.2 Soil Requirement

Considering the needs of various aromatic and medicinal plants with respect to soil makes it clear that these needs vary greatly. Loam and clay loam are better-growing conditions for *Plantago*, *Cassia*, and *Cymbopogon*, which yield larger harvests when grown in light soil. *Vetiveria* is distinctive in that it does not alter its overall oil yield or composition when stressed by soil alkalinity, sporadic flooding, or field waterlogging. Soils with a pH range of 5–6 are preferred for *Cinchona*, *Cephalis*, and *Coptis*. *Aloe*, *Pandanus*, *Urginea*, and *Commiphora*, on the other hand, could thrive well even at a higher pH (Chatterjee, 2002).

25.4.1.3 Photoperiodism

Research has shown that the duration of daylight impacts various aromatic and medicinal plants in diverse manners. Commercial production of *Mentha*, *Glycyrrhiza*, *Coptis*, and *Humulus* requires at least 14 hours of daylight for maximum crop yield. On the other hand, *Pogostemon*, *Pelargonium*, and *Citronella* grow better with shorter day lengths. It is possible to boost essential oil output by about 25% by providing these crops with the light required. *Senna* is a day-neutral plant; thus, the length of the most intense daylight does not affect how well it grows (Chatterjee, 2002).

25.4.1.4 Temperature

It is also clearly evident how air temperature affects how well aromatic and medicinal plants grow. Because of their lower minimum temperatures, which increase the dry matter yield and total pyrethrin contents, southern hills are good places for *Pyrethrum* growth. Medicinal plants synthesize alkaloids and terpenes as secondary metabolites, which are typically produced more efficiently when the temperature is elevated to its optimal range, before reaching levels that cause physiological damage. Lower temperatures encourage the buildup of morphine in opium (Chatterjee, 2002).

25.4.1.5 Nutrient Management

When nutrient management is considered, it is generally indicated that medium levels of N, P, and K are almost optimal for the formation of secondary metabolites. Both biomass and the accumulation of active ingredients may alter as nutrient levels rise. The alkaloid yield of *Rauvolfia* and *Atropa*, diosgenin of *Dioscorea* spp. and *Costus*, sennosides of *Cassia*, and essential oils of *Cymbopogon* grasses, *Vetiveria*, and *Mentha* were all increased by nitrogen fertilizers (Chatterjee, 2002).

25.4.1.6 Water Requirement

Water requirements for plants vary. In the tropical region of *Cinchona*, for example, the high water content during the rainy season prevents the accumulation of secondary metabolites, sometimes leading to the complete lack of alkaloids. On the other hand, where there is more water available, secondary metabolites tend to accumulate. Under specific circumstances, such as moisture stress, the roots, root-shoot ratio, total alkaloid concentrations, and root output of *Rauvolfia* and *Catharanthus* are observed to be amplified. Furthermore, several crops (*Atropa*, *Hyoscyamus*, *Datura*, *Rauvolfia*, and *Catharanthus*) have a profitable rise in alkaloids when subjected to moisture stress (Chatterjee, 2002).

25.4.1.7 Harvesting Schedule

The ideal harvesting time for targeted secondary metabolites varies greatly and is crucial for the recovery of secondary metabolite production. *Rauvolfia* roots are usually taken in the late flowering stage, after about 2.5 years of cultivation. *Catharanthus*, *Datura*, *Solanum*, and *Dioscorea* usually yield active components at their highest when 60–70% of the reproductive phase is reached. *Costus speciosus* exhibits its maximal diosgenin level at the start of the reproductive phase. Because most continuously grown medicinal plant species reach their peak active ingredient attainment after three to four years, this determines the ideal harvest period. Once again, the seventh or eighth year is the ideal window of opportunity for secondary metabolites in Cinchona to recover well (Chatterjee, 2002).

25.4.2 Essentiality of Field Culture of Secondary Metabolites Production

Chemical synthesis can be applied to help with the intense process of secondary metabolite production. However, not all secondary metabolite-producing plants can easily extract and purify biomolecules by chemical synthesis. Furthermore, the synthetic synthesis of these alkaloids is more costly than their extraction from plant sources because of the relative complexity of their chemical structure and their long biosynthetic process. Since the natural chemical was found in plants, biomass could be used to separate enormous amounts of it. The manufacturing of plant secondary metabolites has long been accomplished through the field cultivation of medicinal plants, but chemical synthesis is not cost-effective due to the complicated nature of many secondary metabolites. On the other hand, it might be challenging to cultivate plants from certain biotopes outside of their native environments (Nguyen et al., 2013).

25.5 *IN VITRO* CULTURE FOR SECONDARY METABOLITES PRODUCTION

The use of biotechnological techniques, particularly plant tissue culture, is essential in the quest for substitutes for the extraction of desired medicinal components from plants. *In vitro* plant materials have been refocused as possible factories for secondary metabolites due to the potent and growing need for natural and renewable products in recent demand. This has also opened the door for new research on secondary product expression *in vitro*. When the metabolite of interest is only produced in specialised plant tissues, the manufacture of a wide range of valuable secondary phytochemicals in unorganised callus or suspension cultures is required. Herbal plants like *Hypericum perforatum*, which accumulates the anti-depressant hypericins and hyperforins in foliar glands, have not shown the ability to accumulate phytochemicals in undifferentiated cells. Another example is the biosynthesis of lysine to anabasine in the roots of tobacco plants (*Nicotiana tabacum*), which is then converted to nicotine in the leaves. Because tobacco callus and shoot cultures do not contain the organ-specific chemical anabasine, they are only able to produce negligible levels of nicotine (Karuppusamy, 2009). Cell culture and organ culture are two potential *in vitro* tools for plant secondary metabolite production (Bourgaud, 2001).

25.6 BIOREACTORS

At the dawn of the 1960s, plant cell culture methods were presented as a potential means of both producing and exploring plant secondary metabolites. Numerous studies have been conducted on several approaches that employ *in vitro* systems to enhance the yield of secondary plant chemicals. While hairy roots and other organ cultures have garnered a lot of attention, undifferentiated cell cultures have received the majority of research attention. Particular procedures have been developed to satisfy the demands of organ and plant cell cultures in bioreactors. Plant biotechnologies have yielded relatively few commercial achievements for the generation of valuable secondary chemicals, despite extensive efforts over the past 30 years. In contrast to other biotechnological domains like microbes or mammalian cell cultures, this could be attributed to inadequately designed reactor facilities or a lack of fundamental understanding of biosynthetic pathways. Advancements in recombinant DNA technology have enabled the direct manipulation of gene expression associated with biosynthesis, providing a more recent avenue for potential alteration. The routes leading to secondary plant chemicals can now be altered. The last stage before secondary metabolites from plant cell cultures might potentially be produced commercially is bioreactor research. When cells are grown in big tanks, bioreactor growth is significantly altered, and the generation of cell biomass continues to be a crucial factor in bioreactor productivity. Undifferentiated cell cultures produce a number of secondary metabolites, including taxanes, tropane alkaloids, and indole alkaloids. Generally speaking, organ cultures grown in bioreactors are less sensitive to shear stress, while some hairy root lines such as those of *Catharanthus roseus* have been reported to be susceptible as well. The inhomogeneous nature of biomass non-comparison to thin cell suspensions is a major issue with organ cultures in bioreactors. For hairy root cultures, certain bioreactors have been developed to get around the obstacles that stand in the way of producing secondary metabolites and biomass (Bourgaud, 2001). The bioreactor is a system or device that permits an environment that is biologically active. Pharmaceuticals, agriculture, food processing, environmental cleanup, and the manufacturing of biofuels are just a few of the industries that heavily utilize these systems (Pino et al., 2018). Researchers have also used them as instruments to extract various secondary metabolites. In order to harvest cell biomass for the production of secondary metabolites, bioreactors were initially designed to cultivate microorganisms. Later, they were used to harvest plant cell suspensions and organs such as roots, shoots, and embryos (Baque et al., 2012). In addition to providing the optimum conditions for target materials growth and product extraction, the bioreactor's design must ensure system homogeneity and provide more control over contaminations and process disruptions. Bioreactors can be categorized according to a number of factors, such as the reactor's physical layout, the kind of biological material or organism it uses, and how it operates (Ruíz-Leza et al., 2007).

25.7 MODES OF BIOREACTOR

The paramount importance lies in the efficient production of the desired product or outcome rather than the complexity or sophistication of the bioreactor's instrumentation. Certain bioreactors may benefit from the application of multiple control schemes, or a single control scheme or hybrid control scheme may work well in a specific application like the production of secondary metabolites. There are two primary types of fermentation: continuous and fed-batch. In addition, other types of bioreactors have been evolving continuously.

25.7.1 Batch Bioreactor

The most basic kind of bioreactor is called batch bioreactor because it only has one ewer in which all of the ingredients like nutrients, cells, etc. are added at the outset of the process. The batch reactor is unparalleled in that substrate is not supplied during the batch operation once the reaction has commenced. During the cultivation process, there is no media or gas exchange in batch bioreactors because they are closed systems. Cells grow, waste products build up, and nutrients are consumed until the target endpoint is reached. This implies that in order to keep up nearly ideal process conditions during the batch reaction process, more tangled control algorithms are unexpectedly needed. Throughout this process, the process conditions fluctuate a lot and reproducibility issues arise with each bioreactor when process monitoring is not maintained within close-fitting tolerances. Although this bioreactor is widely used in research labs for small-scale biological material production and short-duration processes, the majority of the procedures are designed with fed-batch and continuous systems in mind (Galvanauskas et al., 2019).

25.7.2 Fed-Batch Bioreactor

Fed-batch bioreactors allow for the controlled addition of nutrients or substrates during the cultivation process, making them comparable to batch reactors. The amenities of fed-batch operations in a plant is that by appropriately directing the feed, the reactors used in this kind of operation can sustain an ideal level of substrate intentness throughout the bioprocess (Sarkar & Modak, 2003). Control strategies adjust feeding rates based on real-time measurements of culture parameters, such as cell density, substrate concentration, and product formation. However, production challenges may sometimes arise because the product inhibits the biocatalyst. The optimal substrate feed rate must then be determined. Generally speaking, a fed-batch operation may be beneficial when the particular rates of growth and/or product development are non-monotonic functions of the substrate concentration. Increasing biomass or metabolite production is the usual objective of fed-batch bioreactor optimization, which has traditionally been carried out with consideration for substrate feed rate (Ozturk & Hu, 2005).

25.7.3 Continuous Bioreactor

It refers to a wider range of bioreactor systems intended for continuous operation, which may include different setups apart from the conventional stirred-tank design. These bioreactors seek to continuously remove the product or waste stream from the system while supplying nutrients and other necessary components. A variety of aspects could have benefits like better productivity, less substrate inhibition, and increased mass transfer. For certain uses, such as high-density cell culture, continuous fermentation, or continuous enzymatic reactions, continuous bioreactors can be tailored (Villadsen et al., 2011). While continuous culture is gaining popularity among researchers as a simple way to recover products and control processes, it still has drawbacks, such as requiring cross-functional teams and process operators to remain constantly alert in order to monitor ongoing operations in a plant facility. One kind of continuous bioreactor is the continuous stirred-tank bioreactor, which consists of a cylindrical vessel with a stirrer to guarantee that the culture is mixed evenly. The goal of both bioreactors is continuous operation; however, productivity and efficiency can be greatly impacted by the configuration of the bioreactor (Mahler et al., 2018).

25.7.4 Perfusion Bioreactor

In perfusion bioreactors, fresh media is continuously added while used media and cell debris are removed to maintain continuous cell culture. When it comes to perfusion, compared to continuous cultures, which involve keeping the cells inside the bioreactor throughout the entire process, the former collects the cells in a cell retention device through alternating tangential flow filtration. This allows the cells to be retained in the ATF filter while the spent media is transferred back into the bioreactor (Zhong, 2010). Because the filter is spun to prevent cell blockage, this type of culture is also known as spin culture. Fresh medium is pumped into the culture vessel while the filtered medium is pumped to a product reservoir. Depending on the concentration of cells in the vessel, the rate of addition and removal can be adjusted (Stanbury et al., 2013).

25.7.5 Solid-State Bioreactor

With its recent track record of success, the solid-state fermentation bioreactor, also known as the submerged fermentation process has drawn attention as a potential replacement. Since the thermal conductivity and heat capacity of water are higher in a continuous aqueous phase, extracting metabolic heat from a particle bed where an aqueous phase occupies the inter-particle phase is preferable to extracting heat from an air-occupied bed. In addition to increasing the surface area available for the reaction to occur by appropriately exposing the substrate-strain combination, the impeller's spinning speed also allows heat to escape, which boosts productivity. In this type, the microorganisms' sensitivity can be regulated by selecting an optimal rotation speed that maintains the equilibrium of the substrate strain mixture and the specific bioreactor's dimensionality (Mitchell et al., 2006).

25.7.6 Trickle Bed Bioreactor

When at least two components, one in gas and the other in liquid phase, need to react through the use of a solid catalyst, these bioreactors have been employed. One of the benefits is that it allows for higher reaction conversions by matching the liquid flow rate to the plug flow rate. These types can also be distinguished by their resistance to mass and heat transfer, the use of tiny catalysts, the absence of moving parts, and their capacity to function at greater pressures and temperatures. Because of their large particle sizes, TBRs have the disadvantage of having lower catalyst activity efficiency in some reactions. The use of viscous and foaming liquids is likewise subject to restrictions (Gianetto & Specchia, 1992).

25.7.7 Micro-Bioreactor

Micro-bioreactors, also known as microfluidic instruments, are rapid, low-cost, and reagent-light methods for high-throughput screening of bioprocess materials, permitting precise control over culture conditions and cellular microenvironments. Molecular biology and microfluidics technology came together to form the new field of micro-bioreactors. As a result of the development of microtechnology, the science of microfluidics examines and absorbs fluids in small quantities through micrometric channels. Managing fluid dynamics and applying it to microsystems can significantly enhance various aspects such as genetic responses to environmental changes, molecular screening, microbial screening, cellular microenvironment simulation, and diagnostics (Prado et al., 2018).

25.7.8 Photo Bioreactor

In most cases, closed, illuminated vessels without direct culture-environment contact are referred to as photo bioreactors because of the gas exchange through sterilized filters that come into contact with the atmosphere, greatly lowering the chance of contamination. In addition, compared to an open system, this setup produces less CO_2 and water loss during the process, and overall productivity is substantially higher. For efficient microalgae cultivation and bio catalytic processes, photo bioreactors are a desirable option. Uniform illumination is a primary concern in this bioreactor. While sunlight is still the primary source of illumination, artificial light sources such as fluorescent lamps, tungsten lamps, and light-emitting diodes (LEDs) have been investigated as alternatives due to fluctuations in light intensity. In addition, high mass transfer rates without causing cell damage and less reactor fouling from film formation are significant contributors to this bioreactor's increased efficiency (Chanquia et al., 2022).

25.8 OPERABLE BIOREACTORS

Based on culture systems, the bioreactor is classified into three main types: microbial bioreactor, plant cell/organ culture bioreactor, and mammalian cell culture bioreactor.

25.8.1 Microbial Bioreactor

Several substances, including growth hormones, antibiotics, and pigments, are examples of secondary metabolites produced by microorganisms. They have been used in food, cosmetics, pharmaceuticals, and biocontrol but are not directly related to microbial growth (Perry et al., 2022). Various microorganisms produce secondary metabolites, and the culture conditions they are grown in have an impact on the physiology, metabolism, and stress responses of these microorganisms. In order to extract different metabolites from microorganisms during the log growth phase of the fermentation process, certain cultivation techniques and the maintenance of optimal culture conditions are prerequisites (Pan et al., 2019). The bioreactor provides the necessary biological environment for the fermentation. Depending on what they are used for, bioreactors come in a variety of shapes and sizes. Operating bioreactors at an industrial scale is made possible by integrated control systems. For the purpose of designing the bioreactors, it is crucial to consider various factors. The majority of the microbial secondary metabolites have been generated in the bioreactors listed in Table 25.1. The design of bioreactors still presents certain difficulties, such as scaling up, integrating all functions into a single bioreactor, or creating a single bioreactor that can manage the majority of fermentation processes or that can hold the majority of microorganisms (Ruíz-Leza et al., 2007).

TABLE 25.1
Types and Functions of Different Microbial Bioreactors

S. no.	Types	Description	Product formation and organisms	References
1.	Stirred Tank Bioreactor	For industrial applications, STB is the most commonly used reactor. Its components include a glass or stainless steel vessel with a motor attached to it, as well as a shaft that has one or more impellers attached. Due to its ability to function in batch, fed-batch, and continuous modes, STB is incredibly versatile. It offers excellent heat transfer, high mass, and mixing. This reactor is perfect for growing microorganisms aerobically.	i. Mycophenolic acid by *Penicillium brevicompactum* ii. Resveratrol by *Vitis labrusca*	Anand & Srivastava, 2022 Chastang et al., 2018
2.	Bubble Column Bioreactor	They often use pneumatic agitation, which involves air traveling through the media as bubbles, in place of mechanical agitation. The rheological characteristics of the fluid must be taken into account, even with the bioreactor's straightforward design. Bubble column bioreactors are mostly used for biological wastewater treatment.	i. PHB production by *Methylocystis hirsuta* ii. DHA production by *Crypthecodinium cohnii*	Rahnama et al., 2012 Hoseinkhani et al., 2021
3	Air-Lift Bioreactor	It has less shear damage to the microbial cells and is constructed simply. Aside from this, the cyclic circulation of liquids is another notable feature. For all these reasons, their most common application is in the production of secondary metabolites.	i. Carotenoids production by *Sporidiobolus pararoseus* ii. Lacic acid production by *Rhizopus oryzae*	Manowattana et al., 2018 Matsumoto & Furuta, 2018
4.	Biofilm Bioreactor	The microorganisms in most biofilm bioreactors are affixed to support materials, form a biofilm, catalyze reactions inside the bioreactor, and then are eliminated when the bio reaction is finished. Many factors, such as the type of support materials used, the availability of nutrients, hydrodynamics, and the thickness of the biofilm, control the formation and performance of biofilms.	i. Oosporin by *Beauveria bassiana* ii. Cephalosporin C by *Cephalosporium acremonium*	Lara-Juache et al., 2021 Srivastava & Kundu., 1999
5.	Solid-State Fermentation (SSF) Bioreactor	The growth of microorganisms in hydrophobic substances is aided by this type of bioreactor, which increases the reactor's overall efficiency and helps concentrate products. Controlling the different parameters, such as temperature, pH, and metabolic heat dissipation, are the main drawbacks.	i. Cephalosporin C by *Acremonium chrysogenum*	Lopez-Calleja et al., 2012
6.	Tray Bioreactor	Thin layers of solid substrate dispersed on horizontal trays with 1–4 cm bed heights are the distinguishing feature of tray reactors. They control a number of parameters and offer effective metabolic heat dissipation and aeration.	i. Lovastatin by *Aspergillus terreus* ii. Various colouring pigments by *Monascus*	Banos et al., 2009 Chaudhary et al., 2022
7.	Stirred and Rotating Drum Bioreactor	Drum that is rotating and stirred, two types of reactors with mixing but no forced aeration are called bioreactors. Usually arranged horizontally, these bioreactors are cylindrical drums that are partially filled with a solid substrate. A current of air is blown through the reactor's headspace to supply oxygen.	i. Rhamnolipids by *P. aeruginosa*	Dabaghi et al., 2023

25.8.2 Plant Cell/Organ Culture Bioreactor

The use of bioreactors has become increasingly important in the process of producing secondary metabolites in plants. Plant *in vitro* cultures have been studied as a possible source of secondary metabolites for a number of decades. It is possible to produce desired chemicals year-round using plant cells or organs grown *in vitro*, independent of the season, climate zone, weather disturbances, and other environmental elements like pests and soil quality. Plant cell suspensions' main benefit is their relative ease of adaptation for large-scale bioreactor cultivation, which is essential for the industrial synthesis of a desired chemical. Maintaining and positively controlling biological reactions while taking into account two areas of parameters, such as reactor constraints for necessary chemical, physical, or macrokinetic systems, is the aim of a bioreactor. In addition to producing high-quality products, bioreactors are implied to be biological factories that offer a number of other benefits like increased nutrient uptake, easier and faster cell harvesting with encouraging multiplication rates, greater concentration of secondary metabolite production, independent plant availability, improved control over environment and growth parameters, and great mechanism stirring mechanism (Sharma et al., 2013). They provide a flexible and effective platform for enhancing the synthesis of target metabolites by enhancing growth conditions, increasing production, and modifying genetic factors. Promising outcomes have been shown in studies that use bioreactors to produce plant metabolites. Pharmaceuticals, nutraceuticals, and specialty chemicals are among the many metabolites whose yields they have successfully increased. Alkaloids, flavonoids, terpenoids, and polyphenols are among the compounds that can be enhanced in bioreactor systems through optimization (Siahsar et al., 2011).

25.9 TYPES OF PLANT CELL/ORGAN CULTURE BIOREACTORS

There are three types of bioreactor systems for plant cell/organ culture – liquid-phase, gas-phase, and hybrid bioreactors – which are discussed here. However, plant cell/organ culture has certain limitations, such as slow growth, difficulty scaling up, sensitive to shear stress, and genetic instability. In view of that, hairy root culture has been adopted and mostly used as an explant because of its excellent biosynthetic ability of secondary metabolites along with other benefits as compared to plant cell/organ culture (Khan et al., 2018).

25.9.1 Liquid-Phase Bioreactors

In a liquid-phase bioreactor, the explant is completely immersed in the liquid medium. Due to the mass transfer of culture media, oxygen availability to the root is a limiting factor here (Sharma & Shahzad, 2013). This category includes the following reactors:

25.9.1.1 Stirred Tank Bioreactor (STB)

STB is a mechanically agitated bioreactor in which air bubbles are sparged into a liquid phase and dispersed uniformly by means of an impeller (Huang & McDonald, 2012). It has a high mass transfer rate. It is primarily used for large-scale biosynthetic production. However, due to the shear force created by the impeller, hairy roots suffer damage and tend to form callus. In the case of *Azadirachta indica*, the hairy root culture failed to grow in STB and showed high phenolic content in the medium as a result of high shear stress (Srivastava & Srivastava, 2013). Hence, a modified system of STB has been developed with a flat blade turbine and hairy roots are placed on a steel cage or nylon mesh to avoid contact with the impeller. While culturing hairy roots of *Artemisia annua* in modified STB, the artemisin content was found significantly higher with a further increase in production by using an elicitor methyl jasmonate (Patra & Srivastava, 2014).

25.9.1.2 Rotating Drum Bioreactor (RDB)

RDB is another mechanically agitated bioreactor. Its surface-to-volume ratio is higher compared to other reactors. Here a drum is rotated through a roller at a speed of 2–6 rpm to protect the hairy root tissues from shear. However, there is still damage to tissues which was overcome by lining polyurethane foam inside the drum wall (Mishra & Ranjan, 2008).

25.9.1.3 Bubble Column Bioreactor (BCB)

BCB is a pneumatically agitated bioreactor. Hairy roots are cultured in a column in which air bubbles are released through a sparger at the bottom to flow upward for the mixing of air into the liquid phase (Georgiev et al., 2012). It causes less shear to the tissues; hence, it is suitable for biosynthetic production. Multiple spargers increase the mass transfer rate. Hairy roots tend to show even distribution inside the column due to low shear stress in *Cichorium intybus* hairy root culture (Pal Bais et al., 2002). *Hyoscyamus niger* hairy root culture produced 1.6, 5.3, and 26.5 mg/g of hyoscyamine, scopolamine, and cuscohygrine, respectively, in BCB when hairy roots were immobilized in mesh support (Jaremicz et al., 2014). A modified BCB was developed by using polypropylene mesh or polyurethane foam support for hairy roots, which resulted in higher biomass and secondary metabolites production (Srivastava & Srivastava, 2013; Thakore et al., 2017).

25.9.1.4 Air Lift Bioreactor (ALB)

ALB is similar to BCB having pneumatically mass transfer of air bubbles in the liquid phase. However, the shear stress to tissues is significantly lower. It contains a column with a central draught tube that controls the circulation of air bubbles. Air bubbles are sparged into the central draught tube and flow upward unidirectionally, which prevents bubbles from coalescing (Kim et al., 2002a). Thus, shear stress caused by air flow is distributed uniformly and minimized. Higher production of puerarin was observed in ALB in the case of *Pueraria phaseoloides* hairy root culture (Kintzios et al., 2004). The use of stainless steel mesh in modified ALB column or stainless steel draught tube enhanced artemisinin (Liu et al., 1998) and cichoric acid (Abbasi et al., 2009) production in *Artemisia annua* and *Echinacea* hairy root cultures, respectively. Besides, cone-type ALB produced the highest betacyanin in *Beta vulgaris* hairy root culture (Shin et al., 2002).

25.9.1.5 Balloon-Type Bubble Bioreactor (BTBB)

Some of the disadvantages of hairy root culture in BCB and ALB are foaming and cell growth at the headspace of their columns. BTBB is designed in such a way that it reduces foaming. It consists of a larger diameter upper part which is balloon-shaped and a concentric tube at the lower part. Besides, a short cylindrical opening on the top section of the BTBB facilitates the harvesting of culture tissues. BTBBs having capacities of 300, 500, and 1000 liters have been used for biosynthetic production in different plant species (Paek et al., 2001). Enhanced production of metabolites has been observed in hairy root cultures of *Hypericum perforatum* (Cui et al., 2010), ginseng (Paek et al., 2005), *Eleutherococcus sessiliflorus* (Abdullah et al., 2005) and *Echinacea purpurea* (Wu et al., 2007) using BTBB.

25.9.1.6 Ebb and Flood Bioreactor (EFB)

It is also known as a periodic immersion system. EFB is similar to BCB and ALB in terms of major equipment required, however, a supporting raft system is built to prevent plant material from being completely submerged in the liquid media. This system pumps liquid medium from a storage tank to the culture vessel. The use of a series of channels ensures even distribution of nutrients to plants, leading to uniform growth. The liquid medium is recycled back to the storage tank after a few minutes which is controlled by a solenoid valve every 3–6 hours, depending upon the type of explant and plant species (Piao et al., 2003).

25.9.1.7 Convective Flow Bioreactor (CFB)

CFB is a combined form of the stirred tank, peristaltic pump, and tubular reactor. Inside the stirred tank, oxygen is sparged into the liquid medium. The peristaltic pump recirculates the medium between the stirred tank and the tubular reactor. In the case of *Hyoscyamus muticus* hairy root culture using CFB, root biomass increased significantly over that of BCB. However, CFB may not be suitable for large-scale systems due to the pressure needed to circulate the culture media at a high enough velocity to overcome the root bed's flow resistance (Carvalho & Curtis, 1998).

25.9.1.8 Turbine Blade Bioreactor (TBB)

This is a combination of ALR and STR. Stainless steel mesh separates the cultivation and agitation spaces, preventing hairy roots from coming into touch with the impeller. Air is introduced from the bottom and disseminated by an eight-blade impeller, stirring the medium. This works well for carrots with bushy roots. Kondo et al. (1989) identified TBB as a superior bioreactor because of its excellent oxygen transfer capabilities. TBB showed the highest growth rate also in carrot root culture.

25.9.2 Gas-Phase Bioreactors

This is a type of bioreactor where explants are cultured in a gas phase not submerged in a liquid medium. Nutrients are supplied in the form of spray, droplets, or mist to the root. Some of the advantages of using this type of reactor are low shear stress and adequate oxygen supply while disadvantages are the non-uniform distribution of roots in the chamber and the need for manual loading (Sharma & Shahzad, 2013).

25.9.2.1 Trickle Bed Bioreactor (TBB)

TBB is a liquid dispersed gas phase bioreactor where nutrient medium in trickled or sprayed on hairy root and recirculated again. It imparts better oxygen mass transfer than liquid phase reactors and low shear stress (Huang et al., 2004). A modified version of TBB called mesh hindrance mist trickling bioreactor (MHMTB) was developed where mesh hindrance allowed lateral roots bridging, thus enhancing the production of L-DOPA in *Stizolobium* culture (Sung & Huang, 2006).

25.9.2.2 Nutrient Mist Bioreactor (NMB)

This system is conceptualized by Weathers and Giles (1988) namely "Mistifier". In NMB, hairy roots are kept on mesh support in the air phase and nutrient mist is applied through an ultrasonic transducer. Some benefits of using NMB are ease of use, low shear stress, better replenishment of nutrients, and removal of toxic byproducts. Intermittent misting for a few minutes with breaks in nutrient supply is better for root growth. NMB system has shown good results in the shoot culture of *Dianthus caryophyllus* also (Woo & Park, 1993). Promising results in potato tuber propagation have been obtained through the NMB system (Kurata et al., 1991; Hao et al., 1998). For hairy root culture, NMB was first used by Dilorio et al. (1992). Comparisons among different bioreactors for hairy root culture in *Nicotiana tabacum* and *Datura stramonium* found that NMB showed better growth than any other bioreactors (Whitney, 1992). A low-cost mist bioreactor (LCMB) has been recently developed with a plastic film chamber that allows hairy root culture under sterile conditions (Mishra & Ranjan, 2008).

25.9.3 Hybrid Bioreactors

Gas phase bioreactors are advantageous over liquid phase bioreactors. However, there is a problem with the non-uniform distribution of roots in the chamber and the need for manual loading. To solve this hybrid bioreactor was developed which is a hybrid of liquid-phase and gas-phase bioreactors. The hairy roots are cultured in the liquid phase first to distribute roots uniformly, then the roots are exposed to the gas phase. Flores and Curtis (1992) suggested that a combination of BCB and TBB is the best combination of hybrid reactors. A BCB and NMB system of hybrid bioreactor has also been used by Kim et al. (2002b, 2003) in the case of the hairy root culture of *Artemisia annua*. Initially, the BCB system was used for a week followed by intermittent NMB. The bubble column-spray bioreactor system developed by Jaremicz et al. (2014) for *Hyoscyamus niger* hairy root to produce tropane alkaloids also followed the same pattern.

25.10 ADVANCED BIOREACTORS

25.10.1 Disposable Bioreactors

These are single-use bioreactors made of ISO 10993–certified and FDA-approved plastic materials, such as polyethylene, polystyrene, polypropylene, etc. The benefit of using a disposable bioreactor over stainless steel or glass bioreactor is no need for sterilization and reuse which reduces the chance of cross-contamination and saves time as well as process cost. Disposable bag bioreactor is the major disposable bioreactor having a flexible bag-like container. This is mostly used for mammalian and insect cell culture; however, plant cell or organ culture is also promising in this setup. Further, it can be classified based on mass and energy transfer and power input, i.e., static bag bioreactor and dynamic bag bioreactor. For static bag bioreactor, the culture of cells is done in a closed system. Therefore there is a limitation in cell growth output due to a restricted closed environment with a lack of modification in it and manual feeding is also problematic as there is a chance of contamination. Dynamic bag bioreactors are of two types based on power supply: mechanically driven and pneumatically driven (Sharma et al., 2022).

25.10.1.1 Stirred Bag Bioreactor

It is a dynamic bag reactor with a sparger ring or microsparger for aeration and a rotating or tumbling impeller for mixing up the medium. Its cubical or cylindrical-shaped bag also contains a gas filter and ports to add sensor probes and line sets. The bag is supported by a steel container with a heating and cooling jacket (Hanson et al., 2009).

25.10.1.2 Vibromixer Bag Bioreactor

This type of bag bioreactor contains a vertically oscillating hollow shaft with a perforated disc heat-sealed in the bag with sensors for dissolved oxygen and pH measurement. An axial flow of culture medium is created and the perforation allows upward or downward flow depending upon its position. Hence, it results in no vortex formation (Hu, 2020).

25.10.1.3 Wave-Mixed Bag Bioreactor

Here the culture bag is kept on a rocking platform that induces a wave inside the culture medium, allowing better mass and energy transfer (Eibl et al., 2010).

25.10.1.4 Hybrid Bag Bioreactor

This has a unique-shaped bag with two propellers and a sparger at the base of the bag for air flow. The air flow happens to be upward and then moves transversely which ensures uniform energy and mass transfer. As it is pneumatically driven, foaming at the top is the major problem which can be reduced by applying pressure at the headspace (Eibl & Eibl, 2009).

25.10.2 Miniature Bioreactor

For studies of a large number of cell lines, miniature versions of large bioreactors have been developed. These miniature bioreactors function based on shaken flask, stirred tank, or bubble column principles (Betts & Baganz, 2006). A novel modular miniature bioreactor as developed by Kloke et al. (2010) involves the stacking of several polycarbonate elements and silicon gaskets with oxygen availability and a mass transfer monitoring system. This is a modular system of reactors with adaptable volumes in milliliter range.

25.11 CONCLUSION

In vitro plant cell/organ culture in medicinal plants for the production of secondary metabolites is a promising approach (Vanisree et al., 2004). Especially, hairy roots, being a great producer of metabolites, are often used for *in vitro* cultures (Mishra & Ranjan, 2008). For example, shear-sensitive hairy roots of *Artemisia annua*, when grown in modified inner loop ALB (Liu et al., 1998), NMB and modified NMB (Liu et al., 1999), modified STB (Patra & Srivastava, 2014), and hybrid bioreactor (Patra & Srivastava, 2016), resulted in higher artemisinin production. In *Hyoscyamus niger* hairy root culture, anisodamine was found higher in the hybrid bioreactor, whereas scopolamine, hyoscyamine, and cuscohygrine were higher in BCB (Jaremicz et al., 2014). Ajmalicine content was significantly higher in modified BCB specifically with polyurethane foam in the case of *Catharanthus roseus* hairy root culture (Thakore et al., 2017). Similar results were obtained in case of shikonin production from *Lithospermum erythrorhizon* in two-phase BCB (Sim & Chang, 1993), hypericin production from *Hypericum perforatum* in BTBB (Cui et al., 2010), azadirachtin production from *Azadirachta indica* in STB, BCB, and NMB (Srivastava & Srivastava, 2012). Different types of conventional bioreactors have shown varied results in terms of scaling up the secondary metabolites production from hairy roots, however, examples are also there of successful production at a commercial scale (Choi et al., 2008). Further advanced bioreactor systems have been developed that have less chance of cross-contamination compared to conventional bioreactors but incur high operating costs due to their disposable nature. Disposable bag bioreactors are in use quite often in the biotechnology sector at a commercial scale (Lehmann et al., 2014). They have certain advantages, including a reduction in sterilization and cleaning costs and less contamination while unit cost is higher for each bag and sensors are expensive. Nevertheless, these advanced bioreactors are going to become the future of commercial biosynthetic machinery. Improvement in cost-effectiveness and easy scaling-up of these reactors will be the way forward for further development in this sector.

REFERENCES

Abbasi, B. H., Liu, R., Saxena, P. K., & Liu, C. Z. (2009). Cichoric acid production from hairy root cultures of *Echinacea purpurea* grown in a modified airlift bioreactor. *Journal of Chemical Technology & Biotechnology*, *84*(11), 1697–1701.

Abdullah, M. S., Chakrabarty, D., Yu, K. W., Hahn, E. J., & Paek, K. Y. (2005). Application of bioreactor system for large scale production of *Eleutherococcus sessilifl* orus somatic embryos in an air-lift bioreactor and the production of eleutherosides. *Journal of Biotechnology*, *120*, 228–236.

Akula, R., & Ravishankar, G. A. (2011). Influence of abiotic stress signals on secondary metabolites in plants. *Plant Signaling & Behavior*, *6*(11), 1720–1731.

Anand, S., & Srivastava, P. (2022). Comparative study for the production of mycophenolic acid using penicillium brevicompactum in batch, fed-batch and continuous fermentation process. *Biointerface Research in Applied Chemistry*, *12*(1), 366–376.

Banos, J. G., Tomasini, A., Szakács, G., & Barrios-González, J. (2009). High lovastatin production by *Aspergillus terreus* in solid-state fermentation on polyurethane foam: An artificial inert support. *Journal of Bioscience and Bioengineering*, *108*(2), 105–110.

Baque, M. A., Elgirban, A., Lee, E. J., & Paek, K. Y. (2012). Sucrose regulated enhanced induction of anthraquinone, phenolics, flavonoids biosynthesis and activities of

antioxidant enzymes in adventitious root suspension cultures of *Morinda citrifolia* (L.). *Acta Physiologiae Plantarum, 34*, 405–415.
Betts, J. I., & Baganz, F. (2006). Miniature bioreactors: Current practices and future opportunities. *Microbial Cell Factories, 5*, 1–14.
Bourgaud, F., Gravot, A., Milesi, S., & Gontier, E. (2001). Production of plant secondary metabolites: A historical perspective. *Plant Science, 161*(5), 839–851.
Broun, P. (2004). Transcription factors as tools for metabolic engineering in plants. *Current Opinion in Plant Biology, 7*(2), 202–209.
Burbulis, I. E., & Winkel-Shirley, B. (1999). Interactions among enzymes of the Arabidopsis flavonoid biosynthetic pathway. *Proceedings of the National Academy of Sciences, 96*(22), 12929–12934.
Canel, C., Lopes-Cardoso, M. I., Whitmer, S., van der Fits, L., Pasquali, G., van der Heijden, R., Hoge J., Hoge, H. C., & Verpoorte, R. (1998). Effects of over-expression of strictosidine synthase and tryptophan decarboxylase on alkaloid production by cell cultures of *Catharanthus roseus*. *Planta, 205*, 414–419. doi: http://dx.doi.org/10.1007/s004250050338.
Cardoso, J. C., Oliveira, M. E., & Cardoso, F. D. C. (2019). Advances and challenges on the in vitro production of secondary metabolites from medicinal plants. *Horticultura Brasileira, 37*, 124–132.
Carvalho, E. B., & Curtis, W. R. (1998). Characterization of fluid-flow resistance in root cultures with a convective flow tubular bioreactor. *Biotechnology and Bioengineering, 60*(3), 375–384.
Chanquia, S. N., Vernet, G., & Kara, S. (2022). Photobioreactors for cultivation and synthesis: Specifications, challenges, and perspectives. *Engineering in Life Sciences, 22*(12), 712–724.
Chastang, T., Pozzobon, V., Taidi, B., Courot, E., Clément, C., & Pareau, D. (2018). Resveratrol production by grapevine cells in fed-batch bioreactor: Experiments and modelling. *Biochemical Engineering Journal, 131*, 9–16.
Chatterjee, S. K. (2002). Cultivation of medicinal and aromatic plants in India – A commercial approach. *Acta Horticutlturae, 576*, 191–202. DOI: 10.17660/ActaHortic.2002.576.28
Chaudhary, V., Katyal, P., Poonia, A. K., Kaur, J., Puniya, A. K., & Panwar, H. (2022). Natural pigment from Monascus: The production and therapeutic significance. *Journal of Applied Microbiology, 133*(1), 18–38.
Chodisetti, B., Rao, K., Gandi, S., & Giri, A. (2013). Improved gymnemic acid production in the suspension cultures of *Gymnema sylvestre* through biotic elicitation. *Plant Biotechnology Reports, 7*, 519–525.
Choi, Y. E., Kim, Y. S., & Paek, K. Y. (2008). Types and designs of bioreactors for hairy root culture. In *Plant tissue culture engineering* (pp. 161–172).
Cui, X. H., Chakrabarty, D., Lee, E. J., & Paek, K. Y. (2010). Production of adventitious roots and secondary metabolites by *Hypericum perforatum* L. in a bioreactor. *Bioresource Technology, 101*(12), 4708–4716.
Dabaghi, S., Ataei, S. A., & Taheri, A. (2023). Performance analysis of a laboratory scale rotating drum bioreactor for production of rhamnolipid in solid-state fermentation using an agro-industrial residue. *Biomass Conversion and Biorefinery, 13*(13), 11513–11520.
DellaPenna, D. (2001). Plant metabolic engineering. *Plant Physiology, 125*(1), 160–163.
DiIorio, A. A., Cheetham, R. D., & Weathers, P. J. (1992). Growth of transformed roots in a nutrient mist bioreactor: Reactor performance and evaluation. *Applied Microbiology and Biotechnology, 37*, 457–462.
Ehrlich, P. R., & Holdren, J. P. (1971). Impact of population growth. *Science, 171*, 1212–1217.
Eibl, R., & Eibl, D. (2009). Application of disposable bag bioreactors in tissue engineering and for the production of therapeutic agents. In *Bioreactor systems for tissue engineering* (Vol. 112, pp. 183–207). Springer.
Eibl, R., Werner, S., & Eibl, D. (2010). Bag bioreactor based on wave-induced motion: Characteristics and applications. In *Disposable bioreactors* (Vol. 115, pp. 55–87). Springer.
Figueredo, A. C., Barroso, J. G., Pedro, L. G., & Scheffer, J. J. (2008). Factors affecting secondary metabolite production in plants: Volatile and essential oils. *Flavour and Fragrance Journal, 23*, 213–226.
Flores, H. E., & Curtis, W. R. (1992). Approaches to understanding and manipulating the biosynthetic potential of plant roots. *Annals of the New York Academy of Sciences, 665*, 188–209.
Galvanauskas, V., Simutis, R., Levišauskas, D., & Urniežius, R. (2019). Practical solutions for specific growth rate control systems in industrial bioreactors. *Processes, 7*(10), 693.
Georgiev, V., Ivanov, I., Berkov, S., Ilieva, M., Georgiev, M., Gocheva, T., & Pavlov, A. (2012). Galanthamine production by *Leucojum aestivum* L. shoot culture in a modified bubble column bioreactor with internal sections. *Engineering in Life Sciences, 12*(5), 534–543.
Gianetto, A., & Specchia, V. (1992). Trickle-bed reactors: State of art and perspectives. *Chemical Engineering Science, 47*(13–14), 3197–3213.
Gómez-Galera, S., Pelacho, A. M., Gené, A., Capell, T., & Christou, P. (2007). The genetic manipulation of medicinal and aromatic plants. *Plant Cell Reports, 26*, 1689–1715. https://doi.org/10.1007/s00299-007-0384-x
Gorelick, J., & Bernstein, N. (2014). Elicitation: An underutilized tool in the development of medicinal plants as a source of therapeutic secondary metabolites. In *Advances in agronomy* (Vol. 124, pp. 201–230). Elsevier.
Hanson, M. A., Brorson, K. A., Moreira, A. R., & Rao, G. (2009). Comparisons of optically monitored small-scale stirred tank vessels to optically controlled disposable bag bioreactors. *Microbial Cell Factories, 8*, 1–7.
Hao, Z., Ouyang, F., Geng, Y., Deng, X., Hu, Z., & Chen, Z. (1998). Propagation of potato tubers in a nutrient mist bioreactor. *Biotechnology Techniques, 12*, 641–644.
Herrmann, K. M., & Weaver, L. M. (1999). The shikimate pathway. *Annual Review of Plant Biology, 50*(1), 473–503.
Hoseinkhani, N., Jalili, H., Ansari, S., & Amrane, A. (2021). Impact of bubble size on docosahexaenoic acid production by *Crypthecodinium cohnii* in bubble column bioreactor. *Biomass Conversion and Biorefinery, 11*, 1137–1144. Springer.
Hu, W. S. (2020). Cell culture bioreactors. In *Cell culture bioprocess engineering* (2nd ed., pp. 279–303). CRC Press.
Huang, S. Y., Hung, C. H., & Chou, S. N. (2004). Innovative strategies for operation of mist trickling reactors for enhanced hairy root proliferation and secondary metabolite productivity. *Enzyme and Microbial Technology, 35*(1), 22–32.
Huang, T. K., & McDonald, K. A. (2012). Bioreactor systems for in vitro production of foreign proteins using plant cell cultures. *Biotechnology Advances, 30*(2), 398–409.
Isah, T., Umar, S., Mujib, A., Sharma, M. P., Rajasekharan, P. E., Zafar, N., & Frukh, A. (2018). Secondary metabolism of pharmaceuticals in the plant in vitro cultures: Strategies, approaches, and limitations to achieving higher yield. *Plant*

Cell, Tissue and Organ Culture (PCTOC), *132*, 239–265. http://doi.org/10.1007/s11240-017-1332-2

Jadaun, J. S., Sangwan, N. S., Narnoliya, L. K., Singh, N., Bansal, S., Mishra, B., & Sangwan, R. S. (2017). Over-expression of DXS gene enhances terpenoidal secondary metabolite accumulation in rose-scented geranium and *Withania somnifera*: Active involvement of plastid isoprenogenic pathway in their biosynthesis. *Physiologia Plantarum*, *159*(4), 381–400. doi: http://dx.doi.org/10.1111/ppl.12507

Jaremicz, Z., Luczkiewicz, M., Kokotkiewicz, A., Krolicka, A., & Sowinski, P. (2014). Production of tropane alkaloids in *Hyoscyamus niger* (black henbane) hairy roots grown in bubble-column and spray bioreactors. *Biotechnology Letters*, *36*, 843–853. Springer.

Jarvis, M. F., & Williams, M. (1989). Direct autoradiographic localization of adenosine A2 receptors in the rat brain using the A2-selective agonist, [3H]. *CGS European Journal of Pharmacology*, *168*, 243–246.

Jung, H. Y., Kang, S. M., Kang, Y. M., Kang, M. J., Yun, D. J., Bahk, J. D., Yang, J. K., & Choi, M. S. (2003). Enhanced production of scopolamine by bacterial elicitors in adventitious hairy root cultures of Scopolia parviflora. *Enzyme and Microbial Technology*, *33*(7), 987–990. https://doi.org/10.1016/S0141-0229(03)00253-9

Kang, S. M., Min, J. Y., Kim, Y. D., Karigar, C. S., Kim, S. W., Goo, G. H., & Choi, M. S. (2009). Effect of biotic elicitors on the accumulation of bilobalide and ginkgolides in Ginkgo biloba cell cultures. *Journal of Biotechnology*, *139*(1), 84–88. https://doi.org/10.1016/j.jbiotec.2008.09.007

Karuppusamy, S. (2009). A review on trends in production of secondary metabolites from higher plants by in vitro tissue, organ and cell cultures. *Journal of Medicinal Plant Research*, *3*(13), 1222–1239.

Khan, S. A., Siddiqui, M. H., & Osama, K. (2018). Bioreactors for hairy roots culture: A review. *Current Biotechnology*, *7*(6), 417–427.

Khani, S., Barar, J., Movafeghi, A., & Omidi, Y. (2012). Production of anticancer secondary metabolites: Impacts of bioprocess engineering. In I. E. Orhan (Ed.), *Biotechnological production of secondary metabolites* (pp. 215–240). Benthan eBooks.

Kim, Y. J., Weathers, P. J., & Wyslouzil, B. E. (2002b). Growth of *Artemisia annua* hairy roots in liquid- and gas-phase reactors. *Biotechnology and Bioengineering*, *80*(4), 454–464.

Kim, Y. J., Weathers, P. J., & Wyslouzil, B. E. (2003). Growth dynamics of *Artemisia annua* hairy roots in three culture systems. *Biotechnology and Bioengineering*, *83*(4), 428–443.

Kim, Y. J., Wyslouzil, B. E., & Weathers, P. J. (2002a). Secondary metabolism of hairy root cultures in bioreactors. *In Vitro Cellular & Developmental Biology-Plant*, *38*, 1–10.

Kintzios, S., Makri, O., Pistola, E., Matakiadis, T., Ping Shi, H., & Economou, A. (2004). Scale-up production of puerarin from hairy roots of *Pueraria phaseoloides* in an airlift bioreactor. *Biotechnology Letters*, *26*, 1057–1059.

Kloke, A., Rubenwolf, S., Bücking, C., Gescher, J., Kerzenmacher, S., Zengerle, R., & von Stetten, F. (2010). A versatile miniature bioreactor and its application to bioelectrochemistry studies. *Biosensors and Bioelectronics*, *25*(12), 2559–2565.

Kondo, O., Honda, H., Taya, M., & Kobayashi, T. (1989). Comparison of growth properties of carrot hairy root in various bioreactors. *Applied Microbiology and Biotechnology*, *32*, 291–294.

Kurata, K., Ibaraki, Y., & Goto, E. (1991). Propagation of potato tubers in a nutrient mist bioreactor. *American Society of Agricultural Engineers*, *34*, 621–624.

Lara-Juache, H. R., Ávila-Hernández, J. G., Rodríguez-Durán, L. V., Michel, M. R., Wong-Paz, J. E., Muñiz-Márquez, D. B., Veana, F., Aguilar-Zárate, M., Ascacio-Valdés, J. A., & Aguilar-Zárate, P. (2021). Characterization of a biofilm bioreactor designed for the single-step production of aerial conidia and oosporein by *Beauveria bassiana* PQ2. *Journal of Fungi*, *7*(8), 582.

Latchman, D. S. (2010). *Eukaryotic transcription factors*. Academic Press.

Lehmann, N., Dittler, I., Lämsä, M., Ritala, A., Rischer, H., Eibl, D., Oksman-Caldentey, K. M., & Eibl, R. (2014). Disposable bioreactors for cultivation of plant cell cultures. *Production of biomass and bioactive compounds using bioreactor technology* (pp. 17–46).

Lessard, P. A., Kulaveerasingam, H., York, G. M., Strong, A., & Sinskey, A. J. (2002). Manipulating gene expression for the metabolic engineering of plants. *Metabolic Engineering*, *4*(1), 67–79. https://doi.org/10.1006/mben.2001.0210

Liang, C., Chen, C., Zhou, P., Xu, L., Zhu, J., Liang, J., Zi, J., & Yu, R. (2018). Effect of *Aspergillus flavus* fungal elicitor on the production of terpenoid indole alkaloids in *Catharanthus roseus* cambial meristematic cells. *Molecules*, *23*(12), 3276. http://doi.org/10.3390/molecules23123276

Liu, C. Z., Wang, Y. C., Zhao, B., Guo, C., Ouyang, F., Ye, H. C., & Li, G. F. (1999). Development of a nutrient mist bioreactor for growth of hairy roots. *In Vitro Cellular & Developmental Biology-Plant*, *35*, 271–274.

Liu, C., Wang, Y., Guo, C., Ouyang, F., Ye, H., & Li, G. (1998). Enhanced production of artemisinin by *Artemisia annua* L hairy root cultures in a modified inner-loop airlift bioreactor. *Bioprocess Engineering*, *19*, 389–392.

Lopez-Calleja, A. C., Cuadra, T., Barrios-González, J., Fierro, F., & Fernandez, F. J. (2012). Solid-state and submerged fermentations show different gene expression profiles in cephalosporin C production by *Acremonium chrysogenum*. *Journal of Molecular Microbiology and Biotechnology*, *22*(2), 126–134.

Maeda, H., & Dudareva, N. (2012). The shikimate pathway and aromatic amino acid biosynthesis in plants. *Annual Review of Plant Biology*, *63*, 73–105.

Mahler, N., Tschirren, S., Pflügl, S., & Herwig, C. (2018). Optimized bioreactor setup for scale-up studies of extreme halophilic cultures. *Biochemical Engineering Journal*, *130*, 39–46.

Mahmoud, S. S., & Croteau, R. B. (2001). Metabolic engineering of essential oil yield and composition in mint by altering expression of deoxyxylulose phosphate reductoisomerase and menthofuran synthase. *Proceedings of the National Academy of Sciences*, *98*(15), 8915–8920. https://doi.org/10.1073/pnas.141237298

Manowattana, A., Techapun, C., Watanabe, M., & Chaiyaso, T. (2018). Bioconversion of biodiesel-derived crude glycerol into lipids and carotenoids by an oleaginous red yeast Sporidiobolus pararoseus KM281507 in an airlift bioreactor. *Journal of Bioscience and Bioengineering*, *125*(1), 59–66.

Matsumoto, M., & Furuta, H. (2018). In situ extractive fermentation of lactic acid by Rhizopus oryzae in an air-lift bioreactor. *Chemical and Biochemical Engineering Quarterly*, *32*(2), 275-280.

Mishra, B. N., & Ranjan, R. (2008). Growth of hairy-root cultures in various bioreactors for the production of secondary metabolites. *Biotechnology and Applied Biochemistry*, *49*(1), 1–10.

Mitchell, D. A., Krieger, N., & Berovic, M. (2006). *Solid-state fermentation bioreactors* (p. 19). Springer.

Murthy, H. N., Lee, E. J., & Paek, K. Y. (2014). Production of secondary metabolites from cell and organ cultures: Strategies and approaches for biomass improvement and metabolite accumulation. *Plant Cell, Tissue and Organ Culture*, *118*, 1–16. https://doi.org/10.1007/s11240-014-0467-7

Namdeo, A., Patil, S., & Fulzele, D. P. (2002). Influence of fungal elicitors on production of ajmalicine by cell cultures of Catharanthus roseus. *Biotechnology Progress*, *18*(1), 159–162. http://doi.org/10.1021/bp0101280

Nguyen, T. K. O., Dauwe, R., Bourgaud, F., & Gontier, E. (2013). From bioreactor to entire plants: Development of production systems for secondary metabolites. In *Advances in botanical research* (Vol. 68, pp. 205–232). Academic Press.

Ochoa-Villarreal, M., Howat, S., Hong, S., Jang, M. O., Jin, Y. W., Lee, E. K., & Loake, G. J. (2016). Plant cell culture strategies for the production of natural products. *BMB Reports*, *49*(3), 149. https://doi.org/10.5483/bmbrep.2016.49.3.264

Ozturk, S., & Hu, W. S. (Eds.). (2005). *Cell culture technology for pharmaceutical and cell-based therapies*. CRC Press.

Paek, K. Y., Chakrabarty, D., & Hahn, E. J. (2005). Application of bioreactor systems for large scale production of horticultural and medicinal plants. *Plant Cell Tissue and Organ Culture*, *81*, 287–300.

Paek, K. Y., Hahn, E. J., & Son, S. H. (2001). Application of bioreactors of large scale micropropagation systems of plants. *In Vitro Cellular and Developmental Biology-Plant*, *37*, 149–157.

Pal Bais, H., Suresh, B., Rachavarao, K. S. M. S., & Ravishankar, G. A. (2002). Performance of hairy root cultures of *Cichorium intybus* L. in bioreactors of different configurations. *In Vitro Cellular & Developmental Biology-Plant*, *38*, 573–580.

Pan, R., Bai, X., Chen, J., Zhang, H., & Wang, H. (2019). Exploring structural diversity of microbe secondary metabolites using OSMAC strategy: A literature review. *Frontiers in Microbiology*, *10*, 418973.

Patra, N., & Srivastava, A. K. (2014). Enhanced production of artemisinin by hairy root cultivation of Artemisia annua in a modified stirred tank reactor. *Applied Biochemistry and Biotechnology*, *174*, 2209–2222.

Patra, N., & Srivastava, A. K. (2016). Artemisinin production by plant hairy root cultures in gas-and liquid-phase bioreactors. *Plant Cell Reports*, *35*, 143–153.

Perry, E. K., Meirelles, L. A., & Newman, D. K. (2022). From the soil to the clinic: The impact of microbial secondary metabolites on antibiotic tolerance and resistance. *Nature Reviews Microbiology*, *20*(3), 129–142.

Piao, X. C., Chakrabarty, D., Hahn, E. J., & Paek, K. Y. (2003). A simple method for mass production of potato microtubers using a bioreactor system. *Current Science*, 1129–1132.

Pino, M. S., Rodríguez-Jasso, R. M., Michelin, M., Flores-Gallegos, A. C., Morales-Rodriguez, R., Teixeira, J. A., & Ruiz, H. A. (2018). Bioreactor design for enzymatic hydrolysis of biomass under the biorefinery concept. *Chemical Engineering Journal*, *347*, 119–136.

Prado, R. C., & Borges, E. R. (2018). Microbioreactors as engineering tools for bioprocess development. *Brazilian Journal of Chemical Engineering*, *35*, 1163–1182.

Rahnama, F., Vasheghani-Farahani, E., Yazdian, F., & Shojaosadati, S. A. (2012). PHB production by *Methylocystis hirsuta* from natural gas in a bubble column and a vertical loop bioreactor. *Biochemical Engineering Journal*, *65*, 51–56.

Ramirez-Estrada, K., Vidal-Limon, H., Hidalgo, D., Moyano, E., Golenioswki, M., Cusidó, R. M., & Palazon, J. (2016). Elicitation, an effective strategy for the biotechnological production of bioactive high-added value compounds in plant cell factories. *Molecules*, *21*(2), 182. https://doi.org/10.3390/molecules21020182

Ruíz-Leza, H. A., Rodríguez-Jasso, R. M., Rodríguez-Herrera, R., Contreras-Esquivel, A. J., & Aguilar, C. N. (2007). Bioreactors desing for solid state fermentation. *Revista Mexicana de Ingeniería Química*, *6*(1), 33–40.

Sanchez, S., & Demain, A. L. (2000). *Secondary metabolites*. American Society of Plant Physiologists.

Sarkar, D., & Modak, J. M. (2003). Optimisation of fed-batch bioreactors using genetic algorithms. *Chemical Engineering Science*, *58*(11), 2283–2296.

Saunders, J. A. (1979). Investigations of vacuoles isolated from tobacco: I. Quantitation of nicotine. *Plant Physiology*, *64*(1), 74–78.

Saxena, M., Saxena, J., Nema, R., Singh, D., & Gupta, A. (2013). Phytochemistry of medicinal plants. *Journal of Pharmacognosy and Phytochemistry*, *1*(6), 168–182.

Sharma, R., Harrison, S. T., & Tai, S. L. (2022). Advances in bioreactor systems for the production of biologicals in mammalian cells. *ChemBioEng Reviews*, *9*(1), 42–62.

Sharma, S., & Shahzad, A. (2013). Bioreactors: A rapid approach for secondary metabolite production. In *Recent trends in biotechnology and therapeutic applications of medicinal plants* (pp. 25–49). Springer Netherlands.

Shin, K. S., Murthy, H. N., Ko, J. Y., & Paek, K. Y. (2002). Growth and betacyanin production by hairy roots of Beta vulgaris in airlift bioreactors. *Biotechnology Letters*, *24*, 2067–2069.

Siahsar, B., Rahimi, M., Tavassoli, A., & Raissi, A. (2011). Application of biotechnology in production of medicinal plants. *American-Eurasian Journal of Agricultural & Environmental Sciences*, *11*(3), 439–444.

Sim, S. J., & Chang, H. N. (1993). Increased shikonin production by hairy roots of *Lithospermum erythrorhizon* in two phase bubble column reactor. *Biotechnology Letters*, *15*, 145–150.

Srivastava, P., & Kundu, S. (1999). Studies on cephalosporin-C production in an air lift reactor using different growth modes of *Cephalosporium acremonium*. *Process Biochemistry*, *34*(4), 329–333.

Srivastava, S., & Srivastava, A. K. (2012). In vitro azadirachtin production by hairy root cultivation of Azadirachta indica in nutrient mist bioreactor. *Applied Biochemistry and Biotechnology*, *166*, 365–378.

Srivastava, S., & Srivastava, A. K. (2013). Production of the biopesticide azadirachtin by hairy root cultivation of *Azadirachta indica* in liquid-phase bioreactors. *Applied Biochemistry and Biotechnology*, *171*, 1351–1361.

Stanbury, P. F., Whitaker, A., & Hall, S. J. (2013). *Principles of fermentation technology*. Elsevier.

St-Pierre, B., Vazquez-Flota, F. A., & De Luca, V. (1999). Multicellular compartmentation of Catharanthus roseus alkaloid biosynthesis predicts intercellular translocation of a pathway intermediate. *The Plant Cell*, *11*(5), 887–900.

Sung, L. S., & Huang, S. Y. (2006). Lateral root bridging as a strategy to enhance l-DOPA production in Stizolobium hassjoo hairy root cultures by using a mesh hindrance mist trickling bioreactor. *Biotechnology and Bioengineering*, *94*(3), 441–447.

Takeuchi, C., Nagatani, K., & Sato, Y. (2013). Chitosan and a fungal elicitor inhibit tracheary element differentiation and promote accumulation of stress lignin-like substance in Zinnia elegans xylogenic culture. *Journal of Plant Research*, *126*, 811–821.

Thakore, D., Srivastava, A. K., & Sinha, A. K. (2017). Mass production of Ajmalicine by bioreactor cultivation of hairy roots of *Catharanthus roseus*. *Biochemical Engineering Journal*, *119*, 84–91.

Twaij, B. M., & Hasan, M. N. (2022). Bioactive secondary metabolites from plant sources: Types, synthesis, and their therapeutic uses. *International Journal of Plant Biology*, *13*(1), 4–14. https://doi.org/10.3390/ijpb13010003

Vanisree, M., Lee, C. Y., Lo, S. F., Nalawade, S. M., Lin, C. Y., & Tsay, H. S. (2004). Studies on the production of some important secondary metabolites from medicinal plants by plant tissue cultures. *Botanical Bulletin of Academia Sinica*, *45*(1), 1–22.

Verpoorte, R., & Alfermann, A. W. (Eds.). (2000). *Metabolic engineering of plant secondary metabolism* (pp. 1–30). Springer Science & Business Media.

Villadsen, J., Nielsen, J., & Lidén, G. (2011). *Bioreaction engineering principles*. Springer Science & Business Media.

Wagner, G. J., & Kroumova, A. B. (2008). The use of RNAi to elucidate and manipulate secondary metabolite synthesis in plants. In *Current perspectives in microRNAs (miRNA)* (431–459). Springer.

Weathers, P. J., & Giles, K. L. (1988). Regeneration of plants using nutrient mists. *In Vitro Cellular and Development Biology*, *24*, 727–732.

Whitney, P. J. (1992). Novel bioreactors for the growth of roots transformed by *Agrobacterium rhizogenes*. *Enzyme and Microbial Technology*, *14*(1), 13–17.

Winkel, B. S. J. (2004). Metabolic channeling in plants. *Annual Review of Plant Biology*, *55*, 85–107.

Woo, S. H., & Park, J. M. (1993). Multiple shoot culture *Dianthus caryophyllus* using mist culture system. *Biotechnology Techniques*, *7*, 697–702.

Wu, C.-H., Murthy, H. N., Hahn, E.-J., & Paek, K.-Y. (2007). Improved production of caffeic acid derivatives in suspension cultures of *Echinacea* purpurea by medium replenishment strategy. *Archives of Pharmacal Research*, *30*(8), 945–949.

Zhang, B., Zheng, L. P., Yi Li, W., & Wen Wang, J. (2013). Stimulation of artemisinin production in *Artemisia annua* hairy roots by Ag-SiO_2 core-shell nanoparticles. *Current Nanoscience*, *9*(3), 363–370. https://doi.org/10.2174/1573413711309030012

Zhang, H., Wang, L., Deroles, S., Bennett, R., & Davies, K. (2006). New insight into the structures and formation of anthocyanic vacuolar inclusions in flower petals. *BMC Plant Biology*, *6*, 1–14. https://doi.org/10.1186/1471-2229-6-29

Zhong, J. J. (2010). Recent advances in bioreactor engineering. *Korean Journal of Chemical Engineering*, *27*, 1035–1041.

26 Unlocking Medicinal Benefits and Augmenting Secondary Metabolite Synthesis in Cultivated and Wild Legumes

Suriya Prakash Eswaran, Santhanalakshmi Balasubramaniam, Anitha Sakthivel, Manojkumar Sampath, Ganeshan Sivanandhan, and Kapildev Gnanajothi

26.1 INTRODUCTION

Fabaceae, or Leguminosae, is a family of flowering plants also recognized as the legume family. One of the diverse groups and economically significant within the Plantae kingdom. Fabaceae is the third-largest angiosperm family, following Orchidaceae and Asteraceae. Nearly 730 genera and over 19,400 species include a wide range from trees to herbaceous plants, characterized by their leguminous root, compounds, and stipulated leaves. Key members include economically important species such as soybean (***Glycine max***), garden pea (***Pisum sativum***), peanut (***Arachis hypogaea***), and alfalfa (***Medicago sativa***), which holds an essential source of nutrients like carbohydrates, proteins, lipids, fatty acids, amino acids, fibers, minerals, and vitamins. Fabaceae is crucial in natural ecosystems with important ecological responsibilities, including symbiotic interactions with nitrogen-fixing bacteria that aid in soil fixation of nitrogen. This family contains several species that have been utilized historically for a variety of reasons, including food, wood, and traditional medicine. (Wink, 2013).

Fabaceae plants are significant in various cultural and economic contexts, serving as traditional medicines, garden ornamentals, dyes, and insecticides. Their bioactive compounds like phenolic acids, flavonoids, lectins, and alkaloids are of pharmacological interest for their potential to combat cancer, inflammation, diabetes and microbial infections, opening avenues for pharmaceutical and cosmetic innovations (Maroyi, 2023) (Table 26.1, Figure 26.1). Secondary metabolites in Fabaceae play crucial roles in plant defense mechanisms and plant signaling pathways. These organic substances obtained through primary metabolism enhance adaptive traits and ecological fitness by defending against herbivores and pathogens, tolerating abiotic and biotic stresses, and interacting with pollinators and root-associated microbes. Unlike primary metabolites, which are vital for cellular functions like growth and reproduction, secondary metabolites provide ecological benefits. The synthesis and storage of secondary metabolites in Fabaceae are essential strategies for defense and communication. These compounds serve as defensive agents and signaling molecules, attracting pollinators, dispersers, and symbiotic bacteria. Fabaceae plants produce a variety of secondary metabolites, including isoflavones, which offer health benefits to mankind. This exploration of Fabaceae aims to reveal the intricate relationship between primary and secondary metabolites, highlighting their ecological, physiological, and pharmacological importance. Understanding these interactions underscores the vital role of Fabaceae in ecosystems and human wellbeing (Erb et al., 2020).

Investigating the phytochemical composition of Fabaceae plants and their possible medicinal uses is an important line of inquiry for science. Anthraquinones, amines, amino acids, flavonoids, isoflavonoids, coumarins, phenylpropanoids, terpenes (diterpenes, sesquiterpenes, triterpenes) protease inhibitors, lectins and cyanogenic glycosides are among the many compounds in this family. Whereas the Caesalpinioideae appear to be paraphyletic, the Papilionoideae and Mimosoideae are monophyletic subfamilies. Moreover, the legume tree does not begin with the Mimosoideae; rather, they are derived from the original Caesalpinoideae. Legumes have two main classes of defense compounds, which are mapped onto the trees to determine if a certain characteristic is consistently present within a monophyletic taxon or whether a specific structural type has developed several times almost invariably (Obistioiu et al., 2021).

26.2 REMARKABLE ROLE OF SECONDARY METABOLITES OF FABACEAE

Secondary metabolites in plants exhibit various biological, pharmacological, and toxicological activities. Many of these compounds, particularly alkaloids, act as neurotoxins or neuromodulators. For instance, In the liver of herbivores,

DOI: 10.1201/9781003475491-26

TABLE 26.1
Pharmacological Properties and Toxicity of the Phyto-Compounds Found in Legumes

S. no.	Compound	Group	Species	Toxicity	Pharmacological properties
1	Bufotenine	Alkaloids/ Amine	*Anadenanthera peregrina*	Psychotropic effects	Psychotropic, hallucinogenic (Wink, 2013; Chao et al., 2021)
2	N, N-Dimethyltry-ptamine	Alkaloids/ Amine	*Mimosa tenuiflora*	Psychotropic effects, mild toxicity at high doses	Psychotropic, hallucinogenic (Wink, 2013; Bezerra et al., 2021)
3	Pinitol	Cyclitol	*Glycine max*, Carob		Anti-diabetic, anti-inflammatory (Sánchez-Hidalgo et al., 2021)
4	Genistein	Isoflavone	*Glycine max*	Neurotoxic at high doses	Anticancer, antioxidant (Tandon and Das, 2018; Sarao et al., 2022)
5	Daidzein	Isoflavone	*Glycine max*		Anticancer, antioxidant (Alshehri et al., 2021; Sarao et al., 2022)
6	Lupanine	Alkaloid	*Lupinus* species	Neurotoxic	Antibacterial, antidiabetic (Wink, 2013)
7	Galegine	Alkaloid	*Galega officinalis*	Toxic at high doses, can cause hypoglycemia	Antidiabetic (Bednarska et al., 2020)
8	Sanggenon C	Flavonoid	*Albizia julibrissin*	Toxic at high doses	Anticancer, anti-inflammatory (Xiao et al., 2017)
9	Corotoxigenin	Cardiac Glycoside	*Coronilla species*	Highly toxic, inhibits Na+/ K+ ATPase	Cardiotonic (Al-Snafi et al., 2016)
10	Lupenone	Terpenoid	*Lupinus species*	Generally safe, potential mild toxicity at high doses	Anticancer, anti-inflammatory (Xu et al., 2018)
11	Vicenin-2	Flavonoid	*Crotalaria pallida*		Antioxidant, anti-inflammatory (de Beer et al., 2023)
12	Medicarpin	Isoflavonoid	*Medicago sativa*	Generally safe, potential mild toxicity at high doses	Antifungal, phytoalexin (Matsouka et al., 2011)
13	Ciceritol	Cyclitols	*Cicer arietinum*		Anti-diabetic (Wang et al., 2021)
14	Astragalin	Flavonoid	*Cassia tora*		Antioxidant, anti-inflammatory (Bratkov et al., 2016)
15	Cassiollin	Flavonoid	*Cassia* species	Mild toxicity at high doses	Antioxidant, antiviral (Yadav et al., 2010)
16	Karanjin	Flavonoid	*Pongamia pinnata*		Antioxidant, anticancer (Gnanaraj et al., 2024)
17	Epinephrine	Neuro-transmitter	*Glycine max*		Neurotransmission, "fight or flight" response (Schenck et al., 2018)
18	Nor-epinephrine	Neurotransmitter	*Glycine max*		Neurotransmission, arousal, attention (Schenck et al., 2018)
19	L-DOPA (Levodopa)	Dopamine Precursor	*Mucuna pruriens*	Side effects include dyskinesias	Treatment of Parkinson's disease (Soares et al., 2014; Schenck et al., 2018)
20	Stachydrine	Alkaloids	*Astragalus membranaceus* and others	Generally safe, potential mild toxicity at high doses	Cardiovascular protection, diuretic, anti-inflammatory (Liao et al., 2023)
21	Rutin	Flavonoid	*Sophora japonica*		Antioxidant, anti-inflammatory (Usman et al., 2022)
22	Quercetin	Flavonoid	*Glycine max*, lentils		Antioxidant, anti-inflammatory (Usman et al., 2022)
23	Luteolin	Flavonoid	Green peas, *Alfalfa*		Antioxidant, anti-inflammatory (Taban et al., 2023)

TABLE 26.1 (*Continued*)
Pharmacological Properties and Toxicity of the Phyto-Compounds Found in Legumes

S. no.	Compound	Group	Species	Toxicity	Pharmacological properties
24	Kaempferol	Flavonoid	Chickpeas, fava beans		Antioxidant, anti-inflammatory (Periferakis et al., 2022)
25	Biochanin A	Isoflavone	Red clover (*Trifolium pratense*)		Anticancer, antioxidant (Anuranjana et al., 2023)
26	Formononetin	Isoflavone	Chickpeas, red clover	Generally safe, potential mild toxicity at high doses	Anticancer, antioxidant (Ding et al., 2024)
27	Glycitein	Isoflavone	*Glycine max*, alfalfa		Anticancer, antioxidant (Bernatoniene et al., 2021)
28	Ellagic Acid	Polyphenol	Various legumes		Antioxidant, anticancer (Gupta et al., 2021)

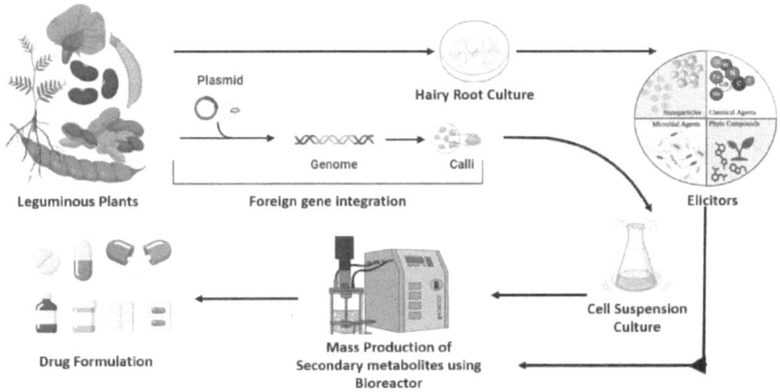

FIGURE 26.1 The chemical structures of the phyto-compounds found in Legumes are mentioned in Table 26.1.

pyrrolizidine alkaloids are activated, which can cause DNA alkylation, mutations, and perhaps cancer in humans and animals. The evolution of these alkaloids is believed to be a defense mechanism against herbivores. Additionally, some alkaloids and amines in legumes, such as bufotenine and N, and N-dimethyltryptamine, have psychotropic effects and have been historically used as drugs that cause hallucination. These toxin substances can either cause immediate side effects or long-term consequences by reducing the survival ability and reproductive fitness of herbivores (Wink, 1992; Wink & Schimmer, 2010).

The primary secondary metabolites found in legumes include polyketides, terpenoids, non-protein amino acids (NPAA), cyanogens, alkaloids, and phenolics. Certain secondary metabolites, including pinitol, triterpenes, and flavonoids are widely distributed, while others are exclusive to a small group of animals. It should be emphasized that certain taxa have not yet been extensively studied and the knowledge of the prevalence and distribution of secondary metabolites in legumes is still inadequate. In other cases, phytochemists were more concerned with publishing the discovery of novel substances than with disclosing the presence of well-known metabolites (Wink, 2013).

26.3 CYANOGENIC GLYCOSIDES

Upon wounding, cyanogenic glycosides in plants release hydrogen cyanide (HCN) through enzymatic hydrolysis. HCN is a potent respiratory poison that inhibits the mitochondrial respiratory chain, making it lethal for most animals. This mechanism provides plants with an effective defense against herbivores (Seigler, 1998).

26.4 NON-PROTEIN AMINO ACIDS (NPAA)

Non-protein amino acids are the analogs of the 20 standard amino acids that can constitute proteins. When incorporated into proteins NPAAs cause the proteins to fold incorrectly, resulting in inactive or malfunctioning proteins. This makes NPAAs toxic antimetabolites affecting not only herbivores but also bacteria, fungi, and viruses (Rosenthal, 2012).

26.5 POLYPHENOLS

Similar to tannins, polyphenols may bind to DNA bases and proteins by forming ionic and hydrogen interactions. This interaction modulates the activity of various proteins, such as enzymes and ion channels, leading to significant pharmacological effects. Polyphenols from legumes are known for their antioxidant, anti-inflammatory, antibacterial, antifungal, and antiviral properties (Wink, 2008b).

26.6 FURANOCOUMARINS (FC)

Furanocoumarins are lipophilic compounds that can easily penetrate cells and intercalate with DNA. When exposed to UV light, they form covalent bonds with pyrimidine bases like cytosine and thymine, causing apoptotic cell death. In the liver, FCs are converted into epoxides that alkylate DNA. Contact with skin can cause severe inflammation similar to sunburn. FCs are mutagenic and exhibit antimicrobial properties against bacteria, fungi, and viruses (Wink & Schimmer, 2010; Wink & Van Wyk, 2008).

26.7 SAPONINS

In legumes, saponins are mostly of the triterpene type and the steroidal saponins are mostly rare. Saponins disrupt biomembranes by forming complexes with cholesterol. The lipophilic core of these molecules intercalates into the membrane while the hydrophilic sugar side chains interact with glycoproteins or glycolipids outside the cells. Saponins can form pores in membranes and lyse cells at higher concentrations. They are bitter-tasting and target herbivores and microbes, especially fungi and viruses (Wink, 2008a).

26.8 CARDIAC GLYCOSIDES (CG)

Cardiac glycosides, a class of steroidal compounds found rarely in legumes, inhibit the Na^+/K^+ ATPase enzyme in animals making them highly toxic. These compounds are produced by a few taxa of the tribe Loteae, such as *Coronilla* and *Securigera*, and synthesize cardenolides like corotoxigenin and glaucotoxigenin. These glycosides have potent cardiotonic effects (Wink & Van Wyk, 2008).

26.9 PHARMACOLOGICAL EFFECTS

Numerous medicinal benefits of the Fabaceae family, such as antibacterial, antifungal, antiviral, antiprotozoal, antiplasmodial, antifeedant, insecticidal, and antiparasitic properties, were found through research. Polymenorrhea, anemia, ulcers, menorrhagia during pregnancy, diarrhea, and kwashiorkor caused by a protein deficit, are treated with these herbs. They work well against hyperglycemia brought on by thyroxine and hypocholesterolemia. Furthermore, plants in the Fabaceae family provide relief against bronchitis, upper respiratory infections, rheumatoid arthritis, gland inflammation, sleeplessness, stress, irregular heartbeats caused by nerves, and melanogenesis inhibition. They also control metabolism and energy expenditure. Anti-osteoporotic, anticancer, anti-atherogenic, anti-inflammatory, antinephritic, antidiabetic, laxative, sedative, diuretic, digestive, hepatoprotective, chemopreventive, neuroprotective, and cytotoxic actions are certain other noteworthy pharmacological qualities.

26.10 ANTIOXIDANT PROPERTIES

Various plants from the Fabaceae family, such as *Albizia lebbeck*, *Cassia fistula*, and *Cicer arietinum*, have demonstrated significant antioxidant activities in their methanol, chloroform and aqueous extracts analyzed through assays of reducing power, Radical scavenging activities such as DPPH. These works varied notably between extracts and assays. *Mucuna pruriens* seeds showed substantial antioxidant potential in methanol extracts, attributed to phenolic compounds, which act as powerful antioxidants. *Acacia saligna* flower extracts exhibited increasing antioxidant activity with concentration, though less potent than standard antioxidants like Butylated hydroxytoluene (BHT). Comparative studies revealed that acetone extract of *Acacia rigidula* displayed the highest anti-peroxidation activity among tested Acacia species, due to higher levels of terpenes and tannins. Additionally, seeds and fruit rinds from six Fabaceae species, the methanolic extracts of *Cajanus indica*, showed high antioxidant activities. Investigations of Turkish fodder plants, including *Vicia villosa*, *Trifolium ochroleucum*, and *Onobrychis altissima*, highlighted antioxidant capacities, underlining their potential as forage with health benefits. The antioxidant properties of Fabaceae plants are primarily attributed to their rich content of secondary metabolites such as phenols, flavonoids, terpenes, and tannins contributing to their therapeutic potential and making them valuable both nutritionally and medicinally (Jadhavar et al., 2022)

26.11 ANTICANCER ACTIVITIES

Metabolites of Fabaceae with natural anticancer potential have gained attention in recent times. Compounds such as beta carotene, alpha-tocopherol, and kaempferol show promising results and some entering clinical trials. The Fabaceae family, known for its legumes, offers anticancer benefits due to being low-fat and high-protein and containing dietary fibers and phytochemicals. Isoflavones like genistein and daidzein inhibit tumor formation, while lectins exhibit antitumor properties. Saponins in legumes such as soybeans and chickpeas possess anticancer activities through mechanisms like apoptosis regulation. Antioxidant-rich phenolic compounds in legumes further support cancer prevention and treatment (Jadhavar et al., 2022). While research is still being conducted to fully understand the potential of some of these substances, others have been the subject of substantial study and are being used globally to combat cancer. These phytochemicals use a variety of strategies to fight cancer, including immune system modulation, apoptosis, oxidative stress reduction, induction of cell cycle arrest, and carcinogen inactivation. However further investigation is required to fully comprehend the anticancer qualities of Fabaceae metabolites, particularly their synergistic actions within a single plant and any potential detrimental physiological consequences over the long term in patients. Furthermore, even though a large number of studies have shown the effectiveness of Fabaceae phytochemicals in experimental settings, additional clinical trials are required to validate these results and develop standardized procedures for evaluating the bioavailability, efficacy, safety, quality, and regulatory needs of these compounds. Reaching global benchmarks in these domains will be essential for propelling this field's future research (Usman et al., 2022)

26.12 ANTI-INFLAMMATORY PROPERTIES

Inflammation, a natural defense mechanism, can lead to adverse effects when dysregulated, contributing to diseases like arthritis, heart disease, and cancer. Conventional treatments often involve non-steroidal anti-inflammatory drugs (NSAIDs), which can cause significant side effects. Hence, exploring natural anti-inflammatory agents from the Fabaceae family has been a focus. Studies have shown that extracts from various Fabaceae plants like *Albizia lebbeck*, *Cassia fistula*, and *Cicer arietinum* exhibit significant antioxidant and anti-inflammatory properties. *Adenantherapavonina* and *Acacia macrostachya* demonstrated dose-dependent anti-inflammatory effects when comparable to standard drugs. Similarly, *Albizia julibrissin* and *Flemingia strobilifera* extracts showed promising results in vitro and in vivo studies, with activities attributed to phenolic compounds and flavonoids. Additionally, *Parkia biglobosa* and *Lathyrus* spp. extracts have shown efficacy in stabilizing membranes and reducing inflammation and pain, suggesting that these plants could be potent natural alternatives for managing inflammation (Jadhavar et al., 2022).

26.13 ANTIVIRAL ACTIVITY

Investigations of the anti–Mayaro virus activity of *Cassia australis* leaf extracts reveal that flavonoids and condensed tannins in the extracts have been proven to activate against the virus. The cytotoxicity and antiviral activities showed 85% cytotoxicity and exhibited over 90% virus inhibition, surpassing the standard antiviral drug ribavirin. This study highlights *C. australis* as a potential source of antiviral compounds (Spindola et al., 2014). *Cassia alata* leaf extract against the dengue virus showed that the butanol fraction demonstrated significant antiviral activity with favorable IC_{50}, CC_{50}, and SI values. Both studies emphasize the promising antiviral properties of *Cassia* species, suggesting further research to understand their mechanisms and potential as antiviral drugs (Angelina et al., 2017).

26.14 METABOLIC OCCURRENCE IN THE FABACEAE FAMILY

26.14.1 Genus – *Crotalaria*

The reason *Crotalaria*, often referred to as rattlepod or rattlebox, gets its name is because of the distinctive noise made by the developing seeds inside the pod. About 600 species belong to this genus, which is found worldwide in tropical and subtropical areas. Generally, species of *Crotalaria* are plants or shrubs with simple or digitately foliate leaves and yellow flowers with anthers that are dimorphic. While many species are used for aesthetics and green manure, some, like *Crotalaria retusa*, have therapeutic use; their seeds are used to treat fevers. Furthermore, *Crotalaria juncea* has characteristics that may be helpful in the fight against diarrhea, cancer, and obesity (Yaquba et al., 2020).

26.14.2 Genus – *Senna*

Several bioactive metabolites, such as flavonoids, anthrones, terpenoids, anthraquinones, tannins, glycosides, steroids, alkaloids, polyphenols, saponins, and essential oils, have been identified by extensive phytochemical analyses of several *Senna* species. Numerous compounds with notable in vitro and in vivo activities were found by this research, opening the door to the development of new and effective medications. Senna is rich in phenolic compounds, which have different chemical compositions according to the extraction technique, solvent polarity, and plant location. Procyanidins, flavones, glycosylated flavonols, and flavan-3-ols are notable phenolic chemicals that have been identified. Major components of *Senna* and anthraquinones comprise derivatives of dimeric anthracene and anthraquinones, which are distinguished by their distinct structures and biological activity. Alkaloids with a variety of pharmacological characteristics have been found, including derivatives of pyridine and piperidine. Stems such as β-sitosterol and stigmasterol, as well as terpenoids, especially triterpenoids and sesquiterpenoids, have also been identified. *Senna* species have therapeutic potential, which is supported by the structural variety and bioactivity of these phytoconstituents, which may be used in herbal medicine and drug development. (Oladeji et al., 2021).

26.14.3 Genus – *Acacia*

The existence of various secondary metabolites in the genus *Acacia* like saponosides, alkaloids, peptides, and a new chromen derivative has been reported. These compounds are primarily isolated from species like *Acacia nilotica* and *Acacia ataxacantha*. Secondary metabolites from *Acacia* species exhibit diverse biological activities, including antifungal, antibacterial, anticancer, antioxidant, antidiabetic, antiparasitic, cytotoxic, and immunomodulatory properties. These activities make them potentially valuable in the treatment of various diseases. For example, triterpenoids, betulinic acid, gallic acid methyl ester, and flavonoids showed significant antibacterial activity against various bacterial strains. They also show promising antimalarial activity against *Plasmodium falciparum* and *Plasmodium berghei*, responsible for causing malaria. The species belonging to *Acacia* are rich sources of polyphenolic compounds with strong antioxidant properties. These compounds help in preventing oxidative stress-related diseases like cardiovascular and neurodegenerative disorders. Several bioactive compounds of this genus demonstrate anticancer properties. Gallic acid, catechins, and other polyphenolic compounds exhibit cytotoxic effects against uveal melanoma cell lines and other cancer cell lines. This genus also shows hypoglycemic effects, making them potentially beneficial in the management of diabetes. Compounds like betulin and polysaccharides from *Acacia tortilis* demonstrate significant reductions in blood glucose levels in diabetic animal models. Although limited, some studies suggest that extracts may influence immune response and cytokine induction, potentially indicating immunomodulatory properties. Certain compounds, such saponins, quercetin, and erythrodiol, show promise as anticancer medicines by exhibiting cytotoxic actions against different cancer cell lines. The other activities include their capacity to reduce nitric oxide production, inhibit acetylcholinesterase activity, and exhibit antimicrobial and antioxidant activities (Amoussa et al., 2020).

26.14.4 Genus – *Hymenaea*

The *Hymenaea* genus belongs to the family Fabaceae. The species in the genus are carefully examined in this study with special attention to their traditional medicinal usage as their bioactive substance holds prospective therapeutic benefits. The principal contributions to the therapeutic capabilities of the genus *Hymenaea* have been identified as flavonoids, terpenoids, and phenolic compounds, which the authors note as being abundant in these secondary metabolites. The pharmacological properties of anti-inflammatory, antibacterial, antioxidant, and anticancer effects are displayed by these substances. Those bioactive substances exert their effects by synthesizing results from frequent investigations. For example, it has been demonstrated that flavonoids found in the genera alter important signaling pathways linked to inflammation and cancer, while terpenoids have strong antibacterial action against a range of diseases. Furthermore, phenolic chemicals are well known for their potent antioxidant capabilities aiding in the mitigation of illnesses linked to oxidative stress. These chemical components emphasize the genus's potential for future pharmacological advancements as well as its medical utility. To confirm the pharmacological potential of *Hymenaea* species and to investigate their bioactive components as possible new therapeutic agents, the study ends with recommendations for future research directions. For those working in the domains of pharmacology and natural product chemistry, this comprehensive analysis of the genus *Hymenaea* is a useful resource (da Silva Oliveira et al., 2018).

26.14.5 Genus – *Lupinus*

Alkaloids, flavonoids, and terpenoids are just a few of the many bioactive substances that phytochemical studies uncover and that support the plant's medicinal potential. Extensive research has been conducted on the biological activity of these substances, revealing the noteworthy anti-inflammatory, antioxidant, antibacterial, and anticancer potential of the *Lupinus* species as a sustainable source of bioactive compounds and in phytoremediation. The pharmacological characteristics and uses in contemporary medicine highlight the significance of additional study to fully realize the plant's therapeutic and commercial potential. The *Lupinus* genus stresses its value in both traditional and current aspects (Ishaq et al., 2022).

26.14.6 Genus – *Ceasalpinieae*

The diverse range of bioactive compounds found in these plants, including flavonoids, terpenoids, and alkaloids, contribute to their significant medicinal potential. Various pharmacological activities such as anti-inflammatory, antimicrobial, antioxidant, and antidiabetic effects, emphasize the importance of these plants in traditional and modern medicine. Specific species like *Caesalpinia bonduc* and *Caesalpinia pulcherrima* are noted for their potent bioactivities; for instance, *C. bonduc* is reported to have anticonvulsant and anti-inflammatory properties, while *C. pulcherrima* shows promise in treating asthma and bronchospasm. The review also addresses the need for further research to isolate and characterize these bioactive compounds and to understand their mechanisms of action (El-Nashar et al., 2015).

26.14.7 Genus – *Dalbergia*

The genus *Dalbergia* comprises over 274 species of woody climbers, lianas, shrubs, and trees. Traditionally, *Dalbergia* species – which are widely dispersed in tropical and subtropical regions – have been used to cure a variety of illnesses including inflammation, bleeding problems, and aphthae. The genus's extensive phytochemical variety is highlighted in the study that also mentions the existence of terpenoids, flavonoids, tannins, and saponins. Numerous pharmacological properties, such as antibacterial, anti-inflammatory, antioxidant, and hepatoprotective effects, are displayed by these substances. The review highlights the importance of *Dalbergia* species in traditional therapeutic methods across many cultures from an ethnomedical perspective. It also highlights the necessity of more clinical and pharmacological research to support conventional usage and investigate novel treatment avenues. Such phytochemical and pharmacological profile reporting offers a useful platform for the development of innovative drugs and therapeutic agents derived from *Dalbergia* species (Saha et al., 2013)

26.14.8 Genus – *Erythrina*

Several species of *Erythrina* are known for their distinctly bioactive alkaloids and great variety. Numerous biological actions, including anti-inflammatory, anticonvulsant, antibacterial, and analgesic effects have been exerted by the alkaloids. Several alkaloids found in distinct *Erythrina* species correlate with their chemical structures and the pharmacological effects reported. It also explores preclinical research that assesses these alkaloids' therapeutic potential in various illness models, presenting encouraging findings that call for greater research. Limited clinical trials are also presented, highlighting the potential medicinal uses of substances produced from this genus. The need for larger-scale clinical research to confirm these substances' safety and effectiveness in human beings. Despite some difficulties in developing natural product-based drugs, including problems with bioavailability, standardization, and possible adverse effects. All things considered, this work offers a comprehensive review of the present status of research on *Erythrina* alkaloids, urging more investigation into their possible therapeutic uses and adding to the expanding corpus of information about ethnopharmacology and natural product research topics (Rambo et al., 2019).

Some plant metabolites are known to be toxic or hallucinogenic because they contain specific alkaloids, terpenoids, or other secondary metabolites that act on neuroreceptors, neurotransmitter-degrading enzymes, ion channels, ion pumps, or cytoskeletal elements in humans or animals. Many of these metabolites are extracted and used in modern medicine for their specific effects shaped by evolutionary molecular modeling. However, most plants used in phytotherapy do not contain the specific compounds listed in pharmacological tables; instead, they are used as complex extracts that contain numerous metabolites from different structural groups, and the bioactivity of these extracts often results from synergistic interactions among multiple metabolites, making it challenging to credit the therapeutic effects to a single compound (Wink, 2015).

Multiple methods are utilized to pinpoint the genes and mutations responsible for the phenotypes associated with legume domestication (Olsen et al., 2013). One such technique is the examination of secondary metabolites, which are often more abundant in their wild counterparts than in their domesticated counterparts because of artificial selection (Gepts, 2017). This selection frequently reduces genetic diversity by favoring features that are advantageous for farming and commercial qualities at the price of advantageous secondary metabolites. However, conventional plant breeding techniques, which have been employed for millennia to create new crop varieties with desired features, may be used to improve farmed legumes by using the genetic diversity of wild populations. In this technique, parents with desirable qualities are crossed, and progenies that show both high yield and the desired trait – such as pest resistance – are chosen (Ku et al., 2020a).

26.15 BIOSYNTHESIS

Leguminous plants are mostly pigmented owing to polyphenols, of which flavonoids like anthocyanins, flavanol glycosides, and proanthocyanidins are important contributors. The production of naringenin chalcone, which is mediated by chalcone synthase (CHS) (Dastmalchi et al., 2014), is the first step in the flavonoid biosynthesis pathway that is conserved among plants that produce seeds. Numerous enzymes are involved in this process, such as flavone 3-hydroxylase (F3H) and chalcone isomerase (CHI). Research on legumes, such as peanuts and soybeans, has revealed important genes and loci linked to the color of the seed coat (Tanaka et al., 2008; Li et al., 2016).

26.16 BIOENGINEERING OF SECONDARY METABOLITES IN LEGUMES

26.16.1 Genetic Engineering

Modifying secondary metabolite contents through plant metabolic engineering involves identifying genes in biosynthesis pathways and altering gene expressions. For example, overexpressing the gene encoding isoflavone reductase (IFR) in soybeans enhances resistance to *Phytophthora sojae* by increasing glyceollin production. Similarly, overexpressing chalcone isomerase in soybeans boosts daidzein accumulation and pathogen resistance (Ku et al., 2020b).

26.17 BIOTECHNOLOGICAL APPROACHES

In alfalfa, engineering genistein glucoside production improves nutritional value. Transgenic approaches such as RNAi-mediated gene silencing in peas, have identified intermediaries in secondary metabolite pathways (Deavours & Dixon, 2005; Kaimoyo et al., 2008). Emerging techniques like biosynthesis in microorganisms and manipulating transcription factors offer additional strategies for optimizing secondary metabolite production.

26.18 METABOLIC MODELS

Genome-scale models help represent the metabolic capabilities of legumes, integrating various omics data to enhance understanding of plant-microbe interactions and design efficient biosynthesis pathways for secondary metabolites. This approach has been applied to crops like alfalfa and soybeans, offering insights into optimizing secondary metabolite production (DiCenzo et al., 2016).

26.19 HOW TO ENHANCE THE SECONDARY METABOLITE PRODUCTION

Nonetheless, biotechnology presents viable pathways to enhance their manufacturing. One strategy is to increase the expression of genes linked to the manufacture of metabolites, albeit this can occasionally be impeded by incomplete genetic information. The text also compares and contrasts yeast and fungus to illustrate how the cell wall and environment interact. There may be a direct or indirect connection between metabolite synthesis and cell wall integrity since mutants with modified cell wall composition have been demonstrated to have modifications in secondary metabolite production. Furthermore, using cell wall inhibitors may cause modifications to the composition of the cell wall, which may affect the synthesis of secondary metabolites. Micronutrients such as silicon (Si) and boron (B) also affect the characteristics of cell walls and the accumulation of secondary metabolites. Secondary metabolite changes have been associated with shortages in boron, but silica has been demonstrated to improve plant vigor and stress tolerance, which may result in enhanced metabolite synthesis. Silicon application has been demonstrated to have beneficial effects on plant tissue culture, encouraging secondary metabolism and cell development. All things considered, comprehending how the cell wall, micronutrients, and secondary metabolism interact offers hope for improving the synthesis of beneficial metabolites in plants by biotechnological treatments (Reshi et al., 2023)

26.20 INTRODUCTION TO IN VITRO CELL CULTURE AND GENETIC TRANSFORMATION USING

26.20.1 *AGROBACTERIUM TUMEFACIENS*

These days, transgenic plant cultivation, genetic variation creation, and clonal multiplication are all accomplished in vitro plant cultures. They support pharmaceutical and environmental biotechnology in the production of plant secondary metabolites, as well as the reduction of ecological dangers and improvement of phytoremediation. Differentiation is employed to develop distinct secondary metabolites from cultures such as calli, cell suspensions, roots, and shoots. Selecting the appropriate culture type is necessary for efficient metabolite synthesis (Hemanthakumar et al., 2019; Kowalczyk et al., 2020).

Discovered more than a century ago, *Agrobacterium tumefaciens* genetically modifies plant genomes by introducing T-DNA into the plant's DNA, which results in tumors. This technique, which was originally applied in 1983 to produce transgenic plants, is now considered a mainstay of plant biotechnology. It permits both temporary and permanent genetic changes, allowing for the introduction of novel features in both dicot and monocot plants. Introducing the soybean isoflavone genistein onto non-legume plants by genetic modification through the *Agrobacterium*-mediated gene Transfer (AMGT) method. This entails using genetic editing techniques to change secondary metabolic pathways. The goal of the research is to increase the production of genistein in non-legume plants by introducing the genes responsible for isoflavone biosynthesis from soybeans. To produce larger yields of genistein – a bioactive substance with substantial promise for use in pharmaceutical and health applications – the study investigates strategies for optimizing these genetic alterations. The study emphasizes the viability and advantages of employing genetic engineering to increase the synthesis of useful plant metabolites beyond what is produced naturally (Liu et al., 2007).

26.21 SECONDARY METABOLITE PRODUCTION THROUGH HAIRY ROOT CULTURES USING ELICITORS

Hairy root cultures, which are genetically and biochemically stable and perfect for the synthesis of secondary metabolites, are produced by *Agrobacterium rhizogenes*. Numerous biotic and abiotic stimulants activate biosynthetic pathways, increasing the yields of metabolites such as alkaloids, terpenoids, and flavonoids. To maximize production, elicitor type, concentration, and exposure period must all be optimized. The pharmaceutical and agricultural industries might benefit financially and commercially from this strategy. Subsequent investigations ought to integrate elicitation with metabolic engineering and systems biology to promote the production of metabolites through this method (Halder et al., 2019). Employing hairy root cultures in pigeon pea crops to increase the yield of the important isoflavonoid isowighteone. The study shows a considerable rise in the synthesis and secretion of isowighteone when numerous elicitors, including salicylic acid, methyl jasmonate, and yeast extract are co-treated with these cells. The study demonstrates how well distinct elicitor combinations may promote secondary metabolite biosynthesis, offering valuable information for improving bioactive chemical manufacturing processes in plant biotechnology (Gajurel et al., 2022).

26.22 EPIGENETIC APPROACHES TO SECONDARY METABOLITE PRODUCTION

Silent biosynthetic gene clusters (BGCs) prevent the production of metabolites, which are important for ecological processes and useful in medicine and nutraceuticals. These clusters can be activated by epigenetic processes like as histone acetylation and DNA methylation, which can result in the creation of new bioactive substances. The variety and quantity of secondary metabolites can be increased using strategies that use tiny chemical modifiers to change the chromatin structure and increase gene expression. These strategies may be used to find novel antibiotics and therapeutic substances. Further research is needed to completely comprehend these epigenetic pathways and create scalable commercial metabolite production (Yang et al., 2014; Bind et al., 2022).

Through precise editing of the genes involved in these processes, CRISPR-Cas technology improves the synthesis of secondary metabolites in plants. Enhancing the synthesis of terpenoids, flavonoids, and alkaloids by targeting transcription factors, biosynthetic genes, or regulatory elements is the advantage of CRISPR-Cas in the fields of biotechnology, agriculture, and medicine. The work demonstrates the usefulness of CRISPR-Cas and how it may revolutionize plant metabolic engineering. (Bhojiya & Joshi, 2024). Numerous genes were targeted those involved in isoflavone synthesis and soybean mosaic virus resistance using CRISPR/Cas9 technology. Important genes that are involved in the production of flavonoids, especially

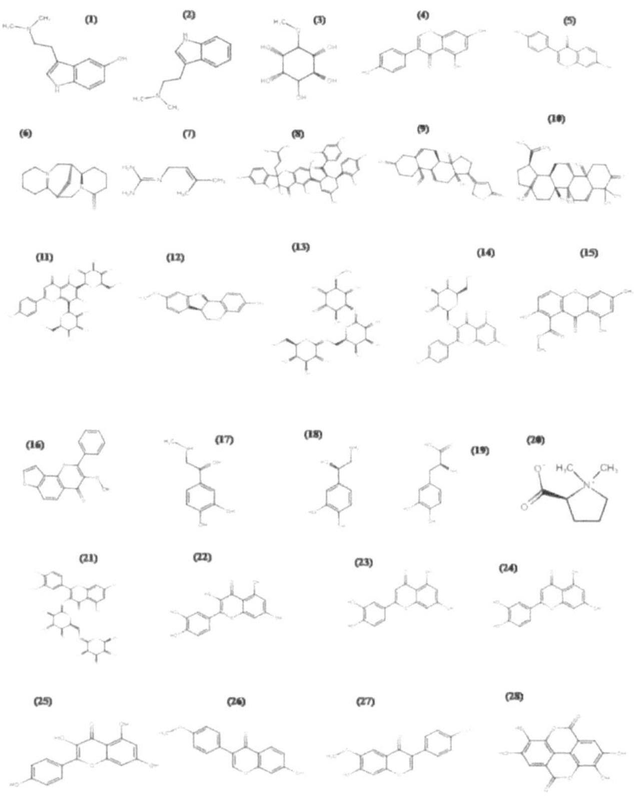

FIGURE 26.2 Schematic representation of the production of secondary metabolites through in vitro cell culture in leguminous plants.

isoflavones such as daidzein, genistein and glycitein which confers viral resistance, were altered. As a consequence, soybeans' resistance to viral infections was enhanced and their levels of isoflavones were increased. This technique highlights the promise of CRISPR/Cas9-mediated metabolic engineering in agricultural applications by strengthening the plant's resistance to disease and introducing a novel strategy for improving the nutritional content of soybeans (Zhang et al., 2020).

26.23 CONCLUSION

Considering a broad array of genera and species, the Fabaceae family has a great deal of medicinal promise because of its abundance of secondary metabolites, which include flavonoids, terpenoids, alkaloids, and saponins. These metabolites have antiviral, anticancer, anti-inflammatory, and antioxidant qualities. Genetic engineering breakthroughs such as microbial biosynthesis, RNA interference, and gene overexpression have increased the yield of these important chemicals. Metabolite synthesis may be further optimized by employing epigenetic procedures such as DNA methylation and histone acetylation, in conjunction with biotechnological approaches like cell wall modification and vitamin supplementation. Plant biotechnology was transformed by Gottlieb Haberlandt's groundbreaking work in 1902 on in vitro cell culture, which made it possible to cultivate plant cells under controlled conditions – a critical capability for applications in agriculture, the environment, and medicine. Genetic transformation has been made possible by *Agrobacterium tumefaciens*, and transgenic plant development has been made possible by the genetic transformation process, which is aided by *Agrobacterium tumefaciens*. Further techniques, such as electroporation and microprojectile bombardment, increase the potential for genetic alteration. Metabolite production is improved by combining metabolic engineering with elicitation approaches in hairy root cultures. These integrated methods emphasize the value of Fabaceae plants for drug development and improving human health, as well as their potential to advance medical research and sustainable healthcare (Figure 26.2).

REFERENCES

Alshehri, M. M., Sharifi-Rad, J., Herrera-Bravo, J., Jara, E. L., Salazar, L. A., Kregiel, D., Uprety, Y., Akram, M., Iqbal, M., Martorell, M., Torrens-Mas, M., Pons, D. G., Daştan, S. D., Cruz-Martins, N., Ozdemir, F. A., Kumar, M., & Cho, W. C. (2021). Therapeutic potential of isoflavones with an emphasis on Daidzein. *Oxidative Medicine and Cellular Longevity, 2021*, 6331630.

Al-Snafi, A. E. (2016). The pharmacological and toxicological effects of *Coronilla varia* and *Coronilla scorpioides*: A review. *The Pharmaceutical and Chemical Journal, 3*(2), 105–114.

Amoussa, A. M. O., Sanni, A., & Lagnika, L. (2020). Chemical diversity and pharmacological properties of genus *Acacia*. *Asian Journal of Applied Sciences, 13*, 40–59.

Angelina, M., Hanafi, M., Suyatna, F. D., Ratnasari, S., & Dewi, B. E. (2017, December). Antiviral effect of sub fraction *Cassia alata* leaves extract to dengue virus Serotype-2 strains new guinea C in human cell line Huh-7 it-1. In *IOP Conference Series: Earth and Environmental Science* (Vol. 101, No. 1, p. 012004). IOP Publishing.

Anuranjana, P. V., Beegum, F., K P, D., George, K. T., Viswanatha, G. L., Nayak, P. G., Kanwal, A., Kishore, A., Shenoy, R. R., & Nandakumar, K. (2023). Mechanisms behind the pharmacological application of biochanin-A: A review. *F1000Research, 12*, 107.

Bednarska, K., Kuś, P., & Fecka, I. (2020). Investigation of the phytochemical composition, antioxidant activity, and methylglyoxal trapping effect of *Galega officinalis* L. Herb in vitro. *Molecules (Basel, Switzerland), 25*(24), 5810.

Bernatoniene, J., Kazlauskaite, J. A., & Kopustinskiene, D. M. (2021). Pleiotropic effects of isoflavones in inflammation and chronic degenerative diseases. *International Journal of Molecular Sciences, 22*(11), 5656.

Bezerra, J. J. L., Pinheiro, A. A. V., & Lucena, R. B. (2021). Phytochemistry and teratogenic potential of Mimosa tenuiflora (willd.) poir.(Fabaceae) in ruminants: A systematic review. *Toxicon, 195*, 78–85.

Bhojiya, A. A., & Joshi, H. (2024). CRISPR gene editing for secondary metabolite production: A review. *Gene editing in plants: CRISPR-Cas and its applications* (pp. 437–475). Springer.

Bind, S., Bind, S., Sharma, A. K., & Chaturvedi, P. (2022). Epigenetic modification: A key tool for secondary metabolite production in microorganisms. *Frontiers in Microbiology, 13*, 784109.

Bratkov, V. M., Shkondrov, A. M., Zdraveva, P. K., & Krasteva, I. N. (2016). Flavonoids from the Genus *Astragalus*: Phytochemistry and biological activity. *Pharmacognosy Reviews*, *10*(19), 11–32.

Chao, Z., Min, C., Shan-Liang, S., Jiao-Jiao, W., Zhong, Y., Huan-Huan, C. H. E. N., He-Min, L. I., Han, X. U., Nian-Guang, L. I., Hong-Yue, M. A., & Xiao-Long, W. (2021). Bufotenine and its derivatives: Synthesis, analgesic effects identification and computational target prediction. *Chinese Journal of Natural Medicines*, *19*(6), 454–463.

da Silva Oliveira, F. G., de Souza Araújo, C., Rolim, L. A., Barbosa-Filho, J. M., & da Silva Almeida, J. R. (2018). The genus *Hymenaea* (Fabaceae): A chemical and pharmacological review. *Studies in Natural Products Chemistry*, *58*, 339–388.

Dastmalchi, M., & Dhaubhadel, S. (2014). Soybean seed isoflavonoids: Biosynthesis and regulation. *Phytochemicals–Biosynthesis, Function and Application*, *44*, 1–21.

de Beer, D., Human, C., & Joubert, E. (2023). Phenolic composition of honeybush and changes during herbal tea and extract production. In J. M. Mérillon, C. Riviere, & G. Lefèvre (Eds.), *Natural products in beverages*. Reference Series in Phytochemistry. Springer. https://doi.org/10.1007/978-3-031-04195-2_219-1

Deavours, B. E., & Dixon, R. A. (2005). Metabolic engineering of isoflavonoid biosynthesis in alfalfa. *Plant Physiology*, *138*(4), 2245–2259.

DiCenzo, G. C., Zamani, M., Milunovic, B., & Finan, T. M. (2016). Genomic resources for identification of the minimal N2-fixing symbiotic genome. *Environmental Microbiology*, *18*(8), 2534–2547.

Ding, M., Bao, Y., Liang, H., Zhang, X., Li, B., Yang, R., & Zeng, N. (2024). Potential mechanisms of formononetin against inflammation and oxidative stress: A review. *Frontiers in Pharmacology*, *15*, 1368765.

El-Nashar, H., Eldahshan, O., & Singab, A. N. (2015). The tribe Caesalpinieae (Fabaceae): An updated review on pharmacological aspects. *Medicinal and Aromatic Plants*, *4*(215), 2167–0412.

Erb, M., & Kliebenstein, D. J. (2020). Plant secondary metabolites as defenses, regulators, and primary metabolites: The blurred functional trichotomy. *Plant Physiology*, *184*(1), 39–52.

Gajurel, G., Nopo-Olazabal, L., Hendrix, E., & Medina-Bolivar, F. (2022). Production and secretion of isowighteone in hairy root cultures of pigeon pea (*Cajanus cajan*) co-treated with multiple elicitors. *Plants*, *11*(6), 834.

Gepts, P. (2017). Genetic aspects of crop domestication. In *Routledge handbook of agricultural biodiversity* (pp. 147–167). Routledge.

Gnanaraj, C., Govendan, M., Loo, C. Y., Yong, Y. S., Sekar, M., Mat Taib, C. N., Subramaniam, S. D., Shah, M. D., & Lee, W. H. (2024). Karanjin: A potential furanoflavonoid for neuroprotection. *Phytochemistry Reviews*, *23*(5), 1351–1375.

Gupta, A., Singh, A. K., Kumar, R., Jamieson, S., Pandey, A. K., & Bishayee, A. (2021). Neuroprotective potential of ellagic acid: A critical review. *Advances in Nutrition (Bethesda, MD)*, *12*(4), 1211–1238.

Halder, M., Sarkar, S., & Jha, S. (2019). Elicitation: A biotechnological tool for enhanced production of secondary metabolites in hairy root cultures. *Engineering in Life Sciences*, *19*(12), 880–895.

Hemanthakumar, A. S., Preetha, T. S., Pillai, P. P., & Seeni, S. (2019). Embryogenesis followed by enhanced micro-multiplication and eco-restoration of *Calamus thwaitesii* Becc.: An economic non-wood forest produces for strengthening agroforestry system. *Agroforestry Systems*, *93*, 1093–1105.

Ishaq, A. R., El-Nashar, H. A., Younis, T., Mangat, M. A., Shahzadi, M., Ul Haq, A. S., & El-Shazly, M. (2022). Genus *Lupinus* (Fabaceae): A review of ethnobotanical, phytochemical and biological studies. *Journal of Pharmacy and Pharmacology*, *74*(12), 1700–1717.

Jadhavar, P., & Deshpande, D. S. (2022). Recent updates on medicinal potentiality of fabaceae family: Critical. *International Journal of Pharmaceutical Sciences*, *13*(3), b32–b41.

Kaimoyo, E., & VanEtten, H. D. (2008). Inactivation of pea genes by RNAi supports the involvement of two similar O-methyltransferases in the biosynthesis of (+)-pisatin and of chiral intermediates with a configuration opposite that found in (+)-pisatin. *Phytochemistry*, *69*(1), 76–87.

Kowalczyk, T., Wieczfinska, J., Skała, E., Śliwiński, T., & Sitarek, P. (2020). Transgenesis as a tool for the efficient production of selected secondary metabolites from plant in vitro cultures. *Plants*, *9*(2), 132.

Ku, Y. S., Contador, C. A., Ng, M. S., Yu, J., Chung, G., & Lam, H. M. (2020a). The effects of domestication on secondary metabolite composition in legumes. *Frontiers in Genetics*, *11*, 581357.

Ku, Y. S., Ng, M. S., Cheng, S. S., Lo, A. W. Y., Xiao, Z., Shin, T. S., Chung, G., & Lam, H. M. (2020b). Understanding the composition, biosynthesis, accumulation and transport of flavonoids in crops for the promotion of crops as healthy sources of flavonoids for human consumption. *Nutrients*, *12*(6), 1717.

Li, Z. R., Li, J., Gu, J. P., Lai, J. Y., Duggan, B. M., Zhang, W. P., Li, Z. L., Li, Y. X., Tong, R. B., Xu, Y., Lin, D. H., Moore, B. S., & Qian, P. Y. (2016). Divergent biosynthesis yields a cytotoxic aminomalonate-containing precolibactin. *Nature Chemical Biology*, *12*(10), 773–775.

Liao, L., Tang, Y., Li, B., Tang, J., Xu, H., Zhao, K., & Zhang, X. (2023). Stachydrine, a potential drug for the treatment of cardiovascular system and central nervous system diseases. *Biomedicine & Pharmacotherapy*, *161*, 114489.

Liu, R., Hu, Y., Li, J., & Lin, Z. (2007). Production of soybean isoflavone genistein in non-legume plants via genetically modified secondary metabolism pathway. *Metabolic Engineering*, *9*(1), 1–7.

Maroyi, A. (2023). Medicinal uses of the Fabaceae family in Zimbabwe: A review. *Plants*, *12*(6), 1255.

Matsouka, I., Beri, D., Chinou, I., Haralampidis, K., & Spyropoulos, C. G. (2011). Metals and selenium induce medicarpin accumulation and excretion from the roots of fenugreek seedlings: A potential detoxification mechanism. *Plant Soil*, *343*, 235–245.

Obistioiu, D., Cocan, I., Tîrziu, E., Herman, V., Negrea, M., Cucerzan, A., & Alexa, E. (2021). Phytochemical profile and microbiological activity of some plants belonging to the Fabaceae family. *Antibiotics*, *10*(6), 662.

Oladeji, O. S., Adelowo, F. E., & Oluyori, A. P. (2021). The genus *Senna* (Fabaceae): A review on its traditional uses, botany, phytochemistry, pharmacology and toxicology. *South African Journal of Botany*, *138*, 1–32.

Olsen, J. E., Hoegh-Andersen, K. H., Casadesús, J., Rosenkrantz, J. T., Chadfield, M. S., & Thomsen, L. E. (2013). The role of flagella and chemotaxis genes in host pathogen interaction of the host adapted *Salmonella enterica* serovar Dublin compared to the broad host range serovar *S. Typhimurium*. *BMC Microbiology*, *13*, 1–11.

Periferakis, A., Periferakis, K., Badarau, I. A., Petran, E. M., Popa, D. C., Caruntu, A., Costache, R. S., Scheau, C., Caruntu, C., & Costache, D. O. (2022). Kaempferol: Antimicrobial properties, sources, clinical, and traditional applications. *International Journal of Molecular Sciences*, *23*(23), 15054.

Rambo, D. F., Biegelmeyer, R., Toson, N. S., Dresch, R. R., Moreno, P. R. H., & Henriques, A. T. (2019). The genus *Erythrina* L.: A review on its alkaloids, preclinical, and clinical studies. *Phytotherapy Research*, *33*(5), 1258–1276.

Reshi, Z. A., Ahmad, W., Lukatkin, A. S., & Javed, S. B. (2023). From nature to lab: A review of secondary metabolite biosynthetic pathways, environmental influences, and in vitro approaches. *Metabolites*, *13*(8), 895.

Rosenthal, G. (2012). *Plant nonprotein amino and imino acids: Biological, biochemical, and toxicological properties*. Elsevier.

Saha, S., Shilpi, J. A., Mondal, H., Hossain, F., Anisuzzman, M., Hasan, M. M., & Cordell, G. A. (2013). Ethnomedicinal, phytochemical, and pharmacological profile of the genus *Dalbergia* L. (Fabaceae). *Phytopharmacology*, *4*(2), 291–346.

Sánchez-Hidalgo, M., León-González, A. J., Gálvez-Peralta, M., González-Mauraza, N. H., & Martin-Cordero, C. (2021). D-Pinitol: A cyclitol with versatile biological and pharmacological activities. *Phytochemistry Reviews*, *20*, 211–224.

Sarao, L., Kaur, S., Malik, T., & Singh, A. (2022). Genistein and daidzein. In *Nutraceuticals and health care* (pp. 331–341). Academic Press.

Schenck, C. A., & Maeda, H. A. (2018). Tyrosine biosynthesis, metabolism, and catabolism in plants. *Phytochemistry*, *149*, 82–102.

Seigler, D. S., & Seigler, D. S. (1998). Phenylpropanoids. In *Plant secondary metabolism* (pp. 106–129). Springer.

Soares, A. R., Marchiosi, R., Siqueira-Soares, R. D. C., Barbosa de Lima, R., Dantas dos Santos, W., & Ferrarese-Filho, O. (2014). The role of L-DOPA in plants. *Plant Signaling & Behavior*, *9*(4), e28275.

Spindola KC, Simas NK, Salles TS, de Meneses MD, Sato A, Ferreira D, Romão W, Kuster RM. (2014). Anti-Mayaro virus activity of Cassia australis extracts (Fabaceae, Leguminosae). *Parasites & Vectors*, *7*, 1–7.

Taban, K., İlhan, M., & Süntar, I. (2023). Luteolin: Advances on resources, biosynthesis pathway, bioavailability, bioactivity, and pharmacology. In J. Xiao (Ed.), *Handbook of dietary flavonoids*. Springer.

Tanaka, Y., Sasaki, N., & Ohmiya, A. (2008). Biosynthesis of plant pigments: Anthocyanins, betalains and carotenoids. *The Plant Journal*, *54*(4), 733–749.

Tandon, V., & Das, B. (2018, June). Genistein: Is the multifarious botanical a natural anthelmintic too? *Journal of Parasitic Diseases*, *42*(2), 151–161.

Usman, M., Khan, W. R., Yousaf, N., Akram, S., Murtaza, G., Kudus, K. A., & Nazre, M. (2022). Exploring the phytochemicals and anti-cancer potential of the members of Fabaceae family: A comprehensive review. *Molecules*, *27*(12), 3863.

Wang, J., Li, Y., Li, A., Liu, R. H., Gao, X., Li, D., & Xue, Z. (2021). Nutritional constituent and health benefits of chickpea (*Cicer arietinum* L.): A review. *Food Research International*, *150*, 110790.

Wink, M. (1992). The role of quinolizidine alkaloids in plant insect interactions. In E. A. Bernays (Ed.), *Insect–plant interactions* (Vol. IV, pp. 133–169). CRC.

Wink, M. (2008a). Plant secondary metabolism: Diversity, function and its evolution. *Natural Products Communications*, *3*, 1934578X0800300801.

Wink, M. (2008b). Evolutionary advantage and molecular modes of action of multi-component mixtures used in phytomedicine. *Current Drug Metabolism*, *9*, 996–1009.

Wink, M. (2013). Evolution of secondary metabolites in legumes (Fabaceae). *South African Journal of Botany*, *89*, 164–175.

Wink, M. (2015). Modes of action of herbal medicines and plant secondary metabolites. *Medicines*, *2*(3), 251–286.

Wink, M., & Schimmer, O. (2010). Molecular modes of action of defensive secondary metabolites. In *Annual plant reviews Vol. 39: Functions and biotechnology of plant secondary metabolites* (pp. 21–161). Blackwell.

Wink, M., & Van Wyk, B.-E. (2008). *Mind-altering and poisonous plants of the world*. Briza Publications.

Xiao, L., Gu, Y., Gao, L., Shangguan, J., Chen, Y., Zhang, Y., & Li, L. (2017). Sanggenon C protects against pressure overload--induced cardiac hypertrophy via the calcineurin/NFAT2 pathway-. *Molecular Medicine Reports*, *16*(4), 5338–5346.

Xu, F., Huang, X., Wu, H., & Wang, X. (2018). Beneficial health effects of lupenone triterpene: A review. *Biomedicine & Pharmacotherapy*, *103*, 198–203.

Yadav, J. P., Arya, V., Yadav, S., Panghal, M., Kumar, S., & Dhankhar, S. (2010). *Cassia occidentalis* L.: A review on its ethnobotany, phytochemical and pharmacological profile. Fitoterapia, *81*(4), 223–230.

Yang, X., Huang, L., & Ruan, X. L. (2014). Epigenetic modifiers alter the secondary metabolite composition of a plant endophytic fungus, Pestalotiopsiscrassiuscula obtained from the leaves of *Fragaria chiloensis*. *Journal of Asian Natural Products Research*, *16*, 412–417.

Yaquba, M. S., Bushra, B., & Aslam, R. (2020). Secondary metabolites of the genus *Crotalaria* (Rattlepods) and their medicinal importance review. *Journal of Horticultural Science and Technology*, *3*(1), 1–7.

Zhang, P., Du, H., Wang, J., Pu, Y., Yang, C., Yan, R., & Yu, D. (2020). Multiplex CRISPR/Cas9-mediated metabolic engineering increases soya bean isoflavone content and resistance to soya bean mosaic virus. *Plant Biotechnology Journal*, *18*(6), 1384–1395.

27 Genetic Engineering and Genome Remodeling for Enhanced Production of Therapeutic Terpenes and Essential Oils of Lamiaceae

Santhanalakshmi Balasubramaniam, Suriya Prakash Eswaran, Anitha Sakthivel, Mayavan Subramani, and Kapildev Gnanajothi

27.1 INTRODUCTION

Plant terpenes are naturally occurring therapeutic compounds with significant uses in traditional and modern medicine. The terpenes and terpenoids are highly active components of essential oils derived from different plant families. In this regard, the species of Lamiaceae are reported to be the potential producers of distinct bioactive essential oils (Cherednichenko et al., 2021). The biotechnological tools for overproduction or enhancing the synthesis of the metabolites within the living system are gaining attention in recent times owing to their advantages. Therapeutically valuable terpenes and terpenoids with diversified and complex structures exhibit potent biological activities; synthesis and extraction of these compounds require advanced biotechnological tools and procedures. Recently, synthetic biology has confined its remarkable efficiency to produce substantial products by engineering on host organisms (Bashor et al., 2010). In this aspect, the approach of synthetic biology with potential methods for improving the yield of therapeutic terpenes and essential oils from the mint family has been reviewed in this chapter (Figure 27.1).

27.2 PLANT TERPENES, TERPENOIDS, AND ESSENTIAL OILS

Plant-derived essential oils (EO) are hydrophobic volatile compounds or secondary metabolites with potent biological activities, consists mixtures of diverse and complex biomolecules, especially terpenes (mostly monoterpenes and sesquiterpenes), terpenoids, and phenylpropanoids. Terpenes are special compounds of simple hydrocarbons spanning isoprene structure while terpenoids are defined as oxygenated hydrocarbons or modified terpenes that are functionally similar to terpenes (Siddiqui et al., 2024). The other major constituent is phenylpropanoids, which are benzene derivates linked to the three-carbon propene (Stevanovic et al., 2020). These components can be colored or colorless; usually aromatic and flavoring; lower in density and soluble in weakly polar or non- polar organic solvents (Stephane & Jules, 2020; Masyita et al., 2022). Essential oils and terpenes are majorly extracted from more than 50 plant families such as Asteraceae, Lamiaceae, Myrtaceae, Myristicaceae, Piperaceae, Rutaceae, Rosaceae, Lauraceae, Umbelliferae, Apiaceae, Poaceae, Zingiberaceae and so (Butnariu & Sarac, 2018; Stephane & Jules, 2020; Roba, 2020). These isolated compounds are still being reported for their widespread pharmacological activities like antimicrobial, antioxidant, antiviral, anti-inflammatory, anthelmintic, immunomodulatory, antinociceptive, anticancer, antidepressant, neuroprotective, etc. (Masyita et al., 2022; Siddiqui et al., 2024). Apart from these therapeutic effects the EO, terpenes and terpenoids have extensive commercial applications in the food industry as fragrances products, natural preservatives and used in culinary as traditional food (Carovic-Stanko et al., 2016); in cosmetics such as perfumery, skin, and hair care products; in agriculture as insecticide and larvicide (Hamilton et al., 2020); and also in aroma and pain therapy (Uritu et al., 2018).

27.3 CLASSIFICATIONS OF TERPENES AND TERPENOIDS

Terpenes and terpenoids are arranged based on their cyclic nature and categorized according to the number of isoprene (five-carbon base unit) structures. This structural diversity in isoprene repetitions, cyclization, and rearrangements are primary reasons for its physical and chemical functions. They are classified as hemiterpenes, with single isoprene unit (C5), monoterpenes (C10), sesquiterpenes (C15), diterpenes (C20), triterpenes (C30), and tetraterpenes (C40) (Ninkuu et al., 2021; Mosquera et al., 2021). The basic classification of terpenes and terpenoids in Lamiaceae are provided in Figure 27.2 and Table 27.1 with specified examples.

DOI: 10.1201/9781003475491-27

27.4 BIOSYNTHETIC PATHWAY OF TERPENE AND TERPENOIDS

The synthetic mechanism of terpenes initiates from isopentenyl pyrophosphate (IPP) and dimethylallyl pyrophosphate (DMAPP) (the five carbon units) driven by two metabolic routes, the mevalonate (MVA) in the cytosol, endoplasmic reticulum and peroxisomes and the methylerythritol phosphate (MEP) pathway in the chloroplast. The mevalonate pathway consisting of six steps starts with the acetyl-CoA, involving a series of reactions to produce IPP. The significant intermediates include mevalonic acid from 3-hydroxy-3-methylglutaryl-CoA (HMG-CoA), which resulted in a reduction of the key enzyme HMG reductase. This IPP on isomerization is inter-converted to DMAPP by IPP isomerase (IDI). The precursors for the biosynthesis of sesquiterpenes, polyprenols, phytosterols, brassinosteroids, and triterpenes were provided from this pathway (Arendt et al., 2016; Tholl, 2015).

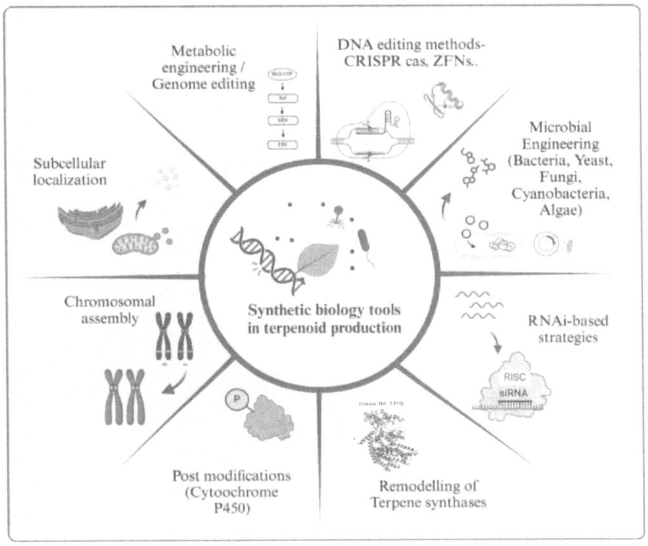

FIGURE 27.1 Synthetic biology tools in terpenoid production.

Unlike the MVA pathway, the MEP mechanism occurs specifically in the plastid region, initiated with pyruvate and glyceraldehyde-3-phosphate (GAP) synthesized from the Calvin cycle. The foremost metabolite of this pathway is the 1-deoxy-D-xylulose 5-phosphate (DXP) catalyzed by DXP synthase. The DXP is the rate-limiting reaction of this pathway. The end products IPP and its isomer DMAPP were simultaneously produced from hydroxy-3-methylbut-2-enyldiphosphate (HMBPP) and IPP was finally converted to DMAPP by IDI. The IPP and DMAPP (C5) are the direct precursors of terpenes synthesizing, Geranyl diphosphate (GPP) monoterpene with 10 carbons by condensation and farnesyl diphosphate (FPP), sesquiterpene with 15 carbons by double condensation. The diterpene and triterpene are further synthesized with a series of reactions as in Figure 27.3, which are catalyzed by specified terpene synthases, the key enzyme for terpenes and terpenoids diversity (Arendt et al., 2016; Roba, 2020).

27.5 ESSENTIAL OIL, TERPENES, AND TERPENOIDS OF LAMIACEAE

Lamiaceae is the aromatic plant family with more than 7000 plant species categorized under 240 genera. It was

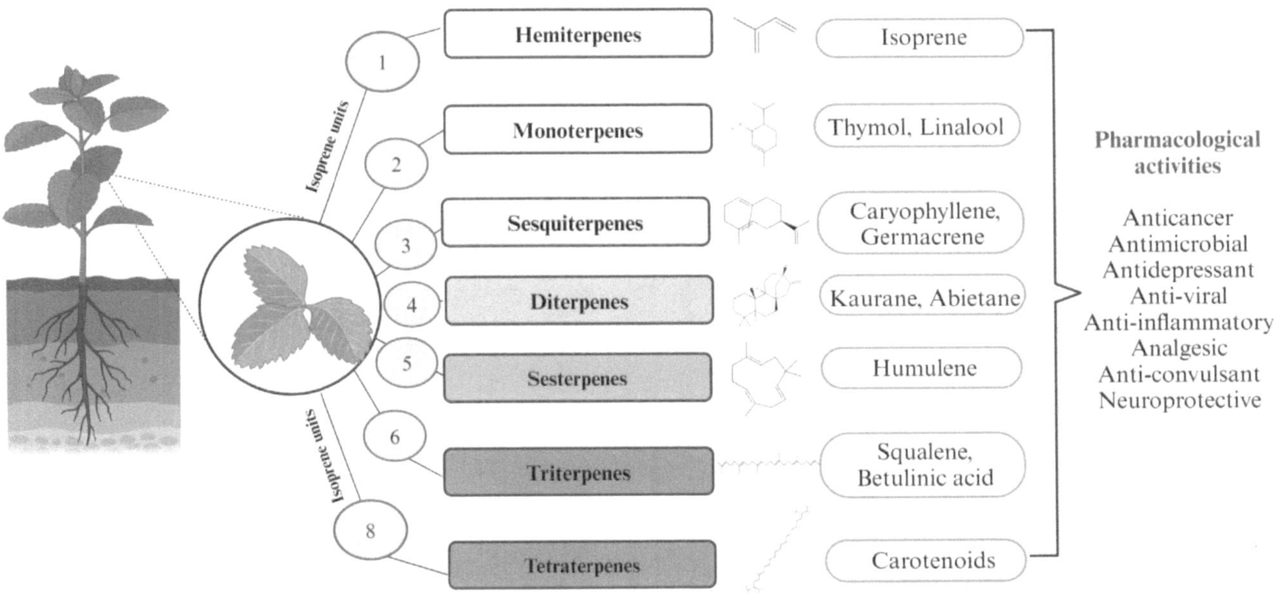

FIGURE 27.2 Terpenoids in Lamiaceae

TABLE 27.1
Terpenes and Terpenoids Classification with Examples from Lamiaceae Species

Classification	Terpenes	Terpenoids
Hemi (C_5)	Hemialboside	
Mono (C_{10})	• Myrcene (acyclic) • Menthane, limonene, *p*-cymene, terpinolene, sabinene, α-phellandrene, β-phellandrene (monocyclic) • Thujane, camphene, α-pinene (bicyclic)	• Linalool, neral, germinol, germinal (acyclic) • α-terpineol, terpinen-4-ol, menthol, isopulegol, thymol, Menthone, pulegone, piperitone (monocyclic) • Thujone, 1,8-cineole, camphor (bicyclic)
Sesqui (C_{15})	• Farnesene, farnesane (acyclic) • α-Humulene, germacrene D (monocyclic) • β-Caryophyllene, δ-selinene, bicyclogermacrene (bicyclic) • Bourbonane, aromadendrene (tricyclic)	• Farnesol (acyclic) • δ-Canadiene, caryophyllene oxide (bicyclic) • Aromadendrene oxide (tricyclic)
Di (C_{20})	• Camphorene, cambrene (monocyclic) • Kaurane, kaur-16-ene (tetracyclic)	• Phytol (acyclic) • 9-Geranyl-α-terpineol (monocyclic) • Manool (bicyclic)
Ses (C_{25})		Spathulenol
Tri (C_{30})	Ursolic acid, lupeol, betulinic acid	Ursolic acid, lupeol, betulinic acid

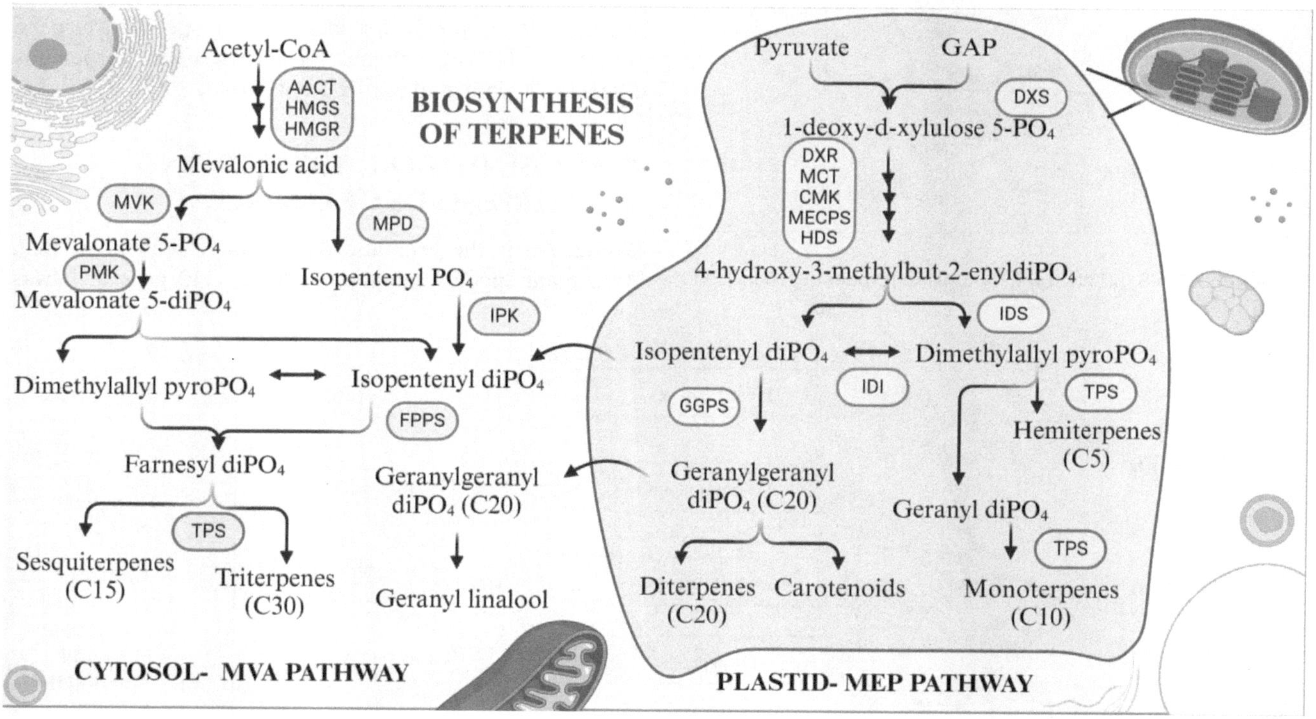

FIGURE 27.3 Biosynthesis of terpenes.

formerly known as Labiatae, owing to its floral structure characterized by bilabiate corolla. Lamiaceae are rich sources of terpene compounds and essential oils possessing numerous applications in the fields of medicine, food, cosmetics, and agriculture conferring great economic value (Ramos da Silva et al., 2021; Uritu et al., 2018). Major genera of Lamiaceae with potential terpene-containing essential oils are *Mentha*, *Ocimum*, *Salvia*, *Clerodendrum*, *Plectranthus*, *Hyptis*, *Satureja*, *Thymus*, *Origanum*, *Rosmarinus*, *Lavandula*, *Scutellaria*, *Marrubium*, *Sideritis*, *Lamium*, *Teucrium*, *Melissa*, *Ziziphora*, *Leonurus*, etc. (Baser & Kirimer, 2018). The terpenes and essential oil compounds detected in the Lamiaceae species are listed in Table 27.2.

TABLE 27.2
Essential Oil Components from Different Species of Lamiaceae

Species	Essential oil components (terpenes and terpenoids)	References
Anisomeles sp.		
Anisomeles indica	Camphene, β-pinene, camphor, borneol, α-thujone, linalyl acetate, anisole, caryophyllene, caryophyllene oxide, Farnesyl acetone, α-bisabolol, trans-phytol	Yogesh & Krishnakant, 2011
Anisomeles malabarica	α-Pinene, camphor, borneol, α-thujone, thymol, terpenyl acetate, eugenol, δ-cadinene, caryophyllene oxide, Farnesyl acetone, α-bisabolol, trans-phytol, 1,3 trans menthol	Yogesh & Krishnakant, 2011
Betonica sp.		
Betonica officinalis	1-octen-3-ol, beta-bourbonene, (E)-caryophyllene, α-pinene, germacrene D	Hajdari et al., 2011
Betonica grandiflora	Ledol, trans-caryophyllene, α-bisabolol, isolongifolol, eudesm-7(11)-en-4-ol	Yousefi et al., 2012
Betonica nivea	Sabinene hydrate, selin-11-en-4-α-ol, β-caryophyllene, δ-cadinene	Morteza-Semnani et al., 2008
Ballota sp.		
Ballota nigra	Hexenal, germacrene D, β-caryophyllene, β-bisabolene	Baser & Kirimer, 2018
Calamintha sp.		
Calamintha nepeta L.	Pulegone, menthone, limonene	Bozovi´c & Ragno, 2017
Clerodendrum sp.		
Clerodendrum polycephalum	β-caryophyllene, α-muurolene, β-pinene, (E)-nerolidol	Ogundajo et al., 2016
Clerodendrum phlomidis	Terpinen-4-ol, caryophyllene, β-bisabolene, phytol	Katekhaye et al., 2012
Clinopodium sp.		
Clinopodium nepeta	Piperitenone, pulegone, menthone, limonene, β-caryophyllene	Baser & Kirimer, 2018
Clinopodium sylvatica	Terpinen-4-ol, limonene, piperitenone	Baser & Kirimer, 2018
Dracocephalum sp.		
Dracocephalum foetidum	Mentha-1,8-dien-10-al, limonene, geranial, and neral	Lee et al., 2007
Dracocephalum kotschyi	β-pinene, (E)-citral, neral, methyl geranate, limonene, campholenal, geraniol, cyclohexylallene	Ghavam et al., 2021
Dorystaechas sp.		
Dorystaechas hastata	1,8-Cineole, borneol, camphor	Baser & Kirimer, 2018
Hyssopus sp.		
Hyssopus officinalis	Pinocamphone, isopinocamphone, hedycaryol, spathulenol	Aebisher et al., 2021
Hyptis sp.		
Hyptis dilatata Benth	Fenchone, 3-carene, α-pinene, ß-caryophyllene, limonene, ß-pinene, camphor	Ramos da Silva et al., 2021
Hyptis martiusii	1,8-cineole, d-carene, limonene, germacrene B	Ramos da Silva et al., 2021
Lamium sp.		
Lamium album	Geranial, terpeniol, linalool, spatulenol, bisabolol, D-germacrene, β-caryophyllene, β-amyrin, neral	Konarska et al., 2021
Lavandula sp.		
Lavandula dentata L.	1,8-cineole, isolimonene, thuj-3-en-10-al, trans-pinocarveol	Ramos da Silva et al., 2021
Lavandula officinalis	Linalool, acetate linalool, monotril	Ramos da Silva et al., 2021
Lavandula angustifolia	Z-ß-ocimene, E-ß-ocimene, linalool, terpinen-4-ol, linalyl acetate, lavandulyl acetate	Giatropoulos et al., 2018
Lavandula x *intermedia*	1.8-Cineole, linalool, camphor, isoborneol, lavandulyl acetate	Popa et al., 2021
Lavandula stoechas	α-Fenchone, 1,8-cineole, camphor, α-thujone	Baser & Kirimer, 2018
Marrubium sp.		
Marrubium persicum	m-Tolualdehyde, acetophenone, germacrene D, β-caryophyllene, β-farnesene, α-pinene	Hamedeyazdan et al., 2013
Marrubium anisodon	(Z)-β-farnesene, β-caryophyllene	Baser & Kirimer, 2018
Marrubium globosum	Spathulenol, β-caryophyllene	Baser & Kirimer, 2018
Melissa sp.		
Melissa officinalis L.	Citral, caryophyllene oxide, citronellal, geraniol, geranyl acetate,	Ramos da Silva et al., 2021
Micromeria sp.		
Micromeria fruticose	Pulegone, isomenthone, piperitenone, β-pinene, βcaryophyllene, piperitone	Baser & Kirimer, 2018
Mentha sp.		
Mentha arvensis L.	Menthol, menthone, isomenthone, citronellal, and nerol	Ramos da Silva et al., 2021
Mentha piperita L.	Menthone, menthol, pulegone and menthyl acetate, β-myrcene, carvone	Ramos da Silva et al., 2021
Mentha pulegium	Menthone, pulegone, piperitenone	Ntalli et al., 2010
Mentha spicata	1,8-Cineole, carvonea, ß-caryophyllene	Giatropoulos et al., 2018

(*Continued*)

TABLE 27.2 (*Continued*)
Essential Oil Components from Different Species of Lamiaceae

Species	Essential oil components (terpenes and terpenoids)	References
Mentha longifolia	Menthone, pulegone, carvone, limonene, 1,8-cineole	Baser & Kirimer, 2018
Mentha suaveolens	Piperitenone, limonene	Baser & Kirimer, 2018
Nepeta sp.		
Nepeta italica	β-Pinene, eucalyptol, longifolene, (E)-β-caryophyllene, (E)-β-farnesene, β-santalene, germacrene D, β-bisabolene,	Celik et al., 2020
Nepeta cataria	Nepetalactone, 1,8-cineole, germacrene D	Baser & Kirimer, 2018
Nepeta conferta	p-Cymene, eucalyptol	Baser & Kirimer, 2018
Nepeta nuda	α-Nepetalactone, germacrene, elemol, β-caryophyllene	Baser & Kirimer, 2018
Ocimum sp.		
Ocimum basilicum L.	Methyl chavicol, linalyl acetate, linalool, eugenol	Ramos da Silva et al., 2021
Ocimum gratissimum L.	1,8-cineole, eugenol, terpineol-4	Rodrigues et al., 2020
Origanum sp.		
Origanum scabrum	Carvacrol, thymol, p-cymene, γ-terpinene	Ramos da Silva et al., 2021
Origanum vulgare	4-Terpineol, sabinene, thymol	Ramos da Silva et al., 2021
Origanum dictamnu	α-Thujene, carvacrol, γ-terpinene, p-cymene, p-cymene	Ntalli et al., 2010
Origanum majorana	Carvacrola, a-terpinene, linalool, terpinen-4-ol,	Giatropoulos et al., 2018
Origanum rotundifolium	Isopulegyl acetate, aromadendrene, limonene oxide, viridiflorol	Baser & Kirimer, 2018
Origanum hypericifolium	p-Cymene, carvacrol, ϒ-terpinene, thymol	Baser & Kirimer, 2018
Origanum acutidens	Aromadendrene, m-cymene, carvacrol, linalool acetate	Baser & Kirimer, 2018
Origanum haussknechtii	Aromadendrene, carvacrol, allo-aromadendrene, α-himachalene	Baser & Kirimer, 2018
Origanum onites	Carvacrol, γ-terpinene, p-cymene, linalool, thymol	Baser & Kirimer, 2018
Phlomis sp.		
Phlomis amanica	Isopimaradien-ol, germacrene D	Baser & Kirimer, 2018
Phlomis grandiflora	β-Eudesmol, β-curcumene, α-pinene, α-cedrene	Sarikaya & Fakir, 2016
Phlomis sieheana	Germacrene D, β-caryophyllene, α-pinene, spathulenol	Ozdemir et al., 2017
Phlomis rigida	β-Caryophyllene, β-selinene, caryophyllene	Baser & Kirimer, 2018
Plectranthus sp.		
Plectranthus amboinicus	Thymol, ß-pinene, γ-terpinene, caryophyllene	Ramos da Silva et al., 2021
Plectranthus ornatus	α-Thujene, pinene, sabinene, 3-carene, 1-octen3-ol, β-ocimene, α-terpinyl acetate, β-caryophyllene, and germacrene D	Passinho-Soares et al., 2017
Plectranthus barbatus	β-Caryophyllene, α-copaene, germacrene	Ramos da Silva et al., 2021
Rosmarinus sp.		
Rosmarinus officinalis	α-Pinene, 1,8-cineole, borneol, a-terpineol	Giatropoulos et al., 2018
Satureja sp.		
Satureja thymbra	Carvacrol, γ-terpinene	Ramos da Silva et al., 2021
Satureja hortensis	Carvacrol, γ-terpinene, o-cymene	Aebisher et al., 2021
Satureja cilicica	p-Cymene, carvacrol, ɣ-terpinene, thymol	Arabaci et al., 2017
Satureja cuneifolia	Carvacrol, ɣ-terpinene, p-cymene, thymol,	Baser & Kirimer, 2018
Salvia sp.		
Salvia officinalis	Viridiflorol, camphene, β-pinene, myrcene, 1,8-cineole, camphor, β-caryophyllene	Ntalli et al., 2010
Salvia lavandulifola	Borneol, manool, 1,8-cineole, camphor and β-pinene, α-humulene β-caryophyllene, limonene	Schmiderer et al., 2008
Salvia aethiopis	α-Copaene, α-cubebene, (-)-spathulenol, germacrene-D, α-limonene, ledol	Goze et al., 2016
Salvia euphratica	cis-Sabinol, myrcenyl acetate, 1,8-cineole	Goze et al., 2016
Salvia divaricata	α-Pinene, camphor, camphene, 1,8-cineole	Temel et al., 2016
Salvia nydeggeri	α-Pinene 16, β-pinene 9, cubebol 6, caryophyllene oxide	Temel et al., 2016
Salvia pachystachys	β-Pinene 24, α-pinene 12, spathulenol	Temel et al., 2016
Salvia russelli	β-Pinene, 1,8-cineole, α-copaene, valeranone	Temel et al., 2016
Salvia sericeo-tomentosa	Sabinyl acetate, α-pinene, trans-sabinol, cumin alcohol	Tan et al., 2017
Salvia ceratophylla	α-Pinene, β-pinene, β-caryophyllene, bornyl acetate, α-terpineol, α-copaene, ɣ-muurolene	Baser & Kirimer, 2018

TABLE 27.2 (*Continued*)
Essential Oil Components from Different Species of Lamiaceae

Species	Essential oil components (terpenes and terpenoids)	References
Salvia cilicica	Spathulenol, caryophyllene oxide	Baser & Kirimer, 2018
Salvia cryptantha	Camphor, 1,8-cineole, borneol, viridiflorol	Baser & Kirimer, 2018
Salvia verticillata	Spathulenol, α-pinene, limonene	Baser & Kirimer, 2018
Salvia virgata	1,8-cineole, α-copaene, germacrene D, camphor	Baser & Kirimer, 2018
Salvia viscosa	α-Copaene, β-caryophyllene, γ-muurolene, δ-cadinene	Baser & Kirimer, 2018
Sidetis sp.		
Sidetis brevibracteata	β-Caryophyllene, germacrene-D, α-cadinene, carvacrol, βpinene	Sagir et al., 2017
Sidetis caesarea	β-Caryophyllene, caryophyllene oxide, spathulenol	Gunbatan et al., 2017
Sidetis cilicica	β-Pinene, α-pinene, β-phellandrene	Baser & Kirimer, 2018
Sidetis erytrantha	α-Pinene, β-caryophyllene, sabinene, β-caryophyllene, α-bisabolo	Baser & Kirimer, 2018
Sidetis trojana	Valeranone, α-bisabolol, β-caryophyllene	Baser & Kirimer, 2018
Sideritis lycia	3-Carene, myrcene, limonene, b-phellandrene, δ-sabinene, δ-cadinene, naphthalene	Baser & Kirimer, 2018
Stachys sp.		
Stachys aetherocalyx	Germacrene D, β-myrcene, linalool, linalyl acetate	Kilic et al., 2017
Stachys annua	(Z)-β-Ocimene, β-pinene, α-pinene, benzaldehyde, βbourbonene	Renda et al., 2017
Stachys balansae	Germacrene D, β-caryophyllene, spathulenol, α-cadinene	Goren et al., 2011
Stachys cretica	α-Cadinol, verbenol, sabinene, α-humulene, germacrene D, β-caryophyllene, α-bisabolol, spathulenol, T-muurolol	Goren et al., 2011
Stachys germanica	Germacrene D, β-caryophyllene, α-copaene, trans-β-farnesene, spathulenol	Goren et al., 2011
Stachys lavandulifolia	α-Bisabolol, bicyclogermacrene, β-phellandrene, α-pinene	Baser & Kirimer, 2018
Stachys mardinensis	Menthyl acetate, isomenthone, pulegone, menthol, spathulenol, caryophyllene, β-bourbonene	Kilic et al., 2017
Stachys pumila	Trans-nerolidol, α-pinene, β-pinene	Baser & Kirimer, 2018
Stachys sylvatica	Limonene, α-ledrene, γ-muurolene 1, (2E)-hexenal	Renda et al., 2017
Stachys tmolea	Germacrene D, β-caryophyllene, valeronone, spathulenol	Goren et al., 2011
Teucrium sp.		
Teucrium alyssifolium	β-Caryophyllene, ar-curcumene, bisabolene, α-humulene	Baser & Kirimer, 2018
Teucrium chamaedrys	β-Caryophyllene, nonacosane, germacrene D, α-pinene	Baser & Kirimer, 2018
Teucrium lamiifolium	β-Caryophyllene, trans-β-bergamotene, germacrene D, (Z)-β-farnesene	Baser & Kirimer, 2018
Teucrium orientale	Nonanal, thuja-2,4(10)-diene, β-cubebene, 2-methyl cumarone, germacrene D	Baser & Kirimer, 2018
Thymbra sp.		
Thymbra spicata	α-Thujene, myrcene, α-terpinene, p-cymene, γ-terpinene, carvacrol	Celik et al., 2020
Thymus sp.		
Thymus cilicicus	Camphene, Sabinene, Linalool, 2-Bornanone, Borneol, α-Terpineol, Citral, Geraniol, Caryophyllene, α-Pinene	Celik et al., 2020
Thymus vulgaris	Thymol, β-caryophyllene, carvacrol, linalool, geraniol	Giatropoulos et al., 2018
Thymus pannonicus	β-Citral, geranial, thymol, carvacrol, p-cymene, γ-terpinene, linalool	Popa et al., 2021
Thymus citriodorus	Geraniol, carvacrol, terpinene	Aebisher et al., 2021
Thymus pseudopulegioides	Thymol, ɣ-terpinene, p-cymene	Baser & Kirimer, 2018
Thymus revolutus	Cymene, ɣ-terpinene, carvacrol, thymol, borneol, α-pinene	Baser & Kirimer, 2018
Thymus sipyleus	α-Pinene, farnesol, 1,8-cineole, limonene, linalool	Maral et al., 2017
Vitex sp.		
Vitex agnus-castus	1,8-Cineole, α-pinene, sabinene, bicyclogermacrene	Zhelev et al., 2022
Vitex negundo	δ-Guaiene, carryophyllene epoxide, α-selinene, germacren-4-ol, and (E)-nerolidol, α-cedrene, germacrene D	Khokra et al., 2008
Ziziphora sp.		
Ziziphora clinopodioides	Eugenol, piperitone, pulegone, terpinen-4-ol, eucalyptol	Celik et al., 2020
Ziziphora persica	Pulegone, limonene, β-pinene, piperitone	Baser & Kirimer, 2018
Ziziphora tenuior	Pulegone, piperitenone, 1,8-cineole	Baser & Kirimer, 2018

27.6 TERPENE AND TERPENOIDS OF LAMIACEAE – PHARMACOLOGICAL EFFECTS

Numerous studies have reported the pharmacological activities of essential oils and their components from plant sources in traditional medicinal and modern therapeutics. The bioactivities of EO components of Lamiaceae are listed in Table 27.3.

27.7 PRINCIPLES OF SYNTHETIC BIOLOGY AND ITS APPROACHES

Synthetic biology can be phenomenally defined as multiple approaches to scientifically engineer living organisms for desirable functioning and synthesis of potential compounds to meet therapeutic and industrial requirements. This system principally needed an in-depth understanding of the entire system for its functioning to eventually alter their biosynthetic networks in a predictable productive way (Neumann & Neumann-Staubitz, 2010; Zhang & Hong, 2020). Fundamentally, this ideology involves the modular decomposition of biological systems into simple parts or sub-compartments which permit the activities exterior to the original environment (Bashor et al., 2010). These artificial biological units such as primary metabolites and macromolecules and their implementation inside the live host set the boundaries between natural and artificial biological systems and point to the future of synthetic biology (Auslander et al., 2017; Tang et al., 2021).

TABLE 27.3
Pharmacological Activities of Reported Terpenes and Terpenoids of Lamiaceae

Terpene/terpenoid	Structure	Activity/property	References
1,8-cineole/eucalyptol	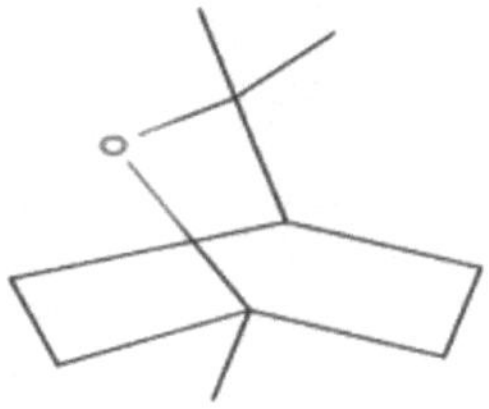	Anti-inflammatory, antioxidant, antimicrobial, bronchodilatory, analgesic and treating neuro-related and respiratory disorders	Hoch et al., 2023
Aromadendrene	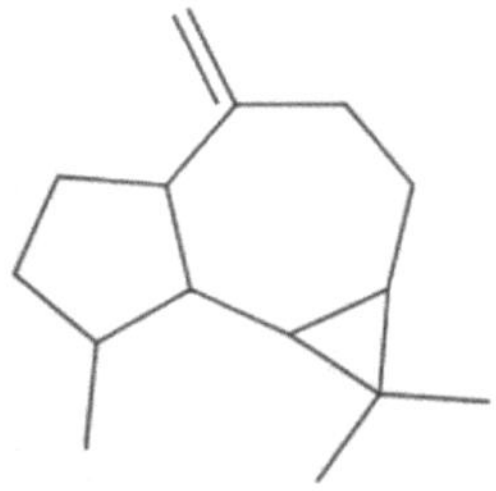	Antimicrobial	Mulyaningsih et al., 2010
Borneol	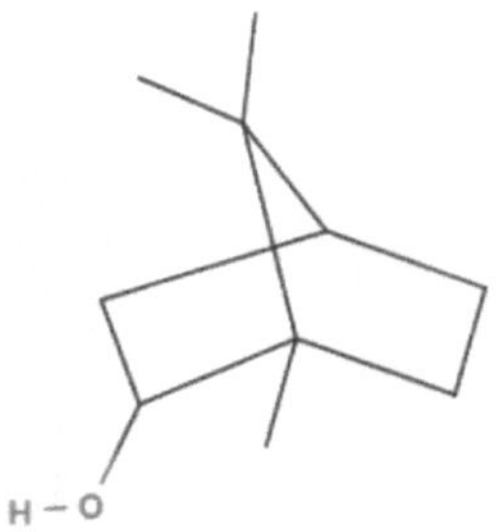	Anti-inflammatory, analgesic, neuroprotective, antiepileptic, antidiabetic, anti-hyperlipidemic, antihypertensive, and anticancer effects	Rajput et al., 2023
Cadinene	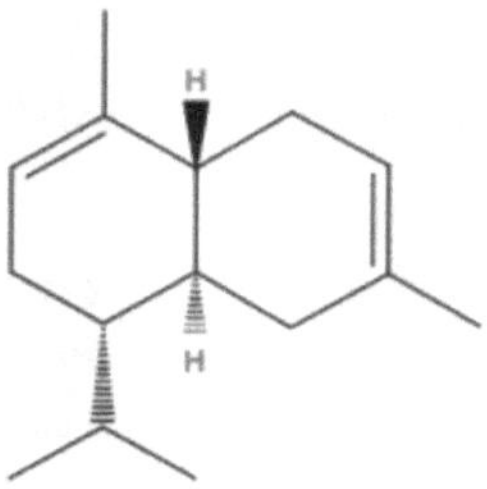	Cytotoxic to ovarian tumor cells, antimicrobial and antiproliferative	de Oliveira et al., 2021

TABLE 27.3 (*Continued*)
Pharmacological Activities of Reported Terpenes and Terpenoids of Lamiaceae

Terpene/terpenoid	Structure	Activity/property	References
Cadinol	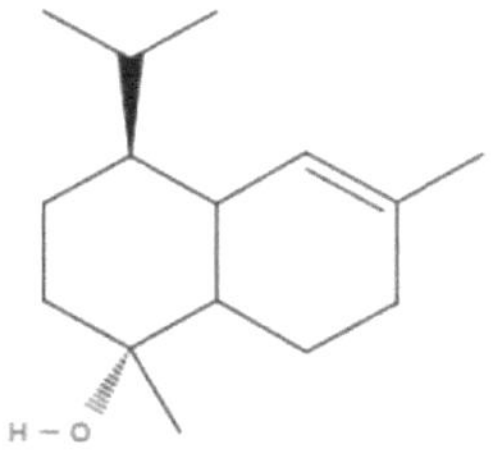	Antimicrobial, antifungal, and anti-inflammatory, cytotoxic activity against MCF7 cells	Ringel et al., 2022
Camphene	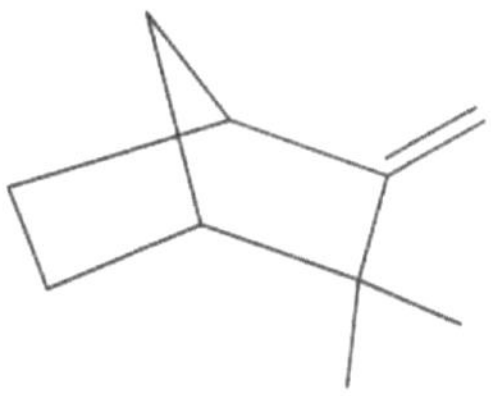	Antidiabetic anticancer, antidiabetic, anti-leishmanial, hepatoprotective, anti-acetylcholinesterase inhibitory activities	Hachlafi et al., 2023
Camphor	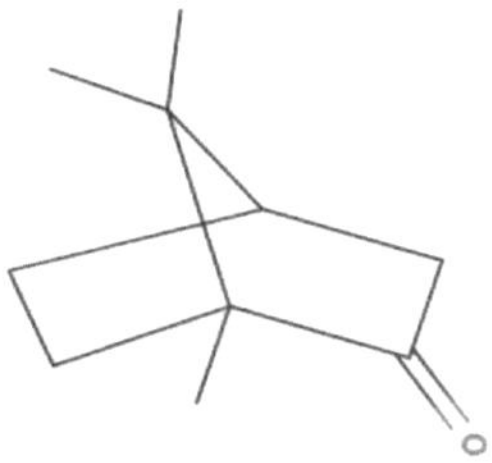	Treating nervine and cerebral, respiratory, and gastrointestinal disorders	Fazmiya et al., 2022
Carryophyllene	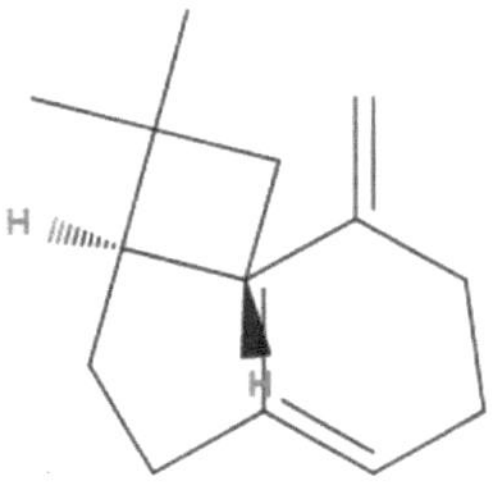	Antibacterial, antioxidant, gastroprotective, anxiolytic, anti-inflammatory, and analgesic properties	Fidyt et al., 2016
Carvacrol	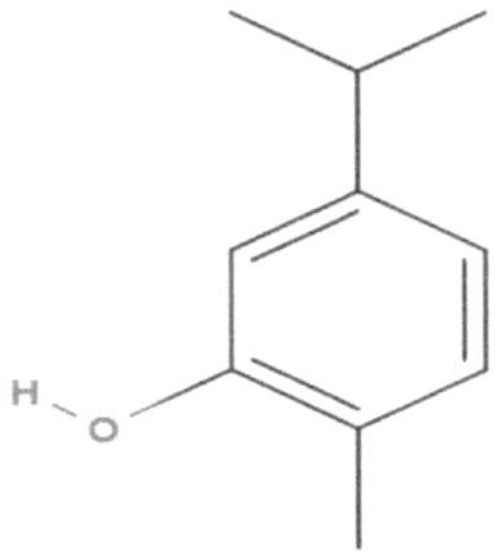	Antimutagenic, antigenotoxic, analgesic, antispasmodic, anti-inflammatory, angiogenic, antiparasitic, antiplatelet, AChe inhibitory, antielastase	Baser, 2008
Cedrene	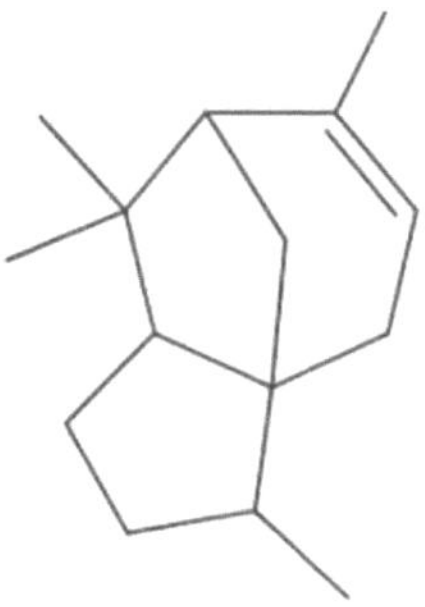	Trypanocidal, antileukemic, antimicrobial, anti-obesity activity	Kim et al., 2015
Citral	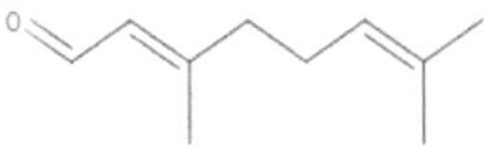	Antimicrobial, antioxidant, anticancer, antidiabetic, and anti-inflammatory	Sharma et al., 2021

(*Continued*)

TABLE 27.3 (*Continued*)
Pharmacological Activities of Reported Terpenes and Terpenoids of Lamiaceae

Terpene/terpenoid	Structure	Activity/property	References
Citronellal	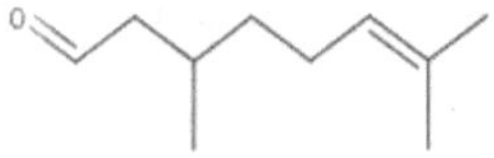	Analgesic, antioxidant, and acetylcholinesterase inhibition	Santos et al., 2019
Copaene	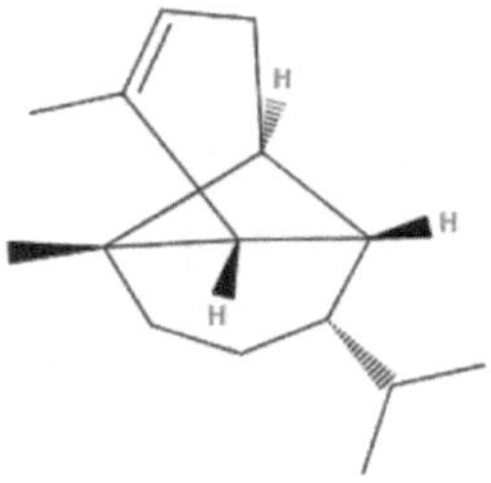	Antigenotoxic, antimicrobial	Turkez et al., 2014
Curcumene	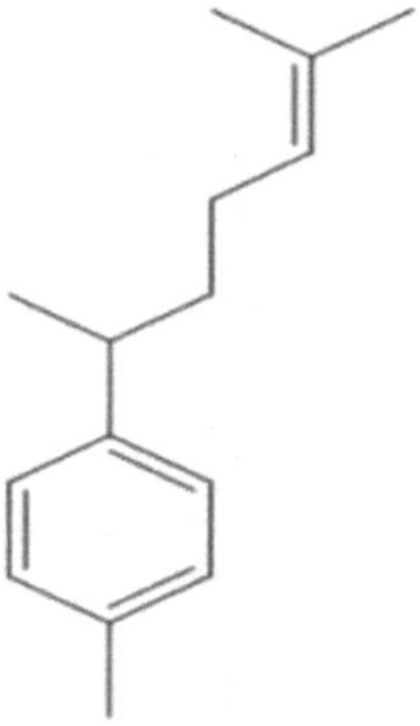	Anti-inflammatory, hypoglycemic, antioxidant, antimicrobial, antiviral, anticancer, neuroprotective	Urosevic et al., 2022
Cymene	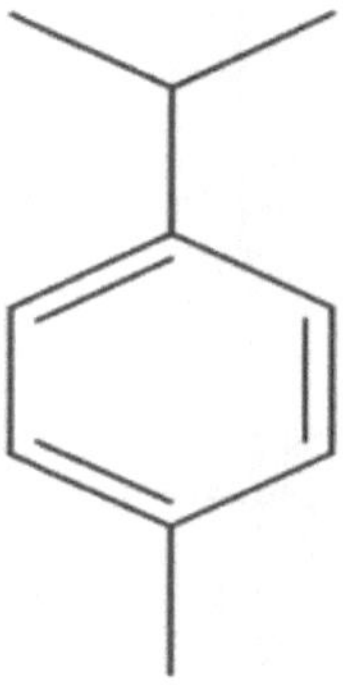	Antioxidant, vasorelaxant, anti-inflammatory, antinociceptive properties	Silva et al., 2015
Eudesmol	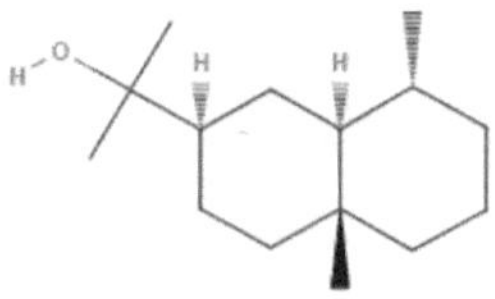	Antitumor and anti-angiogenic	Acharya et al., 2021
Eugenol	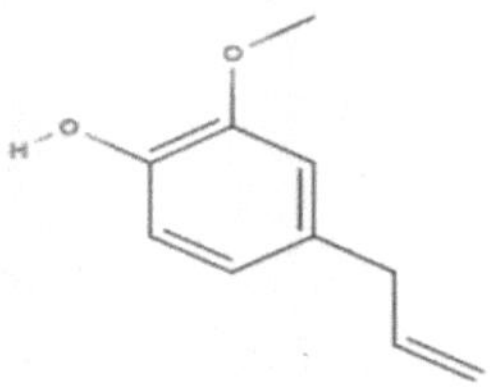	Antiseptic, anesthetic, anti-inflammatory, analgesic	Mohammadi Nejad et al., 2017
Farnesol	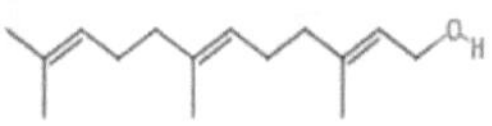	Antitumor, neuroprotective, cardioprotective, reduce pulmonary edema, increase glutathione levels, decrease membrane cytotoxicity	de Araujo Delmondes et al., 2019

TABLE 27.3 (*Continued*)
Pharmacological Activities of Reported Terpenes and Terpenoids of Lamiaceae

Terpene/terpenoid	Structure	Activity/property	References
Geraniol	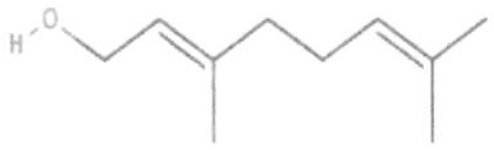	Hepatoprotective, cardioprotective, and neuroprotective effects, antidiabetic, anti-inflammatory	Lei et al., 2019
Germacrene D	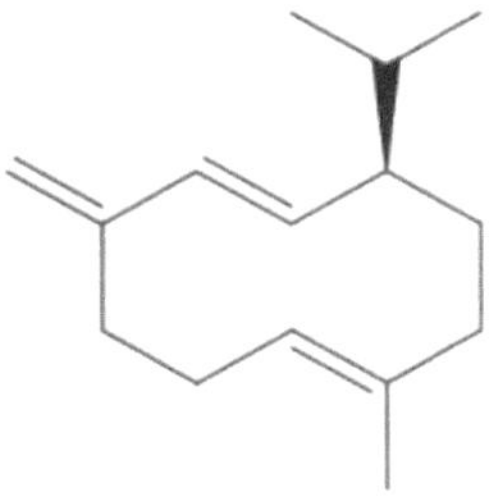	Insecticide, antimicrobial, antioxidant	Noge and Becerra, 2009
Guaiene	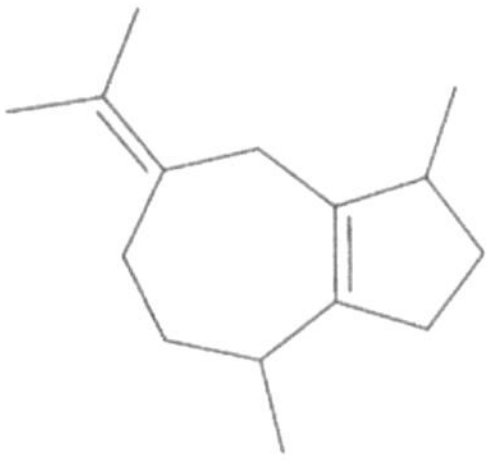	Surfactant and emulsifier; anti-inflammatory activity against platelet activating factor	Widyasanti et al., 2021
Humulene	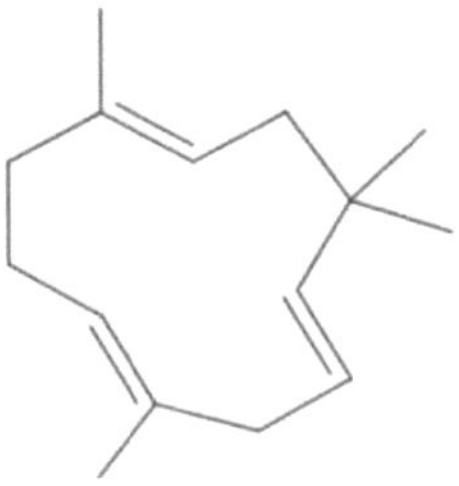	Gastroprotective, cicatrizing, analgesic and antioxidant potential, antitumor and anti-inflammatory	de Lacerda Leite et al., 2021
Lavandulyl acetate	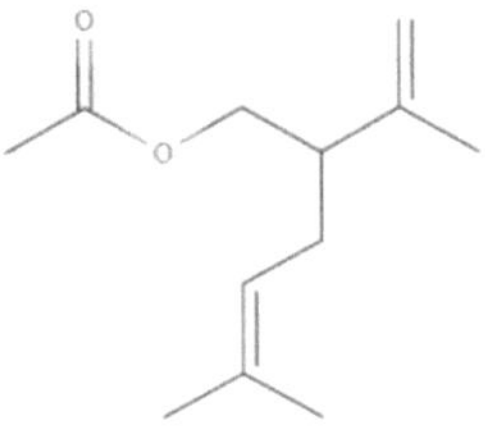	Mosquito larvicidal, anti-genotoxic	Govindarajan and Benelli, 2016
Limonene	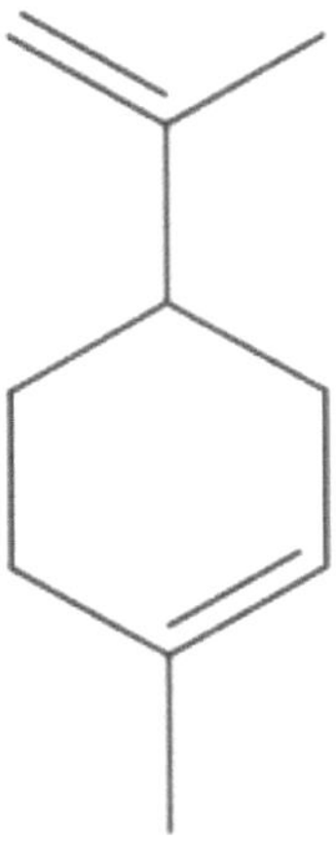	Gastroprotective, hepatoprotective, immune-modulatory, anti-fibrotic, anti-genotoxic	Anandakumar et al., 2021

(*Continued*)

TABLE 27.3 (*Continued*)
Pharmacological Activities of Reported Terpenes and Terpenoids of Lamiaceae

Terpene/terpenoid	Structure	Activity/property	References
Linalool	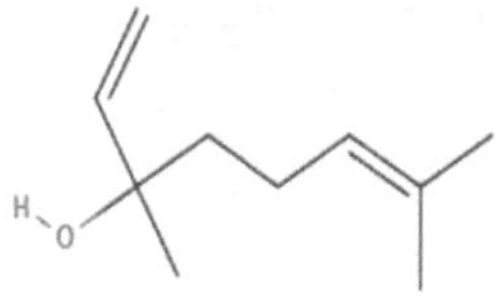	Anxiolytic, antidepressant antinociceptive, analgesic, sedative, anticonvulsant	Dos Santos et al., 2022a
Longifolene	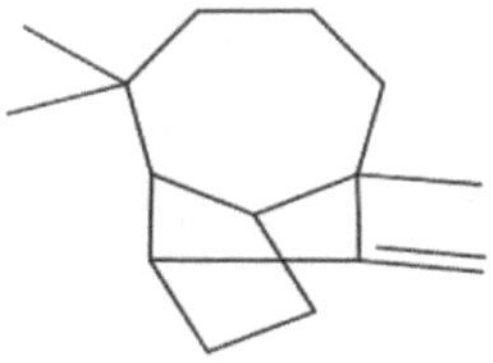	Antifungal, anti-termite, anticancer (prostate, oral, kidney)	Grover et al., 2022
Manool	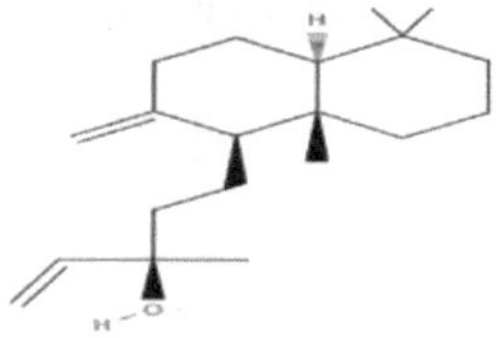	Antigenotoxicity, anticancer anti-inflammatory, antioxidant, and antimicrobial	Nicolella et al., 2014
Menthol	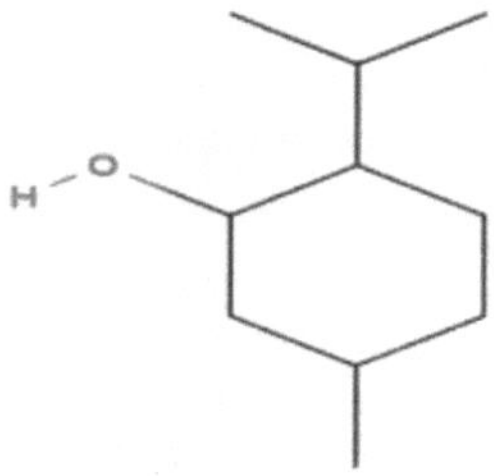	Analgesic, antibacterial, antifungal, anaesthetic, and penetration-enhancing effects, chemopreventive and immunomodulating	Kamatou et al., 2013
Menthone	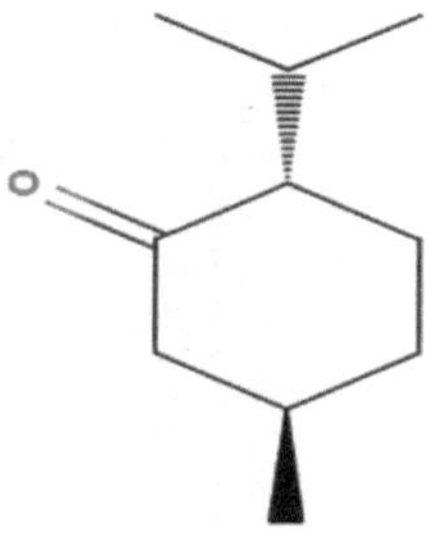	Antibiotic, antibiofilm, and anti-inflammatory	Zhao et al., 2023
Muurolene	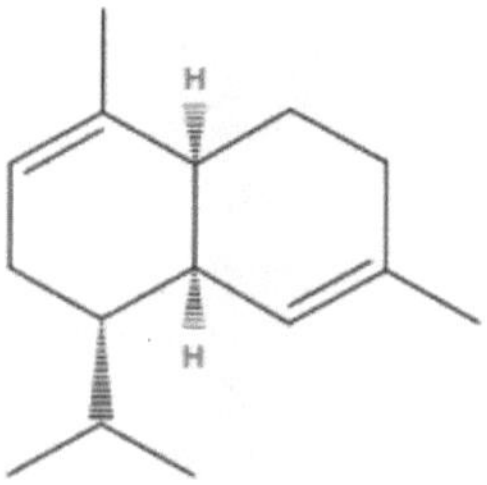	Antioxidant, anti-inflammatory	Martins et al., 2008
Myrcene	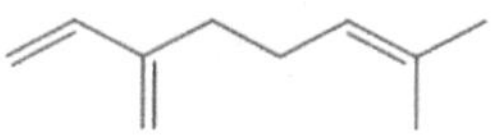	Anxiolytic, antioxidant, anti-aging, anti-inflammatory, analgesic	Surendran et al., 2021
Naphthalene	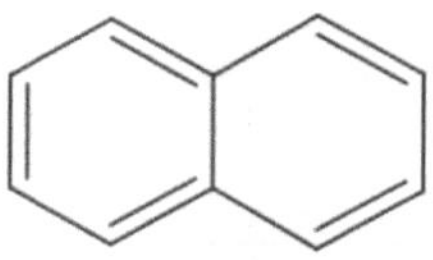	Anticonvulsant, antidepressant, antidiabetic, antihypertensive, anti-inflammatory, antimicrobial, anti-neurodegenerative, antipsychotic, antitubercular	Makar et al., 2019

TABLE 27.3 (*Continued*)
Pharmacological Activities of Reported Terpenes and Terpenoids of Lamiaceae

Terpene/terpenoid	Structure	Activity/property	References
Nepetalactone	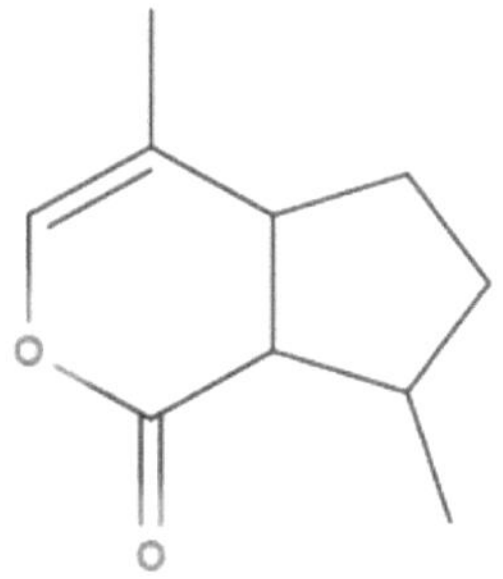	Diuretic, diaphoretic, antitussive, anti-asthmatic, antispasmodic, sedative agent	Shafaghat and Oji, 2010
Nerol	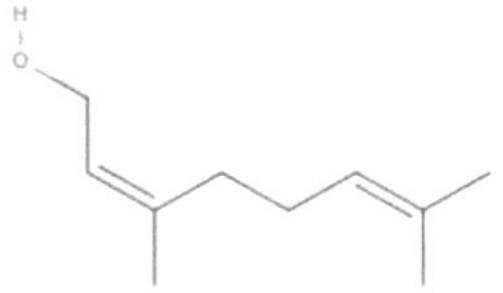	Antispasmodic, anthelmintic, antifungal, and antioxidant	Coelho et al., 2022
Nerolidol	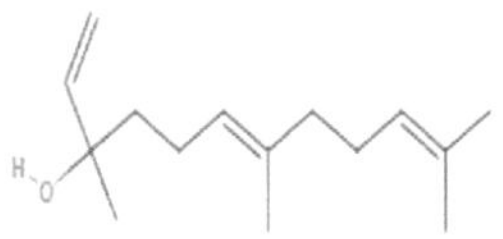	Anti-parasitic, skin-penetration enhancer, skin-repellent, anti-nociceptive, neuroprotective	Chan et al., 2016
Pinene	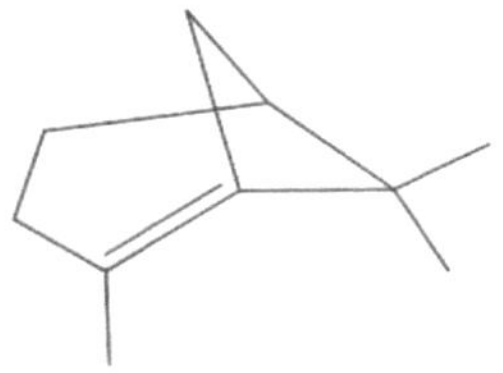	Anticoagulant, anti-inflammatory, anti-leishmania, antimalarial	Park et al., 2021
Piperitone	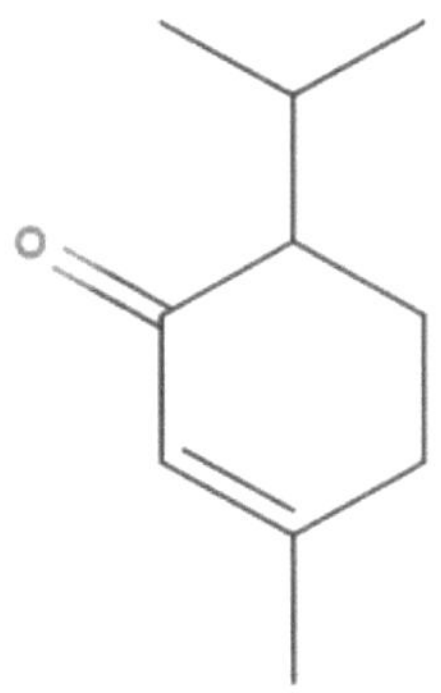	Insecticidal, bactericidal, larvicidal, molluscicidal, cytotoxic, anticholinesterasic, and leishmanicidal activities	Rondon et al., 2016
Pulegone	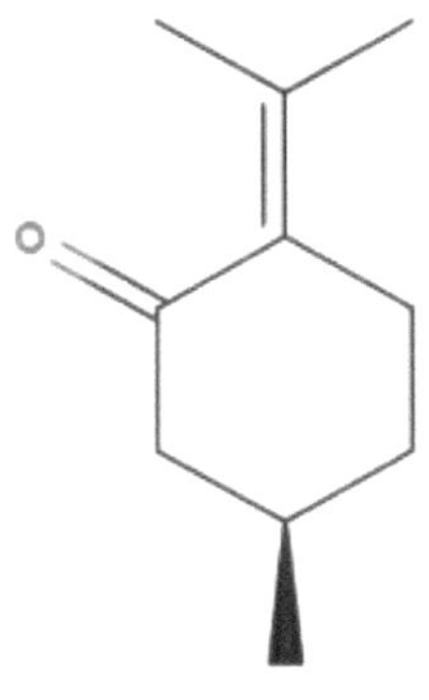	Anti-feeding, antifungal, antiviral, pesticide activities, and abortifacient and emmenagogue.	Dhingra et al., 2023

(*Continued*)

TABLE 27.3 *(Continued)*
Pharmacological Activities of Reported Terpenes and Terpenoids of Lamiaceae

Terpene/terpenoid	Structure	Activity/property	References
Sabinene	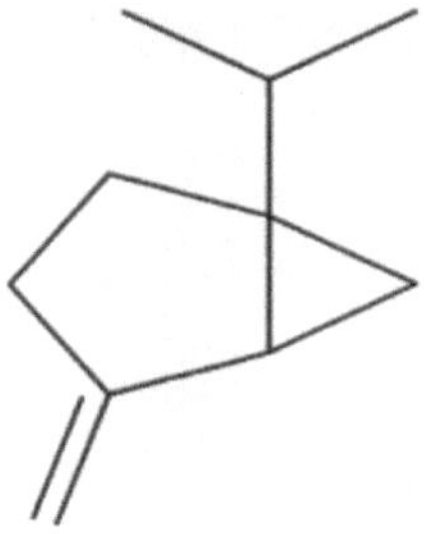	Hepatoprotective, angiostatic, and antiangiogenic effects	Sharma et al., 2019
Selinene	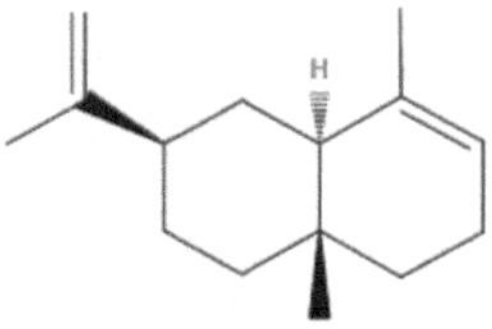	Anti-inflammatory, analgesic, and antipyretic activity	Chandra et al., 2017
Spathulenol	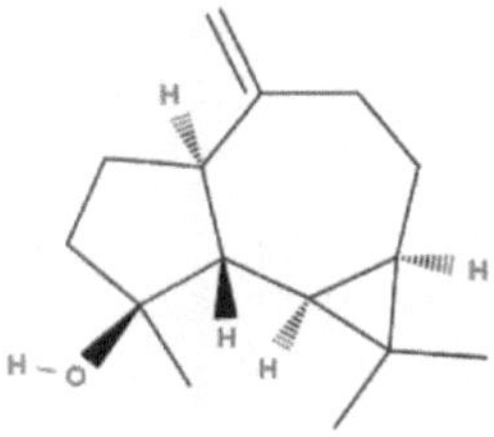	Anti-nociceptive, anxiolytic, and antidepressant	Dos Santos et al., 2022b
Terpinen-4-ol	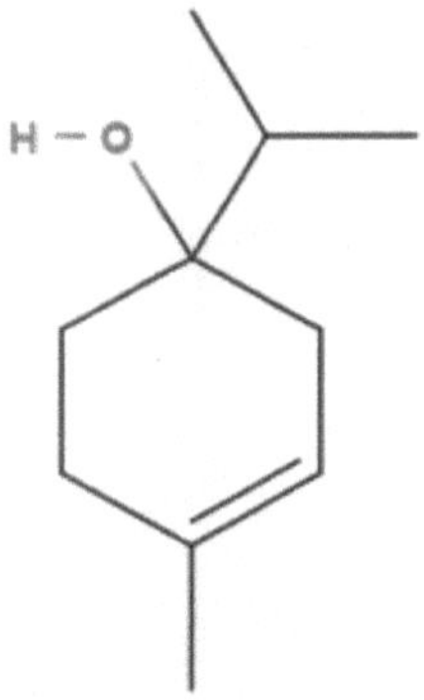	Anti-arthritic, anti-biofilm agent, anti-inflammatory	Aslam et al., 2022
Terpinene	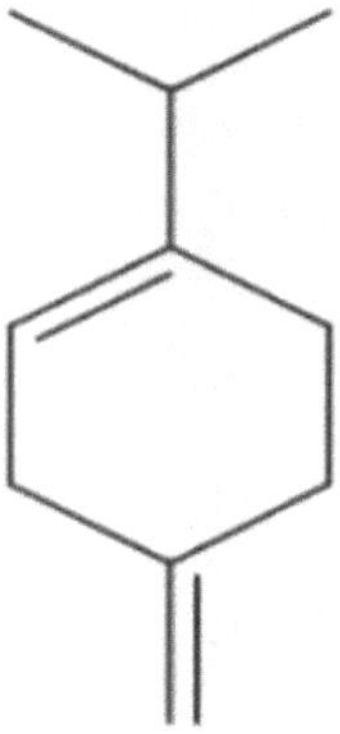	Antioxidant, anti-inflammatory	Ramalho et al., 2015

TABLE 27.3 (*Continued*)
Pharmacological Activities of Reported Terpenes and Terpenoids of Lamiaceae

Terpene/terpenoid	Structure	Activity/property	References
Terpineol	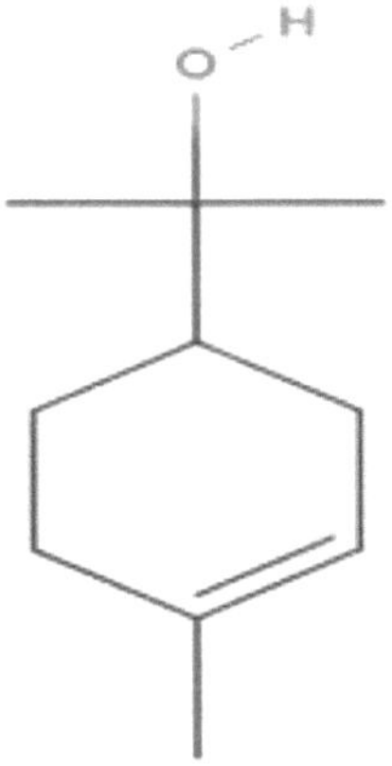	Analgesic, gastroprotective, cardioprotective, neuroprotective, and antidiarrheal effects	Chen et al., 2023
Terpinyl acetate	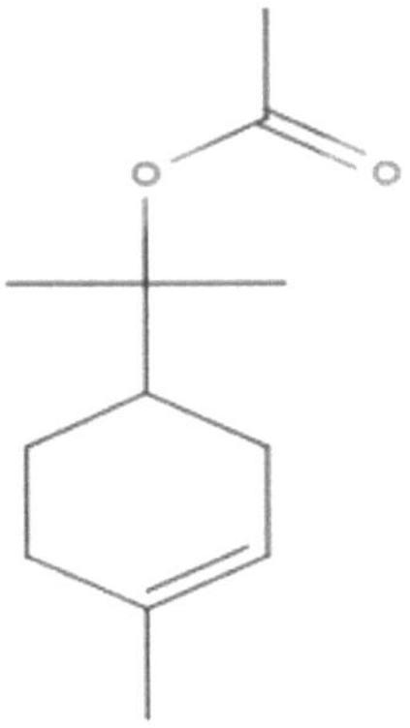	Anticholinesterase activity, anti-aggregation and neuroprotective effect	Chowdhury and Kumar, 2020
Thujone	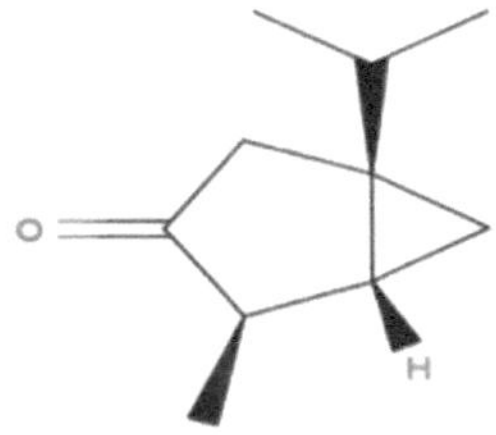	Antidiabetic, antimutagenic, and immune-modulatory effects	Zamborine Nemeth and Thi Nguyen, 2020
Thymol	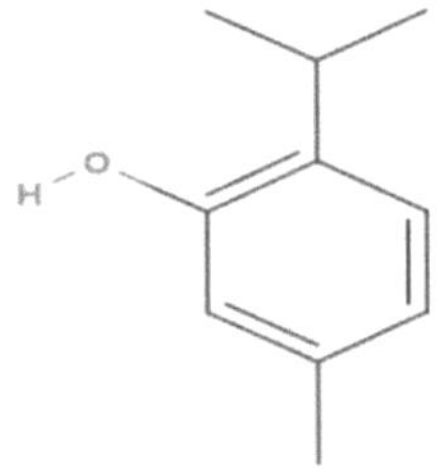	Analgesic, antispasmodic, antiseptic, anticancer, acts against various cardiovascular, neurological, rheumatological, gastrointestinal, diseases	Nagoor Meeran et al., 2017
Verbenol	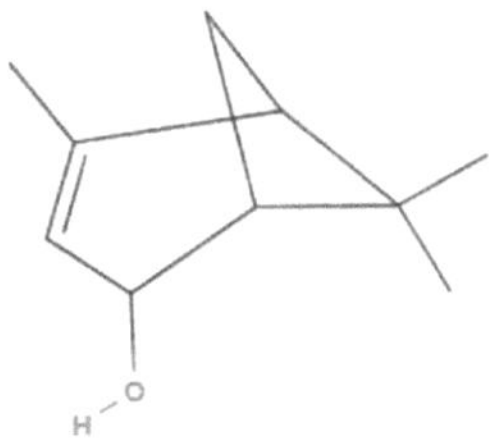	Anti-inflammatory, anti-ischemic activity	Choi et al., 2010
Viridiflorol	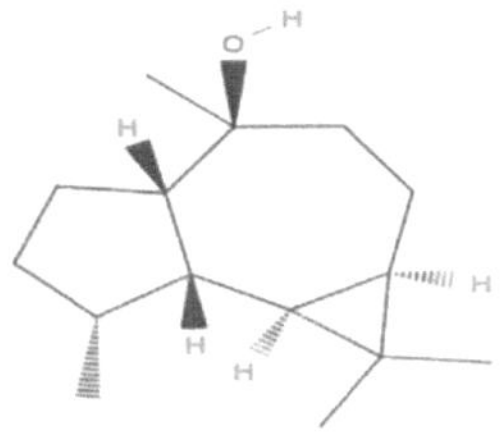	Analgesic, anti-arthritic activity, anti-hyperalgesic, and anti-neoplastic	de Matos Balsalobre et al., 2023

It follows the iterative cycles of design, build, test, and learn to improve the biosynthesis of functional compounds in live cells for industrial applications. This approach globally allows researchers to work in teams across disciplines forming networks to exchange new ideas in various subjects such as metabolic engineering, minimal genomes, regulatory circuits, and biosystems to develop novel products and research tools (Garner, 2021).

27.8 SYNTHETIC BIOLOGY METHODS FOR TERPENE PRODUCTION

Synthetic biological methods are regarded as quite suitable techniques to produce the industrial terpenes of Lamiaceae for application in required fields. Numerous reports on editing or manipulating the system, pathway, genome, and transcriptome for enhanced production of terpene compounds have been reviewed in distinct aspects. Cherednichenko et al. (2021) have provided a general review of Lamiaceae for gene engineering through plant resource modifications, studying the metabolic system, etc. A specific review of the mint family was published by Fuchs et al. (2022), which included studies of genetically manipulating the biosynthetic pathways of metabolites.

27.9 MICROBIAL ENGINEERING

Cao et al. (2019) heterogeneously expressed the codon of longifolene synthase augmenting with farnesyl pyrophosphate (FPP) synthases, which subsequently increased the production of longifolene upto 1.8 fold in *E. coli* under fed-batch conditions. Transformation of viridiflorol synthase (VS) from *Agrocybe aegerita* to *E. coli* was achieved by Shukal et al. (2019). The statistical optimization of the transcriptional pathway along with mutation in translation has improved the yield of the sesquiterpene viridiflorol up to 25.7 g/L in fed-batch fermentation. This standardized methodology applied for amorphadiene improvement has yielded nearly 30 g/L.

Cyanobacteria are a sustainable host organism in biotechnological applications specifically in the metabolic engineering of plant compounds due to their photosynthetic characteristics. Englund et al. (2015) engineered the *Synechocystis* sp. PCC 6803 to produce 13R-manoyl oxide (13R-MO), the precursor of forskolin. This forskolin is naturally seen in the root bark of *Coleus forskohlii* and has therapeutic applications in the treatment of glaucoma. The vector pEERM targeted stable chromosomal integration CfTPS2 and CfTPS3 (diterpene synthases) altered the MEP synthetic pathway.

27.10 GENETIC ENGINEERING IN YEAST – TERPENE SYNTHESIS

Due to the robust and versatile nature of *Saccharomyces cerevisiae*, it has become the most convenient microbe model for metabolic engineering and transformation for synthesizing complex bioactive compounds. Regarding the biosynthesis of industrially important terpenes and terpenoids, yeasts are natural producers of the precursors IPP, DMAPP, GPP, and FPP through the MVA pathway, partially sharing the conserved metabolism from plants (Zhang et al., 2017). A few strategies of terpene production in engineered yeast systems were reported and reviewed.

Amorpha-4,11-diene synthase gene was transformed and cloned in *Saccharomyces cerevisiae* by homologous recombination (Lindahl et al., 2006). Overexpression of the catalytic HMG1 domain has increased the accumulation of cubebol and squalene and down regulation of squalene synthase (ERG9) by replacing its native promoter with the MET3 promoter has improved the production of cubebol (Asadollahi et al., 2010). The MVA metabolic pathway in yeast was modulated for overexpression of ERG9, which improved the productivity of the sesquiterpene, α- santalene (Scalcinati et al., 2012). Similarly, combined repression of ERG9 and overexpression of tHMG1 resulted in mass production of linalool (Amiri et al., 2016). Peng et al. (2017) destabilized the Erg9p protein by endoplasmic-reticulum-associated protein degradation (ERAD) mechanism to reduce squalene accumulation and redirect the synthetic pathway toward sesquiterpene production through upregulation of FPP. This novel action improved nerolidol titer up to 100 mg/L (86%). Along with weakened ERG9 expression, FPP synthase and patchoulol synthase genes were fused, resulting in increased patchoulol production by 7.8-fold (Ma et al., 2019).

27.11 METABOLIC ENGINEERING OR GENOME EDITING

Enhanced production of limonene in *Mentha piperita* was reported on the co-suppressing limonene-3-hydroxylase gene. The yield was increased up to 80% without influencing the expression of limonene synthase (Mahmoud et al., 2004). In *Mentha spicata*, the gene MsYABBY5 (usually expressed in grandular trichomes) silenced using RNAi has improved terpene synthesis. It has been reported that MsYABBY5 transcription factor negatively regulates the terpene biosynthetic pathway and activated the MsWRKY75 gene downregulator of terpene production (Wang et al., 2016). Similarly, RNAi-mediated suppression of the MsMYB, member of R2R3-MYB (negative regulator of terpene biosynthesis suppressing the expression of geranyl diphosphate synthase large subunit gene) gene in spearmint has substantially improved the production of monoterpenes in spearmint (Reddy et al., 2017). The limonene synthase promoter-driven expression of menthone:menthol reductase gene in transgenic peppermint has increased the transcript levels of MMR up to fourfold. This has relatively lowered the synthesis of menthone but increased the menthol production. This study has exemplified the importance of tissue-specific promoters in generating transformants (Vining et al., 2017).

Hwang et al. (2020) have distinctly engineered *M. piperita* f. citrate (orange mint) with the lipid transfer protein of *N. tabacum* (NtLTP1). The overexpression of this gene has resulted in an altered phenotype of trichome head showing increased diameter along with increased levels of limonene (1.6 fold) and 3-pinanone (22 fold). In *Mentha*sp. DXP reductoisomerase was stimulated to improve monoterpene precursor IPP/DAMPP; similarly, in peppermint menthofuran synthase and limonene synthase cDNA were overexpressed for producing menthofuran and limonene respectively. In contrast, silencing the expression of grandular trichomes- specific transcription factors (MsYABBY5, R2R3-MYB) by RNAi relatively increased the levels of terpenes in transgenic spearmint. This study by Mahmoud et al. (2021) emphasized metabolic engineering by overexpression of enzymes involved in isoprenoid synthesis in grandular trichomes transforming mint and lavender with enhanced terpene production

27.12 HAIRY ROOT CULTURES

On investigating the effects of chitosan (CS), methyl jasmonate (MeJA) and β-cyclodextrin (elicitors) on terpenoids production in hairy root cultures of (*Agrobacterium rhizogenes* strain A4) *Scutellaria bornmuelleri* revealed that increased monoterpenes content were increased in cultures with elicited 100 μM methyl jasmonate and the sesquiterpene production were maximum with 100 mg/L of chitosan. While the yield of both terpenes was maximum when elicited with 0.7 and 7 mM of β-cyclodextrin and methyl jasmonate (Gharari et al., 2023). Table 27.4 provides the list of experiments involving *Agrobacterium*-mediated transformations in Lamiaceae.

27.13 TERPENE SYNTHASES – REMODELING

Evolutionary divergence in the family of terpene synthase (TPS) functionally alters the synthetic pathway of terpenoids in plants and is responsible for terpene diversity (Karunanithi & Zerbe, 2019). Terpenes synthases of various families and subfamilies are involved in catalyzing the reaction of diverse terpene production from isoprenoid diphosphate precursor and are responsible for the primary diversification of the complex terpenes. Transgenic *Lavandula latifolia* expressing limonene synthase of spearmint was generated byMunoz-Bertomeu et al. (2008). The mature leaves of transformants contained more than 450% of limonene over the control influencing the other monoterpene profile. Also, this transgenic phenotype was maintained in the progenies inheriting limonene synthase transgene. The specific site of 1,8-cineole synthase from *Salvia fruticosa* acts as a switch to select between two isomers of trans-GPP or cis-NPP as enzyme substrate. This binding of selective substrate influenced the specified terpene synthetic pathway. Applying a similar approach of substrate switching in sabinene synthase from *Salvia pomifera* has yielded sevenfold increase of sabiene, which inturn enabled the production of *trans*-sabin-3-ol by sabinene hydroxylase (Ignea et al., 2019). The terpenoids synthase of *Colquhounia coccinea* var. *mollis* (CcTSP1) is capable of converting isoprenoids to sester-/di-/sesqui-/mono-terpenoids based on its genetic evolution and cellular location. Expression of CcTSP1 in *E. coli* produced three novel terpenoids and four terpenes with cyclohexane skeletons. However, the leaf expression of *CcTSP* in *N. benthamania* and *C. coccinea* revealed its specificity in synthesizing plant diterpenoids and sesquiterpenoids and can be induced by methyl jasmonate (Li et al., 2021).

27.14 CHROMOSOME SCALE ASSEMBLY

A study reported by Hamilton et al. (2020) showed that the chromosome-scale assembly in *Callicarpa americana* L. genome of 506.1 Mb to contig size of 7.5 and scaffold of 29.0 Mb spanning 17 pseudochromosomes by long-read sequencing and hi-C scaffolding. This has developed threewhole genome duplications (WGD) with local tandem repeats resulting in the relative expression of Kolavenyl diphosphate synthase (*camTSP2*) encoding the precursor of callicarpenal, a significant insect-repellent terpenoid. In a similar way, the genes contributing to the terpenoid diversity in lavender showed altered expression on WGD specifically in grandular trichomes. This chromosomal ploidy was

TABLE 27.4
***Agrobacterium*-Mediated Transformations in Lamiaceae Species for Improving terpene production**

Species	Strain	Improved essential oil compounds	References
Mentha piperita	*A. rhizogenes* IFO14554/pBI121	Limonene and Menthol	Inoue et al., 2003
	A. tumefaciens disarmed strain GV3101 (pMP90)	Menthone, menthofuran and pulegone	Krasnyanski et al., 1999
M. arvensis and *M. piperita*	*A. tumefaciens* strain C58pMP90- nptII and 4S-limonene synthase c genes	*M. piperita*- Pinenes (50%); Cineole (70%), Pulegone (18 %)	Diemer et al., 2001
		M. arvensis- Cineole, Isopiperitenone, Peperitenone	
Perovskia abrotanoides	*A. rhizogenes* TR105	cryptotanshinone and IIA tanshonine	Ebrahimi et al., 2017
Salvia runcinata	*Agrobacterium rhizogenes* A4T	Octadecadienoic acid and sugars	Figlan &Makunga, 2017
Trachyspermum ammi	*Agrobacterium tumefaciens* (EHA101) with pBI121-TPS2	Thymol	Nomani & Tohidfar, 2021

found to be lineage-specific duplication in lavender species (Li et al., 2021). The genome assembly of *Schizonepeta tenuifolia*, also known as Japanese catnip, was carried out at the chromosomal level by Liu et al. (2023). This has revealed that the biosynthetic gene cluster (BGC) holding a bipartite structure is involved in the biosynthesis of p-menthane monoterpenoids and encodes for the missing isopiperitenone reductase, evolved from the closely related gene of *Mentha* spp. through convergent evolution. *Thymus mongolicus*genome of 605.2 Mb was assembled with 12 pseudochromosomes resulting in WGD and tandem duplication, which might be involved in species formation. The duplication events were seen in genes of essential regulators of the phenylpropanoid-flavonoid biosynthesis that would influence the terpene production of the plant (Dang et al., 2024).

27.15 ADVANCED STRATEGIES (POST-MODIFICATIONS, CYTOCHROME P450, SUB-CELLULAR LOCALIZATIONS)

Yeasts are more promising organisms than *E. coli* to express modified cytochrome P450s in plant terpene synthesis because of their eukaryotic features like transmembrane system and post-modifications (Zhang & Hong, 2020). The crucial cytochrome P450 enzyme (CYP720B1), is known to catalyze the oxidation of abietadiene as the precursor, structurally resembling manoyl oxide and manool diterpenes. Active site mutagenesis of CYP720B1 altered the specificity of the product by substrate selectivity and improving the catalytic activity. This has encouraged the synthesis of 18- or 19-hydroxy-miltiradiene (twofold increase) (Bureau et al., 2023).

Microbial engineering approaches and systems are more reliable and widely established for complex biosynthetic pathways than the plant host. Sequential transformation and crossing of homozygous transgenic line processes are time-consuming, require intensive laboring, and are unreliable with next-generation transformants (Farre et al., 2014). Expressing polycistronic genes of the complete metabolic pathway can be achieved using a recent approach called the 2A peptide system. This system has been employed in terpene engineering such as amorphadiene in *N. benthamiana* (van Herpen et al., 2010).

Korman et al. (2017) employed cell-free enzyme systems for biosynthesis of monoterpenes from glucose. This was achieved by altering the enzyme system of the glycolysis pathway to provide continuous accumulation of acetyl-CoA, the initiator of the MVA pathway of terpene synthesis. This approach has relatively produced the limonene, pinene, and sabinene titred by 12.5, 14.9, and 15.9 g/L respectively.

Subcellular localization, organelle engineering, or compartmentalization confers the advantage for synthetic biology approaches for terpene synthesis by facilitating significant yield along with the maintenance of redox potential than the endogenous way of biosynthesis. Mitochondrial engineering can be preferred for additional use of the acetyl-CoA pool produced in it. Yuan and Ching (2016) introduced the FPP pathway in mitochondria resulting in the yield of 427 mg/L of amorpha-4,11-diene, and localizing geraniol synthase from *Ocimum basilicum* has produced a twofold increase in geraniol than the wild-type. ER and peroxisomes have been engineered for squalene accumulation with a titer of 634 mg/L and 11 g/L, respectively (Zhang & Hong, 2020). Phosphatidic acid phosphatase-encoding PAH1 gene deletion in ER has improved the β-amyrin triterpenes by 8-fold via ER expansion. Similarly, ER-derived organelles and lipid droplets in yeast are employed for enhancing the lipid synthesis targeted by the terpenoid mechanism and also allow increased production of hydrophilic substances.

27.16 CRISPR-CAS IN THE SYNTHESIS OF TERPENES

CRISPR-Cas system composed of guide RNA and Cas nuclease is an emerging tool in synthetic biology for transcriptional regulation and modulation directing toward to synthesis or blocking of desired products (Jeong et al., 2023). The MVA pathway and the plant-derived terpenoid synthases were harbored in *E. coli* using the CRISPR interference system to modulate the transcriptional blockage of acetoacetyl-CoA thiolase enzyme which substantially increased bisabolol (C15) and lycopene (C40) content. Increased isoprene was achieved by coupling CRISPRi to an endogenous gene (ispA) (Kim et al., 2016). The CRISPR-Cas9 system was established to facilitate the production of terpenoids in *Bacillus subtilis* focusing on the MEP pathway. Amorphadiene was significantly improved to 116 mg/L adopting three combinatorial modules including terpene synthase, branch pathway and central pathway within a single *B. subtilis* strain (Song et al., 2021). Valencene synthase was successfully introduced into *S. cerevisiae* BJ5464 by knocking out *Erg9* gene (responsible for squalene synthesis) in the MVA pathway using a recyclable CRISPR/Cas9 system constructed with Cre/loxP. Valencene production titer reached up to 539.3μg/L (160-fold) compared to the initial production (Chen et al., 2019). In a study conducted by Kim and Lee (2023), 32 nonessential genes in *E. coli* were downregulated by CRISPRi. This has influenced the synthetic pathway of isoprenol yielding up to 4.5 to 5 fold.

27.17 CONCLUSION

Plant terpenes and EO are one of the largest and most diverse classes of naturally occurring therapeutic compounds along with numerous industrial applications. The substantial applications of the essential oil from the mint family have eventually raised the needs in industrial and therapeutic fields. Synthetic biology is an ideal approach involving efficient methods to improve the synthesis and production of the desired terpenes and terpenoids. This phenomenon of producing terpenes via synthetic biology methods specifically involves recombining and editing elements of the MVA pathway (Kitaoka et al., 2015).

From these studies, it can be noted that enormous research reports have been published in the past two decades on various methods for improving terpenes of Lamiaceae in microbial hosts, by transformation and gene silencing. Specifically, silencing or eliminating the expression of *Erg9* gene, which is responsible for squalene synthesis can improve the production of various terpenes and terpenoids by activating the respective terpene synthases. We can also see that the advanced methods of cell-free systems, CRISPR/Cas system are still under research to improve the terpenes and EO of Lamiaceae. Many DNA editing approaches are yet to be applied to this subject of research. The combinatorial approach of utilizing different methods for enhanced production of metabolites is emerging in the field of biotechnology. These studies are expected to be encouraged for metabolic alteration in futuristic research.

REFERENCES

Acharya, B., Chaijaroenkul, W., & Na-Bangchang, K. (2021). Therapeutic potential and pharmacological activities of β-eudesmol. *Chemical Biology & Drug Design*, *97*(4), 984–996. https://doi.org/10.1111/cbdd.13823

Aebisher, D., Cichonski, J., Szpyrka, E., Masjonis, S., & Chrzanowski, G. (2021). Essential oils of seven lamiaceae plants and their antioxidant capacity. *Molecules*, *26*, 3793. https://doi.org/10.3390/molecules2613379

Amiri, P., Shahpiri, A., Asadollahi, M. A., Momenbeik, F., & Partow, S. (2016). Metabolic engineering of *Saccharomyces cerevisiae* for linalool production. *Biotechnology Letters*, *38*(3), 503–508. https://doi.org/10.1007/s10529-015-2000-4

Anandakumar, P., Kamaraj, S., & Vanitha, M. K. (2021). D-limonene: A multifunctional compound with potent therapeutic effects. *Journal of Food Biochemistry*, *45*(1), e13566. https://doi.org/10.1111/jfbc.13566

Arabaci, T., Uzay, G., Kelestemur, U., Karaaslan, M. G., Balcioglu, S., & Ates, B. (2017). Cytotoxicity, radical scavenging, antioxidant properties and chemical composition of the essential oil of *Satureja cilicica* P. H. Davis from Turkey. *Marmara Pharmaceutical Journal*, *21*(3), 500–505. https://doi.org/10.12991/marupj.311814

Arendt, P., Pollier, J., Callewaert, N., & Goossens, A. (2016). Synthetic biology for production of natural and new-to-nature terpenoids in photosynthetic organisms. *The Plant Journal*, *87*(1), 16–37. https://doi.org/10.1111/tpj.13138

Asadollahi, M. A., Maury, J., Schalk, M., Clark, A., & Nielsen, J. (2010). Enhancement of farnesyl diphosphate pool as direct precursor of sesquiterpenes through metabolic engineering of the mevalonate pathway in *Saccharomyces cerevisiae*. *Biotechnology and Bioengineering*, *106*(1), 86–96. https://doi.org/10.1002/bit.22668

Aslam, S., Younis, W., Malik, M. N. H., Jahan, S., Alamgeer Uttra, A. M., Munir, M. U., & Roman, M. (2022). Pharmacological evaluation of anti-arthritic potential of terpinen-4-ol using in vitro and in vivo assays. *Inflammopharmacology*, *30*(3), 945–959. https://doi.org/10.1007/s10787-022-00960-w

Auslander, S., Auslander, D., & Fussenegger, M. (2017). Synthetic biology – the synthesis of biology. *Angewandte Chemie International Edition*, *56*(23), 6396–6419. https://doi.org/10.1002/anie.201609229

Baser, K. H. (2008). Biological and pharmacological activities of carvacrol and carvacrol bearing essential oils. *Current Pharmaceutical Design*, *14*(29), 3106–3119. https://doi.org/10.2174/138161208786404227

Baser, K., & Kirimer, N. (2018). Essential oils of *Anatolian lamiaceae* – an update. *Natural Volatiles and Essential Oils*, *5*(4), 1–28.

Bashor, C. J., Horwitz, A. A., Peisajovich, S. G., & Lim, W. A. (2010). Rewiring cells: Synthetic biology as a tool to interrogate the organizational principles of living systems. *Annual Review of Biophysics*, *39*, 515–537. https://doi.org/10.1146/annurev.biophys.050708.133652

Bozovic, M., & Ragno, R. (2017). *Calamintha nepeta* (L.) Savi and its main essential oil constituent pulegone: Biological activities and chemistry. *Molecules*, *22*(2), 290. https://doi.org/10.3390/molecules22020290

Bureau, J. A., Oliva, M. E., Dong, Y., & Ignea, C. (2023). Engineering yeast for the production of plant terpenoids using synthetic biology approaches. *Natural Product Reports*, *40*, 1822–1848. https://doi.org/10.1039/D3NP00005B

Butnariu, M., & Sarac, I. (2018). Essential oils from plants. *Journal of Biotechnology and Biomedical Science*, *1*(4), 35–43. https://doi.org/10.14302/issn.2576-6694.jbbs-18-2489

Cao, Y., Zhang, R., Liu, W., Zhao, G., Niu, W., Guo, J., Xian, M., & Liu, H. (2019). Manipulation of the precursor supply for high-level production of longifolene by metabolically engineered *Escherichia coli*. *Scientific Reports*, *9*, 95. https://doi.org/10.1038/s41598-018-36495-w

Carovic-Stanko, K., Petek, M., Grdisa, M., Pintar, J., Bedekovic, D., Herak custic, M., & Satovic, Z. (2016). Medicinal plants of the family Lamiaceae as functional foods-a review. *CZECH Journal of Food Sciences*, *34*(5), 377–390. https://doi.org/10.17221/504/2015-CJFS

Celik, G., Kilic, G., Kanbolat, S., Sener, S. O., Karakose, M., Yayli, N., & Karaoglu, S. A. (2020). Biological activity, and volatile and phenolic compounds from five Lamiaceae species. *Flavour and Fragrance Journal*, *36*(2), 223–232. https://doi.org/10.1002/ffj.3636

Chan, W. K., Tan, L. T., Chan, K. G., Lee, L. H., & Goh, B. H. (2016). Nerolidol: A sesquiterpene alcohol with multi-faceted pharmacological and biological activities. *Molecules*, *21*(5), 529. https://doi.org/10.3390/molecules21050529

Chandra, M., Prakash, O., Kumar, R., Bachheti, R. K., Bhushan, B., Kumar, M., & Pant, A. K. (2017). β-Selinene-rich essential oils from the parts of *Callicarpa macrophylla* and their antioxidant and pharmacological activities. *Medicines*, *4*(3), 52. https://doi.org/10.3390/medicines4030052

Chen, H., Zhu, C., Zhu, M., Xiong, J., Ma, H., Zhuo, M., & Li, S. (2019). High production of valencene in *Saccharomyces cerevisiae* through metabolic engineering. *Microbial Cell Factories*, *18*(1), 195. https://doi.org/10.1186/s12934-019-1246-2

Chen, Y., Zhang, L. L., Wang, W., & Wang, G. (2023). Recent updates on bioactive properties of α-terpineol. *Journal of Essential Oil Research*, *35*(3), 274–288. https://doi.org/10.1080/10412905.2023.2196515

Cherednichenko, M. Y., Polivanova, O. B., Khlebnikova, D. A., Sosina, A. V., Belova, M. M., & Kalashnikova, E. A. (2021). Gene engineering of lamiaceae family: Basic and applied research. *Revista Geintec-Gestao Inovacao E Tecnologias*,

11(2), 123–139. https://doi.org/10.47059/revistageintec.v11i2.1648
Choi, I. Y., Lim, J. H., Hwang, S., Lee, J. C., Cho, G. S., & Kim, W. K. (2010). Anti-ischemic and anti-inflammatory activity of (S)-cis-verbenol. *Free Radical Research*, *44*(5), 541–551. https://doi.org/10.3109/10715761003667562
Chowdhury, S., & Kumar, S. (2020). Alpha-terpinyl acetate: A natural monoterpenoid from *Elettaria cardamomum* as multi-target directed ligand in Alzheimer's disease. *Journal of Functional Foods*, *68*, 103892. https://doi.org/10.1016/j.jff.2020.103892
Coelho, M. L., Islam, M. T., Laylson da Silva Oliveira, G., Oliveira Barros de Alencar, M. V., Victor de Oliveira Santos, J., Campinho dos Reis, A., Oliveira Ferreira da Mata, A. M., Correia Jardim Paz, M. F., Docea, A. O., Calina, D., & Sharifi-Rad, J. (2022). Cytotoxic and antioxidant properties of natural bioactive monoterpenes nerol, estragole, and 3, 7-dimethyl-1-octanol. *Advances in Pharmacological and Pharmaceutical Sciences*, *2022*. https://doi.org/10.1155/2022/8002766
Dang, Z., Xu, Y., Zhang, X., Mi, W., Chi, Y., Tian, Y., Liu, Y., & Ren, W. (2024). Chromosome-level genome assembly provides insights into the genome evolution and functional importance of the phenylpropanoid-flavonoid pathway in *Thymus mongolicus*. *BMC Genomics*, *25*(1), 291. https://doi.org/10.1186/s12864-024-10202-8
de Araujo Delmondes, G., Bezerra, D. S., de Queiroz Dias, D., de Souza Borges, A., Araujo, I. M., Lins da Cunha, G., Bandeira, P. F. R., Barbosa, R., Melo Coutinho, H. D., Felipe, C. F. B., Barbosa-Filho, J. M., Alencar de Menezes, I. R., & Kerntopf, M. R. (2019). Toxicological and pharmacologic effects of farnesol ($C_{15}H_{26}O$): A descriptive systematic review. *Food and Chemical Toxicology: An International Journal Published for the British Industrial Biological Research Association*, *129*, 169–200. https://doi.org/10.1016/j.fct.2019.04.037
de Lacerda Leite, G. M., de Oliveira Barbosa, M., Lopes, M. J. P., de Araújo Delmondes, G., Bezerra, D. S., Araújo, I. M., de Alencar, C. D. C., Coutinho, H. D. M., Peixoto, L. R., Barbosa-Filho, J. M., & Felipe, C. F. B. (2021). Pharmacological and toxicological activities of α-humulene and its isomers: A systematic review. *Trends in Food Science & Technology*, *115*, 255–274. https://doi.org/10.1016/j.tifs.2021.06.04
de Matos Balsalobre, N., Dos Santos, E., Dos Santos, S. M., Arena, A. C., Konkiewitz, E. C., Ziff, E. B., Formagio, A. S. N., & Kassuya, C. A. L. (2023). Potential anti-arthritic and analgesic properties of essential oil and viridiflorol obtained from *Allophylus edulis* leaves in mice. *Journal of Ethnopharmacology*, *301*, 115785. https://doi.org/10.1016/j.jep.2022.115785
de Oliveira, J. A. M., Bernardi, D. I., Balbinot, R. B., Cabral, M. R. P., Zanqueta, E. B., Endo, E. H., Dias Filho, B. P., Nakamura, T. U., Figueiredo, M. C., Ruiz, A. L. T. G., & Foglio, M. A. (2021). New cadinene-sesquiterpene from *Chromolaena laevigata* (lam.) RM King & H. Rob (Asteraceae) aerial parts and biological activities. *Natural Product Research*, *35*(21), 3880–3887. https://doi.org/10.1080/14786419.2020.1747456
Dhingra, A. K., & Chopra, B. (2023). Pulegone: An emerging oxygenated cyclic monoterpene ketone scaffold delineating synthesis, chemical reactivity, and biological potential. *Recent Advances in Anti-Infective Drug Discovery*, *18*(1), 16–28. https://doi.org/10.2174/2772434418666221018090507
Diemer, F., Caissard, J. C., Moja, S., Chalchat, J. C., & Jullien, F. (2001). Altered monoterpene composition in transgenic mint following the introduction of 4S-limonene synthase. *Plant Physiology and Biochemistry*, *39*(7–8), 603–614. https://doi.org/10.1016/S0981-9428(01)01273-6
Dos Santos, E. R. Q., Maia, J. G. S., Fontes-Junior, E. A., & do Socorro Ferraz Maia, C. (2022a). Linalool as a therapeutic and medicinal tool in depression treatment: A review. *Current Neuropharmacology*, *20*(6), 1073–1092. https://doi.org/10.2174/1570159X19666210920094504
Dos Santos, E., Radai, J. A. S., do Nascimento, K. F., Formagio, A. S. N., de Matos Balsalobre, N., Ziff, E. B., Castelon Konkiewitz, E., & Kassuya, C. A. L. (2022b). Contribution of spathulenol to the anti-nociceptive effects of *Psidium guineense*. *Nutritional Neuroscience*, *25*(4), 812–822. https://doi.org/10.1080/1028415X.2020.1815330
Ebrahimi, S., Zaker, A., Abrishamchi, P., Bahrami, A. R., Ganjeali, A., & Sodagar, N. (2017). Hairy root induction and secondary metabolite production in *Perovskia abrotanoides* Karel. *Journal of Plant Process and Function*, *6*(20), 17–26.
Englund, E., Andersen-Ranberg, J., Miao, R., Hamberger, B., & Lindberg, P. (2015). Metabolic engineering of *Synechocystis sp.* PCC 6803 for production of the plant diterpenoid manoyl oxide. *ACS Synthetic Biology*, *4*(12), 1270–1278. https://doi.org/10.1021/acssynbio.5b00070
Farre, G., Blancquaert, D., Capell, T., Van Der Straeten, D., Christou, P., & Zhu, C. (2014). Engineering complex metabolic pathways in plants. *Annual Review of Plant Biology*, *65*, 187–223. https://doi.org/10.1146/annurev-arplant-050213-035825
Fazmiya, M. J. A., Sultana, A., Rahman, K., Heyat, M. B. B., Akhtar, F., Khan, S., & Appiah, S. C. Y. (2022). Current insights on bioactive molecules, antioxidant, anti-inflammatory, and other pharmacological activities of *Cinnamomum camphora* Linn. *Oxidative Medicine and Cellular Longevity*, *2022*, 9354555. https://doi.org/10.1155/2022/9354555
Fidyt, K., Fiedorowicz, A., Strządała, L., & Szumny, A. (2016). β-caryophyllene and β-caryophyllene oxide – natural compounds of anticancer and analgesic properties. *Cancer Medicine*, *5*(10), 3007–3017. https://doi.org/10.1002/cam4.816
Figlan, S., &Makunga, N. P. (2017). Genetic transformation of the medicinal plant *Salvia runcinata* L. f. using *Agrobacterium rhizogenes*. *South African Journal of Botany*, *112*, 193–202. https://doi.org/10.1016/j.sajb.2017.05.029
Fuchs, L. K., Holland, A. H., Ludlow, R. A., Coates, R. J., Armstrong, H., Pickett, J. A., Harwood, J. L., & Scofield, S. (2022). Genetic manipulation of biosynthetic pathways in mint. *Frontiers in Plant Science*, *13*, 928178. https://doi.org/10.3389/fpls.2022.928178
Garner, K. L. (2021). Principles of synthetic biology. *Essays in Biochemistry*, *65*(5), 791–811. https://doi.org/10.1042/EBC20200059
Gharari, Z., Bagheri, K., & Sharafi, A. (2023). Enhanced terpenoids production of elicited hairy root cultures of *Scutellaria bornmuelleri*. *Brazilian Archives of Biology and Technology*,*66*, e23210435.https://doi.org/10.1590/1678-4324-2023210435
Ghavam, M., Manconi, M., Manca, M. L., &Bacchetta, G. (2021). Extraction of essential oil from *Dracocephalum kotschyi* Boiss. (Lamiaceae), identification of two active compounds and evaluation of the antimicrobial properties. *Journal of Ethnopharmacology*, *267*, 113513. https://doi.org/10.1016/j.jep.2020.113513
Giatropoulos, A., Kimbaris, A., Michaelakis, A., Papachristos, D. P., & Polissiou, M. G., &Emmanouel, N. (2018). Chemical composition and assessment of larvicidal and repellent capacity of 14 Lamiaceae essential oils against *Aedes*

albopictus. *Parasitology Research*, *117*(6), 1953–1964. https://doi.org/10.1007/s00436-018-5892-9

Goren, A. C., Piozzi, F., Akcicek, E., Kilic, T., Carikci, S., Mozioglu, E., & Setzer, W. N. (2011). Essential oil composition of twenty-two Stachys species (mountain tea) and their biological activities. *Phytochemistry Letters*, *4*(4), 448–453. https://doi.org/10.1016/j.phytol.2011.04.013

Govindarajan, M., & Benelli, G. (2016). Eco-friendly larvicides from Indian plants: Effectiveness of lavandulyl acetate and bicyclogermacrene on malaria, dengue and Japanese encephalitis mosquito vectors. *Ecotoxicology and Environmental Safety*, *133*, 395–402. https://doi.org/10.1016/j.ecoenv.2016.07.035

Goze, I., Vural, N., & Ercan, N. (2016). Characterization of essential oil and antioxidant activities of some species of salvia in Turkey. *Natural Volatiles & Essential Oils*, *3*(4), 1–7.

Grover, M., Behl, T., Virmani, T., Sanduja, M., Makeen, H. A., Albratty, M., Alhazmi, H. A., Meraya, A. M., & Bungau, S. G. (2022). Exploration of cytotoxic potential of longifolene/Junipene isolated from *Chrysopogon zizanioides*. *Molecules*, *27*(18), 5764. https://doi.org/10.3390/molecules27185764

Gunbatan, T., Demirci, B., Gurbuz, I., Demirci, F., & Ozkan, A. M. G. (2017). Comparison of volatiles of *Sideritis caesarea* specimens collected from different localities in Turkey. *Natural Product Communications*, *12*(10), 1639–1642. https://doi.org/10.1177/1934578X1701201029

Hachlafi, N. E., Aanniz, T., Menyiy, N. E., Baaboua, A. E., Omari, N. E., Balahbib, A., Shariati, M. A., Zengin, G., Fikri-Benbrahim, K., & Bouyahya, A. (2023). In vitro and in vivo biological investigations of camphene and its mechanism insights: A review. *Food Reviews International*, *39*(4), 1799–1826.https://doi.org/10.1080/87559129.2021.1936007

Hajdari, A., Mustafa, B., Franz, C., &Novak, J. (2011). Variability of essential oils of *Betonica officinalis* (Lamiaceae) from different wild populations in Kosovo. *Natural Product Communications*, *6*(9), 1343–1346.

Hamedeyazdan, S., Fathiazad, F., & Asnaashari, S. (2013). Chemical composition of the essential oil from *Marrubium persicum* CA Mey (Lamiaceae). *Pharmaceutical Sciences*, *19*(2), 35–38.

Hamilton, J. P., Godden, G. T., Lanier, E., Bhat, W. W., Kinser, T. J., Vaillancourt, B., Wang, H., Wood, J. C., Jiang, J., Soltis, P. S., & Soltis, D. E. (2020). Generation of a chromosome-scale genome assembly of the insect-repellent terpenoid-producing Lamiaceae species, *Callicarpa americana*. *GigaScience*,*9*(9), giaa093. https://doi.org/10.1093/gigascience/giaa093

Hoch, C. C., Petry, J., Griesbaum, L., Weiser, T., Werner, K., Ploch, M., Verschoor, A., Multhoff, G., Dezfouli, A. B., & Wollenberg, B. (2023). 1, 8-cineole (eucalyptol): A versatile phytochemical with therapeutic applications across multiple diseases. *Biomedicine & Pharmacotherapy*, *167*, 115467. https://doi.org/10.1016/j.biopha.2023.115467

Hwang, H.-S., Adhikari, P. B., Jo, H.-J., Han, J. Y., & Choi, Y. E. (2020). Enhanced monoterpene emission in transgenic orange mint (*Mentha × piperita* f. citrata) overexpressing a tobacco lipid transfer protein (NtLTP1). *Planta*, *252*, 44. https://doi.org/10.1007/s00425-020-03447-6

Ignea, C., Raadam, M. H., Motawia, M. S., Makris, A. M., Vickers, C. E., & Kampranis, S. C. (2019). Orthogonal monoterpenoid biosynthesis in yeast constructed on an isomeric substrate. *Nature Communications*, *10*(1), 3799. https://doi.org/10.1038/s41467-019-11290-x

Inoue, F., Sugiura, H., Tabuchi, A., Karasawa, D. E., & Minami, M. (2003). Alteration of essential oil composition in transgenic peppermint (*Mentha piperita*) carrying T-DNA from *Agrobacterium rhizogenes*. *Breeding Science*, *53*(2), 163–167.

Jeong, S. H., Lee, H. J., & Lee, S. J. (2023). Recent advances in CRISPR-Cas technologies for synthetic biology. *Journal of Microbiology*, *61*(1), 13–36. https://doi.org/10.1007/s12275-022-00005-5

Kamatou, G. P., Vermaak, I., Viljoen, A. M., & Lawrence, B. M. (2013). Menthol: A simple monoterpene with remarkable biological properties. *Phytochemistry*, *96*, 15–25. https://doi.org/10.1016/j.phytochem.2013.08.005

Karunanithi, P. S., & Zerbe, P. (2019). Terpene synthases as metabolic gatekeepers in the evolution of plant terpenoid chemical diversity. *Frontiers in Plant Science*, *10*, 1166. https://doi.org/10.3389/fpls.2019.01166

Katekhaye, S. D., Mulik, M. B., & Laddha, K. S. (2012). Identification of constituents of the essential oil isolated from leaves of *Clerodendrum phlomidis* Linn. by GC-MS. *International Journal of Phytothearpy Research*, *2*(1), 14–28.

Khokra, S. L., Prakash, O., Jain, S., Aneja, K. R., & Dhingra, Y. (2008). Essential oil composition and antibacterial studies of *Vitex negundo* Linn. extracts. *Indian Journal of Pharmaceutical Sciences*, *70*(4), 522–526. https://doi.org/10.4103/0250-474X.44610

Kilic, O., Ozdemir, F. A., & Yildirimli, S. (2017). Essential oils and fatty acids of *Stachys* L. taxa, a chemotaxonomic approach. *Progress in Nutrition*, *19*, 49–59.

Kim, J., & Lee, T. S. (2023). Enhancing isoprenol production by systematically tuning metabolic pathways using CRISPR interference in *E. coli*. *Frontiers in Bioengineering and Biotechnology*, *11*, 1296132. https://doi.org/10.3389/fbioe.2023.1296132

Kim, S. K., Han, G. H., Seong, W., Kim, H., Kim, S. W., Lee, D. H., & Lee, S. G. (2016). CRISPR interference-guided balancing of a biosynthetic mevalonate pathway increases terpenoid production. *Metabolic Engineering*, *38*, 228–240. https://doi.org/10.1016/j.ymben.2016.08.006

Kim, T. H., Yoo, S. D., Lee, H. S., Lee, K. M., Seok, S. H., Kim, M. G., Jung, B. H., Kim, M. G.,&Shin, B. S. (2015). In vivo absorption and disposition of α-cedrene, a sesquiterpene constituent of cedarwood oil, in female and male rats. *Drug Metabolism and Pharmacokinetics*, *30*(2), 168–173. https://doi.org/10.1016/j.dmpk.2014.12.003

Kitaoka, N., Lu, X., Yang, B., & Peters, R. J. (2015). The application of synthetic biology to elucidation of plant mono-, sesqui-, and diterpenoid metabolism. *Molecular Plant*, *8*, 6–16. https://doi.org/10.1016/j.molp.2014.12.002

Konarska, A., Weryszko-Chmielewska, E., Matysik-Wozniak, A., Sulborska, A., Polak, B., Dmitruk, M., Piotrowska-Weryszko, K., Stefanczyk, B., & Rejdak, R. (2021). Histochemical and phytochemical analysis of *Lamium albumsubsp.* album *L. Corolla*: Essential oil, triterpenes, and iridoids. *Molecules*, *26*, 4166. https://doi.org/10.3390/molecules26144166

Korman, T. P., Opgenorth, P. H., & Bowie, J. U. (2017). A synthetic biochemistry platform for cell free production of monoterpenes from glucose. *Nature Communications*, *8*, 15526. https://doi.org/10.1038/ncomms15526

Krasnyanski, S., May, R. A., Loskutov, A., Ball, T. M., & Sink, K. C. (1999). Transformation of the limonene synthase gene into peppermint (*Mentha piperita* L.) and preliminary studies on the essential oil profiles of single transgenic plants. TAG. *Theoretical and Applied Genetics. Theoretische und angewandte Genetik*, *99*(3–4), 676–682. https://doi.org/10.1007/s001220051284

Lee, S. B., Cha, K. H., Kim, S. N., Altantsetseg, S., Shatar, S., Sarangerel, O., & Nho, C. W. (2007). The antimicrobial activity of essential oil from *Dracocephalum foetidum* against pathogenic microorganisms. *Journal of Microbiology, 45*(1), 53–57.

Lei, Y., Fu, P., Jun, X., & Cheng, P. (2019). Pharmacological properties of geraniol – a review. *Planta Medica, 85*(1), 48–55. https://doi.org/10.1055/a-0750-6907

Li, J., Wang, Y., Dong, Y., Zhang, W., Wang, D., Bai, H., Li, K., Li, H., & Shi, L. (2021). The chromosome-based lavender genome provides new insights into Lamiaceae evolution and terpenoid biosynthesis. *Horticulture Research, 8*. https://doi.org/10.1038/s41438-021-00490-6

Lindahl, A. L., Olsson, M. E., Mercke, P., Tollbom, O., Schelin, J., Brodelius, M., & Brodelius, P. E. (2006). Production of the artemisinin precursor amorpha-4,11-diene by engineered *Saccharomyces cerevisiae*. *Biotechnology Letters, 28*(8), 571–580. https://doi.org/10.1007/s10529-006-0015-6

Liu, C., Smit, S. J., Dang, J., Zhou, P., Godden, G. T., Jiang, Z., Liu, W., Liu, L., Lin, W., Duan, J., Wu, Q., & Lichman, B. R. (2023). A chromosome-level genome assembly reveals that a bipartite gene cluster formed via an inverted duplication controls monoterpenoid biosynthesis in *Schizonepeta tenuifolia*. *Molecular Plant, 16*, 533–548. https://doi.org/10.1016/j.molp.2023.01.004

Ma, B., Liu, M., Li, Z. H., Tao, X., Wei, D. Z., & Wang, F. Q. (2019). Significantly enhanced production of Patchoulol in metabolically engineered *Saccharomyces cerevisiae*. *Journal of Agricultural and Food Chemistry, 67*(31), 8590–8598. https://doi.org/10.1021/acs.jafc.9b03456

Mahmoud, S. S., Maddock, S., & Adal, A. M. (2021). Isoprenoid metabolism and engineering in glandular trichomes of Lamiaceae. *Frontiers in Plant Science, 12*, 699157. https://doi.org/10.3389/fpls.2021.699157

Mahmoud, S. S., Williams, M., & Croteau, R. (2004). Cosuppression of limonene-3-hydroxylase in peppermint promotes accumulation of limonene in the essential oil. *Phytochemistry, 65*(5), 547–554. https://doi.org/10.1016/j.phytochem.2004.01.005

Makar, S., Saha, T., & Singh, S. K. (2019). Naphthalene, a versatile platform in medicinal chemistry: Sky-high perspective. *European Journal of Medicinal Chemistry, 161*, 252–276. https://doi.org/10.1016/j.ejmech.2018.10.018

Maral, H., Turk, M., Caliskan, T., & Kirici, S. (2017). Chemical composition and antioxidant activity of essential oils of six Lamiaceae plants growing in Southern Turkey. *Natural Volatiles & Essential Oils, 4*(4), 62–68.

Martins, F. T., Doriguetto, A. C., de Souza, T. C., de Souza, K. R., Dos Santos, M. H., Moreira, M. E., & Barbosa, L. C. (2008). Composition, and anti-inflammatory and antioxidant activities of the volatile oil from the fruit peel of *Garcinia brasiliensis*. *Chemistry & Biodiversity, 5*(2), 251–258. https://doi.org/10.1002/cbdv.200890022

Masyita, A., Mustika Sari, R., Dwi Astuti, A., Yasir, B., Rahma Rumata, N., Emran, T. B., Nainu, F., & Simal-Gandara, J. (2022). Terpenes and terpenoids as main bioactive compounds of essential oils, their roles in human health and potential application as natural food preservatives. *Food Chemistry: X. 13*, 100217. https://doi.org/10.1016/j.fochx.2022.100217

Mohammadi Nejad, S., Ozguneş, H., & Başaran, N. (2017). Pharmacological and toxicological properties of Eugenol. *Turkish Journal of Pharmaceutical Sciences, 14*(2), 201–206. https://doi.org/10.4274/tjps.62207

Morteza-Semnani, K., Saeedi, M., & Akbarzadeh, M. (2008). The essential oil composition of *Betonica nivea subsp.* mazandarana. *Acta Botanica Hungarica, 50*(3–4), 385–389.https://doi.org/10.1556/ABot.50.2008.3-4.14

Mosquera, M. E. G. Jimenez, G. Tabernero, V. Vinueza-Vaca, J. García-Estrada, C. Kosalkova, K. Sola-Landa, A. Monje, B. Acosta, C. Alonso, R., & Valera, M. A. (2021). Terpenes and terpenoids: Building blocks to produce biopolymers. *Sustainable Chemistry, 2*,467–492. https://doi.org/10.3390/suschem2030026

Mulyaningsih, S., Sporer, F., Zimmermann, S., Reichling, J., & Wink, M. (2010). Synergistic properties of the terpenoids aromadendrene and 1, 8-cineole from the essential oil of *Eucalyptus globulus* against antibiotic-susceptible and antibiotic-resistant pathogens. *Phytomedicine, 17*(13), 1061–1066. https://doi.org/10.1016/j.phymed.2010.06.018

Munoz-Bertomeu, J., Ros, R., Arrillaga, I., & Segura, J. (2008). Expression of spearmint limonene synthase in transgenic spike lavender results in an altered monoterpene composition in developing leaves. *Metabolic Engineering, 10*(3–4), 166–177. https://doi.org/10.1016/j.ymben.2008.04.002

Nagoor Meeran, M. F., Javed, H., Al Taee, H., Azimullah, S., & Ojha, S. K. (2017). Pharmacological properties and molecular mechanisms of thymol: Prospects for its therapeutic potential and pharmaceutical development. *Frontiers in Pharmacology, 8*, 380. https://doi.org/10.3389/fphar.2017.00380

Neumann, H., & Neumann-Staubitz, P. (2010). Synthetic biology approaches in drug discovery and pharmaceutical biotechnology. *Applied Microbiology and Biotechnology,87*, 75–86. https://doi.org/10.1007/s00253-010-2578-3

Nicolella, H. D., de Oliveira, P. F., Munari, C. C., Costa, G. F., Moreira, M. R., Veneziani, R. C., & Tavares, D. C. (2014). Differential effect of manool-a diterpene from *Salvia officinalis*, on genotoxicity induced by methyl methanesulfonate in V79 and HepG2 cells. *Food and Chemical Toxicology: An International Journal Published for the British Industrial Biological Research Association, 72*, 8–12. https://doi.org/10.1016/j.fct.2014.06.025

Ninkuu, V., Zhang, L., Yan, J., Fu, Z., Yang, T., &Zeng, H. (2021). Biochemistry of terpenes and recent advances in plant protection. *International Journal of Molecular Sciences, 22*(11), 5710. https://doi.org/10.3390/ijms22115710

Noge, K., & Becerra, J. X. (2009). Germacrene D, a common sesquiterpene in the genus Bursera (Burseraceae). *Molecules, 14*(12), 5289–5297.

Nomani, M., & Tohidfar, M. (2021). Plant regeneration and transformation of *Trachyspermum ammi* using *Agrobacterium tumefaciens* and zygotic embryos. *Journal of Genetic Engineering and Biotechnology, 19*(1), 68. https://doi.org/10.1186/s43141-021-00173-8

Ntalli, N. G., Ferrari, F., Giannakou, I., &Menkissoglu-Spiroudi, U. (2010). Phytochemistry and nematicidal activity of the essential oils from 8 Greek Lamiaceae aromatic plants and 13 terpene components. *Journal of Agricultural and Food Chemistry, 58*(13), 7856–7863. https://doi.org/10.1021/jf100797m

Ogundajo, A. L., Owoyele, O. A., Ogunwande, I. A., & Owolabi, M. S. (2016). Chemical composition of essential oil from the leaves of *Clerodendrum polycephalum* Baker growing in Nigeria. *Journal of Essential Oil Bearing Plants, 19*(1), 119–124. https://doi.org/10.1080/0972060X.2014.905759

Ozdemir, F. A., Kilic, O., & Yildirimli, S. (2017). Essential oil composition and antimicrobial activity of endemic *Phlomis sieheana* Rech. From Bingol. *Journal of essential Oil Bearing Plants, 20*(2), 516–523. https://doi.org/10.1080/0972060X.2017.1304833

Park, B. B., An, J. Y., & Park, S. U. (2021). Recent studies on pinene and its biological and pharmacological activities.

EXCLI Journal, *20*, 812–818. https://doi.org/10.17179/excli2021-3714

Passinho-Soares, H. C., David, J. P., de Santana, J. R. F. David, J. M., de Rodrigue, F. M., Mesquita, P. R. R., de Oliveira, F. S., & Bellintani, M. C. (2017). Influence of growth regulators on distribution of trichomes and the production of volatiles in micropropagated plants of *Plectranthus ornatus*. *Revista Brasileira de Farmacognosia*, *27*, 679–680. https://doi.org/10.1016/j.bjp.2017.10.001

Peng, B., Plan, M. R., Chrysanthopoulos, P., Hodson, M. P., Nielsen, L. K., & Vickers, C. E. (2017). A squalene synthase protein degradation method for improved sesquiterpene production in *Saccharomyces cerevisiae*. *Metabolic Engineering*, *39*, 209–219. https://doi.org/10.1016/j.ymben.2016.12.003

Popa, C. L., Lupitu, A., Mot, M. D., Copolovici, L., Moisa, C., & Copolovici, D. M. (2021). Chemical and biochemical characterization of essential oils and their corresponding hydrolats from six species of the Lamiaceae Family. *Plants*, *10*, 2489. https://doi.org/10.3390/plants10112489

Rajput, A., Kasar, A., Thorat, S., &Kulkarni, M. (2023). Borneol: A plant-sourced terpene with a variety of promising pharmacological effects. *The Natural Products Journal*, *13*(1), 13–28.

Ramalho, T. R., Oliveira, M. T., Lima, A. L., Bezerra-Santos, C. R., & Piuvezam, M. R. (2015). Gamma-Terpinene modulates acute inflammatory response in Mice. *Planta Medica*, *81*(14), 1248–1254. https://doi.org/10.1055/s-0035-1546169

Ramos da Silva, L. R., Ferreira, O. O., Cruz, J. N., de Jesus Pereira Franco, C., Oliveira Dos Anjos, T., Cascaes, M. M., Almeida da Costa, W., Helena de Aguiar Andrade, E., & Santana de Oliveira, M. (2021). Lamiaceae essential oils, phytochemical profile, antioxidant, and biological activities. *Evidence-based Complementary and Alternative Medicine: eCAM*, 6748052. https://doi.org/10.1155/2021/6748052

Reddy, V. A., Wang, Q., Dhar, N., Kumar, N., Venkatesh, P. N., Rajan, C., Panicker, D., Sridhar, V., Mao, H. Z., & Sarojam, R. (2017). Spearmint R2R3-MYB transcription factor MsMYB negatively regulates monoterpene production and suppresses the expression of geranyl diphosphate synthase large subunit (MsGPPS.LSU). *Plant Biotechnology Journal*, *15*(9), 1105–1119. https://doi.org/10.1111/pbi.12701

Renda, G., Bektas, N. Y., Korkmaz, B., Celik, G., Sevgi, S., & Yayli, N. (2017). Volatile constituents of three *Stachys* L. species from Turkey. *Marmara Pharmaceutical Journal*, *21*(2), 278–285.

Ringel, M., Dimos, N., Himpich, S., Haack, M., Huber, C., Eisenreich, W., Schenk, G., Loll, B., & Brück, T. (2022). Biotechnological potential and initial characterization of two novel sesquiterpene synthases from Basidiomycota *Coniophora puteana* for heterologous production of δ-cadinol. *Microbial Cell Factories*, *21*(1), 64. https://doi.org/10.1186/s12934-022-01791-8

Roba, K. (2020). The role of terpene (secondary metabolite). *Natural Products Chemistry & Research*,*9*, 411.

Rodrigues, F. C., Almeida-Bezerra, J. W., Fidelis, K. R., Costa, A. R., Souza, M. A. De, Costa, M. H. N. Da, Santos, L. T. Dos, Gonçalo, M. A. B. F., Menezes, S. A. De, Sousa, S. M. De, Araujo, E. K. R. De, Braga, A. V. P., Ribeiro, T. G., Costa, A. R. De S., Santos, M. A. F. Dos, Nunes, R. G. Dos, S., & Rocha, M. I. (2020). Chemical characterization and insecticidal potential of the essential oil of *Ocimum gratissimum* L. (Lamiaceae) against *Nauphoeta cinerea* (Blaberidae). *Research, Society and Development*,*9*(9), e72996412. https://doi.org/10.33448/rsd-v9i9.6412

Rondon, M. E., Velasco, J., Cornejo, X., Fernandez, J., & Morocho, W. (2016). Chemical composition and antibacterial activity of *Piper lenticellosum* C.D.C essential oil collected in Ecuador. *Journal of Applied Pharmaceutical Science*, *6*(08), 156–159. https://doi.org/10.7324/JAPS.2016.60824

Sagir, Z. O., Carikci, S., Kilic, T., & Goren, A. C. (2017). Metabolic profile and biological activity of *Sideritis brevibracteata* P. H. Davis endemic to Turkey. *International Journal of Food Properties*, *20*(12), 2994–3005.

Santos, P. L., Matos, J. P. S., Picot, L., Almeida, J. R., Quintans, J. S., & Quintans-Junior, L. J. (2019). Citronellol, a monoterpene alcohol with promising pharmacological activities-A systematic review. *Food and Chemical Toxicology*, *123*, 459–469. https://doi.org/10.1016/j.fct.2018.11.030

Sarikaya, A. G., & Fakir, H. (2016). The effect of reaping times on volatile components of natural *Phlomis* L. (Lamiaceae) taxa in the lakes district of Turkey. *Applied Ecology and Environmental Research*, *14*(3), 753–772. https://doi.org/10.15666/aeer/1403_753772

Scalcinati, G., Knuf, C., Partow, S., Chen, Y., Maury, J., Schalk, M., Daviet, L., Nielsen, J., & Siewers, V. (2012). Dynamic control of gene expression in *Saccharomyces cerevisiae*engineered for the production of plant sesquitepene α-santalene in a fed-batch mode. *Metabolic Engineering*, *14*(2), 91–103. https://doi.org/10.1016/j.ymben.2012.01.007

Schmiderer, C., Grassi, P., Novak, J., & Franz, C. (2008). Diversity of essential oil glands of Spanish sage (*Salvia lavandulifolia Vahl*, Lamiaceae). *Natural Product Communications*, *3*(7), 1155–1160. https://doi.org/10.1177/1934578X0800300723

Shafaghat, A., & Oji, K. (2010). Nepetalactone content and antibacterial activity of the essential oils from different parts of *Nepeta persica*. *Natural Product Communications*, *5*(4), 625–628. https://doi.org/10.1177/1934578X1000500427

Sharma, S., Gupta, J., Prabhakar, P. K., Gupta, P., Solanki, P., & Rajput, A. (2019). Phytochemical repurposing of natural molecule: Sabinene for identification of novel therapeutic benefits using *In Silico* and *In Vitro* approaches. *Assay and Drug Development Technologies*, *17*(8), 339–351. https://doi.org/10.1089/adt.2019.939

Sharma, S., Habib, S., Sahu, D., & Gupta, J. (2021). Chemical properties and therapeutic potential of citral, a monoterpene isolated from lemongrass. *Medicinal Chemistry*, *17*(1), 2–12. https://doi.org/10.2174/1573406416666191227111106

Shukal, S., Chen, X., & Zhang, C. (2019). Systematic engineering for high-yield production of viridiflorol and amorphadiene in auxotrophic *Escherichia coli*. *Metabolic Engineering*, *55*, 170–178. https://doi.org/10.1016/j.ymben.2019.07.007

Siddiqui, T., Sharma, V., Khan, M. U., & Gupta, K. (2024). Terpenoids in essential oils: Chemistry, classification, and potential impact on human health and industry. *Phytomedicine Plus*, 100549. https://doi.org/10.1016/j.phyplu.2024.100549

Silva, M., Ribeiro, F. P., Medeiros, M. A., Sampaio, P. A., Silva, Y., Silva, M. T., Quintans, J. S., Quintans-Júnior, L. J., & Ribeiro, L. A. (2015). The vasorelaxant effect of p-cymene in rat aorta involves potassium channels. *The Scientific World Journal*, *2015*. https://doi.org/10.1155/2015/458080

Song, Y., He, S., Abdallah, I. I., Jopkiewicz, A., Setroikromo, R., van Merkerk, R., Tepper, P. G., & Quax, W. J. (2021). Engineering of multiple modules to improve amorphadiene production in *Bacillus subtilis* using CRISPR-Cas9. *Journal of Agricultural and Food Chemistry*, *69*(16), 4785–4794. https://doi.org/10.1021/acs.jafc.1c00498

Stephane, F. F. Y., & Jules, B. K. J. (2020). Terpenoids as important bioactive constituents of essential oils. In M. S. de Oliveira, W. A. da Costa, & S. G. Silva (Eds.), *Essential oils – bioactive compounds, new perspectives and applications*. IntechOpen. https://doi.org/10.5772/intechopen.91426

Stevanovic, Z. D., Sieniawska, E., Glowniak, K., Obradovic, N., & Pajic-Lijakovic, I. (2020). Natural macromolecules as carriers for essential oils: From extraction to biomedical application. *Frontiers in Bioengineering and Biotechnology*,*8*. https://doi.org/10.3389/fbioe.2020.00563

Surendran, S., Qassadi, F., Surendran, G., Lilley, D., &Heinrich, M. (2021). Myrcene-what are the potential health benefits of this flavouring and aroma agent? *Frontiers in Nutrition*, *8*, 699666. https://doi.org/10.3389/fnut.2021.699666

Tan, N., Yazici-Tutunis, S., Yesil, Y., Demirci, B., & Tan, E. (2017). Antibacterial activities and composition of the essential oils of *Salvia sericeo-tomentosa* varieties. *Records of Natural Products*, *11*(5), 456–461.

Tang, T. C., An, B., Huang, Y., Vasikaran, S., Wang, Y., Jiang, X., Lu, T. K., & Zhong, C.(2021). Materials design by synthetic biology. *Nature Reviews Materials*,*6*, 332–350. https://doi.org/10.1038/s41578-020-00265-w

Temel, H. E., Demirci, B., Demirci, F., Celep, F., Kahraman, A., Dogan, M., &Baser, K. H. C. (2016). Chemical characterization and anticholinesterase effects of essential oils derived from Salvia species. *Journal of Essential Oil Research*, *28*(4), 322–331.https://doi.org/10.1080/10412905.2016.1159257

Tholl, D. (2015). Biosynthesis and biological functions of terpenoids in plants. *Advances in Biochemical Engineering/Biotechnology*, *148*, 63–106. https://doi.org/10.1007/10_2014_295

Turkez, H., Celik, K., & Togar, B. (2014). Effects of copaene, a tricyclic sesquiterpene, on human lymphocytes cells in vitro. *Cytotechnology*, *66*(4), 597–603. https://doi.org/10.1007/s10616-013-9611-1

Uritu, C. M., Mihai, C. T., Stanciu, G. D., Dodi, G., Alexa-Stratulat, T., Luca, A., Leon-Constantin, M. M., Stefanescu, R., Bild, V., Melnic, S., & Tamba, B. I. (2018). Medicinal plants of the family lamiaceae in pain therapy: A review. *Pain Research & Management*, *2018*, 7801543. https://doi.org/10.1155/2018/7801543

Urosevic, M., Nikolic, L., Gajic, I., Nikolic, V., Dinic, A., & Miljkovic, V. (2022). Curcumin: Biological activities and modern pharmaceutical forms. *Antibiotics*, *11*(2), 135. https://doi.org/10.3390/antibiotics11020135

Van Herpen, T. W. J. M., Cankar, K., Nogueira, M., Bosch, D., Bouwmeester, H. J., & Beekwilder, J. (2010). *Nicotiana benthamiana* as a production platform for artemisinin precursors. *PLoS ONE*, *5*, e14222.

Vining, K. J., Johnson, S. R., Ahkami, A., Lange, I., Parrish, A. N., Trapp, S. C., Croteau, R. B., Straub, S. C., Pandelova, I., & Lange, B. M. (2017). Draft genome sequence of *Mentha longifolia* and development of resources for mint cultivar improvement. *Molecular Plant*, *10*(2), 323–339. http://dx.doi.org/10.1016/j.molp.2016.10.018

Wang, Q., Reddy, V. A., Panicker, D., Mao, H. Z., Kumar, N., Rajan, C., Venkatesh, P. N., Chua, N. H., & Sarojam, R. (2016). Metabolic engineering of terpene biosynthesis in plants using a trichome-specific transcription factor MsYABBY5 from spearmint (*Mentha spicata*). *Plant Biotechnology Journal*, *14*(7), 1619–1632. https://doi.org/10.1111/pbi.12525

Widyasanti, A., Nurjanah, S., Nurhadi, B., & Osman, C. P.(2021). November. Isolation of guaiene from crude and distillate patchouli oil extracted by molecular distillation. In *IOP Conference Series: Earth and Environmental Science*, *924*(1), 012008. https://doi.org/10.1088/1755-1315/924/1/012008

Yogesh, U., & Krishnakant, P. (2011). Chemical composition and antibacterial activity of essential oil from Anisomeles species grown in India. *Pharmacognosy Journal*, *2*(18), 55–59. https://doi.org/10.1016/S0975-3575(11)80026-X

Yousefi, M., Gandomkar, S., &Habibi, Z. (2012). Essential oil from aerial parts of *Betonica grandiflora* Willd. from Iran. *Natural Product Research*, *26*(2), 146–151. https://doi.org/10.1080/14786419.2010.534992

Yuan, J., & Ching, C.-B. (2016). Mitochondrial acetyl-CoA utilization pathway for terpenoid productions. *Metabolic Engineering*, *38*, 303–309. https://doi.org/10.1016/j.ymben.2016.07.008

Zamborine Nemeth, E., & Thi Nguyen, H. (2020). Thujone, a widely debated volatile compound: What do we know about it? *Phytochemistry Reviews*, *19*, 405–423. https://doi.org/10.1007/s11101-020-09671-y

Zhang, C., & Hong, K. (2020). Production of terpenoids by synthetic biology approaches. *Frontiers in Bioengineering and Biotechnology*, *8*, 347. https://doi.org/10.3389/fbioe.2020.00347

Zhang, Y., Nielsen, J., & Liu, Z. (2017). Engineering yeast metabolism for production of terpenoids for use as perfume ingredients, pharmaceuticals and biofuels. *FEMS Yeast Research*, *17*(8). https://doi.org/10.1093/femsyr/fox080

Zhao, W., Yang, C., Zhang, N., Peng, Y., Ma, Y., Gu, K., Liu, X., Liu, X., Liu, X., Liu, Y., Li, S., & Zhao, L. (2023). Menthone exerts its antimicrobial activity against methicillin resistant *Staphylococcus aureus* by affecting cell membrane properties and lipid profile. *Drug Design, Development and Therapy*, *17*, 219–236. https://doi.org/10.2147/DDDT.S384716

Zhelev, I., Petkova, Z., Kostova, I., Damyanova, S., Stoyanova, A., Dimitrova-Dyulgerova, I., Antova, G., Ercisli, S., Assouguem, A., Kara, M., Almeer, R., & Sayed, A. A. (2022). Chemical composition and antimicrobial activity of essential oil of fruits from *Vitex agnus-castus* L., growing in two regions in Bulgaria. *Plants*, *11*, 896. https://doi.org/10.3390/plants11070896

28 Integrated Molecular Strategies for Enhanced Production of Anti-Cancer Compounds from *Abutilon hirtum* (Lam.) Sweet (Malvaceae)

Balasundaram Saraswathy Chithra Devi, S. Tharani, M. R. Elakkiya, M. Krishnasreya, and P. Servin Wesley

28.1 INTRODUCTION

Cancer is a complex and devastating disease characterized by uncontrolled cell growth and proliferation. It remains one of the leading causes of death globally, posing significant challenges to healthcare systems and societies worldwide. Despite advancements in cancer treatment modalities such as chemotherapy, radiation therapy, and surgery, the quest for more effective and less toxic therapies continues. In recent years, there has been increasing interest in exploring natural compounds derived from medicinal plants as potential sources of anti-cancer agents. Among these plants, *Abutilon hirtum* (Lam.) Sweet, belonging to the Malvaceae family, has emerged as a promising candidate due to its rich phytochemical profile and documented medicinal properties.

The term 'cancer' was coined by Hippocrates, the father of medicine, who likened tumors to a crab, using the Greek words 'carcinos' and 'carcinoma' to describe the disease. While Hippocrates may have been the first to give it a name, the history of cancer likely predates his observations (Parkin et al., 2005). Cancer arises from a combination of genetic mutations and environmental factors, leading to abnormal cell growth and tumor formation. Risk factors for cancer include lifestyle choices such as smoking and excessive alcohol consumption, exposure to environmental pollutants, genetic predisposition, and viral infections (Parsa, 2012). Regardless of various therapies to treat cancer, it remains the second most widely spread disease in India next to heart disease. It is accountable for a maximum mortality rate of approximately 0.3 million fatalities per year.

In recent decades, there has been a growing recognition of the therapeutic potential of medicinal plants in cancer treatment. Traditional medicine systems, including Ayurveda, Siddha, Unani, and traditional Chinese medicine, have long utilized plant-based remedies for various ailments, including cancer (Jaiswal et al., 2016). India, with its rich biodiversity, is particularly renowned for its wealth of medicinal plant species, with approximately 43% of known angiosperms possessing medicinal properties (Choudhury et al., 2012). *Abutilon hirtum*, commonly known as Indian mallow or *thuthi* in Ayurveda, is one such plant that has been traditionally used for its medicinal properties (Kailasam, 2015).

Abutilon hirtum is a perennial shrub native to tropical and subtropical regions, including India, where it has been used in traditional medicine for its anti-inflammatory, antioxidant, and antimicrobial properties. Recent scientific investigations have revealed the presence of various bioactive compounds in *Abutilon hirtum*, including alkaloids, flavonoids, phenolic acids, saponins, and tannins, which are known to possess diverse pharmacological activities of particular interest, are its potential anti-cancer properties, which have attracted attention from researchers and pharmaceutical companies seeking novel therapeutic agents (Hamed et al., 2017).

The conventional methods of obtaining bioactive compounds from medicinal plants often involve extraction from plant tissues using organic solvents (Stéphane et al., 2021). However, these methods have limitations, including low yield, environmental concerns, and the potential for solvent residues in the final product (Gullón et al., 2020). Biotechnological approaches offer alternative strategies for the enhanced production of bioactive compounds from plants, including *Abutilon hirtum*. These approaches leverage the principles of plant tissue culture, genetic engineering, and metabolic engineering to manipulate plant metabolism and enhance the production of target compounds.

Plant tissue culture, including techniques such as cell suspension culture and adventitious root culture, has emerged as a powerful tool for the production of secondary metabolites from medicinal plants (Wang et al., 2017; Rahmat & Kang, 2019). By providing a controlled environment for plant growth and metabolism, tissue culture allows for the production of bioactive compounds in large quantities and without the need for land or extensive agricultural

DOI: 10.1201/9781003475491-28

resources. Furthermore, tissue culture can be coupled with elicitation techniques, whereby plants are exposed to specific stimuli or stressors to induce the production of secondary metabolites (Wang et al., 2017; Rahmat & Kang, 2019; Cardoso et al., 2019).

Elicitation techniques have been extensively studied and optimized for various medicinal plants, including *Abutilon hirtum*, with the aim of enhancing the production of specific bioactive compounds. Elicitors can be classified as abiotic or biotic, depending on their nature, and may include substances such as methyl jasmonate, salicylic acid, chitosan, and fungal elicitors (Thakur et al., 2018; Akhtar et al., 2019). These elicitors trigger plant defense mechanisms, leading to the accumulation of secondary metabolites with potential therapeutic properties (Zhao et al., 2005).

In addition to traditional elicitation methods, recent advancements in biotechnology have led to the development of novel approaches for enhancing the production of bioactive compounds from medicinal plants (Niazian et al., 2019). Metabolic engineering, for example, involves the manipulation of plant genes involved in secondary metabolite biosynthesis to increase their expression and yield. This approach has been successfully applied in various plant species to enhance the production of specific compounds, including anti-cancer agents (Sharma et al., 2019).

Another promising biotechnological strategy for enhancing the production of bioactive compounds from medicinal plants is the use of nanotechnology (Majumdar et al., 2020). Nanoparticles can serve as delivery systems for bioactive compounds, improving their stability, bioavailability, and targeted delivery to cancer cells. Furthermore, nanoparticles themselves may possess intrinsic anti-cancer properties, making them attractive candidates for drug delivery and cancer therapy (Rivero-Montejo et al., 2021).

Abutilon hirtum holds great promise as a source of anticancer compounds, and biotechnological strategies offer innovative approaches for enhancing their production. By leveraging techniques such as plant tissue culture, elicitation, metabolic engineering, and nanotechnology, researchers can unlock the full therapeutic potential of this valuable medicinal plant. Continued research and development in this field are essential for advancing cancer treatment and improving patient outcomes.

28.2 BIOACTIVE POTENTIAL OF *ABUTILON HIRTUM*

Abutilon, a genus within the Malvaceae family, encompasses numerous flowering plants with recognized medicinal properties (Das & Islam, 2019). Throughout history, various *Abutilon* species, including *A. hirtum*, have been extensively studied for their therapeutic potential (Rajeshwari & Sevarkodiyone, 2018). These plants have been traditionally utilized to address a spectrum of health issues, ranging from gastrointestinal disorders to inflammatory conditions and wound healing (Sohel, 2015). *Abutilon* species are known to possess antioxidant properties due to the presence of phenolic compounds (Hamed et al., 2017). Antioxidants help neutralize harmful free radicals in the body, potentially reducing oxidative stress and lowering the risk of chronic diseases (Devasagayam et al., 2004). Specifically, *A. indicum*, a closely related species, has been noted for its multifaceted pharmacological activities, including wound healing, immunomodulation, hepatoprotection, hypoglycemia alleviation, pain relief, antimalarial effects, antimicrobial properties, and anti-ulcer activity (Arivoli & Tennyson, 2011; Dashputre & Naikwade, 2011; Mohite et al., 2012).

Some studies suggest that extracts from Abutilon species exhibit anti-inflammatory activity. This property could be beneficial in managing inflammatory conditions like arthritis or inflammatory skin disorders (Gomaa et al., 2018). Within this context, *A. hirtum* also emerges as a valuable resource in traditional medicine, particularly esteemed for its efficacy in managing nervous system disorders (Hamed et al., 2017). The various parts of the plant, such as the stem bark, seeds, leaves, and roots, have been employed for diverse therapeutic purposes. The stem bark, for instance, serves as a febrifuge, anthelmintic, alexiteric, diuretic, and astringent agent. Seeds find applications in addressing bronchitis, piles, and gonorrhea and as a laxative, while leaves are utilized for toothache and lumbago (Das & Islam, 2019). The root is particularly valued for its role in managing conditions like excessive menstrual bleeding and fever (Rajeshwari & Sevarkodiyone, 2018).

Moreover, *A. hirtum*'s pharmacological spectrum extends to include the treatment of liver abnormalities, with the aqueous leaf extract being particularly notable in this regard (Hamed et al., 2017). In Malaysian traditional medicine, *A. hirtum* finds application in alleviating pain associated with kidney stones, often administered topically in combination with sticky rice on ulcers. In Thailand, a decoction of the plant's roots is utilized to manage toothaches, colds, and coughs, while paste made from its leaves or flowers is applied to abscesses (Perumal, 2001; Gomaa et al., 2018).

Notably, *A. hirtum* also serves non-medicinal purposes, such as providing edible fruits and fodder for domestic animals in certain regions of Africa (Lye et al., 2008). In India, the leaves are employed not only in medicinal formulations for dysentery, diarrhea, and digestive disorders but also as demulcents and diuretics. Decoctions derived from the leaves and roots are utilized in the management of bladder inflammations, ulcers, and wounds and also as mouthwashes, reflecting the plant's versatility in traditional healing practices (Kapoor & Kapoor, 1980; Pullaiah, 2002; Chandan et al., 2008). Traditional medicine has used some *Abutilon* species as diuretics, suggesting a potential for managing conditions related to fluid retention (Gomaa et al., 2018).

Certain compounds found in *Abutilon hirtum* may have antimicrobial properties, which could help combat bacterial, fungal, or viral infections. This potential could be explored for the development of new antimicrobial agents. Additionally, alkaloid compounds isolated from the roots of *A. hirtum* further underscore its pharmacological significance. The diverse array of traditional uses across different cultures underscores the plant's broad therapeutic potential and underscores the need for further scientific exploration to elucidate its bioactive constituents and mechanisms of action.

28.3 ISOLATION AND CHARACTERIZATION OF BIOACTIVE COMPOUNDS FROM *ABUTILON HIRTUM*

28.3.1 Isolation Methods

28.3.1.1 Collection and Preparation of Plant Material

Fresh leaves of *Abutilon hirtum* were collected from their natural habitat and carefully washed to remove dirt and debris. The plant material was then dried and ground into a fine powder using a mortar and pestle for further processing (Gomaa et al., 2016).

28.3.1.2 Extraction Techniques

- **Solvent Extraction:**

Solvent extraction is a fundamental technique employed in the isolation of bioactive compounds from plant material. In the case of *Abutilon hirtum*, this method serves as the initial step in the process of obtaining crude extracts enriched with potentially active compounds. The powdered plant material undergoes extraction using a suitable solvent system, such as ethanol, methanol, or dichloromethane, to effectively extract the bioactive constituents present in the plant matrix (Srivastava et al., 2021).

The choice of solvent system is crucial in solvent extraction as it determines the selectivity and efficiency of compound extraction. Commonly used solvents for the extraction of bioactive compounds from plant material include ethanol, methanol, and dichloromethane. The selection of solvent is based on factors such as the polarity of the target compounds, the solubility of compounds in the solvent, and the desired extraction yield (Abubakar & Haque, 2020).

The powdered plant material is then subjected to solvent extraction using the chosen solvent system. This can be achieved through various techniques such as maceration, percolation, or reflux extraction. In maceration, the plant material is soaked in the solvent for a specific duration, allowing the extraction of bioactive compounds by diffusion. Percolation involves the continuous passage of solvent through the plant material, while reflux extraction utilizes heat to accelerate the extraction process (Gokulakrishnan et al., 2023).

Extraction parameters, such as solvent-to-material ratio, extraction time, and temperature, are optimized to maximize the yield and quality of the extract. Factors such as the particle size of the plant material and agitation speed may also influence extraction efficiency and are taken into consideration during optimization (Aboulghazi et al., 2022). After the completion of the extraction process, the solvent-extract mixture is filtered to separate the liquid extract from the solid plant residue. The filtrate containing the extracted bioactive compounds is collected and concentrated under reduced pressure using techniques such as rotary evaporation or vacuum distillation (Goti & Dasgupta, 2023). The concentrated extract obtained from solvent extraction is stored under appropriate conditions to maintain its stability and integrity. This may involve storing the extract in amber-colored vials or containers to protect it from light-induced degradation and storing it at low temperatures to prevent oxidation or degradation of sensitive compounds (Hamed et al., 2017).

- **Soxhlet Extraction:**

Soxhlet extraction is a continuous extraction technique commonly used for the extraction of both polar and non-polar compounds from plant material. It is particularly effective for extracting compounds that are less soluble in cold solvents or present in small quantities (Farooq et al., 2022). In the case of *Abutilon hirtum*, Soxhlet extraction is employed to obtain a crude extract enriched with a wide range of bioactive constituents, including both polar and non-polar compounds. Soxhlet extraction allows for the use of a wide range of solvent systems to extract polar and non-polar compounds. Common solvents include ethanol, methanol, chloroform, dichloromethane, and hexane (Srivastava et al., 2021). The choice of solvent depends on the polarity and solubility of the target compounds.

The Soxhlet apparatus consists of a round-bottom flask containing the solvent, a thimble containing the powdered plant material, a condenser, and a siphon. The extraction process begins with the solvent in the round-bottom flask being heated to reflux. As the solvent vaporizes, it rises through the condenser and drips onto the plant material in the thimble (Abubakar & Haque, 2020). The solvent continuously cycles through the plant material in the thimble, extracting the target compounds as they pass through. The extracted compounds dissolve in the solvent and accumulate in the round-bottom flask. The refluxing action ensures thorough extraction of the bioactive constituents from the plant material (Stéphane et al., 2021).

Soxhlet extraction typically continues for several hours overnight to ensure complete extraction of the target compounds. The duration of extraction may vary depending on factors such as the extraction efficiency desired and the nature of the plant material (Višnjevec et al., 2024). After completion of the extraction process, the solvent-extract mixture is collected in the round-bottom flask. The solvent is then evaporated under reduced pressure using techniques such as rotary evaporation or vacuum distillation to obtain a concentrated extract (de Alba et al., 2023). The resulting crude extract contains a diverse array of bioactive constituents, which can undergo further purification and characterization to identify and elucidate their chemical composition and potential pharmacological activities.

28.3.1.3 Fractionation Techniques

Fractionation techniques play a crucial role in the isolation and purification of bioactive compounds from crude extracts obtained through solvent extraction. Two commonly employed fractionation techniques are column chromatography and thin-layer chromatography (TLC). These techniques enable the separation of compounds based on their chemical properties, allowing for the isolation of specific bioactive constituents from complex mixtures.

- **Column Chromatography:**

Column chromatography is a versatile technique used for the separation and purification of compounds based on their interactions with a stationary phase (typically silica gel or alumina) and a mobile phase (solvent system) (Ramraje et al., 2020). In the case of *Abutilon hirtum*, column chromatography is employed to fractionate the crude extract obtained from solvent extraction, thereby isolating individual bioactive compounds.

Preparation of Silica Gel Column: A glass column is packed with silica gel as the stationary phase. The silica gel is carefully packed to ensure an even bed and optimal flow of solvent through the column.

Loading of Crude Extract: The crude extract obtained from solvent extraction is dissolved in a suitable solvent and loaded onto the top of the silica gel column. The sample is then allowed to adsorb onto the silica gel bed (Yuan et al., 2022).

Elution with Solvent Gradient: Elution of compounds from the column is carried out using a gradient of solvent polarity. Initially, a less polar solvent is used to elute non-polar compounds, followed by a gradual increase in the polarity of the solvent to elute more polar compounds. This gradient elution allows for the separation of compounds based on their chemical properties (Jandera & Churacek, 1985).

Collection of Fractions: As elution progresses, fractions containing individual compounds are collected in test tubes. Each fraction represents a group of compounds with similar polarity and elution characteristics (Erkan, 2012).

Monitoring of Elution: Elution progress is monitored using UV-visible spectroscopy or TLC to identify the fractions containing bioactive compounds. Fractions of interest are pooled together for further analysis and purification (de Souza et al., 2023).

- **Thin-Layer Chromatography (TLC):**

Thin-layer chromatography (TLC) is a rapid and cost-effective technique used for qualitative analysis and preliminary fractionation of compounds in a mixture. In the context of *Abutilon hirtum*, TLC is employed as a preliminary screening tool to monitor the separation of compounds on a thin layer of adsorbent material.

Preparation of TLC Plate: A thin layer of adsorbent material, such as silica gel or cellulose, is coated onto a glass or plastic plate. The sample spots are applied to the TLC plate using a microsyringe or capillary tube (Bele & Khale, 2011).

Development of TLC Plate: The TLC plate is placed in a developing chamber containing a solvent system. As the solvent migrates up the TLC plate via capillary action, it carries the sample components with it. Different compounds in the sample exhibit varying degrees of affinity for the stationary phase and the mobile phase, resulting in separation (Albreht & Vovk, 2016).

Visualization of Separated Compounds: After the TLC plate has been developed, it is removed from the chamber and dried. The separated compounds are visualized using UV light, iodine vapor, or chemical staining reagents. This allows for the identification of distinct spots corresponding to individual compounds (Zarzycki, 2023).

Analysis of TLC Results: The Rf (retention factor) values of the separated compounds are calculated by measuring the distance traveled by the compound spot and the solvent front. These values provide information about the relative polarity and mobility of the compounds (Kumar et al., 2013).

Selection of Fractions: Based on the TLC results, fractions containing bioactive compounds are identified and collected for further analysis and purification using column chromatography or other techniques.

Column chromatography and thin-layer chromatography are indispensable fractionation techniques used in the isolation and purification of bioactive compounds from *Abutilon hirtum*. These techniques enable the separation of complex mixtures into individual compounds based on their chemical properties, facilitating the identification of bioactive constituents with potential pharmacological activities.

28.3.1.4 Purification Techniques

Purification techniques are essential in the isolation of bioactive compounds from crude extracts obtained through extraction and fractionation processes. Two commonly employed purification techniques are high-performance liquid chromatography (HPLC) and preparative thin-layer chromatography (TLC). These techniques enable the isolation of pure compounds from complex mixtures, facilitating further characterization and biological evaluation of the bioactive constituents.

- **High-Performance Liquid Chromatography (HPLC):**

High-performance liquid chromatography (HPLC) is a powerful chromatographic technique used for the separation, identification, and quantification of individual compounds in a mixture (Lozano-Sánchez et al., 2018). In the context of *Abutilon hirtum*, HPLC is employed for further purification of individual compounds obtained from fractionation techniques such as column chromatography or preparative TLC.

The sample containing the mixture of compounds to be purified is dissolved in a suitable solvent and injected into the HPLC system. The sample may undergo filtration or centrifugation to remove particulate matter before injection. HPLC columns packed with stationary phases such as silica, C18, or phenyl are commonly used for compound separation (Qiu et al., 2011). The choice of column depends on the polarity and chemical properties of the compounds to be separated. A suitable mobile phase consisting of a solvent or solvent mixture is selected based on the polarity of the compounds. Gradient elution or isocratic elution can be employed to achieve optimal separation (Sasidharan et al., 2011).

The sample components are separated as they pass through the HPLC column under high pressure. The compounds interact with the stationary phase and mobile phase to varying degrees, leading to their separation based on differences in polarity, size, and chemical structure. Compounds eluting from the HPLC column are detected using a UV-visible detector or other detectors such as diode array detectors (DAD) or mass spectrometers (MS) (Burnaz et al., 2017). The retention times and peak areas of the compounds are recorded for quantification and identification purposes (Hostettmann et al., 2001). Fractions containing individual compounds of interest are collected using a fraction collector. Each fraction corresponds to a specific compound or group of compounds separated by the HPLC system.

- **Preparative Thin-Layer Chromatography (TLC):**

Preparative thin-layer chromatography (TLC) is a chromatographic technique used for the isolation and purification of compounds in larger quantities based on their migration behavior on a TLC plate. In the context of *Abutilon hirtum*, preparative TLC is employed to isolate pure compounds from crude extracts or fractionated mixtures obtained from column chromatography or other fractionation techniques.

The sample containing the mixture of compounds to be purified is applied as a spot at the base of a TLC plate using a microsyringe or capillary tube (Hameed et al., 2023). Multiple sample spots may be applied depending on the number of compounds present in the mixture. The TLC plate is placed in a developing chamber containing a suitable solvent system (Cieśla & Waksmundzka-Hajnos, 2009). As the solvent migrates up the TLC plate, the compounds in the sample are separated based on their affinity for the stationary phase and the mobile phase. After development, the TLC plate is removed from the chamber and dried.

The separated compounds are visualized using UV light or staining reagents. The spots corresponding to individual compounds are marked for collection. The TLC plate is placed under a UV lamp, and the spots corresponding to the desired compounds are scraped off the plate using a sharp blade or spatula (Cai, 2014). The scraped material is eluted from the TLC plate using a suitable solvent to obtain purified compounds. The eluted compounds are concentrated under reduced pressure using techniques such as rotary evaporation or vacuum drying. The purified compounds are then stored in appropriate containers under suitable conditions to maintain their stability and integrity (Zubairi et al., 2014). These techniques enable the separation of pure compounds from complex mixtures, facilitating further characterization and biological evaluation of the bioactive constituents.

28.3.2 Characterization Techniques

28.3.2.1 Spectroscopic Analysis

- **Nuclear Magnetic Resonance (NMR):**

Proton NMR and carbon NMR spectroscopy are powerful techniques used to elucidate the chemical structure of isolated compounds. In the context of *Abutilon hirtum*, proton NMR (^1H-NMR) provides information about the hydrogen atoms present in the molecule, including their chemical environment and connectivity (de Oliveira, 2021; Neri et al., 2001). Carbon NMR (^13C-NMR) spectroscopy, on the other hand, offers insights into the carbon atoms in the molecule, allowing for the determination of carbon connectivity and structural features. By analyzing the NMR spectra, researchers can deduce the molecular framework of bioactive compounds isolated from *Abutilon hirtum*, aiding in their structural characterization (Pradeepkumar & Muthukrishnan, 2022).

- **Infrared Spectroscopy (IR):**

IR spectroscopy is a technique used to identify functional groups present in organic molecules based on their characteristic absorption bands in the infrared region of the electromagnetic spectrum. In the characterization of bioactive compounds from *Abutilon hirtum*, IR spectroscopy provides valuable information about the types of chemical bonds present in the molecules, such as C=O, C—H, O—H, and N-H bonds (Singh et al., 2014). By comparing the IR spectra of isolated compounds with reference spectra or databases, researchers can infer the presence of specific functional groups, thereby contributing to the structural elucidation of the compounds.

- **Mass Spectrometry (MS):**

Mass spectrometry is a versatile analytical technique used to determine the molecular weight and fragmentation pattern of organic compounds. In the characterization of bioactive compounds from *Abutilon hirtum*, MS analysis provides information about the molecular mass of the isolated compounds, allowing for the determination of their empirical formula and molecular formula (Singh et al., 2014). Additionally, MS fragmentation patterns provide insights into the structural features and connectivity of the compound's functional groups. By combining MS data with other spectroscopic techniques, researchers can elucidate the complete structural profile of bioactive compounds isolated from *Abutilon hirtum*.

28.3.2.2 Chromatographic Analysis

- **Gas Chromatography (GC):**

Gas chromatography (GC) analysis is employed to determine the purity and composition of volatile compounds present in extracts obtained from *Abutilon hirtum*. In GC analysis, the sample is vaporized and injected into a chromatographic column, where it undergoes separation based on differences in volatility and affinity for the stationary phase (Hoh & Mastovska, 2008). By comparing the retention times of sample components with those of known standards, researchers can identify and quantify individual volatile compounds present in the extract (De Jager et al., 2008). GC analysis provides valuable information about the chemical composition of volatile constituents in *Abutilon hirtum*, aiding in the characterization and quality assessment of the plant material.

- **Liquid Chromatography (LC):**

Liquid chromatography (LC) analysis is utilized for the separation and quantification of individual compounds present in extracts derived from *Abutilon hirtum* (Shanmugapriyanka, 2021). In LC analysis, the sample is dissolved in a liquid solvent and injected into a chromatographic column, where it undergoes separation based on differences in polarity and interaction with the stationary phase. By monitoring the retention times and peak areas of sample components, researchers can identify and quantify bioactive compounds present in the extract (Czaplicki, 2013). LC analysis offers high sensitivity and selectivity, making it suitable for the characterization of complex mixtures of compounds in *Abutilon hirtum* extracts.

28.3.2.3 Bioassays for Bioactivity Evaluation

- **Antimicrobial Assays:**

The antimicrobial activity of isolated compounds from *Abutilon hirtum* is assessed using various methods such as agar diffusion or broth dilution assays. In agar diffusion assays, discs impregnated with isolated compounds are placed on agar plates inoculated with pathogenic microorganisms. The zones of inhibition around the discs indicate the antimicrobial activity of the compounds against the test microorganisms (gowdu Viswanathan et al., 2012). Broth dilution methods involve determining the minimum inhibitory concentration (MIC) of isolated compounds by serially diluting them in broth media and observing their inhibitory effects on microbial growth (Kuete et al., 2012). These assays provide valuable information about the potential of isolated compounds to combat bacterial, fungal, and other microbial infections.

- **Antioxidant Assays:**

The antioxidant potential of isolated compounds is evaluated using assays such as 2,2-diphenyl-1-picrylhydrazyl (DPPH) or 2,2'-azinobis(3-ethylbenzothiazoline-6-sulfonic acid) (ABTS) radical scavenging assays (Loizzo et al., 2009). In these assays, the ability of compounds to neutralize free radicals is measured by their capacity to reduce the color intensity of DPPH or ABTS radicals (Gupta, 2015). The decrease in absorbance indicates the antioxidant activity of the compounds, with higher scavenging activity correlating to stronger antioxidant potential (Alamed et al., 2009). These assays provide insights into the ability of isolated compounds to protect cells and tissues from oxidative stress and associated diseases.

- **Cytotoxicity Assays:**

The cytotoxic effects of isolated compounds are assessed using cell-based assays such as 3-(4, 5-dimethylthiazol-2-yl)-2,5-diphenyltetrazolium bromide (MTT) or sulforhodamine B (SRB) assays (Gavanji et al., 2022). In MTT assays, isolated compounds are incubated with cultured cancer cells, and cell viability is measured by the reduction of MTT to formazan crystals, which are quantified spectrophotometrically (Awouafack et al., 2013). SRB assays involve staining cellular proteins with sulforhodamine B dye, and the absorbance of the dye is proportional to the number of viable cells. These assays help determine the cytotoxic potency of isolated compounds against cancer cell lines, providing crucial information for the development of anti-cancer therapeutics (Vajrabhaya & Korsuwannawong, 2018). The results obtained from these assays contribute to the understanding of the therapeutic potential of *Abutilon hirtum*–derived compounds and their possible applications in medicine and healthcare.

28.4 ELICITATION OF BIOACTIVE COMPOUNDS FROM CELL SUSPENSION CULTURES

Plant cell cultures have been proven to play a crucial role in biotechnological applications. In the 1950s, researchers recognized the potential of plant cell cultures in producing numerous therapeutic compounds with low molecular weights, prompting the exploration of various methodologies for in vitro propagation (Gamborg, 2002). Cell suspension culture in plant species was first employed by Muir et al. (1954). The growing commercial importance of the plant phytochemical compounds in recent years has resulted in a huge interest, particularly with the chance to alter the secondary metabolite production through in vitro cell culture technologies. The significant benefit of this technology lies in its ability to provide a consistent and reliable source

TABLE 28.1
Isolated Bioactive Compounds from *Abutilon hirtum* and Its Anti-cancer Properties

Compound Name	Medicinal Properties	Anti-cancer Properties	References
Stigmasterol	Lowering blood sugar and cholesterol levels, anti-inflammatory, antioxidant, cytotoxic, anti-tumor, anti-osteoarthritic, antimutagenic activities, central nervous system depressant	Inhibits prostate cancer growth, inhibits carcinoma development, acts as a potent anti-cancer agent	Kaur et al., 2011; Scholtysek et al., 2009.
Gallic acid	Antimicrobial, antioxidant, cytotoxic against prostate cancer and breast cancer cells, scavenge electrophilic mutagens	Possesses cytotoxic effects against liver cancer cells (HepG2)	Karamać et al., 2006; Subramanian et al., 2015.
Quercetin	Anti-inflammatory, antioxidant, antiviral	Inhibits proliferation of cancer cells; induces apoptosis	Duo et al., 2012
Kaempferol	Anti-inflammatory, antioxidant, antimicrobial	Suppresses tumor growth; induces cell cycle arrest and apoptosis	Zhu & Xue, 2019.
β-sitosterol	Anti-inflammatory, lowers cholesterol, immune modulation	Inhibits angiogenesis; induces apoptosis in cancer cells	Sharmila & Sindhu, 2017.
Lupeol	Anti-inflammatory, antioxidant, hepatoprotective	Inhibits cancer cell migration and invasion; induces apoptosis	Liu et al., 2016
Rutin	Antioxidant, anti-inflammatory, venotonic	Anti-proliferative effects on cancer cells; inhibits metastasis	Paudel et al., 2021
Ellagic acid	Antioxidant, anti-inflammatory, antimicrobial	Inhibits tumor initiation and progression; induces apoptosis	Chen et al., 2015

for the production of economically valuable products, along with facilitating large-scale production of plant cell cultures for compound isolation (Rao & Ravishankar, 2002).

Callus, which is undifferentiated tissue, forms the basis for suspension culture. Plant cell suspensions are prepared by the continuous agitation of actively dividing callus tissues in a liquid culture medium containing the hormones in a shaker flask. This leads to the creation of small clusters of delicate callus in controlled laboratory conditions, accompanied by increased levels of secondary metabolites, ideal for producing pharmaceutical compounds of high purity. This process also enables simple downstream processing and cost-effective production (see **Figure 28.1**). Numerous technologies have been effectively utilized in plant *in vitro* systems to enhance the production of valuable phytochemical compounds and to study their biosynthetic and metabolic pathways (Ravishankar & Venkataraman, 1993; Kim et al., 2008; Franconi et al., 2010).

Secondary metabolites secreted by plant cells find extensive applications in commercial and industrial biotechnology. Maintaining in vitro plant cell cultures in sterile conditions ensures the generation of valuable commercial products. Researchers are actively exploring alternatives to intact plants for phytochemical compound production in response to the global market's demand for sustainable and natural products. This shift has spurred novel research into in vitro secondary product expression, with innovative techniques emerging such as differentiated cell and organ cultures, immobilized cell cultures, induction of hairy root cultures via elicitors, tissue engineering, and metabolic engineering (Kumar et al., 2013).

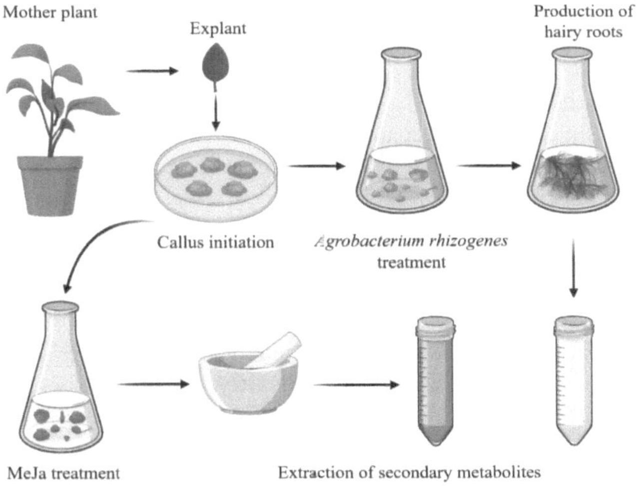

FIGURE 28.1 Elicitation of plant cell cultures for enhancement of secondary metabolites.

Bioactive compound extraction from cell suspension cultures involves the utilization of elicitors, which are substances capable of provoking an intensified reaction within treated plant cells (Babich et al., 2020; Singh & Sharma, 2020). This response triggers increased gene expression, activation of secondary metabolic pathways, and enhanced accumulation of secondary metabolites, leading to greater production of desired compounds in medicinal plant cells and tissue cultures (Ramawat et al., 2020).

Elicitors can be broadly classified into two categories: biotic and abiotic. Biotic elicitors encompass entities such

TABLE 28.2
Types of Elicitors used in Secondary Metabolite Production

Elicitor Type	Description	Examples	References
Abiotic Elicitors	Non-living factors that induce responses	Light (UV, blue, red), pH, temperature, osmotic stress	Samal et al., 2023
Biotic Elicitors	Living organisms or their components	Fungal pathogens, bacterial extracts, insect-derived signals	Zhao et al., 2005
Chemical Elicitors	Synthetic compounds inducing responses	Methyl jasmonate, salicylic acid, chitosan, yeast extract	Bhaskar et al., 2022
Physical Elicitors	Mechanical or physical stimuli	Wounding, ultrasound, electric pulses, magnetic fields	Aguirre-Becerra et al., 2021

as bacteria, glucan polymers, glycoproteins, and fungal cells. Abiotic elicitors, on the other hand, consist of heavy metals, pesticides, organic acids, or physical stressors like cold, salinity, osmotic stress, heat stress and UV exposure (Table 28.2). All of these agents stimulate enzyme activity (Thakur et al., 2018; Akhtar et al., 2019).

The process of elicitation presents an alluring and straightforward strategy for augmenting metabolite productivity in *invitro* culture systems. By eliciting plant cells or tissues, higher yields can be achieved within shorter time frames, owing to the interconnected nature of plant defense mechanisms and metabolite production through secondary metabolism (Sahai & Shuler, 1984). Biotic elicitors activate defense mechanisms, thereby enhancing secondary metabolite accumulation. Plant recognition of elicitors likely involves high-affinity receptors on cell surfaces or intracellular mechanisms, initiating a cascade of signal transduction that triggers characteristic defense responses (Nürnberger, 1999).

Stress-related growth regulators such as jasmonic acid (JA) and methyl jasmonate (MeJA), are commonly employed to induce the biosynthesis of phytochemical compounds by transmitting elicitor signals and promoting the transcription of key enzyme genes, such as phenylalanine ammonia lyase (PAL), which results in elevated enzyme levels in suspension cultures (Moola & Diana, 2019; Ramawat et al., 2020). Jasmonates play a pivotal part in elicitation, stimulating *de novo* transcription, translation, and ultimately, biosynthesis of phytochemical compounds in the *in vitro* cell cultures (Gundlach et al., 1992; Vanisree & Tsay, 2007). Several biotic and abiotic elicitors have successfully elevated the levels of lignins, sterols, steroid glycosides, alkaloids, and triterpenes through elicitation experiments (Almagro et al., 2011).

Given the effectiveness of plant in vitro systems in producing valuable compounds and studying their biosynthesis and metabolism, this study endeavors to enhance secondary metabolite levels in *A. hirtum* suspension cultures using the non-transgenic approach of supplementing with the signaling elicitor molecule, MeJA. When plant cells are treated with elicitor-active substances, there is a swift increase in the production of secondary products, extending beyond defense-related metabolites (Gundlach et al., 1992; Kurosaki, 2012).

Elicitation of bioactive compounds from cell suspension cultures represents a multifaceted process involving the activation of defense mechanisms and biosynthesis of phytochemicals through the application of elicitors. Understanding the intricacies of this process not only facilitates the optimization of metabolite production but also provides insights into the fundamental mechanisms underlying plant responses to environmental stimuli.

A. Secondary Metabolite Production through Hairy Root Cultures

Agrobacterium rhizogenes, a bacterium found in the rhizosphere belonging to the family Rhizobiaceae, initiates the hairy root disease by infecting wounded regions of the roots of higher plants (Veena et al., 2007; Akhtar et al., 2019). This organism transfers a specific DNA fragment called the T-DNA into the genome of the host plant. The T-DNA is flanked by 25-base-pair direct oligonucleotide repeats, from its large Ri (root-inducing) plasmid. This T-DNA segment contains the genes that are responsible for regulating plant hormone levels, such as those involved in cytokinin biosynthesis and auxin control (iaaM, iaaH, ipt), as well as the genes for opines, which are unusual amino acids (Otten, 2018; Pistelli et al., 2010).

The integration of T-DNA alters the hormonal balance, resulting in the development of proliferating roots known as hairy roots at the infection site. Hairy roots display rapid growth that is not influenced by hormone cues, lack the typical response to gravity (geotropism), exhibit lateral branching, and maintain genetic stability. These traits make them valuable for the production of various metabolites typically found in natural roots, leading to extensive research into their potential (Pistelli et al., 2010). Researchers have successfully established long-term sterile cultures of hairy roots from more than 200 higher plant species. These roots possess the capability to produce a diverse range of phytochemical compounds and adjust their metabolic activities in response to environmental stressors (Giri et al., 2000).

Recent advancements have shed light on the molecular mechanisms underlying T-DNA transfer into plants by *A. rhizogenes*, revealing similarities with *A. tumefaciens*. This *A. tumefaciens* manipulates its host cells by transferring a specific DNA fragment from its Ti (tumor-inducing) plasmid to the genome of the host cell. This DNA segment, known as T-DNA, contains oncogenes and genes responsible for opine catabolism. As a result, the transfer leads to abnormal tissue growth and the production of opines in plant cells (Guillon et al., 2006).

The transfer process involves the expression of virulence genes in response to signals from the host plant, leading to excision of T-DNA, formation of a single-stranded T-complex, and exportation to the host cell via a specialized secretion system (Suryadinata et al., 2022). Once the T-DNA enters the host cell nucleus, the T-complex integrates into the plant genome, albeit the exact mechanism remains unclear (Tzfira & Citovsky, 2006; Pistelli et al., 2010). Notably, the transfer of T-DNA is not specific to particular DNA sequences, allowing the replacement of native genes with desired ones in host cells. Different patterns of the integration of T-DNA exist based on the origin of the Ti plasmid, leading to variations in gene expression and phenotype (Tzfira et al., 2004).

The process of hairy root formation involves the identification of specific loci in the T-DNA, known as root loci (rol), namely, A, B, C, and D, which play crucial roles in the neoplastic growth observed in infected plants. These genetically stable neoplastic roots exhibit distinct phenotypic traits, and in some cases, can be regenerated into complete plants (Nilsson et al., 1997). Transformation mediated by *A. rhizogenes* has resulted in transgenic plants across a wide range of taxa, displaying a characteristic HR syndrome characterized by various morphological and physiological alterations. Studying individual rol genes has provided insights into their functions and the phenotypes they induce, either alone or in combination.

Hairy roots formed by infecting wounded higher plants with *A. rhizogenes* display neoplastic growth traits, such as rapid proliferation in hormone-free environments and genetic stability. They maintain a similar phytochemical profile to their wild-type counterparts, rendering them valuable for synthesizing plant secondary metabolites using biotechnological methods. Although elicitation techniques can boost metabolite accumulation, optimizing their cultivation in bioreactors remains a challenge for large-scale production. Moreover, hairy roots can function as biofactories for manufacturing recombinant proteins, offering additional potential for industrial applications. Advances in understanding the molecular mechanisms underlying hairy root development have significantly enhanced their usefulness in various biotechnological processes (Pistelli et al., 2010).

Secondary metabolites are specialized phytocompounds secreted by higher plants. These compounds play important roles in various biological processes, including defense against pathogens and pests (Selwal et al., 2023). Hairy root cultures offer a promising approach for the synthesis of secondary metabolites in *Abutilon hirtum* (Savithramma et al., 2011). These specialized compounds, such as flavonoids, alkaloids, and terpenoids, have potent pharmaceutical and industrial applications due to their diverse biological activities. Understanding the biosynthetic pathways of these phytochemical compounds in hairy root cultures can provide insights into optimizing their production (Velu et al., 2018; Singh & Sharma, 2020).

The establishment of hairy root cultures involves the transformation of plant tissue with *A. rhizogenes*, resulting in the formation of genetically stable and fast-growing roots capable of synthesizing high levels of secondary metabolites. This innovative technique allows for the sustainable and controlled production of valuable compounds while bypassing the need for plant cultivation in the field (Srivastava et al., 2007). Furthermore, elucidating the regulatory mechanisms that govern the biosynthesis of secondary metabolites in *A. hirtum* hairy root cultures is essential for increasing the productivity and quality of these bioactive compounds. By exploring the influence of several factors like media composition, elicitors, and biotic or abiotic stressors, researchers can fine-tune the metabolic pathways to achieve optimal yields of target secondary metabolites (Isah, 2019).

Incorporating advanced analytical techniques, such as metabolomics and transcriptomics, enables comprehensive profiling of the metabolic and gene expression changes associated with the production of secondary metabolites in hairy root cultures (Sharma et al., 2013). This holistic approach facilitates the identification of key biosynthetic genes and regulatory elements, paving the way for metabolic and genetic engineering strategies to improve the biosynthesis of specific bioactive compounds (Yue et al., 2016). Enhancement of secondary metabolites through hairy root cultures includes;

i. **Initiation of Hairy Roots:** Hairy roots are generated from plant explants, typically leaf or stem tissues, through the action of *Agrobacterium rhizogenes*, a soil bacterium capable of gene transfer into plant cells, thereby triggering the formation of these distinct roots. This process typically entails the transfer of T-DNA (transfer DNA) from the bacterium to the plant genome, leading to the development of hairy roots (Halder et al., 2019).
ii. **Growth and Culture Maintenance:** Once hairy roots are established, they are cultured in a suitable growth medium containing all the necessary nutrients, vitamins, and growth regulators. The culture is typically maintained in a controlled environment with appropriate temperature, pH, and light conditions to promote growth (Chandra & Chandra, 2011).
iii. **Enhancement of Secondary Metabolite Production:** Various approaches can be adopted to boost the yield of secondary metabolites in hairy root cultures. These methods encompass optimizing culture parameters like nutrient levels, pH, and temperature, elicitation (which involves inducing stress or introducing elicitors to stimulate secondary metabolite synthesis), genetic engineering (utilizing genetic manipulation techniques to enhance the expression of genes involved in the biosynthesis pathway of target metabolites), and precursor supplementation (supplying precursor molecules to the culture medium to enhance metabolite output) (Wang & Wu, 2013; Akhtar et al., 2019).

Literature commonly recommends introducing an elicitor during the exponential growth phase, characterized by peak enzymatic activity, leading to a more robust response to the elicitor (Vasconsuelo & Boland, 2007). Additionally, the presence of growth regulators in the culture medium significantly influences secondary metabolism elicitation. For instance, studies have demonstrated that *in vitro* carrot cell cultures in a medium devoid of the hormone auxin did not respond to elicitation (Kurosaki et al., 1985), and the biosynthesis of anthraquinone by cell suspensions of *Morinda citrifolia* is affected by various growth regulators (Zenk et al., 1975).

Moreover, the light conditions during culture play a critical role. For instance, in jasmonic acid–induced hypericin synthesis in *Hypericum perforatum* L. cells, increased cell growth and secondary metabolite production were observed in dark conditions compared to light conditions (Walker et al., 2002). However, conflicting results exist, as some studies reveal that light-induced stimulation of secondary metabolite production (Vazquez-Flota & De Luca, 1998). Given the modifications in response to elicitation, factors such as growth phase, growth regulators, and light conditions, optimizing medium composition and culture conditions is crucial in elicitation protocols. This optimization ensures maximal utilization of the elicitor's potential and efficient production of desired secondary metabolites.

iv. **Extraction and Analysis:** Once a sufficient number of secondary metabolites have been produced, they are extracted from the hairy root culture using appropriate extraction techniques. The extracted metabolites undergo analysis using a range of analytical techniques, including chromatography, such as HPLC, LC-MS, and GC-MS, and spectroscopy, such as UV-Vis and NMR. These methods are employed to identify and quantify the desired compounds accurately (Wu et al., 2007).

B. Abiotic Elicitors and Secondary Metabolites Production:

Abiotic elicitors are non-living factors or substances that induce stress in plants, triggering a range of biochemical and molecular responses that often lead to an enhanced synthesis of secondary metabolites with numerous biological activities comprising antioxidant, anti-cancer, and antimicrobial properties (Naik & Al-Khayri, 2016). Chemical elicitors like salicylic acid (SA), methyl jasmonate (MeJA), heavy metals (e.g., cadmium, copper), hydrogen peroxide (H_2O_2), and sodium chloride (NaCl), as well as physical elicitors like UV radiation, temperature stress (heat or cold shock), mechanical damage, and osmotic stress (induced by salts or polyethylene glycol), are among the types of abiotic elicitors studied. These elicitors trigger stress responses in plants by activating signaling pathways and transcriptional cascades that lead to changes in gene expression, often involving the upregulation of genes encoding enzymes involved in secondary metabolite biosynthesis (Zhao et al., 2005). Consequently, the stress-induced metabolic changes redirect resources toward the synthesis of secondary metabolites, which serve various functions, including defense against environmental stressors. Heavy metals have become significant stressors for organisms due to their widespread industrial use, leading to concerns relating to toxicity and bioaccumulation (Cai et al., 2013). These heavy metals disrupt metabolic processes in plants, affecting the synthesis of key components like chlorophyll pigments, proteins, sugars, and non-protein thiols. Furthermore, heavy metals can influence the production of bioactive compounds by altering various aspects of secondary metabolism (Verpoorte et al., 2002). Specific metals such as nickel, silver, iron, and cobalt have been noted to enhance the production of secondary metabolites across various plant species (Zhao et al., 2005). Additionally, different metal ions like lanthanum (La3+), europium (Eu3+), silver (Ag+), cadmium (Cd2+), and oxalate have been shown to affect secondary metabolite synthesis (Marschner, 2011; Menaka Thakur et al., 2019). Salinity exposure triggers diverse physiological and metabolic responses in plants (Bernstein, 2013). These environmental stressors prompt the synthesis of diverse secondary metabolites, including alkaloids, terpenes, and phenols (Haghighi et al., 2012; Selmar, 2008; Winkel-Shirley, 2001).

Drought represents a significant physical abiotic elicitor (Vasconsuelo & Boland, 2007) and ranks among the primary environmental stressors capable of altering the biochemical and physiological characteristics of plant species, leading to increased secondary metabolite concentrations within plant tissues (Zobayed et al., 2007; Menaka Thakur et al., 2019). The capacity for tolerance to drought varies among plant species, with drought stress typically characterized by water deficit conditions, often accompanied by elevated temperatures and solar radiation levels (Xu et al., 2010).

The impact of abiotic elicitors on the production of secondary metabolites varies depending on factors, such as elicitor type, duration of exposure, concentration, and the specific plant species or tissue culture system utilized. While some abiotic elicitors may induce a wide range of secondary metabolites, others may trigger the synthesis of specific compounds. Abiotic elicitation has numerous applications in plant biotechnology, including the production of bioactive compounds for pharmaceuticals, nutraceuticals, and functional foods, as well as enhancing crop resistance to biotic and abiotic stresses, contributing to sustainable agriculture. These elicitors also serve as valuable tools for studying plant stress responses and understanding the molecular mechanisms underlying secondary metabolite biosynthesis. Optimization of elicitor concentration, treatment duration, and timing is crucial for maximizing the production of target compounds, and once optimal conditions are established at the laboratory scale, the process can be scaled up for larger production using bioreactor systems or controlled greenhouse conditions. Secondary metabolites production through abiotic elicitors can include the following.

i. Selection of Abiotic Elicitors:

The selection of abiotic elicitors for enhancing secondary metabolite production in *Abutilon hirtum* involves a systematic approach that considers various factors. Firstly, understanding the plant's physiology and natural habitat, typically tropical and subtropical regions, provides insights into the environmental stressors to which it is naturally exposed. Identification of stress-responsive secondary metabolites, such as phenolic compounds and alkaloids, guides the selection process towards elicitors known to induce the production of these compounds. Reviewing existing literature helps identify abiotic stressors that have been successful in enhancing secondary metabolite production in related plant species, serving as a basis for selection criteria (Yeshi et al., 2022).

Criteria for selecting abiotic elicitors include their ability to induce stress responses without causing irreversible damage to plant health, as well as practical considerations, such as dosage, duration of exposure, and compatibility with growth media. Examples of potential abiotic elicitors include salinity, drought, temperature stress, heavy metals, and UV radiation, each of which can mimic environmental conditions that induce stress responses in plants and subsequently alter secondary metabolism (Guru et al., 2022). Experimental validation through controlled studies is crucial for assessing the efficacy of selected elicitors in enhancing target secondary metabolites. This involves monitoring changes in gene expression, enzyme activity, and metabolite profiles to understand the underlying mechanisms of elicitation.

Following experimental validation, optimization of elicitation protocols is necessary to fine-tune parameters such as elicitor concentration, timing of application, and duration of exposure. Scaling up the elicitation process from laboratory-scale experiments to larger production systems requires ensuring reproducibility and consistency in secondary metabolite yields. Ultimately, the systematic selection and optimization of abiotic elicitors play a critical role in maximizing the synthesis of secondary metabolites with potential medicinal or industrial applications in *Abutilon hirtum* (Narayani & Srivastava, 2017).

ii. Preparation of Elicitor Solutions:

The preparation of elicitor solutions is a crucial step in the elicitation process, ensuring the effective delivery of stress-inducing compounds to the plant system. This preparation involves several key considerations to maximize the efficacy and reproducibility of the elicitation protocol. The choice of solvent or carrier for the elicitor compound is critical. The solvent should be compatible with both the elicitor compound and the plant system, ensuring optimal solubility and stability. Common solvents include water, ethanol, methanol, and dimethyl sulfoxide (DMSO), each with specific advantages and limitations depending on the nature of the elicitor and the experimental setup. The concentration of the elicitor solution must be carefully determined to achieve the desired level of stress without causing undue harm to the plant (Jeong & Park, 2006). Pilot studies and dose-response experiments are typically conducted to establish the optimal concentration range for elicitor application. Furthermore, the method of preparation should ensure uniform distribution and homogeneity of the elicitor solution. Techniques such as stirring, sonication, or gentle heating may be employed to facilitate the dissolution and dispersion of the elicitor compound in the solvent (Szabo et al., 1999).

Moreover, it's essential to take precautions to preserve the stability and integrity of the elicitor solution during storage and handling. Compounds sensitive to light may necessitate preparation and storage in amber or opaque containers to mitigate degradation caused by light exposure. Similarly, temperature-sensitive compounds should be stored under suitable conditions to prevent degradation or loss of activity. Finally, the prepared elicitor solution should be sterilized if necessary to eliminate microbial contamination and ensure the integrity of the plant culture. Sterilization methods such as filtration through sterile membranes or autoclaving may be employed according to the nature of the elicitor and the experimental requirements.

The preparation of elicitor solutions involves careful selection of solvents, determination of optimal concentrations, uniform distribution, and sterilization to ensure effective and reproducible elicitation of stress responses in the plant system. Attention to these details is essential for the success of elicitation experiments and the subsequent increase in the production of secondary metabolite in plants (Wee, 2015).

iii. Treatment of *Abutilon hirtum* Hairy Root Cultures:

The treatment of *Abutilon hirtum* hairy root cultures involves carefully administering the elicitor solutions to the plant tissue culture system to induce stress responses and stimulate secondary metabolite production. This process requires precise timing, dosage, and application methods to ensure optimal efficacy and reproducibility. The hairy root cultures of *Abutilon hirtum* should be established and maintained under sterile conditions using suitable growth media and culture vessels. The cultures should be actively growing and free from contamination before initiating any treatments (Krzemińska et al., 2022; Ramawat et al., 2020).

Next, the elicitor solution prepared in the previous step is added to the culture medium at the desired concentration and timing. The timing of elicitor treatment is crucial and may be different depending upon the factors namely the growth stage of hairy roots and the kinetics of secondary metabolite production. Typically, the elicitor is applied during the exponential phase of growth when the metabolic activity and biosynthetic capacity of the cells are maximal (Sajjalaguddam & Paladugu, 2016).

The method of elicitor application can vary depending on the experimental setup and the specific requirements of the study. Common techniques include direct addition of the elicitor solution to the culture medium, immersion of the hairy roots in the elicitor solution, or foliar application for aerial plant parts. Care should be taken to ensure uniform distribution of the elicitor solution and minimize physical damage to the plant tissue. Incubate the treated and control *Abutilon hirtum* hairy root cultures under controlled conditions in a growth chamber or incubator. Keep the cultures at 25°C ± 2°C, maintaining a photoperiod of 16 hours light and 8 hours dark, with a light intensity of 100 $\mu mol\,m^{-2} \cdot s^{-1}$ (Shankarsingh et al., 2016).

iv. Extraction of Secondary Metabolites:

Secondary metabolites extraction from *Abutilon hirtum* includes the isolation of bioactive compounds from plant tissues or culture medium following elicitor treatment. This process aims to recover and concentrate the target metabolites for further analysis or downstream applications.

Hairy root cultures or plant tissues treated with elicitors are harvested at a specific time point following the treatment. The choice of harvest time may depend on the kinetics of secondary metabolite accumulation and the experimental objectives. The harvested plant material is rinsed with distilled water to remove any adhering debris or contaminants. For hairy root cultures, the roots may be separated from the growth medium and gently blotted dry to remove excess moisture (Hamed et al., 2017).

The plant material undergoes a process of cell disruption to release intracellular contents, including secondary metabolites. Common methods for this include grinding with mortar and pestle, homogenization, or sonication. The selection of solvent for extraction is based on the polarity of the target secondary metabolites. Typically, solvents like methanol, ethanol, acetone, ethyl acetate, and water, alone or in combination, are used. Sequential or gradient systems employing both polar and non-polar solvents can be utilized to extract a wide array of metabolites (Saleem et al., 2020). The disrupted plant material is mixed with the chosen solvent in an appropriate ratio, typically using a solvent-to-sample ratio optimized for efficient extraction. The extraction may be carried out at room temperature or under controlled conditions such as temperature, agitation, or sonication to enhance extraction efficiency. The mixture is then incubated for a specified period, allowing the solvent to penetrate the plant tissue and solubilize the secondary metabolites (Stéphane et al., 2021).

Extraction may be performed once or repeated multiple times to maximize metabolite recovery. The extract is filtered to remove solid debris or particulate matter, and the resulting filtrate containing the dissolved secondary metabolites is collected (Hamed et al., 2017). The extract may then be concentrated using techniques such as rotary evaporation, freeze-drying, or solvent removal under reduced pressure. The concentrated extract is stored under appropriate conditions, such as low temperature and protection from light, to maintain the stability and integrity of the secondary metabolites until further analysis or application (Thakor et al., 2016).

v. Analysis and Quantification:

Analyze the extracted secondary metabolites using analytical techniques such as high-performance liquid chromatography (HPLC) with mass spectrometry (MS) or diode array detection (DAD). Quantify the levels of target compounds (e.g., flavonoids, phenolic acids) based on standard calibration curves and express the results as µg/g dry weight (Bano & Deora, 2020).

vi. Data Analysis and Interpretation:

Statistical methods are employed to assess the significance of differences in metabolite levels between elicitor-treated and control samples. Common statistical tests include analysis of variance (ANOVA), t-tests, or non-parametric tests, depending on the experimental design and distribution of the data. The choice of statistical test should be guided by assumptions about data distribution and sample size (Yang et al., 2015).

Metabolomic data can be subjected to exploratory analysis methods like cluster analysis or multidimensional scaling to reveal inherent structures or trends across treatment groups. Visualization tools such as heat maps or scatter plots can aid in the interpretation of complex datasets and facilitate the identification of outliers or clustering patterns (Granato et al., 2018).

The effects of elicitor treatment are compared to appropriate controls, including untreated samples or samples treated with solvent alone. Changes in metabolite levels, metabolic pathways, or gene expression profiles are evaluated to determine the specific effects of elicitors on secondary metabolism in *Abutilon hirtum*. The interpretation of results involves correlating changes in metabolite levels with the physiological responses of *Abutilon hirtum* to elicitor treatment. Factors such as the timing and dosage of elicitor application, as well as the metabolic pathways involved in secondary metabolite biosynthesis, are considered in interpreting the observed effects.

Validation of findings may involve repeating experiments under controlled conditions to confirm the reproducibility of results. Independent validation by other research groups or complementary analytical techniques can further strengthen the robustness of conclusions drawn from the data (Kim et al., 2013). The interpretation of results should be integrated with the biological context of *Abutilon hirtum*, considering its natural habitat, ecological interactions, and evolutionary history. Insights gained from data analysis may inform future studies on the ecological significance or biotechnological applications of elicitor-induced responses in this species.

vii. Optimization and Scale-Up:

Optimization and scale-up are crucial phases in the elicitation process with *Abutilon hirtum,* aiming to maximize secondary metabolite production and facilitate the transition from laboratory-scale experiments to larger production systems.

Optimization and scale-up of elicitation processes with *Abutilon hirtum* involve fine-tuning elicitation parameters, media composition optimization, bioreactor design and operation, scale-up strategies, process integration and automation, downstream processing optimization, economic and techno-economic analysis, and regulatory considerations (Kumar, 2022). These steps are essential for translating laboratory discoveries into scalable and economically viable production processes for secondary metabolites with potential pharmaceutical, nutraceutical, or industrial applications.

C. Biotic Elicitor and Secondary Metabolites Production:

The strong correlation between the secondary metabolism in plants and defense responses is widely acknowledged, as plants activate defense mechanisms in response to various stressors like pathogens, insects, or environmental factors, leading to the synthesis of secondary metabolites such as phytoalexins and structural defenses like lignin deposition (Thakur et al., 2018; Vasconsuelo & Boland, 2007).

Studies on *Phytophthora megasperma*'s induction of phytoalexin accumulation in soybeans unveiled that pathogen-derived small molecules initiate similar responses in plants as the pathogens themselves, termed elicitors. Subsequent investigations on microbial oligosaccharides demonstrated the potential of elicitors in enhancing the secondary metabolism in plants. Essentially, the term 'elicitors' encompasses chemicals from diverse origins, including biotic, abiotic, and physical factors, capable of prompting responses in living organisms that result in the accumulation of secondary metabolites (Vasconsuelo & Boland, 2007).

Consequently, elicitors serve as valuable tools for augmenting the synthesis of beneficial plant compounds. Biotic elicitors are those compounds that originate from living organisms, such as pathogens or plants themselves, as outlined by Patel and Krishnamurthy in 2013. These substances interact with receptors and modulate enzymatic activity or ion channels. They encompass elicitors from various sources like fungi, bacteria, or herbivores, along with fragments of plant cell walls. Additionally, they comprise chemicals released by plants at the site of attack in response to pathogens or herbivores.

Proteins and enzymes act as biotic elicitors, triggering defense responses in plant cell cultures *in vitro.* Protein elicitors were utilized to investigate the role of ion channels in transmitting signals in response to external stimuli across cell membranes (Menaka Thakur et al., 2019; Thakur et al., 2018). Glycoproteins stimulate the production of phytoalexins in plant cell cultures. Additionally, lectins or agglutinins, carbohydrate-binding proteins, contribute to plant defense against various herbivores, thus influencing the production of secondary metabolites crucial for plant defense mechanisms. For example, elicitins proteins produced by the pathogenic fungus *Phytophthora drechsleri* induce necrosis in tobacco leaves, as observed in a study by Huet et al. in 1992.

Various carbohydrates are utilized in plant tissue culture to stimulate the biosynthesis of plant secondary metabolites. Oligogalacturonides (OGAs), which are derived from pectic polysaccharides present in the cell walls of plants, act as elicitors. These OGAs have been observed to induce the biosynthesis of phytoalexins in *Glycine max* cotyledons and elicit defense responses in *Nicotiana tabacum* plants. Chitin, a component of fungal cell walls, is identified as a potential elicitor signal in numerous plants, as highlighted by Shibuya and Minami in 2001. Since early investigations, carbohydrates have been used to increase the production of secondary metabolites in plant cell cultures *in vitro* (Menaka Thakur et al., 2019).

Numerous studies have demonstrated that pathogenic bacteria and other natural elicitors have the capacity to augment the synthesis of secondary metabolites that possess pharmacological properties. Plant growth promoting rhizobacteria (PGPR) inhabiting the rhizospheres of plants are the bacteria renowned for their ability to improve plant growth through various mechanisms, even under adverse conditions, as noted by Bhattacharya and Jha in 2012. Based on their interaction with plants, they are categorized as extracellular (ePGPR), primarily residing in the rhizosphere or intercellularly within the root cortex, and intracellular (iPGPR), which inhabit specialized structures such as root cell nodules.

Arbuscular mycorrhizal fungi (AMF) area group of fungi that is crucial for promoting the growth and development of the plant. They enhance water absorption in plants and protect them against root-invading pathogenic bacteria, fungi, and parasitic nematodes. Additionally, AMF plays a significant role in stimulating secondary metabolite production in plants. Given the close relationship between plant defense responses and secondary metabolite production, numerous elicitors have been identified to enhance the production of secondary metabolites in plant cell cultures (Zhao et al., 2022; Menaka Thakur et al., 2019; Selwal et al., 2023). Furthermore, research has shown that biotic elicitors can be effectively utilized in plant tissue culture systems to increase the yield of secondary metabolites with pharmacological activity. Overall, biotic elicitors offer promising avenues for improving the synthesis of valuable compounds in medicinal plants, thereby contributing to the development of novel pharmaceuticals and nutraceuticals.

28.5 ELICITOR-MEDIATED SIGNAL TRANSDUCTION IN SECONDARY METABOLISM

The initial stage of elicitor signal transduction commences with signal perception, where plants detect various stimuli, initiating processes such as the activation of kinases, generation of reactive oxygen species, fluxes in ions, and acidification of the cytoplasm. While the concept of elicitors interacting with receptors on plant cell surfaces originated from studies showcasing receptor-like proteins in soybean membranes, recent advancements have deepened our comprehension of elicitor perception. Multiple binding sites for elicitors have been pinpointed, primarily situated in plant plasma membranes (Vasconsuelo & Boland, 2007).

Transmembrane receptor-like protein kinases (RLKs) stand out as primary candidates for pathogen recognition, with the *Arabidopsis* genome housing a diverse array of RLKs, suggesting potential roles in sensing various stimuli (Jose et al., 2020). Notably, the flagellin receptor, a well-explored RLK, belongs to the leucine-rich repeat (LRR) class. Another category of elicitor receptors encompasses plant R-proteins, which recognize race-specific elicitors encoded by avirulence (avr) genes, resulting in specific responses such as phytoalexin biosynthesis. These R-proteins typically feature nucleotide-binding sites and leucine-rich repeat domains (NBS-LRR). Although direct receptor-elicitor interactions have been confirmed, the recognition process likely entails intricate interactions, as evidenced by R-proteins requiring multiple components to trigger resistance (Ferrari S, 2010; Vasconsuelo & Boland, 2007).

Future investigations should focus on elucidating the complexity of these recognition events, including the role of the lipid environment surrounding plant receptors, which may influence their structures and functions similarly to animal cell receptors. These studies will provide insights into the intricate interplay between receptors and elicitors, advancing our understanding of elicitor signal transduction pathways.

D. Enhancement of Secondary Metabolites through Nanomaterials:

Nanotechnology has emerged as a promising strategy in agriculture, providing innovative solutions to enhance plant growth, stress tolerance, and the production of bioactive compounds. The potential of nanomaterials enhances the production of secondary metabolites in *Abutilon hirtum*, a medicinal plant renowned for its pharmacological properties.

Recent research has demonstrated the efficacy of various nanomaterials in enhancing the production of secondary metabolites in plants. Nanoparticles, nanotubes, and nanocomposites have been shown to interact with plant cells, influencing gene expression, enzyme activity, and signaling pathways involved in the biosynthesis of secondary metabolites. For instance, silver nanoparticles were found to upregulate the gene expression of encoding key enzymes in the biosynthetic pathways of secondary metabolites across several plant species.

Several researchers have investigated the use of nanomaterials to improve secondary metabolite production in *Abutilon hirtum*. For instance, Godeto et al. (2023) revealed that treatment with titanium dioxide nanoparticles significantly increased the accumulation of flavonoids and phenolic compounds in *Abutilon hirtum* leaves (Singh & Sharma, 2020). Similarly, Chen et al. (2021) reported that copper oxide nanoparticles stimulated the production of alkaloids and terpenoids in *Abutilon hirtum* roots.

28.6 MECHANISMS OF ACTION

The mechanisms underlying the enhancement of secondary metabolites by nanomaterials involve complex interactions between nanoparticles and plant cells. Nanoparticles can penetrate plant tissues and interact with cellular components, including DNA, proteins, and membranes. These interactions can influence gene expression and enzyme activity, ultimately leading to the upregulation of secondary metabolite biosynthesis pathways (Rivero-Montejo et al., 2021).

28.7 BENEFITS AND FUTURE DIRECTIONS

Utilizing nanomaterials to boost secondary metabolite production in *Abutilon hirtum* presents numerous advantages, such as elevated yields of bioactive compounds and enhanced plant resilience to environmental stressors. Nonetheless, there are challenges to overcome, including nanoparticle toxicity and potential environmental impacts, which necessitate further research and investigation. Future studies should focus on optimizing nanoparticle properties and application methods to maximize their effectiveness while minimizing adverse effects on plant health and the environment (Mahajan et al., 2023).

Nanotechnology holds great promise for enhancing secondary metabolite production in *Abutilon hirtum* and other medicinal plants. By elucidating the mechanisms of action and optimizing application strategies, nanomaterials have the potential to revolutionize plant biotechnology and contribute to the sustainable production of natural products for pharmaceutical and nutraceutical industries (Zhang et al., 2022). Further research in this area is warranted to fully harness the benefits of nanotechnology in agriculture and plant science.

28.8 CONCLUSION

In conclusion, the utilization of integrated biotechnological strategies to enhance the production of anti-cancer compounds from *Abutilon hirtum* holds significant promise for advancing cancer treatment. The implementation of novel biotechnological approaches like plant cell culture, genetic engineering, and metabolic engineering has the potential to markedly amplify the yield of bioactive compounds with anti-cancer properties. This not only presents the prospect of more sustainable and cost-effective production methods but also unveils avenues for discovering novel compounds and formulating new cancer therapies. Continued research and development in this realm could result in substantial strides in combating cancer and enhancing access to potent anti-cancer medications for patients in need. The potential of biotechnological strategies for augmenting the production of anti-cancer compounds from *A. hirtum* is particularly exhilarating due to its broader implications for oncology. Additionally, it underscores the significance of abiotic elicitors, such as salinity, drought, and temperature stress, in eliciting defense responses and accumulating secondary metabolites in *A. hirtum*. Through subjecting the plants to controlled stress conditions, we observed a significant increase in the expression of genes encoding enzymes involved in the production of anti-cancer compounds, resulting in a notable enhancement in their production levels.

Moreover, our exploration of biotic elicitors, including beneficial microbial strains such as *mycorrhizal* fungi and *rhizobacteria*, has revealed their potential as potent inducers of plant defense mechanisms and secondary metabolite biosynthesis. Through symbiotic interactions with *A. hirtum*, these microbes trigger systemic signaling pathways that activate the synthesis of phytochemicals with anti-cancer properties, thereby enhancing the medicinal value of the plant.

Furthermore, the incorporation of nanoparticles into our biotechnological approach has emerged as a groundbreaking strategy for improving the efficiency and efficacy of elicitor-mediated responses in *A. hirtum*. By leveraging the unique physicochemical properties of nanoparticles, such as their high surface area, reactivity, and stability, we were able to enhance the delivery and uptake of elicitors into plant tissues, as well as modulate intracellular signaling pathways involved in secondary metabolite biosynthesis.

Importantly, it underscores the potential of these biotechnological interventions not only in enhancing the production of anti-cancer compounds from *A. hirtum* but also in addressing various challenges associated with traditional cultivation and extraction methods. By harnessing the power of biotic and abiotic elicitors, along with nanotechnology, we can overcome limitations, such as seasonal variability, resource constraints, and environmental degradation, while simultaneously increasing the yield, potency, and sustainability of medicinal plant-based therapies. Furthermore, the sustainable and cost-effective production of these compounds through biotechnological methods has the potential to make anti-cancer drugs more accessible to patients, particularly in resource-limited settings. As research and development in this area continue to advance, the availability of potent anti-cancer drugs may increase, improving the overall landscape of cancer treatment.

Additionally, the potential ecological implications of deploying these biotechnological strategies should be carefully evaluated to ensure their compatibility with existing ecosystems and agricultural practices. Overall, the integration of biotechnological strategies for the production of anti-cancer compounds from *Abutilon hirtum* not only holds promise for enhancing the treatment options for cancer patients but also has the potential to contribute to significant advancements in the fight against cancer on a global scale. By harnessing the power of plant cell culture, genetic engineering, and metabolic engineering, researchers are not only increasing the yield of bioactive compounds but also paving the way for the discovery of novel compounds with potential anti-cancer properties.

REFERENCES

Aboulghazi, A., Bakour, M., Fadil, M., & Lyoussi, B. (2022). Simultaneous optimization of extraction yield, phenolic compounds and antioxidant activity of Moroccan propolis extracts: Improvement of ultrasound-assisted technique using response surface methodology. *Processes*, *10*(2), 297.

Abubakar, A. R., & Haque, M. (2020). Preparation of medicinal plants: Basic extraction and fractionation procedures for experimental purposes. *Journal of Pharmacy and Bioallied Sciences*, *12*(1), 1–10.

Aguirre-Becerra, H., Vazquez-Hernandez, M. C., Saenz de la, O. D., Alvarado-Mariana, A., Guevara-Gonzalez, R. G., Garcia-Trejo, J. F., & Feregrino-Perez, A. A. (2021). Role of stress and defense in plant secondary metabolites production. In *Bioactive natural products for pharmaceutical applications* (pp. 151–195). Springer Nature.

Akhtar, M. S., Swamy, M. K., & Sinniah, U. R. (2019). Natural bio-active compounds. *Natural Bio-Active Compounds*, *1*, 1–608.

Alamed, J., Chaiyasit, W., McClements, D. J., & Decker, E. A. (2009). Relationships between free radical scavenging and antioxidant activity in foods. *Journal of Agricultural and Food Chemistry*, *57*(7), 2969–2976.

Albreht, A., & Vovk, I. (2016). *TLC-MS analysis using solvent elution of compounds from chromatographic media* (pp. 123–140). CRC Press/Taylor & Francis Group.

Almagro, L., Sabater-Jara, A. B., Belchí-Navarro, S., Fernández-Pérez, F., Bru, R., & Pedreño, M. A. (2011). Effect of UV light on secondary metabolite biosynthesis in plant cell cultures elicited with cyclodextrins and methyl jasmonate. *Plants and Environment*, *1*, 115–136.

Arivoli, S., & Tennyson, S. (2011). Larvicidal and adult emergence inhibition activity of *Abutilon indicum* (Linn.)(Malvaceae) leaf extracts against vector mosquitoes (*Diptera: Culicidae*). *Journal of Biopesticides*, *4*(1), 27.

Awouafack, M. D., McGaw, L. J., Gottfried, S., Mbouangouere, R., Tane, P., Spiteller, M., & Eloff, J. N. (2013). Antimicrobial activity and cytotoxicity of the ethanol extract, fractions and eight compounds isolated from *Eriosema robustum* (Fabaceae). *BMC Complementary and Alternative Medicine*, *13*, 1–9.

Babich, O., Sukhikh, S., Pungin, A., Ivanova, S., Asyakina, L., & Prosekov, A. (2020). Modern trends in the in vitro production and use of callus, suspension cells and root cultures of medicinal plants. *Molecules*, *25*(24), 5805.

Bano, I., & Deora, G. S. (2020). Preliminary phytochemical screening and Gc-Ms analysis for identification of bioactive compounds from *Abutilon Fruticosum* Guill and Perr. A rare and endemic plant of Indian Thar desert. *International Journal of Pharmaceutical Sciences*, *11*(6), 2671–2679.

Bele, A. A., & Khale, A. (2011). An overview on thin layer chromatography. *International Journal of Pharmaceutical Sciences and Research*, *2*(2), 256.

Bernstein, N., Eshel, A., & Beeckman, T. (2013). Effects of salinity on root growth. In *Plant roots: The hidden half* (p. 10). CRC Press.

Bhaskar, R., Xavier, L. S. E., Udayakumaran, G., Kumar, D. S., Venkatesh, R., & Nagella, P. (2022). Biotic elicitors: A boon for the in-vitro production of plant secondary metabolites. *Plant Cell, Tissue and Organ Culture (PCTOC)*, *149*(1), 7–24.

Bhattacharyya, P. N., & Jha, D. K. (2012). Plant growth-promoting rhizobacteria (PGPR): Emergence in agriculture. *World Journal of Microbiology and Biotechnology*, *28*, 1327–1350.

Burnaz, N. A., Küçük, M., & Akar, Z. (2017). An on-line HPLC system for detection of antioxidant compounds in some plant extracts by comparing three different methods. *Journal of Chromatography B*, *1052*, 66–72.

Cai, L. (2014). Thin layer chromatography. *Current Protocols Essential Laboratory Techniques*, *8*(1), 6–3.

Cai, Z., Kastell, A., Speiser, C., & Smetanska, I. (2013). Enhanced resveratrol production in *Vitis vinifera* cell suspension cultures by heavy metals without loss of cell viability. *Applied Biochemistry and Biotechnology*, *171*, 330–340.

Cardoso, J. C., Oliveira, M. E. B. D., & Cardoso, F. D. C. (2019). Advances and challenges on the in vitro production of secondary metabolites from medicinal plants. *Horticultura Brasileira*, *37*(2), 124–132.

Chandan, B. K., Saxena, A. K., Shukla, S., Sharma, N., Gupta, D. K., Singh, K., Bhadauria, M., Suri, S., Suri, K. A., Suri, O. P., & Qazi, G. N. (2008). Hepatoprotective activity of *Woodfordia fruticosa* Kurz flowers against carbon tetrachloride induced hepatotoxicity. *Journal of Ethnopharmacology*, *119*(2), 218–224.

Chandra, S., & Chandra, R. (2011). Engineering secondary metabolite production in hairy roots. *Phytochemistry Reviews*, *10*, 371–395.

Chen, H. S., Bai, M. H., Zhang, T., Li, G. D., & Liu, M. (2015). Ellagic acid induces cell cycle arrest and apoptosis through TGF-β/Smad3 signaling pathway in human breast cancer MCF-7 cells. *International Journal of Oncology*, *46*(4), 1730–1738.

Chen, H., Feng, X., Gao, L., Mickymaray, S., Paramasivam, A., Abdulaziz Alfaiz, F., Alnahdi, H. S., Alhardi, S. A., Babiker, A. Y., & Aziz Ibrahim, I. A. (2021). Inhibiting the PI3K/AKT/mTOR signalling pathway with copper oxide nanoparticles from *Houttuynia cordata* plant: Attenuating the proliferation of cervical cancer cells. *Artificial Cells, Nanomedicine, and Biotechnology*, *49*(1), 240–249.

Choudhury, S., Sharma, P., Choudhury, M. D., & Sharma, G. D. (2012). Ethnomedicinal plants used by Chorei tribes of Southern Assam, North eastern India. *Asian Pacific Journal of Tropical Disease*, *2*, S141–S147.

Cieśla, Ł., & Waksmundzka-Hajnos, M. (2009). Two-dimensional thin-layer chromatography in the analysis of secondary plant metabolites. *Journal of Chromatography A*, *1216*(7), 1035–1052.

Czaplicki, S. (2013). Chromatography in bioactivity analysis of compounds. In *Column chromatography* (pp. 100–122). IntechOpen.

Das, U., & Islam, M. S. (2019). A review study on different plants in Malvaceae family and their medicinal uses. *American Journal of Biomedical Science and Research*, *3*(2), 94–97.

Dashputre, N. L., & Naikwade, N. S. (2011). Evaluation of anti-ulcer activity of methanolic extract of *Abutilon indicum* Linn leaves in experimental rats. *International Journal of Pharmaceutical Sciences and Drug Research*, *3*(2), 97–100.

de Alba, S. L., García-González, C., Coronado Ortega, M. A., Ayala Bautista, J. R., Alpírez, G. M., & Montes Núñez, D. G. (2023). Extraction methods and applications of bioactive compounds from neem (*Azadirachta indica*): A mini-review. *Mini-Reviews in Organic Chemistry*, *20*(7), 644–654.

De Jager, L. S., Perfetti, G. A., & Diachenko, G. W. (2008). Comparison of headspace-SPME-GC–MS and LC–MS for the detection and quantification of coumarin, vanillin, and ethyl vanillin in vanilla extract products. *Food Chemistry*, *107*(4), 1701–1709.

de Oliveira, C. R. (2021). *A 1 H NMR network approach for the study of specialized metabolites: Development and applications* (Doctoral dissertation, University of Nevada, Reno).

de Souza, A. L. C., do Rego Pires, A., Moraes, C. A. F., de Matos, C. H. C., dos Santos, K. I. P., Campos e Silva, R., Silva, F. P., Olivera, E. J., Almeida, J. R. G. S., Nascimento, T. G., & dos Santos Araújo, S. (2023). Chromatographic methods for separation and identification of bioactive compounds. In *Drug discovery and design using natural products* (pp. 153–176). Springer Nature Switzerland.

Devasagayam, T. P. A., Tilak, J. C., Boloor, K. K., Sane, K. S., Ghaskadbi, S. S., & Lele, R. D. (2004). Free radicals and antioxidants in human health: Current status and future prospects. *Japi*, *52*(794804), 4.

Duo, J., Ying, G. G., Wang, G. W., & Zhang, L. (2012). Quercetin inhibits human breast cancer cell proliferation and induces apoptosis via Bcl-2 and Bax regulation. *Molecular Medicine Reports*, *5*(6), 1453–1456.

Erkan, N. (2012). Antioxidant activity and phenolic compounds of fractions from *Portulaca oleracea* L. *Food Chemistry*, *133*(3), 775–781.

Farooq, S., Mir, S. A., Shah, M. A., & Manickavasagan, A. (2022). Extraction techniques. In *Plant extracts: Applications in the food industry* (pp. 23–37). Academic Press.

Ferrari, S. (2010). Biological elicitors of plant secondary metabolites: Mode of action and use in the production of nutraceutics. In *Bio-farms for nutraceuticals: Functional food and safety control by biosensors* (pp. 152–166). Springer US.

Franconi, R., Demurtas, O. C., & Massa, S. (2010). Plant-derived vaccines and other therapeutics produced in contained systems. *Expert Review of Vaccines*, *9*(8), 877–892.

Gamborg, O. L. (2002). Plant tissue culture. biotechnology. milestones. *In Vitro Cellular & Developmental Biology-Plant*, *38*, 84–92.

Gavanji, S., Bakhtari, A., Famurewa, A. C., & Othman, E. M. (2023). Cytotoxic activity of herbal medicines as assessed in vitro: A review. *Chemistry & Biodiversity*, *20*(2), e202201098.

Giri, A., & Narasu, M. L. (2000). Transgenic hairy roots: Recent trends and applications. *Biotechnology Advances*, *18*(1), 1–22.

Godeto, Y. G., Ayele, A., Ahmed, I. N., Husen, A., & Bachheti, R. K. (2023). Medicinal plant-based metabolites in nanoparticles synthesis and their cutting-edge applications: An overview. In *Secondary metabolites from medicinal plants* (pp. 1–34). CRC Press.

Gokulakrishnan, S., Vaishnavi, G., & Ilango, K. B. (2023). Standardization and solvent comparative study of *Abutilon hirtum* L. *World Journal of Pharmaceutical Research*, *12*(5), 1192–1215.

Gomaa, A. A. R., Samy, M. N., Desoukey, S. Y., & Kamel, M. S. (2018). Anti-inflammatory, analgesic, antipyretic and anti-diabetic activities of *Abutilon hirtum* (Lam.) Sweet. *Clinical Phytoscience*, *4*, 1–7.

Gomaa, A. A., Samy, M. N., Desoukey, S. Y., & Kamel, M. S. (2016). Pharmacognostical studies of leaf, stem, root and flower of *Abutilon hirtum* (Lam.) Sweet. *International Journal of Pharmacognosy and Phytochemical Research*, *8*(1), 199–216.

Goti, D., & Dasgupta, S. (2023). A comprehensive review of conventional and non-conventional solvent extraction techniques. *Journal of Pharmacognosy and Phytochemistry*, *12*(3), 202–211.

gowdu Viswanathan, M. B., Ananthi, J. D. J., & Kumar, P. S. (2012). Antimicrobial activity of bioactive compounds and leaf extracts in *Jatropha tanjorensis*. *Fitoterapia*, *83*(7), 1153–1159.

Granato, D., Santos, J. S., Escher, G. B., Ferreira, B. L., & Maggio, R. M. (2018). Use of principal component analysis (PCA) and hierarchical cluster analysis (HCA) for multivariate association between bioactive compounds and functional properties in foods: A critical perspective. *Trends in Food Science & Technology*, *72*, 83–90.

Guillon, S., Tremouillaux-Guiller, J., Pati, P. K., Rideau, M., & Gantet, P. (2006). Hairy root research: Recent scenario and exciting prospects. *Current Opinion in Plant Biology*, *9*(3), 341–346.

Gullón, P., Gullón, B., Romaní, A., Rocchetti, G., & Lorenzo, J. M. (2020). Smart advanced solvents for bioactive compounds recovery from agri-food by-products: A review. *Trends in Food Science & Technology*, *101*, 182–197.

Gundlach, H., Müller, M. J., Kutchan, T. M., & Zenk, M. H. (1992). Jasmonic acid is a signal transducer in elicitor-induced plant cell cultures. *Proceedings of the National Academy of Sciences*, *89*(6), 2389–2393.

Gupta, D. (2015). Methods for determination of antioxidant capacity: A review. *International Journal of Pharmaceutical Sciences and Research*, *6*(2), 546.

Guru, A., Dwivedi, P., Kaur, P., & Pandey, D. K. (2022). Exploring the role of elicitors in enhancing medicinal values of plants under in vitro condition. *South African Journal of Botany*, *149*, 1029–1043.

Haghighi, Z., Karimi, N., Modarresi, M., & Mollayi, S. (2012). Enhancement of compatible solute and secondary metabolites production in *Plantago ovata* Forsk. by salinity stress. *Journal of Medical Plant Research*, *6*(12), 1234–1242.

Halder, M., Sarkar, S., & Jha, S. (2019). Elicitation: A biotechnological tool for enhanced production of secondary metabolites in hairy root cultures. *Engineering in Life Sciences*, *19*(12), 880–895.

Hamed, M. M., Refahy, L. A. G., & Abdel-Aziz, M. S. (2017). Assessing the bioactivity and antioxidative properties of some compounds isolated from *Abutilon hirtum* (lam.). *Asian Journal of Pharmaceutical and Clinical Research*, *10*(3), 333–340.

Hameed, K., Khan, M. S., Fatima, A., Shah, S. M., & Abdullah, M. A. (2023). Exploring the world of thin-layer chromatography: A review. *Asian Journal of Applied Chemistry Research*, *14*(3), 23–38.

Hoh, E., & Mastovska, K. (2008). Large volume injection techniques in capillary gas chromatography. *Journal of Chromatography A*, *1186*(1–2), 2–15.

Hostettmann, K., Wolfender, J. L., & Terreaux, C. (2001). Modern screening techniques for plant extracts. *Pharmaceutical Biology*, *39*(Suppl. 1), 18–32.

Huet, J. C., Nespoulous, C., & Pernollet, J. C. (1992). Structures of elicitin isoforms secreted by *Phytophthora drechsleri*. *Phytochemistry*, *31*(5), 1471–1476.

Isah, T. (2019). Stress and defense responses in plant secondary metabolites production. *Biological Research*, *52*.

Jaiswal, Y., Liang, Z., & Zhao, Z. (2016). Botanical drugs in Ayurveda and traditional Chinese medicine. *Journal of Ethnopharmacology*, *194*, 245–259.

Jandera, P., & Churacek, J. (1985). *Gradient elution in column liquid chromatography: Theory and practice*. Elsevier.

Jeong, G. T. A., & Park, D. H. E. (2006). Enhanced secondary metabolite biosynthesis by elicitation in transformed plant root system: Effect of abiotic elicitors. *Applied Biochemistry and Biotechnology*, *130*, 436–446.

Jose, J., Ghantasala, S., & Roy Choudhury, S. (2020). *Arabidopsis* transmembrane receptor-like kinases (RLKs): A bridge between extracellular signal and intracellular regulatory machinery. *International Journal of Molecular Sciences*, *21*(11), 4000.

Kailasam, K. V. (2015). *Abutilon indicum* L (Malvaceae)-medicinal potential review. *Pharmacognosy Journal*, *7*(6).

Kapoor, S. L., & Kapoor, L. D. (1980). Medicinal plant wealth of the Karimnagar district of Andhra Pradesh. *Bulletin of Medical Ethnobotanical Research*, *1*(120).

Karamac, M., Kosińska, A., & Pegg, R. B. (2006). Content of gallic acid in selected plant extracts. *Polish Journal of Food and Nutrition Sciences*, *15*(1), 55.

Kaur, N., Chaudhary, J., Jain, A., & Kishore, L. (2011). Stigmasterol: A comprehensive review. *International Journal of Pharmaceutical Sciences and Research*, *2*(9), 2259.

Kim, J. K., Park, S. Y., Lim, S. H., Yeo, Y., Cho, H. S., & Ha, S. H. (2013). Comparative metabolic profiling of pigmented rice (*Oryza sativa* L.) cultivars reveals primary metabolites are correlated with secondary metabolites. *Journal of Cereal Science*, *57*(1), 14–20.

Kim, T. G., Baek, M. Y., Lee, E. K., Kwon, T. H., & Yang, M. S. (2008). Expression of human growth hormone in transgenic rice cell suspension culture. *Plant Cell Reports*, *27*, 885–891.

Krzemińska, M., Owczarek, A., Gonciarz, W., Chmiela, M., Olszewska, M. A., & Grzegorczyk-Karolak, I. (2022). The antioxidant, cytotoxic and antimicrobial potential of phenolic acids-enriched extract of elicited hairy roots of *Salvia bulleyana*. *Molecules*, *27*(3), 992.

Kuete, V., BetrandTeponno, R., Mbaveng, A. T., Tapondjou, L. A., Meyer, J. J. M., Barboni, L., & Lall, N. (2012). Antibacterial activities of the extracts, fractions and compounds from Dioscorea bulbifera. *BMC Complementary and Alternative Medicine*, *12*, 1–8.

Kumar, N. (Ed.). (2022). *Biotechnology and crop improvement: Tissue culture and transgenic approaches*. CRC Press.

Kumar, S., Jyotirmayee, K., & Sarangi, M. (2013). Thin layer chromatography: A tool of biotechnology for isolation of bioactive compounds from medicinal plants. *International Journal of Pharmaceutical Sciences Review and Research*, *18*(1), 126–132.

Kumar, V., Kumar, C. S., Hari, G., Venugopal, N. K., & Vijendra, P. D. (2013). Homology modeling and docking studies on oxidosqualenecyclases associated with primary and secondary metabolism of *Centella asiatica*. *SpringerPlus*, *2*, 1–12.

Kurosaki, F. (2012). Induction and activation of plant secondary metabolism by external stimuli. In *Drug discovery research in pharmacognosy* (pp. 209–226). IntechOpen.

Kurosaki, F., Futamura, K., & Nishi, A. (1985). Factors affecting phytoalexin production in cultured carrot cells. *Plant and Cell Physiology*, *26*(4), 693–700.

Liu, Y., Bi, T., Shen, G., Li, Z., Wu, G., Wang, Z., Qian, L., & Gao, Q. (2016). RETRACTED ARTICLE: Lupeol induces apoptosis and inhibits invasion in gallbladder carcinoma GBC-SD cells by suppression of EGFR/MMP-9 signaling pathway. *Cytotechnology*, *68*(1), 123–133.

Loizzo, M. R., Said, A., Tundis, R., Hawas, U. W., Rashed, K., Menichini, F., Frega, N. G., & Menichini, F. (2009). Antioxidant and antiproliferative activity of *Diospyros lotus* L. extract and isolated compounds. *Plant Foods for Human Nutrition*, *64*, 264–270.

Lozano-Sánchez, J., Borrás-Linares, I., Sass-Kiss, A., & Segura-Carretero, A. (2018). Chromatographic technique: High-performance liquid chromatography (HPLC). In *Modern techniques for food authentication* (pp. 459–526). Academic Press.

Lye, K. A., Bukenya-Ziraba, R., Tabuti, J. R., & Waako, P. (2008). Botanical-medicinal dictionary for East Africa. In *Makerere Herbarium handbook* (p. 2). Makerere University.

Mahajan, R., Sagar, T., Hussain, S., Nipunta, & Kapoor, N. (2023). Secondary metabolite production in medicinal plants under abiotic stress. In *Medicinal plants: Their response to abiotic stress* (pp. 127–152). Springer Nature Singapore.

Majumdar, M., Shivalkar, S., Pal, A., Verma, M. L., Sahoo, A. K., & Roy, D. N. (2020). Nanotechnology for enhanced bioactivity of bioactive compounds. In *Biotechnological production of bioactive compounds* (pp. 433–466). Elsevier.

Marschner, H. (Ed.). (2011). *Marschner's mineral nutrition of higher plants*. Academic Press.

Menaka Thakur, M. T., Sujata Bhattacharya, S. B., Khosla, P. K., & Sunil Puri, S. P. (2019). Improving production of plant secondary metabolites through biotic and abiotic elicitation. *Journal of Applied Research on Medicinal and Aromatic Plants*, *12*(1), 1–12.

Mohite, M. S., Shelar, P. A., Raje, V. N., Babar, S. J., & Sapkal, R. K. (2012). Review on pharmacological properties of *Abutilon indicum*. *Asian Journal of Pharmaceutical Research*, *2*(4), 156–160.

Moola, A. K., & Diana, R. K. B. (2019). Elicitation as a means for enhanced plant secondary metabolites through hairy root system. *Journal of Advanced Scientific Research*, *10*(03), 6–12.

Muir, W. H., Hildebrandt, A. C., & Riker, A. J. (1954). Plant tissue cultures produced from single isolated cells. *Science*, *119*(3103), 877–878.

Naik, P. M., & Al-Khayri, J. M. (2016). Impact of abiotic elicitors on in vitro production of plant secondary metabolites: A review. *Journal of Advanced Research in Biotechnology*, *1*(2), 1–7.

Narayani, M., & Srivastava, S. (2017). Elicitation: A stimulation of stress in in vitro plant cell/tissue cultures for enhancement of secondary metabolite production. *Phytochemistry Reviews*, *16*, 1227–1252.

Neri, P., Tringali, C., & Bioact, C. (2001). Applications of modern NMR techniques in the structure elucidation of bioactive natural products. In C. Tringali (Ed.), *Bioactive compounds from natural sources* (pp. 69–128). Taylor and Francis.

Niazian, M. (2019). Application of genetics and biotechnology for improving medicinal plants. *Planta*, *249*, 953–973.

Nilsson, O., & Olsson, O. (1997). Getting to the root: The role of the *Agrobacterium rhizogenes* rol genes in the formation of hairy roots. *Physiologia Plantarum*, *100*(3), 463–473.

Nürnberger, T. (1999). Signal perception in plant pathogen defense. *Cellular and Molecular Life Sciences CMLS*, *55*, 167–182.

Otten, L. (2018). The *Agrobacterium* phenotypic plasticity (Plast) genes. In *Agrobacterium biology: From basic science to biotechnology* (pp. 375–419). Springer.

Parkin, D. M., Bray, F., Ferlay, J., & Pisani, P. (2005). Global cancer statistics, 2002. *CA: A Cancer Journal for Clinicians*, *55*(2), 74–108.

Parsa, N. (2012). Environmental factors inducing human cancers. *Iranian Journal of Public Health*, *41*(11), 1.

Paudel, K. R., Wadhwa, R., Tew, X. N., Lau, N. J. X., Madheswaran, T., Panneerselvam, J., Hansbro, P. M., Chellappan, D. K., & Dua, K. (2021). Rutin loaded liquid crystalline nanoparticles inhibit non-small cell lung cancer proliferation and migration in vitro. *Life Sciences*, *276*, 119436.

Perumal, B. (2001). *Abutilon hirtum* (Lamk) Sweet. *Plant Resources of South-East Asia: Medicinal and Poisonous Plants*, *2*, 30–31.

Pistelli, L., Giovannini, A., Ruffoni, B., Bertoli, A., & Pistelli, L. (2010). Hairy root cultures for secondary metabolites production. In *Bio-farms for nutraceuticals: Functional food and safety control by biosensors* (pp. 167–184). Springer Science & Business Media.

Pradeepkumar, S., & Muthukrishnan, S. (2022). Structural prediction of bioactive compound elucidated through 1H and 13C NMR based metabolomics from *Physalis minima*. *Journal of Advanced Materials Science and Engineering*, *2*(1), 1–8.

Pullaiah, T. (2002). *Medicinal plants in Andhra Pradesh, India*. Daya Books.

Qiu, H., Liang, X., Sun, M., & Jiang, S. (2011). Development of silica-based stationary phases for high-performance liquid chromatography. *Analytical and Bioanalytical Chemistry*, *399*, 3307–3322.

Rahmat, E., & Kang, Y. (2019). Adventitious root culture for secondary metabolite production in medicinal plants: A review. *Journal of Plant Biotechnology*, *46*(3), 143–157.

Rajeshwari, S., & Sevarkodiyone, S. P. (2018). Medicinal properties of *Abutilon indicum*. *Open Journal of Plant Science*, *3*(1), 022–025.

Ramawat, K. G., Ekiert, H. M., & Goyal, S. (2020). *Plant cell and tissue differentiation and secondary metabolites*. Springer International Publishing.

Ramraje, G. R., Patil, S. D., Patil, P. H., & Pawar, A. R. (2020). A brief review on: Separation techniques chromatography. *Asian Journal of Pharmaceutical Analysis*, *10*(4), 231–238.

Rao, S. R., & Ravishankar, G. A. (2002). Plant cell cultures: Chemical factories of secondary metabolites. *Biotechnology Advances*, *20*(2), 101–153.

Ravishankar, G. A., & Venkataraman, L. V. (1993). Role of plant cell culture in food biotechnology: Current trends, limitations and future prospects. *Biotechnology Letters*, *15*(3), 189–196.

Rivero-Montejo, S. D. J., Vargas-Hernandez, M., & Torres-Pacheco, I. (2021). Nanoparticles as novel elicitors to improve bioactive compounds in plants. *Agriculture*, *11*(2), 134.

Sahai, O. P., & Shuler, M. L. (1984). Multistage continuous culture to examine secondary metabolite formation in plant cells: Phenolics from *Nicotiana tabacum*. *Biotechnology and Bioengineering*, *26*(1), 27–36.

Sajjalaguddam, R. R., & Paladugu, A. (2016). Quercetin production from hairy root cultures of Indian Mallow (*Abutilon indicum* L.). *World Journal of Pharmaceutical Sciences*, *5*, 956–967.

Saleem, H., Sarfraz, M., Ahsan, H. M., Khurshid, U., Kazmi, S. A. J., Zengin, G., Senan, A. M., Khan, K., Yousaf, Z., & Ahemad, N. (2020). Secondary metabolites profiling, biological activities and computational studies of *Abutilon figarianum* Webb (Malvaceae). *Processes*, *8*(3), 336.

Samal, M., Abass, S., Parveen, R., Ahmad, S., & Iqbal, M. (2023). Effect of abiotic stress on production of secondary metabolites in plants. In *Plants as medicine and aromatics* (pp. 145–172). CRC Press.

Sasidharan, S., Chen, Y., Saravanan, D., Sundram, K. M., & Latha, L. Y. (2011). Extraction, isolation and characterization of bioactive compounds from plants' extracts. *African Journal of Traditional, Complementary and Alternative Medicines*, *8*(1).

Savithramma, N., Rao, M. L., & Suhrulatha, D. (2011). Screening of medicinal plants for secondary metabolites. *Middle-East Journal of Scientific Research*, *8*(3), 579–584.

Scholtysek, C., Krukiewicz, A. A., Sharma, K. P., Sharma, P. C., & Goldmann, W. H. (2009). Characterizing components of the Saw Palmetto Berry Extract (SPBE) on prostate cancer cell growth and traction. *Biochemical and Biophysical Research Communications*, *379*(3), 795–798.

Selmar, D. (2008). Potential of salt and drought stress to increase pharmaceutical significant secondary compounds in plants. *LandbauforschungVolkenrode*, *58*(1/2), 139.

Selwal, N., Rahayu, F., Herwati, A., Latifah, E., Suhara, C., Suastika, I. B. K., Suryadi, H., Maulana, H., Pratama, A. W., Mahendra, A., Handayani, N., Nugraha, R. A., & Wani, A. K. (2023). Enhancing secondary metabolite production in plants: Exploring traditional and modern strategies. *Journal of Agriculture and Food Research*, *14*, 100702.

Shankarsingh, S. M., Reshi, N. A., & Vasanaika, G. H. (2016). In vitro regenerative potentials of the medicinal plant *Abutilon indicum* (L.) Sweet. *African Journal of Biotechnology*, *15*(12), 472–480.

Shanmugapriyanka, S. (2021). *Extraction and quantification of flavonoid from abutilon pannosum flower using LC-ESI-MS and evaluating it's in vitro anti-cancer activity* (Doctoral dissertation, JKK Nattaraja College of Pharmacy, Kumarapalayam).

Sharma, M., Koul, A., Sharma, D., Kaul, S., Swamy, M. K., & Dhar, M. K. (2019). Metabolic engineering strategies for enhancing the production of bio-active compounds from medicinal plants. In *Natural bio-active compounds: Vol. 3: Biotechnology, bioengineering, and molecular approaches* (pp. 287–316). Springer Nature.

Sharma, P., Padh, H., & Shrivastava, N. (2013). Hairy root cultures: A suitable biological system for studying secondary metabolic pathways in plants. *Engineering in Life Sciences*, *13*(1), 62–75.

Sharmila, R., & Sindhu, G. (2017). Modulation of angiogenesis, proliferative response and apoptosis by β-sitosterol in rat model of renal carcinogenesis. *Indian Journal of Clinical Biochemistry*, *32*, 142–152.

Shibuya, N., & Minami, E. (2001). Oligosaccharide signalling for defence responses in plant. *Physiological and Molecular Plant Pathology*, *59*(5), 223–233.

Singh, B., & Sharma, R. A. (2020). *Secondary metabolites of medicinal plants, 4 Volume set: Ethnopharmacological properties, biological activity and production strategies*. John Wiley & Sons.

Singh, K. S., Majik, M. S., & Tilvi, S. (2014). Vibrational spectroscopy for structural characterization of bioactive compounds. In *Comprehensive analytical chemistry* (Vol. 65, pp. 115–148). Elsevier.

Sohel, M. D. (2015). Pharmacologicals and phytochemicals potential of *Abutilon indicum*: A comprehensive review. *American Journal of BioScience*, *3*(2–1), 5–11.

Srivastava, N., Singh, A., Kumari, P., Nishad, J. H., Gautam, V. S., Yadav, M., Gupta, R. K., Tiwari, P., Dubey, S. K., Sinha, P. R., Sharma, M., Agarwal, D. S., Mishra, A. N., & Kharwar, R. N. (2021). Advances in extraction technologies: Isolation and purification of bioactive compounds from biological materials. In *Natural bioactive compounds* (pp. 409–433). Academic Press.

Srivastava, S., & Srivastava, A. K. (2007). Hairy root culture for mass-production of high-value secondary metabolites. *Critical Reviews in Biotechnology*, *27*(1), 29–43.

Stéphane, F. F. Y., Jules, B. K. J., Batiha, G. E., Ali, I., & Bruno, L. N. (2021). *Extraction of bioactive compounds from medicinal plants and herbs*. IntechOpen.

Subramanian, A. P., John, A. A., Vellayappan, M. V., Balaji, A., Jaganathan, S. K., Supriyanto, E., & Yusof, M. (2015). Gallic acid: Prospects and molecular mechanisms of its anticancer activity. *RSC Advances*, *5*(45), 35608–35621.

Suryadinata, R. V., Aditya, D. M. N., Wicaksono, A. S., & Sofyana, I. R. (2022). Antioxidant efficacy of flavonoid in broccoli (*Brassica oleracea* var. italica) to decrease alveolar macrophages in lung tissue due to cigarette smoke exposure. *Teikyo Medical Journal*, *45*(1), 4001–4007.

Szabo, E., Thelen, A., & Petersen, M. (1999). Fungal elicitor preparations and methyl jasmonate enhance rosmarinic acid accumulation in suspension cultures of *Coleus blumei*. *Plant Cell Reports*, *18*, 485–489.

Thakor, P., Mehta, J. B., Patel, R. R., Patel, D. D., Subramanian, R. B., & Thakkar, V. R. (2016). Extraction and purification of phytol from *Abutilon indicum*: Cytotoxic and apoptotic activity. *RSC Advances*, *6*(54), 48336–48345.

Thakur, M., Bhattacharya, S., Khosla, P. K., & Puri, S. (2018). Improving production of plant secondary metabolites through biotic and abiotic elicitation. *Journal of Applied Research on Medicinal and Aromatic Plants*, *12*, 1–12.

Tzfira, T., & Citovsky, V. (2006). Agrobacterium-mediated genetic transformation of plants: Biology and biotechnology. *Current Opinion in Biotechnology*, *17*(2), 147–154.

Tzfira, T., Li, J., Lacroix, B., & Citovsky, V. (2004). *Agrobacterium* T-DNA integration: Molecules and models. *Trends in Genetics*, *20*(8), 375–383.

Vajrabhaya, L. O., & Korsuwannawong, S. (2018). Cytotoxicity evaluation of a Thai herb using tetrazolium (MTT) and sulforhodamine B (SRB) assays. *Journal of Analytical Science and Technology*, *9*(1), 1–6.

Vanisree, M., & Tsay, H. S. (2007). Plant cell cultures: Production of biologically important secondary metabolites from medicinal plants of Taiwan. In *Medicinal plant biotechnology from basic research to industrial application*. Wiley-VCH Verlag GmbH & Co KGaA.

Vasconsuelo, A., & Boland, R. (2007). Molecular aspects of the early stages of elicitation of secondary metabolites in plants. *Plant Science*, *172*(5), 861–875.

Vazquez-Flota, F. A., & De Luca, V. (1998). Developmental and light regulation of desacetoxyvindoline 4-hydroxylase in *Catharanthus roseus* (L.) G. Don. Evidence of a multilevel regulatory mechanism. *Plant Physiology*, *117*(4), 1351–1361.

Veena, V., & Taylor, C. G. (2007). *Agrobacterium rhizogenes*: Recent developments and promising applications. *In Vitro Cellular & Developmental Biology-Plant*, *43*, 383–403.

Velu, G., Palanichamy, V., & Rajan, A. P. (2018). Phytochemical and pharmacological importance of plant secondary metabolites in modern medicine. In *Bioorganic phase in natural food: An overview* (pp. 135–156). Springer.

Verpoorte, R., Contin, A., & Memelink, J. (2002). Biotechnology for the production of plant secondary metabolites. *Phytochemistry Reviews*, *1*, 13–25.

Višnjevec, A. M., Barp, L., Lucci, P., & Moret, S. (2024). Pressurized liquid extraction for the determination of bioactive compounds in plants with emphasis on phenolics. *TrAC Trends in Analytical Chemistry*, 117620.

Walker, T. S., Bais, H. P., & Vivanco, J. M. (2002). Jasmonic acid-induced hypericin production in cell suspension cultures of

Hypericum perforatum L. (St. John's wort). *Phytochemistry*, *60*(3), 289–293.

Wang, J. W., & Wu, J. Y. (2013). Effective elicitors and process strategies for enhancement of secondary metabolite production in hairy root cultures. In *Biotechnology of hairy root systems* (pp. 55–89). Springer.

Wang, J., Li, J. L., Li, J., Li, J. X., Liu, S. J., Huang, L. Q., & Gao, W. Y. (2017). Production of active compounds in medicinal plants: From plant tissue culture to biosynthesis. *Chinese Herbal Medicines*, *9*(2), 115–125.

Wee, S. L. (2015). *The effects of elicitors and precursor on in vitro cultures of Sauropus androgynus for sustainable metabolite production and antioxidant capacity improvement* (Doctoral dissertation, University of Nottingham).

Winkel-Shirley, B. (2001). Flavonoid biosynthesis. A colorful model for genetics, biochemistry, cell biology, and biotechnology. *Plant Physiology*, *126*(2), 485–493.

Wu, J. Y., Ng, J., Shi, M., & Wu, S. J. (2007). Enhanced secondary metabolite (tanshinone) production of *Salvia miltiorrhiza* hairy roots in a novel root–bacteria coculture process. *Applied Microbiology and Biotechnology*, *77*, 543–550.

Xu, Z., Zhou, G., & Shimizu, H. (2010). Plant responses to drought and rewatering. *Plant Signaling & Behavior*, *5*(6), 649–654.

Yang, T., Fang, L., Nopo-Olazabal, C., Condori, J., Nopo-Olazabal, L., Balmaceda, C., & Medina-Bolivar, F. (2015). Enhanced production of resveratrol, piceatannol, arachidin-1, and arachidin-3 in hairy root cultures of peanut co-treated with methyl jasmonate and cyclodextrin. *Journal of Agricultural and Food Chemistry*, *63*(15), 3942–3950.

Yeshi, K., Crayn, D., Ritmejerytė, E., & Wangchuk, P. (2022). Plant secondary metabolites produced in response to abiotic stresses has potential application in pharmaceutical product development. *Molecules*, *27*(1), 313.

Yuan, X., Ni, H., Hou, Y., Lai, M. T., & Hu, S. Q. (2022). Efficient short extraction and purification procedures of kinsenoside from *Anoectochilus roxburghii* with deep eutectic solvent by column chromatographic extraction. *Industrial Crops and Products*, *182*, 114866.

Yue, W., Ming, Q. L., Lin, B., Rahman, K., Zheng, C. J., Han, T., & Qin, L. P. (2016). Medicinal plant cell suspension cultures: Pharmaceutical applications and high-yielding strategies for the desired secondary metabolites. *Critical Reviews in Biotechnology*, *36*(2), 215–232.

Zarzycki, P. K. (2023). Staining and derivatization techniques for visualization in planar chromatography. In *Instrumental thin-layer chromatography* (pp. 213–257). Elsevier.

Zenk, M. H., El Shagi, H., & Schulte, U. (1975). Anthraquinone production by cell suspension cultures of *Morinda citrifolia*. *Planta Medica*, *28*(S 01), 79–101.

Zhang, Y., Qi, G., Yao, L., Huang, L., Wang, J., & Gao, W. (2022). Effects of metal nanoparticles and other preparative materials in the environment on plants: From the perspective of improving secondary metabolites. *Journal of Agricultural and Food Chemistry*, *70*(4), 916–933.

Zhao, J., Davis, L. C., & Verpoorte, R. (2005). Elicitor signal transduction leading to production of plant secondary metabolites. *Biotechnology Advances*, *23*(4), 283–333.

Zhao, Y., Cartabia, A., Lalaymia, I., & Declerck, S. (2022). Arbuscular mycorrhizal fungi and production of secondary metabolites in medicinal plants. *Mycorrhiza*, *32*(3), 221–256.

Zhu, L., & Xue, L. (2019). Kaempferol suppresses proliferation and induces cell cycle arrest, apoptosis, and DNA damage in breast cancer cells. *Oncology Research*, *27*(6), 629.

Zobayed, S. M. A., Afreen, F., & Kozai, T. (2007). Phytochemical and physiological changes in the leaves of St. John's wort plants under a water stress condition. *Environmental and Experimental Botany*, *59*(2), 109–116.

Zubairi, S. I., Sarmidi, M. R., & Aziz, R. A. (2014). Identification of bio-active constituents from Derris elliptica liquid crude extract using liquid chromatographic method coupled with high vacuum pressure. *Advances in Environmental Biology*, 437–441.

29 Network Pharmacology of Ginseng
Modern Insights of a Traditional Medicine

Hien Thi Thu Do, Huy Phu Quang Le, Tram Huyen Ngoc Nguyen, Lam Nguyen Ngoc Pham, Ngoc Diem Thuy Nguyen, Phuong-Hien Nguyen, and Nguyen Phuong Thao

29.1 INTRODUCTION

Ginseng, a traditional herb commonly used in many countries, including the USA, China, and Korea, is renowned for the diverse array of pharmacological effects attributed to its rich content of bioactive compounds[1]. The antioxidant, anti-inflammatory, or anticancer activities of ginseng active constituents have been revealed in terms of chronic disease treatment[2–6]. The study has shown that ginsenoside Rg3, Rh1, Rp1 and compound K inhibit cell proliferation and possess anticancer effects[7]. Lee et al. have determined that the combination of G-Rg3 with cisplatin notably suppressed the proliferation of bladder cancer cells resistant to cisplatin[8]. Another study by Yun et al. investigated that the ethanol extract of wild-type ginseng dose-dependently reduced the blood glucose in mice[9]. Still, the current studies about mechanisms of ginseng in disease treatment have been limited to individual ginseng-related compounds, whereas the interaction between active constituents and treatment responses typically comprises multiple molecular targets and pathways[10].

Network pharmacology emerges as a powerful tool to address these limitations. This approach utilizes network analysis to study interactions between drugs, genes, proteins, and other biological entities and offers immense potential in exploring the comprehensive therapeutic potential of ginseng[11]. By examining the complex interplay between ginseng compounds and biological networks, network pharmacology can identify novel therapeutic targets and provide a better understanding of the detailed mechanisms underlying ginseng's wide effects[12]. Recently, many scientists have shifted toward using network pharmacology to investigate how ginseng's active compounds function in various diseases. Studies employing network pharmacology have successfully deciphered the multi-target interactions of ginseng compounds with key molecular pathways involved in disease development. Given ginseng's multifaceted effects, encompassing anti-inflammatory, antioxidant, neuroprotective, and immunomodulatory properties, this plant emerges as a versatile candidate for integrative healthcare approaches[13–16].

In this chapter, we will summarize findings concerning the relationship between ginseng's active compounds and disease mechanisms, as elucidated by network pharmacology approach. This chapter aims to explore the fundamental steps in conducting network-pharmacology-based research and highlighting the pivotal role of network pharmacology in uncovering ginseng's multi-target mechanisms of action. Through organizing data on ginseng's bioactive components and their interplay within biological networks, we aim to provide insights that can guide the development of novel therapeutic approaches.

29.2 GINSENG CHEMICAL CONSTITUENTS AND PHARMACOLOGICAL PROFILES

Ginseng has been a precious ingredient in ancient medicinal practices in Asian countries for healing and restoration[17], sparking many investigations in ginseng pharmacology associated with the biochemical and molecular properties of ginseng-derived compounds. Recent studies showed that nearly 200 compounds can be isolated from ginseng[18] and ginsenosides represent the major active pharmacological components of ginseng[10]. Ginsenosides (saponin), glycosylated triterpenes, have a four-ring structure with a steroidal structure attached to sugar molecules[19]. Most ginsenosides are named Rx, where R stands for root and x is the alphabet polarity order classified by chromatography[20]. Currently, more than 40 types of ginsenosides have been extracted from *Panax ginseng*[21]. Depending on the carbon-binding position of sugar units, ginsenosides are classified into three main groups: protopanaxadiol (PPD), protopanaxatriol (PPT), and oleanonic acid ginsenosides. Sugar groups of PPD-type ginsenosides attach to dammarane-type triterpenes at position C3, whereas those of PPT ginsenosides bind to carbon-6-position (Figure 29.1). Oleanonic acid–type saponins are pentacyclic triterpene having three different sugar-binding motifs such as 3-monoglycoside, 28-monoglycoside, and 3,28-diglycoside[22]. Previous studies showed that the presence of particular ginsenosides served as a marker to distinguish the ginsenoside genus[23]. For instance, a high ginsenoside Fr/F11 ratio or low ratio between Rb1 and Rg1 indicates *Panax* species[24].

The chemical structural variation of ginsenoside including the type, number, and binding position of sugar motifs

DOI: 10.1201/9781003475491-29

A. PPD Ginsenoside

Ginsenosides	R_1	R_2
Rb1	Glc$^{1\text{-}2}$Glc	Glc$^{1\text{-}6}$Glc
Rb2	Glc$^{1\text{-}2}$Glc	Glc$^{1\text{-}6}$Arap
Rc	Glc$^{1\text{-}2}$Glc	Glc$^{1\text{-}6}$Araf
Rd	Glc$^{1\text{-}2}$Glc	Glc
Rg3	Glc$^{1\text{-}2}$Glc	H
Rh2	Glc	H
Compound K	H	Glc

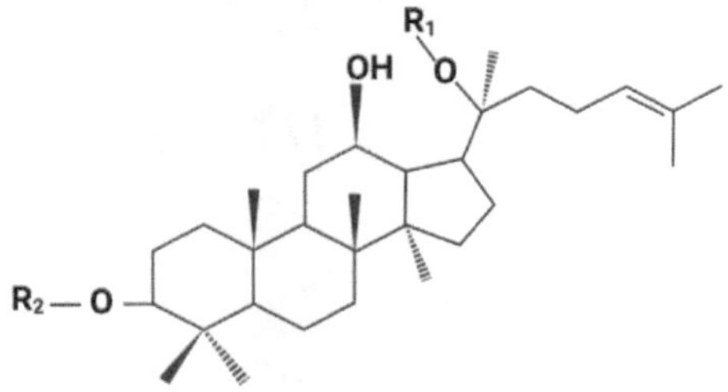

B. PPT Ginsenoside

Ginsenosides	R_3	R_2
Re	Rha$^{1\text{-}2}$Glc	Glc
Rf	Glc$^{1\text{-}2}$Glc	H
Rg1	Glc	Glc
Rg2	Rha$^{1\text{-}2}$Glc	H
Rh1	Glc	H

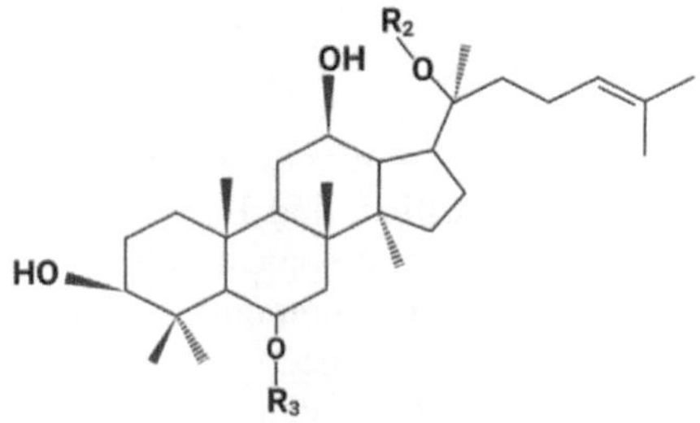

Created in BioRender.com

FIGURE 29.1 Ginsenoside types and chemical structures. A. protopanaxadiol (PPD) ginsenoside, B. protopanaxatriol (PPT) ginsenoside, beta-D-glucopyranosyl (Glc); alpha-L-arabinopyranosyl (Arap); alpha-D-arabinofuranosyl (Araf); alpha-L-rhamnopyranosyl (Rha).

results in a diverse range of pharmacological activities. Recent studies have found that ginsenosides are potential candidates to ameliorate oxidative, inflammatory-related stresses or diseases[25] (Table 29.1). In *in vitro* studies of the pretreatment of Rb2 using H_2O_2-induced stress in H9C2 cells, the damage of reactive oxygen species (ROS) was improved via decreasing the level of lactate dehydrogenase (LDH) and malondialdehyde (MDA) and increasing the activities of superoxide dismutase (SOD)[26]. Recent studies pointed out that ginsenoside Rb3 prevents the accumulation of ROS via triggering nuclear erythroid-2-related factor (Nrf2), protein kinase RNA-like ER kinase (PERK) in hypoxia/reoxygenation induced H9C2 cell[27]. Ginsenoside Rf downregulates the phosphorylation of p38/nuclear factor-κB (p38/NF-κB) on HT-29 and RAW264.7 cells[28]. Furthermore, ginseng plants also contain other non-ginsenoside constituents, including oils, carbohydrates, amino acids, peptides, vitamins, and minerals, contributing to the overall complex pharmacological effects. Gintonin is a glycosylated protein, that can regulate the inflammatory effects via the controlling of calcium influx of ion channels[29]. Ginseng polysaccharides showed protective effects against inflammatory bowel disease (IBD) mouse models via the attenuation of inflammatory-related signaling pathways[30].

29.3 RESEARCH METHODOLOGIES IN NETWORK PHARMACOLOGY OF GINSENG

29.3.1 The Evolution of Network Pharmacology

A modern technique known as network pharmacology merges systems biology, network analysis, and pharmacology to explore drug mechanisms and interactions with multiple molecular targets[51]. This approach provides a simultaneously comprehensive perspective on drug discovery, addressing clinical efficacy, side effects, and toxicity[52]. This modern tool has become a pivotal way for elucidating the relationships between drug formulas and the entire organism[53]. In 1999, Shao Li discovered the interplay between traditional Chinese medicine (TCM) and biological molecule networks, marking the first time network pharmacology gained prominence[54]. This early work suggested that herbal formulations could influence disease gene networks. Subsequently, in 2007, focusing on cold/hot syndrome, Li et al. applied bioinformatics to create the first biomolecular network, revealing the regulatory effects of herbal formulas[55]. Around the same time, Andrew L. Hopkins coined the term 'network pharmacology', which implies that multiple highly effective medications interact

TABLE 29.1
Ginsenoside Pharmacological Activities

No.	Ginsenosides	Experimental model	Function	References
1	Ra1	H/R human umbilical vein endothelial cells	↓ protein tyrosine kinase activation	31
2	Rb1	Obese diabetic db/db C57BLKS/J mice	↓ the hepatic fat formation	32
			↑ insulin sensitivity	
			↓ liver weight	
			↓ hepatic triglyceride contents	
3	Rb2	DIO C57BL/6J mice	↓ body fat weight	33
			↑ glucose metabolism	
			↑ energy expenditure	
4	Rb3	Hypertensive Wistar-Kyoto rats	↓ oxidative stress	34
			↓ endothelial-dependent contractions in renal arteries	
5	Rc	3T3-L1 preadipocytes	↓ lipogenesis	35
			↓ lipid accumulation	
			↑ lipolysis	
			↑ transcription factors of adipogenesis	
6	Rd	LPS-activated N9 microglial cells	↓ LPS-stimulated NF-κB activation	36
			↓ TNF-α expression	
			↓ ERK1/2, JNK, and c-Jun phosphorylation	
7	Compound K	LPS-stimulated BV2 microglial cells	↓ Iba1-positive activated microglia	37
			↓ IL-1β secretion	
			↓ TNF-α production	
			↓ the infarct volume of ischemic brain	
8	Re	LPS-stimulated peritoneal macrophages	↓ the phosphorylation of IKK-β	38
			↓ NF-κB activation	
			↓ proinflammatory cytokines secretion	
			↓ IRAK-1 phosphorylation	
			↓ the degradation of IRAK-1 and IRAK-4	
			↓ the binding of LPS to TLR4	
		TNBS-induced mice	↓ IL-1β secretion	
			↓ TNF-α production	
			↓ NF-κB activation	
			↓ myeloperoxidase activity	
9	Rf	Forced swimming ICR mice	↑ myoblast differentiation	39
			↑ mitochondrial biogenesis	
			↑ AMPK and p38 MAPK signaling pathways	
10	Rg1	IR mice	↓ apoptosis and necrosis of hepatocytes	40
			↑ phosphorylation of NF-κB p65	
11	Rg2	Vascular smooth muscle cells (VSMCs)	↓ PDGF-BB	41
			↓ VSMCs proliferation, migration, and phenotypic transformation	
		Human umbilical vein endothelial cells	↓ NF-κB	
			↓ p-ERK signaling pathway	
12	Rg3	LPS-induced rats	↓ TNF-α, IL-1β, and COX-2 expression in the hippocampus	42
			↓ LPS infusion in lateral ventricle	
13	Rg5	LPS-stimulated mice	↓ LPS binding to macrophages' TLR4	43
			↓ inhibition of NF-κB activation	
			↓ proinflammatory cytokines	
14	Rh1	IFN-γ-stimulated BV2 microglial cells	↓ iNOS gene expression	44
			↓ the phosphorylation of ERK, JAK1, STAT1, and STAT3	
15	Rh2	Breast cancer cells (MDA-MB-15)	↓ m6A RNA methylation	45
		Sarcoma cells (NMFH-1& KYM-1)	↓ KIF26B expression	
		Thyroid cancer cells (TT & 8305C)	↑ ZC3H13/CBLL1 nuclear localization	
		Lung cancer cells (NCIH226)		

(*Continued*)

TABLE 29.1 *(Continued)*
Ginsenoside Pharmacological Activities

No.	Ginsenosides	Experimental model	Function	References
16	Rh3	Cisplatin-induced pig kidney epithelium cells (CL-101)	↓ the phosphorylation of JNK, ERK, MAPKs-caspase-3 apoptotic cascade ↑ apoptosis	46
17	Rh4	LPS-stimulated RAW264.7 macrophage cells	↓(TNF-, IL-6, and IL-1) ↓ PGE2 and NO levels ↓ elevated iNOS and COX-2	47
18	Ro	B16F10 tumor-bearing mice	↓ human umbilical vein endothelial cells tube formation	48
19	Rp1	Human breast cancer cells: MCF-7, MDA-MB-231, and T-47D cells	↓ the stability of IGF-1R in IGF-1R/Akt pathway ↓ cell invasion and metastasis	49
20	Rk1	UVB-irradiated BALB/c nude mouse dorsal skin UVB-irradiated HaCaT keratinocytes	↓ collagen degradation ↓ the expression of matrix metalloproteinases ↓ PI3K/AKT/NF-κB signaling pathways ↓ skin roughness ↓ collagen fiber arrangement disorder ↓ wrinkles	50

Note: 2.4,6-trinitrobenzene sulfonic acid (TNBS), c-Jun NH2-terminal kinase (JNK), diet-induced obese (DIO), extracellular signal-regulated kinase (ERK), hypoxia/reoxygenation (H/R), IL-1 receptor-associated kinases (IRAKs), inhibitor of nuclear factor kappa-B kinase (IKK), insulin-like growth factor 1 receptor (IGF-1R), interleukin (IL), inducible nitric oxide synthase (iNOS), ischemia/reperfusion (IR), lipopolysaccharide (LPS), mitogen-activated protein kinase (MAPKs), mouse strain was named ICR after the initial letters of the Institute of Cancer Research (ICR mice), vascular smooth muscle cells (VSMCs), nuclear factor-κB (NF-κB), phosphatidylinositol 3′-OH kinase (PI3K), platelet-derived growth factor (PDGF), protein kinase B (AKT), signal transducer and activator of transcription (STAT), toll-like receptors (TLRs), tumor necrosis factor (TNF).

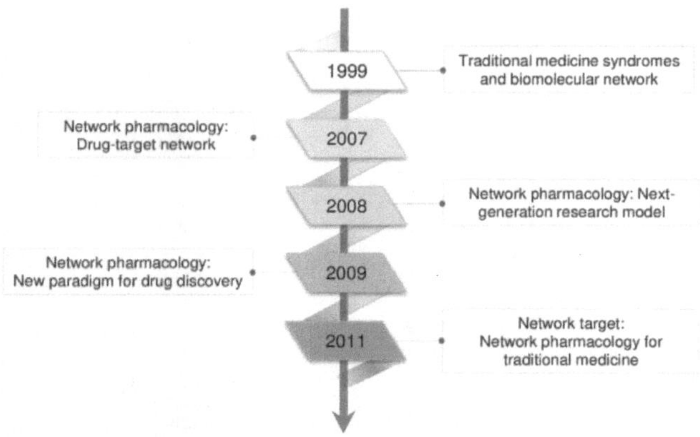

FIGURE 29.2 Timeline of pharmacology network.

with numerous targets rather than exclusively targeting a single one[12]. Overall, the evolution of network pharmacology is associated with the study of TCM, reflecting a symbiotic relationship between traditional medicinal practices and modern pharmacological methodologies. In many ways, network pharmacology represents a modern adaptation of traditional drug discovery methods, offering a promising framework for innovative drug development regardless the challenges of safety, efficacy, and sustainability in the pharmaceutical industry[56].

29.3.2 Strategy to Construct Network Pharmacology

Research strategies in network pharmacology encompass several key steps. Initially, the process involves collecting data on active compounds, disease-associated genes, and other related information. Following this, a protein-protein interaction (PPI) network is constructed to further investigate the mechanisms underlying interactions between compounds and diseases[57]. Subsequently, network analysis is conducted to identify key nodes and predict essential biological pathways. After predicting significant biological targets, enrichment analysis methods are performed to gain further insights into the pharmacological network. Finally, a crucial step is the validation of the findings through other bioinformatics methods or experimental approaches[58–60]. In this chapter, we aim to outline the essential steps involved in conducting a network pharmacology method to evaluate the active compounds of ginseng and their interaction with disease-related pathways.

a. Data Mining

The initial step in constructing a network pharmacology framework involves gathering data on disease genes, bioactive compounds, and target genes from diverse databases. These databases are classified into several categories, including those focused on genes/proteins, herbs/compounds, and

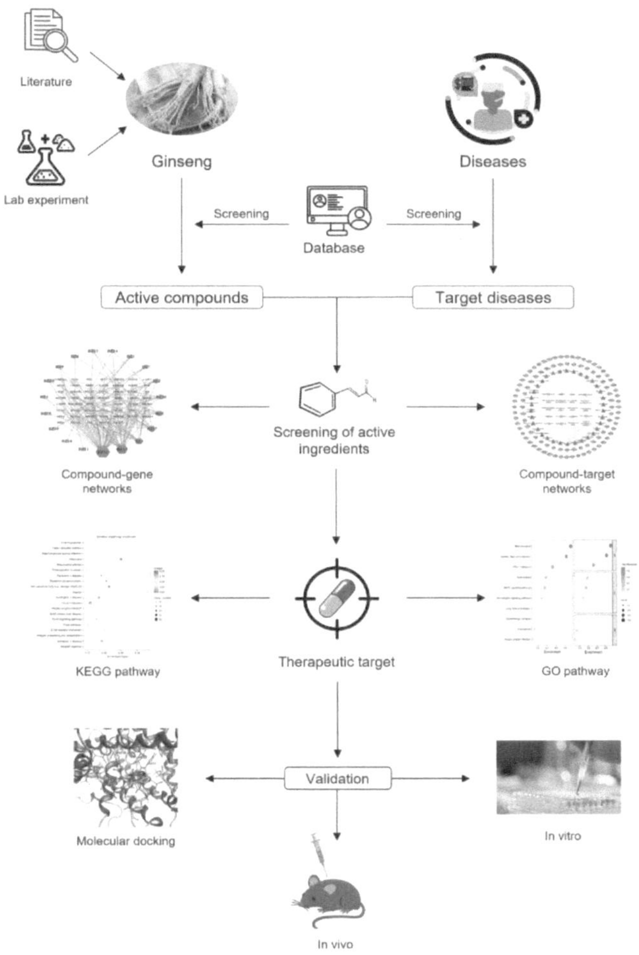

FIGURE 29.3 The workflow in network pharmacology study. Kyoto Encyclopedia of Genes and Genomes (KEGG) pathway, Gene Ontology (GO) analysis.

diseases databases. For network pharmacology research in ginseng, commonly referenced databases include DAVID (the Database for Annotation, Visualization, and Integrated Discovery), Drugbank, GeneCards, KEGG (Kyoto Encyclopedia of Genes and Genomes), Metascape, OMIM (Online Mendelian Inheritance in Man), PubChem, STRING (Search Tool for the Retrieval of Interacting Genes/Proteins), TCM (Traditional Chinese Medicine Information Database), TCMID (Traditional Chinese Medicine Integrated Database), TCMSP (Traditional Chinese Medicines Systems Pharmacology Database and Analysis Platform), and Uniprot (Universal Protein Resource). The information about these databases is listed in the following Table 29.2.

b. Analyzing and Establishing

A typical representation in network pharmacology research is a graphical illustration depicting interactions among disease genes, biological pathways, compound targets, and other relevant components. Networks can be categorized into two main types based on the types of nodes: (1) single-type when all nodes are of the same type (e.g., PPI networks where all nodes represent protein) and (2) multi-type networks, where various components including proteins, genes, pathways constitute the nodes (e.g., compound-target-disease networks and compound-target-pathway networks)[53]. The choice of analysis methods, network topology, and the visualization of results will vary depending on the properties of the components within the network. As researchers progress toward the network construction phase, Venn diagrams play a crucial role in delineating overlapping targets between diseases and compounds[74]. This foundational step sets the stage for identifying shared genes between compounds and diseases, as well as predicting compound-disease-related targets. These shared genes serve as key nodes in the subsequent construction of PPI networks. The subsequent network visualization aims to predict hub genes based on their heightened connectivity within the network. Network visualization involves two main steps: first, adding network nodes, establishing links, and assigning attributes and, second, describing the network and extracting useful features to characterize its structure[75]. Professional software such as Cytoscape is commonly used for network pharmacology visualization on herb-level similarity. Various approaches are employed for network analysis, including structural analysis, functional analysis, and analogy analysis. For example, the integration of Gene Ontology (GO) enrichment analysis and KEGG pathway analysis further enhances the precision of key target features.

c. Validation

The predicted outcomes selected from earlier steps will be validated by various methods to ensure the accuracy and reliability of the results. Molecular docking, a cornerstone in *in silico* validation, allows for the prediction of interactions between ligands and their respective protein ligands when they are bound together[76]. The high docking score and high binding affinity have become primary criteria for confirming reliable interaction between compounds and targets. Another approach for validation is immunoblotting (Western blotting, WB), which serves as an indispensable method for confirming the levels of protein expression in target genes[77]. Researchers frequently perform WB to validate findings obtained from the multi-type interaction network. The precision and trustworthiness of results obtained through WB analysis facilitate significant breakthroughs in drug discovery and development[77]. Scientists also perform various methods and *in vitro* assays to evaluate the consistency of results obtained through network pharmacology, including the 3-[4,5-dimethylthiazol-2-yl]-2,5 diphenyl tetrazolium bromide (MTT) assay, polymerase chain reaction (PCR), and others. These methods have partially demonstrated that the results obtained using network pharmacology are consistent with experimental findings[78–80]. Lastly, *in vivo* analyses are often essential in many studies to scrutinize the obtained results in an intact organism. Various animal models including mouse, dog, and rabbit models

TABLE 29.2
The Used Databases for Network Pharmacology Research

Category	Name	Brief Description	Web Links	Reference
Ginseng- related databases	TCM	Provides comprehensive information on the chemical compounds found in *Panax ginseng*, including their structures and potential pharmacological activities	http://tcm.cmu.edu.tw/	61
	TCMID	Offers information about the chemical constituents of *Panax ginseng*, facilitating the study of pharmacological properties	http://www.megabionet.org/tcmid/	62
	TCMSP	Presents a framework for traditional Chinese medicines, capturing the intricate relationships among drugs and diseases targets.	https://tcmsp-e.com/tcmsp.php	63
	HIT	Acts as a search and curation platform for herbal ingredients and target information, drawing from literature evidence	http://lifecenter.sgst.cn/hit	64
Diseases-related databases	OMIM	Delivers a publicly updated database about human genes, disease-related information, and genetic traits	https://www.omim.org	65
	GAD	Provides a repository of genetic association data linked to complex disorders and diseases	https://geneticassociationdb.nih.gov	66
Gene-/protein-related databases	UniProtKB	Provides public access to a database containing protein sequences and annotations	http://www.uniprot.org/uniprot	67
	STRING	Offers freely accessible information on millions of proteins, aiding researchers in functional enrichment analysis and molecular interactions	http://string-db.org	68
	PubChem	Serves as a comprehensive repository containing details on chemical molecules and their activities within biological contexts	https://pubchem.ncbi.nlm.nih.gov	69
	GeneCards	Provides an extensive range of information encompassing genomic, transcriptomic, proteomic, genetic, and functional aspects of both known and predicted human genes, aggregated from over 150 data sources	https://www.genecards.org	70
Network construction, visualization and analyzing	KEGG	Offers functional annotations and pathway insights specific to Panax Ginseng, facilitating the comprehension of its biological impacts	http://www.genome.jp/kegg	71
	DAVID	Functions as a database tailored for annotating, visualizing, and providing functional insights into ginseng	https://david.ncifcrf.gov	72
	Cytoscape	Serves as a platform dedicated to constructing and visualizing networks	https://cytoscape.org	73

have been used to validate predicted results and foster groundbreaking treatment modalities for lethal diseases in network pharmacology research[81]. Besides *in vitro* and *in vivo* experiments, recent advancements in high-throughput technologies and genomics, single-cell analyses and *in silico* approaches have emerged as valuable tools for result validation. These virtual methods provide a convenient platform for validating predictions, and optimal reliability can be achieved through the integration of all methodologies.

29.3.3 Limitations

Despite recent advances in network pharmacology research throughout the last decade, most studies are still in the preliminary stages, making it challenging to make reliable predictive inferences. Network pharmacology approach largely relies on public databases, so the accuracy, reliability, and completeness of the data significantly effect on the prediction outcomes[82]. The data are gathered from numerous literature and experiment, so there are many false positive and negative interactions, resulting in the unreliability and inaccuracy of the model[83]. Consequently, more sophisticated computational tools are required to analyze and improve more informative interactions. Most network pharmacology research focuses on the qualitative analysis between the component target disease and the efficiency of an herb or natural compound but neglects the side effects of the drugs. Concerns about drug toxicity have developed dramatically over the past years, the mechanism of adverse side effects or the discovery of off-target drugs receives a lot of attention in network pharmacology research. Federico et al. combined network pharmacology to mechanistic and chemocentric approaches to find prominent drug combination in different disease[84].

29.4 APPLICATION OF NETWORK PHARMACOLOGY IN GINSENG IN DISEASES

29.4.1 Cancer

Cancer remains a significant global health concern due to the complex and dynamic nature as a result of the convoluted intrinsic and extrinsic biological mechanisms[85]. Patient heterogeneity, marked by the diverse nature of different cancer entities and therapeutic targets across populations, also poses a significant obstacle in developing effective treatments[86]. Traditional drug discovery, which often focuses on single-target approaches, faces challenges in combating this complexity[87]. By simultaneously targeting multiple proteins and networks involved in malignant progression, network pharmacology emerges as a revolutionary tool, shifting the paradigm toward multifaceted strategies[12]. Moreover, network pharmacology not only fosters deeper understanding of disease mechanisms but also potentiates the discovery of new anticancer therapies[53]. Natural products derived from plants, characterized by low toxicity and little side effects, often exhibit multitargeted activity[77]. Therefore, it is tempting to extent the search for natural substances with anticancer characteristics. Ginseng, which has been used for millennia in traditional medicine, is a promising candidate for cancer treatments[88]. The potential of ginseng resides in the multifaceted mode of action based on many bioactive compounds, coupled with a notably low toxicity profile. Thus, the integration of network pharmacology becomes crucial to unravel the mysteries of ginseng's interactions with various cancers[10]. By meticulously analyzing the network of interactions between ginseng's components, target molecules, and signaling pathways within different cancer cells, network pharmacology has provided a comprehensive overview of its potential therapeutic effects[89]. The exploration of the patterns of interactions between these compounds, cancer cells, and signaling pathways can lead to the development of individualized treatment regimens for complex malignant diseases[90], facilitating more effective prevention and treatment of cancers, and even better survival rates for patients.

a. Lung cancer

Lung cancer represents the most prevalent malignancy worldwide, carrying the highest rates of morbidity and mortality among all tumors[91]. Depending on the disease subtype and stage, therapies range from chemotherapy, radiotherapy, and more recently targeted therapy and immunotherapies. However, current treatments fail to prevent relapse and often cause severe or fatal side effects; thus, exploration of novel therapeutic avenues remains demanding for lung cancer[92, 93]. Recent studies have indicated the potential of ginseng in enhancing lung cancer treatment by influencing cell differentiation, triggering cell cycle arrest, and encouraging apoptosis[94, 95]. Despite the promising effects of ginseng, the exact mechanisms by which this herb impacts lung cancer remain unclear.

A study by Song et al. explored the mechanism of how ginsenoside Rh2 (G-Rh2) inhibited the viability of A549 cells via screening of various databases for potential targets about the relation between this compound and lung cancer[15]. The authors constructed a PPI network that identified 217 targets associated with G-Rh2 against lung cancer. GO enrichment analyses revealed the functional roles of these targets, highlighting their involvement in apoptosis, transcriptional regulation, and cellular processes. Furthermore, KEGG pathway analysis identified 191 important genes and 143 significantly enriched pathways, including those associated with proteoglycans in cancer, focal adhesion, Forkhead box O (FoxO) signaling, apoptosis, and Janus kinase/signal transducers/activators of transcription (JAK/STAT), phosphatidylinositol-3-kinase (PI3K)/AKT signaling pathway[15]. This analysis exhibited phosphatidylinositol-4,5-bisphosphate 3-kinase catalytic subunit alpha (PIK3CA) as a central node, suggesting the crucial involvement in G-Rh2's cancer mechanism. Additionally, regulation of insulin receptor substrate 1 (IRS1), nuclear factor κB subunit 1 (NFKB1), AKT serine/threonine kinase 1 (AKT1), mitogen-activated protein kinase 9 (MAPK9), and DNA polymerase delta 1 catalytic subunit (POLD1) might also contribute to the inhibition of A549 cell viability by G-Rh2. Validation experiments including MTT assay, flow cytometry, and WB analysis have validated the bioinformatics findings and indicated consistent results.

Gao et al. conducted a comprehensive investigation into the therapeutic potential of fat soluble components of ginseng, delineating the underlying molecular mechanisms and identifying crucial targets[96]. An analysis of 410 fat-soluble targets and 1329 differentially expressed genes in lung cancer revealed thirty-three common targets shared between ten ginseng's fat-soluble components (G-FSC) and lung cancer. Enrichment analyses were performed for both function and pathway to unveil the top five biological responses and signaling pathways associated with the key targets of G-FSC for lung cells. These findings suggested that G-FSC compounds exhibit inhibitory effects on lung cancer cell growth and facilitate apoptosis due to the modulation of key signaling molecules including epidermal growth factor receptor (EGFR), kinase insert domain receptor (KDR), and mitogen-activated protein kinase-3 (MAPK3). Thus, G-FSC compounds may exert therapeutic benefits in lung cancer treatment through the regulation of pertinent pathways and target proteins.

Another research performed by Li et al. investigated the anti-lung cancer mechanism of ginseng constituents extracted from leaves[97]. Notably, ginseng leaf extracts possess significant anticancer properties due to their high content of ginsenosides compared to ginseng roots. Ginseng leaves have been extensively studied for their components and potential anticancer properties. However, a comprehensive understanding of how the complex interplay of ginseng's multiple components targeting various pathways contributes to lung cancer treatment via immunomodulation remains lacking attention. The study showed that the eight active component targets identified in ginseng leaves were associated with lung cancer and immune function. As a result, a total of 193 composite targets were selected. The construction of a PPI network revealed key targets, including EGFR, heat shock protein 90 alpha family class A member 1 (HSP90AA1), JUN, and signal transducer and activator of transcription 3 (STAT3)[97]. Functional enrichment analysis of ten key biological processes related to nutrient and oxygen homeostasis, as well as lipopolysaccharide signaling. Molecular function analysis revealed a repertoire of proteins involved in crucial cellular processes, including DNA-binding transcription factors, cytokine receptors, and ubiquitin-like protein ligases. KEGG pathway analysis identified 164 pathways, with prominent involvement in the MAPK, PI3K–AKT, and TNF pathways[97]. The comparison between GO and KEGG analysis in outcomes suggested that target gene of ginseng leaves may exert regulatory effects on critical pathways involved in lung cancer progression and apoptosis. Molecular docking studies further validated the interaction between ginseng leaf components and the top six core targets associated with lung cancer and immunity.

b. Hepatocellular carcinoma

Hepatocellular carcinoma (HCC) stands as the most prevalent primary malignancy of the liver, accounting for approximately 10% of all new cancer diagnoses worldwide[98]. The primary systemic treatments for liver cancer remain conventional approaches like chemotherapy and radiotherapy and are increasingly challenged by the emergence of drug resistance and unsatisfying outcomes[99]. This underscores the critical need for innovative adjuvant therapies to improve the therapeutic landscape for liver cancer.

Ginsenoside Rg5 and Rk1 (G-Rg5 and G-Rk1) were classified as minor ginseng saponins, with high oral bioavailability[100]. The anticancer properties of G-Rg5 and G-Rk1 in HCC have been highlighted in some studies, but a deeper understanding of the underlying mechanisms and interactions with relevant pathways remains to be elucidated[101, 102]. Chen et al. investigated the anticancer properties of G-Rg5 and G-Rk1 in human liver cancer cells using a network pharmacology-based approach[103]. Due to limited information on the structure interaction between ginsenosides and target protein, the researchers applied the similarity in G-Rg5 and G-Rk1 structure to bioinformatic approach. This method led to the identification of 142 potential targets, from which 44 putative genes were pinpointed as common entities shared between G-Rg5 and G-Rk1103. Functional enrichment analysis revealed that these two ginsenosides primarily target genes associated with the PI3K/AKT pathway in cancer cell survival and proliferation. Additionally, a ginsenoside-target-pathway network was constructed, highlighting critical genes that contribute to the anti-tumor activities of these compounds including ESR1, EGFR, ESR2, RELA, and HSP90AA1[103]. Finally, two cell viability assays, including MTT, and mitochondrial membrane potential (MMP) assays were performed to validate the findings in the MHCC-97H human cell line, which align with the outcomes derived from network pharmacology methodologies.

Ginsenoside Rk3 (G-Rk3) displayed anti-tumor activities against HCC[104]. Through the construction of a PPI network, potential target proteins of G-Rk3 were identified along with liver-cancer-related genes. Further analysis using GO analysis and KEGG pathway enrichment analysis demonstrated that G-Rk3 affects cancer-related pathways and biological processes involved in proteolysis and regulation of the apoptosis process, corroborated by the significant inhibition in HCC proliferation[104]. This research underscored G-Rk3 as a promising candidate for the advancement of novel therapeutics targeting the PI3K/AKT pathway in HCC treatment, offering potential advantages in terms of low toxic side effects.

c. Gastric cancer

Gastric cancer has been one of the leading contributors to cancer-related deaths[105]. Ginsenosides found in ginseng have exhibited excellent effectiveness in the treatment of gastric cancer. Recent investigations have indicated ginsenosides exert the potential anti-gastric-cancer activity. Ginsenoside Rb1 (G-Rb1), a primary compound found in ginseng, transforms ginsenoside CK (G-CK) by intestinal flora, exhibiting diverse pharmacological effects[106, 107]. Wan et al. analyzed the fundamental mechanism of G-CK and G-Rb1 in the treatment of gastric cancer[108]. Utilizing the Disgenet, Malacards, OMIM, and HIT databases, 189 targets associated with gastric cancer, along with 24 potential targets of G-Rb1 and G-CK were identified. PPI networks that were built from earlier information indicated several key targets such as NTRK1, FN1, AKT1, and AURKA. GO analysis revealed the correlation between G-Rb1 with biological processes such as cell apoptosis and proliferation, whereas G-CK was linked to the IκB kinase/ NF-κB signaling pathway. KEGG pathway analysis identified common potential pathways, notably the NF-κB, MAPK, TLR, TNF, PI3K/AKT, and WNT signaling pathways, suggesting the suppression of the PI3K/AKT/ NF-κB as a target for gastric cancer therapy[108]. Indeed, these network pharmacology findings are promising because experimental validation, particularly *in vitro* investigations, has been necessarily warranted. The effects of G-CK and G-Rb1 on the cycle-related factors and apoptosis-associated proteins were explored by WB and RT-qPCR. The analyses demonstrated that G-CK effectively inhibited the NF-κB and PI3K/AKT pathways at both transcriptional and translational stages, while G-Rb1 did not exhibit a significant inhibitory effect.

d. Osteosarcoma

Osteosarcoma, a predominant bone cancer, derived from mesenchymal tissue, is characterized by highly malignant features like rapid growth, early metastasis, and dismal prognosis[109]. G-Rg5 has shown potential as a therapeutic agent for sarcoma due to its anti-tumor characteristics[110]. In a study by Liu et al., the mechanism underlying the treatment of osteosarcoma with G-Rg5 was investigated through the application of network pharmacology and molecular techniques[111]. The prediction of the SwissTargetPrediction database and similarity ensemble approach showed that there were 181 identified targets of G-Rg5, and 1322 genes associated with osteosarcoma were gathered from various databases. However, just around 62 key gene targets can be considered as the intersection between G-Rg-5 and osteosarcoma. Subsequent GO and KEGG pathway enrichment analyses revealed key signaling pathways implicated in osteosarcoma, including PI3K-AKT, proteoglycans, lipid metabolism, atherosclerosis, and Kaposi-sarcoma-associated herpesvirus infection. Using PPI analysis, pivotal targets such as PIK3CA. EGFR, MAPK1, proto-oncogene tyrosine-protein kinase (SRC), tumor protein P53 (TP53), and vascular endothelial growth factor A (VEGFA) were selected for further molecular docking investigation. The PPI results demonstrated stable binding energies across all hub targets[111]. Furthermore, survival analysis indicated correlations between the prognosis of sarcoma patients and TP53 and VEGFA expression levels. This study unveils the potential therapeutic mechanisms of G-Rg5 for osteosarcoma treatment and underscores the importance of further clinical exploration.

29.4.2 Diabetes

Diabetes mellitus (DM) poses a significant global health challenge characterized by chronic hyperglycemia and associated complications[112]. DM ranks as the third major disease causing mortality following cardiovascular diseases and tumors[113]. Although current treatments involving peptide and insulin derivatives, alongside dietary control, are effective to some extent, a fundamental therapy for type 2 diabetes stills elusive[114]. The long-term use of these treatments presents challenges due to inconvenience and potential adverse effects, prompting exploration of natural compounds with fewer side effects, for instance traditional herbs like ginseng. The potential of ginseng in managing DM has been demonstrated; however, the precise compounds responsible for therapeutic effects, the targets they interact with, the signaling pathways they influence, and the biological processes involved in DM management are not fully understood. Moreover, challenges including the limited availability of compounds, the difficulty of *in vivo* experiments, and the complexity associated with multiple components and targets emphasize the necessity for innovative research methodologies[115]. Network pharmacology offers a method to investigate pharmacologically active constituents, compounds, and potential molecular network mechanisms within ginseng[116, 117].

Li et al. integrated molecular docking in the network pharmacology studies to explore the mechanism of active components from ginseng to manage DM[14]. Through a comprehensive search of the TCMSP database and literature, the authors identified 28 ginsenosides, which were subsequently analyzed for their target profiles. Notably, the ginsenosides PPD, Rh1, Rh2, Rh4, Rg1, and PPT exhibited a higher number of targets compared to other ginsenosides, suggesting their potential efficacy in DM management. Furthermore, the integration of data from five disease databases enabled the identification of 4341 DM-related targets. After integration, 101 intersection targets were obtained, highlighting the key proteins involved in the pathogenesis of DM. Ninety-nine protein-protein intersection targets were identified in the PPI network and nine core targets – EGFR, MAPK1, STAT3, mTOR, CASP3, IL-β, JUN, TNF-α, and VEGFA – were identified using Cytoscape[14]. GO functional enrichment analysis using R software discovered GO terms linked with molecular functions, cellular components, and biological processes, with the number of data up to 2029, 56, and 153, respectively. GO enrichment analysis suggested that the management of DM by ginsenosides may involve processes such as positive regulation of lipid and reactive oxygen metabolic processes, MAPK activity, and response to a steroid hormone. Additionally, KEGG pathway analysis identified 169 signaling pathways, including the adipocytokine, AGE-RAGE, AMPK, Insulin, PI3K-AKT, and VEGF signaling pathways[14]. Combined with previous reports[118, 119], it has been predicted that B cell receptor, TNF, insulin, and VEGF signaling pathways might be vital in the controlling of DM by ginseng. Molecular dynamics simulation and docking results revealed that most ginsenosides could stably interact with the core targets, primarily through hydrogen and hydrophobic bonds. Overall, the identification of these core targets provides information on the molecular mechanisms underlying DM pathogenesis and highlights prospective treatment targets for further investigations.

Another study by Liu et al. elucidated the potential properties of G-Rk1 and G-Rg5 in type 2 diabetes mellitus (T2DM) treatment[120] via the network pharmacology, first applied to forecast the target of both two ginsenosides. Initial screenings identified 3659 T2DM-related targets, and subsequent intersection with G-Rk1 and G-Rg5 active ingredient targets revealed 250 common gene targets shared between G-Rk1 and G-Rg5 and T2DM. GO enrichment analysis further emphasized the involvement of core targets in insulin signal transduction and related pathways. A network analysis using Cytoscape software identified the top 10 proteins potentially serving as core targets for T2DM treatment. The KEGG and GO enrichment analysis indicated that those ginsenosides enhanced the p53, PPAR, cAMP signaling as well as the insulin resistance[120]. In detail, the study analysis the effect on insulin metabolic pathway such as the insulin receptor, signaling transduction, cellular response and the activation of protein tyrosine kinases. Molecular simulations employing AutoDock 4.2.6 software demonstrated the binding potential of G-Rk1 and G-Rg5 with these core targets, with AKT1 emerging as a promising candidate. Molecular docking analyses revealed intricate binding patterns between G-Rk1, G-Rg5 and AKT1, suggesting potential mechanisms underlying their therapeutic actions in ameliorating T2DM[120]. The study brought valuable insights into the molecular interactions highlighting the therapeutic effects of G-Rk1 + G-Rg5 in T2DM management.

29.4.3 Neurodegenerative Diseases

a. Parkinson's disease

Parkinson's disease (PD) is a neurological ailment posing a substantial threat to the physical and mental health of individuals in middle- and older age groups[121]. This disease progressively impacts the motor cortex, leading to enduring deficits in movement and coordination[122]. Ginseng has been widely used for PD management[123]. Despite the well-established efficacy, the precise mechanism underlying the therapeutic impact of ginseng in PD remains to be elucidated. Recent advancements in network pharmacology have significantly contributed to unraveling the mechanisms associated with the relationship between ginseng and PD. A study was performed by Zhang et al. using network pharmacology to explore the molecular mechanisms of ginseng in PD treatment[124]. The results showed that there were 97 primary effective components from ginseng and 212 target genes from the TCMSP database. GeneCards and OMIM databases established a platform of PD targets, revealing 7772 PD-associated disease targets. Subsequently, an 'effective ingredient-target-disease' interaction network was established from the retrieval STRING database to build a PPI network, analyzing related protein interactions. Network analysis targeted primarily on the regulation of cell metabolism and various other biological mechanisms. Further analysis performed by KEGG and GO pathway enrichment showed that the effects of ginseng on PD treatment involve the regulation of several signaling pathways, including MAPK, AGE, and RAGE[124]. Molecular docking of active compounds with PD targets was performed, demonstrating favorable binding with MAPK3 and MAPK14 targets. These findings suggested that saponins are more likely to block cell death by regulating multiple signaling pathways, including preventing glial cell activation and aggregation, thereby enhancing nerve cell viability.

b. Alzheimer's disease

Alzheimer's disease (AD) illustrates a formidable challenge to global public health, characterized by the insidious onset and progressive decline in cognitive function. Despite extensive research efforts, the etiology and pathogenesis of AD remain elusive, and current treatment options offer only limited symptomatic relief[125]. Notably, emerging evidence suggests a promising role for ginseng in the AD early prevention and treatment[126]. In the study of Li et al., the network pharmacology explored the possible targets of ginseng in AD through modulating ferroptosis, an iron-dependent and lipid peroxidation-driven type of programmed cell death, aiming to offer novel insights for the enhancement of AD-related therapeutics[127]. Ferroptosis relates to neurodegeneration in AD, appears when long-term elevated iron levels and oxidative stress. Through rigorous screening, they identified 16 potentially active ingredients and revealed 94 potential targets of these ginseng active compounds. Subsequently, a comprehensive ginseng-active compound-target network was constructed, incorporating 8699 AD-related molecular targets. Notably, their analysis highlighted kaempferol as a crucial active compound based on the regulation of ferroptosis[128]. Core targets identified in the network included IL1B, TNF, AKT1, JUN, and NFKBIA[127]. Furthermore, through KEGG and GO pathway enrichment analyses, they elucidated TNF, PI3K-AKT, NF-°B, and MAPK signaling pathways as the main enriched pathways. Moreover, by analyzing 260 genes associated with ferroptosis and 84 predicted targets of ginseng active compounds against AD, they hypothesized that ginseng may exert modulatory effects on signaling pathways such as TNF, hypoxia-inducible factor 1 (HIF-1), MAPK, TLR, cAMP, neurotrophic factor, and NOD-like receptor, which modulate ferroptosis and are implicated in the pathogenesis of AD.

29.4.4 Heart Diseases

Cardiovascular diseases (CVDs) including atherosclerosis, hypertension, stroke, heart failure, acute myocardial infarction (AMI), and other conditions are the clinical manifestations forcing enormous global economic and health burdens[129, 130]. The pharmacological potentials of ginseng are early studied such as antiatherosclerosis, antimyocardial ischemia, and antiarrhythmic actions as well as inhibiting ventricular remodeling[131–133]. Yu et al. built a 'components-targets' network, a PPI network, and used a zebrafish disease model to determine the relation between garden ginseng (GG) and mountain-cultivated ginseng (MCG) in CVDs[78]. In a comprehensive analysis, 243 targets in total corresponding to 11 different compounds were identified. The HGDB databases and DrugBank were utilized to gather 4382 targets related to CVD, with gene names obtained from the UniProt database after removing null and repetitive targets. These efforts resulted in the identification of 149 common target genes, which were then integrated into a 'components-targets' network to assess components-targets interaction. This network, comprising 291 edges and 197 nodes, highlighted ginsenoside Re, ginsenoside Rg3, and citric acid as potentially crucial components in the context of CVD[78]. Furthermore, employing the STRING database, a PPI network was constructed and subsequently clustered into five distinct clusters. The top 10 targets within these clusters included tumor protein P53, protooncogene tyrosine-protein kinase SRC, and VEGFA. Additionally, KEGG enrichment analysis yielded 59 pathways, with the top 15 pathways identified at a significance level of $p < 0.05$. Notably, these pathways encompassed critical signaling cascades such as the HIF-1, TNF, and PI3K-AKT signaling pathway, with the PI3K-AKT pathway being particularly enriched[78]. The molecular docking analysis was conducted using the ginsenosides B2, C, Rg3 and Re, along with the key genes P53, NOS2, TNF and STAT3. Notably, the results demonstrated a robust binding affinity of various compounds to these key genes, corroborating the findings obtained through network pharmacology screening. Furthermore, experimental validation of the vascular injury model in zebrafish revealed that both GG and MCG could upregulate the expression of HIF-1α and VEGF, pivotal factors in promoting angiogenesis under pathological conditions. These findings not only validate the therapeutic potential of MCG and GG but also serve as valuable data references for their clinical selection and applications[78].

Heart failure (HF) stands as one of prevalent causes of mortality worldwide. Characterized by symptoms like dyspnea, fluid retention, and reduced exercise tolerance, HF imposes a significant burden on global health, with a low cure rate[134]. Ginsenosides have been demonstrated to exert positive inotropic effects, enhancing myocardial contractility and ameliorating ventricular remodeling, thereby stabilizing hemodynamics in HF patients and improving cardiac structure and function[135]. Nevertheless, the precise active compounds, targets, and the mechanisms of HF-regulatory action remain incompletely understood. Combining network pharmacology with molecular docking technology, Dai et al. predicted the molecular mechanism of processed ginseng in HF treating[136]. The research identified forty ginsenosides by using UPLC-QE-Orbitrap-MS/MS analysis, encompassing notable constituents such as ginsenosides Re, Rh1, Rb1, Rc, and Rd. Subsequently, these identified saponin constituents underwent network pharmacological analysis. PPI analyses of the targets associated with various ginseng preparations for HF treatment were conducted using the STRING database. Through database screening, 124 prospective targets for HF treatment were identified from various derived ginsenosides of ginseng, with 22 main targets identified, including EGFR, STAT3, ESR1, ERBB2, AKT1, TNF, MMP9, PIK3CA, MTOR, and HRAS. The composition-target-disease-pathway network approach revealed that ginsenoside compounds interact with several targets[136]. Based on DAVID database, the KEGG and GO enrichment results indicated that the mechanisms underlying diverse ginseng-derived products in HF treatment primarily involved the PI3K-AKT, RAS, mTOR, and apoptosis pathways. This research highlighted the multi-pathway and multi-target characteristics of ginseng-derived products in HF treatment. Lastly, the major active components of different *Panax ginseng* products were screened, and their core targets were subjected to docking analysis. The molecular docking results demonstrated favorable binding activity of the active components with prominent targets, thereby validating the reliability of the network pharmacology method to a certain extent[136].

29.4.5 Osteoporosis

Osteoporosis is a prevalent systemic bone disorder recognized as bone mass loss and micro-architectural deterioration, leading to bone fragility and fractures. With nearly nine million fractures occurring annually worldwide, predominantly in the hip, distal forearm, spin and proximal humerus, osteoporosis poses a significant global health concern[137]. Current clinical management primarily focuses on fracture relief using bisphosphonates, which come with limitations such as renal failure and musculoskeletal pain, urging the need for safer and more effective treatment options[138]. Ginseng contains pharmacologically active compounds known as ginsenosides, exhibiting diverse therapeutic effects, including anti-osteoporosis properties[139]. Despite its potential, research on ginseng's role in osteoporosis treatment remains limited.

Zhang et al. investigated the therapeutic potential of ginsenosides for osteoporosis, with a focus on G-CK, employing network pharmacology to elucidate its potential targets and underlying mechanisms[80]. Through data mining of public databases, 6590 and 206 targets were identified for osteoporosis and G-CK, respectively, with an intersection analysis revealing 138 co-targets shared between G-CK and osteoporosis. Further analysis involved PPI network analysis to explore the internal interactions and pivotal targets among these co-targets. Notably, nodes with significant degrees, including GRB2, PLCG1, PIK3CA, PIK3R1, SRC, STAT3, and VEGFA, were identified within the PPI network[80]. Additionally, the PPI network data underwent MCODE analysis, revealing five major gene clusters, especially JAK2, MAP2K1, PIK3R1, STAT3, and VEGFA, identified as key genes. GO analysis indicated the involvement of phosphatidylinositol-related biological processes and molecular functions in G-CK's role in osteoporosis treatment. Furthermore, the c-FMS-mediated osteoclast differentiation pathway emerged as a pivotal mechanism for G-CK in osteoporosis treatment[80]. Molecular docking studies provided additional confirmation, demonstrating the low-energy interaction between G-CK and c-FMS protein. In summary, G-CK was found to intervene in the onset and progression of osteoporosis via modulating of c-Fms-mediated PI3K and MAPK signaling pathways, thereby regulating osteoclast differentiation.

On the other hand, Guo et al. explored the potential targets and mechanisms of ginsenosides, key bioactive compounds found in ginseng, in the treatment of osteoporosis[140]. Initially, 190 derivatives of ginsenosides were identified from the TCMSP database, with eight key ginsenosides

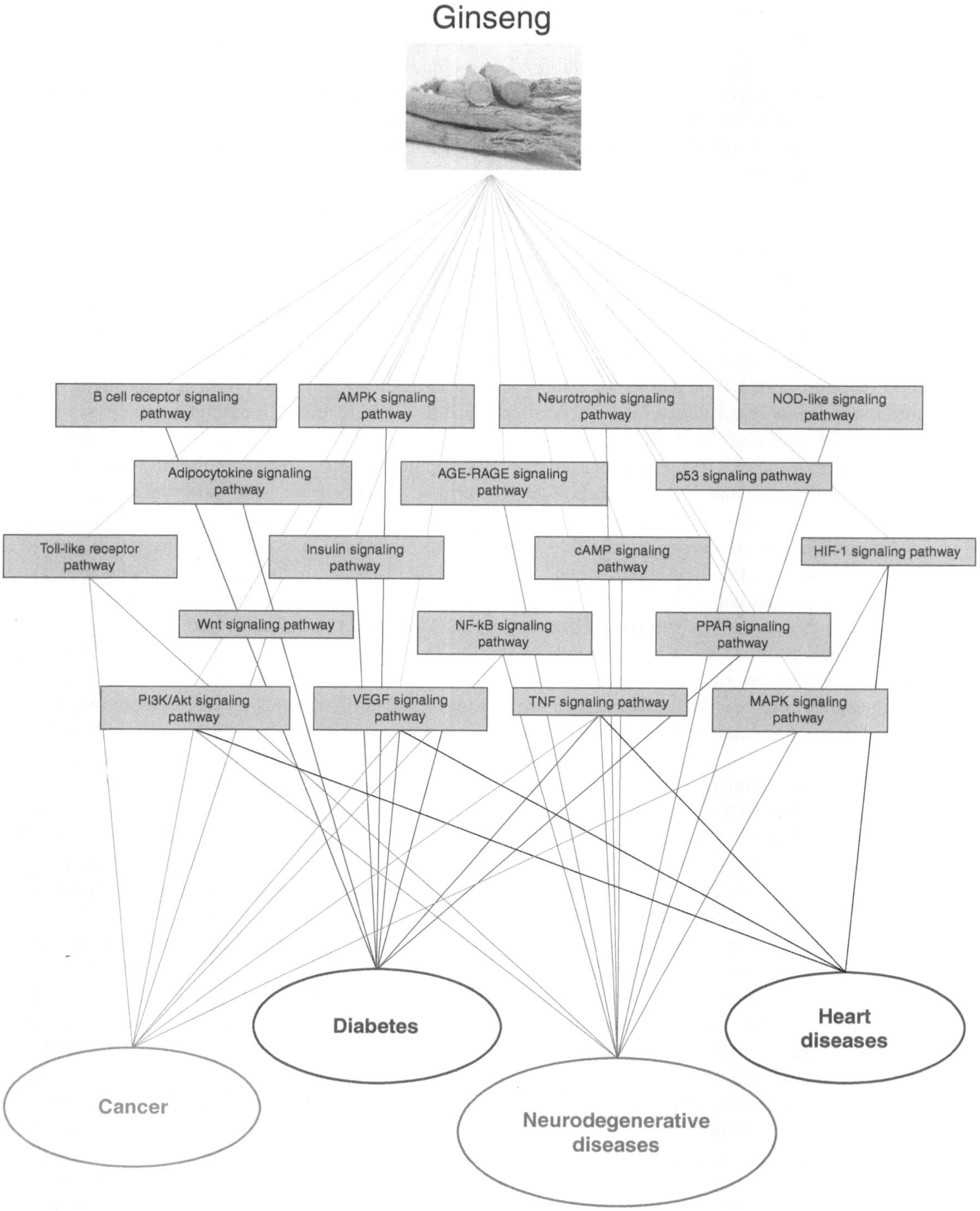

FIGURE 29.4 The interaction of ginseng-derived compounds and disease-related signaling pathways.

selected for further investigation. Target prediction of ginsenosides was performed using both the TCMSP and Pharmmapper databases, resulting in the identification of 80 target proteins associated with ginsenosides. Through cross-validation with osteoporosis-related genes, 17 overlapping genes were identified as prospective targets for ginsenosides in the osteoporosis treatment. Notably, these targets encompassed critical genes involved in osteoclast differentiation and inflammation pathways. Further analysis using the Metascape platform and Cytoscape software unveiled a PPI network comprising 14 key target proteins, including IL1B, TNF, IFNG, and NFKBIA[140]. Among these, TNF and IL1B exhibited the most interactions with other ginsenoside targets, highlighting their significant role in mediating the therapeutic effects of ginsenosides in osteoporosis. Finally, the molecular docking studies database confirmed the binding interactions between ginsenosides and target proteins associated with osteoporosis. Validation through AutoDock software underscored the potential of ginsenosides to interact with key proteins involved in anti-osteoporotic pathways. Overall, this comprehensive analysis provides information on the molecular mechanisms behind the anti-osteoporotic effects of ginsenosides, offering a foundation for further research and the enhancement of novel therapeutic strategies leveraging the therapeutic potential of ginseng-derived compounds[140].

29.5 CONCLUSION

Ginseng has been popular in folk medicines over decades, thanks to its many active compounds, such as ginsenosides, polyphenolics, polysaccharides, and peptides. Some pharmacological activities of ginseng include antioxidant, anti-inflammation, anticancer, and neuroprotection. Ginsenosides are one of the well-known active compounds isolated from ginseng with therapeutic and pharmacological activities. Recently, network pharmacology has represented a novel approach that accelerates the analysis of crucial interactions of ginseng's active compounds with disease-cored pathways. Thus, this chapter focused on the overview of the application of this modern tool on ginsenosides and other compounds, revealing anticancer effects via their involvement in many cancer-related pathways including the NF-κB and PI3K/AKT, TNF, or MAPK signaling pathway. In addition to cancers, the therapeutic potential of ginseng's active constituents in diabetes, neurodegeneration, cardiovascular diseases, and osteoporosis was identified and validated with the PPI, GO, and KEGG enrichment analysis. Through these studies, the underlying mechanisms of ginseng's compounds have been illuminated as complicated networks that have many interactions between ginseng's targets, biological pathways, related genes, and proteins. Thus, this offers further insights into the beneficial effects of ginseng's constituents, especially ginsenosides at molecular levels, and identifies new targets for drug development.

ACKNOWLEDGMENT

This research is funded by Vietnam National University HoChiMinh City (VNU-HCM) under grant number C2024-28-14.

REFERENCES

1. Ratan, Z. A., Haidere, M. F., Hong, Y. H., Park, S. H., Lee, J.-O., Lee, J., & Cho, J. Y. (2021). Pharmacological potential of ginseng and its major component ginsenosides. *Journal of Ginseng Research*, *45*, 199–210.
2. Kim, J.-H. (2018). Pharmacological and medical applications of Panax ginseng and ginsenosides: A review for use in cardiovascular diseases. *Journal of Ginseng Research*, *42*, 264–269.
3. Kim, K. H., Lee, D., Lee, H. L., Kim, C.-E., Jung, K., & Kang, K. S. (2018). Beneficial effects of *Panax ginseng* for the treatment and prevention of neurodegenerative diseases: Past findings and future directions. *Journal of Ginseng Research*, *42*, 239–247.
4. Kim, H.-J., Jung, S.-W., Kim, S.-Y., Cho, I.-H., Kim, H.-C., Rhim, H., Kim, M., & Nah, S.-Y. (2018). *Panax ginseng* as an adjuvant treatment for Alzheimer's disease. *Journal of Ginseng Research*, *42*, 401–411.
5. Kim, J. H., Yi, Y.-S., Kim, M.-Y., & Cho, J. Y. (2017). Role of ginsenosides, the main active components of Panax ginseng, in inflammatory responses and diseases. *Journal of Ginseng Research*, *41*, 435–443.
6. Wang, C.-Z., & Yuan, C.-S. (2008). Potential role of Ginseng in the treatment of colorectal cancer. *The American Journal of Chinese Medicine*, *36*, 1019–1028.
7. Ahuja, A., Kim, J. H., Kim, J.-H., Yi, Y.-S., & Cho, J. Y. (2018). Functional role of ginseng-derived compounds in cancer. *Journal of Ginseng Research*, *42*, 248–254.
8. Lee, Y. J., Lee, S., Ho, J. N., Byun, S.-S., Hong, S. K., Lee, S. E., & Lee, E. (2014). Synergistic antitumor effect of ginsenoside Rg3 and cisplatin in cisplatin-resistant bladder tumor cell line. *Oncol Reports*, *32*, 1803–1808.
9. Wild ginseng prevents the onset of high-fat diet induced hyperglycemia and obesity in ICR mice. *PubMed*. https://pubmed.ncbi.nlm.nih.gov/15357009/
10. Park, S.-Y., Park, J.-H., Kim, H.-S., Lee, C.-Y., Lee, H.-J., Kang, K. S., & Kim, C.-E. (2018). Systems-level mechanisms of action of *Panax ginseng*: A network pharmacological approach. *Journal of Ginseng Research*, *42*, 98–106.
11. He, S., Shi, J., Chai, H., Ma, L., Pei, H., Zhang, P., Shi, D., & Li, H. (2024). Mechanisms with network pharmacology approach of ginsenosides in alzheimer's disease. *Heliyon*, *10*, e26642.
12. Hopkins, A. L. (2008). Network pharmacology: The next paradigm in drug discovery. *Nature Chemical Biology*, *4*, 682–690.
13. Valdés-González, J. A., Sánchez, M., Moratilla-Rivera, I., Iglesias, I., & Gómez-Serranillos, M. P. (2023). Immunomodulatory, anti-inflammatory, and anti-cancer properties of Ginseng: A pharmacological update. *Molecules*, *28*, 3863.
14. Li, M., Jin, M., Hu, R., Tang, S., Li, K., Gong, X.-J., Sun, Y., Wang, Y., Wang, Z., & Li, W. (2023). Exploring the mechanism of active components from ginseng to manage diabetes mellitus based on network pharmacology and molecular docking. *Scientific Reports*, *13*, 793.
15. Song, C., Yuan, Y., Zhou, J., He, Z., Hu, Y., Xie, Y., Liu, N., Wu, L., & Zhang, J. (2022). Network pharmacology-based prediction and verification of ginsenoside Rh2-induced apoptosis of A549 cells via the PI3K/Akt pathway. *Frontiers in Pharmacology*, *13*, 878937.
16. Zhang, W., Chen, J., & Liu, H. (2022). Network pharmacology and molecular docking-based prediction of the molecular targets and signaling pathways of Ginseng in the treatment of Parkinson's disease. https://journals.sagepub.com/doi/full/10.1177/1934578X221102029
17. Lee, C. H., & Kim, J.-H. (2014). A review on the medicinal potentials of ginseng and ginsenosides on cardiovascular diseases. *Journal of Ginseng Research*, *38*, 161–166.
18. Attele, A. S., Wu, J. A., & Yuan, C.-S. (1999). Ginseng pharmacology. *Biochemical Pharmacology*, *58*, 1685–1693.
19. Shehzad, O., Kim, H. P., & Kim, Y. S. (2013). State-of-the-art separation of ginsenosides from Korean white and red

ginseng by countercurrent chromatography. *Analytical and Bioanalytical Chemistry, 405*, 4523–4530.
20. Shibata, S., Fujita, M., Itokawa, H., Tanaka, O., & Ishii, T. (1963). Studies on the constituents of Japanese and Chinese crude drugs. XI. panaxadiol, a sapogenin of Ginseng roots. (1). *Chemical and Pharmaceutical Bulletin, 11*, 759–761.
21. Khan, S., Tosun, A., & Kim, Y. S. (2015). Chapter 31 – Ginsenosides as food supplements and their potential role in immunological and neurodegenerative disorders. In R. R. Watson & V. R. Preedy (Eds.), *Bioactive nutraceuticals and dietary supplements in neurological and brain disease* (pp. 303–309). Academic Press. https://doi.org/10.1016/B978-0-12-411462-3.00031-X
22. Sha, Y., Yan, M.-C., Liu, J., Liu, Y., & Cheng, M.-S. (2008). Facile synthesis of oleanolic acid monoglycosides and diglycosides. *Molecules, 13*, 1472–1486.
23. Hou, M., Wang, R., Zhao, S., & Wang, Z. (2021). Ginsenosides in *Panax* genus and their biosynthesis. *Acta Pharmaceutica Sinica B, 11*, 1813–1834.
24. Li, W., Gu, C., Zhang, H., Awang, D. V., Fitzloff, J. F., Fong, H. H., & van Breemen, R. B. (2000). Use of high-performance liquid chromatography-tandem mass spectrometry to distinguish *Panax ginseng* C. A. Meyer (*Asian ginseng*) and *Panax quinquefolius* L. (North American ginseng). *Analytical Chemistry, 72*, 5417–5422.
25. He, B., Chen, D., Zhang, X., Yang, R., Yang, Y., Chen, P., & Shen, Z. (2022). Oxidative stress and ginsenosides: An update on the molecular mechanisms. *Oxidative Medicine and Cellular Longevity, 2022*, 9299574.
26. Fu, W., Yu, X., Lu, Z., Sun, F., Wang, Y., Zhang, Y., Zhang, Y., Xu, H., & Sui, D. (2016). Protective effects of ginsenoside Rb2 on myocardial ischemia in vivo and in vitro. *International Journal of Clinical and Experimental Medicine, 9*, 9843–9855.
27. Sun, J., Yu, X., Huangpu, H., & Yao, F. (2019). Ginsenoside Rb3 protects cardiomyocytes against hypoxia/reoxygenation injury via activating the antioxidation signaling pathway of PERK/Nrf2/HMOX1. *Biomedicine & Pharmacotherapy, 109*, 254–261.
28. Im, D.-S. (2020). Pro-resolving effect of ginsenosides as an anti-inflammatory mechanism of *Panax ginseng*. *Biomolecules, 10*, 444.
29. Ikram, M., Ullah, R., Khan, A., & Kim, M. O. (2020). Ongoing research on the role of gintonin in the management of neurodegenerative disorders. *Cells, 9*, 1464.
30. Li, Y., Yu, P., Fu, W., Wang, J., Ma, Y., Wu, Y., Cui, H., Zhao, W., Zhang, F., Yu, X., Sui, D., & Xu, H. (2022). Polysaccharides from *Panax ginseng* C. A. Meyer alleviated DSS-induced IBD by inhibiting JAK2/STAT1/NLPR3 inflammasome signalling pathway in mice. *Journal of Functional Foods, 91*, 105013.
31. Dou, D.-Q., Zhang, Y.-W., Zhang, L., Chen, Y.-J., & Yao, X.-S. (2001). The inhibitory effects of ginsenosides on protein tyrosine kinase activated by hypoxia/reoxygenation in cultured human umbilical vein endothelial cells. *Planta Medica, 67*, 19–23.
32. Yu, X., Ye, L., Zhang, H., Zhao, J., Wang, G., Guo, C., & Shang, W. (2015). Ginsenoside Rb1 ameliorates liver fat accumulation by upregulating perilipin expression in adipose tissue of db/db obese mice. *Journal of Ginseng Research, 39*, 199–205.
33. Hong, Y., Lin, Y., Si, Q., Yang, L., Dong, W., & Gu, X. (2019). Ginsenoside Rb2 alleviates obesity by activation of brown fat and induction of browning of white fat. *Frontiers in Endocrinology, 10*.
34. Wang, Y., Dong, J., Liu, P., Lau, C. W., Gao, Z., Zhou, D., Tang, J., Ng, C. F., & Huang, Y. (2014). Ginsenoside Rb3 attenuates oxidative stress and preserves endothelial function in renal arteries from hypertensive rats. *British Journal of Pharmacology, 171*, 3171–3181.
35. Yang, J.-W., & Kim, S. S. (2015). Ginsenoside Rc promotes anti-adipogenic activity on 3T3-L1 adipocytes by down-regulating C/EBPα and PPARγ. *Molecules, 20*, 1293–1303.
36. Wu, C. F., Bi, X. L., Yang, J. Y., Zhan, J. Y., Dong, Y. X., Wang, J. H., Wang, J. M., Zhang, R., & Li, X. (2007). Differential effects of ginsenosides on NO and TNF-α production by LPS-activated N9 microglia. *International Immunopharmacology, 7*, 313–320.
37. Park, J.-S., Shin, J. A., Jung, J.-S., Hyun, J.-W., Van Le, T. K., Kim, D.-H., Park, E.-M., & Kim, H.-S. (2012). Anti-inflammatory mechanism of compound K in activated microglia and its neuroprotective effect on experimental stroke in mice. *Journal of Pharmacology and Experimental Therapeutics, 341*, 59–67.
38. Lee, I.-A., Hyam, S. R., Jang, S.-E., Han, M. J., & Kim, D.-H. (2012). Ginsenoside re ameliorates inflammation by inhibiting the binding of lipopolysaccharide to TLR4 on macrophages. *Journal of Agricultural and Food Chemistry, 60*, 9595–9602.
39. Lim, W.-C., Shin, E. J., Lim, T.-G., Choi, J. W., Song, N.-E., Hong, H.-D., Cho, C.-W., & Rhee, Y. K. (2022). Ginsenoside Rf enhances exercise endurance by stimulating myoblast differentiation and mitochondrial biogenesis in C2C12 myotubes and ICR mice. *Foods, 11*, 1709.
40. Tao, T., Chen, F., Bo, L., Xie, Q., Yi, W., Zou, Y., Hu, B., Li, J., & Deng, X. (2014). Ginsenoside Rg1 protects mouse liver against ischemia–reperfusion injury through anti-inflammatory and anti-apoptosis properties. *Journal of Surgical Research, 191*, 231–238.
41. Xue, Q., Yu, T., Wang, Z., Fu, X., Li, X., Zou, L., Li, M., Cho, J. Y., & Yang, Y. (2023). Protective effect and mechanism of ginsenoside Rg2 on atherosclerosis. *Journal of Ginseng Research, 47*, 237–245.
42. Lee, B., Sur, B., Park, J., Kim, S.-H., Kwon, S., Yeom, M., Shim, I., Lee, H., & Hahm, D.-H. (2013). Ginsenoside Rg3 alleviates lipopolysaccharide-induced learning and memory impairments by anti-inflammatory activity in rats. *Biomolecules and Therapeutics, 21*, 381–390.
43. Kim, T.-W., Joh, E.-H., Kim, B., & Kim, D.-H. (2012). Ginsenoside Rg5 ameliorates lung inflammation in mice by inhibiting the binding of LPS to toll-like receptor-4 on macrophages. *International Immunopharmacology, 12*, 110–116.
44. Jung, J.-S., Kim, D.-H., & Kim, H.-S. (2010). Ginsenoside Rh1 suppresses inducible nitric oxide synthase gene expression in IFN-γ-stimulated microglia via modulation of JAK/STAT and ERK signaling pathways. *Biochemical and Biophysical Research Communications, 397*, 323–328.
45. Hu, C., Yang, L., Wang, Y., Zhou, S., Luo, J., & Gu, Y. (2021). Ginsenoside Rh2 reduces m6A RNA methylation in cancer via the KIF26B-SRF positive feedback loop. *Journal of Ginseng Research, 45*, 734–743.
46. Lee, H. L., & Kang, K. S. (2017). Protective effect of ginsenoside Rh3 against anticancer drug-induced apoptosis in LLC-PK1 kidney cells. *Journal of Ginseng Research, 41*, 227–231.
47. To, K.-I., Zhu, Z.-X., Wang, Y.-N., Li, G.-A., Sun, Y.-M., Li, Y., & Jin, Y.-H. (2022). Integrative network pharmacology and experimental verification to reveal the

anti-inflammatory mechanism of ginsenoside Rh4. *Frontiers in Pharmacology, 13.*
48. Zheng, S., Xiao, S., Wang, J., Hou, W., & Wang, Y. (2019). Inhibitory effects of ginsenoside Ro on the growth of B16F10 melanoma *via* its metabolites. *Molecules, 24*, 2985.
49. Kang, J.-H., Song, K.-H., Woo, J.-K., Park, M. H., Rhee, M. H., Choi, C., & Oh, S. H. (2011). Ginsenoside Rp1 from *Panax ginseng* exhibits anti-cancer activity by down-regulation of the IGF-1R/Akt pathway in breast cancer cells. *Plant Foods for Human Nutrition, 66*, 298–305.
50. Liu, Y., Qu, L., Wan, S., Li, Y., & Fan, D. (2022). Ginsenoside Rk1 prevents UVB irradiation-mediated oxidative stress, inflammatory response, and collagen degradation *via* the PI3K/AKT/NF-κB pathway *In Vitro* and *In Vivo*. *Journal of Agricultural and Food Chemistry, 70*, 15804–15817.
51. Jiao, X., Jin, X., Ma, Y., Yang, Y., Li, J., Liang, L., Liu, R., & Li, Z. (2021). A comprehensive application: Molecular docking and network pharmacology for the prediction of bioactive constituents and elucidation of mechanisms of action in component-based Chinese medicine. *Computational Biology and Chemistry, 90*, 107402.
52. Li, S., & Zhang, B. (2013). Traditional Chinese medicine network pharmacology: Theory, methodology and application. *Chinese Journal of Natural Medicines, 11*, 110–120.
53. Li, L., Yang, L., Yang, L., He, C., He, Y., Chen, L., Dong, Q., Zhang, H., Chen, S., & Li, P. (2023). Network pharmacology: A bright guiding light on the way to explore the personalized precise medication of traditional Chinese medicine. *Chinese Medicine, 18*, 146.
54. Li, S. (1999). Possible relationship between traditional Chinese medicine ZHENG and molecular networks. In *Proceedings of the 1st academic annual meeting of the China Association for Science and Technology.*
55. Li, S. (2007). Framework and practice of network-based studies for Chinese herbal formula. *Zhong Xi Yi Jie He Xue Bao, 5*, 489–493.
56. Chandran, U., Mehendale, N., Patil, S., Chaguturu, R., & Patwardhan, B. (2017). Network pharmacology. *Innovative Approaches in Drug Discovery*, 127–164. https://doi.org/10.1016/B978-0-12-801814-9.00005-2
57. Wu, Y., Bai, Y., Lu, Y., Zhang, Z., Zhao, Y., Huang, S., Tang, L., Liang, Y., Hu, Y., & Xu, C. (2023). Transcriptome sequencing and network pharmacology-based approach to reveal the effect and mechanism of Ji Chuan Jian against Parkinson's disease. *BMC Complementary Medicine and Therapies, 23*, 182.
58. Zuo, H., Zhang, Q., Su, S., Chen, Q., Yang, F., & Hu, Y. (2018). A network pharmacology-based approach to analyse potential targets of traditional herbal formulas: An example of Yu Ping Feng decoction. *Scientific Reports, 8*, 11418.
59. Luo, T., Lu, Y., Yan, S., Xiao, X., Rong, X., & Guo, J. (2020). Network pharmacology in research of Chinese medicine formula: Methodology, application and prospective. *Chinese Journal of Integrative Medicine, 26*, 72–80.
60. A network pharmacology approach to understanding the mechanisms of action of traditional medicine: Bushenhuoxue formula for treatment of chronic kidney disease. *PubMed.* https://pubmed.ncbi.nlm.nih.gov/24598793/
61. Chen, C. Y.-C. (2011). TCM database@Taiwan: The world's largest traditional Chinese medicine database for drug screening in Silico. *PLOS ONE, 6*, e15939.
62. TCMID: Traditional Chinese medicine integrative database for herb molecular mechanism analysis. *Nucleic Acids Research | Oxford Academic.* https://academic.oup.com/nar/article/41/D1/D1089/1057998
63. Ru, J., Li, P., Wang, J., Zhou, W., Li, B., Huang, C., Li, P., Guo, Z., Tao, W., Yang, Y., Xu, X., Li, Y., Wang, Y., & Yang, L. (2014). TCMSP: A database of systems pharmacology for drug discovery from herbal medicines. *Journal of Cheminformatics, 6*, 13.
64. HIT: Linking herbal active ingredients to targets. *Nucleic Acids Research | Oxford Academic.* https://academic.oup.com/nar/article/39/suppl_1/D1055/2507403
65. Online Mendelian Inheritance in Man (OMIM), a knowledgebase of human genes and genetic disorders. *PubMed.* https://pubmed.ncbi.nlm.nih.gov/15608251/
66. The genetic association database. *PubMed.* https://pubmed.ncbi.nlm.nih.gov/15118671/
67. UniProt: The universal protein knowledgebase. *PubMed.* https://pubmed.ncbi.nlm.nih.gov/27899622/
68. STRING: A database of predicted functional associations between proteins. *Nucleic Acids Research | Oxford Academic.* https://academic.oup.com/nar/article/31/1/258/2401231
69. PubChem: A public information system for analyzing bioactivities of small molecules. *PubMed.* https://pubmed.ncbi.nlm.nih.gov/19498078/
70. GeneCards Version 3: The human gene integrator. *Database | Oxford Academic.* https://academic.oup.com/database/article/doi/10.1093/database/baq020/407450
71. KEGG: Kyoto encyclopedia of genes and genomes. *Nucleic Acids Research | Oxford Academic.* https://academic.oup.com/nar/article/28/1/27/2384332
72. DAVID bioinformatics resources: Expanded annotation database and novel algorithms to better extract biology from large gene lists. *Nucleic Acids Research | Oxford Academic.* https://academic.oup.com/nar/article/35/suppl_2/W169/2924156
73. Cytoscape StringApp: Network analysis and visualization of proteomics data. *Journal of Proteome Research.* https://pubs.acs.org/doi/10.1021/acs.jproteome.8b00702
74. Lee, K., Choi, Y.-J., Lim, H.-I., Cho, K. J., Kang, N., & Ko, S.-G. (2024). Network pharmacology study to explore the multiple molecular mechanism of SH003 in the treatment of non-small cell lung cancer. *BMC Complementary Medicine and Therapies, 24*, 70.
75. Wu, L., Wang, Y., & Fan, X. (2011). Tools for network pharmacology study: Network visualization and network analysis. *Zhongguo Zhong Yao Za Zhi, 36*, 2923–2925.
76. Liao, J., Wang, Q., Wu, F., & Huang, Z. (2022). In Silico methods for identification of potential active sites of therapeutic targets. *Molecules, 27*, 7103.
77. Noor, F., Tahir Ul Qamar, M., Ashfaq, U. A., Albutti, A., Alwashmi, A. S. S., & Aljasir, M. A. (2022). Network pharmacology approach for medicinal plants: Review and assessment. *Pharmaceuticals (Basel), 15*, 572.
78. Yu, T., & Zhang, Y.-X. (2022). Investigation of the pharmacological effect and mechanism of mountain-cultivated ginseng and garden ginseng in cardiovascular diseases based on network pharmacology and zebrafish experiments. *Frontiers in Pharmacology, 13.*
79. Ding, Q., Zhu, W., Diao, Y., & Shi, Y. (2021). Elucidation of the mechanism of action of Ginseng against acute lung injury/acute respiratory distress syndrome by a network pharmacology-based strategy. *Frontiers in Pharmacology, 11.*
80. *IJMS | Free full-text | Biochemical targets and molecular mechanism of ginsenoside compound K in treating osteoporosis based on network pharmacology.* https://www.mdpi.com/1422-0067/23/22/13921
81. Li, Q., Huang, Z., Liu, D., Zheng, J., Xie, J., Chen, J., Zeng, H., Su, Z., & Li, Y. (2021). Effect of berberine on hyperuricemia and kidney injury: A network pharmacology analysis

and experimental validation in a mouse model. *DDDT, 15*, 3241–3254.
82. Budovsky, A., & Fraifeld, V. (2012). Medicinal plants growing in the Judea region: Network approach for searching potential therapeutic targets. *Network Biology, 2*, 84–94.
83. Zhang, R., Zhu, X., Bai, H., & Ning, K. (2019). Network pharmacology databases for traditional Chinese medicine: Review and assessment. *Fronties in Pharmacology, 10*.
84. Federico, A., Fratello, M., Scala, G., Möbus, L., Pavel, A., Del Giudice, G., Ceccarelli, M., Costa, V., Ciccodicola, A., Fortino, V., Serra, A., & Greco, D. (2022). Integrated network pharmacology approach for drug combination discovery: A multi-cancer case study. *Cancers, 14*, 2043.
85. Umar, A., Dunn, B. K., & Greenwald, P. (2012). Future directions in cancer prevention. *Nature Review Cancer, 12*, 835–848.
86. Radovich, M., Clare, S. E., Atale, R., Pardo, I., Hancock, B. A., Solzak, J. P., Kassem, N., Mathieson, T., Storniolo, A. M. V., Rufenbarger, C., Lillemoe, H. A., Blosser, R. J., Choi, M. R., Sauder, C. A., Doxey, D., Henry, J. E., Hilligoss, E. E., Sakarya, O., Hyland, F. C., . . . Schneider, B. P. (2014). Characterizing the heterogeneity of triple-negative breast cancers using microdissected normal ductal epithelium and RNA-sequencing. *Breast Cancer Research and Treatment, 143*, 57–68.
87. Wang, X., Wang, Z.-Y., Zheng, J.-H., & Li, S. (2021). TCM network pharmacology: A new trend towards combining computational, experimental and clinical approaches. *Chinese Journal of Natural Medicines, 19*, 1–11.
88. Park, J. Y., Park, W. Y., Song, G., Jung, S. J., Kim, B., Choi, M., Kim, S. H., Park, J., Kwak, H. J., Ahn, K. S., Lee, J. H., & Um, J.-Y. (2023). *Panax ginseng* C.A. Meyer alleviates benign prostatic hyperplasia while preventing finasteride-induced side effects. *Frontiers in Pharmacology, 14*, 1039622.
89. Ding, Q., Zhu, W., Diao, Y., Xu, G., Wang, L., Qu, S., & Shi, Y. (2021). Elucidation of the mechanism of action of *Ginseng* against acute lung injury/acute respiratory distress syndrome by a network pharmacology-based strategy. *Frontiers in Pharmacology, 11*.
90. Chen, L., Wang, L., Ao, W., Chen, Y., Li, S., Huang, Z., Yu, D., Dong, Y., Gu, J., & Zeng, H. (2022). Bioinformatics study of the potential therapeutic effects of ginsenoside Rf in reversing nonalcoholic fatty liver disease. *Biomedicine & Pharmacotherapy, 149*, 112879.
91. Dela Cruz, C. S., Tanoue, L. T., & Matthay, R. A. (2011). Lung cancer: Epidemiology, etiology, and prevention. *Clinics in Chest Medicine, 32*. https://doi.org/10.1016/j.ccm.2011.09.001
92. Li, Y., Yan, B., & He, S. (2023). Advances and challenges in the treatment of lung cancer. *Biomedicine & Pharmacotherapy, 169*, 115891.
93. Araghi, M., Mannani, R., Heidarnejad Maleki, A., Hamidi, A., Rostami, S., Safa, S. H., Faramarzi, F., Khorasani, S., Alimohammadi, M., Tahmasebi, S., & Akhavan-Sigari, R. (2023). Recent advances in non-small cell lung cancer targeted therapy; an update review. *Cancer Cell International, 23*, 162.
94. Li, X., Cao, D., & Sun, S. (2023). Anticancer therapeutic effect of ginsenosides through mediating reactive oxygen species. *Frontiers in Pharmacology, 14*.
95. Majeed, F., Malik, F. Z., Ahmed, Z., Afreen, A., Afzal, M. N., & Khalid, N. (2018). Ginseng phytochemicals as therapeutics in oncology: Recent perspectives. *Biomedicine & Pharmacotherapy, 100*, 52–63.
96. Gao, D., Li, Y., Xiang, S., & Zhang, J. (2023). Key targets and molecular mechanisms of the fat-soluble components of *Ginseng* for lung cancer treatment. *Applied Biochemistry and Biotechnology, 195*, 6495–6515.
97. Li, Z.-H., Yu, D., Huang, N.-N., Wu, J.-K., Du, X.-W., & Wang, X.-J. (2021). Immunoregulatory mechanism studies of ginseng leaves on lung cancer based on network pharmacology and molecular docking. *Scientific Reports, 11*, 18201.
98. Kew, M. C. (2014). Hepatocellular carcinoma: Epidemiology and risk factors. *Journal of Hepatocell Carcinoma, 1*, 115–125.
99. Anwanwan, D., Singh, S. K., Singh, S., Saikam, V., & Singh, R. (2020). Challenges in liver cancer and possible treatment approaches. *Biochimica et Biophysica Acta (BBA) – Reviews on Cancer, 1873*, 188314.
100. Elshafay, A., Tinh, N. X., Salman, S., Shaheen, Y. S., Othman, E. B., Elhady, M. T., Kansakar, A. R., Tran, L., Van, L., Hirayama, K., & Huy, N. T. (2017). Ginsenoside Rk1 bioactivity: A systematic review. *PeerJ, 5*, e3993.
101. Yang, Y., Lee, J., Rhee, M. H., Yu, T., Baek, K.-S., Sung, N. Y., Kim, Y., Yoon, K., Kim, J. H., Kwak, Y.-S., Hong, S., Kim, J.-H., & Cho, J. Y. (2015). Molecular mechanism of protopanaxadiol saponin fraction-mediated anti-inflammatory actions. *Journal of Ginseng Research, 39*, 61–68.
102. Zhang, D., Wang, A., Feng, J., Zhang, Q., Liu, L., & Ren, H. (2019). Ginsenoside Rg5 induces apoptosis in human esophageal cancer cells through the phosphoinositide-3 kinase/protein kinase B signaling pathway. *Molecular Medicine Reports, 19*, 4019–4026.
103. Chen, C., Lv, Q., Li, Y., & Jin, Y.-H. (2021). The anti-tumor effect and underlying apoptotic mechanism of ginsenoside Rk1 and Rg5 in human liver cancer cells. *Molecules, 26*, 3926.
104. Qu, L., Liu, Y., Deng, J., Ma, X., & Fan, D. (2023). Ginsenoside Rk3 is a novel PI3K/AKT-targeting therapeutics agent that regulates autophagy and apoptosis in hepatocellular carcinoma. *Journal of Pharmaceutical Analysis, 13*, 463–482.
105. Iwu, C. D., & Iwu-Jaja, C. J. (2023). Gastric cancer epidemiology: Current trend and future direction. *Hygiene, 3*, 256–268.
106. Yao, H., Wan, J., Zeng, J., Huang, W., Sava-Segal, C., Li, L., Niu, X., Wang, Q., Wang, C., & Yuan, C. (2018). Effects of compound K, an enteric microbiome metabolite of ginseng, in the treatment of inflammation associated colon cancer. *Oncology Letters, 15*, 8339–8348.
107. Wang, C.-Z., Du, G.-J., Zhang, Z., Wen, X.-D., Calway, T., Zhen, Z., Musch, M. W., Bissonnette, M., Chang, E. B., & Yuan, C.-S. (2012). Ginsenoside compound K, not Rb1, possesses potential chemopreventive activities in human colorectal cancer. *International Journal of Oncology, 40*, 1970–1976.
108. Wan, Y., Liu, D., Xia, J., Xu, J.-F., Zhang, L., Yang, Y., Wu, J.-J., & Ao, H. (2022). Ginsenoside, CK., rather than Rb1, possesses potential chemopreventive activities in human gastric cancer via regulating PI3K/AKT/NF-κB signal pathway. *Frontiers in Pharmacology, 13*.
109. Sheng, G., Gao, Y., Yang, Y., & Wu, H. (2021). Osteosarcoma and metastasis. *Frontiers in Oncology, 11*, 780264.
110. Liu, Y., & Fan, D. (2020). The preparation of ginsenoside Rg5, its antitumor activity against breast cancer cells and its targeting of PI3K. *Nutrients, 12*, 246.
111. Liu, M., Jiang, D., Zhao, X., Zhang, L., Zhang, Y., Liu, Z., Liu, R., Li, H., Rong, X., & Gao, Y. (2024). Exploration in the mechanism of Ginsenoside Rg5 for the treatment of osteosarcoma by network pharmacology and molecular docking. *Orthopaedic Surgery, 16*, 462–470.

112. Sugandh, F., Chandio, M., Raveena, F., Kumar, L., Karishma, F., Khuwaja, S., Memon, U. A., Bai, K., Kashif, M., Varrassi, G., Khatri, M., & Kumar, S. (2023). Advances in the management of diabetes mellitus: A focus on personalized medicine. *Cureus*, *15*, e43697.
113. Raghavan, S., Vassy, J. L., Ho, Y., Song, R. J., Gagnon, D. R., Cho, K., Wilson, P. W. F., & Phillips, L. S. (2019). Diabetes mellitus–related all-cause and cardiovascular mortality in a national cohort of adults. *Journal of the American Heart Association*, *8*, e011295.
114. Mishra, V., Nayak, P., Sharma, M., Albutti, A., Alwashmi, A. S. S., Aljasir, M. A., Alsowayeh, N., & Tambuwala, M. M. (2021). Emerging treatment strategies for diabetes mellitus and associated complications: An update. *Pharmaceutics*, *13*, 1568.
115. Su, W.-Y., Li, Y., Chen, X., Li, X., Wei, H., Liu, Z., Shen, Q., Chen, C., Wang, Y.-P., & Li, W. (2021). Ginsenoside Rh1 improves type 2 diabetic nephropathy through AMPK/PI3K/Akt-mediated inflammation and apoptosis signaling pathway. *The American Journal of Chinese Medicine*, *49*, 1215–1233.
116. Li, S., Tang, S.-H., Liu, J.-L., Su, J., & He, F.-Y. (2018). Ginseng prescription rules and molecular mechanism in treating coronary heart disease based on data mining and integrative pharmacology. *Zhongguo Zhong Yao Za Zhi*, *43*, 1303–1309.
117. Jiashuo, W. U., Fangqing, Z., Zhuangzhuang, L. I., Weiyi, J., & Yue, S. (2022). Integration strategy of network pharmacology in Traditional chinese medicine: A narrative review. *Journal of Traditional Chinese Medicine*, *42*, 479–486.
118. Zhang, A., Fang, H., Chen, J., He, L., & Chen, Y. (2020). Role of VEGF-A and LRG1 in abnormal angiogenesis associated with diabetic nephropathy. *Frontiers in Physiology*, *11*.
119. Kutluay, V. M., & Diker, N. Y. (2020). Constitution of a comprehensive phytochemical profile and network pharmacology based investigation to decipher molecular mechanisms of *Teucrium polium* L. in the treatment of type 2 diabetes mellitus. *PeerJ*, *8*, e10111.
120. Liu, Y., Zhang, J., An, C., Liu, C., Zhang, Q., Ding, H., Ma, S., & Xue, W. (2023). Identification of potential mechanisms of Rk1 combination with Rg5 in the treatment of type II diabetes mellitus by integrating network pharmacology and experimental validation. *International Journal of Molecular Sciences*, *24*, 14828.
121. Simon, D. K., Tanner, C. M., & Brundin, P. (2020). Parkinson disease epidemiology, pathology, genetics and pathophysiology. *Clinics in Geriatric Medicine*, *36*, 1–12.
122. Van Kampen, J. M., Baranowski, D. B., Shaw, C. A., & Kay, D. G. (2014). *Panax ginseng* is neuroprotective in a novel progressive model of Parkinson's disease. *Experimental Gerontology*, *50*, 95–105.
123. Cho, I.-H. (2012). Effects of *Panax ginseng* in neurodegenerative diseases. *Journal of Ginseng Research*, *36*, 342–353.
124. Zhang, W., Chen, J., & Liu, H. (2022). Network pharmacology and molecular docking-based prediction of the molecular targets and signaling pathways of ginseng in the treatment of Parkinson's disease. https://journals.sagepub.com/doi/10.1177/1934578X221102029
125. Gong, C.-X., Dai, C.-L., Liu, F., & Iqbal, K. (2022). Multi-targets: An unconventional drug development strategy for Alzheimer's disease. *Frontiers in Aging Neuroscience*, *14*, 837649.
126. Wang, Z., Zhang, Z., Liu, J., Guo, M., & Li, H. (2023). *Panax ginseng* in the treatment of Alzheimer's disease and vascular dementia. *Journal of Ginseng Research*, *47*, 506–514.
127. Li, N., Duan, Y., Chen, L., & Zhang, K. (2023). Iron metabolism: An emerging therapeutic target underlying the anti-Alzheimer's disease effect of ginseng. *Journal of Trace Elements in Medicine and Biology*, *79*, 127252.
128. Majerníková, N., den Dunnen, W. F. A., & Dolga, A. M. (2021). The potential of ferroptosis-targeting therapies for Alzheimer's disease: From mechanism to transcriptomic analysis. *Frontiers in Aging Neuroscience*, *13*.
129. The emerging epidemic of cardiovascular risk factors and atherosclerotic disease in developing countries. *PubMed*. https://pubmed.ncbi.nlm.nih.gov/28232018/
130. Cardiovascular disease in Europe 2014: Epidemiological update. *PubMed*. https://pubmed.ncbi.nlm.nih.gov/25139896/
131. Ginsenosides in vascular remodeling: Cellular and molecular mechanisms of their therapeutic action. *PubMed*. https://pubmed.ncbi.nlm.nih.gov/33964471/
132. Irfan, M., Kwak, Y.-S., Han, C.-K., Hyun, S. H., & Rhee, M. H. (2020). Adaptogenic effects of *Panax ginseng* on modulation of cardiovascular functions. *Journal of Ginseng Research*, *44*, 538–543.
133. Herbal medicine for cardiovascular diseases: Efficacy, mechanisms, and safety. *PMC*. https://www.ncbi.nlm.nih.gov/pmc/articles/PMC7155419/
134. Dokainish, H., Teo, K., Zhu, J., Roy, A., AlHabib, K. F., ElSayed, A., Palileo-Villaneuva, L., Lopez-Jaramillo, P., Karaye, K., Yusoff, K., Orlandini, A., Sliwa, K., Mondo, C., Lanas, F., Prabhakaran, D., Badr, A., Elmaghawry, M., Damasceno, A., Tibazarwa, K., . . . Mondo, C. (2017). Global mortality variations in patients with heart failure: Results from the International Congestive Heart Failure (INTER-CHF) prospective cohort study. *The Lancet Global Health*, *5*, e665–e672.
135. Liu, L., Hu, J., Mao, Q., Liu, C., He, H., Hui, X., Yang, G., Qu, P., Lian, W., Duan, L., Dong, Y., Pan, J., Liu, Y., He, Q., Li, J., & Wang, J. (2022). Functional compounds of ginseng and ginseng-containing medicine for treating cardiovascular diseases. *Frontiers in Pharmacology*, *13*.
136. Dai, T., Gong, J., & Liu, S. (2023). Prediction of molecular mechanism of processed ginseng in the treatment of heart failure based on network pharmacology and molecular docking technology. *Medicine*, *102*, e36576.
137. Eastell, R., O'Neill, T. W., Hofbauer, L. C., Langdahl, B., Reid, I. R., Gold, D. T., & Cummings, S. R. (2016). Postmenopausal osteoporosis. *Nature Reviews Disease Primers*, *2*, 1–16.
138. Gallagher, J. C., & Tella, S. H. (2014). Prevention and treatment of postmenopausal osteoporosis. *The Journal of Steroid Biochemistry and Molecular Biology*, *142*, 155–170.
139. Siddiqi, M. H., Siddiqi, M. Z., Ahn, S., Kang, S., Kim, Y.-J., Sathishkumar, N., Yang, D.-U., & Yang, D.-C. (2013). Ginseng saponins and the treatment of osteoporosis: Mini literature review. *Journal of Ginseng Research*, *37*, 261–268.
140. Guo, L., Zhen, Q., Zhen, X., Cui, Z., Jiang, C., Zhang, Q., Gao, K., Luan, D., & Zhou, X. (2022). A network pharmacology approach to explore and validate the potential targets of ginsenoside on osteoporosis. *International Journal of Immunopathology and Pharmacology*, *36*, 03946320221107239

30 Occurrence, Functional Properties of Plant Vitamins and Biotechnological Approaches for Their Improved Production

P.V. Sreelakshmi, P.T. Vismaya, Srivatsa Udupa, Manoj Kumar, Sachin Ashok Thorat, and Annamalai Muthusamy

30.1 INTRODUCTION

A healthy human diet consists of a profound share of vitamins, which are relatively in highly complex organic compounds especially derived from plants. Most of the vitamins cannot be synthesized by the human body so they are taken up through diet (except some vitamins like vitamin D, B3, and C (Maqbool et al., 2018)). They help in maintaining regular growth, metabolism, and appropriate functioning of the body. A Polish biochemist Casmir Funk coined the word vitamin from *'vitamine'* in 1912 (Funk, 1912). Initially, vitamins were labeled alphabetically as their structural and chemical forms were least known. Once the novel versions of vitamin B started unraveling, naming was a bit difficult in identifying specific structures in between the categories and understanding their chemical functions. As science advanced, they were also named by considering function, occurrence, source, etc. Vitamins are of two types: fat-soluble (A, D, E, and K) and water-soluble (B1, B2, B3, B5, B6, B7, B9, B12, and C family) (Fitzpatrick et al., 2012) all total as thirteen. There are also specific dietary sources that are not true vitamins. Vitamin level is mandatory from the very beginning, even in the fetal stage. A pregnant lady should have a balanced vitamin (mainly vitamin B) intake to avoid defects like neural tube abnormality in the fetus (Wright et al., 2007). Vitamin deficiency can promote many challenging diseases which can even lead to fatal effects at times (Kraemer et al., 2008). The trend shows that developed countries have a marginal effect on vitamin deficiency due to new strategies like imposing government strategies by providing a sufficient amount of food and the addition of vitamins in common food, called fortification.

Plants are the most prominent source of vitamins for humans and animals. Artificially, crop plants are transformed to possess high vitamin sufficiency in them, and the process is termed biofortification (Fang et al., 2019). Such types of transgenic plants like golden rice (a rich source of vitamin A) are introduced in the market very successfully (Storozhenko et al., 2007). The main agenda of plant genome sequencing is to study the occurrence and biosynthesis of vitamins in plants (Boubakri et al., 2016). In fat-soluble vitamins, vitamin A is the most complex one. There are two forms of vitamin A based on sources (plant – provitamin A carotenoids, and animal-retinoids). They are well known for maintaining proper vision, aiding the immune system and inflammatory cells, cell growth and development, cell communication, etc. Daily uptake for male $900\mu g$, female $700\mu g$, children 300 to $400\mu g$. Less intake of vitamin A causes diseases like night blindness and over intake causes hypervitaminosis A. The main food sources are sweet potatoes, carrots, spinach, eggs, etc. (Ross, 2010). Vitamin D is very essential for the human body for skin synthesis with the help of sunlight (as cholecalciferol is produced by converting 7-dehydrocholesterol by the action of ultraviolet B (UVB) wavelength in the skin) and the quantification of the exact level in the body is laborious. This production of vitamin D by skin ranges from person to person depending on their health and the texture of the skin. Cholecalciferol is just a precursor of vitamin D. The complete conversion to vitamin D is done by the kidney and liver. Teenage boys and females should consume 15 micrograms (600 international units) of calcium every day (Khadilkar et al., 2017). Egg yolks, tuna, salmon, sardines, mushrooms, cow milk, soy milk, orange juice, and fortified foods are all good sources of vitamin D.

Vitamin E refers to a group of four tocopherols and four tocotrienols that make up the vitamin. Tocochromanols are the collective name for these eight chemicals. LDL (low-density lipoprotein) cholesterol is destroyed with oxidative damage by free radicals which result in the thickening of blood vessels and arteries. Vitamin E protects LDL cholesterol against oxidative damage produced by free radicals. Vitamin E deficiency has also been linked to heart attacks, cancer, stroke, fibrocystic breast disease, and epilepsy. Vitamin E intake is advised at 15 milligrams per day for adults and 5 milligrams per day for children as per the survey by WHO Foods. Sunflower seeds, spinach, swiss chard,

DOI: 10.1201/9781003475491-30

turnip greens, asparagus, beet greens, mustard greens, chili peppers, almonds, broccoli, and bell peppers are all good sources of vitamin E (Niki & Traber, 2012). Vitamin K is the label on the blood clotting factors since it is derived from the German word 'coagulation', which means 'blood clotting' (Shearer & Newman, 2014). Vitamin K is subdivided into three types: K1, K2, and K3. Blood clotting is both beneficial and also non-beneficial (Hirota et al., 2013). A set of twenty different proteins accompanies the formation of blood clotting in which the four subtypes of Vitamin K have a prime role. Carrying mothers are advised to have more vitamin K as the fetus already gets less of it because there is a preventive barrier of the placenta to reduce intake of vitamin K (Shearer & Newman, 2014). Vitamin K intake should not exceed 80 milligrams per day. WHO Foods recommends a daily dose of 90 micrograms for women and 120 micrograms for men, with 55 micrograms for youngsters.

In water-soluble vitamins, vitamin B1 is important because it serves as a gatekeeper between the carbohydrate breakdown (low energy step), Kreb cycle (high energy phase), and electron transport chain. As a result, this vitamin is essential for energy metabolism, and a deficit in it can cause major problems (Bâ, 2008). Vitamin B1 aids the functioning of the neurological system, stomach, and heart. As the proper working of the brain is highly essential, collection of vitamin B1 is a prior necessity of the brain. Parkinson's, Alzheimer's, and alcohol-related brain diseases are also linked with vitamin B1 deficiency (Keogh et al., 2012). Glutathione is a requisite anti-oxidant for checking unwanted free radical attacks in the body, and the production of glutathione is done with the help of vitamin B1. It is also having a role in the production of iron, which is very necessary for the synthesis of heme in red blood cells. The deficiency of this vitamin can thus lead to anemia. Migraine headaches, iron deficiency anemia, excessive homocysteine, congestive heart failure, Parkinson's disease, cataracts, and hypertension are all linked to a shortage of this vitamin (Rivlin & Pinto, 2007).

Niacin and vitamin B3 are categorized together as niacin and their chemical forms have vitamin like actions too. Nicotinic acid, nicotinamide, and a variety of other enzymatic forms make up vitamin B3. This vitamin is involved in forming a barrier against antioxidants and its deficiency can cause pellagra (Lanska, 2010). Two different types of B3 (Nicotinamide adenine dinucleotide (NAD) and nicotinamide adenine dinucleotide phosphate (NADP)) can produce energy from proteins, lipids, and carbohydrates (Lešková et al., 2006). A deficient level of pantothenic acid can cause deformed hormone preparation and inappropriate working of hormones. Pantothenic acid is incorporated into Coenzyme A (CoA), a key enzyme in energy metabolism. Acne vulgaris, chronic weariness, excessive cholesterol, and diabetes-related foot ulcers are all linked to a lack of this vitamin (Kelly, 2011). B6 is a crucial vitamin in the synthesis of red blood cells, glucose metabolism, liver detoxification, brain and nervous system health. Neurotransmitter synthesis is also maintained by this vitamin by preparing dopamine, Gamma aminobutyric acid (GABA), and serotonin which is essential for message transfer from the brain (Beaudet et al., 2016). Folates play an important role in message transfer. Nerves employ messenger molecules to send messages throughout the body, and these molecules are created. The BH4 cycle (tetrahydrobiopterin) demonstrated that folates and neurotransmitters, particularly serotonin and dopamine have a strong relationship as they use specific methods to break brain barriers and pass through them. BH4 and folates exploited a similar method to pass the blood-brain barrier. Folates helps to maintain the cardiovascular and nervous system (Crider et al., 2011). A specific form of folate (5-methyltetrahydrofolate (5-MTHF)) maintains homocysteine levels; the higher the homocysteine level (hyperhomocysteinemia), the more the risk level for cardiovascular diseases (Hayden & Tyagi, 2004).

Aside from these characteristics, this vitamin is critical for human cardiovascular health. Vitamin B12 is involved in fromation of red blood cells, which is compulsory in oxygen transportation, by producing its elementary unit, succinyl COA. Homocysteine levels in the blood are linked to a variety of disorders, including peripheral vascular disease, stroke, and coronary heart disease. So, homocysteine level is maintained by vitamin B12 which helps in the conversion of homocysteine to methionine (Watanabe, 2007). Biotin, a B-complex member, was previously known as coenzyme R, vitamin H, and vitamin B7, but is now known as biotin which cannot be excluded in sugar and fat synthesis. Once it is deficient the body's sugar levels alter and insulin (a hormone involved in the maintenance of sugar levels) production and function are dismayed. Biotin deficiency can also cause skin rashes because it is required for fat deposition in the skin and also helps in skin moisturizing and suppling, whereas skin that lacks them becomes irritated, flaky, or scaly. Cradle cap symptoms in nursing newborns can be alleviated by increasing biotin consumption in lactating mothers' meals.

30.2 VITAMINS AND HUMAN HEALTH

Vitamins are essential for proper human functioning and the prevention of many illnesses since they are involved in cellular processes. A vitamin deficit can have a number of negative effects on health. The vitamins are also implemented in the treatment of viral infections. In 1996, sub-Saharan Africa was suffering from the widely spread HIV infection (Hunter et al., 1993). It was so severe as up to 105% of the total infection in the world, with middle-aged women in prominence. Clinical trials on vitamin supplements were one main category of treatment method to seal the effect of virus transmission and immune boosting. The Tanzanian women with HIV who gave birth showed a high level of infectious rate (Rosen et al., 1994). The main drug used for HIV is Azidothymidine (AZT), which is very expensive and not so easily available. So, the usage of certain vitamins is a very essential supplement to maintain the course of action of the immune system. The Tanzanian population needed a modular and cheap alternative is much needed. As the clinical trials proved the aid of vitamin supplement efficiency, they tried the usage of vitamins too (O'Connor et al., 2005). Vitamin introduction is increased in mammals (except humans) with an increase in stress levels too. As discussed, the enzymatic functions and antioxidant properties of vitamin C are very well known. The infection of sepsis is vulnerable when vitamin C is less in the body system, which is a prime cause for high production of oxygen species with endothelium spread well which may even cause injury in mitochondria and other cellular compartments (Drouin et al., 2011). This, later on, may even open to organ failures. In certain clinical trials, it was shown that the usage of vitamin C as a nutrient provider during sepsis will help in bringing down the inflammation caused. Shreds of evidence also suggest that corticosteroids, thiamine, etc. with vitamin C will help in the further development of certain drugs to reactivate the damaged organs (Fowler III et al., 2017).

In 2019 December, a few patients were hospitalized in the city of Wuhan, China with symptoms starting from a mild cold, and fever which further developed into severe respiratory tract disorder and even led to death (Jovic et al., 2020). The respiratory syndrome SARS–CoV (coronavirus-like illness) was caused by the RNA virus which is very virulent with high mutation capacities (Duffy et al., 2018). The virus along with its associated mutants caused severe diseases like SARS-CoV, Middle Eastern Respiratory Syndrome Coronavirus (MERS-CoV), and H1N1 influenza A, etc. It was initially spread in the local areas which later led to the mass spreading in the Asian countries and even then, to the Middle Eastern nations, European countries, etc. The fatal rate caused by the virus is 2.3% than expected (Lewis et al., 2019). The symptoms are now grown into unpredictable levels, but still commonly seen ones are, asymptomatic carriage, mild upper respiratory tract infection (URTI), severe viral pneumonia to acute respiratory distress syndrome (ARDS), and death (Junaid et al., 2021). The specification of the virus with the host is still an unveiled mask. There are vaccines modified for the same but still, their efficiency is not yet predictable due to lack of proper clinical trials. The role of vitamins and minerals supplementing or immune-nutrition was studied much with more vitality in this viral season as no effective proper treatment methodology is found. The potentiality of certain vitamins, their antimicrobial property, and their antioxidant roles are studied (Singh, 2020).

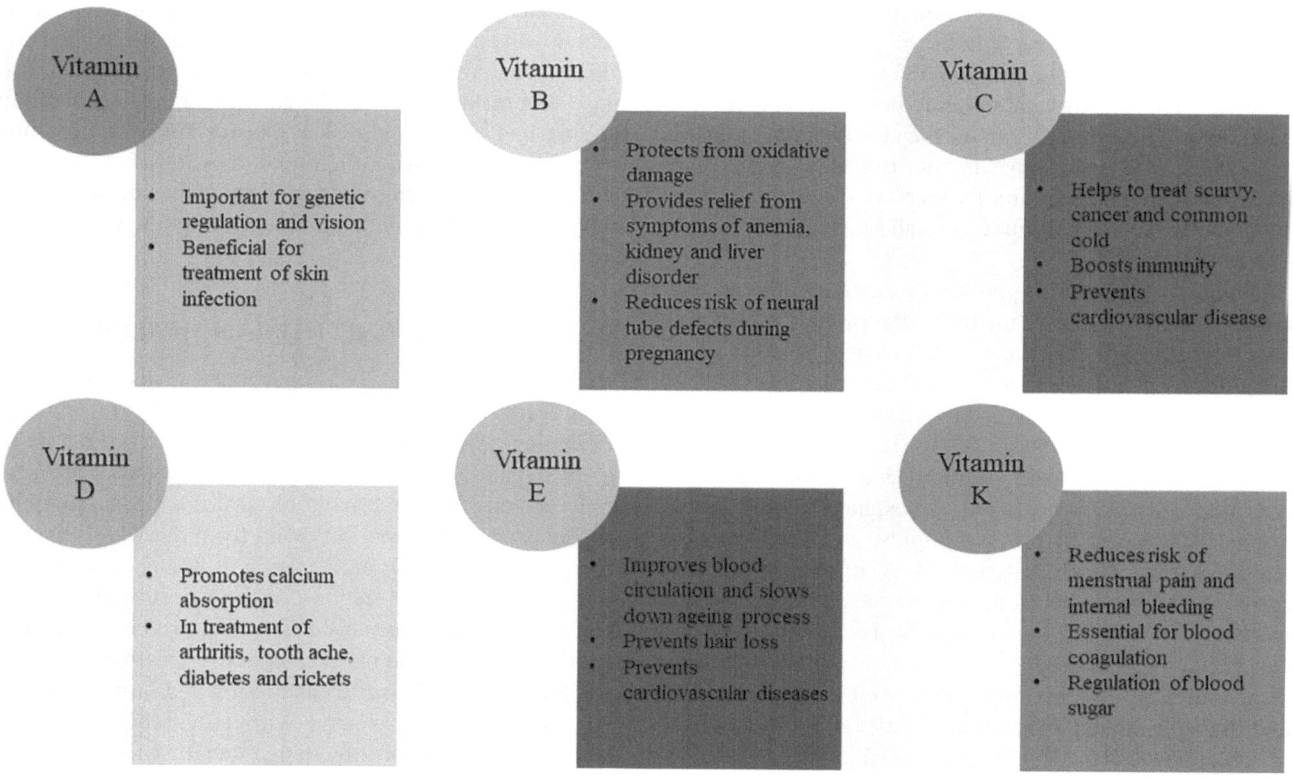

FIGURE 30.1 Benefits of vitamins on human health

It is seen that in certain diseases the vitamin A level lowering may lead to high peak induction in disease. There are certain drug theories based on vitamin B mainly, B12 (An et al., 2019). The complex role of vitamin C in the maintenance of lung health is already studied well (Izquierdo et al., 2019). The Front Line COVID-19 Critical Care (FLCCC) Working Group released a set of strategies to check on the disease by facilitating the usage of drug supplements with vitamin C, D, zinc, iron, etc. The protocol also talks about drugs with major ingredients such as intravenous methylprednisolone, high-dose intravenous ascorbic acid, full-dose low-molecular-weight heparin, and optional treatment components. All the trials and studies show and prove the major role played by each vitamin and their importance in controlling the virus and its after-effects (Singh, 2020). The body system requires immunity from birth till the end of its life span. The neural system has a primary need for vitamins in day-to-day life. The glial cells and neurons work properly only with the help of certain vitamins, namely vitamin D. The reduced intake of vitamin D also leads to neural system failure or viral and bacterial infections which may also lead to many infectious diseases (Fiorino et al., 2020).

30.3 OCCURRENCE OF VITAMINS IN PLANTS

The main food sources of these vitamins are lentils, asparagus, spinach, turnip greens, broccoli, beets, bokchoy, cauliflower, parsley, pinto beans, garbanzo beans, black beans, navy beans, kidney beans, papaya, brussels sprouts, green peas, bell peppers, green beans, celery, cabbage, summer squash, strawberries, tomatoes, leeks, fennel, lima beans, dried peas, avocado, peanuts, sunflower seeds. (Zempleni et al., 2009). Vitamin C is the most well-known nutrient, and citrus fruits are a good source of it and thus it is merely included in the diet. Vitamin C possesses antioxidant capabilities that protect the lens of the eyes, bloodstream molecules, and DNA from free radical damage. It is easily taken up by intestinal walls. Vitamin C is necessary for the formation of collagen, a structural component of the human body. Less dose of vitamin C leads to scurvy. Papaya, bell peppers, broccoli, brussels sprouts, strawberries, pineapple, oranges, kiwifruit, cantaloupe, cauliflower, kale, cabbage, bokchoy, grapefruit, parsley, turnip greens are all good sources of vitamin C (Carr & Maggini, 2017).

Death intensity rate increased up to millions due to severe illnesses. The immune system is being controlled with proper functioning with the aid of several vitamins present in daily dietary regimes. Specialized research and surveys show that vitamins A, B6, B12, C, D, and E are very necessary for maintaining the proper functioning of the human body (Calder et al., 2020). Improper vitamin intake can lead to specific viral diseases like respiratory diseases like bronchitis, HIV infection, sepsis, etc. along with specific neurological problems like Alzheimer's, Parkinson's disease, etc. To avoid the deficiency of vitamins and to have a high nutritional and well-built body, crops are genetically modified. Production of quality protein maize (QPM), breeding strategies in sunflower, improved rice strain (golden rice), etc. are the main genetic modifications being done in plants for improved quality vitamins. This helps in supplementing Recommended Dietary Allowances to maintain safer limits of vitamin intake. Usually, this process is artificially done by the collaboration of public health and government authorities by biofortification (Calder et al., 2020). A distinct set of plants are tagged to study later and modify the micronutrient content in them. These plants are usually consumed by the populace more (can be their staple food) so that the desired amount of daily dietary intake reaches them appropriately (Šramková et al., 2009). Transgenic plants with evolved vitamin synthesis pathways have been very keenly checked in recent times. Still, the development in this field is a long way. The studies reveal that the improvement in the technological aspects of vitamin enrichment in crops by genetic modification leads to a promising steppingstone in eradicating vitamin deficiencies and health issues associated with them. (Herbers, 2003). Along with this, their relevance in pharmaceutical industries due to bioactivity is to be mentioned chiefly.

Plants synthesize a range of vitamins that are crucial not only for human metabolism but also for their growth and development, vitamins play a significant role due to their redox chemistry and as cofactors, and some possess potential antioxidant properties (Rubio et al., 2008). Under stress conditions, vitamins induce an antioxidant response in plants (Tunc-Ozdemir et al., 2009). Vitamins are essential for plant growth and development disruptions in the synthesis of vitamins B1, B2, B3, B5, B6, B7, B9, and E cause embryo-or early-seedling-lethal pheno-types; treatment with the corresponding vitamin can rescue the phenotype (Hanson & Gregory, 2011). Thiamine reduces membrane permeability, increases photo-synthetic pigment accumulation, activates antioxidant enzymes, and promotes maize growth (Kaya et al., 2015). Flowering plants or angiosperms are reported to be an excellent source of various vitamins that are essential for human health (Table 1). Vitamin C or the L- ascorbic acid is the most prevalent water-soluble substance, which is also a vital micronutrient and a key component for the metabolisms in living organisms (Paciolla et al., 2019). Due to its capacity to donate reducing equivalents and the relative stability of the resulting monodehydroascorbate radical, vitamin C (L-ascorbic acid) is a powerful free radical scavenger. However, vitamin C is not only an antioxidant, since it is also a cofactor for numerous enzymes involved in plant and human metabolism. In humans, vitamin C takes part in various physiological processes, such as immune stimulation, collagen synthesis, iron absorption, and epigenetic regulation (Paciolla et al., 2019). Aerial parts of plants such as leaves, flowers and fruits have been reported to contain Vitamin C abundantly. Further, plants namely, *Citrus limon (*Funk, 1912), *Carica papaya* (Prigge et al., 1999), *Ipomoea batatas* (Cartabiano et al., 2020), *Vitis vinifera* (Langová et al., 2020) and *Citrus*

TABLE 30.1
List of plants, types of vitamin and its source from flowering plants.

S. no.	Botanical names of the plants and their families	Type of vitamins	Organ types/whole plant	References
1	*Citrus limon* (Rutaceae)	Vitamin C	Fruits	Funk (1912)
2	*Glycine max* (Fabaceae)	Vitamin B12	Seeds, roots	Mozafar and Oertli (1992)
3	*Elaeis guineensis* (Aracaceae)	Tocotrienols Vitamin E	Fruits, seeds	Ebong et al. (1999)
4	*Spinacia oleracea* (Amaranthaceae)	Phylloquinone/vitamin K1	Leaves	Garber et al. (1999)
5	*Carica papaya* (Caricaceae)	Vitamin C	Fruits	Prigge et al. (1999).
6	*Psidium guajava* (Myrtaceae)	Vitamin C	Fruits, leaves, flowers	Breithaupt and Bamedi (2001)
7	*Brassica oleracea* (Brassicaceae)	Vitamin C	Leaves	Breithaupt and Bamedi (2001)
8	*Cucurbita* sp. (Cucurbitaceae)	α carotene, β cryptoxanthin	Leaves, flowers	Breithaupt and Bamedi (2001)
9	*Daucus carota* (Apiaceae)	α carotene, β carotene, vitamin B1 and niacin	Fruits	Breithaupt and Bamedi (2001); Leclerc et al. (1991)
10	Solanaceae, Cucurbitaceae, Fabaceae (Leguminoseae), and Poaceae species	25[OH]D and 1,25[OH]2D Active form of vitamins D2 and D3	Leaves	Boland (1986); Boland et al. (2003)
11	*Solanum glaucophyllum* (Solanaceae)	Vitamin D3	Leaves	Boland et al. (2003)
12	Green leafy vegetables	Vitamin K	Leaves	Vermeer (2004).
13	*Pisum sativum* (Fabaceae)	Phylloquinone Vitamin K1	Fruits	Damon et al. (2005)
14	*Cucumis sativus* (Cucurbitaceae)	Phylloquinone	Fruits	Damon et al. (2005)
15	*Solanum tuberosum* (Solanaceae)	Phylloquinone	Tubers	Damon et al. (2005)
16	*Triticum aestivum* (Poaceae)	Phylloquinone	Seeds	Damon et al. (2005)
17	*Psidium guajava* (Myrtaceae)	Vitamin C	Fruits	Kamath et al. (2008)
18	*Arabidopsis thaliana* (Brassicaceae)	Vitamin B1	Leaves, flowers	Tunc-Ozdemir et al. (2009)
19	*Solanum lycopersicum* (Solanaceae)	Vitamin E	Fruit	Raiola et al. (2015)
20	*Brassica oleracea* (Brassicaceae)	Vitamin K, C, E	Leaves	Nagraj et al. (2020); Lee et al. (2000)
21	*Malus domestica* (Rosaceae)	Vitamin B12	Fruit	Patocka et al. (2020)
22	*Ipomoea batatas* (Convolvulaceae)	Vitamins C, A		Cartabiano et al. (2020)
23	*Vitis vinifera* (Vitaceae)	Vitamin C	Seeds	Langova et al. (2020)
24	*Nelumbo nucifera* (Nelumbonaceae)	Vitamin C	Flower	Akhter et al. (2021)
25	*Zea mays* (Poaceae)	Vitamins E, A		Das et al. (2021)
26	*Citrus sinensis* (Rutaceae)	Vitamin C	Fruit	Tiencheu et al. (2021)
27	*Citrus limon* (Rutaceae)	Vitamin C	Fruit	Tiencheu et al. (2021)
28	*Anacardium occidentale* (Anacardiaceae)	Vitamin C	Fruit	Assoi et al. (2021)
29	*Cucurbita pepo* (Cucurbitaceae)	Vitamin E	Fruit	Ermis and Yanmaz (2022)
30	*Pisum sativum* (Fabaceae)	Vitamin B	Seeds	İdikut (2023)
31	*Mangifera indica* (Anacardiaceae)	Vitamins A, K, E	Fruit	Rajasekaran and Soundarapandaian (2023)

sinensis (Tiencheu et al., 2021) had been described for their vitamin C content.

Humans are unable to synthesize vitamin A de novo, so the vitamin is metabolically derived from oxidative cleavage of provitamin A carotenoids (Figure 30.2). *Psidium guajava* and *Brassica oleracea* have been excellent sources of carotenoids and β-cryptoxanthin esters, which act as sources of provitamin A and carotenoids in the human diet (Breithaupt & Bamedi, 2001). The carotenoids such as α-carotene, β-carotene and β-cryptoxanthin are referred to as provitamin A as these compounds are the key precursors for vitamin A synthesis. Moreover, β-carotene, α-carotene, and β-cryptoxanthin found mostly in green, leafy, yellow-colored vegetables and orange fruits, can be internally converted to vitamin A (Grune et al., 2010). The leaves, fruits and flowers of the plant *Cucurbita* spp., *Daucus carota* (Breithaupt & Bamedi, 2001), *Ipomoea batatas* (Cartabiano et al., 2020) and *Mangifera indica* (Rajasekaran & Soundarapandian, 2023) was reported as a rich source of vitamin A. It has been studied that, *Zea mays*, *Brassica oleracea*, *Solanum lycopersicum*, *Cucurbita pepo*, and *Elaeis guineensis* are good sources of vitamin E (Ebong et al., 1999; Nagraj et al., 2020; Das et al., 2021; Ermiş & Yanmaz, 2022). Further, vitamin K (VK) is a fat-soluble vitamin that functions as a coenzyme and is involved in the synthesis of several proteins involved in blood clotting and bone metabolism. It may also have a role in osteoporosis and atherosclerosis. Green leafy and flowery vegetables, including broccoli, broccoli

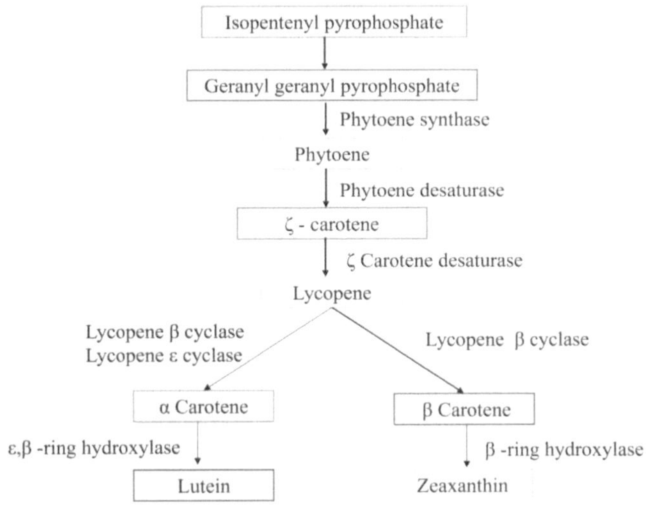

FIGURE 30.2 Biosynthesis of vitamin A in plants (adapted and modified from Srinivasan & Buys, 2019).

raab, spinach, and certain lettuces, have been a great source of vitamin K1 (Damon et al., 2005). Phylloquinone is a prenylated naphthoquinone that plays the role of vitamin K in humans, which are majorly produced by terrestrial plants, aquatic green algae, and some of the blue-green algal (cyanobacteria) members. Vegetable oil and green leafy vegetables (Vermeer, 2004) are important dietary sources of vitamin K (Basset et al., 2017). The various parts of plants like fruit, tuber, seed, and leaves are known to contain vitamin K. *Pisum sativum*, *Cucumis sativus*, *Solanum tuberosum*, *Triticum aestivum* (Damon et al., 2005), *Spinacia oleracea* (Garber et al., 1999), and *Mangifera indica* (Rajasekaran & Soundarapandian, 2023) are rich sources of phylloquinone. Vitamin B12 is mostly obtained from diets derived from animals, while certain plants can also provide significant levels of this vitamin (Watanabe et al., 2014). Mozafar and Oertli (1992) and Patocka et al. (2020) studied the presence of vitamin B12 *in Glycine max* (roots and seeds) and *Malus domestica* (fruit).

30.4 BREEDING STRATEGIES FOR IMPROVED VITAMIN CONTENT FOR TREATING VIRAL DISEASE

The breeding strategies are maintained based on each plant's characteristic features like flowering, hybridization, etc. Reliable techniques are used for crop management to utilize breeding strategies that are useful for farmers. For this, genetic libraries are considered to get the exact genomic sequences for each species, and the desired ones are selected. In African species, high heterozygosity was seen while crossbreeding between the species with the same evolutionary line. The tolerance and susceptibility for certain viruses like cassava mosaic disease (CMD) were also established through this. This was procured with the help of cycle selections and recombination DNA technology. The genetic changes made them achieve the best yield with a high-performance range (Jennings & Iglesias, 2002). The process of biofortification plays a major role in this process. Biofortification is carried out by many plant breeding and transgenic strategies, which develop the characteristic traits and upgrade to the betterment of the crop. The golden development of golden rice is the most vital strategy ever concocted. Development in the technical and scientific aspects has created a new pathway for enriched nutrient diets. These are mainly considered by checking on the effect on juvenile age because the growth curve starts from this group so they should be well nourished, and the effect on them is of prior importance. (Welch & Graham, 2004).

Enhancement of vitamins and other micronutrients with less nutrient values is being done by many strategies where the satisfaction of further demands by consumers was the main aim. The formation of provitamin A in carrots is studied where the production of baby carrots was established. This formed a new pathway for efficient technological development in breeding programs. A usual carrot is notated with 6 carotene sequences including alpha, beta, gamma, zeta, beta zeacarotene, and lycopene. The carrot with more amount of alpha-carotene was suspected with less amount of beta-carotene. The studies by Buishand et al. (1979) suggest that the red-rooted cultivators of Asia have a higher amount of lycopene content than the orange-rooted ones. In the past era, breeding is done with the help of a selection method where the dominant breeds of different interested traits (maturing rate, disease resistance, etc.) are crossed with each other to produce a better version. Carotene was evaluated using the HPLC method where it was seen that in Europe alpha and beta carotene ranged up to 22–49 ppm and lycopene ranged up to 46–103 ppm. Likewise, different germplasms of different continents were collected and subsequently checked on the level of carotenes and the trait formations by studying the society.

The studies also suggested that the color range of carrots ranged in three. This occurred due to polygenicity. When a dominant × dominant cross was conducted based on traits, it was seen that there is significant overdominance but still a slight variant of high expression in beta carotene was seen. The studies suggest that the backcrossing of strawberries for high adaptability and better production with peak nutritional qualities leads to the enrichment of compounds like vitamins too. Backcrossing was done between three generations (BCA1, 2, 3) between species of *Fragaria* × *ananassa* (Fxa) × *Fragaria virginiana glauca*. It increased weight, quality, and nutritional properties. The most important genotype was selected by checking on their total phenolic content, and vitamin C content from spectrophotometric analysis. The results show that the highest level of vitamin C content was seen in backcross 3 and anthocyanin content in Fxa in the F1 generation (Mazzoni et al., 2020). The enhancement of this vitamin is highly important for viral resistance for many human viral diseases like coronavirus. The enhancement of this vitamin also leads to many viral diseases, such as herpes diseases, *Varicella zoster* virus infection, Herpesviridae infections, DNA virus infections, etc.

The quality protein maize (QPM) has a lesser level of vitamin A production, which is being enriched by a specific strategy as by backcrossing *Hp321-1* with desired alleles of *crt RB1* within inbred cell lines CML161 and CML171. The new generations had a development of up to 92% efficiency. This helped in the development of vitamin A in QPM (Liu et al., 2015). Vitamin A increase is also treated by increasing its production in maize by marker-based back cross method (MABB) where beta carotene gene expressing ones are backcrossed for betterment which led to a sevenfold increase (Natesan et al., 2020). Vitamin efficiency can lead to drug development for certain viral diseases like measles, diarrheal diseases, measles-related pneumonia, human immunodeficiency virus infection, and malaria (Semba, 1999).

30.5 PLANT CELL CULTURES AND THEIR COMMON STRATEGIES TO DEVELOP VITAMINS

The specific technology where the plant cells are cultured, multiplied, and grown in an infection-free clean and safe environment in an in vitro condition for ensuring increased cell division and novel genetic structure is plant cell culture (Rao & Ravishankar, 2002). Many tissue culture techniques are being used for vitamin enrichment like somatic hybridization, in-vitro selection, haploid and double haploid selection, encapsulation technology, etc. (Gómez-Lim & Litz, 2004). The in-vitro culture and genetic transformations are the prominent techniques utilized, in which gene transfer techniques may have very little regeneration efficiency, especially in fruit crops. In tissue culture systems, organogenesis (new plant from organs) and somatic embryogenesis (formation of embryo from cells rather than gametic fusion) are mainly used (Rai et al., 2011). Somatic embryogenesis is better than organogenesis because of its high cell division rate, potential augment in bioreactors, etc. (Merkle & Dean, 2000). It was found that certain culture methods resulted in the needed genetically recombined species. Initially, vitamins like vitamin B12 and vitamin E were mainly extracted, but later on, many other vitamins, minerals, and other secondary metabolites were found to have higher efficiency (Pullman & Bucalo, 2014). Additionally, the studies also suggest that embryogenic tissue development initiates at a greater speed when they increase vitamin levels too.

In another method, safflower cells were cultured with respiratory medial, used to maintain the growth of cells as well as vitamin efficiency. The subculture was maintained within three weeks with tobacco basal medium as an extra supplement, casamino acid, kinetin, etc. This method also checks on the preservation of cell organization and its associated chambers. This method showed a high level of efficacy of tocopherols which helps in the prevention of many associated diseases (Takeda et al., 1998). It was found that certain combinations like vitamin C and E will help in processing a double barrier against many diseases like influenza viral infection (Tantcheva et al., 2003). It is also effective in the treatment of novel coronavirus. Normalization of vitamin A deficiency is maintained by organogenesis treatment. In a study by See et al. (2008), vitamin level was increased (250 µg atRA/g diet) in pregnant mice by inducing a higher level of all-trans-retinoic acid (atRA). At the early stage of the experiment, they injected less amount of atRA, (1.5 to 12 µg/g diet) which led to vitamin A deficiency (VAD), resulting in alterations and mutations in Hoxd3 and Hoxb4 genes.

The usage of myo-inositol as a substitute helps in prime treatment for betterment in vitamin synthesis. The vitamin C enrichment is done mainly by escalating its biosynthesis or reducing its degradation techniques. Still, the ascorbate pathway for downregulation is not clear. In Vitaceae species, ascorbate degradation is done by I-idonate to I-thearate where dehydroascorbate plays a key role. The main pathway between ascorbate degradation and oxidation is changed via extracellular pathway degradation check in a cell culture strategy with the help of enzyme 4-O-oxalyl-1-threonate. Some steps may contribute to peroxide generation which may lead to the loosening of the cell wall and its disintegration while cooking which may result in vitamin loss in temperature fluctuations. This cell culture method proved the complete pathway and helped in its inhibition strategy which is useful in industries to make it tolerant to heat (Green & Fry, 2005). Using reasonable stereoisomeric form present in chloroplast containing photosynthetic organism, one can enhance vitamin E. Sunflower cell culture was done with MS basal medium with naphthaleneacetic acid, benzyl aminopurine with casamino acids, and myo-inositol. There was a sudden increase where up to 30% of tocopherol was seen. Given a longer culture interval, it was proved that 78% of the tocopherol production increased (Caretto et al., 2004). Mainly in elderly people, vitamin A plays a crucial role and helps in fighting against viral infections (Rizvi et al., 2014).

Safflower (*Carthamus tinctorius*) cells are used where dissolved oxygen is kept as a limiting factor for the cell culture method. If the culture had an increase in DO concentration, the production of vitamin E would also rise. But this rate was compared with batch cultivation where an exponential loss was seen. But later on, the addition of fresh media increased the confluency rate and dissolved oxygen for better production. Repetition of the batch culture with media change gave 9mg/100g per dry cell, which is comparatively more than usual safflower seeds (Takeda et al., 1993). In *Arabidopsis* species and tobacco *BY2* cell suspension culture, methyl jasmonate addition enriched the level of transcription of two or more methyl jasmonate responsive gene coding enzymes (GDP mannose epimerase and lactodehydrogenase/oxidase) which help in the biosynthesis of vitamin C biosynthesis (Wolucka et al., 2005).

30.6 TRANSGENIC PLANTS FOR IMPROVED VITAMIN CONTENT

To avoid vitamin deficiency, advancement in technology and fields of science, such as biochemistry, was very crucial. The population increase will also increase the demand for nutrient-rich food. As per the predictions, by 2050, there will be a static growth in population which will lead to scarcity of high-rich nutrient food sources (Rani & Usha, 2013). Scientists worked on producing more plants with better adaptability to drastic environmental changes as well as of high nutritional value, which was more preferred by farmers too (Kershen, 2010). The process of modification of DNA is contrived to enhance the genes for a better-versioned trait (which is not present in the plant naturally) by adding new genes from a disparate plant or even species which adds to progressivity in its uses (such as shelf-life improvement, better yield, quality rise, adaptable to climatic conditions, pest, insect, viral tolerant, actively barricading against abiotic stresses, better pharma values, nutrient efficiency, vaccine preparation, bioremediation, etc.) is transgenicity in plants. The genetically modified plants used in vaccine production for humans are found to be cost-efficient too, which leads to the production of vaccines from potatoes and bananas. The antibiotic-resistant tobacco plant introduced in 1983 (China) was a major milestone in the production of transgenic crops. Thereafter, there was a certain leap in the production of transgenic plants in different methods (Rani & Usha, 2013). The most prominent gene transfer methods used are the biolistic/microprojectile bombardment/gene gun method and Agrobacterium-based transformation (usually using *A. tumefaciens*). The gene gun method is important for transgenicity in monocot plants as it is not very effective after *Agrobacterium* transformation (Shrawat & Lörz, 2006). A gold or tungsten-coated DNA molecule with the desired gene of interest is added to the cell or plant by a shot provided the condition of high pressure. This DNA molecule moves into the genomic region and binds to it (Chawla, 2000). The chance for cell disintegration is an important demerit of this method.

The *Agrobacterium* is a soil bacterial species. The tumor-inducing Ti plasmid with specific T DNA region with genes of opine synthesis, auxin, and cytokine production. They are introduced into plant cells with the help of certain factors like *Vir A, G, D1, D2, E,* and *VIP 1,2* genes. The floral dip method of a dipping cluster of inflorescences in culture is also a prominently used technique. The main disadvantage is that for some crops this may lead to bacterial infection (Chawla, 2000). Interkingdom gene transfer can also be done along with interspecies transfer. Some other methods like (a) PEG-mediated transformation, (b) calcium phosphate method, (c) electroporation, and (d) microinjection can also be used, but they are limited in application due to high cost and less reliance (Herrera-Estrella et al., 2005). The issues of transgenic plants are the adversity in human health (Ferber, 1999), natural environment (Losey et al., 1999), wild species genome damage, etc., which are being accessed and maintained by different authorities, such as EPA, USDA, FDA, DHHS, etc. Mainly, transgenic plants with vitamins A, E, and C are seen as most efficient as other vitamins.

The process of biofortification is well celebrated in increasing the efficiency of crops. Instantly, it is done by adding a set of fertilizers adequate for plant growth. However, the downside of this process is differential geographical conditions, climatic changes, plant acceptance level for a particular mineral, etc. The method is also very tedious, laborious, and costly. The typical plant breeding strategies (which have setbacks like high levels of variability in the gene pool and time consumption rate), as well as the ultra-modern biotechnological processes, are used (Hefferon, 2015). The transgenic technologies are better in the sense that specific genes could be altered or modified the phenotypical and genetic changes cost-effectively, whereas in traditional methods there is the chance for improper genetic changes and alterations may not be possible at times. Taking much time, a proper desired trait can be expressed in a plant with the help of transgenecity (Demont & Stein, 2013). The nucleic-acid-based genome editing techniques of biotechnology, such as TALENs and CRISPR (Cas), were a big breakthrough. The exact place is found, and editing is done for the desired sequence for food crops. But such strategies will develop new hybrids replacing the wild varieties. Society was against this replacement which was a major setback for this method. Still, research is going on in balancing the wild and new species (Demont & Stein, 2013). The main agenda behind the production of biofortified crops is to increase micronutrients that are in edible form for organisms and thus can satisfy the neglected micronutrient deficiencies in society.

The drastic leap in biofortification is the modification of wild-type rice varieties into a vitamin-A-rich variety. The eyesight issue in infants is counted up to millions per year, in which half the count dies soon after. Vitamin A is produced by amending its precursor molecule β-carotene, which is naturally absent in cereals. As rice is the common staple food for a large community, deficiency of such an important vitamin may lead to simultaneous ill effects. A transgenic variety, golden-colored golden rice is rich in β-carotene (Tang et al., 2009). It is developed by reconstructing two genes from daffodil and *Erwinia uredovora*, which alters the biosynthetic pathway. The initial variety GR1 was later subjected to different backcross breeding methods, and two different provitamins A pathways, along with a different phytoene synthesizing gene and parallel gene of daffodil (Shumskaya & Wurtzel, 2013), were used to produce a novel variety GR2, which can synthesize up to 35μg of vitamin A than normal rice (Tanumihardjo et al., 2010). The daily dietary intake is up to 100μg of golden rice, which is a stable level for children. Studies show that vitamin A in golden rice is more effective than in the unaltered spinach variety (Haskell, 2012).

Vitamin A in biofortified maize is increased up to 3.1-fold than in normal maize. The usage of biofortified maize porridge has been shown 40–50% of the dietary recommended level in Zimbabwean men (Mugode et al., 2014). A different study where Mongolian gerbils were fed with a β-cryptoxanthin-variety instead of β-carotene showed a faster way of conversion (Jeong & Guerinot, 2008). These studies proved that the usage of biofortified maize is a better option. The production of triple vitamin-fortified maize in the endosperm is the most improved version which produces a large quantity of vitamins A, B12, and C (Mugode et al., 2014). Condensing the level of toxin cyanogen in cassava roots helps in increasing the levels of iron, zinc, vitamin A, and vitamin E. A study was performed on people fed with biofortified cassava and blood samples were taken from participants which showed a sudden shot in β carotene levels (Naqvi et al., 2009). The bio casava project and maize biofortification are a major boon in African countries for reducing multiple micronutrient famine (La Frano et al., 2013).

30.7 GENE TRANSFER STUDIES IN MAPs

Medicinal and aromatic plant (MAP) systems elucidated from the botanical world the actively immense source in the sectors of pharmaceutical, health products, and cosmetic industries. The increased promotion of herbal products peaked the usage, development, and production of medicinal and aromatic plants. Ayurvedic treatment, traditional healthcare practitioners, and even household first aid facilities have a very prime role. The main reason is that MAP cannot be replaced with full effects by any other products. However, the purpose and implementation were certified from the beginning of human life. Seventy-five percent of the whole world is dependent on the properties of MAPs. The modernized implementation brings about the best utility of them with better proceedings, safety measures, and improvised quality assurance. Many diseases are treated with the help of MAP species. Many plants are specifically treated with certain mechanical or synthetic engineering processes for the enrichment of vitamins (Başer, 2003). Krasnyanski et al. (1999) suggest that protoplast-based *Agrobacterium*-mediated and direct gene transfer methods using PEG were the most reliable and stateable method. The Black Mitcham, produced by the cross of *Mentha* and *Piperita*, had a limestone gene insertion. The internodes that are developed were subjected to a very severe method of organogenesis where the shoots had to undergo insertion of *Agrobacterium* species, which has better vector qualities. The addition of disarmed strain *GV3101* was performed. The usage of direct genetic transformation when the integrated protoplast part grows on will aid in the development of the whole plant body system. The promoter CaMV35S is the responsible area for the insertion of pGA643 with *npt2/GUS* genes and was the foremost method to be used. The substitution of *GUS* and the addition of kanamycin as an antibiotic was performed which resulted in a callus culture formed out of protoplast. The efficiency was checked by traditional methods like PCR or Southern blot hybridization. The resultant statistics analysis shows that the *Agrobacterium*-mediated method is more effective than direct DNA uptake. This was finalized by checking on a comparison of different essential species activities and bioactive chemicals like vitamins.

Vitamin production in MAPs is so important because usually in rural areas people prefer more natural plant materials. Now the trend has changed in such a way that the urban population also shows a similar version of inclination toward MAP products (Srivastava et al., 1995). Certain steroidal production helps in the formation of many vitamins like vitamin D. The mulberry leaf extract showed a count of vitamin level in studies (Balakrishnan et al., 2009) on the clonal propagation method. Ganjewala (2008) suggests in their study that lemongrass oil with many pharmaceutical activities is also used in vitamin A synthesis. The carotene synthesis method alteration is adapted by many species of plants for aroma and plant development. Vitamin E is well known for its anti-aging and anticancer properties, which is highly efficient in the pharmaceutical and cosmetic industry (Shahidi & de Camargo, 2016). Studies by Bhat et al. (2002) suggest that the heterozygosity and homozygosity of genes matter when it comes to the enzyme production of genes and matter of expression of proteins for over synthesis of vitamins and other micronutrients. The residual matter obtained after the saponification of certain oils may contain effective materials like different vitamins. Amalaki (*Embelica officinalis*) and bhallataka (*Semicarpus anacardium)* were noted with higher synthesis levels of vitamin C, with many antitumor properties along with many vitamins in them. Gene manipulation for this purpose is mainly done by the micropropagation technique. The cell culture method on the development of hairy roots of Solanaceae family species to override the causal limitations of biosynthesis methods of certain vitamins is checked on. *Datura metel* was modified with the help of *Agrobacterium rhizogenes* with strain A4 in a B5 solid medium as studied by Giri et al. (1997). The studies on hairy root transformation by Giri et al. (1997) and Souret and Weathers (2000) on plants like *Artemisia annua*, *Aconitum heterophyllum*, *Digitalis lanata*, *Papaver somniferum*, *Crocus sativus*, *Eschscholzia californica* (California poppy), *Artemisia annua*, respectively, made a significant growth process (Bajaj, 2012).

30.8 IMPROVEMENT OF VITAMINS VIA GENETIC ENGINEERING

Carotene production is highly necessary in the case of vitamin A synthesis. In a study by Rosati et al. (2000), it was seen that the normal tomatoes with dietary intake level proclaimed as 100 gm which was replaceable by the newly modified crop with lycopene cyclase from *Arabidopsis* species along with the phytoene desaturase promoter as up to 60 gm. Römer et al. (2000) conducted a study where it was proven that the lycopene conversion to carotenoid or the dual expression of *E. uredovora* phytoene desaturase (*crtI*) along with 35s promoter sequence of CaM virus will top up the vitamin A production in transgenic plants up to 50%.

This can also be done by the increase in metabolic flux in the b branch between GGPP and lycopene or the increase sink capacity of carotenoids (Zhu et al., 2009; Farré et al., 2010; Bai et al., 2011).

According to Guiliano et al. (2000), there is a particular strategy where transcription of beta cyclase with its transcript B can lead to an enzyme-like cyclase that forms as the second. When the first and second genes have a better scale of transcription, it will lead to a sudden peak of 1.7-to-2.5-fold level of beta carotene in transgenic plants than the wild species. Rosati et al. (2000) and Römer et al. (2000) found that excluding the phytoene synthase gene and upregulating the rest of *crt 1* genes led to a conflicting situation. Considering Ye et al. (2000), they were able to produce a high level of beta carotene by utilizing phytoene synthase and phytoene reductase mechanisms with simple homology of beta cyclase activity. Further, found that the carotene level changes and its intermediates lead to the higher efficiency of the plant vitamin system. Moving on to Enfissi et al. (2005), the DPX synthase's higher expression level leads to betterment in the carotene precursor.

Welsch et al. (2010) found that carotene level increases up to 34-fold from 21 mg/g (wild type) with the help of *crt B* gene induction from bacterial species. Paine et al. (2005) found that renewal of maize's gene with daffodils produced 371 mg per gram DW carotene. When the same daffodil genes were outshown in maize, they produced almost 601 g of DW carotene as per the studies by Zhu et al. (2009) and Naqvi et al. (2009). The *crt B, Y, I* from bacterial (*Erwinia*) species were transferred to potato species by Diretto et al. (2007), which showed an alteration of beta carotene production and alpha-carotene which produced 471 g/g DW of beta carotene synthesis in total. This was majorly utilized by potato species. Also, Lopez et al. (2008) showed that the cauliflower *Or* genes, which aid in chromoplast formation help in the high deposition system of beta carotene in tubers of the potato, which led to the formation of yellow flesh with ten times beta carotene range than normal species. Certain gene overexpression can lead to the production of a higher amount of vitamin E. It is easier and better to increase the α tocopherol in transgenic plants according to a study by Naqvi et al. (2009) tocopherol in soya increased 14-fold with a change in gene by *Perilla frutescens* γ-TMT. Simultaneously, cDNA clones of two *Arabidopsis* species helped in having a threefold increase in vitamin E production in maize. Plants with a high level of tocochromanol molecules can derive a good yield of vitamin E. By the activation of more energetic isomers or a level change in the content of tocochromanol.

The studies by Tavva et al. (2007) say that there is a sudden peak in alpha-tocopherol and beta-tocopherol in soya when the *Perilla* species *c-TMT* gene was introduced as up to 10.4-fold and 14.9-fold, respectively. The expression of two cDNA by *Arabidopsis* species HPPD and MPBQ MT increase thrice the content of tocopherol in maize as explained by Naqvi et al. (2009). Certain studies also suggest that HPPD from *Arabidopsis* increase in rice will not change the tocopherol content. Chloroplast modification was seen as an important technique where Lee et al. (2016a) showed that incorporating *HPT, MPBQMT,* and *HPPD* genes results in high tocopherol contents in plants. Changing the pathway for better alpha-tocopherol synthesis by incorporating dual gene strategies is also a good method in plants like lettuce. Studies by Yabuta et al. (2013) suggest that the *VTE 1* gene expression increase or analysis by incorporating using new crop will lead to better development of vitamin E synthesis along with other properties like stress tolerance, drought resistance, and other biotic stresses. It was also seen that the knockout of certain genes like *slr1736* and MTI enzyme will alter tocopherol concentrations as suggested by Savidge et al. (2002) and Shintani et al. (2002). *Agrobacterium*-mediated transgenic potatoes had a high content of alpha-tocopherol. The genes like *At-HPT* and *gamma TMT* genes from the CaMV 35S promoter also will reluctantly increase vitamin E content as studied by Upadhyaya et al. (2021).

For ascorbate production, Jain and Nessler (2000) found a method by improvising the biosynthesis method; in lettuce, this led to a sevenfold increase and a twofold increase in potatoes as studied by Hemavathi et al. (2010), as well as the ascorbate recycling (alternate pathway with glutathione as reductant) rate expansion. Naqvi et al. (2009) suggest that the *DHAR* gene from rice, when added to maize, produces a lot of vitamin improvement with high efficiency of vitamin C up to six times. A slight increase in folate level is encountered by pterin fluxing or PABA pathways, which were suggested by de Crécy-Lagard et al. (2007), which showed a double folate production. This method was applied in rice by Storozhenko et al. (2007) suggested that there is a 100-fold increase in vitamin B levels in rice and is suggested to be very powerful. Vitamin B12 biosynthesis levels are increased mainly by translational riboswitch methods. B12 is retrieved by microbial fermentation in *P. denitrificans* and *P. shermanii* by (Fang et al., 2017). As vitamin treatments are available with a few genetic strategies, mainly fermentation and mutagenesis are used. New methods like metabolic engineering in *E. coli* as host were also developed and so on. The compilation of metabolic engineering and synthetic engineering provides a new pathway in the synthesis of enhanced vitamin B12 (Fang et al., 2017).

30.9 MECHANISM OF ACTION OF BIOACTIVE COMPOUNDS IN HUMANS

The biologically active compounds are the compounds that share a major part in the cellular and physiological activity of the living organism. The role played by them may lead to an up- or downregulation in the system based on its dosage, sufficiency, and usage levels (Lee, 2016b). Their efficiency in food and medicinal uptake plays an important part in human health. Plant bioactive compounds help in signaling, competition, defense, etc. (Rai et al., 2011). The providence of certain treatment strategies for lung cancer, prostate cancer, diabetes mellitus, cardiovascular diseases, etc., is maintained with the help of bioactive compounds.

The MAP studies are performed by extracts or purified phytochemicals. Extracts are then carried out in certain preparations in pharmaceutics and nanomaterial photosynthesis for their antioxidant, anti-inflammatory, neuroprotective, and cardiovascular properties. The purified phytoconstituents are used in food additives and natural dyes, which help in antioxidant and antimicrobial activities (Fierascu et al., 2021).

In garlic extraction studies by Shang et al. (2019), the antioxidant property can be improved by maintaining enzyme activations and the Nrf2-ARE pathway. Anti-inflammatory property is maintained by inflammation-inhibiting mediators such as TNF alpha and IL 1. The metabolic processes are customized in a bacterial cell for the antibacterial property. The immune system and cardiovascular problems are checked by the levels of bioactive compounds. The anticancer actions are maintained by CD4=/CD8= ratio maintenance and increase of IL2 and IL4 levels. Gastric issues and respiratory problems are also maintained by the same. Pulses with vitamin B are an important source to degrade breast cancer levels and homocysteine concentration (Patterson et al., 1997). The mitochondrial activity is improved by vitamin D by tabulating its receptors to digest starch (Clarke et al., 2010). The studies on mango suggested that the conditioned bioactive compounds can be maintained by special transformations and genetic changes. Studies by Ribeiro and Schieber (2010) show that the DPPH radical doses are necessary for the maintenance of antioxidant properties in mango. The muscles like *Mytilus edulis* and *Perna canaliculus* show bioactivity when certain processes are done for biotransformation with the help of fermentation activity as studied by Grienke et al. (2013). The inflammation is scored by activation of TF and pro-inflammatory cytokines which help in inflammation as well as the cardiovascular system. Important compounds, such as polyphenols, especially isoflavone, and anthocyanin, but quercetin, catechin, and resveratrol were seen along with alpha-lipoic acid in studies by Ross (2010) where NK–kB plays a significant role. An in silico prediction method (ADMET) by Muhammad et al. (2020) shows that the biochemical compound pathway is necessary for checking breast cancer.

30.10 CONCLUSION

The key role of vitamins in the human diet is to catalyze important chemical reactions in our cells. From the synthesis of cells to the end of apoptosis, every step is modulated by vitamins. They are mainly synthesized by plants. The enriching of micronutrients in plants will help in avoiding the scarcity of vitamins in the coming days. Even treatment of severe viral infection is maintained by specific vitamins. The different methods of breeding for biofortification help in finding new varieties of crops with improved quality at cheap prices. Certain other properties like shelf-life improvement, resistance to climatic conditions, etc. are also studied. Conventional breeding strategies like plant breeding techniques with backcross and crossbreeding also help to enhance vitamins in plant species on a large scale, helping in overproduction and increase in systematization. Genetic engineering methods also help in the study of the whole biosynthesis pathway and lead to novel variety development with the desired gene of interest. The medicinal and aromatic plants are now grown into an irradiated sector of urban and rural life due to their supreme powers. The bioactive compounds with better facilitation help in the production of all essential propertied substances like drugs, cosmetics, etc. The strategic method to utilize them is based on the purpose of use and the method to be used. They rely on the importance of the compound to be formed, and thereby, a specific mechanism is used to produce the subsequent products.

ACKNOWLEDGMENTS

We thank Manipal Academy of Higher Education (MAHE), Manipal, Karnataka, India, TIFAC-CORE and FIST, DST New Delhi, DBT BUILDER grant (BT/INF/22/SP43065/2021), DBT, New Delhi, and K-FIST, VGST, Govt. of Karnataka, for the facilities. The authors (VPT, SU, MK, and SAT) are thankful to MAHE for the T.M.A. Pai Ph.D. scholarship, and SAT is thankful to DST-SERB (CRB/2022/003498 dated 20.06.2023) for JRF, respectively. We are grateful to Dr. B.S. Satish Rao, Professor and Director, Manipal School of Life Sciences, MAHE, for their encouragement and support.

AUTHOR CONTRIBUTIONS

AM and SPV conceived the idea; SPV, VPT, SU, MK, and AM carried out the literature survey and prepared the tables and figures; SPV, VPT, SU, MK, SAT, and AM wrote and edited the chapter, and all authors approved the chapter.

REFERENCES

Akhter, F., Akhter, A., Day, V. W., Nolte, E. D., Bhattacharya, S., & Saeed, M. (2021). *Nelumbo nucifera*. In *Naturally occurring chemicals against alzheimer's disease* (pp. 467–479). Academic Press.

An, Y., Feng, L., Zhang, X., Wang, Y., Wang, Y., Tao, L., Quin, R., & Xiao, R. (2019). Dietary intakes and biomarker patterns of folate, vitamin B 6, and vitamin B 12 can be associated with cognitive impairment by hypermethylation of redox-related genes *NUDT15* and *TXNRD1*. *Clinical Epigenetics*, *11*(1), 1–19.

Assoi, E. K., Bagui, O. K., Kouakou, B. K., Gbogbo, A. Y., Soro, D., & Zoueu, J. T. (2021). Estimating maturity by measuring pH, sugar, dry matter, water, and vitamin C content of cashew apple ('*Anacardium occidentale*') from remote spectral reflectance data using neural network. *Australian Journal of Crop Science*, *15*(7), 1029–1034.

Bâ, A. (2008). Metabolic and structural role of thiamin in nervous tissues. *Cellular and Molecular Neurobiology*, *28*(7), 923.

Bai, J., Huang, L., Yan, D., Wang, Q., Gao, H., Xiao, R., & Huang, C. (2011). Contamination characteristics of heavy metals in wetland soils along a tidal ditch of the Yellow River Estuary,

China. *Stochastic Environmental Research and Risk Assessment*, *25*, 671–676.

Bajaj, Y. P. S. (2012). *Medicinal and aromatic plants I*, *2*(4), 3. Springer Science & Business Media.

Balakrishnan, V., Latha, M. R., Ravindran, K. C., & Robinson, J. P. (2009). Clonal propagation of Morus alba L. through nodal and axillary bud explants. *Botany Research International*, *2*(1), 42–49.

Başer, K. H. C. (2003). New trends in the utilization of medicinal and aromatic plants. *Conservation, Cultivation and Sustainable Use of Medicinal*, *676*(2), 11–23.

Basset, G. J., Latimer, S., Fatihi, A., Soubeyrand, E., & Block, A. (2017). Phylloquinone (vitamin K1): Occurrence, biosynthesis and functions. *Mini Reviews in Medicinal Chemistry*, *17*(12), 1028–1038.

Beaudet, V., Gervais, R., Graulet, B., Nozière, P., Doreau, M., Fanchone, A., Castagnino, D. D. S., & Girard, C. L. (2016). Effects of dietary nitrogen levels and carbohydrate sources on apparent ruminal synthesis of some B vitamins in dairy cows. *Journal of Dairy Science*, *99*(4), 2730–2739.

Bhat, S., Maheshwari, P., Kumar, S., & Kumar, A. (2002). Mentha species: In vitro regeneration and genetic transformation. *Molecular Biology Today*, *3*(1), 11–23.

Boland, R. (1986). Role of vitamin D in skeletal muscle function. *Endocrine Reviews*, *7*(4), 434–448.

Boland, R., Skliar, M., Curino, A., & Milanesi, L. (2003). Vitamin D compounds in plants. *Plant Science*, *164*(3), 357–369.

Boubakri, H., Gargouri, M., Mliki, A., Brini, F., Chong, J., & Jbara, M. (2016). Vitamins for enhancing plant resistance. *Planta*, *244*(3), 529–543.

Breithaupt, D. E., & Bamedi, A. (2001). Carotenoid esters in vegetables and fruits: A screening with emphasis on β-cryptoxanthin esters. *Journal of Agricultural and Food Chemistry*, *49*(4), 2064–2070.

Buishand, J. G., & Gabelman, W. H. (1979). Investigations on the inheritance of color and carotenoid content in phloem and xylem of carrot roots (*Daucus carota* L.). *Euphytica*, *28*, 611–632.

Calder, P. C., Carr, A. C., Gombart, A. F., & Eggersdorfer, M. (2020). Optimal nutritional status for a well-functioning immune system is an important factor to protect against viral infections. *Nutrients*, *12*(4), 1181.

Caretto, S., Speth, E. B., Fachechi, C., Gala, R., Zacheo, G., & Giovinazzo, G. (2004). Enhancement of vitamin E production in sunflower cell cultures. *Plant Cell Reports*, *23*(3), 174–179.

Carr, A. C., & Maggini, S. (2017). Vitamin C and immune function. *Nutrients*, *9*(11), 1211.

Cartabiano-Leite, C. E., Porcu, O. M., & de Casas, A. F. (2020). Sweet potato (*Ipomoea batatas* L. Lam) nutritional potential and social relevance: A review. *History*, *11*, 23–40.

Chawla, H. S. (2000). Introduction to plant biotechnology. *Biotechnology*, *2*, 30.

Clarke, R., Halsey, J., Lewington, S., Lonn, E., Armitage, J., Manson, J. E., & B-Vitamin Treatment Trialists' Collaboration. (2010). Effects of lowering homocysteine levels with B vitamins on cardiovascular disease, cancer, and cause-specific mortality: Meta-analysis of 8 randomized trials involving 37 485 individuals. *Archives of Internal Medicine*, *170*(18), 1622–1631.

Crider, K. S., Bailey, L. B., & Berry, R. J. (2011). Folic acid food fortification – its history, effect, concerns, and future directions. *Nutrients*, *3*(3), 370–384.

Damon, M., Zhang, N. Z., Haytowitz, D. B., & Booth, S. L. (2005). Phylloquinone (vitamin K1) content of vegetables. *Journal of Food Composition and Analysis*, *18*(8), 751–758.

Das, A. K., Gowda, M. M., Muthusamy, V., Zunjare, R. U., Chauhan, H. S., Baveja, A., Bhatt, V., Chand, G., Bhat, J. S., Guleria, S. K., Saha, S., Gupta, H. S., & Hossain, F. (2021). Development of maize hybrids with enhanced vitamin-E, vitamin-A, lysine, and tryptophan through molecular breeding. *Frontiers in Plant Science*, *12*, 659381.

de Crécy-Lagard, V., El Yacoubi, B., de la Garza, R. D., Noiriel, A., & Hanson, A. D. (2007). Comparative genomics of bacterial and plant folate synthesis and salvage: Predictions and validations. *BMC Genomics*, *8*, 1–15.

Demont, M., & Stein, A. J. (2013). Global value of GM rice: A review of expected agronomic and consumer benefits. *New Biotechnology*, *30*(5), 426–436.

Diretto, G., Al-Babili, S., Tavazza, R., Papacchioli, V., Beyer, P., & Giuliano, G. (2007). Metabolic engineering of potato carotenoid content through tuber-specific overexpression of a bacterial mini-pathway. *PLoS One*, *2*(4), e350.

Drouin, G., Godin, J. R., & Pagé, B. (2011). The genetics of vitamin C loss in vertebrates. *Current Genomics*, *12*(5), 371–378.

Duffy, S. K., O'Doherty, J. V., Rajauria, G., Clarke, L. C., Hayes, A., Dowling, K. G., O'Grandy, M. N., Kerry, P. J., Jakebson, J., Cashman, D. K., & Kelly, A. K. (2018). Vitamin D-biofortified beef: A comparison of cholecalciferol with synthetic versus UVB-mushroom-derived ergosterol as feed source. *Food Chemistry*, *256*, 18–24.

Ebong, P. E., Owu, D. U., & Isong, E. U. (1999). Influence of palm oil (*Elaesis guineensis*) on health. *Plant Foods for Human Nutrition*, *53*, 209–222.

Enfissi, E. M., Fraser, P. D., Lois, L. M., Boronat, A., Schuch, W., & Bramley, P. M. (2005). Metabolic engineering of the mevalonate and non-mevalonate isopentenyl diphosphate-forming pathways for the production of health-promoting isoprenoids in tomato. *Plant Biotechnology Journal*, *3*(1), 17–27.

Ermiş, S., & Yanmaz, R. (2022). Comparative Study of The Chemical Composition and Nutritional Value of Pumpkin Seed (*Cucurbita pepo* L.) Grown in the different ecological conditions of Turkey. *Türk Tarım ve Doğa Bilimleri Dergisi*, *9*(2), 413–423.

Fang, C., Luo, J., & Wang, S. (2019). The diversity of nutritional metabolites: Origin, dissection, and application in crop breeding. *Frontiers in Plant Science*, *10*, 1028.

Fang, H., Kang, J., & Zhang, D. (2017). Microbial production of vitamin B 12: A review and future perspectives. *Microbial Cell Factories*, *16*(1), 1–14.

Farré, G., Sanahuja, G., Naqvi, S., Bai, C., Capell, T., Zhu, C., & Christou, P. (2010). Travel advice on the road to carotenoids in plants. *Plant Science*, *179*(1–2), 28–48.

Ferber, D. (1999). Risks and benefits: GM crops in cross hairs. *Science*, *286*(5445), 1662–1666.

Fierascu, R. C., Fierascu, I., Baroi, A. M., & Ortan, A. (2021). Selected aspects related to medicinal and aromatic plants as alternative sources of bioactive compounds. *International Journal of Molecular Sciences*, *22*(4), 1521.

Fiorino, S., Zippi, M., Gallo, C., Sifo, D., Sabbatani, S., Manfredi, R., Rasciti, E., Rasciti, L., Giampieri, E., & de Biase, D. (2020). The rationale for a multi-step therapeutic approach based on antivirals, drugs and nutrients with immunomodulatory activity in patients with coronavirus-SARS2-induced disease of different severities. *British Journal of Nutrition*, *125*(3), 275–293.

Fitzpatrick, T. B., Basset, G. J., Borel, P., Carrari, F., DellaPenna, D., Fraser, P. D., Hellmann, H., Osorio, S., Rothan, C., Valpuesta, V., Caris-Vecrat, C., & Fernie, A. R. (2012). Vitamin deficiencies in humans: Can plant science help?. *The Plant Cell*, *24*(2), 395–414.

Fowler III, A. A., Kim, C., Lepler, L., Malhotra, R., Debesa, O., Natarajan, R., Fisher, J. B., Seyed, A., DeWilde, C., Priday, A., & Kasirajan, V. (2017). Intravenous vitamin C as adjunctive therapy for enterovirus/rhinovirus induced acute respiratory distress syndrome. *World Journal of Critical Care Medicine*, *6*(1), 85.

Funk, C. (1912). The etiology of deficiency diseases. *The Journal of State Medicine (1912–1937)*, *20*(6), 341–368.

Ganjewala, D. (2008). RAPD characterization of three selected cultivars OD-19, GRL-1 and Krishna of East Indian lemongrass (*Cymbopogon flexuosus* Nees ex Steud) Wats. *American-Eurasian Journal of Botany*, *1*(2), 53–57.

Garber, A. K., Binkley, N. C., Krueger, D. C., & Suttie, J. W. (1999). Comparison of phylloquinone bioavailability from food sources or a supplement in human subjects. *The Journal of Nutrition*, *129*(6), 1201–1203.

Giri, S. N., Hyde, D. M., Braun, R. K., Gaarde, W., Harper, J. R., & Pierschbacher, M. D. (1997). Antifibrotic effect of decorin in a bleomycin hamster model of lung fibrosis. *Biochemical Pharmacology*, *54*(11), 1205–1216.

Gómez-Lim, M. A., & Litz, R. E. (2004). Genetic transformation of perennial tropical fruits. *In Vitro Cellular & Developmental Biology-Plant*, *40*(5), 442–449.

Green, M. A., & Fry, S. C. (2005). Vitamin C degradation in plant cells via enzymatic hydrolysis of 4-O-oxalyl-L-threonate. *Nature*, *433*(7021), 83–87.

Grienke, U., Richter, M., Braun, H., Kirchmair, J., von Grafenstein, S., Liedl, K. R., Schmidtke, M., & Rollinger, J. M. (2013). New insights into the anti-influenza activity of licorice constituents. *Planta Medica*, *79*(13), PG3.

Grune, T., Lietz, G., Palou, A., Ross, A. C., Stahl, W., Tang, G., Thurnham, D., Yin, S., & Biesalski, H. K. (2010). β-Carotene is an important vitamin A source for humans. *The Journal of Nutrition*, *140*(12), 2268S–2285S.

Guiliano, M., Boukir, A., Doumenq, P., Mille, G., Crampon, C., Badens, E., & Charbit, G. (2000). Supercritical fluid extraction of bal 150 crude oil asphaltenes. *Energy & Fuels*, *14*(1), 89–94.

Hanson, A. D., & Gregory, J. F. (2011). Folate biosynthesis, turnover, and transport in plants. *Annual Review of Plant Biology*, *62*, 105–125.

Haskell, M. J. (2012). The challenge to reach nutritional adequacy for vitamin A: β-carotene bioavailability and conversion – evidence in humans. *The American Journal of Clinical Nutrition*, *96*(5), 1193S–1203S.

Hayden, M. R., &Tyagi, S. C. (2004). Homocysteine and reactive oxygen species in metabolic syndrome, type 2 diabetes mellitus, and atheroscleropathy: The pleiotropic effects of folate supplementation. *Nutrition Journal*, *3*, 4.

Hefferon, K. L. (2015). Nutritionally enhanced food crops; progress and perspectives. *International Journal of Molecular Sciences*, *16*(2), 3895–3914.

Hemavathi, U. C. P., Akula, N., Young, K. E., Chun, S. C., Kim, D. H., & Park, S. W. (2010). Enhanced ascorbic acid accumulation in transgenic potato confers tolerance to various abiotic stresses. *Biotechnology Letters*, *32*, 321–330.

Herbers, K. (2003). Vitamin production in transgenic plants. *Journal of Plant Physiology*, *160*(7), 821–829.

Herrera-Estrella, L., Simpson, J., & Martínez-Trujillo, M. (2005). Transgenic plants: An historical perspective. In *Transgenic plants: Methods and protocols* (pp. 3–31). Humana Press.

Hirota, Y., Tsugawa, N., Nakagawa, K., Suhara, Y., Tanaka, K., Uchino, Y., Takeuchi, A., Sawada, N., Kamao, M., Wada, A., Oktisu, T., & Okano, T. (2013). Menadione (vitamin K3) is a catabolic product of oral phylloquinone (vitamin K1) in the intestine and a circulating precursor of tissue menaquinone-4 (vitamin K2) in rats. *Journal of Biological Chemistry*, *288*(46), 33071–33080.

Hunter, D. J., Manson, J. E., Colditz, G. A., Stampfer, M. J., Rosner, B., Hennekens, C. H., Speizer, F. E., & Willett, W. C. (1993). A prospective study of the intake of vitamins C, E, and A and the risk of breast cancer. *New England Journal of Medicine*, *329*(4), 234–240.

İdikut, L. (2023). Nutritional importance and agricultural properties of pea (*Pisum sativum* L.). *Advances in Agriculture, Forestry and Aquaculture Sciences*, *110*.

Izquierdo, M., Domínguez, D., Jiménez, J. I., Saleh, R., Hernández-Cruz, C. M., Zamorano, M. J., & Hamre, K. (2019). Interaction between taurine, vitamin E and vitamin C in microdiets for *gilthead seabream* (Sparusaurata) larvae. *Aquaculture*, *498*, 246–253.

Jain, A. K., & Nessler, C. L. (2000). Metabolic engineering of an alternative pathway for ascorbic acid biosynthesis in plants. *Molecular Breeding*, *6*, 73–78.

Jennings, D. L., & Iglesias, C. (2002). Breeding for crop improvement. In *Cassava: Biology, Production and Utilization* (pp. 149–166). Wallingford.

Jeong, J., & Guerinot, M. L. (2008). Biofortified and bioavailable: The gold standard for plant-based diets. *Proceedings of the National Academy of Sciences*, *105*(6), 1777–1778.

Jovic, T. H., Ali, S. R., Ibrahim, N., Jessop, Z. M., Tarassoli, S. P., Dobbs, T. D., Holford, P., Thornton, A. C., & Whitaker, I. S. (2020). Could vitamins help in the fight against COVID-19? *Nutrients*, *12*(9), 2550.

Junaid, K., Qasim, S., Yasmeen, H., Ejaz, H., Alsrhani, A., Ullah, M. I., Fahad, A., & Rehman, A. (2021). Potential inhibitory effect of vitamins against COVID-19. *Computers, Materials & Continua*, *66*(1), 707–714.

Kamath, J. V., Rahul, N., Kumar, C. A., & Lakshmi, S. M. (2008). *Psidium guajava* L: A review. *International Journal of Green Pharmacy (IJGP)*, *2*(1).

Kaya, C., Ashraf, M., Sonmez, O., Tuna, A. L., Polat, T., & Aydemir, S. (2015). Exogenous application of thiamine promotes growth and anti-oxidative defence system at initial phases of development in salt-stressed plants of two maize cultivars differing in salinity tolerance. *Acta Physiologiae Plantarum*, *37*, 1741.

Kelly, G. S. (2011). Pantothenic acid. *Alternative Medicine Review*, *16*(3), 263–275.

Keogh, J. B., Cleanthous, X., Wycherley, T. P., Brinkworth, G. D., Noakes, M., & Clifton, P. M. (2012). Increased thiamine intake may be required to maintain thiamine status during weight loss in patients with type 2 diabetes. *Diabetes Research and Clinical Practice*, *98*(3), 40–42.

Kershen, D. L. (2010). Trade and commerce in improved crops and food: An essay on food security. *New Biotechnology*, *27*(5), 623–627.

Khadilkar, A., Khadilkar, V., Chinnappa, J., Rathi, N., Khadgawat, R., Balasubramanian, S., Praek, B., & Jog, P. (2017). Prevention and treatment of vitamin D and calcium deficiency in children and adolescents: Indian Academy of Pediatrics (IAP) Guidelines. *Indian Pediatrics*, *54*(7), 567–573.

Kraemer, K., Waelti, M., De Pee, S., Moench-Pfanner, R., Hathcock, J. N., Bloem, M. W., & Semba, R. D. (2008). Are low tolerable upper intake levels for vitamin A undermining effective food fortification efforts? *Nutrition Reviews*, *66*(9), 517–525.

Krasnyanski, S., May, R. A., Loskutov, A., Ball, T. M., & Sink, K. C. (1999). Transformation of the limonene synthase gene into peppermint (*Mentha piperita* L.) and preliminary studies on the essential oil profiles of single transgenic plants. *Theoretical and Applied Genetics*, *99*, 676–682.

La Frano, M. R., Woodhouse, L. R., Burnett, D. J., & Burri, B. J. (2013). Biofortified cassava increases β-carotene and vitamin A concentrations in the TAG-rich plasma layer of American women. *British Journal of Nutrition*, *110*(2), 310–320.

Langová, R., Jůzl, M., Cwiková, O., & Kos, I. (2020). Effect of different method of drying of five varieties grapes (*Vitis vinifera* L.) on the bunch stem on physicochemical, microbiological, and sensory quality. *Foods*, *9*(9), 1183.

Lanska, D. J. (2010). historical aspects of the major neurological vitamin deficiency disorders: The water-soluble B vitamins. *Handbook of Clinical Neurology*, *95*, 445–476.

Leclerc, J., Miller, M. L., Joliet, E., & Rocquelin, G. (1991). Vitamin and mineral contents of carrot and celeriac grown under mineral or organic fertilization. *Biological Agriculture & Horticulture*, *7*(4), 339–348.

Lee, C. M. (2016). Fifty years of research and development of cosmeceuticals: A contemporary review. *Journal of Cosmetic Dermatology*, *15*(4), 527–539.

Lee, J., Ye, L., Landen, Jr., W. O., & Eitenmiller, R. R. (2000). Optimization of an extraction procedure for the quantification of vitamin E in tomato and broccoli using response surface methodology. *Journal of Food Composition and Analysis*, *13*(1), 45–57.39–147.

Lee, M. J., Son, K. H., & Oh, M. M. (2016). Increase in biomass and bioactive compounds in lettuce under various ratios of red to far-red LED light supplemented with blue LED light. *Horticulture, Environment, and Biotechnology*, *57*(2), 1.

Lešková, E., Kubíková, J., Kováčiková, E., Košická, M., Porubská, J., & Holčíková, K. (2006). Vitamin losses: Retention during heat treatment and continual changes expressed by mathematical models. *Journal of Food Composition and Analysis*, *19*(4), 252–276.

Lewis, E. D., Meydani, S. N., & Wu, D. (2019). Regulatory role of vitamin E in the immune system and inflammation. *International Union of Biochemistry and Molecular Biology Life*, *71*(4), 487–494.

Liu, L., Jeffers, D., Zhang, Y., Ding, M., Chen, W., Kang, M. S., & Fan, X. (2015). Introgression of the crtRB1 gene into quality protein maize inbred lines using molecular markers. *Molecular Breeding*, *35*(8), 1–12.

Lopez, A. B., Van Eck, J., Conlin, B. J., Paolillo, D. J., O'Neill, J., & Li, L. (2008). Effect of the cauliflower *or* transgene on carotenoid accumulation and chromoplast formation in transgenic potato tubers. *Journal of Experimental Botany*, *59*(2), 213–223.

Losey, J. E., Rayor, L. S., & Carter, M. E. (1999). Transgenic pollen harms monarch larvae. *Nature*, *399*(6733), 214–214.

Maqbool, M. A., Aslam, M., Akbar, W., & Iqbal, Z. (2018). Biological importance of vitamins for human health: A review. *Journal of Agriculture and Basic Sciences*, *2*(3), 50–58.

Mazzoni, L., Di Vittori, L., Balducci, F., Forbes-Hernández, T. Y., Giampieri, F., Battino, M., Mezzeti, B., & Capocasa, F. (2020). Sensorial and nutritional quality of inter and intra – Specific strawberry genotypes selected in resilient conditions. *Scientia Horticulturae*, *261*, 108945.

Merkle, S. A., & Dean, J. F. (2000). Forest tree biotechnology. *Current Opinion in Biotechnology*, *11*(3), 298–302.

Mozafar, A., & Oertli, J. J. (1992). Uptake of a microbially-produced vitamin (B12) by soybean roots. *Plant and Soil*, *139*, 23–30.

Mugode, L., Ha, B., Kaunda, A., Sikombe, T., Phiri, S., Mutale, R., Davis, C., Tanumihardjo, S., & De Moura, F. F. (2014). Carotenoid retention of biofortified provitamin A maize (*Zea mays* L.) after Zambian traditional methods of milling, cooking and storage. *Journal of Agricultural and Food Chemistry*, *62*(27), 6317–6325.

Muhammad, A., Katsayal, B. S., Forcados, G. E., Malami, I., Abubakar, I. B., Kandi, A. I., Idris, A. M., Yufus, S., Musa, S. M., Monday, N., & Umar, Z. W. S. (2020). In silico predictions on the possible mechanism of action of selected bioactive compounds against breast cancer. *In Silico Pharmacology*, *8*, 1–13.

Nagraj, G. S., Chouksey, A., Jaiswal, S., & Jaiswal, A. K. (2020). Broccoli. In *Nutritional composition and antioxidant properties of fruits and vegetables* (pp. 5–17). Academic Press.

Naqvi, S., Zhu, C., Farre, G., Ramessar, K., Bassie, L., Breitenbach, J., Conesa, P. D., Ros, G., Sandmann, G., Capelle, T., & Christou, P. (2009). Transgenic multivitamin corn through biofortification of endosperm with three vitamins representing three distinct metabolic pathways. *Proceedings of the National Academy of Sciences*, *106*(19), 7762–7767.

Natesan, S., Duraisamy, T., Pukalenthy, B., Chandran, S., Nallathambi, J., Adhimoolam, K., Manickam, D., Sampathrajan, V., Muniyandi, S. J., Meitei, L. J., Thirunavukkarasu, N., Nalliyappan, G. K., & Rajasekaran, R. (2020). Enhancing β-carotene concentration in parental lines of CO6 maize hybrid through marker-assisted backcross breeding (MABB). *Frontiers in Nutrition*, *7*, 134.

Niki, E., & Traber, M. G. (2012). History of vitamin E. *Annals of Nutrition and Metabolism*, *61*(3), 207–212.

O'Connor, E. B., Barrett, E., Fitzgerald, G., Hill, C., Stanton, C., & Ross, R. P. (2005). Production of vitamins, exopolysaccharides and bacteriocins by probiotic bacteria. *Probiotic Dairy Products*, 167–194.

Paciolla, C., Fortunato, S., Dipierro, N., Paradiso, A., De Leonardis, S., Mastropasqua, L., & De Pinto, M. C. (2019). Vitamin C in plants: From functions to biofortification. *Antioxidants*, *8*(11), 519.

Paine, J. A., Shipton, C. A., Chaggar, S., Howells, R. M., Kennedy, M. J., Vernon, G., Wright, S. Y., Hinchliffe, E., Adams, J. L., Silverstone, A. L., & Drake, R. (2005). Improving the nutritional value of Golden Rice through increased pro-vitamin A content. *Nature Biotechnology*, *23*(4), 482–487.

Patocka, J., Bhardwaj, K., Klimova, B., Nepovimova, E., Wu, Q., Landi, M., Kuca, K., Valis, M., & Wu, W. (2020). Malus domestica: A review on nutritional features, chemical composition, traditional and medicinal value. *Plants*, *9*(11), 1408.

Patterson, R. E., White, E., Kristal, A. R., Neuhouser, M. L., & Potter, J. D. (1997). Vitamin supplements and cancer risk: The epidemiologic evidence. *Cancer Causes & Control*, *8*, 786–802.

Prigge, S. T., Kolhekar, A. S., Eipper, B. A., Mains, R. E., & Amzel, L. M. (1999). Substrate-mediated electron transfer in peptidylglycine α-hydroxylating monooxygenase. N*ature Structural Biology*, *6*(10), 976–983.

Pullman, G. S., & Bucalo, K. (2014). Pine somatic embryogenesis: Analyses of seed tissue and medium to improve protocol development. *New Forests*, *45*(3), 353–377.

Rai, M. K., Kalia, R. K., Singh, R., Gangola, M. P., & Dhawan, A. K. (2011). Developing stress tolerant plants through *in vitro* selection- An overview of the recent progress. *Environmental and Experimental Botany*, *71*(1), 89–98.

Raiola, A., Tenore, G. C., Barone, A., Frusciante, L., & Rigano, M. M. (2015). Vitamin E content and composition in tomato fruits: Beneficial roles and bio-fortification. *International Journal of Molecular Sciences*, *16*(12), 29250–29264.

Rajasekaran, A., & Soundarapandian, S. (2023). Nutritional and medicinal values of *Mangifera indica* L. fruit. *Trends in Horticulture*, *6*(2), 2949.

Rani, S. J., & Usha, R. (2013). Transgenic plants: Types, benefits, public concerns and future. *Journal of Pharmacy Research*, *6*(8), 879–883.

Rao, S. R., & Ravishankar, G. A. (2002). Plant cell cultures: Chemical factories of secondary metabolites. *Biotechnology Advances*, *20*(2), 101–153.

Ribeiro, S. M. R., & Schieber, A. (2010). Bioactive compounds in mango (*Mangifera indica* L.). In *Bioactive foods in promoting health* (pp. 507–523). Academic Press.

Rivlin, R. S., & Pinto, J. T. (2007). Riboflavin (vitamin B2). *Handbook of Vitamins*, *4*(3), 3–4.

Rizvi, S., Raza, S. T., Faizal Ahmed, A. A., Abbas, S., & Mahdi, F. (2014). The role of vitamin E in human health and some diseases. *Sultan Qaboos University Medical Journal*, *14*(2), e157.

Römer, S., Fraser, P. D., Kiano, J. W., Shipton, C. A., Misawa, N., Schuch, W., & Bramley, P. M. (2000). Elevation of the provitamin A content of transgenic tomato plants. *Nature Biotechnology*, *18*(6), 666–669.

Rosati, C., Aquilani, R., Dharmapuri, S., Pallara, P., Marusic, C., Tavazza, R., Bouvier, F., Camara, B., & Giuliano, G. (2000). Metabolic engineering of beta-carotene and lycopene content in tomato fruit. *The Plant Journal*, *24*(3), 413–420.

Rosen, D. S., Sloan, N. L., Del Rosario, A., & De la Paz, T. C. (1994). Risk factors for vitamin A deficiency in rural areas of the Philippines. *Journal of Tropical Pediatrics*, *40*(2), 82–87.

Ross, A. C. (2010). Vitamin A in bioactive compounds and cancer. *Humana Press*, *4*(7), 33.

Rubio, S., Whitehead, L., Larson, T. R., Graham, I. A., & Rodriguez, P. L. (2008). The coenzyme A biosynthetic enzyme phosphorpantetheine adenylyl transferase plays a crucial role in plant growth, salt/osmotic stress resistance, and seed lipid storage. *Plant Physiology*, *148*, 546–556.

Savidge, B., Weiss, J. D., Wong, Y. H. H., Lassner, M. W., Mitsky, T. A., Shewmaker, C. K., Post-Beittenmiller, D., Valentin, & Valentin, H. E. (2002). Isolation and characterization of homogentisate phytyltransferase genes from *Synechocystis* sp. PCC 6803 and *Arabidopsis*. *Plant Physiology*, *129*(1), 321–332.

See, A. W. M., Kaiser, M. E., White, J. C., & Clagett-Dame, M. (2008). A nutritional model of late embryonic vitamin A deficiency produces defects in organogenesis at a high penetrance and reveals new roles for the vitamin in skeletal development. *Developmental Biology*, *316*(2), 171–190.

Semba, R. D. (1999). Vitamin A and immunity to viral, bacterial and protozoan infections. *Proceedings of the Nutrition Society*, *58*(3), 719–727.

Shahidi, F., & De Camargo, A. C. (2016). Tocopherols and tocotrienols in common and emerging dietary sources: Occurrence, applications, and health benefits. *International Journal of Molecular Sciences*, *17*(10), 1745.

Shang, A., Cao, S. Y., Xu, X. Y., Gan, R. Y., Tang, G. Y., Corke, H., Mavumengwana, V., & Li, H. B. (2019). Bioactive compounds and biological functions of garlic (*Allium sativum* L.). *Foods*, *8*(7), 246.

Shearer, M. J., & Newman, P. (2014). Recent trends in the metabolism and cell biology of vitamin K with special reference to vitamin K cycling and MK-4 biosynthesis. *Journal of Lipid Research*, *55*(3), 345–362.

Shintani, T., Huang, W. P., Stromhaug, P. E., & Klionsky, D. J. (2002). Mechanism of cargo selection in the cytoplasm to vacuole targeting pathway. *Developmental Cell*, *3*(6), 825–837.

Shrawat, A. K., & Lörz, H. (2006). Agrobacterium-mediated transformation of cereals: A promising approach crossing barriers. *Plant Biotechnology Journal*, *4*(6), 575–603.

Shumskaya, M., & Wurtzel, E. T. (2013). The carotenoid biosynthetic pathway: Thinking in all dimensions. *Plant Science*, *208*, 58–63.

Singh, V. (2020). Can vitamins, as epigenetic modifiers, enhance immunity in COVID-19 patients with non-communicable disease? *Current Nutrition Reports*, 1–8.

Souret, F. F., & Weathers, P. J. (2000). The growth of saffron (*Crocus sativus* L.) in aeroponics and hydroponics. *Journal of Herbs, Spices & Medicinal Plants*, *7*(3), 25–35.

Šramková, Z., Gregová, E., & Šturdík, E. (2009). Chemical composition and nutritional quality of wheat grain. *Acta Chimica Slovaca*, *2*(1), 115–138.

Srinivasan, K., & Buys, E. M. (2019). Insights into the role of bacteria in vitamin A biosynthesis: Future research opportunities. *Critical Reviews in Food Science and Nutrition*, *59*(19), 3211–3226.

Srivastava, J., Lambert, J., & Vietmeyer, N. (1995). Medicinal plants: A growing role in development. *Agriculture and Natural Resources Department, Agriculture and Forestry*, *2*(3), 33.

Storozhenko, S., De Brouwer, V., Volckaert, M., Navarrete, O., Blancquaert, D., Zhang, G. F., Lambert, W., & Van Der Straeten, D. (2007). Folate fortification of rice by metabolic engineering. *Nature Biotechnology*, *25*(11), 1277–1279.

Takeda, T., Seki, M., Furusaki, S., & Furuya, T. (1993). Factors affecting vitamin E production using plant cell culture of Carthamus tinctorius. *Journal of Chemical Engineering of Japan*, *26*(5), 470–474.

Takeda, T., Takeuchi, Y., Seki, M., Furusaki, S., & Matsuoka, H. (1998). Kinetic analysis of cell growth and vitamin E production in plant cell culture of *Carthamus tinctorius* using a structured model. *Biochemical Engineering Journal*, *1*(3), 233–242.

Tang, G., Qin, J., Dolnikowski, G. G., Russell, R. M., & Grusak, M. A. (2009). Golden Rice is an effective source of vitamin A. *The American Journal of Clinical Nutrition*, *89*(6), 1776–1783.

Tantcheva, L. P., Stoeva, E. S., Galabov, A. S., Braykova, A. A., Savov, V. M., & Mileva, M. M. (2003). Effect of vitamin E and vitamin C combination on experimental influenza virus infection. *Methods and Findings in Experimental and Clinical Pharmacology*, *25*(4), 259–264.

Tanumihardjo, S. A., Palacios, N., & Pixley, K. V. (2010). Provitamin A carotenoid bioavailability: What really matters? *International Journal for Vitamin and Nutrition Research*, *80*(4), 336.

Tavva, V. S., Kim, Y. H., Kagan, I. A., Dinkins, R. D., Kim, K. H., & Collins, G. B. (2007). Increased α-tocopherol content in soybean seed overexpressing the *Perilla frutescens* γ-tocopherol methyltransferase gene. *Plant Cell Reports*, *26*, 61–70.

Tiencheu, B., Nji, D. N., Achidi, A. U., Egbe, A. C., Tenyang, N., Ngongang, E. F. T., Djikeng, F. T., & Fossi, B. T. (2021). Nutritional, sensory, physico-chemical, phytochemical, microbiological and shelf-life studies of natural fruit juice formulated from orange (*Citrus sinensis*), lemon (*Citrus limon*), Honey and Ginger (*Zingiber officinale*). *Heliyon*, *7*(6).

Tunc-Ozdemir, M., Miller, G., Song, L., Kim, J., Sodek, A., Koussevitzky, S., Misra, A. N., Mittler, R., & Shintani, D. (2009). Thiamin confers enhanced tolerance to oxidative stress in *Arabidopsis*. *Plant Physiology*, *151*(1), 421–432.

Upadhyaya, D. C., Bagri, D. S., Upadhyaya, C. P., Kumar, A., Thiruvengadam, M., & Jain, S. K. (2021). Genetic engineering of potato (*Solanum tuberosum* L.) for enhanced α-tocopherols and abiotic stress tolerance. *Physiologia Plantarum*, *173*(1), 116–128.

Vermeer, C. (2004). Vitamin K and bone health. *Nutritional Aspects of Osteoporosis*, *2*, 79–92.

Watanabe, F. (2007). Vitamin B12 sources and bioavailability. *Experimental Biology and Medicine*, *232*(10), 1266–1274.

Watanabe, F., Yabuta, Y., Bito, T., & Teng, F. (2014). Vitamin B12-containing plant food sources for vegetarians. *Nutrients*, *6*(5), 1861–1873.

Welch, R. M., & Graham, R. D. (2004). Breeding for micronutrients in staple food crops from a human nutrition perspective. *Journal of Experimental Botany*, *55*(396), 353–364.

Welsch, R., Arango, J., Bär, C., Salazar, B., Al-Babili, S., Beltran, J., Chavarriaga, P., Ceballos, H., Tohme, J., & Beyer, P. (2010). Provitamin A accumulation in cassava (*Manihot esculenta*) roots driven by a single nucleotide polymorphism in a phytoene synthase gene. *The Plant Cell*, *22*(10), 3348–3356.

Wolucka, B. A., Goossens, A., & Inze, D. (2005). Methyl jasmonate stimulates the *de novo* biosynthesis of vitamin C in plant cell suspensions. *Journal of Experimental Botany*, *56*(419), 2527–2538.

Wright, A. J., Dainty, J. R., & Finglas, P. M. (2007). Folic acid metabolism in human subjects revisited: Potential implications for proposed mandatory folic acid fortification in the UK. *British Journal of Nutrition*, *98*(4), 667–675.

Yabuta, Y., Tanaka, H., Yoshimura, S., Suzuki, A., Tamoi, M., Maruta, T., & Shigeoka, S. (2013). Improvement of vitamin E quality and quantity in tobacco and lettuce by chloroplast genetic engineering. *Transgenic Research*, *22*(2), 391–402.

Ye, X., Al-Babili, S., Kloti, A., Zhang, J., Lucca, P., Beyer, P., & Potrykus, I. (2000). Engineering the provitamin A (β-carotene) biosynthetic pathway into (carotenoid-free) rice endosperm. *Science*, *287*(5451), 303–305.

Zempleni, J., Wijeratne, S. S., & Hassan, Y. I. (2009). Biotin. *Biofactors*, *35*(1), 36–46.

Zhu, C., Naqvi, S., Capell, T., & Christou, P. (2009). Metabolic engineering of ketocarotenoid biosynthesis in higher plants. *Archives of Biochemistry and Biophysics*, *483*(2), 182–119.

31 Designer Medicinal and Aromatic Crops

Development and Characterization of Improved Bioactive Molecules for Viral Diseases

Shahla Haseena Muthalib, Greeshma P. Kammar, H.S. Likitha Aishwayra, Nikhil Kumar Ramesha, Arya Kaniyassery, and Annamalai Muthusamy

31.1 INTRODUCTION

During prehistoric times, the use of medicinal plants was observed in daily human activities. Plants are the primary source of food for humans, animals, and other living organisms. It is very difficult to prove when and where medicinal plants were used for the first time. Due to the development of civilization, industrialization, and urbanization, people have recognized the importance of using traditional medicine. Traditional medicine is the field of medicine that utilizes the medicinal properties of plants to cure physical and mental disorders. Aromatic plants are those that have aromatic flavors and are used for culinary purposes (Giannenas et al., 2020). Aromatic plants are also known as herbs and spices. They have been used for many years for their preservative and medicinal properties. Natural secondary metabolites produced from aromatic plants are referred to as essential oils and are used in various fields. They are complex, volatile, hydrophobic, and fragile compounds characterized by a strong odor and a mixture of secondary metabolites (Greathead, 2003; Brenes & Roura, 2010).

A wide variety of plants are exploited in traditional medicine as well as for drug synthesis due to their pharmaceutical properties. Treatment of disease by using compounds derived from plants began earlier. Over 50,000 plant species are utilized in medicine, including morphine (analgesic), reserpine (antihypertensive), and artemisinin (antimalarial). The FDA has approved Seven plant-derived chemicals with anticancer potential for clinical application, including vinblastine and vincristine (*Catharanthus roseus*) (Gómez-Galera et al., 2007). Knowledge about medicinal and aromatic plants has improved health and the medicinal field. The selection of these plants is not an easy and reliable task due to various factors, such as climate, region-specific areas, and water requirements (Inoue et al., 2019). The use of these plants has become a crude requirement for the development of integrated medicines, and the discovery of various types of traditional medicines has helped to cure a wide variety of diseases and infections. According to the WHO, herbal medicines are defined as plant-derived materials or products that benefit human health (WHO, 2001). Since ancient times, medicinal and aromatic plants have been utilized to treat a wide range of illnesses and ailments. It is believed that these plants were the first to produce pharmaceutical substances (Giannenas et al., 2009). Herbal or traditional medicines have been practised in many countries, such as India, China, Native America, Sri Lanka, and Tibet. Similarly, traditional medicine has been used in veterinary medicine, which has been passed on from one generation to another (WHO, 2001).

Since the 1920s, the discovery of antibiotics by Alexander Fleming has been the gold standard method for eliminating many bacteria. However, these bacteria are gradually more resistant to antibiotic compounds and have many side effects, such as toxicity, which can damage the organs of the food chain after long consumption (Dhama et al., 2015). The major concern of the WHO is that the extensive use of certain antibiotics leads to infectious diseases such as pneumonia and salmonella, which are very difficult to treat without antibiotics for resistant bacteria (Aidara-Kane et al., 2018). Currently, due to restrictions on the use of certain antibiotics as feed additives in certain countries, a considerable increase in the use of medicinal and aromatic plants as alternatives for certain diseases has been observed (Franz et al., 2010; Stevanović et al., 2018). In India, since ancient times, there has been a strong culture of using medicinal plants as medicines to cure certain diseases. As we all know, Indians believe that our holy books, which are written in Sanskrit, have given us importance and knowledge about medicinal plants for curing diseases (Sumner, 2000). The practice of holistic therapy, which is based on Rig Veda, suggests that the main medicinal plant tulsi (basil) is still used in India by the Hindu community as a holistic and therapeutic plant (Singletary, 2018). Another

DOI: 10.1201/9781003475491-31

plant used for holistic therapy is turmeric, and due to the presence of polyphenolic compounds in curcumin, it has been used worldwide (Henrotin et al., 2013).

The COVID-19 pandemic has increased the importance of traditional medicines and home remedies for the prevention of the disease. AYUSH (Ayurveda, Yoga, Unani, Siddha, and Homeopathy) is an exclusive ministry of India that has brought forth all the benefits and remedies in the prevention of the disease. It basically helps us lead a healthy life with conventional and established ideas for preventing diseases and promoting health (Adhikari et al., 2018). Strong antiviral drugs with immunomodulatory and antiallergic properties have reportedly been used in the Indian medical system (Ahmad et al., 2021). Plants usually produce primary and secondary metabolites that are used in pharmacology. These secondary metabolites play an important role in the metabolism of certain enzymes, hormones, etc. Plant secondary metabolites are phytochemicals that include various types, such as phenolic acid, flavonoids, terpenoids, steroids, and alkaloids (Yang et al., 2020). The secondary metabolites produced by aromatic and medicinal plants provide them with economic significance. There are a wide variety of such metabolites that have pharmaceutical properties. Plants such as *Ipomoea batatas*, *Helianthus annuus*, *Cymbopogon*, and *Rosmarinus officinalis* have anticancer, antioxidant, antimicrobial, anti-inflammatory, and antidiabetic effects (Oliveira et al., 2017; Leite et al., 2010; Bruckner et al., 2014; Ghasemzadeh et al., 2016; Guo et al., 2017). Some plants, such as *Helianthus annuus*, *Cymbopogon*, *Rosmarinus officinalis*, *Pogostemon cablin*, *Plectranthus amboinicus*, and *Pelargonium graveolens*, are used in industry as flavoring agents, perfumes, soaps, cosmetics, and food products (Bouzenna & Krichen, 2013). Therefore, the present review highlights the importance of aromatic and medicinal plants, many biotechnological methods to increase secondary metabolite production, and commercially available bioactive compounds on the market. The findings of this study have shown the current importance of medicinal and aromatic plants for viral diseases.

31.2 OVERVIEW OF SECONDARY METABOLITES IN PLANTS

Plants synthesize a wide range of primary and secondary metabolites that play crucial roles in metabolism, photosynthesis, respiration, growth, and development. Usually, these secondary metabolites include compounds like flavonoids, phenolic acids, terpenoids, and steroids, which play vital roles in the field of pharmacology. Many of these compounds are essential for medicine due to their medicinal properties, such as anticoagulant, anti-inflammatory, antimicrobial, and antiviral properties (Teoh, 2016). These compounds are structurally diverse and are present only in a few species within the plant kingdom (Crozier et al., 2008). Plant secondary metabolites are classified into three major groups – terpenoids, phenolics, and steroids, which are involved in biosynthetic pathways (Harborne, 1999). The main compound in secondary metabolites is phenolic acid, and these compounds are found in many plants. A recent report established that the backbone of pharmaceuticals in Western countries is the use of 25% of plant products to produce a drug. One example is aspirin, which is primarily derived from plant compounds used as a drug for stomach aches (Bourgaud et al., 2001).

Secondary metabolites can be grouped based on their structure and the metabolic pathways responsible for their biosynthesis into three groups: terpenoids, phenolic compounds, and nitrogen-containing compounds (Kabera et al., 2014). Phenolic compounds belong to the largest group of secondary metabolites produced specifically by edible plants (>8000 different structures). The shikimate pathway is responsible for the biosynthesis of phenylpropanoids from which phenolic compounds are derived (Cheynier et al., 2013). Phenolic acids, tannins, flavonoids and are the three classes of phenolic compounds associated with health care, of which flavonoids constitute more than half, phenolic acids constitute one-third of all phenolic compounds, and tannins belong to the third most important group (Cook, 1996). The shikimate pathway combines with the mevalonate pathway for the synthesis of flavonoids. The reactivity of the phenol moiety makes phenolic compounds powerful antioxidants (Fernandez-Panchon et al., 2008). Terpenes and terpenoids also referred to as isoprenoids, are structurally diverse groups of secondary metabolites that are lipid soluble, widely produced in flowers and vegetative tissues, but not often in roots (Dudareva et al., 2004). Through deoxyglucose-5-phosphate or the mevalonate pathway, isopentenyl pyrophosphate (IPP), which is a common precursor for terpenes and terpenoids, is synthesized (Rohmer, 1999). Carotenoids are a class of health-beneficial terpenoids. Alkaloids and glucosinolates are nitrogen-containing secondary metabolites. The nitrogen group of alkaloids is acquired from amino acids, which include lysine, tryptophan, aspartate, and tyrosine, and belongs to a large class of nitrogen-containing compounds (Borrelli & Trono, 2016). Quinine, codeine, morphine, atropine, vincristine, and vinblastine are some alkaloids that have pharmacological importance and therapeutic activities (Schläger & Dräger, 2016). Glucosinolates are diverse but small groups of molecules and, conversely, are of nutritional interest.

31.3 PHENOLICS

These are chemical compounds that are an essential class of aromatic plant secondary metabolites (Malčovská et al., 2014). These compounds consist of a phenol group that is attached to its benzene ring (Achakzai et al., 2009; Ali et al., 2013). Two pathways – the malonic acid pathway and the shikimic pathway – are involved in the manufacture of phenolics. Bacteria, fungi, and plants are all engaged in the shikimic pathway. However, the malonic acid pathway is common in microorganisms, such as fungi and bacteria (Cheynier et al., 2013; Ayoub et al., 2016; Puthur, 2016). Lignin is one of the most complex and common phenolic compounds and is found mainly in plants after cellulose; lignin is the major compound used in various drugs and vaccine development (Taiz et al., 2006; Ziegler et al., 2008). According to Verma et al. (2015), phenolic compounds have been shown to offer numerous therapeutic benefits, including anti-inflammatory and antioxidant capabilities. They also appear to considerably lower the risk of cancer. Phenolic chemicals are essential to health and are found in a wide variety of foods and drinks. The different types of fruits and vegetables contain phenolic compounds, such as in grapes, berries, vegetables, and green vegetables. Therefore, doctors usually recommend following a balanced food diet to accumulate all the essential vitamins and minerals and a few metabolites (Lin et al., 2018).

31.4 TERPENOIDS

Terpenoids are another main class of secondary metabolites that are part of the defence mechanism of plants. Higher plants are the primary source of terpenoids, of which over 23,000 distinct structures have been recognized (Kollner et al., 2004). The biosynthesis pathway of terpenoids involves IPP-opentenyl diphosphate and the DMAPP-allylic isomer dimethylallyl diphosphate, which are usually generated in higher plants to produce terpenoids (Cheng et al., 2007). Terpenoids are considered defense systems in plants against insects, herbivores, and microbes that produce certain chemicals to protect themselves. It has been reported that terpenoids are also found in mangoes, beer, citrus fruits, apples, herbs, spices, etc. Terpenoids are also considered to be important metabolites that have pharmaceutical and medicinal properties, such as antioxidant, antidiabetic, anticancer, and immunomodulatory properties (Srivastava et al., 2021).

31.5 FLAVONOIDS

In general, almost all parts of plants, such as flowers, leaves, seeds, and fruits, are pigmented due to the presence of flavonoids. Normally, all plants possess flavonoids, including legumes and nonleguminous plants (Falcone et al., 2012). Flavonoids are also considered to play a role in defense mechanisms against UV radiation, male fertility, auxin transport, phytopathogens, etc. (Mol et al., 1998; Winkel-Shirley, 2002; Bradshaw et al., 2003). Flavonoids, such as flavanols, flavones, flavanols, anthocyanidins, isoflavones, and chalcones, are usually classified into different types based on their structure (Falcone et al., 2012). Flavonoids are usually synthesized in almost all parts of food plants and are responsible for flavor, fragrance, and pigments in fruits, flowers, vegetables, etc. Important food contains flavonoids, such as carrots, mushrooms, cabbage, strawberries, citrus food, etc. (Ruiz-Cruz et al., 2017).

31.6 RATIONALE FOR IMPROVING SECONDARY METABOLITES FOR VIRAL DISEASES

The unusual and complex properties of secondary metabolite biosynthesis and regulation are described, which shows that synthesizing them in a controllable and duplicable manner is a difficult task. Plants that produce highly valuable compounds are difficult to produce and are at risk due to overharvesting (Nitta et al., 2018). Several synthetic products have been produced that can replace natural plant-derived products. However, natural products derived from plants are much more acceptable to the public than artificial products, similar to the case of drugs (Ochoa-Villarreal et al., 2016). Plant-derived pharmaceuticals obtained from secondary metabolites are often costly due to difficult or burdensome practical chemical synthesis, difficult purification procedures, or lower concentrations, complex structures, and specific stereochemical requirements, sometimes only in rare or endangered species (Kirakosyan & Kaufman, 2009; Urabe et al., 2015).

The genetic variability of plants and the extensive purification processes required to produce pure biologically active chemicals can both lower and raise the cost of the final product (Nitta et al., 2018). Therefore, a variety of methods and approaches are employed to enhance the MAPs' ability to produce pharmaceutically active metabolites. There have been several advances in the development of tools for improving secondary products. The secondary metabolites in MAPs could be enhanced in different novel ways, which leads to the overproduction of these phytochemicals, a decrease in the content of toxic compounds, or the synthesis of new compounds. Secondary metabolites that are normally produced at lower concentrations can be scaled up by methods, such as cell suspension culture, organ culture, transgenesis, and metabolic pathway engineering in aromatic and medicinal plants (Gandhi et al., 2015).

Traditional practitioners have used herbs and medicinal plants to treat infectious and noninfectious diseases. Therefore, this traditional knowledge and development of ideas led to the discovery of pharmacology. In many regions of the world, there is more robust use of medicinal drugs than synthetic drugs due to their side effects; plants produce primary and secondary metabolites that have antiviral, antimicrobial, antifungal, and antibacterial properties (Mukhtar et al., 2008). These plants consist of phytochemicals, such as polyphenols, terpenoids, saponins, flavonoids, peptides, etc., which are very effective against viral diseases due to their structural diversity (Naithani et al., 2008). Therefore, medicinal drugs are being used to treat various kinds of diseases. In the 2019, the coronavirus belongs to a large family called Coronoviridae and is a single-stranded, RNA-encapsulated virus that consists of spike proteins on the surface; it is easier to capture and infects the host cell. This family has a genome size of 23–32 kb, and it is divided into two subfamilies, Coronavirinae and Tororirinae, based on their nucleocapsids.

The outbreak of this virus was primarily observed in Wuhan, China, and was considered to be a pandemic disease declared by the WHO. This viral disease has similar properties and symptoms to pneumonia and is considered to be a highly contagious disease caused by tiny droplets produced by an infected person. Later, the WHO identified two different variants of this virus, Middle East respiratory syndrome coronavirus-2 (MERS-CoV-2) and severe acute respiratory syndrome coronavirus-2 (SARS-CoV-2), which are more infectious than other variants. Therefore, the WHO announced that the public health emergency of international concern (PHEIC) was followed by 113 countries, and the main objective of that rule was social distancing, sanitizing, using gloves and masks, and the healthcare sector's use of PPE kits to protect themselves (Payne, 2017). Similarly, other viruses, such as herpesvirus, which is also known as herpes simplex virus, are type 1 and type 2 viruses (HSV-1 and HSV-2) that cause latent infection in the human body, especially in the dorsal root ganglia (DRG). Reactivation of these DRGs can result in infection. While HSV-2 causes meningoencephalitis in newborns and meningitis in adults, HSV-1 is the cause of encephalitis, blindness, and other nervous system problems. The genome of the virus contains double-stranded DNA molecules that are encapsulated within isopentahedral capsids and consist of 162 capsomers. These viruses mainly consist of four glycoproteins that help in attachment to the host cell, bind to its receptor to invade the host, and cause infection (Steiner et al., 2013).

The other main type of virus is hepatitis B virus (HBV). It is a chronic viral disease that has caused infection around the world, and approximately 200 million people are infected by this virus. It is a major infection that has caused health issues for a large population. It is mainly transmitted through the blood and semen of an infected person. Vaccines have been developed, and the government of India has made it mandatory to administer the vaccine. Therefore, its spread has been minimized (Lee, 1997). It is a circular double-stranded DNA molecule that is 3200 base pairs in length and can replicate independently in the host to cause infection (Miller et al., 1989). Hepatitis C virus is also a major type of virus that causes global health problems due to infection. It is a causative agent for chronic liver infection. HCV mainly causes infection in people aged 39–40 years. The USA and the UK have the greatest number of infections, and Japan has reported more infections among older people.

The genome of this virus consists of small, enveloped RNA molecules that belong to the family Flaviviridae. Single-stranded RNA is positively charged and encodes approximately 3000 amino acids in length (Lauer et al., 2001). It causes infection to hepatocytes by invading various receptors and causing infection. The other type of virus is HIV, a virus that is known worldwide to infect the human immune system. This unique type of virus replicates in the presence of an enzyme called reverse transcriptase, which manipulates the host genome to replicate itself through the lytic cycle. It is a sexually transmitted disease that can also be transmitted through the blood of an infected person. There were an estimated 39.9 million people living with HIV at the end of 2023, 65% of whom are in the WHO African Region (World Health Organization (WHO, 2024)). These are major types of viruses that have threatened people. Therefore, various medicines and vaccines are being developed for various types of viral infections. Modern medicine usually causes many side effects, thus leading to the utilization of compounds extracted from plants.

From the beginning, it has been believed that plants, herbs, vegetables, greens, and spices are the daily medicines that are balanced and consumed through our daily diet. As mentioned, plants contain secondary metabolites that are very useful and have therapeutic properties. The main key elements in secondary metabolites are polyphenols. Therefore, this section describes the importance of polyphenols, which are plant-derived active compounds that have a significant number of reports providing evidence for their mechanism of action against viral diseases, such as coronavirus, influenza virus, herpes simplex, and dengue fever (Montenegro-Landívar et al., 2021). Polyphenols protect plants from biotic stresses, such as fungi, microbes, and weeds, and abiotic stresses, such as UV radiation, temperature, and drought (Tuladhar et al., 2021). Polyphenols are involved only in plant metabolism, reproduction, hormonal regulation, nutrient supplementation, etc., but are not involved in defence mechanisms (Sharma et al., 2019). These compounds are alternative treatments due to their health benefits and are most found in fruits, vegetables, and greens (Montenegro-Landívar et al., 2021). Since frequent eating of fresh fruits and vegetables high in polyphenols can guard against conditions like cancer, diabetes, neurological disorders, and cardiovascular disease, these substances are exploited in the development of new drugs for the benefit of human health (Cheynier, 2012). There are several reports that mention that consuming foods rich in phenolic compounds can lower the risk of disease due to their antioxidant activity. Usually, antioxidants control the rancidity level, inhibit the formation of toxic oxidation products, and are used as food additives because they increase the shelf life of the food (Shahidi et al., 2015). Recently, polyphenols, especially flavonoids, have been reported to be highly effective at different stages of coronavirus entry into host cells, and a unique component of green tea called epigallocatechin-3-gallate (EGCG) inhibits replication of DNA viruses (Steinmann et al., 2013).

Polyphenols are chemical structures that contain one or more hydroxyl groups attached to the benzene ring, and depending upon their chemical structure, they are classified into four families: flavonoids, phenolic acids, lignans and stilbenes (Saurina et al., 2015). Flavonoids are the largest group of polyphenols consumed by approximately 60% of the diet, and there are approximately 6000 flavonoids that are beneficial to human health (Zhang et al., 2020; Panche et al., 2016). It has also been reported that flavonoids can inhibit the replication of type-1 and type-2 herpes viruses and HIV-1 and HIV-2 (Naithani et al., 2008). A variety of polyphenols, such as quercetin, hesperidin, rutin, catechin, apigenin, and morin, are present in various fruits, vegetables, spices, etc., including berries, almonds, beans, citrus species, peppermint, grapefruit, chamomile, mint, etc. The main mechanism involved in SARS-CoV-2's penetration of the host cell (lung epithelium) is via a spike protein (protein S) found in the virus's outer layer. It binds to the host cell using the enzyme angiotensin-converting protein II (ACE2) as a receptor. An endosomal protease called transmembrane cellular serine protease 2 hydrolyzes the protein S, aiding in membrane fusion and being absorbed by the host cell for reproduction. There are 2 main proteases involved in replication: $3CL^{PRO}$ and papain-like protease (PL^{PRO}) (Montenegro-Landívar et al., 2021). Therefore, polyphenols are specifically developed as protease inhibitors, which can be seen in the case of *Broussonetia papyrifera* with potential anti-coronaviral agents that inhibit the PL^{PRO} protease (Park et al., 2016). There are several pieces of evidence indicating that polyphenols, such as hesperidin and naringenin, which are derived from the citrus fruit family, inhibit proteases in coronavirus (Khalifa et al., 2020). Although a wide range of in silico studies and experiments are needed for in vivo and in vitro studies to discover polyphenols for viral diseases, recently, hesperidin, a flavonoid that can bind to SARS-CoV-2 proteases, has been identified as an antiviral drug (Wu et al., 2020). Finally, in vitro analysis of pomegranate peel extract revealed that it contains polyphenols such as punicalin and punicalagin, which can strongly inhibit the binding of the virus to the receptor (Suručić et al., 2021).

Traditional medicines are the best for all age groups of the population, but their concentration should be accurate for them to be effective drugs. Although manufacturing herbal or medicinal drugs is time-consuming, their effects can last for a lifetime and mainly have no side effects. Therefore, there should be increased development of secondary metabolites for treating any kind of disease. Currently, the development of various methodologies, assays, digital machines, etc., has led to greater opportunities in research and development in science, and without these techniques and modern methods, it would be impossible to identify, discover, design, and develop a drug or a vaccine against any kind of disease. Most plants contain secondary metabolites that are essential for the immune system can be consumed to lead a healthy life and are easily available. In certain parts of the world, financial issues cause the unaffordability of basic medicine. Thus, the development of and awareness of medicinal plant drugs should increase.

31.7 BREEDING FOR IMPROVED SECONDARY METABOLITES FOR VIRAL DISEASES

The cultivation of plants in fields or greenhouses under the right set of conditions is advantageous over the cultivation of plants in axenic production systems. The extraction of compounds from plants is difficult because of factors such as climate, sensitivity to pathogens, demand for fertilizers, agrochemicals, and regulatory problems with genetically modified plants. As a result, several methods have been created as complimentary approaches to improve bioactive secondary metabolites. These techniques include in vitro callus and cell suspension culture, root and shoot culture, hairy root culture, whole plant culture, micropropagation, and biotransformation. Genetically transformed cell lines with engineered metabolic pathways involve plants as well as microorganisms, which can lead to increased production (Nitta et al., 2018). This review discusses cell culture methods and UV-B induction for enhancing secondary metabolites in medicinal and aromatic crop species.

At present, there has been a drastic increase in the use of medicinal plants for therapeutic and drug purposes; thus, the use of genetic and biotechnological techniques has accelerated the production of medicinal plants. The classical method is through biotechnological-based breeding methods (BBBMs), in which the main 2 methods are plant tissue culture and artificial polyploidy induction (Niazian, 2019). As plants produce a small number of secondary metabolites, plant breeders can use modern technologies such as next-generation sequencing (NGS), genomics, proteomics, transcriptomics, and other recent technologies to obtain the whole genome and identify specific sites or regions (Zhao et al., 2018). The results obtained from NGS can be utilized in sequence analysis, and DNA barcoding is used to identify the genetic diversity of medicinal plants (Techen et al., 2014). The quick creation of metabolites is aided by the identification of essential genes engaged in metabolic pathways involved in the production of secondary metabolites, which is made possible by the mapping of whole genomes in medicinal plants. Targeted genome editing methods such as TALENS, CRISPR-Cas9, and ZFN have been utilized as genetic tools to produce variant plants with increased metabolites (Pouvreau et al., 2018). To produce a variant plant using genome editing methods, identification of the target gene, which can be possible by using molecular markers, is insufficient.

The complementary binding of molecular markers to target genes is a very important technique used in molecular experiments. Examples of molecular markers include RAPD, RFLP, etc., which are being used, but the only disadvantage of these markers is that they are very time-consuming and expensive and have certain limitations. Therefore, an alternative method used was DNA microarray, which works on the principle of hybridization and binds to the target specifically. Thus, this approach is applicable to medicinal plants because of differences in gene expression under different growing conditions (Ishkanian et al., 2004). However, the most widely used technique is NGS-based transcriptome analysis, which helps in differentiating homologous and paralogous gene copies and identifying genes. This method has been applied to the medicinal plant ajowan (*Trachyspermum ammi* L.), a therapeutic plant with genes essential to the synthesis of monoterpenoids (Howyzeh et al., 2018). Monoterpenoids have anti-inflammatory, antioxidant, and anti-neurodegenerative qualities and have been suggested to be beneficial in the treatment of neurodegenerative illnesses like Alzheimer's (Wojtunik-Kulesza et al., 2021). In addition, all of the mentioned NGS techniques combined with high-throughput sequencing along with omics technologies have improved the identification of specific compounds in medicinal plants (Yamazaki et al., 2018).

31.8 DESIGNER MEDICINAL AND AROMATIC PLANTS (MAPs)

The designer crop is a marker-assisted selection (MAS) method that improves the quality of crops and thus increases yield. In genetically modified crops, foreign genes or DNA from different sources (species) are artificially inserted into the crop. MAS is a fast-forward traditional method for improving crop quality that involves the selection of ideal crop specimens and the repeated crossing of these specimens until a crop with the desired characteristics is obtained. Genes for specific proteins are introduced to plants to encode a specific trait. Human health and culture are intimately linked to medicinal and aromatic crops (MACs). Plant-derived medicines, or the byproducts of secondary metabolism, are widely used in the human healthcare system. Plant genetic engineering can facilitate the engineering of secondary metabolic pathways by increasing the synthesis of secondary metabolites (Gómez-Galera et al., 2007).

Humans have exploited different leaves, flowers, roots, seeds, etc., for medicinal purposes in certain plants. These selected plants have irreplaceable medicinal properties and cannot be modified or produced by any synthetic means without the use of the raw products of these plants. Therefore, people have begun overexploiting medicinal plants, causing red alerts in a few countries. As mentioned earlier, these plants require specific environmental conditions for their growth and survival. Thus, these plants cannot be cultivated, but they are mostly extracted from the wild (Gómez-Galera et al., 2007). Normally, medicinal plants produce a very small number of secondary metabolites, which are not sufficient for large populations around the globe. This has allowed researchers to develop designer medicinal and aromatic plants. These plants are genetically manipulated to express biosynthetic genes or enzymes involved in the production of secondary metabolites (Sato et al., 2001). The evolution of science and technology has led to the development of many new techniques, methods, and arrays, especially in the field of genetic engineering, which has helped in designing various genetically modified plants. Recent ongoing research has focused on cancer, which is a worldwide disease and has been reported to be the number one disease affecting many citizens of the United States. Cancer has a devastating physiological and psychological impact on patients, and there is no proper treatment for cancer due to its proliferative properties (Buyel, 2018). Cancer can be caused by biotic factors such as heredity, epigenetic factors, environment, and types of viruses, whereas abiotic factors include pollution, smoking, stress, alcohol, obesity, etc. The ultimate factor of concern for researchers is the cancer cells that cause mutations in tumor suppressor genes (Bieging et al., 2014).

There are several strategies for curing cancer, such as surgery, chemotherapy, immunotherapy, and radiotherapy, which can be the primary treatments for the first few stages of cancer, but these treatments are temporary. In a few cases, it is expensive and time-consuming, and the major disadvantage is its side effects (Buyel, 2018). Anticancer vaccines are also being developed by exploiting and utilizing knowledge of the immune system (Murphy et al., 2016). By observing all the factors, effects, and criteria of cancer treatment, researchers have explored the use of plant-derived medicinal materials for cancer treatment. The use of genetic engineering, particularly in genetic modification by using gene editing systems such as CRISPR-Cas9, has made it possible to produce seven plant-derived compounds with antiviral activity known as Taxol/paclitaxel, a well-known plant-derived drug that is approved by the FDA (Buyel, 2018; Gómez-Galera et al., 2007). This is the first plant-derived drug that utilizes secondary metabolites for its production. The main source of paclitaxel is the Pacific yew, which is grown only in tropical forests. Large quantities of biomass production have led to the extinction of this plant (Wani et al., 2014). Hence, genetic modification of these plants to produce small molecules of secondary metabolites is important. Similarly, medicinal and aromatic plants have been designed to treat various viral diseases. The design of these plants has not only improved the development of different plants but also increased the market demand for herbal medicines. It has given greater attraction and awareness to the public.

31.9 PLANT TISSUE CULTURE

31.9.1 Production in Plant Cell Culture

Plant cell culture (PCC) is a very promising technique for improving the production of secondary metabolites. PCC is an opportunistic platform for the improvement of metabolites. One of the main drawbacks of this technique is its low yield and variable results. Thus, a few strategies have been used to optimize the level of production (Hussain et al., 2012). The factors for improvement are the selection of high-yield cell lines, transformation methods, immobilization techniques, micropropagation, tissue culture, and transgenic hairy roots. Advancements in plant cell and tissue culture combined with genetic engineering, molecular biology, enzymology, and fermentation techniques have all been combined to increase production (Sarfaraj et al., 2012). Plant cell culture has the potential to produce secondary metabolites. This system was developed to study and produce plant secondary metabolites. This provides practical knowledge and a representation of how the plant cell culture system works (Smetanska, 2008). This method can alter the environmental conditions required for these plants. Therefore, observing these criteria and characteristics has attracted researchers to work on this system (Zhong, 2001). Countries such as Japan, Germany, and the USA have adopted this system to produce a wide range of metabolites (Mulabagal et al., 2004). As mentioned earlier, secondary metabolites are produced only in certain plant kingdoms in low quantities; thus, this system has attracted or attracted the attention of researchers to increase the yield of secondary metabolites. It is also considered to be a robust technique (Smetanska, 2008).

A set of techniques that are designed for the growth and multiplication of cells and tissues under aseptic and controlled environments using nutrient solutions is referred to as plant cell culture (Vargas et al., 2006). The potential for the growth and regeneration of plants through cell culture has limited potential. Suspension cultures are fast-growing, can be used to produce target compounds and are amenable to continuous culturing. Plant cells are totipotent and can produce a wide variety of compounds (chemicals) that are also identified in the parent plant. To increase the desired metabolites in a culture, growth conditions such as aeration, temperature, the intensity of light, hormones, pH, and the composition of the medium are changed according to the specific requirements (Maschke et al., 2015). Most of the secondary metabolites increase during the plant's stationary phase of growth. The cells are cultured under controlled conditions for secondary metabolite production. After multiple subcultures of cells, undifferentiated cultures undergo soma clonal variation (Bourgaud et al., 2001).

In addition to the breeding technologies that have helped us to identify specific targets and their expression levels, the production of secondary metabolites is very low, and therefore, an alternative method of production, plant cell culture, is needed. Tissue culture can be a substitute for the synthesis and enhancement of secondary metabolites on an industrial scale due to (1) the high demand for metabolites of the rivet, (2) the fact that certain plants grow at a lower rate, so a long time is needed; (3) the need for large cultivation areas to reduce land availability; and (4) the ability to endanger wild populations of medicinal plants (Gandhi et al., 2015). As the population increased, the demand for medicinal plants increased, and plant products need to be improved. Plant tissue and cell culture are lifespans that can increase the production of secondary metabolites (Pant, 2014). Plant cell culture has a promising effect on the large-scale production of secondary metabolites. Plant cell culture is one of the most essential systems for increasing yield. This system is not routinely used when it produces a large amount of product that can be detected by using a bioreactor (Roberts & Shuler, 1997). This system has also rapidly advanced the production of secondary metabolites.

The most common technique used to produce secondary metabolites in plant cell culture is the immobilization of plant cells in calcium alginate entrapment. This technique has shown the importance of inducing the production of alkaloids in a tenfold increase, which is also considered to be an elicitor (Curtis et al., 1995). As we can see in the case of a secondary metabolite, terpenoids can be produced in cell culture. Terpenoids are a large group of isoprenoids that include a diverse class of naturally occurring organic chemical derivatives. Plant terpenoids are classified into primary and secondary metabolites; the primary metabolites are involved in plant metabolism and the production of hormones such as gibberellins and carotenoids, but secondary metabolites are involved in medicinal purposes because they have medicinal properties such as antioxidant, anti-inflammatory, and anticancer properties (Roberts, 2007). Alternate methods, such as chemical synthesis, semi-synthesis, and genetic engineering, have also been developed, but these methods have advantages and disadvantages. One of the major advantages of plant cell culture is that it is environmentally friendly, renewable, cost-effective, and easy to maintain, and it is also an alternative for secondary metabolite production. The other advantage is that the culture conditions can be precisely maintained as per the FDA requirements. Changes in environmental conditions are key for metabolic engineering, which enhances selectivity and yield in plant cell culture (Frense, 2007; Roberts, 2007; Tabata, 2006). However, plant cell culture has several drawbacks, such as low yield, slow growth, instability of cell lines, changes in parameters, culture factors, and environmental factors; therefore, various strategies have been employed to enhance plant cell culture (Ravishankar et al., 1993).

There are certain advantages of using plant cell culture over other methods, including the facilitation of high-yield cell selection, increased product harvesting, easy validation, and ease of handling (Nitta et al., 2018). In addition to these advantages, cell culture methods have certain disadvantages. The use of plant cell cultures with lower concentrations of target compounds is expensive. The expression of genes involved in secondary metabolic pathways is low in cell culture due to the higher concentration of phytohormones and carbon sources in culture, which can increase the growth rate of cells. Several other factors, such as competition for low expression of enzymes, regulatory proteins, similar precursors, poor compartmentalization of enzymes, and intermediates, inadequate metabolite storage in vacuoles, nonfunctional plastids, selective proteolysis, and inappropriate functioning function of small RNAs, can impede secondary metabolite production (Pasquali et al., 2006; Matsuura et al., 2013; Gupta et al., 2017). These problems can lead to lower yields of desired metabolites in cell cultures. The cell culture should be flexible enough to scale up, increase the synthesis of metabolites, and increase growth rates (Nitta et al., 2018). Cell culture also leads to biochemical and genetic instability (Ochoa-Villarreal et al., 2016), but success has been achieved in the enhancement of shikonin, Taxol, and ginsenosides.

31.10 SELECTION OF HIGH-YIELDING CELL LINES

For any kind of upgradation, selection is the key element in choosing a plant associated with a high-yield cell line. The screening of heterogeneous populations of variant cell line clones usually yields the desired product. Mutations have been implemented to obtain overproducing cell lines. In this method, a large population of cells is exposed to a cytotoxic agent or environmental stress, so the cells that are resistant to these agents will undergo selection procedures. Different types of agents, such as 5-methyl tryptophan, glyphosphate, and biotin, are used for high-yielding cell lines (Rao et al., 2002). The selection of an appropriate parent plant that has a relatively high content of secondary metabolites plays a major role. The selection of a cell line is the foremost factor to be considered because secondary metabolic accumulation is genotype-specific, and the concentration of metabolites varies across each genotype. For example, the amount of the alkaloid camptothecin varies among different species of camptothecin (Ramesha et al., 2008). The traditional method for the selection of cell lines is based on visual screening if certain secondary metabolites show pigmentation (Yamamoto et al., 1982; John, 1980). However, the present method helps us quantify and select the appropriate cell line and growth kinetics can also be obtained (Liang et al., 2006).

31.11 CHEMICAL FACTORS

Changes in the culture medium play a role in the manipulation of metabolic pathways to produce secondary metabolites. Various types of media are used for the growth of cells with abundant carbon, which plays a significant role in the formation of secondary products. It is also believed that concentrations above 2% stimulate the growth of cells, but an increase or decrease in this concentration will affect production. Thus, manipulation of the concentration of carbon sources has shown little evidence of an increase in yield (Stafford et al., 1986). Nitrogen is one of the major requirements for the primary metabolism of enzymes and hormones and for growth; metabolism is equally as important as a carbon source that helps in growth, and the biochemical synthesis of nucleic acids, lipids, and proteins at slightly higher concentrations will improve the growth of cells (Noguchi et al., 1977). Nitrogen regulates plant cells by regulating the nuclear membrane and has an intermediate effect on growth and cell division in plants. It has also been reported that it enhances flowering in certain plants; increases protein content, membrane stability, biomass, and regulation; and ultimately increases nutrient uptake. Nitrogen is the main component of fertilizers for these reasons (Xin et al., 2014).

The phosphate component in media has a major effect on the production of secondary metabolites. A high concentration leads to an increase in cell growth but can also negatively affect growth and supplement products. Accurate levels of phosphate stimulate the action of hormones; therefore, manipulation of this source has also been shown to increase yield (Rao et al., 2002). Phosphate plays an essential role in plant cells because it maintains the structure of DNA, RNA, proteins, and other components of cells. They are primarily involved in forming high-energy bonds such as ATP. Phosphate is also involved in the formation of macromolecules and nucleic acids that are included in the metabolic pathway through which secondary metabolites are produced (Romero-Rodríguez et al., 2018).

31.12 PHYSICAL FACTORS

The temperature normally required for induction or cell culture is 17–25°C; however, depending on the plant species, a few species will need lower temperatures (19°C), while a few species will need higher temperatures (25–27°C); however, different species are required to have different temperatures for the accumulation of secondary metabolites, as the optimum temperature ranges from 25 to 32°C. There is also another important factor to be noted: the type of culture used for solid culture ranged from 25–30°C, and liquid cultures ranging from 22 to 25 to 28°C were found to be optimal. All these parameters were determined for *P. frutescens* cells (Zhong et al., 1993).

Water is the source of plant survival. It is the main component that functions properly and helps in the growth, development, and supplementation of nutrients from the soil. Therefore, the generation of secondary metabolites is influenced by the water concentration. Studies have indicated that some plants produce more secondary metabolites when there is a greater water content. However, because water is used for main metabolites and only a small number of secondary metabolites are created, water stress is linked to reduced secondary metabolite productivity. According to some reports, when a plant receives the same amounts of sunlight, water, minerals, and nutrients, its production of secondary metabolites increases tenfold (Gershenzon, 1984).

31.13 MAINTENANCE OF pH

The optimum pH is critical for cell growth, and the production of a metabolite ranges from 5–6.6; however, this depends on the plant species used, as few species require an acidic environment for the production of metabolites, whereas other species require a neutral to basic environmental pH. The optimum pH has a role in cellular respiration, cell growth, the production of metabolites, etc., for the production of secondary metabolites. Fluctuations in the pH lead to changes in the concentration of hydrogen ions in a culture, which is directly associated with cell growth and survival. Therefore, investigations on the optimal pH assist in the elucidation of cell growth levels (Butenko et al., 1984). The optimum pH for growth of almost all plants ranges from 5.5 to 6.5. This is the optimum pH required for the availability of nutrients. The optimum pH is a sensitive component of the cell (Hyyryläinen et al., 2001). Slight changes in pH can cause major effects, such as interfering with amino acid functional groups and impairing hydrogen bonding, which results in further changes in protein folding, leading to denaturation, nonfunctional enzymes, and other activities (Msimbira et al., 2020).

31.14 AGITATION AND AERATION

Similarly, as mentioned, the importance of secondary metabolites in the medicinal and pharmaceutical fields has led to most of the interest from academic investigators, and various other fields have led to the intensive production of bioactive components and secondary metabolites using large bioreactors, which has contributed to advancements in technology and reduced the manpower available to produce metabolites. These bioreactors have different shapes for large-scale production. However, there are several factors affecting the growth of cells in a bioreactor, including the gaseous atmosphere, CO2 exchange, pH, minerals, growth concentration, sterilization, agitation, aeration, etc. (Ruffoni et al., 2010).

Agitation modes are classified into two types.

a. Mechanically agitated bioreactors – In these types of bioreactors, agitation is driven by impellers; these impellers are flat-blade turbine impellers that break small bubbles in bioreactors.
b. Pneumatically agitated bioreactors – These are tall and thin bioreactors compared to agitation bioreactors; these are also classified into two types.
1. Bubble column – air is bubbled through the base of the column, and the column is agitated.
b) Airlift gas is sprayed from the top of the column, and the medium flows downward; this process is agitated by using a baffle.

Compared with bubble column bioreactors, airlift bioreactors have more advantages due to their simple construction, lack of high shear, high yields at low input rates, etc. (Ilan et al., 1995; Ziv et al., 1991).

Aeration is the amount of oxygen present in bioreactors and plays a vital role in plant cells. The amount of oxygen present depends on the gas phase and air bubbles in the medium and depends on the plant species used. Compared with microbial cells, plant cells usually have a lower metabolic rate; therefore, they require a lower and more continuous oxygen supply, but at high concentrations, oxygen inhibits cell growth (Smart et al., 1984).

31.15 MICROPROPAGATION

Micropropagation refers to the in vitro multiplication or regeneration of plant cells under aseptic and sterile conditions. It is a conventional plant-breeding technique used to improve the production of medicinal materials. In vitro experiments and tissue culture have tremendous potential for increasing the production of medicinal plants through methods such as micropropagation (Debnath et al., 2006). Micropropagation has many advantages over methods such as vegetative propagation due to its limitations. There has been vast progress and improvement in the production of commercial medicinal plants. This method has a significant role in targeting the biosynthesis of metabolites. The main aim of this technique is to increase production by propagating an existing meristem that can produce a genetically identical plant exactly like the donor plant. This method is widely accepted in the field of biotechnology for enhancing production and for adopting techniques such as in vitro regeneration and genetic transformation. It is favorable and, most importantly, is supplied to selected medicinal plants for the production of secondary metabolites (Kapoor et al., 1994).

31.16 CALLUS AND CELL SUSPENSION

Although plant cell culture has provided an efficient way to produce secondary metabolites under controlled conditions without any interference from environmental restriction, it has limitations. Therefore, researchers have led to the development of calli, and cell suspension methods, which are preferably used in the large-scale production of secondary metabolites due to their rapid growth, high rate of metabolism, and ability to perform controlled monitoring, have been developed. This technique can significantly increase the level of secondary metabolites and is a potential alternative for improving the production of secondary metabolites (Bong et al., 2021).

31.17 HAIRY ROOT SYSTEM

The use of a hairy root system via inoculation with *Agrobacterium rhizogenes* is one of the most efficient methods for producing secondary metabolites in plant roots (Karuppusamy, 2009; Palazon et al., 1997). The secondary metabolite from hairy roots arises from the infection of a bacterium, *Agrobacterium rhizogenes*, which is like the parent root but has a greater yield (Sevón et al., 2002). Using this technique, *Agrobacterium rhizogenes* infection only transmits T-DNA from the Ri plasmid, and the genes in this section are produced by plant cells endogenously (Hussain et al., 2012). Thus, these are the current approaches being used to enhance medicinal plants' ability to produce secondary metabolites.

31.18 ELICITATION AND PRODUCTION OF SECONDARY METABOLITES BY PLANT CELL CULTURE

The biotechnological production of secondary metabolites through plants and organ culture has provided a great alternative, but it has limited commercial success due to the lack of analytical selection of high-yield cell lines, stable culture, lack of understanding of the synthesis of secondary metabolites, etc. (Verpoorte et al., 2002; Sevón et al., 2002). Due to these limitations, researchers have developed various methods and techniques, such as micropropagation, hairy root culture, cell immobilization, large-scale cultivation bioreactors, and biotransformation (Namdeo, 2007). Using cell culture, it has been reported that cells do not produce enough secondary metabolites, which has led to the use of elicitors such as methyl jasmonate, salicylic acid, and heavy metals (DiCosmo et al., 1985; Poulev et al., 2003).

Elicitor: An elicitor is defined as a substance at small concentrations that initiates or improves the biosynthesis of specific compounds in a living cell system.

Elicitation: Elicitation is the induced or enhanced synthesis of secondary metabolites in plants due to their competitiveness, obstinacy, and survival by the addition of small amounts of elicitors. Examples of elicitors include methyl jasmonate, salicylic acid, chitosan, UV, and heavy metals (Radman et al., 2003). The concentration of elicitors, exposure time, nutrient medium, and age of the culture can affect the process of elicitation.

31.19 CLASSIFICATION OF ELICITORS

Elicitors are classified based on their origin: biotic or abiotic.

a. **Abiotic elicitors** are substances of nonbiological origin, such as inorganic salts, such as copper and calcium ions.
b. **Biotic elicitors** are substances of biological origin, such as polysaccharides derived from the plant cell wall and glycoproteins.

The improvement of the production of secondary metabolites from plant cell culture through elicitors has become a new area of interest to researchers and has many economic benefits in the pharmaceutical industry. Several parameters are involved in the utilization of elicitors, such as the elicitor concentration and selectivity, cell line selection, growth regulation, age of culture, and nutrient composition quality of the cell wall (Namdeo, 2007).

A. Concentration of elicitor

The elicitor concentration plays an important role in the elicitation process, which has been reported in the accumulation of ajmalicine in *C. roseus* culture when treated with different concentrations of elicitors from *A. niger* and *F. moniliforme*. The accumulation of ajmalicine was greater in cells treated with a 5% elicitor but was lower in cells treated with a 0.5% elicitor (Nef-Campa et al., 1994). However, higher concentrations of elicitors can lead to cell death, hypersensitivity, etc.

B. Exposure duration elicitor

C. roseus cells were exposed to elicitors extracted from *T. viridae*, *A. niger*, and *F. moniliforme* for 24 h, 48 h, 72 h, or 96 h. There was a drastic increase of approximately threefold in ajmalicine production by *C. roseus* cells exposed to elicitor extracts of *T. viridae* for 48 h. Similarly, there was an increase of twofold in cells treated with elicitors from *A. niger* and *F. moniliforme*. However, a further increase in concentration led to a decrease in ajmalicine content (Rijhwani et al., 1998; Moreno et al., 1993; Negrel et al., 1995).

31.20 CULTURE AGE

Subculture duration is a salient parameter in the production of metabolites by elicitation. This parameter also plays a role in a study of *C. roseus* cells cultured for 20 days, which showed a greater yield of ajmalicine upon elicitation (Namdeo, 2007). Elicitation has shown promising and favorable results in enhancing the production of secondary metabolites, but due to certain drawbacks and limitations, the overall mechanism is not understood; therefore, researchers have focused on the role of elicitation in plant cell culture. Combined efforts from plant sciences, microbiology, biotechnology, and various other fields of science have contributed to reaching this level (Namdeo, 2004). Several studies have shown that elicitation can enhance secondary metabolites in medicinal plants. Elicitation of *C. roseus* with *Trichoderma viride* leads to increased accumulation of ajmalicine in cultures, which is an example of a biotic elicitor (Namdeo et al., 2002). Abiotic elicitors, another class of elicitors of a chemical or physical nature, are produced by biotic elicitors within plants. One example of an abiotic elicitor is UV radiation, which has been widely used to enhance secondary metabolites such as flavonoids.

31.21 TRANSGENIC PLANTS FOR IMPROVED SECONDARY METABOLITES

Recombinant DNA technology uses methods to design plants with new properties by the introduction of genes into the plants. These types of genetically engineered plants are called transgenic plants or genetically modified crops. By using the transgenic technique, one can produce plants with desired traits and increased yields (Rani & Usha, 2013). Plant cells can be grown in culture and manipulated by introducing a gene of interest. There are many methods or methods by which genes can be introduced into plant cells, including physical methods, such as protoplast transformation, electroporation, and microinjection; biolistic or chemical methods, such as the use of PEG, calcium phosphate, and lipofection; and biological methods, such as *Agrobacterium*-mediated transfer or viral vector-mediated transfer. The most common method used for gene transfer is *Agrobacterium*-mediated transfer. In transgenic plants, there is a change in the germ line of the plants, and these changes are stably inherited by the progenies after reproduction. For successful genetic manipulation, there are two criteria: (1) there must be a method to introduce the gene of interest into the target plant, and (2) the molecular genetics of the system in which the gene is introduced should be properly known (Primrose & Twyman, 2006).

Transgenic approaches are used to improve the quality and quantity of secondary metabolites by targeting the

TABLE 31.1
Transgenic Medicinal and Aromatic Plants for Improved Secondary Metabolites

S. no.	Transgenic plant	Gene introduced	Method of transformation	Benefits	Reference
1	*Artemisia annua*	FDS	*A. tumefaciens*	Enhanced the production of artemisinin	(Chen et al., 2000)
2	*Hyoscyamus niger*	PMT, H6H	*A. tumefaciens, A. rhizogenes*	Enhanced scopolamine biosynthesis.	(Zhang et al., 2004)
3	*Papaver somniferum*	PsCor1.1	*Agrobacterium tumefaciens*	Increase in morphinan alkaloid content.	(Larkin et al., 2007)
4	*Salvia miltiorrhiza*	HMGR, DXS, GGPPS	*A. tumefaciens*	Enhanced the production of tanshinone.	(Kai et al., 2011)
5	*Salvia miltiorrhiza*	HMGR	*A. rhizogenes*	Increased enzyme activity and enhanced the production of tanshinones and squalene	(Dai et al., 2011)
6	*Catharanthus roseus*	ORCA3, G10H	*A. tumefaciens*	Enhanced synthesis of MIA (strictosidine, vindoline, catharanthine and ajmalicine)	(Pan et al., 2012)
7	*Hyoscyamus senecionis, H. muticus*	PMT, H6H	*A. tumefaciens, A. rhizogenes*	Enhanced the production of tropane alkaloids	(Dehghan et al., 2017)
8	*Catharanthus roseus*	CrTDC, CrSTR	Sonication-assisted *Agrobacterium* transformation (SAAT)	Enhanced total alkaloid production and increase in vindoline and vinblastine synthesis	(Sharma et al., 2018)
9	*Duboisia leichhardtii*	QPT-RNAi construct	*A. rhizogenes*	Silencing of quinolinic acid phosphoribosyl transferase and enhanced production of scopolamine	(Singh et al., 2018)
10	*Glycyrrhiza glabra*	CYP88D6	*Agrobacterium rhizogenes* (hairy root culture)	Increase in Glycyrrhizin content.	(Shirazi et al., 2018)

rate-limiting step in their biosynthesis by manipulating the expression of individual structural genes. A wide variety of genes encoding enzymes involved in major steps of secondary metabolic pathways have been cloned from medicinal plant species. Such modifications have given new prospects for modifying the composition of secondary metabolites of plants, including pharmaceuticals, nutraceuticals, and defensive phytochemicals (Table 31.1). Using tissue/cell-defined promoter sequence metabolic engineering can also emphasize regulatory, biosynthetic, or transporter genes (Nitta et al., 2018).

Medicinal and aromatic plants (MAPs) have also been linked with mankind and their traditional knowledge. Plant secondary metabolites are highly important in human healthcare and are industrialized in developing countries. Advancements in biotechnological methods such as in vitro plant regeneration have facilitated the engineering of secondary metabolites. Advances in gene cloning, genetic manipulation, omics studies, and high-throughput technology have led to improvements in strategies involved in the enhancement of secondary metabolites (Gómez-Galera et al., 2007). Generally, medicinal plants are not cultivated or extracted from nature; however, an increase in the demand for secondary metabolites has reduced the population of medicinal plants. It has been reported that approximately 4000–10,000 medicinal plants are endangered even when the government provides a warning for continued exploitation of these plants (Canter et al., 2005). The cultivation of medicinal plants is not easy because most medicinal plants are from tropical regions and have different types of requirements for their growth and development, which is difficult. Therefore, despite these difficulties, researchers have developed plant breeding techniques for high yield and better-quality control (Gómez-Galera et al., 2007). At present, through biotechnology, the metabolic pathway of medicinal plants is modified to produce secondary metabolites and desired natural products, lowering the levels of antinutrients, increasing the levels of novel compounds, etc. (Verpoorte et al., 2000; Oksman-Caldentey et al., 2004). However, using certain strategies, such as those mentioned earlier, did not yield satisfactory results. This led to the exploitation of transgenic plants, which are genetically modified plants that have certain useful genes incorporated into their genome by using recombinant DNA technology to express exogenous genes. This technique has yielded promising and promising results. This transgenic technology has made remarkable progress in the expression of engineered genes (Weising et al., 1988). Hairy root culture is a valuable tool for studying the expression of genes, biochemical pathways of metabolic systems, etc.

Hairy root culture or hairy root methods are unique and effective methods for producing transgenic plants (Chilton et al., 1982). This technique has given amazing results and is hence a frequently used method in recombinant DNA technology. The roots, which are transformed, are highly differentiated and produce stable and extensive secondary metabolites. For the successful establishment of a hairy root culture system, certain conditions, such as the use of

TABLE 31.2
List of Designer Medicinal and Aromatic Plants (Monocot) and Screening of Enhanced Bioactive Molecules for Viral Diseases

S. no.	Plant and family	Breeding experiment	Bioactive molecules of viral disease	Disease cured	Reference
1	*Aloe barbadensis miller* (Asphodelaceae)	Tissue engineering scaffolds	Monoterpene, pinene, selinene, tetradecane	Infection	Maenthaisong et. al (2007).
2	*Catharanthus roseus* (Apocynaceae)	Micropropagation through shoot tip and nodal segments	vincamine	Anticancer and antiviral	Özçelik et al (2011).
3	*Acorus calamus* (Aceraceae)	Genotoxicity and mutagenicity of beta and alpha-asarone	Alpha and beta asarone	Anti-inflammatory and immunomodulatory activity, anticancer properties	Rajput et al (2014)
4	*Curcumba longa* (Zingiberaceae)	Nanodrug delivery and tissue engineering	Diacetyl curcumin	For neurodegenerative diseases and SARS COVID	Zorofchian et al. (2014), Singh, Y. D et al. (2020)
5	*Azadirachta indica* (Meliaceae)	In situ reverse transcription polymerase chain reaction (in situ RT-PCR)	Nimbandiol, nimbolide, ascorbic acid	Anti-inflammatory, antimicrobial, etc.	Alzohairy (2016).
7	*Taxus brevifolia* (Pacific yew)	Computational and thermodynamic parameters CRISP-Cas9	Butyl 2-ethylhexyl phthalate and Dactylorhin-A	Anticancer	Buyel, J.F (2018).
8	*Rauwolfia serpentine* (Apocynaceae)	*Agrobacterium*-mediated transformation	Terpenoid indole alkaloids	For snake bite, rheumatism, hypertension, insanity, epilepsy hypertension, high blood pressure, mental illness	Rehman et al. (2021).

appropriate plants, suitable culture media, and plant characteristics, have been highlighted (Hu et al., 2006). This culture system consists of hormone-free conditions for production, but the main edge of this system is fast growth, low doubling time, being essential and economically friendly, easy maintenance, and the ability to synthesize a wide range of metabolites. It has also been used to produce certain phytochemicals in pharmaceutical industries, such as cosmetics and food additives (Table 31.2). Many various types of medicinal plants are transformed for the potential production of secondary metabolites (Sevón et al., 2002). Usually, oxygen deficiency is a major complication in this technique caused by poor mixing of mass transfer conditions. To overcome this, two enzymes, Adh and pyruvate decarboxylase, which are transferred into the hairy root of *Arabidopsis thaliana* L., were used, and positive results were achieved hairy root culture has made this process successful.

31.22 IMPROVEMENT OF SECONDARY METABOLITES VIA GENETIC ENGINEERING

Although the use of general strategies for breeding programs mentioned may not satisfactorily justify the metabolic pathway in medicinal plants, employing recent biotechnological techniques has rationally elucidated the pathways involved. Techniques such as molecular breeding, genetic engineering, expressed sequence tag studies, molecular docking, cloning, and various others have been used to improve secondary metabolites in medicinal and aromatic plants. Compared with conventional breeding techniques, molecular breeding via genetic engineering is the best approach to producing secondary metabolites. Genetic transformation through *Agrobacterium tumefaciens*, as mentioned, has been used effectively in many plants, but genetic engineering has made it possible to increase the levels of various compounds, screen for biological activity, improve quality, etc.

Combining plant cells and tissue culture with genetic engineering, omics studies have led to a wide range of enhancements in production. To improve the understanding of secondary metabolites, we need to understand their metabolic pathways where molecular elucidation has allowed analysis. Molecular elucidation helps us to identify a particular metabolic pathway, enzymes catalyzing biosynthesis, and regulatory factors that control the overall metabolic pathway. The main technique involved in molecular elucidation is the process that begins with identifying genes involved and transcription factors present in that metabolic pathway up to the cloning of desired genes. It has also been shown that sequence-specific DNA binding proteins that bind to the promoter regions of target genes are involved in modifying the initiation of mRNA expression in that metabolic pathway (Bindu et al., 2018). A study in spearmint (*Mentha spicata*) revealed that the sites of secondary metabolite production comprised a tiny structure known as peltate glandular trichomes (PGTs). This project provides a metabolic engineering example. The MsYABB5 gene, which is unique and expressed in this plant, was taken out for this investigation. It was possible to create transgenic plants with this gene either overexpressed or RNA interference-silencing. It was established that the terpene-related genes were expressed, and the findings indicated that a decrease in terpene synthesis was caused by overexpression of the genes, while an increase in terpene production was caused by reduced expression of the genes (Wang et al., 2016).

31.23 POSITIONAL CLONING AND EXPRESSION LIBRARIES

Map-based cloning, another name for positional cloning, is a method that entails locating a defective gene. By producing mutants with defects in the biosynthetic pathway, this approach can be used to identify the desired gene. The typical phenotypes are hybridized and evaluated alongside these mutants. The homologous sequences of two mutants and the restored line, which may exist in the target gene, were used for cloning. It is regarded as a putative sequence in the target of interest if the restored line sequence resembles the anticipated sequence and if the two mutants show any changes in the gene sequence (Bindu et al., 2018).

31.24 DIFFERENTIAL EXPRESSION ANALYSIS AND NEXT-GENERATION SEQUENCING

The emergence of DNA sequence technologies and large-scale proteomics platforms are much better tools in bioinformatics. Differential expression analysis is one such technique that refers to discovering quantitative changes in expression levels under experimental conditions. Generally, medicinal plants contain trichomes, which are plants that produce secondary metabolites; therefore, differential gene expression studies, such as suppression subtractive hybridization (SSH), refer to methods that separate closely related DNA molecules. This very useful method for separating cDNA and genomic DNA is favorable for creating cDNA and genomic DNA libraries (Rebrikov et al., 2004). This method can be used for observing trichome-expressed genes involved in the biosynthesis of metabolic pathways. However, if a genomic resource preexists, a microarray differential study is performed, and differentially expressed RNAs are analyzed to predict the certainty of their involvement in the biosynthesis pathway. Once the full-length cDNA was removed from the genes, the heterologously expressed genes were subjected to in vitro enzyme activity determination. Further gene knockout methods have been used to examine the role of genes in metabolic pathways (Miettinen et al., 2014).

31.25 MECHANISM OF ACTION OF BIOACTIVE MOLECULES IN HUMANS

Currently, several viral diseases, such as COVID-19, HIV, influenza, herpes simplex virus (HSV), chikungunya, hepatitis A, hepatitis B, and hepatitis C, are emerging (Orhan et al., 2009). The objective of this chapter is to highlight the use of bioactive molecules obtained from medicinal and aromatic plants against viral diseases (Table 31.3). As mentioned, treatments have the potential to inhibit viral diseases by binding to cell receptors and inhibiting penetration into host cells, and a few additional mechanisms are involved (Khan et al., 2005; Dikid et al., 2013). First, let us look at the human defense system when a viral infection occurs. The body's defense system recognizes the antigen

TABLE 31.3
List of Designer Medicinal and Aromatic Plants (Dicots) and Screening of Enhanced Bioactive Molecules for Viral Diseases

S. no.	Plant and family	Breeding experiment	Bioactive molecules of viral diseases	Diseases cured	Reference
1	*Citrus limon* (Rutaceae)	Cloning	Luteolin, apigenin,	Dengue	Tang et al. (2012)
2	*Agastache* (Lamiaceae)	In vitro propagation	Phenylpropanoids, tilianin	Anti-nociceptive, anti-inflammatory, antiatherogenic	Zielińska et al. (2014).
3	*Zingiber officinale* Roscoe (Zingiberaceae)	Sequence-characterized amplified region (SCAR)	Fukinolic acid and cimicifugic acids	Allicin	Chavan et al. (2008).
4	*Withania somnifera* (Solanaceae)	*Agrobacterium*-mediated genetic transformation	Withanolides	Produces superior quality of important seconday metabolites withaferin A and withanolide A; COVID-19	Tripathi et al. (2017) Khanal et al. (2022).
5	*Glycyrrhiza glabra* (Favaceae)	Metabolic engineering	Glycyrrhetinic acid and glycyrrhizin	Beta-amyrin 11-oxidase; COVID-19	Shirazi et al. (2018)
6	*Erysimum cheiri* (Brassicaceae)	Plant transformation and genome sequencing		Cardiac glycoside	Züst et al. (2020)

TABLE 31.3 (*Continued*)
List of Designer Medicinal and Aromatic Plants (Dicots) and Screening of Enhanced Bioactive Molecules for Viral Diseases

S. no.	Plant and family	Breeding experiment	Bioactive molecules of viral diseases	Diseases cured	Reference
7	*Asparagus racemosus* Wild (Asparagaceae)	Genetically improved farmed tilapia (GIFT)	Polyphenols, epigallocatechin gallate (EGCG), delphinidin, and curcumin	9,12-Octadecadienoic acid	Parmar et al. (2020)
8	*Emblica officinalis* (Phyllanthaceae)	Molecular docking and molecular dynamics	Amritoside, apigenin-6-C-glucosyl7-O-glucoside, pectolinarin and astragalin	SARS COVID	Murugesan et al. (2021).
9	*Andrographis paniculata* (Burm. f.) (Acanthaceae)	Tissue culture	AD, DAP and DAD	Dengue, HIV, COVID-19	Jiang et al. (2023).

TABLE 31.4
Sources Extracted from Fungi or Plants and Their Activity against Viral Diseases

S. no.	Source of extraction	Target mechanism	Viral disease	Reference
1.	Marine-derived fungus (*Dichotomomyces cejpii* F31-1)	Inhibits the enzymes involved in virus replication	Dengue virus	Wu et al., 2013
2	*Ziziphus jujube*	Inhibits the replication of virus	PEDV porcine epidemic diarrhea virus	Kang et al., 2016
3	*Castanospermum australe*, *Houttuynia cordata*, *Myrtopsis corymbosa*, *Uncaria tomentosa*	Prevents the attachment of terminal glucose residue on N-linked glycans.	Dengue virus	Kaushik et al., 2018
4	*Phaleria macrocarpa*	Inactivates the early steps in the viral life cycle	HSV	Ismaeel et al., 2018
5	*Enantia chlorantha*	Antiviral properties	Malaria, hepatitis A, B, C, and D	Ohemu et al., 2018

through two systems – the adaptive or innate immune system. In the innate immune system, the antigens are recognized by pattern recognition receptors (PRRs), which are present on the surface of the cells; this antigen is further processed by natural killer cells and macrophage-mediated phagocytosis (Medina, 2016; Vaure et al., 2014). In the case of the adaptive immune system, T or B lymphocytes recognize the antigen via a surface-antigen-specific receptor that invades the antigen through a cytotoxic receptor and induces antibody production (Khan et al., 2005). Generally, viruses have high sequence variability, which provides a salient feature for infection or attack of host cells (Dos Santos, 2021). There are several strategies for viruses to escape the antiviral signaling pathway, and altering genetic material decreases the production of cytokines and decreases the expression of PRRs (Zou et al., 2016). One of the unique or smartest ways for viruses to enter the host is through entering a dormant state, after which the viruses reactivate when the immune system is weak and have two types of reproductive cycles – a lytic cycle and a lysogenic cycle – which have a major advantage for upregulating the infection (Phelan et al., 2017). There are several synthetic drugs for viral disease, but their side effects on individuals are the main disadvantage. Therefore, researchers have identified plant-derived molecules that act against viral diseases.

31.26 ANTIVIRAL MOLECULES DERIVED FROM MEDICINAL AND AROMATIC PLANTS

The importance of secondary metabolites is of interest for researchers attempting to develop plant-based medicines due to their effectiveness and lack of side effects. Plants contain secondary metabolites such as flavonoids, terpenoids, lignins, and alkaloids, which are useful in the pharmaceutical field for the development of herbal drugs (Table 31.4) (Kapoor et al., 2017). Among all the secondary metabolite families, phenolic compounds are abundant and most abundant. Several studies on the use of phytochemicals as antiviral agents against various viruses, such as Zika virus, HIV, hepatitis virus, and Japanese encephalitis virus, are ongoing (Oliver et al., 2017; Akram et al., 2018). These viruses have DNA or RNA as their genetic material; based on this, phytochemicals exhibit a mode of action. Various strategies have been used to control viral infection, such as blocking or inhibiting the binding of viruses to hosts, inhibiting transcription, and inhibiting the translation of enzymes (Subudhi et al., 2018). Phytochemicals, for example, epigallocatechin gallate (EGCG), which inhibits viral enzymes that are involved in growth, such as RNA polymerase, proteases, and reverse transcriptase (Ahmad et al., 2015), have several underlying mechanisms.

31.27 FLAVONOIDS

Flavonoids are secondary metabolites, a group of phenolic compounds that are primarily found in fruits, vegetables, greens, tea, wine, bark, stems, leaves, etc., and have various bioactive effects on viral disease (Table 31.5). Many researchers have proven that these are the most effective group of secondary metabolites due to their antioxidant, antiviral, and anticarcinogenic properties (Panche et al., 2016).

31.28 ALKALOIDS

Alkaloids are a type of secondary metabolite that consists of a nitrogen ring in its structure and has various biological properties, such as antimicrobial, anticancer, antiviral, and antiasthmatic properties (Shi et al., 2014). Approximately 43 alkaloids have been reported to exhibit antiviral properties against the influenza virus. Alkaloids enhance the production of interferons and macrophages, thus leading to the early release of antigens, inhibition of growth, and synthesis of viral proteins involved in replication (Moradi et al., 2018) (Table 31.6).

31.29 TERPENES AND TERPENOIDS

Terpenes are the largest category of secondary metabolites and consist of five-carbon isoprene units with simple hydrocarbons. Terpenoids are modified versions of terpenes due to changes in the positional structures of their functional groups. They are also considered to be important phytochemicals in the pharmaceutical industry (Jaeger et al., 2016). Terpenoids exhibit a wide range of antimicrobial activities; usually, they penetrate the cells of microbes, disrupt the cell membrane, and cause microbial death (Table 31.6). There are various types of terpenoids, such as hemiterpenoids, monoterpenoids, iridoids, sesquiterpenoids, diterpenoids, sesterterpenoids, triterpenoids, and tetraterpenoids. These compounds are classified based on their structural differences. Terpenoids are the most important secondary metabolites.

The use of phytochemicals has provided a better option for the development of antiviral medicines. Synergistic studies have been conducted on FDA-approved drugs that can improve the efficacy of antiviral drugs. It has been noted that when nano-encapsulated, phytochemicals provide better delivery and long-lasting antiviral action (Ghildiyal et al., 2020).

TABLE 31.5
Bioactive Components of Flavonoids and Their Mechanism of Action against Viral Diseases

S. no.	Bioactive component	Target mechanism	Viral disease	Reference
1	Quercetin, fisetin	Intracellular replication in early stages of infection	Chikungunya	Lane, 2018
2	Luteolin	Inhibits the viral replication in at post entry of virus	A549 cell lines infected with JEV	Fan et al., 2016
3	EGCG- epigallocatechin gallate in green tea	Inhibits the viral replication	Rotavirus strain SA-11	Lipson et al., 2017
4	Baicalein	Inhibiting different stages of viral replication	Dengue virus and enterovirus A71 in-vitro	Min et al., 2018
5	Pinocembrin	Anti-Zika activity, inhibits the viral replication at post entry of virus	Zika virus	Lee et al., 2019

TABLE 31.6
Bioactive Components of Alkaloids and Terpenoids and Their Target Mechanisms against Various Viral Diseases

S. no.	Bioactive component	Target mechanism	Viral disease	Reference
1	Homonojirimycin- (HNJ)	Inhibits the synthesis of viral proteins during replication	Influenza virus	Zhou et al., 2013
2	Dendrobine	Ability to bind nucleoprotein of virus and restrain virus replication	Influenza A virus	Li & Chen, 2021
3	*Croton mauritianus*	Inhibits viral enzymes involved in synthesis of protein	Chikungunya	Corlay et al., 2014
4	*Marrubium vulgare*	Disrupts the attachment of HSV to cell membrane and interfere in replication of HSV-1 virus	HSV-1	Gaber et al., 2017
5	*Bupleurum kaoi*	Inhibits early life cycle of virus	HCV, hepatitis C virus	Lin et al., 2015
6	*Phyllanthus urinaria*	Hampers the viral attachment to the host cell	HCV	Chung et al., 2018

31.30 COMMERCIALLY AVAILABLE BIOACTIVE MOLECULES IN THE PHARMACEUTICAL MARKET

People observe and experience the effects of commercially available medicines that have harmful effects and are aware of naturally obtained or naturally supplemented components to promote health wellness (Berman et al., 2015). In the present situation, advancements in science and techniques have helped utilize bioactive molecules in certain supplements or medicines to enhance and promote well-being. First, bioactive molecules are small compounds that are isolated or extracted from natural fruits and vegetables that have synergistic and complementary effects on enzymes, minerals, and vitamins (Singh et al., 2020). These compounds are being studied for the prevention of cancer, health diseases, viral diseases, infections, etc. These molecules are usually found in fruits, vegetables, nuts, green vegetables, etc. (Teodoro, 2019). The mode of action of many bioactive molecules is through interactions with proteins or macromolecules, but the underlying mechanism is still unknown. Therefore, computational prediction of bioactive molecules with known targets has made it possible to reduce side effects. Computational predictions, such as datasets and molecular docking, have made it possible to understand the interactions of bioactive molecules (Gfeller et al., 2013). The application of bioinformatics, molecular biology, and pharmacological techniques has made it possible to produce a few commercially available bioactive molecules as supplements, which can be combined with medicines. The following table explains the various bioactive compounds and their respective benefits (Table 31.7).

31.31 CONCLUSION

The importance and knowledge base given in this review about medicinal and aromatic plants have been used as alternative treatments in modern medicine and have also helped in the discovery of new plants and herbs containing desirable compounds and assisted in providing an overview of the bioactive compounds in these plants to cure viruses and other diseases. Medicinal plants have been used since the beginning of the human era and have resulted in positive results; thus, they have continued to be modified to screen various types of bioactive compounds. Therefore, it has been confirmed that medicinal plants are better, more robust, more effective, and more understandable for preventing and curing certain diseases. The various strategies used for improvement are still considered to be the backbone of biotechnological experiments. Thus, secondary metabolites play an important role in human health. These metabolites produced by MACs can thus be exploited in the synthesis of drugs, pharmaceuticals, and economically important products. The demand for these natural products has widely increased in the past few years, and more than 50% of FDA-approved drugs are natural products or their analogues (Vuorela et al., 2004). This review provides insights into the different methods used for enhancing biologically active molecules in medicinal plants and shows that biotechnological tools are important. When employed with other methods, such as elicitation with biotic and abiotic elicitors, cell culture strategies can increase the concentration of secondary metabolites in the culture. Thus, metabolic engineering, cell culture techniques, and genetic engineering can be promising techniques for the characterization and improvement of secondary metabolites. Some of the current studies also show that transcription factors can be used as efficient molecular tools in metabolic engineering. These methods also protect several rare plant species from being endangered.

ACKNOWLEDGMENTS

We thank Manipal Academy of Higher Education (MAHE), Manipal, Karnataka, India, TIFAC-CORE and FIST, DST New Delhi, DBT BUILDER grant (BT/INF/22/SP43065/2021), DBT, New Delhi, and K-FIST, VGST, Govt. of Karnataka for the facilities. The authors (LAHS, NKR, AK) are thankful to MAHE for the T.M.A. Pai Ph.D. scholarship and AK is thankful to the Indian Council of Medical Research (ICMR), New Delhi for financial assistance as Project Associate I through a research grant

TABLE 31.7
List of commercially available bioactive molecules.

S. no.	Bioactive compound	Commercially available compounds	Health benefits	Source of extraction	Reference
1	Polyphenols	Catechin	Heart diseases, diabetes, neurodegenerative diseases	Green tea, black tea, olives, species	Lambert et al., 2007
2	Phytosterols	Beta-siterols, campestrol, stigmasterol	Cholesterol lowering, anti-inflammatory, anticancerous, antioxidant	Legumes, dry fruits, grains, seeds, vegetable oil	Deng et al., 2016
3	Carotenoids	Beta-carotene, lycopene, leutic	Converts provitamin to vitamin A	Vegetables – carrots	Berman et al., 2015
4	Flavonoids	Silbin, galangin, hesperitin	Cures arthritis, early ageing	Nuts, honey, seeds, tea	Bose et al., 2019
5	Phytoestrogens	Daidzein and genstein	Inflammation, post menoapause, cardiovascular disease,	Flaxseeds, chickpea, green beans, diary supplement	Sridevi et al., 2021.

(No. 59/08/2022-TRM/BMS). We are grateful to Dr. B.S. Satish Rao, Professor and Director, Manipal School of Life Sciences, MAHE for their encouragement and support.

AUTHOR CONTRIBUTIONS

AM, SHM and GKP conceived the idea; SHM, GPK, LAHS, NKR, AK, and AM carried out the literature survey and prepared the tables; SHM, GPK, LAHS, NKR, AK, and AM wrote and edited the manuscript; all authors approved the chapter.

REFERENCES

Achakzai, A. K. K., Achakzai, P., Masood, A., Kayani, S. A., & Tareen, R. B. (2009). Response of plant parts and age on the distribution of secondary metabolites on plants found in Quetta. *Pakistan Journal of Botany*, *41*(5), 2129–2135.

Adhikari, P. P., & Paul, S. B. (2018). History of Indian traditional medicine: A medical Inheritance. *History*, *11*(1).

Ahmad, A., Kaleem, M., Ahmed, Z., & Shafiq, H. (2015). Therapeutic potential of flavonoids and their mechanism of action against microbial and viral infections – a review. *Food Research International*, *77*, 221–235.

Ahmad, S., Zahiruddin, S., Parveen, B., Basist, P., Parveen, A., Parveen, R., & Ahmad, M. (2021). Indian medicinal plants and formulations and their potential against COVID-19–preclinical and clinical research. *Frontiers in Pharmacology*, *11*, 2470.

Aidara-Kane, A., Angulo, F. J., Conly, J. M., Minato, Y., Silbergeld, E. K., McEwen, S. A., & Collignon, P. J. (2018). World Health Organization (WHO) guidelines on use of medically important antimicrobials in food-producing animals. *Antimicrobial Resistance & Infection Control*, *7*(1), 1–8.

Akram, M., Tahir, I. M., Shah, S. M. A., Mahmood, Z., Altaf, A., Ahmad, K., Munir, N., Daniyal, M., & Mehboob, H. (2018). Antiviral potential of medicinal plants against HIV, HSV, influenza, hepatitis, and coxsackievirus: A systematic review. *Phytotherapy Research*, *32*(5), 811–822.

Ali, M., & Abbasi, B. H. (2013). Production of commercially important secondary metabolites and antioxidant activity in cell suspension cultures of *Artemisia absinthium* L. *Industrial Crops and Products*, *49*, 400–406.

Alzohairy, M. A. (2016). Therapeutics role of *Azadirachta indica* (Neem) and their active constituents in diseases prevention and treatment. *Evidence-Based Complementary and Alternative Medicine*, *2016*.

Ayoub, M., de Camargo, A. C., & Shahidi, F. (2016). Antioxidants and bioactivities of free, esterified, and insoluble-bound phenolics from berry seed meals. *Food Chemistry*, *197*, 221–232.

Berman, J., Zorrilla-López, U., Farré, G., Zhu, C., Sandmann, G., Twyman, R. M., Capell, T., & Christou, P. (2015). Nutritionally important carotenoids as consumer products. *Phytochemistry Reviews*, *14*(5), 727–743.

Bieging, K. T., Mello, S. S., & Attardi, L. D. (2014). Unravelling mechanisms of p53-mediated tumour suppression. *Nature Reviews Cancer*, *14*(5), 359–370.

Bindu, K. H., Mythili, J. B., & Radhika, R. M. (2018). Genetic engineering in medicinal and aromatic plants. In *Genetic engineering of horticultural crops* (pp. 249–271). Academic Press.

Bong, F. J., Chear, N. J. Y., Ramanathan, S., Mohana-Kumaran, N., Subramaniam, S., & Chew, B. L. (2021). The development of callus and cell suspension cultures of Sabah Snake Grass (*Clinacanthus nutans*) for the production of flavonoids and phenolics. *Biocatalysis and Agricultural Biotechnology*, *33*, 101977.

Borrelli, G., & Trono, D. (2016). Molecular approaches to genetically improve the accumulation of health-promoting secondary metabolites in staple crops – a case study: The lipoxygenase-B1 genes and regulation of the carotenoid content in pasta products. *International Journal of Molecular Sciences*, *17*(7), 1177.

Bose, S. K., Howlader, P., Jia, X., Wang, W., & Yin, H. (2019). Alginate oligosaccharide postharvest treatment preserve fruit quality and increase storage life via Abscisic acid signaling in strawberry. *Food Chemistry*, *283*, 665–674.

Bourgaud, F., Gravot, A., Milesi, S., & Gontier, E. (2001). Production of plant secondary metabolites: A historical perspective. *Plant Science*, *161*(5), 839–851.

Bouzenna, H., & Krichen, L. (2013). Natural product research: Formerly natural product letters *Pelargonium graveolens* L' Her. and *Artemisia arborescens* L. essential oils: Chemical composition, antifungal activity against *Rhizoctonia solani* and insecticidal activity against *Rhysopertha*. *Natural Product Research*, *27*(9), 841–846.

Bradshaw, H. D., & Schemske, D. W. (2003). Allele substitution at a flower colour locus produces a pollinator shift in monkeyflowers. *Nature*, *426*(6963), 176–178.

Brenes, A., & Roura, E. (2010). Essential oils in poultry nutrition: Main effects and modes of action. *Animal Feed Science and Technology*, *158*(1–2), 1–4.

Bruckner, K., Božić, D., Manzano, D., Papaefthimiou, D., Pateraki, I., Scheler, U., & Tissier, A. (2014). Phytochemistry characterization of two genes for the biosynthesis of abietane-type diterpenes in rosemary (*Rosmarinus officinalis*) glandular trichomes. *Phytochemistry*, *101*, 52–64.

Butenko, R. G., Lipsky, A. K., Chernyak, N. D., & Arya, H. C. (1984). Changes in culture medium pH by cell suspension cultures of *Dioscorea deltoidea*. *Plant Science Letters*, *35*(3), 207–212.

Buyel, J. F. (2018). Plants as sources of natural and recombinant anti-cancer agents. *Biotechnology Advances*, *36*(2), 506–520.

Canter, P. H., Thomas, H., & Ernst, E. (2005). Bringing medicinal plants into cultivation: Opportunities and challenges for biotechnology. *Trends in Biotechnology*, *23*(4), 180–185.

Chavan, P., Warude, D., Joshi, K., & Patwardhan, B. (2008). Development of SCAR (sequence-characterized amplified region) markers as a complementary tool for identification of ginger (*Zingiber officinale Roscoe*) from crude drugs and multicomponent formulations. *Biotechnology and Applied Biochemistry*, *50*(1), 61–69.

Chen, D. H., Ye, H. C., & Li, G. F. (2000). Expression of a chimeric farnesyl diphosphate synthase gene in *Artemisia annua* L. transgenic plants via *Agrobacterium tumefaciens*-mediated transformation. *Plant Science*, *155*(2), 179–185.

Cheng, A. X., Lou, Y. G., Mao, Y. B., Lu, S., Wang, L. J., & Chen, X. Y. (2007). Plant terpenoids: Biosynthesis and ecological functions. *Journal of Integrative Plant Biology*, *49*(2), 179–186.

Cheynier, V. (2012). Phenolic compounds: From plants to foods. *Phytochemistry Reviews*, *11*(2), 153–177.

Cheynier, V., Comte, G., Davies, K. M., Lattanzio, V., & Martens, S. (2013). Plant phenolics: Recent advances on their biosynthesis, genetics, and ecophysiology. *Plant Physiology and Biochemistry*, *72*, 1–20.

Chilton, M. D., Tepfer, D. A., Petit, A., David, C., Casse-Delbart, F., & Tempé, J. (1982). *Agrobacterium rhizogenes* inserts T-DNA into the genomes of the host plant root cells. *Nature*, *295*(5848), 432–434.

Chung, Y. S., Choi, J. Y., Han, M. G., Park, K. R., Park, S. J., Lee, H., Jee, Y., & Kang, C. (2018). A large healthcare-associated outbreak of hepatitis C virus genotype 1a in a clinic in Korea. *Journal of Clinical Virology*, *106*, 53–57.

Cook, N. C., Samman, S. (1996). Flavonoids – chemistry, metabolism, cardioprotective effects, and dietary sources. *The Journal of Nutritional Biochemistry*, *7*(2), 66–76.

Corlay, N., Delang, L., Girard-Valenciennes, E., Neyts, J., Clerc, P., Smadja, J., Guéritte, F., Leyssen, H., & Litaudon, M. (2014). Tigliane diterpenes from *Croton mauritianus* as inhibitors of chikungunya virus replication. *Fitoterapia*, *97*, 87–91.

Crozier, A., Clifford, M. N., & Ashihara, H. (Eds.) (2008). *Plant secondary metabolites: Occurrence, structure and role in the human diet*. John Wiley & Sons.

Curtis, W. R., Wang, P., & Humphrey, A. (1995). Role of calcium and differentiation in enhanced sesquiterpene elicitation from calcium alginate-immobilized plant tissue. *Enzyme and Microbial Technology*, *17*(6), 554–557.

Dai, Z., Cui, G., Zhou, S. F., Zhang, X., & Huang, L. (2011). Cloning and characterization of a novel 3-hydroxy-3-methylglutaryl coenzyme A reductase gene from *Salvia miltiorrhiza* involved in diterpenoid tanshinone accumulation. *Journal of Plant Physiology*, *168*(2), 148–157.

Debnath, M., Malik, C. P., & Bisen, P. S. (2006). Micropropagation: A tool for the production of high quality plant-based medicines. *Current Pharmaceutical Biotechnology*, *7*(1), 33–49.

Deeks, S. G., Overbaugh, J., Phillips, A., & Buchbinder, S. (2015). HIV infection. *Nature Reviews Disease Primers*, *1*(1), 1–22.

Dehghan, E., Reed, D. W., Covello, P. S., Hasanpour, Z., Palazon, J., Oksman-Caldentey, K. M., & Ahmadi, F. S. (2017). Genetically engineered hairy root cultures of Hyoscyamus senecionis and H. muticus: Ploidy as a promising parameter in the metabolic engineering of tropane alkaloids. *Plant Cell Reports*, *36*, 1615–1626.

Deng, S., Wei, T., Tan, K., Hu, M., Li, F., Zhai, Y., & Luo, M. (2016). Phytosterol content and the campesterol: Sitosterol ratio influence cotton fiber development: Role of phytosterols in cell elongation. *Science China Life Sciences*, *59*, 183–193.

Dhama, K., Latheef, S. K., Mani, S., Samad, H. A., Karthik, K., Tiwari, R., & Tufarelli, V. (2015). Multiple beneficial applications and modes of action of herbs in poultry health and production-A review. *International Journal of Pharmacology*, *11*(3), 152–176.

DiCosmo, F., & Misawa, M. (1985). Eliciting secondary metabolism in plant cell cultures. *Trends in Biotechnology*, *3*(12), 318–322.

Dikid, T., Jain, S. K., Sharma, A., Kumar, A., & Narain, J. P. (2013). Emerging & re-emerging infections in India: An overview. *The Indian Journal of Medical Research*, *138*(1), 19.

Dos Santos, W. G. (2021). Impact of virus genetic variability and host immunity for the success of COVID-19 vaccines. *Biomedicine & Pharmacotherapy*, 111272.

Dudareva, N., Pichersky, E., & Gershenzon, J. (2004). Biochemistry of plant volatiles. *Plant Physiology*, *135*(4), 1893–1902.

Falcone Ferreyra, M. L., Rius, S., & Casati, P. (2012). Flavonoids: Biosynthesis, biological functions, and biotechnological applications. *Frontiers in Plant Science*, *3*, 222.

Fan, W., Qian, S., Qian, P., & Li, X. (2016). Antiviral activity of luteolin against Japanese encephalitis virus. *Virus Research*, *220*, 112–116.

Fernandez-Panchon, M. S., Villano, D., Troncoso, A. M., & Garcia-Parrilla, M. C. (2008). Antioxidant activity of phenolic compounds: From in vitro results to in vivo a ntioxidant activity of phenolic compounds: From in vitro results to in vivo evidence. *Critical Reviews in Food Science and Nutrition*, *48*, 649–671.

Franz, C., Baser, K. H. C., & Windisch, W. (2010). Essential oils and aromatic plants in animal feeding–a European perspective. A review. *Flavour and Fragrance Journal*, *25*(5), 327–340.

Frense, D. (2007). Taxanes: Perspectives for biotechnological production. *Applied Microbiology and Biotechnology*, *73*(6), 1233–1240.

Gaber, M., Medhat, W., Hany, M., Saher, N., Fang, J. Y., & Elzoghby, A. (2017). Protein-lipid nanohybrids as emerging platforms for drug and gene delivery: Challenges and outcomes. *Journal of Controlled Release*, *254*, 75–91.

Gandhi, S. G., Mahajan, V., & Bedi, Y. S. (2015). Changing trends in biotechnology of secondary metabolism in medicinal and aromatic plants. *Planta*, *241*(2), 303–317.

Gershenzon, J. (1984). Changes in the levels of plant secondary metabolites under water and nutrient stress. In *Phytochemical adaptations to stress* (pp. 273–320). Springer.

Gfeller, D., Michielin, O., & Zoete, V. (2013). Shaping the interaction landscape of bioactive molecules. *Bioinformatics*, *29*(23), 3073–3079.

Ghasemzadeh, A., Talei, D., Jaafar, H. Z., Juraimi, A. S., Mohamed, M. T. M., Puteh, A., & Halim, M. R. A. (2016). Plant-growth regulators alter phytochemical constituents and pharmaceutical quality in sweet potato (*Ipomoea batatas* L.). *BMC Complementary and Alternative Medicine*, *16*, 1–13.

Ghildiyal, R., Prakash, V., Chaudhary, V. K., Gupta, V., & Gabrani, R. (2020). Phytochemicals as antiviral agents: Recent updates. In *Plant-derived bioactives* (pp. 279–295). Springer.

Giannenas, I., Nisianakis, P., Gavriil, A., Kontopidis, G., & Kyriazakis, I. (2009). Trace mineral content of conventional, organic and courtyard eggs analysed by inductively coupled plasma mass spectrometry (ICP-MS), *Food Chemistry*, *114*(2), 706–711.

Giannenas, I., Sidiropoulou, E., Bonos, E., Christaki, E., & Florou-Paneri, P. (2020). The history of herbs, medicinal and aromatic plants, and their extracts: Past, current situation and future perspectives. In *Feed additives* (pp. 1–18). Academic Press.

Gómez-Galera, S., Pelacho, A. M., Gené, A., Capell, T., & Christou, P. (2007). The genetic manipulation of medicinal and aromatic plants. *Plant Cell Reports*, *26*(10), 1689–1715.

Greathead, H. (2003). Plants and plant extracts for improving animal productivity. *Proceedings of the Nutrition Society*, *62*(2), 279–290.

Guo, S., Ge, Y., & Jom, K. N. (2017). A review of phytochemistry, metabolite changes, and medicinal uses of the common sunflower seed and sprouts (*Helianthus annuus* L.) *Chemistry Central Journal*, *11*(1), 95.

Gupta, O. P., Karkute, S. G., Banerjee, S., Meena, N. L., & Dahuja, A. (2017). Contemporary understanding of miRNA-based regulation of secondary metabolites biosynthesis in plants. *Frontiers in Plant Science*, *8*, 374.

Harborne, J. B. (1999). Classes and functions of secondary products from plants. *Chemicals from Plants*, *26*, 1–25.

Henrotin, Y., Priem, F., & Mobasheri, A. (2013). Curcumin: A new paradigm and therapeutic opportunity for the treatment of osteoarthritis: Curcumin for Osteoarthritis Management. *Springer Plus*, *2*(1), 1–9.

Howyzeh, M. S., Noori, S. A. S., Shariati, V., & Amiripour, M. (2018). Comparative transcriptome analysis to identify putative genes involved in thymol biosynthesis pathway in medicinal plant *Trachyspermum ammi* L. *Scientific Reports*, *8*(1), 1–19.

Hu, Z. B., & Du, M. (2006). Hairy root and its application in plant genetic engineering. *Journal of Integrative Plant Biology*, *48*(2), 121–127.

Hussain, M. S., Fareed, S., Saba Ansari, M., Rahman, A., Ahmad, I. Z., & Saeed, M. (2012). Current approaches toward production of secondary plant metabolites. *Journal of Pharmacy & Bioallied Sciences*, *4*(1), 10.

Hyyryläinen, H. L., Bolhuis, A., Darmon, E., Muukkonen, L., Koski, P., Vitikainen, M., Sarvas, M., Pragai, Z., Bron, S., Dijl, J. M. V., & Kontinen, V. P. (2001). A novel two-component regulatory system in Bacillus subtilis for the survival of severe secretion stress. *Molecular Microbiology*, *41*(5), 1159–1172.

Ilan, A., Ziv, M., & Halevy, A. H. (1995). Propagation and corm development of Brodiaea in liquid cultures. *Scientia Horticulturae*, *63*(1–2), 101–112.

Inoue, M., Hayashi, S., & Craker, L. E. (2019). Role of medicinal and aromatic plants: Past, present, and future. *Pharmacognosy-medicinal plants* (pp. 1–13). IntechOpen Limited.

Ishkanian, A. S., Malloff, C. A., Watson, S. K., Ronald, J. D., Chi, B., Coe, B. P., Snijders, A., Albertson, D. G., Pinkel, D., Marra, A. M., Ling, V., & Lam, W. L. (2004). A tiling resolution DNA microarray with complete coverage of the human genome. *Nature Genetics*, *36*(3), 299–303.

Ismaeel, M. Y. Y., Dyari, H. R. E., Yaacob, W. A., & Ibrahim, N. (2018). In vitro antiviral activity of aqueous extract of *Phaleria macrocarpa* fruit against herpes simplex virus type 1. In *AIP Conference Proceedings*, *1940*(1).

Jaeger, R., & Cuny, E. (2016). Terpenoids with special pharmacological significance: A review. *Natural Product Communications*, *11*(9).

Jiang, L., Liu, Y., Su, W., Liu, W., Dong, Z., Long, Y., Luo, L., Jing, Q., Cao, Y., Wu, X., & Di, B. (2023). Epidemiological and genomic analysis of dengue cases in Guangzhou, China, from 2010 to 2019. *Scientific Reports*, *13*(1), 2161.

John, S. E. (1980). *Plant tissue culture as a source of biochemicals*. CRC Press.

Kabera, J. N., Semana, E., Mussa, A. R., & He, X. (2014). Plant secondary metabolites: Biosynthesis, classification, function and pharmacological classification, function and pharmacological properties. *Journal of Pharmacy and Pharmacology*, *2*(7), 377–392.

Kai, G., Xu, H., Zhou, C., Liao, P., Xiao, J., Luo, X., You, L., & Zhang, L. (2011). Metabolic engineering tanshinone biosynthetic pathway in *Salvia miltiorrhiza* hairy root cultures. *Metabolic Engineering*, *13*(3), 319–327.

Kang, K. B., Jang, D. S., Kim, J., & Sung, S. H. (2016). UHPLC-ESI-qTOF-MS analysis of cyclopeptide alkaloids in the seeds of *Ziziphus jujuba* var. spinosa. *Mass Spectrometry Letters*, *7*(2), 45–49.

Kapoor, L. D., Chopra, R. N., Chopra, I. C., & Handa, K. L. (1994). *Indigenous drug of India* (pp. 1–792). Academic Publishers.

Kapoor, R., Sharma, B., & Kanwar, S. S. (2017). Antiviral phytochemicals: An overview. *Biochem Physiol*, 6(2), 7.

Karuppusamy, S. (2009). A review on trends in production of secondary metabolites from higher plants by in vitro tissue, organ and cell cultures. *Journal of Medicinal Plants Research*, *3*(13), 1222–1239.

Kaushik, S., Kaushik, S., Sharma, V., & Yadav, J. (2018). Antiviral and therapeutic uses of medicinal plants and their derivatives against dengue viruses. *Pharmacognosy Reviews*, *12*(24).

Khalifa, I., Nawaz, A., Sobhy, R., Althwab, S. A., & Barakat, H. (2020). Polyacylated anthocyanins constructively network with catalytic dyad residues of 3CLpro of 2019-nCoV than monomeric anthocyanins: A structural-relationship activity study with 10 anthocyanins using in-silico approaches. *Journal of Molecular Graphics and Modelling*, *100*, 107690.

Khan, M. T. H., Ather, A., Thompson, K. D., & Gambari, R. (2005). Extracts and molecules from medicinal plants against herpes simplex viruses. *Antiviral Research*, *67*(2), 107–119.

Khanal, P., Chikhale, R., Dey, Y. N., Pasha, I., Chand, S., Gurav, N., Ayyanar, M., Patil, B. M., & Gurav, S. (2022). Withanolides from *Withania somnifera* as an immunity booster and their therapeutic options against COVID-19. *Journal of Biomolecular Structure and Dynamics*, *40*(12), 5295–5308.

Kirakosyan, A., & Kaufman, P. B. (2009). Recent advances in plant biotechnology. *Recent Advances in Plant Biotechnology*, *8*(January), 1–409.

Kollner, T. G., Schnee, C., Gershenzon, J., & Degenhardt, J. (2004). The variability of sesquiterpenes emitted from two Zea mays cultivars is controlled by allelic variation of two terpene synthase genes encoding stereoselective multiple product enzymes. *The Plant Cell*, *16*(5), 1115–1131.

Lambert, J. D., Sang, S., & Yang, C. S. (2007). Biotransformation of green tea polyphenols and the biological activities of those metabolites. *Molecular Pharmaceutics*, *4*(6), 819–825.

Lane, W. C. (2018). *Temperature-dependent efficacy of the IFN-α/β response against Chikungunya and other arboviruses* (Doctoral dissertation, University of Pittsburgh).

Larkin, P. J., Miller, J. A., Allen, R. S., Chitty, J. A., Gerlach, W. L., Frick, S., Kutchan, T. M., & Fist, A. J. (2007). Increasing morphinan alkaloid production by over-expressing codeinone reductase in transgenic *Papaver somniferum*. *Plant Biotechnology Journal*, *5*(1), 26–37.

Lauer, G. M., & Walker, B. D. (2001). Hepatitis C virus infection. *New England Journal of Medicine*, *345*(1), 41–52.

Le Lee, J., Loe, M. W. C., Lee, R. C. H., & Chu, J. J. H. (2019). Antiviral activity of pinocembrin against Zika virus replication. *Antiviral Research*, *167*, 13–24.

Lee, W. M. (1997). Hepatitis B virus infection. *New England Journal of Medicine*, *337*(24), 1733–1745.

Leite, A. C. R., Araujo, T. G., Carvalho, B. D. M., Maia, M. B. S., & Lima, V. L. D. M. (2010). Characterization of the antidiabetic role of *Parkinsonia aculeata* (Caesalpineaceae). *Evidence-Based Complementary and Alternative Medicine*, *2011*.

Li, C., & Chen, H. (2021). H7N9 influenza virus in China. *Cold Spring Harbor Perspectives in Medicine*, *11*(8), a038349.

Liang, L. F., Keng, C. L., & Lim, B. P. (2006). Selection of cell lines for the production of rosmarinic acid from cell suspension cultures of *Orthosihon stamineus* benth. *In Vitro Cellular & Developmental Biology-Plant*, *42*(6), 538–542.

Lin, L. T., Chung, C. Y., Hsu, W. C., Chang, S. P., Hung, T. C., Shields, J., Russell, S. R., Lin, C. C., Li, C. F., Yen, M. H., Tyrrell, D. L. J., Lin, C. C., & Richardson, C. D. (2015). Saikosaponin b2 is a naturally occurring terpenoid that efficiently inhibits hepatitis C virus entry. *Journal of Hepatology*, *62*(3), 541–548.

Lin, M. H., Moses, D. C., Hsieh, C. H., Cheng, S. C., Chen, Y. H., Sun, C. Y., & Chou, C. Y. (2018). Disulfiram can inhibit MERS and SARS coronavirus papain-like proteases via different modes. *Antiviral Research*, *150*, 155–163.

Lipson, S. M., Karalis, G., Karthikeyan, L., Ozen, F. S., Gordon, R. E., Ponnala, S., Bao, J., Samarrai, W., & Wolfe, E. (2017). Mechanism of anti-rotavirus synergistic activity by epigallocatechin gallate and a proanthocyanidin-containing nutraceutical. *Food and Environmental Virology*, *9*, 434–443.

Maenthaisong, R., Chaiyakunapruk, N., Niruntraporn, S., & Kongkaew, C. (2007). The efficacy of aloe vera used for burn wound healing: A systematic review. *Burns*, *33*(6), 713–718.

Malčovská, S. M., Dučaiová, Z., Maslaňáková, I., & Bačkor, M. (2014). Effect of silicon on growth, photosynthesis, oxidative status, and phenolic compounds of maize (*Zea mays* L.) grown in cadmium excess. *Water, Air, & Soil Pollution*, *225*(8), 1–11.

Maschke, R. W., Geipel, K., & Bley, T. (2015). Modeling of plant in vitro cultures: Overview and estimation of biotechnological processes. *Biotechnology and Bioengineering*, *112*(1), 1–12.

Matsuura, H. N., Rau, M. R., & Fett-Neto, A. G. (2013). Oxidative stress and production of bioactive monoterpene indole alkaloids: Biotechnological implications. *Biotechnology Letters*, *36*(2), 191–200.

Medina, K. L. (2016). Overview of the immune system. In *Handbook of clinical neurology* (Vol. 133, pp. 61–76). Elsevier.

Miettinen, K., Dong, L., Navrot, N., Schneider, T., Burlat, V., Pollier, J., Woittiez, L., Krol, S. V. D., Lugan, R., Ilc, T., Verpoorte, R., Caldentey, O. K. M., Matinio, E., Boumeester, H., Goosens, A., Memlink, J., & Werck-Reichhart, D. (2014). The seco-iridoid pathway from *Catharanthus roseus*. *Nature Communications*, *5*(1), 1–12.

Miller, R. H., Kaneko, S., Chung, C. T., Girones, R., & Purcell, R. H. (1989). Compact organization of the hepatitis B virus genome. *Hepatology*, *9*(2), 322–327.

Min, W., Wu, M., Fang, P., Yu, M., Shi, M., Zhang, Z., & Bo, P. (2018). Effect of baicalein on GLUT4 translocation in adipocytes of diet-induced obese mice. *Cellular Physiology and Biochemistry*, *50*(2), 426–436.

Mol, J., Grotewold, E., & Koes, R. (1998). How genes paint flowers and seeds. *Trends in Plant Science*, *3*(6), 212–217.

Montenegro-Landívar, M. F., Tapia-Quirós, P., Vecino, X., Reig, M., Valderrama, C., Granados, M., Cortina, L., & Saurina, J. (2021). Fruit and vegetable processing wastes as natural sources of antioxidant-rich extracts: Evaluation of advanced extraction technologies by surface response methodology. *Journal of Environmental Chemical Engineering*, *9*(4), 105330.

Moradi, M. T., Karimi, A., & Lorigooini, Z. (2018). Alkaloids as the natural anti-influenza virus agents: A systematic review. *Toxin Reviews*, *37*(1), 11–18.

Moreno, P. R. H., Van der Heijden, R., & Verpoorte, R. (1993). Effect of terpenoid precursor feeding and elicitation on formation of indole alkaloids in cell suspension cultures of *Catharanthus roseus*. *Plant Cell Reports*, *12*(12), 702–705.

Msimbira, L. A., & Smith, D. L. (2020). The roles of plant growth promoting microbes in enhancing plant tolerance to acidity and alkalinity stresses. *Frontiers in Sustainable Food Systems*, *4*, 106.

Mukhtar, M., Arshad, M., Ahmad, M., Pomerantz, R. J., Wigdahl, B., & Parveen, Z. (2008). Antiviral potentials of medicinal plants. *Virus Research*, *131*(2), 111–120.

Mulabagal, V., & Tsay, H. S. (2004). Plant cell cultures-an alternative and efficient source for the production of biologically important secondary metabolites. *International Journal of Applied Science and Engineering*, *2*(1), 29–48.

Murphy, K., & Weaver, C. (2016). Janeway's immunobiology. In *Garland science* (pp. 1–855). Norton & Company, Incorporated, W. W.

Murugesan, K., Jagannathan, P., Pham, T. D., Pandey, S., Bonilla, H. F., Jacobson, K., & Banaei, N. (2021). Interferon-γ release assay for accurate detection of severe acute respiratory syndrome coronavirus 2 T-cell response. *Clinical Infectious Diseases*, *73*(9), e3130–e3132.

Naithani, R., Huma, L. C., Holland, L. E., Shukla, D., McCormick, D. L., Mehta, R. G., & Moriarty, R. M. (2008). Antiviral activity of phytochemicals: A comprehensive review. *Mini Reviews in Medicinal Chemistry*, *8*(11), 1106–1133.

Namdeo, A. G. (2004). *Investigation on pilot scale bioreactor with reference to the synthesis of bioactive compounds from cell suspension cultures of Catharanthus roseus Linn*. Devi Ahilya Vishwavidyalaya.

Namdeo, A. G. (2007). Plant cell elicitation for production of secondary metabolites: A review. *Pharmacognosy Reviews*, *1*(1), 69–79.

Namdeo, A., Patil, S., & Fulzele, D. P. (2002). Influence of fungal elicitors on production of ajmalicine by cell cultures of *Catharanthus roseus*. *Biotechnology Progress*, *18*(1), 159–162.

Nef-Campa, C., Trouslot, M. F., Trouslet, P., & Chrestin, H. (1994). Long-term effect of a Pythium elicitor treatment on the growth and alkaloid production of *Catharanthus roseus* cell suspensions. *Planta Medica*, *60*(02), 149–152.

Negrel, J., & Javelle, F. (1995). Induction of phenylpropanoid and tyramine metabolism in pectinase-or pronase-elicited cell suspension cultures of tobacco (*Nicotiana tabacum*). *Physiologia Plantarum*, *95*(4), 569–574.

Niazian, M. (2019). Application of genetics and biotechnology for improving medicinal plants. *Planta*, *249*(4), 953–973.

Nitta, H., Sonia, M., Fernanda, M., Morteza, D. C., Mohammad, Y., & Mirjalili, H. (2018). Specialized plant metabolism characteristics and impact on target molecule biotechnological production. *Molecular Biotechnology*, *60*(2), 169–183.

Noguchi, M., Matsumoto, T., Hirata, Y., Yamamoto, K., Katsuyama, A., Kato, A., & Kato, K. (1977). Plant tissue culture and its biotechnological application (W. Barz, E. Reinhard, & & M. H. Zenk, p. 85). Springer-Verlag.

Ochoa-Villarreal, M., Howat, S., Hong, S., Jang, M. O., Jin, Y. W., Lee, E. K., & Loake, G. J. (2016). Plant cell culture strategies for the production of natural products. *BMB Reports*, *49*(3), 149–158.

Ohemu, T. L., Agunu, A., Chollom, S. C., Okwori, V. A., Dalen, D. G., & Olotu, P. N. (2018). Preliminary phytochemical screening and antiviral potential of methanol stem bark extract of *Enantia chlorantha Oliver* (Annonaceae) and *Boswellia dalzielii Hutch* (Burseraceae) against Newcastle disease in Ovo. *Pharmacognosy and Traditional Medicine*, *25*(4),1–8.

Oksman-Caldentey, K. M., & Inzé, D. (2004). Plant cell factories in the post-genomic era: New ways to produce designer secondary metabolites. *Trends in Plant Science*, *9*(9), 433–440.

Oliveira, A. F. C. D. S., Teixeira, R. R., Oliveira, A. S. D., Souza, A. P. M. D., Silva, M. L. D., & Paula, S. O. D. (2017). Potential antivirals: Natural products targeting replication enzymes of dengue and chikungunya viruses. *Molecules*, *22*(3), 505.

Oliver, S. E., Hoots, B. E., Paz-Bailey, G., Markowitz, L. E., Meites, E., & NHBS Study Group. (2017). Increasing human papillomavirus vaccine coverage among men who have sex with men – National HIV Behavioral Surveillance, United States, 2014. *JAIDS Journal of Acquired Immune Deficiency Syndromes*, *75*, S370–S374.

Orhan, I., Deliorman-Orhan, D., & Özçelik, B. (2009). Antiviral activity and cytotoxicity of the lipophilic extracts of various edible plants and their fatty acids. *Food Chemistry*, *115*(2), 701–705.

Özçelik, B., Kartal, M., & Orhan, I. (2011). Cytotoxicity, antiviral and antimicrobial activities of alkaloids, flavonoids, and phenolic acids. *Pharmaceutical Biology*, *49*(4), 396–402.

Palazon, J., Pinol, M. T., Cusido, R. M., Morales, C., & Bonfill, M. (1997). Application of transformed root technology to the production of bioactive metabolites. *Recent Research Developments in Plant Physiology*, *1*, 125–143.

Pan, Q., Wang, Q., Yuan, F., Xing, S., Zhao, J., Choi, Y. H., Verpoorte, R., Tian, Y., Wang, Y., & Tang, K. (2012). Overexpression of ORCA3 and G10H in *Catharanthus roseus* plants regulated alkaloid biosynthesis and metabolism revealed by NMR-metabolomics. *PLoS One*, *7*(8), 1–14.

Panche, A. N., Diwan, A. D., & Chandra, S. R. (2016). Flavonoids: An overview. *Journal of Nutritional Science*, *5*.

Pant, B. (2014), Application of plant cell and tissue culture for the production of phytochemicals in medicinal plants. In *Infectious diseases and nanomedicine II* (pp. 25–39). Springer.

Park, J. Y., Ko, J. A., Kim, D. W., Kim, Y. M., Kwon, H. J., Jeong, H. J., Kim, C. Y., Park, K. H., Lee, W. S., & Ryu, Y. B. (2016). Chalcones isolated from Angelica keiskei inhibit cysteine proteases of SARS-CoV. *Journal of Enzyme Inhibition and Medicinal Chemistry*, *31*(1), 23–30.

Parmar, H., Yusufzai, S., Parmar, P., Bajaniya, V., & Chavda, V. (2020). Efficacy of Shatavari supplemented diet on growth performance of genetically improved farmed tilapia (GIFT) fry. *Journal of Entomology and Zoology Studies*, *8*(6), 559–561.

Pasquali, G., Porto, D. D., & Fett-neto, A. G. (2006). Metabolic engineering of cell cultures versus whole plant complexity in production of bioactive monoterpene indole alkaloids: Recent progress related to old dilemma. *Journal of Bioscience and Bioengineering*, *101*(4), 287–296.

Payne, S. (2017). Family *coronaviridae*. *Viruses*, 149.

Phelan, D., Barrozo, E. R., & Bloom, D. C. (2017). HSV1 latent transcription and non-coding RNA: A critical retrospective. *Journal of Neuroimmunology*, *308*, 65–101.

Poulev, A., O'Neal, J. M., Logendra, S., Pouleva, R. B., Timeva, V., Garvey, A. S., Gleba, D., Jenkins, I. S., Halpern, B. T., Kneer, R., Cragg, M. G., & Raskin, I. (2003). Elicitation, a new window into plant chemodiversity and phytochemical drug discovery. *Journal of Medicinal Chemistry*, *46*(12), 2542–2547.

Pouvreau, B., Vanhercke, T., & Singh, S. (2018). From plant metabolic engineering to plant synthetic biology: The evolution of the design/build/test/learn cycle. *Plant Science*, *273*, 3–12.

Primrose, S. B., & Twyman, R. (2006). Gene transfer to plants. In *Principles of gene manipulation and genomics* (7th ed., pp. 274–299). John Wiley & Sons.

Puthur, J. T. (2016). Antioxidants and cellular antioxidation mechanism in plants. *South Indian Journal of Biological Sciences*, *2*(1), 9–13.

Radman, R., Saez, T., Bucke, C., & Keshavarz, T. (2003). Elicitation of plants and microbial cell systems. *Biotechnology and Applied Biochemistry*, *37*(1), 91–102.

Rajput, S. B., Tonge, M. B., & Karuppayil, S. M. (2014). An overview on traditional uses and pharmacological profile of *Acorus calamus Linn.* (Sweet flag) and other Acorus species. *Phytomedicine*, *21*(3), 268–276.

Ramesha, B. T., Amna, T., Ravikanth, G., Gunaga, R. P., Vasudeva, R., Ganeshaiah, K. N., Uma Shaanker, R., Khajuria, R. K., Puri, S. C., & Qazi, G. N. (2008). Prospecting for camptothecines from Nothapodytes nimmoniana in the Western Ghats, South India: identification of high-yielding sources of camptothecin and new families of camptothecines. *Journal of Chromatographic Science*, *46*(4), 362–368.

Rani, S. J., & Usha, R. (2013). Transgenic plants: Types, benefits, public concerns and future. *Journal of Pharmacy Research*, *6*(8), 879–883.

Rao, S. R., & Ravishankar, G. A. (2002). Plant cell cultures: Chemical factories of secondary metabolites. *Biotechnology Advances*, *20*(2), 101–153.

Ravishankar, G. A., & Venkataraman, L. V. (1993). Role of plant cell culture in food biotechnology: Current trends, limitations and future prospects. In *Plant biotechnology: Commercial prospects and problems* (pp. 255–274). R. L. M. Pierikeditors.

Rebrikov, D. V., Desai, S. M., Siebert, P. D., & Lukyanov, S. A. (2004). Suppression subtractive hybridization. In *Gene expression profiling* (pp. 107–134). Humana Press.

Rehman, S., Rehman, I. U., Jan, B., Rashid, I., Reshi, Z. A., & Ganie, A. H. (2021). Genome editing: Applications for medicinal and aromatic plants. In *Medicinal and aromatic plants* (pp. 119–144). Academic Press.

Rijhwani, S. K., & Shanks, J. V. (1998). Effect of elicitor dosage and exposure time on biosynthesis of indole alkaloids by *Catharanthus roseus* hairy root cultures. *Biotechnology Progress*, *14*(3), 442–449.

Roberts, S. C. (2007). Production and engineering of terpenoids in plant cell culture. *Nature Chemical Biology*, *3*(7), 387–395.

Roberts, S. C., & Shuler, M. L. (1997). Large-scale plant cell culture. *Current Opinion in Biotechnology*, *8*(2), 154–159.

Rohmer, M. (1999). The discovery of a mevalonate-independent pathway for isoprenoid biosynthesis in bacteria, algae and higher plants. *Natural Product Reports*, *16*(5), 565–574.

Romero-Rodríguez, A., Maldonado-Carmona, N., Ruiz-Villafán, B., Koirala, N., Rocha, D., & Sánchez, S. (2018). Interplay between carbon, nitrogen and phosphate utilization in the control of secondary metabolite production in *Streptomyces*. *Antonie Van Leeuwenhoek*, *111*(5), 761–781.

Ruffoni, B., Pistelli, L., Bertoli, A., & Pistelli, L. (2010). Plant cell cultures: Bioreactors for industrial production. In *Bio-farms for nutraceuticals* (pp. 203–221). Springer.

Ruiz-Cruz, S., Chaparro-Hernández, S., Hernández-Ruiz, K. L., Cira-Chávez, L. A., Estrada-Alvarado, M. I., Gassos Ortega, L. E., & Lopez Mata, M. A. (2017). Flavonoids: Important biocompounds in food. In *Flavonoids-from biosynthesis to human health* (pp. 353–369). IntechOpen.

Sarfaraj, M., Rahman, M. A., Ahmad, I., & Saeed, M. (2012). Current approaches toward production of secondary plant metabolites. *Journal of Pharmacy and Bioallied Sciences*, *4*(1), 10–20.

Sato, F., Hashimoto, T., Hachiya, A., Tamura, K. I., Choi, K. B., Morishige, T., Fujimoto, H., & Yamada, Y. (2001). Metabolic engineering of plant alkaloid biosynthesis. *Proceedings of the National Academy of Sciences*, *98*(1), 367–372.

Saurina, J., & Sentellas, S. (2015). Determination of phenolic compounds in food matrices: Application to characterization and authentication. In *Fast liquid chromatography–mass spectrometry methods in food and environmental analysis* (pp. 517–547). World Scientific Publishers.

Schläger, S., & Dräger, B. (2016). Exploiting plant alkaloids. *Current Opinion in Biotechnology*, *37*, 155–164.

Sevón, N., & Oksman-Caldentey, K. M. (2002). *Agrobacterium rhizogenes*-mediated transformation: Root cultures as a source of alkaloids. *Planta Medica*, *68*(10), 859–868.

Shahidi, F., & Ambigaipalan, P. (2015). Phenolics and polyphenolics in foods, beverages and spices: Antioxidant activity and health effects–a review. *Journal of Functional Foods*, *18*, 820–897.

Sharma, A., Shahzad, B., Rehman, A., Bhardwaj, R., Landi, M., & Zheng, B. (2019). Response of phenylpropanoid pathway and the role of polyphenols in plants under abiotic stress. *Molecules*, *24*(13), 2452.

Sharma, A., Verma, P., Mathur, A., & Mathur, A. K. (2018). Overexpression of tryptophan decarboxylase and strictosidine synthase enhanced terpenoid indole alkaloid pathway activity and antineoplastic vinblastine biosynthesis in *Catharanthus roseus*. *Protoplasma*, *255*, 1281–1294.

Shi, C., Guo, X., Qu, Q., Tang, Z., Wang, Y., & Zhou, S. (2014). Actively targeted delivery of anticancer drug to tumor cells by redox-responsive star-shaped micelles. *Biomaterials*, *35*(30), 8711–8722.

Shirazi, Z., Aalami, A., Tohidfar, M., & Sohani, M. M. (2018). Metabolic engineering of glycyrrhizin pathway by overexpression of beta-amyrin 11-oxidase in transgenic roots of *Glycyrrhiza glabra*. *Molecular Biotechnology*, *60*, 412–419.

Singh, P., Prasad, R., Tewari, R., Jaidi, M., Kumar, S., Rout, P. K., & Rahman, L. U. (2018). Silencing of quinolinic acid phosphoribosyl transferase (QPT) gene for enhanced production of scopolamine in hairy root culture of *Duboisia leichhardtii*. *Scientific Reports*, *8*(1), 13939.

Singh, Y. D., Jena, B., Ningthoujam, R., Panda, S., Priyadarsini, P., Pattanayak, S., Sing, M. C., & Satapathy, K. B. (2020). Potential bioactive molecules from natural products to combat against coronavirus. *Advances in Traditional Medicine*, 1–12.

Singletary, K. W. (2018). Basil: A brief summary of potential health benefits. *Nutrition Today*, *53*(2), 92–97.

Smart, N. J., & Fowler, M. W. (1984). An airlift column bioreactor suitable for large-scale cultivation of plant cell suspensions. *Journal of Experimental Botany*, *35*(4), 531–537.

Smetanska, I. (2008). Production of secondary metabolites using plant cell cultures. *Food Biotechnology*, 187–228.

Sridevi, V., Naveen, P., Karnam, V. S., Reddy, P. R., & Arifullah, M. (2021). Beneficiary and adverse effects of phytoestrogens: A potential constituent of plant-based diet. *Current Pharmaceutical Design*, *27*(6), 802–815.

Srivastava, M., Maurya, P., & Shanker, K. (2021). *Clerodendrum viscosum*: A critical review on phytochemistry, pharmacology, quality assurance, and safety data. *Medicinal Chemistry Research*, 1–23.

Stafford, A., Morris, P., & Fowler, M. W. (1986). Plant cell biotechnology: A perspective. *Enzyme and Microbial Technology*, *8*(10), 578–587.

Steiner, I., & Benninger, F. (2013). Update on herpes virus infections of the nervous system. *Current Neurology and Neuroscience Reports*, *13*(12), 414.

Steinmann, J., Buer, J., Pietschmann, T., & Steinmann, E. (2013). Anti-infective properties of epigallocatechin-3-gallate (EGCG), a component of green tea. *British Journal of Pharmacology*, *168*(5), 1059–1073.

Stevanović, Z. D., Bošnjak-Neumüller, J., Pajić-Lijaković, I., Raj, J., & Vasiljević, M. (2018). Essential oils as feed additives – future perspectives. *Molecules*, *23*(7), 1717.

Subudhi, B. B., Chattopadhyay, S., Mishra, P., & Kumar, A. (2018). Current strategies for inhibition of chikungunya infection. *Viruses*, *10*(5), 235.

Sumner, J. (2000). *The natural history of medicinal plants*. Timber Press.

Suručić, R., Travar, M., Petković, M., Tubić, B., Stojiljković, M. P., Grabež, M., Savikin, K., Zdunic, G., & Škrbić, R. (2021). Pomegranate peel extract polyphenols attenuate the SARS-CoV-2 S-glycoprotein binding ability to ACE2 Receptor: In silico and in vitro studies. *Bio-Organic Chemistry*, *114*, 105145.

Tabata, H. (2006). Production of paclitaxel and the related taxanes by cell suspension cultures of Taxus species. *Current Drug Targets*, *7*(4), 453–461.

Taiz, L., & Zeiger, E. (2006). Plant physiology/by Lincoln Taiz and Eduardo Zeiger (No. Libro 581.1 T3).

Tang, L. I., Ling, A. P., Koh, R. Y., Chye, S. M., & Voon, K. G. (2012). Screening of anti-dengue activity in methanolic extracts of medicinal plants. *BMC Complementary and Alternative Medicine*, *12*, 1–10.

Techen, N., Parveen, I., Pan, Z., & Khan, I. A. (2014). DNA barcoding of medicinal plant material for identification. *Current Opinion in Biotechnology*, *25*, 103–110.

Teodoro, A. J. (2019). Bioactive compounds of food: Their role in the prevention and treatment of diseases. *Oxidative Medicine and Cell Longitivity*, *2019*, 1–4.

Teoh, E. S. (2016). *Medicinal orchids of Asia* (Vol. 16, No. 4). Springer.

Tripathi, S., Sangwan, R. S., Narnoliya, L. K., Srivastava, Y., Mishra, B., & Sangwan, N. S. (2017). Transcription factor repertoire in Ashwagandha (*Withania somnifera*) through analytics of transcriptomic resources: Insights into regulation of development and withanolide metabolism. *Scientific Reports*, *7*(1), 16649.

Tuladhar, P., Sasidharan, S., & Saudagar, P. (2021). Role of phenols and polyphenols in plant defense response to biotic and abiotic stresses. In *Biocontrol agents and secondary metabolites* (pp. 419–441). Wood Head Publishing.

Urabe, D., Asaba, T., & Inoue, M. (2015). Convergent strategies in total syntheses of complex terpenoids. *Chemical Reviews*, *115*(17), 9207–9231.

Vargas, V. M. L., & Flota, F. V. (2006). An introduction to plant cell culture back to the future. In *Plant cell culture protocols methods in molecular biology* (2nd ed., pp. 3–8). Humana Press Inc.

Vaure, C., & Liu, Y. (2014). A comparative review of toll-like receptor 4 expression and functionality in different animal species. *Frontiers in Immunology*, *5*, 316.

Verma, N., & Shukla, S. (2015). Impact of various factors responsible for fluctuation in plant secondary metabolites. *Journal of Applied Research on Medicinal and Aromatic Plants*, *2*(4), 105–113.

Verpoorte, R., & Memelink, J. (2002). Engineering secondary metabolite production in plants. *Current Opinion in Biotechnology*, *13*(2), 181–187.

Verpoorte, R., van der Heijden, R., & Memelink, J. (2000). Engineering the plant cell factory for secondary metabolite production. *Transgenic Research*, *9*(4), 323–343.

Vuorela, P., Leinonen, M., Saikku, P., Tammela, P., Rauha, J. P., Wennberg, T., & Vuorela, H. (2004). Natural products in the process of finding new drug candidates. *Current Medicinal Chemistry*, *11*(11), 1375–1389.

Wang, Q., Reddy, V. A., Panicker, D., Mao, H. Z., Kumar, N., Rajan, C., Venkatesh, P. N., Cha, N.-H., & Sarojam, R. (2016). Metabolic engineering of terpene biosynthesis in plants using a trichome-specific transcription factor Ms YABBY 5 from spearmint (*Mentha spicata*). *Plant Biotechnology Journal*, *14*(7), 1619–1632.

Wani, M. C., & Horwitz, S. B. (2014). Nature as a remarkable chemist: A personal story of the discovery and development of Taxol®. *Anti-Cancer Drugs*, *25*(5), 482.

Weising, K., Schell, J., & Kahl, G. (1988). Foreign genes in plants: Transfer, structure, expression, and applications. *Annual Review of Genetics*, *22*(1), 421–477.

Winkel-Shirley, B. (2002). Biosynthesis of flavonoids and effects of stress. *Current Opinion in Plant Biology*, *5*(3), 218–223.

Wojtunik-Kulesza, K., Rudkowska, M., Kasprzak-Drozd, K., Oniszczuk, A., & Borowicz-Reutt, K. (2021). Activity of selected group of monoterpenes in alzheimer's disease symptoms in experimental model studies – a non-systematic review. *International Journal of Molecular Sciences*, *22*(14), 7366.

World Health Organization. (2001). *Legal status of traditional medicine and complementary (No. WHO/EDM/TRM/2001.2).* World Health Organization.

World Health Organization (WHO), 2024. https://www.who.int/news-room/fact-sheets/detail/hiv-aids

Wu, C., Liu, Y., Yang, Y., Zhang, P., Zhong, W., Wang, Y., Wang, Q., Yang, X., Li, X., Zheng, M., Chen, L., & Li, H. (2020). Analysis of therapeutic targets for SARS-CoV-2 and discovery of potential drugs by computational methods. *Acta Pharmaceutica Sinica B*, *10*(5), 766–788.

Wu, M. F., Chen, S. T., Yang, A. H., Lin, W. W., Lin, Y. L., Chen, N. J., & Hsieh, S. L. (2013). CLEC5A is critical for dengue virus–induced inflammasome activation in human macrophages. *Blood, the Journal of the American Society of Hematology*, *121*(1), 95–106.

Xin, C., Qing-wei, Y., Jia-lin, S., Shuang, X., Fu-chun, X., & Yajun, C. (2014). Research progress on nitrogen use and plant growth. *Journal of Northeast Agricultural University* (English Edition), *21*(2), 68–74.

Yamamoto, Y., Mizuguchi, R., & Yamada, Y. (1982). Selection of a high and stable pigment-producing strain in cultured *Euphorbia millii* cells. *Theoretical and Applied Genetics*, *61*(2), 113–116.

Yamazaki, M., Rai, A., Yoshimoto, N., & Saito, K. (2018). Perspective: Functional genomics towards new biotechnology in medicinal plants. *Plant Biotechnology Reports*, *12*(2), 69–75.

Yang, X., Yu, Y., Xu, J., Shu, H., Liu, H., Wu, Y., Zhang, L., Yu, Wang, Y., Pan, S., Zou, X., Yuan, S., & Shang, Y. (2020). Clinical course and outcomes of critically ill patients with SARS-CoV-2 pneumonia in Wuhan, China: A single-centered, retrospective, observational study. *The Lancet Respiratory Medicine*, *8*(5), 475–481.

Zhang, H. B., Wang, C., & Zhang, L. H. (2004). The quormone degradation system of Agrobacterium tumefaciens is regulated by starvation signal and stress alarmone (p) ppGpp. *Molecular Microbiology*, *52*(5), 1389–1401.

Zhang, L., McClements, D. J., Wei, Z., Wang, G., Liu, X., & Liu, F. (2020). Delivery of synergistic polyphenol combinations using biopolymer-based systems: Advances in physicochemical properties, stability and bioavailability. *Critical Reviews in Food Science and Nutrition*, *60*(12), 2083–2097.

Zhao, X., Park, S. Y., Yang, D., & Lee, S. Y. (2018). Synthetic biology for natural compounds. *Biochemistry*, *58*(11), 1454–1456.

Zhong, J. J. (2001). Biochemical engineering of the production of plant-specific secondary metabolites by cell suspension cultures. *Plant Cells*, 1–26.

Zhong, J. J., & Yoshida, T. (1993). Effects of temperature on cell growth and anthocyanin production in suspension cultures of *Perilla frutescens*. *Journal of Fermentation and Bioengineering*, *76*(6), 530–531.

Zhou, J., Wang, D., Gao, R., Zhao, B., Song, J., Qi, X., Zhang, Y., Yang, L., Zhu, W., Bai, T., Qin, K., Lan, Y., Zou, S., Guo, J., Dong, J., Dong, L., Zhang, Y., Wei, H., Li, X., Chen, Y., Xu, C., Pei, Y., Yang, Y., Zhang, X., Wang, S., Feng, Z., Han, J., Yang, W., Gao, G., Wu, G., Li, D., Wang, Y., & Shu, Y. (2013). Biological features of novel avian influenza A (H7N9) virus. *Nature*, *499*(7459), 500–503.

Ziegler, J., & Facchini, P. J. (2008). Alkaloid biosynthesis: Metabolism and trafficking. *Annual Review of Plant Biology*, *59*, 735–769.

Zielińska, S., & Matkowski, A. (2014). Phytochemistry and bioactivity of aromatic and medicinal plants from the genus *Agastache* (Lamiaceae). *Phytochemistry Reviews*, *13*, 391–416.

Ziv, M., & Hadar, A. (1991). Morphogenic pattern *of Nephrolepis exaltata* cv. Bostoniensis in agar or liquid cultures: Implications for micropropagation. *Israel Journal of Plant Sciences*, *40*(1), 7–16.

Zorofchian Moghadamtousi, S., Abdul Kadir, H., Hassandarvish, P., Tajik, H., Abubakar, S., & Zandi, K. (2014). A review on antibacterial, antiviral, and antifungal activity of curcumin. *BioMed Research International*, *2014*.

Zou, Z. Q., Wang, L., Wang, K., & Yu, J. G. (2016). Innate immune targets of hepatitis B virus infection. *World Journal of Hepatology*, *8*(17), 716.

Züst, T., Strickler, S. R., Powell, A. F., Mabry, M. E., An, H., Mirzaei, M., York, T., Holland, C. K., Kumar, P., Erb, M., Petschenka, G., Gómez, J. M., Perfectti, F., Muller, C., Pires, C. J., Mueller, L. A., & Jander, G. (2020). Independent evolution of ancestral and novel defenses in a genus of toxic plants (*Erysimum, Brassicaceae*). *Elife*, *9*, e51712.

32 Roles of Polyploidy in the Metabolic and Genetic Improvement of Medicinal Plants

Mostafa Ahmadizadeh and Parviz Heidari

32.1 INTRODUCTION

Polyploidy, the possession of more than two sets of chromosomes, is a ubiquitous phenomenon in the plant kingdom. The process of polyploidy encompasses either the replication of a single genome (autopolyploidy) or the amalgamation of multiple different genomes (allopolyploidy) in order to generate a fresh species (Schiessl et al., 2019; Udall & Wendel, 2006). One possible way for mechanisms to occur is through interspecific hybridization that is then followed by the doubling of chromosomes, or through the crossing of undiminished gametes of diploid species, also two autotetraploid hybridization (Chen et al., 2007; Zaman et al., 2019). Natural polyploid plants are often inaccessible, and synthetic polyploidy, induced using mitotic inhibitors, provides a powerful tool for plant improvement (Šedivá, Mrázková et al., 2019). Polyploids make up roughly 70% of wild grasses and 25% of legumes. Several crucial domesticated crops have been categorized as allopolyploids, including wheat, rapeseed, tobacco, oat, and kiwifruit (Mansouri & Bagheri, 2017; Schiessl et al., 2019). Polyploidy frequently leads to an increase in favorable traits, particularly in terms of biomass and vegetative growth, making it desirable for medicinal plants where these attributes are commercially valuable (Mansouri & Bagheri, 2017). Polyploids may find new niches or boost their fitness by possessing traits like enhanced water deficiency tolerance, apomixis, varied flowering times, some biotic stress resistance, and increased organ size (Chen et al., 2007; Deng et al., 2012; Mansouri & Bagheri, 2017; Ren et al., 2018). The existence of surplus gene copies and alleles may increase allelic heterosis, resulting in hybrid vigor, as well as supplying genetic redundancy (Comai, 2005). The combination of gene dosage and epigenetic factors, along with nuclear enlargement, induces genetic modifications that lead to significant changes in phenotype (Allario et al., 2013; Chen, 2010; Hao et al., 2013; Van Laere et al., 2011).

Both types of polyploids, whether natural or artificial, are essential in shaping plant breeding practices. Utilizing artificial crosses and chromosomal manipulation with polyploidization agents allows for the creation of polyploids, providing insight into the immediate effects of genome duplication, including chromosome restructuring, loss of low-copy DNA sequences, gene activation, retrotransposon activity, and gene silencing (Paterson, 2008). Newly established engineered polyploids (artificial) face genome downsizing, rapid loss of synteny within homoeologous chromosomes, and experience adjustment in epigenetic features (Hancock, 2005). The latest developments have demonstrated a bright future for producing induced polyploidy through cultivation in the control condition. Synthetically tetraploids were successfully derived by somatic embryo induction in diploid cultivars through the application of an in vitro technique within several plant genera (Gao et al., 2002; Koefoed Petersen et al., 2003; Wu & Mooney, 2002). The widespread belief is that induced somaclonal variation plays a crucial role in providing tetraploid genotypes for breeding (Yang et al., 2006). Polyploidy induction in medicinal plants through in vitro methods heavily relies on the selection of an appropriate explant type, encompassing options like zygotic embryos, nodes, seeds, buds, vegetative tissues, and callus (Adaniya & Shirai, 2001; Chung et al., 2017; Dhooghe et al., 2011; Rubuluza, Nikolova et al., 2007; Zhang et al., 2010).

Secondary metabolites from plants exhibit intricate variations, stemming not just from genetic variances but also from the influence of external factors. Various forms of stress can result in fluctuations in both the quality and quantity of secondary metabolites. The chemical composition of flowers, seeds, roots, stems, fruit walls, buds, and leaves often varies, with each part potentially changing at different developmental stages and seasons. The utilization of genome duplication can cause the elevation of plant secondary metabolites by stimulating metabolite biosynthesis through heightened levels of related genus or gene expression under specific conditions (Mansouri & Bagheri, 2017). It is expected that gene silencing will have a significant effect on the regulation of duplicate gene expression through both genetic and epigenetic pathways (Comai, 2000; Prince & Pickett, 2002; Tate et al., 2005; Wendel, 2000). Frequently, epigenetic occurrences are practically characterized exclusively via the lack of nucleotide alteration, as is typically seen in research engaging artificial polyploids generated by existing diploid ancestors (Doyle et al., 2008). Polyploidy induces nuclear enlargement, rearrangement of chromosomes and alterations in epigenetics,

DOI: 10.1201/9781003475491-32

ultimately resulting in the reprogramming of the networks of transcription, protein and metabolite production (Leitch & Leitch, 2008). Despite the benefits of increased flexibility of the evolution and phenotypic variation in fresh polyploids, there are drawbacks, such as higher rates of chromosome pairing errors and reduced mating opportunities (Hegarty & Hiscock, 2008). The creation of polyploid species offers a fresh approach to advancing crop improvement (see Table 32.1). To comprehend the intricacies of polyploid genomes and epigenomes, it is essential to

TABLE 32.1
Summary of the Induced Polyploidy Studies in Various Medicinal Plants

Plant species	Ploidy level	Summary of the obtained results	References
Ispaghul (*Plantago ovata* Forsk.)	Tetraploids	- Larger spike, more height, larger pollen grain, thicker leaf and seed, and more seeds per spike. - The number of chloroplast and content of carotenoids and chlorophyll increased in tetraploids. - Bigger stomata size in tetraploids. - The content of mucilage in seeds was more detected in tetraploids.	(Sabzehzari et al., 2020)
Garden thyme (*Thymus vulgaris* L.)	Tetraploids	- Increased plant height, weight, leaf length. - The production of secondary metabolites has increased significantly. Tetraploids now yield essential oil at a rate of 1.19%, up from 0.81%. Additionally, the content of key components such as thymol and carvacrol has increased by 18.01% and 0.49%, respectively.	(Shmeit et al., 2020)
Drug-type strain of *Cannabis sativa*	Tetraploids	- The size of the leaves was increased, and the stomata were around 30% larger and about half less dense when compared to diploids. - Tetraploid sugar leaves exhibited a 40% increase in trichome density, accompanied by notable alterations in the terpene composition and a noteworthy 9% spike in CBD levels within the buds. - No substantial improvement in the yield of dry bud or THC potency was detected.	(Parsons et al., 2019)
Plantago psyllium	Tetraploids	- Increased dimensions for spike, seed, and pollen grain size along with leaf thickness and plant height. - Elevated number of chloroplasts found in guard cells, along with increased levels of chlorophyll and carotenoids. - Indicates that tetraploidization may enhance seed mucilage content for practical applications.	(Sabzehzari et al., 2019)
Chinese sage (*Salvia miltiorrhiza Bunge*)	Tetraploids	- Enhanced agronomic traits. - In the root extract, there are higher concentrations of dihydrotanshinone I and all tanshinones combined.	(Chen et al., 2018)
Anoectochilus formosanus H.	Tetraploids	- Advancements in different agronomic traits like dry weight, stoma size, fresh weight, leaf width, shoot length, root length, and chloroplast quantity per stoma. - In comparison to the diploid, both the leaf and stem of the plant contain higher levels of total flavonoid and gastrodin.	(Chung et al., 2017)
Cannabis	Tetraploids	- Tetraploid cannabis plants were deemed unsuitable for both fiber production and medical purposes due to their lower levels of fiber and cannabinoids. - The mixoploid plant displayed remarkable features with increased cannabinoid content, protein levels, and sugar concentration.	(Mansouri & Bagheri, 2017)
Tetradenia riparia	Tetraploids	- Tetraploid plants demonstrated the capability to generate essential oil that could help prevent postharvest diseases (against *Geotrichum candidum*).	(Hannweg, Visser et al., 2016)
Thymus persicus	Tetraploids	- Diminished height of plantlets, condensed root length, increased stem girth and darker leaf color. - Reduced stomatal density is observed on both the abaxial and adaxial leaf surfaces, where stomata are longer and wider.	(Tavan et al., 2015)
Egyptian henbane (*Hyoscyamus muticus* L.)	Tetraploids	- Growth rate and alkaloid accumulation were altered due to tetraploidy. - A greater amount of scopolamine was generated by the tetraploid clone compared to its diploid counterpart.	(Dehghan et al., 2012)
Basil (*Ocimum basilicum* L.)	Autotetraploids	- The density of stomata decreases while both the size of the pollen grains and number of chloroplasts within guard cells increase.	(Omidbaigi et al., 2010)
Pimpinella anisum L.	Autotetraploids	- Larger stomata in tetraploid plants. - Tetraploidy increased the contents of DNA, RNA, and total protein.	(Ahmadinia & Heidari, 2023)

sequence numerous polyploid genomes. With each newly sequenced polyploid genome, a more comprehensive picture of the intricate mechanisms and functions governing gene expression and morphological adaptations will emerge. Given the current situation, it may be advantageous to reexamine the recent advancements in polyploidy plants, specifically delving into their metabolic, genetic, genomic restructuring, developmental control, and epigenetic alterations such as histone modifications, DNA methylation, and small RNA involvement. The enhancement of understanding on polyploidy formation and progression in relevant studies focusing on genetic and metabolic enhancement will help plant breeders in creating innovative approaches for crop breeding program.

32.2 CHROMOSOME DUPLICATION AND MEIOTIC RECOMBINATION IN POLYPLOIDS

Plants with polyploidy exhibit multiple copies of chromosomes after experiencing whole-genome duplication (WGD), a process that may be triggered by irregular cell division. It appears that polyploidy is typically a dead end in evolution, with only a few instances managing to survive and establish themselves in the long run (Arrigo & Barker, 2012; Vanneste et al., 2014). Exposing diploid fertilized eggs to environmental factors like temperature extremes and radiation can induce chromosome doubling; however, inhibiting the release of polar bodies during meiosis in diploids can result in the creation of polyploids (Song et al., 2012). Ensuring the stability of polyploidy across generations requires intricate systems to regulate the segregation of multiple homologous chromosomes during meiosis. Multivalents can be formed within meiosis in autopolyploidy when various fusions of homologous chromosomes come together at diverse locations along the chromosomes (Bomblies & Madlung, 2014; Del Pozo & Ramirez-Parra, 2015; Grandont et al., 2013). Neopolyploid descendants exhibit a notable proportion of aneuploids, pseudoeuploids, and homolog-recombinant genotypes, potentially influencing the diverse traits observed in early generation polyploids (Ramsey & Schemske, 2002). The contrast between neopolyploids and established polyploids offers insight into the evolution that has occurred post-origin (Münzbergová, 2017; Pavlíková et al., 2017). As a result, neopolyploids provide a way to assess the rapid impacts of genome duplication and its influence on the development and continuity of the new cytotype, serving as a crucial element in researching the ecological mechanisms implicated in polyploid evolution (Castro et al., 2018). By utilizing flow cytometry, researchers can identify newly developed polyploids (i.e., neopolyploids) within wild populations (Kron et al., 2007; Ramsey, 2011) or induce the creation of polyploids in controlled settings using mutagenic anti-mitotic agents (Husband et al., 2008), offering a distinct chance to measure the immediate effects of genome duplications through experimentation.

32.3 EFFECTS ON ENVIRONMENTAL ROBUSTNESS

Polyploidy exerts a diverse array of impacts on plants, with outcomes varying based on the species, range of ploidy levels, and degree of heterozygosity, and is influenced by mechanisms such as gene dose effects, gene silencing, trait regulation, and genetic factors (Ranney, 2006). Whole genome doubling allows for polyploidy to maintain more than two alleles at each locus, potentially resulting in duplicate genes developing novel functions to better cope with different environmental stresses (Yaghobi & Heidari, 2023). While allopolyploids may be more common in nature, autopolyploids excel in their capacity to harness a wide array of genes and allele variants (Manzoor et al., 2019). Allopolyploidy has historically been linked to alterations in gene expression and epigenetic restructuring, typically leading to a greater adaptive capacity compared to autopolyploidy (Fang & Morrell, 2016; Yoo et al., 2014). Nevertheless, genome replication in autopolyploids can result in rapid alterations in gene activity, physical characteristics, and capacity for adaptation to different environments (Bomblies & Madlung, 2014; Van de Peer et al., 2017). To enhance the agronomic traits of plants, plant breeders carefully choose polyploid specimens that have undergone either autopolyploidization within the same species or allopolyploidization with other species (Rêgo et al., 2011). The implications of genome duplication may lead to increased organ size and enhanced production of secondary compounds, which are often associated with heightened resistance to both living and non-living pressures (Ahmadinia & Heidari, 2023). Besides, plant pigments, along with secondary metabolites, may serve as essential defense mechanisms for plants in the face of environmental stress. Polyploids showing higher levels of chlorophyll may benefit from potentially increased photosynthetic capacity, giving them an advantage (Liao et al., 2016). Moreover, studies have shown that polyploids have a greater antioxidant capacity than their diploid (Deng et al., 2012; Niwa & Sasaki, 2003).

32.4 EVOLUTION AND SPECIATION: GENOMIC DIVERSITY IN POLYPLOID PLANTS

The duplication of entire genomes creates a unique landscape for genetic innovation, offering a wealth of raw material for adaptation and diversification. Plants have greatly benefited from polyploidy as a key mechanism for their evolution and speciation (Mansouri & Bagheri, 2017). It is believed that roughly 70% of present-day angiosperms have encountered one or more occurrences of polyploidy throughout their evolution (Chen et al., 2007; Liu & Wendel, 2002). Recent molecular and genomic studies across various plant systems have started to uncover the secrets of polyploidy by showing that rapid genomic changes occur at multiple levels in plants. These instruments are believed to offer assistance in overcoming obstacles associated with immediate adaptation and establishment in new polyploidy, potentially aiding in their long-term success as emerging species (Adams & Wendel, 2005; Chen & Ni, 2006; Otto, 2007). Allopolyploidy plays a significant role in speciation and is a common occurrence in the evolution of plants, leading to the abundance of copied genes in their genomes (Blanc & Wolfe, 2004). Autopolyploids and allopolyploids frequently experience regulatory incompatibilities, chromosome imbalances, genomic instabilities, and reproductive failures due to elevated gene and genome dosages, resulting in rapid alterations to the expression and structure of the gene and trait development. Through the process of polyploidy, genes may duplicate and either maintain their original function, evolve into diverse protein functions, or become deactivated through mutations or epigenetic modifications (Lavania, 2013). Polyploidy, in all its various forms, is a powerful driver of both speciation and diversification, playing a key role in plant genome evolution and serving as a strong approach for sympatric speciation (Sémon & Wolfe, 2007; Wood et al., 2009).

32.5 POLYPLOIDY AND TRANSCRIPTOME ALTERATIONS

Polyploidy research is experiencing a revival due to advancements in technology and the use of genomic tools. The utilization of diverse methods in thorough investigations is likely to extend the frontiers of new methodologies, enabling the application of achieved knowledge in practical ways. The high-throughput genome-wide techniques for deciphering the genetic and epigenetic effects of polyploidization, along with the phenotyping platforms, will be especially crucial (Aversano et al., 2012). The pairing of high-throughput DNA sequencing with bioinformatics analysis offers comprehensive solutions for studying the genetics of polyploids. Despite this, ploidy poses a major hurdle in the sequencing process of plant genomes. Changes in gene expression throughout the genome are linked to modifications in polyploids. Hence, allopolyploids exhibit significant alterations in their transcriptome, likely stemming from the merging of once-distinct genomes and potential interactions among the diverse genetic elements (Hias et al., 2018; Osborn et al., 2003). On the other hand, alterations in gene activity seen in autopolyploids are not as noticeable as those in allopolyploids since the duplicated genome comes from within the same species (Del Pozo & Ramirez-Parra, 2015).

Plant hormone-related genes are a noteworthy group of genes displaying altered expression levels between autopolyploids and diploids (Mu et al., 2012). Genes that are highly regulated in autopolyploid *Arabidopsis* seedlings control processes related to ethylene and auxin (Yu et al., 2010). The generated autopolyploids in multiple plant species underwent transcriptomic analyses, showing fluctuations in the percentage of differentially expressed genes (Dai et al., 2015; Zhou et al., 2015). The comparison of differentially expressed genes between autopolyploids and diploids reveals unique species-specific and tissue-specific expression (Dai et al., 2015).

Multiple SNP detection methods have been utilized to analyze polyploidy in plant species. Utilizing an Illumina RNA-seq technique, Buggs et al. (Buggs et al., 2010) examined close to 3000 SNP markers in *Tragopogon miscellus* allopolyploids to peruse the differential expression of duplicate homologous genes originating from the diploid parents. The researchers observed differences in expression patterns at tissues in the parental genomes (*T. dubius* and *T. pratensis*) when compared to the natural allopolyploids, along with consistent expression levels in F_1 and first-generation synthetic allopolyploids. Through high-throughput cDNA pyrosequencing, Ha and colleagues analyzed the transcriptome of *Arabidopsis thaliana* allopolyploids, finding evidence that small RNAs may influence the protecting against genomic stress in Arabidopsis allopolyploids and interspecific hybrids (Ha et al., 2009). When exposed to various abiotic stresses, the transcriptomic analyses of *Gossypium hirsutum* allopolyploids revealed distinct expression patterns across each homolog (Liu & Adams, 2007). Similarly, stress conditions in allopolyploid *Brassica napus* lead to the expression of different splice variants compared to its parental species (Zhou et al., 2011). Studies using large-scale microarrays on various polyploid plant species have shown clear evidence that gene expression undergoes significant changes due to polyploidy. It is indicated by these findings that autopolyploidy may result in more subtle alterations in expression compared to allopolyploidy (Hegarty et al., 2006; Wang et al., 2006). Among the 9000 genes tested by Stupar et al. (Stupar et al., 2007) in potatoes, approximately 10% exhibited differing levels of expression between diploids and synthetic autotetraploids, leading to distinct phenotypic variances as well. Large disparities in gene expression within numerous metabolic and signal transduction pathways were discovered in *Isatis indigotica*, a member of the well-established autopolyploid Cruciferae group (Lu et al., 2006). The increased copy number of genes can lead to higher levels of gene transcripts and protein products. However, this effect is often buffered by gene silencing, a mechanism that suppresses the expression of duplicated genes to maintain homeostasis. In summary, it is clear that the influence of some environmental stresses on polyploidy gene expression highlights and is important to plant evolution and improvement.

32.6 ROLE OF REGULATORY ELEMENTS ON TRANSCRIPTIONAL REGULATION IN POLYPLOIDS

A reliable allopolyploid offers a great opportunity to explore *cis-* and *trans-*acting effects due to the shared protein factors in allotetraploid cells. Following the merging of the separate genomes, variations in cis- and trans-regulation play a role in altering the ortholog gene expression, which transforms into homoeologous pairs within the allopolyploid or interspecific nucleus (Wittkopp et al., 2004). In allopolyploids, *cis*-regulatory divergence can directly impact single genes or specific chromatin domains like promoters and enhancers, potentially leading to uneven accumulation of homoeologous transcripts. *Cis*-regulatory alterations are the main factor behind the varying expression of offspring genes in allopolyploids (Lee & Chen, 2001; Wang et al., 2004) and interspecific hybrids (De Meaux et la., 2006), as well as maize diploid hybrids (Stupar & Springer, 2006). Discrepancies in gene expression among identical cells probably stem from distinct allelic or epigenetic variances characteristic of progenitors. However, the variation in *trans*-regulatory hierarchies may lead to two distinct forms of expression shifts. An important variation occurs in the combined amount of homoeologous mRNAs in comparison with the median amount of both parents and nonconventional gene expression (Wang et al., 2006). Besides, an allopolyploid displays a different ratio for homoeolog-encoded mRNAs than the two orthologs in a 1:1 mixture of parental mRNAs (Wang et al., 2006). A model of regulatory hierarchy indicates that allopolyploids are primarily characterized by trans-regulatory differences. *Arabidopsis* allotetraploids display unique expression patterns specific to their species, possibly due to regulatory element sequence divergence over the approximately six million years since their parental species diverged (Chen, 2007). The regulatory interactions within a biological pathway may be influenced by changes in epigenetic modifications and cis/trans-acting regulation of homoeologous genes.

32.7 POLYPLOIDY AND PROTEIN ALTERATIONS

The expression data at the transcriptome level may not be a reliable indicator of proteome-level changes due to the impact of regulatory processes like post-transcriptional modifications and protein interactions. Analyzing protein profiles could offer valuable insights to enhance genomic and transcriptomic information. The two main techniques utilized in proteomic studies are two-dimensional gel electrophoresis and isobaric tags for relative and absolute quantitation (iTRAQ). The results of 2-DE experiments in different polyploidy levels of wheat revealed a relationship between the homeologous protein expression of the hexaploid wheat and interactions within the D, B, and A genomes of the parental plants (Islam et al., 2003). By utilizing protein 2-DE via mass spectrometry measure, researchers found a direct association between ploidy levels and differentially expressed proteins in maize (Yao et al., 2011). Mass spectrometry coupled with isobaric tags for relative and absolute quantitation was utilized to compare autopolyploid and neo-allopolyploid with their diploid in terms of protein changes in leaves (Ng et al., 2012). The analysis of over 1000 proteins revealed nearly 7% with varying levels in *A. thaliana* diploids compared to autotetraploids, and more than an 8% variance between *A. thaliana* and its neo-allopolyploid *Arabidopsis suecica*-like. Autotetraploid citrus displayed distinct protein profiles compared to diploid citrus when subjected to salt stress, with a noticeable prevalence of antioxidant enzymes and ROS scavenging enzymes (Podda et al., 2013). Additionally, the same processes have been outlined in the neo-tetraploid *Dioscorea zingiberensis* as well as in the octoploid *Nicotiana benthamiana* (Zhang et al., 2010). The choice of particular proteins in the ploidy process is implied by these observations.

32.8 EPIGENETIC MODIFICATIONS ASSOCIATED WITH POLYPLOIDY

Comprehending the intricacies of genomes and epigenomes of the polyploidy necessitates the sequencing of numerous polyploid plant genomes. The sequencing of more genomes will lead to uncovering in-depth mechanisms and roles in gene expression and changes in morphology. The impact of epigenetic modifications on genomic diversity may provide additional opportunities for natural selection to operate. Frequently, epigenetic occurrences are typically characterized by a lack of nucleotide alteration, which is frequently seen in research on synthetic polyploids formed from existing diploid ancestors (Doyle et al., 2008). Nuclear expansion, chromosomal rearrangement, and epigenetic modifications triggered by polyploidy cause a transformation of the transcriptome, proteome, and metabolome networks (Song & Chen, 2015). Like structural genomic alterations, changes in DNA methylation patterns induced by allopolyploidy have been discovered in plants (Adams et al., 2003; Madlung et al., 2002). Reports have shown that DNA methylation alterations occur in *Eragrostis curvula* and several *Cymbopogon* species as a result of ploidy (Lavania et al., 2012; Ochogavía et al., 2009). Also, it was reported that epigenetic modifications and shifts in DNA methylation patterns correlate with transcriptomic changes at particular loci in neo-autotetraploid *A. thaliana* (Yu et al., 2010).

The promotion of chromosomal DNA recruitment to heterochromatic chromocenter niches in *A. thaliana* is facilitated by tetraploidy, implying that ploidy plays a role in regulating the integrity of functional heterochromatin (Yu et al., 2010). Epigenetic modifications have the potential to be reversed, allowing polyploid cells to adapt to severe genetic changes during the initial stages of polyploidy development, resulting in diverse genetic alterations. Through meiosis, epigenetic modifications can remain intact, exhibiting a lasting legacy over multiple generations (Del Pozo & Ramirez-Parra, 2015).

32.9 DNA METHYLATION CHANGES IN POLYPLOIDS

The top investigated epigenetic event in polyploids is known to be methylation. Methylation-related changes have been investigated using molecular markers in plants. The difference in DNA methylation among allopolyploids, as well as their offspring had been recorded in numerous plant species. In the initial three generations of synthetic *Gossypium* allohexaploids and allotetraploids, no nonadditive banding patterns were observed among the 22,000 AFLP (fragment length polymorphism) bands (Liu et la.,

2001). Despite this, nonadditivity was detected in 20% of bands across both wild and resynthesized allohexaploid wheat (Dong et al., 2005). It was found that in *Spartina anglica*, 30% of parental methylation patterns were nonadditive in hybrids and allopolyploids, implying that the alters in methylation were a result of hybridity (Salmon et al., 2005). In *Arabidopsis* allotetraploids, a study using methylation-sensitive amplification polymorphism has shown that 8.3% of the tested fragments display varying levels of methylation among F_3 synthetic allotetraploids and their progenitors, *A. thaliana* and *A. arenosa*. Homoeologous genes in allopolyploids may be deactivated by DNA methylation. For example, hypermethylation leads to the repression of *A. thaliana* homoeologs of teosinte-branched1 and cycloidea3 (TCP3) and neighboring genes in *Arabidopsis* allotetraploids, which can be reversed by blocking DNA methylation (Lee & Chen, 2001). Moreover, directly after hybridization, changes in methylation affect the transposable element (TE) ratio in allopolyploid genomes. Changes were markedly more common near TE insertions in contrast to random DNA sequences, predominantly influencing the maternal subgenome and implicating TEs in driving genetic modifications during genome merging (Parisod, Alix et al., 2010; Parisod, Holderegger et al., 2010).

Immunofluorescence analysis of 5-methylcytosine indicates increased DNA methylation across the entire genome in autotetraploids of *Cymbopogon* when compared to diploid plants, suggesting a potential role in controlling natural secondary compounds and physical dimensions (Lavania et al., 2012). Following tetraploidization, autopolyploids exhibit a peculiar epiallele of a transgene locus due to unique epigenetic regulation. Hygromycin resistance remains steady in the diploid *A. thaliana* following transformation by the HPT gene, lasting through multiple generations derived from self-pollination or outcrossing of wild-type plants (Mittelsten Scheid et al., 2003). Nevertheless, breeding with tetraploid plants results in decreased HPT activity, while offspring from autotetraploid descendants have suppressed HPT. The term polyploidy-associated transcriptional gene silencing (paTGS) describes this phenomenon, and additional research reveals that paTGS is controlled by both DNA methylation and histone methylation (Baubec et al., 2010).

32.10 HISTONE MODIFICATIONS IN POLYPLOIDS

Various modifications, such as acetylation, methylation, and ubiquitination, can covalently alter core histones (H_2A, H_2B, H_3, and H_4) at various positions of amino-terminal tails (Liu et al., 2010). Diverse changes to histones can result in different outcomes for gene activity. Studies using genome-wide analysis and biochemical techniques have demonstrated that key circadian clock genes (CCA1, LHY, TOC1, and GI) are responsible for regulating the expression of other genes in allotetraploids to increase chlorophyll and starch production (Ni et al., 2009). In allotetraploids, decreased levels of H_3K_9 acetylation and H_3K_4 methylation in promoters coincide with the repression of CCA1 and LHY, while increased levels of these epigenetic marks are linked to the upregulation of TOC1 and GI (Song & Chen, 2015). The findings demonstrate that histone modifications are influential in regulating circadian rhythms on a daily basis, subsequently amplifying the expression of genes involved in starch metabolism and photosynthesis, ultimately contributing to enhanced vigor in allopolyploids and hybrids.

32.11 THE miRNAs FUNCTION IN THE POLYPLOID EVOLUTION

Small RNAs known as miRNAs are key players in regulating gene expression through the cleavage or suppression of mRNA at the level of post-transcription (Khraiwesh et al., 2012). miRNAs are crucial players in a range of physiological and developmental mechanisms such as auxin signaling, organ separation, leaf growth, and reproduction, exhibiting an ability to adapt to both biotic and abiotic stresses (Kumar, 2014; Li et al., 2017; Meng et al., 2010). Ancestral copies of miRNA families are significantly expanded through the presence of allopolyploids. A diverse range of small RNAs contribute to the design and operation of genomes in polyploid species. By analyzing Arabidopsis hybrids and allopolyploids, scientists observed a discrepancy in miRNA content between natural and artificially produced allopolyploids, providing further evidence for the essential function of miRNAs in shaping the evolution of these plants (Ha et al., 2009). In the analysis of miRNA expression levels between diploid and autotetraploid *Paulownia fortunei* after salt treatment, it was found that eight miRNAs showed significant differences in expression, suggesting that tetraploid *P. fortunei* may have a greater tolerance to salt than diploid *P. fortune* (Fan et al., 2016). The differentially expressed microRNAs are not only affected by whole-genome duplication in the natural autopolyploid in comparison with the diploid, but they are also influenced via evolutionary changes.

32.12 CONCLUSION AND FUTURE PERSPECTIVE

Evaluations of plants are a key outcome of polyploidy, a process often linked to enhanced cell size, organ enlargement, biomass, and significant alterations in the secondary metabolism of medicinal plants. Artificial polyploidy typically boosts the concentration of secondary metabolites, altering the chemical profile by modifying biochemical synthesis, introducing new components through gene activation, and potentially eliminating others through gene repression. Advancements in high-throughput molecular genetics and DNA sequencing have deepened our understanding of the molecular and cytological mechanisms driving ploidy changes, paving the way for genetic enhancements in polyploid crops. Epigenetic remodeling, specifically through DNA methylation, is crucial in controlling gene expression and selectively silencing duplicated genes in polyploids. This process is essential for maintaining genetic stability and the proper functioning of cells in organisms with multiple sets of chromosomes. Metabolic, genetic, genomic rearrangement, and epigenetic changes, including histone modifications, DNA methylation, and small RNAs, are essential for improving polyploid crops. The enhanced understanding of polyploid formation, maintenance, and post-duplication divergence across different genomic levels is invaluable for plant biologists seeking to unravel the complexities of plant evolution and diversification. Additionally, this knowledge provides crucial support for plant breeders as they strive to develop cutting-edge strategies for enhancing crop productivity and quality. While the success of polyploid breeding is evident in increased plant growth, vigor, and metabolic efficiency, challenges such as reduced gametic and seed fertility may constrain the feasibility of improvement programs. Future comparative analyses of specific proteomic, metabolomics, transcriptomic profile and epigenetics can release additional pathways and cellular signaling that may be responsible for the modified growth vigor, adaptive behavior, and phenotypes correlated with polyploidization in medicinal plants.

REFERENCES

Adams, K. L., & Wendel, J. F. (2005). Novel patterns of gene expression in polyploid plants. *TRENDS in Genetics*, *21*(10), 539–543.

Adams, K. L., Cronn, R., Percifield, R., & Wendel, J. F. (2003). Genes duplicated by polyploidy show unequal contributions to the transcriptome and organ-specific reciprocal silencing. *Proceedings of the National Academy of Sciences*, *100*(8), 4649–4654.

Adaniya, S., & Shirai, D. (2001). In vitro induction of tetraploid ginger (*Zingiber officinale* Roscoe) and its pollen fertility and germinability. *Scientia Horticulturae*, *88*(4), 277–287.

Ahmadinia, H., & Heidari, P. (2023). Effects of polyploidy induction on the performance of anise (*Pimpinella anisum* L.). *Journal of Plant Molecular Breeding*, *11*(2), 17–30.

Allario, T., Brumos, J., Colmenero-Flores, J. M., Iglesias, D. J., Pina, J. A., Navarro, L., Talon M., Ollitrault, P., & Morillon, R. (2013). Tetraploid Rangpur lime rootstock increases drought tolerance via enhanced constitutive root abscisic acid production. *Plant, Cell & Environment*, *36*(4), 856–868.

Arrigo, N., & Barker, M. S. (2012). Rarely successful polyploids and their legacy in plant genomes. *Current Opinion in Plant Biology*, *15*(2), 140–146.

Aversano, R., Ercolano, M. R., Caruso, I., Fasano, C., Rosellini, D., & Carputo, D. (2012). Molecular tools for exploring polyploid genomes in plants. *International Journal of Molecular Sciences*, *13*(8), 10316–10335.

Baubec, T., Dinh, H. Q., Pecinka, A., Rakic, B., Rozhon, W., Wohlrab, B., von Haeseler, A., & Mittelsten Scheid, O. (2010). Cooperation of multiple chromatin modifications can generate unanticipated stability of epigenetic states in Arabidopsis. *The Plant Cell*, *22*(1), 34–47.

Blanc, G., & Wolfe, K. H. (2004). Widespread paleopolyploidy in model plant species inferred from age distributions of duplicate genes. *The Plant Cell*, *16*(7), 1667–1678.

Bomblies, K., & Madlung, A. (2014). Polyploidy in the Arabidopsis genus. *Chromosome Research*, *22*, 117–134.

Buggs, R. J. A., Chamala, S., Wu, W. E. I., Gao, L. U., May, G. D., Schnable, P. S., Soltis, D. E., Soltis, P. S., & Barbazuk, W. B. (2010). Characterization of duplicate gene evolution in the recent natural allopolyploid *Tragopogon miscellus* by next generation sequencing and Sequenom iPLEX MassARRAY genotyping. *Molecular Ecology*, *19*, 132–146.

Castro, M., Castro, S., & Loureiro, J. (2018). Production of synthetic tetraploids as a tool for polyploid research. *Web Ecology*, *18*(2), 129–141.

Chen, E. G., Tsai, K.-L., Chung, H.-H., & Chen, J.-T. (2018). Chromosome doubling-enhanced biomass and dihydrotanshinone I production in *Salvia miltiorrhiza*, a traditional Chinese medicinal plant. *Molecules*, *23*(12), 3106.

Chen, Z. J. (2007). Genetic and epigenetic mechanisms for gene expression and phenotypic variation in plant polyploids. *Annual Review of Plant Biology*, *58*, 377–406.

Chen, Z. J. (2010). Molecular mechanisms of polyploidy and hybrid vigor. *Trends in Plant Science*, *15*(2), 57–71.

Chen, Z. J., & Ni, Z. (2006). Mechanisms of genomic rearrangements and gene expression changes in plant polyploids. *Bioessays*, *28*(3), 240–252.

Chen, Z. J., Ha, M., & Soltis, D. (2007). Polyploidy: Genome obesity and its consequences: Polyploidy workshop: Plant and animal genome XV conference, San Diego, CA, USA, January 2007. *The New Phytologist*, *174*(4), 717.

Chung, H.-H., Shi, S.-K., Huang, B., & Chen, J.-T. (2017). Enhanced agronomic traits and medicinal constituents of autotetraploids in *Anoectochilus formosanus* Hayata, a top-grade medicinal orchid. *Molecules*, *22*(11), 1907.

Comai, L. (2000). Genetic and epigenetic interactions in allopolyploid plants. In *Plant gene silencing* (pp. 267–279). Springer.

Comai, L. (2005). The advantages and disadvantages of being polyploid. *Nature Reviews Genetics*, *6*(11), 836–846.

Dai, F., Wang, Z., Luo, G., & Tang, C. (2015). Phenotypic and transcriptomic analyses of autotetraploid and diploid mulberry (*Morus alba* L.). *International Journal of Molecular Sciences*, *16*(9), 22938–22956.

De Meaux, J., Pop, A., & Mitchell-Olds, T. (2006). Cis-regulatory evolution of chalcone-synthase expression in the genus Arabidopsis. *Genetics*, *174*(4), 2181–2202.

Dehghan, E., Häkkinen, S. T., Oksman-Caldentey, K.-M., & Shahriari Ahmadi, F. (2012). Production of tropane alkaloids in diploid and tetraploid plants and in vitro hairy root cultures of Egyptian henbane (*Hyoscyamus muticus* L.). *Plant Cell, Tissue and Organ Culture (PCTOC)*, *110*, 35–44.

Del Pozo, J. C., & Ramirez-Parra, E. (2015). Whole genome duplications in plants: An overview from Arabidopsis. *Journal of Experimental Botany*, *66*(22), 6991–7003.

Deng, B., Du, W., Liu, C., Sun, W., Tian, S., & Dong, H. (2012). Antioxidant response to drought, cold and nutrient stress in two ploidy levels of tobacco plants: Low resource requirement confers polytolerance in polyploids? *Plant Growth Regulation*, *66*, 37–47.

Dhooghe, E., Van Laere, K., Eeckhaut, T., Leus, L., & Van Huylenbroeck, J. (2011). Mitotic chromosome doubling of plant tissues in vitro. *Plant Cell, Tissue and Organ Culture (PCTOC)*, *104*(3), 359–373.

Dong, Y. Z., Liu, Z. L., Shan, X. H., Qiu, T., He, M. Y., & Liu, B. (2005). Allopolyploidy in wheat induces rapid and heritable alterations in DNA methylation patterns of cellular genes and mobile elements. *Russian Journal of Genetics*, *41*, 890–896.

Doyle, J. J., Flagel, L. E., Paterson, A. H., Rapp, R. A., Soltis, D. E., Soltis, P. S., & Wendel, J. F. (2008). Evolutionary genetics of genome merger and doubling in plants. *Annual Review of Genetics*, *42*, 443–461.

Fan, G., Li, X., Deng, M., Zhao, Z., & Yang, L. (2016). Comparative analysis and identification of miRNAs and their target genes responsive to salt stress in diploid and tetraploid Paulownia fortunei seedlings. *PloS One*, *11*(2), e0149617.

Fang, Z., & Morrell, P. L. (2016). Domestication: Polyploidy boosts domestication. *Nature Plants*, *2*(8), 1–2.

Gao, S. L., Chen, B. J., & Zhu, D. N. (2002). In vitro production and identification of autotetraploids of Scutellaria baicalensis. *Plant Cell, Tissue and Organ Culture*, *70*, 289–293.

Grandont, L., Jenczewski, E., & Lloyd, A. (2013). Meiosis and its deviations in polyploid plants. *Cytogenetic and Genome Research*, *140*(2–4), 171–184.

Ha, M., Lu, J., Tian, L., Ramachandran, V., Kasschau, K. D., Chapman, E. J., Carrington, J. C., Chen, X., Wang, X. J., & Chen, Z. J. (2009). Small RNAs serve as a genetic buffer against genomic shock in Arabidopsis interspecific hybrids and allopolyploids. *Proceedings of the National Academy of Sciences*, *106*(42), 17835–17840.

Hancock, J. F. (2005). Contributions of domesticated plant studies to our understanding of plant evolution. *Annals of Botany*, *96*(6), 953–963.

Hannweg, K., Visser, G., De Jager, K., & Bertling, I. (2016). In vitro-induced polyploidy and its effect on horticultural characteristics, essential oil composition and bioactivity of Tetradenia riparia. *South African Journal of Botany*, *106*, 186–191.

Hao, G., Lucero, M. E., Sanderson, S. C., Zacharias, E. H., & Holbrook, N. M. (2013). Polyploidy enhances the occupation of heterogeneous environments through hydraulic related trade-offs in *Atriplex canescens* (Chenopodiaceae). *New Phytologist*, *197*(3), 970–978.

Hegarty, M. J., & Hiscock, S. J. (2008). Genomic clues to the evolutionary success of polyploid plants. *Current Biology*, *18*(10), R435–R444.

Hegarty, M. J., Barker, G. L., Wilson, I. D., Abbott, R. J., Edwards, K. J., & Hiscock, S. J. (2006). Transcriptome shock after interspecific hybridization in Senecio is ameliorated by genome duplication. *Current Biology*, *16*(16), 1652–1659.

Hias, N., Svara, A., & Keulemans, J. W. (2018). Effect of polyploidisation on the response of apple (Malus× domestica Borkh.) to *Venturia inaequalis* infection. *European Journal of Plant Pathology*, *151*(2), 515–526.

Husband, B. C., Ozimec, B., Martin, S. L., & Pollock, L. (2008). Mating consequences of polyploid evolution in flowering plants: Current trends and insights from synthetic polyploids. *International Journal of Plant Sciences*, *169*(1), 195–206.

Islam, N., Tsujimoto, H., & Hirano, H. (2003). Proteome analysis of diploid, tetraploid and hexaploid wheat: Towards understanding genome interaction in protein expression. *Proteomics*, *3*(4), 549–557.

Khraiwesh, B., Zhu, J.-K., & Zhu, J. (2012). Role of miRNAs and siRNAs in biotic and abiotic stress responses of plants. *Biochimica et Biophysica Acta (BBA)-Gene Regulatory Mechanisms*, *1819*(2), 137–148.

Koefoed Petersen, K., Hagberg, P., & Kristiansen, K. (2003). Colchicine and oryzalin mediated chromosome doubling in different genotypes of *Miscanthus sinensis*. *Plant Cell, Tissue and Organ Culture*, *73*, 137–146.

Kron, P., Suda, J., & Husband, B. C. (2007). Applications of flow cytometry to evolutionary and population biology. *Annual Review of Ecology, Evolution, and Systematics*, *38*, 847–876.

Kumar, R. (2014). Role of microRNAs in biotic and abiotic stress responses in crop plants. *Applied Biochemistry and Biotechnology*, *174*, 93–115.

Lavania, U. C. (2013). Polyploidy, body size, and opportunities for genetic enhancement and fixation of heterozygosity in plants. *The Nucleus*, *56*(1), 1–6.

Lavania, U. C., Srivastava, S., Lavania, S., Basu, S., Misra, N. K., & Mukai, Y. (2012). Autopolyploidy differentially influences body size in plants, but facilitates enhanced accumulation of secondary metabolites, causing increased cytosine methylation. *The Plant Journal*, *71*(4), 539–549.

Lee, H.-S., & Chen, Z. J. (2001). Protein-coding genes are epigenetically regulated in Arabidopsis polyploids. *Proceedings of the National Academy of Sciences*, *98*(12), 6753–6758.

Leitch, A. R., & Leitch, I. J. (2008). Genomic plasticity and the diversity of polyploid plants. *Science*, *320*(5875), 481–483.

Li, S., Castillo-González, C., Yu, B., & Zhang, X. (2017). The functions of plant small RNA s in development and in stress responses. *The Plant Journal*, *90*(4), 654–670.

Liao, T., Cheng, S., Zhu, X., Min, Y., & Kang, X. (2016). Effects of triploid status on growth, photosynthesis, and leaf area in Populus. *Trees*, *30*(4), 1137–1147.

Liu, B., Brubaker, C. L., Mergeai, G., Cronn, R. C., & Wendel, J. F. (2001). Polyploid formation in cotton is not accompanied by rapid genomic changes. *Genome*, *44*(3), 321–330.

Liu, B., & Wendel, J. F. (2002). Non-Mendelian phenomena in allopolyploid genome evolution. *Current Genomics*, *3*(6), 489–505.

Liu, C., Lu, F., Cui, X., & Cao, X. (2010). Histone methylation in higher plants. *Annual Review of Plant Biology*, *61*, 395–420.

Liu, Z., & Adams, K. L. (2007). Expression partitioning between genes duplicated by polyploidy under abiotic stress and during organ development. *Current Biology*, *17*(19), 1669–1674.

Lu, B., Ding, R., Zhang, L., Yu, X., Huang, B., & Chen, W. (2006). Molecular cloning and characterization of a novel calcium-dependent protein kinase gene IiCPK2 responsive to polyploidy from tetraploid Isatis indigotica. *BMB Reports*, *39*(5), 607–617.

Madlung, A., Masuelli, R. W., Watson, B., Reynolds, S. H., Davison, J., & Comai, L. (2002). Remodeling of DNA methylation and phenotypic and transcriptional changes in synthetic Arabidopsis allotetraploids. *Plant Physiology*, *129*(2), 733–746.

Mansouri, H., & Bagheri, M. (2017). Induction of polyploidy and its effect on *Cannabis sativa* L. In *Cannabis sativa L. – Botany and biotechnology* (pp. 365–383). Springer.

Manzoor, A., Ahmad, T., Bashir, M. A., Hafiz, I. A., & Silvestri, C. (2019). Studies on colchicine induced chromosome doubling for enhancement of quality traits in ornamental plants. *Plants*, *8*(7), 194.

Meng, Y., Chen, D., Ma, X., Mao, C., Cao, J., Wu, P., & Chen, M. (2010). Mechanisms of microRNA-mediated auxin signaling inferred from the rice mutant osaxr. *Plant Signaling & Behavior*, *5*(3), 252–254.

Mittelsten Scheid, O., Afsar, K., & Paszkowski, J. (2003). Formation of stable epialleles and their paramutation-like interaction in tetraploid *Arabidopsis thaliana*. *Nature Genetics*, *34*(4), 450–454.

Mu, H.-Z., Liu, Z.-J., Lin, L., Li, H.-Y., Jiang, J., & Liu, G.-F. (2012). Transcriptomic analysis of phenotypic changes in birch (*Betula platyphylla*) autotetraploids. *International Journal of Molecular Sciences*, *13*(10), 13012–13029.

Münzbergová, Z. (2017). Colchicine application significantly affects plant performance in the second generation of synthetic polyploids and its effects vary between populations. *Annals of Botany*, *120*(2), 329–339.

Ng, D. W. K., Zhang, C., Miller, M., Shen, Z., Briggs, S. P., & Chen, Z. J. (2012). Proteomic divergence in Arabidopsis autopolyploids and allopolyploids and their progenitors. *Heredity*, *108*(4), 419–430.

Ni, Z., Kim, E.-D., Ha, M., Lackey, E., Liu, J., Zhang, Y., Sun, Q., & Chen, Z. J. (2009). Altered circadian rhythms regulate growth vigour in hybrids and allopolyploids. *Nature*, *457*(7227), 327–331.

Niwa, Y., & Sasaki, Y. (2003). Plant self-defense mechanisms against oxidative injury and protection of the forest by planting trees of triploids and tetraploids. *Ecotoxicology and Environmental Safety*, *55*(1), 70–81.

Ochogavía, A. C., Cervigni, G., Selva, J. P., Echenique, V. C., & Pessino, S. C. (2009). Variation in cytosine methylation patterns during ploidy level conversions in *Eragrostis curvula*. *Plant Molecular Biology*, *70*, 17–29.

Omidbaigi, R., Mirzaei, M., & Moghadam, M. S. (2010). Difference of growth traits, essential oil content and compositions between diploid and iinduced tetraploid plants of basil (*Ocimum basilicum* L.). *Journal of Essential Oil Bearing Plants*, *13*(5), 579–587.

Osborn, T. C., Pires, J. C., Birchler, J. A., Auger, D. L., Chen, Z. J., Lee, H.-S., Comai, L., Madlung, A., Doerge, R. W., & Colot, V. (2003). Understanding mechanisms of novel gene expression in polyploids. *Trends in Genetics*, *19*(3), 141–147.

Otto, S. P. (2007). The evolutionary consequences of polyploidy. *Cell*, *131*(3), 452–462.

Parisod, C., Alix, K., Just, J., Petit, M., Sarilar, V., Mhiri, C., Ainouche, M., Chalhoub, B., & Grandbastien, M. (2010). Impact of transposable elements on the organization and function of allopolyploid genomes. *New Phytologist*, *186*(1), 37–45.

Parisod, C., Holderegger, R., & Brochmann, C. (2010). Evolutionary consequences of autopolyploidy. *New Phytologist*, *186*(1), 5–17.

Parsons, J. L., Martin, S. L., James, T., Golenia, G., Boudko, E. A., & Hepworth, S. R. (2019). Polyploidization for the genetic improvement of *Cannabis sativa*. *Frontiers in Plant Science*, *10*, 449166.

Paterson, A. H. (2008). Paleopolyploidy and its impact on the structure and function of modern plant genomes. In *Plant genomes* (Vol. 4, pp. 1–12). Karger Publishers.

Pavlíková, Z., Paštová, L., & Münzbergová, Z. (2017). Synthetic polyploids in Vicia cracca: Methodology, effects on plant performance and aneuploidy. *Plant Systematics and Evolution*, *303*, 827–839.

Podda, A., Checcucci, G., Mouhaya, W., Centeno, D., Rofidal, V., Del Carratore, R., Luro, F., Morillon, R., Ollitrault, P., & Maserti, B. E. (2013). Salt-stress induced changes in the leaf proteome of diploid and tetraploid mandarins with contrasting Na+ and Cl– accumulation behaviour. *Journal of Plant Physiology*, *170*(12), 1101–1112.

Prince, V. E., & Pickett, F. B. (2002). Splitting pairs: The diverging fates of duplicated genes. *Nature Reviews Genetics*, *3*(11), 827–837.

Ramsey, J. (2011). Polyploidy and ecological adaptation in wild yarrow. *Proceedings of the National Academy of Sciences*, *108*(17), 7096–7101.

Ramsey, J., & Schemske, D. W. (2002). Neopolyploidy in flowering plants. *Annual Review of Ecology and Systematics*, *33*(1), 589–639.

Ranney, T. G. (2006). Polyploidy: From evolution to new plant development. *Combined Proceedings International Plant Propagators' Society*, *56*, 137–142.

Rêgo, M. M. do, Rêgo, E. R., Bruckner, C. H., Finger, F. L., & Otoni, W. C. (2011). In vitro induction of autotetraploids from diploid yellow passion fruit mediated by colchicine and oryzalin. *Plant Cell, Tissue and Organ Culture (PCTOC)*, *107*(3), 451–459.

Ren, J., Wu, X., Song, C., Liang, Y., Gao, W., & Wang, Y. (2018). Induction of polyploid tillered onion using colchicine and pendimethalin. *Sains Malaysiana*, *47*(11), 2617–2624.

Rubuluza, T., Nikolova, R. V., Smith, M. T., & Hannweg, K. (2007). In vitro induction of tetraploids in Colophospermum mopane by colchicine. *South African Journal of Botany*, *73*(2), 259–261.

Sabzehzari, M., Hoveidamanesh, S., Modarresi, M., & Mohammadi, V. (2019). Morphological, anatomical, physiological, and cytological studies in diploid and tetraploid plants of *Plantago psyllium*. *Plant Cell, Tissue and Organ Culture (PCTOC)*, *139*, 131–137.

Sabzehzari, M., Hoveidamanesh, S., Modarresi, M., & Mohammadi, V. (2020). Morphological, anatomical, physiological, and cytological studies in diploid and tetraploid plants of Ispaghul (*Plantago ovata* Forsk.). *Genetic Resources and Crop Evolution*, *67*(1), 129–137.

Salmon, A., Ainouche, M. L., & Wendel, J. F. (2005). Genetic and epigenetic consequences of recent hybridization and polyploidy in Spartina (Poaceae). *Molecular Ecology*, *14*(4), 1163–1175.

Schiessl, S.-V., Katche, E., Ihien, E., Chawla, H. S., & Mason, A. S. (2019). The role of genomic structural variation in the genetic improvement of polyploid crops. *The Crop Journal*, *7*(2), 127–140.

Šedivá, J., Mrázková, M., Zahumenická, P., Cusimamani, E. F., & Zahradník, D. (2019). Identification of Phytophthora tolerance in the *Anemone sylvestris* tetraploid. *Scientia Horticulturae*, *256*, 108579.

Sémon, M., & Wolfe, K. H. (2007). Consequences of genome duplication. *Current Opinion in Genetics & Development*, *17*(6), 505–512.

Shmeit, Y. H., Fernandez, E., Novy, P., Kloucek, P., Orosz, M., & Kokoska, L. (2020). Autopolyploidy effect on morphological variation and essential oil content in *Thymus vulgaris* L. *Scientia Horticulturae*, *263*, 109095.

Song, C., Liu, S., Xiao, J., He, W., Zhou, Y., Qin, Q., Zhang, Ch., & Liu, Y. (2012). Polyploid organisms. *Science China Life Sciences*, *55*, 301–311.

Song, Q., & Chen, Z. J. (2015). Epigenetic and developmental regulation in plant polyploids. *Current Opinion in Plant Biology*, *24*, 101–109.

Stupar, R. M., & Springer, N. M. (2006). Cis-transcriptional variation in maize inbred lines B73 and Mo17 leads to additive expression patterns in the F_1 hybrid. *Genetics*, *173*(4), 2199–2210.

Stupar, R. M., Bhaskar, P. B., Yandell, B. S., Rensink, W. A., Hart, A. L., Ouyang, S., Veilleux, R. E., Busse, J. S., Erhardt, R. J., Buell, C. R.,.& Buell, C. R. (2007). Phenotypic and transcriptomic changes associated with potato autopolyploidization. *Genetics*, *176*(4), 2055–2067.

Tate, J. A., Soltis, D. E., & Soltis, P. S. (2005). Polyploidy in plants. In *The evolution of the genome* (pp. 371–426). Elsevier.

Tavan, M., Mirjalili, M. H., & Karimzadeh, G. (2015). In vitro polyploidy induction: Changes in morphological, anatomical and phytochemical characteristics of *Thymus persicus* (Lamiaceae). *Plant Cell, Tissue and Organ Culture (PCTOC)*, *122*(3), 573–583.

Udall, J. A., & Wendel, J. F. (2006). Polyploidy and crop improvement. *Crop Science*, *46*, S-3.

Van de Peer, Y., Mizrachi, E., & Marchal, K. (2017). The evolutionary significance of polyploidy. *Nature Reviews Genetics*, *18*(7), 411–424.

Van Laere, K., França, S. C., Vansteenkiste, H., Van Huylenbroeck, J., Steppe, K., & Van Labeke, M.-C. (2011). Influence of ploidy level on morphology, growth and drought susceptibility in *Spathiphyllum wallisii*. *Acta Physiologiae Plantarum*, *33*, 1149–1156.

Vanneste, K., Maere, S., & Van de Peer, Y. (2014). Tangled up in two: A burst of genome duplications at the end of the Cretaceous and the consequences for plant evolution. *Philosophical Transactions of the Royal Society B: Biological Sciences*, *369*(1648), 20130353.

Wang, J., Tian, L., Lee, H.-S., & Chen, Z. J. (2006). Nonadditive regulation of FRI and FLC loci mediates flowering-time variation in Arabidopsis allopolyploids. *Genetics*, *173*(2), 965–974.

Wang, J., Tian, L., Lee, H.-S., Wei, N. E., Jiang, H., Watson, B., Madlung, A., Osborn, T. C., Doerge, R. W., Comai, L., & Chen, Z. J. (2006). Genomewide nonadditive gene regulation in Arabidopsis allotetraploids. *Genetics*, *172*(1), 507–517.

Wang, J., Tian, L., Madlung, A., Lee, H.-S., Chen, M., Lee, J. J., Watson, B., Kagochi, T., Comai, L., & Chen, Z. J. (2004). Stochastic and epigenetic changes of gene expression in Arabidopsis polyploids. *Genetics*, *167*(4), 1961–1973.

Wendel, J. F. (2000). Genome evolution in polyploids. In *Plant molecular evolution* (pp. 225–249). Springer.

Wittkopp, P. J., Haerum, B. K., & Clark, A. G. (2004). Evolutionary changes in cis and trans gene regulation. *Nature*, *430*(6995), 85–88.

Wood, T. E., Takebayashi, N., Barker, M. S., Mayrose, I., Greenspoon, P. B., & Rieseberg, L. H. (2009). The frequency of polyploid speciation in vascular plants. *Proceedings of the National Academy of Sciences*, *106*(33), 13875–13879.

Wu, J.-H., & Mooney, P. (2002). Autotetraploid tangor plant regeneration from in vitro Citrus somatic embryogenic callus treated with colchicine. *Plant Cell, Tissue and Organ Culture*, *70*(1), 99–104.

Yaghobi, M., & Heidari, P. (2023). Genome-wide analysis of aquaporin gene family in *Triticum turgidum* and its expression profile in response to salt stress. *Genes*, *14*(1), 202.

Yang, X. M., Cao, Z. Y., An, L. Z., Wang, Y. M., & Fang, X. W. (2006). In vitro tetraploid induction via colchicine treatment from diploid somatic embryos in grapevine (*Vitis vinifera* L.). *Euphytica*, *152*, 217–224.

Yao, H., Kato, A., Mooney, B., & Birchler, J. A. (2011). Phenotypic and gene expression analyses of a ploidy series of maize inbred Oh43. *Plant Molecular Biology*, *75*, 237–251.

Yoo, M.-J., Liu, X., Pires, J. C., Soltis, P. S., & Soltis, D. E. (2014). Nonadditive gene expression in polyploids. *Annual Review of Genetics*, *48*, 485–517.

Yu, Z., Haberer, G., Matthes, M., Rattei, T., Mayer, K. F. X., Gierl, A., & Torres-Ruiz, R. A. (2010). Impact of natural genetic variation on the transcriptome of autotetraploid *Arabidopsis thaliana*. *Proceedings of the National Academy of Sciences*, *107*(41), 17809–17814.

Zaman, Q. U., Li, C., Cheng, H., & Hu, Q. (2019). Genome editing opens a new era of genetic improvement in polyploid crops. *The Crop Journal*, *7*(2), 141–150.

Zhang, X.-Y., Hu, C.-G., & Yao, J.-L. (2010). Tetraploidization of diploid Dioscorea results in activation of the antioxidant defense system and increased heat tolerance. *Journal of Plant Physiology*, *167*(2), 88–94.

Zhou, R., Moshgabadi, N., & Adams, K. L. (2011). Extensive changes to alternative splicing patterns following allopolyploidy in natural and resynthesized polyploids. *Proceedings of the National Academy of Sciences*, *108*(38), 16122–16127.

Zhou, Y., Kang, L., Liao, S., Pan, Q., Ge, X., & Li, Z. (2015). Transcriptomic analysis reveals differential gene expressions for cell growth and functional secondary metabolites in induced autotetraploid of Chinese woad (*Isatis indigotica* Fort.). *PLoS One*, *10*(3), e0116392.

Index

Note: Page numbers in *italics* indicate a figure and page numbers in **bold** indicate a table on the corresponding page.

H

I

J

K

L

M

N

O

T

U

V

W

X

Y

Z